Gastrointestinal Physiology

Menizibeya Osain Welcome

Gastrointestinal Physiology

Development, Principles and Mechanisms of Regulation

Springer

Menizibeya Osain Welcome
Department of Physiology, Faculty of Basic Medical Sciences
College of Health Sciences, Nile University of Nigeria
FCT-Abuja
Nigeria

The Figures in this book were designed with help of Motifolio Biomedical Drawing Toolkit Suite (Motifolio Inc., Ellicott City, USA).

ISBN 978-3-030-08165-2 ISBN 978-3-319-91056-7 (eBook)
https://doi.org/10.1007/978-3-319-91056-7

Softcover re-print of the Hardcover 1st edition 2018

Printed on acid-free paper

This Springer imprint is published by the registered company Springer International Publishing AG part of Springer Nature
The registered company address is: Gewerbestrasse 11, 6330 Cham, Switzerland

Preface

The digestive system is responsible for about 60–90% of diseases that affect humans, making the digestive system one of the most important systems in life processes. This book is a review of key findings of the last two and present millennia on the area of digestion. This text is a leading textbook and most comprehensive review ever written in the field of gastrointestinal (GI) physiology.

This book is written to address the gaps in other texts. This book provides key information, yet robust, required to have a detailed and contemporary understanding of GI physiology. This text incorporates key concepts of translational physiology by systematically examining pertinent areas of the GI system, including anatomy, embryology, histology, biochemistry, pharmacology, biophysics, behavioral science, bioinformatics, pathophysiology, public health, genetics, epigenetics, and therapeutics, in accordance with physiology. The text provides crucial information on the molecular, cellular, tissue, organ, and system levels of functioning of the GI tract in health and disease.

This book thoroughly explains the normal functioning of the digestive system in humans, relates the concepts to how diseases develop, and unravels the mechanisms and basis of medical approach to treatment of the different ailments of the GI tract.

This text apart from incorporating historical information on developmental course of GI physiology from antiquity to the contemporary era also outlines contemporary trends and gives a comprehensive description of developmental path that determined the study of digestive functions in the present-day world.

New data that have accumulated over the past decades on the functioning of the digestive system are systematically reviewed, and emphases are made on breakthrough studies. This book incorporates latest information on functional communication network between the gut and other organs and tissues of the body such as the brain, lungs, kidney, heart, pancreas, skin, bone, and adipose tissue. New information on the roles of the gut as endocrine, exocrine, and neural organ is not pretermitted. History of over 60 hormones and neurotransmitters currently discovered in the gut alone as well as their functional aspects are discussed. This book also provides detailed historical and functional information on all digestive enzymes. New information on mechanisms of enzymatic breakdown of food substances is also discussed. Recently discovered enzymes of the GI tract identified to play useful role in digestion are also reviewed.

The text strategically highlights key functions of the gut microbiota. Both traditional and emerging roles of *H. pylori* in gastric physiology are discussed.

This book is carefully designed for biomedical, medical, and health science students, scientists, and researchers. It also serves as an inevitable reference text for clinicians and other medical, health, and allied professionals.

For beneficial comprehension, the book is systematically divided into topics and subtopics. There are also numerous color illustrations. Recommended readings are separated into original articles, review articles, guidelines, books, and Nobel lectures.

The book contains special in-text references on some high-quality publications. Key information or exceptional discoveries of global significance are systematically outlined as "Spotlights." Concepts traditionally used in science, originating from historians or other areas of science other than physiology, are briefly described as "Reference Note." This is needed to provide an adequate and broader understanding of the information applied in physiology.

To aid comprehension of the association between the physiological concepts, principles, and clinical presentations, clinical examples such as pathologies that link basic science with clinical practice are outlined in special sections "Clinical Correlates." Contemporary approaches to the basis of treatment of some GI tract diseases are systematically outlined.

In addition to providing an adequate and broader understanding of the information applied in GI physiology, this approach addresses the challenges of translational physiology. It also provides the necessary background for application of basic science information to medical practice, as well as utilization of bedside clinical data and application to produce a solid knowledge base in physiology. This approach represents a high-quality evidence-based delivery of physiological information to the learner and allows the learner to appreciate the value and usefulness of physiology to nature and human existence.

Thus, the book applies the basic concepts of translational physiology. This design of the text is aimed at closing the gap between basic science and its application (such as in the clinics, public health), which is largely due to the reductionist approach, rather than an integrative in addressing human maladies.

Finally, I would like to take this opportunity to express my sincere gratitude to Prof. Vladimir Alexeevich Pereverzev MD, Ph.D., DSc, Head of Department of Normal Physiology of the Belarusian State Medical University, Minsk, Belarus, and the editorial team of Springer for all their support and encouragement.

FCT-Abuja, Nigeria Menizibeya Osain Welcome
March 2018

Key Features

- Most comprehensive, up-to-date text ever written in the field of gastrointestinal physiology in the world
- Basic and reference text for medical and allied health science students, as well as practicing doctors and other health professionals
- Provides a detailed analysis of the trend of development of knowledge on all aspects of gastrointestinal physiology from antiquity to the contemporary world
- Provides detailed mechanisms of regulation of gastrointestinal functioning in normal and pathology
- Outlines groundbreaking studies of the past centuries in the field of gastrointestinal physiology and provides contemporary information on the direction of future investigations
- Numerous color illustrations.

Target Groups

- Undergraduates and graduates of medicine, dentistry, pharmacy, nursing, human biology, science, and other allied health professions
- Interns, residents, and other practicing medical doctors as well as health professionals
- Academicians
- Researchers and scientists
- Policy makers.

Contents

About the Author

Dr. Menizibeya Osain Welcome, MD, Ph.D. is a Senior Lecturer in the Department of Physiology, Faculty of Basic Medical Sciences, College of Health Sciences, Nile University of Nigeria, FCT-Abuja. He earned the Doctor of Medicine in 2011 and his Ph.D. in Medicine (Physiology) in 2013. He is the author or co-author of over 200 papers and holds several awards.

Chapter 1
History of Development of Gastrointestinal Physiology: From Antiquity to Modern Period and the Birth of Modern Digestive Physiology

Abstract Curiosity and the quest for addressing human maladies or problems of nature are possibly the key driving forces for discoveries in science. Apart from shaping the reputation of the scientists and historically inscribing his or her name in the memorable plate, outstanding discoveries in science gain varying levels of recognition from the local to the global stage. The Nobel Prize is the most respected honor and highest level of recognition given to any individual on planet Earth for exceptional contribution to solving key problems of nature and mankind. Since 1901, this highest accolade started recognizing extraordinary endeavor of scientists in addressing glaring issues of existence and nature. Traditionally, the prize is given in literature, physics, chemistry, and physiology or medicine. Over the past years, award recipients, probably due to the interchangeability and relationship between disciplines in science, scientists who are physicians and researchers in physiology or medicine, for example, have been awarded the prize not only in physiology or medicine, but also in chemistry. Chemistry and physiology or medicine have had a close relationship as regards discoveries that have necessitated the award of the Nobel Prize. Except in rare cases where the prize was awarded to a discovery that happened by chance, most awards were made to discoveries that had solid roots in the previous works of other scientists. Therefore, it is imperative to have a basic knowledge of the historic timeline of major developmental events and achievements to allow for meaningful future investigations that could fetch the world results for the betterment of life. Since its inception, out of 211 Nobel laureates in Physiology or Medicine, awarded between 1901 and 2016 (107 times of Nobel Prize awards), five laureates have received the prestigious award in the area of the gastrointestinal physiology. All five prizes were won by scientists whose works were rooted in previous works of other scientists. The research works of the scientists whose ideas laid the basis for breakthrough studies are worth considering. The first Nobel laureate in the area of gastrointestinal (GI) physiology, Ivan Petrovich Pavlov (1849–1936), who set the stage for breakthrough discoveries in GI physiology, is a physician, physiologist, and pharmacologist. He won the prize in 1904, making him the fourth scientist in the list of Nobel Prize Winners since the inception of the award in 1901. Apart from the Nobel Prize, there are other scientific prizes with

M. O. Welcome, *Gastrointestinal Physiology*,
https://doi.org/10.1007/978-3-319-91056-7_1

near equivalent rating to the Nobel Prize. Some of the scientific prizes are awarded specifically to young scientists. This chapter provides contemporary information on the historical timeline of events that transpired on GI physiology beginning from antiquity to the contemporary period.

Keywords Developmental history · Evolution · Nobel Prize · Antiquity Renaissance · Modern era · Modern digestive physiology · Gastrointestinal physiology · Gastric fistula · Digestion · Sham feeding · Pepsis Nervism · Heidenhain pouch · Pavlov's pouch · Translational physiology Translational medicine · Translational research · Reductionist · Integrative Physiome project · Giome Project · Hippocrates · Aristotle · Plato Herophilus (Herofilos) of Alexandria · Praxagoras · Andreas Vesalius Erasistratos of Keos · Susruta (Sushruta) Samhita and Charaka Samhita Claudius Galen of Pergamum · Magnus of Nisibis (or Emesa) · Alexandrias Byzantine physicians · Caelius Aurelianus · Soranus of Ephesus Avicenna · Norman Guitmund · Theodoric Borgognoni · Guido Lanfranchi (Lanfranc of Milan) · Johannes Guttenberg · Alessandro Benedetti Jacopo Berengario da Carpi · Leonardo da Vinci · Paracelsus · Johan Thölde René Descartes · Giovanni Alphonso Alexander Mikhailovich Ugolev Arne Dahlqvist · Borelli · Herman Boerhaave · Regnier de Graaf Friedrich Tiedemann · Leopold Gmelin · Jöns Jakob Berzelius Andrés Laguna de Segovia · Christopher Columbus · Jan Baptiste van Helmont Andreas Vesalius · Franciscus de la Boë Sylvius · Viridet · William Harvey Caspar Bauhin · Johann Georg Wirsung · Francis Glisson · Johann Conrad Peyer Johann Conrad Brunner · Johann Nathanael Lieberkühn · René de Réaumur Edward Stevens · Johann Nepomuk Eberle · Erhard Friedrich Leuchs Anselme Payen · Jean-François Persoz · Wilhelm Friedrich Kühne Andreas Sigismund Marggraf · Jean Baptiste André Dumas · Louis-Nicolas Vauquelin · Pierre Jean Robiquet · William Hyde Wollaston · Karl Axel Hampus Mörner · Jons Jakob Berzelius · Gerardus Johannes Mulder · Gabriel G. Valentina Apollinaire Bouchardat · Claude Marie Sandras · Louis Mialhe Albrecht von Haller · Fredericus Bernardus Albinus · Bernhard Siegfried Albinus Hermann Boerhaave · Albinus · Andreas Bonn · Abbe Lazzaro Spallanzani John Richardson Young · Guillaume Dupuytren · William Prout Camillo Golgi · Theodor Schwann · Johannes Peter Müller · John Newport Langley · John Sydney Edkins · John Howard Northrop · Roger Moss Herriott François Magendie · Sir Charles Bell · Theodor Schwann · Karl (Carl) Ludwig Ernst von Brücke · William Bayliss · Nikolai Konstantinovich Kulchitsky Karl Hugo Kronecker · Samuel James Meltzer · Saturnin Arloing Ivan Mikhailovich Sechenov · Edwin Burket Twitmyer · Sir Charles Scott Sherrington · Edgar Adrian · Johanes Andreas Fibiger · Henry Hallet Dale George Palade · James Whyte Black · Robin Warren · Barry Marshall Basov Vasiliy Alexandrovich · Sergey Petrovich Botkin · André Latarjet Lester Reynold Dragstedt · William Beaumont · Claude Bernard Jan Evangelista Purkyně · Rudolph Heidenhain · Ivan Petrovich Pavlov

Abbreviations

AD	Anno Domini
BC	Before Christ
BCE	Before Common Era
ca (syn. c., c, cir., circ., cca.)	Circa
CE	Common Era
DNA	Deoxyribonucleic acid
GI	Gastrointestinal
HCl	Hydrochloric acid
mRNA	Messenger ribonucleic acid
NASA	National Aeronautics and Space Administration
α	Alpha
β	Beta
γ	Gamma

1.1 Introduction

Before elaborating the history of the development of gastrointestinal (GI) physiology, it is imperative to explain the meaning of physiology briefly and mention the characters that made substantial contributions to the development of the subject. Physiology is a word formed from the Greek words "physis" meaning nature and "logos" meaning study. Therefore, physiology can be defined as the study of nature and the essence of life processes. In a broader view, physiology means the study of life processes and functions and their dynamics at the molecular, cellular, tissue, organ, system, and organismal levels. It is the science of the life of an organism as a whole, its interactions with the environment, and the dynamics of life processes. Physiology can be delineated further as the science of life functions, its structures, and mechanisms of their realization and the principles of regulations [1].

The word "physiology" was first used by the Greek philosophers around sixth century BC to mean the inquiry into the nature of things. This view of physiology continued for several centuries. However, during and after the end of the Renaissance, the use of "physiology" included functions of human organs and systems (see below). There were different schools of thought on the subject of physiology across centuries. But the philosophical root and reasoning enmeshed in physiology and as defined by the ancient Greek philosophers was unavoidably transferred across generations until the nineteenth–twentieth centuries. According to the German physician, physiologist, and philosopher, professor Wilhelm Maximilian Wundt (1832–1920), "physiology is concerned with all the phenomena of life that present themselves to humans in sense of perception as bodily processes, and accordingly form part of that total environment which we name the external world." The increase in the quest for knowledge and curiosity of humans substantially enhanced the development of physiology in later centuries. This is unarguably true; however, the philosophical view of physiology was subject to unending discussion in the scientific community [1].

It is important to bear in mind that physiology is older than man is. It is man's inquiry into nature that has continued to unravel the principles and mechanisms of body functioning. Key personalities that made exceptional contributions to the fundamental development of physiology as a scientific discipline include Hippocrates (460–377 BC) who, among other things, observed that differences in the composition and movement of liquid of the body are the result of different psychological makeup of humans. Others include René Descartes (1596–1650), William Harvey (1578–1657), Daniel Bernoulli 1700–1782), Antoine-Laurent de Lavoisier (1743–1794), Mikhail Vasilyevich Lomonosov (1711–1765), Claude Bernard (1813–1878), Rudolph Heidenhain (1834–1897), Emil du Bois-Reymond (1818–1896), Karl Landsteiner (1868–1943), Santiago Ramón y Cajal (1852–1934), Camillo Golgi (1843–1926), Ivan Mikhailovich Sechenov (1829–1905), Alexei Alexeevich Ukhtomsky (1875–1942), Ilya Ilyich Mechnikov (1845–1916), Verigo Bronislav Fortunatovich (1860–1925), Schack August Steenberg Krogh (1874–1949), Carl Friedrich Wilhelm Ludwig (1816–1895), Christian Harald Lauritz Peter Emil Bohr (1855–1911), John Scott Haldane (1860–1936), Nikolai Evgenevich Vvedensky (1852–1922), Ivan Petrovich Pavlov (1849–1936), Pyotr Kuzmich Anokhin (1898–1974), Walter Bradford Cannon (1871–1945), David E. Goldman (1910–1998), Alan Lloyd Hodgkin (1914–1998), Andrew Fielding Huxley (1917–2012), Bernard Katz (1911–2003), Johannes Peter Müller (1801–1858), Justus Freiherr von Liebig (1803–1873), Carl Friedrich Wilhelm Ludwig (1816–1895), Sir Michael Foster (1836–1907), François Magendie (1783–1855), Silas Weir Mitchell (1829–1914), Henry Pickering Bowditch (1840–1911), Eduard Friedrich Wilhelm Pflüger (1829–1910), Sámuel Rácz (1744–1807), Elias Cyon (1843–1912), Henry Newell Martin (1848–1896), Oscar Langendorff (1853–1908), Frederick Gowland Hopkins (1861–1947), Walter Clement Alvarez (1884–1978), Curt Paul Richter (1894–1988), Alexander Mikhailovich Ugolev (1926–1991), and Arne Dahlqvist (1909–1995) [2, 3]. A couple of other scientists have contributed significantly to physiology; some of these scientists can be found on the Web site of the Nobel Prize (www.nobelprize.org), American Physiological Society (www.the-aps.org), International Union of Physiological Sciences (http://iups.org), and Physiological Society (http://www.physoc.org).

As a tradition, the study of physiology has been carried out at the molecular, cellular, tissue, organ, and system levels, and hence, physiology is divided into different subdisciplines: molecular physiology, cellular physiology, organ physiology (e.g., liver, lungs, or stomach physiology), system physiology (physiology of the nervous system or neurophysiology, physiology of digestive system or gastrointestinal physiology, physiology of the circulatory system or cardiovascular physiology, physiology of the respiratory system or respiratory physiology, physiology of the urinary system, and so on). Other areas of physiology that are taught as courses in higher institution include developmental physiology, geriatric physiology, physiology of labor, aviation and cosmologic physiology, ecological or environmental physiology, evolutional and comparative physiology. Also, to distinguish differences in functioning of different living things, physiology is classified as human physiology, animal physiology, plant physiology, fungal physiology, and microbial (viral, bacterial) physiology. The study of physiology combines information from other disciplines, including histology, embryology, anatomy, biochemistry, molecular and cellular biology, pharmacology, internal medicine, therapeutics. The study of the mechanisms of occurrence, course, and outcome of pathological processes and diseases is called pathological physiology (pathophysiology). Both physiology and pathological physiology are core disciplines in medicine and clinical practice, providing firsthand information on ways in addressing pathological conditions in the contemporary world.

This book deals specifically with "GI physiology." In this book, physiology of digestion and GI physiology are used interchangeably. Digestion is the process involving the initiation of secretions in preparation for the mechanochemical breakdown of food substances, intake of food, proper biomechanical breakdown from polymers to monomers with varying intensity at different levels of the digestive tract as well as the de novo formation of certain biomolecules in the colon and the absorption and transport of final products of mechanochemical breakdown or synthesis into the body fluids that subsequently serve for plastic and energy functions, and the removal of undigested products from the anus. The discipline that is concerned with this area of science is called GI physiology. It should be emphasized, however, that though the definition of digestion given above is far encompassing, it is not exactly complete, which we will see in later part of this book. This is because accumulating evidences indicate that not only monomers are absorbed in the intestine but also dimers or oligomers of not only peptides but also saccharides.

The hollow structure through which digestion occurs is the digestive tract (or alimentary canal). It is made up of mouth, pharynx, esophagus, stomach, small and large intestine, and the anus (Fig. 1.1). The GI system comprises of the digestive tract and accessory organs such as teeth, tongue, salivary glands, pancreas, liver, and gallbladder (Fig. 1.1). Digestion is regulated by a complex network of neural, hormonal, auto-para-juxa-crine, and electromechanical systems/factors.

1.2 Organization of the GI System, Periods of Development of Knowledge on the Digestive Physiology

Organization of the GI system is the arrangement of the constituent parts of the digestive system that ensure execution of its functions.

The first level of organization of the digestive system is called the chemical level. In this level, the system is viewed as being composed of molecules which are formed from atoms. In the second level, several molecules come together to form the cell, which is the smallest structural and functional basis of living things.

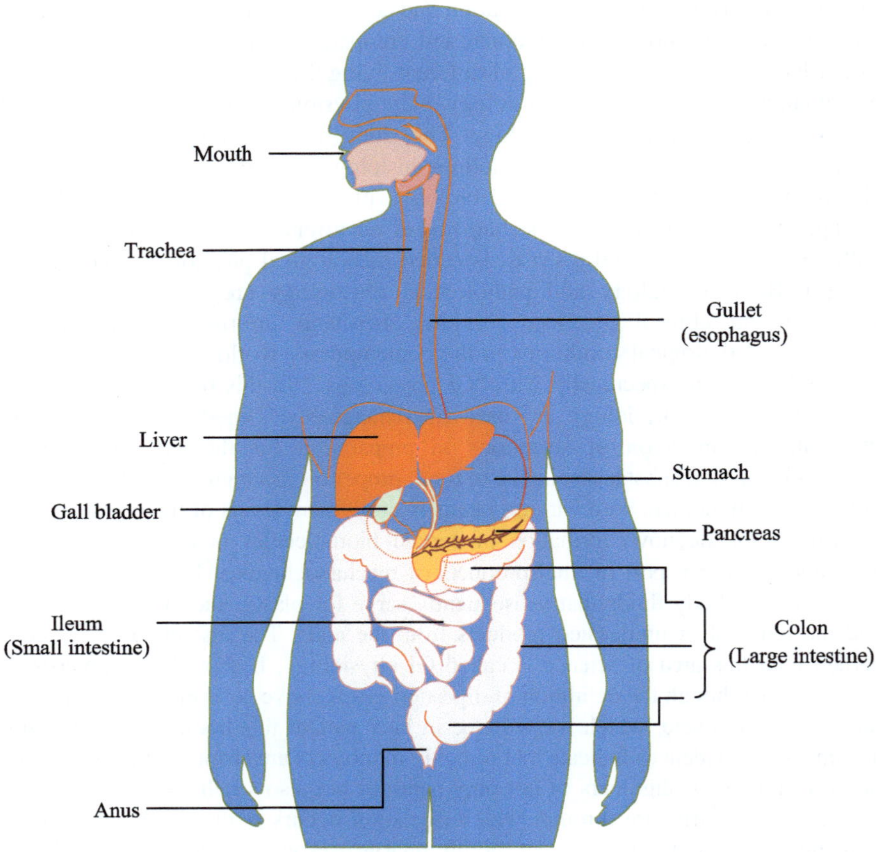

Fig. 1.1 Human digestive system. The digestive system comprises the digestive or GI tract and accessory organs. The tract starts from the mouth and extends from the pharynx down to the esophagus, and then to the stomach (a hollow muscular organ), which leads to the intestine (small and large intestine) and finally to the anus, the exit point of undigested products. The accessory organs of digestion are teeth, tongue, salivary glands, liver, gallbladder, and pancreas. Note that some of the terminologies/words used today in GI physiology and many other areas were formed from Latin and Greek words. For instance, the rectum was formed from the Latin "rectum intestinum," meaning straight intestine. The origin of other words used in GI physiology is discussed in the text

A group of similar cells forms tissue. A group of tissues forms an organ. An organ such as the intestine is formed from epithelial, neural, smooth muscle, and connective tissues. Different organs work closely together to form organ system level of organization. For instance, the GI system is composed of digestive tract (which itself contains organs) and the accessory organs. The organismal level of organization is comprised of many organ systems. The rise of knowledge in different areas of GI physiology may not be separated with precision since one period overlapped with other periods (Fig. 1.2).

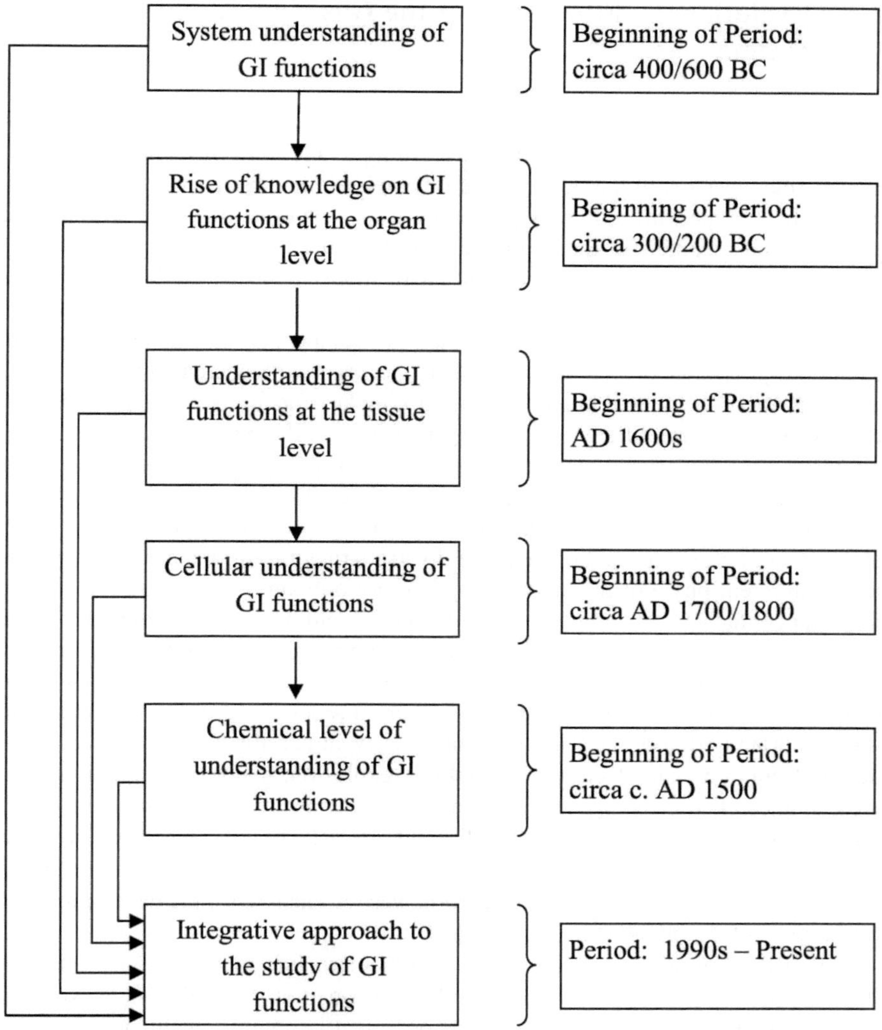

Fig. 1.2 Periods of development of knowledge on different levels of organization of GI physiology. Physiology evolved from initial understanding of the system functioning. This system understanding expanded subsequently to the study of separate organs, which also expanded to tissue and cellular level investigations. Chemical quest for the nature of physiological systems developed at a relatively fast rate and occurred almost simultaneously with cellular and molecular levels of development. All these levels of development, however, are interlaced, and no strong demarcation by time actually exists. The timeline provided in this chart is estimation. It should be pointed that the present trend in development of physiology is directed toward integrative approach, though relatively small research group investigators still study physiological systems at the cellular, molecular, and chemical levels

1.3 Evolutionary Emergence of the Gastrointestinal System

Digestion as a central role of the GI system was predetermined evolutionally, and this physiological process is executed according to the laws of nature. Because it serves other systems of the body, it could have been the first system to evolve and acquire higher specialization during the course of evolution. Digestion is a function present in all members of the animal kingdom ranging from single-cell organisms (with single body cavity) to mammals (with more sophisticated body cavity or tube). The emergence of the GI system is an important evolutionary feat that was required to support the future complexity of members of the animal kingdom. Importantly, complexity of the GI system increases with increase in complexity across phyla of the animal kingdom. The striking key functional similarities despite the diverse structural differences in GI system across different classes of the animal kingdom remain a crucial evidence of the emergence of the GI system during the Cambrian explosion, a period of heightened diversification of animals that may have been triggered or accelerated by environmental or biological factors. The striking similarities in functions despite diverse structural differences in GI system across different phyla of the animal kingdom not only are a testament, but also an important evidence of the emergence of the GI system during the Cambrian period. This suggests that the ancestral descendants of the primitive GI system sequentially rather than inorderly evolved, possibly, during the Cambrian explosion [4]. The Cambrian explosion is believed to have occurred around 542 million years ago in the Cambrian Period of the Paleozoic era. This event is estimated to have lasted for about 20 million years. Before the Cambrian explosion, majority of the organisms were single-celled or composed of colonies. Following this explosion, there was a significant acceleration of diversification of animal species, and over the next millions of years, the environment and its inhabitants gradually began to resemble that of today. Thus, the Cambrian explosion was responsible for the development of a true gut in the animal kingdom [5–8].

The pattern of emergence of the GI system's structure determined its functioning. The digestive process of single-cell organism is intracellular. However, the digestive process of simple multicellular organism is both intracellular and extracellular; thus, food is digested outside the cells in a digestive cavity or tract and then absorbed into the body. The emergence of both intracellular and extracellular digestion in multicellular organisms indicates an adaptive measure acquired during the course of evolution, which was required for enhanced provision of energy and survival. Recent investigations show similarities in digestive system of earliest animals, indicating that GI evolved from the simplest to more complex structure [4].

Across the phyla of the animal kingdom, there is a marked increase in complexity. A tube-like or sack-like structure was the first digestive apparatus to evolve in the animal kingdom comprising of nematodes (roundworms), annelids (earthworms), mollusks, arthropods (insects, crustaceans, arachnids, myriapods), echinoderms, and vertebrates (class fish, amphibians, reptiles, birds, mammals). But the simplest form of the digestive tube first evolved in the nematodes. The first digestive tubes in these animals may have occurred about 3.5–4 billion years ago. Nematodes were the first animals to have a true extracellular digestion; they have the most primitive digestive tract lined with epithelial cells. The tract starts from the mouth and extends through the pharynx down to intestine and to the anus. Food is taken in from the mouth and passes out from the anus. There is a marked gradual increase in complexity of the tube-like structure of digestion across the animal kingdom. There is also gradual increase in regional specialization of digestive functions and processes: chemical and mechanical digestion. Thus, the increase in complexity of digestive tract continuously increased with the evolution of a storage organ coupled with mechanical breakdown of food particles with enzymatic breakdown that previously evolved in plants and microorganisms (bacteria) [4–8].

The functional diversity of body processes coupled with increased search for foods that could compensate the increase in complexity of functions of multicellular organisms may have led to increase in complexity of GI tract in mammals. Unicellular organisms that inhabit higher organisms such as Homo sapiens are a testament to the evolutionary sophistication of digestion. The symbiotic relationship on which humans cannot live without certain population of microorganisms is a key surviving strategy that associates the simplest form of digestion with the most complex ones. Therefore, it may be deduced that different processes in digestion in mammals evolved in different stages in evolution and progressively increase in functional and structural complexity across the animal kingdom [4–8].

1.4 Digestive Physiology in Antiquity

The history of digestive physiology is related to human history as advancement in both knowledge and technology significantly and positively affected the development of physiology as a whole. In GI physiology, in particular, different mystical and philosophical views about digestion spanned across centuries before scientists could have a precise knowledge of the subject. The views and conceptions were mixed with misconceptions and philosophical reasoning. It should be mentioned that progress in digestive physiology was in part stimulated by the human strive to alleviate the sufferings of their fellows. Since antiquity, symptoms of GI disorders such as heartburn, belching, and epigastric pains were observed, and recommendations were made on feeding to ease these symptoms. This indicated that people at the time realized that effective digestion (via feeding with special "good" foods) was necessary to ensure adequate life processes of humans, though, at the time, no such word as "digestion" was known [9].

Digestion as a process had been in existence before humans became inquisitive on why food intake was necessary, the course of the swallowed food, and why feces were passed out. The ancient men wondered why the feeling to pass out feces following some hours of food intake was beyond the volition of man. It is necessary to emphasize, however, that before man's existence, other species had a functional digestive system. However, how they learned to gather food for energy and plastic functions could not be exactly ascertained. In the same line of view, human digestion may not have a precise origin due to the issues of human cognition and record keeping. Thus, history of digestion might not accurately define the timeline of events of human knowledge on the subject. Notwithstanding basic facts on digestion and the trend of development could be deduced based on available data and knowledge.

After humans had come into existence, they were at one time or the other hungry and became weak as they could not move easily from place to place and unable to carry out their daily activities, so they needed to satisfy their urge for movement and other activities. Hence, they tried to consume herbs and other substances. The initial results were, probably, both negative and positive as some herbs or substances could have led to constipation, diarrhea, and even death. Thanks to the physiological senses that would have helped man in distinguishing between bad and good food or substances, after experimenting with different food types and substances over a period. The experimental use of some of the herbs and substances obviously provided man with the required energy to move from place to place and carry out basic activities of life. These events probably led to man thinking that some spiritual forces were involved in food digestion [(they never knew that any bodily function or process like digestion existed, even though they were aware that food must be taken into the mouth (if necessary chewed), swallowed to feel strong and faces passed out after a given period of food intake)] [10]. Almost in every culture and society, the result of consumption of food was tied to one form of spirituality since in some cases consumption of certain food led to sick people regaining health and strength, people who were apparently not sick suddenly developing sickness or even dying subsequently. At one time, the Egyptian papyruses were thought to have suffered from spiritual forces when they developed diseases related to digestion. Though highly learned men as at the time made suggestions on the physiology of digestion, the course of ingested food remained a mystery that further fueled the concept of spiritism. In ancient Greece, Rome, Chinese, and India, philosophers and scientists made numerous proposals on digestion myth and how it occurred in humans.

Reference Note 1.1

Meaning of the Historical Designations—AD, BC, CE, BCE

Confusion usually arises when historical dates are used in research or academic texts. Religious historians were probably the first to have proposed various systems of identifying dates. The widely recognized ones are the "AD, BC, CE, and BCE." The abbreviation "AD"—Anno Domini, meaning "in the year of our Lord" or Anno Domini Nostri Iesu (Jesu) Christi, meaning "in the Year of Our Lord Jesus Christ," is a little over two millennia ago; "BC"—Before Christ; "CE"—Common Era; "BCE"—Before Common Era. The historical designation system "AD/BC" is similar to "BCE/CE" (i.e., AD is equivalent to BCE and BC is equivalent to CE). Therefore, 210 CE is the same as AD 210. These designations were used since the middle period. Remember that there is no zero year in these systems. The AD/BC system is based on the Julian and Gregorian calendars. The Latin abbreviation "AD" is placed before the year number, while "BC" is put after the year number (e.g., AD 2000, 210 BC). The numbers can be used as century or millennium. For example, nineteenth century AD is the same as second millennium AD. The BCE/CE system was developed not only by religious heads but also by scientists. Some scientists designated the period after the Year of Our Lord as "Vulgar Era" and became increasingly popularly used. During that time "Vulgar" means "common." The date of birth and death or reign of an individual may have different historical dates due to inadequate record keeping and different views and research data acquired by various historians. Notice that Galen's year of death is represented with two separate years (216/210). This is because the period represented is not exact, as with many other historical dates. To avoid the cumbersomeness of two separate dates, only one date may be represented using the "circa" attached. Circa is a Latin word meaning "about," i.e., approximately, and usually refers to a date. It is used when the date of an event of somebody is not exactly known. The word is used before the date that is not precisely known. Circa may be used in full or abbreviated as ca, ca., c., c, cir., circ., cca. Thus, Galen's existence on Earth could be written as 129–c. 210.

The Helenians theorized that food when taken is changed in the stomach to chyme and then to basic systemic fluids—blood, mucous, lymph, and bile [10, 11]. The father of medicine, Hippocrates (460–377 BC) who is reported to have introduced the term "pepsis" to mean digestion, may have advanced the view on digestion that set the stage for initial scientific exploits. He performed series of experiments on different parts of the organism, including the esophagus [12]. Other philosophers and scientists that lived during the period also made significant contributions to the subject of digestion. The Greek philosopher and scientist, Aristotle (384–322 BCE), born in the Macedonian city of Stagirus, contributed substantially to the understanding of digestion. He wrote several books on science to mark his contributions: History of Animals, Parts of Animals, Generation of Animals, Motion of Animals, Progression of Animals, Parva Naturalia, and De Anima [13, 14]. The numerous works produced by Aristotle generally on the life of the animal (and man) indicated that he was a fountain of knowledge. Aristotle made mention of an organ that connected with breathing, referring to the gullet and stomach, and suggested that they were necessary for food digestion. Unfortunately, at one time, he rejected the role of the stomach wall in digestion. Aristotle's vast knowledge in science is no surprise as his father Nicomachus was a physician, who worked with King Amyntas of Macedon. Thus, Aristotle must have learned and heard from his father speak about digestion. In addition, the fact that Aristotle studied under his mentor, Plato (ca. 428–347 BCE), a classical Greek philosopher and mathematician, could suggest that Aristotle must have been taught some basic concepts of life, a lesson that might have involved digestion. Similarly, Plato, who was a student of Socrates (ca. 470–399 BC), a classical Greek philosopher, could have listened to lessons on life from his mentor. He rightly identified the location of the liver and the approximate topography of the intestines. Although different philosophers and scientists had their views on science and life generally at the time, some of the concepts raised by Aristotle, Socrates, and Plato must have set the stage for other philosophers and scientists for the initial development of physiology of digestion. Unfortunately, many of their views were also tied to spiritism and entangled with misconceptions [2, 3, 13].

In the early days of the reign of science and philosophy, in antiquity, the Greeks continued to actively investigate the digestive system and terminologies were been identified for specific parts of the GI tract by the pioneer Greek philosophers. Herophilus (Herofilos) of Alexandria (ca. 335–255 BC) described the salivary glands (organs located around the mouth mostly involved in secretion of fluid of varying composition for the purpose of effective digestion) and named the first part of the small intestine following the stomach, duodenum because its length was 12 Greek measures. Herofilos was actually born in the Greek town of Chalcedon, Asia Minor, and schooled under the guidance of Praxagoras, a renowned anatomist and physician who taught at the Hippocratean medical school on the island of Cos (Kos). After completing his education, Herophilus moved to the city of Alexandria to practice medicine, during the reign of the first two Ptolemaic Pharaohs [15, 16]. Following his return to Alexandria, surprisingly, Herophilus abandoned the principles of his teacher Praxagoras. He rightly rejected cardio-centrism and followed the Hippocratic teachings on the soul and body when he proposed that primary parts of the human body are perceptible by the senses. Obviously, he was a proponent of the Hippocratic encephalo-centrism [17]. It was in Alexandria that Herofilos became well acclaimed for his discoveries. It was Herofilos who became the first to accurately describe the liver and its structure (The liver is a large gland located in the upper right quadrant of the abdomen, just beneath the diaphragm, and takes part in digestive and non-digestive functions). He was also the first philosopher and scientist to have investigated the pancreas, a glandular organ of digestion having endocrine functions. Herofilos might have been positively influenced by the city he lived. The city of Alexandria, Egypt, at the time of Herofilos, had the largest book repository in the world. Being one of the great physicians of Antiquity and regarded as the father of anatomy, Herofilos made phenomenal discoveries not only in anatomy, but also in physiology. His close rival in terms of intellectual capabilities may be Andreas Vesalius (1514–ca. 1564) who is regarded as the founder of modern human anatomy (see contributions of Vesalius below) [15, 16].

Another Greek philosopher and scientist who reigned during the time of Herofilos, Erasistratos of Keos (ca. 310–250 BC), proposed that after food intake, some processes occur in vivo because of the sounds of peristalsis during active digestion of food substances. Besides, he noted that food substances pass into the blood, and they could get to other parts of the body. This genius suggestion by Erasistratos may have been the first on food absorption in the GI tract [2, 18].

The ancient Chinese tied their belief to the tastes of herbs, which indicated the value man placed on food quality and digestion of food. However, food, and if at all they knew anything about digestion, it was associated with religion and spirituality as was inherent in other cultures. However, during the Zhou dynasty (ca. 1000–256 BC), the value of physiological regulation of digestion (and physiological systems in general) must have increasingly received recognition. As part of that physiological regulation, which was, though, still widely believed to be a misery would have been how food was eaten and how it transformed, whether to the body fluid or it formed waste or spirits. During the time, organs of digestion were known, for instance, the liver; however, the Chinese philosophers of Antiquity believed that these organs were present in humans on mystical reasons. Rather than the liver being the organ necessary for digestion, they asserted the process of digestion to the functions of the spleen. In addition to the stomach, other organs such as the gallbladder, intestines, and urinary bladder were all considered to be involved in digestion. Importantly, it was believed that digestion was necessary for providing nutrients to the body and delivering waste products out of the body. Interestingly, ancient Chinese associated good nutrition to healthy living as they warned against overeating or drinking, and ingestion of spicy foods [19]. The Chinese were probably the first to have associated GI dysfunctions with excessive emotions. The Chinese believed that ingestion of herbs was necessary to treat different diseases related to digestion. Unfortunately, during the time, like in many cultures, the ancient Chinese had poor knowledge of anatomy and physiology of the digestive system, which led to the death of numerous people as some procedures that could have saved life were still considered taboo and dangerous to the existence of the spirits [19].

Widely acknowledged personalities in Indian physiology or medicine are Susruta (Sushruta) Samhita and Charaka Samhita. The Vedic philosophical teachings of these personalities formed the basis of the ayurvedic medicine (Indian traditional medicine), which is considered to be one of the oldest known systems of medicine in the world today. Globally acknowledged oldest texts in medicine were introduced by Susruta and Charaka. Susruta, a ca. 600 BC physician, provided important anatomical information in the sixth century BCE; however during this era, in their vicinity, more attention was given to anatomy, rather than regulation of physiological systems [20]. Nonetheless, Susruta advocated moderate exercise for ancient Indians because it increased digestion, suggesting that digestion, they acknowledged, was important for health [21]. Charaka, a physician and physiologist, relied heavily on physical examination and direct observation to address human maladies, indicating that Charaka had a relatively good background of body systems and their functioning. In fact, it was Charaka that introduced the concept of digestion and metabolism in India. He emphasized the importance of not only cleanliness, exercise, and lifestyle in disease treatment, but also a healthy diet, which indicates the value and above all the knowledge background of the people at the time on digestion [22].

A review of physiology of antiquity will not be complete if the name and contributions of one of the Great physiologists, Claudius Galen (AD 129–216/210) of Pergamum (now Bergama, Turkey), are not mentioned. This is particularly important because, besides, his contributions, the period of antiquity ends with the death of Galen in circa 210 CE. In addition, Galen's knowledge and school of thought formed the basis of the Galenical school of physiology and medicine, which was the leading academic and research institution at the time. Unfortunately, the death of Galen marked the end of Galenical school of physiology and medicine. This renowned Greco-Roman physician and physiologist pointed out that digestion started in the stomach from where food substances were passed into the intestines (bowel). Further, Galen proposed that in the bowel, these food substances are decomposed and are transported to the liver through blood. This view of digestion was obviously from a genius. As we now know today, Galen was very close to the truth even though there was no useful experimental tool at the time that could provide proof of his hypothesis. He later, however, wrongly attributed that some mythical forces (such as animal heat), based on the initial Hippocratic view, were involved as a driving force in digestion. Another false attribution to digestion made by Galen was that the content of food eaten determined the volume of blood. I suppose Galen would have proposed that food substances consumed increased the concentration or level of corresponding nutrients in the blood. Of course, it would have been difficult for anybody to propose, for instance, that carbohydrate food increased the blood level of sugar. Philosophers and scientists at the time knew nothing about food nutrient types (such as sugar, protein, lipids) and blood sugar have not been discovered at the time. However, Galen's proposals laid the background for productive studies into the process of digestion [10]. Galen also proposed that food digested to produce heat—referred to as Galen's thermal theory of digestion. Some teachings of Galen lasted for about 1300 years until the Renaissance [23].

1.5 Digestive Physiology During the Renaissance

The Renaissance period (Renaissance is a French word meaning new birth) was marked with changes in culture, politics, diplomacy, increased scientific experiments in Europe. The period spanned from the fourteenth to the seventeenth century. Scholars of history suggest that this period was a connecting period between the middle and modern periods. The modern history (period) started after the postclassical era (middle period or middle ages) and after Renaissance period. The period must have started during the sixteenth–seventeenth centuries. The modern period is sometimes divided into the early modern period and the late modern period after the French Revolution and the Industrial Revolution [24, 25].

The Renaissance period opened the way for everybody to learn, invent, and borne new ideas for societal benefits, as opposed to the middle period when only a few (especially religious heads and societal leaders) were privileged to learn. Although there is no agreed consensus on the specific name or duration of this period, it may represent the period that starts after the fall of Rome in 476.

Renaissance must have closed with the discovery of America by Columbus or other notable events that took place in various parts of the world, including after the fall of Constantinople around 1453 [24, 25].

Contrary to antiquity, some people during Renaissance thought that digestive system did not have innate spiritual forces since it was filled with impunity. The period was marked by outright rejection of some of the teachings of Galen. For instance, Galen's century-old humoral and thermal theories on the functioning of the human body were rejected by evolving scientists. In some cases, Galen's view on digestion was completely debunked and considered a thing not worthy of discussion among the religious people, whereas in other situations, anybody who challenged Galen's theory on digestion was worthy of facing a tribunal for punishment and possibly sentencing to death. Some of the heads of the religious faith did not believe in the new science that was emerging during the period. Norman Guitmund (c. 999–1095), Bishop of Aversa, a Benedictine monk who was an opponent of the teachings of Berengar of Tours and never accepted Galen's theory of digestion. Guitmund fervently argued that religious ceremony such as the Lord's Supper performed at the alter is not subject to the laws of bodily digestion with respect to normal bread and wine. He completely rejected the notion that the sacraments can corrupt or decay as he argued that Christ can never corrupt nor decay, for Christ is the food of eternal life [26]. Like other religious groups, the Muslims outrightly punished individuals who proclaimed ideas of authorities such as Galen. Thus among the people of faith, only those teachings of their spiritual leaders were allowed and anybody seen proclaiming self or the teachings of any philosopher or scientist not of their faith was considered an evildoer, liable for punishment or death. In addition, the series of wars that occurred in different parts of the globe also hindered scientific developments as many works were destroyed and scientists killed. So, history of digestive system would not be complete if these events that caught short the lives of scientists or that must have wasted valuable works are not mentioned.

Progress in science and technology generally was greatly hindered during AD 200–1500 because of different cultural and religious ideologies. In fact, many writings of scientists were destroyed or burnt, and the life of some was truncated. These scientists died because of heresy as their opinions were at variance with the official or orthodox teachings [27]. Both scientists and physicians suffered a lot from the religious sector and government. Some scientists were imprisoned, sent into exile, guillotined, banished, blinded, beaten, shot, burned, for opposing the teachings of top authorities [28]. The great anatomist and physician from Spain, Andreas Vesalius (1514–1564), who discovered pulmonary circulation of blood (also known as lesser circulation) together with other scientists were tortured, and some copies of their books burnt to dust and charcoal. Although saliva was widely known to be a secretion of the mouth, the secretions of stomach and intestines were only been speculated upon during the early days of scientific endeavors, which was probably due to the difficulties involved in researching the gastric content since technical knowledge at the time did not allow investigating gastric secretion in vivo. Subsequently, Andreas Vesalius (1514–1564) conducted one of the most notable dissections that exposed the human body structure; chyme was noted to be present in this dissection. In his book, *De Humani Corporis Fabrica*, published in Basel in 1543, which formed the basis of modern anatomical research and medicine, also had numerous details and was accurate conserving the encephalocentric teachings of Hippocrates. The chyme was thought to be formed from ingested food and gut secretion. However, he did not report on the secretions of other regions of the gut. Marcello Malpighi of Bologna (1628–1694), a globally acclaimed anatomist and leading pioneer in liver, kidney, and spleen research, was subjected to physical assault by his foes and people who strongly disbelieved his scientific findings. It should be noted, however, that scientist received tribulations even from their colleagues. For instance, Johann G. Wirsung (1600–1643), a Bavarian monk and physician, who discovered the excretory duct of the pancreas in 1642, was tortured and later killed by his university professors and fellow physicians [2, 28].

While Galenical school of physiology and medicine crumbled in certain parts of the world, some scientists even leading scientists continued to retain the school of thought of this most renowned Greco-Roman physiologist of antiquity. Surprisingly, however, a few scientists who rather than proclaiming, opposed the teachings of Galen, were punished and even killed [28]. Thus, the teachings of Galen, arguably, were considered to be sacred and received various levels of acceptance across the globe during the Renaissance.

The Italian physician and surgeon, Guido Lanfranchi (1252–1315), also known as Lanfranc of Milan (ca. 1250–1306), Archbishop of Canterbury and student of the Italian surgeon and cleric Guglielmo da Saliceto (also called Guilielmus de Saliceto, William of Saliceto or Guillaume de Salicet, ca. 1210–1277), was outspoken critic of Galen school of thought on the thermal theory of digestion. Lanfranc was among the first scientists that received torture due to differences in political ideologies. He escaped persecution in Italy and traveled to Paris to settle in 1295, where he earned respect as the best surgeon in the country at the time [2, 28].

In Asia, pro-Galenic and anti-Galenic scientists and philosophers formed two strong opposing groups across the continent. Anti-Galenist scholars were mostly the religious heads, whereas the pro-Galenists were more of the scientists and physicians. The well-acclaimed Persian physician Avicenna (980–1037) was in concordance with Galen's view on digestion. Avicenna was a leading scientist in Asia and became renowned throughout the world. He made numerous contributions generally in medicine [29–31].

The theory of digestion flourished in Alexandria (Egypt) during the fourth–fifth centuries. More importantly, a crucial link between digestion and extragut organ (precisely the kidney) functioning was documented for the first time. Alexandria, being a city of academicians that housed some of the biggest libraries in the world during the time, produced renowned scientists on kidney functions and diseases, who taught digestion and urinary physiology (and medicine in general). The ancient nephrologist, Magnus of Nisibis (or Emesa), who taught medicine in Alexandria, at one time, may have pointed the link between food intake and urination [32]. This link will later be called the gut–kidney axis, which is discussed in Chap. 15 of this book. As globally recognized as the city of knowledge at the time, where numerous learned men lived, Alexandria had numerous resources for intellectual excellence. During the reign of the Alexandrias, Byzantine physicians (fourth–twelfth centuries AD) also made some observations on digestion [33]. The Byzantines were mostly interested in craniomaxillofacial medicine and surgery; they studied the mouth, the first region where mechanical and chemical digestion took place. They also studied the salivary glands as an accessory organ of digestion [33].

As a noted trend in the development of knowledge, many findings in the subject of digestion were made by physicians and scientists at the time as they searched for the causes of digestive problems while trying to provide solutions to sick people. The physician, Caelius Aurelianus (ca. AD fifth century) of Sicca Veneria, Numidia (now in Algeria), wrote many volumes on "*De morbis acutis et chronicis*" ("Concerning acute and chronic diseases") [34]. However, his works are largely adapted from those of Soranus of Ephesus (ca. 98–138 AD), a lucid Greco-Roman physician and second-century leader of the Methodist school of medicine. Ephesus was an ancient city in Ionian (Greek) Asia Minor, now western Turkey [35]. Another intersystem functional relationship was thought to exist between digestion and mental consciousness (central nervous system) by the Greek scientist. Although epilepsy (a disease involving the central nervous system due to disorders in neural pathways characterized by seizures and loss of conscious) was noted by the Assyrians and Babylonians circa 2000 BC, Hippocrates became the first to dispute that the disease was a divine malady or demonic possession, and notably, Caelius Aurelianus and Soranus of Ephesus genius proposed that patients should be treated with remedies that help digestion, pointed to a possible link between digestive and brain functions, even though it was not completely known at the time that epilepsy was a disease of the brain [34].

Religious heads that had a vast knowledge of science used scientific experimentations to strengthen their spiritual teachings and doctrines, which made their followers not only to have a better understanding, but also believed in the teachings of their leaders. The Italian surgeon, Theodoric Borgognoni (1205–1296), who later became Bishop of Bitonto in 1262 and then Bishop of Cervia appears not only to have made contributions to blood circulation. He was reported to have made illustrations that included the structure of the digestive tract of humans. His physiological understanding of the human body and ability to provide illustrated answers helped his followers had better understanding of his teachings and doctrines [36].

The mode of communication and information dissemination significantly hampered the growth of science during this period as knowledge was spreading slowly during this time, and results of experimentation in one region did not usually get to the people in other regions of the world. However, with a gradual increase in the number of travelers and number of people interested in science, there was a marked improvement in understanding of the human bodily functioning. Unlike nowadays where information can get to anyone located in any part of the world in seconds or milliseconds, information was disseminated during the time especially through traders and people who sailed around the world to find out what was happening in other regions. Movable-type printing was only introduced in 1450 by the German,

Johannes Guttenberg (1395–1468) [28], which significantly increase dissemination of information by the publication of numerous texts in arts, science, and medicine. The first medical book was printed in 1478 and contains the works of many renowned physicians and scientists. This exclusive discovery of printing in Germany would give an opportunity for the printing of the first physiology textbook in the world, which will also aid better understanding of the digestive system.

The period was also marked by freedom of expression and fast advancement of science, especially in Europe. Some acts that were previously considered a taboo could be executed on the premise of scientific exploits and advancement. The first public postmortem dissection of two human bodies in 1315 in Bologna, Italy, significantly improved hypothetical understanding of the human body, including the digestive system and related organs [28].

As philosophers and scientists became more aware of the structure of the digestive tract, numerous proposals were made on the functions of the tract and associated organs. Unfortunately, scientists in those days had a very little or no chance in studying the functions of digestive tract as technological progress could not allow for groundbreaking studies in science. However, the possibility of increased need for experimentation grew as proofs were required for the numerous hypotheses made on bodily functions. In 1497, Alessandro Benedetti (circa 1450–1512), an Italian anatomist and physician, thought that the digestive tract, described to comprise of the stomach and intestines, was impure and as a result did not have connection with the seat of mind. He placed huge importance to experimentation, rather than valuing the words of authorities such as Aristotle or Galen. At this time of history, precise knowledge on the structure of the digestive tract has not been formed. Obviously, scientists needed to first understand the structure of the digestive tract before achieving meaningful understanding on the functions of the tract. Further information on the structure of the digestive tract was provided by the Italian physician, Jacopo Berengario da Carpi (also known as Jacobus Berengarius Carpensis or Jacopo Barigazzi or Giacomo Berengario da Carpi) (1460–1530). During the fifteenth century, Jacopo Berengario da Carpi noted that the anus was linked to the intestines and the mouth was linked to the gullet [2, 37]. His suggestion provided one of the first precise anatomical overviews of the GI tract. Thus, by the description of da Carpus, the structure of the "gut" was near complete and now known as a tract that connects the mouth to the anus.

A well-pronounced personality at the time, Leonardo da Vinci (1452–1519), was an Italian polymath, painter, sculptor, architect, musician, mathematician, engineer, inventor, anatomist, geologist, botanist, and writer, who also contributed to the anatomical understanding of the GI tract. His numerous contributions to anatomy and the drawings that show different structures, organs, and systems also included the digestive system. Leonardo da Vinci suggested that the digestive system helped respiratory functions. This genius suggestion, as we now know is the gut–lungs axis, which is useful in respiration (and more importantly, the two organs functionally support each other). This unique relationship will be briefly discussed in later part of this book. Interestingly, da Vinci argued that digestion involved not only the organs but also the abdominal muscles because they contracted and relaxed. He was probably referring to the role of abdominal muscles in respiration. Unfortunately, he wrongly attributed it to the digestive system. It is possible that rather than mentioning the abdominal muscles, he meant to say stomach muscles, when he was talking about the role of the muscles in digestion [38–40].

The Renaissance period may have ended with the rise of new concepts of understanding of the digestive functions; notably, iatrochemical and iatromechanical views of digestion were borne during this period (ca. 1500). The German physician and alchemist, Paracelsus, born Philippus Aureolus Theophrastus Bombastus von Hohenheim (1493–1541), was keen to observation and could trace that some diseases including GI-related malfunctions were related to psychological dysfunctions. As a pioneer of physiological chemistry, Paracelsus thought that the stomach was functioning as a chemical laboratory within the body. Thus, he tried to use alchemical theory to explain digestion. Paracelsus noted that the content of the stomach was acidic. But he incorrectly reasoned that the acidity of the stomach was due to drinking of acidic spare water, which he called "acetosum ensurinum," meaning "hungry acid." Paracelsus further mentioned that the stomach acid was catalytic in action, necessary to prevent the formation of precipitations. Paracelsus and the work of other alchemists laid the foundations for discovering the exact chemical nature of this hungry acid. Around the same period, the German Alchemist, Johann Thölde (1565–1614), was probably the first to identify the hungry acid as hydrochloric acid (HCl) [9]. The iatrochemists also found other types of fluid in the GI tract. The Dutch physician, anatomist, and physiologist, Regnier de Graaf (1641–1673), was a classical iatrochemist, who studied the chemistry of digestion and the action of pancreatic juice. In 1664, De Graaf suggested that the juice obtained from pancreas was alkaline in nature [41]. However, the nature of GI juice remained to be fairly understood. In 1826, the German physician, anatomist, and physiologist, Friedrich Tiedemann (1781–1861), and the German chemist, Leopold Gmelin (1788–1853), became the first scientists to demonstrate the alkalinity of pancreatic juice, which was necessary for adequate digestion in the intestine. At age 35 years, Tiedmann became a professor of physiology and anatomy. Gmelin, a German chemist professor, conducted research not only in chemistry, but also in physiology. He worked on the nature of gastric acid by conducting several in vitro experiments. Furthermore, Tiedemann and Gmelin showed that the mixing of pancreatic juice and bile enhanced digestion of proteins. In 1835, the Swedish Uppsala University-trained physician and chemist, Jöns Jakob Berzelius (1779–1848), introduced the concept of catalysis as a necessary process for the formation of biomolecules. For Tiedemann, Gmelin, and Berzilius, there was no place for a vital force or spirits in digestion [2, 42–44].

As opposed to the alchemists, the iatromathematicians (also called iatromechanists) made some contributions to the understanding of mechanical aspects of digestion. They reasoned that digestion is an act of trituration (i.e., reduction of substances to smaller pieces by mechanical grinding/mixing). Their principles of digestion and medicine as a whole were based on mechanics, as they paid little or no attention to the role of chemistry in physiology or medicine. Pioneer iatromathematicians were the French philosopher and scientist, René Descartes (Latin—

Renatus Cartesius) (1596–1650), the Italian physiologist, physicist, and mathematician, Giovanni Alphonso Borelli (1608–1679), and the Dutch physician and physiologist, Herman Boerhaave (1668–1738), as well as Andreas Vesalius (1514–1564). Vesalius notably described digestion as a mechanical process and debunked Paracelsus's suggestion that digestion was a chemical process. So, at this point, at least three types of digestion were recognized: chemical, mechanical, and mixed (mechanochemical) [45–49]. While the iatromechanists exclusively believed that bodily functions could be explained with mechanical concepts alone, another school of thought believed that bodily functions including digestion could be explained by both iatrochemical and mechanical theories and not one alone. These iatromechanochemists were the British scientists. These different schools of thought on digestion spanned from the sixteenth to the early decades of the eighteenth century [50].

At this point in history, the digestive system and the possible processes occurring in vivo during food ingestion were becoming clearer. It should be borne in mind that the numerous views on the digestive system did not change; however, fundamental concept or level of knowledge on digestion gradually improved over time. At a certain period, there was relatively huge spread of information on digestion and science and art in general. Although knowledge on digestion spreads during time of Christopher Columbus (1450–1506), particularly due to commerce, colonization, and exploration, the information was accompanied with misconceptions and mainly composed of philosophical reasoning or mythology [51].

1.6 Digestive Physiology in Modern History

Understanding of digestion in the modern history did not mean that humans had modern views of digestion. It only depicts the period of the history of human development. The start of modern era was accompanied with substantial development in physiology and science in general. The first part of the seventeenth century emerged with increasing emphasis on hypotheses and experimental data [10, 27]. This period is regarded by some authors as the beginning of modern physiology [10, 27]. During this period—the seventeenth century—a Flemish chemist, physiologist, physician, and follower of Paracelsus, Jan Baptiste Van Helmont (1580–1644), provided the first chemical explanation of digestion, representing a key transition from alchemy to experimental chemistry. He not only showed that digestive agent in the stomach was a specific type of an acid, as opposed to the hungry acid, but also proposed certain chemical agents in the stomach juice were necessary for digestion. According to one of his theories on fermentation, Van Helmont proposed that ferments were chemical substances responsible for digestion and all physiological processes in the body. However, as it was later known, many works of other scientists on physiological systems did not at all confirm the reductionist thinking of Van Helmont. Consequently, Van Helmont's theory was ignored at the time. However, accumulating research evidences in the contemporary world over the past decades have undoubtedly shown the presence of fermentation in intestines by resident microbiota—population of microbial commensals in the intestines that are capable of digesting food substances where the organism's enzymes cannot digest. Van Helmont is also credited for discoveries in other fields. He introduced the word "gas" and also identified carbon dioxide (gas silvestre), carbon monoxide, nitrous oxide, and methane. Given his contributions to physiology and particularly digestion, the theories of Baptiste Van Helmont must have marked the beginning of digestive physiology [52].

A proponent to the fermentation theory of digestion, Franciscus de la Boë Sylvius (1614–1672), a renowned professor of medicine in Paris and Leida (France), during the seventeenth century, agreed with postulator of the modern theory of intestinal fermentation; however, he did not agree with Baptiste Van Helmont that all physiological processes took place under fermentation. A key contribution of Professor Franciscus Sylvius was that he became the first to describe the digestive ability of saliva and pointed that oral cavity is the first place of digestion. The liquid "saliva" was known before this French scientist, but its digestive function was not previously known [53, 54].

As knowledge increased, there was also continuous progress in all areas of science and digestive physiology had a fair share of the progress in knowledge. As the chemical understanding of the nature of digestion increased, different aspects of digestion including mechanism of movement of food in the intestine and the anatomical structures that regulate this movement became increasingly investigated. The understanding of scientists also grew in the aspect of tissue-level organization of the digestive tract and their functioning. For the first time in 1605, the Swiss anatomist, botanist, and physician, Caspar Bauhin (1560–1624), described the ileocolic valve and correctly explained its function of preventing the intestinal contents from coming back to the small intestine from the colon. So, valves (also called sphincters) were known as anatomical structures that regulate the movement of food substances and release of juice into the intestine in the different regions of the GI tract. Several other valves were also identified in the GI tract. During the first half of the 1600s, the German anatomist, Johann Georg Wirsung (1600–1643), discovered the pancreatic duct in man and showed that it was necessary for the transport of pancreatic juice into the intestine, though details on the anatomy and physiology of the hepatopancreatoduodenal axis were not known [2, 51, 54–56].

In 1654, the Cambridge-trained physician, anatomist, physiologist, pathologist, and colleague of Willaim Harvey, Francis Glisson (1597–1677), reported outstanding findings on one of the large organs of the GI system. After completing his medical degree at the University of Cambridge, and following a period of scientific research and teaching, Glisson was appointed professor of physics in the same university, where he conducted several experiments on the nature of things. His deep interest in medicine made him conduct many experimental studies on the human body, particularly on the organ of the GI system—the liver. His investigation on the liver is still considered as one of the most detailed studies of the human liver in the modern period. Hence, the liver fibrous sack is named after Glisson [2, 51].

Apart from the large glands of the gut (liver, pancreas, gallbladder, salivary glands), scientists began to identify glands that were much smaller in size and located in the mucosa of the intestinal tract. Duodenal glands were identified and described by the Swiss physician, anatomist, and physiologist, Johann Conrad Brunner (1653–1727). These glands were later named after the discoverer—Brunner's glands. Brunner's glands are comprised of tubular submucosal glands found in the duodenum, precisely the part, which is above the hepatopancreatic sphincter (sphincter of Oddi). Following the discovery of one of the first minor glands in the duodenal mucosa, another gland type was identified in the intestinal mucosa. In the middle eighteenth century, Johann Nathanael Lieberkühn (1711–1756) reported the observation of mucous glands in the small intestine, which were later called intestinal glands or crypt of Lieberkühn [10, 51]. The intestinal glands were found to have similar functions with duodenal glands. Further details on these glands are discussed in later part of this book.

In the second half of the seventeenth century, the Swiss anatomist, Johann Conrad Peyer (1653–1712), conducted studies that identified groups of intestinal cells, now called Peyer's patches—aggregates of gut-associated lymphoid tissue usually found in the lowest portion of the small intestinal mucosa. These patches will be later identified to be responsible for gut immunomodulatory functions [51]. Details on the immunomodulatory functions of Payer's patches are discussed in Chap. 10.

The foremost contributor to the tissue level of organization of the GI tract was William Harvey (1578–1657). Born in England as the eldest of seven sons of a farmer, Harvey studied arts and medicine at Cambridge University, where he received a Bachelor of Arts degree in 1597 and in 1602 earned his medical degree from the medical school at Padua, Italy. His younger brothers became London merchants. On receiving his medical degree, William Harvey returned to London to practice medicine and also conducted research [57, 58]. He wrote on human organs in his widely renowned book "Lectures on the Whole of Anatomy," published in 1653. Even though the world knows him mainly on his outstanding work on cardiovascular physiology, this far-famed physiologist was the first to describe the layers of the intestines as tunics, comprising of fibers, flesh, parenchyma, veins, arteries, mesenteries, and mucous. His estimate showed that the intestines were about six times the length of the human body [57, 58].

Pioneer fields (e.g., iatrochemistry, iatromechanics) that directed the flow of knowledge in chemical and mechanical levels of functioning in GI physiology received substantial progress as a new discipline, iatrophysics, which was applied to further explain the mechanisms of GI functioning. Outstanding iatrophysicist scientist in the emerging field of digestive science included René-Antoine Ferchault de Réaumur (or just René de Réaumur) (1683–1757), a French physiologist who showed that digestive chemicals played a huge role on the processes of digestion. De Réaumur isolated gastric juice and demonstrated that digestive secretions could digest meat outside of the body. This discovery not only confirmed the work of Eberle, but also opened the way for future nineteenth-century scientists to investigate the protein digestibility in the digestive tract, which will later lead to the discovery of digestive enzymes. De Réaumur's book on digestion in birds printed in 1752 unraveled many secretes of digestion. As an iatrophysicist, De Réaumur believed that digestion was the result of churning process induced mechanically by the muscles of the stomach wall [51, 59]. Rather than explaining digestion in chemical terms alone, De Réaumur viewed the process of digestion as physical and chemical.

Albrecht von Haller (1708–1777), an eminent Swiss anatomist, physiologist, and student of Herman Boerhaave, published the first textbook of physiology in 1747 in the city of Berne, Switzerland. He later published some volumes of *Elementa Physiologiae Corporis Humani* (Elements of Human Physiology). The text also contained a description of digestion incorporating previously known concepts [2, 51, 60]. As a supporter of iatrochemical school of thought, he wrote in his book that acid in the stomach was produced from the degradation of food. He also spelt out the fact that bile was produced from the liver. Further, von Haller mentioned that bile was important for fat digestion [61]. Even though von Haller realized the importance of secretory activity of the GI tract, he was intellectually handicapped to discuss the matter [62, 63]. Besides digestion, von Haller made contributions to future understanding of the nervous, circulatory, and respiratory systems. He acknowledged the tendency of muscle fibers to contract when the attached nerve is stimulated. In the absence of nerve stimulation, no contraction took place. Thus, he showed that only nerves can transmit sensation [60]. In this same year that von Haller's Elementa Physiologiae was printed, Fredericus Bernardus Albinus (1747–1770), the younger brother of Professor Bernhard Siegfried Albinus (1721–1745), showed that all swallowed substances are transported into the stomach via the esophagus [12], a confirmation of views that was previously identified by other scientists.

The teacher and mentor of von Haller, Hermann Boerhaave (1668–1738), a Dutch physician and pathologist, was one of the most famous physicians of the eighteenth century, teacher and modernizer of medical education and a follower of the school of thought of Hippocrates [64–66]. Boerhaave in 1724 was the first to describe life-threatening non-iatrogenic spontaneous esophageal perforation and the rupture which usually occurs in the left postero-lateral wall of the lower one-third of the esophagus that subsequently formed mediastinal sepsis. His observations were based on his personal clinical and autopsy investigation. This pathological condition is now called Boerhaave's syndrome [64, 67]. Boerhaave's syndrome is barogenic rupture and results from a rapid rise in intraluminal pressure in the distal esophagus. The syndrome is characterized by forceful vomiting, pain, dyspnea and shock. Bleeding occurs but is not usually massive. Barogenic rupture is one of the causes of esophageal perforation. About nine out of ten cases of esophageal rupture usually occur in the anatomically weakest point of the esophagus—the lower one-third and in the left lateral position. The average length of tear is usually about 2.2 cm, lying 3–6 cm above the diaphragm [64]. Some authors even argue that Boerhaave's investigations represent the birth of pathological physiology, the branch of physiology or medicine that deals with the mechanisms of disease development [62]. In 2006, Adams and coworkers [68] published a work and proposed that the name of the Dutch countryman, Baron Jan Gerrit van Wassenaer heer van Rosenberg (1672–1723), from whom the first report of esophageal rupture was made, be incorporated into Boerhaave's syndrome (Boerhaave-van Wassenaer's syndrome). This suggestion may be arguably correct and, though, yet to receive wide recognition.

This era of history was marked by substantial advancement of science as certain procedures that were previously feared and even considered a taboo, which led to the loss of numerous lives, were performed with great success. The modern period had progressive knowledge improvement compared to the period of Renaissance. Many ailments of GI tract were treated with success. Notably, during the eighteenth century in Greece, GI stomas (artificial connections of the gut to the skin; stoma is derived from the Greek word, "stomat," meaning "mouth"), colostomies, designed to relieve digestive obstruction were largely used in surgical operations to treat inflammatory bowel, hepatobiliary obstructions among other conditions. Gastric stomas were introduced to decompress the gut or to deliver food substances [69].

By the mid-end of eighteenth century, the Italian philosopher, Abbe Lazzaro Spallanzani (1729–1799), who became a chemical physiologist and pioneer in modern experimental biology, had produced many diagrams of human digestive system, in which he accurately showed the digestive tract and associated organs. In the middle eighteenth century, he published many scientific papers. Spallanzani was a reductionist, as he believed that many phenomena in nature (including humans) could better be explained in physical or chemical terms. Apart from other areas of science, he notably contributed to respiration. To test his numerous hypotheses that he had formulated, at one time he used animals and another time conducted experiment on himself by swallowing sponges, porous bags, and tubes (note that nowadays there are ethical principles formulated by Research/Ethics Committee of an established institution, based on guiding principles of the Helsinki Declaration that regulate experiments on animals or humans or their tissues). Spallanzani was fortunate to have recovered some of the items he had previously swallowed from his feces or vomitus. Spallanzani reportedly documented the action of saliva in digestion. He correctly showed using his "classical experiment" that gastric juice was responsible for the chemical breakdown of food substances. This renowned Italian scientist showed that gastric juice contains acid that was produced by the stomach itself and not from drinking excess acidic or spar water and did not flow from any other organ or system. As a great physiologist at the time, he proposed that the action of stomach juice on food was due to acid fermentation, not putrefaction or vinous fermentation. Thus, Spallanzani's suggestion was somewhat in conformity with Van Helmont's earlier proposal on the chemical physiology of digestion; however, it later became clearer that Spallanzani's views were incorrect. Spallanzani suggestion that fermentation occurred in the stomach and that it was due to the presence of an acid was not correct as the stomach could have not been necessary for the fermentation process. However, he rightly reasoned and showed that rather than being a process of putrefaction, digestion was a fermentation process. Frankly, during the process of digestion, some processes of fermentation also take place especially in the large intestine and it is carried out by the gut microbiota. Thus, he was able to show experimentally with results indicating that digestion was a chemical process. Another important discovery led by Spallanzani revealed that gastric juice with HCl was responsible for coagulating milk to form "curdled milk." However, it was not certain whether rennet (rennin), pepsin, the acid itself, or another substance in the juice was the agent responsible for the milk coagulation. Following the work of Spallanzani, numerous scientists and experimental researchers started to confirm the results of stomach acid investigations on animal models [9, 59]. Rennin (also known as chymosin) is a member of the aspartic protease family of enzymes. Rennin is an enzyme complex produced by the fourth stomach chamber of cows, called abomasum, gastric chief cells of infants, *Escherichia coli* (*E. coli*), *Kluyveromyces* (formerly *Saccharomyces*) *lactis* (*K. lactis*), and *Aspergillus niger var awamori*. In the presence of pepsin, rennin functions to curdle or coagulate the casein in milk and break the milk into a liquid or semisolid substance, allowing a longer residence in the bowels and consequently better absorption. Rennin is produced in its inactive form known as prorennin. Prorennin is activated by pepsin 1 or 2 [70]. The pH also plays a critical role in this activation. Different components of rennin have their varying activity levels at varying pH [71–73]. Because of the milk clotting and proteolytic activity of rennin, some components of rennin have been successfully produced in the industries to be used in the production of cheeses [73]. There are now available recombinant chymosin and pepsin for cheese making [74]. Milk is the emulsion of fat globules and suspension of casein micelles in water. It contains certain substances (casein) that stimulates, on ingestion, HCl secretion in the stomach, which activates prorennin to convert to rennin. Milk proteinases plasmin and cathepsin D are also bound into micelles. Milk contains the caseinogen protein that includes kappa-casein, beta-casein, alpha-casein, and gamma-casein molecules [75, 76]. Caseins (beta-casein, alpha-casein) are involved in catalyzing the interaction between calcium and milk which is necessary for precipitation and the subsequent absorption of milk. Kappa-casein affects the precipitation of milk. Kappa-casein is the substrate for chymosin. The enzyme, chymosin, cleaves the peptide bond between amino acid residues 105 (phenylalanine) and 106 (methionine) to produce calcium phosphocaseinate. When the specific linkage between the hydrophobic (para-casein) and hydrophilic (acidic glycopeptide) groups of casein is broken, the hydrophobic groups unite and form a network that traps the aqueous phase of the milk. Chymosin acts to extensively precipitate and curdle milk [77, 78].

In 1777, Edward Stevens (ca. 1755–1834), then a medical student in Edinburgh, who arrived from West Indies, performed experimental investigations as part of the criteria for award of the degree and title "physician." His thesis for the degree award involved an experimental description of the isolation of gastric juice from a dog's stomach and the subsequent action of the juice on food substances. In one of his experiments, Stevens placed meat into the gastric juice and after several hours he found that the meat had softened, whereas the meat did not soften when it was placed in water for the same number of hours. He had discovered that gastric juice can digest meat. However, the agent and mechanisms involved in these actions of the gastric juice will continue to remain a challenge for physicians and scientists for some years to come [79].

In 1832, Johann Nepomuk Eberle (1798–1834), a physician from Würzburg, Germany, found that an acid extract of gastric juice that he collected from mammals dissolved protein foods—egg white and meat. This finding was documented in his new book. At this point, it was clear that the agents necessary for digestion were present in the gastric juice. So, there was confusion about the actual agent in the juice that was able to digest the protein substances. The acidity of the gastric juice was clearly known; so many scientists thought that the digestive properties of the juice may be due to the acid content. Because many scientists did not believe in the theory of ferments, and the theory gradually faded away. Obviously, Eberle was not a follower of the theory of ferments. In a further experimental observation made by Eberle in the same year, he noted that HCl in vitro did not digest proteinous substance, but natural gastric juice did similar to the digestive properties of an acidified extract of gastric mucosa. This observation obviously indicted that a non-acid digestive component is present in the juice [80, 81]. But the task was to identify this non-acid component of gastric juice that was responsible for digestion. In a subsequent experiment, Eberle produced a concentrated solution of pepsin using the technique of alcohol extraction of the mucosa. By this method of enzyme extraction, Eberle was able to identify the substance in gastric juice that softens meat or dissolved egg white. Importantly, he also extracted pancreatic protease, which he utilized in his experiments. This German scientist could be regarded as the father of modern chemical physiology of digestion. Interestingly, he further observed using fatty substances that pancreatic juice was responsible for turning fat into a very finely divided state "known as emulsion." This pioneer scientist in GI mucosa physiology, Eberle, believed pancreatic juice aided digestion of fat and its absorption in the intestine. By this contribution, Eberle singled himself out as a modern physiologist who not only using techniques that were considered contemporary at the time, to validate the digestibility of gut juices, initially claimed to be due to the presence of ferments [2, 51, 80, 81].

In the previous year, before the publication of Eberle's book, Erhard Friedrich Leuchs (1800–1837) had first reported the starch digesting properties of saliva. By 1833, two French chemists, Anselme Payen (1795–1871) and Jean-François Persoz (1805–1868), working in a French sugar factory, discovered that alcohol precipitate of malt extract converted starch into sugar. They named the substance responsible for the conversion "diastase" and suggested it to be an enzyme. The two scientists later suggested that the last three letters of diastase "ase" be appended to the root that indicated which substance was been acted upon by the enzyme. The name "diastase" is derived from the Greek "diastasis," meaning a separation. This is because when beer mash is heated, the enzyme causes the starch in the barley seed to transform into soluble sugars and hence resulting in the separation of the husk from the rest of the seed. Diastase refers to a group of enzymes that hydrolyzes carbohydrates; they include α-, β-, and γ-amylases. The term "enzyme" was not introduced at the time of its discovery. The term was later coined in 1876 by Wilhelm Friedrich Kühne (1837–1900) [82, 83]. In a paper which Kühne presented to the Heidelberger Naturhistorischen und Medizinischen Verein (Natural History and Medical Association of Heidelberg, Germany) on the February 4, 1876, Kühne suggested that non-organized ferments called enzymes were responsible for the proteolytic activities of the intestinal (pancreatic) juice [84]. The word "enzyme" is derived from the Greek for "in yeast" or "leavened." The term "enzyme" was later applied to all ferments, not only those of yeast or other unicellular organisms, but also ferments of higher organisms, as their ferments were very similar.

At this time, even though it was increasingly known that GI juice softened starch, fat, or meat, nobody knew how it came about, and besides, while glucose was known to be probably a product of starch digestion, it was not clear the products of action of gastric juice on meat or the products of action of intestinal juice on fats. Glucose was discovered by the German chemist Andreas Sigismund Marggraf (1709–1782) in 1747, but the name "glucose" was actually given by the French chemist, Jean Baptiste André Dumas (1800–1884) in 1838 [2, 3, 51]. Before the discovery of the enzyme that break down starchy food, glucose was yet to be discovered, so it was difficult to think that the sweetened products obtained following addition of saliva or intestinal juice into starch were the hexoses of which glucose was a probable candidate.

The same situation applies to proteins (meat) and their cleavage products—amino acids. While it was reported by the German physiologist Eberle as early as 1832 that gastric juice dissolved protein foods such as egg white and meat, the intermediate or end products of the reaction were yet to be identified. Not many scientists were aware of these monomers. Moreover, it was quite difficult at the time for some researchers in the area of digestive physiology even to suspect that digestion of proteins resulted in monomers such as those already discovered. The first amino acid was discovered as early as 1806 by the French chemists Louis-Nicolas Vauquelin (1763–1829) and Pierre Jean Robiquet (1780–1840) who successfully isolated the compound in asparagus; hence, they named it "asparagine," the name of the amino acid. It should be noted that at the time there was not even a fair knowledge of amino acids in the scientific community. Not many scientists even knew about protein or the composite units of this macromolecule. The term "protein" (from Greek "proteios," meaning "primaries," i.e., the primitive or principal or the first position or rank as it appears as the most important nutrient needed by herbivores) was introduced just in 1838 by the Swedish chemist, Jons Jakob Berzelius (1779–1848). Though a few scientists including the Dutch chemist, Gerardus Johannes Mulder (1802–1880), must have had a fair knowledge of this macromolecule, at this time of history, virtually nothing was known concerning its building or composite blocks. Berzelius reportedly had series of communication with Mulder, which suggested that Mulder had a fair understanding of what proteins were during the time. Moreover, Berzelius's letter to Mulder meant that the Swedish chemist must have known of his Dutch counterpart earlier and probably read one of his papers on protein. Furthermore, in one of the reply to Berzelius, Mulder noted that proteins were important for the maintenance of chemical metabolism. So, arguably, protein was discovered by both Berzelius and Mulder. Subsequently, series of investigations carried out by different laboratories in the world resulted in the isolation of proteins and discovery of proteins as principal sources of the animal body [85]. Following the discovery of asparagine, other amino acids were continuously added to the list. One of such amino acids includes cystine, discovered in bladder stones (urinary calculus) in 1810 by the English chemist, William Hyde Wollaston (1766–1828). The dimer, cysteine, was discovered in 1884 by Baumann E. Cystine is the amino acid formed by the oxidation of two cysteine molecules that covalently link via a disulfide bond. Cysteine was discovered in proteins by Karl Axel Hampus Mörner (1854–1917) in 1899. The structure of cysteine was first defined in 1902 and then confirmed in 1903 [86–88]. By 1935, all the 20 amino acids that are contained in proteins have been discovered. William Cumming Rose (1887–1985) discovered threonine, the last of the 20 amino acids universally present in proteins to be identified [89]. Following the discovery of the genetic codes, two other natural amino acids have been discovered. For a review of the history of the discovery of other amino acids used in making proteins of our body, see Wieland and Bodanszky (2012) [90]. These amino acids are primarily used for making the proteins accounting for about 20% of the body weight in humans. It became a serious issue for science to deal with, during the beginning of this millennium, when it was reported in 2009 by NASA (http://www.nasa.gov/), following analysis, that they have discovered traces of amino acids in extraterrestrial bodies (meteorites, comets) that entered into the Earth from outer space. Many amino acids including glycine and alanine that make up protein of humans were discovered in fossils and remains of comets that entered our Earth. This raised a dozen of questions whether or not there is life in extraterrestrial bodies other than the Earth or in other galaxies in our Milky Way. The branch of science that has been dedicated to solving this scientific problem is "astrobiology." This area of science studies the basic building blocks of life and their presence in extraterrestrial bodies. Using imaging techniques with satellite-equipped devices, some traces of elemental building blocks of life have been discovered in extraterrestrial bodies, although some scientists and the religious sect have been opponents to this discovery as identification of elemental blocks of life might not mean that life exists there.

The great Jan Evangelista Purkyně (also called Johannes Evangelist Purkinje) (1787–1869), a Czech anatomist, physiologist, and physician, became one of the first to identify the end product of food digestion in the GI tract. Purkyně observed that a mixture of fat with pancreatic juice together with bile not only leads to emulsification but also splits the emulsified fatty molecules. By this discovery, he had set a positive prospect for other scientists to investigate the end products of digestion of other food substances. Purkinje also observed in 1836 that pancreatic extracts possess proteolytic properties. Purkinje made enormous contributions to physiology in general and was very famous in and outside Europe. He was the one who named the fluid content of the cytoplasm of the cell "protoplasm" in 1839. The tissues located in the inner ventricular walls of the heart, just beneath the endocardium in a space called the subendocardium, are named after Purkinje "Purkinje fibers." These fibers are specialized in conducting action potential originating from the sinoatrial node to the muscles of the heart and allow the electrical conduction system of the heart to create synchronized contractions of its ventricles. The synchronized contractions are essential for maintaining a consistent heart rhythm. These fibers do not set heart rate but are influenced by electrical discharge from the sinoatrial node. Purkinje fibers carry the electrical impulse from the left and right bundle branches to the myocardium of the ventricles during ventricular contraction portion of the cardiac cycle. This spread of impulse causes the muscle tissue of the ventricles to contract to produce a force required to eject blood out of the heart. If embarrassment of upstream conduction or pacemaking ability occurs, Purkinje fibers can fire impulse at a rate of 15–40 beats per minute. Purkinje is accredited for discovering dark adaptation that was later called Purkinje effect in 1819. Dark adaptation is the tendency for the peak luminance sensitivity of the human eye to shift toward the blue end of the color spectrum at low illumination levels [2, 3, 51, 91]. In administrative sphere, this very great nineteenth-century German physician and physiologist, Purkyne, founded the first physiological institute in Germany. Many renowned physiologists including Johannes Müller, Karl Ludwig, and Herman Helmholtz attended this institute [2, 3, 51].

In 1844, Gabriel G. Valentina (1810–1883) printed his physiology book in which he treated the mechanisms of digestive process by describing diastatic properties of pancreatic juice. He reported that starch (potato) when exposed to the juice became sweetened. Valentina later identified the sweetened substance as glucose, thanks to the earlier discovery of glucose. A year later, a similar observation was made by Apollinaire Bouchardat (1806–1869) and Claude Marie Sandras (1802–1856) [92]. Bouchardat was a great pharmacist, physician, and biochemist and is regarded as the founder of the field of diabetology. He was a specialist who provided advice on diets for the management of diabetes mellitus. More importantly, he made plausible emphasis on patient education and self-monitoring as a requirement for the management of diabetes mellitus [93].

In 1845, Louis Mialhe (1807–1886) identified diastase from saliva and he called it salivary diastase—had similar properties with the diastase of Payen and Persoz. Mialhe was a French physician, chemist, physiologist, and pharmacist who investigated many physiological phenomena in animals and humans. Salivary diastase was latter called ptyalin. Mialhe studied the mechanism of absorption of carbohydrates, fats, and protein substances. Mialhe also investigated the changes accompanying food changes in the presence of different gut fluids and substances—saliva, gastric fluid, and pepsin [80, 94, 95]. Salivary diastase of Mialhe, which was similar to that of Payen and Persoz, was an amylase enzyme. Amylase is an enzyme that catalyzes the hydrolysis of starch into sugars. It is the predominant enzyme in saliva that begins the chemical process of digestion of starch in the mouth. The enzyme is produced by the salivary gland as well as the pancreas and plants and some microbes. The enzyme in human is alpha-amylase that hydrolyzes dietary starch into monosaccharides, disaccharides, and trisaccharides. The final product of alpha-amylase hydrolysis is glucose. Glucose is the major energy source of the body. Detailed information on this is provided in Chap. 12 of this book.

The contributions of Alexander Mikhailovich Ugolev (1926–1991), Arne Dahlqvist (1909–1995), among others to the catalysis of food in the GI lumen are discussed in Chap. 12.

Tissue-level understanding of the gut structure and functions was further strengthened by Albinus, who admitted the existence of a covering that continued from the epidermis to the tongue—the outer layer of cells on the muscular organ. Building on the words of Albinus, Andreas Bonn (1738–1817) was able to trace this covering to the entire buccal cavity and the pharynx. Bonn suggested that all the mucous membranes are only a duplicate of the skin in the interior of the animal. Hence, he furthered that this covering must have a kind of cuticle, or epithelium. However, nobody knew whether or not this cuticle or epithelium extended beyond the pharynx [96]. This cuticle theory set forth by Bonn represented a turning point on the previous views of some scientists and philosophers who thought that the surfaces of the mucous membranes, in the interior of the body, are covered by a structure, which was unknown at the time. Epithelium (plural, epithelia) is a layer of membranous tissue of interconnected cells lining the surfaces of hollow organs as well as cavities of organs. This tissue layer is composed of a layer of one or two cells interconnected anatomically and functionally. The functions of the epithelium include secretion, selective absorption, protection, transcellular shunt and sensation, and chemosensation detection. The word "epithelium" is formed from the Greek "*epi*" meaning "on or upon," and "*thēlē*" meaning "nipple." Epithelium was chosen as the name as it was originally used to describe the translucent covering of small "nipples" of tissue on the lip.

1.7 The Beginning of Modern Digestive Physiology

On the physiology of digestion, in particular, outstanding pioneer scientists who made enormous contributions to the development of modern GI physiology include William Beaumont, Claude Bernard, Jan Evangelista Purkyne, Rudolph Heidenhain, and Ivan Pavlov. These scientists were modern physiologists that made tremendous experimental and developmental impact in physiology or medicine or more precisely, radically changed and laid fundamental concepts that constitute the cornerstone of our contemporary understanding of GI physiology. Interestingly, they also made immense managerial contributions to the development of the subject.

The study of gastric acid by the pioneer American experimental physiologist, John Richardson Young (1782–1804), at the University of Pennsylvania led to him falsely conclude that gastric acid was phosphoric acid. Thirty years later (in 1803) following the description of the process of acid formation in gastric digestion by the American bioscientist, Richardson [59], another American scientist and surgeon, William Beaumont (1785–1853), published series of experimental investigations on gastric juices and the physiology of digestion based on his observations of a patient suffering from a gastric fistula. (Gastric fistula is a patho-developmental link between the stomach and another organ in the body.) As he worked in the US Army, when treating wounded soldiers, with the permission of the patients, Beaumont took samples of gastric juice, which he used for experimentation purposes [97].

Beaumont is believed to be a pioneer in modern digestive physiology, and he is referred to the father of gastric physiology [98]. During his period of existence, there was a marked increase in clinical investigations on the stomach functions. In the month of July 1822, William Beaumont started the first modern experiments of the physiology of digestion. He was able to isolate HCl from the stomach juice. He was the first to discover the connection between the stomach secretion and emotional changes and gave a description of the stomach motility [99, 100]. Besides, Beaumont William, several scientists contributed to the development of gastric physiology during the time.

In 1823, William Prout (1785–1850), a physician of remarkable fame, proved beyond doubt on the exact nature of gastric acid. Even though he initially thought that the acid of the stomach was phosphoric, he later identified HCl as the stomach acid and noted that it is present in many species. Second, he was able to quantify the free and total HCl and chloride present in the stomach juice. The experimental investigation on the true HCl origin of gastric acid was confirmed by neutralization reaction and titration [9, 61, 101]. Some authors argue that the branch of medicine, modern gastroenterology, began with the investigation findings of Prout on the HCl origin of gastric acid [102]. One of Prout's works that earned him numerous accolades all over the globe was that he became the first to divide food substances into carbohydrate, protein, and fats. He hypothesized that the atomic weights of all the elements were multiples of the atomic weight of hydrogen atom—later called Prout's hypothesis. He realized that the elements were formed by a condensation or grouping of hydrogen atoms [59].

The source of gastric acid remained delusive since the stomach was fairly a large organ with many components and layers as it was known at the time. Further insight into the origin of gastric acid was made by Purkinje and Camillo Golgi (1843–1926), an Italian physician, anatomist, physiologist, and Nobel laureate. Golgi was a pioneer in the study of many cellular structures and phenomena including Golgi apparatus, the Golgi tendon organ, and the Golgi tendon reflex. During the mid-to-late nineteenth century, Purkinje and Golgi conducted series of experiments which showed that gastric glands were the source of gastric acid. They also showed that gastric acid level changed upon digestive stimulus [103].

In 1836, Theodor Schwann (1810–1882) (who discovered animal cells and Schwann cell of the nervous system), a German physiologist and student of the famous Johannes Peter Müller (1801–1858), discovered that a substance in the stomach was activated by the HCl. He identified this substance as pepsin. Schwann also discovered the proteolytic action of isolated gastric mucosa and called the substance responsible for the proteolytic action, pepsin—the substance he had previously identified to be activated by HCl. Furthermore, around the mid-1800s, Schwann showed that bile was an essential fluid in the digestive process in the intestine [59].

Scientists soon realized that pepsin was produced as a zymogen, not really in its active form "pepsin." It was during the 1880s, the English physiologists, John Newport Langley (1852–1925) and John Sydney Edkins (1863–1940), working at Cambridge conducted experiments to identify the site and nature of pepsin secretion. Langley and Edkins through a series of experiments concluded that dilute sodium carbonate (an alkaline solution) readily inactivated pepsin alkaline solution, and the precursor pepsinogen was relatively stable when placed in the same solution. In 1881, Langley and Edkins became the first to discover mechanisms of secretion of pepsinogen from the gastric mucosa. The first crystallization of pepsin (from swine) was achieved by John Howard Northrop (1891–1987), while working at Rockefeller Institute for Medical Research. However, it was Roger Moss Herriott (1908–1992) who isolated and crystallized pepsinogen, the inactive form of the enzyme [80]. Reports also indicate that the enzyme in its pure form was isolated by Ernst von Brücke (1819–1892) [104, 105]. Rockefeller Institute for Medical Research, now called Rockefeller University (New York City, USA) founded in 1901 by John D. Rockefeller Sr., one of the foremost biomedical research centers in the USA and world at large. This institute is known to have produced 24 Nobel laureates. It represents a key research institute where many leading scientists work [106].

François Magendie (1783–1855), a French physician, physiologist, and teacher of Claude Bernard, initially worked in the area of digestive physiology; however, he later switched to other areas of investigations. In digestion, he studied deglutition (the act of swallowing) and the mechanics of the digestive tract. His later investigation, which he shares with the Scottish surgeon, anatomist, neurologist, and theologian, Sir Charles Bell (1774–1842), is widely recognized throughout the world—is the work on the division of anterior and posterior spinal roots. He demonstrated that the posterior root stimulation elicited pain (sensory function) while this stimulation of the anterior root produced motor effects (motor function). Magendie described the apertura medialis ventriculi quarte—the foramen of Magendie. This median aperture drains cerebrospinal fluid from the fourth ventricle in the brain into the cisterna magna (for further review, see nervous system section in modern Anatomy texts). There are other openings in the fourth ventricle (they are two): lateral apertures (foramina of Luschka or foramina Luschkae), one on the left and another on the right, which drain cerebrospinal fluid into the cerebellopontine angle cistern. Topographically, the foramen of Magendie is posterior to the pons and anterior to the caudal cerebellum. It is surrounded by the obex and gracile tubercles of the medulla, tela choroidea of the fourth ventricle, and its plexus choroideus, which is attached to the cerebellar vermis. There is also a clinical sign called "Magendie sign," occurring during downward and inward rotation of the eye due to a lesion in the cerebellum [107].

Claude Bernard (1813–1878), a French physician, well-acclaimed physiology professor and student of François Magendie, known as one of the pioneers of modern physiology, also used fistulas (inch-wide openings) to study digestive process in laboratory animals. He was able to determine correctly that the major site of digestion was in the small intestine. In 1848, he also reported that pancreatic secretions were important digestive agents, particularly for fat digestion [108]. Bernard was also involved, elucidating the fat-splitting function of the pancreatic secretions, which was in part due to the presence of the enzyme lipase [2, 51, 108]. The mechanism of fat digestion involves many factors and enzymes, details of which are discussed in Chap. 12.

Worthy mentioning is the invention of the kymograph ("wave writer") in 1847 by the German physician and physiologist, Carl Ludwig (1816–1895). At about 30 years of age, Carl Ludwig had become a professor of physiology, and later professor of anatomy. The wave writer was intended to record changes in arterial blood pressure. However, it turned out to be useful in other areas of physiological investigations including respiration and digestion. A kymograph is a device that gives a graphical representation of spatial position over time in which a spatial axis represents time. It consists of a revolving drum wrapped with a sheet of paper on which a stylus moves back and forth recording perceived changes of phenomena such as motion or pressure. Modern kymographs represent a tremendous improvement on the nineteenth-century kymographs. There are different types of developments over the last centuries. Further information about kymograph could be retrieved from http://www.physiologyinfo.org. The first use of kymograph to measure activity of the digestive tract was carried out in 1883 by the German physiologist, Karl Hugo Kronecker (1839–1914), and his student, Samuel James Meltzer (1851–1920), with a couple of other scientists [for review, see ref. 109]. However, the first to attempt a study of esophageal motility using graphic method of recording swallowing in mammals was Saturnin Arloing (1846–1911). His results were presented in 1877 to the Faculty in Sorbonne, Paris [12]. Kronecker and his students used the balloon kymograph to investigate esophageal motility in humans [12, 110, 111].

The American physiologist, professor and one-time chair of the Department of Physiology at Harvard Medical School, Harvard University, Walter Bradford Cannon (1871–1945) [112, 113], made numerous contributions to the study of physiology. He studied digestion, precisely, the mechanism of swallowing, and the motility of the stomach, using the newly discovered at the time, X-ray technology. Cannon was a pioneer contributor to the publication of the American Journal of Physiology in 1898. In 1898, the Harvard physiologist reportedly used a kymograph attached to an X-ray (X-ray kymography) to register the shape, position, and size of the different parts of the digestive tract and some physiological functions in normal and disease. The height, shape, and a number of waves indicated peristalsis and smooth muscle tone [114]. Thus, Cannon reported the first use of roentgen rays (X-rays) combined with kymograph in the study of esophageal motility. He also investigated stress and was credited for coining "fight-or-flight" response to fear in the year 1915 [111].

In 1897, the Russian anatomist, histologist, and the last Minister of Education of the Russian Empire, Nikolai Konstantinovich Kulchitsky or just Kulchitsky Nicholas (1856–1925), described the endocrine cells of the small intestine which now bear his name "Kulchitsky cells." These cells are enterochromaffin (abbreviated "EC") cells, a type of enteroendocrine and neuroendocrine cell occurring in the epithelial lining the lumen of the digestive tract (gullet, stomach, intestine) and the respiratory tract. It is now believed that these cells secrete tens of hormones, neurotransmitters, and peptides that regulate the functions of the GI system and other organs and tissues of the body. Malfunction of these cells can result in diseases such as cancers [115].

The German physiologist and surgeon Rudolph Heidenhain (1834–1897) carried out numerous investigations on the GI tract and one of the large digestive glands—the pancreas. In 1878, Heidenhain differentiated three gastric cell types based on their secretions in mucosa of the stomach: chief or zymogenic cells (pepsin), parietal cells (HCl), and epithelial (mucous) cells [104, 105]. During the time, there was increased interest in investigating intestinal cells.

The British physiologists, William Bayliss (1860–1924) and Ernest Starling (1866–1927), discovered internal secretions of the gut—hormones. Detailed information on the course of discovery of hormones of the gut and their mechanisms of action are discussed in Chap. 8.

One of the most renowned scientists of the last millennium, Ivan Pavlov (1849–1936), a Russian physician, physiologist, pharmacologist, and Nobel laureate, made a tremendous impact on physiology by studying the mechanisms underlying some functions of the digestive system in dogs. Pavlov worked to unveil the secrets of the digestive system and also studied the signals that triggered secretory phenomena in the GI tract. In one faithful day, Pavlov realized that saliva began to drool when a dog sights food. Moreover, at times when no proper or adequate stimuli were present, no sighting of food was involved; the dog was still noticed to drool saliva. Pavlov became inquisitive of this phenomenon. It was later hypothesized that in the absence of proper stimuli, the dog may be reacting to the person's appearance, probably the laboratory coat since at each food serving time; the same appearance was seen by the dog. Consequently, the dog must have learnt the situation involved when food was served. So, Pavlov designed an experiment in an attempt to investigate and reproduce this development in an experimental model. In his model, he struck a bell each time the dog was provided with food. After some days, the dog started responding to the sound of the bell by drooling if it was mealtime even though the meal was not given the dog. This was the origin of reflexes, though was initially mentioned long before Pavlov by another Russian scientist, Ivan Mikhailovich Sechenov (1829–1905), as a mechanism in brain functioning. Pavlov started actively investigating the reflex basis of the nervous system, which he called higher neural functions or higher brain activity [116, 117]. Two types of stimulus and two types of responses were identified: unconditioned and conditioned stimuli or responses. Unconditioned stimulus is a stimulus that can produce the response without any learning. An example of unconditioned stimulus is the taste, sighting of food, or talks about food. Conditioned stimulus is a neutral stimulus that can produce a response if it is paired (or associated) with the unconditioned stimulus. An example of conditioned stimulus is the sound of a bell. The sound of a bell cannot elicit salivary flow on its own. However, when a bell rings a few minutes before the sighting of food, and this process occurs over a given period, only the sound of a bell can cause salivary flow. This process is called classical conditioning (or conditioned reflex or Pavlovian conditioning). Classical conditioning is a process in which a neutral (indifferent) stimulus when repeatedly paired with an adequate (potent physiological) stimulus becomes expressed in response to the indifferent one. There are principles for achieving a desired response in such stimulus pairing. Like the stimulus, the response is either unconditioned response or conditioned response. Unconditioned response is the unlearned or inborn reaction to the unconditioned stimulus. The conditioned response occurs when a conditioned stimulus elicits a response. An example of both responses includes salivation. On the basis of the type and mode of stimuli combination yielding different types of responses, Pavlov observed certain trends, which are known as the laws of classical conditioning (laws of conditioned reflex or Pavlovian conditioning). The laws of conditioned reflex comprise four different entities—extinction, acquisition, generalization, and discrimination. Acquisition is the process of learning the association between conditioned and unconditioned stimuli. This is the first stage of conditioning. The pairing time interval may vary or may occur simultaneously until the response is observed. When the conditioned stimulus is continuously presented with an unconditioned stimulus, at a particular point, conditioned stimulus will no longer elicit a response. This is called extinction. Extinction may be due to loss of reinforcement or disinhibition process of the stimulus. Sometimes, a different stimulus other than the one being presented, but similar to the conditioned stimulus may produce a response. This is called generalization. Discrimination is the opposite of generalization. It is the ability to differentiate between similar stimuli [118–121]. These processes and stages of conditioning are important strategies allowing mammals to adapt to changes in their environment and feeding behavior [122, 123]. The American psychologist, professor Edwin Burket Twitmyer (1873–1943), must have made similar contribution to the concept of classical conditioning as did Pavlov, which was reported almost the same time in 1904 [124].

The mechanism of saliva secretion and innervation of the salivary glands were partly unraveled by the German Karl Ludwig (1816–1895) [104, 105]. However, Pavlov made a greater contribution to unraveling the mechanisms and phases of salivary secretion.

In Pavlov's conditioning experiment, he used the secretory rate of salivary gland as a quantitative measure of the psychical activity of the animal, in order to emphasize the advantage of objective physiological measures of higher neural activity. Interestingly, saliva and some of its components are still used for analyzing the activity of the nervous system. Synonymously, the higher neural functions were later called integrative functions of the brain or higher integrative functions of the brain because it was later known that as impulses (signals) are conveyed to brain cells, they integrate, amplify, and relay it to other higher areas for final processing, storage, and decision making. A key contribution to this regards was made by Sir Charles Scott Sherrington (1857–1952), an English physiologist who conducted series of experiments that laid the foundation for understanding the integrative functions of the nervous system. Sherrington, who was quite conversant with Pavlov's experiments, excised part of the cerebral hemisphere of mammals (dogs, etc.) and observed that they lost responses to signals. He, therefore, suggested that response to an impulse, constituting the reflex arc of Pavlov, was the result of integrated activities of brain cells. In his book titled "The Integrative Action of the Nervous System" published in 1906, Sherrington distinguished three main groups of sense organs based on the location and perception (reception) of signals from external or internal environment: exteroceptive; interoceptive; and proprioceptive. In later chapters of this book, you will notice that the second group "interoceptive," nowadays, might include not only traditional organs like the tongue but also upper respiratory tract, pancreas, some blood cells, parts of the brain. Sherrington, in a bid to strengthen his observation on "nervous integration," during the last few years of the nineteenth century, demonstrated "reciprocal innervation" of muscles, which was later referred to "Sherrington's law." According to this law, when one set of muscles is stimulated, muscles opposing the action of the first are simultaneously inhibited. Sherrington demonstrated that spinal reflex is composed of integrated actions of the nervous system involving such complex components as the excitation and inhibition of many nerves, induction (i.e., the increase or decrease of inhibition brought on by previous excitation), and the irradiation of nerve impulses to many nerve centers [2, 51, 125].

After reviewing Sherrington's concept on nervous system, Pavlov classified nervous system functions at the cortical and subcortical level and also made other contributions to the study of neurophysiology and GI physiology. Professor Sherrington shared the 1932 Nobel Prize for Physiology or Medicine with another English physiologist and physician, Professor Edgar Adrian (1889–1977), for their work on nervous system functions. Adrian was a leading specialist in electrophysiology (precisely on electroencephalogram or EEG). He carried out investigations on functions of sensory organs, brain, and provided evidence for the "all-or-none" law of nerves. The all-or-none law states that any single nerve or muscle fiber response to a stimulus above a threshold level is maximal and independent of the intensity of the stimulus [125].

Spotlight 1.1

Nobel Prize Winners in Physiology or Medicine (In the Area of Gastrointestinal Physiology/Gastroenterology)

The highest honor accorded a scientist is the Nobel Prize. The honor was founded with the aim to accomplishing the vision of Alfred Nobel. Visit www.nobelprize.org for further details.

The first Nobel laureate in the field of digestive physiology, Ivan Pavlov (1849–1936) was awarded the 1904 Nobel Prize for his investigations on the digestive processes, even though he was better known for his research on physiology of higher nervous functions.

The second Nobel laureate in the field of digestive physiology, Johanes Andreas Fibiger (1867–1928) received his award in 1926 for discovering the causative agent of stomach cancer. He reported inflammatory and degenerative lesions after feeding rats with cockroach infected with larva of *Spiroptera neoplastica.*

The third Nobel laureate in the field of digestive physiology, Henry Hallet Dale (1875–1968) was awarded the Nobel Prize in 1936 for isolation of acetylcholine and determining its role in transmission of stimuli between neurons and effectors.

The fourth Nobel laureate in the field of digestive physiology, George Emil Palade (1912–2008) was awarded the 1974 Nobel Prize for his description of the biochemical steps in protein synthesis, segregation, transport, storage and secretion in the exocrine pancreatic cell, and the ultrastructure of these processes.

The fifth Nobel laureate in the field of digestive physiology, James Whyte Black (1924–) was awarded the 1988 Nobel Prize for research on histamine (H_1) receptor blockers and for discovery of a specific substance inhibiting hydrochloride acid secretion by parietal cells by antagonism of H_2 receptors in the stomach. He established the fact that specific chemical transmitter in the duodenal mucosa can stimulate secretion of the pancreas.

The sixth Nobel laureate in the field of digestive physiology, Robin Warren (1937–present) and Barry Marshall (1951–present) were awarded the 2005 Nobel Prize for their discovery of *Helicobacter pylori* (spiral bacteria) and its role in the development of gastritis and peptic ulcer disease and duodenal ulcer.

Although Pavlov is most famous for his works on nervous system, he obtained his Nobel Prize in digestive physiology, awarded in the year 1904 for invention of the vagally innervated gastric pouch ("Pavlov's pouch") that secretes gastric acid in similar manner as the main stomach from which it was prepared. This honor came only three years after the beginning of the Nobel Prize award. In his book, "Lectures on the Work of Major Digestive Glands," published in 1887 [116, 126], major digestive glands were referred to hepar (*Latin* for liver), large salivary glands, vesica biliaris (*Latin* for gallbladder), and pancreas. The minor (or small) digestive glands are located in multiple areas of the digestive system, especially in the mucosa and submucosa [127]. Pavlov's success was in part due to the fact that he modified the experimental protocol of widely known researchers at the time. The extraction of gastric juice by surgical techniques was fairly known at the time, even though details were as yet to be unraveled. Pioneers in extraction of gastric juice by a fistula technique were William Beaumont and Basov Vasiliy Alexandrovich (1812–1879). Beaumont used a fistula to collect gastric juice in 1822, and Basov applied the technique in 1842. A gastric fistula is an artificially generated link between the stomach and the external environment. In such experimental animals, stomach juice can be obtained for different types of investigations. Unfortunately, gastric juice extraction via fistula technique often results in collection of contaminated juice. Thus, there was a need to modify the fistula technique of gastric juice collection, which was made by the German scientist, Rudolph Heidenhain (1834–1897). Heidenhain obtained pure gastric juice after overlaying gastric fistula with additional simultaneous surgery referred to esophagotomy, an operation that involves surgical excision of the esophagus. In Heidenhain's experimental model, food can no longer pass into the stomach. If food is delivered to the animal, it escapes through the part of excised esophagus, a type of feeding called "sham feeding." Importantly, in this situation, when the animal is been fed, the stomach secretes juice, and its content can be obtained as pure juice. Also, the technique enables the researcher to investigate the influence of different reflexes, activation or inhibition of oral or pharyngeal receptors on the activity of gastric glands [2, 51].

Sham feeding is a procedure that mimics normal food consumption, but where the substance been swallowed is not digested or absorbed. In such experiment, a tube is inserted into the esophagus or stomach so that the swallowed substance does not enter the stomach but passes from an esophageal fistula surgically created in the neck region. Animals used in this kind of experiment cannot be satiated as they continuously masticate and swallow any available food. The technique is used to study the psychological influence on digestion or cephalic phase of gastric secretion, mechanisms of hunger, feeding behavior, digestion, and eating disorder. Before swallowing, during food mastication, there is increased GI secretion via vagal pathways [128, 129]. A different technique is used to study sham feeding in humans, referred to as modified sham feeding. Modified sham feeding includes the use of smelling, tasting, and gum chewing and then spitting it out. Sham feeding stimulates digestive system nerves which trigger the release of GI hormones and also increase the production of GI secretions, especially saliva and pancreatic juice [128, 129].

In humans, besides smelling, tasting, and gum chewing, and then spitting it out, sham feeding may be studied during vagotomy performed in chronic ulcer disease patients. The amount of gastric acid secretion in sham feeding following vagotomy is a useful measure in defining the effectiveness of vagotomy [128].

Alas, the Heidenhain technique provides possibilities to investigate the influence of food and food substances on the secretions of gastric glands; however, the model gives inaccurate result. Pavlov had nurtured interest in these techniques used to study gastric secretion. Interestingly, he observed that the stomach in Heidenhain's experiment was denervated (denervation—removal of nerves). Hence, only the humoral phase of gastric secretion can be studied. He had made a first-class conclusion that will take him to another level of research. So, Pavlov proposed a new technique for isolating small stomach, which allowed obtaining pure gastric juice throughout the period of digestion [51].

Pavlov, a superlative scientist, then devised an operation to prepare a miniature stomach (also called pouch). He isolated the stomach while preserving its vagal nerve supply. The surgical procedure enabled him to study the GI secretions in the animal over its life span. This experimental model was referred to the "small Pavlov stomach." This model with an esophageal fistula enabled him to observe the secretion of the gastric juice under the influence of taste or odor stimuli, which was named the reflex or nervous phase of secretion (now called cephalic phase). Pavlov was able to differentiate between the types of secretions in the cephalic phase and when the food was in the stomach or mouth—chemical phase of secretion (now called gastric phase) [51].

While the mechanisms of action of major gastric enzymes were fairly understood, how food was further broken down into smaller pieces and the mechanism of activation of intestinal enzymes were poorly known. It was in Pavlov's laboratory that the mechanisms of activation of pancreatic proteolytic enzymes were discovered. In 1899, he introduced the name "enterokinase" for the enzyme of the intestinal juice that was responsible for activating proteolytic agents of the pancreatic juice [51].

Pavlov's investigations of the mechanisms of digestive secretions laid the experimental evidence for foundations of the theory of nervism. Proponents of this theory believe that the nervous system plays the dominant role in the regulation of the physiological functions and processes that occur in animals and man [130]. The theory of nervism was based on the initial works of Sechenov and Pavlov. However, it was Sergey Petrovich Botkin (1832–1889) that made enormous contributions to defining "nervism"; hence, the theory is sometimes called Botkin's nervism. Botkin was a famous Russian physician, activist, and one of the founders of modern Russian medical science and education, who introduced pathological anatomy in Russia. This branch of anatomy was key in postmortem diagnostics in Russian medical practice [131–133]. Following the birth of nervism, proponents thought that neural mechanisms govern all activities of humans. However, the discovery of genes and the cracking of the genetic codes sent a strong message to the world on the determinants of human activities. It was heralded throughout the world that genes are the molecular basis of human activities and functions. More recently, however, it is increasingly been acknowledged that the epigenome, rather than the genome, is the determinant of physiological functions in humans and other animals [134, 135].

Together with expanding research in digestion, scientists worked in different areas. As nervism flourished, Pavlov's initial experimental findings on vagus nerve physiology regarding how the nerve affected secretions of the stomach interested another group of scientists. Key figures widely recognized in making enormous contributions to the field were André Latarjet (1877–1947) and Lester Reynold Dragstedt (1893–1975). Even though Galen had described the anatomy of the vagus nerve in the second century, and Pavlov made exciting findings on the physiological basis of stomach secretions, certain aspects of vagus physiology appeared to be unknown, as it was later identified by Latarjet and Dragstedt [136]. Latarjet explored the effect of vagal denervation on stomach secretions and how it could be used to treat patients suffering from gastric problems such as peptic ulcer. In certain cases of peptic ulcer, the application of therapeutics does not adequately treat the conditions. This was the first time physiological understanding of stomach secretions will be applied to the surgical practice of treating such group of patients. In 1943, Dragstedt, after keenly studying the physiology of vagal denervation, proposed that it might be a favorably technique to use in ulcer disease (peptic and duodenal). After documenting the procedures and techniques of vagal denervation (vagotomy) and pyloroplasty—Dragstedt operation—the procedure became widely used in cases of acute bleeding and obstruction. There are currently different techniques and procedures with differing rates of success and mortality for different types of vagal denervation. The part of the vagus innervating the parietal cells can be denervated to study the influence of this cell type on digestion [136–138].

However, a better control of stomach acid secretion for treating gastric or peptic ulcer disease was achieved through a better understanding of the cellular and molecular basis of the GI functioning. Details on this are provided in Chap. 11. It should be noted, however, that the incidence of vagotomy has significantly reduced following a better control of stomach acid secretions with pharmacological agents and understanding of the role of *Helicobacter pylori* in peptic ulcer disease [138].

Other aspects of the history of development of GI physiology are systematically discussed in the various chapters.

1.8 Current and Emerging Trends in the Study of Physiology

The reductionist approach in addressing human problems led to narrowing of knowledge on integrative functioning of the whole organism. Thus, it was necessary to address these issues. Two directions of research that seems to be addressing the reductionist approach or preventing narrowing of knowledge in physiology are translational research and integrative approach.

1.8.1 The Pendulum of Investigation of Physiological Systems Is Swinging Toward Integrative Approach. The Physiome and Giome Projects

With progress in technology, scientists were more committed to conducting research with new techniques, thus making them investigate the cellular, molecular, and genetic basis of body functioning, rather than considering functioning of the systemic organism. Beginning from the twentieth century, there was massive increase in investigations that led to significant development in physiology and rise in knowledge at the cellular and molecular levels. This trend in research actually led to a better understanding of the cellular and molecular basis of bodily processes and functioning. However, much attention was not paid to the integration of the molecular and cellular understanding to the functioning of the whole organism. This unmatched increase in cellular and molecular inquiry into the basis of body functioning resulted in narrowing of knowledge on integrative functioning of the whole organism. In order word, the organism's molecular registry does not function as an entity; rather they function as an integrative unit with the whole organism. Obviously, there was need for more investigations that could increase our understanding of the organism as an integrative entity. This gradual shift toward integrative method of investigating physiological systems led to the introduction of the Physiome Project (for news and latest information, visit: http://physiomeproject.org or http://www.physiome.org).

The Physiome (from "physio" meaning "nature, life" and "ome"—"as a whole") Project is a global concrescence of research crusade toward the mapping of the intact organism functional dynamics in health and disease, based on collaborative data on structure–function relationship at genomic, proteomic, and morphomic levels and aims to beef up the fabric of international scientific coaction, political, and diplomatic allegiance cross-wise. In mapping the intact organism, statistical, mathematical, and computational methods are used. The genome is the total genetic composition of an organism. The proteome is the total protein composition that makes up an organism (information on the gene and protein of an organism is provided in Chap. 3). The morpheme is the quantitative description of anatomical structure, chemical and biochemical composition, and material properties of an intact organism. The morpheme may include the genome, proteome, cells, tissues, and organ structures of the whole intact organism [139–142]. The Physiome Project was initiated in 1993 at the 32nd World Congress in Glasgow by the Commission on Bioengineering in Physiology of the International Union of Physiological Sciences. In 1997, the first meeting on designing the Project was held. A few years later, precisely in 2001 the first global congress on the Physiome Project was successfully held. As part of the congress, the aims and objectives of the Project were outlined. The Physiome Project is aimed at providing comprehensive databases and an integrative method to the study of physiology or medicine [140, 142]. The Physiome Project will provide an integrative quantitative and descriptive modeling of how functions of the whole organism are related to its structural and functional composition. The Physiome will also produce the virtual physiological human which will aid not only in the analysis of integrative physiological functions, but also in the provision of a system for hypothesis testing, a model in drug design, prediction of the results of various treatments and interventions [141, 143, 144].

The Giome (from "GI" meaning "gastrointestinal" and "ome"—"as a whole") Project is an integrative approach to the study of GI functions, which was borne out from the lunch of the Physiome Project. The Giome Project is the GI system part of the Physiome. The Project is aimed at obtaining an integrated understanding of the human GI tract based on bioengineering principles and models [144]. The Giome

Project will provide quantitative and qualitative information on the structure, function, biomechanics, geometry of GI system in health and disease [144].

Giome Project incorporates all aspects of digestive functioning, which include GI motility, secretion, absorption, metabolism, enteric nervous/endocrine system activities, gut-associated lymphoid system activities, gut fluid (blood, lymph, water) regulation and circulation, and the interrelationship among these factors that determine gut activity. For news and latest information on Giome Project, visit: http://www.giome.com.

1.8.2 Bedside-to-Bench or Bench-to-Bedside Physiology—The Cornerstone of Translational Physiology or Medicine

The area of digestion is one of the most important to human life. Although not widely acknowledged, the digestive system is responsible for, depending on the type of food regimen, about 60–90% diseases that affect humans, making the digestive system one of the most important systems in life processes. Unfortunately, over the last decades, there has been a marked deepening in the gap between basic science research and its application (such as in the clinics, public health, industries, or society). Thus, the importance of basic science such as physiology was not really felt especially among new learners in their respective disciplines. Physiology, which is supposed to serve as a building block for knowledge in medicine and allied disciplines, bridging the basic sciences with application, has not been devotionally addressed in many learning environments in accordance with the trends in scientific and technological development [1, 51, 129, 145–147].

According to the paper published in the Journal of American Physiological Society, translational research is "the transfer of knowledge gained from basic research to new and improved methods of preventing, diagnosing, or treating disease, as well as the transfer of clinical insights into hypotheses that can be tested and validated in the basic research laboratory." This definition means that translational research is bidirectional—from bench to bedside and from the bedside back to bench [146]. Translational physiology or medicine therefore is the discipline that studies life process with the aim to "translate" experimental results in basic research into practice and meaningful health or societal outcomes or translate clinical insights into hypothesis for validation, necessary for beneficial use by the society.

Translational research is fairly a new area that was introduced approximately two decades and hopes to solve the widening gap between basic science and application [147, 148]. It holds promise for enhancing societal impact of basic science research and also solidifies the role of physiology in the research enterprise [148].

1.9 Conclusion

Research in GI physiology as a prime area of physiology can provide valuable information for the advancement of science and especially human living. The brief historical perspectives presented in this chapter provide the basis of furthering our knowledge on GI functioning. The initial works and endeavor by different philosophers and scientists had a key bearing on the future development and understanding of digestive functions. Recognition of exceptional contributions of scientists to scientific progress precisely in the area of physiology or medicine gives them a huge sense of belonging to science rather than humiliating, punishing, and killing scientists for their selfless research service to the progress of the human race. The Nobel Prize will continue to provide more opportunities for acknowledging outstanding contributions to science and addressing key issues of nature and human existence.

Each notable development in course of studying digestive functions was important for subsequent advancement on the subject. Research activities resulting in the advancement in cellular and molecular basis of GI functions were aided by improvement in technologies and provided a better means for studying living systems in general and addressing human diseases.

Basic concepts in GI functioning were developed through the combined efforts of several scientists spanning at least two millennia.

Current and emerging efforts in GI functions are the results of the collaborative work of several scientists throughout the globe.

Recommended Readings

Original Articles

1. Bing FC (1971) The history of the word 'metabolism'. J Hist Med Allied Sci 26 (2):158–180
2. Gritti I, Banfi G, Roi GS (2000) Pepsinogens: physiology, pharmacology pathophysiology and exercise. Pharmacol Res 41(3):265–281
3. Kisch B (1954) Jacob Helm's observations and experiments on human digestion: a one hundred and fiftieth anniversary. J Hist Med Allied Sci 9(3):311–328
4. Murray CJ, Vos T, Lozano R, Naghavi M, Flaxman AD, Michaud C, et al (2012) Disability-adjusted life years (DALYs) for 291 diseases and injuries in 21 regions, 1990–2010: a systematic analysis for the Global Burden of Disease Study 2010. Lancet 380(9859):2197–2223
5. Pagel W (1958) The Position of Harvey and Van Helmont in the history of European thought: to commemorate H. E. Sigerist's essay on Harvey (1928). J Hist Med Allied Sci 13(2):186–199
6. Worth EJ (1996) The medical properties of food in the eighteenth century. J Hist Med Allied Sci 51(2):127–154

Review Articles

1. Barbosa AJA (2013) Superficial tumors of the gastrointestinal mucosa: the problem of the histological diagnosis of intravascular infiltration in tissues from endoscopic submucosal dissection. J Brasil Patol Med Lab 49(4):238–240
2. Cheng LK, O'Grady G, Du P, Egbuji JU, Windsor JA, Pullan AJ (2010). Gastrointestinal system. Wiley Interdiscip Rev Syst Biol Med 2(1):65–79
3. Conroy MS (2012) The imperial laboratory: experimental physiology and clinical medicine in Post-Crimean Russia (review). J Hist Med Allied Sci 67 (3):504–507
4. Dockray GJ (2009) The versatility of the vagus. Physiol Behav 97(5):531–536
5. Dockray GJ (2013) Enteroendocrine cell signalling via the vagus nerve. Curr Opin Pharmacol 13(6):954–958
6. Dockray GJ (2014) Gastrointestinal hormones and the dialogue between gut and brain. J Physiol 592(Pt 14):2927–2941
7. Goetz M, Kiesslich R (2010) Advances of endomicroscopy for gastrointestinal physiology and diseases. Am J Physiol Gastrointest Liver Physiol 298(6): G797–806
8. Goetz M, Watson A, Kiesslich R (2011) Confocal laser endomicroscopy in gastrointestinal diseases. J Biophotonics 4(7–8):498–508
9. Hirst GDS, Ward SM (2003) Interstitial cells: involvement in rhythmicity and neural control of gut smooth muscle. J Physiol 550(2):337–346
10. Hoetker MS, Goetz M (2013) Molecular imaging in endoscopy. United European Gastroenterol J 1(2):84–92
11. Horowitz B, Ward SM, Sanders KM (1999) Cellular and molecular basis for electrical rhythmicity in gastrointestinal muscles. Ann Rev Physiol (61):19–43
12. Konturek PC, Konturek SJ (2003) The history of gastrointestinal hormones and the polish contribution to elucidation of their biology and relation to nervous system. J Physiol Pharmacol 54(S3):83–98
13. Konturek SJ (2003) Gastric secretion—from pavlov's nervism to popielski's histamine as direct secretagogue of oxyntic glands. J Physiol Pharmacol 54 (S3):43–68
14. Konturek SJ, Konturek JW, Pawlik T, Brzozowski T (2004) Brain-gut axis and its role in the control of food intake. J Physiol Pharmacol 55(1 Pt 2):137–154
15. Raybould HE (2007) Mechanisms of CCK signaling from gut to brain. Curr Opin Pharmacol 7(6):570–574
16. Ryan KL (2013) APS at 125: a look back at the founding of the American Physiological Society. Adv Physiol Educ 37:10–14
17. San Roman AK, Shivdasani RA (2011) Boundaries, junctions and transitions in the gastrointestinal tract. Exp Cell Res 317(19):2711–2718
18. SRódka A (2003) The short history of gastroenterology. J Physiol Pharmacol 54(S3):9–21

Books

1. Farquhar FJ (1930) Selected readings in the history of physiology. C.C. Thomas in Springfield, Baltimore, MD
2. Foster M (2012) Lectures on the history of physiology during the sixteenth, seventeenth and eighteenth centuries. Cambridge University Press, UK
3. Franklin KJ (1949) A short history of physiology. Staples Press, London, UK
4. Fulton JF, Wilson LG (1966) Selected readings in the history of physiology, 2nd edn. CC Thomas in Springfield, Baltimore, MD
5. Johnson LR, Barret KE, Gishan FK, Merchant JL, Said HM, Wood JD (eds) (2006) Physiology of the gastrointestinal tract, 4th edn. Elsevier Academic Press, Burlington, San Diego and London
6. Kelly K (2009) The history of medicine—early civilizations: prehistoric times to 500 C.E. Infobase Publishing, NY
7. Kelly K (2009) The history of medicine. The middle ages: 500–1450. Infobase Publishing, NY
8. Mark J (2011) The Oxford handbook of the history of medicine. Oxford University Press, UK
9. Rothschuh KE (1973) History of physiology. RE Krieger Pub Co, Huntington, NY
10. Schultz SG, Field M, Frizzell RA, Rauner BB (eds) (1991) Handbook of physiology. Oxford University Press, USA
11. Strathern P (2005) A brief history of medicine: from Hippocrates to gene therapy. Robinson, London, UK

Lectures of Nobel Laureates

1. Ehrlich P (1908) Nobel lecture on partial functions of the cell. In: Himmelweit F, Marquardt M, Dale HH (eds) The collected papers of P. Ehrlich. Pergamon Press, Oxford
2. Frangsmyr T, Lindsten J (eds) (1993) Physiology or Medicine 1981–1990: Nobel lectures: including presentation speeches and Laureates' biographies. World Scientific Pub Co Inc, Singapore
3. Jornvall H (ed) (2003) Nobel Lectures: Physiology or Medicine 1996–2000. World Scientific Publishing, Singapore
4. Palade GE (1975) Intracellular aspects of the process of protein synthesis. Science 189:347–358

Additional Texts and Websites for Review

1. Bunick CG (2014) Perspectives on physiology and medicine from Nobel prize winners: The 64th Lindau Nobel Laureate meeting. J Invest Dermatol 134, 2853–2855
2. Eppendorf. Eppendorf Award for Young European Investigators. http://corporate.eppendorf.com/
3. European Students' Conference (2011) Promising medical scientists willing to look beyond. Eur J Med Res 16(Suppl 1):1–178. http://www.esc-berlin.com/

4. Hellström PM (2005) Spotlight on gastroenterology—the nobel prize laureates in physiology or medicine 2005: John Robin Warren and Barry James Marshall. Scand J Gastroenterol 40(12):1383–1385
5. Queralt AV, Fragou D, Panoutsopoulou K, Anguizola J (2013) Bioanalysis young investigator: award 2013. Bioanalysis 5(11):1341–1345
6. Roberto M, Kash TL, Mulholland PJ, Marty VN, Gilpin NW, Walker BM (2012) Introduction to young investigator award symposium: symposium XII: young investigator award. Alcohol 46(4):301–302
7. Seeberger PH (2010) 2010 Tetrahedron young investigator award for medicinal and bioorganic chemistry. Bioorg Med Chem 18(11):3642

Bibliography

1. Sprung L (2001) Wundt, Wilhelm Maximilian (1832–1920). In: Smelser NJ, Baltes PB (eds) International encyclopedia of the social & behavioral sciences. Pergamon Press (Elsevier), NY, USA
2. Koertge N (ed) (2007) New dictionary of scientific biography, 1st edn. Charles Scribners and Sons, Detroit, Michigan
3. Bourgoin SM (2012) Encyclopedia of world biography, 2nd edn. Gale Research, Detroit, Michigan
4. Vannier J, Liu J, Lerosey-Aubril R, Vinther J, Daley AC (2014) Sophisticated digestive systems in early arthropods. Nat Comm 5:3641
5. Maloof AC, Porter SM, Moore JL, Dudas FO, Bowring SA, Higgins JA, Fike DA, Eddy MP (2010) The earliest Cambrian record of animals and ocean geochemical change. Geol Soc Am Bull 122(11–12):1731–1774
6. Valentine JW, Jablonski D, Erwin DH (1999) Fossils, molecules and embryos: new perspectives on the Cambrian explosion. Development 126(5):851–859
7. Graham B (2013) At the origin of animals: the revolutionary Cambrian fossil record. Curr Genom 14(6):344–354
8. Erwin DH, Laflamme M, Tweedt SM, Sperling EA, Pisani D, Peterson KJ (2011) The Cambrian conundrum: early divergence and later ecological success in the early history of animals. Science 334:1091–1097
9. Kousoulis AA, Tsoucalas G, Armenis I, Marineli F, Karamanou M, Androutsos G (2012) From the hungry acid to pepsinogen: a journey through time in quest for the stomachs secretion. Ann Gastroenterol 25(2):119–122
10. SRódka A (2003) The short history of gastroenterology. J Physiol Pharmacol 54(S3):9–21
11. Riva MA, Riva E, Spicci M, Strazzabosco M, Giovannini M, Cesana G (2011) "The city of Hepar": rituals, gastronomy, and politics at the origins of the modern names for the liver. J Hepatol 55(5):1132–1136
12. Pelemans W, Texter EC Jr (1974) The manometric examination of the esophagus. Dis Esophagus 3:235–245
13. Zalta EN (ed) (2012) The Stanford encyclopedia of philosophy. Stanford University, Stanford, CA
14. Blits KC (1999) Aristotle: form, function, and comparative anatomy. Anat Record 257 (2):58–63
15. Bay NS-Y, Bay B-H (2010) Greek anatomist Herophilus: the father of anatomy. Anat Cell Biol 43(4):280–283

16. Wiltse LL, Pait TG (1998) Herophilus of Alexandria (325–255 BC). The father of anatomy. Spine (Phila Pa 1976) 23(17):1904–1914
17. Imai M (2011) Herophilus of Chalcedon and the Hippocratic tradition in early Alexandrian medicine. Hist Sci (Tokyo) 21(2):103–122
18. Tomey MI, Komotar RJ, Mocco J (2007) Herophilus, Erasistratus, Aretaeus, and Galen: ancient roots of the Bell-Magendie Law. Neurosurg Focus 23(1):E12
19. Hong FF (2004) History of medicine in China. McGill J Med 8:79–84
20. Loukas M, Lanteri A, Ferrauiola J et al (2010) Anatomy in ancient India: a focus on the SusrutaSamhita. J Anat 217:646–650
21. Tipton CM (2008) Susruta of India, an unrecognized contributor to the history of exercise physiology. J Appl Physiol 104:1553–1556
22. Singh J, Desai MS, Pandav CS, Desai SP (2012) Contributions of ancient Indian physicians —implications for modern times. J Postgrad Med 58:73–78
23. Baas JH (1971) Outlines of the history of medicine and the medical profession. Krieger, Huntington, NY
24. Marcel D (1964) Larousse encyclopedia of modern history, from 1500 to the present day. Harper & Row, New York
25. Editors of the American Heritage Dictionaries (2011) The American Heritage Dictionary of the English Language, 4th edn. Houghton Mifflin Harcourt Trade, Boston, Massachusetts, USA
26. Vaillancourt MG (2009) On the body and blood of the Lord on the truth of the body and blood of Christ in the Eucharist. Catholic university of America press, Washington, DC, USA
27. West JB (2014) Galen and the beginnings of Western physiology. Am J Physiol Lung Cell Mol Physiol 307(2):L121–128
28. Hajdu SI (2011) A note from history: landmarks in history of cancer, part 2. Cancer 117 (12):2811–2820
29. Dunn PM (1997) Avicenna (AD 980–1037) and Arabic perinatal medicine. Arch Dis Child Fetal Neonatal Ed 77(1):F75–76
30. Subbarayappa BV (2001) The roots of ancient medicine: an historical outline. J Biosci 26 (2):135–143
31. Zargaran A, Mehdizadeh A, Zarshenas MM, Mohagheghzadeh A (2012) Avicenna (980–1037 AD). J Neurol 259(2):389–390
32. Iorio L, Lamagna M (2011) Byzantine doctrines on uroscopy in the Liber Orinalibus of Hermogenes (codex 69 Montecassino). J Nephrol 24(S17):S103–S107
33. Mylonas AI, Poulakou-Rebelakou E-F, Androutsos GI, Seggas I, Skouteris CA, Papadopoulou EC (2014) Oral and cranio-maxillofacial surgery in Byzantium. J Cranio-Maxillofacial Surg 42(2):159–168
34. Magiorkinis E, Sidiropoulou K, Diamantis A (2010) Hallmarks in the history of epilepsy: Epilepsy in antiquity. Epilepsy Behav 17:103–108
35. Langslow D (2012) Caelius Aurelianus. In: Bagnall RS, Brodersen K, Champion CB, Erskine A, Huebner SR (eds) The encyclopedia of ancient history. Wiley-Blackwell, Hoboken, New Jersey, USA
36. Prioreschi P (2001) A history of medicine. Horatius Press, Omaha, Nebraska, USA
37. Clarke E (Reviewer) (1960) Jacopo Berengario da Carpi: a short introduction to anatomy (Isagogae breves). Med Hist 4(4): 362–364
38. Burke P (1998) The European renaissance: centre and peripheries. Wiley, Hoboken, New Jersey, USA
39. Strathern P (2003) The Medici: godfathers of the renaissance. Jonathan Cape, Florence, Italy
40. Harris MH (1999) History of libraries in the Western world. Scarecrow Press, Lanham, Md., Maryland, USA
41. Houtzager HL (1981) Reinier de Graaf. Eur J Obst Gynecol Reprod Biol 12(6):385–387
42. Mann RJ (1976) Regnier de Graaf, 1641–1673, investigator. Fertil Steril 27(4):466–468

43. Modlin IM (2000) Regnier de Graaf: Paris, purging, and the pancreas. J Clin Gastroenterol 30(2):109–113
44. Prout W (1974) On the nature of acid and saline matters usually existing in stomach of animals. Gut 15(1):45.1–49.1
45. Russell GA (2013) Vesalius and the emergence of veridical representation in Renaissance anatomy. Prog Brain Res 203:3–32
46. Van Gijn J (2005) René Descartes (1596–1650). J Neurol 252(2):241–242
47. Jeune B (2004) Descartes and medicine. Dan Medicinhist Arbog 75–117
48. Boschiero L (ed) (2007) Giovanni Alfonso Borelli (1608–1679). Experiment and natural philosophy in seventeenth-century Tuscany: the history of the academia del Cimento. Springer, Netherlands
49. Pope MH (2005) Giovanni Alfonso Borelli—the father of biomechanics. Spine (Phila Pa 1976) 30(20):2350–5235
50. Clericuzio A (2012) Chemical and mechanical theories of digestion in early modern medicine. Stud Hist Philos Biol Biomed Sci 43(2):329–337
51. Encyclopædia Britannica (2003) Encyclopaedia Britannica, Inc, Chicago, IL
52. Partington JR (1936) Joan Baptista Van Helmont. Ann Sci 1:359–384
53. van Gijn J (2001) Franciscus Sylvius (1614–1672). J Neurol 248(10):915–916
54. Tubbs RS, Linganna S, Loukas M (2007) Franciscus Sylvius (1614–1672): a historical review. Childs Nerv Syst 23(1):1–2
55. Howard JM, Hess W, Traverso W (1998) Johann Georg Wirsüng (1589–1643) and the pancreatic duct: the prosector of Padua, Italy. J Am Coll Surg 187(2):201–211
56. Bergmann M, Wendler D (1986) Caspar Bauhin (1560-1624). Gegenbaurs Morphol Jahrb 132(2):173–181
57. Harvey W (2012) Lectures on the whole of anatomy: an annotated translation of prelectiones anatomiae universalis (trans: O'Malley CD, Poynter FNL). Literary Licensing, LLC
58. Harvey W, O'Malley CD, Poynter FNL, Russell KF (1961) Lectures on the whole of anatomy: an annotated translation of prelectiones anatomiae universalis. University of California Press, Berkeley, CA
59. Richmond C (1986) Two and a quarter centuries of digestive juices. Trends Biochem Sci 11 (12):528–530
60. von Haller A (1801) First lines of physiology. Ad Neill & Co, Edinburgh
61. Konturek PC, Konturek SJ (2003) The history of gastrointestinal hormones and the polish contribution to elucidation of their biology and relation to nervous system. J Physiol Pharmacol 54(S3):83–98
62. Buess H (1959) Albrecht von Haller and his 'Elementa Physiologiae' as the beginning of pathological physiology. Med Hist 3(2):123–131
63. Wong PY (1989) Secretory mechanisms of salivary glands. Ann R Australas Coll Dent Surg 10:72–83
64. de Schipper JP, Pull ter Gunne AF, Oostvogel HJM, van Laarhoven CJHM (2009) Spontaneous rupture of the oesophagus: Boerhaave's syndrome in 2008: literature review and treatment algorithm. Dig Surg 26:1–6
65. Bloch H (1995) Hermann Boerhaave (1668-1738): life, teacher, practitioner, and activator of hippocratic clinical teaching. Heart Lung 24(2):91–93
66. Underwood EA (1968) Boerhaave after three hundred years. Br Med J 4(5634):820–825
67. Hecser L, Siklodi KP, Csiki G, Lungu MF, Jung H, Buda O (2011) Boerhaave syndrome: a case report. Rom J Leg Med 19:283–286
68. Adams BD, Sebastian BM, Carter J (2006) Honoring the Admiral: Boerhaave-van Wassenaer's syndrome. Dis Esophagus 19(3):146–151
69. Fazio VW, Church JM, Wu JS (eds) (2012) Atlas of intestinal stomas. Springer, US
70. Kitamura N, Tanimoto A, Hondo E, Andrén A, Cottrell DF, Sasaki M, Yamada J (2001) Immunohistochemical study of the ontogeny of prochymosin—and pepsinogen-producing cells in the abomasum of sheep. Anat Histol Embryol 30(4):231–235

71. Rand AG, Ernstrom CA (1968) Activation of prorennin by pepsin1 and 2. J Dairy Sci 51 (11):1756–1760
72. Mickelsen R, Ernstrom CA (1972) Effect of pH on the stability of rennin-porcine pepsin blends. J Dairy Sci 55(3):294–297
73. Kumar A, Grover S, Sharma J, Batish VK (2010) Chymosin and other milk coagulants: sources and biotechnological interventions. Crit Rev Biotechnol 30(4):243–258
74. Rogelj I, Perko B, Francky A, Penca V, Pungercar J (2001) Recombinant lamb chymosin as an alternative coagulating enzyme in cheese production. J Dairy Sci 84(5):1020–1026
75. Roncada P, Gaviraghi A, Liberatori S, Canas B, Bini L, Greppi GF (2002) Identification of caseins in goat milk. Proteomics 2(6):723–726
76. Groves ML, Kiddy CA (1970) γ-Caseins isolated from milk samples typed β-casein A^1 and A^3. J Diary Sci 53(7):931–933
77. Garnot P, Valles E, Thapon J-L, Toullec R, Tomassone R, Ribadeau-Dumas B (1974) Influence of dietary proteins on rennin and pepsin content of preruminant calf vell. J Dairy Res 41(1):19–23
78. Wilde CD (1986) Pseudogenes. CRC Crit Rev Biochem 19(4):323–352
79. Day SB (ed) (1968) Edward Stevens: gastric physiologist, physician and American statesman, with a complete translation of his inaugural dissertation De alimentorum concoctione and interpretive notes on gastric digestion along with certain other selected and diplomatic papers. Cultural and Educational Productions, Montreal, Cincinnati
80. Fruton JS (2002) A history of pepsin and related enzymes. Quart Rev Biol 77(2):127–147
81. Davenport HW (1992) A history of gastric secretion and digestion: experimental studies to 1975. Springer, New York
82. Arago F, Berthelot M, Haller A (2011) Annales de Chimie Et de Physique. Nabu Press, LLC, USA
83. Tipton K, Boyce S (2000) History of the enzyme nomenclature system. Bioinformatics 16 (1):34–40
84. Gutfreund H (1976) Wilhelm Friedrich Kühne: an appreciation. FEBS Lett 62(1):E1–E2
85. Sullivan MX (1926) A distinctive test for cysteine. Public Health Reports (1896–1970) 41 (22):1030–1056
86. Stephens AD (1989) Cystinuria and its treatment: 25 years experience at St. Bartholomew's hospital. J Inher Metab Dis 12(2):197–209
87. Torchinskii YM (1974) Introduction. Sulfhydryl and disulfide groups of proteins, Springer, US
88. Arunachalam C, Woywodt A (2010) Turbid urine and beef-eating rabbits: Claude Bernard (1813–78)—a founder of modern physiology. NDT Plus 3(4):335–337
89. Simoni RD, Hill RL, Vaughan M (2002) The discovery of the amino acid threonine: the work of William C Rose. J Biol Chem 277:e25
90. Wieland Theodor, Bodanszky Miklos (2012) The world of peptides: a brief history of peptide chemistry. Spinger, Berlin, Germany
91. Shaheen NJ, Hansen RA, Morgan DR et al (2006) The burden of gastrointestinal and liver diseases, 2006. Am J Gastroenterol 101:2128–2138
92. Howard JM, Hess WR (2012) History of the pancreas mysteries of a hidden organ. Springer, New York
93. Karamanou M, Koutsilieris M, Laios K, Marineli F, Androutsos G (2014) Apollinaire Bouchardat (1806–1886): founder of modern diabetology. Hormones 13(2):296–300
94. Wisniak J (2015) Louis Mialhe. Revista CENIC Ciencias Biológicas 46(1):90–101
95. Green JR (1901) The soluble ferments and fermentation. University Press, Cambridge
96. Lelut M (1828) Anatomical researches on the epithelium, or internal epidermis. The Lancet 9 (228):556–558
97. Brodman E (1964) Scientific and editorial relationships between Joseph Lovell and William Beaumont. Bull Hist Med 38:127–132
98. Dubois A, Johnson LF (1985) William Beaumont: frontier physician and founding father of gastric physiology. J Clin Gastroenterol 7(6):472–474

99. Mai FM (1988) Beaumont's contribution to gastric psychophysiology: a reappraisal. Can J Psychiatr 33(7):650–653
100. Beaumont W (1838) Experiments and observations on the gastric juice and the physiology of digestion. Maclachlan and Stewart, Edinburgh
101. Prout W (1824) On the nature of acid saline matters usually existing in the stomach of animals. Phil Trans R Soc Lond 1:45–49
102. Konturek SJ, Konturek PC, Konturek JW, Plonka M, Czesnikiewicz-Guzik M, Brzozowski T, Bielanski W (2006) Helicobacter pylori and its involvement in gastritis and peptic ulcer formation. J Physiol Pharmacol 57(3):29–50
103. Banić M, Malfertheiner P, Babić Z, Ostojić R, Kujundzic M, Fatović-Ferenčić S et al (2011) Historical impact to drive research in peptic ulcer disease. Dig Dis 29(5):444–453
104. Solcia E, Rindi G, Buffa R, Fiocca R, Capella C (2000) Gastric endocrine cells: types, function and growth. Regul Peptides 93(1–3):31–35
105. Modlin IM, Kidd M, Malfertheiner MV, Gustafsson BI (2009) Gastric neuroendocrine neoplasia. In: Wang TC, Fox JG, Giraud AS (eds) Biology of gastric cancers. Springer, New York
106. Hersey SJ, Norris SH, Gibert AJ (1984) Cellular control of pepsinogen secretion. Ann Rev Physiol 46:393–402
107. Ciolkowski M, Sharifi M, Tarka S, Ciszek B (2011) Median aperture of the fourth ventricle revisited. Folia Morphol 70(2):84–90
108. Cziko G (2000) A biological perspective on purpose: the physiology of Bernard and Cannon. In: The things we do: using the lessons of Bernard and Darwin to understand the what, how, and why of our behavior. MIT Press, Cambridge, Massachusetts
109. Miller F, Sherrington C (1916) Some observations on the bucco-pharyngeal stage of reflex deglutition in the cat. Q J Exp Physiol 9:147–186
110. Meltzer A (2000) Samuel James Meltzer, M.D. March 22, 1851—November 7, 1920. Proc Soc Exp Biol Med 223:114–117
111. Texter EC Jr (1963) Motility in the gastrointestinal tract. JAMA 184(8):640–647
112. Benison S, Barger AC, Wolfe EL (1987) Walter B. Cannon: the life and times of a young scientist. Belknap Press of Harvard University Press, Massachusetts
113. Dale HH (1947) Walter Bradford Cannon. 1871–1945. Obituary notices of fellows of the royal society 5(15):407–426
114. Popov MM, Klymov KP (1963) An X-ray kymograph as a method for studying the physiology of the gastro-intestinal tract. Bull Exp Biol Med 53(3):352–355
115. Kulchitsky N (Emeritus Professor) (1925) Br Med J 1(3346):340
116. Lange KA, Ovsiannikov VI (1996) I. P. Pavlov at the Imperial Institute of Experimental Medicine and his contribution to digestive physiology. Vestn Ross Akad Med Nauk 1:11–16
117. Todes DP (1997) Pavlov's physiology factory. Isis 88(2):204–246
118. Dunsmoor JE, Mitroff SR, LaBar KS (2009) Generalization of conditioned fear along a dimension of increasing fear intensity. Learn Mem 16(7):460–469
119. Baeyens F, Vansteenwegen D, Beckers T, Hermans D, Kerkhof I, De Ceulaer A (2005) Extinction and renewal of Pavlovian modulation in human sequential feature positive discrimination learning. Learn. Mem 12:178–192
120. Clark RE (2004) The classical origins of Pavlov's conditioning. Integr Physiol Behav Sci 39 (4):279–294
121. Mackintosh NJ (2003) Pavlov and associationism. Span J Psychol 6(2):177–184
122. García-Hoz V (2003) Signalization and stimulus-substitution in Pavlov's theory of conditioning. Span J Psychol 6(2):168–176
123. Rescorla RA (2003) Contemporary study of Pavlovian conditioning. Span J Psychol 6 (2):185–195
124. Coon DJ (1982) Eponymy, obscurity, Twitmyer, and Pavlov. J Hist Behav Sci 18(3):255–262
125. Lectures Nobel (1965) Physiology or Medicine 1922–1941. Elsevier Publishing Company, Amsterdam

126. Pavlov IP, Thompson WH (1902, 2007) The work of the digestive glands: lectures. Griffin, London
127. Sapin MR, Nikitiuk DB, Shadlinsky VB (2001) Small glands of the digestive and respiratory systems. Dzhangar, Elista, Russia
128. Feldman M, Richardson CT, Fordtran JS (1980) Experience with sham feeding as a test for vagotomy. Gastroenterology 79(5 Pt 1):792–795
129. Barshteyn V (2011) Physiology of digestion: pages of history. Probl Nutr 22(5):75–81
130. Halsey WD (ed) (1979) Great Soviet Encyclopedia. Gale, Michigan, USA
131. Gurevich ES, Figurina MM (1957) Research and service activities of S. P. Botkin's hospital in Leningrad; 125th anniversary of S. P. Botkin's birth. Klin Med (Mosk) 35(8):74–81
132. Beliaeva VS (2007) Sergeĭ Petrovich Botkin—founder of physiological school in Russian medicine (to 175th anniversary of birthday). Exp Clin Gastroenterol 5:152–154
133. Konturek SJ (2003) Gastric secretion—from Pavlov's nervism to Popielski's histamine as direct secretagogue of oxyntic glands. J Physiol Pharmacol 54(S3):43–68
134. Holliday R (2006) Epigenetics: a historical overview. Epigenetics 1(2):76–80
135. Felsenfeld G (2014) A brief history of epigenetics. Cold Spring Harb Perspect Biol 6: a018200
136. Woodward ER (1987) The history of vagotomy. Am J Surg 153(1):9–17
137. Modlin IM, Darr U (1994) The centenary of Lester Dragstedt—fifty years of therapeutic vagotomy. Yale J Biol Med 67(3–4):63–80
138. Lagoo J, Pappas TN, Perez A (2014) A relic or still relevant: the narrowing role for vagotomy in the treatment of peptic ulcer disease. Am J Surg 207(1):120–126
139. Alessio AM, Butterworth E, Caldwell JH, Bassingthwaighte JB (2010) Quantitative imaging of coronary blood flow. Nano Rev 1(5110):1–13
140. Hunter JP, Borg TK (2003) Integration from proteins to organs: the Physiome Project. Nat Rev Mol Cell Biol 4(3):237–243
141. Bassingthwaighte JB (2000) Strategies for the Physiome Project. Ann Biomed Eng 28 (8):1043–1058
142. Bassingthwaighte JB, Raymond GM, Butterworth E, Alessio A, Caldwell JH (2010) Multiscale modeling of metabolism, flows, and exchanges in heterogeneous organs. Ann New York Acad Sci 1188:111–120
143. Beard DA, Neal ML, Tabesh-Saleki N, Thompson CT, Bassingthwaighte JB, Shimoyama M, Carlson BE (2012) Multi-scale modeling and data integration in the Virtual Physiological Rat Project. Ann Biomed Eng 40(11):2365–2378
144. Gregersen H (2006) The Giome project. Neurogastroenterol Motil 18(5):401–402
145. Thomas S, Wolstencroft K, de Bono B, Hunter PJ (2016) A physiome interoperability roadmap for personalized drug development. Interface Focus 6(2):20150094
146. Woolf SH (2008) The meaning of translational research and why it matters. JAMA 299:211–213
147. Hall JE (2002) The promise of translational physiology. Am J Physiol Lung Cell Mol Physiol 283:235–236
148. Seals DR (2013) Translational physiology: from molecules to public health. J Physiol 591 (14):3457–3469

Chapter 2
Structural and Functional Organization of the Gastrointestinal Tract

Abstract Gastrointestinal (GI) tract is organized into regions and layers with structural and functional peculiarities. The tract which begins from the mouth, extending to the anus, is comprised of different layers and tissues. The tissues, composed of different cells, play diverse roles and functions and they constitute the chief determinants of the state of GI functioning. Accessory organs of the GI tract (including the mesentery, which is currently regarded as an organ as well as its structural and functional unit) are discussed. In this chapter, the structural and functional organization of the entire GI tract is presented.

Keywords Gut functions · GI functions · GI structure · GI tract layers
Alimentary canal · Gut epithelium · Columnar epithelium · Stratified epithelium
GI tract circulation · GI perfusion · GI lymphatic drainage

Abbreviations

CGRP	Calcitonin gene-related peptide
EA	Esophageal atresia
GI	Gastrointestinal
ICCs	Interstitial cells of Cajal
MAP	Mitogen-activated protein
TEF	Tracheoesophageal fistula

2.1 Introduction

Gastrointestinal (GI) tract (also called alimentary canal) is an open-ended or hollow-like tube, organized into regions and layers with each having peculiar features in structure and functions. The tract begins at the mouth (oral cavity) extending to the esophagus, then to the stomach, from where it runs to the small and large intestines and finally to the anus [1–3]. The lumen of the alimentary canal is characterized by a space, which is bordered by polarized epithelial cells and overlaid with

M. O. Welcome, *Gastrointestinal Physiology*,
https://doi.org/10.1007/978-3-319-91056-7_2

mucus in a healthy person [3]. The physiological functions of each region are determined by the structural characteristics of the layers constituting the region of the GI tract. Each layer possesses a peculiar cellular network and tissue organization [4]. The GI tract is designed to perform a variety of physiological functions which include food processing for energy, defense, secretion of water and a pretty huge quantity of biomolecules, and other body processes. About 60–90% of diseases of humans is either due to or leads to alteration of the structural, physical, chemical, secretory, and neuroendocrine properties of the epithelial cells, resident microbiota, and blood-borne effector cells [3]. In this chapter, the basic structure of the regions and layers of the GI tract and their corresponding functions are discussed.

2.2 Structural Architecture of the Gastrointestinal Tract

Why is the normal architecture of the GI tract important? After review of Clinical Correlate 2.1, you will find out that certain structural features of a given region of the GI tract may lead to delay or inability to process food, which is required for normal functioning of the organism as a whole. In essence, if a malformation occurs in a region of the GI tract during the process of ontogenesis, resulting in a deviation from the normal structure of the tract, food digestion will be inevitably impeded. Thus, it is essential to examine the normal architecture of the GI tract, on which the GI and other bodily functions depend.

Clinical Correlate 2.1

Esophageal Atresia—A Case Where Structure Meets Function

Introduction: Esophageal atresia (EA) is a congenital malformation of the GI tract characterized by an interruption of the continuity of the esophagus with or without a persistent communication with the trachea. In EA, the esophagus ends up in a blind pouch rather than connecting with the stomach as in a normal situation (Fig. 2.1). The esophagus may be linked to one or more fistulae between the interrupted esophagus and the trachea—called tracheoesophageal fistula (TEF). However, TEF may occur separately from the blind-ended esophagus [5].

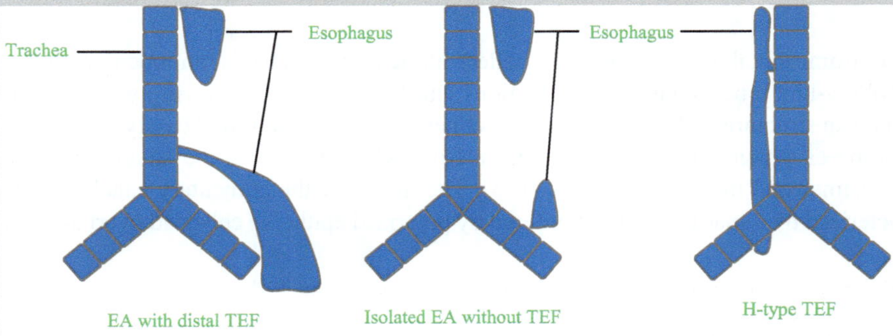

Fig. 2.1 Major structural types of EA

Epidemiology: The incidence of EA is approximately 1 in 2500–3500 live births [5]. Although the cause of this congenital disease is largely unknown, it is believed that the etiology is multifactorial. Accumulating evidences indicate that disorders in gene expression and multiple signaling pathways are responsible for tissue and organ morphogenesis, suggesting a crucial role of genetic mutations and signaling pathways in the etiology of the disease [5–7].

Risk factors: The risk factors that predispose an individual to the development of EA are not completely known, but the following may likely contribute to the development of the disease—maternal polyhydramnios, previous history of congenital anomalies, previous intrauterine deaths, maternal age less than 25 years, and consanguinity [5–8]. Generally, high-risk pregnancy (a condition where the mother or the developing fetus or both are at an increased risk for complications during or after pregnancy and birth) is a risk of EA development for the baby. Also, paternal behavioral contingencies such as medication intake, smoking, drinking alcohol in the periconceptional period may also predispose the fetus to developing EA. Occupational exposure of both parents to toxic solvents, pesticides, or welding fumes during the periconceptional period is a predisposing factor for EA [8–10]. In the baby, combined factors such as low birth weight and the presence of major cardiac anomalies may indicate the presence of EA [11].

Classification of EA: There are currently different classification systems of EA. They include the Vogt, Ladd, Gross, Spitz, and Montreal classifications [5, 11]. Waterston's risk categories are also used for the classification of EA [12]. Some classifications have been revised over time. For instance, the revised Spitz classification identifies three groups of patients with EA and divides them into four classes [11]:

1. Class I. This is the low-risk group comprising of patients with birth weight greater than 2000 g and without major cardiac defects.
2. Class II. This is the moderate-risk group comprising of patients with birth weight less than 2000 g and without major cardiac defects.
3. Class III. This is the relatively high-risk group comprising of patients with birth weight greater than 2000 g and with major cardiac anomalies.
4. Class IV. This is a high-risk group comprising of patients with birth weight less than 2000 g.

Major types of EA include [5, 13]:

1. EA with distal TEF (incidence of 86%).
2. Isolated EA without TEF (incidence of 7%).
3. TEF without atresia (also called H-type TEF) (incidence of 4%).
4. EA with proximal TEF (2–3%).
5. EA with proximal and distal TEF (usually less than 1%).

Additional anomalies are identified in over 50% of individuals with EA [5, 14]. Majority of the associated anomalies involve one or more of the

VACTERL association. First described in 1972 by Quan and Smith, VACTERL (an acronym for "vertebral, anorectal, cardiac, tracheoesophageal, renal, and limb defects") is a random association of the listed malformations occurring due to a genetic defect during the process of ontogenesis [15–17].

Another case of atresia, Feingold syndrome (also called oculodigitoe-sophagoduodenal syndrome), is an autosomal dominant hereditary disorder caused by mutations in the MYCN gene (neuroblastoma-derived V-myc avian myelocytomatosis viral-related oncogene), located on chromosome 2p23-p24. Feingold syndrome is named after Murray Feingold (1930) [18]. This inborn disorder, also known as syndromic EA and TEF, is characterized by hand and foot defects, short palpebral fissures, esophageal, and duodenal atresia with or without fistula, plus microcephaly with or without cognitive or learning impairment [18].

The presence of intestinal fistula in EA is an indication of associated disorders involving other systems [19, 20]. The signs and symptoms depend on the degree of genetic mutation involved. Following surgical repair of the defects and when adequately managed, sufferers of these conditions can reach adulthood. However, they usually experience such conditions as esophageal dysmotility, esophagitis, gastroesophageal reflux, neoplastic growth, and cancer of the GI tract [21, 22]. Other examples of developmental disorders including esophageal stenosis and congenital hiatal hernia also compromise the structural and functional integrity of the gullet [23–25].

Pathogenesis: There is no consensus on the mechanisms of EA and TEF as it is not exactly known how the early foregut differentiates into the respiratory tract and the intestinal tract. But it is generally accepted that these defects are due to disorders in the period of differentiation of the primitive foregut [26–28]. Notwithstanding, however, emerging data have indicated the role of environmental and genetic factors as the possible causes of the disease [29]. Research results from independent laboratories have implicated the sonic hedgehog, bone morphogenetic protein, and fibroblast growth factor signaling pathway in the etiology of the disease [30, 31]. For instance, defects in IIIb-isoform of fibroblast growth factor receptor-2 signaling may be associated with blind-ended esophagus [29–33]. Other signaling pathways such as Wnt and Notch also play important role in normal morphogenesis and organogenesis of the GI tract in the embryo [34]. Details on these signaling pathways are discussed in Chap. 5

Clinical manifestations: If the defect was not diagnosed during intrauterine period, neonates with EA will present usually soon after birth with copious oral secretions, coughing, gagging, cyanosis, vomiting, and/or respiratory distress, inability to swallow saliva, and excessive salivation [5].

Diagnosis: The defect may be observed during antennal visit to a health facility or immediately after birth of the baby. If the defect was not diagnosed during the period of pregnancy, the disease could easily be suspected when certain signs and symptoms such as drooling, which may be associated with choking, coughing, and sneezing; regurgitation of ingested food and fluid through the mouth and nose after normal swallowing are noticed. The returning food and fluid may enter into the respiratory tract and cause decrease in breathing rate or even lead to arrest of breathing. Hence, the neonate becomes cyanotic [34–37].

Passage of suctioning tube into a child immediately after delivery may help to refute or establish the diagnosis of EA. If the condition is present, the tube may not progress beyond 10 cm from the mouth. The suspected condition is confirmed with radiological or surgical techniques such as plain X-ray of the chest and abdomen, magnetic resonance imaging, and endoscopy [5, 35–37]. However, the condition can be detected during routine antenatal visit of the pregnant mother to clinics. At around 18 weeks gestation ultrasonography may show a small or absent stomach bubble. If EA is present ultrasonography of the fetal neck may reveal a blind-ending upper pouch [5, 35–37]. However, rare cases of EA may be diagnosed by preoperative endoscopy involving the use of bronchoscopy or esophagoscopy [5, 35–37].

Complications: The most likely complication of EA is aspiration pneumonia as food accumulates in the blind pouch and can overflow into the trachea and lungs. In addition, a fistula connecting the lower esophagus to trachea may allow stomach acid to flow into the lungs and cause pulmonary disease [13, 38–41].

Prevention: Maternal medications such as folic acid are known to have positive protective effects on the fetus and prevent possible congenital malformation including EA. Furthermore, pregnant women and both parents during the periconception period should avoid contact with occupational hazards as these hazards may predispose conceptus to developing EA [8–10].

Treatment: Treatments for the disease vary depending on its severity. The most immediate, definitive, and effective treatment in the majority of cases is a surgical repair to close the fistula/fistulae and reconnect the two ends of the esophagus to each other with primary anastomosis of the esophagus. This reconstructive surgery restores physiological functions in majority of cases [42]. Specific type of reconstructive surgery is required and depends on the gap between the ends of the esophagus. However, in very rare occasions, esophageal replacement may be required [5]. The surgical procedure referred to as esophageal dilation (Savary bougienage or balloon dilation) may be performed under general anesthesia in some occasions [43]. Endoscopic dilation has also been used as a treatment option. However, it should be pointed that while surgery is the definitive treatment option, no technique is superior to the other. The success of some of the techniques or procedures depends on experience of the surgeon or preference [43].

Postoperative complications that may arise include development of esophageal stricture, resulting from esophageal dysmotility, making it difficult to swallow food or fluid. Other complications are heartburn, esophageal perforation, and dysphagia. In persistent esophageal structure due to esophageal dysmotility, for instance, additional (repeat) surgical treatment may be required [43].

There are other cases of structural disorders comprising of loss of integrity of GI tissue, which can result in physiological dysfunctions. For instance, esophagitis, a clinical condition, characterized by inflammation of the epithelium of the gullet, and esophageal stenosis (narrowing of the lumen of the esophagus), which may occur independently can result to considerable functional disorder such as dysphagia [44–46].

Annoted Review Texts on Congenital Diseases of the GI Tract

Normal development of the gut and the associated defects are discussed in Chap. 6. For further details on congenital malformations of the GI tract, review the following publications.

1. Berrocal T, Lamas M, Gutieérrez J, Torres I, Prieto C, del Hoyo ML (1999) Congenital anomalies of the small intestine, colon, and rectum. Radiographics 19(5):1219–1236
2. Berrocal T, Torres I, Gutiérrez J, Prieto C, del Hoyo ML, Lamas M (1999) Congenital anomalies of the upper gastrointestinal tract. Radiographics 19(4):855–872
3. Chesley PM, Melzer L, Bradford MC, Avansino JR (2015) Association of anorectal malformation and intestinal malrotation. Am J Surg 209 (5):907–912
4. Gupta AK, Guglani B (2005) Imaging of congenital anomalies of the gastrointestinal tract. Indian J Pediatr 72(5):403–414
5. Haak BW, Bodewitz ST, Kuijper CF, de Widt-Levert LM (2014) Intestinal malrotation and volvulus in adult life. Int J Surg Case Rep 5 (5):259–261
6. Karnak I, Ocal T, Senocak ME, Tanyel FC, Büyükpamukçu N (2000) Alimentary tract duplications in children: report of 26 years' experience. Turk J Pediatr 42(2):118–125
7. Klein MD, Hertzler JH (1981) Congenital defects of the abdominal wall. Surg Gynecol Obstet 152(6):805–808
8. Kumar P, Singh S, Rawat JD, Singh S (2016) Postoperative ileocolic intussusception in a neonate with anorectal malformation. APSP J Case Rep 7(5):40
9. Lentze M (2014) Congenital diseases of the gastrointestinal tract. Georgian Med News 230):46–53
10. Martin V, Shaw-Smith C (2010) Review of genetic factors in intestinal malrotation. Pediatr Surg Int 26(8): 769–781
11. Morris G, Kennedy A Jr, Cochran W (2016) Small bowel congenital anomalies: a review and update. Curr Gastroenterol Rep 18(4):16

12. Murruste M, Rajaste G, Kase K (2014) Torsion of Meckel's diverticulum as a cause of small bowel obstruction: A case report. World J Gastrointest Surg 6(10):204–207
13. Podda M, Atzeni J, Messina Campanella A, Saba A, Pisanu A (2016) Syncope with surprise: an unexpected finding of huge gastric diverticulum. Case Rep Surg 2016:1941293
14. Rattan KN, Singh J, Dalal P (2016) Neonatal duodenal obstruction: a 15-year experience. J Neonatal Surg 5(2):13
15. Sasikumar K, Noonavath RN, Sreenath GS, Maroju NK (2013) Axial torsion of gangrenous meckel's diverticulum causing small bowel obstruction. J Surg Tech Case Rep 5(2):103–105
16. Singal R, Garg LN, Singal RP, Gupta S, Shahi SR, Singal S, Singh B (2011) Omphalocele and gastroschisis associated with multiple congenital abnormalities. J Med Life 4(3):295–296
17. Sumer A, Kemik O, Olmez A, Dulger AC, Hasirci I, Iliklerden U, Kisli E, Kotan C (2010) Small bowel obstruction due to mesodiverticular band of meckel's diverticulum: a case report. Case Rep Med 2010. pii:901456
18. Vaos G, Misiakos EP (2010) Congenital anomalies of the gastrointestinal tract diagnosed in adulthood—diagnosis and management. J Gastrointest Surg 14(5):916–925
19. Wanjari AK, Deshmukh AJ, Tayde PS, Lonkar Y (2012) Midgut malrotation with chronic abdominal pain. N Am J Med Sci 4(4):196–198
20. Zheng Z-H, Huang J-L, Wei H-B, Liu J-P, Huang Y (2014) Adult congenital intestinal malrotation accompanied by midgut volvulus: report of eight cases. Int J Clin Exp Med 7(6):1614–1618

The GI tract is the site where digestion takes place. It is in this hallow structure that the mechanical and chemical breakdown of food particles, absorption and transport of digested substances occur. These processes are accomplished not only by the GI tract alone but also accessory organs and all the structures that regulate their functioning. The GI tract and the accessory organs are referred to as the GI system [47, 48].

2.2.1 Regions of the Gastrointestinal Tract

The GI tract is composed of the mouth, pharynx, esophagus, stomach, small intestine, large intestine, rectum, and anus (Fig. 2.2). The mouth, pharynx, esophagus, and stomach belong to the upper GI tract. The structure and functions of

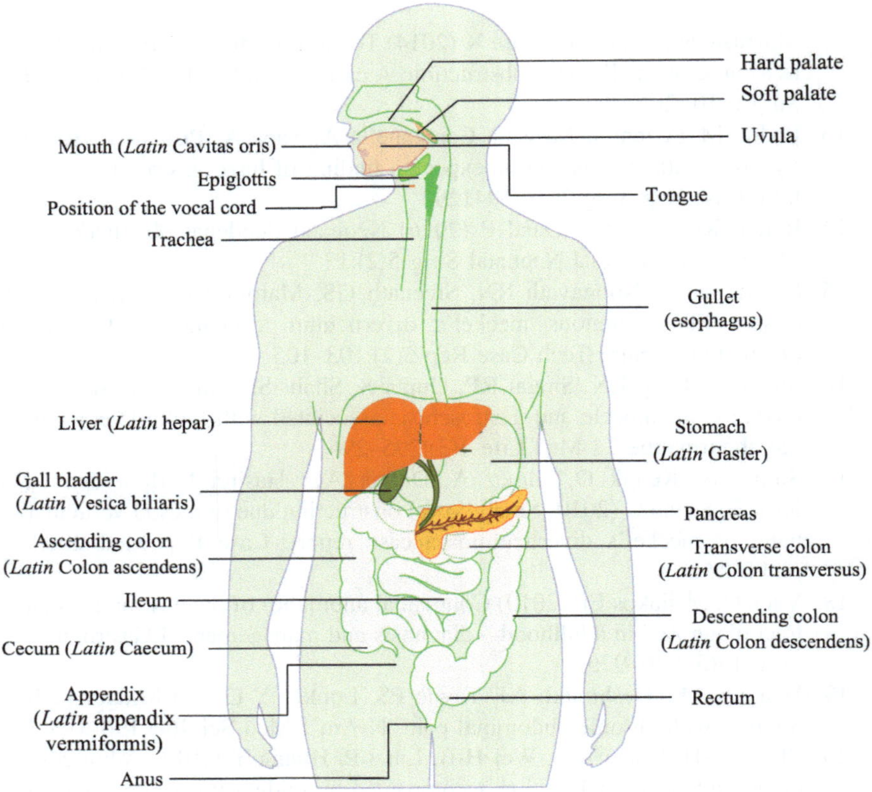

Fig. 2.2 Digestive system of an adult human showing the various regions and associated organs

these regions will be discussed in their respective chapters (Chaps. 7–12). The small and large intestine, rectum, and anus make up the lower GI tract. It should be noted that the anatomical demarcation between the lower and upper GI tract exist precisely at the duodenojejunal (D-J) junction (ligament of Treitz), a thin suspensory muscle of duodenum connecting the junction between the duodenum, jejunum, and duodenojejunal flexure to the connective tissue surrounding the superior mesenteric artery and celiac artery [49–51]. The ligament of Treitz has important clinical application. For instance, it is used to demarcate bleeding occurring in the upper and lower GI tract. Some neoplasms are known to affect this region in particular [49, 50]. More importantly, recent evidences point to the role of this region in regulating evacuation of duodenal contents into the jejunum and is thought to be due to the presence of a physiological sphincter in the region [52–54]. In addition to the duodenum and jejunum, the small intestine is composed of the ileum. The duodenum, a C-shaped segment of the initial part of the small intestine receives chyme from the stomach and secretions from the pancreas that contribute digestive enzymes and bicarbonate as well as other ions that aid in the breakdown of food particles and neutralization of acid chyme. The jejunum, which constitutes the middle small intestinal segment, is involved in secretive and absorptive functions of water, ions, and certain ingested substances. The ileum, which is the terminal end of the small intestine, is responsible for the absorption of nutrients and other molecules [55, 56]. The large intestine is composed of ascending, transverse, descending colon, sigmoid colon, and rectum. The cecum, first part of the large intestine, is a blind-ended sac that leads into the ascending (proximal) colon, then the transverse colon, descending (distal) colon, and rectum. The cecum contains numerous resident gut microbes involved in the fermentative digestion of certain carbohydrates that could not be digested by the intestinal brush border enzymes [57–59]. In addition to other physiological functions (such as absorption of nutrients, water, electrolytes; synthesis of numerous biomolecules including short-chain fatty acids, certain vitamins by colonic resident microbes, production of secondary bile acids required for colonic salvage), these regions of the colon are involved in mechanical transport of undigested products aborally to the rectum for temporary storage, and subsequent evacuation of the formed fecal mass [60]. Table 2.1 shows the length of

Table 2.1 Length of the different region of the digestive tract in an adult human [102, 263, 264]

Region/ organ	Length (cm)	Comments
Tongue	10	Tongue length from the tip to the oropharynx
Gullet	18–25	
Stomach	25	
Duodenum	25–32	Length of the small intestine may substantially vary among different people; it may be in range of 5–8.6 m (50–86 cm) or slightly less
Jejunum	22–25	
Ileum	33–36	
Appendix	2.5	In some individuals the length may reach ~18 cm
Cecum	6.5	
Ascending colon	13	
Transverse colon	38	
Descending colon	25	
Sigmoid colon	25–38	
Rectum	13	
Anus	4	

Note The length of the different portions of the GI tract shown here represents the average length usually occurring in healthy adults. The length of the adult GI tract (i.e., the oroanal length) is around 5–9 m. Two-third of this length is considered the length of the small intestine. The inner diameter of different regions of the GI tract significantly varies. In the small intestine, the average tract diameter is ~2.5 cm and that of the large intestine is ~4.8 cm [102, 228] the regions of the GI tract. The digestive system is a cardinal system connecting the other ten systems of the body (nervous, endocrine, circulatory, respiratory, urinary, reproductive, muscular, lymphatic, skeletal, and integumentary). The GI tract and associated organs are formed from the major tissue types: epithelial, muscle, nervous, connective tissue. These tissues are formed from association of different cells that perform a specific function.

2.2.2 Layers of the Gastrointestinal Tract

Tunica Mucosa (Mucous Layer)

This layer is the side that lies proximal to the lumen (i.e., the passage) of the intestine. The tunica mucosa contains a thick layer of mucous covering the epithelial cells and an underlying layer called the lamina propria (Fig. 2.3) (The cells of the epithelium are linked together by intercellular linkages) (Fig. 2.4). The lamina propria is a thin layer of areolar connective tissue composed of cells and an extracellular matrix made of ground substances (water; glycosaminoglycans mainly hyaluronan and also heparin and heparin sulfate; glycoproteins; proteoglycans) and fibers (collagen and elastin) with numerous lymph nodules, minute blood vessels, small mucous glands, and sensory nerve endings. The lamina propria forms the core of the connective tissue. In most areas of the digestive tract, the lamina propria also contains smooth muscle cells, which make up a layer of tissue called the muscularis mucosae (singular mucosa). This muscle layer produces local movements of the mucosa and does not really move food through the tract [61–65].

Tunica Submucosa (Submucous Layer)

This is a relatively thick and highly vascularized layer of connective tissue that serves the mucosal and submucosal structures. The submucosa is a denser connective tissue than the lamina propria of the mucosa. Absorbed molecules that pass through the columnar epithelial cells of the mucosa enter into blood and lymphatic vessels of the submucosa. The submucosa also contains glands and nerve plexus (submucosal or Meissner's plexus named after the German scientist, Georg Meissner, 1829–1905), which not only innervate submucosal/mucosal glands but also provide nerve supply to the muscularis mucosae of the small and large intestines. GI glands located in the submucosa, pass through the sublayers and open into the lumen via ducts. The arteries and veins of the submucous layer extend from the mesentery. The layer also contains lymphatic vessels [61–64, 66].

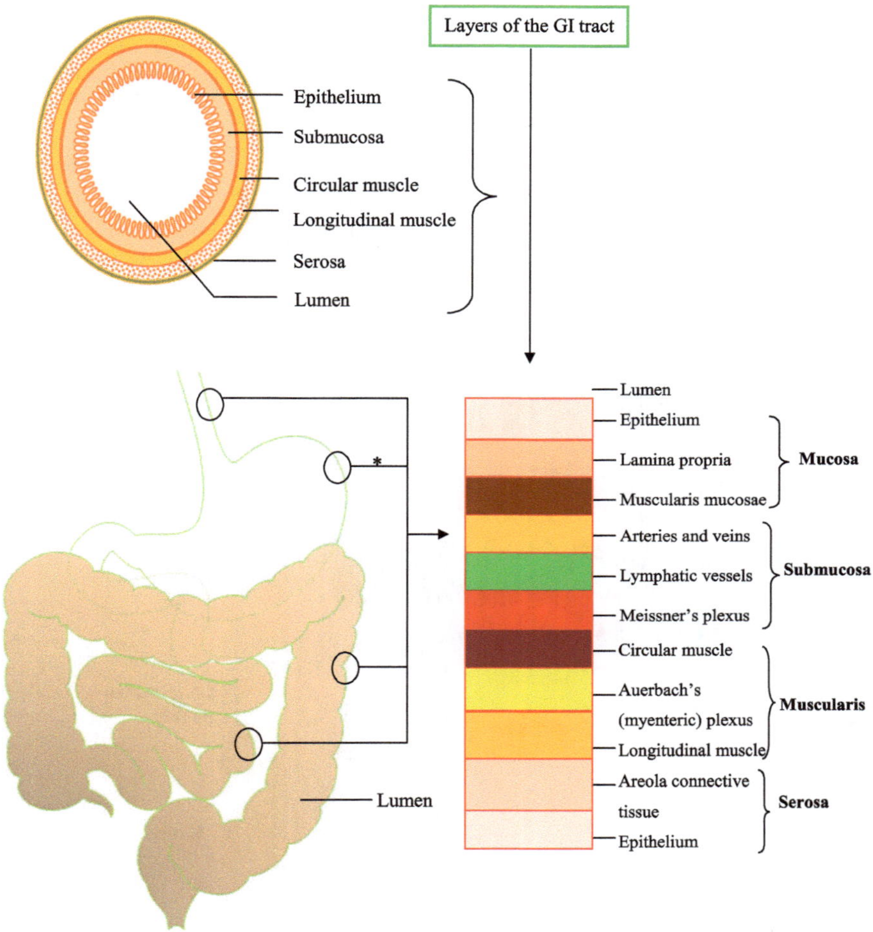

Fig. 2.3 Basic structure of the layers of the GI tract. *The stomach additionally contains an oblique layer of muscles. The muscles of the stomach from the lumen to the serosa are located in the order: oblique (inner layer), circular (middle layer), and longitudinal (outer layer)

Tunica Muscularis (Muscle Layer)

The muscular layer comprises two layers of muscle, the inner and outer layers of smooth muscle cells. The inner layer is arranged in circular rings around the GI tract, and it is called the circular muscle layer. The outer layer is arranged longitudinally, and it is called the longitudinal muscle layer. The stomach has an additional muscle layer, oblique muscle layer, located in the inner side of the organ; it participates in churning the partially digested food. The muscle layers are called *tunica muscularis externa* (external muscle layer; also called the *muscularis propria*), which are oriented smooth muscles of the GI tract responsible for segmental

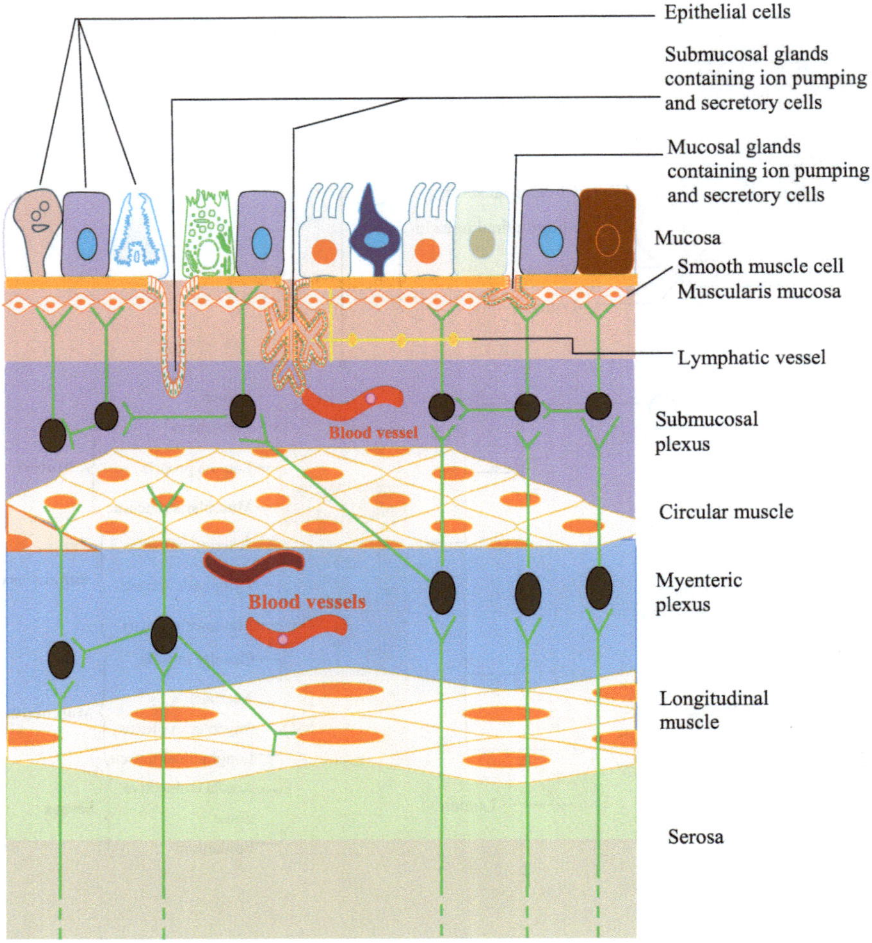

Fig. 2.4 A simplified schematic representation of GI wall showing the intrinsic nervous system and associated cells

contractions and peristaltic movement required for mechanical breaking down of food particles, mixing them with digestive enzymes, and propulsion of food particles along the tract aborally. This motor function (motility or contractility) of the GI tract is regulated by neurons of the sympathetic and parasympathetic divisions of the autonomic nervous system and other factors, which will be discussed in later part of this book [61–63].

In the colon, the outer longitudinal layer forms three discontinuous longitudinal bands, known as *taeniae* coli (meaning "bands of the colon." *Taenia*, plural *taeniae*, is a Latin word meaning ribbon or band). The *taeniae coli* are key features that help to distinguish between the large and small intestine. Another morphological characteristic of the colon is the presence of *haustra* (singular *haustrum*). *Haustra* are formed because the *taenia coli* are shorter than the length of colon, so that the colon becomes sacculated (sac formation) between the *teniae coli*, thereby forming the *haustra* [60, 67]. Contraction of *haustra* is slow segmenting occurring periodically, and it is important in movement of intraluminal contents to the rectum for disposal. The segmentation of the colon into *haustra* facilitates mixing, retention of luminal residues, and formation of solid stools. The reservoir function of the colon is particularly due to the presence of the *haustra*, which have high distendability property [60].

Part of the circular and longitudinal muscular layers, in some junctions of the GI tract, is thickened, forming sphincter, which regulate the movement of food particles and fluid from one region of the GI tract to the other. The thickened circular and longitudinal muscular layers that can be distinguished structurally are called anatomical sphincters. These sphincters are present in many parts of the GI tract. For example, the gastric pylorus has a thickened portion of the inner circular layer, referred to as the pyloric sphincter, which regulate the movement of partially digested food (chyme) from the stomach into the duodenum. The upper esophageal sphincter regulates the movement of substance between the pharynx and the esophagus; lower esophageal sphincter regulates the movement of substances between the esophagus and the stomach. It is believed that the circumferential narrowing by a fold of muscle tissue at the ceco-ascending colon junction is also due to the presence of an anatomical sphincter [60, 68–70]. Physiological sphincters may be formed due to reflex stimulation of the autonomic nerve fibers which innervate area of the tract. Some physiological sphincters of the GI tract are discussed by Gagliardi et al. [71] and Saranović et al. [72].

The property of the smooth muscles to develop force at a given length (called motility or contractility) of the GI tract is regulated by neurons of the sympathetic and parasympathetic divisions of the autonomic nervous system [61–63]. However, this motor function of the smooth muscles is also regulated by the intrinsic neurons of the gut, which are capable of functioning without influence from the autonomic nervous system. The intrinsic neurons of the gut are located in two plexuses—the submucosal plexus and the myenteric or Auerbach's plexus. The former has been discussed in the previous subsection. The latter is named after the German scientist, Leopold Auerbach (1828–1897); it is located between the inner and outer muscular layers and provides innervation to smooth muscle cells of the GI tract. In strict consideration, however, the plexuses include fibers and ganglia from both the sympathetic and parasympathetic divisions of the autonomic nervous system as well as the intrinsic neurons of the gut [73–77]. The autonomic and the intrinsic nervous systems regulate the tone of smooth muscles of the GI tract (Tone is the state of relaxation or contraction of smooth muscle) [60].

The coordinated contractions of the GI wall resulting from the activities of the smooth muscles are initiated by the pacemaker cells called interstitial cells of Cajal (ICC). ICCs are widely distributed within the submucosal, intra- and intermuscular layers of the GI tract beginning from the esophagus to the internal anal sphincter. But most ICCs are present around the circumference of the myenteric plexus [78–80].

The muscle layer and its mechanisms of functioning are discussed in detail in Chap. 7.

Tunica Serosa/Adventitia (Serous/Adventitial Layer)

The *tunica serosa* (also called serous layer) is the outer layer of the GI tract, formed by simple squamous epithelium (called mesothelium) and a connective tissue. The serosa helps to hold the GI tract and its structures in place. The epithelium of the serosa reduces frictional forces during movement of the walls of the GI tract. The simple squamous epithelium doubles as a continuation of the visceral peritoneum. The serous layer is present in some regions (or organs) of the GI system and it protrudes into the peritoneal cavity. Such organs or regions of the GI tract are referred to as intraperitoneal. Thus, intraperitoneal organs (suspended by the peritoneum) are covered by serosa. The intraperitoneal regions of the GI tract include most parts of the stomach, the entire small intestine (except the distal part of the duodenum). The initial portion of the duodenum, cecum, appendix, transverse colon, and sigmoid colon are intraperitoneal. The mesentery is connected to these parts of the tract. In other organs or regions of the GI tract, only the connective tissue portion of the serosa (called adventitia) is present (Fig. 2.3). Such organs or regions of the GI tract are devoid of serosa and are referred to as retroperitoneal [56, 81, 82]. Thus, retroperitoneal organs/regions of the GI tract are covered with adventitia. The retroperitoneal organ or regions include the pancreas, oral cavity, esophagus, pylorus of the stomach, distal duodenum, ascending colon, descending colon, rectum, and anal canal [81, 83–85]. For the purpose of emphasis, retroperitoneal organs lie in the posterior part of the peritoneum, the membrane that lines the abdominal cavity, covering most of the viscera (viscus).

2.2.3 *Epithelium of the Gastrointestinal Tract*

GI epithelial cellular network comprises a network of structurally and functionally interrelated cells of the GI tract. The epithelium covers the whole of the GI tract. The epithelium of the GI tract is either stratified or simple columnar [48].

Figure 2.5 is a typical schematic representation of the simple columnar epithelium. Simple columnar epithelium is found in regions that have high secretion and absorption such as the stomach and small intestine. The cells of simple columnar epithelium possess extensions on the luminal side, referred to as microvilli in the small intestine, or cilia found in the female reproductive tract and respiratory tract. Thus, the epithelium of the stomach and intestines is not only simple columnar but also non-ciliated [86–94].

Stratified epithelium differs from simple epithelium in that it is multilayered. It is therefore found where body linings have to withstand mechanical or chemical insult such that layers can be abraded and lost without exposing subepithelial structures.

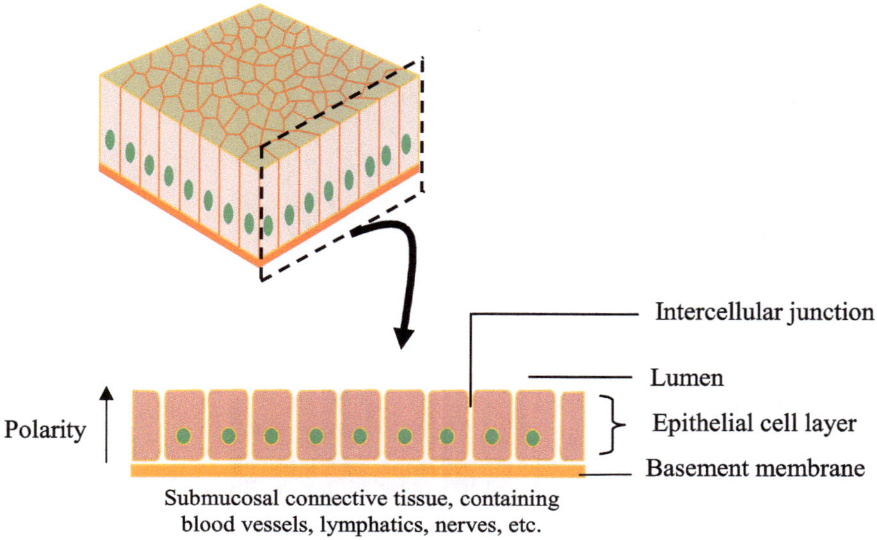

Fig. 2.5 Schematic representation of simple columnar epithelium. The epithelial cell layer contains different types of cells. Intercellular junction comprises the tight junction, adherens junction, desmosome, and gap junction

(Regions of the GI tract where stratified epithelium is present include the oral cavity, pharynx, and esophagus). The cells of stratified epithelium flatten as the layers become more apical, though in their most basal layers the cells may be squamous, cuboidal, or columnar. Stratified cuboidal epithelium contains cuboidal cells, which, as the name implies, are roughly cuboidal in shape, with each cell having a nucleus in the center [86, 88, 95]. The ducts of pancreas and salivary glands are composed of cuboidal epithelium [96–99].

Stratified squamous epithelium contains squamous cells having the appearance of thin plates, closely fitted together in tissues (Fig. 2.6). The shape of the nucleus usually corresponds to the cell form, usually having horizontally flattened and elliptical. Stratified squamous epithelium helps to reduce friction on the surfaces through which fluid flows. The stratified squamous epithelium lines the esophagus and mouth. In addition, the epithelium in these regions is non-keratinized [86, 88, 92]. The anal valves consist of stratified squamous non-keratinized epithelium, whereas the anus comprises stratified squamous keratinized [86].

The proximal part of the anal canal is composed of cuboidal epithelium. Figure 2.7 is a schematic diagram of a typical cuboidal epithelium [86].

Other tissues of the gut, besides the epithelia, include the nervous, connective, and muscle tissues [100, 101]. Details on these tissues and their cell compositions as well as their physiology are discussed in later part of this book.

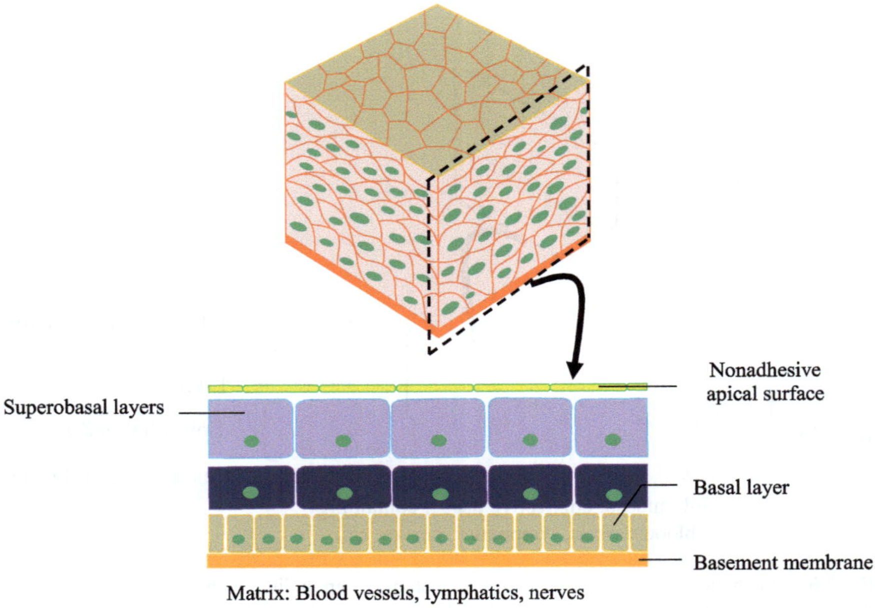

Fig. 2.6 Schematic representation of stratified squamous epithelium

Fig. 2.7 Schematic representation of stratified cuboidal epithelium. The composition of basic components of stratified cuboidal epithelium is the same as in Fig. 2.6

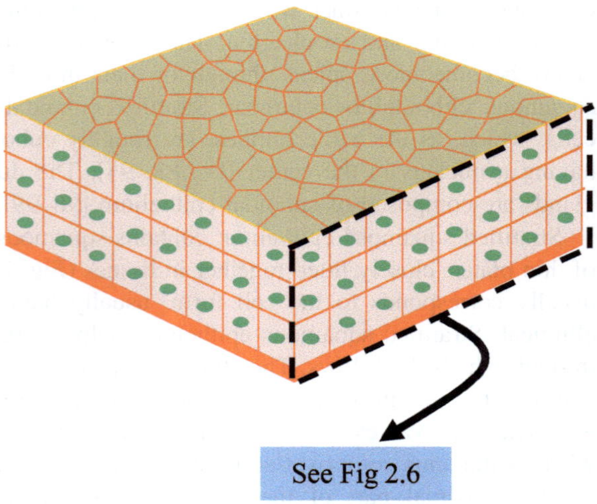

2.2.4 Microarchitecture of the Mucosa

As previously noted, the mucosa comprises the epithelium, the lamina propria, and the muscularis mucosae. Here, we shall discuss the structure of the mucosa, which is critical to digestion, in detail.

Recent estimations indicate that the surface area of an adult human intestinal epithelium exceeds the surface area of a tennis court and more closely corresponds with half a badminton court, suggesting an enormous reserve capacity of the human epithelium for digestive processes [87, 102]. The vast surface area of the intestinal epithelium is due to its morphological characteristics. The epithelium of the intestine possesses invaginations called crypts of Lieberkuhn and finger-like projections called villi. In addition to intestinal villi, the surface of the intestinal lumen contains plicae circulares, and microvilli, which collectively increase the surface area of the lumen by 400–600-fold (this estimate depends on the method used in the evaluation of the mucosal length and diameter). Figure 2.8 is a typical representation of the morphological structures of the intestines. These special morphological structures substantially increase absorption of the products of digestion in the walls of the small intestine [103]. The degree of amplification of surface area for absorption by different intestinal structures depends on the region of the tract. For instance, the colon contributes only $\sim 2\ m^2$ to the total mucosal surface of the GI tract ($\sim 32\ m^2$) (The surface area of the human gut mucosa is not 260–300 m^2 as previously thought, but $\sim 32\ m^2$) [102].

Villi were discovered in the small intestine as early as 1861 by the Swiss anatomist and physiologist, Rudolf Albert von Kölliker (1817–1905). There are no villi in the large intestine. In embryonic life, however, both the small and the proximal large intestine possess villi, but in many species, (including human but not in chick) the embryonic colonic villi continue to flatten until fetal life when they are completely lost following birth. Human colon has a relatively flat epithelium separated regularly by crypts. Thus, all parts of the colon (including the initial part—the cecum) are devoid of villi. The colonic mucosa consists mainly of crypts, which are smaller compared to the small intestinal crypts; the epithelium is flattened [57, 104–106].

Villi are permanent finger-like extensions of the lamina propria and epithelium, covered by a surface epithelium that has an extensive brush border comprising the microvilli, which contain digestive enzymes and possess in their surrounding plasma cells, lymphocytes, fibroblasts, mast cells, smooth muscle cells, capillary loops, and a blind-ended lymphatic capillary, called lacteal. The lacteal is responsible for absorption of digested products of fats and oils (further details on lacteal are presented below). The villi increase the surface area tenfold. The mucosa of the finger-like extensions has a large population of simple columnar epithelium cells (absorptive cells), which in turn possess microvilli on their luminal surfaces [1, 57, 102, 106, 107]. The villi contain the majority of the absorptive cells, possessing a peculiar morphology which increases the surface area of the intestine for absorption. The villi large surface area decreases the average distance traveled by nutrients, which increases the effectiveness of diffusion [64, 108]. The base of the villi is made up of tubular glands called crypts of Lieberkuhn (also called intestinal glands), which secrete various substances that regulate the functions of the GI tract [81, 109–112]. Such glands in the duodenum are called Brunner's glands (or duodenal glands) which produce alkaline fluid rich in mucus and numerous biologically active molecules [56]. In the large intestine, such glands, formed as

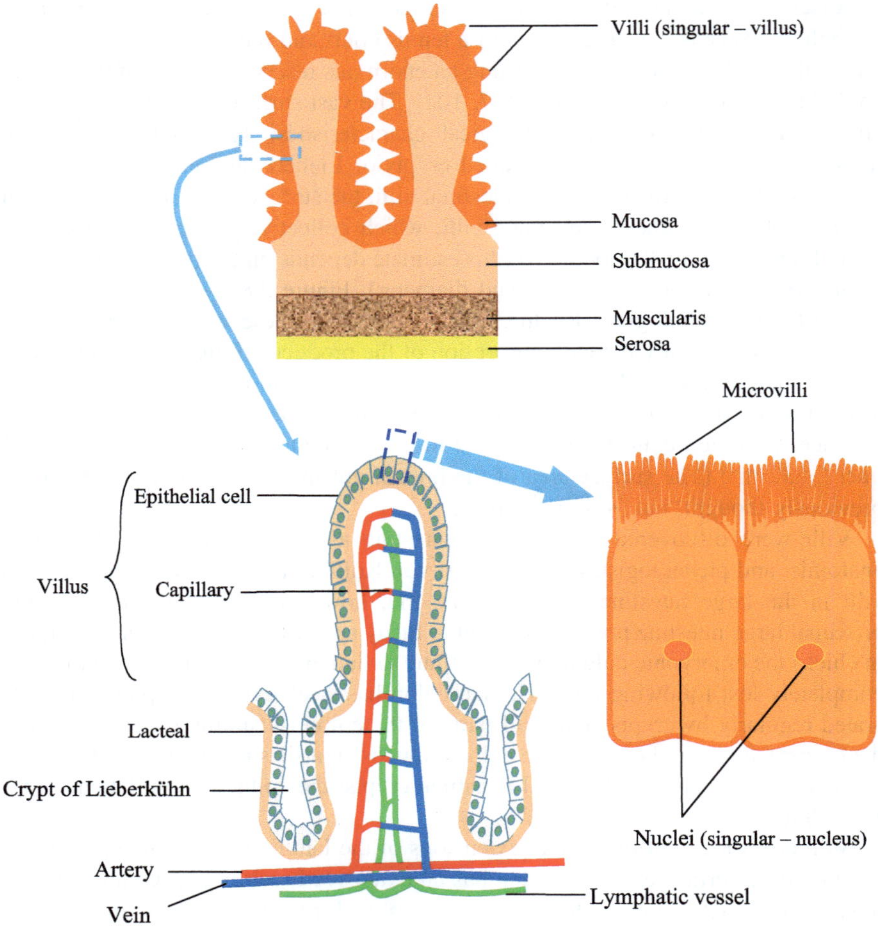

Fig. 2.8 Structure of the villus. A villus is about 1 µm in length, 0.5–1.6 mm in height, and a diameter of ~0.1–0.3 mm. Each villus is covered by a single layer of columnar epithelial cells, mainly goblet cells [228, 259]. Each cell contains series of microvilli, which consist of a cylindrical membrane protrusion of about 1–2 µm long with a diameter of approximately ~100 nm. The apical microvilli play a role in the maintenance of epithelial cell polarization. The use of stereological techniques in studying the morphology of striated brush border of intestinal absorptive epithelial cells showed that microvilli are adaptive in different physiological states [259, 260]. For instance, compared with control, fasting induces hypoproliferation, and streptozotocin-induced diabetes in animal model results in hyperproliferation. It is believed that cell number is crucial in determining villous and microvillous surface areas as the mentioned condition or state do not necessarily alter average length, diameter, or packing density of microvilli [260]. In the perspectives of molecular physiology, different signaling pathways maintain the structure of the microvilli, and hence the entire epithelia of the GI tract. The microvillus length is maintained under complex mechanisms involving several signaling pathways notably, MAP (mitogen-activated protein) kinase, a type of protein kinase enzyme present in cells. The MAP-kinase pathway controls the length of the microvilli by phosphorylation/dephosphorylation reactions. Phosphorylation results in elongation of microvilli length [261]

invaginations of the colonic epithelium, responsible for fluid and mucin secretion are generally called colonic pits (or colonic glands) [65].

The microvilli are finger-like projections of the epithelia, measuring about 1 μm in length and 1 micron in height, extending into the lumen of the intestinal tract. The microvilli constitute the brush border (or striated brush border or just striated border) of intestinal columnar absorptive cells. Each microvillus contains bundle of 30 actin filaments running longitudinally through the core and extending into a zone of intersecting filaments in the apical cytoplasm. Interconnected cytoskeletal filaments include special types of myosin and intermediate filaments. Apart from the absorptive columnar cells, the mucosa also contains a number of different cells (these cell types of the GI mucosa are discussed in Chap. 3) [81, 109–112]. The microvilli are very unstable structurally which is possibly due to high level of plasticity and thereby constantly undergoing dynamic changes in the cytoskeleton [1, 81, 102, 107, 109]. The greatest contribution to the surface area of the intestine for absorption comes from the microvilli.

In 1670, the Dutch physician, Kerckring Theodor (1640–1693) became the first to describe permanent circular folds of the luminal surface of the small intestine, which now bear his name—Kerckring folds or valves. These folds (also called Plicae circulares or valvulae conniventes) are represented by permanent spiral folds of the mucosa and submucosa projecting into the lumen of the small intestine and arranged perpendicular to the long axis. Compared to the villi and microvilli, they contribute to minimum increase (about ~1.5–2-fold) in the surface area of the intestine. The villi lie on the plicae circulares [1, 56, 102, 107, 109].

While the intestinal structures listed here primarily help in absorption of nutrients, water and ions and other molecules, these substances directly affect the absorptive functions of these structures themselves. It has long been established that enteral nutrients maintain small intestinal structure and functions [113]. The growth of mucosal structures, height and area, as well as digestive and absorptive functions can be significantly enhanced by certain nutrients [114]. Thus, unhealthy nutrition, fasting, and parenteral nutrition can induce atrophy and functional compromise of the intestinal mucosa. Surprisingly, surgical intervention such as partial rejection of the distal part of the intestine referred to as partial distal enterectomy confers on the intestine, adaptive function, which, in part, is due to the stimulatory effect of specific nutrient components, a phenomenon referred to as functional workload. The nutrients stimulate increased secretions from the pancreatobiliary system and intestinal glands [113]. This area of experimental research is a promising one as molecules whose secretions are enhanced following partial distal enterectomy could be produced in vitro for possible use in cases of intestinal atrophy.

2.2.5 *Accessory Organs of Digestion*

The activities of certain organs associated with the digestive tract help in digestion of food substances and thus are integral in normal GI functioning. These organs are collectively called accessory organs of digestion. Accessory organs of digestion are the teeth, tongue, salivary glands, pancreas, gallbladder, and liver. Each organ is adapted to carrying out specific functions [115–117]. Disorders of these organs can result in dysregulation of digestive processes [115, 116]. The mesentery was recently considered as an organ in a recent study [118]. The mesentery is a sheet of serous membrane that connects the parietal peritoneum to the visceral peritoneum. Between the parietal peritoneum (part of the peritoneum that lines the abdominal wall) and visceral peritoneum (part of the peritoneum that covers the abdominal organs) are spaces where serous fluid is found. The serous fluid provides lubrication and maintenance of homeostasis for the viscera [119–122]. The mesentery is responsible for suspending some parts of the digestive tract within the peritoneal cavity. The mesentery provides a passage for blood vessels, nerves, and lymphatic vessels in the GI tract [120–122]. The mesentery aids in digestive functions by providing protection for some organs/regions of the digestive system, maintaining transport of fluid and cells across the serosal layers, and control of inflammation and tissue repair in the GI tract [123]. Other physiological and structural aspects of the mesentery are discussed below.

2.3 Gastrointestinal Circulation

GI blood circulation can be grouped based on different criteria. Like any other part of the circulatory system, the GI tract circulation can be divided into arterial blood and venous blood circulation. However, some authors consider classification of GI tract blood circulation at the organ level [124, 125]. GI tract circulation belongs to the broader splanchnic (from Greek "*splanchnikos*" meaning "inwards") circulation, composed of gastric, small intestinal, large intestinal, hepatic, pancreatic, and splenic circulations, arranged in parallel. Splanchnic circulation is the type of circulation that occurs in organs that are supplied by the celiac, superior mesenteric, and inferior mesenteric arteries. The blood supplying the GI tract (from the three branches of the abdominal artery mentioned above) is returned to the portal vein. From here, the fairly deoxygenated blood is transported to the liver [126–129].

2.3.1 Gastrointestinal Venous Circulation

Figure 2.9 shows the veins associated with GI tract circulation. Of particular importance to digestion is the portal vein—a valveless, low pressure (5–10 mm Hg), and low resistance system that play an immense role in the delivery of nutrients from the gut to the liver. The blood vessels, through which absorbed substances are transported, extend from the core of the villi through the submucosa into the serosa and pass through the mesentery and empties into the portal vein from the branches of the mesenteric veins. The portal vein is formed not only from the mesenteric veins

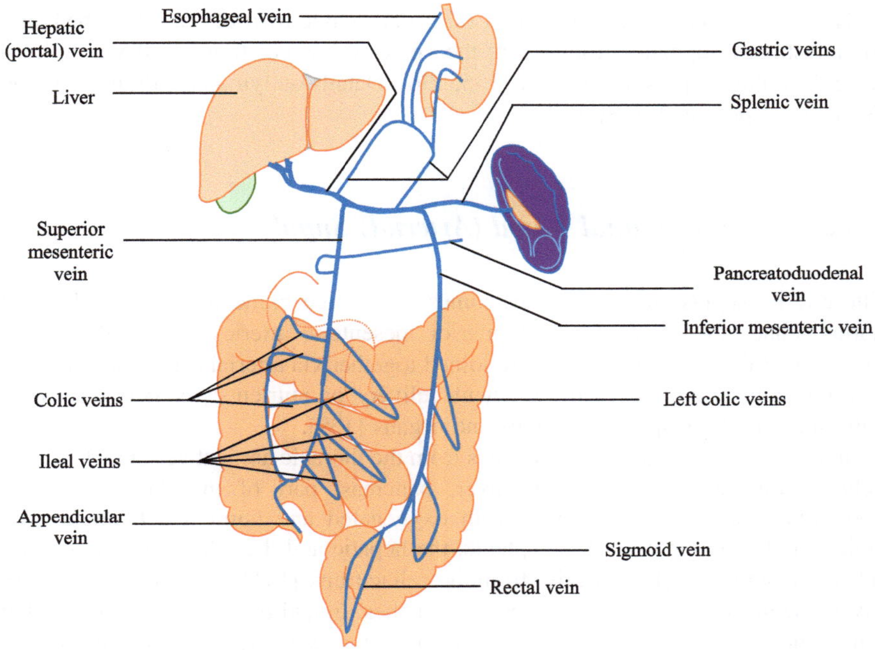

Fig. 2.9 A schematic representation of the venous system of the GI tract

(superior and inferior mesenteric veins), but also the union of all veins draining the intraabdominal organs—spleen, stomach, intestines, gallbladder, and pancreas to the hepatic sinusoid, a type of capillary present in the liver [130–132]. Overall, about 40% of the mesenteric circulation is contained in venules alone, suggesting the integral role veins play in regulating GI circulation [129]. It is estimated that about 75–80% of hepatic blood is supplied by the portal vein alone. The total volume of blood flow in the portal vein is estimated to be about 700–850 ml/min. Blood flow rate in the portal vein is affected by many factors including inflow rate from mesenteric arteries; velocity of venous drainage from the spleen, pancreas, intestines, stomach, and omentum; and the intrahepatic vascular resistance [133, 134]. Further details on the factors that generally affect GI tract blood flow are provided in Sect. 2.3.3.

The portal vein delivers nutrient-rich blood to the liver [130–132]. About 50–80% of substances absorbed in the GI tract are stored by the reticuloendothelial and hepatic cells of the hepatic sinusoids. The hepatic vein transports deoxygenated venous blood from the liver to the inferior vena cava and then to the right atrium from where it is transported, first to the lungs for oxygenation and later to the heart to be sent to all parts of the body including the GI tract [135–138].

However, synthesized chylomicra (singular "chylomicron") are absorbed into the intestinal lymphatics from where they enter the circulating blood via the thoracic duct, thus bypassing the liver [135, 136]. Details on lymphatic drainage to the GI tract are provided in Sect. 2.4.

2.3.2 Gastrointestinal Blood (Arterial) Supply

The arterial supply of the GI tract comes from three branches of the abdominal aorta: celiac artery, superior and inferior mesenteric arteries [139, 140]. These branches of the abdominal artery are joined together via arterial trunk anastomoses, forming collaterals that ensure efficient delivery of nutrients, oxygen, and other substances to the respective regions and organs [139].

In humans, the celiac trunk branches from the thoracic aorta at vertebra T12. The celiac trunk feeds the stomach, liver, pancreas, part of the duodenum, and abdominal part of the esophagus with oxygenated and nutrient-rich blood. The celiac trunk is composed of the splenic, the hepatic and the left gastric artery. But other collateral vessels may arise from the celiac trunk [141]. The left gastric artery gives rise to the esophageal and gastric branches. In addition, the esophagus is fed with little arteries branching off the thoracic aorta (esophageal branch of the thoracic aorta). The common hepatic artery produces the following branches—proper hepatic artery, right gastric artery, gastroduodenal artery, which feed their respective organs and regions of the GI tract. The splenic artery gives rise to numerous branches that supply the pancreas, omentum, and spleen [142]. Overall, 25% of hepatic blood is supplied by the hepatic artery (recall that about 75% is supplied by the portal vein) [133]. The mean pressure of blood in the hepatic artery is approximately equal to the pressure of blood in the aorta [134].

The superior mesenteric artery supplies the right colon, part of transverse colon, and small intestine, while the inferior mesenteric artery supplies the left colon [126]. There are two main components of arterial blood supply of the GI tract—intramural and extramural components. The intramural component of GI tract blood supply involves arterial vascular system that is located inside the walls of the regions and organs of the GI tract such as the intestines and liver. The extramural component indicates the type of vascular distribution seen in the esophagus [140].

It should be noted that numerous anatomical variations on GI tract blood vessels have been reported in the literature. Most of these variations do not impose health problems to the individual [142–144] (Fig. 2. 10).

2.3.3 Gastrointestinal Perfusion

Perfusion may be defined as the volume of blood that flows through a given quantity of tissue per unit time. It is usually expressed in ml of blood per minute per

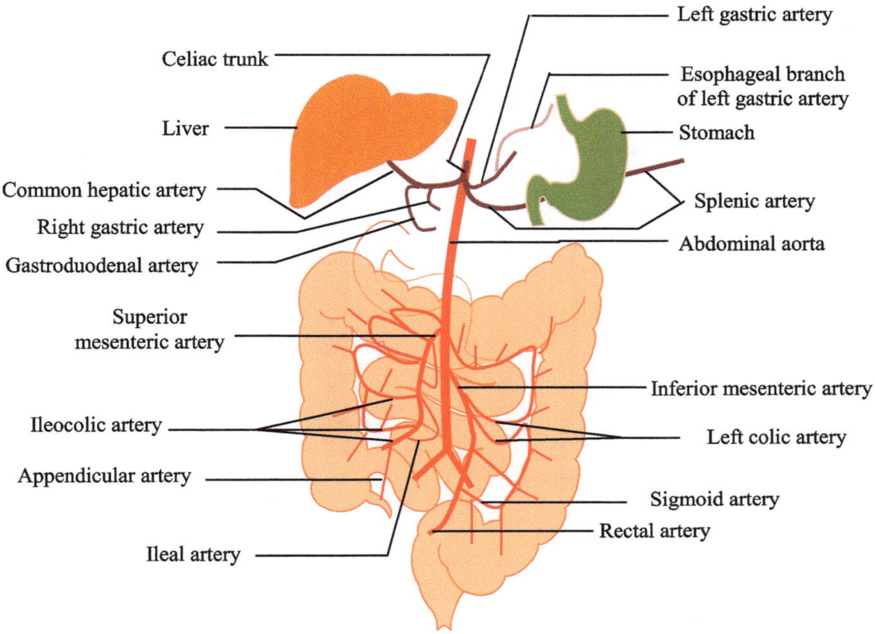

Fig. 2.10 A schematic representation of the arterial system of the GI tract

100 g of tissue [145]. GI perfusion substantially varies in different regions or organs of the tract. Perfusion is 100–120 ml/100 g per min for the pancreas [146], and 100–130 ml/min per 100 g for the liver (or 30 ml/min per kilogram of body weight, which is equivalent to about 800–1200 ml/min [134]. This indicates a rich blood supply in the liver—it is only 2.5% of the total body weight, but receives nearly 25% of the cardiac output [134]. The blood flow rates of different layers of the GI tract also differ. For instance, about 75% of total blood flow is distributed to the mucosa and submucosa, whereas the muscle layer gets only 25% of the total blood flow [147]. Even in different parts of a given GI tract organ or region, blood flow substantially varies. In the colon, for instance, blood flow for the mucosal–submucosal layers may be as high as 100–120 ml per minute per 100 g tissue, whereas it is only 8–28 ml per minute per 100 g tissue for the muscle layer. In the ileum, out of the total blood flow rate of about 21–42 ml per minute per 100 g tissue, estimates for the muscle layer show a blood flow rate of 19–24 ml per minute per 100 g, whereas it is 30–57 ml per minute per 100 g tissue for the mucosal–submucosal layers [147, 148] (Fig. 2. 11).

Overall the GI tract receives about 20–25% of the total cardiac output during periods of fasting [149], and this value may increase to 25–200% (depending on the blood vessel type, calorie content, and type of nutrients as well as the physiological state of the individual) reaching a maximum value at about 20–40 min following feeding. The increased blood flow can remain for about 1.5–2 h and up to 7 h.

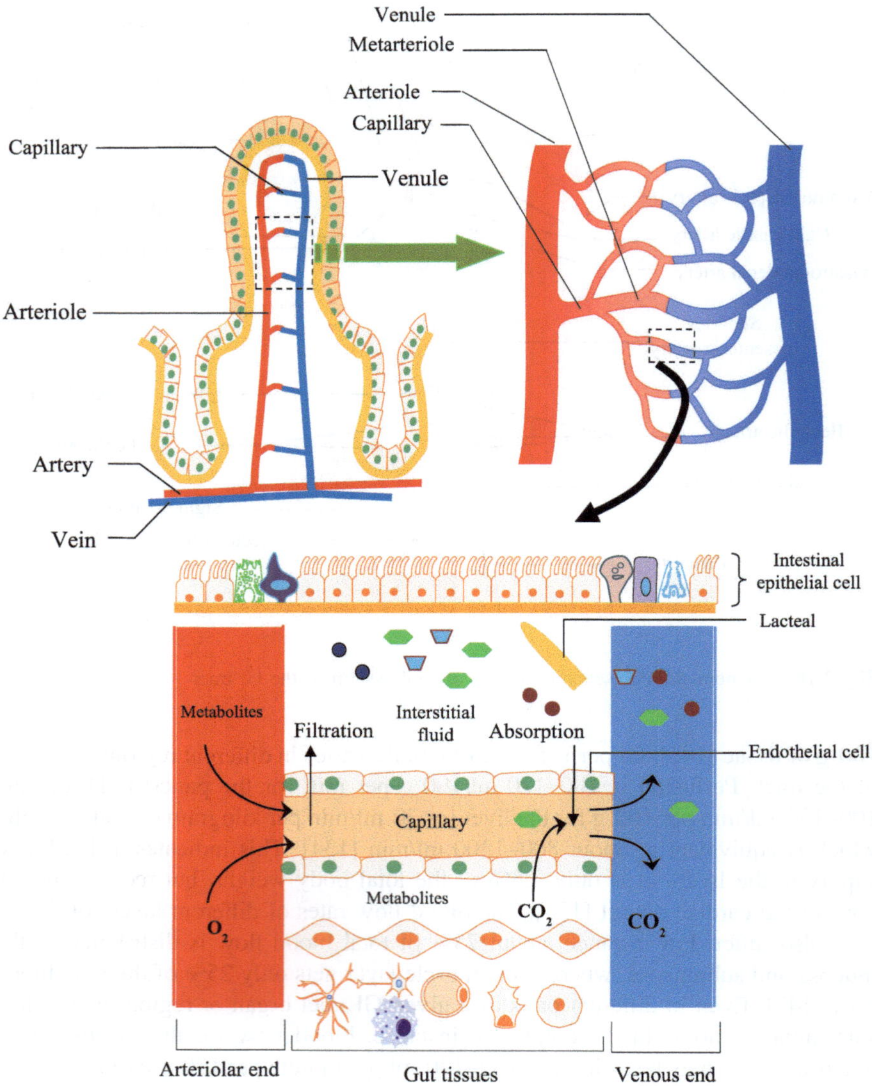

Fig. 2.11 Simplified scheme of exchange of substances between the gut and blood

This increase GI blood flow is due to increased capillary recruitment in response to ingested nutrients—proteins, carbohydrates, and fats [129, 130, 150, 151]. During high-intensity exercise, blood flow in the gut reduces, which is one of the main factors responsible for GI-related symptoms experienced by elite athletes [152, 153]. Depending on certain factors (e.g., method of analysis, duration of exercise) reduction in GI blood flow may reach 80% and is required to ensure adequate supply of blood to the working skeletal muscle during high-intensity exercise [152].

The oxygen consumption by the GI tract is estimated to be 20–35% of total body oxygen consumption. Research indicates that when blood flow decreases in the GI tract, the percentage of oxygen consumption is maintained, due to compensatory increase in oxygen extraction, suggesting a considerable reserve in the GI tract. This reserve is due to the adaptation by the microvascular beds, as well as the architecture of the extensive collateral capillary beds serving as additional conduit during periods of decreased oxygen delivery. The effect of oxygen decrease on the GI tract may only become evident at about 30–50% reduction in oxygen supply to the microvascular beds. Oxygen consumption seems not to be affected during the postprandial state. For similar reasons (adaptive compensatory extraction of oxygen), low to moderate intensity exercise may not affect the oxygen consumption rate in healthy individuals [149, 154, 155]. Thus, even at relatively lower level of oxygen supply, the mucosa and submucosa continue to be nourished with the required substances. In fact, the total blood flow that passes to the mucosal and submucosal layers remain maintained at the physiological range of 60–80% of the total intramural blood flow. The remaining 20% is fed to the muscular layer [130, 149]. Maintenance of GI perfusion is the main role of splanchnic circulation, which is required to ensure adequate circulation especially in the mucosa and submucosa to ensure tissue integrity and functions [129]. Decrease in GI perfusion may contribute to some pathologies of the GI tract. Such condition may occur in necrotizing enterocolitis in neonates, which may be caused, in part, by substantial decrease in intestinal perfusion [156].

Efficient exchange of materials occurs in the microvascular beds and depends particularly on the structure of the capillaries. Capillaries are composed of endothelial cells coupled by tight junctions and anchored to a basal membrane. The interendothelial junctions formed between endothelial cells can be opened or closed. The basement membrane is the abluminal side of the capillary endothelium, whereas the opposite side—the luminal side is covered by a glycocalyx [65, 157]. The permeability of capillaries is controlled by the exchange segment of the vascular tree, comprising the glycocalyx and basement membranes from capillaries, pre- and post-capillary sphincters—these are responsible for controlling the exchange of molecules between the blood and the interstitial fluid, while maintaining blood and tissue homeostasis. Transport of substances across the capillaries may occur by movement of substances in between the cells via interendothelial junctions (paracellular transport) or by movement of the substances across the cells through solute transporters or vesicular carriers (e.g., caveolae) or pore-like subcellular structures (e.g., fenestrae and channels). Movement of substances by solute transporters or carriers as well as transcytosis is called transcellular transport [158]. The fenestrae, caveolae, transendothelial channels, and interendothelial junctions participate in regulating the permeability of capillaries to different substances [65, 157–159].

The microvascular bed of the GI tract is composed of different types of capillaries—discontinuous and fenestrated [160]. In addition to these two types of capillaries, initial electron microscopic investigation into the structure of the capillary endothelium also identified another type of blood vessel capillary—

continuous capillary. This classification is based on the differences in intercellular junctions between the endothelial cells of the capillaries [161]. Currently, there are more complex classifications of capillaries. One of the widely recognized classifications is based on the presence or absence of fenestrae (singular fenestra—from Latin meaning "window") as well as presence or absence of diaphragms [159, 162].

Fenestrated capillaries are found in intestines (ileum and colon), pancreas, adrenal cortex, and peritubular renal tissue [65, 157, 158, 163–165]. These capillaries are characterized by the presence of fenestrae. Fenestrae are round or ovoid transcellular pores measuring about 60–70 nm in diameter. Fenestrae are found in the endothelium of organs where a higher rate of exchange between intra- and extravascular compartments is required [159]. The type of fenestrae discussed above is the most predominant type found in humans [159, 165]. Fenestrae of GI tract capillaries are traversed by diaphragms, measuring 3–5 nm [65, 157, 158]. The other types of fenestrae are found in the glomeruli and liver; they differ from each other by the size of fenestrations and absence of diaphragms [159, 166, 167].

In discontinuous capillaries, the endothelium lacks a continuous basal membrane and it is characterized by the presence of large fenestrations measuring about 100–175 nm, but without diaphragm, exposing the underlying extracellular matrix [166, 167]. These wide fenestrae may allow passage of larger macromolecules including lipid particles and cellular debris [159]. Discontinuous capillaries occur in hepatic sinusoids, spleen, and bone marrow [159, 160, 165].

The continuous capillaries are the exchange blood vessels in which their endothelium does not contain fenestrae, thus all endothelial cells are closely and continuously connected to each other with interendothelial (particularly tight) junctions [165]. Continuous capillaries are found in brain, skin, lung, heart, muscle, retina, and testis [160].

Regulation of Gastrointestinal Perfusion

GI tract blood flow is controlled by numerous factors, which may be classified as extrinsic and intrinsic. The extrinsic factors include general hemodynamic conditions of the cardiovascular system, tone of the autonomic nervous system, and circulating neurohumoral substances. The intrinsic factors include intrinsic neurons, special properties of the microvascular beds, local metabolites, paracrine substances, and local hormones or neurotransmitters [126, 129]. When overall blood volume reduces (known as hypovolemia), the sympathetic nervous system is activated, thus causing the release of norepinephrine (note that the role of epinephrine in the GI tract in this case is almost negligible). This hormone causes vasoconstriction of main arterioles, and to lesser extent venules in the microvascular beds of the GI tract. The resultant effect is the redistribution of venous blood away from peripheral tissues toward the intrathoracic region. Together with these changes, the hypothalamic–pituitary hormone arginine vasopressin is simultaneously released with subsequent activation of the renin–angiotensin system, resulting in the formation of several angiotensin isoforms including angiotensin II [149].

The hormone arginine vasopressin plays a crucial role in retention of water in the renal tubules. Angiotensin II acts as a vasoconstrictor. The overall activities of these hormones aid in the maintenance of blood pressure by counteracting hypovolemia [168, 169].

GI tract blood flow is largely regulated by resistance arterioles. Blood flow in the GI tract capillary beds is inversely proportional to resistance at constant hydrostatic pressure. The tone of the capacitance vessels is crucial to regulating flow, which is determined by pre- and post-capillary sphincters. These sphincters and the resistance arterioles are the factors responsible for intraorgan redistribution of blood [129].

The dynamic balance between vasoconstriction and vasodilation in the GI tract determines the state of GI perfusion. The vasoconstrictors and vasodilators may be endogenous (e.g., neural, hormonal factors) or exogenous (e.g., certain pharmacological drugs) substances [129, 149]. Examples of vasodilators include nitric oxide, prostaglandin I2, GI (including pancreatic) hormones, neurotransmitters, and autacoids. The most potent vasodilator in the human body is calcitonin gene-related peptide (CGRP), followed by nitric oxide. The vasodilatory action of CGRP by several times exceeds that of nitric oxide. Endothelins, sympathetic nervous system, and epinephrine are potent vasoconstrictors. Other factors including arterial baro- and chemo-receptors, cardiopulmonary receptors also affect GI tract blood flow and blood volume [149, 170].

For further details on control of blood flow in microvascular beds, review the following articles:

1. Fry BC, Roy TK, Secomb TW (2013) Capillary recruitment in a theoretical model for blood flow regulation in heterogeneous microvessel networks. Physiol Rep 1(3):e00050.
2. Hudlicka O (2011) Microcirculation in skeletal muscle. Muscles Ligaments Tendons J 1(1):3–11.
3. Itoh Y, Suzuki N (2012) Control of brain capillary blood flow. J Cereb Blood Flow Metab 32(7):1167–1176.
4. Jacob M, Chappell D, Becker BF (2016) Regulation of blood flow and volume exchange across the microcirculation. Crit Care 20:319.
5. Popel AS, Johnson PC (2005) Microcirculation and Hemorheology. Annu Rev Fluid Mech 37:43–69.
6. Secomb TW (2008) Theoretical models for regulation of blood flow. Microcirculation 15(8):765–775.
7. Segal SS (2005) Regulation of blood flow in the microcirculation. Microcirculation 12(1):33–45.

Factors Affecting Gastrointestinal Perfusion

GI tract perfusion is affected by age, feeding, exercise, stress, and different disease conditions including hemorrhage and septic shock [129, 149].

GI perfusion is known to decrease with age [129]. So increasing the efficiency of microcirculation in the GI tract may help to improve GI health and the overall well-being of geriatric patients.

Food ingestion substantially increases GI perfusion. Minutes before food ingestion, increase in heart rate and cardiac output is observed. These influences occur through cephalic mechanism. About 15 min after meal, the mesenteric artery blood flow increases from 500 to 1000 ml/min depending on the type of food ingested. The cardiac responses may cease after 30 min following meal intake. However, the splanchnic responses may be sustained for over 30 min after a meal. In contrast, fasting, even for a few days, leads to mucosal atrophy [129].

Acute (intense) exercise is associated with decrease in GI tract perfusion, but with substantial blood flow increase in cardiac and active skeletal muscle tissues. Acute exercise is also associated with reduction of blood flow in other tissues such as the skin and kidneys. In acute exercise, GI tract perfusion may reduce by 25–50%. The intensity of exercise is proportional to the reduction in GI perfusion [129]. For instance, blood flow gradually decreases from low intensity to high-intensity exercise. Strenuous exercise is associated with substantial reduction in GI perfusion such that some regions of the gut may even develop ischemia. Recovery from exercise is marked by restoration of perfusion (also called reperfusion) to GI regions and organs [125].

Some pharmacological agents are known to affect GI perfusion. Epoprostenol increases GI blood flow [171]. Splanchnic vasoconstrictor drugs such as vasopressin and terlipressin decrease portal blood flow thereby reducing pressure in the portal system. The drug terlipressin selectively stimulates V1 receptors, resulting in arteriolar vasoconstriction in the splanchnic bed. This in turn leads to a shift in blood flow from splanchnic to systemic circulation. In addition, terlipressin increases renal perfusion by increasing both effective blood volume and mean arterial pressure [172].

GI perfusion may be measured by different techniques, which include radiological and non-radiological. The methods of GI perfusion can either be invasive or noninvasive (the invasive methods allow determining GI perfusion through limited access or incisions. In noninvasive methods, no access or incisions are made [173]). For instance, gastric perfusion can be measured with tonometry, a noninvasive technique used to monitor early signs of hypoperfusion [174, 175]. Other noninvasive methods of measuring GI perfusion include esophageal Doppler monitoring, thoracic bioimpedance monitoring, and sublingual capnometry [176, 177]. Examples of invasive or minimally invasive methods include near-infrared fluorescence imaging angiography, possibly in combination with indocyanine green fluorescence dye, or laser-assisted fluorescent-dye angiography [178–180]. Blood flow rate generally can be measured by laser Doppler flowmetry, reflectance spectrophotometry, duplex sonography, pulsatile or perfusion index [129, 181–184].

2.4 Gastrointestinal Lymphatic Drainage

2.4.1 *Brief Historical Background*

"Lymphatic (from Greek meaning "white blood") vessels" were discovered independently at different times by the ancient Greeks and Egyptians. For instance, Hippocrates (460–370 BC) opined that certain structures drain fluid in some parts of the body and he named these structures "lymphatic glands." The ancient scientists and philosophers, Herophilus (335–280 BC), Erasistratus (310–250 BC) and Ruphus (98–117 AD) at one time also reasoned that the structures initially observed by Hippocrates may be present in the GI tract. However, no considerable understanding about the lymphatic vessels or glands was made until the era of Claudius Galen (c. 129–c. 199 AD) [103]. Galen reasoned that certain minute vessels existed in proximity with veins and they were much smaller compared to the vessels that transported deoxygenated blood to the heart. Unfortunately, no one knew precisely the anatomy or physiology of these minute vessels. It is interesting to state that Galen may have been the first to describe the blind-ended lymphatic vessels in the villus (which will later be called lacteals). Up to this point in history, however, there was still no precise understanding of the structure or functions of lymphatic vessels. The lack of precise anatomical and physiological information on the general circulation in man may have hindered the advancement of knowledge in the areas of lymphatic circulation. After centuries of speculations, the Italian physician and physiologist, Gasparo Asellius or Aselli (1581–1626), in 1622, accurately described the anatomy and physiology of these minute vessels (later called lymphatic vessels) in the mesentery of a dog. It was Asellius who named the vessels "milky veins." Further, he showed that fluid (called chyle) was transported along the lacteals down to the lymphatic vessels [185]. The work of Asellius was posthumously published in 1627. Recall from the history of circulation that this period was marked by groundbreaking studies in physiology and anatomy of blood circulation. In the same decade, the English William Harvey provided doubtless evidences on the correct structure and functions of blood circulation. Evidences on blood circulation were widely available at the time and this may have helped the Swedish Olof Rudbeck (1630–1678) to identify the link between lymphatic and blood circulation. Rudbeck discovered the thoracic duct, lymphatic vessels of some organs and regions of the GI tract [103].

2.4.2 *Anatomical Architecture of the Lymphatic System*

The lymphatic system is a network of vessels through which special "milky" fluid (referred to as lymph) circulates and also contains immune cells and numerous substances. The lymphatic system comprises lymph, lymphatic vessels, lymphatic tissues, lymph nodes, and red bone marrow. The system is organized into blind-ended initial lymphatic vessels, precollectors, prenodal collecting lymphatic vessels, lymph nodes, postnodal collecting lymphatic vessels, and the lymphatic trunks (thoracic duct and right lymphatic duct) that subsequently connect to the subclavian veins [186, 187]. The lymph nodes act as a filter for foreign bodies such as germs (bacteria, viruses). The initial lymphatic vessels participate in passive drainage of lymph. The collecting lymphatic vessels are segmented vessels with unidirectional valves that participate in active lymph drainage (Figs. 2.12 and 2.13) [186, 187]. The unidirectional semilunar valves ensure that the lymphatic vessels

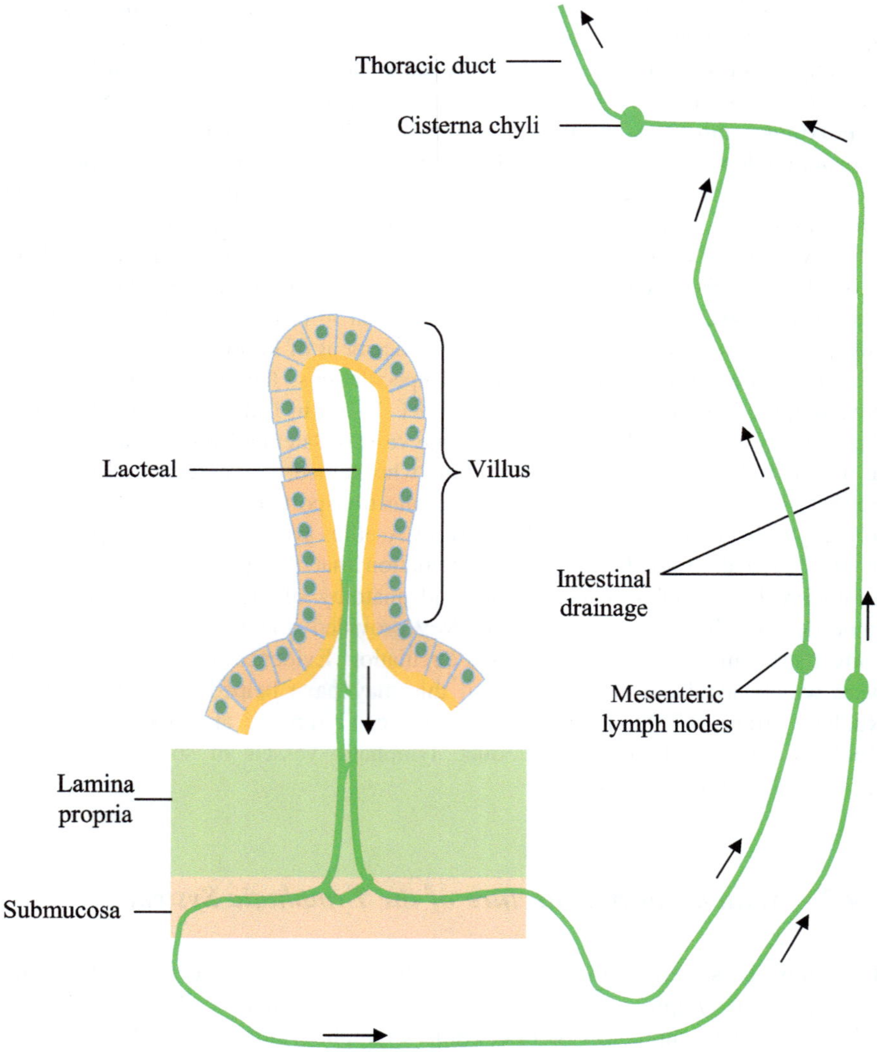

Fig. 2.12 Secretion of lacteal components into lymphatic vessels

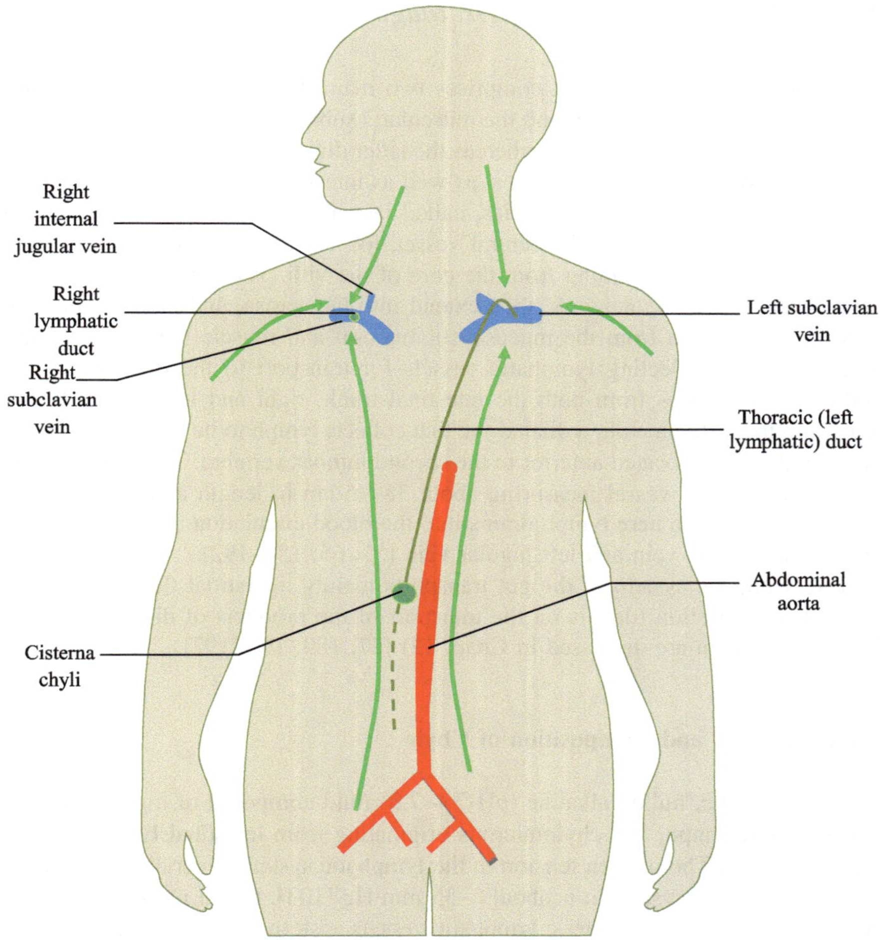

Fig. 2.13 Schematic representation of lymphatic drainage in humans. The normal lymphatic system is unidirectional and begins with the uptake of interstitial fluids via the lacteals or lymph plexuses in the intradermal space. From the lacteals or plexuses, lymph travels through lymphatic vessels via the action of contractile lymphangions to the lymph nodes. From the lymph nodes, the lymph is propelled via the thoracic and the right lymphatic ducts to the subclavian veins for entry into the blood circulatory system. The green arrows indicate the direction of flow of lymph [188, 191]

drain lymph in one-way, thus allowing excess interstitial fluids and proteins to return to the general circulation via the subclavian veins. To carry substances from the different body parts, lymph (containing amino acids, immunoglobulins, and products of digestion) is transported, first, from lymphatic capillaries into lymphatic vessels and via lymph nodes into lymph trunks (Figs. 2.12 and 2.13) [57, 188–190].

2.4.3 *Pattern of Lymphatic Drainage of the Gut*

The GI tract lymphatic system comprises two main divisions—the mucosal–submucosal lymphatic subsystem and the muscularis subsystem. The former drains the mucosal and submucosal layers, whereas the latter drains the muscle layer. Luminal contents absorbed by the enterocytes, as well as interstitial fluid of the GI tract are delivered to the lymphatic capillaries, called lacteals—the origin of the intestinal lymphatic system, formed as a central vessel, located approximately 50 μm from the epithelium, and extending from the core of the villi. The lacteals connect to a submucosal lymphatic network, then extend into the serosa, and pass through the mesenteries. Lymph from the mucosal/submucosal and muscle lymphatic vessels accumulates in collecting lymphatic vessels for transport to the draining lymph nodes. Lymph nodes from both the intestinal trunk, right and left lumbar trunks drain into the cisterna chili, a dilated sac that collects lymph to be transported to the thoracic duct; it is located anterior to the second lumbar vertebra. The thoracic duct is a relatively long vessel measuring about 38–45 cm in length and begins at the cisterna chyli. From here lymph drains into the blood circulation at the junction of the left subclavian vein and left jugular vein [57, 65, 188–192].

The lymphatic system of the gut transports mainly interstitial fluid and certain products of digestion (details on the transport of the products of digestion by the lymphatic system are discussed in Chap. 12) [57, 189, 190, 192].

Characteristics and Composition of Chyle

Chyle is a sterile, milky, alkaline (pH 7.4–7.8) fluid composed of high number of lymphocytes, lymph, and chylomicrons originating from intestinal lymphatic vessels [193, 194]. The oxygen tension in the lymphatic system generally is similar to that in the venous system; it is about 8–35 mm Hg [103]. Under normal condition, chyle is found in the mesenteric lymphatic vessels, cisterna chyli, and thoracic duct [193]. In some disease conditions, however, chyle may leak into the surrounding tissues; it may be found together with ascitic fluid as observed in chylous ascites [195]. Table 2.2 provides information on the composition of chyle.

The essential functions of chyle include transport and regulation of immune cells, maintenance of interstitial fluid volume, transport of absorbed water and chylomicra, which are all essential for the maintenance of homeostasis [65, 103]. These functions of chyle are executed by the lymphatic system. The lymphatic system generally has been implicated in a couple of disease conditions including inflammation, cancer cell dissemination, and lymphedema [187].

Table 2.2 Composition of chyle [193, 194, 265, 266]

Components	Concentration
Total protein	20–40 g/l
Albumin	10–30 g/l
Globulin	10–15 g/l
Fibrinogen	150–250 mg/l
Total fat	10–60 g/l
Triglycerides > plasma level (pleural/plasma ratio >1)	
Cholesterol < plasma values (pleural/plasma ratio <1)	
Cholesterol/triglyceride ratio	<1
Glucose	2–11 mmol/l
Urea	1–3 mmol/l
Electrolytes	
Potassium	3.5–5.0 mmol/l
Chloride	96–106 mmol/l
Sodium	130–145 mmol/l
Blood cells	
Lymphocytes	400–7000 mm^{-3}

Note Chyle electrolyte concentration is equal to plasma values except that calcium level is low

Chyle Flow Rate and the Lymphatic Pump

Estimates indicate that about 1–4 l of lymph is produced in an adult human per day [103]. Over 50% of the total lymph circulating the lymphatic system originates from the GI tract. This fact obviously points to the immense importance of digestion and maintenance of fluid allostasis in the body [57, 103]. The flow rate of chyle expressed in ml per minute per 100 g of tissue substantially varies in different regions and organs of the GI tract. Lymph flow rate in the stomach is about 0.06 ml/min/100 g; small intestine—0.045 ml/min/100 g, large intestine—0.015 ml/min/100 g; salivary gland—0.14 ml/min/100 g, pancreas—0.009 ml/min/100 g; liver—0.05 ml/min/100 g [103]. The flow rate of chyle also varies in different regions of the lymphatic vascular tree. Under normal circumstances, flow rate of chyle in the thoracic duct of an adult is about 1500–2500 ml/day. But daily chyle output can be as low as 240–360 ml during periods of immobility, starvation, and continuous gastric suction. Flow rate of chyle substantially increases after a meal rich in long-chain triglycerides. Thus, the small intestines contribute the greater proportion to the flow rate of chyle following intake of meal rich, especially, in long-chain fatty acids. In periods of resting/absence of feeding, the liver contributes about 75% to lymph flow in the thoracic duct [193] (Fig. 2.14).

The lymphatic pump is the contractile unit of lymphatic system, consisting of a lymphatic segment bounded by two unidirectional valves and driven by contraction of specialized smooth muscle, which pumps lymph against pressure gradient in the body. This pump is called lymphangion. Lymphangions are vessel subunits

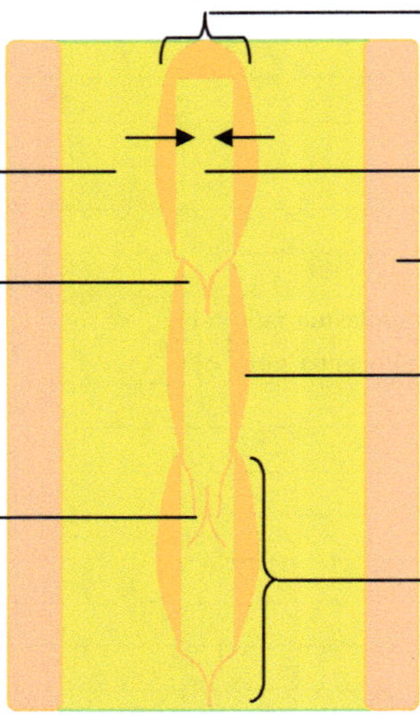

Fig. 2.14 Diagram showing lymphatic capillary and lymphangion (functional unit of lymphatic system). The arrows indicate direction lymph flow. Each lymphangion comprises a semilunar valve plus the part of the lymphatic segment extending to the next valve [187, 262]

bounded by valves that prevent backflow of lymph, propelling lymph forward, and ensuring a unidirectional flow in healthy lymphatics. These functional units of the lymphatic system propel lymph by mechanism of peristaltic and rhythmic muscular contractions. These muscular contractions are due to the intrinsic "pacemaker" activity of smooth muscles contained in the lymphangion wall [103, 196]. The pacemaker activity of the smooth muscle cells of the lymphangion wall makes it contract at a rate of 1–15 cycles per minute. (the initial lymphatic vessels do not contain smooth muscles; they are passive structures) [186, 187, 197]. The interstitial fluid is transported into initial lymphatics by oscillating pressure gradients, which also propel lymph centripetally [186, 187]. In some situations, however, the pressure gradient that modulates the pacemaking of the lymphangions allowing the peristaltic and rhythmic contraction of this morpho-functional units, may reverse, thus forcing the valves to remain continuously open, resulting in passive lymph flow down the pressure gradient [197]. Such reversal of pressure gradient may occur in edema, external compression (e.g., with compression stocking) and limb elevation [197]. The lymphagions are also composed of lymphatic endothelial cells, which are also responsible for modulating several functions of the lymphatics [186, 187].

Numerous factors affect chyle flow rate, which may determine the quantity of daily chyle output. The quantity of chyle output depends on the level of activity, bowel function, fat content of the diet, activities of neurotransmitters and peptides, hormones, free radicals, and disease conditions such as inflammation [193]. These factors may be grouped as extrinsic and intrinsic factors; or as factors that decrease and those that increase lymph flow. Lymph flow rate is increased by feeding, acute and chronic inflammation, intestinal obstruction, cholecystokinin, glucagon, endothelin, bradykinin, angiotensin II, substance-P, serotonin, histamine, adrenaline, dopamine, prostaglandins, AMP, ATP, and lipopolysaccharide [103, 198, 199], whereas the flow rate is decreased by fasting, vasopressin, adrenomedullin, vasoactive intestinal peptide, atrial natriuretic factor, prostacyclin, acetylcholine, and reactive oxygen species [103, 193, 198, 200]. The extrinsic factors responsible for lymph pumping include GI peristalsis, arterial pulsation, skeletal muscle contraction, fluctuation of central venous pressure, and respiration. The intrinsic factors include the properties of the lymphatic valve, lymphatic smooth muscles [187, 201, 202].

2.5 Nerve Supply of the Gastrointestinal Tract

The nerves of the GI tract are supplied by the autonomic nervous system. Both the parasympathetic and sympathetic divisions of the autonomic nervous system are involved in nervous regulation of GI system. The parasympathetic nervous system of the gut originates from facial (VII), glossopharyngeal (IX), and vagus (X) nerves —contains both sensory and motor neurons as well as pelvic splanchnic nerves (S2–4), that is why the parasympathetic nervous system is sometimes termed craniosacral. The sympathetic nerves that innervate the GI tract arise from the celiac, superior and inferior mesenteric ganglia (the preganglionic neurons arise from the intermediolateral columns of T1–L2/L3); hence, the sympathetic nervous system is sometimes termed thoracolumbar [203–207]. While the nerves that supply the gut come from branches of the autonomic nervous system, it should be remembered that the gut walls are extensively innervated by intrinsic neurons of the GI tract (Auerbach's and Meissner's plexuses). The intrinsic nervous system of the gut is capable of independent existence and its activities may only be modulated by the autonomic nervous system. Details of the innervation of the GI tract and the physiological implications are discussed in Chap. 9.

2.6 Functions of the Gastrointestinal Tract

The processes involved in digestion include the following steps: ingestion, mechanical breakdown (mechanical digestion), chemical breakdown of food substances (chemical digestion), absorption, and defecation. Except defecation that is characterized specifically by the distal end of the large intestine, all other processes in the digestive tract are interwoven. Nevertheless, one or two processes might predominate in one region of the tract. This peculiarity in digestive processes is also dependent on the types of food or substance consumed. In addition to mechanical breakdown (motor function of GI tract), chemical breakdown of food substances, absorption, defecation, other functions of the GI tract include secretion, transport, defense, metabolism, excretion, and regulation of extraenteric tissues. Mechanical and chemical breakdowns of food are also considered as major types of digestion. The third type of digestion called symbiotic digestion, though it predominantly takes place in the large intestine, is controlled by the intestinal microflora (gut microbiota). Symbiotic digestion involves the cleavage of non-digestible components of food by the gut microbiota to produce substances essential for the normal functioning of both the gut microbes and the host cells [47, 48, 208] (Fig. 2.15).

Chemical Function: Chemical function of the gut (chemical digestion in particular) entails the action of enzymes, bile, and HCl on food substances, breaking them down to smaller molecules are absorbable by the intestines. Enzymes that participate in chemical digestion are released from all regions of the GI tract—mouth, stomach, intestines, and pancreas. Bile is released from gallbladder into the intestinal lumen in response to adequate stimuli. HCl is released from the stomach [209–211]. Details on chemical digestion are provided in Chap. 12.

Motor Function: Motor function of the GI tract is characterized by periodic waves of contraction and relaxation of the muscles that make up the digestive tract. The contraction and relaxation of the muscles not only allow the mixing and aboral movement of food, but also aid to regulate the activities of the whole GI tract. Motor function is a key characteristic of every region of the gut and involves chewing, swallowing, grinding, peristaltic, and other types of movement of food substances/luminal contents in the GI tract [209, 212].

Secretory Function: The GI tract cells continuously secrete different substances into the lumen, extracellular space of the epithelial cells that regulate a host of functions including absorption and defense. The secretions include mucin, ions (e.g., bicarbonate), enzymes, hormones, neurotransmitters, and other biomolecules that regulate the activities of the gut and other functionally associated tissues or organs [213–220].

Absorption: Ingested food and other substances as well as molecules produced in vivo in the GI tract are absorbed in the intestine. This allows the utilization of these molecules and substances for energy production, plastic function, and other cellular activities [221]. The intestine of a healthy human adult absorbs about 1 kg of nutrients together with 9 l of fluid per day [222]. Inadequate absorption of the required nutrients from ingested food can lead to serious health problems. The condition may be due to disorder in the mucosa, extensively eroded epithelium, extremely short gut possibly resulting from surgery. This condition is termed malabsorption. Depending on the region of the tract affected, absorption of one or more nutrients may be affected [223, 224]. Malabsorptive situations do occur in diseases such as inflammatory bowel diseases (Crohn's disease and ulcerative colitis) [225], diarrhea [226], gastric ulcers, and necrotizing enterocolitis [227].

Transport: The gut also serves as a medium for transport of biomolecules and ions for the maintenance of life process. The gut selectively allows the passage of molecules from the lumen to the basolateral side of the epithelial cells from where absorbed molecules are moved to their various destinations [228]. Water, ions, and

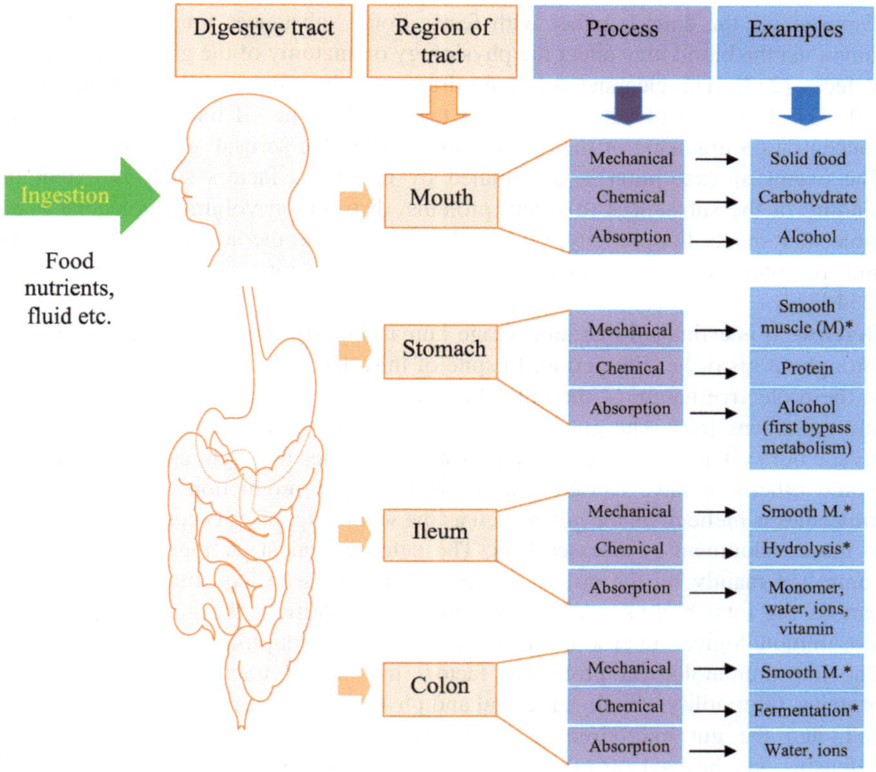

Fig. 2.15 Schematic representation of the processes involved in digestion. The diagram shows the interwoven processes and steps that take place in the mechanochemical conveyor. The major process occurring in the mouth is mechanical breakdown (of solid food); extent of chemical breakdown of the food (e.g., carbohydrates) depends on the time of localization of food in the mouth. Chemical functions of oral secretions are diverse, but they also participate in the formation of bolus. A small percentage of some liquid consumed (e.g., alcohol) may be absorbed in the mouth. Some drugs that dissolve in the mouth are also absorbed here. Liquid such as alcohol, if swallowed, may be absorbed in the stomach. The absorption stages in the oral cavity and stomach bypass the usual metabolic pathway required for the catabolism of alcohol in the liver and it is called first bypass metabolism. The major process that takes place in the ileum involves the hydrolytic cleavage of macromolecules of food substances to form simpler molecules, primarily monomers, which are subsequently absorbed and transported into blood. But dimers and oligomers are also absorbed. In the colon, the fermentative action of microbes produces many useful substances that are absorbed together with water, sodium, potassium, etc. [47, 48, 208]

nutrients are transported through certain channels or receptors to the basolateral side of the enterocyte from where they pass into the bloodstream [229–232].

Exsorptive function: Exsorption is the movement of substances from the blood into the lumen of the gut. Besides transport of substances into the bloodstream from the luminal side, the gut actively participates in moving substances including drugs and toxins from the blood to the intestines, from where these substances are disposed via the anus together with feces. Some substances transported into the lumen via the blood may affect the physiology or anatomy of the gut if not excreted in feces [233]. The clearance of some drugs may be enhanced through this route [40, 41]. Transport of substances from the blood to the GI tract depends on the concentration gradients of the substances between the serosal and mucosal sides. The extent of exsorption is determined by numerous factors such as extent of binding of the substances to serum proteins, distribution volume, molecular size, molecular shape, lipophilicity, GI blood flow rate, and the activity of transporters and receptors including organic anion/cation transporters and P-glycoprotein [234–237].

Defense: It is estimated that an average human ingests about 700 tons of antigens throughout his or her life period. In spite of this, diseases may not develop, thanks to the protective nature of the gut. The gut actively prevents the uptake of toxins and pathogens [87]. The gut is a source of defense against microorganisms that invade our systems. It produces a number of molecules that fight against germs that would otherwise have caused disease. However, if production of such defense molecules is ineffective, the gut as well as the whole organism becomes predisposed to the development of diseases [57]. The defense against pathogens in the gut is controlled mainly by the gut-associated lymphoid tissue, constituting an integral part of the total body's immune system [3]. In addition, the gut also possesses non-immunological defense mechanisms, which confers protection to the GI tract. The non-immunological protective factors include GI tract juice, antimicrobial proteins, GI motility, certain structural and physiological properties of the epithelial cell, and the gut microflora [238]. Numerous antimicrobial peptides, such as defensins, synthesized by GI epithelial cells, are secreted together with GI tract juices. The antimicrobial peptides are important in controlling inflammatory processes and the invasion by pathogens [239, 240].

Metabolic Function: Like any other system of the body, the gut dispenses energy to carry out its activities. However, most importantly, the gut remains an integral source of precursors for key metabolic process particularly gluconeogenesis. The gut feeds the body with gluconeogenetic precursors that constantly circulate between the gut and the general circulation. Some of the metabolic substances, particularly short-chain fatty acids, produced by the gut microbiota, are implicated in health and metabolic disease [241–243].

Microbiota Function: The gut harbors millions of microbes, which may be potentially harmful in disease condition. However, under normal conditions, the gut microflora serves a protective role against the invasion by pathogenic microbes, and as a store house for synthesis of many substances of immense benefit to the whole body functioning. The gut microbiota takes part in de novo synthesis of many biomolecules including vitamins and short-chain fatty acids [103]. Dysfunctions of the gut microbiota have been implicated in several diseases of humans including obesity and diabetes [244, 245]. Further details on the gut microbiota are discussed in Chap. 15 of this book and in the book "Gastrointestinal Physiology: Contemporary Trends, Methods and Models" by the same author.

Excretory Function: The end products of metabolism and toxic molecules as well as excess ion (e.g., calcium, zinc, potassium) are continuously excreted from the gut [246–248].

Endocrine Function: The gut is an endocrine organ that synthesizes numerous hormones and peptides that regulate digestion and generally gut and extragut activities [249, 250]. The hormones of the gut participate in a range of metabolic processes regulating weight balance and availability of nutrients for the entire body's processes [251].

Sensory Function: The GI tract possesses numerous sensory receptors that translate the conditions of the immediate gut environment to nerve impulses that are transmitted to the central nervous system or analyzed in gut local ganglia or plexuses. The coordinated signaling of the visceral sensory system of the gut is important in food digestion and absorption. Sensory coding of information in the gut environment is carried out by receptors. The activities of the transient receptor potential channel (a type of membrane receptors) have been implicated in sensory transduction pathway of visceral sensory neurons of the esophagus, stomach, and intestines. Noxious stimuli (pain) originating from the GI tract are transmitted, first, by the sensory receptors to the central nervous system, where they are analyzed and the result relayed back to the GI tract via the motor system. Other signals such as taste of different foods are sensed and coded by sensory receptors [252–255]. Sensory functions of the GI tract also involve constant evaluation of the concentration of various nutrients in the GI tract, and degree of stretch produced by movement of food particles [256].

Regulation of Extragut Tissues and Organs: The gut provides a useful functional link that regulates the activities of almost all organs and tissues of the body. For instance, the gut plays a considerable role in cognitive functioning. The gut feeds the brain with numerous hormones and neurotransmitters that participate either directly in memory and cognition or enhance the structural–functional integrity of systems responsible for memory and cognitive processes. Emerging evidences from research using state-of-the-art diagnostic techniques indicate that many brain pathologies are due to disorders in the sensory and generally gut nervous system functioning. Surprisingly, some of these brain disorders could be identified earlier by state-of-the-art diagnostic evaluation of the gut intrinsic nervous functioning; thus, these diseases could be averted or treated earlier, even before the symptoms are observed [257, 258]. Similar functional associations are present in other systems of the body. Further information on gut-extragut relationship is discussed in Chap. 15.

2.7 Conclusion

The GI tract is a relatively long tube running from the mouth to the anus, structurally organized into regions with each region having a peculiar characteristic in its layers, and executing a host of physiological functions in a functionally coordinated manner, required to ensure maintenance of life processes.

Recommended Readings

Original Articles

1. Aharinejad S, Lametschwandtner A, Franz P, Firbas W (1991) The vascularization of the digestive tract studied by scanning electron microscopy with special emphasis on the teeth, esophagus, stomach, small and large intestine, pancreas, and liver. Scan Microsc 5(3):811–849
2. Beasley DE, Koltz AM, Lambert JE, Fierer N, Dunn RR (2015) The evolution of stomach acidity and its relevance to the human microbiome. PLoS One 10(7): e0134116
3. Burdan F, Rozylo-Kalinowska I, Szumilo J, Zinkiewicz K, Dworzanski W, Krupski W, Dabrowski A (2012) Anatomical classification of the shape and topography of the stomach. Surg Radiol Anat 34(2):171–178
4. He G, Shankar RA, Chzhan M, Samouilov A, Kuppusamy P, Zweier JL (1999) Noninvasive measurement of anatomic structure and intraluminal oxygenation in the gastrointestinal tract of living mice with spatial and spectral EPR imaging. Proc Natl Acad Sci USA 96(8):4586–4591
5. Hollister EB, Riehle K, Luna RA, Weidler EM, Rubio-Gonzales M, Mistretta T-A, Raza S, Doddapaneni HV, Metcalf GA, Muzny DM, Gibbs RA, Petrosino JF, Shulman RJ, Versalovic J (2015) Structure and function of the healthy pre-adolescent pediatric gut microbiome. Microbiome 3:36
6. Ursic M, Ravnik D, Hribernik M, Pecar J, Butinar J, Fazarinc G (2007) Gross anatomy of the portal vein and hepatic artery ramifications in dogs: corrosion cast study. Anat Histol Embryol 36(2):83–87
7. Vdoviaková K, Petrovová E, Maloveská M, Krešáková L, Teleky J, Elias MZJ, Petrášová D (2016) Surgical anatomy of the gastrointestinal tract and its vasculature in the laboratory rat. Gastroenterol Res Pract 2016:2632368

Review Articles

1. Alexander JS, Ganta VC, Jordan PA, Witte MH (2010) Gastrointestinal lymphatics in health and disease. Pathophysiology 17(4):315–335
2. Bennett DG (1992) Medical examination of the digestive system relevant to purchase. Vet Clin North Am Equine Pract 8(2):387–393
3. Borghei P, Sokhandon F, Shirkhoda A, Morgan DE (2013) Anomalies, anatomic variants, and sources of diagnostic pitfalls in pancreatic imaging. Radiology 266(1):28–36

4. Chait MM (2010) Lower gastrointestinal bleeding in the elderly. World J Gastrointest Endosc 2(5):147–154
5. Cheng LK, O'Grady G, Du P, Egbuji JU, Windsor JA, Pullan AJ (2010) Gastrointestinal system. Wiley Interdiscip Rev Syst Biol Med 2(1):65–79
6. Ghassemi KA, Jensen DM (2013). Lower GI bleeding: epidemiology and management. Curr Gastroenterol Rep 15:333
7. Hansen MB (2003) The enteric nervous system I: organisation and classification. Pharmacol Toxicol 92(3):105–113
8. Holcomb SS (2008) Acute abdomen: what a pain! Nursing 38(9):34–41
9. House JK, Smith BP, Fecteau G, VanMetre DC (1992) Assessment of the ruminant digestive system. Vet Clin North Am Food Anim Pract 8(2):189–202
10. House JK, Smith BP, VanMetre DC, Fecteau G, Craychee T, Neves J (1992) Ancillary tests for assessment of the ruminant digestive system. Vet Clin North Am Food Anim Pract 8(2):203–232
11. Johansson MEV, Sjövall H, Hansson GC (2013) The gastrointestinal mucus system in health and disease. Nat Rev Gastroenterol Hepatol 10(6):352–361
12. Karasov WH, Douglas AE (2013) Comparative digestive physiology. Compr Physiol 3(2):741–783
13. Kim BSM, Li BT, Engel A, Samra JS, Clarke S, Norton ID, Li AE (2014) Diagnosis of gastrointestinal bleeding: a practical guide for clinicians. World J Gastrointest Pathophysiol 5(4):467–478
14. Lackner AA, Mohan M, Veazey RS (2009) The gastrointestinal tract and AIDS pathogenesis. Gastroenterology 136(6):1965–1978
15. Lam YY, Ravussin E (2016) Analysis of energy metabolism in humans: a review of methodologies. Mol Metab 5(11):1057–1071
16. Manocha M, Khan WI (2012) Serotonin and GI disorders: an update on clinical and experimental studies. Clin Transl Gastroenterol 3:e13
17. Miller MJ, Newberry RD (2010) Microanatomy of the intestinal lymphatic system. Ann N Y Acad Sci 1207(Suppl 1):E21–E28
18. Miller SK, Alpert PT (2006) Assessment and differential diagnosis of abdominal pain. Nurse Pract 31(7):38–45, 47
19. Mudie DM, Amidon GL, Amidon GE (2010) Physiological parameters for oral delivery and in vitro testing. Mol Pharm 7(5):1388–1405
20. O'Toole MT (1990) Advanced assessment of the abdomen and gastrointestinal problems. Nurs Clin North Am 25(4):771–776
21. Navuluri R, Kang L, Patel J, Van Ha T (2012) Acute lower gastrointestinal bleeding. Semin Intervent Radiol 29(3):178–186
22. Raphaeli T, Menon R (2012) Current treatment of lower gastrointestinal hemorrhage. Clin Colon Rectal Surg 25(4):219–227
23. Sehgal R, Coffey JC (2014) Historical development of mesenteric anatomy provides a universally applicable anatomic paradigm for complete/total mesocolic excision. Gastroenterol Rep (Oxf) 2(4):245–250
24. Shen L (2009) Functional morphology of the gastrointestinal tract. Curr Top Microbiol Immunol 337:1–35

25. Sreedharan R, Mehta DI (2004) Gastrointestinal tract. Pediatrics 113(4 Suppl):1044–1050
26. Vighi G, Marcucci F, Sensi L, Di Cara G, Frati F (2008) Allergy and the gastrointestinal system. Clin Exp Immunol 153(Suppl 1):3–6
27. Wilkins T, Khan N, Nabh A, Schade RR (2012) Diagnosis and management of upper gastrointestinal bleeding. Am Fam Physician 85(5):469–476

Books and Book Chapters

1. De Martino RR (2015) Normal and variant mesenteric anatomy. In: Oderich GS (ed) Mesenteric vascular disease—current therapy. Springer, New York
2. Ferguson CM (1990) An overview of the gastrointestinal system. In: Walker HK, Hall WD, Hurst JW (eds) Clinical methods: the history, physical, and laboratory examinations, 3rd edn. Butterworths, Boston
3. Ferguson CM (1990) Inspection, auscultation, palpation, and percussion of the abdomen. In: Walker HK, Hall WD, Hurst JW (eds) Clinical methods: the history, physical, and laboratory examinations, 3rd edn. Butterworths, Boston
4. Kvietys PR (2010) The gastrointestinal circulation. Morgan & Claypool, San Rafael, CA
5. Meerveld BG-V (eds) (2017) Gastrointestinal pharmacology. Handbook of experimental pharmacology, Springer International Publishing AG, Cham, Switzerland
6. Pereverzev VA, Lobanok LM (2014) Physiology of digestion. In: Kubarko AI (ed) Normal physiology. In Two parts. Part 1. Visheishaya Shkola, Minsk, Belarus

Anatomy and Physiology Apps and Models

1. Anatomy Warehouse. https://www.anatomywarehouse.com
2. Virtual Physiology. http://virtual-physiology.com
3. Visible body. http://www.visiblebody.com

Bibliography

1. Kararli TT (1995) Comparison of the gastrointestinal anatomy, physiology, and biochemistry of humans and commonly used laboratory animals. Biopharm Drug Dispos 16(5): 351–380
2. Van de Graaff KM (1986) Anatomy and physiology of the gastrointestinal tract. Pediatr Infect Dis 5(1):S11–S16
3. Gelberg HB (2014) Comparative anatomy, physiology, and mechanisms of disease production of the esophagus, stomach, and small intestine. Toxicol Pathol 42(1):54–66
4. Kimmey MB, Martin RW, Haggitt RC, Wang KY, Franklin DW, Silverstein FE (1989) Histologic correlates of gastrointestinal ultrasound images. Gastroenterology 96(2 Pt 1): 433–441
5. Spitz L (2007) Oesophageal atresia. Orphanet J Rare Dis 2:24

6. Sunitha T, Prasoona KR, Kumari TM, Srinadh B, Deepika MLN, Aruna R, Jyothy A (2017) Risk factors for congenital anomalies in high risk pregnant women: a large study from South India. Egypt J Med Hum Genet 18(1):79–85
7. Sarkar S, Patra C, Dasgupta MK, Nayek K, Karmakar PR (2013) Prevalence of congenital anomalies in neonates and associated risk factors in a tertiary care hospital in Eastern India. J Clin Neonatol 2(3):131–134
8. Jiang X, Xu G, Shen L, Wu J, Chen H, Wang Y (2014) Influential factors on congenital gastrointestinal malformation: a hospital-based case-control study. Zhonghua Liu Xing Bing Xue Za Zhi 35(1):81–84
9. Mashuda F, Zuechner A, Chalya PL, Kidenya BR, Manyama M (2014) Pattern and factors associated with congenital anomalies among young infants admitted at Bugando medical centre, Mwanza Tanzania. BMC Res Notes 7:195
10. El-Helaly M, Abdel-Elah K, Haussein A, Shalaby H (2011) Paternal occupational exposures and the risk of congenital malformations—a case-control study. Int J Occup Med Environ Health 24(2):218–227
11. Okamoto T, Takamizawa S, Arai H, Bitoh Y, Nakao M, Yokoi A, Nishijima E (2009) Esophageal atresia: prognostic classification revisited. Surgery 145(6):675–681
12. Poenaru D, Laberge JM, Neilson IR, Guttman FM (1993) A new prognostic classification for esophageal atresia. Surgery 113(4):426–432
13. Pinheiro PFM, e Silva ACS, Pereira RM (2012) Current knowledge on esophageal atresia. World J Gastroenterol 18(28):3662–3672
14. Shaw-Smith C (2006) Oesophageal atresia, tracheo-oesophageal fistula, and the VACTERL association: review of genetics and epidemiology. J Med Genet 43:545–554
15. Morgan RD, O'Callaghan JM, Wagener S, Grant HW, Lakhoo K (2012) Surgical correction of tracheo-oesophageal fistula and oesophageal atresia in infants with VACTERL association: a retrospective case-control study. Pediatr Surg Int 28(10):967–970
16. Carli D, Garagnani L, Lando M, Fairplay T, Bernasconi S, Landi A, Percesepe A (2014) VACTERL (vertebral defects, anal atresia, tracheoesophageal fistula with esophageal atresia, cardiac defects, renal and limb anomalies) association: disease spectrum in 25 patients ascertained for their upper limb involvement. J Pediatr 164(3):458–462 (e1–e2)
17. La Placa S, Giuffrè M, Gangemi A, Di Noto S, Matina F, Nociforo F et al (2013) Esophageal atresia in newborns: a wide spectrum from the isolated forms to a full VACTERL phenotype? Ital J Pediatr 39:45
18. Koçak H, Ozaydin E, Köse G, Marcelis CL, Kamsteeg EJ, Ceylaner S (2009) A Feingold syndrome case with previously undescribed features and a new mutation. Genet Couns 20 (3):261–267
19. Büttiker V, Wojtulewicz J, Wilson M (2000) Imperforate anus in Feingold syndrome. Am J Med Genet 92(3):166–169
20. Layman-Pleet L, Jackson CC, Chou S, Boycott KM (2007) Feingold syndome: a rare but important cause of syndromic tracheoesophageal fistula. J Pediatr Surg 42(9):E1–E3
21. Versteegh HP, van Rooij IALM, Levitt MA, Sloots CEJ, Wijnen RMH, de Blaauw I (2013) Long-term follow-up of functional outcome in patients with a cloacal malformation: a systematic review. J Pediatr Surg 48:2343–2350
22. Rintala RJ, Sistonen S, Pakarinen MP (2009) Outcome of esophageal atresia beyond childhood. Sem Pediatr Surg 18(1):50–56
23. Terui K, Saito T, Mitsunaga T, Nakata M, Yoshida H (2015) Endoscopic management for congenital esophageal stenosis: a systematic review. World J Gastrointest Endosc 7(3): 183–191
24. Stylopoulos N, Rattner DW (2005) The history of hiatal hernia surgery: from Bowditch to laparoscopy. Ann Surg 241(1):185–193
25. Lord RV, DeMeester SR, Peters JH, Hagen JA, Elyssnia D, Sheth CT, DeMeester TR (2009) Hiatal hernia, lower esophageal sphincter incompetence, and effectiveness of Nissen fundoplication in the spectrum of gastroesophageal reflux disease. J Gastrointest Surg 13(4): 602–610

26. Kluth D, Fiegel H (2003) The embryology of the foregut. Semin Pediatr Surg 12(1):3–9
27. Kluth D, Steding G, Seidl W (1987) The embryology of foregut malformations. J Pediatr Surg 22(5):389–393
28. Metzger R, Wachowiak R, Kluth D (2011) Embryology of the early foregut. Sem Pediatr Surg 20(3):136–144
29. Felix JF, de Jong EM, Torfs CP, de Klein A, Rottier RJ, Tibboel D (2009) Genetic and environmental factors in the etiology of esophageal atresia and/or tracheoesophageal fistula: an overview of the current concepts. Birth Defects Res A Clin Mol Teratol 85(9):747–754
30. Spilde TL, Bhatia AM, Mehta S, Ostlie DJ, Hembree MJ, Preuett BL et al (2003) Defective sonic hedgehog signaling in esophageal atresia with tracheoesophageal fistula. Surgery 134 (2):345–350
31. Spilde TL, Bhatia AM, Marosky JK, Preuett B, Kobayashi H, Hembree MJ et al (2003) Fibroblast growth factor signaling in the developing tracheoesophageal fistula. J Pediatr Surg 38(3):474–477
32. Chuong C-M, Patel N, Lin J, Jung H-S, Widelitz RB (2000) Sonic hedgehog signaling pathway in vertebrate epithelial appendage morphogenesis: perspectives in development and evolution. Cell Mol Life Sci 57(12):1672–1681
33. Warburton D, Bellusci S, De Langhe S, Del Moral P-M, Fleury V, Mailleux A et al (2005) Molecular mechanisms of early lung specification and branching morphogenesis. Pediatr Res 57:26R–37R
34. Jacobs IJ, Que J (2013) Genetic and cellular mechanisms of the formation of esophageal atresia and tracheoesophageal fistula. Dis Esophagus 26(4):356–358
35. Kunisaki SM, Bruch SW, Hirschl RB, Mychaliska GB, Treadwell MC, Coran AG (2014) The diagnosis of fetal esophageal atresia and its implications on perinatal outcome. Pediatr Surg Int 30(10):971–977
36. Langer JC, Hussain H, Khan A, Minkes RK, Gray D, Siegel M, Ryan G (2001) Prenatal diagnosis of esophageal atresia using sonography and magnetic resonance imaging. J Pediatr Surg 36(5):804–807
37. Véronique H-D, Julien B (2011) Ultrasound and MRI prenatal diagnosis of esophageal atresia: effect on management. J Pediatr Gastroenterol Nutr 52:S9–S11
38. Chang EY, Chang HK, Han SJ, Choi SH, Hwang EH, Oh J-T (2012) Clinical characteristics and treatment of esophageal atresia: a single institutional experience. J Korean Surg Soc 83 (1):43–49
39. Kovesi T, Rubin S (2004) Long-term complications of congenital esophageal atresia and/or tracheoesophageal fistula. Chest 126(3):915–925
40. Kovesi T (2013) Long-term respiratory complications of congenital esophageal atresia with or without tracheoesophageal fistula: an update. Dis Esophagus 26(4):413–416
41. Shah R, Varjavandi V, Krishnan U (2015) Predictive factors for complications in children with esophageal atresia and tracheoesophageal fistula. Dis Esophagus 28(3):216–223
42. Spitz L, Kiely E, Brereton RJ, Drake D (1993) Management of esophageal atresia. World J Surg 17(3):296–300
43. Michaud L, Coutenier F, Podevin G, Bonnard A, Becmeur F, Khen-Dunlop N et al (2013) Characteristics and management of congenital esophageal stenosis: findings from a multicenter study. Orphanet J Rare Dis 8:186
44. Serrao E, Santos A, Gaivao A, Tavares A, Ferreira S (2010) Congenital esophageal stenosis: a rare case of dysphagia. J Radiol Case Rep 4:8–14
45. Rastogi R, Majid A, Singh VP, Joon P, Gupta Y (2016) Congenital esophageal stenosis: a rare case of childhood dysphasia. J Gastrointest Dig Syst 6:2
46. Yousef MM, Yantiss RK, Baker SP, Banner BF (2006) Duodenal intraepithelial lymphocytes in inflammatory disorders of the esophagus and stomach. Clin Gastroenterol Hepatol 4(5):631–634
47. Pereverzev VA, Lobanok LM (2014) Physiology of digestion. In: Kubarko AI (ed) Normal physiology. In two parts. Part 1. Visheishaya Shkola, Minsk, Belarus

48. Nikitina OS, Welcome MO, Pereverzev VA (2016) Human anatomy and physiology. In: 2 parts. Part 1. Belarusian State Medical University Press, Minsk
49. Kim SK, Cho CD, Wojtowycz AR (2008) The ligament of treitz (the suspensory ligament of the duodenum): anatomic and radiographic correlation. Abdom Imaging 33(4):395–397
50. Jabbour SK, Hashem SA, Bosch W, Kim TK, Finkelstein SE, Anderson BM et al (2014) Upper abdominal normal organ contouring guidelines and atlas: a radiation therapy oncology group consensus. Pract Radiat Oncol 4(2):82–89
51. Meyers MA (1995) Treitz redux: the ligament of treitz revisited. Abdom Imaging 20(5): 421–424
52. Shafik A, El Sibai O, Shafik AA, Shafik IA (2006) Demonstration of a physiologic sphincter at duodeno-jejunal junction. Front Biosci 11:2790–2794
53. Saenko VF, Virchenko SB, Kucherenko TL, Ettinger AP, Belianskiĭ LS, Llu Markulan (1990) The role of the duodenojejunal junction in regulating the evacuatory function of the duodenum. Fiziol Zh 36(4):56–63
54. Virchenko SB, Sayenko VF, Kucherenko TL, Tsedik NN, Elbrønd H, Djurhuus JC, Funch-Jensen P (1993) The duodenojejunal junction and treitz ligament in the regulation of duodenal emptying. Scand J Gastroenterol 28(9):753–759
55. Hamilton KL, Butt AG (2013) Glucose transport into everted sacs of the small intestine of mice. Adv Physiol Educ 37(4):415–426
56. Matsumoto S, Miyatani H, Yoshida Y (2015) Future directions of duodenal endoscopic submucosal dissection. World J Gastrointest Endosc 7(4):389–395
57. Mowat AM, Agace WW (2014) Regional specialization within the intestinal immune system. Nat Rev Immunol 14:667–685
58. Rodriguez-Bigas MA, Lin EH, Crane CH (2003) Surgical anatomy. In: Kufe DW, Pollock RE, Weichselbaum RR, Bast RC Jr, Gansler TS, Holland JF, Frei E III (eds) Holland-frei cancer medicine, 6th edn. Hamilton, ON, BC Decker
59. Sadahiro S, Ohmura T, Yamada Y, Saito T, Taki Y (1992) Analysis of length and surface area of each segment of the large intestine according to age, sex and physique. Surg Radiol Anat 14(3):251–257
60. Camilleri M, Ford MJ (1998) Colonic sensorimotor physiology in health, and its alteration in constipation and diarrhoeal disorders. Aliment Pharmacol Ther 12:287–302
61. Liao D-H, Zhao J-B, Gregersen H (2009) Gastrointestinal tract modelling in health and disease. World J Gastroenterol 15(2):169–176
62. Liao D, Frøkjaer JB, Yang J, Zhao J, Drewes AM, Gilja OH, Gregersen H (2006) Three-dimensional surface model analysis in the gastrointestinal tract. World J Gastroenterol 12(18):2870–2875
63. Shen L (2009) Functional morphology of the gastrointestinal tract. Curr Top Microbiol Immunol 337:1–35
64. Balbi V, Ciarletta P (2013) Morpho-elasticity of intestinal villi. J R Soc Interface 10(82): 20130109
65. Kvietys PR, Granger DN (2010) Role of intestinal lymphatics in interstitial volume regulation and transmucosal water transport. Ann N Y Acad Sci 1207(1):E29–E43
66. Howard ER (2013) Joseph Lister: his contributions to early experimental physiology. Notes Rec R Soc Lond 67(3):191–198
67. Spencer NJ, Dinning PG, Brookes SJ, Costa M (2016) Insights into the mechanisms underlying colonic motor patterns. J Physiol 594(15):4099–4116
68. Eastwood PR, Katagiri S, Shepherd KL, Hillman DR (2007) Modulation of upper and lower esophageal sphincter tone during sleep. Sleep Med 8(2):135–143
69. Lang IM, Shaker R (1997) Anatomy and physiology of the upper esophageal sphincter. Am J Med 103(5A):50S–55S
70. Mittal RK, Liu J (2005) Flow across the gastro-esophageal junction: lessons from the sleeve sensor on the nature of anti-reflux barrier. Neurogastroenterol Motil 17(2):187–190
71. Dj Saranović, Ilić N, Stafanović Z, Kasiković M (1991) Colonic sphincters—Hirsch's sphincter. Srp Arh Celok Lek 119(5–6):166–168

72. Gagliardi JA, Radvany MG, Kilkenny TE, Russo RD Jr (1994) Colonic sphincters revisited: simulators of organic disease. Hawaii Med J 53(10):278–282
73. Costa M, Brookes S, Hennig G (2000) Anatomy and physiology of the enteric nervous system. Gut 47(4):iv15–iv19
74. Odegaard S, Nesje LB, Hoff DA, Gilja OH, Gregersen H (2006) Morphology and motor function of the gastrointestinal tract examined with endosonography. World J Gastroenterol 12(18):2858–2863
75. Goyal RK (2000) Targets of enteric motor neurones: smooth muscle cells. Gut 47(4):iv38–39; discussion iv52
76. Liao D, Lelic D, Gao F, Drewes AM, Gregersen H (2008) Biomechanical functional and sensory modelling of the gastrointestinal tract. Philos Trans A Math Phys Eng Sci 366 (1879):3281–3299
77. Dahm R (2010) From discovering to understanding. EMBO Rep 11(3):153–160
78. Takaki M (2003) Gut pacemaker cells: the interstitial cells of Cajal (ICC). J Smooth Muscle Res 39(5):137–161
79. Al-Shboul OA (2013) The importance of interstitial cells of Cajal in the gastrointestinal tract. Saudi J Gastroenterol 19(1):3–15
80. Huizinga JD, Chen J-H, Mikkelsen HB, Wang X-Y, Parsons SP, Zhu YF (2013) Interstitial cells of Cajal, from structure to function. Front Neurosci 7:43
81. Pakurar AS, Bigbee JW (2004) Digital histology: an interactive CD atlas with review text. Wiley, Hoboken, New Jersey
82. Hosseini SV, Abbasi HR, Rezvani H, Vasei M, Ashraf MJ (2009) Comparison between gallbladder serosal and mucosal patch in duodenal injuries repair in dogs. J Invest Surg 22(2):148–153
83. Oliphant R, Gardiner S, Reid R, McPeake J, Porteous C (2007) Glomus tumour of the ascending colon. J Clin Pathol 60(7):846
84. American Joint Committee on Cancer (2010) Colon and rectum. In: AJCC cancer staging manual, 7th edn. Springer, New York
85. Amrani S, Polcino M, Rodriguez-Bigas M, Chu QD (2015) Colon cancer. In: Chu Q, Gibbs J, Zibari G (eds) Surgical oncology. Springer, New York
86. Gartner LP, Hiatt JL, Strum JM (2003) Board review series. Cell biology and histology, 4th edn. Lippincott Williams & Wilkins, Philadelphia, PA
87. Barrett KE (2012) Epithelial biology in the gastrointestinal system: insights into normal physiology and disease pathogenesis. J Physiol 590(Pt 3):419–420
88. Nagpal R, Patel A, Gibson MC (2008) Epithelial topology. BioEssays 30(3):260–266
89. Bellmann S, Carlander D, Fasano A, Momcilovic D, Scimeca JA, Waldman WJ et al (2015) Mammalian gastrointestinal tract parameters modulating the integrity, surface properties, and absorption of food-relevant nanomaterials. WIREs Nanomed Nanobiotechnol 7:609–622
90. Bullen TF, Forrest S, Campbell F, Dodson AR, Hershman MJ, Pritchard DM et al (2006) Characterization of epithelial cell shedding from human small intestine. Lab Invest 86:1052–1063
91. Rao JN, Wang J-Y (2010) Regulation of gastrointestinal mucosal growth. Morgan & Claypool Life Sciences, San Rafael, CA
92. San Roman AK, Shivdasani RA (2011) Boundaries, junctions and transitions in the gastrointestinal tract. Exp Cell Res 317(19):2711–2718
93. Shoemark A, Burgoyne T, Dixon M, Luther P, Hogg C (2012) Three dimensional ultrastructure of human respiratory cilia in health and disease. Cilia 1(1):47
94. Chilvers M, Rutman A, O'Callaghan C (2003) Functional analysis of cilia and ciliated epithelial ultrastructure in healthy children and young adults. Thorax 58(4):333–338
95. Roffler B, Fäh A, Sauter SN, Hammon HM, Gallmann P, Brem G, Blum JW (2003) Intestinal morphology, epithelial cell proliferation, and absorptive capacity in neonatal calves fed milk-born insulin-like growth factor-I or a colostrum extract. J Dairy Sci 86 (5):1797–1806

96. Villasenor A, Chong DC, Henkemeyer M, Cleaver O (2010) Epithelial dynamics of pancreatic branching morphogenesis. Development 137:4295–4305
97. Amano O, Mizobe K, Ya Bando, Sakiyama K (2012) Anatomy and histology of rodent and human major salivary glands—overview of the Japan salivary gland society-sponsored workshop. Acta Histochem Cytochem 45(5):241–250
98. Redman RS (2008) On approaches to the functional restoration of salivary glands damaged by radiation therapy for head and neck cancer, with a review of related aspects of salivary gland morphology and development. Biotech Histochem 83(3):103–130
99. Nolte T, Brander-Weber P, Dangler C, Deschl U, Elwell MR, Greaves P et al (2016) Nonproliferative and proliferative lesions of the gastrointestinal tract, pancreas and salivary glands of the rat and mouse. J Toxicol Pathol 29(1):1S–125S
100. Bohórquez DV, Liddle RA (2015) The gut connectome: making sense of what you eat. J Clin Invest 125(3):888–890
101. Thelen K, Coboeken K, Willmann S, Burghaus R, Dressman JB, Lippert J (2011) Evolution of a detailed physiological model to simulate the gastrointestinal transit and absorption process in humans, Part 1: oral solutions. J Pharm Sci 100(12):5324–5345
102. Helander HF, Fändriks L (2014) Surface area of the digestive tract—revisited. Scand J Gastroenterol 49(6):681–689
103. Alexander JS, Ganta VC, Jordan PA, Witte MH (2010) Gastrointestinal lymphatics in health and disease. Pathophysiology 17(4):315–335
104. Lim S-S, Low FN (1988) Morphological changes in the developing alimentary canal: a review by scanning electron microscopy. In: Motta PM, Fujita H, Correr S (eds) Ultrastructure of the digestive tract. Electron microscopy in biology and medicine (current topics in ultrastructural research), vol 4. Springer, Boston, MA
105. Von Kölliker A (1968/1817–1905) Würzburger histologist. JAMA 206(9):2111–2112
106. De Santa PB, Van Den Brink GR, Roberts DJ (2003) Development and differentiation of the intestinal epithelium. Cell Mol Life Sci 60(7):1322–1332
107. Madara JL, Wolf JL, Trier JS (1982) Structural features of the rat small intestinal microvillus membrane in acute experimental diabetes. Digest Dis Sci 27(9):801–806
108. Walton KD, Kolterud Å, Czerwinski MJ, Bell MJ, Prakash A, Kushwaha J et al (2012) Hedgehog-responsive mesenchymal clusters direct patterning and emergence of intestinal villi. PNAS 109(39):15817–15822
109. Vieira WA, Pretorius E (2010) The impact of asthma on the gastrointestinal tract (GIT). J Asthma Allergy 3:123–130
110. Yen TH, Wright NA (2006) The gastrointestinal tract stem cell niche. Stem Cell Rev 2(3): 203–212
111. Fennerty MB (1999) Tissue staining (chromoscopy) of the gastrointestinal tract. Can J Gastroenterol 13(5):423–429
112. Wong WM, Wright NA (1999) Cell proliferation in gastrointestinal mucosa. J Clin Pathol 52 (5):321–333
113. Tappenden KA (2006) Mechanisms of enteral nutrient-enhanced intestinal adaptation. Gastroenterology 130(2):S93–S99
114. Claus R, Günthner D, Letzguß H (2007) Effects of feeding fat-coated butyrate on mucosal morphology and function in the small intestine of the pig. J Anim Physiol Anim Nutr 91 (7–8):312–318
115. Wang J (2012) Pathologic aspects of tumors of gastrointestinal tract in the era of personalized medicine. J Gastrointest Oncol 3(3):151–152
116. Rubin E, Farber JL (1990) Environmental diseases of the digestive system. Med Clin North Am 74(2):413–424
117. JimÉnez RM, Connolly S, Blickman JG, Hoogeveen YL (2008) Accessory organs of digestion. In: Devos AS, Blickman JG (eds) Radiological imaging of the digestive tract in infants and children. Part of the series medical radiology. Springer, Berlin, Heidelberg
118. Coffey JC, O'Leary DP (2016) The mesentery: structure, function, and role in disease. Lancet Gastroenterol Hepatol 1(3):238–247

119. Sundler F (2004) GI tract, general anatomy (cells). Encyclopedia of Endocrine Diseases
120. Soffers JH, Hikspoors JP, Mekonen HK, Koehler SE, Lamers WH (2015) The growth pattern of the human intestine and its mesentery. BMC Dev Biol 15:31
121. Kachlik D, Baca V, Stingl J (2010) The spatial arrangement of the human large intestinal wall blood circulation. J Anat 216(3):335–343
122. Jedrzejewski KS, Cendrowska I, Okraszewska E (2002) Evaluation of several methods used in anatomical investigations of the blood and lymphatic vessels. Folia Morphol (Warsz) 61 (2):63–69
123. Mutsaers SE (2004) The mesothelial cell. Int J Biochem Cell Biol 36(1):9–16
124. Jansson L, Barbu A, Bodin B, Carlsson P-O (2016) Pancreatic islet blood flow and its measurement. Upsala J Med Sci 121(2):1–15
125. Granger DN, Holm L, Kvietys P (2015) The gastrointestinal circulation: physiology and pathophysiology. Compr Physiol 5(3):1541–1583
126. Parks DA, Jacobson ED (1985) Physiology of the splanchnic circulation. Arch Intern Med 145(7):1278–1281
127. Clemmesen O (2002) Splanchnic circulation and metabolism in patients with acute liver failure. Dan Med Bull 49(3):177–193
128. Mukhtar A, Dabbous H (2016) Modulation of splanchnic circulation: role in perioperative management of liver transplant patients. World J Gastroenterol 22(4):1582–1592
129. Ackland G, Grocott MPW, Mythen MG (2000) Understanding gastrointestinal perfusion in critical care: so near, and yet so far. Crit Care 4(5):269–281
130. Kvietys PR (2010) The gastrointestinal circulation. Morgan & Claypool Life Sciences, San Rafael, CA
131. Walker TG (2009) Mesenteric vasculature and collateral pathways. Semin Intervent Radiol 26(3):167–174
132. Graf O, Boland GW, Kaufman JA, Warshaw AL, Fernandez del Castillo C, Mueller PR (1997) Anatomic variants of mesenteric veins: depiction with helical CT venography. AJR Am J Roentgenol 168(5):1209–1213
133. Kim PTW, Klintmalm GB (2016) Importance of hepatic flows in liver transplantation. J Hepatol Gastroint Dis 2:2
134. Eipel C, Ke Abshagen, Vollmar B (2010) Regulation of hepatic blood flow: the hepatic arterial buffer response revisited. World J Gastroenterol 16(48):6046–6057
135. Matsuura T, Gad MZ, Harrison EH, Ross AC (1997) Lecithin: retinol acyltransferase and retinyl ester hydrolase activities are differentially regulated by retinoids and have distinct distributions between hepatocyte and nonparenchymal cell fractions of rat liver. J Nutr 127 (2):218–224
136. Kumar A, Tripathi A, Jain S (2011) Extracorporeal bioartificial liver for treating acute liver diseases. J Extra Corpor Technol 43(4):195–206
137. Tabata K, Yamaoka K, Fukuyama T, Nakagawa T (1995) Evaluation of intestinal absorption into the portal system in enterohepatic circulation by measuring the difference in portal-venous blood concentrations of diclofenac. Pharm Res 12(6):880–883
138. Ponziani FR, Zocco MA, Campanale C, Rinninella E, Tortora A, Di Maurizio L et al (2010) Portal vein thrombosis: insight into physiopathology, diagnosis, and treatment. World J Gastroenterol 16(2):143–155
139. Douard R, Chevallier JM, Delmas V, Cugnenc PH (2006) Clinical interest of digestive arterial trunk anastomoses. Surg Radiol Anat 28(3):219–227
140. Geboes K, Geboes KP, Maleux G (2001) Vascular anatomy of the gastrointestinal tract. Best Pract Res Clin Gastroenterol 15(1):1–14
141. Vandamme JP, Bonte J (1985) The branches of the celiac trunk. Acta Anat (Basel) 122 (2):110–114
142. Nayak SR, Prabhu LV, Krishnamurthy A, Ganesh Kumar C, Ramanathan LA, Acharya A, Prasad SA (2008) Additional branches of celiac trunk and its clinical significance. Rom J Morphol Embryol 49(2):247–249

143. Lovisetto F, De Lorenzi FG, Stancampiano P, Corradini C, De Cesare F, Geraci O, Manzi M, Arceci F (2012) Thrombosis of celiacomesenteric trunk: report of a case. World J Gastroenterol 18(29):3917–3920
144. Yan J, Nagasawa Y, Nakano M, Hitomi J (2014) Origin of the celiac and superior mesenteric arteries in a common trunk: description of a rare vessel variation of the celiacomesenteric trunk with a literature review. Okajimas Folia Anat Jpn 91(2):45–48
145. He H, Long Y, Liu D, Wang X, Zhou X (2015) Clinical classification of tissue perfusion based on the central venous oxygen saturation and the peripheral perfusion index. Crit Care 19(1):330
146. Pieńkowska J, Gwoździewicz K, Skrobisz-Balandowska K, Marek I, Kostro J, Szurowska E et al (2016) Perfusion-CT—can we predict acute pancreatitis outcome within the first 24 hours from the onset of symptoms? PLoS ONE 11(1):e0146965
147. Hulten L, Jodal M, Lindhagen J, Lundgren O (1976) Blood flow in the small intestine of cat and man as analyzed by an inert gas washout technique. Gastroenterology 70(1):45–51
148. Hultén L, Lindhagen J, Lundgren O, Fasth S, Ahrén C (1977) Regional intestinal blood flow in ulcerative colitis and Crohn's disease. Gastroenterology 72(3):388–396
149. Hasibeder W (2010) Gastrointestinal microcirculation: still a mystery? Br J Anaesth 105(4): 393–396
150. Someya N, Endo MY, Fukuba Y, Hayashi N (2008) Blood flow responses in celiac and superior mesenteric arteries in the initial phase of digestion. Am J Physiol Regul Integr Comp Physiol 294(6):R1790–R1796
151. Matheson PJ, Wilson MA, Garrison RN (2000) Regulation of intestinal blood flow. J Surg Res 93:182–196
152. de Oliveira EP, Burini RC, Jeukendrup A (2014) Gastrointestinal complaints during exercise: prevalence, etiology, and nutritional recommendations. Sports Med 44:79–85
153. van Wijck K, Lenaerts K, van Loon LJC, Peters WHM, Buurman WA, Dejong CHC (2011) Exercise-induced splanchnic hypoperfusion results in gut dysfunction in healthy men. PLoS ONE 6(7):e22366
154. Schumacker PT, Samsel RW (1989) Oxygen delivery and uptake by peripheral tissues: physiology and pathophysiology. Crit Care Clin 5(2):255–269
155. Guzman JA, Lacoma FJ, Kruse JA (1998) Relationship between systemic oxygen supply dependency and gastric intramucosal PCO2 during progressive hemorrhage. J Trauma 44:696–700
156. Sabnis A, Carrasco R, Liu SX, Yan X, Managlia E, Chou PM et al (2015) Intestinal vascular endothelial growth factor is decreased in necrotizing enterocolitis. Neonatology 107(3): 191–198
157. Hu D, Cai D, Rangan AV (2012) Blood vessel adaptation with fluctuations in capillary flow distribution. PLoS ONE 7(9):e45444
158. Stan RV, Tse D, Deharvengt SJ, Smits NC, Xu Y, Luciano MR et al (2012) The diaphragms of fenestrated endothelia—gatekeepers of vascular permeability and blood composition. Dev Cell 23(6):1203–1218
159. Satchell SC, Braet F (2009) Glomerular endothelial cell fenestrations: an integral component of the glomerular filtration barrier. Am J Physiol Renal Physiol 296(5):F947–F956
160. Hallmann R, Horn N, Selg M, Wendler O, Pausch F, Sorokin LM (2005) Expression and function of laminins in the embryonic and mature vasculature. Physiol Rev 85(3):979–1000
161. Benett HS, Luft JH, Hampton JC (1959) Morphological classifications of vertebrate blood capillaries. Am J Physiol 196:381–390
162. Sarin H (2010) Physiologic upper limits of pore size of different blood capillary types and another perspective on the dual pore theory of microvascular permeability. J Angiogenes Res 2:14
163. Henderson JR, Moss MC (1985) A morphometric study of the endocrine and exocrine capillaries of the pancreas. Q J Exp Physiol 70(3):347–356
164. Aharinejad S, Böck P (1994) Identification of fenestrated capillary segments in microvascular corrosion casts of the rat exocrine pancreas. Scanning 16(4):209–214

165. Strell C, Entschladen F (2008) Extravasation of leukocytes in comparison to tumor cells. Cell Commun Signal 6:10
166. Aird WC (2007) Phenotypic heterogeneity of the endothelium: I. Structure, function and mechanisms. Circ Res 100:158–173
167. Pries AR, Kuebler WM (2006) Normal endothelium. In: Moncada S, Higgs A (eds) Handbook of pharmacology: the vascular endothelium
168. Aoyagi T, Koshimizu T, Tanoue A (2009) Vasopressin regulation of blood pressure and volume: findings from V1a receptor–deficient mice. Kidney Int 76(10):1035–1039
169. Gordan R, Gwathmey JK, Xie L-H (2015) Autonomic and endocrine control of cardiovascular function. World J Cardiol 7(4):204–214
170. Richardson PD (1982) Physiological regulation of the hepatic circulation. Fed Proc 41 (6):2111–2116
171. Carlton LD, Pollack GM, Brouwer KL (1996) Physiologic pharmacokinetic modeling of gastrointestinal blood flow as a rate-limiting step in the oral absorption of digoxin: implications for patients with congestive heart failure receiving epoprostenol. J Pharm Sci 85 (5):473–477
172. Mukhtar A, Dabbous H (2016) Modulation of splanchnic circulation: role in perioperative management of liver transplant patients. World J Gastroenterol 22(4):1582–1592
173. Balaram SK, Markham J, DeRose JJ (2008) Minimally invasive perfusion techniques. In: Mongero LB, Beck JR (eds) On bypass: current cardiac surgery. Humana Press, NY
174. Ruffolo DC (1998) Gastric tonometry: early warning of tissue hypoperfusion. Crit Care Nurs Q 21(3):26–32
175. Marshall AP, West SH (2004) Gastric tonometry and monitoring gastrointestinal perfusion: using research to support nursing practice. Nurs Crit Care 9(3):123–133
176. Eachempati SR, Barie PS (1999) Minimally invasive and noninvasive diagnosis and therapy in critically ill and injured patients. Arch Surg 134(11):1189–1196
177. Lima A, Bakker J (2006) Noninvasive monitoring of peripheral perfusion. In: Applied physiology in intensive care medicine. Springer, Berlin, Heidelberg
178. Sarkaria IS, Bains MS, Finley DJ, Adusumilli PS, Huang J, Rusch VW et al (2014) Intraoperative near-infrared fluorescence imaging as an adjunct to robotic assisted minimally invasive esophagectomy. Innovations (Phila) 9(5):391–393
179. Fikfak V, Gaur P, Kima MP (2016) Endoscopic evaluation of gastric conduit perfusion in minimally invasive Ivor Lewis esophagectomy. Int J Surg Case Rep 19:112–114
180. Zehetner J, DeMeester SR, Alicuben ET, Oh DS, Lipham JC, Hagen JA, DeMeester TR (2015) Intraoperative assessment of perfusion of the gastric graft and correlation with anastomotic leaks after esophagectomy. Ann Surg 262(1):74–78
181. Otte JA, Geelkerken RH, Huisman AB, Kolkman JJ (2007) What Is the best diagnostic approach for chronic gastrointestinal ischemia [quest]. Am J Gastroenterol 102:2005–2010
182. Atef HM, Fattah SA, Gaffer MEA, Al Rahman AA (2013) Perfusion index versus non-invasive hemodynamic parameters during insertion of i-gel, classic laryngeal mask airway and endotracheal tube. Indian J Anaesth 57(2):156–162
183. Jeng EI, Gravenstein N, Klodell CT (2017) Perfusion index: an indicator of success during endoscopic thoracic sympathectomy for hyperhidrosis. Ann Thorac Surg 104(2):426–430
184. Lima A, Bakker J (2005) Noninvasive monitoring of peripheral perfusion. Intensive Care Med 31(10):1316–1326
185. Rovenská E, Rovenský J (2011) Lymphatic vessels: structure and function. IMAJ 13: 762–768
186. Breslin JW (2014) Mechanical forces and lymphatic transport. Microvasc Res 96:46–54
187. Gashev AA (2002) Physiologic aspects of lymphatic contractile function: current perspectives. Ann N Y Acad Sci 979:178–196
188. Al-Busafi SA, Ghali P, Deschênes M, Wong P (2014) Chylous ascites: evaluation and management. ISRN Hepatol 240473:1–10
189. Miller MJ, McDole JR, Newberry RD (2010) Microanatomy of the intestinal lymphatic system. Ann N Y Acad Sci 1207:E21–E28

190. Choe K, Jang JY, Park I, Kim Y, Ahn S, Park DY et al (2015) Intravital imaging of intestinal lacteals unveils lipid drainage through contractility. J Clin Invest 125(11):4042–4052
191. Rasmussen JC, Tan I-C, Marshall MV, Fife CE, Sevick-Muraca EM (2009) Lymphatic imaging in humans with near-infrared fluorescence. Curr Opin Biotechnol 20(1):74–82
192. Zawieja DC (2009) Contractile physiology of lymphatics. Lymphat Res Biol 7(2):87–96
193. Mallick A, Bodenham AR (2003) Disorders of the lymph circulation: their relevance to anaesthesia and intensive care. Br J Anaesth 91:265–272
194. Thompson MD, Carr AP (2002) Hyponatremia and hyperkalemia associated with chylous pleural and peritoneal effusion in a cat. Can Vet J 43(8):610–613
195. Steven BR, Carey S (2015) Nutritional management in patients with chyle leakage: a systematic review. Eur J Clin Nutr 69:776–780
196. Borisov AV (2005) Functional anatomy of lymphangion. Morfologiia 128(6):18–27
197. Quick CM, Venugopal AM, Gashev AA, Zawieja DC, Stewart RH (2007) Intrinsic pump-conduit behavior of lymphangions. Am J Physiol—Regul, Integr Comp Physiol 292 (4):R1510–R1518
198. Brace RA, Andres RL (1991) Left thoracic duct lymph flow responses to angiotensin II or atrial natriuretic factor infusion in the ovine fetus. Am J Obstet Gynecol 165(6 Pt 1):1607–1613
199. Lachance P-A, Hazen A, Sevick-Muraca EM (2013) Lymphatic vascular response to acute inflammation. PLoS ONE 8(9):e76078
200. Dunworth WP, Fritz-Six KL, Caron KM (2008) Adrenomedullin stabilizes the lymphatic endothelial barrier in vitro and in vivo. Peptides 29(12):2243–2249
201. Schmid-Schönbein GW (1990) Microlymphatics and lymph flow. Physiol Rev 70(4): 987–1028
202. Reddy NP (1986) Lymph circulation: physiology, pharmacology, and biomechanics. Crit Rev Biomed Eng 14(1):45–91
203. Browning KN, Travagli RA (2014) Central nervous system control of gastrointestinal motility and secretion and modulation of gastrointestinal functions. Compr Physiol 4 (4):1339–1368
204. Phillips RJ, Powley TL (2007) Innervation of the gastrointestinal tract: patterns of aging. Auton Neurosci 136(1–2):1–19
205. Straub RH, Wiest R, Strauch UG, Härle P, Schölmerich J (2006) The role of the sympathetic nervous system in intestinal inflammation. Gut 55(11):1640–1649
206. Furness JB, Callaghan BP, Rivera LR, Cho HJ (2014) The enteric nervous system and gastrointestinal innervation: integrated local and central control. Adv Exp Med Biol 817:39–71
207. Margolis KG, Gershon MD, Bogunovic M (2016) Cellular organization of neuroimmune interactions in the gastrointestinal tract. Trends Immunol 37(7):487–501
208. Pereverzev VA, Welcome MO, Melnichuk VI, Razvodovsky YE, Pereverzeva EV (2016) Fasting normoglycemia: classification, parameters, maintenance mechanisms. Healthcare 4:18–27
209. Duke GE (1986) Alimentary canal: secretion and digestion, special digestive functions, and absorption. In: Sturkie PD (ed) Avian physiology. Springer, New York
210. Karasov WH, Douglas AE (2013) Comparative digestive physiology. Compr Physiol 3(2): 741–783
211. Furness JB, Cottrell JJ, Bravo DM (2015) Comparative gut physiology symposium: comparative physiology of digestion. J Anim Sci 93(2):485–491
212. Egorov VI, Schastlivtsev IV, Prut EV, Baranov AO, Turusov RA (2002) Mechanical properties of the human gastrointestinal tract. J Biomech 35(10):1417–1425
213. Kauffman GL Jr (1981) Gastric mucus and bicarbonate secretion in relation to mucosal protection. J Clin Gastroenterol 3(2):45–50
214. Agarwal R, Afzalpurkar R, Fordtran JS (1994) Pathophysiology of potassium absorption and secretion by the human intestine. Gastroenterology 107(2):548–571
215. O'Connor A, O'Moráin C (2014) Digestive function of the stomach. Dig Dis 32(3):186–191
216. Chu S, Schubert ML (2012) Gastric secretion. Curr Opin Gastroenterol 28(6):587–593

217. Schubert ML (2014) Gastric secretion. Curr Opin Gastroenterol 30(6):578–582
218. Schubert ML (2009) Gastric exocrine and endocrine secretion. Curr Opin Gastroenterol 25(6):529–536
219. Ménard D (2004) Functional development of the human gastrointestinal tract: hormone-and growth factor-mediated regulatory mechanisms. Can J Gastroenterol 18(1):39–44
220. Eckmann L, Stenson WF, Savidge TC, Lowe DC, Barrett KE, Fierer J et al (1997) Role of intestinal epithelial cells in the host secretory response to infection by invasive bacteria. Bacterial entry induces epithelial prostaglandin h synthase-2 expression and prostaglandin E2 and F2alpha production. J Clin Invest 100(2):296–309
221. Bronner F (2003) Mechanisms of intestinal calcium absorption. J Cell Biochem 88(2): 387–393
222. Ward JB, Keely SJ, Keely SJ (2014) Oxygen in the regulation of intestinal epithelial transport. J Physiol 592(12):2473–2489
223. Pizzorno L (2016) Bariatric surgery: bad to the bone, part 1. Integr Med (Encinitas) 15(1): 48–54
224. Terra X, Auguet T, Guiu-Jurado E, Berlanga A, Orellana-Gavaldà JM, Hernández M et al (2013) Long-term changes in leptin, chemerin and ghrelin levels following different bariatric surgery procedures: Roux-en-Y gastric bypass and sleeve gastrectomy. Obes Surg 23(11): 1790–1798
225. Ghishan FK, Kiela PR (2014) Epithelial transport in inflammatory bowel diseases. Inflamm Bowel Dis 20(6):1099–1109
226. Field M (2003) Intestinal ion transport and the pathophysiology of diarrhea. J Clin Invest 111(7):931–943
227. Kiel JW, Shepherd AP (1990) Gastrointestinal blood flow. In: Shepherd AP, Åke Öberg P (eds) Laser-doppler blood flowmetry. Developments in cardiovascular medicine. Springer Science + Business Media, New York
228. Kiela PR, Ghishan FK (2009) Ion transport in the intestine. Curr Opin Gastroenterol 25(2): 87–91
229. Marks J, Debnam ES, Unwin RJ (2013) The role of the gastrointestinal tract in phosphate homeostasis in health and chronic kidney disease. Curr Opin Nephrol Hypertens 22(4): 481–487
230. Reining SC, Liesegang A, Betz H, Biber J, Murer H, Hernando N (2010) Expression of renal and intestinal Na/Pi cotransporters in the absence of GABARAP. Pflugers Arch 460(1): 207–217
231. Laforenza U (2012) Water channel proteins in the gastrointestinal tract. Mol Aspects Med 33 (5–6):642–650
232. Zhu C, Chen Z, Jiang Z (2016) Expression, distribution and role of aquaporin water channels in human and animal stomach and intestines. Int J Mol Sci 17(9). pii:E1399
233. Page DA, Carlson GP (1994) The role of the intestinal tract in the elimination of carbon tetrachloride. Toxicol Appl Pharmacol 124(2):268–274
234. Arimori K, Nakano M (1998) Drug exsorption from blood into the gastrointestinal tract. Pharm Res 15(3):371–376
235. Hollander D, Dadufalza VD (1983) Intestinal exsorption of oleic acid: influence of aging, bile, pH and ethanol. J Nutr 113(3):511–518
236. Leu BL, Huang J (1995) Inhibition of intestinal P-glycoprotein and effects on etoposide absorption. Cancer Chemother Pharmacol 35(5):432–436
237. Carr RA, Pasutto FM, Foster RT (1996) Renal, biliary and intestinal clearance of sotalol enantiomers in rat model: evidence of intestinal exsorption. Biopharm Drug Dispos 17(8): 725–735
238. Sarker SA, Gyr K (1992) Non-immunological defence mechanisms of the gut. Gut 33(7): 987–993
239. Dommett R, Zilbauer M, George JT, Bajaj-Elliott M (2005) Innate immune defence in the human gastrointestinal tract. Mol Immunol 42(8):903–912

240. Hansson GC (2012) Role of mucus layers in gut infection and inflammation. Curr Opin Microbiol 15(1):57–62
241. Janssen AW, Kersten S (2015) The role of the gut microbiota in metabolic health. FASEB J 29(8):3111–3123
242. Fukuda S, Ohno H (2014) Gut microbiome and metabolic diseases. Semin Immunopathol 36(1):103–114
243. Rubino F, R'bibo SL, del Genio F, Mazumdar M, McGraw TE (2010) Metabolic surgery: the role of the gastrointestinal tract in diabetes mellitus. Nat Rev Endocrinol 6(2):102–109
244. Devaraj S, Hemarajata P, Versalovic J (2013) The human gut microbiome and body metabolism: implications for obesity and diabetes. Clin Chem 59(4):617–628
245. Turnbaugh PJ, Gordon JI (2009) The core gut microbiome, energy balance and obesity. J Physiol 587(Pt 17):4153–4158
246. Krebs NF (2000) Overview of zinc absorption and excretion in the human gastrointestinal tract. J Nutr 130(5S):1374S–1377S
247. Klevay LM, Bogden JD, Aladjem M, Sandstead HH, Kemp FW, Li W et al (2007) Renal and gastrointestinal potassium excretion in humans: new insight based on new data and review and analysis of published studies. J Am Coll Nutr 26(2):103–110
248. Cheng LK, O'Grady G, Du P, Egbuji JU, Windsor JA, Pullan AJ (2010) Gastrointestinal system. Wiley Interdiscip Rev Syst Biol Med 2(1):65–79
249. Gunawardene AR, Corfe BM, Staton CA (2011) Classification and functions of enteroendocrine cells of the lower gastrointestinal tract. Int J Exp Pathol 92(4):219–231
250. Arnold R (2005) Endocrine tumours of the gastrointestinal tract. Introduction: definition, historical aspects, classification, staging, prognosis and therapeutic options. Best Pract Res Clin Gastroenterol 19(4):491–505
251. Maggs D, MacDonald I, Nauck MA (2008) Glucose homeostasis and the gastrointestinal tract: insights into the treatment of diabetes. Diabetes Obes Metab 10(1):18–33
252. Ewart WR (1985) Sensation in the gastrointestinal tract. Comp Biochem Physiol, A: Comp Physiol 82(3):489–493
253. Yu X, Yu M, Liu Y, Yu S (2016) TRP channel functions in the gastrointestinal tract. Semin Immunopathol 38(3):385–396
254. Blackshaw LA, Brookes SJ, Grundy D, Schemann M (2007) Sensory transmission in the gastrointestinal tract. Neurogastroenterol Motil 19(1):1–19
255. Brock C, Arendt-Nielsen L, Wilder-Smith O, Drewes AM (2009) Sensory testing of the human gastrointestinal tract. World J Gastroenterol 15(2):151–159
256. Rayner CK, Samsom M, Jones KL, Horowitz M (2001) Relationships of upper gastrointestinal motor and sensory function with glycemic control. Diabetes Care 24(2): 371–381
257. Yi C-X, Tschöp MH (2012) Brain–gut–adipose-tissue communication pathways at a glance. Dis Model Mech 5(5):583–587
258. Carabotti M, Scirocco A, Maselli MA, Severi C (2015) The gut-brain axis: interactions between enteric microbiota, central and enteric nervous systems. Ann Gastroenterol 28(2): 203–209
259. van der Flier LG, Clevers H (2009) Stem cells, self-renewal, and differentiation in the intestinal epithelium. Annu Physiol 71:241–260
260. Mayhew TM (1990) Striated brush border of intestinal absorptive epithelial cells: stereological studies on microvillous morphology in different adaptive states. J Electron Microsc Tech 16(1):45–55
261. Lan M, Kojima T, Murata M, Osanai M, Takano K, Chiba H, Sawada N (2006) Phosphorylation of ezrin enhances microvillus length via a p38 MAP-kinase pathway in an immortalized mouse hepatic cell line. Exp Cell Res 312(2):111–120
262. Venugopal AM, Stewart RH, Laine GA, Dongaonkar RM, Quick CM (2007) Lymphangion coordination minimally affects mean flow in lymphatic vessels. Am J Physiol Heart Circ Physiol 293(2):H1183–H1189

263. Dudek RW (2010) High-yield systems gastrointestinal tract. Lippincott Williams & Wilkins, Philadelphia, PA
264. Kuo B, Urma D (2006) Esophagus—anatomy and development. GI Motil Online. https://doi.org/10.1038/gimo6
265. Gahutu JB, Wane J (2006) Reference values for serum protein and electrolyte study from Rwanda. East Afr Med J 83(2):64–67
266. Lolekha PH, Vanavanan S, Teerakarnjana N, Chaichanajarernkul U (2001) Reference ranges of electrolyte and anion gap on the beckman E4A, beckman synchron CX5, nova CRT, and nova stat profile ultra. Clin Chim Acta 307(1–2):87–93

Chapter 3
Cellular Organization of the Gastrointestinal Tract

Abstract The gastrointestinal (GI) tract is composed of basic tissue types (epithelia, connective, blood, lymphatic, muscle, and nerve tissue) formed from different structural and functional units that determine fundamental life processes. This chapter provides detailed and contemporary information on the structural and functional characteristics of the cells that constitute the GI tract.

Keywords Gut cells · Epithelial cells · GI epithelial cells · Absorptive columnar enterocyte · Mucin-secreting · Enteroendocrine or neuroendocrine cells
Epithelial enteroimmune response cells · GI epithelial cellular network
Structure and function · Cellular and molecular basis of function
Plasma membrane · Cell · Cytoplasm · Organelles · Polarity

Abbreviations

5-HT	5-Hydroxytryptamine
ABC	ATP-binding cassette
AQP	Aquaporin
ATP	Adenosine triphosphate
BiP	Binding immunoglobulin protein
CCAAT	Cytidine-cytidine-adenosine-adenosine-thymidine
COP	Coat protein complex
DAA	Derivatives of arachidonic acid
DHA	Docosahexaenoic acid
DNA	Deoxyribonucleic acid
EC	Enterochromaffin cell
ER	Endoplasmic reticulum
ERM	Ezrin, radixin, moesin
GI	Gastrointestinal
GIP	Gastric inhibitory peptide
GLP-1 and GLP-2	Glucagon-like peptide 1 & 2
GLUT1	Glucose transporter type 1
GM2	Ganglioside monosialic type 2
GPI	Glycosylphosphatidylinositol

M. O. Welcome, *Gastrointestinal Physiology*,
https://doi.org/10.1007/978-3-319-91056-7_3

GTP	Guanosine triphosphate
lncRNAs	Long noncoding RNAs
miRNAs	MicroRNAs
mRNA	Messenger RNA
MRP	Multidrug resistance-associated protein
mtDNA	Mitochondrial DNA
MTOC	Microtubule organizing center
NANA	*N*-acetylneuraminic acid
ncRNA	Noncoding RNA
NF-H	Neurofilament-heavy
NF-L	Neurofilament-light
NF-M	Neurofilament-medium
OXM	Oxyntomodulin
PYY	Peptide YY
RM	Membrane resistance
RNA	Ribonucleic acid
rRNA	Ribosomal RNA
snRNA	Small nuclear RNA
tRNA	Transfer RNA

3.1 Introduction

Understanding the behavior of the microscopic units that constitute the gastrointestinal (GI) tract provides better chance for addressing contemporary problems related to the GI tract. The behavior of the functional and structural entities of a living system possibly renders a somewhat robust picture on the functioning of the whole organism in normal and disease states. The structural and functional integrity of cellular and subcellular components as well as the molecular signatures of the cell form the basis of functioning of the GI system. The GI tract comprises a network of different cell types that work in a coordinated manner to carry out digestive and associated functions. The integrity of the cells comprising the layers is integral to determining the functional state of the GI tract. The pattern of communication among the cells of the GI tract (and with the cells of extraenteric tissues and organs) is integral to the physiology of the gut [1–5]. This chapter examines the structural and functional characteristics of the different cell types that constitute the GI tract.

3.2 Brief Historical Background

The origin and evolution of cells generally are unknown, but available data indicate that cells may have emerged on Earth at least 3.5 billion years ago. Several theories have been formulated to explain how cells evolved. The exobiologist hypothesis of existence suggests that cell may have originated on Earth from the debris of extraplanetary bodies that accidentally entered the Earth from outer space. The "primordial soup" hypothesis provides another view of the evolution of cells—which may have occurred in an earliest water environment present at the very early period of the Earth's existence and had all molecules required for the commencement of life. The common ancestor gave rise to all cells present in its unicellular (e.g., bacteria) and multicellular forms (e.g., plants and animals, including the total number of cells contained in humans, which is currently estimated to be 3.0×10^{13}) [6–9]. But how these minute life units will be seen with the human eye proved to be a huge challenge that took several hundreds of years of human civilization.

Until technological progress was achieved in magnifying objects, man did not have any idea about the existence of minute life forms. From about 1000 BC to the early sixteenth century, merchants and scientists had attempted to magnify objects. However, no success was recorded until Dutchman Zacharius Jansen (1585–c. 1638) and his father Hans invented the first microscope around the 1590s. This microscope was not used in magnifying minute life forms until its advancement in the second half of the seventeenth century. In 1665 using a compound microscope, the English Robert Hooke (1635–1703) became the first to identify the smallest functional unit of life in cork, then in living plant tissue, and coined the word "cell" (from Latin "cella," meaning "small room" or "cubicle"). The thin slices of cork he viewed under the microscope were seen as tiny rows like honeycomb, which were somewhat similar to the small rooms monks lived in. The results of Hook's observations were published in his book "Micrographia" [10–16].

Advancement into the cellular basis of gut functioning was made by many scientists notably Anton van Leeuwenhoek (1632–1723), Andreas Bonn (1738–1817), Johann Conrad Peyer (1653–1712), Johann Conrad Brunner (1653–1727), Johann Nathanael Lieberkühn (1711–1756), Johannes Evangelist Purkinje (1787–1869), Theodor Schwann (1810–1882) among others [10, 14, 17, 18].

The Dutch scientist, Leeuwenhoek, though a businessman who never had basic scientific knowledge as at the time, developed interest in microscopy. In the 1660s, he started making lenses and refining microscopes. Leeuwenhoek discovered living "animalcules," bacteria in the oral mucosa fluid and protozoa from pond water. In 1684, Leeuwenhoek published the first drawings of bacteria as seen under the compound microscope, which he had invented at the time [10, 17, 19]. The discovery of the compound microscope and observation of small animalcules must have set the stage for cellular understanding of digestive functions. Following the work of Leeuwenhoek, cellular studies into secretes of the digestive system were enhanced. The discovery of microscopic units under a microscope was a useful step forward in the development and understanding of physiological basis of digestive functioning at the cellular level.

Bonn, also a Dutch scientist, is credited for discovering that the GI tract is covered by epithelium [17]. This discovery was useful to future scientists who studied the composition of the layers of the GI tract as well as the epithelium in particular. The epithelium is the main site of absorption, synthesis, and secretion of different molecules in the GI tract.

The Swiss scientist Peyer in 1677 observed groups of cells, which is now named after him "Peyer's patch"—isolated or aggregated lymphoid follicles of the gut-associated lymphoid tissue—the largest lymphoid organ, comprising approximately 70% of the body's immune cells. Around the same period, a Swiss anatomist, Brunner, discovered duodenal glands. Small intestinal glands were discovered by the German physician, Lieberkühn, in the mid-eighteenth century and were later named after him—"crypt of Lieberkühn." These intestinal glands are composed of both minor mucosa and submucosal secretory cells that release alkaline molecules, fluid, and enzymes into the lumen of the GI tract [18, 20].

The Czech experimental physiologist, Purkinje, identified gastric glands as the source of HCl. He was also instrumental in identifying the fluidity of the internal cellular environ "cytoplasm" in 1839. The German physiologist Schwann discovered pepsin as a major product of chief cells and also discovered animal cells in 1836 [21]. Schwann, together with the German botanist, Matthias Schleiden (1804–1881), and the German physician and biologist Rudolf Virchow (1821–1902), developed the cell theory, proposing that the cell is the basic structural and functional units of life; all organisms are made up of cells; all cells arise from preexisting cells (*Omnis cellula e cellula*) [22, 23]. Other scientists including Robert Remak (1815–1865) and Albert von Kölliker (1817–1905) immensely contributed to the advancement of the cell theory. Remak was a Jewish Polish/German physiologist, who discovered that the division of preexisting cells was the origin of cell types. The Swiss anatomist and physiologist, Kölliker, became one of the pioneers to interpret tissue structure in terms of cellular elements. The modern theory of the cell is built on technological advancement and further proposes that all vital functions of an organism occur within cells; all cells contain the hereditary information necessary for regulating cell functions and for transmitting information to the next generation of cells [14, 23, 24].

Ivan Petrovich Pavlov (1849–1936), the first Nobel Laureate in the area of digestive physiology, conducted many experiments on the GI tract to ascertain the origin of its secretory functions [25–27]. Pavlov is also regarded as the first Noble Laureate in the area of neuroscience—the scientific study of the structure and functions as well as pathological conditions of the nervous system. Pavlov studied the psychological processes involved in, as well as the phases of gastric secretion [26].

The quest into the structure of cells was important in the development of both cellular and molecular physiology. Evidently, the period between the seventeenth and nineteenth centuries marked a turning point in discoveries that made tremendous advancement of knowledge in cellular and molecular functioning of the GI tract. This period represented a substantial shift from spiritism to realities, founded on the background of research and experimental results. Galenic teachings, during the second century, about the existence of "vital spirits" in the body fluids that regulated human bodily functions were gradually becoming obsolete, as certain phenomena initially thought, were completely controlled and dictated by spirits, were experimentally confirmed [28, 29]. Further developments on digestive physiology at the molecular and cellular levels were enhanced in the computer age with the emergence of state-of-the-art technology [30–35].

3.3 Types of Gastrointestinal Tract Cells

The GI tract is composed of cells of all corresponding tissue types—epithelia, connective, blood and lymphatic, muscle, and nerve tissues. The structural–functional network of the gut tissues is composed of different cell types. Thus, cells of the GI tract include GI epithelial cells, cells of the gut nervous system (and plexuses), muscle cells, connective tissue cells, cells of accessory organs of the GI tract, and accessory cells of the gut [36–41].

The cells of the GI tract are formed from a single lineage, which is thought to be located at the base of the intestinal crypt from where epithelial cells of the GI tract are periodically renewed. These cells generally include absorptive columnar enterocyte, mucin-secreting, enteroendocrine or neuroendocrine cells, and epithelial enteroimmune response cells. It should be mentioned that this functional classification of gut epithelial cells may be incorrect in strict terms, based on recent evidences showing that traditional definition of endocrine cells as cells that secrete their substances into blood may not be special feature for enteroendocrine or neuroendocrine cells, but also non-endocrine cells. Thus, classification of these groups of cells becomes quite difficult [37–39, 42, 43]. Notwithstanding, however, the GI tract cells can be divided on the basis of their physiological role. The majority of epithelial cells are classified on the basis of their role—secretory, absorptive cells, and so on.

3.3.1 Muscle Cells of the Gut

The muscle tissue of the gut is composed of two cell types—smooth and striated muscle types (Figs. 3.1 and 3.2). The main differences between these cells are striations, number of nuclei and their location, as well as the size of the cells [44, 45].

The smooth muscle is found in the walls of hollow organs (except the heart). This type of muscle is found in the digestive tract; it is involuntary and non-striated (Fig. 3.1). The name "smooth" is derived from its non-striations unlike in cardiac and skeletal muscles that are striated. Smooth muscle is either single unit (also

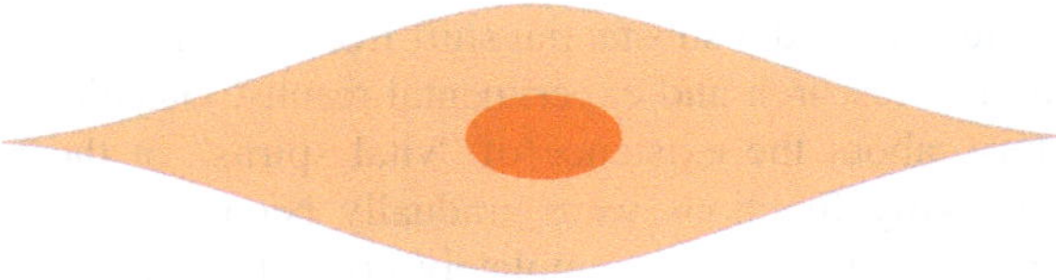

Fig. 3.1 Smooth (involuntary) muscle cell. The size of this cell is about 15–150 μm [45]

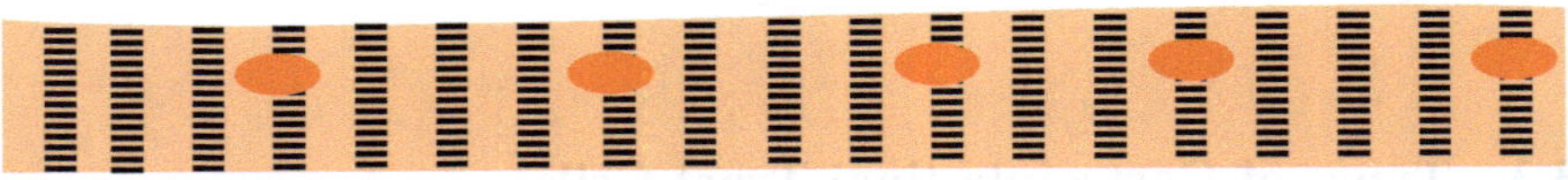

Fig. 3.2 Skeletal (voluntary) muscle cell. The size of symplast of skeletal muscle is about 2000–200,000 μm [44, 45]

called unitary) or multiunit. Single-unit muscle cells in a bundle are innervated by varicosities of autonomic nerve fiber. The bundle of single-unit muscle cells contracts as a syncytium—contracting in a coordinated fashion. This property of smooth muscle cells is due to the presence of numerous intercellular junctions called gap junctions. Thus when one muscle cell depolarizes, waves of calcium ions within a very short interval spread to all muscle cells of the bundle causing it to contract as one unit. Correspondingly, single-unit smooth muscle can contract regularly without input from a motor neuron. Moreover, some cells within the unitary smooth muscle cells may behave as pacemaker cells, generating rhythmic action potentials due to their intrinsic electrical activity. The response of single-unit visceral smooth muscle cells is thus referred to as myogenic. Single-unit muscle is found in the GI tract [46–49]. In multiunit smooth muscle tissues, individual muscle cell is innervated by an autonomic fiber. Thus, each cell behaves (contract and relax) independently. In contrast to single-unit muscle cell, multiunit smooth muscle contraction must be initiated by a nerve fiber of the autonomic nervous system and thus it expresses a substantially lower number of gap junctions. Multiunit muscle cells are therefore referred to as neurogenic. Multiunit smooth muscle behaves like skeletal muscle as they allow for gradual responses to stimuli. Multiunit smooth muscle is present in such locations of the body as major breathing pipe leading the lungs, large arteries, and internal eye muscles [46–49].

Smooth muscles may be classified as rhythmic (also called phasic) and tonic [50–52]. This classification is based on their mechanical response to increase in intracellular calcium level following the action of a stimulus. Phasic smooth muscle is the type of smooth muscle that contracts only for discrete periods of time [53, 54]. The underlying mechanism of phasic contraction is the electrical pacing characterized by rhythmic changes in membrane potential (or slow waves) which gradually increases to the threshold level of depolarization, resulting in opening of voltage-gated membrane calcium channels with substantial calcium influx into the muscle cells. This calcium spike current initiates a cascade of events, which subsequently leads to contraction of the muscle cells [55]. Phasic smooth muscles are present in the ureter, lymphatics, portal vein, and uterus. Tonic smooth muscles are found in the urethra. Tonic smooth muscle is continuously active (i.e., contraction is sustained) [53, 54]. Certain agents can induce increase in intracellular calcium by the activation of intracellular stores (e.g., endoplasmic reticulum). Increase in intracellular calcium may become so prolonged, causing the muscle to tonically contract [55]. Both types of smooth muscles are found in the GI tract [54]. However, specific contraction type dominates in certain regions of the GI tract [56]. In the GI tract, for instance, phasic contraction of sphincters can result from movement of chyme or fluid along the GI tract, whereas the same sphincters can contract tonically if they need to remain closed for a considerably long period of time [50, 51].

The striated (skeletal muscles) of digestion include masticatory muscles and muscles located in some regions of the gullet. The gullet contains skeletal muscles in the upper one-third and a mixture of skeletal and smooth muscles in the middle part of the gullet. The lower portion of the gullet and other regions of the GI tract contains only smooth muscles [57]. It is believed that the skeletal muscles of the gullet are formed from transdifferentiation of the progenitors of smooth muscle cells present around that region of the gullet [58]. The muscles of the gullet also exhibit the motor pattern of the GI tract. Thus, they are responsible for the mechanical processing of food and its peristaltic movement along the digestive tract [55, 57, 59–62].

3.3.2 Neurons and Glial Cells of the Gut

Neural and glial cells, grouped into plexuses, make up the enteric or intrinsic nervous system. Thus, these two groups of cells of the nervous system belong to the peripheral nervous system. There are different types of glial cells and neurons of the enteric nervous system distributed mostly in two intramuscular plexuses that extend along the entire length of the gut forming a complex lattice network in the layers of the gut [63–65] (Fig. 3.3).

The complexity in the arrangement of enteric neurons, in contrast to other peripheral organs, confers the GI tract a unique ability to mediate intrinsic reflexes. In addition, the enteric nervous system consists of pacemaker cells which are able to generate impulses without external input. These are some of the reasons why the enteric nervous system is sometimes called the second brain or gut–brain [66].

3.3.3 Connective Tissue Cells

Connective tissues bind structures together to form a supportive framework for organs and the body as a whole; they help to store fat, transport substances, and take

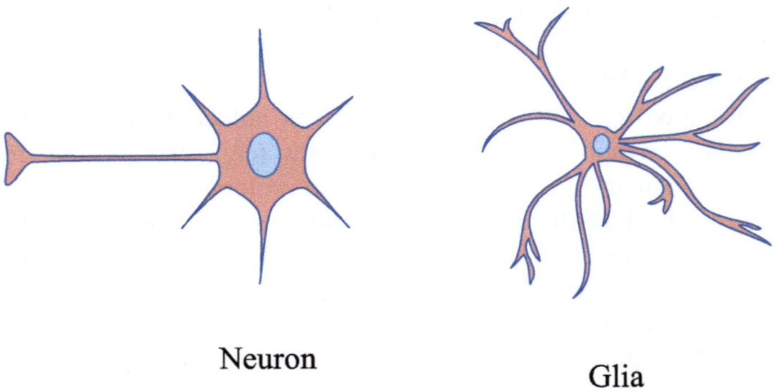

Fig. 3.3 Neural and glial cells

part in tissue reparative and regenerative processes as well as cellular signaling processes. The connective tissues help to maintain the integrity of the structures of the GI system. Certain diseases of the GI tract such as cancers can cause disorders in the composition of the connective tissues of the affected region, consequently, resulting in disorders in GI tract functioning [66].

The different classification types of connective tissues include loose connective tissue, adipose tissue, dense fibrous connective tissue, elastic connective tissue, cartilage, and bone [67]. The most common cells found in connective tissues are fibroblasts, macrophages, and mast cells. Connective tissue also contains adipocytes and leukocytes (Fig. 3.4) [66, 67].

Connective tissues are composed of fibers and ground substance, except for blood and lymph [67]. Elastic, collagenous, and reticular fibers are the major types of connective tissue fibers. Elastic fibers are essential extracellular matrix macromolecules comprising an elastin core surrounded by a mantle of microfibrils with characteristic elasticity [66–70]. Elastic fibers aid in sustaining mechanical stress in tissues. They resume their original length upon removal of the force acting on the

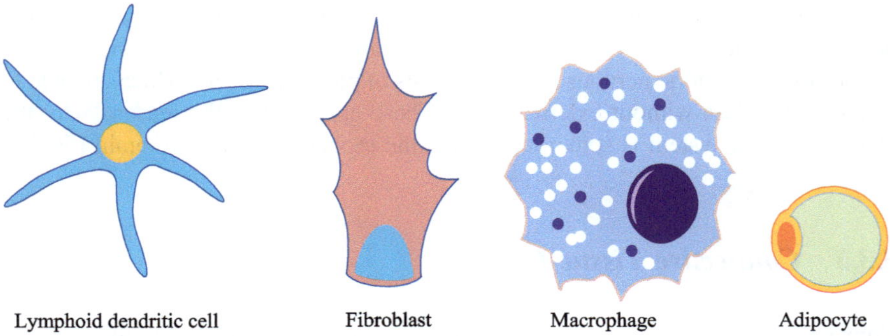

Fig. 3.4 Some cells of the connective tissue

fibers [71]. This property of elastic fibers is required for the maintenance of tissue anatomy and homeostasis [66, 67]. The principal structural component of microfibril mantle is the protein fibrillin. (Fibrillin-1 is the main fibrillin present in microfibrils. Mutation of fibrillin-1 is associated with the heritable disease, Marfan syndrome, characterized by aortic, ocular, and skeletal defects.) But substances such as glycoproteins and collagen are also present in microfibrils. It should be mentioned that microfibrils rich in fibrillin may be also found in extracellular matrix molecules of non-elastic tissues. Elastic fibers are composed of elastin, elaunin, and oxytalan, which are some of the major proteins of extracellular matrix of connective tissue produced by fibroblasts and other cells [66–70].

Collagenous fibers are extracellular matrix macroaggregates that impart structural integrity to the connective tissue, providing an additional strength and supporting tissue–cellular framework, regulating signaling pathways that dictate organ patterning, morphogenesis, and growth. Collagen fibers are the dominant types of fibers present in most connective tissues. Collagen is abundant in the dermis, mucous membranes, nerves, blood vessels, and organs [72–78].

Reticular fibers are a type of connective tissue fiber of the extracellular matrix, providing a certain level of support to the cells of the connective tissue. These fibers form a fine network of collagen fibers acting as a supporting mesh in tissues and organs such as the liver and the lymphatic system [66, 67, 79, 80]. Reticular fibers are composed of type III collagen, synthesized by reticular cell, which is a type of fibroblasts. Reticular fibers connect collagen fibers with the basal lamina of epithelial, muscle, and adipose cells to maintain structural competency of tissues and organs [66, 67].

Ground substance is an amorphous gel-like substance surrounding connective tissue cells, mainly composed of water, glycosaminoglycans (mostly hyaluronan), proteoglycans, and glycoproteins [77, 81, 82]. To test the role of different cells in various connective tissues, specific markers are used. For instance, transcription factor 4 (TCF4) is used as a marker for fibroblasts. Ablation of TCF4 can result in connective tissue disorder and atrophy [67].

3.3.4 Accessory Cells of the Gastrointestinal Tract

Accessory cells of the GI tract aid several functions of the gut, maintaining and preserving the functions of the GI tract and accessory organs. Such cells include but are not limited to lymphocytes and the resident microbes. The microbes are mostly concentrated in the mucus layer (including the crypts of the different regions) of the GI tract (Fig. 3.5) (see Chap. 15).

The commensal microbes of the gut, generally referred to as the gut microbiota, are becoming increasingly recognized as key players in health and disease [83, 84]. Studies have indicated that dysfunctional gut microbiota composition and signaling represent a major cause of a range of behavioral disorders observed in neuropsychiatric diseases such as autism and schizophrenia. The link between the gut

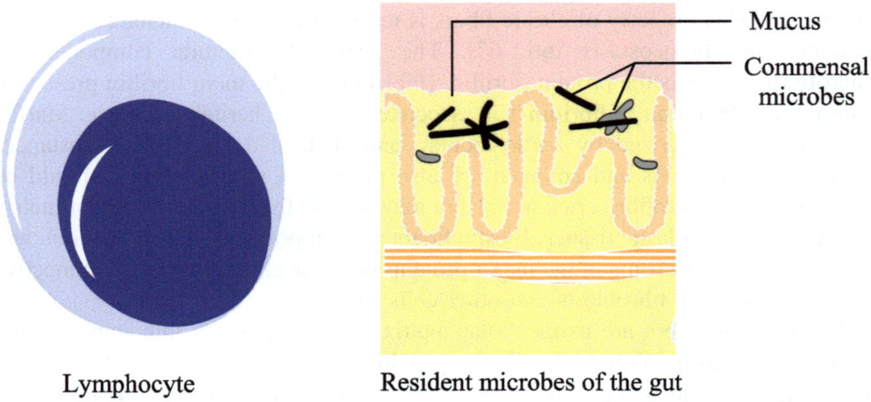

Fig. 3.5 Accessory cells of the GI tract

microbiota and brain (known as gut microbiota–brain axis) is a bidirectional communication that mainly involves the activities of several signaling molecules released by the resident microbes, vagus nerve, immune mediators, neuroendocrine molecules, which modulate peripheral-to-central neurotransmission and consequently affect behavior and cognition. Dysfunctions of this signaling pathway have been implicated in anxiety, depression, and visceral pain [83–85]. This way, the gut microbiota affects the physiology of the central nervous system.

3.3.5 Cells of the Accessory Organs of the Digestive System

The accessory organs of digestion (teeth, tongue, salivary glands, pancreas, gallbladder, and liver) are made up of cell populations specific to each organ. But some of the organs may have one or more cell types in common [45].

The teeth are comprised of such cells as enameloblast, dentinoblast (also called odontoblast), cementoblast, epithelial rests of Malassez (named after Louis-Charles Malassez, 1842–1909, and these cells are part of the periodontal ligament cells that surround each tooth). These cells of the teeth take part in regeneration and renewal of worn-out cells in various parts of the teeth [86–91].

In addition to muscle cells, the tongue is composed of numerous taste cells that can sense the physicochemical properties of different fluids and food types [92, 93].

The salivary glands and pancreas are composed of cells in common—acinar and ductal cells, which are the major cells responsible for the secretory activities of these glands [94–96]. The pancreas, in addition, contains endocrine part with numerous types of cells (see Chap. 8). After review of Chap. 11, you will also notice that the salivary gland cells secrete a couple of endocrine substances into blood, participating in modulation of a host of physiological processes in the body. The salivary glands also contain myoepithelial cells [94].

The gallbladder contains epithelial cells mainly cholangiocytes lining the intra- and extrahepatic ducts of the biliary tree. The main function of cholangiocytes is the modification of hepatocyte-derived bile, a process regulated by hormones, neurotransmitters, neuropeptides, and other substances via signaling cascades that lead to the formation of final bile that is released into the intestine upon activation by an adequate stimulus. It should be noted, however, that the gallbladder is composed of different layers with each layer having specific groups of cells: mucosal (comprises the sublayers—surface or mucosal epithelium and lamina propria), muscle (composes smooth muscle cells), perimuscular subserosal, and serosal layers. Unlike the walls of the GI tract, the wall of the gallbladder does not contain the muscularis mucosae and submucosal layers [97–100].

The liver is composed of parenchymal and non-parenchymal cells [101, 102]. The main representative of hepatic parenchymal cells is the hepatocyte, constituting approximately 70% of the total hepatic cell population. The hepatocytes are basically involved in metabolism of lipids, drugs, etc., as well as secretion of coagulation and complement factors. The hepatic non-parenchymal cells include hepatic stellate cells, Kupffer cells, and hepatic sinusoidal endothelial cells—altogether constitute the remaining ~30% of the total hepatic cell population. The hepatic stellate cells comprise about 5% of the non-parenchymal hepatic cells; they play a central role in lipid metabolism and vitamin A storage. The Kupffer cells represent 10% of the hepatic non-parenchymal cells, and function as immune sensors in the liver. The hepatic sinusoidal endothelial cells comprise half of the total hepatic non-parenchymal cell population, function in filtration of substances from the blood to the hepatic tissue or vice versa, and also secrete numerous substances into blood; they separate the hepatocytes from the lumen of the sinusoids [103, 104]. But the liver also contains a certain proportion of progenitor–stem cells (oval cells—intrahepatic precursor) and hepatocyte precursors, which are responsible for the unique capacity of the liver to regulate its mass (the liver can rapidly regenerate after partial hepatectomy involving resection of more than 50% of its mass). Although it is believed that this growth takes place mainly by compensatory hyperplasia rather than true regeneration [105–107], the role of hepatic precursor cells in such a rapid growth (and regeneration) cannot be completely excluded [107].

Other organs or regions of the GI tract also contain several precursor cells that participate in its growth and regenerative processes. For instance, the teeth, in addition to the cells described above, also harbor a host of mesenchymal stem cells. The stem cell niches in the teeth are mainly responsible for regeneration and renewal of worn-out cells [108].

3.3.6 Major Types of Gastrointestinal Epithelial Cells

There are four main cell lineages in the intestinal epithelium which are differentiated on the basis of their morphology or functional property: columnar (enterocyte), mucin-secreting (goblet), (entero) endocrine (chromaffin-like cells, etc.), and epithelial enteroimmune response (Paneth) cells (Fig. 3.6). There are other less abundant cell lineages of the epithelium, such as caveolated cells and motilin (M) cells (note that a type of endocrine cell in the GI epithelial is also classified as "M cell"—see Chap. 10 for details) [109–114]. It should be noted that nowadays the functional classification of epithelial cells is not strictly universal as one epithelial cell may secrete more than one type of endocrine substance and participate in a range of processes. More so, even non-endocrine cell has been reported to produce and release certain substances by endocrine mechanism. At the same time, some endocrine cells have been reported to release their contents into the GI tract lumen as if they were exocrine glands. These cells that secrete substances in an endocrine manner and can also release substances into the lumen by exocrine mechanism are called amphicrine cells. Examples of amphicrine cells include gastric mucous–endocrine and parietal–endocrine cells. While such cells have been reported in normal functioning of the gastric mucosa, amphicrine cells are usually present in pathological states such as tumors [109–114].

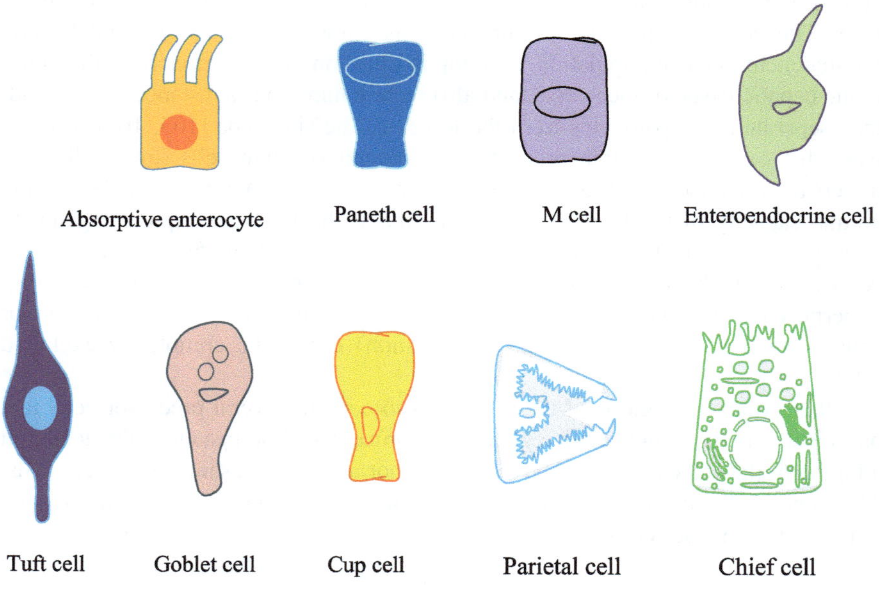

Fig. 3.6 Epithelial cells of the GI tract. The cells shown here are the major cell types currently known to be present in different parts of the GI tract; some of the cells form the major cells found in the Brunner glands and the crypt of Lieberkühn. The functions of some of these cells are not completely known

Absorptive Columnar Enterocytes

The columnar cells are the most abundant cell types in the intestine that is why they are sometimes called enterocytes in the small intestine and colonocytes in the colon. Enterocytes are polarized cells with a basal membrane (also called basolateral membrane) and an apical brush border bearing the glycocalyx as well as numerous intracellular organelles. The enterocytes constitute about 95, 94, and 89% of the epithelial cells in duodenum, jejunum, and ileum, respectively [109].

The absorptive columnar cells are continually renewed in the intestine by crypt stem cells. The turnover time of intestinal epithelial cells on the basis of experimental results obtained using continuous 3H-thymidine infusion is about three days [115]. Thymidine (deoxythymidine) is a pyrimidine deoxynucleoside that constitutes part of the nucleic acid "DNA." Thymidine pairs with deoxyadenosine in double-stranded DNA. In an experiment involving 3H-thymidine infusion, radioactive 3H-thymidine is incorporated into DNA following each cell division. Thus, the greater the number of cell divisions the higher the rate of radioactive thymidine incorporation in the cell, correspondingly, the higher the radioactivity. The rate of radioactivity is also a measure of cell proliferation. The technique can be used to measure the rate of cell migration, differentiation, function, and survival. The amount of radioactivity is counted in a scintillation or gamma counter [116–118]. Enterocytes are structurally linked to neighboring cells via intercellular linkages (details about these linkages are discussed in Chap. 4).

Substances (water, ions, nutrients, drugs) transported from the lumen of the GI tract pass mainly via transporters and channels of the enterocyte apical brush border or are moved via paracellular pathways between two epithelial cells [119–121]. Besides their absorptive role, emerging studies indicate that enterocytes may play a role in GI immune response [109]; notwithstanding, however, the main functions of these columnar cells are absorption of food nutrients and other substances as well as secretion of certain resynthesized inside the cytoplasm metabolic substances into the lymph [115].

Mucin-Secreting Cells

The mucin-secreting cells of the GI tract are usually referred to as goblet cells because of their bowl-shaped resemblance to the drinking glass with a base and stem (Fig. 3.6) [122]. Goblet cells were known already during the 1800s [123]. Goblet cells are scattered among the surface epithelial cells covering the villi and other regions of the GI tract. The cytoplasm of these cells contains mucous granules (also called mucigen or mucin granules), which are secreted onto the lumen as intestinal mucus. Mucus is a complex carbohydrate molecule, responsible for protecting and lubricating the mucosa of the GI tract [109]. There are different types of mucins, which may be classified as neutral, acidic (e.g., sialomucins and sulfomucins), O-acetylated sialomucins [124, 125].

There are other epithelial cells that secrete mucins, collectively referred to as non-goblet columnar cells [125]. The gastric foveolar mucous cell contains numerous mucigens in the cytoplasm. Foveolar mucous cell differs from intestinal goblet cells in that they do not have true theca. Another mucous-secreting cell called mucous neck cell is found in the neck and isthmus of the gastric gland. Mucous neck cells possess apical secretory granules of mucin which are released by exocytosis into the lumen [109]. The mucin-producing cells become more numerous down the GI tract [126]. In pathological conditions such as cystic fibrosis, Crohn's disease, benign and malignant lesions of GI tract, mucin production in these cells is usually impaired [95, 124, 127].

Enzyme-Secreting Cells

The catalytic agents (enzymes) responsible for digesting food particles are secreted from glands or cells located in the mouth, stomach, intestine, and pancreas. These catalytic agents perform the chemical breakdown of bonds in food particles. Enzymes are usually secreted in their inactive form and then activated by certain substances in the surrounding where it is transported to carry out their activities [128, 129].

Both minor and major salivary glands in the mouth secrete enzymes mainly amylase and lingual lipase to initiate the process of chemical breakdown of food [130–132].

Pepsin is secreted in its inactive form "pepsinogen" by pepsin-secreting cells called chief cells (also called peptic cells or gastric zymogenic cells) (Fig. 3.6). However, this cell releases not only pepsinogen, but also gastric prolipase (the inactive form of the lipid-digesting enzyme) and prochymosin (the inactive form of the enzyme responsible for clotting or coagulating milk). The chief cells are located in the base of the glands of the gastric corpus and fundus. The cytoplasm of these cells is packed with secretory zymogenic granules or vesicles which contain pepsinogen. The vesicles are released to the exterior in response to various stimuli. The secretion of inactive pepsinogen is dependent on some factors such as level of cholinergic activity and increased gastric acidity. This means that cells of the GI tract work cooperatively to achieve a set goal. Thus, the chief cells work in concert with parietal cell. The latter releases the gastric acid that converts pepsinogen into pepsin [128, 129].

The pancreatic acinar cells are an integral source of enzymes for continuation of food digestion in the intestine (duodenum). Pancreatic cells secrete enzymes that continue the digestion of carbohydrates (pancreatic α-amylase), lipids (pancreatic lipase, pancreatic lipase-related protein-1 and -2, carboxyl ester lipase, phospholipase A2), proteins (trypsinogen-1, -2, and -3, chymotrypsinogen, proelastase-1 and 2, protease E, kallikreinogen, procarboxypeptidase-A1, -A2, -B1, and -B2) [133–137].

The brush border of the intestine produces numerous enzymes such as lactase, sucrase, maltase, which are involved in final chemical processing of food for absorption by the enterocytes [133].

The secretion of some of these enzymes may be dysfunctional in some diseases of the pancreas (e.g., pancreatic insufficiency), lactose intolerance, cystic fibrosis, or other inherited enzymopathies. Thus, enzyme therapy by supplementation with digestive enzymes is done to aid nutrient digestion and absorption in patients with these diseases [134].

Acid-Secreting Cells

Gastric acid (hydrochloric acid) is secreted by the parietal cells (also called oxyntic or delomorphous cells) of the stomach. These cells are also known to secrete gastric juice in considerable quantity. Parietal cells are localized mainly in the fundus and corpus of the stomach. The surface of the parietal cell is made of numerous infoldings forming secretory canaliculi, a complex network of canals extending nearly to the cell base (Fig. 3.6).

Gastric acid is produced in response to activation by histamine released from oxyntic enterochromaffin-like cells (stimulates histamine H2-receptors), acetylcholine released from antral and oxyntic intramural neurons (stimulates muscarinic-sensitive M3-receptors), and gastrin released from antral G cells (stimulates gastrin receptors) [109, 138–140]. GI hormones such as ghrelin and motilin, as well as GI-derived neurotransmitter (such as hydrogen sulfide), act as stimulants of HCl secretion. Inhibitors of gastric acid secretion include glucagon-like peptide-1 and somatostatin. The latter is the main inhibitor of acid secretion; it is released from oxyntic and antral D cells [141].

The mechanism of acid secretion involves active transport of ions into the gastric lumen. Many ion channels and transporters are responsible for normal secretion of chloride and hydrogen ions in the stomach [109, 138]. Gastric acid has a bactericidal effect and modulates the gut microbiota. It also assists in digestion of protein and facilitates absorption of intestinal ions and elements (such as calcium and iron), as well as vitamin B12 [141]. Details on the mechanism of gastric acid secretion are discussed in Chap. 11.

Neuro-(Entero)endocrine Cells

The neuroendocrine or enteroendocrine cells are endoderm-derived epithelial cells, located in the mucosa of the GI tract [142]. The cytoplasm of these cells contains numerous secretory granules or vesicles which contain peptide hormones, secreted in the basolateral pole of the cell and normally acting in an endocrine manner. The enteroendocrine cell may be regarded as neuroendocrine cell. But not all neuroendocrine cells may be called enteroendocrine. Although in the gut, in reality, there may be no clear-cut definition for the differences between these two cells, neuroendocrine cells may be found in other regions of the body, besides the gut, whereas enteroendocrine cells, as the name implies, are only found in the gut [109, 143]. Further discussion on the differences between these cells is provided in Chap. 8.

Enteroendocrine (or just "endocrine" for short) cells are located throughout the epithelium of the gut and may be found between enterocytes, goblet, or Paneth cells [109, 143]. Endocrine cells possess receptors for different nutrients as they could sense their environment for the specific type of available nutrients or other adequate stimuli such as mechanical distension. When receptors of these endocrine cells are stimulated, many signaling pathways are activated which include secretory pathways that lead to the release of secretory granules from the basal side of the cell membrane. The released peptide hormones may directly stimulate afferent nerves, neighboring cells through paracrine mechanism or in an endocrine manner and enter the bloodstream to stimulate the target cells. These peptide hormones are involved in regulation of many functions of the GI system including digestion, appetite and satiety, body weight and may even take part in modulation of memory formation, storage, and retrieval via the gut–brain axis [84, 85, 109, 143, 144].

There are over ten types of endocrine cells currently identified in the GI tract [144]. These endocrine cells may be classified according to their localization in the gut: esophageal, gastric (antral, fundic, corporal, and pyloric), postpyloric, intestinal (ileal, colonic) endocrine cells. Even though enterochromaffin-like cells, which produce histamine and other substances, are mainly located in gastric fundus and corpus, there are other enterochromaffin-like cells in other regions of the gut, and thus the location can be somewhat useful in differentiating between two sources of the hormone. In the same vein, gastrin cells are mainly found in the antral mucosa (antral gastrin-producing cells). On the basis of their structural differences as regards to the relationship between microvilli and the intestinal lumen, endocrine cells can also be divided into two types: open type or close type. Open-type endocrine cells are those ones whose microvilli of the apical membrane extend into the lumen of the intestine. Examples of the open type include secretin- and gastrin-secreting cells. Closed-type are those endocrine cells that are separated by epithelial tight junctions. Examples of the close type include enterochromaffin-like cells that secrete histamine and serotonin. In the literature, however, endocrine cells in the gut are usually classified based on the major peptide been secreted: G cells secrete gastrin; N cells secrete neurotensin; D cells secrete somatostatin; S cells secrete secretin; P/D1-like cells or X/A-like cells secrete acyl ghrelin, des-acyl ghrelin, obestatin, and nesfatin-1; I cells secrete CCK; K cells secrete gastric inhibitory peptide GIP; EC cells secrete histamine and serotonin; L cells secrete glucagon-like peptide 1 & 2 (GLP-1 and GLP-2), peptide YY (PYY), oxyntomodulin (OXM), and glicentin [42, 109, 143, 145–148]. Around the 1960s, many terminologies for designating endocrine cells of the gut were invented by independent groups of scientists around the world. This problem was solved in October 1969 in the Merck Conference on the "origin, chemistry, physiology, and pathophysiology of the GI hormones," held in Wiesbaden, Germany. A common terminology for naming endocrine cells of the gut was agreed in this conference. This terminology is popularly known as the Wiesbaden nomenclature in which the enteroendocrine cell types were described by names and letters [149]. It should be noted that because a particular enetroendocrine cell secreting a specific type of hormone or mediator may also release other types of hormones or mediators [109, 143], designating a cell as "D", "G", "I", or "S" may be considered not completely correct in line with current knowledge.

Gastrointestinal Opioid-Secreting Cells

Opioids may be produced by more than one cell type in the gut. An example of opioid-producing cell in the gut is tuft cell. The presence of this cell type was discovered in rat respiratory tract by Rhodin and Dalhamn in 1956 [150]. In same year, Jarvi and Keyrilainen found similar cells in the gastric glands of experimental animals [151]. The presence of tuft cells was later confirmed in the GI tract of many mammals [152, 153] including human [154].

Depending on the morphological criterion used, tuft cells are named as peculiar, fibrillovesicular, caveolated, and brush cells. Tuft cells are epithelial cells having a unique tubulovesicular system and apical bundle of microfilaments connected to a tuft of long and thick microvilli protruding into the lumen of the intestine. Tuft cells may exhibit different functions depending on their location. Even though the functions of tuft cells remain completely not clear, it is widely known that this cell type is a type of secretory cell that produces opioids in the GI tract [143, 155, 156]. More recently, tuft "brush" cells have been identified to play a role in immune responses [157].

Gastrointestinal Epithelial Immune Sensor Cell: Paneth Cell

Paneth cell was discovered in the nineteenth century by Joseph Paneth (1857–1890), an Austrian physiologist who conducted initial studies on this cell [13, 158]. Paneth cells are specialized type of epithelial cells found in the mucous membrane lining of the small intestine at the base of tubelike depressions known as Lieberkühn glands and of the appendix. But Paneth cells are rare in the cecum. Paneth cells contain large secretory granules and express a number of proteins, including lysozyme, tumor necrosis factor, and small molecular weight peptides called cryptins (related to the antimicrobial peptides "defensins"), which have antibacterial activity. Paneth cells also possess phagocytic properties. These characteristics of the Paneth cell indicate a role in the maintenance of a relatively sterile environment in the crypt, preventing the colonization of the gut microbes by foreign invaders. While Paneth cells are usually confined to the small intestine, they may be seen in the proximal colon and in some inflammatory conditions in the colon—referred to as Paneth cell metaplasia. Similarly, the secretory columnar epithelium of digestive glands may be replaced with stratified squamous epithelium (see Section "Metaplasia and Dysplasia") [126, 159–161].

Metaplasia and Dysplasia

Under normal conditions, certain cells are usually confined to specific location in the GI tract. In pathological conditions, however, these cells (e.g., Paneth cells, goblet cells) may occur in an unusual location of the tract. Similar situations may occur with the GI epithelium, in which the normal columnar epithelium is replaced with other epithelia. This pathological condition is called dysplasia. In some cases, the cells may revert to the normal type—referred to as metaplasia. In contrast to dysplasia, metaplasia is a physiological response of a fully differentiated cell to abnormal stimulus characterized by the change of the original cell type to another. This cell usually returns to the original cell when environmental conditions return to normal. Thus, metaplasia is an adaptive mechanism of the cell to unfavorable environmental conditions. Because metaplasia usually occurs in chronic inflammation, it can be referred to as a precancerous condition. If a metaplastic cell becomes irreversible, the result is dysplasia. Interestingly, accumulating research data indicate that metaplastic cell may gradually progress into dysplastic ones [159–161].

3.4 Structural Composition and Functions of the Cell

Structural composition of epithelial cells is the architectural design of these cells that is required to ensure survival of the cell. All epithelial cells have a basic structural design which will be discussed in this section.

Like any other human cell, the epithelial cell is made up of a cytoplasm enclosed within a membrane. The cell membrane precisely called the plasma membrane is fluidic and viscous and contains numerous biomolecules such as proteins, lipids, and carbohydrates. These biomolecules are major constituents of the plasma membrane and cytoplasmic constituents. In addition to these biomolecules, the cytoplasm contains many membrane-bound organelles such as nucleus, centrosomes/centrioles, mitochondria, ribosomes, endoplasmic reticulum, Golgi complex, lysosomes, and peroxisomes (Fig. 3.7). These components of the cell together function to ensure metabolism (including nutrients uptake, breakdown, and waste product excretion), respiration, growth, motility, response to both internal and external stimuli. Cells use unique signaling mechanisms and DNA/protein synthesis to regulate growth and size [162].

3.4.1 Plasma Membrane

The plasma membrane is a physicobiological membrane that separates the interior of the cell from its external milieu. It is composed of phospholipids and proteins, cholesterol, and carbohydrates, which form the major components of the plasma

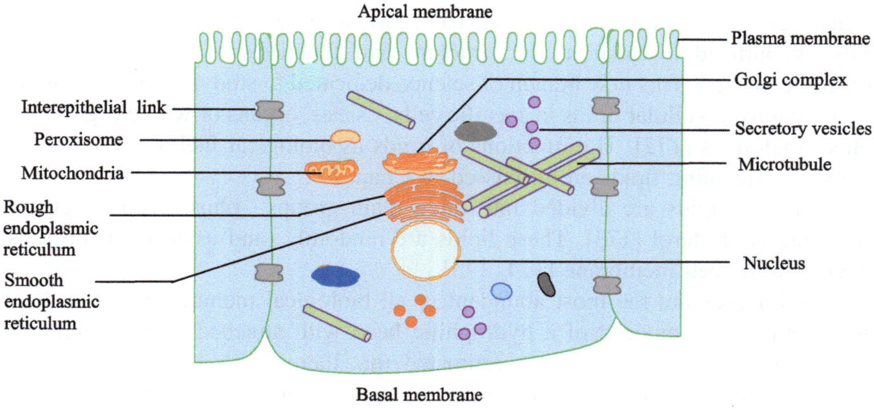

Fig. 3.7 A schematic representation of the structure and composition of a typical GI epithelial cell

membrane. Of the components of the membrane, lipids and proteins (~40 and ~50%, respectively) are the most abundant. The remaining ~10% is the proportion of carbohydrates, which occur as glycolipids and glycoproteins [163, 164]. It should be noted that even though the proportion of proteins and lipids may be approximately the same in the plasma membrane of some cells, in reality, the plasma membrane of majority of cells contains a higher proportion of lipid molecules compared to the protein molecules. This is because lipids are much smaller in size compared to protein molecules [163, 165–167]. The proportion of these components in different cells substantially varies. For instance, the myelin cell membrane has a very low protein content (about 18%) and high lipid content, required for effective insulation of neurons [168]. Disorder in the composition of plasma membrane of cells can alter the physiology of the cells. For instance, dysfunctions in the structural architecture of the enterocyte plasma membrane can result in impairment in the binding and transport of nutrients and drugs in the GI tract [169]. It is important to indicate that the membrane components (lipids, carbohydrates, proteins) are also present in the cytoplasm as well as membranes of organelles [170, 171].

Components of the Plasma Membrane

Lipids

Lipids are water-insoluble biomolecules that serve as biological fuel for cellular processes and form an integral component of the plasma membrane and signaling molecules, regulating the temporal and spatial orientations of the interacting proteins [164, 165]. Our discussion here borders on membrane lipids, which are estimated to be approximately 10^9 molecules in the plasma membrane of a typical animal cell of ~20 μm in size [164–166]. Advancement in technology substantially encouraged research in the structure and functions of biomolecules that constitute the cell. This new branch of science dedicated to studying the distribution and functions of cellular lipids in specific cells, tissues, organs or whole organism is called lipidomics [172]. The functions of lipids as biological fuel are discussed in Chap. 12. Signaling lipids are discussed in Chap. 5.

Membrane lipids are divided into three main groups—phospholipids, glycolipids, and cholesterol [173]. These lipids are randomly (and asymmetrically) distributed in the cell membrane [174, 175].

Phospholipids are the most abundant in all biological membranes [165]. Each phospholipid is composed of a hydrophilic head with attached hydrophobic acyl (aliphatic) chains (Fig. 3.8). It is estimated that there are hundreds of different species of phospholipids which result from variations in the heads and acyl chains [176]. The major membrane phospholipids widely distributed in the plasma membrane include phosphatidylcholine, phosphatidylethanolamine, phosphatidylserine, and sphingomyelin. Figure 3.8 represents a typical phospholipid. The outer leaflet of the plasma membrane mainly consists of phosphatidylcholine and sphingomyelin, whereas phosphatidylethanolamine and phosphatidylserine are the predominant phospholipids of the inner leaflet. Phosphatidylinositol is also localized to the inner half of the plasma membrane and plays a substantial role in cellular signaling. The negative charge of the cytoplasmic face of the plasma membrane is mainly due to the negative charges of the heads of phosphatidylinositol and phosphatidylserine [164, 173, 177–183]. Of all phospholipids, phosphatidylcholine and phosphatidylethanolamine constitute about three-quarters of phospholipids of the animal plasma membrane [176]. These major membrane lipids, phosphatidylcholine and phosphatidylethanolamine, are the main constituents of membrane sites required for vesicle budding and fusion, which are

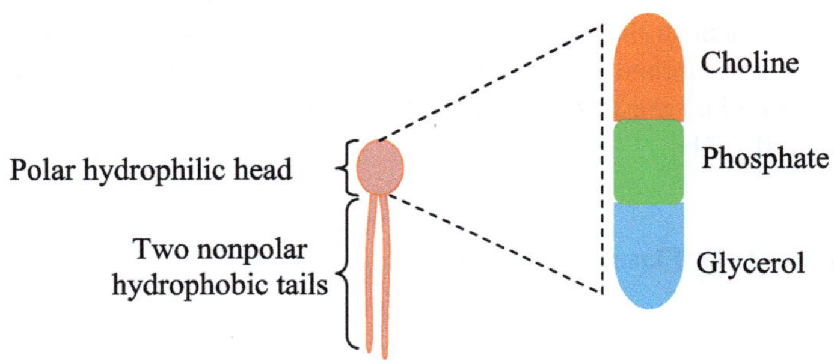

Fig. 3.8 Structure of a phospholipid molecule (phosphatidylcholine—the most abundant phospholipid, constituting about 50% of this category of lipids). The phospholipid tails are composed of fatty acids, with varying lengths of carbon atoms (usually around 14–24 carbon atoms), and are linked to the phospholipid molecule by esterification reaction. One of the tails usually has one or more *cis*-double bonds (i.e., unsaturated) [165, 184]. The head is made up of three different molecules, which are related to the phospholipid by esterification. The polar head group (choline in the case of phosphatidylcholine) determines their classifications [164]

important processes in exo-endocytosis coupling [184]. The remaining proportion is due to other of phospholipids. The plasma membrane also contains other phospholipids such as cardiolipin, and glycosphingolipids [176].

Sphingomyelin is a type of sphingolipids that form a key component of the plasma membrane [183, 185, 186]. Sphingolipids are characterized by the presence of a sphingoid backbone [164]. Other sphingolipids include cerebrosides and gangliosides [187].

Glycolipids are lipids to which sugar residues are conjugated. But they constitute a quantitatively minor portion of the lipids comprising the plasma membrane (about 2% of the membrane mass). Examples of glycolipids are glycoglycerolipids and glycosphingolipids [183, 185, 186].

Cholesterol is a neutral lipid carrying no charged group and one of the most abundant lipids in plasma membrane (Fig. 3.9). It belongs to the group of biomolecules called "sterol." Its hydroxyl group interacts with the phosphate head of phospholipids, whereas the steroid portion interacts with phospholipid tails. Plasma membrane cholesterol quantity constitutes about 50% of the total membrane phospholipids [164]. Cholesterol substantially modulates the fluidity of plasma membrane by limiting or increasing the motion of fatty acid chains of phospholipids. This modulatory effect of cholesterol on membrane fluidity is dependent on

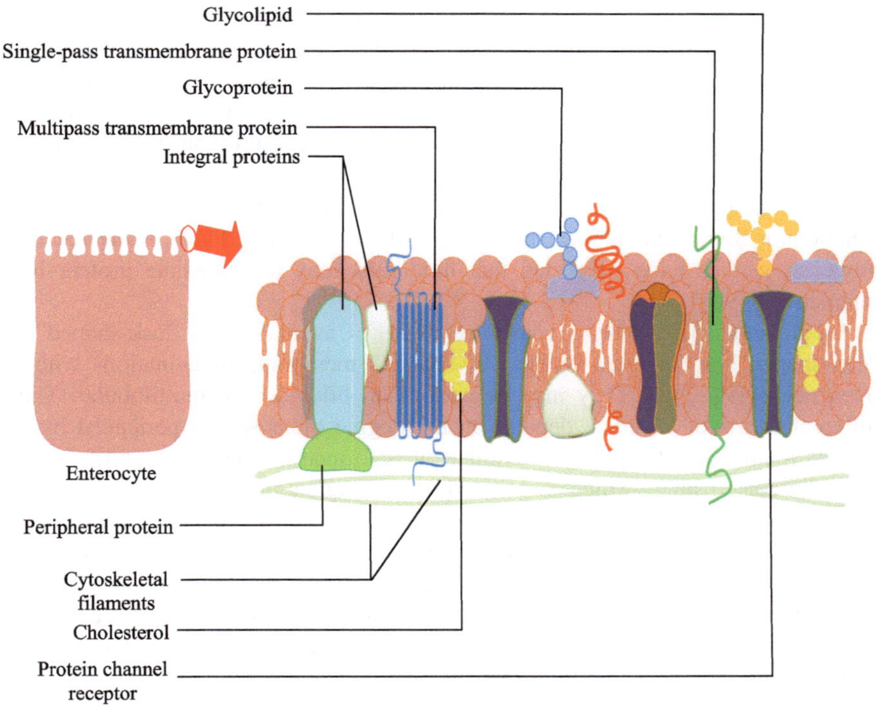

Fig. 3.9 Structure of the plasma membrane showing integral and peripheral membrane proteins

temperature. So, at low temperatures, cholesterol prevents cell freezing by increasing the membrane fluidity. In contrast, at high temperature, cholesterol reduces membrane fluidity to prevent melting of cellular components. Consequently, cholesterol has been described as a "temperature buffer" for the lipid bilayer [187]. Maintenance of membrane fluidity is required to ensure adequate functioning of the cell, movement of ions and molecules to and from the extracellular space [187, 188]. Recent work indicates that not only cholesterol, but also phospholipids (e.g., phosphatidylethanolamine) take part in the modulation of fluidity of plasma membrane [188].

Certain membrane components occur as aggregates. Cholesterol and sphingolipids are organized into tiny, dynamic, and ordered domains in the plasma membranes and are called lipid or membrane rafts, which are currently believed to play numerous roles in cellular signaling and regulation of cellular homeostasis [170, 171, 183]. In 2006, the Keystone Symposia on Molecular and Cellular Biology held in Cambridge, UK, resolved to define "lipid rafts" as small (10–200 nm in diameter), heterogeneous, highly dynamic, sterol- and sphingolipid-rich domains that compartmentalize cellular processes. Lipid rafts may regulate lateral diffusion of membrane proteins in response to receptor activation, shear stress, electrical conductance, and nutrient demand [171]. Lipid rafts form scaffolds for many molecular entities, including signaling molecules and receptors that relay extracellular information to the intracellular milieu. It is believed that lipid rafts regulate cellular polarity, adherence to the extracellular matrix, growth and migration, cellular mechanotransduction, lymphocyte activation, and other cellular signaling activities, and also serve as sites of cellular entry of certain pathogens, toxins, and nanoparticles [171]. The lipid raft hypothesis was first put forward by the Finnish professor of biochemistry and cell biology, Kai Simons (1938–) and his colleague, Elina Ikonen [189], and posits that cholesterol and sphingolipid clusters modulate the protein–protein interactions that form a basic requirement for cellular functions including membrane trafficking and cell signaling [170]. More recently, however, studies have suggested that lipid rafts may also mediate protein–lipid interaction [190].

Lipid rafts may exist as caveolae [171]. Caveolae are "flask-shaped" or "cup-shaped" or "inverted omega (Ω)-shaped" membrane invaginations with an open neck [191–193]. They may be unilobular, bilobular, or multilobular. These different forms of caveolae depend on the degree of tension experienced by the plasma membrane [194, 195]. A similar pit is formed in membrane and is termed "clathrin-coated pit"—which is a membrane invagination, lined (i.e., coated) by clathrin protein. This invagination is one of the initial processes of endocytosis, resulting in the formation of vesicles. In reality, though many proteins are involved in the coating process, clathrin proteins appear to be the most investigated group of polypeptides involved in pit coating [196]. Caveolae were first observed by the Romanian-American cell biologist George Emil Palade (1912–2008) in 1953 [197] and were rediscovered and described in microvilli membrane of the mouse gallbladder epithelium in 1955 by Yamada [198]. The proteins that constitute caveolae were discovered in the 1990s as caveolin-1, caveolin-2, and caveolin-3, which are integral membrane proteins that constitute the structural backbone of caveolae and play an integral role in receptor-mediated endocytosis [191, 199–202]. Caveolin-1 was discovered in 1992 by a team of researchers including the American cell biologist Richard G.W. Anderson (1940–2011) [199]. Caveolin-2 and caveolin-3 were discovered through different experimental methods. Caveolin-2 was identified by Scherer et al. [203] and caveolin-3 (also known as M-caveolin) by Tang et al. [204], and Way and Parton [205]. Caveolae are found in the GI tract epithelial cells as well as cells and tissues of other systems including the cardiovascular system (endothelium and smooth muscle cells of blood vessels, cardiac muscle cells, macrophages, and fibroblasts) [198, 206]. In smooth muscle cells of the GI tract and vascular system, caveolin-1, -2, and -3 are all involved in caveolae formation. Cardiac (and skeletal) muscle cells express only caveolin-3. Epithelial cells and other cells of the GI tract and cardiovascular system express caveolin-1 and -2 [193]. Caveolae are known to contain over 100 different molecules. But of integral importance to the functioning of this cholesterol-sphingolipid-rich invagination, are the proteins that were later discovered in the twentieth century [207]. These proteins are termed cavins, and they are often referred to as support or adaptor proteins that regulate the formation of caveolae by modulating the functions of caveolins and may be involved in membrane remodeling. There are currently four types—designated cavin-1, -2, -3, and 4 [202].

Caveolae play a variety of structural and functional roles in different cells. They participate in cellular remodeling by actively sensing environmental stress [208]. These mini bulb-like pits are involved in the regulation of macromolecular transport across the plasma membrane (e.g., cholesterol deposition). They also regulate the volume of cell and homeostasis of calcium [192, 209]. Emerging studies indicate that the caveolae–caveolin–cavin system plays a critical role in mechanoprotection of skeletal muscle [194, 195]. Importantly, this mechanoprotective role of caveolae also extends to other muscle types. Dysfunction in caveolae formation has been implicated in a host of human diseases including cancers and muscular dystrophies, heart failure, myocardial infarction, pulmonary hypertension, and familial hypertrophic cardiomyopathy [208, 210]. Some of these diseases may be due to mutations in the caveolae–caveolin–cavin system. These diseases are generally referred to as caveolinopathies. Emerging studies indicate that there are dozens of mutations identified for caveolins [211].

Proteins

The plasma membrane is composed of a plethora of different protein molecules which may either be embedded within the phospholipid matrix (called integral or transmembrane proteins) or loosely associated with the cytoplasmic face of the plasma membrane (called peripheral proteins) [212].

The entire protein composition of the cells of an organism is called the proteome. Proteomics is a large-scale study of the cell's proteome of the organism [213]. The field of proteomics is currently subdivided into subfields such as clinical proteomics and functional proteomics [214, 215]. On the basis of their chemical composition, proteins are either simple or complex. The proteins in the plasma membrane are complex proteins, containing more than one amino acid. Some proteins may be conjugated, as they contain a nonprotein part (e.g., carbohydrate residue). Such proteins with carbohydrate residue are termed glycoproteins. This type of conjugated proteins constitutes about 5% of the mass of the plasma membrane [183, 216, 217]. Glycoproteins usually consist of 2–20 sugar residues, mostly glucose, galactose, mannose, arabinose, fucose (6-deoxygalactose), *N*-acetylglucosamine, and *N*-acetylneuraminic acid (NANA—an acetylated derivative of the carbohydrate sialic acid). Examples of glycoprotein include glycophorin, most blood plasma proteins, except albumin, and immunoglobulins. Some membrane proteins are soluble, whereas others are insoluble in fluids surrounding the membrane. Protein can also be classified on the basis of their structure as fibrous or globular. The latter is widely expressed in the plasma membrane. Globular proteins act as enzymes, hormones, transporters, and receptors. The former mainly provides support to the cells. Types of fibrous proteins include collagen and elastin. They have not been found in the plasma membrane. Proteins can also be classified on the basis of their functions—receptors (e.g., ion channels, etc.), transporters (e.g., membrane carriers, fatty acid-binding proteins, plasma lipoproteins, etc.), enzymes, storage proteins, motor function (e.g., actin, myosin etc.), and mechanical support (e.g., elastin and collagen) proteins. Proteins may be classified according to their subcellular locations as plasma membrane proteins, cytoplasmic proteins, organellar (endoplasmic reticulum, Golgi apparatus, lysosome, mitochondria, nucleus, and peroxisome) proteins [165, 168, 218–223]. In this subsection, only integral and peripheral proteins will be discussed [168].

It is estimated that mammalian cells contain one billion individual protein molecules [127]. Of these, approximately 30% of all cellular proteins are localized in the plasma membrane [128, 190]. A protein molecule is found in every 50–100 molecules of plasma membrane lipid [183, 216, 217]. Recent analysis has revealed that there are altogether 6,718 membrane proteins in a single human cell [224]. Of these, 1352 are receptors, 817 are transporters, and 533 are enzymes. The remaining 697 integral membrane proteins are classified as miscellaneous [224]. Of the total membrane proteins, scientists have decoded the structure of more than 1000 proteins [225]. Membrane proteins perform multiple and crucial functions in the cell [226]. Membrane proteins interact with intracellular as well as extracellular molecules and receive signals from both sides of the plasma membrane, and thus membrane proteins can translate extracellular and intracellular cues into meaningful signals and recognize foreign bodies, involved in ferrying of molecules and nutrients across the plasma membrane [190, 227, 228]. Of all proteins in the cell, about 10% are involved in signal transduction, suggesting the important roles proteins play in cell communication [229]. The wide range of drugs that act by targeting this group of biomolecules in the plasma membrane underpins the importance of membrane proteins to the organism. Interestingly, about 50% of prescription drugs are known to relay their signal into the cell via membrane proteins [227, 228].

The integral proteins (also called intrinsic proteins) are membrane-spanning proteins that have one or more portions of their structures permanently embedded in the plasma membrane [164, 168]. These proteins may transverse the plasma membrane once—called single-pass transmembrane protein. If integral protein transverses the membrane multiple times, such protein is called multipass transmembrane protein [168]. However, certain proteins may just transverse the membrane by about 50% of the membrane thickness and remain embedded. Integral proteins act as transporters, receptor, and enzymes [168, 224]. It should be pointed that while some proteins may not be anchored to lipid molecules (and may just be located in the membrane by non-covalent interactions), others may be anchored to such membrane component as glycosylphosphatidylinositol (GPI). Protein molecules possess both polar and nonpolar side chains. The polar hydrophilic linkages are localized toward the outer side of the membrane. The hydrophobic linkages are folded inside or used to establish connections with hydrophobic part of the lipids, specifically the fatty aliphatic chains of the phospholipids [168]. These proteins provide a medium for ion movement across the membrane, and act as signaling molecules. Membrane adhesion proteins, peripheral and integral proteins are linked to internal cytoskeleton. Proteins can interact with other membrane components through electrostatic and hydrophobic interactions, van der Waals forces, and hydrogen bonding [164, 218, 219]. Some integral membrane proteins function as enzymes. Approximately 30 different kinds of enzymes including phosphatases, protein kinases, ATPases, esterases, and nucleases are found in plasma membranes [168, 183, 216, 217].

Peripheral (extrinsic or integral monotopic) membrane proteins are proteins, associated with one leaflet of the phospholipid bilayer and not possessing any hydrophobic transmembrane domain [190]. Peripheral membrane proteins do not interact with the hydrophobic portion of membrane lipid; they are connected to the plasma membrane either by interaction with polar heads of lipid molecules or integral membrane proteins. Examples of peripheral membrane proteins include G proteins, GTPases, Ras family proteins, Wnt proteins, prostaglandin H2 synthase, fatty acid amide hydrolase, and some protein kinases [164, 190]. Peripheral proteins are usually localized to the cytoplasmic face of the membrane. But these proteins shuttle between the cytoplasm and inner half of the membrane, playing a crucial role in signal transduction [168]. Peripheral membrane proteins which also serve to anchor the plasma membrane to the cytoskeleton include actin, spectrin, ankyrin, and band 4.1. In erythrocyte, for example, ankyrin protein serves as the principal link between the plasma membrane and the cytoskeleton by binding to both spectrin and the integral membrane protein “band 3.” The protein band 4.1 links the membrane to the cytoskeletal proteins “spectrin, actin, and glycophorin” [183, 216, 217]. Peripheral membrane proteins are key players in cellular signaling cascades that determine numerous life outcomes [190, 230].

In 2006, a group of researchers, using the yeast *Saccharomyces cerevisiae* as an experimental model, discovered that certain areas of the cell membrane containing heterodimeric, immobile protein complexes mark the site of endocytosis in eukaryotic cell [231]. These complexes were called eisosomes (from Greek “eis” meaning portal and "soma" meaning body). Eisosomes are organelle-like structures occurring in some domains of the plasma membrane, responsible for concentrating lipids, transporters, and signaling molecules. The discovery of these complexes has sparked an increasing interest on their role in health and disease [232].

Carbohydrates

Carbohydrates are the third class of information-encoding biological macromolecules that form integral component of the cell [233]. The extracellular face of the plasma membrane is covered by a carbohydrate-rich layer, described as fuzzy, and often called the glycocalyx [183, 234]. With a thickness of 6–500 nm, the glycocalyx comprises glycoproteins, collagen proteins, and mucopolysaccharides (also known as glycosaminoglycans). The most prominent glycocalyx is found in GI epithelial and endothelial cells, where the cells have evolved to withstand a relatively high level of stress, modulating the flow of macromolecules and fluids [166, 235–237].

The glycocalyx is connected to the epithelium or endothelium through backbone molecules such as proteoglycans and glycoproteins [183, 234]. The glycocalyx is also connected to the cell cytoplasm either through direct contact or indirect contact with proteoglycans or glycosaminoglycans. Thus, proteoglycans and glycosaminoglycans also serve as "backbone" molecules of the glycocalyx. The most important backbone molecules, however, are the proteoglycans, which are comprised of a core protein to which one or more glycosaminoglycan chains are attached [234, 237–239]. The glycoproteins are oligosaccharide chain (approximately 2–9, but may reach 15 sugar residues), which may be branched or unbranched, and are usually covalently conjugated to a protein molecule [183, 237, 240]. The oligosaccharide chain gives the glycocalyx its characteristic negative charge, which is usually due to the presence of anionic oligosaccharides (e.g., heparan sulfate) [217, 240]. This fuzzy coat serves as the attachment surface for several cellular structures such as cytoplasmic cytoskeletal materials as well as extracellular anchors [183, 234]. The glycocalyx provides for cell support, recognition, and cell-to-cell signaling [183, 234].

The complete repertoire of the cell's carbohydrates is termed "glycome." It is the composition of the carbohydrates and their conjugates of the cell, tissue, or organism under specific conditions in time and space [241, 242]. The study of the cells glycome is referred to as glycomics [242]. Glycomics, proteomics, and lipidomics are newly developing branches that aim to characterize all the classes of information-encoding biological macromolecules.

Turnover Rate of Plasma Membrane Components

The components of the cell are continuously removed and replaced by new ones via numerous mechanisms. This turnover allows renewing cell components so that damaged ones are continuously replaced. The turnover rate (half-life) of cell components varies. The turnover rate for some membrane phospholipids is ~3 h; plasma membrane cholesterol of erythrocyte is about 2 h at ambient temperature. Protein turnover rate may range from several minutes to several years, but a typical protein has a half-life of about 55–78 h [166, 243]. The turnover rate also substantially varies among different sugars. The half-life of L-fucose is 12.5 h, whereas for *N*-acetylneuraminic acid, it is 33 h. The half-life of galactose when found as a component of plasma membrane is 20 h [243]. Numerous molecular mechanisms are responsible for and regulate the quantity of plasma membrane component turnover rate. One of the mechanisms involves degradation of old proteins by cytoplasmic and lysosomal protease enzymes [244].

Synthesis of Plasma Membrane Components

The majority of the components (lipids, proteins, glycoproteins, and glycolipids) that constitute the plasma membrane are synthesized in the endoplasmic reticulum of the cell, refined in the Golgi complex, and then distributed to different locations of the cell. The synthesis is regulated by different molecular signaling cascades and is required to create and maintain structure and functions of the plasma membrane [163, 245, 246].

Models of Plasma Membrane Structure

Several types of models have been put forward to explain the structure of the plasma membrane [247–251]. The widely known ones are the phospholipid bilayer (bilipid layer) and the fluid mosaic models (Figs. 3.8 and 3.9) [252, 253].

The Bilipid Layer Model of Membrane Structure: The phospholipid bilayer model of membrane structure, introduced by the Dutch physiologists Evert Gorter (1881–1954) and François Grendel (1897–1969) [254], is based on certain behavioral characteristics of the major constituents (particularly phospholipids) of the plasma membrane (Fig. 3.10). Recall that the phospholipid molecule is composed of a head, which is polar hydrophilic (i.e., water loving), and two fatty acid tails, which are nonpolar hydrophobic (i.e., water repelling). Thus, the phospholipid molecule is termed amphipathic. Consequently, when present in biological fluids the molecules become arranged such that the heads are directed into the fluid medium, while the tails extend to the opposite direction. [Remember that in biological environ, both the cell exterior and the interior are composed of high quantity of water, and thus one layer of the phospholipid heads is directed toward the exterior, while the phospholipid heads of the other layer, proximal to the first one, are directed toward the interior (cytoplasm).] The tails of each layer are located in close proximity to each other, so that only one layer of lipid is formed. But in reality, it is a bilayer, composed of inner (cytoplasmic) and outer (extracellular) leaflets of phospholipids. The result of such behavior of phospholipids in biological

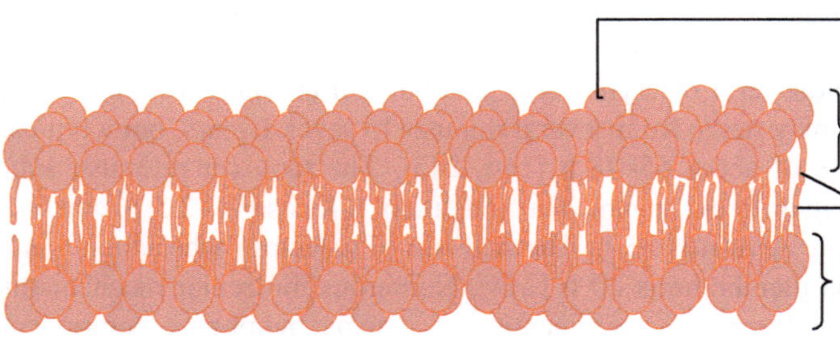

Fig. 3.10 Structure of the lipid bilayer. The heads of the phospholipids are directed toward the interior (cytosolic) part and exterior (extracellular) part of the cell. The fatty acid tails of each phospholipid are directed toward each other. It should be mentioned that some plasma membrane lipids (such as phosphatidylethanolamine and cardiolipin) do not form bilayers and are called non-bilayer lipids [174]. In most membranes, the thickness of each leaflet is 3–4 nm [168]. But the thickness of the plasma membrane in different cell types significantly varies and may reach 10 nm or more [602]

medium is the formation of a continuous spherical lipid bilayer. The formation of the lipid bilayer is controlled by molecular forces including van der Waals forces, hydrogen bonds, hydrophobic, electrostatic, and non-covalent interactions [183].

The Fluid Mosaic Model of Membrane Structure: This model of the plasma membrane structure was proposed in 1972 by the American biologist, Seymour Jonathan Singer (1924–), and the American biochemist, Garth L. Nicolson (1943–). The fluid mosaic model is now generally accepted as the basic paradigm for the organization of all biological membranes. In this model, membranes are viewed as two-dimensional fluids in which proteins are inserted into the lipid bilayers (known as integral or intrinsic proteins) [255]. The other type of membrane protein is called peripheral protein. (Details on integral and peripheral membrane proteins have been discussed in the previous subsection.) The protein molecules are bound to the membrane by protein–protein or protein–lipid interaction [256, 257].

The lipid bilayer behavior is somewhat similar to a layer of oil on a water surface. This simple analogy provides an idea of how lipids behave in biological fluids. Some of the phospholipids in each leaflet of the bilayer may wiggle, and thus the phospholipid bilayer possesses a fluidic property. The to and fro movement in some plasma membrane is catalyzed by the membrane proteins generally called lipid translocators (e.g., flippases, floppases, and scramblases). Lipid flippases are proteins that translocate lipids from the outer leaflet of the plasma membrane to the inner (cytoplasmic) leaflet, thereby maintaining membrane lipid asymmetry [258–260]. The activity of lipid flippases is dependent on cellular energy [260]. These energy-dependent membrane enzymes use ATP to drive the translocation of membrane lipids, and they belong to the P4 subfamily of P-type ATPases [258, 259]. Using the energy from ATP hydrolysis, lipid flippases mediate the translocation of aminophospholipids (phosphatidylserine and phosphatidylethanolamine) from the outer leaflet to the inner one against a concentration gradient; hence, they are also called aminophospholipid translocases [181, 182, 261]. Another

ATP-dependent membrane lipid translocator is called floppase. Lipid floppases use energy from ATP dissociation to transport membrane lipids such as phosphatidylcholine, sphingolipid, and cholesterol from the inner to the outer leaflet against concentration gradients [181, 261]. The third membrane lipid translocating protein, scramblase, is ATP-independent, but dependent on Ca^{2+} ions. This bidirectional ATP-independent membrane lipid translocator facilitates the transportation (scrambling) of lipids between the two leaflets along concentration gradients [181, 261]. Lipid scramblase transports negatively charged phospholipids from the inner monolayer to the outer monolayer and vice versa, thus destroying membrane lipid asymmetry [262, 263]. These lipid translocators may have numerous other functions which are yet to be fully understood [262, 264]. The asymmetric distribution of phospholipids is controlled by the three phospholipidic pumps—flippases and floppases, and scramblases [181]. The transport of membrane lipid molecules mediated by the different lipid translocators is termed "flip-flop" [260]. The flip-flop movement between the inner and outer leaflets is referred to as transverse diffusion. The rate of this diffusion is estimated to be about one movement per hour [166].

The molecular organization of the lipid translocases or translocators determines the absorption and metabolism and tolerance of some drugs as well as the pathophysiology of some diseases of the GI system. The multidrug resistance-associated protein (MRP) may function as a floppase in human hepatocytes and other cells (including erythrocyte) translocate phosphatidylcholine from inner leaflet to outer one. This way lipid translocators function as plasma membrane efflux pumps, transporting various molecules including xenobiotics (such as anticancer drugs) out of cell cytoplasm [265, 266]. Malfunction of the MRPs [(including MRP-1 (P-glycoprotein), MRP-2, MRP-3 belonging to ATP-binding cassette—ABC transporter)] has been shown to be related to disorder in phospholipid secretion into bile resulting in cholestasis, a disease characterized by obstruction or decrease in the secretion or flow of bile into the lumen of the GI tract. The MRP-3 and other variants translocate phosphatidylcholine from the inner to the outer leaflet of the canalicular membrane of the hepatocyte. This makes the phospholipids available for extraction into the canalicular lumen by bile salts [174, 267–269]. The biliary phospholipids are associated with bile salts and cholesterol in mixed micelles, thereby reducing the detergent activity and cytotoxicity of bile salts [269]. Phosphatidylcholine is the major component of mammalian bile [267]. A significant decrease or the absence of this phospholipid leads to dysfunction in the protective role of bile against the detergent effect of bile salts. This can result in biliary epithelial damage, which predisposes the liver to different pathological conditions [267]. Mutations in the lipid translocators can lead to different disease states [269]. Mutation of the liver-specific membrane transporter of phosphatidylcholine, MRP-3, can result in progressive familial intrahepatic cholestasis type 3 [267].

While transverse diffusion is a pretty infrequent phenomenon, the most frequent type of movement of lipids occurring in the phospholipid bilayer is lateral diffusion. Lateral diffusion is the movement of lipids within a monolayer, considered a relatively faster rate, compared to transverse diffusion [166]. At ambient temperature, a lipid molecule exchanges location with its neighbors in a monolayer approximately 107 times per second, diffusing several microns per second. A typical lipid can diffuse the entire length of a typical animal cell in approximately 20 s [168]. The lateral diffusion constant of plasma membrane lipids is about 10–8 cm^2/s [165, 168].

Proteins can also move in the plasma membrane [270, 271]. The majority of mobile plasma membrane proteins are the integral proteins. Depending on the cell type, approximately 30–90% of all integral proteins are mobile within the plasma membrane. On the contrary, proteins that do not change position, i.e., immobile, are usually permanently attached to the cytoskeleton [168]. The lateral diffusion of a mobile membrane protein decreased linearly with increasing concentration of the protein in the membrane [270]. This relationship is true when the protein-to-lipid ratios are up to 3000 μm^{-2} (recall that the normal protein-to-lipid ratio in most membranes is about 10–100 proteins per μm^2 of the membrane surface) [270].

The mobility of membrane proteins and lipids confers on the plasma membrane a property of fluidity of the lipid bilayer. This lateral movement of these membrane components was first reported in 1970 by Larry Frye and Michael Edidin while working at Johns Hopkins University [270]. Several types of proteins are embedded in the fluid matrix of the membrane phospholipids similar to a "mosaic," hence the name "fluid mosaic" [270]. The phospholipids move laterally in the membrane at a speed of about 2 μm per second. The rate of phospholipid motion occurs in reaction to surrounding extracellular and intracellular fluid media [270].

The fluid mosaic model of Singer and Nicolson was invented by combining the experimental evidences from a number of previously conducted studies. The membrane in this model is viewed as a quasi-fluid and not rigid. In the Singer and Nicolson model, lipids and proteins are uniformly inserted into the lipid bilayer [183, 272, 273]. The key features of the Singer–Nicolson model are "fluidity," "diffusion," and "mosaicism," which predict the interspersion of proteins and lipids and their ability to undergo dynamic rearrangement via Brownian motion [274]. The fluid mosaic model predicts lateral and rotational freedom and random distribution of the components in the viscous phospholipid bilayer [274].

Is the Fluid Mosaic Model of Membrane Structure Still Relevant Today? The Plasma Membrane is More Fluidic than Mosaic. On the basis of the experimental results of advanced technological studies, there was apparent necessity for an improvement on the fluid mosaic model of membrane structure. So, the dynamically structured fluid mosaic model was suggested to account for the shift in emphasis from fluid to mosaic [275, 276]. The technological advancements in research utilizing scanning confocal microscopy and photobleaching technique among others provided further insight into the dynamism of membrane structure and functions and gave a better understanding of the composition, molecular structure, mobility, co-mobility, and molecular composition of the plasma membrane. The studies on membrane structure and functions conducted with advanced technologies resulted in a shift from the less dynamic model of Singer and Nicolson to a more dynamic one. Dynamism of the plasma membrane is the change in functions of mobility and proximity relationships of lipid, protein molecules in the plasma membrane, resulting in substantial effects on different aspects of cellular signaling including ligand–receptor interaction and functions [274, 277]. This shift from fluidity to mosaicism indicates a non-random distribution of specific membrane protein and lipid clusters that form islands or microdomains via lipid–lipid, protein–protein, and protein–lipid interactions, as well as sub- and supramembrane (cytoskeletal and extracellular matrix) interactions [272, 274]. Membrane fluidity, according to Vereb et al. [274], is the permissiveness of the architecture to continuous and dynamic restructuring of the clusters on the basis of the cell requirements modulated by environmental parameters. In the current view, membranes are inhomogeneous fluids in which distinct lipids and proteins are non-randomly dispersed into dynamic domains [274].

It is believed that the dynamism of membrane structure is marked by patches—segregated regions of the membrane comprising of tightly packed lipids and proteins with varying thickness and composition, which limit exposure of the membrane portion to the adjacent fluid regions [278]. The patches are due to membrane-associated cytoskeletal fences and fenceposts that maintain the non-random mosaicism of the plasma membrane [272]. The dynamic but persistent patchiness is produced via combination of several factors including lateral diffusion, inhibitors of lateral diffusion, and vesicle traffic to and from the plasma membrane. On the average, the patchiness is determined by the pattern of organization of lipids and proteins into domains, producing a composition different from the mean composition of the entire membrane [279, 280]. Not only patches, but also lipid rafts as well as extracellular matrix regulate membrane dynamics and limit the lateral diffusion of membrane components [272, 281].

The term "mosaic" in the modified fluid mosaic model (i.e., the dynamically structured fluid mosaic model) is used with the understanding that membrane proteins are organized as molecular (small scale) and island (large scale) clusters. In this modification of the fluid mosaic model, membrane fluidity reflects dynamism where the membrane can restructure integral and peripheral protein clusters on the basis of cellular requirements. The dynamism of this model, in part, is due to the genetic influences as well as the numerous interactions among cellular components. It is important to note that lateral movement of proteins and phospholipid molecules occurs, but not in chaos [274, 282]. However, the mechanism by which order is gained from possible chaotic motion of membrane components is not completely understood. An experimental evidence of non-random distribution of proteins was provided by the discovery of co-capping, a phenomenon in which a protein links one cytoplasmic protein to another protein located in the plasma membrane. Co-capping proteins include adducin, ezrin, radixin, moesin, which are intracellular proteins that link actin filament with transmembrane proteins [274, 283, 284]. These proteins function as cross-linkers between the cytoskeleton and the plasma membrane of cells of the microvilli (and other regions of the body) and with adherens junctions [285–287].

Modifications of the Plasma Membrane—The Microvilli

The plasma membrane of some cells may be modified to form microvilli, cilia, lamellipodia, filopodia, or junctional complexes. Though modifications of the plasma membrane are sometimes referred to as organelles of the cell, they constitute part of the plasma membrane. Junctional complexes (e.g., gap junction, tight junctions) are not regarded as organelles, but they form the basis of intercellular linkages and signaling between cells [288–293].

The modified plasma membrane structures are supported by a web mesh of microfilaments containing motor proteins such as actin, myosin, tropomyosin, spectrin, ankyrin (Fig. 3.11) [288–293]. The filaments that constitute the core of the microvillar form a network that is closely associated with the membrane of the microvilli and also form extensions that link adjacent cells together via communicating linkages such as tight junctions, zonula adherens, and desmosomes. This region of the cell is termed the terminal web. It is composed of bands of interwoven actin and myosin as well as cross-linker proteins. It characterizes the apical zone of the epithelial cell with high level of plasticity and is required for the normal functioning of the epithelium of the GI tract (e.g., absorption) [294–297].

Functions of the Plasma Membrane

Cellular Support and Attachment

Plasma membrane contains proteins that function to join the cell to a neighboring one as attachment to the extracellular matrix, required for modulation of signaling pattern and maintenance of the structure and the functions of the cell [298–300]. The plasma membrane serves as an attachment surface for several extracellular matrix structures such as glycocalyx—a carbohydrate-rich layer lining the apical aspect of epithelial and endothelial cells and providing for cell support, recognition, and cell-to-cell communication. The glycocalyx is connected to the epithelium or endothelium through "backbone" molecules such as proteoglycans and glycoproteins [234]. The glycocalyx is also connected to cell cytoplasm either via direct contact or indirect contact via proteoglycans or glycosaminoglycans. Thus, proteoglycans and glycosaminoglycans also serve as "backbone" molecules of the glycocalyx. The most important backbone molecules, proteoglycans, are composed of a core protein to which one or more glycosaminoglycan chains are linked [234, 237–239]. The core protein groups of syndecans, glypicans, mimecan, perlecan, and biglycan form a firm connection to the cell membrane via a membrane-spanning domain (syndecans) [237]. Membrane glycoproteins are also regarded as "backbone" molecules, connecting the glycocalyx to the membrane of the epithelial or endothelial cell. These glycoproteins are oligosaccharide chains (around 2–9, but may reach 15 carbohydrate residues) that may be branched or unbranched, and in most cases, covalently conjugated to a membrane protein molecule, but they may exist in the membrane freely by non-covalent interactions

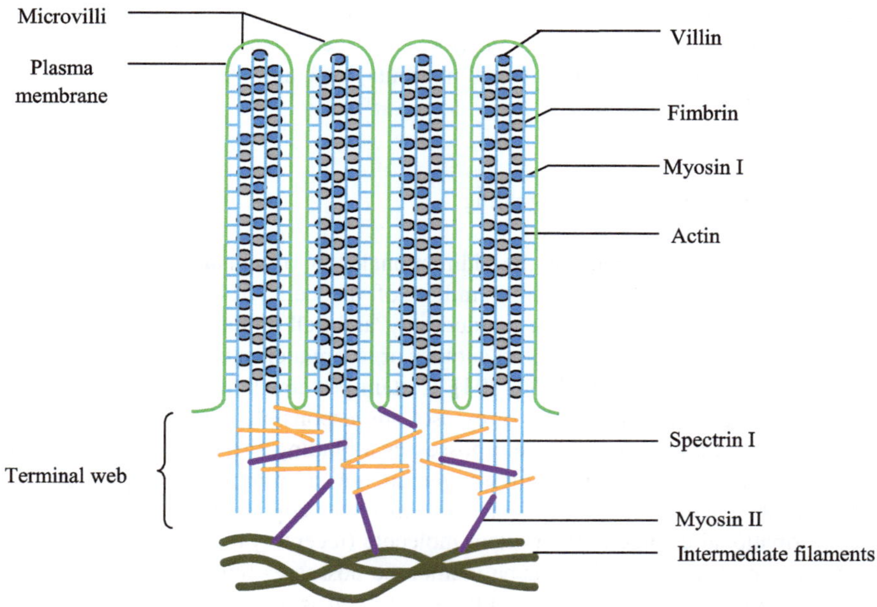

Fig. 3.11 Molecular structure of the microvilli. The cytoskeletal filaments actin and myosin in microvilli are the non-muscle type, but they do have contractile functions [297, 603, 604]. Microvilli are fingerlike evaginations of 0.6–0.8 μm length and 0.1 μm diameter forming fold in the cell membrane that increases the surface area for absorption by hundreds of times. More so, the spaces in between microvilli take part in pinocytosis. There are thousands of microvilli in the apical membrane of each enterocyte and hepatic cells [605, 606]. Microvilli are also found in mesothelial cells [607, 608], uriniferous tubules (proximal tubule, intermediate or thin tubule, distal tubule, and collecting duct) [609–611]. Microvilli consist of a network of motor proteins attached to the plasma membrane of the cell. The motor proteins include actin, myosin, spectrin, intermediate filaments. These motor proteins are linked to the plasma membrane via such proteins as ezrin, radixin, moesin (ERM) proteins. But the proteins villin and fimbrin have also been observed to form cross-linkages with actin filaments. These actin-bundling proteins induce tightly packed actin [612, 613]. Villin, fimbrin (plastin-1), and epsin together with calmodulin belong to calcium-binding proteins of the microvilli core filament and play a crucial role in microvillar structure and signaling [614, 615]. The apical microvilli and associated proteins reorganize and migrate during damage to the intestinal epithelium, thereby enhancing healing process [616]

[237, 240, 298]. The oligosaccharide chain gives the glycocalyx its characteristic negative charge, which is usually due to the presence of anionic oligosaccharides (such as heparan sulfate) [240].

Membrane lipids also serve as attachment site or support for proteins. A classic example is glycosylphosphatidylinositol anchor (glypicans) that helps to attach protein molecules to the plasma membrane, thus maintaining membrane architecture [237, 301]. The glycocalyx also comprises the glycolipids which are membrane lipids attached to a carbohydrate skeleton [185, 186]. The plasma membrane also provides an attachment for the motor proteins of the cytoplasm and the cytoskeleton in general (discussed in the previous subsection on "microvilli"). Thus, the plasma membrane not only protects the interior of the cell but also controls cellular homeostasis, regulating the transport of nutrients, ions, and water across the plasma membrane [169, 302, 303].

Cellular Communication and Signaling

The plasma membrane controls the bidirectional flow of signals (ions and biomolecules) in and out of the internal milieu of the cell [234, 304]. The plasma membrane also provides for ion conductivity and cell signaling by ensuring the flow of information from one cell to another or to organelles or between organelles of the same or different cells [234, 304]. Certain proteins that reside at the plasma membrane, organellar membrane, or cytosol are responsible for the transduction of signal. Usually, a membrane protein possesses a binding site with a specific shape that fits the shape of a chemical messenger, such as a hormone, neuropeptide. The external messenger (also called signal) recognizes its cognate receptor, resulting in a conformational change in the protein molecule (receptor) that relays the message to the inside of the cell. The plasma membrane possesses numerous receptors that control cell cargo traffic. It is this differential expression of cell-surface receptors that produces differences in membrane charge that determine many activities of excitable cells [305]. Details on membrane receptors are discussed in Chap. 5.

Cellular Transport Function

Plasma membrane has protein molecules responsible for different modes of movement of cargoes across the membrane. A protein that spans the membrane may provide a hydrophilic channel across the membrane that is selective for a particular solute. Other transport proteins shuttle a substance from one side to the other by changing the shape. These proteins are termed transporters or carriers [306–308]. Some of these proteins use ATP as an energy source to actively pump substances across the membrane—they are called pumps. For example, pumps that use ATP to actively pump calcium ions are called calcium pump [309].

Transport of substances across the lipid bilayer is highly regulated and depends on the nature of the substance been transported [183, 254, 310]. The phenomenon was first studied around 1900 by the German physician and pharmacologist Hans Horst Meyer (1853–1939) and the British physiologist and biologist and pioneer of the theory of the cell membrane, Charles Ernest Overton (1865–1933). Following a series of investigations independently performed by Meyer and Overton, who observed that lipid-soluble compounds easily travelled through a biological membrane with a relatively high velocity, whereas water-soluble compounds only moved slowly, it was believed that the membrane of the cell was selective and probably took part in movement of substances in and out of the cell. The findings led these pioneer researchers to believe that the cell membrane possesses properties similar to those of lipids and reasoned that cell membrane was composed of certain types of lipids [311, 312]. The lipid bilayer is impermeable to ions and polar molecules. The arrangement of heads and tails of the phospholipid bilayer is necessary to prevent the diffusion of polar molecules (such as amino acids, nucleic acids, carbohydrates, proteins, and ions) across the membrane. However, hydrophobic molecules may diffuse passively across the membrane [163, 165, 168, 313, 314].

There are different types of cellular transport, which will be discussed one after the other.

Passive transport: This is a mode of membrane transport that occurs down the concentration gradient and hence does not involve cellular energy expenditure. The rate of passive transport depends on the permeability of the cell membrane, which, in turn, depends on the architectural organization and characteristics of the membrane lipids and proteins. Passive diffusion can occur either directly through the phospholipid matrix or with the help of channels. Passive transport is characterized by downhill movement of ions or molecules across the plasma membrane [168, 315]. Molecules such as water, carbon dioxide, oxygen can diffuse through the membrane via channels downhill along their concentration gradient [315]. However contrary to the previous views that gases freely penetrate the phospholipid matrix, accumulating studies have documented the existence of "gas" protein channels, which facilitate the transport of gases, such as carbon dioxide, nitric oxide, and ammonia across the plasma membrane [316]. Among other substances, aquaporins (AQPs) are traditional water channels known to transport gases such as oxygen and carbon dioxide [317]. Specifically, AQP1, AQP4, AQP9, and Rh (Rhesus protein channel) have been implicated in gas transport across the plasma membrane [317–323]. Pathological functioning of these channels not only predisposes an individual to the development of edema in their respective location, but also oxygen deficit especially during periods of high oxygen demand [324]. However, the results of some of these studies are in contrast to the well-known Meyer–Overton rule [325]. Around 1900, Meyer and Overton observed that lipid-soluble substances pass rapidly across the cell membrane on the basis of their concentration gradient [311, 312]. Their observation was the basis for the formulation of the well-known Meyer–Overton rule, which states that transport of a molecule across the phospholipid matrix into a cell is controlled by its lipid solubility [326]. The rule is used to predict membrane permeability. Recent studies, however, have shown that there are some exceptions to the Meyer–Overton rule [326]. Accordingly, the membrane permeability (PM) of a molecule increases with its hydrophobicity [327]. Thus, the higher the lipophilicity of a molecule the greater the molecule will penetrate the plasma membrane. But every substance moving across the plasma membrane has a certain level of resistance resulting from the properties of the membrane and the fluid compartments. The membrane resistance (RM = 1/PM) of near-membrane unstirred layers is estimated at approximately $1/6.1 \times 10^{-5}$ cm/s [325, 327]. The Meyer–Overton rule does not account for transport processes, mediated by membrane channels, carriers, and pumps. Accumulating evidences also indicate that both hydrophilic and hydrophobic substances can be transported by carrier proteins

[325]. Substances that do not obey this rule include salicylic acid (an aromatic weak acid belonging to the family of aromatic carboxylic acids; O_2, CO_2, ammonia (and possibly NO, H_2S) when transported by aquaporin channels; short-chain carboxylic acids when transported by membrane fatty acid transporters). Contrary to the Meyer–Overton rule, the transport velocity of a more hydrophilic molecule across the membrane is slower than a lesser hydrophilic molecule [328]. Thus, of the four carboxylic acids, hexanoic acid is expected to diffuse across the membrane faster than valeric. Butanoic and acetic acids are the slowest (the higher the carbon chain length of carboxylic acids the higher the hydrophobicity [327]).

Types of passive transport include osmosis (also called simple diffusion), facilitated diffusion, and filtration [315]. **Osmosis** is the movement of water across a semipermeable membrane (i.e., plasma membrane). Plasma membrane acts as a differential membrane permitting the movement of water molecules in and out of the cell. Osmosis can be defined as the net transport of water or solvent molecules through plasma membrane from low osmotic pressure region to high osmotic pressure region (region of high solvent concentration). The effect of various concentrations of solutes on cell shape is called tonicity. On the basis of tonicity, three types of solution can be distinguished—isotonic, hypertonic, and hypotonic solution. Isotonic solution is that which has the same solute concentration with the cell interior; hypotonic solution has a low solute concentration in relation to the cell interior; hypertonic solution has a high solute concentration in relation to the cell interior. Water molecules move from the region of hypotonic solution to the hypertonic side [329–332]. The difference in the concentration between the cytoplasm and extracellular fluid is referred to as concentration gradient. Transport of water continuously takes place across the cell membrane, and it is carried out by transporters called "aquaporins." As noted earlier, these water channels not only transport water across the cell membrane, but also move other substances including oxygen, carbon dioxide, hydrogen sulfide, hydrogen peroxide, urea, ammonia, glycerol, polyols, carbamides, purines, pyrimidines, glycine, and lactate. That is why aquaporins are now termed multifunctional channels [319, 321–323, 333, 334]. Other channels also transport water across the cell membrane. **Facilitated diffusion or transport**: This is the transport of molecules via specific transmembrane integral proteins along a concentration gradient without the expenditure of chemical energy. The integral proteins through which facilitated diffusion occurs are called carrier proteins. This is why facilitated diffusion is sometimes called **carrier-mediated transport**. In strict terms, this transport type can be considered to involve some sort of energy (electrochemical energy produced by ion gradients). This type of diffusion is accomplished in a stepwise order. First, the membrane-permeating molecule recognizes its cognate carrier protein, associating with it to form a complex with corresponding conformational change. This change in conformation of the receptor (i.e., the carrier protein) allows the movement of the permeating molecule or the cargo across the plasma membrane. Usually, conformational change in a receptor decreases its affinity for its ligands such that no similar substance may be attached to the same carrier protein receptor until the molecule is released. In some cases, however, the attachment of the diffusing molecule to the receptor can increase the transport capacity of the carrier protein receptor. Following release of the membrane-permeating molecule, the carrier protein resumes its original conformation, ready to bind another molecule. Crucial features of facilitated diffusion include the chemical specificity and spatial orientation of the membrane-permeating molecule. Molecules to be transported via carrier proteins must possess the requisite chemical structure and spatial orientation to bind to the receptor. Thus, facilitated diffusion is stereo-specific. This means that only a specific isomer of the molecule (L-isomer for "levo rotatory" or D-isomer for "dextra" rotatory, meaning rotation to the left and right, respectively) is transported and not both. In addition, carrier protein receptors through which facilitated diffusion occurs possess saturability property (which means that the carrier stops moving the substrate across the membrane when specific transport rate is attained). The functions of some carriers may depend on the pH of the cell [335, 336]. Membrane voltage dependency of substrate transport by some carriers is well known [337]. For example, at least one phase of transitions of the activity of the loaded Na^+/glucose co-transporter may be voltage-dependent [338]. For some carriers, the membrane voltage or conductance for a particular ion may either decrease or increase flux of the permeating molecule across the plasma membrane [339]. Facilitated transport is also temperature-dependent [339, 340]. Examples of facilitated diffusion are the transport of glucose from the lumen of the intestine into the cytoplasm of enterocyte; movement of glucose into the red blood cell; transport of free fatty acid from blood through the endothelial cell and the interstitial tissue to the sarcoplasm of the cardiomyocyte via special sarcolemmal transporter called fatty acid-binding protein [341–343]. **Filtration** is the movement of solutes and water across the cell membrane via carrier proteins down the concentration gradient without energy expenditure, due to hydrostatic pressure generated by the cardiovascular system within the physiological range. Membrane transport by filtration depends on the sizes of the membrane pores that allow passage of solutes. For instance, cells of the liver have relatively large membrane pores which allow the passage of large molecules. In contrast, the membrane pores of Bowman's capsule in the kidneys are very small, such that only extremely small proteins can be filtered [165, 344].

Active transport: This is a type of carrier-mediated movement of molecules across the plasma membrane against concentration gradient (chemical concentration or electrochemical gradient) with the expenditure of energy. This form of transport involves the uphill movement of ions and molecules by loss of energy derived from ATP hydrolysis [315]. In active transport, the rate of transport increases with solute concentration till a maximum is reached. Beyond this maximum value, the rate of membrane transport does not increase indicating that it is mediated by means of carrier proteins. The carrier proteins are ATPases—enzymes that catalyze the hydrolysis of ATP. The carrier may exhibit increase in binding to the substrate if the ligands are already present on the same carrier (this is known as cooperative binding). The extent of cooperativity is measured by the Hill coefficient, named after the English physiologist Archibald Vivian Hill (1886–1977) who in 1910 described the sigmoidal O_2 binding curve of hemoglobin. The Hill coefficient describes the fraction of the receptor saturated by the substrate as a function of the substrate concentration [345, 346]. There are basically two types of active transport—primary and secondary active transport. **Primary active transport** involves the transport of substrates across the membrane with loss of chemical energy. In primary active transport, ATP directly provides the energy for the work of the protein carriers, which have an ATP-binding site. ATP transfers its phosphate to the transport protein, thus changing the protein's conformation so that the target substrate can be moved across the membrane. These carrier proteins are called primary pumps—membrane transporters that hydrolyze ATP with corresponding uphill transport of the substrate (usually an ion) [315]. The sodium–potassium pump (Na^+/K^+ pump, also called Na^+/K^+-ATPase), which maintains the appropriate Na^+/K^+ ion balance in animal cells, is an example of primary active transport [347]. This transporter, also acting as an enzyme, is responsible for the consumption of about 40% of ATP produced in the body. The carrier protein "H^+/K^+-ATPase" is localized in many cells including the gastric mucosal cells and renal tubules, and is responsible for shuttling hydrogen and potassium ions across the cell membrane [315, 339, 340, 348]. **Secondary active transport** is the simultaneous movement of two molecules through the same carrier protein referred to as electrogenic pumps against their concentration gradient. The energy used to move the target molecule is provided by the electrochemical gradient, not directly by ATP. However, as noted above, ATP is used to pump the ions across the membrane to generate the electrochemical gradient [349–351]. The electrogenic pumps are termed secondary pumps, which utilize the energy stored in electrochemical gradients of ions by coupling downhill movements of the ions to drive the uphill transport of another substrate. On the basis of the directions of the coupled downhill and uphill ion flux, secondary pumps are categorized as co-transporters and exchangers (also known as countertransporters) [315]. Examples of secondary pumps include the D-glucose/Na^+ transporter in the brush border membrane of the enterocyte [352], electrogenic co-transporter Na–HCO_3, which transports three HCO_3^- and one Na^+ [337]. The transport of Na^+ by this co-transporter is voltage-independent, whereas bicarbonate transport is voltage-dependent [337]. In active transport, the carrier protein possesses multiple binding sites both on the cytoplasmic and extracytoplasmic sides of the plasma membrane. The molecule to be transported binds with the carrier to form a complex with corresponding change in the carrier conformation. This change in conformation allows the molecule to pass via the membrane into the cytoplasm with corresponding loss of energy. Depending on the type of carrier protein involved, the process may be accompanied by transport of another molecule from the interior of the cell to the outside. Types of carrier proteins include uniporters, symporters, and antiporters [353]. **Uniporters** are carrier proteins that transport only one molecule in one direction across the membrane [354]. *Uniporters activity can be driven by ion gradients*. The uniporters move substrates across the membrane by facilitated transport. An example of a uniporter is the integral, transmembrane protein, GLUT1—glucose transporter type 1 [165, 168]. **Symporters** are carrier proteins that transport two molecules in the same direction at a time. Monosaccharides such as D-glucose and amino acids are transported from the intestinal lumen into the enterocyte cytoplasm through symporters [165, 168, 352, 354]. **Antiporters** are carriers that exchange one substance for another in opposite directions. The target substance to be moved binds to the opposite side of the coupled transport protein as the gradient of the coupling ion provides the force to move the target substance [165, 168]. Examples of antiporters include the sodium–potassium exchange pump, which hydrolyzes one ATP molecule to pump three Na^+ outside and two K^+ ions into the cell [354]. In Cl^-/H^+ antiporter (chloride–proton exchanger), the flow of Cl^- in one direction against the electrochemical gradient is coupled to the movement of proton downhill in the opposite direction or vice versa. The Cl^-/H^+ exchangers are implicated in the acidification of the intracellular vesicles and compartments such as secretory vesicles endosomes and lysosomes [355, 356]. The Na^+/Ca^{2+} antiporter (or exchanger) transports 3 molecules of Na^+ across the plasma membrane down its concentration gradient into the extracellular space in exchange for one Ca^{2+}, which is transported in the opposite direction, by exploiting the energy stored in the electrochemical gradient of sodium. However, in certain situations, this transporter may contribute to Ca^{2+} influx (called the reverse Na^+/Ca^{2+} transporter). The Na^+/Ca^{2+} antiporter is the principal mechanism responsible for calcium efflux in cardiomyocytes, but this transporter is also expressed in other cells such as neurons, epithelial cells of the renal tubules, smooth and skeletal muscle cells [357]. The Na^+/H^+ antiporter is an electroneutral pump that regulates the movement of sodium ions and protons in and out of the cell [352, 358]. The activities of Na^+/H^+ antiporter control intracellular saline concentration, cell volume, and pH homeostasis, which are all crucial to the physiology of the cell [359, 360]. Like other transporters, the mutation of this transporter leads to disorder in the transport of the ions [361]. Both symporters and antiporters are also known as coupled transporters or co-transporters or biporters. The transport systems in the inner membrane of the mitochondria occur by active transport [352].

Bulk transport: This type of transport involves the movement of solid particles across the membrane by invagination or evagination of the plasma membrane. Bulk transport is useful in carrying large molecules which otherwise would have not been transported through the cell membrane normally. Bulk transport is accomplished by endocytosis and exocytosis [165, 168, 362]. **Endocytosis** is the process of engulfing large-sized particles of substances (e.g., nutrients, foreign substances, plasma membrane receptors, and other cargoes) by membrane invagination, subsequently releasing the contents into the cytoplasm (Fig. 3.12) [363–366]. Endocytosis occurs on plasma membrane sites or hotspots—eisosomes. These endocytotic spots are mainly composed of the cytoplasmic proteins, Pil1 and Lsp1. The expression or the absence of these proteins directs the site of endocytosis [231]. **Exocytosis** (also called emeicytosis or cell vomiting) is the process of exudating the secretory products or contents of the intracellular vesicles to the outside of the cell cytoplasm (Fig. 3.13). In cells of pancreas, the intracellular vesicles containing enzymes move from the interior of the cytoplasm to the plasma membrane, where they fuse or merge with plasma membrane, leading to the formation of a fusion pore and discharge of the contents to the extracellular space [367]. According to the size of the pore and time of contact between the vesicles with the plasma membrane, transient and full-fusion exocytosis can be distinguished. In full-fusion exocytosis,

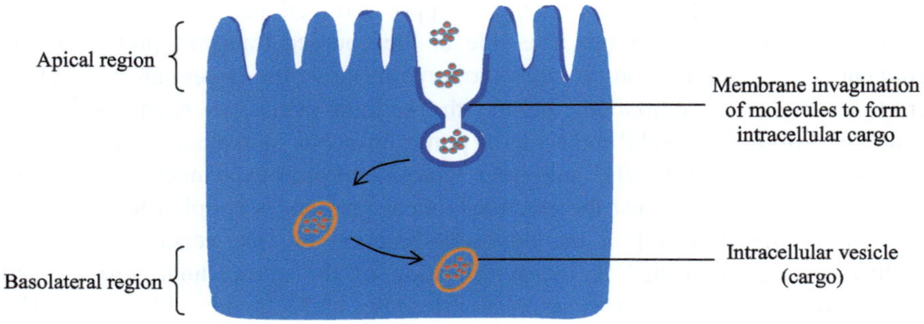

Fig. 3.12 Endocytosis in the apical side of the enterocyte. Endocytosis occurring in the enterocyte apical membrane leads to the loss of microvilli [617–619]. In certain diseases of the GI tract (e.g., cystic fibrosis), apical endocytosis is dysfunctional and delayed [620]. But endocytosis can also occur in the basolateral zone [621]

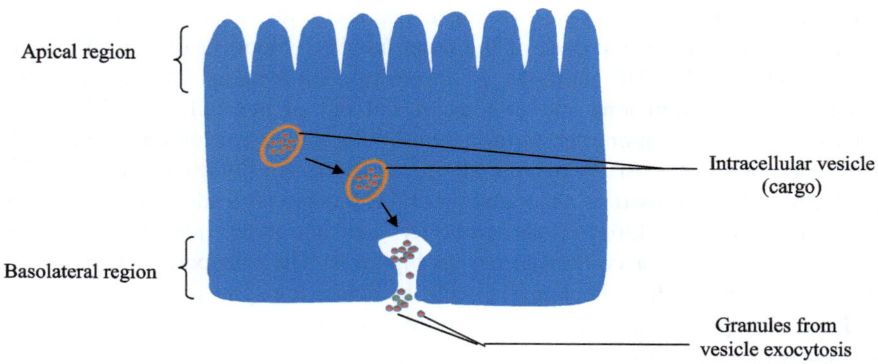

Fig. 3.13 Exocytosis in the basolateral side of the enterocyte. Exocytosis of intracellular cargo is controlled by multiple signaling systems involving motor and associated proteins, membrane lipids, and calcium [622, 623]. Two types of vesicles are shed from the mammalian cell—exosomes and ectosomes [624]. The former are vesicles measuring 50–100 nm in diameter released constitutively from the intracellular compartment, whereas the later, also called microvesicles, are vesicles measuring 100–1000 nm in diameter shed directly from the plasma membrane in a regulated manner [624]. Exocytosis can occur in a constitutive or regulated manner. In constitutive exocytosis, secretory vesicles from the trans-Golgi network (further explained below in the text) move down the cytoskeletal network to directly fuse with the plasma membrane, release vesicular substances into the exterior continuously. This type of exocytosis occurs all the time in the cell and may be required to maintain the basal secretion in the cell [625, 626]. Constitutive exocytosis occurs in all cells [627]. In regulated exocytosis, released cargo from the trans-Golgi network first replenishes a pool of secretory vesicles located in close proximity to the plasma membrane, and then vesicles from the pool fuse with the plasma membrane and are subsequently released into the extracellular space when the cell is stimulated [628]. Regulated exocytosis occurs in specialized cells such as exocrine, endocrine, and neuronal cells [627]

the pore size is relatively large so that almost all contents of the cargo are poured into the exterior. The vesicular cargo remains fused to the plasma membrane for relatively longer interval. In transient exocytosis, however, the intracellular cargo for a very short interval with small pore size forms contact with the plasma membrane. Consequently, only a few contents of the vesicles are released into the exterior [368].

3.4.2 Intracellular Components

The components in the interior of the cell, referred to as intracellular components, include the cytoplasm and all its structures. The cytoplasm is a gelatinous solution comprising the cytoskeleton of the cells as well as the organelles. It is composed of about 75–95% water, large quantity of carbohydrate, fat, and protein molecules. The fluid portion of the cytoplasm is called the cytosol. The cytoplasm is a site for chemical reactions (anaerobic energy metabolism); synthesis of cellular molecules; and packaging of chemicals for export [306, 369]. Organelles are specialized structures of the cell, having characteristic shape and specific functions. They are crucial for a variety of physiological roles including growth, repair, and control of cellular functions. Examples of organelles include the endoplasmic reticulum, nucleus, ribosomes, Golgi complex, mitochondria, lysosomes, peroxisomes, centrosomes/centrioles [370–373].

Endoplasmic Reticulum

The endoplasmic reticulum (ER) is a dynamic membrane-bound synthesis, transport, and quality control organelle that is an extension of the nuclear envelope and plays a variety of roles including protein and lipid synthesis and metabolism as well as calcium depot. The ER is made up of flattened sacs and branching tubules that are interconnected forming a continuous space inside the organelle (Fig. 3.14). This space is called the ER lumen (also known as the ER cisternal space). The lumen comprises about ten percent of the entire volume of the cell (Fig. 3.12) [374–376]. In addition, some regions of the cytoplasmic surface of the ER are covered with ribosomes (Fig. 3.13). The presence of ribosomes on the ER gives it a bumpy appearance when viewed through an electron microscope. ER with ribosomes on its surface is called granular or rough ER, whereas ER without ribosomes is called agranular or smooth ER (Fig. 3.15). The two distinct types of ER are connected to each other, but differ in structure and function. The rough ER is the site of protein processing, whereas the agranular or smooth ER represents the site of lipid metabolism (e.g., cholesterol and steroid hormone synthesis) [163, 374–376]. Smooth ER is abundantly present in cells specialized for lipid synthesis. However in the majority of cells, smooth regions are scanty, are usually partly smooth and rough, and are referred to as transitional ER. The transitional ER is a region of the

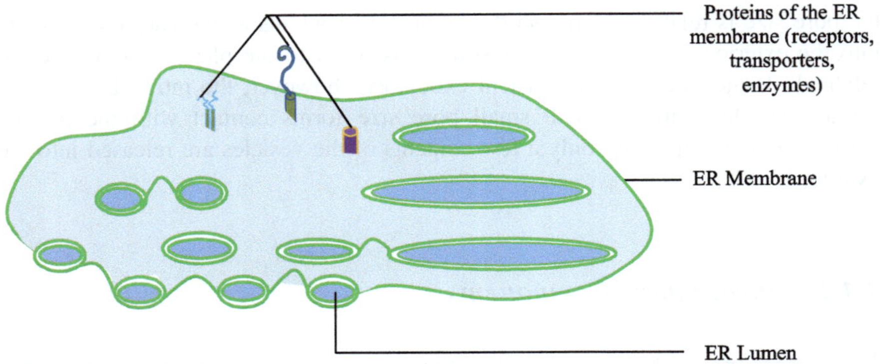

Fig. 3.14 Structure of the ER

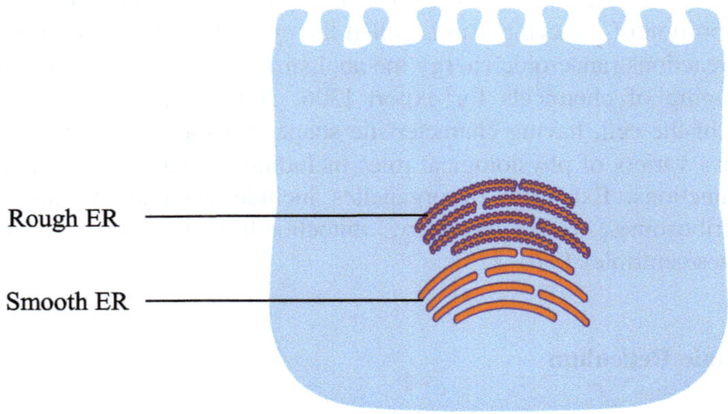

Fig. 3.15 Rough and smooth ER of the epithelial enterocyte

ER where secretory proteins are exported to the Golgi complex and thus represent quality control sites in the secretory cargo proteins. Although the mechanisms of ER functions are not completely understood, studies suggest that the peculiarities of transitional ER functions are characterized by the presence of the coat protein complex II (COPII) proteins, which are highly expressed in the region [165, 377, 378].

The ER membrane allows molecules to be selectively transferred between the lumen and the cytoplasm, and since it is connected to the nuclear envelope, it provides a channel between the nucleus and the cytoplasm. The ER has a central role in synthesis, processing, and trafficking biomolecules for use inside and outside the cell. The ER is the site of production of all the proteins (including transmembrane proteins) and lipids for most of the plasma membrane and organelles, including the ER itself, the Golgi apparatus, lysosomes, endosomes, peroxisomes, and secretory vesicles. Furthermore, almost all of the proteins that will exit the cell, plus those destined for the lumen of the ER, Golgi apparatus, and lysosomes, are originally delivered to the ER lumen. Consequently, many of the proteins found in the cisternal space of the ER lumen are temporarily located there as they transit to other locations. Proteins that do not leave the ER remain in the lumen of the ER, are inserted into the ER membrane—called resident proteins of the ER. These resident proteins contain a special retention signal that enables them to be retained in the ER [374, 375, 379]. An example of an ER resident protein is the chaperone protein "BiP (binding immunoglobulin protein, also known as heat shock protein-70 or glucose-regulated protein-78)." The resident ER proteins (such as BiP) are molecular chaperonins that aid in protein folding, the process by which a newly synthesized protein from the mRNA attains its three-dimensional conformation or native structure required to carry out a specific function at a particular location. The native protein is held together by hydrogen bonds [380, 381]. The ER resident proteins can recognize newly synthesized proteins that have been synthesized or processed and keep them from being sent to their final destinations. Similarly, any protein leaving the ER to another organelle or plasma membrane will possess the signal that dictates its final destination or the protein is sent to the lysosomal degradation pathway [382–384]. The ER regulates cellular stress by sensing intracellular and extracellular stress levels to maintain homeostasis. Consequently, the ER forms an important nexus between environment and the cell function. One mechanism by which ER senses stress level is through the amount of unfolded proteins. ER stress encourages protein unfolding, thus stimulating the ER to work more to bring the protein to its functional "folded" state—which is required to alleviate the stress. Unfortunately, chronic ER stress can induce cell death by activating several signaling pathways implicated in the pathophysiology of several diseases. The pathways include the glycogen synthase kinase 3/3β, c-Jun N-terminal kinase, CCAAT (cytidine-cytidine-adenosine-adenosine-thymidine)/ enhancer binding protein homologous protein, and caspase-12, PKR (double-stranded RNA-dependent protein kinase)-like endoplasmic reticulum kinase, inositol-requiring 1 and activating transcription factor 6 [376, 385–388]. The malfunctioning of this organelle is involved in a range of diseases and conditions including renal fibrosis, inflammation- or osmolar-induced renal injury, ischemia-reperfusion cardiac injury, proteinuria, atherosclerosis, diabetic nephropathy, type 2 diabetes, cancer, preeclampsia, some liver and brain diseases [385, 387, 389]. The ER functions as the intracellular Ca^{2+} reservoir of the cell, maintaining the concentration of Ca^{2+} in cytosol, which is required for regulated signaling in the cell and with its neighbors [374].

Golgi Complex

The Golgi apparatus (also known as the Golgi body or the Golgi complex) is a central station of the cell for protein modification, sorting and secretion or degradation via the lysosomal pathway. The Golgi apparatus is composed of membrane-bound compartments (flattened cisternae), which are interconnected sacs in the form of a stack or ribbon in the perinuclear area (Fig. 3.16). The stacks contain different enzymes (e.g., glycosyltransferases, sulfatases, phosphatases) and other polypeptide molecules [390–394]. Vesicles destined for further processing are sent from the ER to the Golgi apparatus via one side of the stack referred to as cis face—the closest portion of the Golgi complex to the ER (Fig. 3.16). Proteins and lipids transiting the Golgi complex are modified by the sequential action of enzymes of the individual cisternae. The opposite end of the Golgi apparatus, trans face, is where the modified molecules leave. The trans face is usually facing the plasma membrane, where most of the substances the Golgi apparatus modifies are sent. These vesicles may contain various proteins or protein subunits meant for the replenishment of the receptor, transport, and enzymatic protein of the plasma membrane or other organelles [390, 391]. The Golgi complex is the central site of protein modification, removing and substituting sugar monomers to produce a variety of oligosaccharide-associated protein. Upon completion of the modification process, the Golgi complex sorts the final products of its processing. Molecular labels or tags are added by the resident enzymes of the Golgi complex to help in the proper identification, indicating the correct destination of the protein. The Golgi apparatus sends off the final product to various parts of the cell by budding vesicles from its trans face. Some modified proteins may be sent to lysosomes for degradation [390, 391].

Proteins already discharged through the Golgi complex may be affected by other factors and conditions of the cytosol microenvironment on their journey to their final destination.

The mechanisms of vesicle transport in the Golgi complex are still debated. But two models of Golgi vesicle transport have been proposed. They are the vesicular

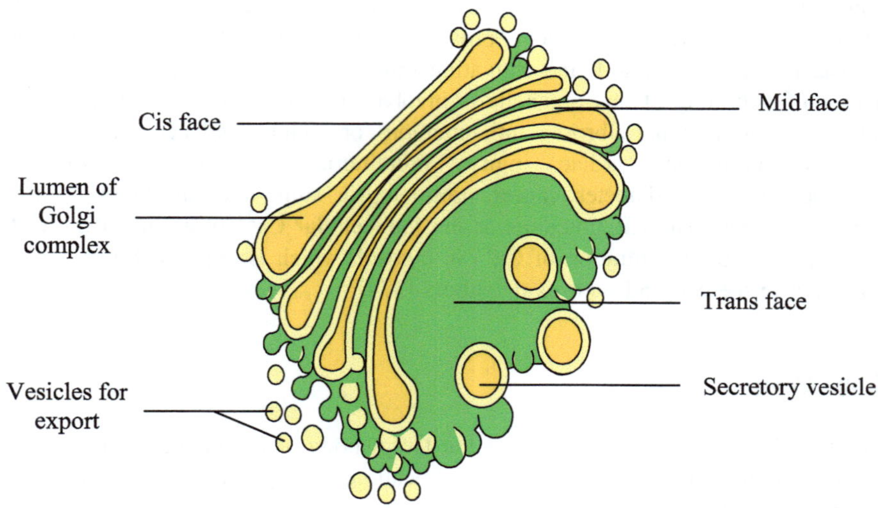

Fig. 3.16 Structure of the Golgi apparatus

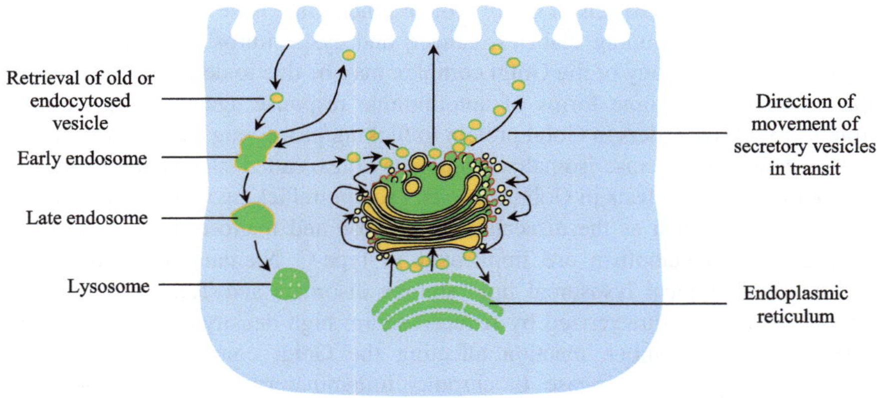

Fig. 3.17 Trafficking functions of the Golgi apparatus and other organelles. The arrows indicate the direction of movement of vesicles. The Golgi apparatus is often found in close proximity to the ER in cells. Protein cargo moves from the ER to the Golgi complex, where it is modified, sorted, and then sent to various destinations including the plasma membrane, endosome, and lysosomes. But secretory vesicles already sent to the Golgi complex can be retrieved back to the ER [390, 397]

transport and cisternal maturation models (see Fig. 3.17). According to the **vesicular transport model**, the Golgi compartments are relatively stable with each region having a peculiar expression of resident enzymes that function to modify proteins. Vesicle traffic is believed to be anterograde from the cis to the trans face. Cargo proteins transported to the Golgi complex arrive at each compartment and are modified by the resident enzymes located within that compartment. Vesicle transport takes place by sequential budding of vesicles containing modified at each compartment cargo proteins from the subsequent cisterna to the next one. So, vesicles transported directly from the ER first arrive at the cis face, where resident enzymes modify the proteins, and then they bud off this compartment to the medial (or middle) face, where similar modification takes place but with different enzymes and to the trans face—the final phase of protein modification [395]. Thus, according to the vesicular transport model, cisternae do not change their relative position in the stack [395]. **In the cisternal maturation model**, it is believed that cargo proteins progress through the Golgi stack by cisternal maturation, which is balanced by a backward flow of Golgi resident enzymes-coated vesicles [396]. According to this model, the transport vesicles function as vehicles for Golgi resident enzymes and not for the cargo proteins [395]. The formation of a new cis cisterna is accompanied subsequently by its maturation, accumulating medial, then trans resident enzymes via retrograde (backward) transport of vesicles to younger or earlier cisterna as opposed to the anterograde transport in the vesicular transport model [390, 395, 397]. Consequently, each newly formed cis cisterna matures to form a medial cisterna and then to a trans cisterna, from where vesicles bud off to their final destinations [395].

Defects in various aspects of Golgi protein functioning have been implicated in a row of disorders including diabetes, cancer, and cystic fibrosis [390]. These disorders in the physiology of the Golgi complex may be due to defects in protein and lipid trafficking or some forms of autoimmune response against Golgi resident proteins [398]. Disorders in Golgi protein trafficking/processing can lead to diseases such as congenital sucrase-isomaltase deficiency (also called sucrose intolerance), a disorder involving defects in Golgi complex protein trafficking resulting in inability to digest sugars such as the disaccharides sucrose and maltose. Defects in Golgi complex lipid metabolism are implicated in type-C Niemann–Pick disease (a neurovisceral atypical lysosomal lipid storage disorder) and Tangier disease (an inherited disorder characterized by reduced serum high-density lipoprotein levels) [398, 399]. Autoimmune reaction affecting the Golgi complex is particularly observed in Sjogren's disease (a chronic, inflammatory systemic autoimmune exocrinopathy characterized by lymphocytic infiltration of the exocrine glands in multiple sites, particularly, lacrimal and salivary glands with the symptomatic triad (dryness of the buccal mucosa, fatigue, and diffuse pain)). The hallmark of the disease is keratoconjunctivitis sicca and xerostomia and may occur as primary or secondary disorder in association with other autoimmune diseases [400, 401] and systemic lupus erythematosus (a clinically heterogeneous, multisystem autoimmune disease, characterized by the presence of autoantibodies directed against nuclear antigens, and in which patients present in different ways, with an estimated incidence of approximately 1 in every 1000, but more frequently occurring among females than males by tenfold) [398, 402, 403]. In Sjogren's syndrome, Golgi autoantibodies are produced against the trans-Golgi complex hydrophilic protein p230 (also called golgin-245) found in the cytosol and associated with the Golgi membrane and are responsible for vesicles budding off the trans cisterna [402, 404, 405]. Golgin-95 and golgin-160 are implicated in the pathophysiology of systemic lupus erythematosus [398, 399]. Other Golgi resident proteins include golgin-45, golgin-67, p115, p200, golgin-84, golgin-95, golgin-97, golgin-160, golgin-245, and giantin/macrogolgin [406–408]. An example of the medial cisternae protein of the Golgi complex is golgin-45 [391]. Golgi proteins associated with the cis-Golgi network include golgin 84, Golgi matrix protein GM130, giantin, Golgi microtubule-associated protein 210 (GMAP-210) [402, 409–411]. Malfunctioning of these proteins in the respective compartment can lead to disruption of the cisternae structure and functions resulting in a dys-polarized Golgi network [391]. These Golgi associated proteins, though exhibit specific functions, are generally involved in vesicle budding and trafficking and are altogether called golgins—a family of coiled-coil protein, associated with the Golgi complex that are believed to be involved in the tethering of vesicles and the stacking of cisternae, and in several other functions including structural supports for Golgi cisternae, mediation of Golgi membrane–cytoskeleton, and Golgi membrane–membrane interactions [394, 407, 408, 412]. There are about 100 Golgi complex proteins discovered [402]. The golgins, though form an integral architecture and functions of the Golgi complex, constitute the target of autoantibodies produced by the immune system as in humoral or cell-mediated immune response. These golgins are therefore termed

Golgi autoantigens [391]. For instance, the autoantigen golgin-245 reacts with autoantibodies formed against the golgin to form a complex that is implicated in the pathophysiology of Sjogren's syndrome [413].

Peroxisomes

Peroxisomes, previously called microbodies, are ubiquitous, roughly spherical, and single membrane-bound organelles, mediating a wide variety of biosynthetic and biodegradative reactions in the cytoplasm. They are often found in the liver and kidney because of their high metabolic levels [414–416]. It is not exactly clear how peroxisomes are formed, but research suggests that these organelles are either formed from preexisting peroxisomes by organelle division or they are formed in the absence of preexisting peroxisomes possibly by fission and fusion of homotypic organelles or budding from heterotypic organelles [417]. Peroxisomes do not possess an organelle genome. Peroxisomal proteins are encoded in the nuclear genome and translated in the cytoplasm by the rough ER. Upon completion of the protein translation process with subsequent modification and sorting, the proteins are imported from the cytoplasm into the organelle matrix and membrane by specific signal sequences that direct their final destinations [415, 418]. For further information on the history of discovery of this organelle, its evolution, and functions, review the publication by Gabaldón [415].

Peroxisomes are synthesis machinery that produces key components of the cell. This organelle harbors enzyme required for the synthesis of biomolecules such as ether phospholipids, bile acids, and docosahexaenoic acid [414, 415]. Ether phospholipids (or ether-linked phospholipids) are lipids characterized by an ether linkage between the glycerol backbone and one or both fatty acid side chains (usually the sn-1 position) as opposed to an ester linkage in the phospholipids discussed in earlier section of this chapter. Examples of ether-linked phospholipids include ethanolamine ether phospholipid (ether-linked phosphatidylethanolamine), choline ether phospholipid (ether-linked phosphatidylcholine) [172, 419], and plasmalogens. Plasmalogens are characterized by a vinyl ether linkage at the sn-1 position with an ester linkage at the sn-2 position. Ether-linked phospholipids are vital biomembrane protective molecules present in several cells of the body. They can function as free radical scavengers and as reservoirs for arachidonic acid in the plasma membrane. The Schwann cells that ensheathe peripheral neurons express plasmalogens and ether-linked phospholipids. The absence or malfunction of these molecules can lead to neurological disorder [172, 419]. Bile acids, derived from cholesterol, are important molecules involved in digestion through their ability to emulsify fats [416]. Docosahexaenoic acid (DHA) is a highly polyunsaturated long-chain omega-3 fatty acid that is a primary structural component of the human brain, retina, etc. [420, 421]. Peroxisomally synthesized DHA is used as pivotal precursors of maresins (mediator produced in macrophage whose role is to resolve inflammation) and protectins (molecules whose roles are to protect the cell from damage). These molecules are potent anti-inflammatory agents mediating response of the body cells and tissues to the damaging effects of injury or other stimuli [416]. DHA is produced by peroxisomes, but in minimal amounts, and thus, must be taken in as food or supplement. In the form of supplement, DHA is used in a variety of diseases characterized by cognitive decline (including dementia, Alzheimer's disease), cancer, cerebrovascular, cardiovascular, and inflammatory diseases [422–426].

Peroxisomes perform a variety of degradative reactions according to the physiological requirements of the cell and its environs [418]. Peroxisomal enzymes found in the lumen of the organelle are important for beta-oxidation of fatty acids, detoxification of reactive oxygen species, alcohol, and other xenobiotics [414, 415] and also mediate oxidative stress response [415]. The peroxisomes shuttle metabolites for continued processing or anaplerotic metabolism [416]. The oxidative reactions occurring within the organelle produce reactive oxygen species such as hydrogen peroxide, which is instantaneously processed by catalase, a peroxisomal resident enzyme [416, 427]. Several other molecules are degraded—some amino acids, polyamines, glyoxylate, alcohol, and other xenobiotics [414, 415]. Peroxisomes also help to degrade derivatives of arachidonic acid (DAA)—eicosanoids. DAA also includes prostaglandins, thromboxanes, leukotrienes, and prostacyclins. These molecules form critical nexus between the damaging effect of extracellular and intracellular stimuli and mediate cellular signaling that exert control over inflammatory reactions [416].

Dysfunctions in peroxisomal synthesis or degradation processes are implicated in a range of human diseases [416]. For instance, Zellweger syndrome (named after the Swiss-American physician Hans Zellweger (1909–1990), who conducted groundbreaking studies on the disease), a congenital disorder, is characterized by peroxisome dys-biogenesis and single peroxisomal enzyme deficiency, which may be caused by defective trafficking of peroxisomal membrane and matrix proteins [417, 428–431]. The disorder is characterized by hypotonia, hepatomegaly, severe neurodevelopmental delay, renal cysts, retinal dysfunction, sensorineural deafness, and facial dysmorphism in newborns [427].

Lysosomes

Lysosomes are acidic and dynamic organelles that mediate the degradation of extracellular and intracellular particles from endocytic, autophagic, phagocytic, and secretory pathways [432–434]. They were discovered in 1955 by Christian De Duve (1917–2013), a Belgian physiological chemist, cytologist, and biochemist who shared the 1974 Nobel Prize in Physiology or Medicine with Albert Claude (1899–1983) and George Emil Palade for their discoveries concerning the structural and functional organization of the cell. De Duve made numerous contributions in cell biology. He also discovered peroxisome in 1965. Furthermore, he investigated subcellular organization of biochemical pathways in synthesis and degradation of biomolecules [435–437]. Albert Claude was one of the founders of modern cell biology, who provided one of the first most accurate diagrams of cell structure in

1945. Palade accurately described the biochemical steps in protein synthesis, segregation, transport, storage, and secretion, and the ultrastructural units related to each process in the exocrine pancreatic cell [437]. The lysosomes together with secretory organelles constitute the endomembrane system of the eukaryotic cell [438].

The main function of lysosomes is digestion of particles engulfed by the cell or intracellular particles, helping to recycle the components of worn-out cell [432]. The degradative functions of lysosomes are due to the expression of hydrolytic enzymes (acid hydrolases) in this organelle. Over 60 types of such enzymes are present in lysosomes. Lysosomal enzymes include nucleases (e.g., DNase, RNase), acid phosphatases (e.g., ester phosphatase), proteinases or proteases (e.g., cathepsins), carbohydrases and glycosidases (e.g., beta-glucosidase, hexosaminidase A, alpha-mannosidase, alpha-fucosidase, sialidase), lipases or lipid hydrolases (e.g., sphingomyelinase, esterases), and sulfatases (e.g., cerebroside sulfatase) [433–444]. The enzymes mediate the degradation of extracellular particles from endocytosis and of intracellular components from autophagy, degraded macromolecules (proteins, polysaccharides, and complex lipids) from the secretory, endocytic, autophagic, and phagocytic vesicles into building-block molecules such as amino acids, monosaccharides, free fatty acids, or nucleotides [432–434]. To carry out its functions effectively, the aqueous environ inside the organelle is maintained at an acidic range of pH of 5.0. Recall that the pH in the cytosol is much higher, and thus lysosomal enzyme leakage will make them ineffective in carrying out their functions. However, excessive leakage of lysosomal enzymes can lead to cell autodigestion. Vesicles or particles (e.g., cellular waste products or debris) destined for degradation in the cell fuse with lysosomes resulting in the release of the hydrolytic enzymes into the vesicles/particles with subsequent digestion of the vesicles/particles to sugars, amino acids, and other molecules. These products of intracellular digestion return to the cytosol for reuse. Obsolete organelles are also recycled by the lysosomes hydrolytic cleavage via a process called autophagy. Lysosomes can degrade the entire cell through a process called autolysis [432, 433, 441, 443]. This can happen in some cases of cell injury, when lysosomes fuse with the plasma membrane [432]. The mechanism of autophagy was discovered by Yoshinori Ohsumi, Japanese physiologist who won the 2016 Nobel Prize in Physiology or Medicine "for his discoveries of mechanisms for autophagy." The process of autophagy is controlled by autophagy-related genes (ATGs) first identified in the yeast *S. cerevisiae*. The products of these genes (Atg8) initiate or control biogenesis of autophagosome via multistep conjugation and finally lipidation that culminates in the formation of autophagocytotic vesicle [445]. Upon stimulation of its active sites, Atg12 is an ubiquitin-like protein that conjugates to Atg5, forming a stable Atg12–Atg5 complex. The process occurs via the activating enzyme Atg7. The Atg12–Atg5 complex interacts with Atg16 to form Atg16–Atg12–Atg5 complex. Under a second activating enzyme, Atg10, Atg8 is conjugated to phosphatidylethanolamine with the formation of the vesicle [445].

The normal functions of lysosome are perturbed in some conditions involving mutations of the gene responsible for the synthesis of its enzymes or non-enzymatic biogenic lysosomal proteins—a condition generally called lysosomal storage diseases or lipidoses. These disorders belong to a subgroup of the broad category of inborn errors of metabolism, which lead to abnormal storage of macromolecules [446, 447]. In these disorders, macromolecules could not be degraded by the acid hydrolases of the endolysosomal system resulting in accumulation of these non-degraded macromolecules in the lysosome and progressively spreading to other intracellular regions [448]. This leads to disorder in functions of the cell and may turn out as a systemic manifestation. Over 50 different types of inherited metabolic diseases have been identified as lysosomal storage disorders [447]. The first vivid description of individuals with lysosomal storage disorders was reported by Warren Tay in 1881. The clinical condition was later named after Warren Tay and Bernard Sachs who in 1887 conducted studies to describing the condition—Tay-Sachs disease [449]. The link between lysosomal storage disorders and enzyme deficiency came only in 1963 [450]. Tay-Sachs disease is a rare, an autosomal recessive inherited disease affecting the central nervous system and results from mutations in the gene encoding the alpha subunit of beta-hexosaminidase A and the accumulation of GM2 ganglioside in neurons. Beta-hexosaminidase A is a lysosomal enzyme composed of alpha and beta polypeptides whose loss of function mutation impairs its folding or trafficking process [451–453]. GM2 (ganglioside monosialic type 2) is a family of the complex lipids "glycosphingolipids (ceramide and oligosaccharide)" having one or more sialic acids (e.g., NANA) linked on the sugar chain abundantly expressed in neurons [454]. The name was first used in 1942 by the German Ernst Klenk (1896–1971) to refer to newly discovered lipid molecules from ganglion cells of the brain. In each disorder, a virtually identical course of neurodegeneration begins in infancy and leads to death generally by 4–6 years of age. The child becomes blind, deaf, and unable to swallow. Muscles begin to waste away, and paralysis sets in [452, 455].

In lysosomes of certain cells (e.g., macrophages), disorder in lipid metabolism may lead to accumulation of cholesterol, and this accumulation may subsequently spread to other regions of this cell, occupying a substantial portion of the volume of the cell, producing a foamy appearance in the cell. That is why such cell is referred to as foam cell [456]. Foam cells are integral in the pathophysiology of atherosclerosis, a progressive disease that leads to the disorder of vascular function predisposing the sufferer to the development of cardiovascular and cerebrovascular accidents, which is often due to rupture of atheromatous plaque [456]. Rupture of plague is frequent in regions with high level of foam cells. Though the etiopathogenesis of plaque is multifactorial, accumulation of cholesterol in macrophages is one cause of the plague development [456]. The diagnosis of lysosomal storage disease is done by detailed patient's history, clinical evaluation including investigation of some urinary metabolites with the aim of detecting specific enzymatic deficiency [447]. For further information on lysosomal storage disease, review Kaye [455], and Filocamo and Morrone [447].

Mitochondria

Mitochondrion (singular) is a complex oblong-shaped organelle that is found in the cytoplasm of all eukaryotic cells, referred to as the power-generating house of the cell (Fig. 3.18). Mitochondrion possesses its own genetic machinery (DNA) and is maternally inherited. Mitochondrial DNA (mtDNA) codes for mitochondrial ribosomal and messenger RNAs, and some of the mitochondrial proteins/enzymes/complexes [457, 458]. Human mtDNA encodes a total of 13 proteins, which are essential for oxidative process in this organelle. The mitochondrial ribosomes are the sites where these proteins are translated [459]. The remaining proteins of this organelle are nuclear gene products that are translated on cytoplasmic ribosomes and imported into the mitochondrion where they perform their functions [459].

This power generating organelle converts oxygen and nutrients into energy required for the continued activity of the cell [457, 460]. Mitochondrion is the major site of ATP synthesis in aerobic cells. ATP, the energy currency of the cell, is synthesized by oxidative phosphorylation requiring several enzymatic steps. Mitochondrion possesses an intraorganelle space (matrix and intermembrane space) containing the enzymes of the citrate cycle and beta-oxidation, enclosed by an inner membrane containing the four complexes of the electron transport chain, ATP synthase, and specific carriers for metabolites. The membranes of the mitochondria are relatively permeable to the transport of molecules. Thus, molecules synthesized in the mitochondrion can be released into the cytosol, however, in a regulated manner [457, 460–462].

Disorders of the mitochondrial functioning can result from mutations of mitochondrial DNA or nuclear genes that encode mitochondrial proteins [463]. These mutations are inherited on the maternal line. Most mutations of the mitochondrial DNA or nuclear genes that encode mitochondrial proteins are multisystemic, but some may occur in specific tissues [464]. Most mitochondrial diseases affect the proteins of the respiratory chain, which usually leads to reduced energy production. Diseases that involve mitochondrial dysfunctions include mitochondrial myopathy,

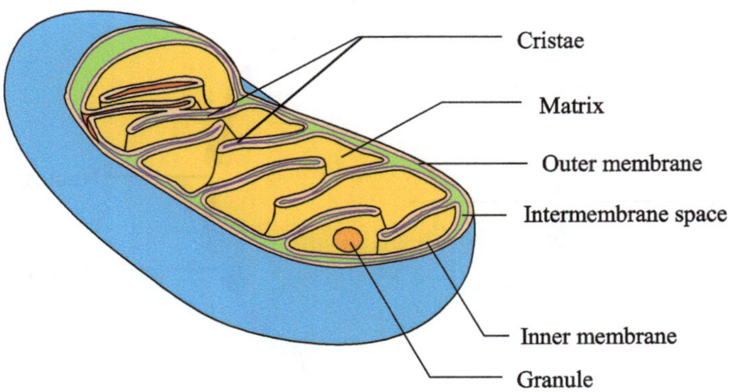

Fig. 3.18 Structure of the mitochondrion

lactic acidosis, some encephalopathies, myoclonus epilepsy, ragged red fibers, optic neuropathy, sensorineural deafness, and Kearns–Sayre syndrome—retinitis pigmentosa, progressive external ophthalmoplegia, ataxia, and heart conduction defects [462, 464].

Nucleus

The nucleus is a highly specialized dynamic membrane-bound organelle that harbors the genetic material of the cell and coordinates the activities of the cell [465, 466]. The nucleus is bounded by a nuclear membrane, which contains pores through which materials are exchanged between the cytoplasm and the nucleoplasm, inner environment of the nucleus (Fig. 3.19). The central part of the nucleus contains a dense material called the nucleolus (Fig. 3.19) [465–467]. The nucleus is important for intermediary metabolism, protein synthesis, RNA synthesis, and associated functions including growth and reproduction [465, 466]. It should be noted, however, that while RNA synthesis takes place in the nucleus itself, protein synthesis and metabolism, though controlled by the nucleus, occur in the cytoplasm.

The genetic materials of the nucleus generally called nucleic acids are deoxyribonucleic acid (DNA) and ribonucleic acid (RNA)—responsible for the transmission of genetic information from one generation to another, control of gene expression and protein synthesis (Fig. 3.19). The nucleus contains tightly packed chromosomes with histone proteins. The chromosomes are the short nuclear rodlike materials that harbor the DNA, often observed when the cell is dividing. In a nondividing cell, the DNA is found in the nuclear structure called chromatin, which is threadlike in appearance. During gene expression, the information present in DNA is transcribed into the RNA, which is then translocated into the cytoplasm. In the cytoplasm, the information present in RNA is finally translated into protein

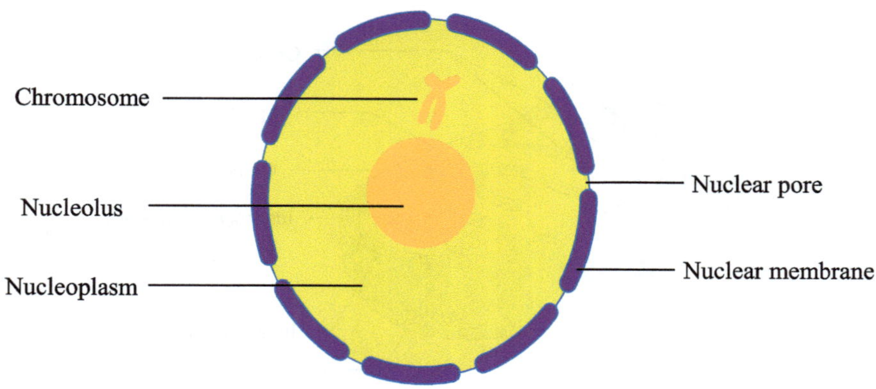

Fig. 3.19 Nucleus and its component parts

molecule. This process of information retrieval from the DNA to produce protein molecule with specific functions forms the cornerstone (central dogma) of molecular biology. Although it was known that the hereditary material resided in the DNA, precise structure and specifications of DNA were not exactly understood until around 1951–1953 when the British molecular biologist Francis Crick (1916–2004) and the American molecular biologist and geneticist James Watson (1928–) reported the results of their experiments on the structure of nucleic acids [468–471]. Crick and Watson together with the New Zealand-born British physicist and molecular biologist Maurice Wilkins (1916–2004) shared the 1962 Nobel Prize in Physiology or Medicine "for their discoveries concerning the molecular structure of nucleic acids and its significance for information transfer in living material." This led to the formulation of the "central dogma" of molecular biology, which states that "DNA makes RNA, RNA makes protein" [472]. But the discovery of reverse transcription (a process in which the enzyme reverse transcriptase transcribes RNA into DNA) questioned this agelong central dogma of molecular biology [473]. Over the past decades, accumulating research results from different laboratories in different parts of the world have added substantial information to the structure and functions of nucleic acids [474–480].

The DNA strands are wrapped around proteins called histones, which function to modulate the activities of the genetic materials (Fig. 3.18) [465, 466]. The RNA comprises three main types—messenger RNA (mRNA), ribosomal RNA (rRNA), and transfer RNA (tRNA). mRNA is produced from transcription of the polypeptide codes contained in a DNA and subsequently transported into the cytoplasm, where ribosomes bind to [481]. Thus, mRNA is used to convey genetic information that directs synthesis of specific proteins. rRNA is used to direct the assembly of proteins on ribosomes linking amino acids together to form proteins. tRNA sorts, delivers amino acids to the ribosome–mRNA complex, and binds to the mRNA. The attached amino acids are then linked together by another part of the ribosome. Upon formation of the protein, it folds to produce a functional three-dimensional structure. The ribosome–mRNA complex splits apart as soon as the ribosome finishes reading the information in the mRNA molecule [481–486]. Further details on the physiology of the ribosomes are discussed in the "subsection on Ribosomes" (see below). In addition to mRNA, tRNA, rRNA, physiologically important RNA includes noncoding RNA (ncRNA), small nuclear RNA (snRNA), microRNAs (miRNAs), long noncoding RNAs (lncRNAs). miRNAs are a class of small RNA molecules that are found within the eukaryotic nucleus, composed of 19–22 nucleotide sequences, and are involved in posttranscriptional regulation of gene expression in health and disease conditions. NcRNA is an RNA molecule that is not translated into a protein [487–490]. LncRNAs are composed of a minimum length of 200 nucleotides which are not translated into protein but involved in development and disease [491, 492]. These RNAs molecules found in the nucleus are integral in the regulation of the genome and epigenome [492] (Fig. 3.20).

The nucleolus is the site of ribosome synthesis. It is the nuclear domain where ribosomal RNAs and ribosomal proteins are synthesized, processed, and assembled for export [493].

The nuclear membrane: The nuclear membrane (also called nuclear envelope or nucleolemma or karyotheca) is the structure that surrounds the nucleus, separating its contents from the cytoplasm and the nucleoplasm. The nuclear membrane is a lipid bilayer with embedded proteins, which is composed of inner nuclear membrane and the outer nuclear membrane [494–497]. Between the two membranes is a space referred to as the perinuclear luminal space. This space is connected with ER lumen [271, 299]. The inner membrane is lined by a meshwork containing a polymer of the laminin proteins, called the nuclear lamina with

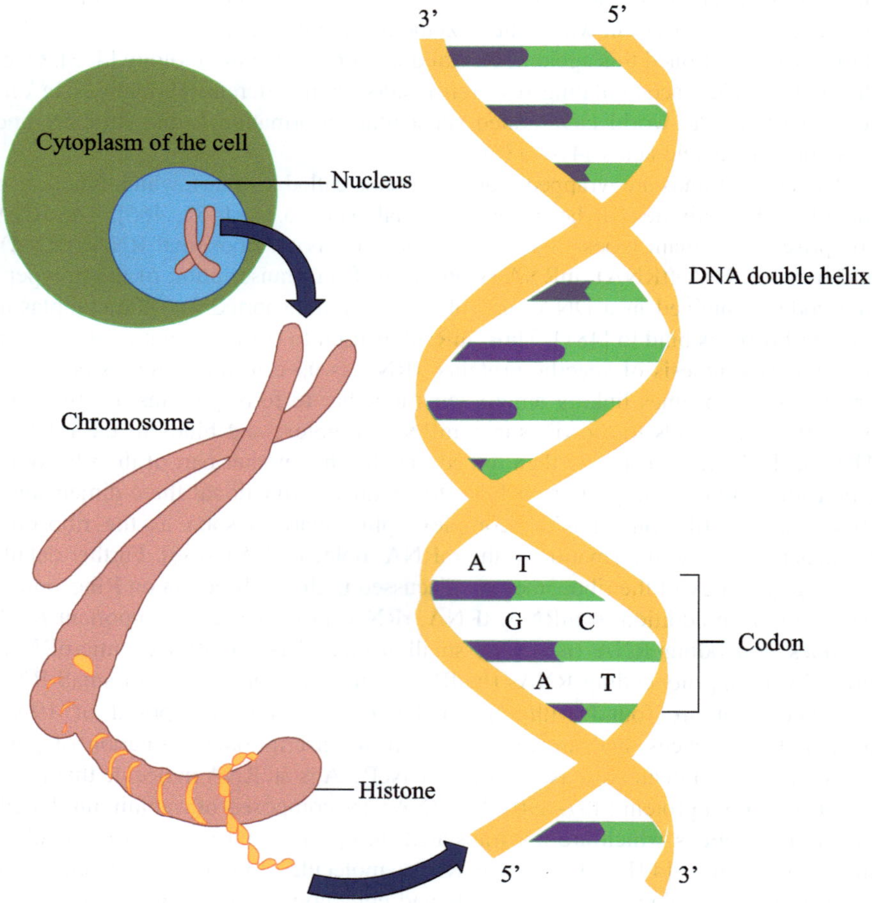

◂**Fig. 3.20** Genetic material of the cell. The codon (three nitrogenous bases) is the genetic information that represents a specific amino acid [163, 168]. The nitrogenous bases in DNA are cytosine (C), guanine (G), adenine (A), and thymine (T). The DNA strands are composed of monomer units called nucleotides. Each nucleotide in DNA is composed of a five-carbon sugar (deoxyribose), nitrogenous base and a phosphate group. Without the phosphate, the structure is called deoxyribonucleoside (deoxyribose joined only to the nitrogenous base). An example of deoxyribonucleoside is deoxyadenosine. The addition of the phosphate to it makes it a nucleotide. In the case of deoxyadenosine, the structure will be called deoxyadenosine monophosphate. The nucleotides are joined to one another in a chain by covalent bonds between the sugar of one nucleotide and the phosphate of the next, resulting in an alternating sugar phosphate backbone. One strand is linked to the other by hydrogen bonds according to base pairing rules (A with T, and C with G) to make double-stranded DNA—the predominant form of DNA [629–631]. One strand of the DNA runs in the direction $3' \rightarrow 5'$, while the other strand runs in the direction $5' \rightarrow 3'$. The DNA strands are synthesized in a $5' \rightarrow 3'$ direction. In DNA, 3′ (three prime) refers to the third carbon of the deoxyribose which is linked to a hydroxyl group and 5′ (five prime) which is linked to a triple phosphate group. There is functional implication of the structure of DNA molecules. The double helix confers a substantial level of protection against physical and chemical damage and thus conserves the information coding in the cell. It should be noted that the DNA is able to adopt non-canonical structures, which include multistranded DNA helices through folding of one of the two strands or association of two, three, or four strands [630, 632–637]. Three-stranded or triple helical DNA is called triplex DNA—a complex of three oligonucleotide strands oriented in the parallel and antiparallel direction [633]. The basic pairing rules in the triplet helix are as follows: T–C, G–A, and G–T. The G–T can bind in both orientations to the duplex, while T–C binds parallel and G–A binds antiparallel to the purine [638]. Four-stranded DNA is called quadruplex [632]. A segment or locus of the DNA which is composed of nucleotides that provides the instructions for synthesis of RNA, which is subsequently translated into a functional protein, is referred to as a gene—the basic functional unit of heredity. Gerstein et al. [639] defined the gene as "a union of genomic sequences encoding a coherent set of potentially overlapping functional products." Genes are composed of DNA. Genes are found in chromosome. The nuclear genome of human cells contains 3×10^9 base pairs that comprise 30,000–40,000 genes [459]. RNA consists of nitrogenous bases, a ribose sugar and a phosphate group. In RNA molecule, the bases are C, G, A, and U which are bound together by hydrogen bonds according to the pairing rule. Adenine and guanine are purines; cytosine and uracil are pyrimidines [484–486]. RNA is usually single stranded and runs in the direction $5' \rightarrow 3'$. In RNA, 3′ refers to the third carbon of the ribose which is linked to a hydroxyl group and 5′ which is linked to a triple phosphate group. Like DNA, RNA is able to adopt non-canonical structures, which include multistranded, double-stranded helices [637, 640]. The different types of RNA have been implicated in a wide variety of physiological and pathological functions [637, 640]

associated nuclear integral proteins and chromatin as well as the nuclear pore complexes [271, 498]. The laminins are members of the intermediate filament family of proteins that form an interwoven mesh at the nuclear periphery underlying the nuclear membrane [499–501]. These proteins are similar to their counterpart cytoplasmic intermediate filament [501] and constitute the major structural proteins of the nucleus. The main constituents of the lamina are the type V intermediate filament proteins [498]. There are four major types of laminins in a mammalian nucleus: laminins A, C, B1, and B2 [271, 495, 502, 503]. The laminins together with actin form the main constituent parts of the nucleoskeleton [504]. The lamina is involved in maintenance of nuclear structure, providing a platform for the binding of structural and functional proteins, protecting the genome from mechanical stress, and regulation of gene expression [271, 495, 498, 504, 505].

The membrane of the nucleus is made up of tiny gaps or membrane perforations referred to as nuclear pore complexes that allow material to move in and out of the nucleus [299]. Nuclear pore complexes are structures with octagonal rotational symmetry. These structures are ~90 nm in length and are ~50–200 nm in diameter [506]. The estimated mass of the nuclear pore complexes in an animal cell is ~60–120 MDa formed by more than 30–50 different types of nuclear pore proteins—nucleoporins [165]. The nuclear pores are highly efficient at selectively allowing the passage of materials to and from the nucleus [499, 500]. There are estimated 3000–4000 pore complexes in the nuclear envelope of a typical animal cell [165, 507]. A nuclear pore complex can transport about 5 to several millions of molecules to and out of the nucleoplasm per second [165, 508]. Relatively small molecules usually less than 20–40 kD are transported through the pore complex by passive diffusion or signal-independent transport [506, 508]. Larger molecules require transport receptor (importin/exportin proteins) for transit through the pore complex by a process called carrier-mediated transport (e.g., facilitated translocation or signal-dependent transport) [506, 508].

Several abnormalities of the nucleus have been implicated in diseases [501]. Some neurodegenerative diseases and striated muscular diseases can result from disorders in nuclear functions [509, 510]. For instance, mutations in gene that codes for laminins result in a group of diseases, called laminopathies [503]. Laminopathies can involve not only striated muscle, but also adipocytes and peripheral nerves. Such diseases include lipoatrophy, muscular dystrophies including cardiomyopathy [510, 511]. Mutation in laminin A gene is associated with the development of Hutchinson–Gilford progeria syndrome, a disease characterized by premature cellular senescence in which children die of severe atherosclerosis at an average age of 13 years [512]. The initial physical signs include severe failure to thrive, complete hair loss, hypopigmentation, hard skin, severe lipoatrophy, bony abnormalities, small/beaked nose, and receding mandible. Children with Hutchinson–Gilford progeria syndrome are seen as physically being many decades older that they really are. Progression of the disease can lead to increase rate of formation of plagues in blood vessels, consequently leading to cardiac failure and stroke [495].

Ribosomes

Ribosomes are small granular organelles that constitute the macromolecular machinery for translating the genetic code into functional proteins; they function as sites of protein synthesis [473, 513, 514]. Ribosomes are complexes composed of about 60% RNA (rRNA) and 40% protein [514, 515]. Ribosomes are sometimes called ribozymes because the catalytic peptidyl transferase (aminoacyltransferase) activity that links amino acids together is performed by the large subunit of the rRNA [165, 516]. All living cells contain ribosomes. In eukaryotic cell, ribosomes are made of four strands of RNA, whereas prokaryotic ribosome consists of three strands of RNA [513, 517, 518]. In mammals, there are two main types of ribosomes—cytoplasmic ribosomes (cytoribosomes) and mitochondrial ribosomes (mitoribosomes). Cytoribosomes are embedded in the intercellular membranes of the rough ER. Mitoribosomes are produced by the mitochondrial DNA, are located in the matrix, and synthesize majority of the proteins found in the inner mitochondrial membrane. Like the cytoplasmic ribosomes, mitoribosomes also consist of small and large subunits [459, 519, 520].

Ribosomes of animal cells are the 80S ribosomes composed of two subunits. The small subunit has a sedimentation value of 40S, whereas that of the large subunit is 60S (Fig. 3.21). The large subunit is composed of varying lengths of different RNA molecules with about 49 different types of proteins. The small subunit has a fewer RNA with much smaller length of RNA chain and about 33 proteins [521, 522]. The small subunit of the ribosome reads the RNA, while the large subunit links the amino acid to form a polypeptide chain in the order specified by mRNA [513, 517, 518]. Each ribosomal subunit is composed of one or more ribosomal RNA (rRNA) molecules and a variety of proteins [521].

In addition to their involvement in protein synthesis, ribosomes (their protein components) are crucial in DNA repair and programmed cell death [521].

Disorders of ribosomal (ribosomal protein and RNA) functioning resulting from mutations in ribosomal genes have been associated with a variety of hematologic disorders including congenital anemias (Diamond–Blackfan anemia and

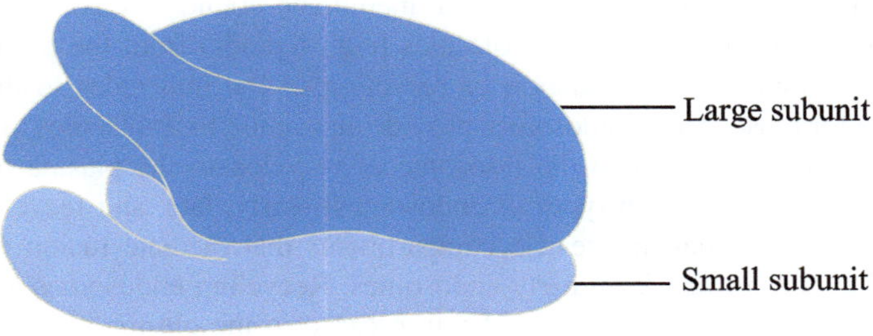

Fig. 3.21 Subunits of the ribosome of an animal cell. The ribosome in animal cell is the 80S, which is composed of two subunits—large (60S) and small (40S) subunits. The eukaryotic ribosome is much larger than the prokaryotic one [513]

Shwachman–Diamond syndrome). Some cancers of the liver, colon, prostate, etc., are characterized by overexpression of specific ribosomal proteins suggesting their interactions with the p53 tumor suppressor pathway and some oncogenes [521]. Some ribosomal diseases are due to disorder of biogenesis of ribosomes [523].

Intracellular Vesicles

Vesicles are small membrane-enclosed transport units that can transfer molecules between different cellular compartments. Most vesicles transfer the cargo components assembled in the ER to the Golgi apparatus, then from the Golgi apparatus to various destinations. A vesicle budding off a membrane contains specific signal (amino acid sequence) on its cytosolic surface that directs its destination. Once the vesicle finds the membrane, they fuse, initiating a series of transformation/reaction processes. Types of vesicles include clathrin-coated, COP-I-coated, and COPII-coated vesicles. Each vesicle performs different functions in the cell. For example, clathrin-coated vesicles transport substances between the Golgi complex and the plasma membrane. COPI- and COPII-coated vesicles are often used for the transport of cargoes between the ER and the Golgi complex [438–440]. Some vesicles are endocytotic (vesicles formed by endocytosis), whereas others are exocytotic (vesicles formed by exocytosis) [524, 525]. Some vesicles are composed of nutrients being transported to the respective sites. For instance, in the enterocyte, just after feeding, the formation of vesicles increases in the cytoplasm of the intestinal cell [526].

The GI tract epithelial cells such as goblet cells contain a cytoplasm with dense mucin granules, which are continuously transported to the apical side of the plasma membrane for exocytosis to replenish the mucus layer of the GI epithelium. The cytoplasmic vesicles are constantly replenished by regulated signaling [527].

Some intracellular vesicles are relatively large and are formed by autophagy. Autophagic vacuoles contain different kinds of materials including hydrolytic enzymes (e.g., acid phosphatase). These relatively large vesicles can originate from the lysosomes [528]. Another large vesicle, which represents the endocytic membrane transport pathway originating from the membrane of the trans-Golgi cisterna, is referred to as endosome. This is a membrane-bound compartment that is responsible for internalization of molecules (e.g., ligands) from the plasma membrane to lysosomes for degradation or recycling of the molecules to the plasma membrane for reuse. Thus, endosomes provide an avenue for molecules to be sorted before they are either sent to the lysosome or recycled to the plasma membrane [529]. There are three main types of endosomes—early, late, and recycling endosomes. Late endosomes are vesicle compartments that mediate fusion with lysosomes. They are formed from early endosomes. Recycling endosomes are distinct compartments of tubular vesicles that return cargo to the plasma membrane; they act as machinery for reuse of cell components (e.g., protein subunits) to build receptors or replenish worn-out regions of the plasma membrane. However, in some cells, materials can be recycled to the plasma membrane by peripheral early endosomes [530–532]. The movement of materials through the endosome to the plasma membrane for fusion is controlled by coat proteins, tethering factors, Rabs, and SNARE proteins. These proteins determine the destination of the vesicle or cargo [533]. The cytoskeleton plays an essential role in this process. In spatially and temporally regulated manner, microtubule facilitates sorting of vesicles, vesicle fusion and fission, delivery to lysosomes or peroxisomes, cytosolic dispersal, and nuclear uptake [534]. Similar mechanisms are involved in the motor protein-mediated transport of internalized extracellular material [534, 535].

Cytoskeleton

The cytoskeleton is described as a dynamic interconnected network of filamentous polymers, tubules, and regulatory proteins, ubiquitously located throughout the cell, extending throughout the cytosol, having both structural and functional implications on the cells and tissues of the body [536, 537]. Cytoskeletal monomers have the ability to self-assemble into linear polymers which are organized with cytoskeletal-associated proteins forming higher order structures [538]. The cytoskeleton plays an important role in cellular transport, cell growth and division, cell differentiation, signal transduction, gene transcription, cell and organelle motility, cell shape maintenance, innate immunity, and cellular self-defense [536, 537, 539, 540]. For the purpose of emphasis, in a couple of cells, the cytoskeletal proteins comprise about 80% or more of the cellular proteome, indicating the huge role the cytoskeleton plays in these cells [541]. The three widely recognized principal components of the cytoskeleton are microfilaments (also called actin filaments), microtubules, and intermediate filaments [538, 539, 542]. Accumulating evidences indicate that a fourth component of cytoskeleton is septin filament [539] (Fig. 3.22).

Intermediate filaments: These are a class of fibrous proteins that play a crucial role as structural elements of the cytoskeleton that confer mechanical strength to the cell. Intermediate filaments form highly viscoelastic, nonpolar, smooth, and flexible filament networks functioning as tension-bearing elements, which help to maintain

Fig. 3.22 Principal components of the cytoskeleton. To visualize the cytoskeleton, a video is available at Martys et al. [641]

cell shape [542–545]. Intermediate filaments have a diameter of 8–12 nm (average of 10 nm), which is intermediate between the diameters of the microfilaments (5–8 nm, usually about 7 nm) and microtubules (23–25 nm, usually about 25 nm) [163, 543, 544]. There are two categories of intermediate filaments—cytoplasmic and nuclear intermediate filaments [538, 539]. (Nuclear intermediate filaments have been discussed in the previous subsection "Nucleus.") It is important to note that nuclear intermediate filaments are coupled to the cytoplasmic intermediate filaments through peptide linkages. One of the proteins linking these two categories of intermediate filaments is called plectin. Cytoplasmic intermediate filaments are generally divided into four types I–IV (recall that type V intermediate filament is found in the nucleus) [546]. During cell division, the nuclear intermediate filaments are disassembled through the action of mitotic kinases, whereas the cytoplasmic intermediate filaments are depolymerized. These changes occur only in some cells [542]. The major protein of the cytoplasmic intermediate filaments is keratin [546–548]. The mammalian intestinal epithelium, for instance, in region-dependent manner, differentially, expresses type I keratin (K18, K19, and K20) and type II keratin (K8). These are the major intermediate filament cytoskeletal proteins found in the enterocytes and other epithelial cells of the gut [549]. The intermediate filament is produced by heterodimerization reaction of types I and II keratins—to form a parallel heterodimer of keratin I and II. The formed parallel heterodimer is joined together by hydrophobic forces. The parallel heterodimer interacts with antiparallel heterodimer to form a tetramer. It is this tetramer that forms the building unit of intermediate filaments. Altogether, eight tetramers are required, via lateral interactions, to form a unit length filament, which then attach longitudinally to form the intermediate filament [550]. Depending on the cell type, however, other proteins expressed in intermediate filaments may include vimentin (mesenchymal, endothelial cells), desmin, synemin, and syncoilin (in myoblasts), peripherin, alpha-internexin, NF-L (neurofilament-light with a molecular weight of ~61–70 kDa), NF-M (neurofilament-medium with a molecular weight of ~95–150 kDa), NF-H (neurofilament-heavy with a molecular weight of ~110–210 kDa) (neurons) [542, 546, 551]. Intermediate filament proteins interact with structures that link neighboring cells together (desmosomes, extracellular matrix structures) by way of adapter proteins [541, 552, 553]. Thus, intermediate filaments anchor the cell to the extracellular matrix as well as its neighbors to maintain the integrity of not only the cells themselves but also the tissues of the body [541].

Microfilaments: These components of the cytoskeleton are solid polar rods made up of globular proteins called actin, formed by polymerization of actin monomers. Microfilament is composed of two intertwined actin strands. These filaments have a wide range of functions, but they are primarily structural in function and are an important component of the cytoskeleton [554–556]. Actin can exist as either filamentous actin (F-actin) or globular actin (G-actin) in cells. G-actin polymerizes to form F-actin, a process that is controlled by several factors including the activities of actin-binding proteins (see below). About 80% of total actin proportion in smooth muscle cells, for instance, exists as F-actin. The percentage of F-actin in muscle cells can increase above 90% upon stimulation by neurotransmitters or hormones [557]. Several proteins are associated with actin referred to as actin-binding proteins, and they include cofilin, profilin, gelsolin, thymosin beta4, DNase I, CapZ, tropomodulin, and Arp2/3). They play a variety of roles in maintaining the structure and functions of actin filaments [558]. Like other cytoskeletal proteins, microfilament proteins can connect to the plasma membrane by way of junctional proteins (such as vinculin) [559].

Microtubules: These components are dynamic polar hollow cylinders and carry out a variety of functions, ranging from transport to structural support structures that undergo continual assembly and disassembly within the cell. Their functions include maintenance of cell shape, cell movement, intracellular transport of organelles and cargoes, and the separation of chromosomes during cell division [560, 561]. Tubulin is a globular protein made up of α- and β-tubulin polypeptides [541]. Thus, tubulin is a dimer protein molecule. (However, a third type of tubulin (γ-tubulin) exists, but this polypeptide is localized to the centrosome, where it is involved in the initiation of microtubule assembly.) Tubulin dimers polymerize to form microtubules, which generally consist of 13 linear protofilaments (or columns of tubulin heterodimer backbone) assembled around a hollow core. This microtubule backbone is linked with microtubule-associated proteins. The protofilaments are arranged in parallel. Like actin, tubulin not only polymerizes but also depolymerizes to disassemble. This reaction is regulated by guanine triphosphate, GTP. Thus, like actin filaments, microtubule is dynamically instable, which allows for turnover of microtubules [560, 562].

Septin filaments: These components of the cytoskeleton are rodlike filament formed by the polymerization of the GTP-binding proteins, septins [563]. There are different types of septins which include septin-2, septin-4, septin-7, septin-9, septin-14 [564]. Septin-9 and septin-14 are expressed in the stomach and large intestine [565]. How septins polymerize to form filaments and rings are not completely understood and research is ongoing to understand the mechanism of septin formation [566]. Septins are known to exhibit a range of functions which include maintenance of cell structure and shape, cell division, cellular transport, and signaling [564].

The cytoskeleton is involved in a variety of diseases affecting humans [548, 567]. Over 100 different types of diseases have been associated with the cytoskeleton [568, 569]. Mutations of the cytoskeletal proteins can lead to several diseases of the cardiovascular system, nervous system, GI system, integumentary system, and pulmonary system. A wide range of cancers affect the cytoskeleton. The cytoskeleton can be affected by the body's own immune system by recognizing the cytoskeletal proteins as foreign [541]. For example, research indicates that an immune response against actin filaments as well as extracellular matrix structures have a role to play in the etiopathogenesis of celiac disease, an autoimmune disease characterized by sensitivity of the lining of the small intestine to gluten-containing food [570]. Mutations in the epithelial keratin K8 were identified in patients with inflammatory bowel disease (Crohn's disease or ulcerative colitis) [571]. Abnormal expression of K7 and K20 has been noted in colitis-associated dysplasia and cancers. Low expression of K20 is found in sporadic colorectal cancers [547].

Mutations of the cytokeratins of the GI epithelium have also been implicated in liver cirrhosis and pancreatitis [571]. Some diseases of the nervous system (e.g., Charcot–Marie–Tooth disease and amyotrophic lateral sclerosis, frontotemporal dementia with Parkinson's disease, Alzheimer's disease, and neuropsychiatric systemic lupus erythematosus) have been linked to some mutations in the intermediate filament cytoskeletal proteins [567]. For example, the hallmark of Alzheimer's disease is the expression, phosphorylation, and aggregation of tau, which is a microtubule-associated protein [560].

Mutations in cytoskeletal proteins have also been associated with some types of cancer. Septin mutations, for example, have been implicated in tumors of the skin, endometrium, stomach, and large intestine [565]. To this end, cytoskeletal depolymerizing compounds (anticytoskeletal or microtubule-targeted antimitotic drugs) are successfully used in cancer chemotherapy [479]. These drugs also inhibit cell division, thereby reducing the proliferation of cancer cells [541]. The drug, cytochalasin B, causes disruption of the microfilaments. Colchicine is another drug used to disrupt microtubules [572]. They are termed microtubule destabilizing drugs as they inhibit microtubule polymerization. The other group of anticytoskeletal drugs is called microtubule-stabilizing drugs because they stimulate microtubule polymerization. Examples of microtubule-stabilizing drugs include paclitaxel, docetaxel, and ixabepilone. There are several drugs used experimentally on animals and applied in humans for addressing various maladies of mankind related to cytoskeletal disorders [573–577].

Centrosomes/Centrioles

The centrosome (from the Latin "centrum" meaning "center" and Greek "soma" meaning "body") is a dense area of the cytoplasm located near the nucleus. It is a membraneless organelle, the core of which consists of a pair of orthogonally arranged (perpendicularly oriented) barrel-shaped cylindrical microtubule-based structures called centrioles, surrounded in a network of proteins called pericentriolar material [578–581]. The microtubules are anchored to the centrosome by their minus-ends [579]. The pericentriolar material contains proteins such as γ-tubulin, pericentrin, dynactin, and ninein, which are responsible for microtubule nucleation, organization, and anchoring [582, 583].

The centrosome is the microtubule organizing center (MTOC) of the cell, aids to organize the microtubules that form the mitotic spindle in dividing cells, and is also involved in a range of cell processes including cell motility, cell cycle, signaling, adhesion, trafficking of biomolecules, and polarity [578, 584]. As the MTOC of the cell, the organelle functions as the initiation site for microtubule assembly (microtubule-nucleating function), which grows outward from the centrosome to the periphery of the cell. This microtubule nucleation is carried out by γ-tubulin molecules of the pericentriolar material. The outward growth of microtubules takes place by the progressive addition of tubulin to the plus end [578–580, 585]. The centrosome regulates cell cycle arrest and repair in response to stress [586]. This organelle plays a crucial role in the regulation of the cell cycle, a highly ordered series of events (phases) required for the duplication of one cell into two daughter cells—DNA replication or synthesis (S), and mitosis (M). Cells begin the cell cycle with one centrosome per cell. In interphase (the stage in the life cycle of a cell between two successive cell divisions), the centrosome is located near the nucleus and microtubules extend to the cell periphery of the cell [578, 581]. The centrosome replicates during the S phase (is a priori generally agreed on the starting point in a cell cycle) of the interphase to serve as the spindle poles during mitosis. The mitotic spindle forms between the two centrosomes. During prophase (the first phase of cell division), centrosomes migrate to opposite poles of the cell. In mitosis, the nuclear membrane breaks down and the nucleated microtubules interact with the chromosomes to build the mitotic spindle. Upon division, each daughter cell receives one centrosome [168, 579, 587].

The expression of centrosomes is dependent on many factors including age. The number of centrosome is greater in differentiating small intestinal crypt cells of embryos and newborn compared with adult [588]. Similarly, the centrosome of young hepatocytes is more active than a well-differentiated adult hepatocyte [589]. The regulation of the centrosome during cell cycle is controlled by mitosis-specific kinases [579]. Correspondingly, the intestinal cells of a younger mammal have higher potential for replication required to replace the continuously shedding crypt cells [590, 591].

Centrosome dysfunctions have been implicated in numerous diseases including neurodegenerative diseases, cancer, infertility, and developmental disorders [581, 585]. Malfunctions of centrosomal proteins or their overexpression can lead to the development of cancer. For example, the mitosis-specific kinases called aurora kinases regulate the activity of centrosome during the cell cycle. The heightened activity of these kinases can lead to amplification of centrosomal proteins and cellular transformation, possibly resulting in the development of cancer or other diseases [578]. Relatively, recently, mutations in centrosomal proteins have been linked to microcephaly and dwarfism. This evidently highlights the role centrosomal proteins play in development [592].

Centrioles: These are a pair of self-replicating barrel-shaped cylindrical organelles located within the centrosome that serves as the center of chromosome movement during cell division [580]. Each centriole consists of nine triplets of microtubules assembled in a spoke-like arrangement at the end of the centriole referred to as cartwheel [578, 580, 581, 586]. Centrioles are essential for the organization of centrosomes [592]. Although centrioles are required, they are not exclusively inevitable, for the formation of centrosome. This is because destruction of centrioles from cells does inhibit the formation of bipolar spindles, suggesting that the cell uses an alternative self-organization pathway to carry out cell division [580].

Centrioles are essential for the formation of cilia and flagella [592]. A centriole that forms a cilium is called a basal body. Centrioles form the basal bodies required for the formation of cilia and flagella [580, 592]. Centrioles can perform multiple functions on behalf of cilia. The microtubule doublets of cilia grow as a direct outgrowth of the microtubule triplets of the centriole [580].

Disorder in the expression of the centrosome can lead to instability of the genome and promotion of tumorigenesis [592].

3.5 Polarity of the Epithelial Cell

One feature that characterizes many cells including epithelial cell as well as all motile cells is polarity. This means that certain part of the cell or structures is always found at the frontal part whereas others are localized to the rear side [168, 593]. Polarity is a ubiquitous property of epithelial cells that characterize their apical (luminal) and basolateral locations and ensures regulated absorption, transport, secretion, and maintained morphogensis [593, 594]. This property of epithelial cells is crucial for regulated functioning of the gut, liver, and exocrine glands as well as other extragut organs including the kidney and genital organs [594]. In the intestinal epithelium, for example, the apical membrane of the intestinal epithelial cells always faces inward to the lumen, whereas the basolateral membrane (the surface of the plasma membrane that forms its basal and lateral surfaces) is located to the opposite side of the cell, facing the interstitium. Although usually called the basolateral membrane, there is no consensus on whether or not this side of the membrane should be called basal and the lateral membranes. In epithelial cells, the two membranes are identical and have similar functional roles. Substances can move between the two closely related membranes. For instance, there is possibility of migration of phospholipids and proteins from the lateral to basal side and vice versa. However, such movements are restricted between the apical and basolateral membranes due to the presence of junctional complexes which will be discussed in later part of this book. The localization and maintenance of the membrane proteins which determine in part the polarity of plasma membrane of the epithelial cell are due to the sharp structural and physiological asymmetry of epithelial cells. Specific pores, channels, carriers, pumps, channels, transporters are peculiar to the apical pole of the epithelia. In the basolateral membrane, the structural types and functions of such pores, channels, carriers, and pumps are essentially different. This property of the epithelia is crucial for its vectorial functions in which the amount of a given substance that crosses in one direction either from the lumen to blood or blood to lumen is usually orders of magnitude larger than in the opposite direction [593, 595, 596]. Thus, disorders in the localization of proteins specific to the polar side of the epithelia could result in serious consequences. In the intestine, such dysfunctions have been associated with cancer and some chronic illnesses. The localization of proteins to the plasma membrane is controlled by several factors of which clathrin adaptor protein complex is an important one [597, 598]. One protein known to determine polarity of the epithelia is band 3. This protein may be apical at certain locations of the cell, whereas at other locations, it may be localized to the basolateral side of the membrane [599]. Cell polarization involves many signaling pathways, but the cytoskeleton appears to play a central role in polarization. The components of the cytoskeleton, microtubules, and microfilaments are oriented along the apicobasal axis of the cells in which the minus-end faces the apical membrane and the plus end is directed to the basal side [594, 600, 601]. The molecular mechanisms of orientation of microtubules, in part, are believed to be related to the calmodulin-regulated-spectrin-associated protein 3, a microtubule minus-end binding protein, which plays an integral role in orienting the apical-to-basal polarity of microtubules in epithelial cells [600].

3.6 Conclusion

The GI tract is characterized by a wide range of cell and tissue diversity that determine the varying functions of the gut and the accessory as well as associated organs. The information discussed in this chapter provides a fundamental and contemporary understanding of the cellular basis of GI functioning.

Recommended readings

Original articles

Brügger B, Erben G, Sandhoff R, Wieland FT, Lehmann WD (1997) Quantitative analysis of biological membrane lipids at the low picomole level by nano-electrospray ionization tandem mass spectrometry. Proc Natl Acad Sci USA 94(6):2339–2344

Review articles

1. Almén MS, Nordström KJV, Fredriksson R, Schiöth HB (2009) Mapping the human membrane proteome: a majority of the human membrane proteins can be classified according to function and evolutionary origin. BMC Biol 7:50
2. Catalá A (2012) Lipid peroxidation modifies the picture of membranes from the "Fluid Mosaic Model" to the "Lipid Whisker Model". Biochimie 94(1):101–109
3. Escribá PV, González-Ros JM, Goñi FM, Kinnunen PKJ, Vigh L, Sánchez-Magraner L et al (2008) Membranes: a meeting point for lipids, proteins and therapies. J Cell Mol Med 12(3):829–875
4. Fahy E, Cotter D, Sud M, Subramaniam S (2011) Lipid classification, structures and tools. Biochim Biophys Acta 1811(11):637–647
5. Fahy E, Subramaniam S, Murphy RC, Nishijima M, Raetz CR, Shimizu T et al (2009) Update of the LIPID MAPS comprehensive classification system for lipids. J Lipid Res 50(Suppl):S9–S14
6. Fahy E, Sud M, Cotter D, Subramaniam S (2007) LIPID MAPS online tools for lipid research. Nucleic Acids Res 35:W606–W612

7. Greenberg ME, Li XM, Gugiu BG, Gu X, Qin J, Salomon RG et al (2008) The lipid whisker model of the structure of oxidized cell membranes. J Biol Chem 283:2385–2396
8. Ray S, Kassan A, Busija AR, Rangamani P, Patel HH (2016) The plasma membrane as a capacitor for energy and metabolism. Am J Physiol Cell Physiol 310(3): C181–C192
9. Sahoo S, Aurich MK, Jonsson JJ, Thiele I (2014) Membrane transporters in a human genome-scale metabolic knowledgebase and their implications for disease. Front Physiol 5:91
10. Tabaei SR, Jackman JA, Kim M, Yorulmaz S, Vafaei S, Cho NJ (2015) Biomembrane fabrication by the solvent-assisted lipid bilayer (SALB) method. J Vis Exp (106)
11. van Meer G, Voelker DR, Feigenson GW (2008) Membrane lipids: where they are and how they behave. Nat Rev Mol Cell Biol 9(2):112–124
12. Vyšniauskas A, Qurashi M, Kuimova MK (2016) A molecular rotor that measures dynamic changes of lipid bilayer viscosity caused by oxidative stress. Chemistry 22(37):13210–13217
13. Zhou EH, Martinez FD, Fredberg JJ (2013) Cell rheology—mush rather than machine. Nat Mater 12(3). https://doi.org/10.1038/nmat3574

Books

1. Alberts B, Johnson A, Lewis J, Raff M, Roberts K, Walter P (2002) Molecular biology of the cell, 4th edn. Garland Science, New York, USA
2. Berg JM, Tymoczko JL, Stryer L (2002) Biochemistry, 5th edn. WH Freeman, New York, USA
3. Cooper GM (2000) The cell: a molecular approach, 2nd edn. Sinauer Associates, Sunderland, MA
4. Lodish H, Berk A, Zipursky SL, Matsudaira P, Baltimore D, Darnell J (2000) Molecular cell biology, 4th edn. WH Freeman, New York

Methods and Protocols in Cellular Biology

1. Baggerman G, Husson S, Wegener C, Verhaert P (2014) Peptidomics state-of-the-art. EuPA Open Proteomics 3(C):76–77
2. Ding S (ed) (2010) Cellular programming and reprogramming. Springer, UK, USA
3. Farwanah H, Kolter T (2012) Lipidomics of glycosphingolipids. Metabolites 2 (1):134–164
4. Fenn LS, McLean JA (2013) Structural separations by ion mobility-MS for glycomics and glycoproteomics. In: Kohler JJ, Patrie SM (eds) Mass spectrometry of glycoproteins. Springer, NY, USA
5. Griffiths WJ, Wang Y (2009) The importance of steroidomics in the study of neurodegenerative disease and ageing. Comb Chem High Throughput Screen 12 (2):212–228

6. Jamal M (ed) (2017) The CRISPR/Cas system: emerging technology and application. Caister Academic Press
7. Lindström S, Andersson-Svahn H (eds) (2012) Single-cell analysis. Springer, NY, USA
8. Verma M, Barh D, de Carvalho VAA, Hassan SS (2014) Molecular biomarkers: overview, technologies, and strategies. In: Rapley R, Whitehouse D (eds) Molecular biology and biotechnology, 6th edn. Royal Society of Chemistry, UK
9. Zhang X, Li L, Wei D, Yap Y (2007) Moving cancer diagnostics from bench to bedside. Trends Biotechnol 25(4):166–173

Bibliography

1. Langkamp-Henken B, Glezer JA, Kudsk KA (1992) Immunologic structure and function of the gastrointestinal tract. Nutr Clin Pract 7(3):100–108
2. Vigneulle RM, Rao S, Fasano A, MacVittie TJ (2002) Structural and functional alterations of the gastrointestinal tract following radiation-induced injury in the rhesus monkey. Dig Dis Sci 47(7):1480–1491
3. Inoue H, Kudo S, Shiokawa A (2005) Technology insight: laser-scanning confocal microscopy and endocytoscopy for cellular observation of the gastrointestinal tract. Nat Rev Gastroenterol Hepatol 2:31–37
4. Pacifici R, Zuccaro P, Farré M, Pichini S, Di Carlo S, Roset PN et al (2002) Cell-mediated immune response in MDMA users after repeated dose administration: studies in controlled versus noncontrolled settings. Ann N Y Acad Sci 965:421–433
5. Ferguson CM (1990) An overview of the gastrointestinal system. In: Walker HK, Hall WD, Hurst JW (eds) Clinical methods: the history, physical, and laboratory examinations, 3rd edn. Butterworths, Boston
6. Bernhardt HS, Tate WP (2012) Primordial soup or vinaigrette: did the RNA world evolve at acidic pH? Biol Direct 7:4
7. McNichol J (2008) Primordial soup, fool's gold, and spontaneous generation. Biochem Mol Biol Educ 36(4):255–261
8. Forterre P, Gribaldo S (2007) The origin of modern terrestrial life. HFSP J 1(3):156–168
9. Pross A, Pascal R (2013) The origin of life: what we know, what we can know and what we will never know. Open Biol 3(3):120190
10. Wollman AJ, Nudd R, Hedlund EG, Leake MC (2015) From animaculum to single molecules: 300 years of the light microscope. Open Biol 5(4):150019
11. Koertge N (ed) (2007) New dictionary of scientific biography (Dictionary of scientific biography (8 vols, 1st ed). Charles Scribners & Sons, Detroit
12. Encyclopædia Britannica (2003) Encyclopædia Britannica, Inc, Chicago
13. Bourgoin SM (2012) Encyclopedia of world biography, 2nd edn. Gale Research, Detroit
14. Gest H (2004) The discovery of microorganisms by Robert Hooke and Antoni Van Leeuwenhoek, fellows of the Royal Society. Not Rec Royal Soc Lond 58(2):187–201
15. Sepel LMN, Loreto ELS, Rocha JBT (2009) Using a replica of Leeuwenhoek's microscope to teach the history of science and to motivate students to discover the vision and the contributions of the first microscopists. CBE Life Sci Educ 8(4):338–343
16. Haseloff J (2003) Old botanical techniques for new microscopes. Biotechniques 34(6):1174–1182

17. Lelut M (1828) Anatomical researches on the epithelium, or internal epidermis. Lancet 9 (228):556–558
18. SRódka A (2003) The short history of gastroenterology. J Physiol Pharmacol 54(S3):9–21
19. Karamanou M, Poulakou-Rebelakou E, Tzetis M, Androutsos G (2010) Anton van Leeuwenhoek (1632–1723): father of micromorphology and discoverer of spermatozoa. Rev Argent Microbiol 42(4):31–34
20. Jung C, Hugot JP, Barreau F (2010) Peyer's patches: the immune sensors of the intestine. Int J Inflam 2010:823710
21. Richmond C (1986) Two and a quarter centuries of digestive juices. Trends Biochem Sci 11 (12):528–530
22. Schultz M (2008) Rudolf Virchow. Emerg Infect Dis 14(9):1480–1481
23. Mazzarello P (1999) A unifying concept: the history of cell theory. Nat Cell Biol 1(1):E13–E15
24. Tavassoli M (1980) The cell theory: a foundation to the edifice of biology. Am J Pathol 98 (1):44
25. Smith GP (2000) Pavlov and integrative physiology. Am J Physiol Regul Integr Comp Physiol 279(3):R743–R755
26. Samoilov VO (2007) Ivan Petrovich Pavlov (1849–1936). J Hist Neurosci 16(1–2):74–89
27. Haas L (1999) Ivan Petrovich Pavlov (1849–1936). J Neurol Neurosurg Psychiatry 67 (3):299
28. Drozdov I, Modlin IM, Kidd M, Goloubinov VV (2009) From Leningrad to London: the saga of Kulchitsky and the legacy of the enterochromaffin cell. Neuroendocrinol 89(1):1–12
29. Drozdov I, Modlin IM, Kidd M, Goloubinov VV (2009) Nikolai Konstantinovich Kulchitsky (1856–1925). J Med Biogr 17(1):47–54
30. Smith TF, Gruskin K, Tolman S, Faulkner D (1986) The molecular biology computer research resource. Nucleic Acids Res 14(1):25–29
31. Fang FC, Casadevall A (2011) Reductionistic and holistic science. Infect Immun 79 (4):1401–1404
32. Pandolfino JE, Kahrilas PJ (2009) New technologies in the gastrointestinal clinic and research: impedance and high-resolution manometry. World J Gastroenterol 15(2):131–138
33. Zhu B, Wang Y, Yan G, Jiang P, Liu Z (2014) A gastrointestinal electrical stimulation system based on transcutaneous power transmission technology. Gastroenterol Res Pract 2014:728572
34. Zhang JG, Liu HF (2011) Functional imaging and endoscopy. World J Gastroenterol 17 (38):4277–4282
35. Peck ML (1992) The future of nursing in a technological age: computers, robots, and TLC. J Holist Nurs 10(2):183–191
36. Furness JB, Rivera LR, Cho HJ, Bravo DM, Callaghan B (2013) The gut as a sensory organ. Nat Rev Gastroenterol Hepatol 10:729–740
37. Yen TH, Wright NA (2006) The gastrointestinal tract stem cell niche. Stem Cell Rev 2 (3):203–212
38. Kennelly R, Conneely JB, Bouchier-Hayes D, Winter DC (2011) Mast cells in tissue healing: from skin to the gastrointestinal tract. Curr Pharm Des 17(34):3772–3775
39. Bitar KN, Raghavan S (2012) Intestinal tissue engineering: current concepts and future vision of regenerative medicine in the gut. Neurogastroenterol Motil 24(1):7–19
40. Bitar KN (2003) Function of gastrointestinal smooth muscle: from signaling to contractile proteins. Am J Med 115(Suppl 3A):15S–23S
41. Xu AT, Tong JL, Ran ZH (2016) Organoids derived from digestive tract, liver, and pancreas. J Dig Dis 17(1):3–10
42. Gunawardene AR, Corfe BM, Staton CA (2011) Classification and functions of enteroendocrine cells of the lower gastrointestinal tract. Int J Exp Pathol 92(4):219–231
43. Langlands AJ, Almet AA, Appleton PL, Newton IP, Osborne JM, Näthke IS (2016) Paneth cell-rich regions separated by a cluster of lgr5+ cells initiate crypt fission in the intestinal stem cell niche. PLoS Biol 14(6):e1002491

44. Shenkman BS, Turtikova OV, Nemirovskaya TL, Grigoriev AI (2010) Skeletal muscle activity and the fate of myonuclei. Acta Naturae 2(2):59–66
45. Nikitina OS, Welcome MO, Pereverzev VA (2016) Human anatomy and physiology. In: 2 parts. Part 1. Belarusian State Medical University Press, Minsk
46. Andersson KE, Arner A (2004) Urinary bladder contraction and relaxation: physiology and pathophysiology. Physiol Rev 84(3):935–986
47. Fisher SA (2010) Vascular smooth muscle phenotypic diversity and function. Physiol Genomics 42A(3):169–187
48. Ikeda T, Murai H, Kaneko S, Usui S, Kobayashi D, Nakano M et al (2012) Augmented single-unit muscle sympathetic nerve activity in heart failure with chronic atrial fibrillation. J Physiol 590(3):509–517
49. Macefield VG, Wallin BG (1999) Firing properties of single vasoconstrictor neurones in human subjects with high levels of muscle sympathetic activity. J Physiol 516(1):293–301
50. He WQ, Qiao YN, Peng YJ, Zha JM, Zhang CH, Chen C et al (2013) Altered contractile phenotypes of intestinal smooth muscle in mice deficient in myosin phosphatase target subunit 1. Gastroenterol 144(7):1456–1465
51. Thornbury KD (1999) Tonic and phasic activity in smooth muscle. Ir J Med Sci 168(3):201–207
52. Mandrek K, Golenhofen K (1990) Phasic-rhythmical and tonic components in gastrointestinal motility. Prog Clin Biol Res 327:463–481
53. McHale N, Hollywood M, Sergeant G, Thornbury K (2006) Origin of spontaneous rhythmicity in smooth muscle. J Physiol 570(Pt 1):23–28
54. Zhao JB, Chen PM, Gregersen H (2013) Changes of phasic and tonic smooth muscle function of jejunum in type 2 diabetic Goto-Kakizaki rats. World J Diab 4(6):339–348
55. Fiorenza V, Yee YS, Zfass AM (1987) Small intestinal motility: normal and abnormal function. Am J Gastroenterol 82(11):1111–1114
56. Golenhofen K, Mandrek K (1991) Phasic and tonic contraction processes in the gastrointestinal tract. Dig Dis 9(6):341–346
57. Gartner LP, Hiatt JL, Strum JM (2003) Board review series. Cell biology and histology, 4th edn. Lippincott Williams & Wilkins, Philadelphia
58. Bates MD (2002) Development of the enteric nervous system. Clin Perinatol 29(1):97–114
59. Kuo B, Urma D (2006) Esophagus—anatomy and development. Part 1: Oral cavity, pharynx and esophagus. GI Motility Online. https://doi.org/10.1038/gimo6
60. Mittal RK (2011) Motor function of the pharynx, esophagus, and its sphincters. Morgan & Claypool Life Sciences, San Rafael
61. Mittal RK (2013) Longitudinal muscle of the esophagus: its role in esophageal health and disease. Curr Opin Gastroenterol 29(4):421–430
62. Lang IM, Medda BK, Jadcherla SR, Shaker R (2016) Characterization and mechanisms of the pharyngeal swallow activated by stimulation of the esophagus. Am J Physiol Gastrointest Liver Physiol 311(5):G827–G837
63. Hanani M (1993) Neurons and glial cells of the enteric nervous system: studies in tissue culture. J Basic Clin Physiol Pharmacol 4(3):157–179
64. Rühl A (2005) Glial cells in the gut. Neurogastroenterol Motil 17(6):777–790
65. Bassotti G, Villanacci V, Antonelli E, Morelli A, Salerni B (2007) Enteric glial cells: new players in gastrointestinal motility? Lab Invest 87:628–632
66. Estofolete CF, Botelho-Machado C, Taboga SR, Zucoloto S, Polli-Lopes AC, Gil CD (2010) Effects of myenteric denervation on extracellular matrix fibers and mast cell distribution in normal stomach and gastric lesions. Cancer Cell Int 10:18
67. Murphy MM, Lawson JA, Mathew SJ, Hutcheson DA, Kardon G (2011) Satellite cells, connective tissue fibroblasts and their interactions are crucial for muscle regeneration. Development 138(17):3625–3637
68. Cain SA, Morgan A, Sherratt MJ, Ball SG, Shuttleworth CA, Kielty CM (2006) Proteomic analysis of fibrillin-rich microfibrils. Proteomics 6(1):111–122

69. Kielty CM, Sherratt MJ, Marson A, Baldock C (2005) Fibrillin microfibrils. Adv Protein Chem 70:405–436
70. Hayakawa M, Kobayashi M, Hoshino T (1990) Microfibrils: a constitutive component of reticular fibers in the mouse lymph node. Cell Tissue Res 262(1):199–201
71. Barros EM, Rodrigues CJ, Rodrigues NR, Oliveira RP, Barros TE, Rodrigues AJ Jr (2002) Aging of the elastic and collagen fibers in the human cervical interspinous ligaments. Spine J 2(1):57–62
72. Doillon CJ, Dunn MG, Bender E, Silver FH (1985) Collagen fiber formation in repair tissue: development of strength and toughness. Coll Relat Res 5(6):481–492
73. Chvapil M, Speer DP, Holubec H, Chvapil TA, King DH (1993) Collagen fibers as a temporary scaffold for replacement of ACL in goats. J Biomed Mater Res 27(3):313–325
74. Ushiki T (2002) Collagen fibers, reticular fibers and elastic fibers. A comprehensive understanding from a morphological viewpoint. Arch Histol Cytol 65(2):109–126
75. Montes GS (1996) Structural biology of the fibres of the collagenous and elastic systems. Cell Biol Int 20(1):15–27
76. Kielty CM, Sherratt MJ, Shuttleworth CA (2002) Elastic fibres. J Cell Sci 115(Pt 14):2817–2828
77. Sherratt MJ, Wess TJ, Baldock C, Ashworth J, Purslow PP, Shuttleworth CA, Kielty CM (2001) Fibrillin-rich microfibrils of the extracellular matrix: ultrastructure and assembly. Micron 32(2):185–200
78. Ramirez F, Sakai LY, Dietz HC, Rifkin DB (2004) Fibrillin microfibrils: multipurpose extracellular networks in organismal physiology. Physiol Genomics 19(2):151–154
79. Wen SL, Feng S, Tang SH, Gao JH, Zhang LH, Tong H et al (2016) Collapsed reticular network and its possible mechanism during the initiation and/or progression of hepatic fibrosis. Sci Rep 6:35426. https://doi.org/10.1038/srep35426
80. Raica M, Dema E, Iacovliev M, Alexa A, Mederle O (1996) Reticular fibers in the stroma of the thymus. Rom J Morphol Embryol 42(3–4):141–145
81. Tanzer ML (2006) Current concepts of extracellular matrix. J Orthop Sci 11(3):326–331
82. Pupa SM, Ménard S, Forti S, Tagliabue E (2002) New insights into the role of extracellular matrix during tumor onset and progression. J Cell Physiol 192(3):259–267
83. Sherwin E, Sandhu KV, Dinan TG, Cryan JF (2016) May the force be with you: the light and dark sides of the microbiota-gut-brain axis in neuropsychiatry. CNS Drugs 30 (11):1019–1041
84. Kennedy PJ, Cryan JF, Dinan TG, Clarke G (2017) Kynurenine pathway metabolism and the microbiota-gut-brain axis. Neuropharmacol 112(Pt B):399–412
85. Sandhu KV, Sherwin E, Schellekens H, Stanton C, Dinan TG, Cryan JF (2017) Feeding the microbiota-gut-brain axis: diet, microbiome, and neuropsychiatry. Transl Res 179:223–244
86. Chen Z, Li W, Wang H, Wan C, Luo D, Deng S et al (2016) Klf10 regulates odontoblast differentiation and mineralization via promoting expression of dentin matrix protein 1 and dentin sialophosphoprotein genes. Cell Tissue Res 363(2):385–398
87. Saito MT, Salmon CR, Amorim BR, Ambrosano GM, Casati MZ, Sallum EA et al (2014) Characterization of highly osteoblast/cementoblast cell clones from a CD105-enriched periodontal ligament progenitor cell population. J Periodontol 85(6):e205–e211
88. Rodríguez-Lozano FJ, Insausti CL, Iniesta F, Blanquer M, Ramírez MDC, Luis M et al (2012) Mesenchymal dental stem cells in regenerative dentistry. Med Oral Patol Oral Cir Bucal 17(6):e1062–e1067
89. Edwards PC, Mason JM (2006) Gene-enhanced tissue engineering for dental hard tissue regeneration: (2) dentin-pulp and periodontal regeneration. Head Face Med 2:16
90. Xiong J, Gronthos S, Bartold PM (2013) Role of the epithelial cell rests of Malassez in the development, maintenance and regeneration of periodontal ligament tissues. Periodontol 63 (1):217–233
91. Volponi AA, Kawasaki M, Sharpe PT (2013) Adult human gingival epithelial cells as a source for whole-tooth bioengineering. J Dent Res 92(4):329–334

92. Roper SD (2013) Taste buds as peripheral chemosensory processors. Semin Cell Dev Biol 24(1):71–79
93. Breslin PA, Huang L (2006) Human taste: peripheral anatomy, taste transduction, and coding. Adv Otorhinolaryngol 63:152–190
94. Delporte C, Bryla A, Perret J (2016) Aquaporins in salivary glands: from basic research to clinical applications. Int J Mol Sci 17(2):166
95. Park HW, Lee MG (2012) Transepithelial bicarbonate secretion: lessons from the pancreas. Cold Spring Harb Perspect Med 2(10):a009571
96. Amano O, Mizobe K, Bando Y, Sakiyama K (2012) Anatomy and histology of rodent and human major salivary glands—overview of the japan salivary gland society-sponsored workshop. Acta Histochem Cytochem 45(5):241–250
97. Weiss H, Brockmann W, Bersch W, Fiedler L (1993) The layers of the gallbladder wall: an ultrasound-anatomic comparative study. Bildgebung 60(3):125–130
98. Lee SP, Savard CE, Kuver R (2009) Gallbladder epithelial cells that engraft in mouse liver can differentiate into hepatocyte-like cells. Am J Pathol 174(3):842–853
99. Yoo KS, Lim WT, Choi HS (2016) Biology of cholangiocytes: from bench to bedside. Gut Liver 10(5):687–698
100. Tabibian JH, Masyuk AI, Masyuk TV, O'Hara SP, LaRusso NF (2013) Physiology of cholangiocytes. Compr Physiol 3(1). https://doi.org/10.1002/cphy.c120019
101. Kmieć Z (2001) Cooperation of liver cells in health and disease. Adv Anat Embryol Cell Biol 161(III–XIII):1–151
102. Abdel-Misih SR, Bloomston M (2010) Liver anatomy. Surg Clin North Am 90(4):643–653
103. Ding C, Li Y, Guo F, Jiang Y, Ying W, Li D et al (2016) A cell-type-resolved Liver Proteome. Mol Cell Proteomics 15(10):3190–3202
104. Davenport M, Hadzic N (2016) Pediatric hepatic physiology. In: Puri P (ed) Pediatric surgery. Springer, Berlin
105. Ghosheh N, Küppers-Munther B, Asplund A, Edsbagge J, Ulfenborg B, Andersson TB et al (2017) Comparative transcriptomics of hepatic differentiation of human pluripotent stem cells and adult human liver tissue. Physiol Genomics. https://doi.org/10.1152/physiolgenomics.00007.2017
106. Aloia L, McKie MA, Huch M (2016) Cellular plasticity in the adult liver and stomach. J Physiol 594(17):4815–4825
107. Fausto N, Campbell JS (2003) The role of hepatocytes and oval cells in liver regeneration and repopulation. Mech Dev 120(1):117–130
108. Otsu K, Kumakami-Sakano M, Fujiwara N, Kikuchi K, Keller L, Lesot H, Harada H (2014) Stem cell sources for tooth regeneration: current status and future prospects. Front Physiol 5:36
109. Wright NA (2000) Epithelial stem cell repertoire in the gut: clues to the origin of cell lineages, proliferative units and cancer. Int J Exp Path 81:117–143
110. Merendino N, Dwinell MB, Varki N, Eckmann L, Kagnoff MF (1999) Human intestinal epithelial cells express receptors for platelet-activating factor. Am J Physiol Gastrointest Liver Physiol 277(4):G810–G818
111. Peterson LW, Artis D (2014) Intestinal epithelial cells: regulators of barrier function and immune homeostasis. Nat Rev Immunol 14(3):141–153
112. Lee KZ, Lestradet M, Socha C, Schirmeier S, Schmitz A, Spenlé C et al (2016) Enterocyte purge and rapid recovery is a resilience reaction of the gut epithelium to pore-forming toxin attack. Cell Host Microbe 20:716–730
113. Peck BCE, Mah AT, Pitman WA, Ding S, Lund PK, Sethupathy P (2016) Functional transcriptomics in diverse intestinal epithelial cell types reveals robust gut microbial sensitivity of microRNAs in intestinal stem cells. BioRxiv. https://doi.org/10.1101/087882
114. Al-Ghadban S, Kaissi S, Homaidan FR, Naim HY, El-Sabban ME (2016) Cross-talk between intestinal epithelial cells and immune cells in inflammatory bowel disease. Sci Rep 6:29783

115. Cheng H, Leblond CP (1974) Origin, differentiation and renewal of the four main epithelial cell types in the mouse small intestine I. Columnar cell. Am J Anat 141(4):461–479
116. Murat JC, Gamet L, Cazenave Y, Trocheris V (1990) Questions about the use of [3H] thymidine incorporation as a reliable method to estimate cell proliferation rate. Biochem J 270(2):563–564
117. Duque A, Rakic P (2011) Different effects of BrdU and 3H-thymidine incorporation into DNA on cell proliferation, position and fate. J Neurosci 31(42):15205–15217
118. Hu VW, Black GE, Torres-Duarte A, Abramson FP (2002) 3H-thymidine is a defective tool with which to measure rates of DNA synthesis. FASEB J 16(11):1456–1457
119. Kodama N, Iwao T, Katano T, Ohta K, Yuasa H, Matsunaga T (2016) Characteristic analysis of intestinal transport in enterocyte-like cells differentiated from human induced pluripotent stem cells. Drug Metab Dispos 44(10)
120. Chen L, Tuo B, Dong H (2016) Regulation of intestinal glucose absorption by ion channels and transporters. Nutrients 8(1):43
121. Pouokam E, Althaus M (2016) Epithelial electrolyte transport physiology and the gasotransmitter hydrogen sulfide. Oxid Med Cell Longev 2016:4723416
122. Maccallum JB (1906) Factors influencing secretion. J Biol Chem 1:335–344
123. Dwight T (1877) Recent progress in anatomy. Boston Med Surg J 97:270–273
124. Prasanna LC (2016) Analysis of the distribution of mucins in adult human gastric mucosa and its functional significance. J Clin Diagn Res 10(2):AC01–AC04
125. Chen YY, Wang HH, Antonioli DA, Spechler SJ, Zeroogian JM, Goyal R et al (1999) Significance of acid-mucin-positive nongoblet columnar cells in the distal esophagus and gastroesophageal junction. Hum Pathol 30(12):1488–1495
126. Mowat AM, Agace WW (2014) Regional specialization within the intestinal immune system. Nat Rev Immunol 14:667–685
127. Wang K, Wu LY, Dou CZ, Guan X, Wu HG, Liu HR (2016) Research advance in intestinal mucosal barrier and pathogenesis of Crohn's disease. Gastroenterol Res Pract 2016:9686238
128. Ugolev AM, Iezuitova NN, Timofeeva NM, Egorova VV, Nikitina AA, Gordova LA (1992) The digestive enzymes in the gastrointestinal tract, kidney, liver and spleen in different functional states. Fiziol Zh SSSR Im I M Sechenova 78(9):76–83
129. Pandol SJ (2010) The exocrine pancreas. Morgan & Claypool Life Sciences, San Rafael
130. Hamosh M (1990) Lingual and gastric lipases. Nutrition 6(6):421–428
131. Brandtzaeg P (2013) Secretory immunity with special reference to the oral cavity. J Oral Microbiol 5. https://doi.org/10.3402/jom.v5i0.20401
132. Hand AR, Pathmanathan D, Field RB (1999) Morphological features of the minor salivary glands. Arch Oral Biol 44(Suppl 1):S3–10
133. Ianiro G, Pecere S, Giorgio V, Gasbarrini A, Cammarota G (2016) Digestive enzyme supplementation in gastrointestinal diseases. Curr Drug Metab 17(2):187–193
134. Fieker A, Philpott J, Armand M (2011) Enzyme replacement therapy for pancreatic insufficiency: present and future. Clin Exp Gastroenterol 4:55–73
135. Miled N, Canaan S, Dupuis L, Roussel A, Rivière M, Carrière F et al (2000) Digestive lipases: from three-dimensional structure to physiology. Biochimie 82(11):973–986
136. Sternby B, Nilsson A, Melin T, Borgström B (1991) Pancreatic lipolytic enzymes in human duodenal contents. radioimmunoassay compared with enzyme activity. Scand J Gastroenterol 26(8):859–866
137. Sternby B, Nilsson A (1997) Carboxyl ester lipase (bile salt-stimulated lipase), colipase, lipase, and phospholipase A2 levels in pancreatic enzyme supplements. Scand J Gastroenterol 32(3):261–267
138. Chew CS, Nakamura K, Ljungström M (1992) Calcium signaling mechanisms in the gastric parietal cell. Yale J Biol Med 65(6):561–623
139. Waldum HL, Kleveland PM, Sørdal ØF (2016) Helicobacter pylori and gastric acid: an intimate and reciprocal relationship. Therap Adv Gastroenterol 9(6):836–844
140. Helander HF, Keeling DJ (1993) Cell biology of gastric acid secretion. Baillieres Clin Gastroenterol 7(1):1–21

141. Schubert ML (2016) Gastric acid secretion. Curr Opin Gastroenterol 32(6):452–460
142. Latorre R, Sternini C, De Giorgio R, Greenwood-Van Meerveld B (2016) Enteroendocrine cells: a review of their role in brain-gut communication. Neurogastroenterol Motil 28 (5):620–630
143. Posovszky C, Wabitsch M (2015) Regulation of appetite, satiation, and body weight by enteroendocrine cells. Part 1: characteristics of enteroendocrine cells and their capability of weight regulation. Horm Res Paediatr 83:1–10
144. El-Salhy M, Mazzawi T, Hausken T, Jan Hatlebakk G (2016) Interaction between diet and gastrointestinal endocrine cells. Biomed Rep 4(6):651–656
145. Sun FP, Song YG, Cheng W, Zhao T, Yao YL (2002) Gastrin, somatostatin, G and D cells of gastric ulcer in rats. World J Gastroenterol 8:375–378
146. Margolis KG, Gershon MD, Bogunovic M (2016) Cellular organization of neuroimmune interactions in the gastrointestinal tract. Trends Immunol 37(7):487–501
147. Furness JB (2016) Integrated neural and endocrine control of gastrointestinal function. Adv Exp Med Biol 891:159–173
148. Holst JJ (1997) Enteroglucagon. Annu Rev Physiol 59:257–271
149. Pearse AGE, Coulling I, Weavers B, Friesen S (1970) The endocrine polypeptide cells of the human stomach, duodenum, and jejunum. Gut 11(8):649–658
150. Rhodin J, Dalhamn T (1956) Electron microscopy of the tracheal ciliated mucosa in rat. Z Zellforsch Mikrosk Anat 44(4):345–412
151. Jarvi O, Keyrilainen O (1956) On the cellular structures of the epithelial invasions in the glandular stomach of mice caused by intramural application of 20-methylcholantren. Acta Pathol Microbiol Scand Suppl 39(Suppl 111):72–73
152. Luciano L, Reale E, Ruska H (1968) On a glycogen containing brusc cell in the rectum of the rat. Z Zellforsch Mikrosk Anat 91(1):153–158
153. Isomaki AM (1973) A new cell type (tuft cell) in the gastrointestinal mucosa of the rat. A transmission and scanning electron microscopic study. Acta Pathol Microbiol Scand A Suppl 240:235–241
154. Reid L, Meyrick B, Antony VB, Chang LY, Crapo JD, Reynolds HY (2005) The mysterious pulmonary brush cell: a cell in search of a function. Am J Respir Crit Care Med 172(1):136–139
155. Galligan JJ, Akbarali HI (2014) Molecular physiology of enteric opioid receptors. Am J Gastroenterol Suppl 2:17–21
156. Holzer P (2009) Opioid receptors in the gastrointestinal tract. Regul Pept 155(1–3):11–17
157. Gerbe F, Jay P (2016) Intestinal tuft cells: epithelial sentinels linking luminal cues to the immune system. Mucosal Immunol 9:1353–1359
158. Gillispie CC, Holmes FL, Koertge N (eds) (2008) Complete dictionary of scientific biography. Charles Scribner's Sons, Detroit
159. Gutierrez-Gonzalez L, Graham TA, Rodriguez-Justo M, Leedham SJ, Novelli MR, Gay LJ et al (2011) The clonal origins of dysplasia from intestinal metaplasia in the human stomach. Gastroenterol 140(4):1251–1260
160. Colby JKL, Klein RD, McArthur MJ, Conti CJ, Kiguchi K, Kawamoto T et al (2008) Progressive metaplastic and dysplastic changes in mouse pancreas induced by cyclooxygenase-2 overexpression. Neoplasia 10(8):782–796
161. Camilo V, Garrido M, Valente P, Ricardo S, Amaral AL, Barros R et al (2015) Differentiation reprogramming in gastric intestinal metaplasia and dysplasia: role of SOX2 and CDX2. Histopathol 66(3):343–350
162. Dolznig H, Grebien F, Sauer T, Beug H, Müllner EW (2004) Evidence for a size-sensing mechanism in animal cells. Nat Cell Biol 6:899–905
163. Cooper GM (2000) The cell: a molecular approach, 2nd edn. Sinauer Associates, Sunderland
164. Escribá PV, González-Ros JM, Goñi FM, Kinnunen PKJ, Vigh L, Sánchez-Magraner L et al (2008) Membranes: a meeting point for lipids, proteins and therapies. J Cell Mol Med 12 (3):829–875

165. Alberts B, Johnson A, Lewis J, Raff M, Roberts K, Walter P (2002) Molecular biology of the cell, 4th edn. Garland Science, New York
166. Freitas RA Jr (1999) Nanomedicine: basic capabilities, vol 1. Landes Bioscience, Georgetown
167. Lee AG (2004) How lipids affect the activities of integral membrane proteins. BBA—Biomembranes 1666(1–2):62–87
168. Lodish H, Berk A, Zipursky SL, Matsudaira P, Baltimore D, Darnell J (2000) Molecular cell biology, 4th edn. WH Freeman, New York
169. Kararli TT (1995) Comparison of the gastrointestinal anatomy, physiology, and biochemistry of humans and commonly used laboratory animals. Biopharm Drug Dispos 16(5):351–380
170. Kraft ML (2013) Plasma membrane organization and function: moving past lipid rafts. Mol Biol Cell 24:2765–2768
171. Head BP, Patel HH, Insel PA (2014) Interaction of membrane/lipid rafts with the cytoskeleton: impact on signaling and function: membrane/lipid rafts, mediators of cytoskeletal arrangement and cell signaling. Biochim Biophys Acta 1838(2):532–545
172. Donovan EL, Pettine SM, Hickey MS, Hamilton KL, Miller BF (2013) Lipidomic analysis of human plasma reveals ether-linked lipids that are elevated in morbidly obese humans compared to lean. Diabetol Metab Syndr 5:24
173. Berg JM, Tymoczko JL, Stryer L (2002) Biochemistry, 5th edn. WH Freeman, New York
174. van Meer G, Voelker DR, Feigenson GW (2008) Membrane lipids: where they are and how they behave. Nat Rev Mol Cell Biol 9(2):112–124
175. Fadeel B, Xue D (2009) The ins and outs of phospholipid asymmetry in the plasma membrane: roles in health and disease. Crit Rev Biochem Mol Biol 44(5):264–277
176. Naudí A, Jové M, Ayala V, Cabré R, Portero-Otín M, Pamplona R (2013) Non-enzymatic modification of aminophospholipids by carbonyl-amine reactions. Int J Mol Sci 14(2):3285–3313
177. Jain SK (1985) In vivo externalization of phosphatidylserine and phosphatidylethanolamine in the membrane bilayer and hypercoagulability by the lipid peroxidation of erythrocytes in rats. J Clin Invest 76(1):281–286
178. Saheki Y, Bian X, Schauder CM, Sawaki Y, Surma MA, Klose C et al (2016) Control of plasma membrane lipid homeostasis by the extended synaptotagmins. Nat Cell Biol 18:504–515
179. Okeke E, Dingsdale H, Parker T, Voronina S, Tepikin AV (2016) Endoplasmic reticulum-plasma membrane junctions: structure, function and dynamics. J Physiol 594 (11):2837–2847
180. Watson H (2015) Biological membranes. Essays Biochem 59:43–69
181. Jaiswal R, Luk F, Gong J, Mathys JM, Grau GER, Bebawy M (2012) Microparticle conferred microRNA profiles—implications in the transfer and dominance of cancer traits. Mol Cancer 11:37
182. Bevers EM, Williamson PL (2010) Phospholipid scramblase: an update. FEBS Lett 584 (13):2724–2730
183. Doherty GJ, McMahon HT (2008) Mediation, modulation and consequences of membrane-cytoskeleton interactions. Annu Rev Biophys 37:65–95
184. van Meer G, Halter D, Sprong H, Somerharju P, Egmond MR (2006) ABC lipid transporters: extruders, flippases, or flopless activators? FEBS Lett 580(4):1171–1177
185. Holgersson J, Jovall PÅ, Breimer ME (1991) Glycosphingolipids of human large intestine: detailed structural characterization with special reference to blood group compounds and bacterial receptor structures. J Biochem 110(1):120–131
186. Breimer ME, Hansson GC, Karlsson KA, Larson G, Leffler H (2012) Glycosphingolipid composition of epithelial cells isolated along the villus axis of small intestine of a single human individual. Glycobiology 22(12):1721–1730
187. Khan MS, Dosoky NS, Williams JD (2013) Engineering lipid bilayer membranes for protein studies. Int J Mol Sci 14(11):21561–21597

188. Dawaliby R, Trubbia C, Delporte C, Noyon C, Ruysschaert JM, Van Antwerpen P, Govaerts C (2016) Phosphatidylethanolamine is a key regulator of membrane fluidity in eukaryotic cells. J Biol Chem 291(7):3658–3667
189. Simons K, Ikonen E (1997) Functional rafts in cell membranes. Nature 119:569–572
190. Morozova D, Guigas G, Weiss M (2011) Dynamic structure formation of peripheral membrane proteins. PLoS Comput Biol 7(6):e1002067
191. Bastiani M, Parton RG (2010) Caveolae at a glance. J Cell Sci 123:3831–3836
192. Taggart MJ (2001) Smooth muscle excitation-contraction coupling: a role for caveolae and caveolins? Physiology 16(2):61–65
193. Echarri A, Del Pozo MA (2012) Caveolae. Curr Biol 22(4):R114–R116
194. Lo HP, Nixon SJ, Hall TE, Cowling BS, Ferguson C, Morgan GP et al (2015) The caveolin-cavin system plays a conserved and critical role in mechanoprotection of skeletal muscle. J Cell Biol 210(5):833–849
195. Lo HP, Hall TE, Parton RG (2016) Mechanoprotection by skeletal muscle caveolae. Bioarchitecture 6(1):22–27
196. Merrifield CJ, Perrais D, Zenisek D (2005) Coupling between clathrin-coated-pit invagination, cortactin recruitment, and membrane scission observed in live cells. Cell 121(4):593–606
197. Palade GE (1953) Fine structure of blood capillaries. J Appl Phys 24:1424
198. Yamada E (1955) The fine structure of the gall bladder epithelium of the mouse. J Biophys Biochem Cytol 1(5):445–458
199. Rothberg KG, Heuser JE, Donzell WC, Ying YS, Glenney JR, Anderson RGW (1992) Caveolin, a protein component of caveolae membrane coats. Cell 68(4):673–682
200. Glenney JR Jr, Soppet D (1992) Sequence and expression of caveolin, a protein component of caveolae plasma membrane domains phosphorylated on tyrosine in Rous sarcoma virus-transformed fibroblasts. Proc Natl Acad Sci USA 89(21):10517–10521
201. Brown MS, Goldstein JL (2011) Richard G.W. Anderson (1940–2011) and the birth of receptor-mediated endocytosis. J Cell Biol 193(4):601–603
202. Chidlow JH Jr, Sessa WC (2010) Caveolae, caveolins, and cavins: complex control of cellular signalling and inflammation. Cardiovasc Res 86(2):219–225
203. Scherer PE, Okamoto T, Chun M, Nishimoto I, Lodish HF, Lisanti MP (1996) Identification, sequence, and expression of caveolin-2 defines a caveolin gene family. Proc Natl Acad Sci USA 93(1):131–135
204. Tang Z, Scherer PE, Okamoto T, Song K, Chu C, Kohtz DS et al (1996) Molecular cloning of caveolin-3, a novel member of the caveolin gene family expressed predominantly in muscle. J Biol Chem 271:2255–2261
205. Way M, Parton R (1995) M-caveolin: a muscle-specific caveolin-related protein. FEBS Lett 376:108–112
206. Sowa G (2011) Caveolae, caveolins, cavins, and endothelial cell function: new insights. Front Physiol 2:120
207. Vinten J, Johnsen AH, Roepstorff P, Harpoth J, Tranum-Jensen J (2005) Identification of a major protein on the cytosolic face of caveolae. Biochim Biophys Acta 1717:34–40
208. Parton RG, del Pozo MA (2013) Caveolae as plasma membrane sensors, protectors and organizers. Nat Rev Mol Cell Biol 14:98–112
209. Liu P, Li WP, Machleidt T, Anderson RGW (1999) Identification of caveolin-1 in lipoprotein particles secreted by exocrine cells. Nat Cell Biol 1:369–375
210. Fridolfsson HN, Patel HH (2013) Caveolin and caveolae in age associated cardiovascular disease. J Geriatr Cardiol 10(1):66–74
211. Gazzerro E, Sotgia F, Bruno C, Lisanti MP, Minetti C (2010) Caveolinopathies: from the biology of caveolin-3 to human diseases. Eur J Hum Genet 18(2):137–145
212. Fletcher A (2013) The cell membrane and receptors. Anaest Intens Care Med 14(9):417–421
213. Graves PR, Haystead TAJ (2002) Molecular biologist's guide to proteomics. Microbiol Mol Biol Rev 66(1):39–63

214. Mischak H, Apweiler R, Banks RE, Conaway M, Coon J, Dominiczak A et al (2007) Clinical proteomics: a need to define the field and to begin to set adequate standards. Proteomics Clin Appl 1(2):148–156
215. Monti M, Orrù S, Pagnozzi D, Pucci P (2005) Functional proteomics. Clin Chim Acta 357 (2):140–150
216. Vermeer BJ (1987) Plasma membrane receptors. J Invest Dermatol 88:529–531
217. Wilson J, Hunt T (2008) Molecular biology of the cell: the problems book, 5th edn. Garland Science, New York
218. Chou KC, Elrod DW (1999) Prediction of membrane protein types and subcellular locations. Proteins 34(1):137–153
219. Chou KC, Elrod DW (1999) Protein subcellular location prediction. Protein Eng 12(2):107–118
220. Ouzounis CA, Coulson RM, Enright AJ, Kunin V, Pereira-Leal JB (2003) Classification schemes for protein structure and function. Nat Rev Genet 4(7):508–519
221. Sunde M, Blake CC (1998) From the globular to the fibrous state: protein structure and structural conversion in amyloid formation. Q Rev Biophys 31(1):1–39
222. Thiriet M (2014) Mechanotransduction and vascular resistance. In: Lanzer P (ed) Pan vascular medicine. Springer, Berlin
223. Banfalvi G (2016) Biological membranes. In: Permeability of biological membranes. Springer, Cham
224. Almén MS, Nordström KJV, Fredriksson R, Schiöth HB (2009) Mapping the human membrane proteome: a majority of the human membrane proteins can be classified according to function and evolutionary origin. BMC Biol 7:50
225. Stansfeld PJ, Goose JE, Caffrey M, Carpenter EP, Parker JL, Newstead S, Sansom MS (2015) MemProtMD: automated insertion of membrane protein structures into explicit lipid membranes. Structure 23(7):1350–1361
226. Ali F, Hayat M (2015) Classification of membrane protein types using voting feature interval in combination with Chou's Pseudo amino acid composition. J Theor Biol 384:78–83
227. Butt AH, Rasool N, Khan YD (2017) A treatise to computational approaches towards prediction of membrane protein and its subtypes. J Membr Biol 250(1):55–76
228. Josic D, Clifton JG (2007) Mammalian plasma membrane proteomics. Proteomics 7 (16):3010–3029
229. Good MC, Zalatan JG, Lim WA (2011) Scaffold proteins: hubs for controlling the flow of cellular information. Science 332(6030):680–686
230. Ziemba BP, Falke JJ (2013) Lateral diffusion of peripheral membrane proteins on supported lipid bilayers is controlled by the additive frictional drags of 1) bound lipids and 2) protein domains penetrating into the bilayer hydrocarbon core. Chem Phys Lipids 172–173:67–77
231. Walther TC, Brickner JH, Aguilar PS, Bernales S, Pantoja C, Walter P (2006) Eisosomes mark static sites of endocytosis. Nature 439:998–1003
232. Olivera-Couto A, Salzman V, Mailhos M, Digman MA, Gratton E, Aguilar PS (2015) Eisosomes are dynamic plasma membrane domains showing pil1-lsp1 heteroligomer binding equilibrium. Biophys J 108(7):1633–1644
233. Aoki-Kinoshita KF (2008) An introduction to bioinformatics for glycomics research. PLoS Comput Biol 4(5):e1000075
234. Egberts HJ, Koninkx JF, van Dijk JE, Mouwen JM (1984) Biological and pathobiological aspects of the glycocalyx of the small intestinal epithelium—a review. Vet Q 6(4):186–199
235. Roseman S (2001) Reflections on glycobiology. J Biol Chem 276:41527–41542
236. Weinbaum S, Tarbell JM, Damiano ER (2007) The structure and function of the endothelial glycocalyx layer. Annu Rev Biomed Eng 9:121–167
237. Reitsma S, Slaaf DW, Vink H, van Zandvoort MAMJ, Oude Egbrink MGA (2007) The endothelial glycocalyx: composition, functions, and visualization. Pflugers Arch 454 (3):345–359
238. Frey A, Giannasca KT, Weltzin P, Giannasca PJ, Reggio H, Lencer WI, Neutra MR (1996) Role of the glycocalyx in regulating access of microparticles to apical plasma membranes of

intestinal epithelial cells: implications for microbial attachment and oral vaccine targeting. Exp Med 184:1045–1059
239. Artis D (2008) Epithelial-cell recognition of commensal bacteria and maintenance of immune homeostasis in the gut. Nat Rev Immunol 8:411–420
240. Cruz-Chu ER, Malafeev A, Pajarskas T, Pivkin IV, Koumoutsakos P (2014) Structure and response to flow of the glycocalyx layer. Biophys J 106(1):232–243
241. Hart GW, Copeland RJ (2010) Glycomics hits the big time. Cell 143(5):672–676
242. Soo EC, Hui JP (2010) Metabolomics in glycomics. Methods Mol Biol 600:175–186
243. Kreisel W, Volk BA, Büchsel R, Reutter W (1980) Different half-lives of the carbohydrate and protein moieties of a 110,000-dalton glycoprotein isolated from plasma membranes of rat liver. Proc Natl Acad Sci USA 77(4):1828–1831
244. Murray RK, Granner DK, Mayes PA, Rodwell VW (1993) Harper's biochemistry, 23rd edn. Appleton & Lange, Norwalk
245. Hankins HM, Baldridge RD, Xu P, Graham TR (2015) Role of flippases, scramblases and transfer proteins in phosphatidylserine subcellular distribution. Traffic 16(1):35–47
246. Farquhar MG, Palade GE (1998) The Golgi apparatus: 100 years of progress and controversy. Trends Cell Biol 8(1):2–10
247. Greenberg ME, Li XM, Gugiu BG, Gu X, Qin J, Salomon RG et al (2008) The lipid whisker model of the structure of oxidized cell membranes. J Biol Chem 283:2385–2396
248. Catalá A (2012) Lipid peroxidation modifies the picture of membranes from the "Fluid Mosaic Model" to the "Lipid Whisker Model". Biochimie 94(1):101–109
249. Vyšniauskas A, Qurashi M, Kuimova MK (2016) A molecular rotor that measures dynamic changes of lipid bilayer viscosity caused by oxidative stress. Chemistry 22(37):13210–13217
250. Israelachvili JN (1977) Refinement of the fluid-mosaic model of membrane structure. Biochim Biophys Acta 469(2):221–225
251. Bagatolli LA, Mouritsen OG (2013) Is the fluid mosaic (and the accompanying raft hypothesis) a suitable model to describe fundamental features of biological membranes? What may be missing? Front Plant Sci 4:457
252. Lombard J (2014) Once upon a time the cell membranes: 175 years of cell boundary research. Biol Direct 9:32
253. Hoff HE, Geddes LA (1957) The rheotome and its prehistory: a study in the historical interrelation of electrophysiology and electromechanics. Bull Hist Med 31(4):327–347
254. Gorter E, Grendel F (1925) On bimolecular layers of lipoids on the chromocytes of the blood. J Exp Med 41(4):439–443
255. Singer SJ, Nicolson GL (1972) The fluid mosaic model of the structure of cell membranes. Science 175(4023):720–731
256. Dowhan W, Bogdanov M (2011) Lipid-protein interactions as determinants of membrane protein structure and function. Biochem Soc Trans 39(3):767–774
257. Kahya N (2010) Protein-protein and protein-lipid interactions in domain-assembly: lessons from giant unilamellar vesicles. Biochim Biophys Acta 1798(7):1392–1398
258. Lopez-Marques RL, Theorin L, Palmgren MG, Pomorski TG (2014) P4-ATPases: lipid flippases in cell membranes. Pflugers Arch 466(7):1227–1240
259. Andersen JP, Vestergaard AL, Mikkelsen SA, Mogensen LS, Chalat M, Molday RS (2016) P4-ATPases as phospholipid flippases-structure, function, and enigmas. Front Physiol 7:275
260. Pomorski T, Menon AK (2006) Lipid flippases and their biological functions. Cell Mol Life Sci 63(24):2908–2921
261. Clark MR (2011) Flippin' lipids. Nat Immunol 12:373–375
262. Sahu SK, Gummadi SN, Manoj N, Aradhyam GK (2007) Phospholipid scramblases: an overview. Arch Biochem Biophys 462(1):103–114
263. Prasad R, Rawal MK (2014) Efflux pump proteins in antifungal resistance. Front Pharmacol 5:202
264. Williamson P (2015) Phospholipid scramblases. Lipid insights 8(Suppl 1):41–44

265. Huang Z, Chang X, Riordan JR, Huang Y (2004) Fluorescent modified phosphatidylcholine floppase activity of reconstituted multidrug resistance-associated protein MRP1. Biochim Biophys Acta 1660(1–2):155–163
266. Kamp D, Haest CW (1998) Evidence for a role of the multidrug resistance protein (MRP) in the outward translocation of NBD-phospholipids in the erythrocyte membrane. Biochim Biophys Acta 1372(1):91–101
267. Degiorgio D, Colombo C, Seia M, Porcaro L, Costantino L, Zazzeron L et al (2007) Molecular characterization and structural implications of 25 new ABCB4 mutations in progressive familial intrahepatic cholestasis type 3 (PFIC3). Eur J Hum Genet 15:1230–1238
268. Choi CH (2005) ABC transporters as multidrug resistance mechanisms and the development of chemosensitizers for their reversal. Cancer Cell Int 5:30
269. Morita S, Terada T (2014) Molecular mechanisms for biliary phospholipid and drug efflux mediated by abcb4 and bile salts. Biomed Res Int 954781:1–11
270. Ramadurai S, Holt A, Krasnikov V, van den Bogaart G, Killian JA, Poolman B (2009) Lateral diffusion of membrane proteins. J Am Chem Soc 131(35):12650–12656
271. Chen IHB, Huber M, Guan T, Bubeck A, Gerace L (2006) Nuclear envelope transmembrane proteins (NETs) that are up-regulated during myogenesis. BMC Cell Biol 7:38
272. Nicolson GL (2014) The fluid-mosaic model of membrane structure: still relevant to understanding the structure, function and dynamics of biological membranes after more than 40 years. Biochim Biophys Acta 1838(6):1451–1466
273. Morange M (2013) What history tells us XXX. The emergence of the fluid mosaic model of membranes. J Biosci 38(1):3–7
274. Vereb G, Szöllosi J, Matkó J, Nagy P, Farkas T, Vigh L et al (2003) Dynamic, yet structured: the cell membrane three decades after the Singer-Nicolson model. PNAS 100 (14):8053–8058
275. Edidin M (2015) Light and life in Baltimore and beyond. Biophys J 108(3):466–470
276. de la Serna JB, Schütz GJ, Eggeling C, Cebecauer M (2016) There is no simple model of the plasma membrane organization. Front Cell Dev Biol 4:106
277. Marguet D, Lenne PF, Rigneault H, He HT (2006) Dynamics in the plasma membrane: how to combine fluidity and order. EMBO J 25(15):3446–3457
278. Engelman DM (2005) Membranes are more mosaic than fluid. Nature 438:578–580
279. Gheber LA, Edidin M (1999) A model for membrane patchiness: lateral diffusion in the presence of barriers and vesicle traffic. Biophys J 77(6):3163–3175
280. Edidin M (1993) Patches, posts and fences: proteins and plasma membrane domains. Trends Cell Biol 2(12):376–380
281. Contreras FX, Ernst AM, Wieland F, Brügger B (2011) Specificity of intramembrane protein–lipid interactions. Cold Spring Harb Perspect Biol 3(6):a004705
282. Edidin M (2003) Lipids on the frontier: a century of cell-membrane bilayers. Nat Rev Mol Cell Biol 4:414–418
283. Zhang M, Bohlson SS, Dy M, Tenner AJ (2005) Modulated interaction of the ERM protein, moesin, with CD93. Immunology 115(1):63–73
284. Xu K, Zhong G, Zhuang X (2013) Actin, spectrin, and associated proteins form a periodic cytoskeletal structure in axons. Science 339(6118):452–456
285. Hirao M, Sato N, Kondo T, Yonemura S, Monden M, Sasaki T et al (1996) Regulation mechanism of ERM (ezrin/radixin/moesin) protein/plasma membrane association: possible involvement of phosphatidylinositol turnover and Rho-dependent signaling pathway. J Cell Biol 135(1):37–51
286. Niggli V, Rossy J (2008) Ezrin/radixin/moesin: versatile controllers of signaling molecules and of the cortical cytoskeleton. Int J Biochem Cell Biol 40(3):344–349
287. Kondo T, Takeuchi K, Doi Y, Yonemura S, Nagata S, Tsukita S (1997) ERM (ezrin/radixin/moesin)-based molecular mechanism of microvillar breakdown at an early stage of apoptosis. J Cell Biol 139(3):749–758

288. Machnicka B, Czogalla A, Hryniewicz-Jankowska A, Bogusławska DM, Grochowalska R, Heger E, Sikorski AF (2014) Spectrins: a structural platform for stabilization and activation of membrane channels, receptors and transporters. Biochim Biophys Acta 1838(2):620–634
289. Machnicka B, Grochowalska R, Bogusławska DM, Sikorski AF, Lecomte MC (2012) Spectrin-based skeleton as an actor in cell signaling. Cell Mol Life Sci 69(2):191–201
290. Cohen CM, Tyler JM, Branton D (1980) Spectrin-actin associations studied by electron microscopy of shadowed preparations. Cell 21(3):875–883
291. Coleman TR, Fishkind DJ, Mooseker MS, Morrow JS (1989) Contributions of the beta-subunit to spectrin structure and function. Cell Motil Cytoskeleton 12(4):248–263
292. Tilney LG (1975) The role of actin in nonmuscle cell motility. Soc Gen Physiol Ser 30:339–388
293. Fowler VM, Davis JQ, Bennett V (1985) Human erythrocyte myosin: identification and purification. J Cell Biol 100(1):47–55
294. Bjorkman DJ, Allan CH, Hagen SJ, Trier JS (1986) Structural features of absorptive cell and microvillus membrane preparations from rat small intestine. Gastroenterol 91(6):1401–1414
295. Hull BE, Staehelin LA (1979) The terminal web. A reevaluation of its structure and function. J Cell Biol 81(1):67–82
296. Walton KD, Freddo AM, Wang S, Gumucio DL (2016) Generation of intestinal surface: an absorbing tale. Development 143(13):2261–2272
297. Leeson TS (1982) The terminal web of the duodenal enterocyte. J Anat 134(4):653–666
298. Xiao H, Wu R (2017) Quantitative investigation of human cell surface N-glycoprotein dynamics. Chem Sci 8:268–277
299. Christ L, Raiborg C, Wenzel EM, Campsteijn C, Stenmark H (2017) Cellular functions and molecular mechanisms of the ESCRT membrane-scission machinery. Trends Biochem Sci 42(1):42–56
300. Triffo SB, Huang HH, Smith AW, Chou ET, Groves JT (2012) Monitoring lipid anchor organization in cell membranes by PIE-FCCS. J Am Chem Soc 134(26):10833–10842
301. Paulick MG, Bertozzi CR (2008) The glycosylphosphatidylinositol anchor: a complex membrane-anchoring structure for proteins. Biochem 47(27):6991–7000
302. Smith HW (1962) The plasma membrane, with notes on the history of botany. Circulation 26:987–1012
303. Kraft ML (2013) Plasma membrane organization and function: moving past lipid rafts. Mol Biol Cell 24(18):2765–2768
304. Syrovatkina V, Alegre KO, Dey R, Huang XY (2016) Regulation, signaling, and physiological functions of G-proteins. J Mol Biol 428(19):3850–3868
305. Wu W, Shi X, Xu C (2016) Regulation of T cell signalling by membrane lipids. Nat Rev Immunol 16:690–701
306. Loganathan SK, Lukowski CM, Casey JR (2016) The cytoplasmic domain is essential for transport function of the integral membrane transport protein SLC4A11. Am J Physiol Cell Physiol 310(2):C161–C174
307. Levin ER (2002) Cellular functions of plasma membrane estrogen receptors. Steroids 67 (6):471–475
308. Bonifacino JS, Lippincott-Schwartz J (2003) Coat proteins: shaping membrane transport. Nat Rev Mol Cell Biol 4:409–414
309. Shull GE, Okunade G, Liu LH, Kozel P, Periasamy M, Lorenz JN, Prasad V (2003) Physiological functions of plasma membrane and intracellular Ca^{2+} pumps revealed by analysis of null mutants. Ann N Y Acad Sci 986:453–460
310. Seelig J (1993) Phospholipid headgroups as sensors of electric charge. New developments in lipid-protein interactions and receptor function. NATO ASI Series 246:241–248
311. Meyer H (1899) Zur Theorie der Alkoholnarkose. Erste Mittheilung. Welche Eigenschaft der Anasthetica bedingt ihre narkotis che Wirkung? (On the theory of alcohol narcosis. What property of anesthetics produces their anesthetic action?) Arch Exp Pathol Pharmakol 425:109–118

312. Overton CE (1991) Studien uber die Narkose, zugleich ein Beitrag zur allgemeiner Pharmakologie (Studies of narcosis). Verlag Gustav Fischer, Jena
313. Li S, Hu PC, Malmstadt N (2011) Imaging molecular transport across lipid bilayers. Biophys J 101(3):700–708
314. Chakrabarti AC, Deamer DW (1992) Permeability of lipid bilayers to amino acids and phosphate. Biochim Biophys Acta 1111(2):171–177
315. Gadsby DC (2009) Ion channels versus ion pumps: the principal difference, in principle. Mol Cell Biol 10:344–352
316. Itel F, Al-Samir S, Öberg F, Chami M, Kumar M, Supuran CT et al (2012) CO_2 permeability of cell membranes is regulated by membrane cholesterol and protein gas channels. FASEB J 26(12):5182–5191
317. Wang Y, Cohen J, Boron WF, Schulten K, Tajkhorshid E (2007) Exploring gas permeability of cellular membranes and membrane channels with molecular dynamics. J Struct Biol 157 (3):534–544
318. Ripoche P, Goossens D, Devuyst O, Gane P, Colin Y, Verkman AS, Cartron JP (2006) Role of RhAG and AQP1 in NH_3 and CO_2 gas transport in red cell ghosts: a stopped-flow analysis. Transfus Clin Biol 13(1–2):117–122
319. Peng J, Huang CH (2006) Rh proteins vs Amt proteins: an organismal and phylogenetic perspective on CO_2 and NH_3 gas channels. Transfus Clin Biol 13(1–2):85–94
320. Hub JS, de Groot BL (2006) Does CO_2 permeate through aquaporin-1? Biophys J 91 (3):842–848
321. Hub JS, de Groot BL (2008) Mechanism of selectivity in aquaporins and aquaglyceroporins. Proc Natl Acad Sci USA 105(4):1198–1203
322. Yu L, Villarreal OD, Chen LL, Chen LY (2016) 1,3-Propanediol binds inside the water-conducting pore of aquaporin 4: does this efficacious inhibitor have sufficient potency? J Syst Integr Neurosci 2(1):91–98
323. Assentoft M, Kaptan S, Schneider HP, Deitmer JW, de Groot BL, MacAulay N (2016) Aquaporin 4 as a NH_3 Channel. J Biol Chem 291(36):19184–19195
324. Al-Samir S, Goossens D, Cartron JP, Nielsen S, Scherbarth F, Steinlechner S et al (2016) Maximal oxygen consumption is reduced in aquaporin-1 knockout mice. Front Physiol 7:347
325. Missner A, Pohl P (2009) 110 years of the Meyer-Overton rule: predicting membrane permeability of gases and other small compounds. ChemPhysChem 10(9–10):1405–1414
326. Al-Awqati Q (1999) One hundred years of membrane permeability: does Overton still rule? Nat Cell Biol 1:E201–E202
327. Missner A, Kügler P, Antonenko YN, Pohl P (2008) Passive transport across bilayer lipid membranes: overton continues to rule. PNAS 105(52):E123
328. Grime JMA, Edwards MA, Rudd NC, Unwin PR (2008) Quantitative visualization of passive transport across bilayer lipid membranes. Proc Natl Acad Sci USA 105:14277–14282
329. Farinas J, Kneen M, Moore M, Verkman AS (1997) Plasma membrane water permeability of cultured cells and epithelia measured by light microscopy with spatial filtering. J Gen Physiol 110(3):283–296
330. Falath W, Sabir A, Jacob KI (2017) Novel reverse osmosis membranes composed of modified PVA/Gum Arabic conjugates: Biofouling mitigation and chlorine resistance enhancement. Carbohydr Polym 155:28–39
331. Shannon MA (2010) Water desalination: fresh for less. Nat Nanotechnol 5:248–250
332. Xie M, Gray SR (2017) Silica scaling in forward osmosis: from solution to membrane interface. Water Res 108:232–239
333. Conde A, Diallinas G, Chaumont F, Chaves M, Gerós H (2010) Transporters, channels, or simple diffusion? Dogmas, atypical roles and complexity in transport systems. Int J Biochem Cell Biol 42(6):857–868
334. Madeira A, Moura TF, Soveral G (2016) Detecting aquaporin function and regulation. Front Chem 4:3

335. Kido Y, Tamai I, Okamoto M, Suzuki F, Tsuji A (2000) Functional clarification of MCT1-mediated transport of monocarboxylic acids at the blood-brain barrier using in vitro cultured cells and in vivo BUI studies. Pharm Res 17(1):55–62
336. Takanaga H, Tamai I, Tsuji A (1994) pH-dependent and carrier-mediated transport of salicylic acid across Caco-2 cells. J Pharm Pharmacol 46(7):567–570
337. Gross E, Hopfer U (1998) Voltage and cosubstrate dependence of the Na-HCO_3 cotransporter kinetics in renal proximal tubule cells. Biophys J 75(2):810–824
338. Bennett E, Kimmich GA (1996) The molecular mechanism and potential dependence of the Na^+/glucose cotransporter. Biophys J 70(4):1676–1688
339. King RB (1996) Modeling membrane transport. Adv Food Nutr Res 40:243–262
340. Arcizet D, Meier B, Sackmann E, Rädler JO, Heinrich D (2008) Temporal analysis of active and passive transport in living cells. Phys Rev Lett 101:248103
341. Carruthers A (1991) Mechanisms for the facilitated diffusion of substrates across cell membranes. Biochem 30(16):3898–3906
342. Barta E, Sideman S, Bassingthwaighte JB (2000) Facilitated diffusion and membrane permeation of fatty acid in albumin solutions. Ann Biomed Eng 28(3):331–345
343. Yang NJ, Hinner MJ (2015) Getting across the cell membrane: an overview for small molecules, peptides, and proteins. Methods Mol Biol 1266:29–53
344. Kvietys PR, Granger DN (2010) Role of intestinal lymphatics in interstitial volume regulation and transmucosal water transport. Ann NY Acad Sci 1207(Suppl 1):E29–E43
345. Hill AV (1910) The possible effects of the aggregation of the molecules of haemoglobin on its dissociation curves. J Physiol 40 (Suppl):4–7
346. Holt J, Ackers GK (2009) The hill coefficient: inadequate resolution of cooperativity in human hemoglobin. Method Enzymol 455:193–212
347. Mandel LJ (1986) Primary active sodium transport, oxygen consumption, and ATP: coupling and regulation. Kidney Int 29(1):3–9
348. Zeuthen T (1991) Secondary active transport of water across ventricular cell membrane of choroid plexus epithelium of Necturus maculosus. J Physiol 444:153–173
349. Shi Y (2013) Common folds and transport mechanisms of secondary active transporters. Annu Rev Biophys 42:51–72
350. Quistgaard EM, Löw C, Guettou F, Nordlund P (2016) Understanding transport by the major facilitator superfamily (MFS): structures pave the way. Nat Rev Mol Cell Biol 17:123–132
351. Gayen A, Leninger M, Traaseth NJ (2016) Protonation of a glutamate residue modulates the dynamics of the drug transporter EmrE. Nat Chem Biol 12:141–145
352. Shechter E (1986) Secondary active transport. Biochimie 68(3):357–365
353. Goswitz VC, Brooker RJ (1995) Structural features of the uniporter/symporter/antiporter superfamily. Protein Sci 4(3):534–537
354. Wolfersberger MG (1994) Uniporters, symporters and antiporters. J Exp Biol 196:5–6
355. Zhang Y, Voth GA (2011) The coupled proton transport in the ClC-ec1 Cl^-/H^+ Antiporter. Biophys J 101(10):L47–L49
356. Tarek M (2014) Prerequisites to proton transport in the bacterial ClC-ec1 Cl^-/H^+ exchanger. Proc Natl Acad Sci USA 111(5):1668–1669
357. Reeves JP, Condrescu M, Chernaya G, Gardner JP (1994) Na^+/Ca^{2+} antiport in the mammalian heart. J Exp Biol 196:375–388
358. Dudeja PK, Foster ES, Brasitus TA (1989) Na^+-H^+ antiporter of rat colonic basolateral membrane vesicles. Am J Physiol 257(4 Pt 1):G624–G632
359. Hunte C, Screpanti E, Venturi M, Rimon A, Padan E, Michel H (2005) Structure of a Na^+/H^+ antiporter and insights into mechanism of action and regulation by pH. Nature 435 (7046):1197–1202
360. Arkin IT, Xu H, Jensen MØ, Arbely E, Bennett ER, Bowers KJ et al (2007) Mechanism of Na^+/H^+ antiporting. Science 317(5839):799–5803
361. Krulwich TA (1983) Na^+/H^+ antiporters. Biochim Biophys Acta 726(4):245–264

362. Geng J, Kim K, Zhang J, Escalada A, Tunuguntla R, Comolli LR et al (2014) Stochastic transport through carbon nanotubes in lipid bilayers and live cell membranes. Nature 514:612–615
363. Du Toit A (2015) Endocytosis—a new gateway into cells. Nat Rev Mol Cell Biol 16:68
364. Haucke V (2015) Cell biology: on the endocytosis rollercoaster. Nature 517:446–447
365. Houy S, Croisé P, Gubar O, Chasserot-Golaz S, Tryoen-Tóth P, Bailly Y et al (2013) Exocytosis and endocytosis in neuroendocrine cells: inseparable membranes! Front Endocrinol (Lausanne) 4:135
366. Frazier MN, Jackson LP (2017) Watching real-time endocytosis in living cells. J Cell Biol 216(1):9–11
367. Lang T (2008) Imaging Ca^{2+} -triggered exocytosis of single secretory granules on plasma membrane lawns from neuroendocrine cells. Methods Mol Biol 440:51–59
368. Kreft M, Jorgačevski J, Vardjan N, Zorec R (2016) Unproductive exocytosis. J Neurochem 137(6):880–889
369. Charras GT, Mitchison TJ, Mahadevan L (2009) Animal cell hydraulics. J Cell Sci 122(Pt 18):3233–3241
370. Ziviani E, Scorrano L (2016) Cell biology: the organelle replication connection. Nature 538:326–327
371. Lu Q, Haragopal H, Slepchenko KG, Stork C, Li YV (2016) Intracellular zinc distribution in mitochondria, ER and the Golgi apparatus. Int J Physiol Pathophysiol Pharmacol 8(1):35–43
372. Suarez C, Kovar DR (2016) Internetwork competition for monomers governs actin cytoskeleton organization. Nat Rev Mol Cell Biol 17:799–810
373. Strzyz P (2016) Cytoskeleton—a new face of the centrosome. Nat Rev Mol Cell Biol 17:66–67
374. Schwarz DS, Blower MD (2016) The endoplasmic reticulum: structure, function and response to cellular signaling. Cell Mol Life Sci 73(1):79–94
375. Phillips MJ, Voeltz GK (2016) Structure and function of ER membrane contact sites with other organelles. Nat Rev Mol Cell Biol 17:69–82
376. Liu D, Zhang M, Yin H (2013) Signaling pathways involved in endoplasmic reticulum stress-induced neuronal apoptosis. Int J Neurosci 123(3):155–162
377. Mezzacasa A, Helenius A (2002) The transitional ER defines a boundary for quality control in the secretion of tsO45 VSV glycoprotein. Traffic 3(11):833–849
378. Shindiapina P, Barlowe C (2010) Requirements for transitional endoplasmic reticulum site structure and function in *Saccharomyces cerevisiae*. Mol Biol Cell 21(9):1530–1545
379. Iurlaro R, Muñoz-Pinedo C (2016) Cell death induced by endoplasmic reticulum stress. FEBS J 283(14):2640–2652
380. Valastyan JS, Lindquist S (2014) Mechanisms of protein-folding diseases at a glance. Dis Models Mech 7:9–14
381. Hartl FU, Hayer-Hartl M (2009) Converging concepts of protein folding in vitro and in vivo. Nat Struct Mol Biol 16:574–581
382. Haas IG (1994) BiP (GRP78), an essential hsp70 resident protein in the endoplasmic reticulum. Experientiam 50(11):1012–1020
383. Chung KT, Shen Y, Hendershot LM (2002) BAP, a mammalian BiP-associated protein, is a nucleotide exchange factor that regulates the ATPase activity of BiP. J Biol Chem 277 (49):47557–47563
384. Ledford BE, Leno GH (1994) ADP-ribosylation of the molecular chaperone GRP78/BiP. Mol Cell Biochem 138(1–2):141–148
385. Fu J, Zhao L, Wang L, Zhu X (2015) Expression of markers of endoplasmic reticulum stress-induced apoptosis in the placenta of women with early and late onset severe pre-eclampsia. Taiwan J Obstetr Gynecol 54(1):19–23
386. Su J, Zhou L, Kong X, Yang X, Xiang X, Zhang Y et al (2013) Endoplasmic reticulum is at the crossroads of autophagy, inflammation, and apoptosis signaling pathways and participates in the pathogenesis of diabetes mellitus. J Diabetes Res 2013:193461

387. Taniguchi M, Yoshida H (2015) Endoplasmic reticulum stress in kidney function and disease. Curr Opin Nephrol Hypertens 24(4):345–350
388. Teske BF, Wek SA, Bunpo P, Cundiff JK, McClintick JN, Anthony TG, Wek RC (2011) The eIF2 kinase PERK and the integrated stress response facilitate activation of ATF6 during endoplasmic reticulum stress. Mol Biol Cell 22(22):4390–4405
389. Ozcan L, Tabas I (2012) Role of endoplasmic reticulum stress in metabolic disease and other disorders. Annu Rev Med 63:317–328
390. Connerly PL (2010) How do proteins move through the golgi apparatus? Nat Educ 3(9):60
391. Pfeffer SR (2001) Constructing a Golgi complex. J Cell Biol 155(6):873–876
392. Nakano A (2015) Cell biology: polarized transport in the Golgi apparatus. Nature 521 (7553):427–428
393. Andreeva AV, Kutuzov MA, Evans DE, Hawes CR (1998) The structure and function of the Golgi apparatus: a hundred years of questions. J Exp Bot 49(325):1281–1291
394. De Matteis MA, Mironov AA, Beznoussenko GV (2008) The Golgi ribbon and the function of the Golgins. In: Mironov AA, Pavelka M (eds) The Golgi apparatus. Springer, Vienna
395. Malhotra V, Mayor S (2006) Cell biology: the Golgi grows up. Nature 441:939–940
396. Pelham HR (2001) Traffic through the Golgi apparatus. J Cell Biol 155(7):1099–1102
397. Storrie B (2005) Maintenance of Golgi apparatus structure in the face of continuous protein recycling to the endoplasmic reticulum: making ends meet. Int Rev Cytol 244:69–94
398. Gonatas NK (1997) The Golgi apparatus in disease. In: Berger EG, Roth J (eds) The Golgi apparatus. Birkhäuser, Basel
399. Vanier MT (2010) Niemann-pick disease type C. Orphanet J Rare Dis 5:16
400. Jadhav S, Jadhav A, Thopte S, Marathe S, Vhathakar P, Chivte P, Jamkhande A (2015) Sjögren's syndrome: a case study. J Int Oral Health 7(3):72–74
401. Di Ngo YJ, Thomson WM, Nolan A, Ferguson S (2016) The lived experience of Sjögren's syndrome. BMC Oral Health 16:7
402. Nozawa K, Fritzler MJ, von Mühlen CA, Chan EKL (2004) Giantin is the major Golgi autoantigen in human anti-Golgi complex sera. Arthritis Res Ther 6(2):R95–R102
403. Manson JJ, Rahman A (2006) Systemic lupus erythematosus. Orphanet J Rare Dis 1:6
404. Williams PH, Cobb BL, Namjou B, Scofield RH, Sawalha AH, Harley JB (2007) Horizons in Sjögren's syndrome genetics. Clin Rev Allergy Immunol 32(3):201–209
405. Gleeson PA, Anderson TJ, Stow JL, Griffiths G, Toh BH, Matheson F (1996) p230 is associated with vesicles budding from the trans-Golgi network. J Cell Sci 109(Pt 12):2811–2821
406. Fridmann-Sirkis Y, Siniossoglou S, Pelham HRB (2004) TMF is a golgin that binds Rab6 and influences Golgi morphology. BMC Cell Biol 5:18
407. Barr FA, Short B (2003) Golgins in the structure and dynamics of the Golgi apparatus. Curr Opin Cell Biol 15(4):405–413
408. Narula N, Stow JL (1995) Distinct coated vesicles labeled for p200 bud from trans-Golgi network membranes. Proc Natl Acad Sci USA 92(7):2874–2878
409. Weide T, Bayer M, Köster M, Siebrasse JP, Peters R, Barnekow A (2001) The Golgi matrix protein GM130: a specific interacting partner of the small GTPase rab1b. EMBO Rep 2:336–341
410. Diao A, Rahman D, Pappin DJC, Lucocq J, Lowe M (2003) The coiled-coil membrane protein golgin-84 is a novel rab effector required for Golgi ribbon formation. J Cell Biol 160 (2):201–212
411. Infante C, Ramos-Morales F, Fedriani C, Bornens M, Rios RM (1999) GMAP-210, A cis-Golgi network-associated protein, is a minus end microtubule-binding protein. J Cell Biol 145(1):83–98
412. Sohda M, Misumi Y, Ogata S, Sakisaka S, Hirose S, Ikehara Y, Oda K (2015) Trans-Golgi protein p 230/golgin-245 is involved in phagophore formation. Biochem Biophys Res Comm 456(1, 2):275–281

413. Lu L, Tai G, Hong W (2004) Autoantigen Golgin-97, an effector of Arl1 GTPase, participates in traffic from the endosome to the trans-Golgi network. Mol Biol Cell 15 (10):4426–4443
414. Roels F, Espeel M, Pauwels M, De Craemer D, Egberts HJ, van der Spek P (1991) Different types of peroxisomes in human duodenal epithelium. Gut 32(8):858–865
415. Gabaldón T (2010) Peroxisome diversity and evolution. Philos Trans R Soc Lond B Biol Sci 365(1541):765–773
416. Terlecky SR, Terlecky LJ, Giordano CR (2012) Peroxisomes, oxidative stress, and inflammation. World J Biol Chem 3(5):93–97
417. South ST, Gould SJ (1999) Peroxisome synthesis in the absence of preexisting peroxisomes. J Cell Biol 144(2):255–266
418. Smith JJ, Aitchison JD (2013) Peroxisomes take shape. Nat Rev Mol Cell Biol 14:803–817
419. Mawatari S, Hazeyama S, Fujino T (2016) Measurement of ether phospholipids in human plasma with HPLC-ELSD and LC/ESI-MS after hydrolysis of plasma with phospholipase A1. Lipids 51(8):997–1006
420. Hashimoto M, Hossain S, Al Mamun A, Matsuzaki K, Arai H (2017) Docosahexaenoic acid: one molecule diverse functions. Crit Rev Biotechnol 37(5):579–597
421. Marszalek JR, Lodish HF (2005) Docosahexaenoic acid, fatty acid-interacting proteins, and neuronal function: breastmilk and fish are good for you. Annu Rev Cell Dev Biol 21:633–657
422. Siddiqui RA, Harvey KA, Xu Z, Bammerlin EM, Walker C, Altenburg JD (2011) Docosahexaenoic acid: a natural powerful adjuvant that improves efficacy for anticancer treatment with no adverse effects. BioFactors 37(6):399–412
423. Terano T, Fujishiro S, Ban T, Yamamoto K, Tanaka T, Noguchi Y et al (1999) Docosahexaenoic acid supplementation improves the moderately severe dementia from thrombotic cerebrovascular diseases. Lipids 34(Suppl):S345–S346
424. Holub BJ (2009) Docosahexaenoic acid (DHA) and cardiovascular disease risk factors. Prostaglandins Leukot Essent Fatty Acids 81(2–3):199–204
425. Allaire J, Couture P, Leclerc M, Charest A, Marin J, Lépine MC et al (2016) A randomized, crossover, head-to-head comparison of eicosapentaenoic acid and docosahexaenoic acid supplementation to reduce inflammation markers in men and women: the Comparing EPA to DHA (ComparED) Study. Am J Clin Nutr 104(2):280–287
426. Horrocks LA, Yeo YK (1999) Health benefits of docosahexaenoic acid (DHA). Pharmacol Res 40(3):211–225
427. FitzPatrick DR (1996) Zellweger syndrome and associated phenotypes. J Med Genet 33 (10):863–868
428. Brul S, Westerveld A, Strijland A, Wanders R, Schram A, Heymans H et al (1988) Genetic heterogeneity in the cerebrohepatorenal (Zellweger) syndrome and other inherited disorders with a generalized impairment of peroxisomal functions. A study using complementation analysis. J Clin Invest 81(6):1710–1715
429. Wiedemann HR (1991) Hans-Ulrich Zellweger (1909-1990). Eur J Pediatr 150(7):451
430. Rosewich H, Waterham H, Poll-The BT, Ohlenbusch A, Gärtner J (2015) Clinical utility gene card for: Zellweger syndrome spectrum. Eur J Hum Genet 23(8):1111
431. Steinberg SJ, Raymond GV, Braverman NE, Moser AB (2003) Zellweger spectrum disorder. In: Adam MP, Ardinger HH, Pagon RA, Wallace SE, Bean LJH, Mefford HC et al (eds) GeneReviews. Seattle
432. Luzio JP, Pryor PR, Bright NA (2007) Lysosomes: fusion and function. Nat Rev Mol Cell Biol 8:622–632
433. Xu H, Ren D (2015) Lysosomal physiology. Annu Rev Physiol 77:57–80
434. Samie MA, Xu H (2014) Lysosomal exocytosis and lipid storage disorders. J Lipid Res 55 (6):995–1009
435. de Duve C (2005) The lysosome turns fifty. Nat Cell Biol 7:847–849
436. Sabatini DD, Adesnik M (2013) Christian de Duve: explorer of the cell who discovered new organelles by using a centrifuge. Proc Natl Acad Sci USA 110(33):13234–13235

437. Opperdoes F (2013) A feeling for the cell: Christian de Duve (1917–2013). PLoS Biol 11 (10):e1001671
438. Burd CG (2011) Physiology and pathology of endosome-to-Golgi retrograde sorting. Traffic 12(8):948–955
439. Bharadwaj R, Cunningham KM, Zhang K, Lloyd TE (2016) FIG4 regulates lysosome membrane homeostasis independent of phosphatase function. Hum Mol Genet 25(4):681–692
440. Kornfeld S (1986) Trafficking of lysosomal enzymes in normal and disease states. J Clin Invest 77(1):1–6
441. Kornfeld S (1987) Trafficking of lysosomal enzymes. FASEB J 1(6):462–468
442. von Figura K, Hasilik A (1986) Lysosomal enzymes and their receptors. Annu Rev Biochem 55:167–193
443. Jaishy B, Abel ED (2016) Lipids, lysosomes, and autophagy. J Lipid Res 57(9):1619–1635
444. Sinha D, Valapala M, Shang P, Hose S, Grebe R, Lutty GA et al (2016) Lysosomes: Regulators of autophagy in the retinal pigmented epithelium. Exp Eye Res 144:46–53
445. Amar N, Lustig G, Ichimura Y, Ohsumi Y, Elazar Z (2006) Two newly identified sites in the ubiquitin-like protein Atg8 are essential for autophagy. EMBO Rep 7:635–642
446. Platt FM, Boland B, van der Spoel AC (2012) Lysosomal storage disorders: the cellular impact of lysosomal dysfunction. J Cell Biol 199(5):723–734
447. Filocamo M, Morrone A (2011) Lysosomal storage disorders: molecular basis and laboratory testing. Hum Genomics 5:156
448. Schulze H, Sandhoff K (2011) Lysosomal lipid storage diseases. Cold Spring Harb Perspect Biol 3(6):a004804
449. Vellodi A (2005) Lysosomal storage disorders. Br J Haematol 128(4):413–431
450. Hers HG (1963) α-Glucosidase deficiency in generalized glycogenstorage disease (Pompe's disease). Biochem Jl 86:11–16
451. Myerowitz R (1997) Tay-Sachs disease-causing mutations and neutral polymorphisms in the Hex A gene. Hum Mutat 9(3):195–208
452. Myerowitz R, Lawson D, Mizukami H, Mi Y, Tifft CJ, Proia RL (2002) Molecular pathophysiology in Tay-Sachs and Sandhoff diseases as revealed by gene expression profiling. Hum Mol Genet 11(11):1343–1350
453. Dersh D, Iwamoto Y, Argon Y (2016) Tay-Sachs disease mutations in HEXA target the α chain of hexosaminidase A to endoplasmic reticulum-associated degradation. Mol Biol Cell 27(24):3813–3827
454. Mocchetti I (2005) Exogenous gangliosides, neuronal plasticity and repair, and the neurotrophins. Cell Mol Life Sci 62(19–20):2283–2294
455. Kaye EM (2001) Lysosomal storage diseases. Curr Treat Options Neurol 3(3):249–256
456. Jerome GW (2010) Lysosomes, cholesterol and atherosclerosis. Clin Lipidol 5(6):853–865
457. Sherratt HS (1991) Mitochondria: structure and function. Rev Neurol (Paris) 147(6–7):417–430
458. Picard M, Taivassalo T, Gouspillou G, Hepple RT (2011) Mitochondria: isolation, structure and function. J Physiol 589(Pt 18):4413–4421
459. Sylvester JE, Fischel-Ghodsian N, Mougey EB, O'brien TW (2004) Mitochondrial ribosomal proteins: candidate genes for mitochondrial disease. Genet Med 6:73–80
460. Mourier A, Devin A, Rigoulet M (2010) Active proton leak in mitochondria: a new way to regulate substrate oxidation. Biochim Biophys Acta 1797(2):255–261
461. Rigoulet M, Mourier A, Galinier A, Casteilla L, Devin A (2010) Electron competition process in respiratory chain: regulatory mechanisms and physiological functions. Biochim Biophys Acta 1797(6–7):671–677
462. Lee YM (2012) Mitochondrial diseases. J Epilepsy Res 2(1):1–4
463. Parikh S, Goldstein A, Koenig MK, Scaglia F, Enns GM, Saneto R et al (2015) Diagnosis and management of mitochondrial disease: a consensus statement from the Mitochondrial Medicine Society. Genet Med 17(9):689–701
464. Schapira AHV (2012) Mitochondrial diseases. Lancet 379(9828):1825–1834

465. Lamond AI, Earnshaw WC (1998) Structure and function in the nucleus. Science 280 (5363):547–553
466. Cardoso MC, Sporbert A, Leonhardt H (1999) Structure and function in the nucleus: subnuclear trafficking of DNA replication factors. J Cell Biochem 32–33:15–23
467. Newport JW, Forbes DJ (1987) The nucleus: structure, function, and dynamics. Annu Rev Biochem 56:535–565
468. Crick F (1970) Central dogma of molecular biology. Nature 227(5258):561–563
469. Wilkins MHF, Randall JT (1953) Crystallinity in sperm heads: molecular structure of nucleoprotein in vivo. Biochem Biophys Acta 10:192–193
470. Wilkins MH, Stokes AR, Wilson HR (1953) Molecular structure of deoxypentose nucleic acids. Nature 171(4356):738–740
471. Watson JD, Crick FH (1953) Molecular structure of nucleic acids; a structure for deoxyribose nucleic acid. Nature 171(4356):737–738
472. Shapiro JA (2009) Revisiting the central dogma in the 21st century. Ann N Y Acad Sci 1178:6–28
473. Nierhaus KH (1982) Structure, assembly, and function of ribosomes. Curr Topics Microbiol Immunol 97:81–155
474. Koonin EV (2012) Does the central dogma still stand? Biol Direct 7:27
475. Travers A, Muskhelishvili G (2015) DNA structure and function. FEBS J 282(12):2279–2295
476. Chouard T (2011) Structural biology: breaking the protein rules. Nature 471:151–153
477. Rigden DJ, Fernández-Suárez XM, Galperin MY (2016) The 2016 database issue of nucleic acids research and an updated molecular biology database collection. Nucleic Acids Res 44 (Database issue):D1–D6
478. Varshney GK, Burgess SM (2016) DNA-guided genome editing using structure-guided endonucleases. Genome Biol 17:187
479. Palazzo AF, Gregory TR (2014) The Case for Junk DNA. PLoS Genet 10(5):e1004351
480. Bacolla A, Wang G, Vasquez KM (2015) New perspectives on DNA and RNA triplexes as effectors of biological activity. PLoS Genet 11(12):e1005696
481. Leonhardt H, Rahn HP, Cardoso MC (1999) Functional links between nuclear structure, gene expression, DNA replication, and methylation. Crit Rev Eukaryot Gene Expr 9(3–4):345–351
482. Wang B, Yanez A, Novina CD (2008) MicroRNA-repressed mRNAs contain 40S but not 60S components. Proc Natl Acad Sci USA 105(14):5343–5348
483. Krokowski D, Gaccioli F, Majumder M, Mullins MR, Yuan CL, Barbara P et al (2011) Characterization of hibernating ribosomes in mammalian cells. Cell Cycle 10(16):2691–2702
484. Auxilien S, Rasmussen A, Rose S, Brochier-Armanet C, Husson C, Fourmy D et al (2011) Specificity shifts in the rRNA and tRNA nucleotide targets of archaeal and bacterial m5U methyltransferases. RNA 17(1):45–53
485. Neumann S, Petfalski E, Brügger B, Großhans H, Wieland F, Tollervey D, Hurt E (2003) Formation and nuclear export of tRNA, rRNA and mRNA is regulated by the ubiquitin ligase Rsp5p. EMBO Rep 4(12):1156–1162
486. Abraham M, Dror O, Nussinov R, Wolfson HJ (2008) Analysis and classification of RNA tertiary structures. RNA 14(11):2274–2289
487. Pritchard CC, Cheng HH, Tewari M (2012) MicroRNA profiling: approaches and considerations. Nat Rev Genet 13:358–369
488. Panwar B, Arora A, Raghava GPS (2014) Prediction and classification of ncRNAs using structural information. BMC Genom 15:127
489. Moreno-Moya JM, Vilella F, Simón C (2014) MicroRNA: key gene expression regulators. Fertil Steril 101(6):1516–1523
490. Peskoe SB, Barber JR, Zheng Q, Meeker AK, De Marzo AM, Platz EA, Lupold SE (2017) Differential long-term stability of microRNAs and RNU6B snRNA in 12–20 year old archived formalin-fixed paraffin-embedded specimens. BMC Cancer 17:32

491. Hung T, Chang HY (2010) Long noncoding RNA in genome regulation: prospects and mechanisms. RNA Biol 7(5):582–585
492. McFadden EJ, Hargrove AE (2016) Biochemical methods to investigate lncRNA and the influence of lncRNA: protein complexes on chromatin. Biochem 55(11):1615–1630
493. Hernandez-Verdun D, Roussel P, Thiry M, Sirri V, Lafontaine DL (2010) The nucleolus: structure/function relationship in RNA metabolism. Wiley Interdiscip Rev RNA 1(3):415–431
494. Enyedi B, Jelcic M, Niethammer P (2016) The cell nucleus serves as a mechanotransducer of tissue damage-induced inflammation. Cell 165(5):1160–1170
495. Scaffidi P, Gordon L, Misteli T (2005) The cell nucleus and aging: tantalizing clues and hopeful promises. PLoS Biol 3(11):e395
496. Wiener J, Spiro D, Loewenstein WR (1965) Ultrastructure and permeability of nuclear membranes. J Cell Biol 27(1):107–117
497. Gay H (1956) Chromosome-nuclear membrane-cytoplasmic interrelations in drosophila. J Biophys Biochem Cytol 2(4):407–414
498. Dechat T, Pfleghaar K, Sengupta K, Shimi T, Shumaker DK, Solimando L, Goldman RD (2008) Nuclear lamins: major factors in the structural organization and function of the nucleus and chromatin. Genes Dev 22(7):832–853
499. Hoelz A, Glavy JS, Beck M (2016) Toward the atomic structure of the nuclear pore complex: when top down meets bottom up. Nat Struct Mol Biol 23:624–630
500. Beck M, Hurt E (2017) The nuclear pore complex: understanding its function through structural insight. Nat Rev Mol Cell Biol 18:73–89
501. Worman HJ (2012) Nuclear lamins and laminopathies. J Pathol 226:316–325
502. Lund E, Collas P (2013) Nuclear lamins—making contacts with promoters. Nucleus 4 (6):424–430
503. Burke B, Stewart CL (2013) The nuclear lamins: flexibility in function. Nat Rev Mol Cell Biol 14:13–24
504. Shumaker DK, Kuczmarski ER, Goldman RD (2003) The nucleoskeleton: lamins and actin are major players in essential nuclear functions. Curr Opin Cell Biol 15(3):358–366
505. Dittmer TA, Misteli T (2011) The lamin protein family. Genome Biol 12(5):222
506. Yang W, Musser SM (2006) Nuclear import time and transport efficiency depend on importin β concentration. J Cell Biol 174(7):951–961
507. Wozniak RW, Lusk CP (2003) Nuclear pore complexes. Curr Biol 13(5):R169
508. Ribbeck K, Görlich D (2001) Kinetic analysis of translocation through nuclear pore complexes. EMBO J 20(6):1320–1330
509. Woulfe JM (2007) Abnormalities of the nucleus and nuclear inclusions in neurodegenerative disease: a work in progress. Neuropathol Appl Neurobiol 33(1):2–42
510. Worman HJ, Courvalin JC (2004) How do mutations in lamins A and C cause disease? J Clin Invest 113(3):349–351
511. Gruenbaum Y, Medalia O (2015) Lamins: the structure and protein complexes. Curr Opin Cell Biol 32:7–12
512. Butin-Israeli V, Adam SA, Goldman AE, Goldman RD (2012) Nuclear lamin functions and disease. Trends Genet 28(9):464–471
513. Wilson DN, Cate JHD (2012) The structure and function of the eukaryotic ribosome. Cold Spring Harb Perspect Biol 4(5):a011536
514. Earnest TM, Cole JA, Peterson JR, Hallock MJ, Kuhlman TE, Luthey-Schulten Z (2016) Ribosome biogenesis in replicating cells: integration of experiment and theory. Biopolymers 105(10):735–751
515. Sanchez CG, Teixeira FK, Czech B, Preall JB, Zamparini AL, Seifert JR et al (2016) Regulation of ribosome biogenesis and protein synthesis controls germline stem cell differentiation. Cell Stem Cell 18(2):276–290
516. Zhang B, Cech TR (1998) Peptidyl-transferase ribozymes: trans reactions, structural characterization and ribosomal RNA-like features. Chem Biol 5(10):539–553
517. Oeffinger M (2016) Structural biology: moulding the ribosome. Nature 537:38–40

518. Brandman O, Hegde RS (2016) Ribosome-associated protein quality control. Nat Struct Mol Biol 23:7–15
519. Englmeier R, Pfeffer S, Förster F (2017) Structure of the human mitochondrial ribosome studied in situ by cryoelectron tomography. Structure 25(10):1574–1581
520. Richman TR, Rackham O, Filipovska A (2014) Mitochondria: unusual features of the mammalian mitoribosome. Int J Biochem Cell Biol 53:115–120
521. Shenoy N, Kessel R, Bhagat TD, Bhattacharyya S, Yu Y, Mcmahon C, Verma A (2012) Alterations in the ribosomal machinery in cancer and hematologic disorders. J Hematol Oncol 5:32
522. Korobeinikova AV, Garber MB, Gongadze GM (2012) Ribosomal proteins: structure, function, and evolution. Biochem (Mosc) 77(6):562–574
523. Freed EF, Bleichert F, Dutca LM, Baserga SJ (2010) When ribosomes go bad: diseases of ribosome biogenesis. Mol BioSyst 6(3):481–493
524. Stein K, Chiang HL (2014) Exocytosis and endocytosis of small vesicles across the plasma membrane in *Saccharomyces cerevisiae*. Membranes (Basel) 4(3):608–629
525. Wu LG, Hamid E, Shin l, Chiang HC (2014) Exocytosis and endocytosis: modes, functions, and coupling mechanisms. Annu Rev Physiol 76:301–331
526. Visentin M, Diop-Bove N, Zhao R, Goldman ID (2014) The intestinal absorption of folates. Annu Rev Physiol 76:251–274
527. Pelaseyed T, Bergström JH, Gustafsson JK, Ermund A, Birchenough GM, Schütte A et al (2014) The mucus and mucins of the goblet cells and enterocytes provide the first defense line of the gastrointestinal tract and interact with the immune system. Immunol Rev 260 (1):8–20
528. Yamauchi K, Tarachai P (2000) Changes in intestinal villi, cell area and intracellular autophagic vacuoles related to intestinal function in chickens. Br Poult Sci 41(4):416–423
529. Wj Luo, Chang A (2000) An endosome-to-plasma membrane pathway involved in trafficking of a mutant plasma membrane ATPase in yeast. Mol Biol Cell 11(2):579–592
530. van IJzendoorn SCD (2006) Recycling endosomes. J Cell Sci 119:1679–1681
531. Mainou BA, Dermody TS (2012) Transport to late endosomes is required for efficient reovirus infection. J Virol 86(16):8346–8358
532. Bonifacino JS, Rojas R (2006) Retrograde transport from endosomes to the trans-Golgi network. Nat Rev Mol Cell Biol 7:568–579
533. Cai H, Reinisch K, Ferro-Novick S (2007) Coats, tethers, Rabs, and SNAREs work together to mediate the intracellular destination of a transport vesicle. Dev Cell 12(5):671–682
534. Murray JW, Wolkoff AW (2003) Roles of the cytoskeleton and motor proteins in endocytic sorting. Adv Drug Deliv Rev 55(11):1385–1403
535. Rogers SL, Gelfand VI (2000) Membrane trafficking, organelle transport, and the cytoskeleton. Curr Opin Cell Biol 12(1):57–62
536. Damania D, Subramanian H, Tiwari AK, Stypula Y, Kunte D, Pradhan P et al (2010) Role of cytoskeleton in controlling the disorder strength of cellular nanoscale architecture. Biophys J 99(3):989–996
537. Fletcher DA, Mullins RD (2010) Cell mechanics and the cytoskeleton. Nature 463 (7280):485–492
538. Moseley JB (2013) An expanded view of the eukaryotic cytoskeleton. Mol Biol Cell 24 (11):1615–1618
539. Mostowy S, Shenoy AR (2015) The cytoskeleton in cell-autonomous immunity: structural determinants of host defence. Nat Rev Immunol 15(9):559–573
540. Szymanski WG, Zauber H, Erban A, Gorka M, Wu XN, Schulze WX (2015) Cytoskeletal components define protein location to membrane microdomains. Mol Cell Proteomics 14 (9):2493–2509
541. Ramaekers FCS, Bosman FT (2004) The cytoskeleton and disease. J Pathol 204:351–354
542. Herrmann H, Strelkov SV, Burkhard P, Aebi U (2009) Intermediate filaments: primary determinants of cell architecture and plasticity. J Clin Invest 119(7):1772–1783

543. Steinert PM, Jones JC, Goldman RD (1984) Intermediate filaments. J Cell Biol 99(1 Pt 2):22s–27s
544. Fuchs E, Weber K (1994) Intermediate filaments: structure, dynamics, function and disease. Annu Rev Biochem 63:345–382
545. Köster S, Weitz D, Goldman RD, Aebi U, Herrmann H (2015) Intermediate filament mechanics in vitro and in the cell: from coiled coils to filaments, fibers and networks. Curr Opin Cell Biol 32:82–91
546. Xiao S, McLean J, Robertson J (2006) Neuronal intermediate filaments and ALS: a new look at an old question. Biochim Biophys Acta 1762(11–12):1001–1012
547. Majumdar D, Tiernan JP, Lobo AJ, Evans CA, Corfe BM (2012) Keratins in colorectal epithelial function and disease. Int J Exp Pathol 93(5):305–318
548. Fuchs E (1996) The cytoskeleton and disease: genetic disorders of intermediate filaments. Annu Rev Genet 30:1–702
549. Kucharzik T, Lugering N, Schmid K, Schmidt M, Stoll R, Domschke W (1998) Human intestinal M cells exhibit enterocyte-like intermediate filaments. Gut 42(1):54–62
550. Coch RA, Leube RE (2016) Intermediate filaments and polarization in the intestinal epithelium. Cells 5(3):32
551. Lu XY, Chen XX, Huang LD, Zhu CQ, Gu YY, Ye S (2010) Anti-alpha-internexin autoantibody from neuropsychiatric lupus induce cognitive damage via inhibiting axonal elongation and promote neuron apoptosis. PLoS ONE 5(6):e11124
552. Herrmann H, Bär H, Kreplak L, Strelkov SV, Aebi U (2007) Intermediate filaments: from cell architecture to nanomechanics. Nat Rev Mol Cell Biol 8(7):562–573
553. Wang Q, Tolstonog GV, Shoeman R, Traub P (2001) Sites of nucleic acid binding in type I-IV intermediate filament subunit proteins. Biochem 40(34):10342–10349
554. Hill TL (1981) Microfilament or microtubule assembly or disassembly against a force. Proc Natl Acad Sci USA 78(9):5613–5617
555. Yang X, Salas PJI, Pham TV, Wasserlauf BJ, Smets MJD, Myerburg RJ et al (2002) Cytoskeletal actin microfilaments and the transient outward potassium current in hypertrophied rat ventriculocytes. J Physiol 541(Pt 2):411–421
556. Connolly JA, Kalnins VI, Barber BH (1981) Microtubules and microfilaments during cell spreading and colony formation in PK 15 epithelial cells. Proc Natl Acad Sci USA 78 (11):6922–6926
557. Kim HR, Appel S, Vetterkind S, Gangopadhyay SS, Morgan KG (2008) Smooth muscle signalling pathways in health and disease. J Cell Mol Med 12(6a):2165–2180
558. dos Remedios CG, Chhabra D, Kekic M, Dedova IV, Tsubakihara M, Berry DA, Nosworthy NJ (2003) Actin binding proteins: regulation of cytoskeletal microfilaments. Physiol Rev 83(2):433–473
559. Huang SC, Soong HK, Kowal VO (1991) Microfilaments, microtubules, and intermediate filaments in cultured corneal fibroblasts. Cornea 10(2):110–120
560. Oz S, Gozes I (2013) The cytoskeleton as a pharmacological target in neurodegenerative diseases. In: Dermietzel R (ed) The cytoskeleton Volume 79 of the series neuromethods. Springer, New York
561. Oriolo AS, Wald FA, Ramsauer VP, Salas PJI (2007) Intermediate filaments: a role in epithelial polarity. Exp Cell Res 313(10):2255–2264
562. Kleele T, Marinković P, Williams PR, Stern S, Weigand EE, Engerer P et al (2014) An assay to image neuronal microtubule dynamics in mice. Nat Commun 5:4827
563. Füchtbauer A, Lassen LB, Jensen AB, Howard J, Quiroga Ade S, Warming S et al (2011) Septin9 is involved in septin filament formation and cellular stability. Biol Chem 392(8–9):769–777
564. McMurray MA, Bertin A, Garcia G 3rd, Lam L, Nogales E, Thorner J (2011) Septin filament formation is essential in budding yeast. Dev Cell 20(4):540–549
565. Angelis D, Spiliotis ET (2016) Septin mutations in human cancers. Front Cell Dev Biol 4:122

566. Mostowy S, Cossart P (2012) Septins: the fourth component of the cytoskeleton. Nat Rev Mol Cell Biol 13:183–194
567. Cairns NJ, Lee VMY, Trojanowski JQ (2004) The cytoskeleton in neurodegenerative diseases. J Pathol 204(4):438–449
568. Leube RE, Schwarz N (2016) Intermediate filaments. In: Bradshaw RA, Stahl PD (eds) Encyclopedia of cell biology, vol 2. Elsevier, Amsterdam
569. Omary MB (2009) IF-pathies: A broad spectrum of intermediate filament-associated diseases. J Clin Invest 119:1756–1762
570. Clemente M, Musu M, Frau F, Brusco G, Sole G, Corazza G, De Virgiliis S (2000) Immune reaction against the cytoskeleton in coeliac disease. Gut 47(4):520–526
571. Owens DW, Lane EB (2004) Keratin mutations and intestinal pathology. J Pathol 204 (4):377–385
572. Spooner BS, Yamada KM, Wessells NK (1971) Microfilaments and cell locomotion. J Cell Biol 49(3):595–613
573. Zhou J, Giannakakou P (2005) Targeting microtubules for cancer chemotherapy. Curr Med Chem Anticancer Agents 5(1):65–71
574. Stanton RA, Gernert KM, Nettles JH, Aneja R (2011) Drugs that target dynamic microtubules: a new molecular perspective. Med Res Rev 31(3):443–481
575. Mukhtar E, Adhami VM, Mukhtar H (2014) Targeting microtubules by natural agents for cancer therapy. Mol Cancer Ther 13(2):275–284
576. Dumontet C, Jordan MA (2010) Microtubule-binding agents: a dynamic field of cancer therapeutics. Nat Rev Drug Discov 9(10):790–803
577. Jordan MA, Wilson L (2004) Microtubules as a target for anticancer drugs. Nat Rev Cancer 4:253–265
578. Nigg EA, Stearns T (2011) The centrosome cycle: centriole biogenesis, duplication and inherent asymmetries. Nat Cell Biol 13(10):1154–1160
579. Lukasiewicz KB, Lingle WL (2009) Aurora A, centrosome structure, and the centrosome cycle. Environ Mol Mutagen 50(8):602–619
580. Marshall WF (2009) Centriole evolution. Curr Opin Cell Biol 21(1):14–19
581. Badano JL, Teslovich TM, Katsanis N (2005) The centrosome in human genetic disease. Nat Rev Genet 6:194–205
582. Dammermann A, Merdes A (2002) Assembly of centrosomal proteins and microtubule organization depends on PCM-1. J Cell Biol 159(2):255–266
583. Didier C, Merdes A, Gairin JE, Jabrane-Ferrat N (2008) Inhibition of proteasome activity impairs centrosome-dependent microtubule nucleation and organization. Mol Biol Cell 19 (3):1220–1229
584. Bettencourt-Dias M (2013) Q&A: who needs a centrosome? BMC Biol 11:28
585. Schatten H, Sun QY (2015) Centrosome–microtubule interactions in health, disease, and disorders. In: Schatten H (ed) The cytoskeleton in health and disease. Springer, New York
586. Löffler H, Lukas J, Bartek J, Krämer A (2006) Structure meets function–centrosomes, genome maintenance and the DNA damage response. Exp Cell Res 312(14):2633–2640
587. Rieder CL, Faruki S, Khodjakov A (2001) The centrosome in vertebrates: more than a microtubule-organizing center. Trends Cell Biol 11(10):413–419
588. Komarova IA, Vorob'ev IA (1994) The ultrastructure of the cell center in the enterocytes of mouse embryos and newborn mice. Ontogenez 25(2):76–88
589. Sysoeva VY, Onishchenko GE (2006) Centrosome and Golgi complex during differentiation of hepatocytes in early postnatal development of mice. Russ J Dev Biol 37(3):163–172
590. Nalapareddy K, Nattamai KJ, Kumar RS, Karns R, Wikenheiser-Brokamp KA, Sampson LL et al (2017) Canonical Wnt signaling ameliorates aging of intestinal stem cells. Cell Rep 18 (11):2608–2621
591. McDonald SAC, Graham TA, Humphries A, Wright NA, Preston SL, Brittan M, Direkze NC (2009) Stem cells in the gastrointestinal tract. In: Lanza R, Gearhart J, Hogan B, Melton D, Pedersen R, Thomas ED et al (eds) essentials of stem cell biology, 2nd edn. Elsevier, San Diego

592. Nigg EA, Raff JW (2009) Centrioles, centrosomes, and cilia in health and disease. Cell 139 (4):663–678
593. Rodriguez-Boulan E, Macara IG (2014) Organization and execution of the epithelial polarity programme. Nat Rev Mol Cell Biol 15(4):225–242
594. Oriolo AS, Wald FA, Canessa G, Salas PJI (2007) GCP6 binds to intermediate filaments: a novel function of keratins in the organization of microtubules in epithelial cells. Mol Biol Cell 18(3):781–794
595. Cereijido M, Contreras RG, García-Villegas MR, González-Mariscal L, Valdés J (1996) Epithelial polarity. In: Wills NK, Reuss L, Lewis SA (eds) Epithelial transport—a guide to methods and experimental analysis. Springer, The Netherlands
596. Wilson PD (1997) Epithelial cell polarity and disease. Ame J Physiol—Renal Physiol 272 (4):F434–F442
597. Ozdamar B, Bose R, Barrios-Rodiles M, Wang HR, Zhang Y, Wrana JL (2005) Regulation of the polarity protein Par6 by TGFbeta receptors controls epithelial cell plasticity. Science 307(5715):1603–1609
598. Hase K, Nakatsu F, Ohmae M, Sugihara K, Shioda N, Takahashi D et al (2013) AP-1B-mediated protein sorting regulates polarity and proliferation of intestinal epithelial cells in mice. Gastroenterol 145(3):625–635
599. van Adelsberg J, Edwards JC, Takito J, Kiss B, Al-Awqati Q (1994) An induced extracellular matrix protein reverses the polarity of band 3 in intercalated epithelial cells. Cell 76(6):1053–1061
600. Toya M, Kobayashi S, Kawasaki M, Shioi G, Kaneko M, Ishiuchi T et al (2016) CAMSAP3 orients the apical-to-basal polarity of microtubule arrays in epithelial cells. Proc Natl Acad Sci USA 113(2):332–337
601. Massey-Harroche D (2000) Epithelial cell polarity as reflected in enterocytes. Microsc Res Tech 49(4):353–362
602. Morone N, Fujiwara T, Murase K, Kasai RS, Ike H, Yuasa S, Usukura J, Kusumi A (2006) Three-dimensional reconstruction of the membrane skeleton at the plasma membrane interface by electron tomography. J Cell Biol 174(6):851–862
603. Clarke M, Spudich JA (1977) Nonmuscle contractile proteins: the role of actin and myosin in cell motility and shape determination. Annu Rev Biochem 46:797–822
604. Sauvanet C, Wayt J, Pelaseyed T, Bretscher A (2015) Structure, regulation, and functional diversity of microvilli on the apical domain of epithelial cells. Annu Rev Cell Dev Biol 31:593–621
605. Gissen P, Arias IM (2015) Structural and functional hepatocyte polarity and liver disease. J Hepatol 63(4):1023–1037
606. Mayhew TM (1990) Striated brush border of intestinal absorptive epithelial cells: stereological studies on microvillous morphology in different adaptive states. J Electron Microsc Tech 16(1):45–55
607. Mutsaers SE (2004) The mesothelial cell. Int J Biochem Cell Biol 36(1):9–16
608. Mutsaers SE, Herrick SE (2006) Mesothelial cells. In: Laurent GJ, Shapiro SD (eds) Encyclopedia of respiratory medicine. Academic Press, San Diego
609. Christensen EI, Wagner CA, Kaissling B (2012) Uriniferous tubule: structural and functional organization. Compr Physiol 2(2):805–861
610. Roy A, Al-bataineh MM, Pastor-Soler NM (2015) Collecting duct intercalated cell function and regulation. Clin J Am Soc Nephrol 10(2):305–324
611. Mount DB (2014) Thick ascending limb of the loop of henle. Clin J Am Soc Nephrol 9 (11):1974–1986
612. Fath KR, Burgess DR (1995) Microvillus assembly: not actin alone. Current Biol 5(6):591–593
613. Glenney JR Jr, Kaulfus P, Matsudaira P, Weber K (1981) F-actin binding and bundling properties of fimbrin, a major cytoskeletal protein of microvillus core filaments. J Biol Chem 256(17):9283–9288

614. Hampton CM, Liu J, Taylor DW, DeRosier DJ, Taylor KA (2008) The 3D structure of villin as a unique f-actin crosslinker. Structure 16(12):1882–1891
615. Revenu C, Ubelmann F, Hurbain I, El-Marjou F, Dingli F, Loew D et al (2012) A new role for the architecture of microvillar actin bundles in apical retention of membrane proteins. Mol Biol Cell 23(2):324–336
616. Ubelmann F, Chamaillard M, El-Marjou F, Simon A, Netter J, Vignjevic D et al (2013) Enterocyte loss of polarity and gut wound healing rely upon the F-actin–severing function of villin. Proc Natl Acad Sci USA 110(15):E1380–E1389
617. Danielsen EM (2015) Probing endocytosis from the enterocyte brush border using fluorescent lipophilic dyes: lipid sorting at the apical cell surface. Histochem Cell Biol 143(5):545–556
618. Rodríguez-Fraticelli AE, Bagwell J, Bosch-Fortea M, Boncompain G, Reglero-Real N, García-León MJ et al (2015) Developmental regulation of apical endocytosis controls epithelial patterning in vertebrate tubular organs. Nat Cell Biol 17(3):241–250
619. Sauvanet C, Wayt J, Pelaseyed T, Bretscher A (2015) Structure, regulation, and functional diversity of microvilli on the apical domain of epithelial cells. Annu Rev Cell Dev Biol 31:1–805
620. Kravtsov DV, Ameen NA (2013) Molecular motors and apical CFTR traffic in epithelia. Int J Mol Sci 14(5):9628–9642
621. Weisz OA, Rodriguez-Boulan E (2009) Apical trafficking in epithelial cells: signals, clusters and motors. J Cell Sci 122:4253–4266
622. Miyake K, McNeil PL (1995) Vesicle accumulation and exocytosis at sites of plasma membrane disruption. J Cell Biol 131(6 Pt 2):1737–1745
623. Vogel GF, Klee KM, Janecke AR, Müller T, Hess MW, Huber LA (2015) Cargo-selective apical exocytosis in epithelial cells is conducted by Myo5B, Slp4a, Vamp7, and Syntaxin 3. J Cell Biol 211(3):587–604
624. Choi DS, Kim DK, Kim YK, Gho YS (2013) Proteomics, transcriptomics and lipidomics of exosomes and ectosomes. Proteomics 13(10–11):1554–1571
625. Jahn R, Lang T (2008) Exocytosis. In: Offermanns S, Rosenthal W (eds) Encyclopedia of molecular pharmacology. Springer, Berlin
626. Schmoranzer J, Goulian M, Axelrod D, Simon SM (2000) Imaging constitutive exocytosis with total internal reflection fluorescence microscopy. J Cell Biol 149(1):23–32
627. Morgan A (1995) Exocytosis. Essays Biochem 30:77–95
628. Chieregatti E, Meldolesi J (2005) Regulated exocytosis: new organelles for non-secretory purposes. Nat Rev Mol Cell Biol 6:181–187
629. Wang W, Liu J, Zhou X (2014) Identification of single-stranded and double-stranded DNA binding proteins based on protein structure. BMC Bioinform 15(Suppl 12):S4
630. Richard DJ, Bolderson E, Khanna KK (2009) Multiple human single-stranded DNA binding proteins function in genome maintenance: structural, biochemical and functional analysis. Crit Rev Biochem Mol Biol 44(2–3):98–116
631. Arnott S (2006) DNA polymorphism and the early history of the double helix. Trends in Biochem Sci 31(6):349–354
632. David AP, Margarit E, Domizi P, Banchio C, Armas P, Calcaterra NB (2016) G-quadruplexes as novel cis-elements controlling transcription during embryonic development. Nucleic Acids Res 44(9):4163–4173
633. Jain A, Wang G, Vasquez KM (2008) DNA triple helices: biological consequences and therapeutic potential. Biochimie 90(8):1117–1130
634. Griffiths AJF, Gelbart WM, Miller JH, Lewontin RC (1999) Modern genetic analysis. WH Freeman, New York
635. Buske FA, Bauer DC, Mattick JS, Bailey TL (2012) Triplexator: detecting nucleic acid triple helices in genomic and transcriptomic data. Genome Res 22(7):1372–1381
636. Duca M, Vekhoff P, Oussedik K, Halby L, Arimondo PB (2008) The triple helix: 50 years later, the outcome. Nucleic Acids Res 36(16):5123–5138

637. Escudé C, François JC, Sun JS, Ott G, Sprinzl M, Garestier T, Hélène C (1993) Stability of triple helices containing RNA and DNA strands: experimental and molecular modeling studies. Nucleic Acids Res 21(24):5547–5553
638. Buske FA, Mattick JS, Bailey TL (2011) Potential in vivo roles of nucleic acid triple-helices. RNA Biol 8(3):427–439
639. Gerstein MB, Bruce C, Rozowsky JS, Zheng D, Du J, Korbel JO et al (2007) What is a gene, post-ENCODE? History and updated definition. Genome Res 17(6):669–681
640. Gupta P, Muse O, Rozners E (2012) Recognition of double stranded RNA by guanidine-modified peptide nucleic acids (GPNA). Biochem 51(1):63–73
641. Martys JL, Ho CL, Liem RKH, Gundersen GG (1999) Intermediate filaments in motion: observations of intermediate filaments in cells using green fluorescent protein-vimentin. Mol Biol Cell 10(5):1289–1295

Chapter 4
Intercellular Network of Junctions of the Gastrointestinal Tract

Abstract The peculiar organization of cells of the gastrointestinal tract, in part, is due to the integrity of the anatomical architecture of the linkages or junctions between the neighboring cells. In this chapter, the structural and functional characteristics of these junctions are discussed.

Keywords Epithelium · Epithelial cells · Intercellular junction
Intercellular network · Intercellular linkage · Gap junction (*macula communicans*)
Nexus · Connexin · Pannexin · Innexin · Tight junction (*zonula occludens*)
Occludin · Cadherins · Intermediate junction · Adherens junction (*zonula adherens*) · Catenin · Plakins (plakophilin, plakoglobin) · Desmoglein
Desmocollin · Desmosome (*macula adhaerens)*

Abbreviations

ADP	Adenosine diphosphate
AMP	Adenosine monophosphate
ATP	Adenosine triphosphate
Bves	Blood vessel epicardial substance
cAMP	Cyclic adenosine monophosphate
CAR	Coxsackievirus and adenovirus receptor
CaSR	Ca^{2+}-sensing receptor
cGMP	Cyclic guanosine monophosphate
Cx	Connexin
GK	Guanylate kinase
GTP	Guanosine triphosphate
Inx	Innexin
IP3	Inositol 1,4,5-triphosphate
JAM	Junctional adhesion molecule
MAGUK	Membrane-associated guanylate kinase
MARVEL	MAL (myelin and lymphocyte) and related proteins for vesicle trafficking and membrane link
NAD	Nicotinamide adenine dinucleotide
Panx	Pannexin

M. O. Welcome, *Gastrointestinal Physiology*,
https://doi.org/10.1007/978-3-319-91056-7_4

PDZ	Postsynaptic density, disk-large, ZO
PKC	Protein kinase type C
Popdc	Popeye domain-containing gene family of proteins
SH3	Src (sarcoma) homology-3
ZO	Zona occludens protein

4.1 Introduction

Cells of the GI tract are structurally and functionally linked by special anatomical structures that connect the plasma membrane of neighboring cells. These structures are present in all the cells of the GI tract and aid in the regulation of the activities of the GI tract and accessory organs [1–5]. Of particular importance are those structures that join neighboring epithelial cells. Interepithelial cellular linkages or junctions of the GI tract comprise a network of plasma membrane-associated protein-complex linkages between neighboring cells that maintain the environment required to execute normal functioning of the gut and accessory organs [4, 6–8]. This chapter deals with the structural and functional aspects of these junctions as key regulators of GI activities.

4.2 Brief Historical Background

The structural linkages between epithelial cells of the GI tract were first extensively researched by the Romanian-born American cell biologist and Nobel Laureate George Emil Palade (1912–2008) around the middle of the twentieth century. In later part of his life, he worked with the renowned cell biologist, Marilyn Gist Farquhar (1928), who he married a year later following the death of his first wife Irina Malaxa in 1969. Farquhar, a Distinguished Professor of cellular and molecular medicine and pathology at the University of California, San Diego, La Jolla, California, actively collaborated with his husband, Palade, especially in the areas of cell-to-cell contact and secretory mechanisms in tissues (including the GI tract) [9, 10].

Palade and Farquhar proposed a concept of tripartite junctional complex as the architectural basis of cell-to-cell linkages. They investigated intercellular junctions in the epithelium and differentiated the various types of cell-to-cell junctions—gap junction, tight junction, and desmosome [11] (Fig. 4.1). These intercellular complexes were found to join adjacent cells, especially in the mucosal epithelia of the GI tract (stomach, intestine, gallbladder), ductal epithelia of the GI glands (pancreas, liver, salivary glands), glandular epithelia of endocrine organs. The pioneer researchers also identified the cell-to-cell linkages in the reproductive tract,

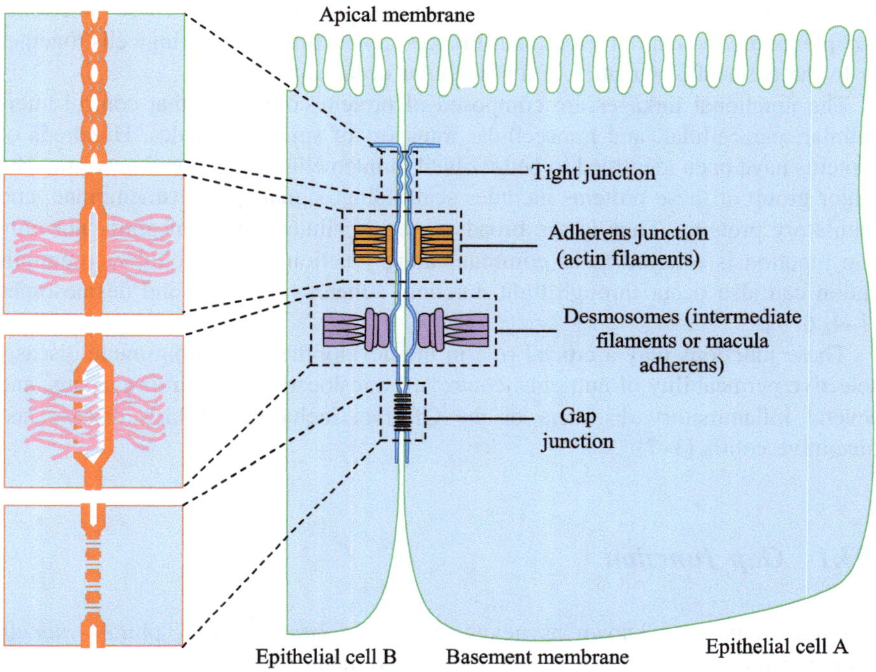

Fig. 4.1 Types of intercellular junctions

proximal and distal convoluted and collecting ducts of the nephron. In hollow organs, Palade and Farquhar noted that cellular complexes of the junctions have a distinct location, succeedingly arranged in the order from the apical to basal side of the epithelial cell [10].

In addition to the structural characteristics, first identified by Farquhar and Palade, it was also observed that the junctional complexes formed a selective diffusion barrier or seal allowing the passage of very small molecules but not large molecules [10]. Today, state-of-the-art research has identified other types of intercellular junctional complexes, as well as their mechanisms of functioning. The latest members of the intercellular junctions will also be covered in this chapter.

4.3 Gastrointestinal Epithelial Cells Are Structurally and Functionally Attached to Each Other via Intercellular Network of Junctions

Like other cells of the body, epithelial cells are joined together by connecting network through which they form cell population and communicate with each other. The intercellular communicating network of junctions are gap junctions, tight junction, adherens junction, and desmosomes (Fig. 4.1) [1–3]. These junctional complexes are the main intercellular linkages connecting neighboring cells together or to the extracellular matrix in multicellular organisms [4].

The junctional linkages are composed of protein complexes that control intercellular, paracellular, and transcellular transport of small molecules. Hundreds of proteins have been identified to be involved in intercellular junctional network. The major group of these proteins includes scaffolding, signaling, transmembrane, and regulatory proteins. Thus in the broad sense of cellular communication, not only gap junction is referred to as communicating junction as intercellular communication can also occur through tight junction, adherens junction, and desmosomes [1–4, 6, 7].

These junctions play a crucial role in the development of autoimmune disease, selective permeability of nutrients, controlled translocation of microorganisms, and several inflammatory disorders of the GI tract including Crohn's disease and ulcerative colitis [1–7].

4.3.1 Gap Junction

Gap junctions (also known as nexus or *macula communicans*, plural—*macula communicantes*) are aggregates of specialized intercellular channels connecting neighboring cells that regulate flow of small molecules such as ions (K^+, Mg^{2+}, Ca^{2+}), second messengers (cAMP, cGMP, and inositol 1,4,5-triphosphate), metabolites (glucose, glutamate, adenosine, AMP, ADP, ATP, and glutathione), as well as chemical and electrical coupling between adjacent cells. There are numerous gap junctions in one cell [12].

Each gap junction channel comprises two hemichannels, joined together from adjacent cells. A hemichannel is a connexon that connects the intracellular space with the extracellular environ and also allows the flow of small molecules. The majority of hemichannels is closed under physiological conditions. However, certain stimuli may increase their permeability to flow of molecules that may act in autocrine or paracrine manner. Pathological conditions such as inflammation, ischemia, and genetic mutations resulting in deafness or cataracts have been associated with increase in hemichannel activity or the so-called leaky hemichannels [13]. Substances that may be transported across hemichannels include ATP, glutamate, NAD^+, as well as other signaling peptides [14].

One connexon (hemichannel) is formed from hexamer of connexins (six connexins), the protein that forms the gap junction in vertebrates. Each connexin protein has four transmembrane domains (Figs. 4.2 and 4.3) [12]. In invertebrates, the protein forming gap (communicating) junctions is called innexins (Inxs). A relatively recent discovery has included another family of gap junction protein "pannexins" expressed in vertebrates. Thus, the vertebral connexins and pannexins are the equivalent of innexins in invertebrates [12, 14–16]. A comparative structure in higher plants is the plasmodesmata [14].

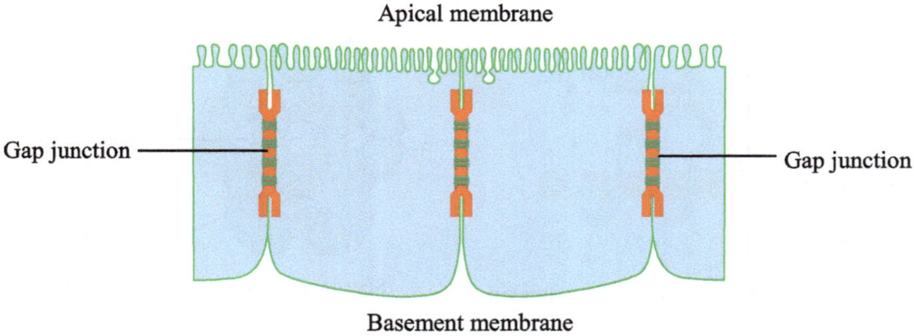

Fig. 4.2 Gut epithelial cells are joined together via gap junctions. Gap junctions may connect two nonidentical cells

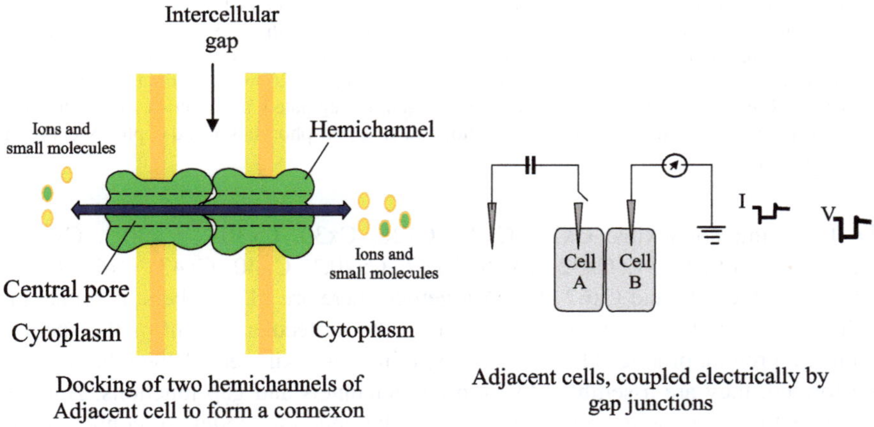

Fig. 4.3 Formation of gap junction from two hemichannels. The gap channels are formed by close end-to-end docking of hexameric hemichannels of tetraspan integral membrane proteins. The close membrane apposition achieved after the docking allows for the formation of a gap measuring about 2–4 nm connecting the cytoplasm of adjacent cells. The gap produced allows direct diffusion of ions and small molecules between adjacent cells. Gap junction varies not only in size but also in polarity. For instance, a negatively polar gap junction will repel negatively charged molecules, allowing only positively charged molecules of comparable size to flow through it [12]

Pannexins channels apart from allowing the diffusion of small molecules also serve as ATP-release channels. Other gap junctions may as well perform similar functions as ATP-release channels [15, 17] (Fig. 4.4).

There are different types of gap junctions, which may be grouped as homotypic and heterotypic. A homotypic gap junction is one formed from the same type of hemichannels, whereas heterotypic gap junction is formed from one homomeric connexon and one heteromeric connexon. Heteromeric connexon is formed from different types of connexins. The connexin (Cx) gene family of humans produces

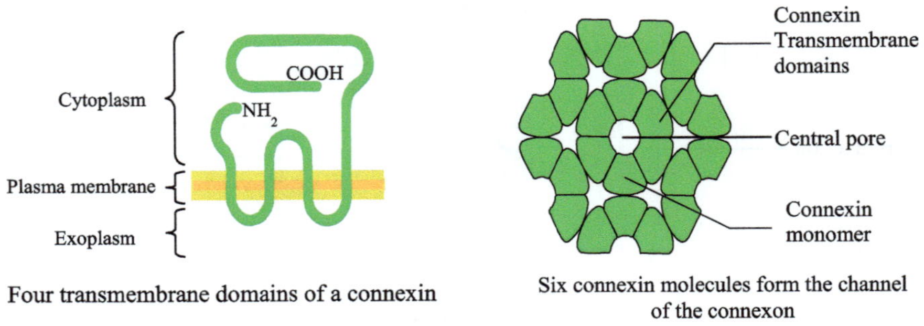

Fig. 4.4 Molecular structure of a gap junction. Gap junction is composed of connexin protein which has four transmembrane domains—two extracellular loops, one cytoplasmic loop, and two cytoplasmic N and C termini. The oligomerization of connexins results in the formation of hemichannels called "connexons." One connexon each from adjacent cell aligns to form a gap junction. The structure of connexins is maintained via numerous interacting bonds and molecular forces. Notably, intraconnexin disulfide bonds are formed from three cysteine residues in each loop. The C terminus of the cytoplasmic loop varies in terms of sequence and length across connexin family members. The shorter C terminus is found in Cx26, whereas the longest is found for Cx50. The functionality of connexons is constantly regulated by a series of biochemical reactions involving posttranslational modifications such as phosphorylation and dephosphorylation reactions [14]

the following connexins: Cx23, Cx25, Cx26, Cx30, Cx30.2, Cx30.3, Cx31.1, Cx31.3, Cx31.9, Cx31, Cx32, Cx36, Cx37, Cx40.1, Cx40, Cx43, Cx45, Cx46, Cx47, Cx50, Cx59, and Cx62 [16]. Altogether, there are 21 members of connexins in humans. This naming system is based on the molecular weight of the specific connexin protein, produced by the corresponding connexin gene. When the proteins are formed, they are assembled to form hemichannels and gap junctions. Another nomenclature uses the abbreviation "GJ" for the gene that codes for connexin. On the basis of gene sequence similarity, gap junction genes have been classified as α(GJA), β(GJB), γ(GJC), δ(GJD), and ε(GJE) according to the order of their discovery. For instance, a member of GJA1 or GJα1 is Cx43; GJB1—Cx32, GJC1—Cx45, and GJE1—Cx30.2 (Cx31.3) [18–20] (Fig. 4.5).

Gap junctions that are expressed in the gut include Cx26, Cx32, Cx36, Cx40, Cx43, Cx45, and some members of the pannexins family. Out of the three paralogs of pannexins (Panx1, Panx2, and Panx3), Panx1 and Panx2 have been implicated in gut physiology and pathology. They modulate numerous biochemical couplings in the cell. Panx 1 is also regarded as an ATP channel. Panx2 is involved in differentiation of neurons. Both Panx 1 and 2 are expressed in the enteric nervous system. Interestingly, emerging results indicate that pannexin hemichannels may even be more important in maintaining gut physiology than pannexin gap junction itself [21, 22].

Gap junctions are selectively distributed in different proportions in various tissues of the gut. For instance, Cx43 and Cx26 are expression in the circular layer of muscularis externa of the colon [23]. These gap junctions play immense role in intestinal pacemaking and neurotransmission [24]. The results of Wang and Daniel

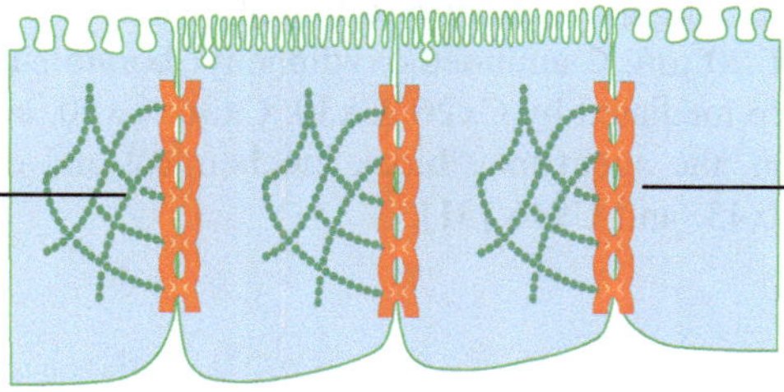

Fig. 4.5 Intestinal epithelial cells joined together via tight junctions. Tight junctions are the most apical component of the epithelial junctional complex and are connected to the actin cytoskeleton [41]

(2001) indicate that the protein Cx43 is present throughout the gut, especially at the site of gap junction formation [24]. The gap junction protein Cx40 is widely distributed in the circular muscle of the lower esophageal sphincter, stomach, and ileum. Gap junction proteins Cx40 and Cx45 are scantly distributed in the interstitial cells of Cajal networks of the myenteric, deep muscular, and submuscular plexuses. More recently, investigation by Nagy et al. (2014) has showed that Cx36 is localized at appositions between the smooth muscle cells and the somata of enteric neurons secreting nitric oxide [25]. This indicates that gap junctions are formed between different types of cells [25].

Gap junction serves as one of the pathways through which cells discharge their contents into another cell or into the lumen or selectively absorb certain molecules. Gap junctions mediate the transport of molecules across the endothelium of the cell (transcytotic pathway of material secretion). Transport of molecules by transcytosis selectively allows passage of certain size of molecules (usually less 10 nm) across the endothelial or epithelial cells. Gap junctions possess the property of selective permeability (permselectivity)—which regulates the passage, conductance, and gating to specific molecules across the cell. While Cx32 hemichannels are permeable to cAMP and cGMP, heteromeric hemichannels of Cx26 and Cx32 have reduced permeability to cAMP only and not cGMP. The permeability of Cx32 gap junctions to adenosine is about tenfold higher compared with Cx43 gap junctions [14, 26, 27].

To study the physiology of gap junction, researchers use gap junction-blocking agents such as 18-β-glycyrrhetinic acid or carbenoxolone (3β-hydroxy-11-oxoolean-12-en-30-oic acid 3-hemisuccinate). Carbenoxolone is a glycyrrhetinic acid derivative with a steroid-like structure and is one of the most widely water-soluble gap junction blockers used in cellular physiology [28]. Another gap junction blocker is the drug quinine [29]. Gap junction blockers are now gaining attention as new frontiers for the treatment of many human maladies including arrhythmia, defibrillation, seizure, and cancer [30, 31]. The gap junction blockers 16-doxyl-stearic acid and 1-heptanol have been shown to significantly reduce defibrillation [30]. Another gap junction blocker, 2-aminoethoxydiphenyl borate, selectively blocks different gap junctions at different concentrations. At a relatively high concentration of 20 μm, 2-aminoethoxydiphenyl borate substantially blocks junctional conductance mediated by Cx26, Cx30, Cx36, Cx40, and Cx45, whereas at lower concentration, the agent may block biochemical and electrical coupling mediated by Cx32, Cx43, and Cx46 [31].

4.3.2 Tight Junction

Tight junction (also called occluding junction or *zonula occludens*, plural—*zonulae occludentes*) is the most apical component of intercellular network of junctions formed by the association of the membranes of adjacent epithelial cells to produce a structure that acts as a paracellular barrier regulating the movement of water and solutes between epithelial layers [32]. Tight junctions form a multifunctional complex of continuous branching belt-like, highly dynamic intermembrane network at the boundary between the apical and the basolateral membrane domains in both epithelial and endothelial cells [8, 10, 33, 34]. More precisely, mammalian tight junctions are observed at the boundary between the apical and lateral surfaces of the outer leaflet of plasma membrane [7]. Data from transmission electron microscopy indicate that tight junctions appear as very close membrane appositions of adjacent cells [Hideki Chiba, et al. 2008]. These close oppositions of adjacent membranes have been termed "kissing points" or "kisses" of adjacent cells [8, 32, 35]. Tight junctions are highly selective for solutes on the basis of their size, charge, polarity, and transepithelial resistance. The corresponding structure in invertebrates (e.g., Drosophila and crustaceans) is the septate junction. Tight and septate junctions are what seal cells together. Tight junctions are present in endothelial cells, mesothelial cells, Schwann cells, oligodendrocytes, and Sertoli cells. They are abundantly expressed in the GI epithelium [32, 35]. The gut–immune barrier in the GI tract is formed by tight junction. These junctions are also responsible for forming the glia–blood–brain barrier in both invertebrates and vertebrates [4, 36–40]. In the kidney, tight junctions create a paracellular transport pathway for reabsorption of various ions and substances into the blood [33].

The property of tight junctions acting like a "gate" to the paracellular transport of ions, solutes, water, and microbes and as a "fence" that separates the apical and basolateral membranes to form a diffusion barrier in the outer leaflet of the plasma membrane is crucial to the functioning of the epithelium as a whole [11, 32, 41, 42]. Thus, tight junctions act as both physical and chemical barriers. The epithelium of the GI tract prevents the transport of certain bacterial toxins across the epithelial cells due to its barrier functions. However, the same epithelium allows the selective movement of water, solutes, and ions across the epithelial cells. Similar functions of tight junctions are seen in the skin epidermis, kidney, urinary bladder, and the reproductive tract [6, 8] (Fig. 4.6).

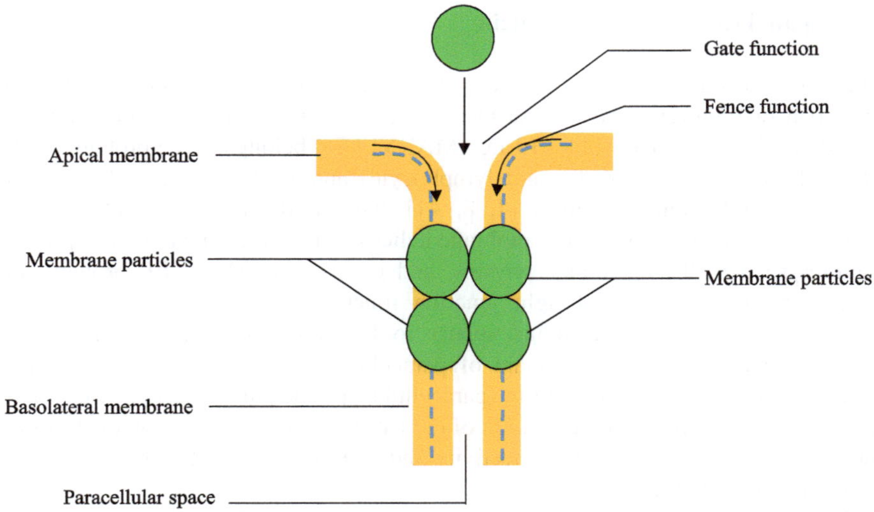

Fig. 4.6 Gate and fence functions of tight junctions. The gate function allows selective transport of ions, water, and some solutes between cells. The fence function of tight junctions prevents the diffusion of solutes or other substances between the apical and the basolateral membranes [7]. This fence function allows the tight junction to mechanically restrict diffusion of proteins and lipids within the plane of the lipid bilayer [8]. This function of tight junctions slows these biomolecules to be transported via the transcellular pathway [8]

Tight junctions also contribute to signal transduction, cell differentiation, proliferation, and polarity [32, 34].

The functions of tight junctions are disrupted in many pathologies, affecting humans including cancer, infectious (including sexually transmitted diseases) and some inherited diseases [5, 6, 32, 34, 41, 43]. For example in inflammatory bowel disease, vasogenic edema, the barrier functions of tight junctions are disrupted resulting not only in paracellular transport irregularities, but also dysfunctional cell polarity, differentiation, and proliferation [32, 43]. Pathogenic microbes such as the enterohemorrhagic *Escherichia coli* can cause intestinal lesions by release of their effector toxins which are capable of disrupting the tight junction architecture of the GI epithelium resulting in effacement of the brush border microvilli. One of the mechanisms by which these pathogens disrupt the tight junctions is by interfering with the polymerization/depolymerization of actin filament, and integral cytoskeletal material in the maintenance of tight junction integrity [4].

The functions and properties of tight junction are due to the proteins from which it is formed. Tight junctions are formed from a network of different types of transmembrane proteins which include single-span transmembrane, triple-span transmembrane, and tetraspan transmembrane proteins [8]. However, the major types of tight junction transmembrane proteins are the single- and tetraspan transmembrane proteins [4, 41].

Tetraspan Transmembrane Protein

The tetraspan transmembrane proteins of the tight junctions are integral proteins that span or transverse the plasma membrane four times. They include claudins, occludin, tricellulin, and MARVELD3 (MARVELD3 belongs to the protein family of MARVEL—MAL (myelin and lymphocyte) and related proteins for vesicle trafficking and membrane link) [4, 8, 41]. The MARVELs are proteins with M-shaped topology—four transmembrane helices that are identified in some protein families such as the occludins, physins, and gyrins. The MARVEL proteins are crucial in the regulation of tight junction functions [44, 45]. Of all tetraspan transmembrane proteins, claudins appear to be most abundant and important, functioning as a major determinant of paracellular permeability [4, 8, 41]. The tetraspan proteins not only form paracellular permeability barrier, but more importantly, the type and composition of tetraspan transmembrane proteins forming the tight junction determine the selectivity and rate of the paracellular ion transport between two cells [4, 41].

Claudins (from Latin *claudere*, meaning "to close"): These tetraspan proteins consist of a family of about 26 members with molecular mass ranging from 20 to 28 kDa. Claudin proteins possess two extracellular domains, amino and carboxy terminus cytoplasmic domains, which possess different functions (Fig. 4.7). For example, the carboxy-terminal contains a PDZ-binding motif, which interacts with the cytoplasmic scaffolding proteins ZO-1, -2, -3, etc. (ZO means zonula occludens, also called occludins)—this is important for correct localization of the components of tight junctions. The acronym "PDZ" is derived from the names of the first proteins in which the domain was observed "Postsynaptic density-95/disk-large tumor suppressor lethal/Zonula occludens-1" [48, 49]. Paracellular selective diffusion of negatively and positively charged ions between cells is due to the first

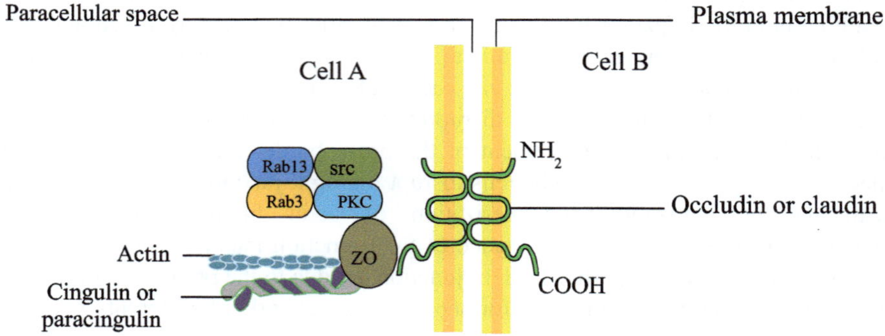

Fig. 4.7 Molecular structure of tight junction. The tight junction integral membrane proteins (claudin and occluding) are linked to ZO, which interact with a wide range of cytoplasmic molecules such as small GTPase (via atypical PKC), actin filaments, cingulin, and paracingulin proteins. The small GTPases (e.g., Rab3, Rab13) and PKC help to maintain the length and functionality of the intercellular belt of tight junctions [46, 47]

extracellular loop of claudins. Claudins that increase paracellular ion transport are generally called pore-forming claudins, whereas those that reduce paracellular transport are termed sealing claudins. In addition, certain regions of the extracellular loop mediate entry of microbes into the cells. Example of pore-forming claudins includes claudin-2 [50]. The expression of claudin-2 in a cell makes the tissues highly permeable to ions, described as "leaky." For example, epithelial tissue is described as "leaky" if this claudin is expressed in it. Leaky epithelium is abundant in the proximal tubule of the nephron, gallbladder, small intestine, and choroid plexus [8, 11, 50]. Sealing claudins make the barrier between cells tighter, and they include claudin type 1, 3, 4, 5, 7, 8, 11, 14, 15, 16, 18, and 19. Sealing claudins are abundantly expressed in the distal convoluted tubule and collecting duct of the nephron, urinary bladder, stomach, ducts of salivary and sweat glands [11]. It should be noted, however, that the functions of some of these leaky claudins depend on the tissue where it is expressed. For example, claudin-5 in endothelium of the blood–brain barrier is associated with a tight paracellular pathway, whereas expression of this claudin type in the epithelium of the breathing pipe and alveolar is associated with a leaky paracellular pathway between the epithelial cells [50]. The degree of leakiness and tightness of tight junctions depends on a variety of factors which include pore size, transepithelial electrical resistance, charge, and size of the claudins expressed in the junction. All these factors are important determinants of paracellular transport. For instance, even if a negatively charged molecule has the right radius to pass through the pore formed by a claudin that is negatively charged, transport of the molecule may not be possible due to the charge. However, solute with slightly higher radius may be transported across a paracellular pathway if it possesses the opposite charge. Positive ions such as Na^+ and Mg^{++} may easily pass through paracellular pathways formed by claudin-2, -15, -16, and -19. Negatively charged ions such as Cl^- easily pass through paracellular pathways formed by claudin-10 [32, 43, 51–53]. Research has shown that the majority of leaky epithelium shows a preference for Na^+ over Cl^- paracellular transport. This allows paracellular transport of Na^+ to follow secreted Cl^-, which is required to maintain a saline environment [32, 43, 51–53]. In the GI tract, the colon has the highest transepithelial resistance, which is required to reabsorb NaCl and water to form a semisolid mass [32, 43, 51–54].

In addition to paracellular mediation of water, solute, and ion transport, claudins also regulate signal transduction via the extracellular G protein-coupled Ca^{2+}-sensing receptor (CaSR). Claudins interact with CaSR to maintain proper Ca^{2+} homeostasis in the cell [55, 56].

In certain abnormal conditions, the barrier functions of claudins may be disrupted. Such conditions may occur in oxidative stress possibly induced by chronic alcohol abuse and other disease conditions [50]. Diseases associated with a loss of barrier function include inflammatory bowel disease, infectious diseases, and cancers [32, 57]. One mechanism by which cholera toxin causes its diarrheal effects is by upregulation of pore-forming claudins in the GI epithelium, resulting in widening of intercellular spaces, which leads to uncontrolled passage of water and ions [58]. The GI epithelium of patients with Crohn's disease has reduced and broken tight junction strands with increased expression of pore-forming claudins and decreased expression of non-pore-forming claudins [59]. Patients with ulcerative colitis also have increased expression of pore-forming claudins, caused by the increased activity of the cytokine “interleukin-13,” which is also responsible for intestinal epithelial apoptosis, resulting in microerosions [60, 61]. Defects in intestinal tight junctions can result from the activity of certain *E. coli* strains which can gain pathologic relevance in response to signaling by proinflammatory cytokines in conditions of inflammation in the GI tract [62].

Mutations in the genes encoding the tight junction proteins can cause inherited human diseases. For instance, claudin-16 and -19 mutations can result in renal familial hypomagnesemia, hypercalciuria and nephrocalcinosis, and hypertension [33, 63].

Claudin-like proteins are expressed in many regions of the body and mediate normal functioning of the organ. Mammalian proteins that are homologous to claudins include the membrane protein of eye lens-20; epithelial membrane protein 1, 2, 3; and peripheral myelin protein 22. Each of these homologs is crucial for the formation of tight junctions in their respective cellular and tissue environment [33, 50, 64].

Occludins (from Latin *occludere* meaning “restrict passage”): Also called zonula occludens (ZO), occludins represent a family of tight junction-associated proteins that function as cross-linkers, anchoring the tight junction transmembrane proteins to the actin cytoskeleton [65–67]. ZO proteins can be described as scaffolding proteins that provide a structural basis for the assembly of protein complexes at the cytoplasmic surface of intercellular junctions. These proteins provide a link between the integral membrane proteins and the filamentous cytoskeleton [64]. ZO proteins contribute to the “cytoplasmic plaque” [64]. The cytoplasmic plaque is formed by a complex network of adaptors, scaffolding, and cytoskeletal proteins that cross-link junctional membrane proteins and connect tight junctions to the actin cytoskeleton [41, 68]. The protein occludin is also called MARVELD1 [69].

There are three types of ZO proteins (ZO-1, -2, and -3), which belong to the protein family of MAGUK (membrane-associated guanylate kinase homologs) [64]. MAGUKs are characterized by multiple protein-binding domains, which include Src-homology 3 (SH3), PDZ, and guanylate kinase (GK) domains. This family of proteins aids in the assembly of cytoskeletal materials, receptors, and signaling molecules [64]. Src is abbreviation for sarcoma, a proto-oncogene encoding a non-receptor tyrosine kinase. Src comprises an SH2 domain, an SH3 domain, and a tyrosine kinase domain. It was originally discovered by the American scientists John Michael Bishop (1936–) and Harold Eliot Varmus (1939–). Both were awarded the 1989 Nobel Prize in Physiology or Medicine.

The ZO proteins have several protein-binding domains—three PDZ domains, one SH3 domain, one GK domain, and an F-actin-binding domain. ZO-1 interacts with claudins through its first PDZ domain, while the second PDZ domain is used for interaction with ZO-2 or ZO-3. SH3 of ZO provides a binding site to several signaling partners [8, 41, 64, 68, 70, 71]. The ZO proteins can connect to the actin cytoskeleton via alpha-actinin [7]. ZO proteins also bind to alpha-catenin, p120 catenin, and afadin [64]. Afadin is a nectin-interacting protein that acts as a scaffold, maintaining cell shape and barrier function not only in tight junctions but also in adherens junctions. Increase in the level of muscle contractility has been associated with increased expression of junctional afadin [72]. The cytoplasmic proteins cingulin and paracingulin are also associated with ZO (Fig. 4.7). Cingulin and paracingulin connect the tight junctions to the microtubules of the cell [4].

The ZO proteins participate in the regulation of cell growth, proliferation, gene transcription, and signal transduction [7, 64]. ZO proteins regulate the transepithelial migration of neutrophils and the paracellular diffusion of small hydrophilic tracers [64].

ZO protein dysfunction may be associated with chronic inflammation, and hyperplasia of the GI epithelium [33]. The translocation of pathogenic microbes (e.g., hepatitis C virus) through the GI epithelium is made possible due to its action on occludin, which is used in the initial steps of their internalization. The causative organism of the parasitic disease "amebiasis," *Entamoeba histolytica,* produces a cysteine protease, which induces a proinflammatory response that results in upregulation of pore-forming claudins, and may destabilize occludin functions. The causative agent of peptic ulcer disease, *Helicobacter pylori,* injects the CagA protein into the cytoplasm of gastric epithelial cells via the type 4 secretion system. This protein associates with ZO to alter the composition and function of the apical junctional complex which may subsequently result in progressive erosion of the epithelium [4].

Tricellulin: Also known as MARVELD2, tricellulin is a tetraspan transmembrane protein formed as an integral component of tricellular (meeting points of three cells) as well as bicellular (meeting point of two cells) tight junctions. Tricellular junctions form vertical strands, whereas bicellular tight junctions form horizontal band of strands [8, 73, 74]. The protein tricellulin is involved in modulation of major functions of the tight junction such as barrier, pore, and fence functions [8, 74]. Low expression of tricellulin in tricellular tight junctions provides a pathway for the transport of macromolecules via the central tube formed by these proteins. High expression of tricellulin in tricellular tight junctions leads to the formation of a barrier against macromolecules alone, whereas in bicellular tight junctions, paracellular transport of both solutes and macromolecules is hindered [74]. The paracellular ion permeability is decreased when tricellulin is overexpressed [73].

Mutations of tricellulin have been associated with degeneration of hair cells and integral component of the cochlea—the sensory organ responsible for hearing. This mutation progressively results in a non-syndromic form of deafness by a mechanism which is yet to be completely understood. However, the mechanism may be related to tricellulin involvement in the regulation of ion homeostasis around cochlear hair cells, a required condition for normal physiological functioning of the hair cells [73–75].

Triple Transmembrane Protein

Certain tight junction proteins span through the plasma membrane three times forming three transmembrane domains that are involved in cell–cell interaction/ adhesion, and cell motility. An example of triple-span protein of tight junctions is the blood vessel epicardial substance (Bves) [8, 76]. Bves is abundantly expressed in epithelial, smooth, cardiac, and skeletal muscle cells [76–78]. Bves co-localizes with ZO proteins in the epithelium of the intestine [79]. Bves is the first member of the Popdc (Popeye domain-containing) gene family of proteins [79]. The name "*Popeye*" was introduced by Andrée et al. [80] because of the abundant expression of the protein in muscle tissue, which is in reference to the hero of the comic strip "Thimble Theatre" "Popeye the sailor man," a cartoon fictional character created by Elzie Crisler Segar (1894–1938). This fictional character is most famous for his supernatural muscle strength [81]. In addition to *Popdc1*, the Popdc genes encode other two membrane proteins—*Popdc2* and *Popdc3*—all three proteins possess triple-span transmembrane helices [82]. The domain of this protein that is conserved in the family members resides on the cytoplasmic side of the cell, a carboxy-terminal of the protein—called the Popeye domain. The gene for Popdc proteins is expressed in cardiac, smooth, and skeletal muscle cells as well as neuronal cells (in some areas of the nervous system) and in epithelial cells [76, 83].

Bves plays a role in cell–cell adhesion and motility [76]. This protein, according to new research, is involved in rhythmicity of muscle cells [82]. This function of Bves is in part, due to its ability to bind cAMP and modulate the expression of ion channels such as potassium channel [82].

The Popdc protein is involved in the pathogenesis of various types of epithelial cell cancers [81].

Single Transmembrane Protein

The single transmembrane proteins include junctional adhesion molecule (JAM), crumbs protein homolog 3, coxsackievirus and adenovirus receptor (CAR), CAR-like membrane protein, and the endothelial cell-selective adhesion molecule [8]. These single-span transmembrane proteins interact with adaptors of the cytoplasmic plaque. Of the single-span proteins, the widely investigated is JAM—a family of glycoproteins characterized by two immunoglobulin-like domains [8]. Several isoforms have been identified in JAMs: JAM-A, -B, -C and -D. JAMs act as key regulators of leukocyte–endothelial cell adhesion and leukocyte/platelet–endothelium interactions. These cell adhesion proteins are also responsible for the migration of leukocytes across the endothelium through the paracellular pathway. JAMs are also involved in the regulation of cell polarity [4, 33].

4.3.3 Adherens Junction

Adherens junctions (*Zonula adherens*, adhesion belt or intermediate junction) are protein complexes characterized by close apposition of plasma membranes of neighboring cells (about 10–20 nm), with intercellular rod-shaped molecules, and condensed actin filaments in the cytoplasmic side of the junctions (Fig. 4.8). These junctional complexes appear either as bands surrounding the cell (referred to as *Z. adherens*) or spots of attachment to the extracellular materials (referred to as adhesion plaques). Though both adherens junctions and tight junctions help to seal the space between cells, adherens junctions are usually more basal than tight junctions (tight junctions are more apically located) [7, 10, 84, 85]. The adherens junctions almost totally enclose the cells together with the F-[filamentous]-actin (also called actin filament or microfilament) lining to form what is called the circumferential actin belt, which is required especially for apical epithelial cell contraction or relaxation and morphogenesis [47, 72, 86].

The *Z. adherens* is formed as a ring of the transmembrane glycoprotein cadherins linked to an underlying actomyosin filament via catenin proteins [72]. There are about 20 members in the cadherin family. Type I classical cadherins and the related type II cadherins are single-span membrane proteins that possess both extracellular and cytoplasmic domains [87, 88]. However, some members of the cadherins are usually located in the intercellular space, providing connections between the plasma membrane of neighboring cells. Examples of such cadherins include E-(epithelial type) and N-(neuronal type) cadherins. These extracellular cadherins may be chemically cross-linked as E-N heterodimer and as hemihomotypic dimers. They may also be expressed as homodimers in tight junctions [87–89]. In the majority of situations, cadherins are linked together by the principles of homophilic cell–cell adhesion of cell-type-specific, densely clustered cadherins. In this regard, epithelial cells are usually linked to neighboring epithelial cells by E-cadherin, whereas neurons are usually linked to neighboring neurons by N-cadherin in adherens junctions (Fig. 4.9) [89]. The adhesion functions of these classical cadherins are dependent on binding of their extracellular domains to Ca^{2+} [88]. The interaction of Ca^{2+} with the extracellular domain functions as an "off" and

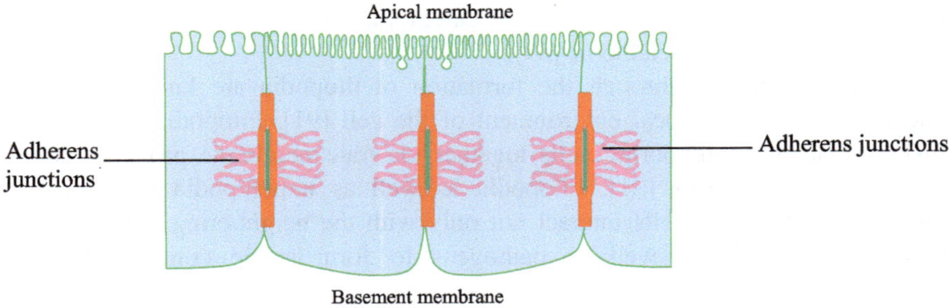

Fig. 4.8 Intestinal epithelial cells joined together via adherens junctions

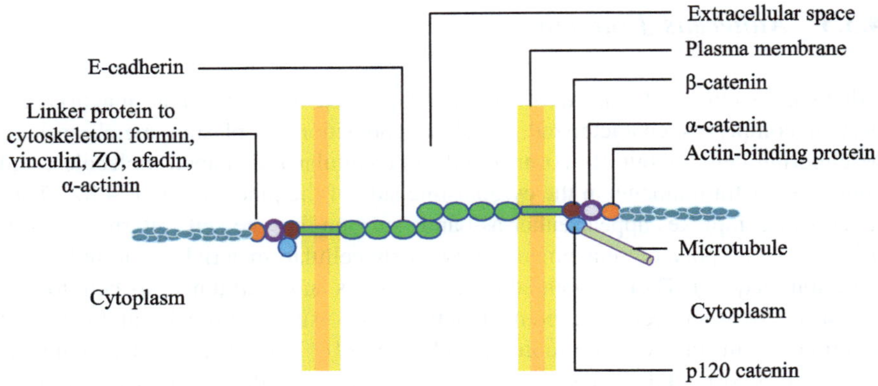

Fig. 4.9 Molecular structure of the adherens junction

"on" switch of adherens junctions [86, 87]. The cytoplasmic domain of classical cadherins (which contains the carboxy-terminal) also regulates the functions of tight junctions via numerous mechanisms including phosphorylation reactions as well as binding to the armadillo proteins. These so-called armadillo proteins (α-, β-, γ-, and p120-catenins) provide anchorage of cytoskeletal materials such as actin filaments and microtubules [86, 87, 89]. Moreover, the cytoskeletal material, in particular, microtubules, can associate with β-catenin by the activities of the protein dynein, which binds with β-catenin to form dynein–β-catenin complex [86]. The α-catenin links cadherin with actin filament by indirectly binding via β-catenin or γ-catenin (plakoglobin) [88]. The indirect link is provided by the actin-binding proteins such as formin and vinculin [72, 86]. The β-catenin binds with cadherins (such as E-cadherin) via the cytoplasmic carboxy-terminal of the classical cadherins [86, 88].

Adherens junctions are expressed not only in epithelial, but also in mesenchymal, muscular, and neural cells [89]. Adherens junctions play a variety of physiological roles including control of plasticity, polarity, cell division, cell death, cell movement, cell shape, cell signaling, and gene transcription [86–88, 90]. These junctions are responsible for maintenance of the gut–immune barrier as well as blood–brain and blood–nerve barriers [87]. Adherens junction has been implicated in some forms of cancers. For example, carcinoma progression and metastasis are related to decrease in E-cadherin [89].

Adherens junctions through the formation of filopodia are known to act as crucial sensors of the local environment of the cell [91]. Filopodia are cell protrusions that link neighboring cells together via transmembrane proteins such as cadherins and integrins [91]. Filopodia as well as lamellipodia are structures through which certain cells interact not only with the neighboring cells, but also extracellular matrix as well as pathogens to form stable contacts [91, 92] (Fig. 4.10).

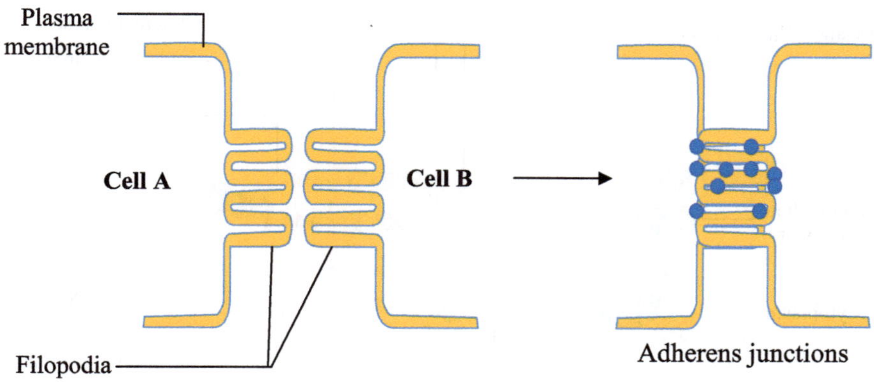

Fig. 4.10 Formation of adherens junction from close apposition of filopodia of neighboring cells

4.3.4 Desmosomes

Desmosomes (macula adherens) is a dynamic patch-like specialized supra-adhesive discontinuous, button-like complex that is formed by linkage of the cadherin transmembrane proteins (desmocollin and desmoglein) via desmoplakins (plakophilin and plakoglobin) to intermediate filaments, which help to maintain shear forces and mechanical stress [7, 10, 93, 94] (Fig. 4.11).

Desmosomal proteins aid in anchoring intermediate filaments to cytoplasmic plaques, forming a complex structure that stabilizes mechanical forces around the cells. These junctional complexes are also involved in other physiological functions including morphogenesis, cell proliferation, differentiation, control of intracellular signaling possibly through calcium waves [93, 95–97]. It should be noted that desmosomes do not determine paracellular permeability [4] (Fig. 4.12).

Desmosomes are abundantly distributed in epithelia and also in several non-epithelial tissues such as cardiac muscle, meninges, and epidermal tissues [93,

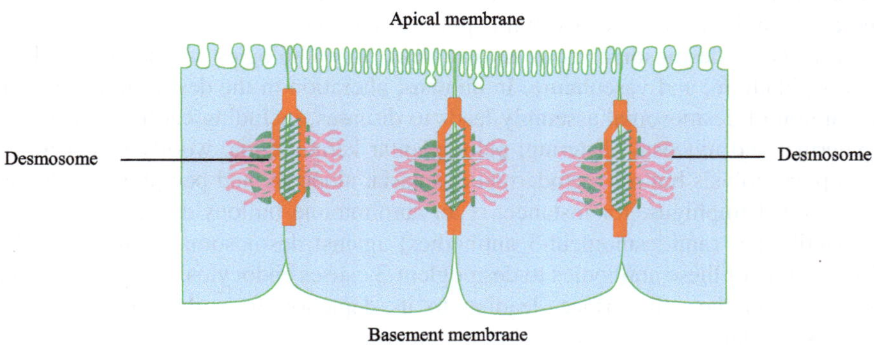

Fig. 4.11 Intestinal epithelial cells joined together via desmosomes

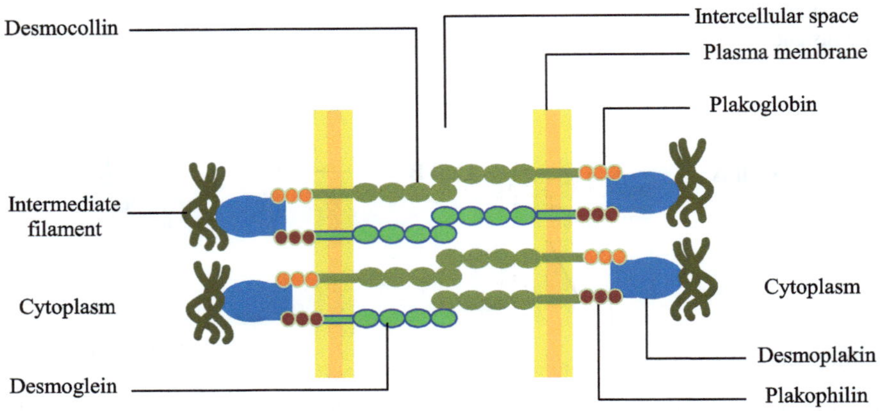

Fig. 4.12 Molecular structure of desmosomes. Desmosomes are formed by desmosomal cadherins on neighboring cells—desmogleins and desmocollins. The linkage may be homo- or heterophilic formed by the two cadherin types. The cytoplasmic tail of each cadherin is linked to proteins such as plakoglobin. Plakoglobin then binds the adaptors "plakin protein family" such as desmoplakin, plakophilin, which associate with the intermediate filament, e.g., keratin filament [94, 97–99]

96, 100–102]. Studies have shown that desmosomal proteins such as desmoplakins are expressed in lymphatic and vascular endothelial cells. The presence of these desmosomal proteins may enhance the association between the intermediate filaments and the plasma membrane in epithelial cells [103].

The GI tract epithelium is usually under constant forces from peristaltic contraction and shear stress from luminal contents. Under physiological conditions, however, this does not cause diseases, thanks to the cytoskeletal organization of the terminal web [98]. Integral components of the microvilli terminal web cytoskeleton are keratins and intermediate filament which is tethered to desmosome to maintain the organization of the apical region of the GI epithelial cells. The desmosomal protein, desmoplakin, may be involved in maintaining the shape and length of the cytoskeletal material of the microvilli [98]. Certain filaments of the basal region of the cell may be also associated with spot desmosomes [104].

Defects in desmosomal proteins can lead to pathologies in the heart, skin, neuroepithelium, and vasculature. In humans, alteration in the desmosome gene or disruption of desmosomal assembly leads to diseases, including cardiomyopathies, epidermal and mucosal blistering, palmoplantar keratoderma, woolly hair, keratosis, epidermolysis bullosa, ectodermal dysplasia, alopecia, and pemphigus vulgaris [98, 105]. Pemphigus, for instance, is an autoimmune bullous disease caused by autoantibodies (antidesmoglein-3 antibodies) against desmosomal cadherins [94]. The binding of these antibodies to desmoglein-3 causes endocytosis of this cadherin protein from the cell surface, leading to its depletion from desmosomes. This process is related to acantholysis in the epidermis [97].

4.4 Conclusion

Intercellular linkages or junctions of the GI tract provide a means of maintaining the structural integrity of the cells and tissues as well as functionality of the GI tract, required to ensure adequate digestive, absorptive, secretive, and transport functions in the GI tract.

Recommended readings

Original Articles

1. Li C, Gao M, Zhang W, Chen C, Zhou F, Hu Z, Zeng C (2016) Zonulin regulates intestinal permeability and facilitates enteric bacteria permeation in coronary artery disease. Sci Rep 6: 29142
2. Navabi N, McGuckin MA, Lindén SK (2013) Gastrointestinal cell lines form polarized epithelia with an adherent mucus layer when cultured in semi-wet interfaces with mechanical stimulation. PLoS One 8(7): e68761
3. Song H-L, Lv S, Liu P (2009) The roles of tumor necrosis factor-alpha in colon tight junction protein expression and intestinal mucosa structure in a mouse model of acute liver failure. BMC Gastroenterol 9:70

Review Articles

1. Arrieta MC, Bistritz L, Meddings JB (2006) Alterations in intestinal permeability. Gut 55(10): 1512–1520
2. Binder HJ (2009) Mechanisms of diarrhea in inflammatory bowel diseases. Ann N Y Acad Sci 1165:285–93
3. Bischoff SC, Barbara G, Buurman W, Ockhuizen T, Schulzke J-D, Serino M, et al. (2014) Intestinal permeability—a new target for disease prevention and therapy. BMC Gastroenterol 14: 189
4. Chopra DP, Dombkowski AA, Stemmer PM, Parker GC (2010) Intestinal epithelial cells in vitro. Stem Cells Dev 19(1): 131–141
5. Groschwitz KR, Hogan SP (2009) Intestinal barrier function: molecular regulation and disease pathogenesis. J Allergy Clin Immunol 124(1): 3–22
6. Khailova L, Dvorak K, Arganbright KM, Williams CS, Halpern MD, Dvorak B (2009) Changes in hepatic cell junctions structure during experimental necrotizing enterocolitis: effect of EGF treatment. Pediatr Res 66(2): 140
7. Lagaud G, Davies KP, Venkateswarlu K, Christ GJ (2002) The physiology, pathophysiology and therapeutic potential of gap junctions in smooth muscle. Curr Drug Targets 3(6):427–40
8. Landy J, Ronde E, English N, Clark SK, Hart AL, Knight SC, et al. (2016) Tight junctions in inflammatory bowel diseases and inflammatory bowel disease associated colorectal cancer. World J Gastroenterol 22(11): 3117–3126
9. Maes M, Yanguas SC, Willebrords J, Cogliati B, Vinken M (2015) Connexin and pannexin signaling in gastrointestinal and liver disease. Transl Res 166(4): 332–343

10. Odijk M, van der Meer AD, Levner D, Kim HJ, van der Helm MW, Segerink LI, et al. (2015) Measuring direct current trans-epithelial electrical resistance in organ-on-a-chip microsystems. Lab Chip 15(3):745–52
11. Srinivasan B, Kolli AR, Esch MB, Abaci HE, Shuler ML, Hickman JJ (2015) TEER measurement techniques for in vitro barrier model systems. J Lab Autom 20(2): 107–126
12. Tse G, Lai ET, Yeo JM, Tse V, Wong SH (2016) Mechanisms of electrical activation and conduction in the gastrointestinal system: lessons from cardiac electrophysiology. Front Physiol 7:182
13. Van Patten K, Parkash V, Jain D (2010) Cadherin expression in gastrointestinal tract endometriosis: possible role in deep tissue invasion and development of malignancy. Modern Pathol 23: 38–44
14. Wang YF, Daniel EE (2001) Gap junctions in gastrointestinal muscle contain multiple connexins. Am J Physiol Gastrointest Liver Physiol 281(2):G533–G543
15. Yu YB, Li YQ (2014) Enteric glial cells and their role in the intestinal epithelial barrier. World J Gastroenterol 20(32):11273–80

Books

1. Alberts B, Johnson A, Lewis J, Raff M, Roberts K, Walter P (2002) Molecular biology of the cell. 4th edition. Garland Science, New York, USA
2. Lodish H, Berk A, Zipursky SL, Matsudaira P, Baltimore D, Darnell J (2000) Molecular cell biology. 4th edition. WH Freeman, New York, USA
3. Nikitina OS, Welcome MO, Pereverzev VA (2016) Human anatomy and physiology. In 2 parts. Part 1. Belarusian State Medical University Press, Minsk, Belarus
4. Offermanns S, Rosenthal W eds (2004) Encyclopedic reference of molecular pharmacology. Springer, Berlin Heidelberg, Germany

Methods and Protocols in Cellular Biology

1. Graves CL, Harden SW, LaPato M, Nelson M, Amador B, Sorenson H, et al. (2014) A method for high purity intestinal epithelial cell culture from adult human and murine tissues for the investigation of innate immune function. J Immunol Methods 414: 20–31
2. Mahe MM, Sundaram N, Watson CL, Shroyer NF, Helmrath MA (2015) Establishment of human epithelial enteroids and colonoids from whole tissue and biopsy. J Vis Exp 97. https://doi.org/10.3791/52483
3. Mahé MM, Aihara E, Schumacher MA, Zavros Y, Montrose MH, Helmrath MA, et al. (2013) Establishment of gastrointestinal epithelial organoids. Curr Protoc Mouse Biol 3: 217–240

Bibliography

1. Fasano A (2012) Zonulin, regulation of tight junctions, and autoimmune diseases. Ann N Y Acad Sci 1258(1):25–33
2. Lerner A, Matthias T (2015) Changes in intestinal tight junction permeability associated with industrial food additives explain the rising incidence of autoimmune disease. Autoimm Rev 14(6):479–489
3. Lee SH (2015) Intestinal permeability regulation by tight junction: implication on inflammatory bowel diseases. Intest Res 13(1):11–18
4. Ugalde-Silva P, Gonzalez-Lugo O, Navarro-Garcia F (2016) Tight junction disruption induced by type 3 secretion system effectors injected by enteropathogenic and enterohemorrhagic *Escherichia coli*. Front Cell Infect Microbiol 6:87
5. Sharma R, Young C, Neu J (2010) Molecular modulation of intestinal epithelial barrier: contribution of microbiota. J Biomed Biotechnol 2010:305879
6. Blaskewicz CD, Pudney J, Anderson DJ (2011) Structure and function of intercellular junctions in human cervical and vaginal mucosal epithelia. Biol Reprod 85(1):97–104
7. Giepmans BNG, van IJzendoorn SCD (2009) Epithelial cell–cell junctions and plasma membrane domains. BBA—Biomembr 1788(4):820–831
8. Günzel D, Yu ASL (2013) Claudins and the modulation of tight junction permeability. Physiol Rev 93(2):525–569
9. Palade GE (1974) Intracellular aspects of the process of protein secretion. In: Nobel Lecture physiology or medicine. Almqvist & Wiksell International, Stockholm
10. Farquhar MG, Palade GE (1963) Junctional complexes in various epithelia. J Cell Biol 17:375–412
11. Diamond JM (1977) Twenty-first Bowditch Lecture. The epithelial junction: Bridge, gate, and fence. Physiol 20(1):10–18
12. Goodenough DA, Paul DL (2009) Gap Junctions. Cold Spring Harb Symp Quant Biol 1(1): a002576
13. Retamal MA, Reyes EP, García IE, Pinto B, Martínez AD, González C (2015) Diseases associated with leaky hemichannels. Front Cell Neurosci 9:267
14. Mee G, Richard G, White TW (2007) Gap junctions: basic structure and function. J Invest Dermatol 127:2516–2524
15. Scemes E, Spray DC, Meda P (2009) Connexins, pannexins, innexins: novel roles of "hemi-channels". Pflugers Arch 457(6):1207–1226
16. Söhl G, Willecke K (2004) Gap junctions and the connexin protein family. Cardiovasc Res 62:228–232
17. Dahl G, Locovei S (2006) Pannexin: to gap or not to gap, is that a question? IUBMB Life 58 (7):409–419
18. Frinchi M, Di Liberto V, Turimella S, D'Antoni F, Theis M, Belluardo N, Mudò G (2013) Connexin36 (Cx36) expression and protein detection in the mouse carotid body and myenteric plexus. Acta Histochem 115(3):252–256
19. Kumar NM, Gilula NB (1996) The gap junction communication channel. Cell 84(3):381–388
20. Sohl G, Willecke K (2003) An update on connexin genes and their nomenclature in mouse and man. Cell Commun Adhes 10:173–180
21. Le Vasseur M, Lelowski J, Bechberger JF, Sin W-C, Naus CC (2014) Pannexin 2 protein expression is not restricted to the CNS. Front Cell Neurosci 8:392
22. Diezmos EF, Sandow SL, Markus I, Perera DS, Lubowski DZ, King DW et al (2013) Expression and localization of pannexin-1 hemichannels in human colon in health and disease. Neurogastroenterol Motil 25(6):e395–e405
23. Kanczuga-Koda L, Sulkowski S, Koda M, Sobaniec-Lotowska M, Sulkowska M (2004) Expression of connexins 26, 32 and 43 in the human colon–an immunohistochemical study. Folia Histochem Cytobiol 42(4):203–217

24. Wang YF, Daniel EE (2001) Gap junctions in gastrointestinal muscle contain multiple connexins. Am J Physiol Gastrointest Liver Physiol 281(2):G533–G543
25. Nagy JI, Urena-Ramirez V, Ghia JE (2014) Functional alterations in gut contractility after connexin36 ablation and evidence for gap junctions forming electrical synapses between nitrergic enteric neurons. FEBS Lett 588(8):1480–1490
26. Palade GE (1975) Intracellular aspects of the process of protein synthesis. Sci 189:347–358
27. Saucan L, Palade GE (1994) Membrane and secretory proteins are transported from the Golgi complex to the sinusoidal plasmalemma of hepatocytes by distinct vesicular carriers. J Cell Biol 125:733–741
28. Connors BW (2012) Tales of a dirty drug: carbenoxolone, gap junctions, and seizures. Epilepsy Curr 12(2):66–68
29. Srinivas M, Hopperstad MG, Spray DC (2001) Quinine blocks specific gap junction channel subtypes. PNAS 98(19):10942–10947
30. Qi X, Varma P, Newman D, Dorian P (2001) Gap junction blockers decrease defibrillation thresholds without changes in ventricular refractoriness in isolated rabbit hearts. Circ 104:1544–1549
31. Bai D, del Corsso C, Srinivas M, Spray DC (2006) Block of specific gap junction channel subtypes by 2-aminoethoxydiphenyl borate (2-APB). JPET 319(3):1452–1458
32. Chiba H, Osanai M, Murata M, Kojima T, Sawada N (2008) Transmembrane proteins of tight junctions. BBA Biomembr 1778(3):588–600
33. Hou J (2014) The kidney tight junction (Review). Int J Mol Med 34(6):1451–1457
34. Förster C (2008) Tight junctions and the modulation of barrier function in disease. Histochem Cell Biol 130(1):55–70
35. Anderson JM, Van Itallie CM (2009) Physiology and function of the tight junction. Cold Spring Harb Perspect Biol 1(2):a002584
36. Baumgartner S, Littleton JT, Broadie K, Bhat MA, Harbecke R, Lengyel JA et al (1996) A Drosophila neurexin is required for septate junction and blood-nerve barrier formation and function. Cell 87(6):1059–1068
37. Oshima K, Fehon RG (2011) Analysis of protein dynamics within the septate junction reveals a highly stable core protein complex that does not include the basolateral polarity protein Discs large. J Cell Sci 124(Pt 16):2861–2871
38. Juang JL, Carlson SD (1994) Analog of vertebrate anionic sites in blood-brain interface of larval Drosophila. Cell Tissue Res 277(1):87–95
39. Abbott NJ, Rönnbäck L, Hansson E (2006) Astrocyte–endothelial interactions at the blood–brain barrier. Nat Rev Neurosci 7:41–53
40. Willis CL (2011) Glia-induced reversible disruption of blood–brain barrier integrity and neuropathological response of the neurovascular unit. Toxicol Patholvol 39(1):172–185
41. Balda MS, Matter K (2008) Tight junctions at a glance. J Cell Sci 121:3677–3682
42. Mandel LJ, Bacallao R, Zampighi G (1993) Uncoupling of the molecular 'fence' and paracellular 'gate' functions in epithelial tight junctions. Nat 361(6412):552–555
43. Landy J, Ronde E, English N, Clark SK, Hart AL, Knight SC et al (2016) Tight junctions in inflammatory bowel diseases and inflammatory bowel disease associated colorectal cancer. World J Gastroenterol 22(11):3117–3126
44. Sánchez-Pulido L, Martín-Belmonte F, Valencia A, Alonso MA (2002) MARVEL: a conserved domain involved in membrane apposition events. Trends Biochem Sci 27 (12):599–601
45. Raleigh DR, Marchiando AM, Zhang Y, Shen L, Sasaki H, Wang Y et al (2010) Tight junction–associated MARVEL proteins MarvelD3, tricellulin, and occludin have distinct but overlapping functions. Mol Biol Cell 21(7):1200–1213
46. Wu Y-C, Liu C-Y, Chen Y-H, Chen R-F, Huang C-J, Wang I-J (2012) Blood vessel epicardial substance (Bves) regulates epidermal tight junction integrity through atypical protein kinase C. J Biol Chem 287:39887–39897
47. Abe K, Takeichi M (2008) EPLIN mediates linkage of the cadherin—catenin complex to F-actin and stabilizes the circumferential actin belt. PNAS 105(1):13–19

48. Cho KO, Hunt CA, Kennedy MB (1992) The rat brain postsynaptic density fraction contains a homolog of the Drosophila discs-large tumor suppressor protein. Neuron 9(5):929–942
49. Woods DF, Bryant PJ (1993) ZO-1, DlgA, PSD-95/SAP90: homologous proteins in tight, septate and synaptic cell junctions. Mech Dev 44:85–89
50. Overgaard CE, Daugherty BL, Mitchell LA, Koval M (2011) Claudins: control of barrier function and regulation in response to oxidant stress. Antioxid Redox Signal 15(5):1179–1193
51. Shen L, Weber CR, Raleigh DR, Yu D, Turner JR (2011) Tight junction pore and leak pathways: a dynamic duo. Annu Rev Physiol 73:283–309
52. Pei L (2016) Paracellular epithelial transport maximizes energy efficiency in the kidney. University of Kansas, Kansas, USA
53. Anderson JM, Van Itallie CM (2009) Physiology and function of the tight junction. Cold Spring Harb Perspect Biol 1(2):a002584
54. Markov AG, Amasheh S (2011) Barrier properties and tight junction protein expression along the longitudinal axis of rat intestine. Ross Fiziol Zh Im I M Sechenova 97(10):1066–1083
55. Hou J (2012) Lecture: new light on the role of claudins in the kidney. Organogenesis 8(1):1–9
56. Gong Y, Hou J (2014) Claudin-14 underlies Ca^{++}-sensing receptor-mediated Ca^{++} metabolism via NFAT-microRNA-based mechanisms. J Am Soc Nephrol 25(4):745–760
57. Amasheh S, Fromm M, Günzel D (2011) Claudins of intestine and nephron—a correlation of molecular tight junction structure and barrier function. Acta Physiol (Oxf) 201(1):133–140
58. Markov AG, Falchuk EL, Kruglova NM, Rybalchenko OV, Fromm M, Amasheh S (2014) Comparative analysis of theophylline and cholera toxin in rat colon reveals an induction of sealing tight junction proteins. Pflugers Arch 466(11):2059–2065
59. Zeissig S, Bürgel N, Günzel D, Richter J, Mankertz J, Wahnschaffe U et al (2007) Changes in expression and distribution of claudin 2, 5 and 8 lead to discontinuous tight junctions and barrier dysfunction in active Crohn's disease. Gut 56(1):61–72
60. Mankertz J, Schulzke JD (2007) Altered permeability in inflammatory bowel disease: pathophysiology and clinical implications. Curr Opin Gastroenterol 23(4):379–383
61. Heller F, Florian P, Bojarski C, Richter J, Christ M, Hillenbrand B et al (2005) Interleukin-13 is the key effector Th2 cytokine in ulcerative colitis that affects epithelial tight junctions, apoptosis, and cell restitution. Gastroenterol 129(2):550–564
62. Schulzke JD, Ploeger S, Amasheh M, Fromm A, Zeissig S, Troeger H et al (2009) Epithelial tight junctions in intestinal inflammation. Ann N Y Acad Sci 1165:294–300
63. Hou J, Goodenough DA (2010) Claudin-16 and claudin-19 function in the thick ascending limb. Curr Opin Nephrol Hypertens 19(5):483–488
64. Bauer H, Zweimueller-Mayer J, Steinbacher P, Lametschwandtner A, Bauer HC (2010) The dual role of zonula occludens (ZO) proteins. J Biomed Biotechnol 2010:402593
65. Itoh M, Nagafuchi A, Moroi S, Tsukita S (1997) Involvement of ZO-1 in cadherin-based cell adhesion through its direct binding to alpha catenin and actin filaments. J Cell Biol 138 (1):181–192
66. Itoh M, Furuse M, Morita K, Kubota K, Saitou M, Tsukita S (1999) Direct binding of three tight junction-associated MAGUKs, ZO-1, ZO-2, and ZO-3, with the COOH termini of claudins. J Cell Biol 147(6):1351–1363
67. Woods DF, Bryant PJ (1993) ZO-1, DlgA and PSD-95/SAP90: homologous proteins in tight, septate and synaptic cell junctions. Mech Dev 44(2–3):85–89
68. Guillemot L, Paschoud S, Pulimeno P, Foglia A, Citi S (2008) The cytoplasmic plaque of tight junctions: a scaffolding and signalling center. BBA Biomembranes 1778(3):601–613
69. Mariano C, Sasaki H, Brites D, Brito MA (2011) A look at tricellulin and its role in tight junction formation and maintenance. Eur J Cell Biol 90(10):787–796

70. Fanning AS, Jameson BJ, Jesaitis LA, Anderson JM (1998) The tight junction protein ZO-1 establishes a link between the transmembrane protein occludin and the actin cytoskeleton. J Biol Chem 273(45):29745–29753
71. Kale G, Naren AP, Sheth P, Rao RK (2003) Tyrosine phosphorylation of occludin attenuates its interactions with ZO-1, ZO-2, and ZO-3. Biochem Biophys Res Commun 302(2):324–329
72. Choi W, Acharya BR, Peyret G, Fardin M-A, Mège R-M, Ladoux B et al (2016) Remodeling the zonula adherens in response to tension and the role of afadin in this response. J Cell Biol 213(2):243–260
73. Kamitani T, Sakaguchi H, Tamura A, Miyashita T, Yamazaki Y, Tokumasu R et al (2015) Deletion of tricellulin causes progressive hearing loss associated with degeneration of cochlear hair cells. Sci Rep 5:18402
74. Krug SM, Amasheh S, Richter JF, Milatz S, Günzel D, Westphal JK et al (2009) Tricellulin forms a barrier to macromolecules in tricellular tight junctions without affecting ion permeability. Mol Biol Cell 20(16):3713–3724
75. Riazuddin S, Ahmed ZM, Fanning AS, Lagziel A, Kitajiri S-I, Ramzan K et al (2006) Tricellulin is a tight-junction protein necessary for hearing. Am J Hum Genet 79(6):1040–1051
76. Brand T (2005) The Popeye domain-containing gene family. Cell Biochem Biophys 43 (1):95–103
77. Osler ME, Smith TK, Bader DM (2006) Bves, a member of the Popeye domain-containing gene family. Dev Dynam 235(3):586–593
78. Hager HA, Bader DM (2009) Bves: ten years after. Histol Histopathol 24(6):777–787
79. Wu Y-C, Chen R-F, Liu C-Y, Hu FR, Huang C-J, Wang I-J (2014) Knockdown of Zebrafish blood vessel epicardial substance results in incomplete retinal lamination. Sci World J 2014:803718
80. Andrée B, Hillemann T, Kessler-icekson G, Schmitt-john T, Jockusch H, Arnold HH, Brand T (2000) Isolation and characterization of the novel popeye gene family expressed in skeletal muscle and heart. Dev Biol 223(2):371–382
81. Schindler RFR, Brand T (2016) The Popeye domain containing protein family—a novel class of cAMP effectors with important functions in multiple tissues. Prog Biophys Mol Biol 120(1–3):28–36
82. Froese A, Breher SS, Waldeyer C, Schindler RF, Nikolaev VO, Rinné S et al (2012) Popeye domain containing proteins are essential for stress-mediated modulation of cardiac pacemaking in mice. J Clin Invest 122(3):1119–1130
83. Smith TK, Bader DM (2006) Characterization of Bves expression during mouse development using newly generated immunoreagents. Dev Dynam 235:1701–1708
84. Pardo JV, Craig SW (1979) alpha-Actinin localization in the junctional complex of intestinal epithelial cells. J Cell Biol 80:203–210
85. Renyong G, Hiroshi S, Shigeki S (2006) Endothelial cell motility is compatible with junctional integrity. J Cell Physiol 211:327–335
86. Meng W, Takeichi M (2009) Adherens junction: molecular architecture and regulation. Cold Spring Harb Perspect Biol 1(6):a002899
87. Harrison OJ, Jin X, Hong S, Bahna F, Ahlsen G, Brasch J et al (2011) The extracellular architecture of adherens junctions revealed by crystal structures of type I cadherins. Struct 19 (2):244–256
88. Hartsock A, Nelson WJ (2008) Adherens and tight junctions: structure, function and connections to the actin cytoskeleton. Biochim Biophys Acta 1778(3):660–669
89. Straub BK, Rickelt S, Zimbelmann R, Grund C, Kuhn C, Iken M et al (2011) E-N-cadherin heterodimers define novel adherens junctions connecting endoderm-derived cells. J Cell Biol 195(5):873–887
90. Coopman P, Djiane A (2016) Adherens junction and E-Cadherin complex regulation by epithelial polarity. Cell Mol Life Sci 73(18):3535–3553

91. Hoffmann B, Schäfer C (2010) Filopodial focal complexes direct adhesion and force generation towards filopodia outgrowth. Cell Adh Migr 4(2):190–193
92. Vasioukhin V, Fuchs E (2001) Actin dynamics and cell-cell adhesion in epithelia. Curr Opin Cell Biol 13(1):76–84
93. Green KJ, Simpson CL (2007) Desmosomes: new perspectives on a classic. J Invest Dermatol 127(11):2499–2515
94. Waschke J (2008) The desmosome and pemphigus. Histochem Cell Biol 130(1):21–54
95. Green KJ, Gaudry CA (2000) Are desmosomes more than tethers for intermediate filaments? Nat Rev Mol Cell Biol 1:208–216
96. Garrod D, Chidgey M (2008) Desmosome structure, composition and function. Biochim Biophys Acta 1778(3):572–587
97. Kitajima Y (2013) New insights into desmosome regulation and pemphigus blistering as a desmosome-remodeling disease. Kaohsiung J Med Sci 29(1):1–13
98. Sumigray KD, Lechler T (2012) Desmoplakin controls microvilli length but not cell adhesion or keratin organization in the intestinal epithelium. Mol Biol Cell 23(5):792–799
99. Acehan D, Petzold C, Gumper I, Sabatini DD, Müller EJ, Cowin P, Stokes DL (2008) Plakoglobin is required for effective intermediate filament anchorage to desmosomes. J Invest Dermatol 128(11):2665–2675
100. Bornslaeger EA, Godsel LM, Corcoran CM, Park JK, Hatzfeld M, Kowalczyk AP, Green KJ (2001) Plakophilin 1 interferes with plakoglobin binding to desmoplakin, yet together with plakoglobin promotes clustering of desmosomal plaque complexes at cell-cell borders. J Cell Sci 114(Pt 4):727–738
101. Holthöfer B, Windoffer R, Troyanovsky S, Leube RE (2007) Structure and function of desmosomes. Int Rev Cytol 264:65–163
102. Windoffer R, Borchert-Stuhlträger M, Leube RE (2002) Desmosomes: interconnected calcium-dependent structures of remarkable stability with significant integral membrane protein turnover. J Cell Sci 115(Pt 8):1717–1732
103. Kowalczyk AP, Navarro P, Dejana E, Bornslaeger EA, Green KJ, Kopp DS, Borgwardt JE (1998) VE-cadherin and desmoplakin are assembled into dermal microvascular endothelial intercellular junctions: a pivotal role for plakoglobin in the recruitment of desmoplakin to intercellular junctions. J Cell Sci 111(Pt 20):3045–3057
104. Hull BE, Staehelin LA (1979) The terminal web. A reevaluation of its structure and function. J Cell Biol 81(1):67–82
105. Brooke MA, Nitoiu D, Kelsell DP (2012) Cell-cell connectivity: desmosomes and disease. J Pathol 226(2):158–171

Chapter 5
Molecular Mechanisms of Gastrointestinal Signaling

Abstract The epithelial cells of the gastrointestinal (GI) tract communicate with each other and with cells of other organs via a complex network of highly regulated movement of ions and biomolecules. The molecules ensure regulated activity of cells, tissues, and organs of the GI system and the body as whole. The regulated movement and subsequent activities of the biomolecules released from one cell to the target are made possible by receptive substances (receptors) localized on the membrane of the target cells or intracellular organelles, or in the cytosol. This process, which is referred to as cell-to-cell communication or cellular signaling, ensures the regulated functioning of the cells and tissues of the GI system and the whole organism. This chapter is dedicated to the mechanism of cell-to-cell communication and signaling in normal and relates it to how disease develops. Basic mechanisms of GI epithelial cell signaling and gut nutrient receptor sensing (GI chemosensation) are discussed.

Keywords Cellular communication · Cell-to-cell communication
Cell signaling · Steroid receptors · G protein-coupled receptor · Ion channels
Ion channel receptor · Channelopathies · Tyrphostins · Catalytic receptors
Enzyme-linked receptor · Guanylate cyclase · Receptor serine/threonine kinase
Tyrosine receptor kinase · Morphogens · Morphogen receptor · Notch
Hedgehog · Wingless/wnt · Cytokines · Interleukins · Cytokine receptor
JAK/STAT · Integrin receptors · Cell-surface adhesion receptors
Gastrointestinal chemosensation · Gut nutrient sensing · Receptor nutrient sensing
Amino acid sensors · Fatty acid/lipid sensors · Glucose sensors

Abbreviations

5-HT	5-hydroxytryptamine type
7TM	Seven transmembrane
ADAM	A disintegrin and metalloproteinase protein
Akt	"Ak" (mouse bred) that developed thymoma ("t")
AMPA	Alpha-amino-3-hydroxy-5-methyl-4-isoxazolepropionic acid
ANP	Atrial natriuretic peptide
APC	Adenomatous polyposis coli

M. O. Welcome, *Gastrointestinal Physiology*,
https://doi.org/10.1007/978-3-319-91056-7_5

ATF3	Activating transcription factor 3
BMP	Bone morphogenic protein
BNP	Brain natriuretic peptide
CaM	Calmodulin
CaMK	Calcium/calmodulin-dependent protein kinases
cAMP	Cyclic adenosine monophosphate
CaSR	Calcium-sensing receptor of aromatic amino acid
CBPs	Ca^{2+}-binding proteins
CCK	Cholecystokinin
CDK	Cyclin-dependent kinase
CFTR	Cystic fibrosis transmembrane conductance regulator
cGMP	Cyclic guanosine monophosphate
CLK	CDK-like kinase
CNP	C-type natriuretic peptide
CO	Carbon monoxide
CRAC	Ca^{2+} release-activated Ca^{2+}
CREB	CAMP response element-binding protein
CSL	(CBF1/Su(H)/Lag-1) C promoter Binding Factor-1, Suppressor of Hairless, Longevity-Assurance Gene
DAG	Diacylglycerol
Dhh	Desert Hedgehog
Dll-1-4	Delta/Delta-like
Dvl or Dsh	Disheveled
EGF	Epidermal growth factor
EPAC	Cyclic nucleotide-gated ion channels and exchange proteins activated by cAMP
ER	Endoplasmic reticulum
ErbB	Erythroblastic leukemia viral oncogene
ERK	Extracellular receptor kinase
FABP	Fatty acid-binding proteins
FAK	Focal adhesion kinase
FDA	Food and Drug Administration
FFAR	Fatty Acid Transport Protein
FGF	Fibroblast growth factor
FRAP1	FK506-binding protein 12-rapamycin-associated protein 1
Fzd	Frizzled
GABA	Gamma-aminobutyric acid
GAG	Glycosaminoglycan
GAPs	GTPase-activating proteins
GBP	GSK-3-binding proteins
GC	Guanylate cyclase
GDI	Guanine nucleotide dissociation inhibitor
GDNF	Glial-derived neurotrophic factor
GDP	Guanosine 5′-diphosphate

GEFs	Guanine nucleotide exchange factors
GIP	Glucose-dependent insulinotropic polypeptide
GLP-1	Glucagon-like peptide-1
GPBAR-1	G protein-coupled bile acid receptor 1
GPCR	G protein (guanine nucleotide-binding protein) coupled receptor
GRKs	GPCR kinases
GSK3	Glycolgen synthase kinase 3
GSK-3β	Glycogen synthase kinase-3β
GTP	Guanosine 5′-triphosphate
HGF	Hepatocyte growth factor
Hh	Hedgehog
HMG	High mobility group
HS	Heparan sulfate
HSPGs	Heparan sulfate proteoglycans
ICAM	Intercellular adhesion molecules
ICD	Intracellular domain of the notch receptor
Ig	Immunoglobulin
IGF1R	Insulin-like growth factor
IGF-2	Insulin-like growth factor-2
Ihh	Indian Hedgehog
IL	Interleukins
ILK	Integrin-linked kinase
IP_3	Inositol 1,4,5-trisphosphate
IP_3R	IP_3 receptor
IRAG	Nitric oxide
IRS1	Insulin receptor substrate 1
Jag-1, Jag-2	Jagged
JAK	Janus kinase
LCFA	Long-chain fatty acid
LEF	Lymphoid enhancer factor
MAML	Mastermind-like
MAPK	Mitogen-activated protein kinase
M-BAR	Membrane-type receptor for bile acids
MCT	Medium chain triglyceride
mTOR	Mechanistic or mammalian target of rapamycin
NAADP	Cyclic ADP-ribose and nicotinic acid adenine dinucleotide phosphate
nAChRs	Nicotinic acetylcholine receptors
NAD^+	Nicotinamide adenine dinucleotide
N-CAM	Neural cell adhesion molecule
NFAT	Nuclear factor of activated T cell
NGF	Neuronal growth factor

NMDA	*N*-methyl-D-aspartate
NO	IP_3R-associated cGMP kinase substrate
NO	Nitric oxide
NT-3	Neurotrophin-3
P2X receptors	ATP-gated channels
PDGF	Platelet-derived growth factor
PI-3-K	Phosphoinositide-3-kinase
PIP_2	Phosphatidylinositol 4,5-bisphosphate
PKA	Protein kinase A
PKC	Protein kinase C
PKG	Protein kinase G
PLC	Phospholipase C
PPAR	Peroxisome proliferator-activated receptors
PTB	Phosphotyrosine binding
Ras	Rat sarcoma
RBPJ	Jκ immunoglobulin gene
RSTK	Receptor serine/threonine kinase
RTK	Receptor tyrosine kinase
SARAF	Store-operated calcium entry-associated regulatory factor
SCF (also called *c-kit*)	Stem cell factor
SCFA	Short chain fatty acid
SGLT1	Sodium glucose cotransporter type 1
SH2	Src homology 2
Shh	Sonic hedgehog
Shh	Sonic Hedgehog
SIM	Stromal interacting molecule (previous name for STIMI1)
SOCE	Store-operated calcium channel
STa	Bacterial heat-stable enterotoxins
STAT	Signal transducers and activators of transcription
STIM1	Stromal interaction molecule 1
TCF	T cell factor
TGF-beta	Transforming growth factor beta
TKI	Tyrosine kinase inhibitors
TMEM66	Transmembrane protein 66
TNFs	Tumor necrosis factors
TRP	Transient Receptor Potential
TRPV1	Transient Receptor Potential Valinoid type 1
TSC1-TSC2	Tuberous sclerosis complex subunit 1 and 2
VCAM	Vascular cell adhesion molecule
VEGF	Vascular endothelial growth factor
Wif-1 & 2	Wnt inhibitory factor 1 and 2
Wnt	Wingless-related integration site

5.1 Introduction

The structural and functional architecture of the gastrointestinal (GI) tract is maintained by a wide array of signals impinging on the different cells of this digestive apparatus. These signals are responsible for initiating, modulating, or regulating the activities of the entire GI system via multiple cellular signaling pathways. Cellular signaling is a phenomenon involving the initiation or modulation of specific function(s) by the activities of an extracellular signal on a cognate receptor molecule located usually on the membrane of the cell, transducing the information carried by the extracellular signal to the effectors, required to achieve the desired physiological response. (It is estimated that a single human cell harbors over 300 different receptor types). Some signals originating from inside the cell (intracellular signal) or from outside the cell (extracellular signal) may directly act on surface receptors, intracellular receptors, localized inside the cytoplasm, inside the nucleus, on the nuclear membrane, on membrane of other organelles, or inside the organelles to initiate or modulate cellular activities. The extracellular signals are called the first messengers. First messengers include peptides (peptide hormones, neuropeptides, and growth factors), steroids, ions, products of metabolism, gases, chemical agent, physical agents (light, mechanical stress). These messengers initiate downstream activity, producing another messenger in the intracellular milieu. This intracellular messenger is called the second messenger. To date, over 20 second messengers have been discovered. Examples of second messengers include calcium, inositol 1,4,5-trisphosphate (IP3), cyclic adenosine monophosphate (cAMP), cyclic guanosine monophosphate (cGMP), diacylglycerol (DAG). These signaling molecules constitute the major second messengers of GI cell signaling [1–5].

The GI tract possesses special sensors (receptor molecules) that translate the chemical properties of food substances or extracellular signal into intracellular signal that culminate in secretion of hormones and neurotransmitters as well as exocrine (lumencrine) release of substances. The release of hormones and neurotransmitters is mediated, usually, by chemical, electrical, and mechanical stimulation of the cell. Chemical stimulation involves the recognition of receptor sensors of GI cells (such as neurons, endocrine, and other epithelial cells) by signaling molecules, resulting in subsequent release of hormones, neurotransmitters or neuromodulators. Mechanical signaling is usually initiated via specific tension-sensing receptors on the cell membrane. Upon activation, these receptors respond to changes in membrane tension produced by local movement of an organ or region of the organ. Electrical stimulation by nervous impulses represents an integral aspect of gut signaling and plays a key role in regulation of GI tract activities. All signaling types work in a coordinated manner to ensure regulated signaling and functioning of the cells and tissues of the GI tract [1, 6–10].

Receptor nutrient sensor signaling in the GI tract (also referred to as GI chemosensation) is a regulated signaling initiated by the interaction between specific receptor group (sensors) localized on the apical membrane of the gut epithelial cell and nutrient or biomolecule in the lumen, resulting in the release of hormone, neurotransmitter, or paracrine factor that not only possesses local, but also considerable systemic significance. Different types of GI epithelial cell taste-sensing receptors are expressed in the entire GI tract. These taste receptors (five types of taste receptors) are involved in encoding the chemical property of food into sensory impulses that are relayed to the central nervous system for further analysis. The taste sensory system also modulates several properties of the GI cells by its endocrine and neural secretions [11–13].

The ways by which the cells of the GI tract interact among themselves are numerous. Figures 5.1 and 5.2 only show some of the possible ways. Also bear in mind that the same cell can signal with itself by autocrine mechanism.

5.2 Brief Historical Background

The first class of receptors to be discovered was the ion channels. The rise of knowledge about ion channels dates as far back to the 1700–1800s with the work of the Italian physiologist Aloysio Luigi Galvani (1737–1798), Italian physicist, Alessandro Volta (1745–1827), Italian physicist and neurophysiologist Carlo Matteucci (1811–1868), and German physician and physiologist Emil du Bois-Reymond (1818–1896) on electricity concept in living tissue. Though these scientists did not opine that ion channels were present in living cells, their works directed the course of discovery of these pore-forming receptors in living cells. It is widely acknowledged that the science of receptor signaling began in the first decade of the twentieth century when the English physiologist John N. Langley (1852–1925) introduced the concept of "receptive molecules," which were required to determine the action of a specific chemical agent on the nerve cell. The idea about "receptive molecules" stemmed originally from the work of the German scientist Paul Ehrlich (1854–1915) whose experimental results indicated that drugs and dyes bind to specific sites or receptors on cell surfaces. The concept of "receptive molecules" was further developed by Sir Henry Dale [14]. Further information on the contributions of these individuals is provided in Chaps. 7 and 11. The introduction of the squid giant neuron as an experimental model in 1937 by the English zoologist and neurophysiologist John Zachary Young (1907–1997) to study different properties of neurons gave promise to early neurophysiology research and ion channel activity investigations. The squid giant neuron is convenient model because of its size compared to a typical mammalian neuron. It is hundreds of times larger than an animal neuron [15, 16]. This made it easier to work with the giant neuron and this may have helped the invention of the first instrument to measure ionic currents from these cells. In 1946–1949, the device for accurately measuring the electrical potentials of living cells was developed by the American neurophysiologist Ralph Waldo Gerard (1900–1974), Judith Graham Pool (1919–1975) and cell physiologist and biochemist Gilbert Ning Ling (1919–) [17–20]. This device, called microelectrode, uses minute glass electrodes whose diameter was much smaller than the size of a neuron. The electrodes were filled with a reference

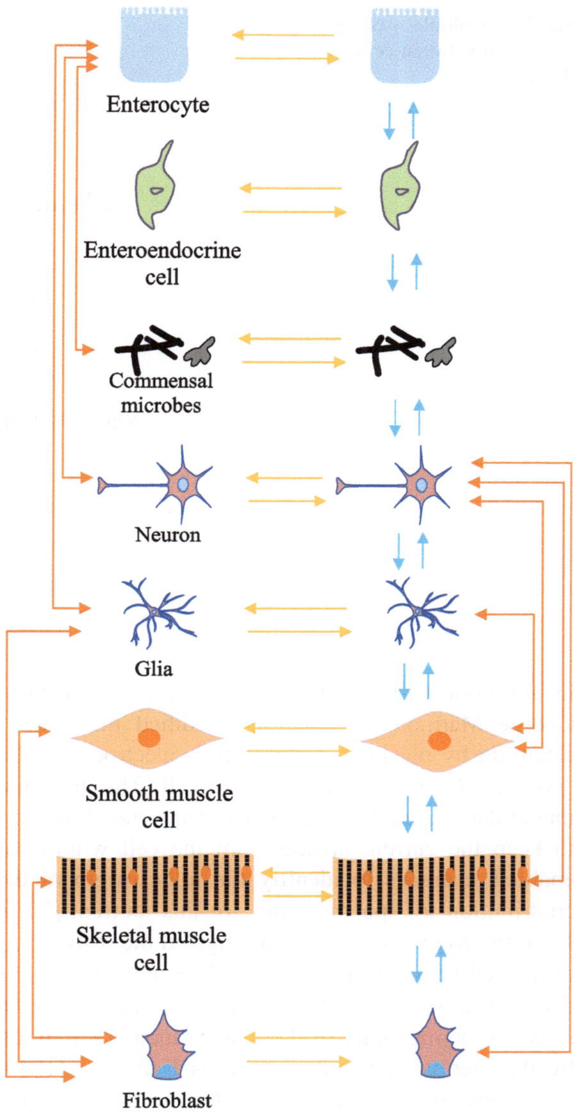

Fig. 5.1 Possible ways of cell-to-cell communication and signaling in the GI tract. The arrows indicate the possible ways of interaction between the cells

saline solution. This solution was used to detect current changes upon placement of the electrode inside a living cell by piercing the sharp end through the plasma membrane. The current changes in the cell were detected by the microelectrode device connected to a screen to visualize the electrical phenomenon occurring in the cell. At this time, cell researchers were now able to measure directly and quantitatively the current present in living cells—the result of differences in concentration of ions inside and outside the cell [21, 22]. In the same decade, following the

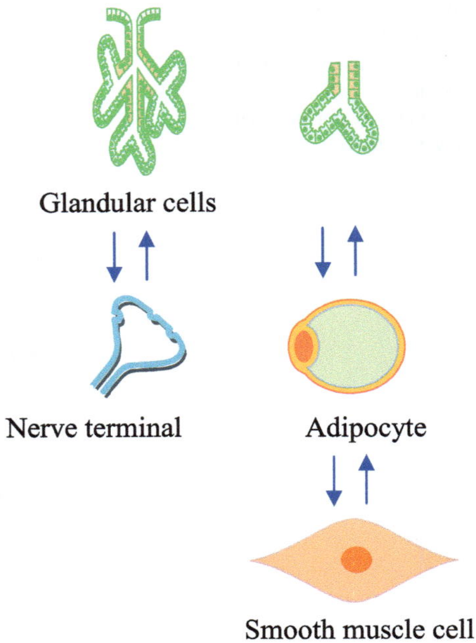

Fig. 5.2 Multiple ways of interaction between cells of GI tract

invention of the microelectrode technique, Kenneth Stewart Cole (1900–1984) and George Marmont successfully modified the initial Graham-Gerard-Ling microelectrode by placing a second glass electrode inside the cell, which was used to "voltage clamp" the inside of the cell. This technique allowed to experimentally manipulate the cell potential at varying membrane potential. Thus, it was possible to keep the current arising from the cell within constant range. This technique enabled scientists to identify the ionic origin of the membrane current and the contribution of specific ions to this current [23–26]. Together with Howard J. Curtis, Kenneth S. Cole investigated the resting as well as depolarizing potential of the cell [27]. During the late 1940s and early 1950s, it was observed that during depolarization the membrane was still selectively permeable to some ions [28]. This selective permeability of the plasma membrane to ions was first proposed in 1902 by the German physiologist Julius Bernstein (1839–1917). On the basis of his electrophysiological studies and the electrochemical theory earlier propounded by the German physicist, Walther Hermann Nernst (1864–1941), Bernstein formulated the membrane theory, postulating that resting potential of an excitable cell is formed as a diffusion potential gradient due to the tendency of positively charged ions (e.g., K^+ ions) to diffuse from a region of high ion concentration (cytoplasm) to a region of low ion concentration (extracellular fluid), thus resulting in the development of a more negatively charged cytoplasmic surface in relation to the extracellular surface of the plasma membrane of the cell [29–31]. The modern interpretation of theory postulates that the plasma membrane of excitable cells is selectively permeable to the movement of ions and that the electrochemical gradient dictates the movement of specific ions at different concentrations of ions inside and outside the cell [32, 33]. However, the mechanisms for this phenomenon remained unclear at the time. The German physiologist Ludimar Hermann (1838–1914) had initially proposed a "theory of local circuit," which at that time served the basis for explaining the flow of current across the nerve or muscle fiber. Hermann's theory indicated that flow of current along the nerve or muscle fiber may depend on the local flow of current. Hermann showed that the whole surface of an uninjured muscle was electrically isopotential (equipotential), meaning that every point of the entire surface had same potential [29]. Upon stimulation of the cell, there is reversal in potential gradient. In both cases, the diffusion or membrane permeability changes to different ions [29–31].

The series of advancement in technology to study the electrical activities in living cells enabled researchers to unravel some aspects of cell signaling. The uncovering of plasma membrane structure around the 1920 s also gave way to the discovery of proteins in the plasma membrane that act as channels or receptors. Of special note, the English physiologists and biophysicists Sir Andrew Fielding Huxley (1917–2012) and Sir Alan Lloyd Hodgkin (1914–1998) and the German-born biophysicist Katz Bernard (1911–2003), using the giant squid neuron, in a series of experiment, resolved that changes in ion concentration (in particular sodium and potassium ions) inside and outside the cell were due to the formation of action potential [34]. The modern history of ion channel receptors began with the works of Huxley and Hodgkin, who reported their experimental findings of the action potentials of ion channel activities by using a voltage clamp technique [34–36]. On the basis of their experimental data, Hodgkin and Huxley developed a mathematical model of ionic conductance to demonstrate the role of ion channels in the electrical behavior of nerve cells [37]. Hodgkin and Katz may have discovered special types of ion channels (now called leaky channels) when they noted a background conductance that is always present even though all known channels have been blocked [38]. Huxley, Hodgkin, and the Australian neurophysiologist Sir John Carew Eccles (1903–1997) won the 1963 Nobel Prize in Physiology or Medicine for their works on the ionic mechanisms of neurotransmission. The trio discovered the ionic mechanisms connected with excitation and inhibition in nerve fibers. Hodgkin and Huxley became the first to provide a quantitative description of membrane current and applied it to conduction and excitation in nerve cells. Eccles was recognized for his work on the synapse [34, 36, 39]. There are currently many modifications of the initial microelectrode invented for scientific investigation [40–42].

On the basis of his experimental results using the squid giant axon, Mullins LJ became one of the earliest scientists in 1959 to have used the term "channel" referring to the conductance changes occurring in the squid giant axon due to differences in ion concentration on either side of the membrane [43]. In the late 1950s (precisely in 1957–1959), Furshpan and Potter (1957, 1959) identified that electrical transmission was possible across a synapse [44, 45]. Thus, they discovered electrical synapse. Even though it was still unclear whether or not special integral proteins in the plasma membrane aided the movement of ions or charges (cellular electricity), there were increasing speculations that differences in ion gradients in the cell were due to movement of ions across the membrane. The discovery of the mechanism of action of tetrodotoxin, a neurotoxin isolated in 1909 by Japanese scientist Yoshizumi Tahara from puffer fish (Tetraodon) provided a clue to whether or not the cell membrane possesses ion pore-forming proteins or channels through which the ions can permeate. The mechanism of action of tetrodotoxin first identified in 1964 by Toshio Narahashi (1927–2013), John W. Moore, and William R. Scott found to exclusively block voltage-gated sodium channels which results to the inhibition of action potentials in nerve cell membranes. This discovery further intensified belief that really, there exists an ion preforming integral protein or channel, specific for ions, in this case, sodium ions [46, 47]. Identification in the 1960s that tetraethylammonium ion blocks specific potassium ion flow suggested that membrane channels or pores specific for potassium ions are also present in the plasma membrane [48–51]. Discovery of tetrodotoxin and tetramethylammonium and their mechanisms of action indicated that there were specific membrane channels, in particular, for sodium and potassium ions [48, 51]. But there were yet no all-encompassing experimental evidences to unravel specific ion channel activity.

A huge disadvantage for the experiment on squid giant axon was that the technique including its modifications cannot be used for cells much smaller than the squid neuron. Thus, the voltage clamp technique could not be used for smaller cells. A breakthrough was made in 1976 when Bert Sakmann and Erwin Neher developed the patch clamp technique that allowed investigating the voltage of a single ion channel. Further, in 1981, Sakmann, Neher, and colleagues developed the Gigaseal which significantly improved the signal-to-noise ratio of patch clamp recording. With the Gigaseal, smaller current can be recorded through an ion channel [25, 52–54]. They were awarded the 1991 Nobel Prize in Physiology and Medicine. There are presently numerous improvements on the original patch clamp technique of Sakmann and Neher. There are different types of patch clamp ranging from the manual to automated, highly resolved, high throughput, and to patch clamp on a chip. By the 1980s, numerous papers using the words "ion channels" have been published [55–58]. For further review, see Káradóttir and Attwell [59] and Dunlop et al. [60].

Patch clamping is a widely used electrophysiological technique for the study of ion channels, membrane proteins that regulate the flow of ions across cellular membranes. The technique allows to also record the electrical activity of the whole cell. To study ion channel, an isolated piece of cell membrane or whole is clamped to determine the current flowing through it or the level of electrical activity. The current measured here is the one provided by the patch clamp amplifier keeping the voltage steady (clamped). Through this technique, current in the picoampere (pA) range (10^{-12} A) or lower can be measured [61–65]. The different concentrations of ions in the cytoplasm and extracellular fluid of the cell are shown in Table 5.1.

By this time it was widely acknowledged, on the basis of electrophysiological experiments, that specific ion channels exist to regulate flow of ions across the

Table 5.1 Intracellular and extracellular concentration of the main ions found in animal fluids [66]

Types of ion	Intracellular concentration (mM)	Extracellular concentration (mM)
Na^+	5–20	130–160 (145)
K^+	130–160 (140)	4–8 (5)
Mg^{2+}	10–20 (0.5)	1–5
Ca^{2+}	50–1000 nM (10^{-4})	1.2–4
H^+	$10^{-7.2}$, pH = 7.2	$10^{-7.4}$, pH = 7.4
Cl^-	1–60	100–140
HCO_3^-	1–3	20–30

plasma membrane; however, the structure of ion channels remained unknown for decades not until the Roderick MacKinnon (1956–) of Rockefeller University started unraveling the architecture of different pore-forming membrane integral proteins [25]. Roderick MacKinnon together with the American Peter C. Agre (1949–) won the 2003 Nobel Prize in Chemistry for their work on the structure and operation of ion channels [67]. Peter Agre was awarded the prize for the discovery of water channels in 1988 from human erythrocytes. The water channels were initially called by Agre and coworkers as channel-forming integral membrane protein-28 kDa and later renamed in 1992 as water channel. A year later Agre's group called it aquaporin. The water channel identified by Agre and associates was later called aquaporin-1 [28, 68–71]. Aquaporins are transmembrane proteins that mainly function to transport water across biological membranes. Following the award of the Nobel Prize to Agre, a series of reports emerged pointing that the first water channel was actually first discovered by the Romanian research group, Gheorghe Benga (1944–) and coworkers, in 1985 and not by Agre [70, 72–74]. Benga's group discovered the first protein water channel in erythrocyte and they published their work in 1986 in the Journals Biochemistry and the European Journal of Cell Biology [75, 76].

The importance of ion channels in human health cannot be overemphasized. A variety of humans diseases are now believed to be caused by mutations in the subunits of specific ion channels, generally called channelopathies [77–79].

Around the 1950–1960s, the American pharmacologist and biochemist Earl Wilbur Sutherland Jr. (1915–1974) and coworkers discovered that some hormones (such as glucagon, epinephrine) can stimulate the production of cyclic AMP (cAMP) in the cytoplasm [80–85]. cAMP was discovered in 1956 as an intracellular signal (second messenger) induced by the action of hormones that control a range of cellular functions including hormone secretion, ion channel permeability, gene expression, lipolysis, and glycogenolysis [86]. Sutherland won the 1971 Nobel Prize for Physiology or Medicine "for his discoveries concerning the mechanisms of the action of hormones" [87, 88]. However, nobody knew how this increased production of intracellular cAMP came about. Following a series of speculations and experimentations, it became apparent that intracellular production of this second messenger may require an intermediary intracellular protein, which was later identified as the heterotrimeric guanine nucleotide-binding protein (G protein) [89]. The structure and functions of the subunits of the G protein were discovered by the American pharmacologist and biochemist Alfred Goodman Gilman (1941–2015). Gilman and the American biochemist and molecular endocrinologist Martin Rodbell (1925–1998) shared the 1994 Nobel Prize for Physiology or Medicine "for their discovery of G proteins and the role of these proteins in signal transduction" [89–92].

G protein is coupled to a major membrane receptor called G protein-coupled receptor (GPCR). The structure of this receptor only became known following the discovery of rhodopsin (a receptor exclusively located in rod photoreceptor cells of the retina and activated by light photons) and the complete sequencing of this photoreceptor protein in the early 1980s. The molecular investigation led to the discovery that rhodopsin is a GPCR, a receptor with seven-transmembrane α-helical segments embedded in the plasma membrane. This prototypical GPCR, rhodopsin, became the first GPCR for which the structure was unraveled [92, 93]. Independent groups of laboratories around the world played crucial roles in the discovery of the structure of rhodopsin. In particular, the works of Ovchinnikov Yuri Anatolevich (1934–1988) [94, 95] and Paul A. Hargrave [96, 97] provided considerable insight to the structure and functions of rhodopsin. But it was the discovery of the Polish Biochemist Krzysztof Palczewski, et al. that undoubtedly showed that rhodopsin is a GPCR [98].

Technological advancement led to the discovery of other receptors with similar structure as rhodopsin. Around the same time with the discovery that rhodopsin is a GPCR (1970–1980s), beta-2-adrenergic receptor and other adrenoceptors were identified to have GPCR topology. This receptor in particular interested the American physiologist Brian Kent Kobilka (1955–), a former postdoctoral fellow of the American physician and biochemist Robert Joseph Lefkowitz (1943–) [99, 100]. Kobilka and Lefkowitz determined the structure of the agonist-bound β2-adrenoreceptor-G protein stimulatory subunit complex [99, 100]. The researchers discovered a universal mechanism by which the G protein is regulated. The two medical doctors and scientists discovered novel mechanisms by which receptor endocytosis and activation/deactivation of a range of signaling pathways take place through the activities of the intracellular proteins GPCR kinases and arrestins [101]. Kobilka and Lefkowitz won the 2012 Nobel Prize in Chemistry for discoveries that reveal the mechanism of functioning of this very important family of GPCR [102].

The middle and the second half of the twentieth century were marked by increased interest in research on how viruses cause cancers. In particular, the avian sarcoma virus previously discovered in 1911 by the American virologist Francis Peyton Rous (1879–1970) as the factor responsible for causing cancer [103]. The 1966 Nobel Prize in Physiology or Medicine was shared by Peyton Rous "for his discovery of tumor-inducing viruses" and the Canadian-American physician and physiologist Charles Brenton Huggins (1901–1997) "for his discoveries concerning hormonal treatment of prostatic cancer" [35, 104].

In 1979, it was discovered by chance by Raymond L. Erikson (1936–) and coworkers that the gene product of this avian sarcoma virus, Src (from "sarcoma") acts as a protein kinase that phosphorylates mainly tyrosine residues on proteins [105]. In the same year, Tony Hunter and colleagues identified a tyrosine kinase activity in the polyomavirus middle T-transforming protein [106]. This tyrosine kinase activity was confirmed in Src by Tony Hunter's group in 1980 [107, 108]. The findings of Michael Bishop and colleagues were similar to the results of the Erikson's and Hunter's groups [109, 110]. These studies not only identified that protein phosphorylation is important in the transformation process of normal cells to pathological ones, but also led to the identification of a new receptor that functions by phosphorylating proteins. This was widely regarded as the year of tyrosine phosphorylation discovery that marked the identification of a new type of protein modification [110]. One of the first proteins with similar properties was discovered long before this time but its mechanism of functioning only became known following the works of the groups of independent researchers led by Erikson, Hunter, and Bishop in different laboratories [105, 106, 109, 110]. Interestingly, in 1962 Stanley Cohen discovered the epidermal growth factor (EGF) receptor [111] and later showed that it has a tyrosine kinase activity [112]. The receptors with tyrosine kinase activity had a peculiar property of ligand-binding-induced autophosphorylation. This special property of this type of receptor was crucial for growth, proliferation in normal and pathology [108]. Other receptors with protein kinase activity were subsequently identified [113].

Advancement in science enabled the identification of other receptor types, which will be discussed in later part of this chapter.

5.3 Classification of Receptors and Their Signal Transduction Mechanisms

Receptors can be classified according to their localization: plasma membrane receptors, and intracellular receptors. Plasma membrane receptors are localized as integral proteins on the plasma membrane [114]. These receptors are distributed throughout the body including all cell types of the GI tract. The receptors may be generally classified as sensory and non-sensory [115]. Intracellular receptors are receptors localized in the cytosol, cytoplasmic organelles, or nucleus (Fig. 5.3). Receptors of the cytosol are usually protein in nature. Cellular organelles possess numerous receptor types that serve to modulate their functions. Receptors on the membrane of intracellular organelles act to direct protein and other macromolecular trafficking inside the cell. Thus, protein synthesized by the endoplasmic reticulum can be directed to the correct location on the organelle by localizing its cognate receptor. Nuclear receptors are usually named on the basis of the activating ligand. For instance, cortisol activates the cortisol receptors; estrogen activates estrogen receptors; progesterone—progesterone receptors; vitamin D—vitamin D receptor;

thyroid hormone—thyroid hormone receptor; and retinoic acid—retinoic acid receptors (Fig. 5.3) [116–120]. The corresponding ligands of the receptors are able to diffuse through the plasma membrane and bind to their specific receptors in the nucleus, where they regulate the expression of specific genes (Fig. 5.3) [116, 121].

In this chapter, we shall concentrate on the different types of membrane receptors and their mechanisms of signal transduction. Research on the theory of receptive substances, spanning over a century identified three major receptor types: ion channels, GPCR, receptor tyrosine kinase, RTK, also called enzyme-linked receptor. Any of the three receptor types may be located in the plasma membrane, cytoplasm, and nucleus of the cell. These receptors are widely distributed in the body including all layers and tissues of the GI tract.

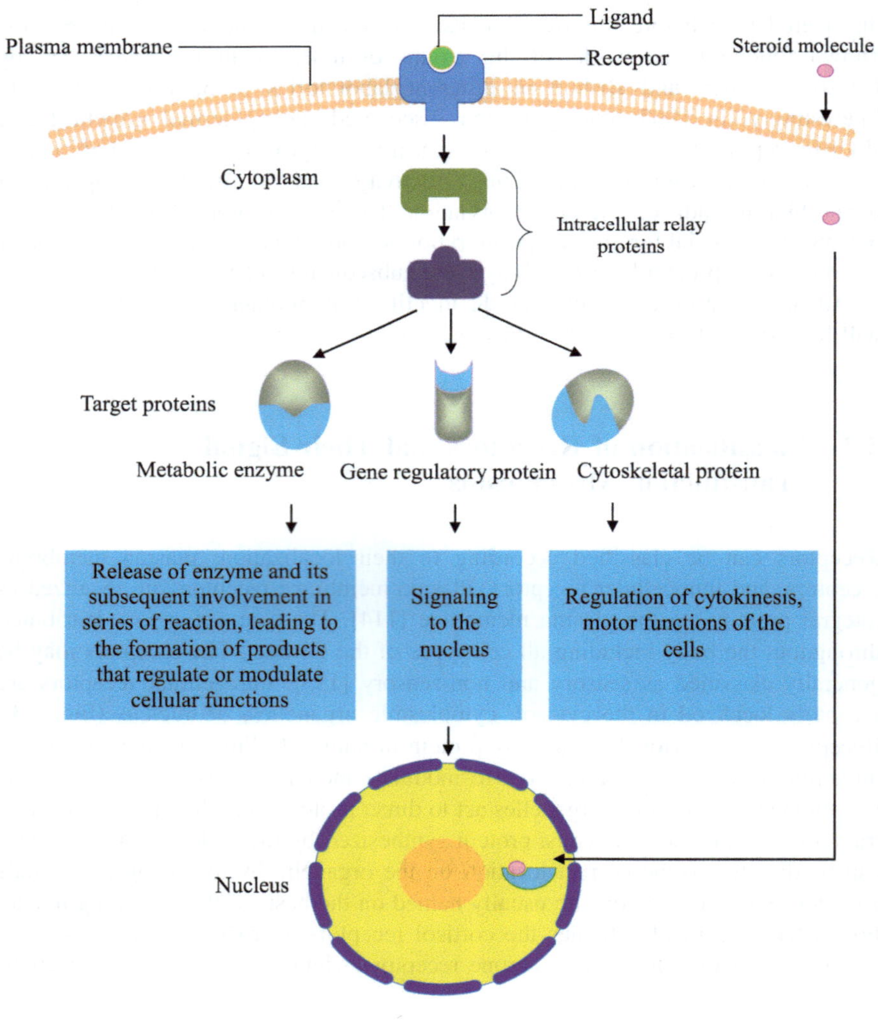

◀**Fig. 5.3** A simple depiction of downstream signaling following interaction between a ligand and a receptor on the plasma membrane and translocation of ligand (steroid molecule) to the nucleus. The binding of the ligand to the plasma membrane receptor leads to downstream signaling of several molecules and may involve cross-signaling or cross talks with other pathways. The intracellular receptors include steroid hormone receptors. However, plasma membrane receptors for steroid hormones have been identified. In the classical model of steroid receptor action, before binding of ligands, the receptors (nuclear receptors) are associated with the heat-shock proteins Hsp70 and Hsp90. The binding of the ligand (hormone) to its cognate receptor induces a conformational change that leads to dissociation of Hsp70 and Hsp90. This initial dissociation promotes dimerization, phosphorylation, and binding to the hormone response elements located in the promoter region of target genes. The binding sites in the genes are called DNA-binding domain, which is specific for a given molecule. Upon activation, the receptor modulates transcription by recruiting components of the transcriptional machinery and gene expression including enzymes acetyltransferases, ligases, ATPases, methylases, cell cycle regulators, RNA helicases, and docking proteins to bridge to basal transcription factors [122]. Some of the molecules may suppress signaling by the steroid molecule. Apart from this classical view, steroid hormones also have plasma membrane receptors through which they relay their signal into the interior of the cell. For instance, progesterone interacts with its membrane receptors to mediate signaling via GPCR pathways, with subsequent activation of Ca^{2+} channel by opening membrane Ca^{2+} channels and also stimulating the MAPK, mitogen-activated protein kinase/ERK (extracellular receptor kinase), and other kinases. Similarly, testosterone can depolarize its target cell by stimulation of its membrane receptors that lead to activation of Ca^{2+} channel by inhibition of K_{ATP} channels, activation of ERK kinase/MAPK pathway, and other pathways. Androgen may also act via their GPCRs leading to changes in ion channel fluxes, cyclic AMP and cyclic GMP production, MAPKs, tyrosine kinases, and lipid kinases. Estradiol may act via the transmembrane GPCR (GPR30) to activate the Src kinase that phosphorylates the EGF receptor, releasing metalloproteases, which in turn trigger the release of EGF ligand from heparin [122]. Steroid hormones may signal via their traditional germ cells, tumor cells, vascular and immune cells [122]. Steroid hormones are mainly synthesised by the gonads, adrenal gland, and placenta and are released into the bloodstream to act both in peripheral target tissues and the central nervous system. However, certain steroid transmitters may be produced by the GI cells. Steroids are important in homeostasis, metabolism, reproduction, development, differentiation, cell proliferation, and inflammation [122]. Estradiol, for instance, inhibits gastric emptying. In contrast, progesterone increases gastric emptying [123]. But progesterone has been reported to inhibit intestinal motility [124]. Testosterone has no known role in GI functions [123]. The female sex hormone estrogen stimulates intestinal enzyme levels and facilitates nutrient absorption [124]. Administration of exogenous glucocorticoids leads to increase brush border enzyme levels. However, excessive production and action of glucocorticoids may lead to decrease in GI functions. Also, high doses of glucocorticoids in the stomach may result in inhibition of prostaglandin synthesis, which predisposes to the development of gastric ulcers [124]

Table 5.2 Signal transduction databases [2, 125]

Name of databases	Website
Signal transduction classification database	http://bibiserv.techfak.uni-bielefeld.de/stcdb/
Simple modular architecture research tool	http://smart.embl-heidelberg.de

The mechanisms of signal transduction for various receptors will be discussed in their respective subsections. Further information on these mechanisms can be reviewed in details in the following databases (Table 5.2).

5.3.1 Ion Channel Receptors

Ion channel receptors (or ion channels) are pore-forming transmembrane proteins that selectively allow ions to flow across the plasma membrane according to electrochemical gradients [126]. Ion channels selectively conduct ions across the cell membrane—which means that ion channel receptors allow the passage of ions from the inside to the outside of a semipermeable membrane or vice versa [37]. The flux of ions through ion channels drives electrical and biochemical processes in cells and plays a critical role in shaping the secretory and sensory transduction processes in the cells [37, 126]. There are hundreds of genes in humans that code for ion channel receptors [126]. The mammalian brain alone is known to express about 500 ion channel genes [127]. Table 5.3 gives a list of databases in which currently known ion channels are embedded.

Ion channel receptors that become functional upon activation to allow the passage of ion are called gated ion channels [114, 132, 133]. Some ion channels are highly selective for a given ion, whereas others may freely allow the passage of ions. It should be noted, however, that some channels classified under one group may be activated by a range of signals. So the classification of ion channels is not strictly rigid. If a signal that activates a channel is taken out of the medium, the receptor returns to a deactivated phase, which is the original conformation of the receptor, provided that there are no other activating signals in the immediate cellular environment. This receptor is said to be at rest. Ion channels are the effectors in a number of regulatory pathways [134].

Ion channels can be classified according to their structure and functions. Structural classification basically considers the three-dimensional structure of the ion channel. Functional classification of ion channels is based on the ion types that are permeable to the membrane and the stimulus type that triggers the gating of the ion channels. According to the ion-gated classification, ion channels include chloride, sodium, and potassium channels. According to the stimulus-gated classification, ion channels include potential- (voltage-), ligand-, temperature-, light-, mechanically (membrane tension or stretch) gated ion channels [126].

Table 5.3 Ion channel databases

Types of database	Website
International union of basic and clinical pharmacology database [128]	http://www.iuphar-db.org
Predictor from Sequence of ION channels (PSION) [129]	https://sourceforge.net/projects/psion
Voltage-gated K^+ channel database [130]	http://vkcdb.biology.ualberta.ca
Ligand-gated ion channel database [131]	http://www.ebi.ac.uk/compneur-srv/LGICdb http://lenoverelab.org/LGICdb/LGICdb.php?LGICdb/index.html
Channelpedia [37]	http://channelpedia.net

Leaky Ion Channels

Some ion channels are always at all times open to the passage of ions. These channels do not close and may allow any of Ca^{2+}, Na^{+}, or K^{+} to leak down their concentration gradients [135, 136]. The degree of leak in these ion channels is not equal. At rest, cell membranes are considerably more permeable to K^{+} than to Na^{+} because they have many more K^{+} than Na^{+} leaky channels. Example of leaky ion channel is the hyperpolarization-activated cation channel, which at rest allows Na^{+} conduction. This background Na^{+} conductance is probably aimed at balancing the K^{+} leak to set the resting membrane potential following action of a stimulus, which keeps the membrane around the resting membrane potential. It should be noted, however, that some channels, though selective, may also function as leaky ion channels. For instance, the previously known to be non-selective/ voltage-independent, but now selective ion channel, NALCN channel (sodium leak channel, non-selective), is permeable to Na^{+}, K^{+} and Ca^{2+} [136, 137]. In theory, a leak of Ca^{2+}, Mg^{2+}, or H^{+} can achieve the same goal of maintaining the membrane potential, but excessive leak of these ions can be damaging to the cells. High levels of these ions in the cell substantially affect the metabolism and other functions of the cell. As previously noted, the leaky channels can affect the membrane potential of the cell. In practice, cells have different resting membrane potentials. Even among same group of cells, the resting membrane potential is not the same. The differences in membrane potential, in part, are due to the difference in the types of different sodium and potassium channels with different conductance, thus the basal permeability of Na^{+} and K^{+} also vary [38]. For neurons, the resting membrane potential is in the range −50 to −80 mV. The main contributor to this potential is K^{+} channels, however, basal Ca^{2+} in addition to Na^{+} are all important in maintaining the resting membrane potential [136]. The most abundant anion in the cell can also influence the resting membrane potential [136].

Voltage-Gated Ion Channel

Voltage-gated ion channels are activated by changes in membrane potential. The changes in membrane potential are constantly evaluated by voltage sensor of the voltage-gated ion channel. The flow of ions across an ion channel depends on the current of the ion channel and the charge of the ion as well as the transmembrane potential. According to Ohm's law, current flowing through an open channel can be defined as $I = \gamma \cdot Vm$, provided that the concentration of the ion on both sides of the membrane is the same. Vm is the membrane potential; γ is the constant defining the conductivity of the channel. A channel with high conductivity will allow the flow of large current, whereas channels with low conductivity will allow the flow of low current. Conductivity is expressed in picosiemens, transmembrane potential in millivolt, and current in ampere [132, 133]. Figure 5.4 shows different types of voltage-gated ion channels. From the figure, types of voltage-gated ion channels include voltage-gated sodium, calcium, potassium, anion (e.g., chloride) channels

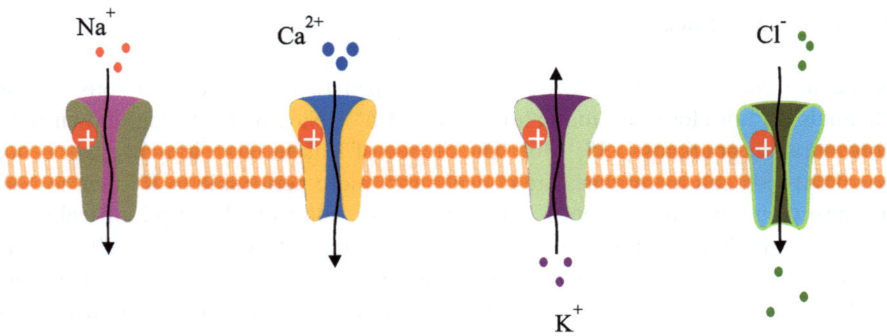

Fig. 5.4 Voltage-gated channels. These channels are named according to the ion to which they are permeable upon change in membrane potential, resulting in changes in electrical potential of integral protein ion channel. Different voltage-gated ion channels open at different potential levels—some open to the passage of ions at very low potential, whereas others open at very high potential

[131]. The proton channel and transient receptor potential (TRP) channels are also considered as types of voltage-gated ion channels. The hyperpolarization-activated cyclic nucleotide-gated channel discussed as ligand-gated channel is also a type of voltage-gated ion channel [129]. Voltage-sensitive dye imaging has emerged as a powerful tool to detect cell activity (including changes in membrane potential, and intracellular signaling of ions). The activity of enteric neurons or other cell types can be reliably recorded with voltage-sensitive dyes [133].

Except leaky ion channels, ion channels generally have one or two gates. But keep in mind that many channels that are classified as voltage or ligand-gated have a basal level of ion conductance. The position of a gate is not specific for all ion channels, thus the location of a gate may vary across different types of ion channels. The flow of ion across the ion channel is controlled by the gate. When the gate is closed, theoretically the ion does not translocate through it. When the gate is open, the ion can flow through it [132, 133, 138, 139]. The diagrams (Figs. 5.5 and 5.6) show hypothetical gating mechanism of ion channels. Apart from the gate and voltage sensor, an ion channel may have other integral parts like specificity pocket, activation, or deactivation portion. The activity of these different parts of the ion channel is associated with different conformations of the channel that either allow or prevent the translocation of ion across the plasma membrane. For further details on gating mechanism, review Gadsby [139].

The quantity of ion translocated through the membrane depends on the sign of the charge of the ions, magnitude, and direction of the transmembrane potential. Ion flow through a membrane may be either uphill or downhill. The net flow of ions down their electrochemical gradient is referred to as downhill ion flow. Uphill flow occurs in the presence of cellular energy by the movement of ions against electrochemical gradient. A single ion channel can translocate some hundreds or millions of ions per second. Estimates show that most ion channel can transport 1,000,000–100,000,000 ions per second. The change in potential resulting from the

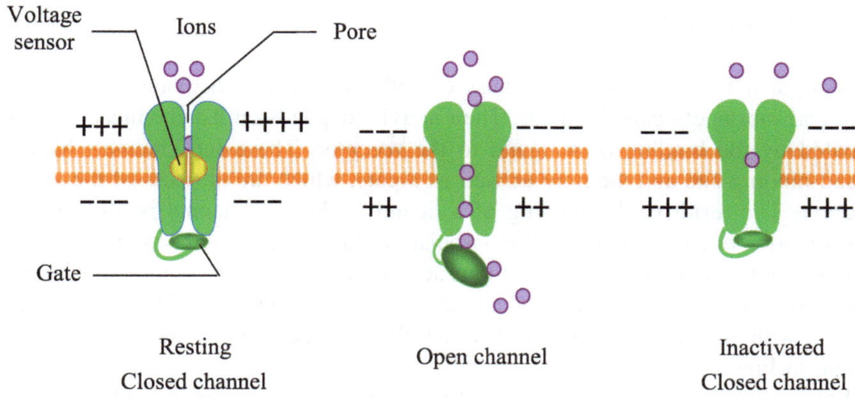

Fig. 5.5 One-way gating mechanism of ion channels

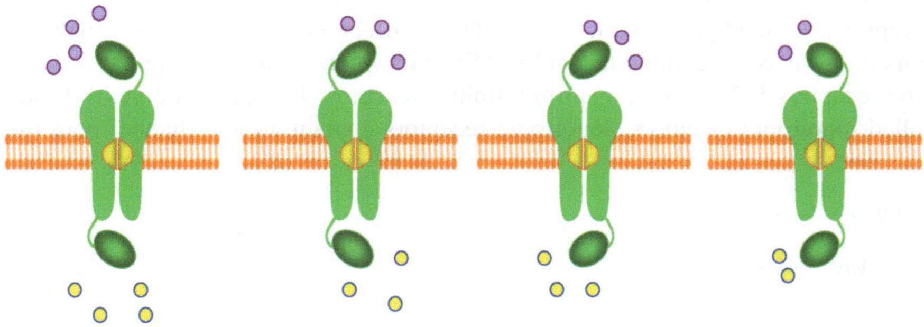

Fig. 5.6 Two-way gating mechanism of ion channels

movement of this large number of ion across one ion channel could be recorded by very sensitive amplifiers, for example, by using the patch clamp technique [139, 140]. Such measurements reveal that ion flow via a single ion channel can create as much electrical current on the order of 10^{-12}–10^{-10} A. Under normal condition, potassium ion concentration is higher in the cell interior than in the exterior. The exterior of the cell has a higher concentration of sodium, chloride, and calcium ions [140]. If an ion channel is activated, ion flow occurs according to the gradient of concentration. Thus, calcium and sodium ions will flow into the cell interior via their corresponding channels. The result is the inversion of the negativity of the inner membrane of the cell resulting in depolarization—the cell is activated. Supposedly, an impinging signal on the cell leads to activation of other channel receptors, in particular, chloride and potassium channels. The response is efflux of potassium and influx of chloride ions which subsequently leads to an increase in the negative value of interior of the cell—the cell is deactivated or hyperpolarized [140].

Ligand-Gated Ion Channels

Ligand-gated ion channels are a class of cellular receptors activated by various chemical substances called ligands. Their activity depends on the chemical reaction initiated by the binding of a ligand to the receptor. These ion channels have a ligand-binding site and the pore as well as a gate, which are all parts of the single receptor. The result of the binding of a ligand to the ion channel is a change in conformation due to ligand-receptor association. In essence, the receptor assumes a conformation that allows the flow of ions across the pore into the membrane interior [138]. The opening and closing of ligand-gated ion channels depend on the interactions with specific ligands [129]. Examples of ligand-gated ion channels are shown in Figs. 5.7 and 5.8.

Ligand-gated ion channels are found in a wide range of cells, including enteric neurons, and muscle cells. These channels include nicotinic acetylcholine receptors (nAChRs), ATP-gated channels (P2X receptors or purinoceptors), 5-hydroxytryptamine type 3 (5-HT3) receptors, gamma-aminobutyric acid ($GABA_A$) receptors, *N*-methyl-D-aspartate (NMDA) receptors, alpha-amino-3-hydroxy-5-methyl-4-isoxazolepropionic acid (AMPA) receptors, kainate receptors, and glycine receptors [138]. These neurotransmitter receptors belong to the general class called ionotropic receptors. Ionotropic receptors contain an ion channel pore that

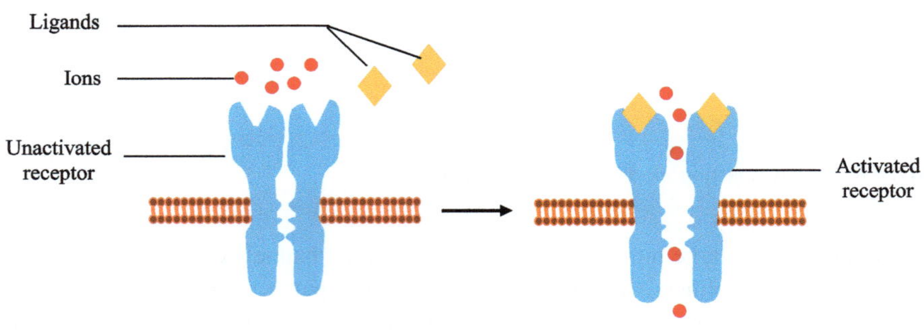

Fig. 5.7 Ligand-gated ion channel receptor. This receptor is activated by a ligand leading to the opening of the channel for the passage of the respective ions

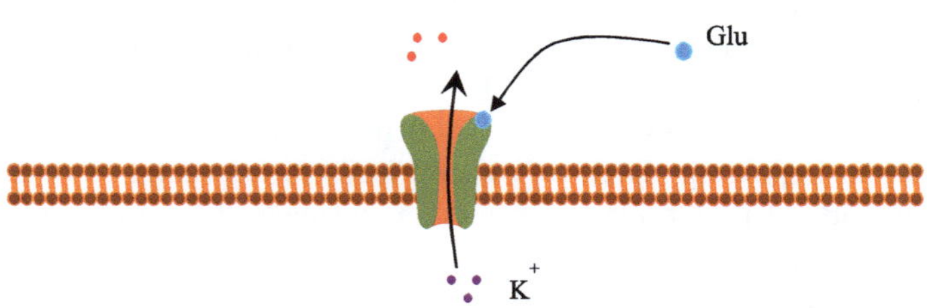

Fig. 5.8 Ion channel gated by glutamate (Glu). The receptor is activated by the binding of the neurotransmitter "glutamate"

open or close upon binding of a chemical messenger or molecule [142–145]. These ion channels that open upon binding of a neurotransmitter are also called receptor-operated channels [146]. NMDA, AMPA, and kainate receptors are activated by the binding of glutamate and are generally termed ionotropic glutamate receptors. In the GABA receptor, binding of GABA molecules results in the opening of an intrinsic Cl^- conducting pore [147, 148]. Acetylcholine is activated by the binding of acetylcholine neurotransmitter (Fig. 5.9). Some of these receptors are also classified as cys-loop receptors, named after a characteristic loop formed by a disulfide bond between two cysteine residues in the N-terminal extracellular domain. The cys-loop receptors are four-pass membrane-spanning pentameric ligand-gated ion channels responsible for fast inhibitory or excitatory synaptic transmission [149]. A functional cys-loop receptor is formed by the assembly of five similar (homomer) or different (heteromer) protein subunits. Each subunit consists of a large extracellular N-terminal domain, a short extracellular C-terminal domain, and four

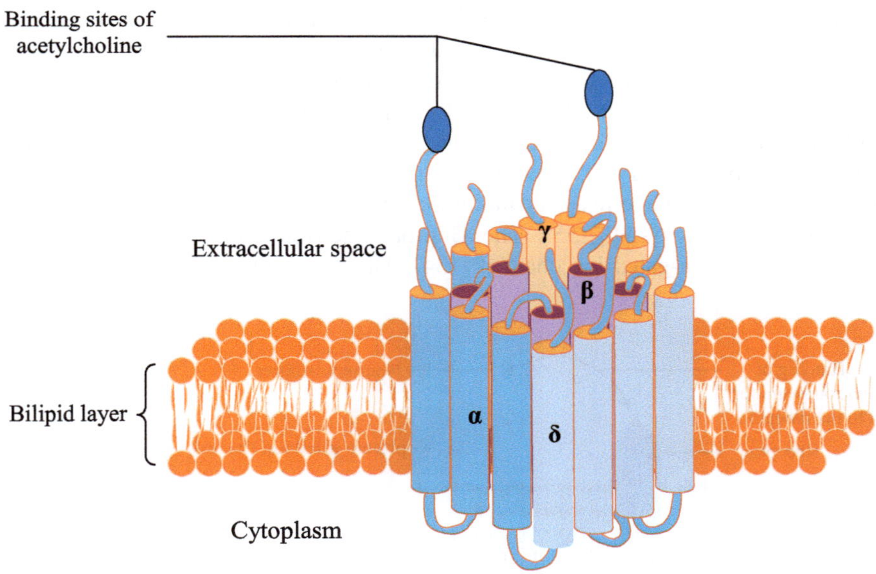

Fig. 5.9 Three-dimensional representation of the acetylcholine receptor. Acetylcholine receptors are present in many cells and tissues of the body (they are found in neurons and smooth muscles of the GI system). Acetylcholine acts as a ligand that binds to the acetylcholine receptor. Different types of acetylcholine receptors exist. For instance, the muscle-type nAChR is a heteropentamer consisting of four subunits organized around a central pore in the plasma membrane. Each receptor has two α-subunits and one each of β, δ, and γ subunits. The binding of the neurotransmitter to the receptor activates the receptor leading to changes in receptor conformation in such a way that the portion that blocks the pore moves away. This leads to changes in ion flux across the cell and membrane potential. As a result, different types of cellular activity are modulated. Upon activation, the opening of the pore allows increase in ion conductance across the membrane. The skeletal muscle and neuronal type allows Na^+ ions to flow into the cytosol down their electrochemical gradient [141]

membrane-spanning helices [150]. The extracellular domain contains the ligand-binding sites; intracellular domain controls the passage of ions. These cys-loop receptors are generally divided into anionic and cationic cys-loop receptors [151]. Anionic cys-loop receptors include $GABA_A$, glycine receptors, while cationic cys-loop receptors include some serotonin, nicotinic acetylcholine (nAChR), and zinc-activated ion channel cys-loop receptors [150, 152, 153]. The cys-loop receptors may also be classified as inhibitory or excitatory. The excitatory cys-loop receptors include some acetylcholine, 5-HT, GABA, and zinc receptors, while the inhibitory receptors include some acetylcholine, 5-HT, GABA, glycine, glutamate, histamine, and even pH receptors [150]. The activation of GABA and glycine receptors will result to increased chloride channel activity [150]. The conformational changes resulting from the binding of the agonist-chemical messenger to the receptors culminates in the opening of a central ion pore [152]. Many drugs are known to activate ligand-gated ion channels. For instance, the anthelmintic drug, ivermectin inhibits neuronal activity, and muscular contractility in nematodes by activating glutamate-gated chloride channels [150].

Some ion channels are activated or gated by intracellular molecules (Figs. 5.10 and 5.11). Ion channels that are gated or activated by G protein subunits are called G protein-gated ion channels, a family of transmembrane ion channels that is highly expressed in excitable cells (Fig. 5.10). Research has shown that different subunits of G protein modulate different ion channel activity in various ways. Some calcium channels are modulated by G protein beta/gamma and alpha subunits. For instance, Gαo mediates a voltage-resistant inhibition of N-type calcium channel by neurotransmitter binding [154, 155]. The G protein-activated K^+ channel is another example G protein-gated ion channel [134]. The specificity and magnitude of the

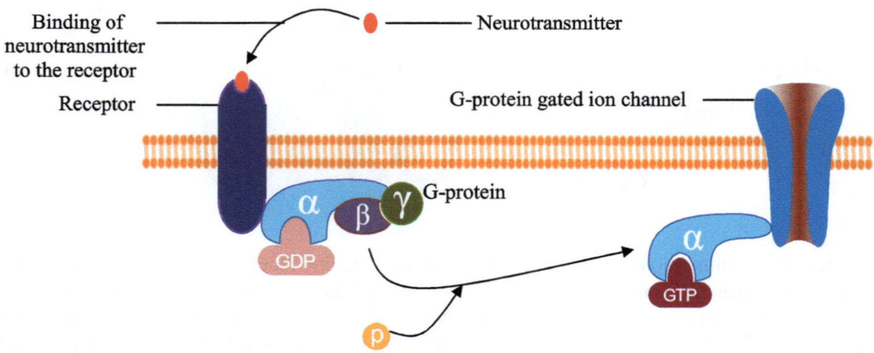

Fig. 5.10 Ion channel activation via G protein-coupled receptor (G protein-gated ion channel). The receptor is activated by a neurotransmitter, which is linked to a G protein. Upon binding of the neurotransmitter the G protein is activated leading to separation of the GTP bound α-subunit from the βγ-complex. The activated α-subunit directly activates the ion channel resulting in changes in ion fluxes across the membrane. However, depending on the type of ion channel, and the isoform of the activating/inhibiting G protein subunit, βγ-complex may serve as an activator of ion channel [154]

overall regulation of physiological functions are determined by the intricate relationships between ion channel receptor and G protein subunit activity [134]. Not only ligand-gated ion channels, but also voltage-gated ion channels may be regulated by neurotransmitter receptors coupled to the heterotrimeric G proteins [154]. The modulation of ion channels by G proteins may be indirect (via second messengers and protein kinases) or direct, via physical interactions between G protein subunits and the channel protein (Fig. 5.11) [134, 155]. Such types of G protein involved in the modulation of ion channels by second messenger system are generally called metabotropic receptors. The metabotropic receptors are indirectly linked with ion channels via signal transduction mechanism that activate G protein with dissociation of its subunits, which in turn activate downstream targets producing a secondary messenger. The second messenger can bind to ion channels and modulates its activity. Examples of metabotropic receptors include the metabotropic glutamatergic, muscarinic sensitive cholinergic, $GABA_B$, most serotoninergic, adrenergic histaminergic, dopaminergic, and peptidergic receptors [156–158].

Certain channels may be activated by intracellular binding of Ca^{2+} to the receptors. The Ca^{2+}-activated K^+ channel is an example of a channel that is activated by intracellular Ca^{2+} (Fig. 5.11). Instead of Ca^{2+} ions, these channels may be gated by increase in intracellular ATP. An example of ATP-gated channel is the inwardly rectifying K^+ channel, which allows K^+ to move more easily into rather than out of the cell [162]. Depending on the cell type, there are different types of inwardly rectifying K^+ channel, which include G protein-gated inwardly rectifying K^+ channel, and ATP-sensitive K^+ channel [163].

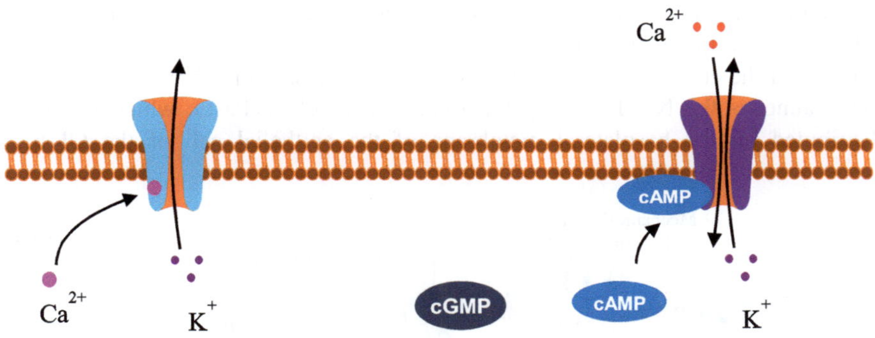

Fig. 5.11 Calcium- and cAMP- (second messenger)-gated ion channels. Calcium-gated ion channel is activated by calcium leading to the opening of the potassium ion channel. That is why the receptor is termed Ca^{2+}-activated K^+ channel [159]. cAMP-gated ion channel is a cyclic nucleotide-gated channel. The channel is activated by cyclic nucleotides, which results to changes in ion fluxes across the membrane [160]. The movement of the ions shown in the diagram is a simplified scheme. Cyclic nucleotide-gated channels are very complex in their functioning (and depends on the type of cell), permeable not only to K^+, Ca^{2+} ions but also to Na^+ and Mg^{2+} [161]. The channels that are stimulated by the second messengers (such as Ca^{2+}, cAMP, and cGMP) are generally called second messenger-gated channels [146]

Ion channel receptors can be activated by lipid molecules such as phosphatidylinositol 4,5-bisphosphate (PIP_2) [164]. PIP_2 is a plasma membrane lipid that regulates the activities of numerous channels including the inwardly rectifying K^+ channel, voltage-gated K^+ channel, hyperpolarization-activated channel, cyclic nucleotide-regulated channel, and voltage-gated Ca^{2+} channel [165]. The action of PIP_2 on ion channel activity may be positive or negative, but this lipid usually mediates a bidirectional flow of ions across the plasma membrane. PIP_2 allows a bidirectional flow of ions across voltage-gated calcium channels. PIP_2 has been shown to enhance the opening of hyperpolarization-activated cyclic nucleotide-gated channels by shifting their voltage-dependent activation toward depolarized potentials [164, 166].

Mechanically Gated Ion Channels

Mechanically gated channels are those channels that open when activated by membrane tension, stretch, or distension (Fig. 5.12). A mechanosensitive cell is activated by mechanical distension or force acting on the membrane of the cell. If the mechanical signal is strong enough to cause depolarization up to the critical level, then action potential develops, which culminates in secretion of molecules by the cell. In some cells, the excitation is propagated to neighboring cells [115].

Ion Pump

Ion pumps are ion channel receptors that have intrinsic enzymatic activity for ATP. Upon splitting of ATP by this channel, the associated change leads to the opening of the channel to transport ions across the plasma membrane. One of such ion channel is the Na^+/K^+ pump (also called Na^+/K^+ ATPase), which is widely distributed on the basolateral membrane of the epithelial cell of the GI tract

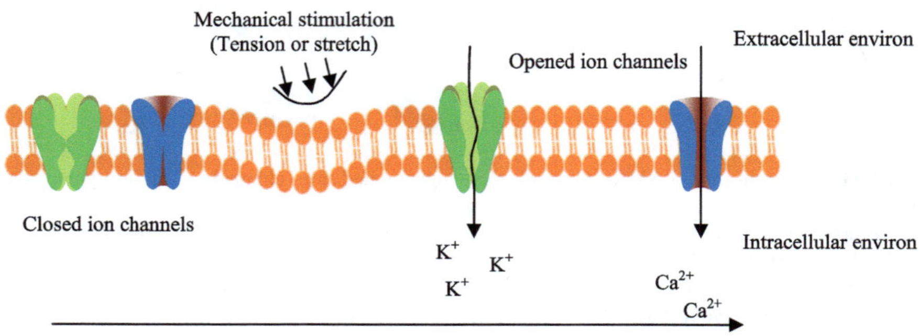

Fig. 5.12 Activation of mechanically gated ion channel. Mechanical stimulation leading to membrane distension or distortion may stimulate certain types of ion channels (which are usually non-selective) to increase in ion fluxes

(Fig. 5.13). The channel import molecule of potassium from the extracellular environment for every three molecules of sodium ions exported into the intracellular environment with loss of energy in the form of ATP. Hydrolysis of one ATP molecule produces some joules of energy required for the activity of this ion carrier. ATP is a metabolic currency, regenerated by a transport ATPase working in reverse, i.e., the ATP synthase. The reactions involving the cleavage of ATP are shown in Figs. 5.14, 5.15 and 5.16 [167, 168]. The ATP-gated sodium/potassium channel functions by a two-way gating mechanism which is diagrammatically presented in Fig. 5.13.

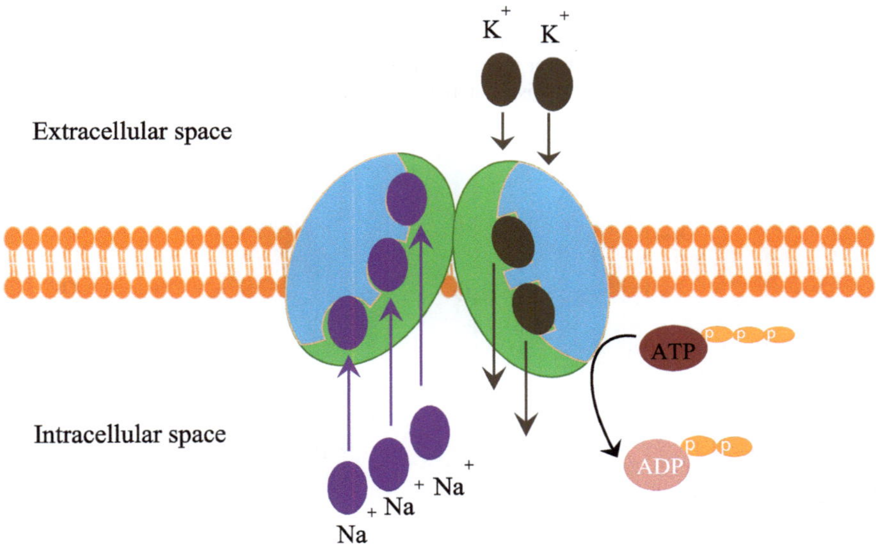

Fig. 5.13 Sodium–potassium ion channel (Na^+/K^+ pump)

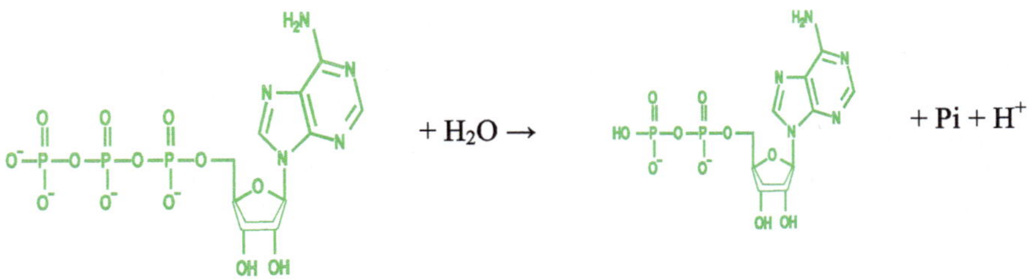

Ap~p~p (ATP) Ap~p (ADP)

Fig. 5.14 Hydrolysis of ATP to ADP. The high-energy bonds in ATP (two phosphoanhydride bonds) are hydrolyzed to produce energy for various cellular processes such as electromotive force for the movement of ions across the plasma membrane. In the hydrolysis of ATP to ADP, the first phosphoanhydride bond is cleaved with the production of 7.3 kcal/mol

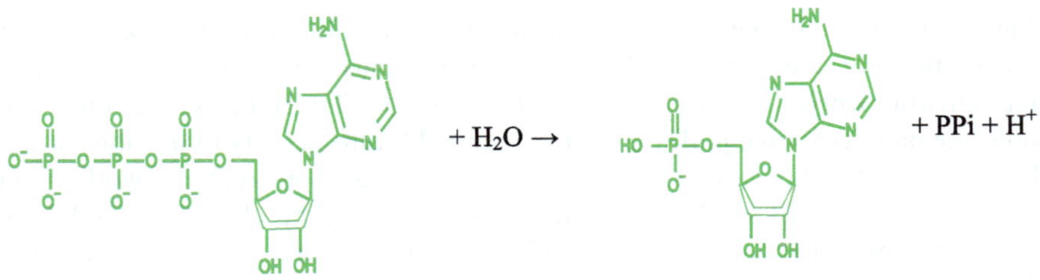

Ap~p~p (ATP) Ap (AMP)

Fig. 5.15 Hydrolysis of ATP to AMP. The high-energy bonds in ATP (two phosphoanhydride bonds) are hydrolyzed to produce energy. The phosphoanhydride bond is a covalent bond, but it releases about 7.3 kcal/mol of free energy when it is broken. In contrast, hydrolysis of the phosphoester bond in AMP, forming inorganic phosphate and adenosine, releases only about 2 kcal/mol of free energy. P_i stands for inorganic phosphate and PP_i for inorganic pyrophosphate, two phosphate groups linked by a phosphoanhydride bond

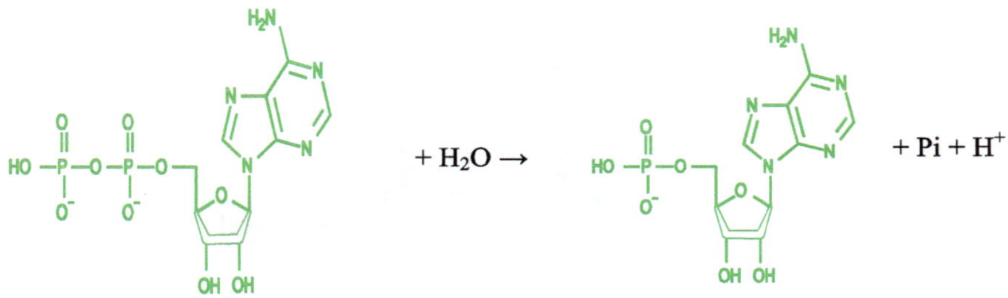

Ap~p (ADP) Ap (AMP)

Fig. 5.16 Hydrolysis of ADP to AMP

There are other types of pump, which include calcium pump (Ca-ATPase), and proton pump (H^+/K^+ ATPase) [169, 170]. Proton pump is found in plasma membrane and intracellular vesicles of many cells and plays a key role in shuttling hydrogen ions across the membrane thus regulating the pH of the immediate environment of the cell or organelle. Ca-ATPase is expressed in many cells including the GI tract cells. This pump is involved in the regulation of calcium homeostasis in the cell. These ATPases belong to the class "P" of transport ATPases. There are different classes of P-ATPases, which transport specific types of ion: H^+, Na^+, K^+, Mg^{2+}, Ag^+, Ag^{2+}, Zn^{2+}, Co^{2+}, Pb^{2+}, Ni^{2+}, Cd^{2+}, Cu^+, and Cu^{2+}. P-ATPases (also called E1–E2 ATPases) comprises of two polypeptide chains, which usually assume two main conformations, called E1 and E2, hence the name E1–E2 ATPases [170–172]. Other members of class of transport ATPases are V, F, and ABC types [167, 168, 173]. It should be noted that P-ATPases transport not only ions but also phospholipids, across the membrane. Malfunction of some of these ATPases has been implicated in a number of diseases that affect different

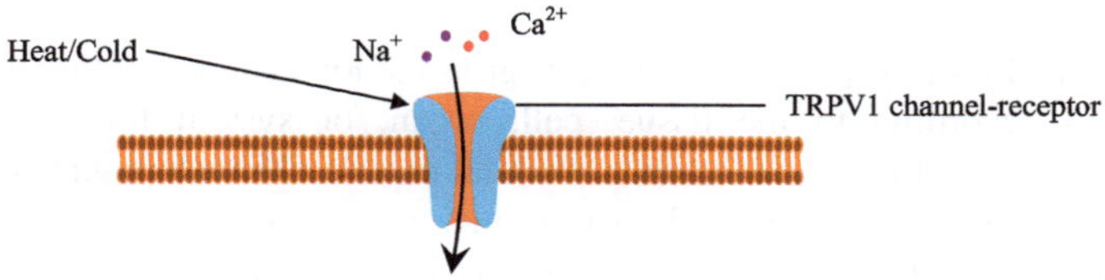

Fig. 5.17 Temperature-gated ion channel. The transient receptor potential valinoid type 1 (TRPV1) is activated by increase or decrease in environmental temperature which results to increase in ion fluxes across the plasma membrane

organs including the pancreas, stomach, and salivary glands. Malfunctions of ABC type ATPases have been implicated in the pathogenesis of gastric ulcers [167, 168].

Temperature-Gated Ion Channels

Temperature-gated ion channels also called thermosensitive ion channels are ion channels that open following a change in the temperature of the environment. These channels sense hot and cold environment and react appropriately by changes in ion flux across the membrane [126]. Temperature-gated ion channels belong to the TRP channel family (Fig. 5.17) [115].

TRPV1 and TRPM8 are examples of TRP channels that sense and respond to changes in ambient temperature. The mechanisms of the activities of these channels are yet to be fully understood. But they may involve the intracellular association with lipid and other biomolecules [115].

Clinical Correlate 5.1
Channelopathies

Introduction: Channelopathies are disorders resulting from a defect in ion channels located in the membranes or organelles of cells. The prevalence of these disorders is about 1 in every 10,000. Malfunctions of ion channel have been linked to many diseases of humans [60, 174–180]. Consistent with the distribution of ion channels throughout the human body, ion channel defects have been implicated in a range of diseases, including epilepsy, muscle paralysis migraine, blindness, deafness, diabetes, hypertension, cardiac arrhythmia, pain syndromes, asthma, irritable bowel syndrome, and cancers [176]. Thus, channelopathies affect almost all organs or systems of the body [126].

Classification: Channelopathies can be classified according to different criteria [181]. On the basis of the activating signal of the ion channel, channelopathies are grouped as voltage-, ligand-gated ion channelopathies. These diseases can also be classified according to the type of ion for which the channel is permeable to. Thus, there are sodium, potassium, zinc channelopathies [135]. Furthermore, this group of ion channel diseases can be classified according to the tissue, cell, organ, or system for which the pathology is identified. For instance, skeletal muscle channelopathies include periodic paralysis, malignant hyperthermia, paramyotonia congenita, and myotonia congenita. The central nervous system channelopathies are generalized epilepsy with febrile seizures, familial hemiplegic migraine, episodic ataxia, hyperkalemic, and hypokalemic periodic paralysis. Cardiovascular system channelopathies include long QT syndrome, short QT syndrome, Brugada syndrome, and catecholaminergic polymorphic ventricular tachycardia. Respiratory system channelopathies include cystic fibrosis. Endocrine system channelopathies are neonatal diabetes mellitus, familial hyperaldosteronism, pseudohypoaldosteronism type 1, Liddle syndrome, familial hyperinsulinemic hypoglycemia, persistent hyperinsulinemic hypoglycemia of infancy, Dent's disease, hypomagnesemia with secondary hypocalcemia, and thyrotoxic hypokalemic periodic paralysis. Urinary system channelopathies are Bartter syndrome, nephrogenic diabetes insipidus, autosomal-dominant polycystic kidney disease, and hypomagnesemia with secondary hypocalcemia. Immune system channelopathies are myasthenia gravis, neuromyelitis optica, Isaac syndrome, and anti-NMDA [*N*-methyl-D-aspartate] receptor encephalitis. GI tract channelopathies include irritable bowel syndrome and certain epithelial cell cancers. Channelopathies can be classified according to the type of mutations in the ion channel receptor [176, 182–187]. Channelopathies can also be classified according to the organelle where the channel is mutated, as mitochondrial, nuclear, and plasma membrane channelopathies [188].

Etiopathogenesis: Mutations in genes encoding ion channels, which impair channel function, are the most common cause of channelopathies [175, 176, 189]. The first mutation in human ion channel was discovered by Ptacek and coworkers in 1991 [185–187]. There are over 160 different types of diseases resulting from mutations of ion channels [126]. Though channelopathies resulting from mutations of ion channels may manifest in a multitude of ways depending on the tissue specificity of the channel that is affected [181], mutation of a single gene can result to different channelopathies due to the overlap of gene functions. It should be noted that silent mutations of ion channel gene also occur and such individuals are termed silent mutation carriers [190, 191]. Some channelopathies can be acquired if changes in channel functions are capable of compromising the integrity of the channel associated proteins [188].

Mutations of ion channel lead to varying degree of ligand affinity (measure of how tightly the ligand can bind to the ion channel receptor) and efficacy (the ability of the channel to open upon establishment of the ligand-channel complex) [60, 174, 175, 189]. Loss of functions or mutations of the water channel, for instance, can lead to edema. In this case, depending on the degree of malfunction, water molecules are poorly transported via the channel pore.

Water channels are located in different tissues of the body and mediate fluid secretion in various organs [68, 69]. Mutation by deletion of phenylalanine at amino acid position 508 (delta F508) in the membrane protein, cystic fibrosis transmembrane conductance regulator (CFTR) leads to the development of cystic fibrosis, a disease characterized by a defect in epithelial chloride ion transport. Under normal condition, CFTR functions as a cyclicAMP-sensitive Cl-channel. However, mutation of this channel gene disrupts the maturation process of the glycoprotein as well as its translocation process to the correct location of the cell where it is required to carry out its functions [192].

In addition to ion conductance and electrical signal transduction that are affected upon mutation of ion channel gene, other aspects of ion channel functions, which include involvement in neurotransmitter and hormonal secretion, trans-epithelial transport, cell motility, growth, cell volume regulation, regulation of acidification of lysosomal compartments, regulation of cytoplasmic or vesicular ion concentration and pH, and second messenger signaling are affected in channelopathies [182, 193].

Diagnosis: A suspicion of channelopathy may be evidenced by the pattern of presentation of the illness as well as a recorded history of development of similar condition in the family. When there is high degree of suspicion, clinical investigation, and laboratory analysis are required. Polymerase chain reaction (PCR), multiple ligation probe analysis, gene sequencing are genetic techniques that can be used for the laboratory diagnosis of channelopathies [181]. In addition, basic serum, and urine tests are performed [194, 195]. Laboratory diagnosis of the condition requires a state-of-the-art tool and is not available in many laboratories. Diagnostic laboratories of channelopathies in different parts of the world can be assessed at www.cardiogenetica.nl; www.centogene.com; www.digenom.de; www.mangen.co.uk; www.pgxhealth.com; www.genedx.com; www.correlagen.com; www.hpcgg.org.

Treatments: The treatment of channelopathies is based on the clinical laboratory findings and clinical presentations. Treatment is based on curbing the symptoms arising from specific channelopathies, depends on the severity of the condition, and is individualized. The disorders usually require lifelong treatment. Patients following adequate evaluation may require intravenous KCl, glucose (with insulin), calcium gluconate, diuretics, and beta blockers. Patients may require changes in exercise and diet. It is advised that patients reduce or completely restrain from stressful situation or likely factors that trigger the onset or worsen their condition. For identified group of patients, specific ion channel activator drugs may be used. The drug cromakalim is a known activator of potassium channel [176, 189]. Carbonic anhydrase inhibitors (dichlorphenamide and acetazolamide) have been shown to improve some cases of channelopathies involving hypokalemia (i.e., reduced potassium ion concentration) [196]. Some of the drugs are currently undergoing clinical trials [196, 197]. For more information on some of the channelopathy drugs that are undergoing clinical trials, visit clinicaltrials.gov. It should be mentioned that gene therapy is also a promising area of application for the treatment of channelopathies. Other lines of treatment may be considered on the basis of disease presentation such as the use of methylprednisolone, phenytoin etc. [198, 199].

Annoted Review Texts on Channelopathies

1. Ptacek LJ (1997) Channelopathies: ion channel disorders of muscle as a paradigm for paroxysmal disorders of the nervous system. Neuromuscul Disord 7:250–255
2. Zhou P, Wang J (2010) Genetic testing for channelopathies, more than ten years progress and remaining challenges. J Cardiovasc Dis Res 1(2): 47–49

5.3.2 G Protein-Coupled Receptor

GPCRs or GPRs, also called seven-transmembrane Serpentine (7TM or 7TMS) receptors, are membrane-bound proteins that transverse the cell membrane seven times. GPCRs are activated by extracellular ligands such as hormones, neurotransmitters, chemokines, light, odorants, tastants and these receptors transduce signals downstream via interactions with G proteins. It is estimated that there are over 800 types of GPCRs in the human genome [200]. The GPCR is the most predominant and highly investigated cell membrane receptors. More importantly, about 50% of all prescription drugs function through the GPCRs or use it as their docking site. Impairment in GPCRs functions may lead to a range of diseases [158].

The binding of a ligand to GPCR leads to conformational change of the receptor, which results in GDP–GTP exchange in the G protein α-subunit, followed by dissociation of the G protein into α-GTP and βγ complexes (Figs. 5.18 and 5.19). In the inactive state, G protein "switch" is bound to guanosine 5′-diphosphate (GDP). When the G protein is activated, guanosine 5′-triphosphate (GTP) is exchanged for GDP within the α-subunit of the G protein with dissociation of the G protein into α-GTP and βγ complexes. This "on" and "off" switch of the G protein is due to the intrinsic GTPase hydrolytic activity of the α-subunit. The exchange of GTP for GDP promotes the re-association of the three subunits, thus returning the G protein to its inactive form. The cellular response depends on the type of G protein subunit isoform activated. G protein complex is named after the α-subunit. The α-subunits of G proteins are classified according to their function and structure in terms of amino acid similarity. On the basis of this classification, there are over 15 types of G proteins: Gαs, Gαi/o, $G_{\alpha q}$, etc. Gαs stimulates adenylate cyclase, Gαi/o inhibits adenylate cyclase, Gαq stimulates phospholipase C [158, 201]. The Gα12/13

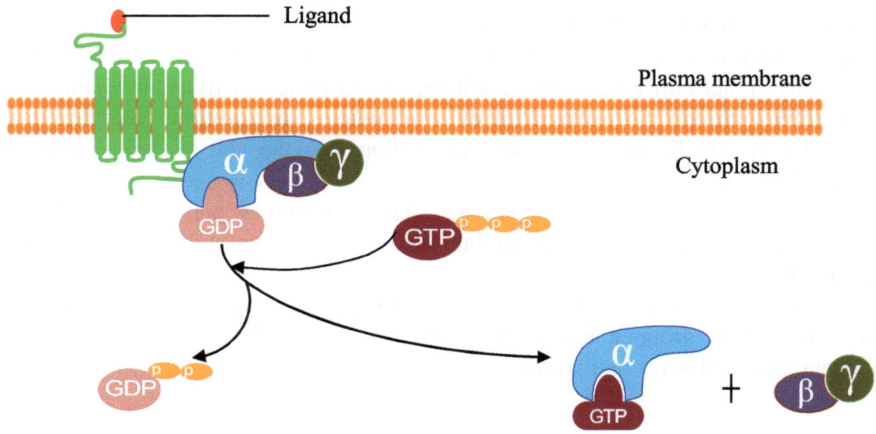

Fig. 5.18 Activation of G protein and the dissociation of the α- and βγ-subunits. The different types of GPCRs are classified according to the following classes. Class A is termed rhodopsin-like GPCRs—this class contains the majority of GPCRs, including receptors for biogenic amines, nucleotides, peptides, and glycoprotein hormones. The majority of classical neurotransmitter conveys their signal through GPCR class A. Class B (secretin-like GPCRs) contains purely peptide receptors. Class C (metabotropic glutamate family receptors) contains metabotropic glutamate and GABA-B receptors, some taste receptors, calcium-sensing receptors, and pheromone receptors [158]. Class D contains the fungal pheromone receptors. Class E contains the cAMP receptors of Dictyostelium. The remaining is the Frizzled/Smoothened class of GPCRs. There are also a number of putative classes of newly discovered GPCRs [201]. Further details on GPCR classification can be found in Kristiansen [204], Horn et al. [203], and Fredriksson et al. [202]

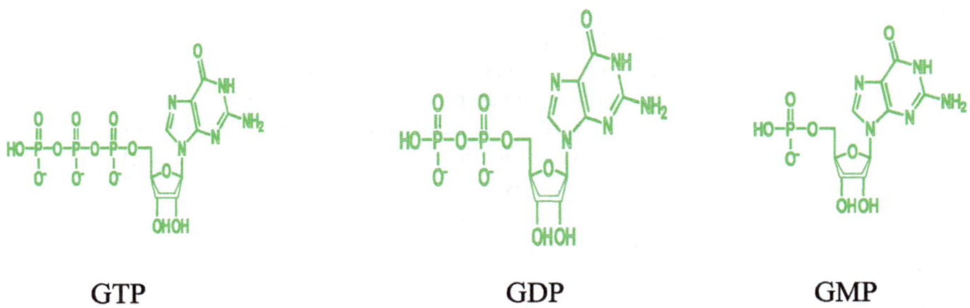

Fig. 5.19 Guanine based ribonucleotides

activates the Na^+/H^+ exchanger. Studies indicate that some GPCR functions by association not with the G proteins, but with small G proteins such as ADP ribosylation factor (Arf) [158, 202–204]. Small G proteins are discussed in later part of this chapter.

The dissociated subunits of the G protein can activate or inhibit several effectors including adenylyl cyclases, phospholipases (PLCβ, PLC-δ, PLC-γ, PLC-ε, PLC-ζ, and PLC-η), phosphodiesterases, phosphoinositide-3-kinase (PI-3-K), GPCR kinases, ion channels, MAPK (extracellular kinases), tyrosine kinases, and transcription factors [158, 201, 205]. These proteins have different isoforms. The impinging signal either activates or inhibits the G protein activity, mediating the interaction with specific isoforms of effector proteins. In addition, emerging data indicate that GPCR may transduce information into the cell via non-canonical means [201, 205].

In addition to the natural endogenous ligands (hormones, peptides, neurotransmitters), GPCR can be activated or inhibited by infectious and pharmacological agents (Table 5.4).

According to the type of intracellular substrate activated, G protein signaling pathway can be grouped into cAMP-dependent or cAMP-independent, inositol phospholipid pathways and so on.

Table 5.4 Action of some agents (infectious, non-infectious, and pharmacological) on cellular signaling [89, 206–217]

Agents	Characteristics	Mechanism of action
Cholera toxin	An exotoxin produced by *Vibrio cholerae*, the bacterium that causes cholera. This toxin is a protein consisting of a larger A (required for intracellular toxicity) of molecular weight 28 kDa and B (required for binding to the animal cell receptor) of 11 kDa subunits	Acts as a mono-ADP-ribosyltransferase to increase intracellular cAMP. This enzyme catalyzes the transfer of the ADP-ribose of nicotinamide adenine dinucleotide (NAD) to a protein substrate
Pertussis toxin	An exotoxin produced by *Bordetella pertussis*, the organism that causes whooping cough. This organism produces not only pertussis toxin, but also adenylate cyclase toxin, a dermo-necrotic toxin, and tracheal cytotoxin, which exert a combined damaging effect on the cell	Acts as a mono-ADP-ribosyltransferase to increase cAMP levels by inhibiting G protein inhibitory subunit. Thus, negative regulation of adenylyl cyclase is inhibited. This will lead to an increase in adenylyl cyclase activity, thereby increasing cAMP levels
Forskolin	A diterpene produced by the Indian Coleus plant (*Plectranthus barbatus*)	Activates adenylyl cyclase to increase cAMP level
Caffeine (1,3,7-trimethylxanthine)	A methylxanthine alkaloid found in the seeds, nuts, and leaves of a couple of plants	Inhibits cAMP phosphodiesterase, which degrades cAMP. Thus, the level of cAMP increases. It also antagonizes the adenosine receptor (and prevents the binding of adenosine to its

(continued)

Table 5.4 (continued)

Agents	Characteristics	Mechanism of action
		cognate receptor) to stimulate the central nervous system (psychotropic effect). It exerts anti-inflammatory effects on a number of tissues of the body
Theophylline (1,3-dimethylxanthine)	A methylxanthine alkaloid that can be derived from tea	Similar to the mechanism of action of caffeine. It acts as a phosphodiesterase inhibitor, adenosine receptor blocker, and histone deacetylase activator. It can function as a smooth muscle relaxant, and causes bronchial dilation
Theobromide (3,7-dimethylxanthine)	A bitter, primary methylxanthine alkaloid of the cacao plant (*Theobroma cacao*). It is also found in chocolate, and a range of other food substances, including the leaves of the tea plant, and kola nut	Mechanism of action is similar to that of the other xanthines
Bucladesine (dibutyryl cAMP)	A synthetic organic compound	Inhibits phosphodiesterase. Reduces cAMP levels by dephosphorylating cAMP to AMP

cAMP–PKA Pathway

The G protein α-subunit upon dissociation activates the plasma membrane enzyme, adenylate cyclase, which in turn catalyzes the production of cAMP in the presence of ATP (Figs. 5.20 and 5.21). cAMP is a second messenger that target different intracellular proteins. One of the widely studied actions of this messenger occurs via its interaction with the enzyme protein kinase A (PKA). This enzyme in the inactive state has four binding sites for cAMP. Upon activation by cAMP, PKA is ready to phosphorylate different intracellular (cytoplasmic and nucleoplasmic) acceptors including enzymes of glycogenolysis, enzymes, and proteins involved in smooth or cardiac muscle contraction, secretion of fluids, digestion, and transcription factors [222–225]. There are different types of protein kinases. PKA is a member of the family of AGC protein kinases, which include PKC (protein kinase C) and PKG (protein kinase G). Other families of protein kinases are: tyrosine kinases; calcium/calmodulin-dependent protein kinases; casein kinase-1; CDK (cyclin-dependent kinase), MAPK, GSK3 (glycogen synthase kinase 3), and CLK (CDK-like kinase) etc. [226, 227].

Increase in cAMP also activates ion channels (including cyclic nucleotide-gated ion channels and exchange proteins activated by cAMP, EPAC) resulting in

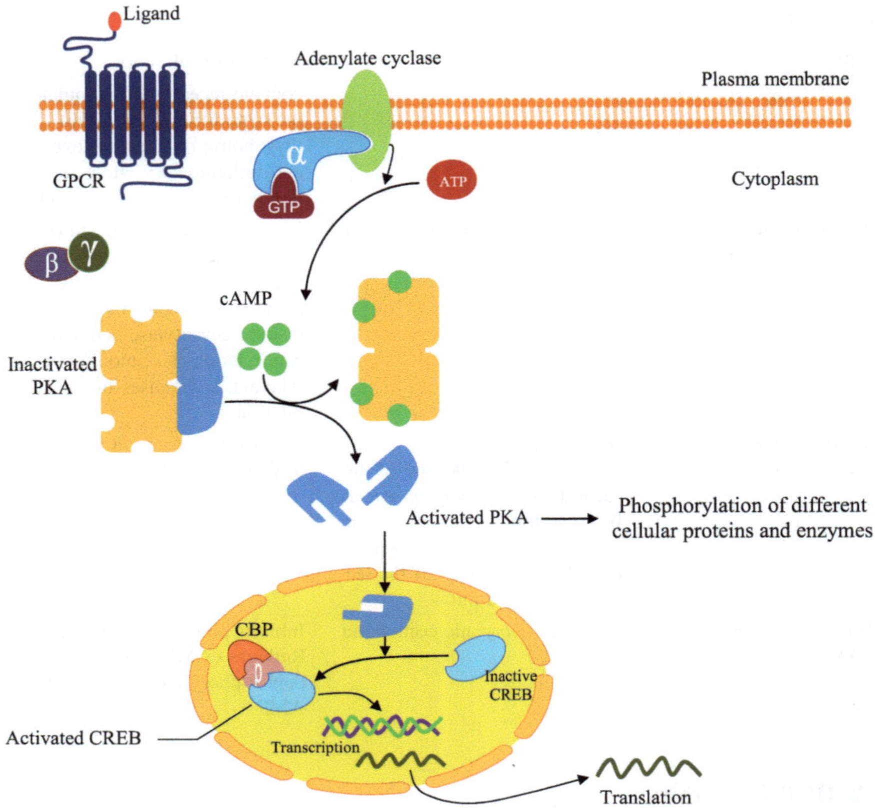

Fig. 5.20 Activation of cAMP pathway and PKA signaling. Each GPCR is activated by a specific ligand stimulus that ranges in size from small-molecule catecholamines, lipids, neurotransmitters to large protein hormones. When a GPCR is activated by its extracellular ligand, a conformational change is induced in the receptor that is transmitted to an attached intracellular heterotrimeric G protein complex. The G_s alpha subunit of the stimulated G protein complex exchanges GDP for GTP and dissociates from the beta–gamma complex. In a cAMP-dependent pathway, the activated G_s alpha subunit binds to and activates an enzyme called adenylyl cyclase, which, in turn, catalyzes the conversion of ATP into cyclic adenosine monophosphate (cAMP). Increase in concentration of the second messenger cAMP may lead to the activation of ion channels and intracellular proteins and enzymes and bind to their nuclear receptors. cAMP can activate cAMP—dependent protein kinase, which in turn phosphorylates different targets in the cell to effect physiological response [218–221]

changes in the activities of ion channels (e.g., calcium handling), and several pivotal cellular processes, including proliferation, survival, differentiation, polarization, motility, adhesion, and secretion [228–232].

There are currently a couple of databases of GPCR that provides structural and functional information on hundreds of GPCR discovered to date. Some of these databases are shown in Table 5.5.

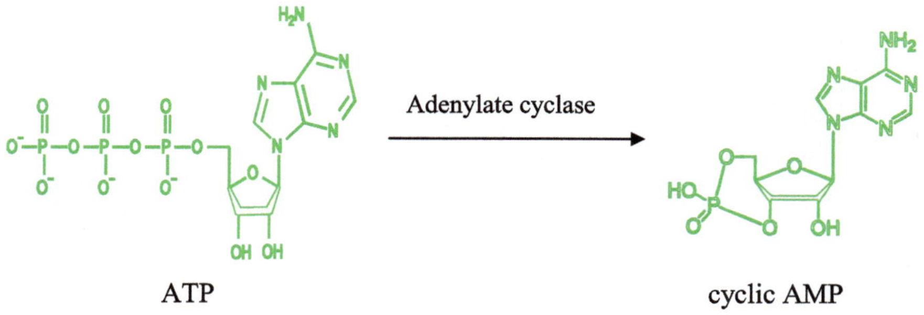

Fig. 5.21 Production of AMP from ATP by adenylate cyclase

Table 5.5 GPCR databases

Name of databases	Website
G protein DataBase (GpDB) [201]	http://bioinformatics.biol.uoa.gr/gpDB
GPCR-LIgand DAtabase (GLIDA) [233]	http://gdds.pharm.kyoto-u.ac.jp:8081/glida
GPCR DataBase (GPCRdb) [233]	http://gpcrdb.org
International Union of Basic and Clinical Pharmacology (IUPHAR) [128]	http://iuphar-db.org/iuphar-rd/index.html; http://www.guidetopharmacology.org
GPCR-Sequence-Structure-Feature-Extractor (SSFE) database (GPCR-SSFE) [234]	http://www.ssfa-7tmr.de/ssfe

PLC–IP_3 Pathway

Binding of an agonist (hormone) to its cognate GPCR recruits $G_{\alpha q}$-subunit of the G protein, which activates the phospholipase C enzyme isozyme-β (PLC-β) located in the inner leaflet of the cell membrane (Fig. 5.22). The activated PLC-β catalyzes the hydrolysis of phosphatidylinositol 4,5-bisphosphate (PIP_2) to produce DAG (1,2-diacylglycerol) and $InsP_3$ or IP_3 (inositol 1,4,5-trisphosphate). IP_3 is a soluble second messenger that is capable of diffusing through the cytoplasm to the endoplasmic reticulum (ER), where it binds to a membrane receptor of IP_3 (IP_3R) on a ligand-gated Ca^{2+} channel located on the ER membrane. The binding of IP_3 to IP_3R forms IP_3-IP_3R complex that changes the conformation of the channel—this triggers the opening of the Ca^{2+} channel, and thus results to efflux of calcium ions into the cytosol. PKC can be directly activated by the lipid/insoluble second messenger DAG. PKC in turn phosphorylates several protein molecules [158, 201, 235, 236, 247, 248]. Not only hormones function through the Gαq-subunit of the G protein but also some neurotransmitters. Certain growth factors can bind to GPCR that leads to the recruitment of the beta–gamma subunits of the Gi/o family, which in turn activate the γ-isozyme of PLC through receptors that become autophosphorylated due to their stimulated tyrosine kinase activity and provide binding sites for the Src homology domains of the isozyme [249].

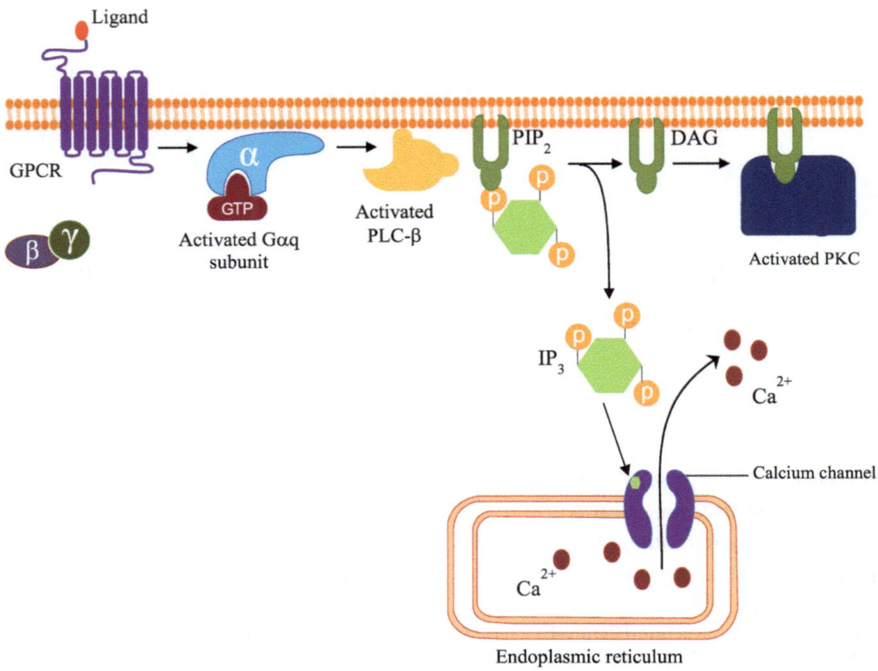

Fig. 5.22 Phospholipase C–inositol phospholipid pathway. The binding of the ligand triggers the dissociation of the αq-subunit of the G protein from the βγ-subunit. The αq-subunit activates the β-isozyme of the PLC, which in turn hydrolyzes the membrane phospholipid PIP_2 to form the second messengers DAG and IP_3. DAG stimulates another enzyme PKC located in the cytosol. IP_3 binds to its receptor located on the calcium channel on the ER membrane. The conformational change of this channel resulting in binding of the IP_3 to its receptor triggers the opening of the channel pore that subsequently leads to calcium efflux into the cytosol [235, 236]. The PKC enzyme can also be activated by calcium. Upon activation, PKC phosphorylates hydroxyl groups of serine and threonine residues of proteins to affect a wide range of physiological responses [237]. PKC has different isozymes that are activated by different agents. PKC isozymes include PKC-α, -β, -γ, -δ, -ε, -η, -θ, -ι, -ζ. PKC-α, -β, -γ are often called classical PKCs and they require Ca^{2+}, DAG, and a phospholipid such as phosphatidylserine for activation. PKC δ, ε, η, and θ are novel PKCs that require DAG, but not Ca^{2+} for activation. The isozymes ζ and ι/λ are atypical PKCs that do not need Ca^{2+} or DAG for activation [238–240]. PKCζ can be activated by lipid components, such as phosphatidylinositol, phosphatidic acid, arachidonic acid, and ceramide. Both phosphatidylinositol (3,4,5)-trisphosphate and PDK1 are necessary for the complete and stable activation of PKCζ [241]. Other families of protein kinases include PKG, PKD, and PK-N [242–246]

Cellular Calcium Handling

The activation of GPCR signaling pathways results in substantial changes in the intracellular Ca^{2+} concentration ($[Ca^{2+}]$) [114]. At rest the $[Ca^{2+}]$ gradient across the plasma membrane is about ~10,000-fold (Fig. 5.23). This gradient plus the hyperpolarized resting membrane potential produces an enormous electrochemical

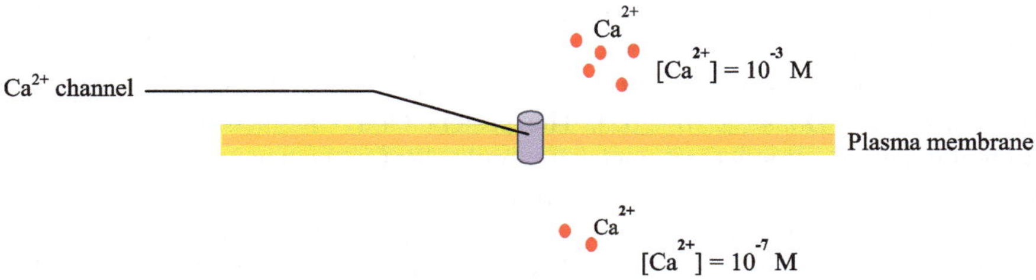

Fig. 5.23 Concentration of calcium inside and outside the cell

driving force in favor of Ca^{2+} influx. In most resting cells, even a small increase in permeability of Ca^{2+} can lead to significant increase in Ca^{2+} influx [146]. Such changes in the intracellular free [Ca^{2+}] constitute one of the main ways by which extracellular information or signals are transferred to the interior of the cell to modulate cellular activities including secretion, development, proliferation, differentiation, and apoptosis. These intracellular Ca^{2+} signals are waves of Ca^{2+} flow characterized by magnitude and duration in terms of spatiotemporal parameters [250–255].

The influx of Ca^{2+} across the plasma membrane is termed "store-operated calcium entry," which is also known as "capacitative calcium entry" [250–252]. Capacitative calcium entry is a process by which the depletion of intracellular Ca^{2+} stores activates Ca^{2+} influx across the plasma membrane [256]. This process indicates reciprocal linkage between intracellular Ca^{2+} concentration (especially in the ER and mitochondria) and Ca^{2+} influx across the plasma membrane [146, 256]. There are different Ca^{2+} channels in the plasma membrane of cells. As discussed in the previous subsection on ion channels, Ca^{2+} channels include voltage-gated, ligand-gated (usually second messenger-gated), and receptor-operated ion channels. Excitable cells such as nerve and muscle cells have high expression of voltage-gated and receptor-operated Ca^{2+} channels, but may have lower expression of second messenger-operated channels. Non-excitable cells such as glandular cells exhibit relatively higher expression of ligand-gated ion channel receptors including the second messenger-operated channels [146]. Off these channels, the best characterized is the Ca^{2+} release-activated Ca^{2+} (CRAC) channels, which are specialized plasma membrane Ca^{2+} ion channels, activated to replenish the level of Ca^{2+} in the ER upon depletion. The channel is present in almost all cell types. CRAC is believed to form a dynamic linkage between intracellular Ca^{2+} store and Ca^{2+} influx across the plasma membrane [146]. The mechanism involved in this sensing is gradually been unraveled. The ER membrane is equipped with a calcium sensor called stromal interaction molecule 1 (STIM1). It was previously abbreviated "SIM" (stromal interacting molecule) on account of its role in mediating the binding of lymphocytes to stromal cells [257]. However, the name later changed to STIM when the protein was identified as Ca^{2+} sensor protein in store-operated Ca^{2+} influx across the plasma membrane [258, 259]. STIM1 is a single transmembrane protein located in the ER, but also associated with the plasma membrane [260, 261]. The main function of this protein is to sense the concentration of Ca^{2+} inside the ER, activate Orai, which forms the pore of the plasma membrane CRAC channel. Orai was identified as membrane proteins and named after the Greek keepers of Heaven's gate "Orais" [256, 262]. The protein Orai1 interacts with the STIM1 protein in periods of low Ca^{2+} concentration in ER. On the hand, in periods of high Ca^{2+} concentration in ER, the protein SARAF (store-operated calcium entry-associated regulatory factor) is activated to interact with STIM1, which results to inactivation of the store-operated calcium channel [250–252]. SARAF (also called transmembrane protein 66, TMEM 66) is an ER membrane resident protein that associates with STIM1, functioning as a negative regulator of the store-operated calcium channel (SOCE) thereby facilitating the inactivation of SOCE [251].

In addition to the Ca^{2+} influx into the cytoplasm, release of stored Ca^{2+} in organelles (endoplasmic reticulum and mitochondria) also contributes to the changes in intracellular Ca^{2+} (Fig. 5.24). In most cases, both the organellar Ca^{2+} stores and extracellular influx contribute to the Ca^{2+} changes in the cell [146].

Some cytoplasmic proteins also regulate intracellular concentration of Ca^{2+}. Such proteins are generally termed cytoplasmic Ca^{2+}-binding proteins (CBPs). The protein calmodulin (CaM) is one of the most inbiquitous and crucial of the CBPs [256, 272]. CaM is composed of two lobes, each of which binds two Ca^{2+} ions [272]. Though the expression of CaM varies from cell to cell, their role as calcium responsive protein is conserved across different phyla of living things. In neurons, however, the main calcium-sensing protein in the cytoplasm appears to be synaptotagmin. But neuronal CaM also plays a role in cytoplasmic Ca^{2+} binding [272]. The activity of some cytoplasmic enzymes that regulate a range of cellular processes is also dependent on CaM. The enzymes are activated in the presence of CaM and are called CaM-dependent protein kinases (CaMK) (Figs. 5.25 and 5.26). There are different types of CaMK, which include the twitchin kinase, titin kinase,

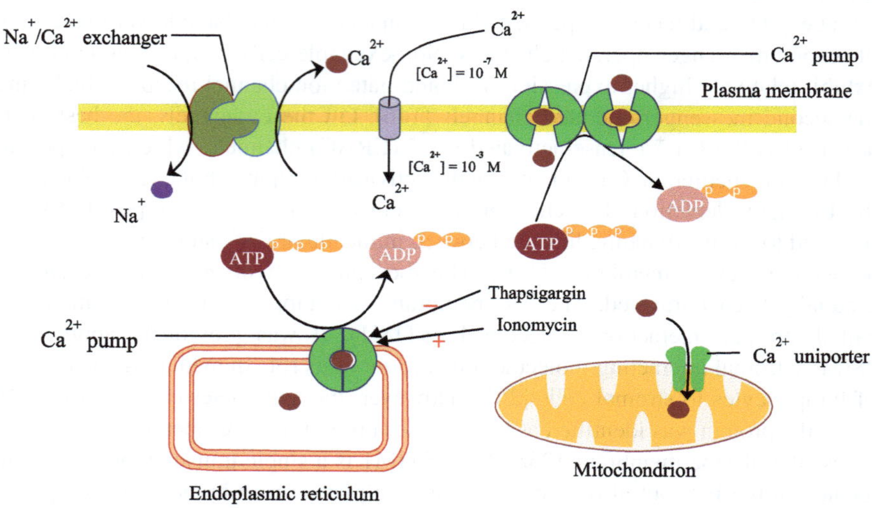

◂**Fig. 5.24** Regulation of intracellular calcium. The sign "–" indicates inhibition and the "+ sign"—activation. The plasma membrane receptors that regulate the concentration of Ca^{2+} in the cell are channels such as Na^{+}/Ca^{2+}, and Ca^{2+} ATPase. Upon stimulation of the cell, Ca^{2+} channel allows influx of Ca^{2+} according to gradient of concentration. Eventually, the high concentration of Ca^{2+} in the cytosol is decreased through the activities of Na^{+}/Ca^{2+} and Ca^{2+} ATPase. The release of calcium ion in the cytosol from the ER occurs via the PLC—IP_3 signaling pathway [256]. The ER is a Ca^{2+}store and sink organelle [146]. Apart from IP_3, second messengers such as cyclic ADP-ribose and nicotinic acid adenine dinucleotide phosphate (NAADP) have been implicated in agonist mediated calcium release from intracellular stores [146]. High level of Ca^{2+} in the cytosol activates both the intracellular stores and the plasma membrane calcium channels resulting in the resequestration of calcium into the ER and mitochondria, and extrusion of calcium ions across the plasma membrane via Na^{+}/Ca^{2+} exchangers and Ca^{2+}-ATPase [146]. Influx of Ca^{2+} into the ER is mediated by sarcoendoplasmic/ER—Ca^{2+} channel and Ca^{2+} ATPase. The release of Ca^{2+} into the cytosol is mediated by IP_3R and ryanodine receptors [263]. Ryanodine receptors are mainly expressed in muscle cells. Most cells express IP_3R, which is a tetrameric channel with each subunit having a binding site for IP_3. Thus, four molecules of IP_3 are required to bind the receptor [264]. Like any other receptor, the IP_3 receptor is regulated by different cellular molecules. Under physiological conditions, the main modulators of this receptor are ATP and Ca^{2+}. The transport of Ca^{2+} is believed to occur at maximum rate at optimal ATP level. But at varying ATP level, different subunits exhibit varying degrees of contribution to Ca^{2+} transport [265]. Again, calmodulin binds to the IP_3 receptor and activates NO/cGMP/cGMP kinase I signaling pathway to negatively regulate calcium stores. cGMP kinase I can phosphorylate the IP_3 receptor and the IP_3R-associated cGMP kinase substrate (IRAG) to inhibit calcium release from ER stores. The association IP_3R-cGMP kinase I, IP_3R-associated cGMP kinase substrate (IRAG), NO, and cGMP inhibit calcium release from ER stores [266]. Different pharmacological agents have varying roles on the ER calcium stores. To this end, the calcium ionophore, ionomycin depletes ER calcium stores by interacting with the ER membrane channel that is responsible for calcium fluxes. The agent thapsigargin inhibits ER calcium ion pump. In an experimental condition, intracellular Ca^{2+} changes are investigated through calcium imaging using fluorescent or chelating indicators with FRET based devices [256]. The mitochondria also play a considerable role in the regulating intracellular Ca^{2+} levels [267]. This organelle regulates cytosolic Ca^{2+} levels by direct and indirect pathways. In indirect pathway, mitochondria, through changes in the concentration of metabolites (NADPH, ATP, pyruvate, and reactive oxygen species), buffer cytosolic Ca^{2+} via release from internal stores and influx across the plasma membrane. Plasma membrane Ca^{2+} influx has been discussed. In the direct pathway, the mitochondria influence the cytosolic Ca^{2+} levels by the mitochondrial channel receptors. The mitochondrial Na^{+}/Ca^{2+} exchanger is a Na^{+}-dependent exporter of Ca^{2+} that functions to return one Ca^{2+} ion back to the cytosol in exchange for three Na^{+} ions. The mitochondrial antiporter H^{+}/Ca^{2+} also functions by returning Ca^{2+} to the cytosol in exchange for H^{+} in a 1:1 ratio [268]. The mitochondrial Ca^{2+} uniporter functions at high cytosolic Ca^{2+} concentration, importing Ca^{2+} into the matrix [269–271]. The Ca^{2+} signaling of the mitochondria is mainly influenced by the state of cellular metabolism. Also, the level of calcium in the cytosol affects mitochondrial metabolic rate. In high calcium concentration, the mitochondria increases cellular energy output (but may trigger cell death), whereas in low calcium levels, it decreases cellular energy output and buffers Ca^{2+} level [271]

myosin light chain kinase, phosphorylase kinase, and calcium/calmodulin-dependent kinase (CaMK) [273]. CaMK is probably the most studied member of the CaM-dependent kinases. Types of CaMK include CaMK I, CaMK II, CaMK III (also called EF-2 kinase), and CaMK IV [274, 275]. Other CaM kinases identified in nature include CaMK-V, -VI, -VII, -VIII, and calmodulin-like protein kinase [276, 277]. Each of these types also has different

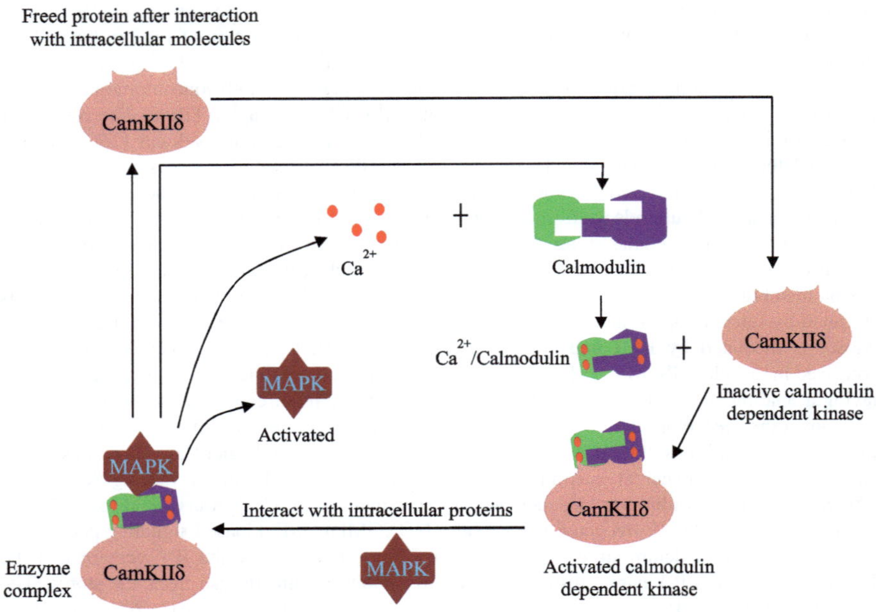

Fig. 5.25 Activation loop of calmodulin

isoforms [272, 278]. CaMKs phosphorylate different intracellular proteins to affect physiological responses. CaM is highly expressed in smooth muscle cells and plays a crucial role in contraction–relaxation cycle. CaM is also involved in regulation of ion channel activity and assembly of microtubules. The regulation of physiological processes by CBPs, in part, depends on their spatiotemporal organization in the cell. Other CBPs include caldesmon, protein S-100 (a vitamin D-dependent CBP), calcineurin, calbindin-D, and annexin VI (Ca^{2+}-dependent phospholipid binding protein)—these CBPs do not possess the classical Ca^{2+}-binding site as in CaM, rather they (except calcineurin) have a single Ca^{2+}-binding site [279]. The spatiotemporal localization of the CBPs in the cytosol or different subcellular compartments determines the role of Ca^{2+} in the cell [279]. Calcineurin is a Ca^{2+}- and CaM-dependent serine/threonine protein phosphatase that dephosphorylates intracellular or membrane proteins (including ion channel) inactivating it to limit Ca^{2+} influx across the plasma membrane [275]. Calcineurin was first identified in the mammalian brain and has been shown to be present in a range of tissues including cardiac and smooth muscles [280]. Increased activity of calcineurin is associated with a range of diseases including hypertension and cardiac hypertrophy [281]. An important line of therapy in these disorders involving increased activity of this CBP is to administer calcineurin inhibitors. Calcineurin inhibitors are immunosuppressive drugs that include cyclosporin, pimecrolimus, and tacrolimus [280, 282–284]. The nuclear effects of calcineurin are achieved by activation of nuclear factor of activated T cell (NFAT), a cytoplasmic transcription factor activated

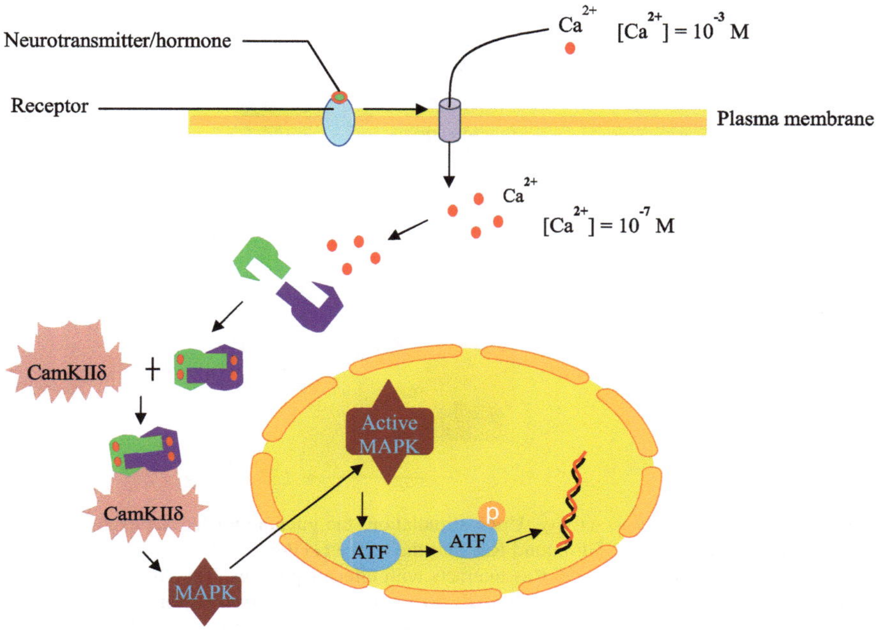

Fig. 5.26 CaM kinase activation and signaling downstream nuclear effectors. Upon release of calcium either from intracellular stores or from receptor-operated calcium channel, calmodulin-dependent protein kinase II (CaMKII) is recruited to associate with 4 calcium ions forming calcium/CaMKII complex, which then activates intracellular proteins such as MAPK, and also interact with other proteins of the cytoplasm to modulate metabolism or hormone signaling. The activated MAPK can be translocated into the nucleus to interact with certain genes; it can phosphorylate specific regions of the FoxO1, ATF3 (activating transcription factor 3) an ATF/CREB (cAMP response element-binding protein) family of transcription factors [290–292]. The expression of ATF3 gene is induced by a wide range of signals, including stress. MAPK kinase 6 is an isozyme of MAPK, a protein kinase upstream of p38 that induce the activation of the p38 pathway resulting in activation of gene expression mediated by ATF3 [291]

(dephosphorylated) by calcineurin. The activated NFAT then translocates into the nucleus, where it upregulates interleukin-2 expression to stimulate differentiation and growth of lymphocytes. Calcineurin is also involved in neuronal and muscle development, as well as morphogenesis of cardiac valves [285]. Caldesmon is a major CaM-binding protein that plays essential role in actomyosin interaction in smooth muscle and non-muscle cells [286]. For further information of CBPs, review Yáñez et al. [287], Weinman [288], and Heizmann [289].

Activation/Deactivation Cycle of G Protein-Coupled Receptor

G protein is inactivated when the α-subunit-bound GTP is hydrolyzed to GDP—this leads to the reassembly of the heterotrimeric G protein [158]. The hydrolysis of

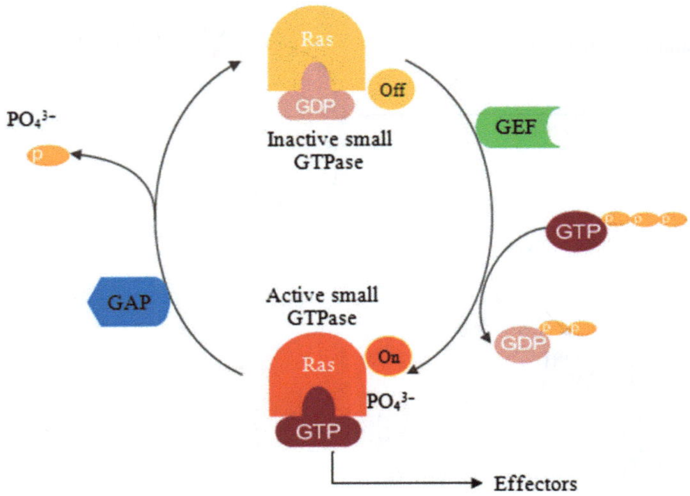

Fig. 5.27 Regulation of Ras activity. Upon stimulation, the guanine nucleotide exchange factor (GEF) converts the inactive GDP-bound Ras into the active GTP-bound Ras (switch is on) in the presence of GTP. In the "on" state, Ras interacts with effector proteins. The "on" state of Ras is regulated by GTPase-activating protein (GAP)—This protein is responsible for converting active Ras to inactive Ras. This is because GTP-bound Ras by its own accord is incapable for switching to the "off" state due to its weak GTPase activity, thus, the active Ras requires a special protein (GAP) for the hydrolysis of GTP to GDP. This is how small GTPases cycle to affect cellular processes. However, these small GTPases do not cycle continuously, but may be removed from cycling between the "on" and "off" states by regulated signaling coordinated by guanine nucleotide dissociation inhibitor (GDI) located in the cytosol. GDI is responsible for removing GTPases from the "on" and "off" cycling, thus regulating the number of available small GTPases required for cellular processes [293]

GTP is catalyzed by GTPase activating proteins (GAPs), while GTP exchange is catalyzed by guanine nucleotide exchange factors (GEFs) (Fig. 5.27). Activation of a GEF correspondingly activates its cognate G protein, while activation of a GAP results in inactivation of the cognate G protein.

Small GTPases are monomeric guanine nucleotide-binding molecular-switch proteins that function as enzymes to hydrolyze GTP to form GDP. These GTPases are a type of G protein found in the cytosol and are related to the α-subunit of heterotrimeric G proteins [253, 294, 295]. These GTPases are usually membrane-bound, but can cycle between the cytosol and the inner leaflet of the plasma membrane [293].

All small GTPases belong to Ras superfamily because the initially identified members were encoded by Ras genes. The name "Ras" is derived from "Rat sarcoma," which reflects the way the first members of these proteins were discovered [293]. To date, more than a hundred and fifty proteins in the Ras superfamily have been discovered. Based on structure, sequence, and function, the Ras superfamily is divided into five main families (Ras, Rho, Ran, Rab, and Arf GTPases). The Ras family itself is further divided into six subfamilies: Ras, Ral,

Rit, Rap "Ras-related proteins," Rheb "Ras homolog enriched in brain," and Rad. These proteins can function upstream and downstream to regulate multiple effectors [296]. They are involved in a host of cellular processes including cell differentiation, proliferation, metabolism. Malfunctions of the signaling of these proteins can result in serious human diseases, including cancer as well as developmental disorders [295].

Regulation of GPCR Activity

Receptor activity is constantly regulated by their ligand concentration. The number and activity of receptors are also regulated by chemical reactions such as phosphorylation, desensitization, and a host of other modifications [162, 297, 298]. The GPCR, in particular, is usually regulated by desensitization—which is responsible for its deactivation. This desensitization process is basically mediated by GPCR kinases (GRKs) and proteins called arrestins. GRKs are serine/threonine protein kinases that bind to the agonist-receptor complex, thereby promoting receptor phosphorylation (inactivation), initiating the uncoupling of the receptor from heterotrimeric G proteins, which in turn leads to arrestin binding. The binding of arrestin prevents G protein interaction with GPCR resulting in functional desensitization. The desensitized GPCRs are then removed from the plasma membrane via endocytosis regulated by the clathrins [298]. There are different types of GRKs, which include GRK types-2, -3, beta-adrenergic receptor kinases. The GRKs are peculiar in their substrate specificities and functions. The GRKs mentioned above associated with beta–gamma subunits of G protein, released upon receptor activation [299]. Arrestins occur as α and β types and each of these types also has several variants [300, 301].

5.3.3 Receptors with Intrinsic Enzymatic Activities

Receptors with intrinsic enzymatic activities, also referred to as catalytic receptors, are nodal points in the signaling cascades that translate extracellular information (usually growth factors and cytokines) through intrinsic catalytic activity of the receptors' intracellular domain. These receptors constitute the second largest type of cell-surface receptor. The extracellular signals, usually, are mediators that are normally present at a very low concentration ($\sim 10^{-9}$–10^{-11} M) but sufficient to initiate receptor signaling [302, 303]. These receptors regulate a wide range of cellular processes including growth, development, metabolism, migration, survival, immune defense, adhesion, motility, proliferation, cell cycle control, and cell death [304, 305]. This functional diversity of the tyrosine kinases is due to its ability to phosphorylate diverse target molecules—covalently attaching a phosphate group to the target. There are 525 kinases in the human genome (referred to as "kinome") divided into eight major groups [306]. Receptors with intrinsic enzymatic activities can be classified according to the type of amino acid that constitutes the catalytic site of the receptor [307]. The majority of the catalytic receptors are usually tyrosine kinases [302, 303]. Receptors with intrinsic enzymatic activities include receptor tyrosine kinase (RTK), receptor serine/threonine kinase, RSTK, (e.g., bone morphogenetic protein receptor, BMPR, and transforming growth factor beta, TGF-β), and guanylate cyclase (GC) [302, 303]. The Nomenclature Committee of the IUPHAR further identifies a third and fourth group of receptors with intrinsic catalytic activities—extrinsic protein tyrosine kinase receptors (EPTKRs) and receptor tyrosine phosphatase (RTP) [308]. The EPTKR has a ligand-binding site and a catalytic site located on separate proteins of the heterodimer. Thus, ligand binding to one subunit triggers signaling via tyrosine phosphorylation in the other subunit. GDNF and ErbB receptors are examples of EPTKR [308]. The RTP currently has no known ligand; nevertheless, it is known that signaling in these receptors is triggered by cell-to-cell contact. RTP signaling plays a crucial role in hematopoietic, immune, and muscle tissue morphogenesis [308]. This chapter is mainly concerned with RTK, RSTK, and GC.

Receptor Serine/Threonine Kinase

RSTK is a kinase enzyme that phosphorylates serine or threonine residues on proteins. The serine–threonine kinases are present in both prokaryotes and eukaryotes. In prokaryotes (e.g., *Escherichia coli*, *Salmonella enterica*, *Mycobacterium tuberculosis*), this group of catalytic receptors confer virulence [309, 310]. Some drug-susceptible and drug-resistant microbes may act through the RSTK. In eukaryotes, this group of receptors plays an essential role in signal transduction triggered by certain group of hormones and is also involved in a range of diseases. Raf-1 is an example of a serine-threonine kinase conserved in multicellular organisms [311, 312]. The receptors—TGF-β type I and type II receptors—are examples of RSTK. TGF-β represents the largest and most versatile cytokine family known in metazoans. (Cytokines are a large family of more than 100 low molecular weight proteins secreted by the immune cells that usually act as mediators at short range between neighboring cells. Examples of cytokines include tumor necrosis factors (TNFs), interleukins (IL), growth factors (e.g., hematopoietic growth factors—colony stimulating factors), and interferons, among others. These small protein molecules are involved in cell proliferation, migration, repair, angiogenesis, immunity, and inflammation [313, 314]). TGF-β binds to their cognate receptors on target cells through the formation of heteromeric RSTK complexes [315–317]. The bone morphogenetic proteins also act through RSTK [317].

Guanylate Cyclase

Guanylate cyclase, also called guanylyl cyclase, is an enzyme that catalyzes the formation of the second messenger cGMP and pyrophosphate from GTP. GC exists in both the soluble (cytosolic) and insoluble (membrane) forms in cells [318]. cGMP participates in a range of cellular processes such as platelet aggregation, neurotransmission, and gut peristalsis.

In the soluble form, GC exists as a heme-containing enzyme consisting of two subunits—alpha and beta subunits that form the heterodimer [318]. The soluble GC is a receptor for nitric oxide (NO) and carbon monoxide (CO). These gaseous ligands bind to the heme moiety thus initiating the formation of the second messenger cGMP [319]. However, the affinity of NO for soluble GC is about 200 times greater compared with CO [320]. The NO donor, sodium nitroprusside can stimulate soluble GC activity resulting in increased cGMP levels [321]. GC can also be activated by bicarbonate and guanylyl cyclase activating proteins [322].

The membrane GC is non-heme single-membrane-spanning protein having a single subunit, with extracellular ligand-binding domain and cytoplasmic-catalytic domains that act as a cell-surface receptor for natriuretic peptides (atrial natriuretic peptide, ANP; brain natriuretic peptide, BNP, and C-type natriuretic peptide, CNP), STa (the bacterial heat-stable enterotoxins that causes traveler's diarrhea) and the endogenous peptides uroguanylin and guanylin, which are highly expressed in the small and large intestines [323–325]. The catalytic domain activity is due to the protein tyrosine phosphatase [318, 326–329]. Ligand binding to the membrane GC, for instance, peptide hormone ANP binding to ANP receptors, activates the catalytic sites of the receptor, resulting in stimulation of cGMP formation which in turn induce diuresis, natriuresis, and vasorelaxation. Known isozymes of GC include GC-A, GC-B, GC-C, GC-D, GC-E, GC-F, and GC-G [326–329]. It was recently reported that thermoreceptors of a subset of neurons respond to cold stimulus by dimerization/oligomerization of GC-G [330].

Receptor Tyrosine Kinase

RTK is a receptor with intrinsic tyrosine catalytic sites, which are capable of transferring phosphate groups from ATP to tyrosine residues of cytoplasmic proteins or the intracellular domains of transmembrane receptors. Thus, RTK can function in an "on" or "off" state [304, 305]. RTKs comprise 58 of the 90 tyrosine kinases in the human genome, representing 10–15% of total protein kinase genes found only in metazoans. Of the 58 RTKs, four lack catalytic activity (e.g. ErbB3) [305–307, 331]. To date, about 20 types of tyrosine kinase receptors have been identified [306, 335]. RTK consists of an N-terminal extracellular domain which binds the ligand; a single transmembrane domain, and a C-terminal intracellular domain with the catalytic sites [103, 305].

In addition to RTK, the tyrosine kinase family also consists of cytoplasmic protein tyrosine kinase, which is a non-receptor tyrosine kinase. But this non-receptor tyrosine kinase aids numerous cellular processes in the cytoplasm [335]. To date, about 32 non-receptor tyrosine kinases have been discovered [332].

Activation of Receptor Tyrosine Kinase

The activation of RTK occurs following binding of the cognate ligands. RTK ligands include epidermal growth factor (EGF), insulin, insulin-like growth factor-1 (IGF-1), fibroblast growth factor (FGF), platelet-derived growth factor (PDGF), vascular endothelial growth factor (VEGF), stem cell factor (SCF) (also called *c-kit*), ErbB (erythroblastic leukemia viral oncogene—the name of a viral oncogene to which the molecule is homologous), neuronal growth factor (NGF), glial-derived neurotrophic factor (GDNF), and neurotrophin-3 (NT-3). These ligands bind to their corresponding receptors—EGFR, insulin receptor, IGF1R, FGFR, PDGFR, VEGFR, SCFR (also called *c-kit* receptor), ErbB receptor, NGFR, GDNFR, and NT-3R, respectively [245, 246, 333–337]. RTKs are numerous in the human tissue and cells and they also include Toll-like receptors (TLRs) 1-10 (the ligands for the TLRs are microbial cell wall components) [303, 337].

Recall that RTK is a monomeric transmembrane protein, however, upon binding of the ligand, the conformation of the receptor changes in such a way that results to receptor dimerization (Fig. 5.28). RTK ligands bind in dimers, where each ligand forms a complex with each RTK monomer. The binding of the dimeric ligands leads to RTK dimerization. The change in conformation resulting from binding of the ligand stimulates the kinase activity of each subunit of the dimeric receptor, phosphorylating tyrosine residues near the catalytic site in the other subunit (referred to as transphosphorylation of tyrosine in the activation loop or juxtamembrane domain) [103, 245, 246]. Then tyrosine residues in other parts of the cytosolic domain are autophosphorylated. The phosphorylated tyrosines also called phosphotyrosines in the cytosolic domain act as docking sites for other intracellular proteins [103]. This way, signal is transduced into the intracellular domain of the cell. These receptors may exist as tetramers as well as oligomers even in the absence of ligands. However, a cognate ligand must be present for autophosphorylation to occur [108, 245, 246].

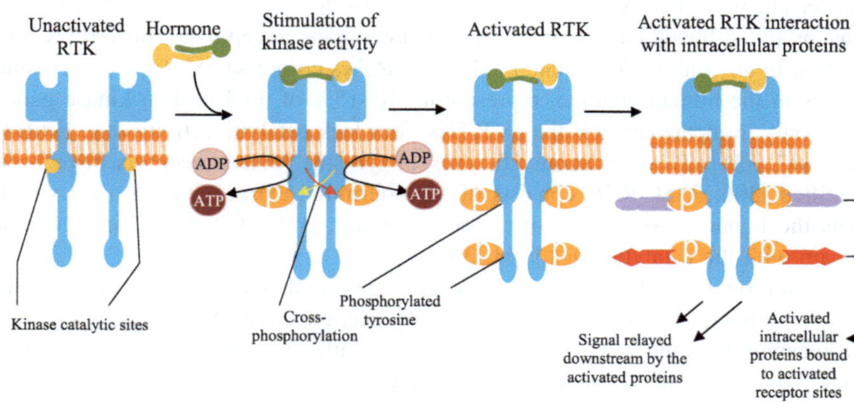

Fig. 5.28 Activation process of RTKs

Receptor Tyrosine Kinase Signaling Downstream

How does RTK initiate intracellular signaling? The phosphotyrosine residues on activated RTKs are recognized by a phospho-dependent-binding domain, the SH2 (Src homology 2) or PTB (phosphotyrosine binding) domains of other proteins. The recruitment of SH2 domain proteins to autophosphorylated RTKs at the plasma membrane is integral for signaling downstream effector molecules [108]. For instance, activation of RTK by hormones such as EGF leads to intracellular activation that recruits the SH2 domain of PLCγ to bind with the phosphotyrosine on the EGFR [103]. Similarly the SH2 domain of the protein tyrosine phosphatase Shp2, GAP, and the adaptor proteins Grb2, Crk, Nck can bind to specific phosphotyrosines of the RTKs. Through these signaling molecules, extracellular information is relayed downstream, activating small GTPases and other proteins including MAPK, Akt, and JAK/STAT (Fig. 5.29) [103].

So the PTB domain of proteins can form complexes with RTKs. Proteins with such domain include insulin receptor substrate 1 (IRS1) (Fig. 5.30). This docking protein binds with its PTB domain to the insulin receptor resulting in phosphorylation of tyrosine residues. The phosphorylated tyrosines can serve as binding sites for other cytoplasmic proteins using the SH2 domains [103]. The insulin receptor is an integral membrane glycoprotein composed of two alpha and two beta subunits linked by disulfide bonds. The alpha subunits contain the insulin binding site, while the beta subunits possess tyrosine kinase activity [343]. The insulin receptors have a dual cellular function: cellular signaling and endocytosis of insulin. The binding of insulin to its plasma membrane receptor induces transfer of extracellular signal to the catalytic cytoplasmic receptor domain that leads to phosphorylation of the tyrosine residues on the beta subunit initiating downstream signaling. In addition, the insulin receptor complex is internalized leading to intracellular proteolysis of insulin (by insulin-degrading enzyme, insulysin, also known as insulinase), while the receptors are recycled back to the plasma membrane [344–347]. The insulin receptor kinase activity can be reversed in vivo through dephosphorylation by alkaline phosphatase [348].

Databases that provide extensive information on RTKs can be assessed in the websites provided in Table 5.6.

Clinical Correlate 5.2

Receptor Tyrosine Kinase Inhibitors (Tyrphostins), Cancer, and Other Human Diseases

While the phosphorylation of proteins by receptor and non-receptor kinases is an important step in communicating signals within into the cell, required for cellular processes (such as cell proliferation, and cell growth), malfunctions in the activities of these kinases can result in serious diseases including cancers, immune disorders, inflammatory diseases such as rheumatoid arthritis, inflammatory bowel disease, and psoriasis. Malfunctions of protein kinases can be due to mutations that make them perpetually switched in the "on" state resulting in uncontrolled gene expression that compromise the normal cellular processes [306, 361–364]. At least 50 different types of tyrosine kinases are implicated in the etiopathogenesis of cancer [362–367] (Also see annoted review texts on tyrphostins and human diseases). Tyrosine kinase inhibitors (TKI) have been successfully used in controlling malfunctions of tyrosine kinase activities resulting in various human ailments. TKIs are a pharmaceutical class of small-molecule kinase inhibitors, orally available [306, 365]. At least 28 TKIs have been approved by the United States Food and Drug Administration (FDA) [366]. The US FDA-approved kinase inhibitors include imatinib, dasatinib, nilotinib, gefitinib, cetuximab, panitumumab, bevacizumab. Imatinib was the first to be introduced into clinical oncology, and it was followed by dasatinib, gefitinib, erlotinib, sorafenib, and sunitinib [304, 367]. TKIs can be further categorized into specific inhibitors such as Janus kinase inhibitors (Jakinibs), TNF inhibitors etc. [304]. The Jakinibs include tofacitinib, ruxolitinib, and baracitinib [304]. Examples of TNF inhibitors are etanercept, adalimumab, golimumab, adalimumab, certolizumab, etanercept, and infliximab [304].

Though TKIs can cause skin toxicity (such as folliculitis, in more than 50% of patients), and other side effects, the clinical efficacy of these drugs have been proven for various human diseases including cancers [304, 367].

Annoted Review Texts on Tyrphostins and Human Diseases

1. Callejas-Rubio JL, López-Pérez L, Ortego-Centeno N (2008) Tumor necrosis factor-alpha inhibitor treatment for sarcoidosis. Ther Clin Risk Manag 4(6):1305–1313
2. Crawford M, Curtis JR (2008) Tumor necrosis factor inhibitors and infection complications. Curr Rheumatol Rep 10(5):383–389
3. Dhillon S (2015) Palbociclib: first global approval. Drugs 75(5):543–51
4. Famenini S, Wu JJ (2013) Combination therapy with tumor necrosis factor inhibitors in psoriasis treatment. Cutis 92(3):140–7
5. Gentile C, Martorana A, Lauria A, Bonsignore R (2017) Kinase inhibitors in multitargeted cancer therapy. Curr Med Chem
6. Gomez-Puerta JA, Mócsai A (2013) Tyrosine kinase inhibitors for the treatment of rheumatoid arthritis. Curr Top Med Chem 13:760–773
7. Jacobsson LTH, Turesson C, Nilsson J-Å, Petersson IF, Lindqvist E, Saxne T, Geborek P (2007) Treatment with TNF blockers and mortality risk in patients with rheumatoid arthritis. Ann Rheum Dis 66(5):670–675
8. Kontzias A, Laurence A, Gadina M, O'Shea JJ (2012) Kinase inhibitors in the treatment of immune-mediated disease. F1000 Med Reports 4:5
9. No Authors (2016) Drug and Device News. P T 41(7):408–412, 415, 422
10. Wu P, Nielsen TE, Clausen MH (2016) Small-molecule kinase inhibitors: an analysis of FDA-approved drugs. Drug Discov Today 21(1):5–10
11. Yamaoka K (2016) Janus kinase inhibitors for rheumatoid arthritis. Curr Opin Chem Biol 32:29–33

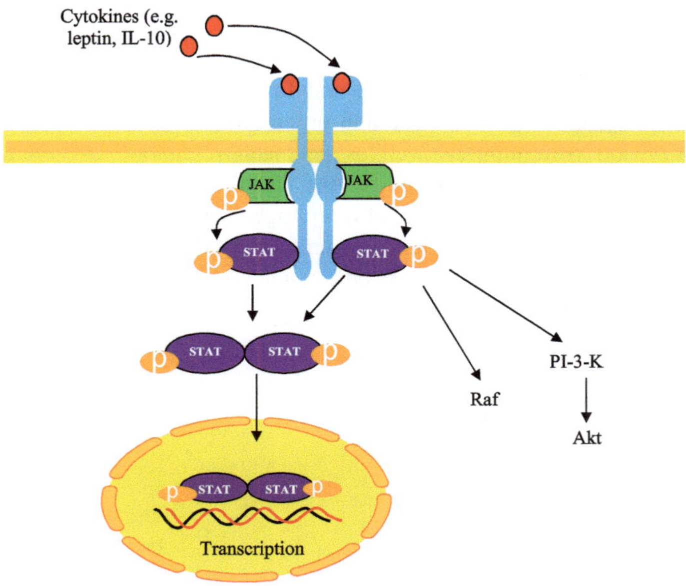

Fig. 5.29 Cytokine receptor signaling via the JAK/STAT pathway. Ligand binding to cytokine receptors (type I/II cytokine receptors) induces their dimerization. But ligand binding can also lead to oligomerization. Named from the two-faced Roman god of the beginning and the end, "Janus," the Janus kinase (JAK) is a non-receptor tyrosine kinase protein attached to cytokine receptors that possesses intrinsic protein tyrosine kinase activity. Ligand binding leads to conformational change in the receptor, which stimulates kinase catalytic sites of JAK resulting in transphosphorylation. JAKs subsequently phosphorylate a range of intracellular proteins including the SH2 domain-containing signal transducers and activators of transcription (STAT). The activated STATs dimerize, functioning as a transcription factor, and translocate to the nucleus, where they activate or repress target gene promoters [306, 338]. STATs can also be directly activated by Src kinases. The phosphorylated JAK can serve as sites for recruitment of adaptors that link the receptor to MAPK, PI-3-K/Akt, among other signaling molecules [339–341]. Akt is derived from a temporary classification name "Ak" for a mouse bred that developed spontaneous thymic lymphomas. The "t" in Akt is derived from the first letter in "thymoma" on the account that a transforming retrovirus was isolated from the Ak strain. Akt (also referred to as PKB) is a serine/threonine-specific protein kinase that plays a key role in multiple cellular processes such as glucose metabolism, cell proliferation, migration, transcription, and apoptosis [342]. Some cytokines such as hematopoietic growth factors (e.g., Erythropoietin, EPO), interferon, thrombopoietin, leptin, and IL-10 signal through the JAK/STAT pathway. This pathway has been implicated in some diseases including cancer. To this end, antagonists of STAT3, a key signaling molecule in this pathway, have shown promise in the treatment of cancers. The action of tyrosine kinases is opposed (regulated) by about 108 protein phosphatases that function to remove phosphate from phosphotyrosines [108, 306]

5.3.4 Morphogen Receptors

Morphogen receptors are the structural and functional units of the signal transduction pathways that regulate tissue and organ morphogenesis during embryonic development and patterning as well as postnatal regulation of cell and tissue positioning [368, 369]. Morphogens are the signaling molecules that govern the process of morphogenesis. The effect of morphogens on cells and tissues is believed to be concentration dependent. The concentration gradient of morphogens is considered a major driving force for early embryonic development. This concentration gradient is regulated by the morphogen receptors as well as extracellular matrix

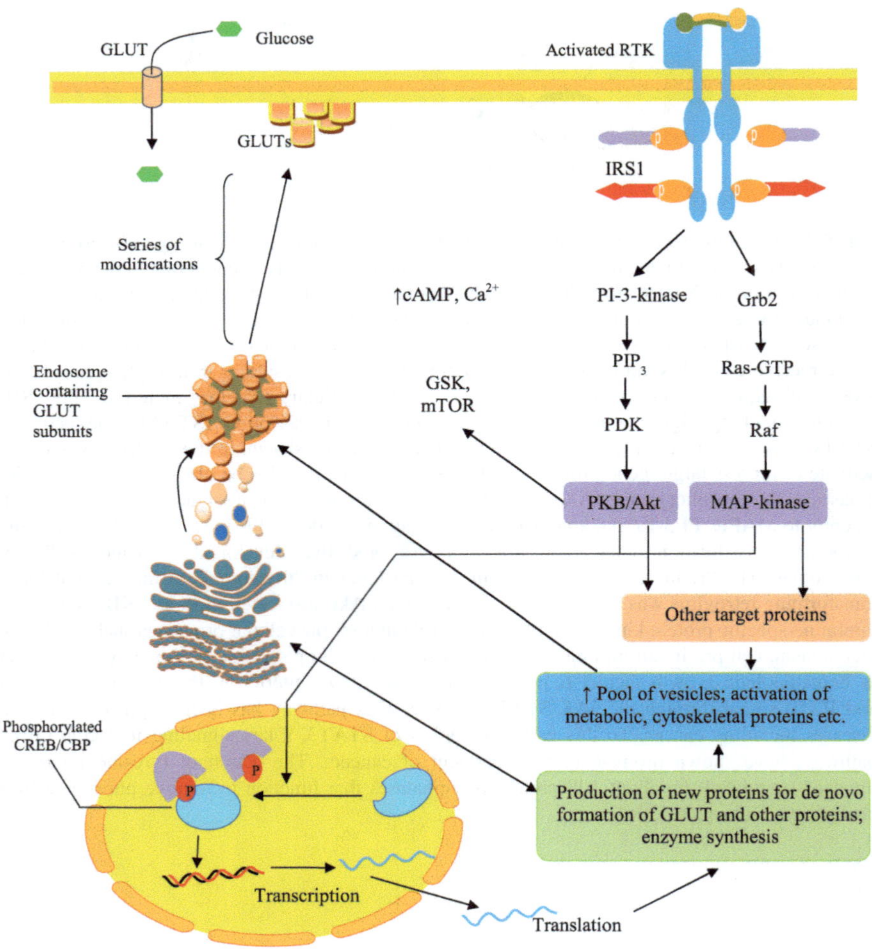

◂**Fig. 5.30** Mechanisms of action of insulin. Insulin receptor is a hetero-tetrametric complex consisting of two α and two β-subunits. The two α extracellular subunit of the insulin receptor binds with two molecules of insulin. This leads to transphosphorylation of specific tyrosine residues on the β-subunits in the activation loop, which in turn leads to increased catalytic kinase activity. The receptor undergoes autophosphorylation of the tyrosine residues in the juxtamembrane regions and intracellular tail. The activated insulin receptor continues to phosphorylate tyrosine residues on intracellular substrates, including the insulin receptor substrate (IRS), IRS-5/DOK-4 (Docking protein), IRS-6/DOK-5, Gab-1 (GRB2-associated-binding protein 1) (GRB2—Growth factor receptor-bound protein 2), Cbl (Casitas B-lineage Lymphoma), APS (Adapter protein with a PH and SH2 domain). One of the enzymes that is activated by IRS is PI-3-K, which signals downstream kinase Akt. The activated insulin receptor also activates certain small G proteins [349, 350]. The insulin receptor signaling is also associated with GSK-3β (glycogen synthase kinase-3 beta) and mTOR (mechanistic or mammalian target of rapamycin) pathways. The mTOR, also known as FRAP1 (FK506-binding protein 12-rapamycin-associated protein 1), is a serine/threonine protein kinase that regulates cell metabolism, growth, proliferation, differentiation, protein synthesis, gene transcription and motility. mTOR belongs to the family of PI-3-K, and plays an important role in insulin-mediated glucose transport by activation of S6 K [347, 351]. S6 K, which is the underlying target kinase in the mTOR pathway, is also involved in insulin resistance [352, 353]. Of the two mTOR complexes, the first complex (mTORC1) regulates glucose metabolism. The signal directed to mTORC1, first assemble at the small G protein—Rheb. In the active state (GTP-bound Rheb), this small G protein activates mTORC1. However, this state of Rheb can change via regulation of complex TSC1–TSC2 (Tuberous sclerosis complex subunit 1 and 2, which also have GTPase activity) [354, 355]. Phosphorylation of TSC1–TSC2 leads to removal of their inhibitory effects on mTORC1 and Rheb [354, 355]. GSK-3 is an enzyme that is inhibited by mTOR via phosphorylation of TSC2. The expression of GSK-3β results in a decrease in absorption of glucose. The effects of GSK-3 on absorption of glucose and expression of GLUT1 occur through mTOR by phosphorylation of TSC2. Without the TSC2, there is increase in glucose entry into the cell, as well as the expression of GLUT1 [356, 357]

Table 5.6 RTK databases

Name of databases	Website
Receptor Tyrosine Kinase database (RTKdb) [305]	http://pbil.univ-lyon1.fr/RTKdb/
Nuclear receptors (NR) and Receptor tyrosine kinase (RTK) (NR-RTK database) [358]	http://www.bioinfo-cbs.org/NR-RTK/
HUGO Gene Nomenclature Committee (HGNC[a]) database HGNC database [359, 360]	http://www.genenames.org/cgi-bin/genefamilies/set/321; http://www.genenames.org/
International Union of Basic and Clinical Pharmacology (IUPHAR) [170]	http://www.guidetopharmacology.org/

[a]HGNC—human gene name and symbol database

substances such as heparan sulfate proteoglycans (HSPGs) [368, 369]. The integrity of the transcription process of the genes that take part in gut formation in an embryo and postnatal regulation of cell and tissue positioning and localization is crucial to growth and development. The genes that control the following pathways have been implicated in morphogenesis: sonic hedgehog (Shh), Wnt family member Wingless, and Notch. Other genes involved in TGF-β and BMP pathways also have crucial role to play in morphogenesis. These later molecules (TGF-β and BMP) are known to signal through RTKs and RSTKs [370–372].

Morphogenesis is a very complex process that involves numerous signaling pathways and growth factors including EGF, insulin-like growth factor-2 (IGF-2), and hepatocyte growth factor (HGF). This subsection deals with the mechanism of Shh, Wnt, and Notch signaling as well as their roles in morphogenesis. Accumulating studies have identified the involvement of other genes including the homeobox genes in tissue morphogenesis. These genes and signaling pathways are key determinant of the formation of the gut cells, migration of the cells to their specific regions, and organogenesis [370–372].

The concentration gradient of morphogens is a major determinant of morphogenesis and is thought to be under controlled diffusion, mediated by exosomes, cytonemes, argosomes, and tunneling nanotubes [373]. These microvesicles serve as vehicles transporting morphogens and RNA across long-range distances (>100 μm) from one cell to another [374–376]. They form a crucial nexus in cell–cell communication [377]. Exosomes are extracellular microvesicles measuring about 30–100 nm in diameter, formed from exocytic vesicles, and are involved transport of morphogens and other cellular substances across long-range distances [375, 378]. The actin-based close-ended filopodial-like protrusions or bridges, cytonemes are dynamic structures that connect cells together and control the spatiotemporal diffusion of morphogens (morphogen gradient) to establish a genetically predetermined structure of the tissue or organ [379, 380]. Importantly, these microvesicles are not only involved in normal morphogenesis but have been implicated in a range of pathological conditions including cancers [381]. Argosomes are membrane exovesicles that shuttle some proteins (including ligands) to distant cells [382, 383]. Tunneling nanotubes are thin open-ended actin tubes that connect cells over long distances, facilitating the intercellular transport of various cellular components including organelles [381, 384, 385].

In addition to the microvesicles, some substances of the extracellular matrix or cell-surface molecules also mediate morphogen diffusion gradient [386]. Heparan sulfate proteoglycans (HSPGs) are polysaccharides (glycoproteins) that occur in close proximity to cell surface or extracellular matrix proteins and they interact with a host of cell receptors thus modulating their functions. They also interact with morphogens. HSPGs have common characteristics of containing one or more covalently attached heparan sulfate (HS) chains. HS is a type of glycosaminoglycan (GAG) family of carbohydrates and is structurally closely related to heparin. Both GAG and heparin (including the other members—hyaluronic acid or hyaluronan, keratan sulfate, chondroitin sulfate, dermatan sulfate) are polymers consisting of a variably sulfated repeating disaccharide units of D-glucuronic or L-iduronic acid, and either *N*-acetylglucosamine or *N*-acetylgalactosamine. There are about 17 types of HSPGs, which are categorized into three groups on the basis of their location. They include membrane HSPGs (syndecans; glycosylphosphatidylinositol-anchored proteoglycans or glypicans), secreted extracellular matrix HSPGs (agrin, perlecan, type XVIII collagen), and the secretory vesicle proteoglycan, serglycin.

Serglycin plays a role in packaging granular contents, maintaining proteases in an active state, and regulating various biological activities after secretion such as coagulation, host defense, and wound repair. (Syndecans are single transmembrane domain proteins that are thought to act as coreceptors, especially for GPCR) [386–389].

Wnt Signaling Pathway

The Wnt signaling pathway was discovered in 1982 from mouse breast tumors [390]. In 1982, Roel Nusse (1950–present) and Harold Eliot Varmus (1939–present) identified a new mouse proto-oncogene (gene that forms oncogene when mutated; oncogene is a cancer-causing gene) that they named int1 (integration 1). Further research into this gene revealed that it was the same gene in *Drosophila melanogaster* called "Wingless" already characterized by Christiane Nüsslein-Volhard (1942–present) and Eric Francis Wieschaus (1947–present) (both shared the Nobel Prize in Physiology or Medicine in 1995 with Edward B. Lewis (1918–2004) for their discoveries concerning the "genetic control of early embryonic development"). The functions of the Wingless gene were known as segment polarity gene involved in the formation of the body axis during embryonic development. This gene and its functions are conserved throughout the animal kingdom. The gene "int/Wingless" was renamed "Wnt"—a combination of "int" and "Wg" and it stands for "Wingless-related integration site" [391–394]. The *Wnt* family consists of over 15 genes that encode secreted glycoproteins and are all implicated in developmental processes [395]. The Wnt pathway is responsible for embryonic and postnatal development and also plays a useful role in adult tissue homeostasis. The pathway is important in the regulation of cell fate determination, proliferation, differentiation, adhesion, and polarity. A disorder in this signaling pathway results in a series of malformations and dysfunctions that include developmental defects, cancers, and metabolic disorders [396, 397].

The Wnt pathway is subdivided into the following types—canonical (Wnt/β-catenin pathway), planar cell polarity, and Wnt/Ca^{2+} pathways [398]. In the canonical pathway, Wnt ligands (lipoglycoproteins such as Wnt 1, 3a, and 8) [396]) attach to the Wnt receptor, a dimeric cell-surface receptor composed of seven-transmembrane frizzled (fzd) receptor (type of GPCR: 10 Fzd receptors have been identified) and the transmembrane low-density lipoprotein receptor-related protein 5 and 6 (LRP5 and 6). But LRP5 and 6 can act as coreceptors with the Fzd receptor and function via receptor-mediated endocytosis. However, receptors like receptor tyrosine kinase and receptor tyrosine kinase-like orphan receptor 2 are considered as possible candidates for Wnt signaling [396]. There are at least 19 Wnt ligands currently identified [399, 400]. The binding of Wnt ligands to the Wnt receptor causes phosphorylation of the cytoplasmic protein disheveled (Dvl or Dsh) leading to its activation. Dvl is a cytoplasmic phosphoprotein that acts downstream fzd receptors in the canonical and non-canonical Wnt signaling pathways. The Dvl protein is encoded by a gene whose mutation results to disordered orientation of body hairs of fruit flies hence the name [401–404]. The Dvl protein is made up of 500–600 amino acids and it is an essential component of canonical Wnt, planar cell polarity and Wnt/Ca^{2+} signaling pathways [405].

Apart from Dvl, other key molecules of this pathway include the serine/threonine kinase GSK-3β, β-catenin, adenomatous polyposis coli (APC), Axin, Axil, and conductin [406]. The GSK-3β usually exists in association with proteins such as Axin—which enables the phosphorylation (degradation) of β-catenin by proteolysis. Activation of Dvl inhibits GSK-3β by inducing its dissociation from Axin. Thus, the degradation of β-catenin is inhibited and the molecule associates with the high mobility group (HMG)-domain proteins, T cell factor (TCF) and lymphoid enhancer factor (LEF) family transcription factors (Fig. 5.31) [407]. The formed complex freely translocates into the nucleus to initiate Wnt gene expression [395–398, 408]. Genes that are activated in the process include c-myc, c-jun, fra-1, and cyclin D1, Axin, Axil, Ccnd1, peroxisome proliferator-activated receptor (PPARδ) [406]. Studies indicate that β-catenin is also regulated by APC protein, a 300 kDa protein associated with β-catenin, GSK-3β and a homolog of the Drosophila discs large tumor suppressor protein. GSK-3β was first identified as an inhibitor of glycogen synthase hence the name. A part from being a major component of the Wnt pathway, studies have shown that this enzyme is involved in EGF and insulin signaling downstream processes including metabolism, gene expression, proliferation and differentiation, and also interacts with numerous cellular proteins including cell adhesion molecule, synapsin I, tau, a couple of transcription factors, and cytoskeletal proteins [409, 410].

Axin and related proteins (Axin-1, -2) interact with APC, β-catenin, GSK-3β, and protein phosphatase 2. Axin, axil, and conductin have similar functions as they all negatively regulates the Wnt signaling pathway through control of the level of β-catenin and also stabilize GSK-3. They belong to the family of peptides called

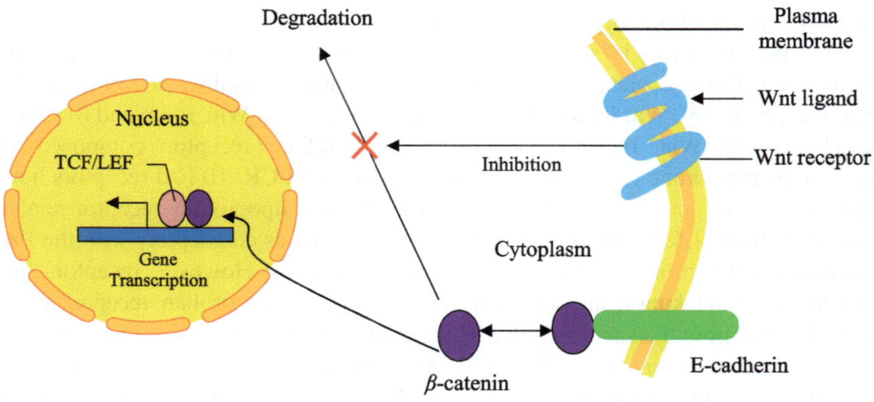

Fig. 5.31 The Wnt signaling pathway. Activation of Wnt receptor leads to the inhibition of β-catenin degradation in the cytoplasm. β-catenin then translocates into the nucleus to initiate or inhibit gene expression

GSK-3 binding proteins (GBP) expressed in fruit fly. Axil binds to β-catenin and GSK-3β; however, phosphorylation of β-catenin is enhanced by the binding of axil to GSK-3β. This suggests that since axil is also phosphorylated by GSK-3β, axil may readily bind to β-catenin to form axil-β-catenin—a complex that is phosphorylated first from the axil terminal, thereby exposing the β-catenin molecule to further phosphorylation [408, 411].

Another crucial regulator of the Wnt pathway is GBP which functions to increase the level of cytosolic β-catenin by interacting with GSK-3β. Thus there are negative and positive regulators of GSK-3. Axin functions as a scaffolding protein that directly binds both GSK-3 and its substrate β-catenin and greatly enhances the ability of GSK-3β to phosphorylate β-catenin [395, 396, 408]. The negative regulation (inhibition) of Wnt pathway can occur at the extracellular or intracellular domains. The negative regulators of Wnt at the extracellular domain are mostly secreted factors that interact with the Wnt receptors or ligands. Such factors include Wnt inhibitory factor-1 and -2, sclerostin, and secreted fzd-related proteins. The negative regulators at the intracellular domain are proteins that inhibit the activity of β catenin and Tcf/Lef-1. For instance, Chibby is a protein that inhibits the interaction between β catenin and Tcf/Lef-1 in the nucleus [396, 412].

The non-canonical pathway is yet to be well defined. But research shows that in the non-canonical Wnt pathway, ligands, such as Wnt 5a and Wnt 11 activate their respective Wnt receptors to initiate downstream cellular processes. The receptors of this pathway have coreceptors which include but not limited to Knypek, Ror2, and Cripto [413]. The non-canonical pathway includes calcium-dependent pathway and the planar cell polarity pathway. The Wnt calcium-dependent pathway is initiated by Wnt ligands binding to their Fzd receptors followed by the activation of G protein and a signal cascade releasing intracellular calcium from the ER. The released calcium activates downstream mediators, such as PLC, PKC, calcineurin, and calcium/CaMK II, which in turn activates transcription factors such as cAMP response element-binding protein (CREB), nuclear factor κB, and nuclear factor of activated T cells. The planar cell polarity pathway is important in cell polarity, migration, motility, and division as well as cytoskeleton maintenance. In this pathway, fzd can phosphorylate Dsh without co-receptor recruitment via the activation of small GTPases such as rho and rac, ultimately leading to activation of downstream kinases such as c-jun NH2 kinase. This Wnt signaling pathway may inhibit gene transcription via the recruitment of β-catenin [396, 398, 413].

It has been shown that the canonical pathway interacts with the non-canonical pathways, thus allowing for cross talks between different Wnt sub-pathways. The ligand Wnt 5a of the non-canonical pathway can activate the canonical β-catenin pathway in the presence of fzd receptor 5 or 4 and LRP 5 [396].

This signaling pathway has been associated with numerous types of cancer including colon carcinoma, hepatocellular carcinoma, hepatoblastomas, neuroblastomas, and ovarian adenocarcinomas [395].

Notch Signaling Pathway

The "notch" signaling pathway derives its name from the notch-like appearance of the wings of the fruit fly caused by mutation in specific groups of gene referred to as notch genes. Notch signaling is a type of contact-dependent signal transduction used to specify cell identities and regulate cell behavior throughout the process of animal development. Thus, the notch signaling determines how groups of cells are organized to form larger and complex structures such as tissues and organs. To ensure this, the notch pathway control lateral intercellular signaling between neighboring cells and regulate the transition from undifferentiated to committed (fully differentiated) cells [414–416]. Thus, components of the notch pathway play a crucial role in cell fate determination. The cells of the GI tract including the enteric neurons and glial cells express notch ligands and their cognate receptors. The notch signaling pathway is essential in intestinal crypt and esophageal basal layer stem cell differentiation [417, 418]. The signaling pathway is involved in the maintenance of transit-amplifying cells through the control of transcription factors in many tissues of the developing embryo as well as postnatal period [418–420]. The notch signaling is dysregulated in many diseases including cancers, multiple sclerosis, and tetralogy of fallot [421].

The notch pathway is initiated by the binding of notch ligands to their cognate receptors. Notch ligands are membrane-bound ligands called Delta/Delta-like (Dll-1-4) and Jagged (Jag-1, Jag-2) on the surface of neighboring cells. Other ligands (atypical) such as contactin can also bind notch receptors. The notch receptor is a single transmembrane protein composed of one extracellular and one intracellular domain. Altogether, there are four notch receptors (Notch-1, -2, -3, -4) identified in mammals [422–424].

There are two types of notch signaling: canonical and non-canonical. The canonical signaling pathway involves ligand-induced cleavage of notch for transcriptional regulation. In the non-cannonical pathway, notch functions independently of ligand and transcription factors (Fig. 5.32) [439].

Hedgehog Signaling Pathway

The hedgehog (Hh) family of morphogens includes three members: Sonic Hedgehog (Shh), Indian Hedgehog (Ihh), and Desert Hedgehog (Dhh). The sonic hedgehog was named after Sega's video game character "Sonic the Hedgehog." Sega is a Japanese multinational video game developer and publisher with headquarter located in Tokyo. The desert hedgehog and Indian hedgehog were named for species of hedgehogs (porcupine), a type of mammal covered with both hair and protective spines [427, 440, 441].

Hh is widely expressed in the GI tract including the muscle cells, enteric neurons, glial cells as well as their progenitors [418]. Hh is crucial in embryonic development and tissue patterning events, shaping cell and tissue compartment border. Dysfunctions in Hh signaling have been implicated in many human diseases

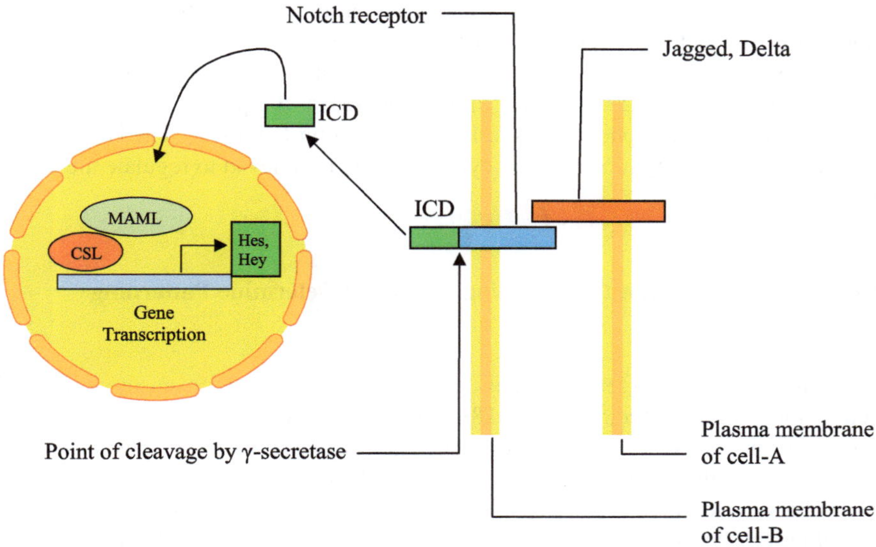

Fig. 5.32 Notch signaling pathway. The notch receptor is activated by direct cell-to-cell contact. Ligand proteins binding to the extracellular domain induce proteolytic cleavage in two steps. The first cleavage is mediated by ADAM (a disintegrin and metalloproteinase protein)—a family of transmembrane proteins known to control membrane fusion, development, cell fate determination, and cytokine/growth factor shedding because of their protease action and are thus referred to as sheddases—as they shed the extracellular part of transmembrane proteins [425]. ADAM proteins are involved in the regulation of notch signaling by receptor proteolysis of the extracellular domain [426]. Following the first cleavage, the enzyme γ-secretase "cuts off" the intracellular domain of the notch receptor (ICD) from the membrane resulting in the subsequent release of the intracellular domain, which enters the nucleus to bind with the DNA-binding proteins such as CSL (C promoter binding factor—1, suppressor of hairless, longevity-assurance gene (CBF1/Su(H)/Lag-1)), recombination signal binding protein of the Jκ immunoglobulin gene (RBPJ). If CSL is activated, then it recruits a positive regulator of notch called mastermind-like (MAML), which in turn initiates the transcription of various target genes including hairy and E (spl)-1 (Hes) and Hey. There are different isoforms of Hes and Hey transcription factors and thus activation or inhibition of gene expression depends on the isoform of transcription factor that is activated. For instance, Hes1 acts as a transcriptional repressor [407, 413, 418, 427–433]. This notch pathway (CSL-ICD-Mastermind complex) induced by typical ligands is called the canonical pathway. The ICD can interact with Deltex proteins to inhibit the activity of transcription factors through the same pathway that recruits ICD [419, 434–438]

[377, 418]. Absence of Hh has been associated with anomalies of development. Malfunctions of Hh signaling can lead to reduction in the thickness of the GI tract smooth muscle layer, anorectal malformations [427], and can result to Hirschsprung's disease (see Chap. 7 for details on this disease) [418].

The receptor of Hh is called patched. Two types are known—patched-1 and -2. In the inactive state (i.e., in the absence of Hh ligand), patched inhibits smoothened, a 7TM receptor. Following activation by Hh ligand, the inhibitory action of patched on this 7TM is relieved, resulting in the activation of the pathway via the Gli (glioblastoma) proteins (Gli1, Gli2, and Gli3 proteins), a family of zinc-finger transcriptional factors [427]. Gli1 and Gli2 function as transcriptional activators, whereas Gli3 is a repressor [418]. For further details on this signaling pathway, review Liu and Ngan (2014) [418]. In addition to Hh, the protein decapentaplegic also functions as a morphogen in similar way as Hh protein [377]. The Hh pathway is controlled in part by receptor endocytosis, which is believed to regulate the signal gradient [427].

Cooperativity Between the Signaling Pathways Determine Patterning and Development

Hh, wnt, and notch signaling pathways work cooperatively to ensure regulated functioning during development [418]. These pathways are also associated with other morphogen receptor signaling. The pathways are strongly interconnected, which ensures their regulated activity [413, 427]. For instance, Hh signaling regulates the development of gut in part by inhibiting notch activity [418]. Cooperative signaling can be investigated in details in database of signaling pathway cross talk "XTalkDB" (http://www.xtalkdb.org) [442].

5.3.5 Integrin Receptor

The name "integrin" was first used in 1986 to indicate an integral membrane protein [443]. The receptor for this integral protein was first identified the following year by Hynes [444]. Integrins are membrane-spanning heterodimer consisting of one α chain and one β chain, which are non-covalently linked. Integrins are restricted to the metazoan. In vertebrates, there are 18 α- and 8 β-subunits that combine to form 24 currently identified receptors [445–448]. Both subunits contain extracellular and cytoplasmic domains. The extracellular portion resembles a large "head" that is continuous with two "legs" extending into the cytoplasmic domain (Fig. 5.33) [446]. Integrins receptors are the best-characterized members of the cell-surface adhesion receptors, which include cadherins, immunoglobulin (Ig) family, and selectins (Fig. 5.33) [448–450].

Integrin receptors play numerous roles in the organism. These integral protein receptors are necessary for cell adhesion to extracellular matrix proteins; transmembrane connections to the cytoskeleton (anchorage function); tissue organization, maintenance, and repair as well as cell development; intracellular signaling; providing information on the cell's location; host defense; hemostasis; and act as receptors for microbes [445–448, 451, 452]. In addition to these functions, integrins of the GI tract are involved in a variety of protective roles. The GI tract integrin $\alpha4\beta7$ is involved in regulating the migration (homing) of lymphocyte to the intestine through interaction with the mucosal cell adhesion molecule-1, expressed primarily in the lamina propria of the GI tract and venules of the gut-associated

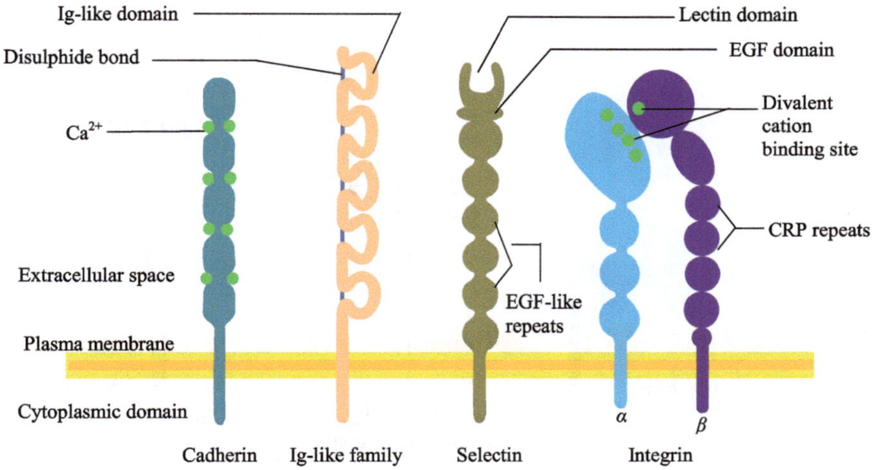

Fig. 5.33 The cell-surface adhesion receptors. Cadherins are cell-surface adhesion receptors involved in a range of cellular processes including morphogenesis, cell adhesion, and cell-to-cell signaling. Both ions (e.g., calcium) and proteins (e.g., internalin) can bind to cadherins to initiate signaling process [458–460]. Immunoglobulin (Ig) family consists of cell-surface adhesion receptors that play a crucial role in inflammatory process. Ig family of cell-surface adhesion receptors include intercellular adhesion molecules (ICAM-1, -2, and -3), neural cell adhesion molecule (N-CAM), vascular cell adhesion molecule (VCAM), and T cell receptors (CD3, CD4, etc.)—they are also involved in inflammatory response [461–463]. Selectins are a type of cell-surface adhesion receptors that bind to carbohydrate residue through their lectin-domain. Selectins regulate homing and adhesion of platelet (P-selectin), leucocyte (L-selectin) [464–466]

lymphoid tissue such as Peyer's patches and gut mesenteric lymph nodes [450, 453]. The integrin α4β7 is considered as a receptor for lymphocytes [454]. Absence of integrin β7 in the gut has been implicated in lymphocyte deficit resulting in the loss of Peyer's patches [445]. Lymphocytes as well as other cells (dendritic cells, macrophages, fibroblasts, and endothelial cells) that express integrin receptors are involved in gut immune responses [450, 455]. The regulation of tissue composition and structure by stem cells is dependent on their interaction with integrin-associated adhesion substances [456]. The prevention of cell death, at least, in part, is due to the ability of integrin to regulate signaling through PI-3-K/Akt and ERK and other pathways [445]. The structure and functions of certain integrins in GI tract in inflammatory diseases of the gut (e.g., Crohn's disease, ulcerative colitis) are compromised. Consequently, anti-α4 antibody has been shown to be effective in treating inflammatory bowel diseases [457].

Integrins do not have intrinsic catalytic activity and can be activated upon binding of a cognate ligand to extracellular or intracellular domain (outside and inside receptor activation) (Fig. 5.34). The intercellular adhesion molecule is an example of integrin ligand [467]. The ligand can bind to the extracellular or intracellular domain to initiate signaling from outside to in or inside to out, respectively [446]. Extracellularly, ICAM-4 can bind with the integrin α4β1,

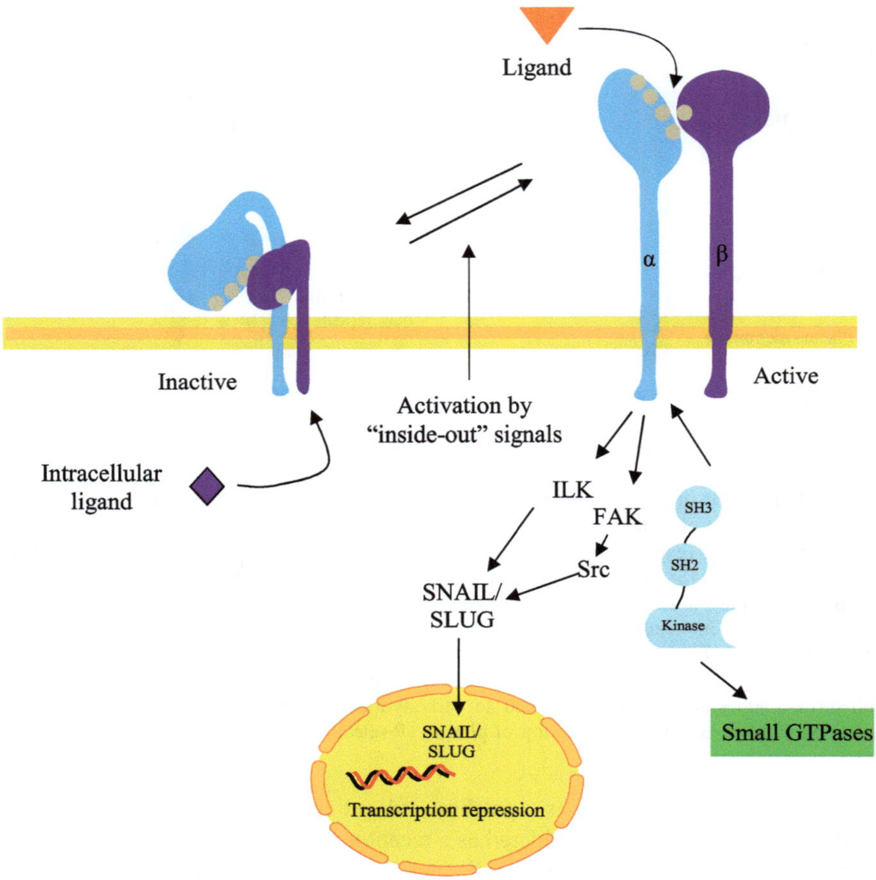

Fig. 5.34 Integrin signaling

connective tissue growth factor binds with αVβ3 or αIIbβ3, VEGF binds with α9β1 or αVβ3 [445]. The binding of a ligand to integrins depends on the quantity of divalent cations such as Ca^{2+}, Mg^{2+}, and Mn^{2+} [446]. The extracellular domain of each subunit contributes to the binding of ligands to induce conformational change that results in the interaction of the cytoplasmic domain with several signaling molecules [457]. The cytoplasmic tail of the β-subunit can form a hub of interactions with several proteins [446]. So, receptor activation initially leads to receptor clustering with subsequent conformational change that allows for interactions with a range of proteins. Proteins that associate with the cytoplasmic domain of integrins include focal adhesion kinase (FAK), Src, small GTPases (e.g., Ras), integrin-linked kinase (ILK), and adaptors (e.g., paxillin). First, the enzymes associate with the activated receptor, followed by attachment by other proteins that mediate interactions with the cytoskeleton or with transcription factors that regulate gene expression. The activation of the zinc-finger transcription factors SNAIL/SLUG results to translocation into the nucleus—these transcription factors are responsible for gene transcription repression [446, 451, 457]. The SNAIL signaling pathway controls numerous functions of the cell including apical-basal polarity, and cell fate determination [468, 469]. Proteins that may form linkages with downstream components of the cytoplasm include talin, ezrin, radixin, moesin, vinculin, actin-binding proteins that link the cytoskeleton with integrin receptors [292, 451, 457].

Integrins can be activated intracellularly by signals from intracellular proteins (inside-out signaling). The GPCR signaling can initiate the phosphorylation of the cytoplasmic domain of the β-subunit of the integrin receptor, which in turn modulates a range of signaling pathways [457].

5.4 Gut Nutrient Sensing and Nutrient Receptor Signaling (Gastrointestinal Chemosensation)

The GI tract possesses special receptor-sensors that translate the various chemical content of food into signal that culminate in secretion of hormones and neurotransmitters as well exocrine (lumencrine) release of substances. Though the release of hormones and neurotransmitters is mediated by chemical, electrical and mechanical stimulation, the chemically mediated stimulation is crucial to the functioning of the entire GI tract and involves the identification of different nutrients by receptor-sensors located on the membrane of endocrine cells that subsequently culminate in the secretion of hormones, neurotransmitters or paracrine factors [13, 115, 470].

Receptor nutrient sensor signaling in the GI tract (also referred to as GI chemosensation) is a regulated signaling initiated by the interaction between specific receptor group (sensors) on the apical membrane of the gut epithelial cell and nutrients or biomolecules in the lumen, resulting in the release of hormones, neurotransmitters, or paracrine factors that possess not only local, but also considerable systemic significance. Gut receptors that sense the conformational change in neighboring gut epithelial receptors due to translocation of luminal nutrient are also considered as gut chemosensors [11–13].

Some of the GI tract receptor sensors are called taste receptors as they recognize and respond to the presence of different tastes by initiation of cellular signaling. There are five taste receptors: sweet, sour, salty, bitter, and umami (savory) tastes [115]. However, emerging studies indicate that there is a sixth modality of taste concerned with perception of dietary lipid or fat. The proposed term for the sixth type of taste is "oleogustus" [471]. The CD36 and GPR120 are the receptors involved in perception of lipid or fat taste [472, 473]. The CD36 receptor is a fatty acid transporter (glycoprotein) that is mainly expressed in the lingual circumvallate and foliate papillae taste buds [473]. The GPR120 is mainly expressed in circumvallate, fungiform papillae taste buds, but also in extralingual epithelial cells

[474]. Evidences mounting indicate that GPR40 is also involved in fat taste perception and is predominantly expressed in foliate and fungiform taste buds [473, 474]. The G protein-associated taste receptors are activated by fatty acids signaling downstream with subsequent release of Ca^{2+}, which in turn activates the cation channel TRP channel type M5 [475]. This signaling results to physiological responses that affect food ingestion and digestive processes [472, 473]. The different components of fat appear to have varying effects on the taste receptors mentioned above. Oleogustus is mainly due to the presence of non-esterified medium and long-chain fatty acids [471]. The perception of taste initiated by short-chain fatty acids is like sour taste [471–473]. Disordered signaling of CD36, GPR120, and GPR40 receptors underlie many feeding behaviors involved in the development of metabolic diseases like obesity [473, 476]. These receptors are present almost in every region of the GI tract beginning from the tongue. Taste receptors belong to the class of ion channel receptors and GPCR. For instance, the sweet taste receptor (T1R2/T1R3), umami taste receptor (T1R1/T1R3), and bitter taste receptors (T2R) are GPCR. However, sweet and bitter taste receptor signaling is coupled to the activities of the ion channel—TRP channel through G protein gustducin activation. The α-subunit of the G protein activates PLC resulting in the production of IP3 and DAG from PIP2. DAG activates PKC, while IP3 activates 1P3R of the ER membrane. PKC can phosphorylate several effectors. The activation of IP3R leads to efflux of Ca^{2+} ions into the cytosol. Ca^{2+} ions can activate Na^+ channels with influx of Na^+ ions resulting depolarization of the cell membrane. Ca^{2+} ions can also initiate the release of mediators or neurotransmitters by recruiting the pool secretory vesicles [115]. These hormones, neurotransmitters can signal with the vagus nerve, with the hypothalamus to regulate a range of behaviors (either via direct contact or transport via the bloodstream) [470]. The sour taste is due to the direct sensing of Na^+, Ca^{2+}, or H^+ ion flux through specific channels [115, 477]. The channels responsible for sour taste include hyperpolarization-activated cyclic nucleotide-gated channel, Na^+/H^+ exchangers, TRP channels—PKD2L1 (polycystic kidney disease-2-like 1)-PKD1L3 ion channel complex, acid-sensing ion channels, amiloride-sensitive H^+-activated cation (Na^+, Ca^{2+}, K^+) channels, and the two-pore domain potassium channels that contribute to a steady outward leak of K^+ ions [478–480]. Further details on taste receptor signaling are discussed in Chap. 9. Fat, amino acids, and carbohydrates are sensed by different receptors on the epithelial membrane of the GI tract (discussed below) [470].

5.4.1 Carbohydrate Sensing in the Gastrointestinal Tract

Carbohydrate sensors of the GI tract include cotransporter Na^+/glucose (SGLT3), GLUT2, TIR—which are highly expressed in many tissues of the body (especially epithelial cells). The human SGLT3 does not transport hexoses but generates Na^+ current which extends into the cell, which in turn depolarizes the membrane potential. SGLT3 is found in the nervous system of the intestine, brain, etc. SGLT3 has functional significance in neuroendocrine cells in the synthesis of incretins (glucagon-like peptide-1, GLP-1; and glucose-dependent insulinotropic polypeptide, GIP) (Fig. 5.35) [481–484]. Since the neurons and nerve endings of the GI tract also express the hexose (e.g., glucose) sensors, the digestive tract afferent nerve endings of the mucosa can also identify change in nutrient concentrations to directly initiate signaling process resulting in generation of nerve impulse.

5.4.2 Amino Acid Sensing

Amino acid sensors include CaSR (calcium-sensing receptor of aromatic amino acid), GPRC6A (G protein receptor that senses the basic amino acids), GPR92 (G protein receptor that senses peptones) [470, 488]. Amino acid sensors are distributed beginning from the oropharyngeal cavity to the small intestine. For example, umami taste receptor (T1R1–T1R3) of the lingual epithelium has been implicated in sensing of L-amino acids such as phenylalanine, leucine, glutamate, and tryptophan. This sensing is associated with the secretion of gut peptides including cholecystokinin from enteroendocrine cells [488]. GI tract amino acid sensing is involved in the modulation of endocrine and lumencrine secretion as well

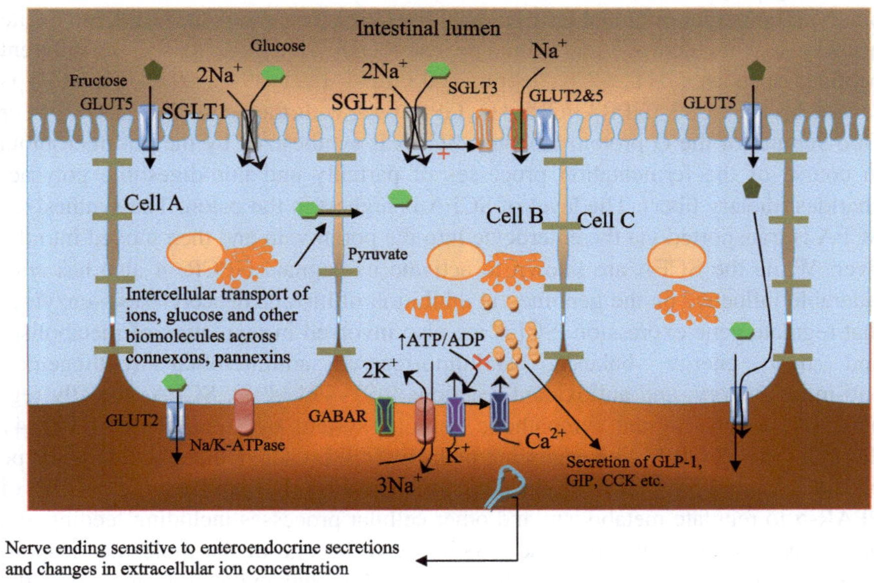

Fig. 5.35 GI tract epithelial cell nutrient receptor signaling. A and C—enterocytes; B—enteroendocrine cell. The absorptive enterocytes comprise more than 96% of all epithelial cells in the GI tract, while the enteroendocrine cells are just about 1–3% of all GI epithelial cells. Current data indicate that both types of cells contain hexose sensors, however, only the enteroendocrine cells have been largely implicated in the secretion of gut peptides upon activation by luminal nutrients [485–487]

as protein digestion and metabolism. It also plays certain role in ensuring host cell defense of the GI mucosa [489].

5.4.3 *Lipid (Fatty Acid) Sensing*

Free fatty acid sensors include nuclear receptors peroxisome proliferator-activated receptors (PPARs) and fatty acid-binding proteins (FABPs) as well as fatty acid transport protein (FFAR). These fatty acid sensors differentially sense LCFA (long-chain fatty acid), SCFA (short-chain fatty acid), and MCT (medium-chain triglyceride), etc., to produce physiological responses. For example, MCFA/LCFA (C6-C22) is sensed by GPR120/FFAR1 via the activation of Gq/11 that regulates the secretion of glucagon-like peptide-1 in the intestine and also facilitates glucose-stimulated insulin secretion from pancreatic β cells. However, recent studies have shown that GPR120 acts as a physiological receptor of ω3 fatty acids in macrophages and adipocytes, which are involved in mediating anti-inflammatory and insulin-sensitizing effects [490–495]. The GPR84 is a free fatty acid sensor that is activated by MCFA [496]. The cation channel TRP ankyrin 1 (TRPA1) is a fatty acid sensor that is expressed in sensory neurons and gut tissues and is also involved in sensing polyunsaturated fatty acids (≥ 18 carbon atoms and three unsaturated bonds). The polyunsaturated fatty acids of the GI tract can activate TRPA1 to excite primary sensory neurons and stimulate enteroendocrine cells to secrete different peptides such as 5-HT, CCK, leptin, and GLP-1 [491–494]. SCFA (C2–C4) is sensed by FFAR2/3 (GPR43/GPR41), GPR109A with the activation of Gq/11- or Gi/o-subunit of the G protein receptor. SCFA is synthesized by the gut microbiota in course of the fermentation processes of partially and non-digestible polysaccharides (dietary fiber). The level of SCFA is highest in the colon. The synthesized SCFA is transported via the enterocyte into the portal vein and then moved into the liver. While the SCFA are known to activate its cognate GPCR, it also has considerable influence on the genome via inhibition of histone deacetylases—enzymes that regulate gene expression. SCFA are also involved in regulation of metabolism and host energy balance, antitumorigenic, antimicrobial (pathogenic), anti-inflammatory, and antioxidant reactions [491–494, 497]. SCFAs directly regulate the activity of the sympathetic nervous system by stimulation of GPR41 through Gβγ-PLCβ-MAPK signaling pathway [498]. Oleoylethanolamide is a type of endogenous lipid that is produced in the intestines. It stimulates the activity of PPAR-α to regulate metabolism and other cellular processes including feeding and body weight. Oleoylethanolamide is believed to be one of the molecules responsible for the feeling of satiety after meals. Oleoylethanolamide signal via the GPR119. The interaction with this receptor can trigger the release of hormones and neurotransmitters including GLP-1 [470, 499–502].

Bile acid sensors include G protein-coupled bile acid receptor 1 (GPBAR-1), also known as TGR5, G protein-coupled receptor 19 (GPCR19), or membrane-type receptor for bile acids (M-BAR) [503, 504]; vitamin D receptor, [505]; and farnesoid X receptor [506]. Vitamin D receptor and farnesoid X receptor are nuclear receptors. These receptors are expressed in endocrine glands (e.g., pancreatic β cells), adipocytes (e.g., brown adipose tissue), muscles, immune organs, spinal cord, and the enteric nervous system (e.g., enteric neurons, enteroendocrine cells). Thus, their activation will lead to, depending on the organ or cell involved, hormone secretion, increase in energy expenditure, etc. [504, 507]. The stimulation of TGR5, for instance, by bile acids influences energy homeostasis, intestinal motility, and immunomodulatory functions. The receptor is also involved in the pathogenesis of metabolic disorders including diabetes and obesity [508]. Bile acids via paracrine and endocrine mechanisms are involved in the regulation of cholesterol homeostasis, lipid, and carbohydrate metabolism [505]. The bile acid sensor and receptor, farnesoid X receptor, is required for the regulation of bile acid biosynthesis [505]. The mechanism for this process is based on the activation of farnesoid X receptor by primary bile acid (cholic and chenodeoxycholic acid)—which then mediates the feedback suppression by bile acids of cholesterol 7α-hydroxylase, the rate-limiting enzyme in bile acid biosynthesis from cholesterol [506]. The vitamin D receptor is a nuclear receptor responsible for the action of calcemic vitamin D hormone 1,25-dihydroxyvitamin D_3 that regulate gene expression. This receptor also acts as intestinal bile acid sensor [505]. Vitamin D receptor can also act as a receptor for the secondary bile acid lithocholic acid, which is hepatotoxic and potentially carcinogenic. The binding of this bile acid to this receptor-induced expression of CYP3A, a cytochrome P450 enzyme that detoxifies lithocholic acid in the liver and intestine [509]. Thus, disordered signaling of vitamin D receptor or excessive production of this secondary bile acid can predispose the individual to the development of liver and intestinal cancer [510–513].

5.5 Conclusion

GI signaling involves an array of multiple interactions at various levels between different cells of the GI system, mediated by several extracellular and intracellular signaling molecules. GI signaling represents a key way in which functions of the GI system are regulated. The study of receptors, referred to as "receptorology," is a wide growing field of science that has obviously brought a considerable level of help to addressing human maladies.

Recommended Readings

Original Articles

1. Bahlawane C, Schmitz M, Letellier E, Arumugam K, Nicot N, Nazarov PV, Haan S (2017) Insights into ligand stimulation effects on gastro-intestinal stromal tumors signaling. Cell Signal 29:138–149

2. Baumann K (2013) Cell signalling: Inside GPCR signaling. Nat Rev Mol Cell Biol 14:267
3. Teng L, Zhao J, Wang F, Ma L, Pei G (2010) A GPCR/secretase complex regulates β- and γ-secretase specificity for Aβ production and contributes to AD pathogenesis. Cell Res 20:138–153

Review Articles

1. Almén MS, Nordström KJV, Fredriksson R, Schiöth HB (2009) Mapping the human membrane proteome: a majority of the human membrane proteins can be classified according to function and evolutionary origin. BMC Biol 7:50
2. An SS, Liggett SB (2018) Taste and smell GPCRs in the lung: Evidence for a previously unrecognized widespread chemosensory system. Cell Signal 41: 82–88
3. Catterall WA (2012) Voltage-gated sodium channels at 60: structure, function and pathophysiology. J Physiol 590(Pt 11):2577–2589
4. Kobilka BK (2007) G protein coupled receptor structure and activation. Biochim Biophys Acta 1768(4):794–807
5. Lemmon MA, Schlessinger J (2010) Cell signaling by receptor tyrosine kinases. Cell 141(7):1117–1134
6. Liu JA, Ngan ES-W (2014) Hedgehog and notch signaling in enteric nervous system development. Neurosignals 22:1–13
7. Parsons SJ, Parsons JT (2004) Src family kinases, key regulators of signal transduction. Oncogene 23:7906–7909
8. Ramírez-Weber F-A, Kornberg TB (1999) Cytonemes: cellular processes that project to the principal signaling center in *Drosophila* imaginal discs. Cell 97 (5):599–607
9. Uings IJ, Farrow SN (2000) Cell receptors and cell signaling. Mol Pathol 53 (6):295–299
10. Weiss RA, Vogt PK (2011) 100 years of Rous sarcoma virus. J Exp Med 208 (12):2351–2355

Books

1. Bradshaw RA, Dennis EA (eds) (2009) Handbook of cell signaling. In: Cell Biology, 2nd edn. Academic Press, San Diego, USA
2. Gomperts BD, Kramer IM, Tatham PER (2009) Signal transduction, 2nd edn. Academic Press, San Diego, USA
3. Hancock JT (2016) Cell signaling, 4th edn. Oxford University Press, New York, USA
4. Lim W, Mayer B, Pawson T (2014) Cell signaling—principles and mechanisms. Garland Science (Taylor & Francis Group), New York, USA
5. Tsien RW, Barrett CF (2000–2013) A brief history of calcium channel discovery. Madame Curie Bioscience Database. Landes Bioscience, Texas, USA

Nobel Lectures

1. Eccles JC (1943) The ionic mechanism of postsynaptic inhibition. Nobel Lecture. https://www.nobelprize.org/nobel_prizes/medicine/laureates/1963/eccles-lecture.pdf
2. Lefkowitz RJ (2013) A brief history of G-protein coupled receptors (Nobel Lecture). Angew Chem Int Ed Engl 52(25):6366–78
3. Sutherland EW (1971) Studies on the mechanism of hormone action. Nobel Lecture. http://www.nobelprize.org/nobel_prizes/medicine/laureates/1971/sutherland-lecture.pdf

Websites for Review

1. Cell Signaling. https://www.cellsignal.com/
2. QIAGEN GeneGlobe. https://www.qiagen.com/ng/geneglobe/

Bibliography

1. Banghart MR, Volgraf M, Trauner D (2006) Engineering light-gated ion channels. Biochemistry 45(51):15129–15141
2. Chen M, Lin S, Hofestaedt R (2004) STCDB: signal transduction classification database. Nucleic Acids Res 32:D456–D458
3. Lim W, Mayer B, Pawson T (2014) Cell signaling—principles and mechnaisms. Garland Science, New York, USA
4. Hancock JT (2016) Cell signaling, 4th edn. Oxford University Press, New York, USA
5. Mooren FC (2016) Intercellular signaling. In: Mooren F (ed) Encyclopedia of exercise medicine in health and disease. Springer, Heidelberg, USA
6. Jobst EE, Enriori PJ, Cowley MA (2004) The electrophysiology of feeding circuits. Trends Endocrinol Metab 15(10):488–499
7. Nuche-Berenguer B, Jensen RT (2015) Gastrointestinal hormones/neurotransmitters and growth factors can activate P21 activated kinase 2 in pancreatic acinar cells by novel mechanisms. Biochim Biophys Acta 1853(10):2371–2382
8. Nezami BG, Srinivasan S (2010) Enteric nervous system in the small intestine: pathophysiology and clinical implications. Curr Gastroenterol Rep 12(5):358–365
9. Mace OJ, Tehan B, Marshall F (2015) Pharmacology and physiology of gastrointestinal enteroendocrine cells. Pharmacol Res Perspect 3(4):e00155
10. Cummings DE, Overduin J (2007) Gastrointestinal regulation of food intake. J Clin Invest 117(1):13–23
11. Psichas A, Reimann F, Gribble FM (2015) Gut chemosensing mechanisms. J Clin Invest 125 (3):908–917
12. Gribble FM, Reimann F (2016) Enteroendocrine cells: chemosensors in the intestinal epithelium. Annu Rev Physiol 78:277–299
13. Nøhr MK, Pedersen MH, Gille A, Egerod KL, Engelstoft MS, Husted AS et al (2013) GPR41/FFAR3 and GPR43/FFAR2 as cosensors for short-chain fatty acids in enteroendocrine cells versus FFAR3 in enteric neurons and FFAR2 in enteric leukocytes. Endocrinology 154(10):3552–3564
14. Hill SJ (2006) G-protein-coupled receptors: past, present and future. Br J Pharmacol 147 (Suppl 1):S27–S37

15. Boycott BB (1998) John Zachary Young. 18 March 1907–4 July 1997. Biogr Mem Fellows R Soc 44:487–509
16. Young JZ (1992) Sources of discovery in neuroscience. In: Worden FG, Swazey JP, Adelman G (eds) The neurosciences: paths of discovery, I. Birkhäuser, Boston, USA
17. Ling G, Gerard RW (1949) The normal membrane potential of frog sartorius fibers. J Cell Physiol 34(3):383–396
18. Ling G, Gerard RW (1949) The membrane potential and metabolism of muscle fibers. J cellular Physiol 34(3):413–438
19. Ling G, Gerard RW (1950) External potassium and the membrane potential of single muscle fibres. Nature 165:113–114
20. Edwards C (1983) Who invented the intracellular microelectrode? Trends Neurosci 6:44
21. Verkhratsky A, Parpura V (2014) History of electrophysiology and the patch clamp. Methods Mol Biol 1183:1–19
22. Rall JA (2014) Excitation-contraction coupling and the role of calcium in contraction and relaxation in the 1950s and 1960s. Mechanism of muscular contraction part of the series perspectives in physiology. Springer, New York
23. Cole KS (1979) Mostly membranes. Annu Rev Physiol 41:1–24
24. Huxley A (1996) Kenneth Stewart Cole: July 10, 1900–April 18, 1984. Biogr Mem Natl Acad Sci 70:25–45
25. Schwiening CJ (2012) A brief historical perspective: Hodgkin and Huxley. J Physiol 590(Pt 11):2571–2575
26. Cole KS, Moore JW (1960) Potassium ion current in the squid giant axon: dynamic characteristic. Biophys J 1:I
27. Curtis HJ, Cole KS (1942) Membrane resting and action potentials from the squid giant axon. J Cell Comp Physiol 19:135–144
28. Vandenberg JI, Waxman SG (2012) Hodgkin and Huxley and the basis for electrical signalling: a remarkable legacy still going strong. J Physiol 590(Pt 11):2569–2570
29. De Palma A, Pareti G (2011) Bernstein's long path to membrane theory: radical change and conservation in nineteenth-century German electrophysiology. J Hist Neurosci 20(4):306–337
30. Grundfest H (1965) Julius Bernstein, Ludimar Hermann and the discovery of the overshoot of the axon spike. Arch Ital Biol 103(3):483–490
31. Seyfarth E-A (2006) Julius Bernstein (1839–1917): pioneer neurobiologist and biophysicist. Biol Cybern 94(1):2–8
32. From J (2014) The discovery of the nerve impulse. A chapter in the history of physiology. Dan Medicinhist Arbog 42:81–98
33. Finkelstein G (2006) Emil du Bois-Reymond versus Ludimar Hermann. Comptes Rendus Biol 329(5–6):340–347
34. Hodgkin AL, Huxley AF, Katz B (1952) Measurement of current-voltage relations in the membrane of the giant axon of Loligo. J Physiol 116(4):424–448
35. Raju TN (1999) The Nobel chronicles. 1963: Sir Alan Lloyd Hodgkin (1914–98), Sir Andrew Fielding Huxley (b 1917), and Sir John Carew Eccles (1903–97). Lancet 354 (9174):263
36. Brock LG, Coombs JS, Eccles JC (I952) The recording of potentials from motoneurones with an intracellular electrode. J Physiol II7:43I–460
37. Ranjan R, Khazen G, Gambazzi L, Ramaswamy S, Hill SL, Schürmann F, Markram H (2011) Channelpedia: an integrative and interactive database for ion channels. Front Neuroinform 5:36
38. Hodgkin AL, Katz B (1949) The effect of sodium ions on the electrical activity of the giant axon of the squid. J Physiol 108:37–77
39. Raju TN (1999) The Nobel chronicles. 1971: Earl Wilbur Sutherland, Jr. (1915–74). Lancet 354(9182):961

40. Cao H, Yu F, Zhao Y, Zhang X, Tai J, Lee J et al (2014) Wearable multi-channel microelectrode membranes for elucidating electrophysiological phenotypes of injured myocardium. Integr Biol (Camb) 6(8):789–795
41. Serpe MJ, Zhang X (2007) The principles, development and application of microelectrodes for the in vivo determination of nitric oxide. In: Michael AC, Borland LM (eds) Electrochemical methods for neuroscience. CRC Press/Taylor & Francis, Boca Raton, Florida, USA
42. Wijdenes P, Ali H, Armstrong R, Zaidi W, Dalton C, Syed NI (2016) A novel bio-mimicking, planar nano-edge microelectrode enables enhanced long-term neural recording. Sci Rep 6:34553
43. Mullins LJ (1959) An analysis of conductance changes in squid axon. J Gen Physiol 42:1013
44. Furshpan EJ, Potter DD (1957) Mechanism of nerve-impulse transmission at a crayfish synapse. Nature 180:342–343
45. Furshpan EJ, Potter DD (1959) Transmission at the giant motor synapses of the crayfish. J Physiol 145(2):289–325
46. Narahashi T, Moore JW, Scott WR (1964) Tetrodotoxin blockage of sodium conductance increase in lobster giant axons. J Gen Physiol 47(5):965–974
47. Narahashi T (2008) Tetrodotoxin—A brief history. Proc Jpn Acad Ser B Phys Biol Sci 84 (5):147–154
48. Armstrong CM (1969) Inactivation of the potassium conductance and related phenomena caused by quaternary ammonium ion injection in squid axons. J Gen Physiol 54(5):553–575
49. Armstrong CM, Binstock L (1965) Anomalous rectification in the squid giant axon injected with tetraethylammonium chloride. J Gen Physiol 48:859–872
50. Armstrong CM (1971) Interaction of tetraethylammonium ion derivatives with the potassium channels of giant axons. J Gen Physiol 58:413–437
51. Binstock L, Lecar H (1969) Ammonium ion currents in the squid giant axon. J Gen Physiol 53(3):342–361
52. Hamill OP, Marty A, Neher E, Sakmann B, Sigworth FJ (1981) Improved patchclamp techniques for high-resolution current recording from cells and cell-free membrane patches. Pflugers Arch 391:85–100
53. Neher E, Sakmann B (1976) Single-channel currents recorded from membrane of denervated frog muscle fibres. Nature 260:799–802
54. Neher E, Sakmann B, Steinbach JH (1978) The extracellular patch clamp: a method for resolving currents through individual open channels in biological membranes. Pflugers Arch 375:219–228
55. Hille B (1975) Ionic selectivity, saturation, and block in sodium channels. A four-barrier model. J Gen Physiol 66:535–560
56. Geck P, Heinz E (1986) The Na-K-2Cl cotransport system. J Membr Biol 91(2):97–105
57. Kostyuk PG (1984) Metabolic control of ionic channels in the neuronal membrane. Neuroscience 13(4):983–989
58. Beale R, Dutton GR, Currie DN (1980) An ion flux assay of action potential sodium channels in neuron- and glia-enriched cultures of cells dissociated from rat cerebellum. Brain Res 183(1):241–246
59. Káradóttir R, Attwell D (2006) Combining patch-clamping of cells in brain slices with immunocytochemical labelling to define cell type and developmental stage. Nat Protoc 1 (4):1977–1986
60. Dunlop J, Bowlby M, Peri R, Vasilyev D, Arias R (2008) High-throughput electrophysiology: an emerging paradigm for ion-channel screening and physiology. Nat Rev Drug Discov 7:358–368
61. Aldea M, Jun K, Shin HS, Andrés-Mateos E, Solís-Garrido LM, Montiel C et al (2002) A perforated patch-clamp study of calcium currents and exocytosis in chromaffin cells of wild-type and alpha(1A) knockout mice. J Neurochem 81(5):911–921
62. Brüggemann A, Farre C, Haarmann C, Haythornthwaite A, Kreir M, Stoelzle S et al (2008) Planar patch clamp: advances in electrophysiology. Methods Mol Biol 491:165–176

63. Gardner P (1990) Patch clamp studies of lymphocyte activation. Annu Rev Immunol 8: 231–252
64. Palmer LG (1986) Patch-clamp technique in renal physiology. Am J Physiol 250(3 Pt 2): F379–F385
65. Long Y, Li Z (2012) Drug screening and drug safety evaluation by patch clamp technique. In: Kaneez FS (ed) patch clamp technique. InTech, Croatia
66. Molleman A (2003) Patch clamping: an introductory guide to patch clamp electrophysiology. John Wiley & Sons, New York
67. Vandenberg JI, Kuchel PW (2003) Nobel Prizes for magnetic resonance imaging and channel proteins. Med J Aust 179(11):611–613
68. Rothman PB (2016) Introduction of Peter Agre. J Clin Invest 126(12):4735–4741
69. Knepper MA, Nielsen S (2004) Peter Agre, 2003 Nobel Prize winner in chemistry. J Am Soc Nephrol 15(4):1093–1095
70. Benga G (2006) Water channel proteins: from their discovery in 1985 in Cluj-Napoca, Romania, to the 2003 Nobel Prize in Chemistry. Cell Mol Biol (Noisy-le-grand) 52(7):10–9
71. Benga G (2004) The first water channel protein (later called aquaporin 1) was first discovered in Cluj-Napoca. Romania. Rom J Physiol 41(1–2):3–20
72. Kuchel PW (2006) The story of the discovery of aquaporins: convergent evolution of ideas–but who got there first? Cell Mol Biol (Noisy-le-grand) 52(7):2–5
73. Agre P (2009) The 2009 Lindau Nobel Laureate meeting: Peter Agre, Chemistry 2003. J Vis Exp 34: pii: 1565
74. Cucuianu M (2006) The discovery by Gh. Benga of the first water channel protein in 1985 in Cluj-Napoca, Romania, A few years before P. Agre (2003 Nobel Prize in Chemistry). Rom J Intern Med 44(3):323–334
75. Benga G, Popescu O, Pop VI, Holmes RP (1986) p-(Chloromercuri)benzenesulfonate binding by membrane proteins and the inhibition of water transport in human erythrocytes. Biochemistry 25(7):1535–1538
76. Benga G, Popescu O, Borza V, Pop VI, Muresan A, Mocsy I et al (1986) Water permeability of human erythrocytes. Identification of membrane proteins involved in water transport. Eur J Cell Biol 41(2):252–262
77. Hatta S, Sakamoto J, Horio Y (2002) Ion channels and diseases. Med Electron Microsc 35 (3):117–126
78. Enkvetchakul D (2010) Genetic disorders of ion channels. Mo Med 107(4):270–275
79. Sanguinetti MC, Spector PS (1997) Potassium channelopathies. Neuropharmacology 36 (6):755–762
80. Rall TW, Sutherland EW (1958) Formation of a cyclic adenine ribonucleotide by tissue particles. J Biol Chem 232:1065–1076
81. Sutherland EW, Rall TW (1958) Fractionation and characterization of a cyclic adenine ribonucleotide formed by tissue particles. J Biol Chem 232:1077–1092
82. Exton JH, Robison GA, Sutherland EW, Park CR (1971) Studies on the role of adenosine 3′,5′-monophosphate in the hepatic actions of glucagon and catecholamines. J Biol Chem 246(20):6166–6177
83. Jefferson LS, Exton JH, Butcher RW, Sutherland EW, Park CR (1968) Role of adenosine 3′,5′-monophosphate in the effects of insulin and anti-insulin serum on liver metabolism. J Biol Chem 243(5):1031–1038
84. Butcher RW, Ho RJ, Meng HC, Sutherland EW (1965) Adenosine 3′,5′-monophosphate in biological materials. II. The measurement of adenosine 3′,5′-monophosphate in tissues and the role of the cyclic nucleotide in the lipolytic response of fat to epinephrine. J Biol Chem 240(11):4515–4523
85. Sutherland Earl W, Wosilait WD (1956) The relationship of epinephrine and glucagon to liver phosphorylase. I. Liver phosphorylase; preparation and properties. J Biol Chem 218:459–468
86. Blumenthal SA (2012) Earl Sutherland (1915–1974) and the discovery of cyclic AMP. Perspect Biol Med 55(2):236–249

87. Kresge N, Simoni RD, Hill RL (2005) Earl W. Sutherland's discovery of cyclic adenine monophosphate and the second messenger system. J Biol Chem 280:e39
88. Raju TN (1999) The Nobel chronicles. 1966: Francis Peyton Rous (1879–1970) and Charles Brenton Huggins (1901–97). Lancet 354(9177):520
89. Milligan G, Kostenis E (2006) Heterotrimeric G-proteins: a short history. Br J Pharmacol 147(Suppl 1):S46–S55
90. Bourne HR (2016) Alfred Gilman: intrepid, committed scientist. Proc Natl Acad Sci USA 113(13):3414–3416
91. Birnbaumer L (2007) The discovery of signal transduction by G proteins. A personal account and an overview of the initial findings and contributions that led to our present understanding. Biochim Biophys Acta 1768(4):756–771
92. Hofmann L, Palczewski K (2015) The G protein-coupled receptor rhodopsin: a historical perspective. Methods Mol Biol 1271:3–18
93. Costanzi S, Siegel J, Tikhonova IG, Jacobson KA (2009) Rhodopsin and the others: a historical perspective on structural studies of G protein-coupled receptors. Curr Pharm Des 15(35):3994–4002
94. Ovchinnikov YA (1982) Rhodopsin and bacteriorhodopsin: structure-function relationship. FEBS Lett 148:179–191
95. Ovchinnikov YA, Abdulaev NG, Feigina MY, Artamonov ID, Zolotarev AS, Kostina MB et al (1982) The complete amino acid sequence of visual rhodopsin. Bioorg Khim 8:1011–1014
96. Schertler GFX, Hargrave PA (1995) Projection structure of frog rhodopsin in two crystal forms. PNAS 192:11578–11582
97. Unger VM, Hargrave PA, Baldwin JM, Schertler GFX (1997) Arrangement of rhodopsin transmembrane alpha-helices. Nature 389:203–206
98. Palczewski K, Kumasaka T, Hori T, Behnke CA, Motoshima H, Fox BA et al (2000) Crystal structure of rhodopsin: a G protein-coupled receptor. Science 289(5480):739–745
99. Clark RB (2013) Profile of Brian K. Kobilka and Robert J. Lefkowitz, 2012 Nobel Laureates in chemistry. Proc Natl Acad Sci USA 110(14):5274–5275
100. Kobilka B, Jamal Azouz H (2014) Q&A: Brian Kobilka. Stuck on structure. Nature 514 (7522):S12–S13
101. Lefkowitz RJ (2004) Historical review: a brief history and personal retrospective of seven-transmembrane receptors. Trends Pharmacol Sci 25(8):413–422
102. Bourne HR (2006) G-proteins and GPCrs: from the beginning. Ernst Schering Found Symp Proc 2:1–21
103. Schlessinger J (2014) Kinases: legacy of the first two decades. Cold Spring Harb Perspect Biol 6(3). pii: a008912
104. Becsei-Kilborn E (2010) Scientific discovery and scientific reputation: the reception of Peyton Rous' discovery of the chicken sarcoma virus. J Hist Biol 43(1):111–157
105. Collett MS, Erikson RL (1978) Protein kinase activity associated with the avian sarcoma virus src gene product. Proc Natl Acad Sci USA 75:2021–2024
106. Eckhart W, Hutchinson MA, Hunter T (1979) An activity phosphorylating tyrosine in polyoma T antigen immunoprecipitates. Cell 18:925–933
107. Hunter T, Sefton BM (1980) Transforming gene product of Rous sarcoma virus phosphorylates tyrosine. Proc Natl Acad Sci USA 77:1311–1315
108. Hunter T (2015) Discovering the first tyrosine kinase. Proc Natl Acad Sci USA 112 (26):7877–7882
109. Stehelin D, Varmus HE, Bishop JM, Vogt PK (1976) DNA related to the transforming gene (s) of avian sarcoma viruses is present in normal avian DNA. Nature 260:170–173
110. Levinson AD, Oppermann H, Levintow L, Varmus HE, Bishop JM (1978) Evidence that the transforming gene of avian sarcoma virus encodes a protein kinase associated with a phosphoprotein. Cell 15:561–572
111. Cohen S (1962) Isolation of a mouse submaxillary gland protein accelerating incisor eruption and eyelid opening in the newborn animal. J Biol Chem 237:1555–1562

112. Cohen S, Carpenter G, King L Jr (1980) Epidermal growth factor-receptor-protein kinase interactions: co-purification of receptor and epidermal growth factor-enhanced phosphorylation activity. J Biol Chem 255:4834–4842
113. Cohen S (2008) Origins of growth factors: NGF and EGF. J Biol Chem 283(49):33793–33797
114. Galligan JJ (2002) Ligand-gated ion channels in the enteric nervous system. Neurogastroenterol Motil 14(6):611–623
115. Julius D, Nathans J (2012) Signaling by sensory receptors. Cold Spring Harb Perspect Biol 4 (1):a005991
116. Garcia-Villalba P, Jimenez-Lara AM, Aranda A (1996) Vitamin D interferes with transactivation of the growth hormone gene by thyroid hormone and retinoic acid. Mol Cell Biol 16(1):318–327
117. Weigel NL (1996) Steroid hormone receptors and their regulation by phosphorylation. Biochem J 319(Pt 3):657–667
118. Lazar MA, Chin WW (1990) Nuclear thyroid hormone receptors. J Clin Invest 86(6):1777–1782
119. Kumar R, Thompson EB (1999) The structure of the nuclear hormone receptors. Steroids 64 (5):310–319
120. Castillo AI, Sánchez-Martínez R, Moreno JL, Martínez-Iglesias OA, Palacios D, Aranda A (2004) A permissive retinoid X receptor/thyroid hormone receptor heterodimer allows stimulation of prolactin gene transcription by thyroid hormone and 9-cis-retinoic acid. Mol Cell Biol 24(2):502–513
121. Weigel NL, Zhang Y (1998) Ligand-independent activation of steroid hormone receptors. J Mol Med (Berl) 76(7):469–479
122. García-Gómez E, González-Pedrajo B, Camacho-Arroyo I (2013) Role of sex steroid hormones in bacterial-host interactions. Biomed Res Int 2013:928290
123. Chen TS, Doong ML, Chang FY, Lee SD, Wang PS (1995) Effects of sex steroid hormones on gastric emptying and gastrointestinal transit in rats. Am J Physiol 268(1 Pt 1):G171–G176
124. Black HE (1988) The effects of steroids upon the gastrointestinal tract. Toxicol Pathol 16 (2):213–222
125. Küntzer J, Backes C, Blum T, Gerasch A, Kaufmann M, Kohlbacher O, Lenhof HP (2007) BNDB—the Biochemical Network Database. BMC Bioinformatics 8:367
126. Hinard V, Britan A, Rougier JS, Bairoch A, Abriel H, Gaudet P (2016) ICEPO: the ion channel electrophysiology ontology. Database (Oxford) 2016: baw017
127. Ashburner M, Ball CA, Blake JA, Botstein D, Butler H, Cherry JM et al (2000) Gene ontology: tool for the unification of biology. The gene ontology consortium. Nat Genet 25 (1):25–29
128. Mpamhanga CP, Sharman JL, Harmar AJ (2012) How to use the IUPHAR receptor database to navigate pharmacological data. Methods Mol Biol 897:15–29
129. Gao J, Cui W, Sheng Y, Ruan J, Kurgan L (2016) PSIONplus: accurate sequence-based predictor of ion channels and their types. PLoS ONE 11(4):e0152964
130. Gallin WJ, Boutet PA (2011) VKCDB: voltage-gated K+ channel database updated and upgraded. Nucleic Acids Res 39:D362–D366
131. Donizelli M, Djite MA, Le Novère N (2006) LGICdb: a manually curated sequence database after the genomes. Nucleic Acids Res 34:D267–D269
132. Michel K, Michaelis M, Mazzuoli G, Mueller K, Vanden Berghe P, Schemann M (2011) Fast calcium and voltage-sensitive dye imaging in enteric neurones reveal calcium peaks associated with single action potential discharge. J Physiol 589(Pt 24):5941–5947
133. Vignali S, Peter N, Ceyhan G, Demir IE, Zeller F, Senseman D et al (2010) Recordings from human myenteric neurons using voltage-sensitive dyes. J Neurosci Methods 192(2):240–248
134. Dascal N (2001) Ion-channel regulation by G proteins. Trends Endocrinol Metab 12(9):391–398

135. George AL Jr (2005) Inherited disorders of voltage-gated sodium channels. J Clin Invest 115 (8):1990–1999
136. Ren D (2011) Sodium leak channels in neuronal excitability and rhythmic behaviors. Neuron 72(6):899–911
137. Swayne LA, Mezghrani A, Varrault A, Chemin J, Bertrand G, Dalle S et al (2009) The NALCN ion channel is activated by M3 muscarinic receptors in a pancreatic β-cell line. EMBO Rep 10:873–880
138. Galligan JJ (2009) Cannabinoid signalling in the enteric nervous system. Neurogastroenterol Motil 21(9):899–902
139. Gadsby DC (2009) Ion channels versus ion pumps: the principal difference, in principle. Nat Rev Mol Cell Biol 10:344–352
140. Cooper EC, Jan LY (1999) Ion channel genes and human neurological disease: recent progress, prospects, and challenges. Proc Natl Acad Sci USA 96(9):4759–4766
141. Albuquerque EX, Pereira EFR, Alkondon M, Rogers SW (2009) Mammalian nicotinic acetylcholine receptors: from structure to function. Physiol Rev 89(1):73–120
142. McGehee DS, Role LW (1996) Presynaptic ionotropic receptors. Curr Opin Neurobiol 6 (3):342–349
143. Draguhn A, Axmacher N, Kolbaev S (2008) Presynaptic ionotropic GABA receptors. Results Probl Cell Differ 44:69–85
144. Bardoni R, Takazawa T, Tong CK, Choudhury P, Scherrer G, Macdermott AB (2013) Pre- and postsynaptic inhibitory control in the spinal cord dorsal horn. Ann NY Acad Sci 1279:90–96
145. Forostyak O, Butenko O, Anderova M, Forostyak S, Sykova E, Verkhratsky A, Dayanithi G (2016) Specific profiles of ion channels and ionotropic receptors define adipose- and bone marrow derived stromal cells. Stem Cell Res 16(3):622–634
146. Parekh AB, Putney JW Jr (2005) Store-operated calcium channels. Physiol Rev 85(2):757–810
147. MacDermott AB, Role LW, Siegelbaum SA (1999) Presynaptic ionotropic receptors and the control of transmitter release. Annu Rev Neurosci 22:443–485
148. Bowie D (2008) Ionotropic glutamate receptors & CNS disorders. CNS Neurol Disord: Drug Targets 7(2):129–143
149. Nys M, Wijckmans E, Farinha A, Yoluk Ö, Andersson M, Brams M et al (2016) Allosteric binding site in a Cys-loop receptor ligand-binding domain unveiled in the crystal structure of ELIC in complex with chlorpromazine. Proc Natl Acad Sci USA 113(43):E6696–E6703
150. Lynagh T, Lynch JW (2012) Molecular mechanisms of Cys-loop ion channel receptor modulation by ivermectin. Front Mol Neurosci 5:60
151. Kozuska JL, Paulsen IM (2012) The Cys-loop pentameric ligand-gated ion channel receptors: 50 years on. Can J Physiol Pharmacol 90(6):771–782
152. Thompson AJ, Lester HA, Lummis SC (2010) The structural basis of function in Cys-loop receptors. Q Rev Biophys 43(4):449–499
153. Connolly CN, Wafford KA (2004) The Cys-loop superfamily of ligand-gated ion channels: the impact of receptor structure on function. Biochem Soc Trans 32(Pt3):529–534
154. Kaneko S, Akaike A, Satoh M (1999) Receptor-mediated modulation of voltage-dependent Ca2+ channels via heterotrimeric G-proteins in neurons. Jpn J Pharmacol 81(4):324–331
155. Herlitze S, Garcia DE, Mackie K, Hille B, Scheuer T, Catterall WA (1996) Modulation of Ca2+ channels by G-protein beta gamma subunits. Nature 380(6571):258–262
156. Karim F, Bhave G, Gereau RW 4th (2001) Metabotropic glutamate receptors on peripheral sensory neuron terminals as targets for the development of novel analgesics. Mol Psychiatry 6(6):615–617
157. Lin W, Kinnamon SC (1999) Physiological evidence for ionotropic and metabotropic glutamate receptors in rat taste cells. J Neurophysiol 82(5):2061–2069
158. Niswender CM, Conn PJ (2010) Metabotropic glutamate receptors: physiology, pharmacology, and disease. Annu Rev Pharmacol Toxicol 50:295–322

159. Ledoux J, Werner ME, Brayden JE, Nelson MT (2006) Calcium-activated potassium channels and the regulation of vascular tone. Physiology 21(1):69–78
160. Baker EH (2000) Ion channels and the control of blood pressure. Br J Clin Pharmacol 49 (3):185–198
161. Lane BR, Strassmaier T, Brady JD, Karpen JW (2006) The pharmacology of cyclic nucleotide-gated channels: emerging from the darkness. Curr Pharm Des 12(28):3597–3613
162. Wickman KD, Clapham DE (1995) G-protein regulation of ion channels. Curr Opin Neurobiol 5(3):278–285
163. Hibino H, Inanobe A, Furutani K, Murakami S, Findlay I, Kurachi Y (2010) Inwardly rectifying potassium channels: their structure, function, and physiological roles. Physiol Rev 90(1):291–366
164. de la Cruz L, Puente EI, Reyes-Vaca A, Arenas I, Garduño J, Bravo-Martínez J, Garcia DE (2016) PIP2 in pancreatic β-cells regulates voltage-gated calcium channels by a voltage-independent pathway. Am J Physiol Cell Physiol 311(4):C630–C640
165. Suh B-C, Hille B (2008) PIP2 is a necessary cofactor for ion channel function: How and why? Annu Rev Biophys 37:175–195
166. Rodríguez-Menchaca AA, Adney SK, Zhou L, Logothetis DE (2012) Dual regulation of voltage-sensitive ion channels by PIP2. Front Pharmacol 3:170
167. Pedersen PL (2005) Transport ATPases: structure, motors, mechanism and medicine: a brief overview. J Bioenerg Biomembr 37(6):349–357
168. Pedersen PL (2007) Transport ATPases into the year 2008: a brief overview related to types, structures, functions and roles in health and disease. J Bioenerg Biomembr 39(5–6):349–355
169. Lopina OD (2000) Na+/K+ -ATPase: structure, mechanism, and regulation. Membr Cell Biol 13(6):721–744
170. Alexander SPH, Kelly E, Marrion N, Peters JA, Benson HE, Faccenda E et al (2015) The concise guide to pharmacology 2015/16: transporters. Br J Pharmacol 172:6110–6202
171. Rappas M, Niwa H, Zhang X (2004) Mechanisms of ATPases—a multi-disciplinary approach. Curr Protein Pep Sci 5(2):89–105
172. Axelsen KB, Palmgren MG (1998) Evolution of substrate specificities in the P-type ATPase superfamily. J Mol Evol 46(1):84–101
173. Futai M, Sun-Wada GH, Wada Y (2004) Proton pumping ATPases and diverse inside-acidic compartments. Yakugaku Zasshi 124(5):243–260
174. Kaupp UB, Seifert R (2002) Cyclic nucleotide-gated ion channels. Physiol Rev 82(3):769–824
175. Jackson MB (1995) Presynaptic excitability. Int Rev Neurobiol 38:201–251
176. Kim J-B (2014) Channelopathies. Korean. J Pediatr 57(1):1–18
177. Lehmann-Horn F, Jurkat-Rott K (eds) (2000) Channelopathies. Elsevier, New York
178. Wilders R (2015) A note on the prevalence of cardiac ion channelopathies in the sudden infant death syndrome. Europace 17(11):1739
179. Abriel H, Zaklyazminskaya EV (2013) Cardiac channelopathies: genetic and molecular mechanisms. Gene 517(1):1–11
180. Horga A, Raja Rayan DL, Matthews E, Sud R, Fialho D, Durran SC et al (2013) Prevalence study of genetically defined skeletal muscle channelopathies in England. Neurology 80 (16):1472–1475
181. Davies NP, Hanna MG (1999) Neurological channelopathies: diagnosis and therapy in the new millennium. Ann Med 31(6):406–420
182. Rolim AL, Lindsey SC, Kunii IS, Fujikawa AM, Soares FA, Chiamolera MI et al (2010) Ion channelopathies in endocrinology: recent genetic findings and pathophysiological insights. Arq Bras Endocrinol Metabol 54(8):673–681
183. Tester DJ, Ackerman MJ (2011) Genetic testing for potentially lethal, highly treatable inherited cardiomyopathies/channelopathies in clinical practice. Circulation 123(9):1021–1037
184. Lehnart SE, Ackerman MJ, Benson DW Jr, Brugada R, Clancy CE, Donahue JK et al (2007) Inherited arrhythmias: a National Heart, Lung, and Blood Institute and Office of Rare

Diseases workshop consensus report about the diagnosis, phenotyping, molecular mechanisms, and therapeutic approaches for primary cardiomyopathies of gene mutations affecting ion channel function. Circulation 116(20):2325–2345

185. Ptáĉek LJ, Tawil R, Griggs RC, Meola G, McManis P, Barohn RJ et al (1994) Sodium channel mutations in acetazolamide-responsive myotonia congenita, paramyotonia congenita, and hyperkalemic periodic paralysis. Neurology 44(8):1500–1503
186. Ptácek LJ, George AL Jr, Griggs RC, Tawil R, Kallen RG, Barchi RL et al (1991) Identification of a mutation in the gene causing hyperkalemic periodic paralysis. Cell 67:1021–1027
187. Ptacek LJ, Trimmer JS, Agnew WS, Roberts JW, Petajan JH, Leppert M (1991) Paramyotonia congenita and hyperkalemic periodic paralysis map to the same sodium channel gene locus. Am J Hum Genet 49:851–854
188. Pi Y, Goldenthal MJ, Marín-García J (2007) Mitochondrial channelopathies in aging. J Mol Med (Berl) 85(9):937–951
189. Camerino DC, Desaphy JF, Tricarico D, Pierno S, Liantonio A (2008) Therapeutic approaches to ion channel diseases. Adv Genet 64:81–145
190. Zhou P, Wang J (2010) Genetic testing for channelopathies, more than ten years progress and remaining challenges. J Cardiovasc Dis Res 1(2):47–49
191. Schwartz PJ, Crotti L, Insolia R (2012) Long QT syndrome: from genetics to management. Circ Arrhythm Electrophysiol 5(4):868–877
192. Dalemans W, Barbry P, Champigny G, Jallat S, Dott K, Dreyer D et al (1991) Altered chloride ion channel kinetics associated with the delta F508 cystic fibrosis mutation. Nature 354(6354):526–528
193. Hübner CA, Jentsch TJ (2002) Ion channel diseases. Hum Mol Genet 11(20):2435–2445
194. Ackerman MJ, Marcou CA, Tester DJ (2013) Personalized medicine: genetic diagnosis for inherited cardiomyopathies/channelopathies. Rev Esp Cardiol (Engl Ed) 66(4):298–307
195. Ackerman MJ, Priori SG, Willems S, Berul C, Brugada R, Calkins H et al (2011) HRS/EHRA expert consensus statement on the state of genetic testing for the channelopathies and cardiomyopathies. Europace 13(8):1077–1109
196. Cleland JC, Griggs RC (2008) Treatment of neuromuscular channelopathies: current concepts and future prospects. Neurother: J Amer Soc Exp NeuroTher 5:607–612
197. Cleland JC, Griggs RC (2008) Treatment of neuromuscular channelopathies: current concepts and future prospects. Neurotherapeutics 5(4):607–612
198. Beck RW, Cleary PA, Anderson MM Jr, Keltner JL, Shults WT, Kaufman DI et al (1992) A randomized, controlled trial of corticosteroids in the treatment of acute optic neuritis. The Optic Neuritis Study Group. N Engl J Med 326(9):581–588
199. Imbrici P, Liantonio A, Camerino GM, De Bellis M, Camerino C, Mele A et al (2016) Therapeutic approaches to genetic ion channelopathies and perspectives in drug discovery. Front Pharmacol 7:121
200. Premont RT, Gainetdinov RR (2007) Physiological roles of G protein-coupled receptor kinases and arrestins. Annu Rev Physiol 69:511–534
201. Elefsinioti AL, Bagos PG, Spyropoulos IC, Hamodrakas SJ (2004) A database for G proteins and their interaction with GPCRs. BMC Bioinformatics 5:208
202. Fredriksson R, Lagerström MC, Lundin LG, Schiöth HB (2003) The G-protein-coupled receptors in the human genome form five main families. Phylogenetic analysis, paralogon groups, and fingerprints. Mol Pharmacol 63:1256–1272
203. Horn F, Bettler E, Oliveira L, Campagne F, Cohen FE, Vriend G (2003) GPCRDB information system for G protein-coupled receptors. Nucleic Acids Res 31(1):294–297
204. Kristiansen K (2004) Molecular mechanisms of ligand binding, signaling, and regulation within the superfamily of G-protein-coupled receptors: molecular modelingand mutagenesis approaches to receptor structure and function. Pharmacol Ther 103:21–80
205. Rhee SG, Choi KD (1992) Multiple forms of phospholipase C isozymes and their activation mechanisms. Adv Second Messenger Phosphoprotein Res 26:35–61

206. Page CP, Costello J (2000) Theophylline and new generation phosphodiesterase inhibitors in the treatment of asthma. In: Giembycz MA, O'Connor BJ (eds) Asthma: epidemiology, anti-inflammatory therapy and future trends part of the series respiratory pharmacology and pharmacotherapy. Birkhäuser, Basel
207. Li X, Mumby SM, Greenwood A, Jope RS (1995) Pertussis toxin-sensitive G protein alpha-subunits: production of monoclonal antibodies and detection of differential increases on differentiation of PC12 and LA-N-5 cells. J Neurochem 64(3):1107–1117
208. Gunther EC, von Bartheld CS, Goodman LJ, Johnson JE, Bothwell M (2000) The G-protein inhibitor, pertussis toxin, inhibits the secretion of brain-derived neurotrophic factor. Neuroscience 100(3):569–579
209. Mangmool S, Kurose H (2011) Gi/o protein-dependent and -independent actions of Pertussis Toxin (PTX). Toxins (Basel) 3(7):884–899
210. Akram F, Nasiruddin M, Ahmad Z, Khan RA (2012) Doxofylline and theophylline: a comparative clinical study. J Clin Diagn Res 6(10):1681–1684
211. Feneck R (2007) Phosphodiesterase inhibitors and the cardiovascular system. Contin Educ Anaesth Crit Care Pain 7(6):203–207
212. Banner KH, Page CP (1995) Theophylline and selective phosphodiesterase inhibitors as anti-inflammatory drugs in the treatment of bronchial asthma. Eur Respir J 8:996–1000
213. Institute of Medicine (US) (2001) Committee on military nutrition research. Caffeine for the sustainment of mental task performance: formulations for military operations. National Academies Press (US), Washington DC, USA
214. Jacobson KA, Gao Z-G (2006) Adenosine receptors as therapeutic targets. Nat Rev Drug Discov 5(3):247–264
215. Smit HJ (2011) Theobromine and the pharmacology of cocoa. Handb Exp Pharmacol 200:201–234
216. Smit HJ, Gaffan EA, Rogers PJ (2004) Methylxanthines are the psycho-pharmacologically active constituents of chocolate. Psychopharmacology 176(3–4):412–419
217. Shively CA, Tarka SM Jr (1984) Methylxanthine composition and consumption patterns of cocoa and chocolate products. Prog Clin Biol Res 158:149–178
218. Cheng Z, Garvin D, Paguio A, Stecha P, Wood K, Fan F (2010) Luciferase reporter assay system for deciphering GPCR pathways. Curr Chem Genomics 4:84–91
219. Schou KB, Pedersen LB, Christensen ST (2015) Ins and outs of GPCR signaling in primary cilia. EMBO Rep 16:1099–1113
220. McCudden CR, Hains MD, Kimple RJ, Siderovski DP, Willard FS (2005) G-protein signaling: back to the future. Cell Mol Life Sci 62(5):551–577
221. Tuteja N (2009) Signaling through G protein coupled receptors. Plant Signal Behav 4 (10):942–947
222. Alberts B, Johnson A, Lewis J, Raff M, Bray D, Hopkin K et al (2004) Essential cell biology, 2nd edn. Garland Science, New York
223. Hanoune J, Defer N (2001) Regulation and role of adenylyl cyclase isoforms. Annu Rev Pharmacol Toxicol 41:145–174
224. Meinkoth JL, Alberts AS, Went W, Fantozzi D, Taylor SS, Hagiwara M et al (1993) Signal transduction through the cAMP-dependent protein kinase. Mol Cell Biochem 127–128:179–186
225. Walsh DA, Van Patten SM (1994) Multiple pathway signal transduction by the cAMP-dependent protein kinase. FASEB J 8(15):1227–1236
226. Park G, Servin JA, Turner GE, Altamirano L, Colot HV, Collopy P et al (2011) Global analysis of serine-threonine protein kinase genes in Neurospora crassa. Eukaryot Cell 10 (11):1553–1564
227. Cohen P (2002) Protein kinases—the major drug targets of the twenty-first century? Nat Rev Drug Discov 1(4):309–315
228. Roscioni SS, Elzinga CR, Schmidt M (2008) Epac: effectors and biological functions. Naunyn Schmiedebergs Arch Pharmacol 377(4–6):345–357

229. Ster J, De Bock F, Guérineau NC, Janossy A, Barrère-Lemaire S, Bos JL et al (2006) Exchange protein activated by cAMP (Epac) mediates cAMP activation of p38 MAPK and modulation of Ca2+ -dependent K+ channels in cerebellar neurons. PNAS 104(7):2519–2524
230. Bos JL (2006) Epac proteins: multi-purpose cAMP targets. Trends in Biochem Sci 31 (12):680–686
231. Gloerich M, Bos JL (2010) Epac: defining a new mechanism for cAMP action. Annu Rev Pharmacol Toxicol 50:355–375
232. Jensen JK, Dolmer K, Schar C, Gettins PG (2009) Receptor-associated protein (RAP) has two high-affinity binding sites for the low-density lipoprotein receptor-related protein (LRP): consequences for the chaperone functions of RAP. Biochem J 421(2):273–282
233. Okuno Y, Yang J, Taneishi K, Yabuuchi H, Tsujimoto G (2006) GLIDA: GPCR-ligand database for chemical genomic drug discovery. Nucleic Acids Res 34:D673–D677
234. Worth' CL, Kreuchwig A, Kleinau G, Krause G (2011) GPCR-SSFE: a comprehensive database of G-protein-coupled receptor template predictions and homology models. BMC Bioinform 12:185
235. Gumbleton M, Kerr WG (2013) Role of inositol phospholipid signaling in natural killer cell biology. Front Immunol 4:47
236. Kerr WG, Colucci F (2011) Inositol phospholipid signaling and the biology of natural killer cells. J Innate Immun 3(3):249–257
237. Asaoka Y, Nakamura S-I, Yoshida K, Nishizuka Y (1992) Protein kinase C, calcium and phospholipid degradation. Trends in Biochem Sci 17(10):414–417
238. Cejas PJ, Carlson LM, Zhang J, Padmanabhan S, Kolonias D, Lindner I et al (2005) Protein kinase C βII plays an essential role in dendritic cell differentiation and autoregulates its own expression. J Biol Chem 280:28412–28423
239. Wu SC, Solaro RJ (2007) Protein kinase C ζ: a novel regulator of both phosphorylation and de-phosphorylation of cardiac sarcomeric proteins. J Biol Chem 282(42):30691–30698
240. Shin OS, Behera AK, Bronson RT, Hu LT (2007) Role of novel protein kinase C (PKC) isoforms in lyme arthritis. Cell Microbiol 9(8):1987–1996
241. Xiao H, Liu M (2013) Atypical protein kinase C in cell motility. Cell Mol Life Sci 70 (17):3057–3066
242. Watanabe G, Saito Y, Madaule P, Ishizaki T, Fujisawa K, Morii N et al (1996) Protein kinase N (PKN) and PKN-related protein rhophilin as targets of small GTPase Rho. Science 271(5249):645–648
243. Browning DD (2008) Protein kinase G as a therapeutic target for the treatment of metastatic colorectal cancer. Expert Opin Ther Targets 12(3):367–376
244. Rozengurt E (2011) Protein kinase D signaling: multiple biological functions in health and disease. Physiol (Bethesda) 26(1):23–33
245. Hanks SK, Quinn AM (1991) Protein kinase catalytic domain sequence database: identification of conserved features of primary structure and classification of family members. Methods Enzymol 200:38–62
246. Shigeki H, Hidehiko I, Chie M, Miki H, Hirofumi I, Natsuki M (2008) Membrane-anchored growth factors, the epidermal growth factor family: beyond receptor ligands. Cancer Sci 99 (2):214–220
247. Wood JD, Kirchgessner A (2004) Slow excitatory metabotropic signal transmission in the enteric nervous system. Neurogastroenterol Motil 16(Suppl 1):71–80
248. Sternweis PC, Smrcka AV (1992) Regulation of phospholipase C by G proteins. Trends Biochem Sci 17(12):502–506
249. Exton JH (1996) Regulation of phosphoinositide phospholipases by hormones, neurotransmitters, and other agonists linked to G proteins. Annu Rev Pharmacol Toxicol 36:481–509
250. Feske S (2010) CRAC channelopathies. Pflügers Archiv Eur J Physiol 460(2):417–435
251. Palty R, Raveh A, Kaminsky I, Meller R, Reuveny E (2012) SARAF inactivates the store operated calcium entry machinery to prevent excess calcium refilling. Cell 149:425–438

252. Wilson CH, Ali ES, Scrimgeour N, Martin AM, Hua J, Tallis GA et al (2015) Steatosis inhibits liver cell store-operated Ca(2)(+) entry and reduces ER Ca(2)(+) through a protein kinase C-dependent mechanism. Biochem J 466:379–390
253. Vallejo M (2009) PACAP signaling to DREAM: a cAMP-dependent pathway that regulates cortical astrogliogenesis. Mol Neurobiol 39(2):90–100
254. Vogalis F, Harvey JR, Neylon CB, Furness JB (2002) Regulation of K+ channels underlying the slow afterhyperpolarization in enteric afterhyperpolarization-generating myenteric neurons: role of calcium and phosphorylation. Clin Exp Pharmacol Physiol 29(10):935–943
255. Boesmans W, Ameloot K, van den Abbeel V, Tack J, Vanden Berghe P (2009) Cannabinoid receptor 1 signalling dampens activity and mitochondrial transport in networks of enteric neurones. Neurogastroenterol Motil 21(9):958–e77
256. Putney JW (2010) Pharmacology of store-operated calcium channels. Mol Interv 10(4):209–218
257. Oritani K, Kincade PW (1996) Identification of stromal cell products that interact with pre-B cells. J Cell Biol 134:771–782
258. Williams RT, Manji SS, Parker NJ, Hancock MS, Van Stekelenburg L, Eid JP et al (2001) Identification and characterization of the STIM (stromal interaction molecule) gene family: coding for a novel class of transmembrane proteins. Biochem J 357:673–685
259. Cai X (2007) Molecular evolution and functional divergence of the Ca2+ sensor protein in store-operated Ca2+ entry: stromal interaction molecule. PLoS ONE 2:e609
260. FreWilliams RT, Senior PV, Van Stekelenburg L, Layton JE, Smith PJ, Dziadek MA (2002) Stromal interaction molecule 1 (STIM1), a transmembrane protein with growth suppressor activity, contains an extracellular SAM domain modified by N-linked glycosylation. Biochim Biophys Acta 1596(1):131–137
261. Soboloff J, Rothberg BS, Madesh M, Gill DL (2012) STIM proteins: dynamic calcium signal transducers. Nat Rev Mol Cell Biol 13(9):549–565
262. Beech DJ (2012) Orai1 calcium channels in the vasculature. Pflugers Arch 463(5):635–647
263. Taylor CW, Konieczny V (2016) IP3 receptors: take four IP3 to open. Sci Signal 9(422):pe1
264. Alzayady KJ, Wang L, Chandrasekhar R, Wagner LE, Van Petegem F, Yule DI (2016) Defining the stoichiometry of inositol 1,4,5-trisphosphate binding required to initiate Ca2 + release. Sci Signal 9(422):ra35
265. Chandrasekhar R, Alzayady KJ, Wagner LE 2nd, Yule DI (2016) Unique Regulatory Properties of heterotetrameric inositol 1,4,5-trisphosphate receptors revealed by studying concatenated receptor constructs. J Biol Chem 291(10):4846–4860
266. Schlossmann J, Ammendola A, Ashman K, Zong X, Huber A, Neubauer G et al (2000) Regulation of intracellular calcium by a signalling complex of IRAG, IP3 receptor and cGMP kinase Ibeta. Nature 404(6774):197–201
267. Berghe VP (2004) Fluorescent molecules as tools to study Ca2+ signaling, mitochondrial dynamics and synaptic function in enteric neurons. Verh K Acad Geneeskd Belg 66(5–6):407–425
268. Smith JB (1996) Calcium homeostasis in smooth muscle cells. New Horiz 4(1):2–18
269. Uzhachenko R, Shanker A, Yarbrough WG, Ivanova AV (2015) Mitochondria, calcium, and tumor suppressor Fus1: at the crossroad of cancer, inflammation, and autoimmunity. Oncotarget 6(25):20754–20772
270. Walsh C, Barrow S, Voronina S, Chvanov M, Petersen OH, Tepikin A (2009) Modulation of calcium signalling by mitochondria. Biochim Biophys Acta Bioenergetics 1787(11):1374–1382
271. Santo-Domingo J, Demaurex N (2010) Calcium uptake mechanisms of mitochondria. Biochim Biophys Acta Bioenergetics 1797(6–7):907–912
272. Pang ZP, Cao P, Xu W, Südhof TC (2010) Calmodulin controls synaptic strength via presynaptic activation of calmodulin kinase II. J Neurosci 30(11):4132–4142
273. Wilmann M, Gautel M, Mayans O (2000) Activation of calcium/calmodulin regulated kinases. Cell Mol Biol (Noisy-le-grand) 46(5):883–94

274. Pfisterer SG, Mauthe M, Codogno P, Proikas-Cezanne T (2011) Ca2+/calmodulin-dependent kinase (CaMK) signaling via CaMKI and AMP-activated protein kinase contributes to the regulation of WIPI-1 at the onset of autophagy. Mol Pharmacol 80 (6):1066–1075
275. Nairn AC, Picciotto MR (1994) Calcium/calmodulin-dependent protein kinases. Semin Cancer Biol 5(4):295–303
276. Mohanta TK, Kumar P, Bae H (2017) Genomics and evolutionary aspect of calcium signaling event in calmodulin and calmodulin-like proteins in plants. BMC Plant Biol 17:38
277. Zhu X, Dunand C, Snedden W, Galaud J-P (2015) CaM and CML emergence in the green lineage. Trends Plant Sci 20(8):483–489
278. Means AR, Cruzalegui F, Lemagueresse B, Needleman DS, Slaughter GR, Ono T (1991) A novel Ca2+/calmodulin-dependent protein kinase and a male germ cell-specific calmodulin-binding protein are derived from the same gene. Mol Cell Biol 11(8):3960–3971
279. Tsunoda T, Yamakawa M, Takahashi T (1999) Differential expression of Ca2+ -binding proteins on follicular dendritic cells in non-neoplastic and neoplastic lymphoid follicles. Am J Pathol 155(3):805–814
280. Armstrong DL (1989) Calcium channel regulation by calcineurin, a Ca2+ -activated phosphatase in mammalian brain. Trends Neurosci 12(3):117–122
281. Lim HW, De Windt LJ, Steinberg L, Taigen T, Witt SA, Kimball TR, Molkentin JD (2000) Calcineurin expression, activation, and function in cardiac pressure-overload hypertrophy. Circulation 101:2431–2437
282. Song YH, Cai GY, Xiao YF, Wang YP, Yuan BS, Xia YY et al (2017) Efficacy and safety of calcineurin inhibitor treatment for IgA nephropathy: a meta-analysis. BMC Nephrol 18(1):61
283. Naesens M, Kuypers DRJ, Sarwal M (2009) Calcineurin inhibitor nephrotoxicity. CJASN 4 (2):481–508
284. Qiu TT, Zhang C, Zhao HW, Zhou JW (2017) Calcineurin inhibitors versus cyclophosphamide for idiopathic membranous nephropathy: a systematic review and meta-analysis of 21 clinical trials. Autoimmun Rev 16(2):136–145
285. Crabtree GR (2001) Calcium, calcineurin, and the control of transcription. J Biol Chem 276:2313–2316
286. Wang C-LA (2008) Caldesmon and the regulation of cytoskeletal functions. Adv Exp Med Biol 644:250–272
287. Yáñez M, Gil-Longo J, Campos-Toimil M (2012) Calcium binding proteins. Adv Exp Med Biol 740:461–482
288. Weinman S (1991) Calcium-binding proteins: an overview. J Biol Buccale 19(1):90–98
289. Heizmann CW (1992) Calcium-binding proteins: basic concepts and clinical implications. Gen Physiol Biophys 11(5):411–425
290. Ozcan L, de Souza JC, Harari AA, Backs J, Olson EN, Tabas I (2013) Activation of calcium/calmodulin-dependent protein kinase II in obesity mediates suppression of hepatic insulin signaling. Cell Metab 18(6):803–815
291. Lu D, Chen J, Hai T (2007) The regulation of ATF3 gene expression by mitogen-activated protein kinases. Biochem J 401(Pt 2):559–567
292. Kim HR, Appel S, Vetterkind S, Gangopadhyay SS, Morgan KG (2008) Smooth muscle signalling pathways in health and disease. J Cell Mol Med 12(6a):2165–2180
293. Yang Z (2002) Small GTPases—versatile signaling switches in plants. Plant Cell 14(Suppl): s375–s388
294. Wennerberg K, Rossman KL, Der CJ (2005) The Ras superfamily at a glance. J Cell Sci 118 (Pt 5):843–846
295. Goitre L, Trapani E, Trabalzini L, Retta SF (2014) The Ras superfamily of small GTPases: the unlocked secrets. Methods Mol Biol 1120:1–18
296. Manser E (2002) Small GTPases take the stage. Dev Cell 3(3):323–328
297. Ivanova-Nikolova TT, Breitwieser GE (1997) Effector contributions to G beta gamma-mediated signaling as revealed by muscarinic potassium channel gating. J Gen Physiol 109(2):245–253

298. Krupnick JG, Benovic JL (1998) The role of receptor kinases and arrestins in G protein-coupled receptor regulation. Annu Rev Pharmacol Toxicol 38:289–319
299. Premont RT, Inglese J, Lefkowitz RJ (1995) Protein kinases that phosphorylate activated G protein-coupled receptors. FASEB J 9(2):175–182
300. Palczewski K (1994) Structure and functions of arrestins. Protein Sci 3(9):1355–1361
301. Puca L, Chastagner P, Meas-Yedid V, Israël A, Brou C (2013) A-arrestin 1 (ARRDC1) and β-arrestins cooperate to mediate Notch degradation in mammals. J Cell Sci 126(Pt 19):4457–4468
302. Alberts B, Johnson A, Lewis J, Raff M, Roberts K, Walter P (2002) Molecular biology of the cell, 4th edn. Garland Science, New York, USA
303. Alexander SP, Mathie A, Peters JA (2007) Catalytic receptors. Br J Pharmacol 150 Suppl 1 (S1):S122–S127
304. Gomez-Puerta JA, Mócsai A (2013) Tyrosine kinase inhibitors for the treatment of rheumatoid arthritis. Curr Top Med Chem 13(6):760–773
305. Grassot J, Mouchiroud G, Perrière G (2003) RTKdb: database of receptor tyrosine kinase. Nucleic Acids Res 31(1):353–358
306. Kontzias A, Laurence A, Gadina M, O'Shea JJ (2012) Kinase inhibitors in the treatment of immune-mediated disease. F1000 Med Rep 4:5
307. Martin J, Anamika K, Srinivasan N (2010) Classification of protein kinases on the basis of both kinase and non-kinase regions. PLoS ONE 5(9):e12460
308. Alexander SPH, Fabbro D, Kelly E, Marrion N, Peters JA, Benson HE et al (2015) The concise guide to pharmacology 2015/16: catalytic receptors. Br J Pharmacol 172:5979–6023
309. Tyagi N, Anamika K, Srinivasan N (2010) A framework for classification of prokaryotic protein kinases. PLoS ONE 5(5):e10608
310. Prisic S, Husson RN (2014) Mycobacterium tuberculosis serine/threonine protein kinases. Microbiol Spectr 2(5):1–26
311. Magnuson NS, Beck T, Vahidi H, Hahn H, Smola U, Rapp UR (1994) The Raf-1 serine/threonine protein kinase. Semin Cancer Biol 5(4):247–253
312. Taylor SS, Radzio-Andzelm E, Hunter T (1995) How do protein kinases discriminate between serine/threonine and tyrosine? Structural insights from the insulin receptor protein-tyrosine kinase. FASEB J 9(13):1255–1266
313. Feldmann M (2008) Many cytokines are very useful therapeutic targets in disease. J Clin Invest 118(11):3533–3536
314. Vilček J, Feldmann M (2004) Historical review: cytokines as therapeutics and targets of therapeutics. Trends Pharmacol Sci 25(4):201–209
315. Massagué J, Weis-Garcia F (1996) Serine/threonine kinase receptors: mediators of transforming growth factor beta family signals. Cancer Surv 27:41–64
316. ten Dijke P, Franzén P, Yamashita H, Ichijo H, Heldin CH, Miyazono K (1994) Serine/threonine kinase receptors. Prog Growth Factor Res 5(1):55–72
317. Josso N, di Clemente N (1997) Serine/threonine kinase receptors and ligands. Curr Opin Genet Dev 7(3):371–377
318. Schulz S, Chinkers M, Garbers DL (1989) The guanylate cyclase/receptor family of proteins. FASEB J 3(9):2026–2035
319. Martin E, Berka V, Tsai AL, Murad F (2005) Soluble guanylyl cyclase: the nitric oxide receptor. Methods Enzymol 396:478–492
320. Ma X, Sayed N, Beuve A, van den Akker F (2007) NO and CO differentially activate soluble guanylyl cyclase via a heme pivot-bend mechanism. EMBO J 26(2):578–588
321. Murthy KS (2004) Modulation of soluble guanylate cyclase activity by phosphorylation. Neurochem Int 45(6):845–851
322. Potter LR (2011) Guanylyl cyclase structure, function and regulation. Cell Signal 23 (12):1921–1926
323. Qian X, Prabhakar S, Nandi A, Visweswariah SS, Goy MF (2000) Expression of GC-C, a receptor-guanylate cyclase, and its endogenous ligands uroguanylin and guanylin along the rostrocaudal axis of the intestine. Endocrinology 141(9):3210–3224

324. Sharma RK, Duda T, Goraczniak R, Sitaramayya A (1997) Membrane guanylate cyclase signal transduction system. Indian J Biochem Biophys 34(1–2):40–49
325. Garbers DL (1991) Guanylyl cyclase-linked receptors. Pharmacol Ther 50(3):337–345
326. Garbers DL (1989) Guanylate cyclase, a cell surface receptor. J Biol Chem 264(16):9103–9106
327. Chinkers M, Garbers DL, Chang MS, Lowe DG, Chin HM, Goeddel DV, Schulz S (1989) A membrane form of guanylate cyclase is an atrial natriuretic peptide receptor. Nature 338 (6210):78–83
328. Thorpe DS, Garbers DL (1989) The membrane form of guanylate cyclase. Homology with a subunit of the cytoplasmic form of the enzyme. J Biol Chem 264(11):6545–6549
329. Singh S, Lowe DG, Thorpe DS, Rodriguez H, Kuang WJ, Dangott LJ et al (1988) Membrane guanylate cyclase is a cell-surface receptor with homology to protein kinases. Nature 334(6184):708–712
330. Chao YC, Chen CC, Lin YC, Breer H, Fleischer J, Yang RB (2015) Receptor guanylyl cyclase-G is a novel thermosensory protein activated by cool temperatures. EMBO J 34 (3):294–306
331. Manning G, Whyte DB, Martinez R, Hunter T, Sudarsanam S (2002) The protein kinase complement of the human genome. Science 298:1912–1934
332. Robinson SDR, Wu Y-M, Lin S-F (2000) The protein tyrosine kinase family of the human genome. Oncogene 19:5548–5557
333. Bublil EM, Yarden Y (2007) The EGF receptor family: spearheading a merger of signaling and therapeutics. Curr Opin Cell Biol 19(2):124–134
334. Ribchester RR, Thomson D, Haddow LJ, Ushkaryov YA (1998) Enhancement of spontaneous transmitter release at neonatal mouse neuromuscular junctions by the glial cell line-derived neurotrophic factor (GDNF). J Physiol 512(3):635–664
335. Mead B, Logan A, Berry M, Leadbeater W, Scheven BA (2014) Paracrine-mediated neuroprotection and neuritogenesis of axotomised retinal ganglion cells by human dental pulp stem cells: comparison with human bone marrow and adipose-derived mesenchymal stem cells. PLoS ONE 9(10):e109305
336. Hollis ER II, Tuszynski MH (2011) Neurotrophins: potential therapeutic tools for the treatment of spinal cord injury. Neurotherapeutics 8(4):694–703
337. Blume-Jensen P, Hunter T (2001) Oncogenic kinase signaling. Nature 411:355–365
338. Seidel HM, Lamb P, Rosen J (2000) Pharmaceutical intervention in the JAK/STAT signaling pathway. Oncogene 19(21):2645–2656
339. Sansone P, Bromberg J (2012) Targeting the interleukin-6/Jak/stat pathway in human malignancies. J Clin Oncol 30(9):1005–1014
340. Vainchenker W, Constantinescu SN (2013) JAK/STAT signaling in hematological malignancies. Oncogene 32(21):2601–2613
341. Jatiani SS, Baker SJ, Silverman LR, Reddy EP (2010) JAK/STAT pathways in cytokine signaling and myeloproliferative disorders. Genes Cancer 1(10):979–993
342. Song G, Ouyang G, Bao S (2005) The activation of Akt/PKB signaling pathway and cell survival. J Cell Mol Med 9(1):59–71
343. Gammeltoft S, Van Obberghen E (1986) Protein kinase activity of the insulin receptor. Biochem J 235(1):1–11
344. Zick Y (1989) The insulin receptor: structure and function. Crit Rev Biochem Mol Biol 24 (3):217–269
345. Rosen OM (1987) After insulin binds. Science 237(4821):1452–1458
346. Kasuga M, Izumi T, Tobe K, Shiba T, Momomura K, Tashiro-Hashimoto Y, Kadowaki T (1990) Substrates for insulin-receptor kinase. Diabet Care 13(3):317–326
347. Hay N, Sonenberg N (2004) Upstream and downstream of mTOR. Genes Dev 18(16):1926–1945
348. Morris DL, Cho KW, Zhou Y, Rui L (2009) Enhances insulin sensitivity by both stimulating the insulin receptor and inhibiting tyrosine dephosphorylation of insulin receptor substrate proteins. Diabetes 58(9):2039–2047

349. Watson RT, Pessin JE (2001) Subcellular compartmentalization and trafficking of theinsulin-responsive glucose transporter, GLUT4. Exp Cell Res 271:75–83
350. Huang S, Czech MP (2007) The GLUT4 glucose transporter. Cell Metab 5(4):237–252
351. Takai H, Wang RC, Takai KK, Yang H, de Lange T (2007) Tel2 regulates the stability of PI3 K-related protein kinases. Cell 131:1248–1259
352. Krebs M, Brunmair B, Brehm A, Artwohl M, Szendroedi J, Nowotny P et al (2007) The mammalian target of rapamycin pathway regulates nutrient-sensitive glucose uptake in man. Diabetes 56(6):1600–1607
353. Smith GC, Ong WK, Costa JL, Watson M, Cornish J, Grey A et al (2013) Extended treatment with selective phosphatidylinositol 3-kinase and mTOR inhibitors has effects on metabolism, growth, behaviour and bone strength. FEBS J 280(21):5337–5349
354. Yecies JL, Manning BD (2011) Transcriptional control of cellular metabolism by mTOR signaling. Cancer Res 71(8):2815–2820
355. Huang J, Manning BD (2009) A complex interplay between Akt, TSC2 and the two mTOR complexes. Biochem Soc Trans 37(Pt 1):217–222
356. Buller CL, Loberg RD, Fan MH, Zhu Q, Park JL, Vesely E et al (2008) A GSK-3/TSC2/ mTOR pathway regulates glucose uptake and GLUT1 glucose transporter expression. Am J Physiol Cell Physiol 295(3):C836–C843
357. Feng Z, Levine AJ (2010) The regulation of energy metabolism and the IGF-1/mTOR pathways by the p53 protein. Trends Cell Biol 20(7):427–434
358. Choura M, Rebaï A (2010) Application of computational approaches to study signalling networks of nuclear and tyrosine kinase receptors. Biol Direct 5:58
359. Gray KA, Seal RL, Tweedie S, Wright MW, Bruford EA (2016) A review of the new HGNC gene family resource. Hum Genomics 10:6
360. Yates B, Braschi B, Gray KA, Seal RL, Tweedie S, Bruford EA (2017) Genenames.org: the HGNC and VGNC resources in 2017. Nucleic Acids Res 45(D1):D619–D625
361. Brown JE, Krodel M, Pazos M, Lai C, Prieto AL (2012) Cross-phosphorylation, signaling and proliferative functions of the Tyro3 and Axl receptors in Rat2 cells. PLoS ONE 7(5): e36800
362. Levitzki A, Gazit A, Osherov N, Posner I, Gilon C (1991) Inhibition of protein-tyrosine kinases by tyrphostins. Methods Enzymol 201:347–361
363. Gillespie J, Dye JF, Schachter M, Guillou PJ (1993) Inhibition of pancreatic cancer cell growth in vitro by the tyrphostin group of tyrosine kinase inhibitors. Br J Cancer 68 (6):1122–1126
364. Zlotnik Y, Patya M, Vanichkin A, Novogrodsky A (2005) Tyrphostins reduce chemotherapy-induced intestinal injury in mice: assessment by a biochemical assay. Br J Cancer 92(2):294–297
365. Natoli C, Perrucci B, Perrotti F, Falchi L, Iacobelli S (2010) Consorzio Interuniversitario Nazionale per Bio-Oncologia (CINBO), tyrosine kinase inhibitors. Curr Cancer Drug Targets 10(5):462–483
366. Wu P, Nielsen TE, Clausen MH (2016) Small-molecule kinase inhibitors: an analysis of FDA-approved drugs. Drug Discov Today 21(1):5–10
367. Hartmann JT, Haap M, Kopp HG, Lipp HP (2009) Tyrosine kinase inhibitors—a review on pharmacology, metabolism and side effects. Curr Drug Metab 10(5):470–481
368. Kaplan FS, Groppe J, Pignolo RJ, Shore EM (2007) Morphogen receptor genes and metamorphogenes: skeleton keys to metamorphosis. Ann NY Acad Sci 1116:113–133
369. Christian JL (2012) Morphogen gradients in development: from form to function. Wiley Interdiscip Rev Dev Biol 1(1):3–15
370. Artells R, Navarro A, Diaz T, Monzó M (2011) Ultrastructural and immunohistochemical analysis of intestinal myofibroblasts during the early organogenesis of the human small intestine. Anat Record 294(3):462–471
371. Lévy E, Delvin E, Ménard D, Beaulieu J-F (2009) Functional development of human fetal gastrointestinal tract. Human Embryogenesis: Methods Mol Biol 550:205–224

372. Drozdowski LA, Clandinin T, Thomson ABR (2010) Ontogeny, growth and development of the small intestine: understanding pediatric gastroenterology. World J Gastroenterol 16 (7):787–799
373. Tabata T, Takei Y (2004) Morphogens, their identification and regulation Development 131:703–712
374. Janas T, Janas MM, Sapoń K, Janas T (2015) Mechanisms of RNA loading into exosomes. FEBS Lett 589(13):1391–1398
375. Lakkaraju A, Rodriguez-Boulan E (2008) Itinerant exosomes: emerging roles in cell and tissue polarity. Trends Cell Biol 18(5):199–209
376. Vincent J-P, Magee T (2002) Argosomes: membrane fragments on the run. Trends Cell Biol 12(2):57–60
377. Ramírez-Weber F-A, Kornberg TB (1999) Cytonemes: cellular processes that project to the principal signaling center in Drosophila imaginal discs. Cell 97(5):599–607
378. Zhang L, Wrana JL (2014) The emerging role of exosomes in Wnt secretion and transport. Curr Opin Genet Dev 27:14–19
379. Bischoff M, Gradilla AC, Seijo I, Andrés G, Rodríguez-Navas C, González-Méndez L, Guerrero I (2013) Cytonemes are required for the establishment of a normal Hedgehog morphogen gradient in Drosophila epithelia. Nat Cell Biol 15(11):1269–1281
380. Gradilla A-C, González E, Seijo I, Andrés G, Bischoff M, González-Mendez L et al (2014) Exosomes as Hedgehog carriers in cytoneme-mediated transport and secretion. Nat Commun 5:5649
381. Lim YS, Tang BL (2012) Intercellular organelle trafficking by membranous nanotube connections: a possible new role in cellular rejuvenation? Cell Commun Adhes 19(3–4):39–44
382. Shilo B-Z (2001) The organizer and beyond. Cell 106(1):17–22
383. Greco V, Hannus M, Eaton S (2001) Argosomes: a potential vehicle for the spread of morphogens through epithelia. Cell 106(5):633–645
384. Austefjord MW, Gerdes H-H, Wang X (2014) Tunneling nanotubes: diversity in morphology and structure. Commun Integr Biol 7:e27934
385. Gurke S, Barroso JFV, Gerdes H-H (2008) The art of cellular communication: tunneling nanotubes bridge the divide. Histochem Cell Biol 129(5):539–550
386. Sarrazin S, Lamanna WC, Esko JD (2011) Heparan sulfate proteoglycans. Cold Spring Harb Perspect Biol 3(7):a004952
387. Rodgers KD, San Antonio JD, Jacenko O (2008) Heparan sulfate proteoglycans: a GAGgle of skeletal-hematopoietic regulators. Dev Dyn 237(10):2622–2642
388. Fisher MC, Li Y, Seghatoleslami MR, Dealy CN, Kosher RA (2006) Heparan sulfate proteoglycans including syndecan-3 modulate BMP activity during limb cartilage differentiation. Matrix Biol 25(1):27–39
389. Li JP, Spillmann D (2012) Heparan sulfate proteoglycans as multifunctional cell regulators: cell surface receptors. Methods Mol Biol 836:239–255
390. Nusse R, Varmus HE (1982) Many tumors induced by the mouse mammary tumor virus contain a provirus integrated in the same region of the host genome. Cell 31(1):99–109
391. Roel N, Harold V (2012) Three decades of Wnts: a personal perspective on how a scientific field developed. EMBO J 31(12):2670–2684
392. Nusse R, van Ooyen A, Cox D, Fung YK, Varmus H (1984) Mode of proviral activation of a putative mammary oncogene (int-1) on mouse chromosome 15. Nature 307(5947):131–136
393. Klaus A, Birchmeier W (2008) Wnt signaling and its impact on development and cancer. Nat Rev Cancer 8(5):387–398
394. Nüsslein-Volhard C, Wieschaus E (1980) Mutations affecting segment number and polarity in Drosophila. Nature 287(5785):795–801
395. Nakamura T, Hamada F, Ishidate T, Anai K, Kawahara K, Toyoshima K, Akiyama T (1998) Axin, an inhibitor of the Wnt signalling pathway, interacts with β-catenin, GSK-3β and APC and reduces the β-catenin level. Genes Cells 3(6):395–403

396. Kim JH, Liu X, Wang J, Chen X, Zhang H, Kim SH et al (2013) Wnt signaling in bone formation and its therapeutic potential for bone diseases. Ther Adv Musculoskel Dis 5 (1):13–31
397. Yao H, Ashihara E, Maekawa T (2011) Targeting the Wnt/β-catenin signaling pathway in human cancers. Expert Opin Ther Targets 15(7):873–887
398. Kühl M, Sheldahl LC, Park M, Miller JR, Moon RT (2000) The Wnt/Ca2+ pathway: a new vertebrate Wnt signaling pathway takes shape. Trends Genet 16(7):279–283
399. Bryja V, Andersson ER, Schambony A, Esner M, Bryjová L, Biris KK et al (2009) The extracellular domain of Lrp5/6 inhibits noncanonical wnt signaling in vivo. Mol Biol Cell 20 (3):924–936
400. Ren D, Chen J, Li Z, Yan H, Yin Y, Wo D et al (2014) LRP5/6 directly bind to Frizzled and prevent Frizzled-regulated tumour metastasis. Nat Commun 6:6906
401. Malbon CC (2004) Frizzleds: new members of the superfamily of G-protein-coupled receptors. Front Biosci 9:1048–1058
402. Huang HC, Klein PS (2004) The Frizzled family: receptors for multiple signal transduction pathways. Genome Biol 5(7):234
403. Sear RP (2007) Dishevelled: a protein that functions in living cells by phase separating. Soft Matter 3:680–684
404. Pulvirenti T, Van Der Heijden M, Droms LA, Huse JT, Tabar V, Hall A (2011) Dishevelled 2 signaling promotes self-renewal and tumorigenicity in human gliomas. Cancer Res 71 (23):7280–7290
405. Wallingford JB, Habas R (2005) The developmental biology of Dishevelled: an enigmatic protein governing cell fate and cell polarity. Development 132:4421–4436
406. Kikuchi A (1999) Modulation of Wnt signaling by Axin and Axil. Cytokine Growth Factor Rev 10(3–4):255–265
407. Li KK-W, Lau K-M, Ng H-K (2013) Signaling pathway and molecular subgroups of medulloblastoma. Int J Clin Exp Pathol 6(7):1211–1222
408. Farr GH III, Ferkey DM, Yost C, Pierce SB, Weaver C, Kimelman D (2000) Interaction among Gsk-3, Gbp, Axin, and APC in Xenopus axis specification. J Cell Biol 148(4):691–702
409. Hedgepeth CM, Deardorff MA, Rankin K, Klein PS (1999) Regulation of glycogen synthase kinase 3β and downstream wnt signaling by Axin. Mol Cell Biol 19(10):7147–7157
410. Ikeda S, Kishida S, Yamamoto H, Murai H, Koyama S, Kikuchi A (1998) Axin, a negative regulator of the Wnt signaling pathway, forms a complex with GSK-3β and β-catenin and promotes GSK-3β-dependent phosphorylation of β-catenin. The EMBO J 17:1371–1384
411. Yamamoto H, Kishida S, Uochi T, Ikeda S, Koyama S, Asashima M, Kikuchi A (1998) Axil, a member of the Axin family, interacts with both glycogen synthase kinase 3beta and beta-catenin and inhibits axis formation of Xenopus embryos. Mol Cell Biol 18(5):2867–2875
412. Takemaru K-I, Yamaguchi S, Lee YS, Zhang Y, Carthew RW, Moon RT (2003) Chibby, a nuclear β-catenin-associated antagonist of the Wnt/Wingless pathway. Nature 422:905–909
413. Lin GL, Hankenson KD (2012) Integration of BMP, Wnt, and Notch signaling pathways in osteoblast differentiation. J Cell Biochem 112(12):3491–3501
414. Djabrayan NJV, Dudley NR, Sommermann EM, Rothman JH (2012) Essential role for Notch signaling in restricting developmental plasticity. Genes Dev 26(21):2386–2391
415. Huang T, Zhou Y, Cheng ASL, Yu J, To KF, Kang W (2016) Notch receptors in gastric and other gastrointestinal cancers: oncogenes or tumor suppressors? Mol Cancer 15:80
416. Demitrack ES, Samuelson LC (2016) Notch regulation of gastrointestinal stem cells. J Physiol 594(17):4791–4803
417. Kim TH, Kim BM, Mao J, Rowan S, Shivdasani RA (2011) Endodermal Hedgehog signals modulate Notch pathway activity in the developing digestive tract mesenchyme. Development 138(15):3225–3233
418. Liu JA, Ngan ES-W (2014) Hedgehog and Notch signaling in enteric nervous system development. Neurosignals 22:1–13

419. Katoh M, Katoh M (2007) Notch signaling in gastrointestinal tract (review). Int J Oncol 30 (1):247–251
420. Fernandez-Valdivia R, Takeuchi H, Samarghandi A, Lopez M, Leonardi J, Haltiwanger RS, Jafar-Nejad H (2011) Regulation of mammalian Notch signaling and embryonic development by the protein O-glucosyltransferase Rumi. Development 138(10):1925–1934
421. Ungerbäck J, Elander N, Grünberg J, Sigvardsson M, Söderkvist P (2011) The Notch-2 gene is regulated by Wnt signaling in cultured colorectal cancer cells. PLoS ONE 6(3):e17957
422. Kangsamaksin T, Murtomaki A, Kofler NM, Cuervo H, Chaudhri RA, Tattersall IW et al (2015) Notch decoys that selectively block Dll/Notch or Jagged/Notch disrupt angiogenesis by unique mechanisms to inhibit tumor growth. Cancer Discov 5(2):182–197
423. Van de Walle I, De Smet G, Gärtner M, De Smedt M, Waegemans E, Vandekerckhove B et al (2011) Jagged2 acts as a Delta-like Notch ligand during early hematopoietic cell fate decisions. Blood 117(17):4449–4459
424. Chillakuri CR, Sheppard D, Lea SM, Handford PA (2012) Notch receptor–ligand binding and activation: insights from molecular studies. Semin Cell Dev Biol 23(4):421–428
425. Seals DF, Courtneidge SA (2003) The ADAMs family of metalloproteases: multidomain proteins with multiple functions. Genes Dev 17:7–30
426. Tsai YH, VanDussen KL, Sawey ET, Wade AW, Kasper C, Rakshit S et al (2014) ADAM10 regulates Notch function in intestinal stem cells of mice. Gastroenterology 147(4):822–834. e13
427. De Santa BP, Van Den Brink GR, Roberts DJ (2003) Development and differentiation of the intestinal epithelium. Cell Mol Life Sci 60(7):1322–1332
428. D'mello NP, Childress AM, Franklin DS, Kale SP, Pinswasdi C, Jazwinski SM (1994) Cloning and characterization of LAG1, a longevity-assurance gene in yeast. J Biol Chem 269(22):15451–15459
429. Teufel A, Maass T, Galle PR, Malik N (2009) The longevity assurance homologue of yeast lag1 (Lass) gene family (review). Int J Mol Med 23(2):135–140
430. Lehman RF, Jimenez W, Dietrich U, Campos-Ortega JA (1983) On the phenotype and development of mutants of early neurogenesis in Drosophila melanogaster. Wilhelm Roux's Arch Dev Biol 192:62–74
431. Smoller D, Friedel C, Schmid A, Bettler D, Lam L, Yedvobnick B (1990) The Drosophila neurogenic locus mastermind encodes a nuclear protein unusually rich in amino acid homopolymers. Genes Dev 4:1688–1700
432. Artavanis-Tsakonas S, Matsuno K, Fortini ME (1995) Notch signaling. Science 268:225–232
433. Kopan R, Ilagan MX (2009) The canonical Notch signaling pathway: unfolding the activation mechanism. Cell 137(2):216–233
434. Yamamoto N, Yamamoto S, Inagaki F, Kawaichi M, Fukamizu A, Kishi N et al (2001) Role of Deltex-1 as a transcriptional regulator downstream of the Notch receptor. J Biol Chem 276(48):45031–45040
435. Matsuno K, Eastman D, Mitsiades T, Quinn AM, Carcanciu ML, Ordentlich P et al (1998) Human deltex is a conserved regulator of Notch signalling. Nat Gene 19(1):74–78
436. Matsuno K, Diederich RJ, Go MJ, Blaumueller CM, Artavanis-Tsakonas S (1995) Deltex acts as a positive regulator of Notch signaling through interactions with the Notch ankyrin repeats. Development 121(8):2633–2644
437. Deregowski V, Gazzerro E, Priest L, Rydziel S, Canalis E (2006) Role of the Ram domain and ankyrin repeats on notch signalingand activity in cells of osteoblastic lineage. J Bone Miner Res 21(8):1317–1326
438. Panelos J, Massi D (2009) Emerging role of Notch signaling in epidermal differentiation and skin cancer. Cancer Biol Ther 8(21):1986–1993
439. Andersen P, Uosaki H, Shenje LT, Kwon C (2012) Non-canonical Notch signaling: emerging role and mechanism. Trends Cell Biol 22(5):257–265
440. Heretsch P, Tzagkaroulaki L, Giannis A (2010) Modulators of the hedgehog signaling pathway. Bioorg Med Chem 18(18):6613–6624

441. Zhou J, Wei X, Wei L (2014) Indian Hedgehog, a critical modulator in osteoarthritis, could be a potential therapeutic target for attenuating cartilage degeneration disease. Connect Tissue Res 55(4):257–261
442. Sam SA, Teel J, Tegge AN, Bharadwaj A, Murali TM (2017) XTalkDB: a database of signaling pathway crosstalk. Nucleic Acids Res 45(D1):D432–D439
443. Tamkun JW, DeSimone DW, Fonda D, Patel RS, Buck C, Horwitz AF, Hynes RO (1986) Structure of integrin, a glycoprotein involved in the transmembrane linkage between fibronectin and actin. Cell 46:271–282
444. Hynes RO (1987) Integrins: a family of cell surface receptors. Cell 48:549–554
445. Humphries JD, Byron A, Humphries MJ (2006) Integrin ligands. J Cell Sci 119(Pt 19):3901–3903
446. Campbell ID, Humphries MJ (2011) Integrin structure, activation, and interactions. Cold Spring Harb Perspect Biol 3(3):a004994
447. Hynes RO (2002) Integrins: bidirectional, allosteric signaling machines. Cell 110(6):673–687
448. Akiyama SK (1996) Integrins in cell adhesion and signaling. Hum Cell 9(3):181–186
449. Wagner G (1994) Cell surface adhesion receptors. Curr Opin Struct Biol 4(6):841–851
450. Gorfu G, Rivera-Nieves J, Ley K (2009) Role of β7 integrins in intestinal lymphocyte homing and retention. Curr Mol Med 9(7):836–850
451. Harburger DS, Calderwood DA (2009) Integrin signalling at a glance. J Cell Sci 122:159–163
452. Lee JW, Juliano RL (2000) α5β1 integrin protects intestinal epithelial cells from apoptosis through a phosphatidylinositol 3-kinase and protein kinase B–dependent pathway. Mol Biol Cell 11(6):1973–1987
453. Wang X, Xu H, Gill AF, Pahar B, Kempf D, Rasmussen T et al (2009) Monitoring α4 β7 integrin expression on circulating CD4+ T cells as a surrogate marker for tracking intestinal CD4+ T-cell loss in SIV infection. Mucosal Immunol 2:518–526
454. Liu Y-X, Yoshino T, Ohara N, Oka T, Jin Z-S, Hayashi K, Akagi T (2001) Loss of expression of α4 β7 integrin and L-selectin is associated with high-grade progression of low-grade MALT lymphoma. Mod Pathol 14(8):798–805
455. Heino J (2000) The collagen receptor integrins have distinct ligand recognition and signaling functions. Matrix Biol 19(4):319–323
456. Teixidó J, Hemler ME, Greenberger JS, Anklesaria P (1992) Role of beta 1 and beta 2 integrins in the adhesion of human CD34hi stem cells to bone marrow stroma. J Clin Invest 90(2):358–367
457. Takada Y, Ye X, Simon S (2007) The integrins. Genome Biol 8(5):215
458. Mengaud J, Ohayon H, Gounon P, Mege R-M, Cossart P (1996) E-cadherin is the receptor for internalin, a surface protein required for entry of L. monocytogenes into epithelial cells. Cell 84(6):923–932
459. Niessen CM, Leckband D, Yap AS (2011) Tissue organization by cadherin adhesion molecules: dynamic molecular and cellular mechanisms of morphogenetic regulation. Physiol Rev 91(2):691–731
460. Takeichi M (1991) Cadherin cell adhesion receptors as a morphogenetic regulator. Science 251(5000):1451–1455
461. Shimono Y, Rikitake Y, Mandai K, Mori M, Takai Y (2012) Immunoglobulin superfamily receptors and adherens junctions. Subcell Biochem 60:137–170
462. Geissmann F, Ruskoné-Fourmestraux A, Hermine O, Bourquelot P, Belanger C, Audouin J et al (1998) Homing receptor α4β7 integrin expression predicts digestive tract involvement in mantle cell lymphoma. Am J Pathol 153(6):1701–1705
463. Baruch-Morgenstern NB, Shik D, Moshkovits I, Itan M, Karo-Atar D, Bouffi C et al (2014) Paired immunoglobulin-like receptor A is an intrinsic, self-limiting suppressor of IL-5-induced eosinophil development. Nat Immunol 15(1):36–44
464. Michail S, Mezoff E, Abernathy F (2005) Role of selectins in the intestinal epithelial migration of eosinophils. Pediatr Res 58:644–647

465. Kawamura YI, Kawashima R, Fukunaga R, Hirai K, Toyama-Sorimachi N, Tokuhara M et al (2005) Introduction of Sda carbohydrate antigen in gastrointestinal cancer cells eliminates selectin ligands and inhibits metastasis. Cancer Res 65(14):6220–6227
466. Trinchera M, Aronica A, Dall'Olio F (2017) Selectin ligands sialyl-lewis a and sialyl-lewis x in gastrointestinal cancers. Biology 6(16):1–18
467. Hermand P, Gane P, Callebaut I, Kieffer N, Cartron JP, Bailly P (2004) Integrin receptor specificity for human red cell ICAM-4 ligand. Critical residues for alphaIIbeta3 binding. Eur J Biochem 271(18):3729–3740
468. Kajita M, McClinic KN, Wade PA (2004) Aberrant expression of the transcription factors Snail and Slug alters the response to genotoxic stress. Mol Cell Biol 24(17):7559–7566
469. Weng M, Wieschaus E (2016) Myosin-dependent remodeling of adherens junctions protects junctions from Snail-dependent disassembly. J Cell Biol 212(2):219–229
470. Janssen S, Depoortere I (2013) Nutrient sensing in the gut: new roads to therapeutics? Trends Endocrin Metab 24(2):92–100
471. Running CA, Craig BA, Mattes RD (2015) Oleogustus: the unique taste of fat. Chem Senses 40(7):507–516
472. Besnard P, Passilly-Degrace P, Khan NA (2016) Taste of fat: a sixth taste modality? Physiol Rev 96(1):151–176
473. Dramane G, Akpona S, Simonin AM, Besnard P, Khan NA (2011) Cell signaling mechanisms of gustatory perception of lipids: can the taste cells be the target of anti-obesity agents? Curr Med Chem 18(22):3417–3422
474. Liu D, Archer N, Duesing K, Hannan G, Keast R (2016) Mechanism of fat taste perception: association with diet and obesity. Prog Lipid Res 63:41–49
475. Keast RSJ, Costanzo A (2015) Is fat the sixth taste primary? Evidence and implications. Flavour 4:5
476. Degrace-Passilly P, Besnard P (2012) CD36 and taste of fat. Curr Opin Clin Nutr Metab Care 15(2):107–111
477. Richter TA, Caicedo A, Roper SD (2003) Sour taste stimuli evoke Ca2+ and pH responses in mouse taste cells. J Physiol 547(Pt 2):475–483
478. Chandrashekar J, Hoon MA, Ryba NJP, Zuker CS (2006) The receptors and cells for mammalian taste. Nature 444:288–294
479. Liman ER, Zhang YV, Craig M (2014) Peripheral coding of taste. Neuron 81(5):984–1000
480. Kawaguchi H, Yamanaka A, Uchida K, Shibasaki K, Sokabe T, Maruyama Y et al (2010) Activation of polycystic kidney disease-2-like 1 (PKD2L1)-PKD1L3 complex by acid in mouse taste cells. J Biol Chem 285(23):17277–17281
481. Wright EM, Turk E (2004) The sodium/glucose cotransport family SLC5. Pflugers Arch 447 (5):510–518
482. Jung H (2001) Towards the molecular mechanism of Na+/solute symport in prokaryotes. Biochim Biophys Acta Bioenerg 1505(1):131–143
483. Harada N, Inagaki N (2012) Role of sodium-glucose transporters in glucose uptake of the intestine and kidney. J Diabetes Invest 3(4):352–353
484. Takata K, Hirano H, Kasahara M (1997) Transport of glucose across the blood-tissue barriers. Int Rev Cytol 172:1–53
485. Weale AR, Edwards AG, Bailey M, Lear PA (2005) Intestinal adaptation after massive intestinal resection. Postgrad Med J 81:178–184
486. Bertrand PP (2009) The cornucopia of intestinal chemosensory transduction. Front Neurosci 3:48
487. Sykaras AG, Demenis C, Case RM, McLaughlin JT, Smith CP (2012) Duodenal enteroendocrine I-cells contain mRNA transcripts encoding key endocannabinoid and fatty acid receptors. PLoS ONE 7(8):e42373
488. Daly K, Al-Rammahi M, Moran A, Marcello M, Ninomiya Y, Shirazi-Beechey SP (2013) Sensing of amino acids by the gut-expressed taste receptor T1R1-T1R3 stimulates CCK secretion. Am J Physiol Gastrointest Liver Physiol 304(3):G271–G282

489. Gabriel AS, Uneyama H (2013) Amino acid sensing in the gastrointestinal tract. Amino Acids 45(3):451–461
490. Talukdar S, Olefsky JM, Osborn O (2011) Targeting GPR120 and other fatty acid sensing GPCRs ameliorates insulin resistance and inflammatory diseases. Trends Pharmacol Sci 32 (9):543–550
491. Hara T, Hirasawa A, Ichimura A, Kimura I, Tsujimoto G (2011) Free fatty acid receptors FFAR1 and GPR120 as novel therapeutic targets for metabolic disorders. J Pharm Sci 100 (9):3594–3601
492. Ichimura A, Hirasawa A, Hara T, Tsujimoto G (2009) Free fatty acid receptors act as nutrient sensors to regulate energy homeostasis. Prostaglandins Other Lipid Mediat 89(3–4):82–88
493. Mohamed HA, Khaled DM, Mohamed HA (2014) Homology modeling and explicit membrane molecular dynamics simulation to delineate the mode of binding of thiazolidinediones into FFAR1 and the mechanism of receptor activation. Bioorg Med Chem Lett 24 (22):5330–5336
494. Mace OJ, Lister N, Morgan E, Shepherd E, Affleck J, Helliwell P et al (2009) An energy supply network of nutrient absorption coordinated by calcium and T1R taste receptors in rat small intestine. J Physiol 587:195–210
495. Li T, Holmstrom SR, Kir S, Umetani M, Schmidt DR, Kliewer SA, Mangelsdorf DJ (2011) The G protein-coupled bile acid receptor, TGR5, stimulates gallbladder filling. Mol Endocrinol 25(6):1066–1071
496. Motter AL, Ahern GP (2012) TRPA1 is a polyunsaturated fatty acid sensor in mammals. PLoS ONE 7(6):e38439
497. Tan J, McKenzie C, Potamitis M, Thorburn AN, Mackay CR, Macia L (2014) The role of short-chain fatty acids in health and disease. Adv Immunol 121:91–119
498. Kimura I, Inoue D, Maeda T, Hara T, Ichimura A, Miyauchi S et al (2011) Short-chain fatty acids and ketones directly regulate sympathetic nervous system via G protein-coupled receptor 41 (GPR41). Proc Natl Acad Sci USA 108(19):8030–8035
499. Gaetani S, Oveisi F, Piomelli D (2003) Modulation of meal pattern in the rat by the anorexic lipid mediator oleoylethanolamine. Neuropsychopharmacology 28(7):1311–1316
500. Criado AS, Pavon-Moron FJ, de Arco I, de Fonseca FR (2006) Oleoylethanolamide reverses changes in both fatty acid composition and desaturase mRNA expression in a new model of liver steatosis. Obes Metab 2(3):155–164
501. Kim CH, Park J, Kim M (2014) Gut microbiota-derived short-chain fatty acids, T cells, and inflammation. Immune Netw 14(6):277–288
502. Eberle JA-M, Widmayer P, Breer H (2014) Receptors for short-chain fatty acids in brush cells at the "gastric groove". Front Physiol 5:152
503. Thomas C, Gioiello A, Noriega L, Strehle A, Oury J, Rizzo G et al (2009) TGR5-mediated bile acid sensing controls glucose homeostasis. Cell Metab 10(3):167–177
504. Kumar DP, Rajagopal S, Mahavadi S, Mirshahi F, Grider JR, Murthy KS, Sanyal AJ (2012) Activation of transmembrane bile acid receptor TGR5 stimulates insulin secretion in pancreatic β cells. Biochem Biophys Res Commun 427(3):600–605
505. Tiwari A, Maiti P (2009) TGR5: an emerging bile acid G-protein-coupled receptor target for the potential treatment of metabolic disorders. Drug Discov Today 14(9–10):523–530
506. Tu H, Okamoto AY, Shan B (2000) FXR, a bile acid receptor and biological sensor. Trends Cardiovasc Med 10(1):30–35
507. Kawamata Y, Fujii R, Hosoya M, Harada M, Yoshida H, Miwa M et al (2003) A G protein-coupled receptor responsive to bile acids. J Biol Chem 278(11):9435–9440
508. Duboc H, Taché Y, Hofmann AF (2014) The bile acid TGR5 membrane receptor: from basic research to clinical application. Int J Gatroenterol Hepatol 46(4):302–312
509. Makishima M, Lu TT, Xie W, Whitfield GK, Domoto H, Evans RM et al (2002) Vitamin D receptor as an intestinal bile acid sensor. Science 296(5571):1313–1316
510. Ajouz H, Mukherji D, Shamseddine A (2014) Secondary bile acids: an underrecognized cause of colon cancer. World J Surg Oncol 12:164

511. Bernstein H, Bernstein C, Payne CM, Dvorakova K, Garewal H (2005) Bile acids as carcinogens in human gastrointestinal cancers. Mutat Res/Rev Mutat Res 589(1):47–65
512. Vuolo L, Di Somma C, Faggiano A, Colao A (2012) Vitamin D and cancer. Front Endocrinol (Lausanne) 3:58
513. Konstantakis C, Tselekouni P, Kalafateli M, Triantos C (2016) Vitamin D deficiency in patients with liver cirrhosis. Ann Gastroenterol 29(3):297–306

Chapter 6
Gastrointestinal Growth and Development: From Embryo to Adult. The Aging Gut

Abstract The gastrointestinal (GI) system is one of the first systems to develop during the process of ontogenesis. Any disorder of GI tract during this period can result to serious consequences on the organism. Therefore, it is important to address the processes of normal development of the GI tract. This chapter provides data on the origin as well as structure and functions of the developing GI tract at the molecular, cellular, tissue, organ, and system levels at different stages of ontogenesis. The chapter also provides the fundamental concepts and principles governing the developing GI tract beginning from the formation of the primitive gut to maturity and also discusses the processes of aging of the gut up to old age.

Keywords GI growth · GI development · Primitive gut · GI aging
Fetal gut · Newborn gut · Infant · Geriatrics · Old age · Gut microbes

Abbreviations

EGF	Epidermal growth factor
Epo	Erythropoietin
FGF	Fibroblast growth factor
GDNF	Glial cell-derived neurotrophic factor
GFRα1	GDNF family receptor-α1
GH	Growth hormone
GI	Gastrointestinal
GLP-2	Glucagon-like peptide-2
Hand2	Heart And Neural Crest Derivatives-expressed Protein 2
HGF	Hepatocyte growth factor
ICC	Interstitial cells of Cajal
IGF-1	Insulin-like growth factor 1
KGF	Keratinocyte growth factor
NCCs	Neural crest cells
NEC	Necrotizing enterocolitis

M. O. Welcome, *Gastrointestinal Physiology*,
https://doi.org/10.1007/978-3-319-91056-7_6

Phox2b	Paired-like homeobox 2b
RET	Rearrangement during transformation
Sox10	SRY (sex-determining region Y)-box 10
TGFβ	Transforming Growth Factor beta

6.1 Introduction

One of the characteristics of mammals is that they reproduce like ones through a series of processes involving fertilization, development of the zygote to embryo and fetus, and subsequently, birth of an organism normally weighing ~3 kg. This process occurs within a period of about 38–40 weeks (266–280 days) following fertilization and is evolutionarily preprogrammed and genetically determined [1–9].

The digestive system is one of the first systems to develop in the embryo and thus draws attention to its degree of importance to other developing organs and systems of the body and as a pathway for migration of cells. The physiology and development of the gut from embryo to maturity are environmentally and genetically determined by thousands of genes and controlled by signaling pathways that are dependent on the activities of morphogens and growth factors (signaling pathways that dictate the pattern of growth and development during ontogenesis and postnatal period have been discussed in Chap. 5) [1, 10]. After fertilization, a zygote (measuring about 0.1 mm in diameter and weighing less than 10^{-9} g) is formed (Fig. 6.1). Mitotic cell division of the zygote results in the formation of a solid mass called morula, on the third day. Morula forms a blastocyst, which contains two layers: an inner layer of cells (embryoblast) and an outer layer (trophoblast). The blastocyst forms approximately 24 h following morula formation by development of blastocele, which is an inner fluid-filled cavity. The blastocyst is a good source of embryonic stem cells, which are located in the inner portion and subsequently form the preimplantation epiblast and hypoblast. The outer portion is called the trophectoderm. The stem cells derived from the embryo have been used for a variety of purposes. Human embryo with all germ layers and genetic composition similar to the one that develops in vivo has been successfully grown in vitro from human embryonic stem cells [1–14].

Further, the blastocyst to be implanted breaks apart from the zona pellucida by alternating expansion and contraction [13, 15, 16]. (The zona pellucida is the glycoprotein-rich membrane enveloping the egg cell, through which the sperm penetrates during fertilization.) Several sequential processes are required for implantation of the blastocyst. These processes include apposition, attachment, penetration, and trophoblast invasion [13]. The trophoblast of the blastocyte contacts with the epithelium of the endometrium forming syncytiotrophoblast on the sixth–seventh day. Initially, the inner layer of cell has two layers: the hypoblast and the epiblast. However, continued cell division leads to the formation of a bilaminar

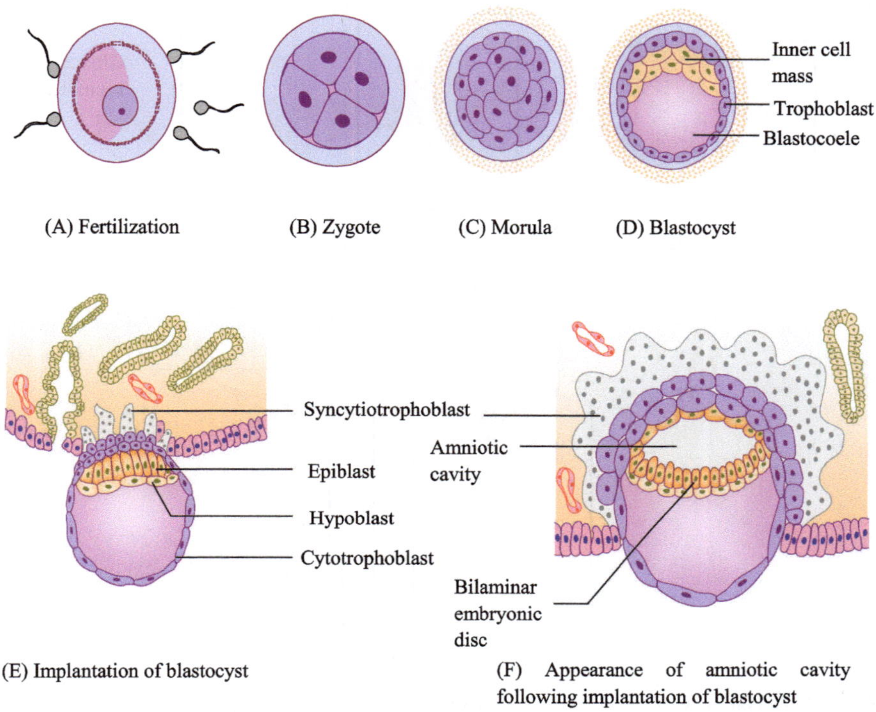

Fig. 6.1 Processes involved in fertilization and implantation of the product of fertilization. The fusion of the sperm (with 23 chromosomes) and the oocyte (with 23 chromosomes) at fertilization [A] results in a single-cell zygote [B]. The number of chromosomes in the zygote is 46, which is the total number of chromosomes in a human. Single-cell zygote continuously divides to form multicellular morula [C]. Further cell division of the morula results in the formation of a blastocyst [D], which is transported to the endometrium for implantation [E]. This stage begins from about 7 days of gestation following fertilization. Amniotic cavity appears on the eighth day [F]. From fertilization to the eighth week of gestation is the embryonic period of development. Fetal physiology starts from the ninth week of gestation

embryonic disk, which is the amniotic cavity with fluid around the cells of epiblast that have now formed embryoblast. The implantation of the product of fertilization begins from about seventh–eighth day of gestation (Fig. 6.1) [17–19].

Following implantation, the cells of the embryo continue to divide until about 14 days of gestation; a primitive streak is formed between the epiblast and extraembryonic tissue on the posterior side of the embryo and the site of ingression (Fig. 6.2). (The extraembryonic tissues are the yolk sac and placenta, which are formed from the hypoblast and trophectoderm [14].) The primitive streak is the structure that determines the process of gastrulation. Gastrulation can be defined as the development of a gastrula from a blastula through the inward migration of cells. A gastrula is an embryo with two layers of cells—the inner and outer layers. The inner layer of cells forms the mesoderm and endoderm, while the outer layer of cells

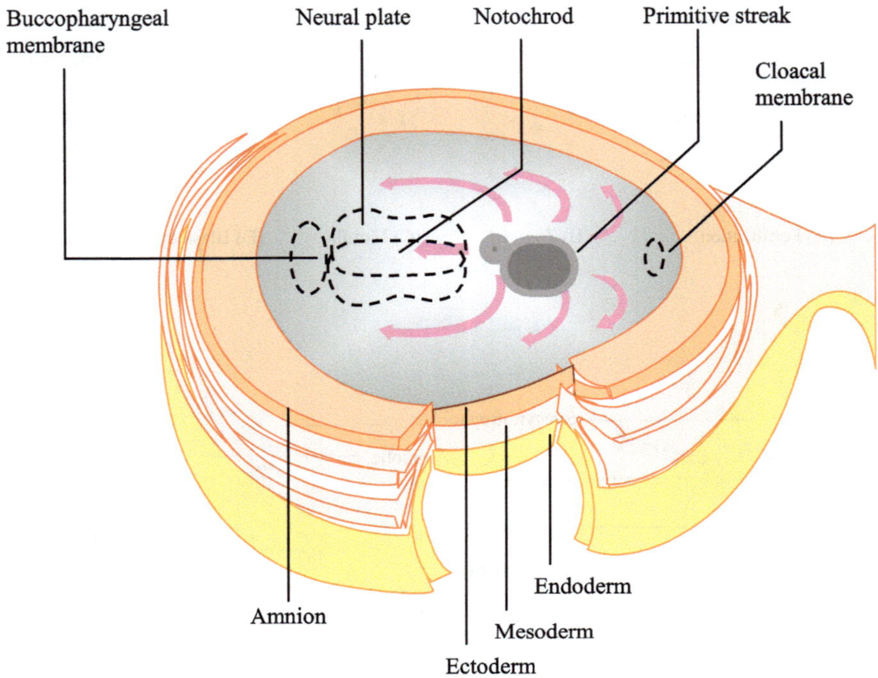

Fig. 6.2 Dorsal view of the embryonic disk showing the germs layers, associated structures, and the primitive gut. The three germ layers are formed by 9–17 days following embryonic implantation

forms the ectoderm (these layers are defined below). The epithelia of the epiblast undergo an epithelial to mesenchymal transition, losing their initial characteristics and moving through the primitive streak by ingression to form a new cell layer called the endoderm [12, 20–22]. In essence, the epiblast forms the embryo proper [14]. The formation of endoderm pushes the hypoblast out, thereby forming the amnion. As the extension process of the epiblast continues, another cell layer, the mesoderm is formed from the inductive interactions between the other two germ layers. The layer at the top is called the ectoderm [22, 23].

The terms "ectoderm" and "endoderm" were coined in 1853 by George James Allman (1812–1898), while "mesoderm" was introduced in 1871 by Thomas Henry Huxley (1825–1895). These layers are the three germ layers of gastrulation which have been known for over a century. They represent the fundamental embryonic layers from which tissues and organs arise and are specified genetically during the development of the egg. However, this triploblastic model of vertebrate development is presently contended with; the neural crest is sometimes considered the fourth germ layer [24, 25]. During the end of the last century, Brian Hall (1941–) suggested a quadriploblastic model of vertebrate development, in which he opined that like the mesoderm, the neural crest which is the dorsal part of the neural fold, is a secondary layer, though, formed through the inductive interactions between neural and epidermal portion of the ectoderm. This secondary formation has some common features, which include diversity of cells produced following the migration from the midline and both forms embryonic mesenchyme [24]. Emerging evidences also suggest that the endoderm and mesoderm arise from common precursor cells called the mesendoderm. In course of gastrulation, the progenitor cells of the mesendoderm give rise to either endoderm or mesoderm [26, 27].

6.2 Development of the Gastrointestinal System: From Embryo to Fetus

6.2.1 The Primitive Gut, the Mesenchyme–Epithelial Transition and the Derivatives of the Muscle Layers

The endoderm and the mesoderm form the primitive gut in early stages of development of the embryo at about 4 weeks of gestation. The primitive gut represents a tube extending from the buccopharyngeal membrane to the cloacal membrane, and it is divided into the foregut, midgut, and hindgut (Fig. 6.3). The epithelial lining of the gut tube progressively divides, obliterating the lumen. However, the lumen is developed later by recanalization. The precursors of the future small intestine are formed from the mesoderm and endoderm around the seventh week of embryonic development. The lumen of the intestine becomes visible from the ninth week of

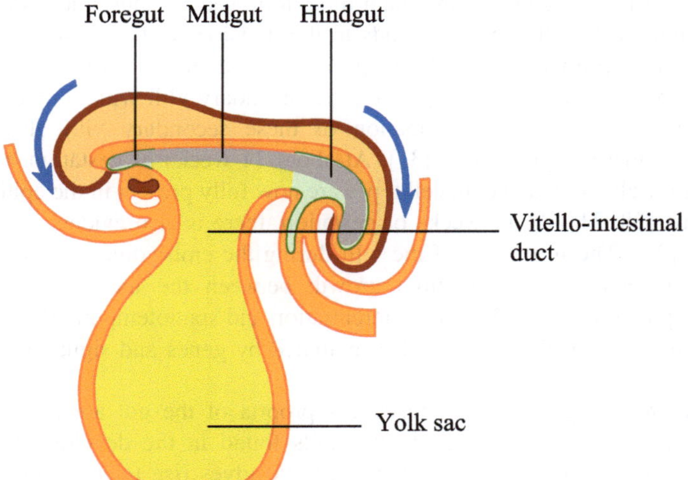

Fig. 6.3 Sagittal section showing the process of longitudinal folding by day 25 of gestation

gestation. During this period, the intestinal lumen is covered by cells of mesenchymal origin, having signs of epithelial differentiation with the beginning of development of the basement membrane [27–29]. Continuous epithelial differentiation and the development of the basement membrane produce evaginations into the epithelium that form the precursors of the lamina propria and the intestinal crypts. The mesenchymal cells also have a role to play in the formation of these structures [28].

The mesenchyme is a group of undifferentiated embryonic cells of the mesoderm that give rise to structures such as blood and lymphatic vessels as well as connective tissues. The cells of the mesenchyme are closely located with the basement membrane of the epithelial cells, fibroblasts, vascular endothelial cells, and proto-myofibroblasts. These cell types interact with each other via cross-signaling. For instance, the epithelial cells strongly interact with the mesenchyme to regulate the formation of different gut tissues. The fibroblasts, vascular endothelial cells, and collagen fibers, which are formed by the ninth week, are not an exception to the extensive signaling cross talks. The proto-myofibroblasts, which are formed already around the seventh week of gestation, interact via cross talks with the mesenchyme and also with the epithelium [28, 30]. The proto-myofibroblasts can differentiate into myofibroblasts in response to specific factors [12, 28]. Like the fibroblasts, the proto-myofibroblasts and the myofibroblasts play an integral role in the development of the villi, lamina propria, and the muscularis propria of the gut [28, 30].

By the tenth week, several longitudinal ridges have been formed by an elongation of the luminal surfaces of individual epithelial cells [31]. During formation of the ridges, the bases of some of the ridges lie along with the mesenchyme to form longitudinal mucosal folds. Thereafter, small cavities develop within the epithelium near to its base. These cavities are not continuous with the main gut tube, but gradually begin to extend into the main tube, followed by subsequent shedding of non-functional cells. This process leads to the division of the intestinal folds into structures called primary villi [31]. The primary villi continue to divide and enlarge, extending into the lumen. Subsequently, the secondary villi with simple columnar epithelium are formed. The formation of these secondary villi also requires upgrowth of mesenchymal cells [31]. At about 11 weeks of gestation, the crypts and villi are well-formed and all cell types are now fully present in the epithelium of the GI tract [28, 31]. At 11 weeks of gestation, there is also evidence of vascular structures [28]. The formation of the villi during the embryonic stage of development is triggered by a differential growth between the mesenchyme and the intestinal epithelium [32, 33]. The differentiation and spatiotemporal distribution of the epithelial cells of the villi are also regulated by genes and signaling pathways [34, 35].

The lamina propria and the muscularis propria of the gut are formed by the myofibroblasts, a type of mesenchymal cells found in the developing gut [28]. Thus, the intestinal mesenchyme of the embryo gives rise to the smooth muscle cells of the GI tract. By the 11th–14th week, the muscularis layers have been formed in the foregut, midgut, and hindgut [12, 36]. Figure 6.4 shows the GI tract of a fetus at about 11 weeks of gestation. In addition, a group of pacemaker cells,

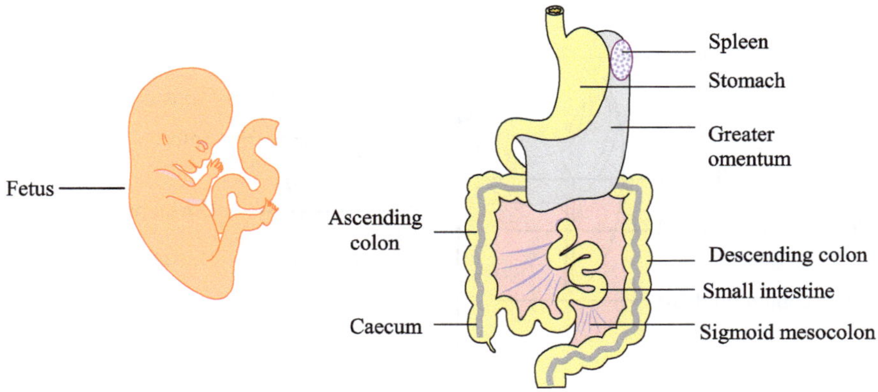

Fig. 6.4 Fetus at about 11 weeks of gestation. At this stage of development, the gut (including the mesenteries) is fully formed and continues to undergo maturation process *in utero*

called interstitial cells of Cajal (ICC)—pacemaker cells that coordinate muscular contraction in the GI tract, are also derived from the intestinal mesenchyme [37]. The ICC of the GI tract is localized in the ganglion plexus initially and later moves to the periphery [12, 36].

Continued development of the gut and the maturation process are regulated by intestinal stem cells present in the gut. The stem cells are present in the gut already around the 10th–11th week of gestation and form the intestinal stem cell niche responsible for renewing the intestinal epithelium [38–40]. It is believed that these stem cells are located in the crypt; however, the precise location of the stem cells responsible for the maintenance and differentiation of the villus and crypt epithelial lining and other regions of the gut are still under serious debate [28]. Details on the mechanisms of differentiation of intestinal stem cells and replacement of epithelial cells of the GI tract are discussed in Chap. 9 of Gastrointestinal Physiology: Contemporary Trends, Methods and Models.

6.2.2 *Development of Gastrointestinal Organs and Supportive Tissues*

Supportive organs of the GI tract as well as organs linking the tract develop in various stages of embryonic and fetal development (Fig. 6.5; Table 6.1). The development of the stomach is observed around 5–10 weeks of gestation. The accessory organs of digestion begin the process of development during the same time of development with other parts of the gut. Pancreatobiliary system develops at about 4–6 weeks. The liver starts to form at about 5 weeks of gestation [12, 41].

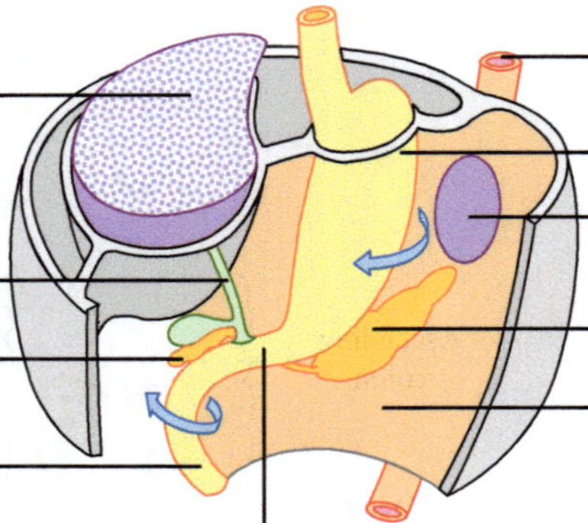

Fig. 6.5 Development of the structures of the GI tract, including the duodenum and mesenteries at 5 weeks of gestation

Table 6.1 Embryonic derivative of germ layers showing GI tract and associated system [26, 27, 42–50]

Germ layers	Brief description	Derivatives
Endoderm	This is the innermost of the primitive germ layers of the embryo and forms the midgut or archenteron—primitive gut. The term endoderm is synonymous with entoderm. This germ layer is concerned primarily with metabolism and homeostasis.	Gut tube
		Epithelia of the GI tract and associated glands
		Liver
		Pancreas
		Biliary system
		Thyroid, thymus, and lungs
Ectoderm[a]	This is the outermost of the germ layers that form the foregut and hindgut. The ectoderm interacts with the environment	Brain
		Spinal cord
		Peripheral nervous system
		Enamel of the tooth
Neural crest[a]	Together with the neural tube, the neural crest develops from the ectoderm	Peripheral nervous system
		Dentin of tooth
Mesoderm[a]	This is the middle layer of the primitive germ layers. A cavity, called coelom, is formed from the mesoderm. The coelom forms organs that move, grow, and develop independently. The primary role of the mesoderm is structural organization of the body	Blood and lymph vessels
		Serous membranes lining the body cavities
		Connective tissue
		Striated and smooth muscles

[a]The neurochord makes neuroectoderm from the ectoderm. The neuroectoderm forms CNS neurons, astrocytes, oligodendrocytes. The neural tube of the ectoderm develops into: brain, spinal cord. Neural crest cells are formed from the neural tube [51, 52]

6.2.3 Development of Gastrointestinal Nervous System

The nerves of the mesentery are connected to the intrinsic neural cells of the gut (which forms the enteric neurons). The nervous system of the GI tract is made up of intrinsic (enteric neurons) and extrinsic (peripheral) nerves (details on GI nervous system are discussed in Chap. 9). The enteric nervous system is formed from cells of the neural crest (NCC) that migrates into and along the gut, leading to the formation of a complex network of neurons and glial cells that regulates motility, secretion, and blood flow [30, 37]. Before the fourth week of gestation, NCCs begin their linear rostrocaudal migration to foregut and then down the bowel (Fig. 6.6). By the seventh week of gestation, this migration is expected to be complete. The majority of the NCCs that migrate to the gut come from the hindbrain (vagal) NCCs. However, a number of cells also migrate from the sacral region to the hindgut [30, 53, 54]. During the process of migration, some NCCs differentiate to form neuroblasts, which are multipotent stem cells and precursors of the enteric nervous system [55–57].

Some of the migrating NCCs differentiate to form clusters called ganglia, which produces a network of neurons (plexuses) throughout the bowel (Fig. 6.7) [54]. The enteric neurons are found in two major plexuses: myenteric and submucosal. Both plexuses are formed by neuroblasts. The myenteric plexus is formed by neuroblasts which were distributed to the GI tract by cranio-caudal migration during the 5th–12th week of gestation. The submucous plexus is formed by neuroblasts migrating from the myenteric plexus into the submucous layer. The origin of the myenteric neuroblasts is believed to be conveyed in part via vagus trunks from the central nervous system to the esophagus, and then into the rest of the GI tract [12, 36].

The migrating neuroblasts further differentiate into glial cells and neurons which can be detected in the submucosa of the foregut and midgut by the 12th week of gestation. The enteric glia can also arise from Schwann cells that enter the gut from the extrinsic nerves [54, 55, 58]. The different subtypes of neurons and glia are formed on the basis of influence of specific signaling/transcription and guidance factors as well as morphogens. The formation of a functional nervous system depends on these factors. Signals that direct the journey of these cells to the gut

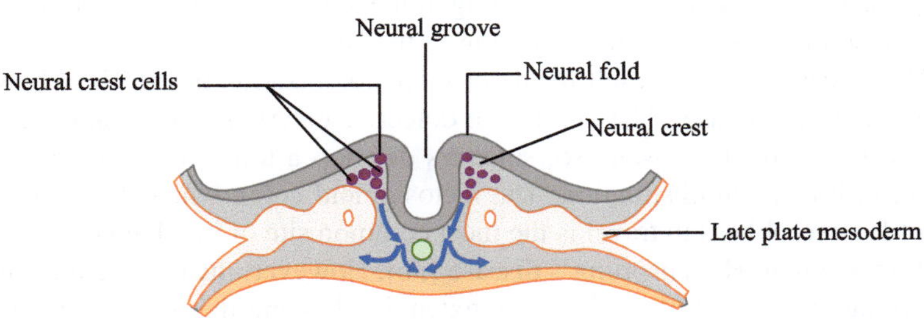

Fig. 6.6 Transverse section of the embryo showing the origin and migration of cells of the neural crest to form the enteric ganglion and other neural structures at 19 days of gestation

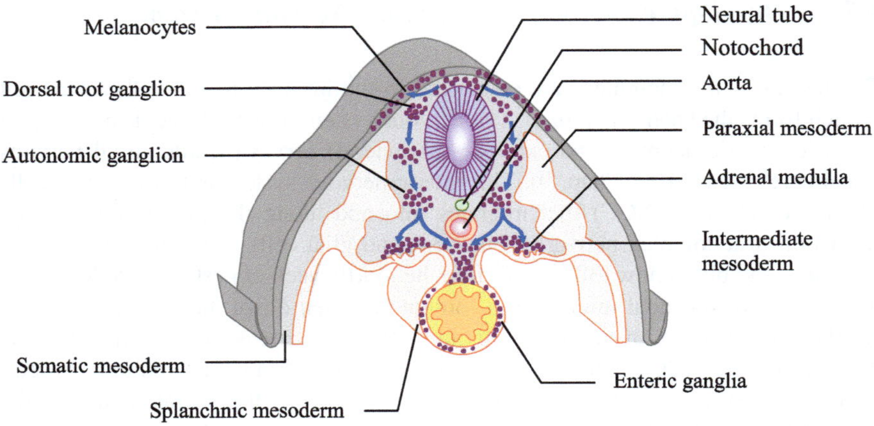

Fig. 6.7 Transverse section of the embryo showing the origin and migration of neural crest cells. Continued migration of these cells forms enteric and other ganglia at 21 days of gestation

include glial cell-derived neurotrophic factor, endothelin-3, etc., and their receptors as well as transcription factors such as Sox10 ((SRY (sex-determining region Y)-box 10)) and Phox2b (paired-like homeobox 2b). The migration of these cells is controlled by semaphorin 3A, cell adhesion molecules, morphogens, and small GTPases [54]. Transcription factors such as Mash1 also known as achaete-scute complex-like 1 (a member of the basic helix–loop–helix family of transcription factors), rearrangement during transformation (RET) and GFRα1 (GDNF family receptor-α1) and growth factors including GDNF, bone morphogenic protein, Hand2 (Heart And Neural Crest Derivatives-expressed Protein 2), and retinoic acid control the differentiation of the neuroblasts into different neuronal and glial sub-types [30, 53, 54, 59, 60]. The NCCs pass through different stages of migration, differentiation, proliferation, and survival to form a functional nervous system of the gut [54, 58].

In some conditions, the formation of the neural plexuses is disordered. Hirschsprung's disease (aganglionosis) is a developmental anomaly that results from dysfunction of intramural plexus development before the 12th week of gestation [54]. The disease is a life-threatening non-Mendelian genetic (developmental) disorder resulting from cessation or failure of the NCCs to colonize the distal bowel [54]. The disease can also result from defective colonization of the distal bowel by the migrating neuroblasts. The missing or defective enteric neurons cannot actively relax intestinal smooth muscle, so, sufferers develop a tonically contracted bowel resulting in functional obstruction. The rectosigmoid colon which is the most distant part for neuroblasts to travel is the most common site of involvement—referred to as short-segment Hirschsprung's disease [36]. However, the disease may affect a relatively large segment of the intestine extending beyond the rectum and sigmoid colon—long-segment Hirschsprung's disease. Hirschsprung's disease has an incidence rate of about 1 in 5000 live births. The majority of reported cases of the disease occur as an isolated condition; however, about 20% may be inherited. The reported cases of the disease more frequently tend to occur among male infants by about fourfold than among the females, especially in short-segment disease [54]. Symptoms of the disease include constipation, vomiting, abdominal pain, and growth failure. Children may die if intervention is not carried out in a timely manner [54].

6.2.4 *Separation of Larynx and Trachea from the Pharynx and Esophagus*

From Fig. 6.8, it is obvious that the gut (pharynx and esophagus) and the respiratory tract (larynx and trachea) develop from a single tube. Though it is not exactly clear when these two systems separate during ontogenesis, it is acknowledged that malfunctions during the separation process can lead to congenital disorders [61]. Any genetic malformation caused by signaling disorders vis-à-vis may result in serious structural malformations of the gut, which subsequently result in physiological dysfunctions that are evident in the newborn after birth. Drugs including alcohol, tobacco, nicotine, cocaine, amphetamine, ecstasy, and opiates (among others) as well as socioeconomic factors can affect the intrauterine development of the fetus, which may compromise the development of the GI tract and possibly other organs of the fetus [62–65]. Clinical correlate 6.1 gives brief information on some congenital diseases of the gut.

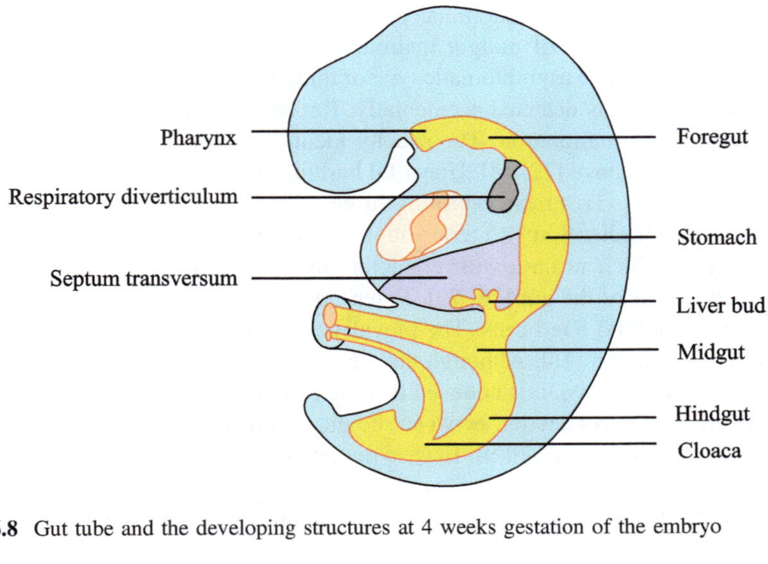

Fig. 6.8 Gut tube and the developing structures at 4 weeks gestation of the embryo

Clinical Correlate 6.1
Some Congenital Anomalies of the Gut

There are many disorders that can arise as a result of malfunctions during the stages of development of the embryo and fetus. These disorders, also called congenital diseases, include esophageal atresia, gut malrotation, duplication, diverticulum, annular pancreas, duodenal web, Brunner gland hyperplasia, among others (see below). Esophageal atresia has been discussed in Chap. 2.

Midgut Malrotation: Also called intestinal malrotation, midgut malrotation simply refers to an impaired rotation of the midgut and is characterized by intestinal obstruction [66, 67]. It has an incidence rate of 1 per 500 live births, but symptoms occur in one out of every 6000 live births [68, 69]. The etiopathogenesis of the disease has not been completely unraveled, but studies indicate that it is due to a malfunction in rotation of the intestinal loop around the axis of the superior mesenteric artery around tenth week of gestation. In normal development of the embryo gut, the intestine protrudes into the base of the umbilical cord. This is followed by elongation of the bowel and its return to the abdominal cavity. During this return process, the bowel turns counterclockwise through an angle of about 270° (i.e., often referred to as midgut rotation). The rotation leads to the formation of the duodenojejunal junction (also called duodenojejunal flexure) normally found on the left aspect of the midline, at the level of L1 vertebra. These processes lead to the formation of a wide mesentery extending from the duodenojejunal junction to the cecum. A defect resulting during the rotation process or formation of the mesentery can cause a disorder in (or prevent) rotation around the superior mesenteric artery. In symptomatic cases, infants with midgut malrotation experience vomiting (usually bilious), and volvulus, which may result in necrosis. There may be abdominal pain, melena, and weight loss [70, 71]. There are three types of midgut malrotation: non-rotation, reverse rotation, and malrotation. The asymptomatic type of midgut malrotation is described as non-rotation as it is detected accidentally. Reverse rotation is an uncommon form of midgut malrotation. The usually identified form—malrotation—has been described above [72–74]. Upper GI barium study is used for detection of malrotation [72]. Treatment option is surgery [75].

Enteric Duplication: Enteric duplication refers to a developmental anomaly in which a mass with epithelial and muscle layers is structurally attached to part of the gut [76, 77]. Enteric duplication has an incidence rate of about 1 in 4500 live births. The prevalence of the anomaly is 0.2% in the children population [78]. Although the mechanisms of enteric duplication are not completely known, it is believed that the condition results from a disorder in the process of connection between the gut and neural tube or canalization in the developing embryo [79]. Enteric duplication varies in size and location [80]. It may occur in the oral cavity (such as in sublingual duplication), esophagus, ileum, and colon. However, the ileum is the most common site of duplication [81–83]. Duplication can either occur within or outside the wall of the GI tract such as thoracoabdominal (intrathoracic and intraabdominal) duplications [81–83]. Enteric duplication may be cystic or tubular depending on the basis of whether or not it is spherical or has a communicating channel with the GI tract [81, 84]. The condition may be asymptomatic or symptomatic [79]. In symptomatic cases, presentation may mimic similar disorders with majority of cases observed around the first two years of life [78, 79]. The symptoms of enteric duplication include difficulty in breathing, dysphagia, dyspepsia, nausea, vomiting, abdominal distention, and chronic constipation. In some cases, an abdominal mass may be palpable [78, 79, 83–85]. However, the symptoms depend on the location and size of the duplications [84]. The complications include obstruction, perforation, acute abdomen, malignancy, and bleeding [79, 84]. Diagnosis of the condition is made with ultrasonography, barium meal ultrasonography, computed tomography, magnetic resonance imaging, and Technetium-99m pertechnetate scintigraphy [72, 76, 79, 82, 86]. Treatment is usually surgical and involves complete excision [77, 83].

Volvulus of the Intestines: The term "volvulus" is derived from the Latin "*volvere*"; meaning to roll. Volvulus of the intestine, sometimes called midgut volvulus, is a condition characterized by fixation of a segment or loop of the intestine to the wall of the gut via fibrous bands called Ladd's bands, due to midgut malrotation, combined with malfixation of the mesentery [87–89]. The small proportion of volvulus can develop due to torsion of the whole ileum, but with normal fixation of the mesentery [89]. The twisting of a segment of intestine around itself is the main cause of intestinal obstruction, which results in bilious vomiting [87, 90–92]. Twisting or malfixation of the mesentery that supports the affected segment of the intestine can compress the blood vessels in the affected region resulting in ischemia (loss of local blood circulation) and necrosis (death of cells) of the affected region [87, 91, 92]. Volvulus of the intestine usually occurs around the sigmoid colon and cecum, which are the regions affected in adults, whereas in children the commonly affected region is the ileum [90–93]. As previously noted, certain congenital disorders predispose the individual to the development of intestinal volvulus. The risk factors are chronic constipation, intestinal malrotation, Hirschsprung's disease, abdominal adhesions, possibly due to previous laparotomy [90–93]. Intestinal volvulus may be asymptomatic or symptomatic [90]. The signs and symptoms include abdominal pain, bilious vomiting, bloating, constipation, bloody stool, and abdominal distention [93]. The condition is diagnosed with abdominopelvic X-rays, color Doppler ultrasonography, and CT scan [89–94]. However, the gold standard is abdominopelvic CT scan [93]. The treatment option is surgery and involves exploratory laparotomy with bowel resection [91, 92].

Situs Ambiguus (heterotaxy), Situs Inversus, and Situs Solitus: Situs ambiguus or situs ambiguous, also known as heterotaxy or heterotaxia or visceral malposition, is a rare congenital defect in which the visceral organs are distributed abnormally within the thorax and abdomen. Heterotaxy can be referred to as a disorder of left–right laterality and arrangement of the visceral organs [95, 96]. Situs ambiguus may have components of situs solitus and situs inversus [97]. Situs solitus is the normal anatomy of the thoracoabdominal visceral organs, whereas situs inversus is the mirror image of situs solitus [95, 98]. Individuals with heterotaxy can have a variety of disorders including asplenia, polysplenia, heterotaxy subtype with malposition of the right or left atrial appendages [97]. The incidence of heterotaxy is estimated to be approximately 1 per 8000–25,000 live births [95]. The pathogenesis is not fully understood, but studies have indicated that heterotaxy may be due to some sort of visceral organ rotation disorder during embryogenesis. The left–right axis is defined around day 15 of embryogenesis. Around the 5th–6th week of gestation, the pancreas, liver, spleen, duodenum, and stomach are involved in some form of rotation [98, 99]. A disorder in the process of axis determination or rotation of these visceral organs or during this period of development can result in heterotaxy. Several genes have been identified in normal development of the right–left axis. Genes as well as their signaling pathways implicated in this syndrome include ZIC3 gene (encodes for the first zinc finger transcription factor) and genes that encode proteins in the TGF-beta pathway [100–104]. The use of echocardiogram and chest X-ray is essential for diagnosis. Treatment depends on the nature and severity of the disorders [97].

Omphalocele: Omphalocele, also known as exomphalos, is a congenital anterior abdominal wall defect in which the abdominal viscera (mostly intestines, liver, stomach, spleen, and in rare cases other visceral organs—urinary bladder or gonads) herniated through a central defect at the site of the umbilical ring into a membranous sac [87, 105, 106]. The sac can rupture in a few cases especially in utero or during childbirth [105]. Omphalocele is due to failure of normal return of the viscera to the abdominal cavity around the ninth–tenth week of gestation [107]. The condition is usually associated with chromosomal abnormalities and other congenital disorders involving the heart and neural tube defect, ectopia vesicae (also known as exstrophy of bladder, a condition characterized by herniation of the urinary bladder through a defect in the anterior abdominal wall), and Beckwith–Wiedemann syndrome (an overgrowth disorder which is susceptible to tumorigenesis) [87, 106, 108–110]. The incidence of omphalocele substantially varies across regions of the world, but generally occurs in 2.5 per 10,000 births [105].

Omphaloceles vary in size and therefore can be classified as minor or major [105, 106, 111]. The minor omphaloceles, also known as small omphaloceles, have an average size of defect of approximately 2.5–5 cm. The major omphaloceles, also known as large omphaloceles, have an average size of defect of approximately 6–12 cm [105, 111]. Minor omphaloceles are more common than the major ones [105].

Conservative treatment of omphalocoele is the primary care for this condition and is done with the aid of topical agents (e.g., escharotics). The use of Acacia nilotica paste and povidone-iodine solution is also used in omphalocoele treatment (but povidone-iodine is associated with thyrotoxicosis and hypothyroidism in children) [112]. The combination of povidone-iodine with powdered triple antibiotics (Bacitracin, Neomycin, and Polymyxin) has been shown to be associated with a faster process of escharification and epithelialization of the omphalocele and also minimizes the incidence of thyrotoxicosis and hypothyroidism [113]. The application of mercurochrome and alcohol is not considered safe in omphalocoele management [112]. The surgical treatment is to repair the defect and depends on defect size and associated congenital anomalies [114, 115].

Diverticulum: This congenital anomaly can be described as an outpouching of any portion of the GI tract. This disorder usually occurs in the colon [72, 116, 117]. The disease entity that characterizes the presence of diverticula (plural) is referred to as diverticulosis. It should be noted that some diverticula are acquired. Acquired diverticula are known as extraluminal diverticula, occurring in the duodenum, jejunum, ileum proper [72, 116]. The congenital types of diverticula are the intraluminal and Meckel's diverticula [116]. Meckel's diverticulum can be defined as a true intestinal diverticulum that results from the failure of the vitelline duct to obliterate or to fully regress around the fifth–eighth week of gestation [117]. Meckel's diverticulum can occur as an ectopic gastric or pancreatic tissue that continuously secretes juice, acid, and digestive enzymes. This can cause injury to the surrounding tissues during intrauterine life. During the first few days of life GI bleeding or excessive output of melena, abdominal pain due to tissue injury may be noticed, but bowel obstruction may occur in the next few years of development [87]. Meckel's diverticulum is the most prevalent congenital anomaly of the GI tract with a prevalence rate of about 2–4% in the population, whereas the prevalence of diverticula generally is estimated at 0.01–0.11% in the entire population [117–119]. Diverticula may be either symptomatic or asymptomatic. The symptoms of diverticula may be diverse and mimic other diseases such as peptic ulcer disease, Crohn's disease, or other acquired disorders of the gut [117]. Symptoms of diverticula include dyspepsia, pain (may occur anywhere in the abdomen), nausea, vomiting, and bloating [116–119]. Meckel's diverticulum is symptomatic diverticula, while diverticula of the ileum and colon may be asymptomatic [116]. There are several complications associated with diverticulosis. Major complications include diverticulitis (inflammation of diverticula), GI bleeding (e.g., upper or lower GI bleeding), perforation, bowel obstruction, abscess, volvulus, small intestinal bacterial overgrowth, intussusception, and pancreatobiliary diseases due to duodenal diverticula [116, 117, 119, 120]. Both radiological and surgical methods such as technetium-99m pertechnetate scan, X-ray with barium enema, and diagnostic laparoscopy are used for diagnosis of the condition [116, 117, 119]. The treatment for diverticulosis depends on the history and presentation. The ultimate treatment option is surgical resection of the diverticula [117]. For further review on this congenital anomaly, see Boynton and Floch [121], Weizman and Nguyen [122], Mantas et al. [123], and Ferreira-Aparicio et al. [116].

Annular Pancreas: This is a congenital anomaly, characterized by a ring of pancreatic tissue that surrounds the descending portion (i.e., second part) of the duodenum [124, 125]. The anomaly is due to a defect of rotation of the ventral bud with the duodenum [72, 126, 127]. Annular pancreas may be associated with Down syndrome, esophageal atresia, duodenal atresia, Meckel's diverticulum, and imperforate anus. Pregnant mothers with polyhydramnios (i.e., excess amniotic fluid) are likely to give birth to a child with annular pancreas compared to a normal pregnancy [72, 124–127]. The symptoms of the condition usually arise due to intestinal obstruction. This obstruction results from the narrowing of the intestinal lumen by the ring of pancreatic tissue that encircles the duodenum. The degree of obstruction determines the time of identification of the condition. Annular pancreas may be identified at any age. This anomaly can be identified by radiological imaging with barium salt, CT scan, and MRI [72, 124–128]. Other inborn diseases of the pancreas include pancreatic divisum, hypoplasia, agenesis, ectopia, heterotopia with a pancreatic head invasion. Congenital pancreatic anomalies are identified in about 10% of children [126, 127, 129].

Intestinal Atresia and Stenosis: This congenital disease is characterized by complete closure or absence of the lumen of the intestine. The condition is due to a failure of the lumen to canalize during intrauterine development [130]. The anomaly can occur anywhere along the GI tract, but ileal atresia is the most common accounting for over 90% of bowel obstruction in newborns. Intestinal atresia may be single or multiple. The former is the most prevalent [130–134]. Intestinal atresia is associated with other congenital anomalies of the gut such as gastroschisis, intrauterine growth retardation, maternal polyhydramnios, Down syndrome, annular pancreas, midgut malrotation, and meconium ileus [132, 133, 135]. Signs and symptoms include abdominal distension, bilious vomiting, jaundice, and inability to pass meconium within 24 h following birth [130]. Routine ultrasonography may be helpful in diagnosis of the condition [136]. Treatment of the anomaly is surgery [135, 137].

Intestinal stenosis refers to the incomplete obstruction of the lumen. It may occur anywhere along the GI tract and usually secondary to a web or other congenital pathologies of the gut [138]. Intestinal stenosis may be intrinsic or extrinsic. Examples of intrinsic stenosis include duodenal atresia and windsock deformity of the duodenum (intraluminal diverticulum of the duodenum). Examples of extrinsic stenosis include midgut malrotation, annular pancreas, and choledochal cyst [139]. The condition can be identified at any age depending on the time of onset of and severity of symptoms [140].

Gastroschisis: This congenital anomaly can be defined as a full-thickness abdominal wall defect mostly located on the right side of the umbilicus (navel) with intestinal prolapse but may also involve other abdominal visceral organs [141–143]. Treatment for this condition involves surgical repair of the abdominal wall [141, 142].

Duodenal Web: This is a congenital anomaly that presents as different entities such as windsock deformity or web, imperforate web (also known as complete duodenal atresia), non-concentric or centric web. Duodenal web usually occurs around the ampulla [72]. The occurrence of web can lead to stenosis and is rare causes of bowel obstruction [67, 144–146]. Symptoms of this anomaly include vomiting and abdominal distention, but there may be gastric outlet obstruction especially in infants [144, 146]. Diagnostic tests for identification of duodenal web include upper GI tract barium, CT scan, and upper GI endoscopy [72, 146]. The treatment of duodenal web is surgery [144].

Windsock Deformity: This is an extremely rare congenital anomaly involving a balloon-like dilation observed on a barium enema study in newborns with membranous intestinal atresia. The commonest site of windsock deformity is the descending portion of the duodenum [147, 148]. Symptoms of this condition include partial bowel (duodenal) obstruction [139, 147].

Brunner Gland Hyperplasia: This is a benign tumor of the duodenum that is usually asymptomatic [72, 129]. The lesions may be solitary or multiple and usually not more than 5 mm in diameter observed in the second portion of the duodenum [72, 149]. The disease, though, usually asymptomatic [72], may present with symptoms relating to those of upper GI obstruction or narrowing of the lumen. Brunner gland hyperplasia may be complicated by lower GI bleeding as evidenced by passage of black tarry stools [129, 149]. Diagnostic investigations include upper GI tract barium study, abdominal CT scan, upper GI endoscopy with histology [72, 129, 149]. Surgery is the treatment option for the condition [149].

Anorectal Malformation: Anorectal malformations refer to a spectrum of congenital disorders that involve the rectum and distal anus as well as the genitourinary tracts. The malformations vary and are usually associated with other congenital diseases [150]. The incidence of anorectal malformations is approximately 1 in 2000–5000 live births [151]. The etiopathogenesis is not fully understood, but multiple factors are likely to contribute to the development of the condition. Genes such as HLXB9 and Shh located on chromosome 7q39 have been implicated in the development of anorectal malformations [151].

There are currently several classification systems of anorectal malformations. The traditional classification uses “low,” “intermediate,” and “high” defects to depict various types of the malformations [150, 151]. Anorectal malformations may be associated with rectovaginal, rectovesical, rectobladder neck, and rectourethral fistulae. It should be pointed that some of these associated conditions may be misdiagnosed and could in fact be cloaca as in the case of, for instance, rectovaginal fistula [150, 152]. Furthermore, ectopic anal openings may be seen in some cases [152].

Symptoms of the condition include constipation, soiling (fecal incontinence), and urinary incontinence [153]. Radiological investigations include ultrasonography, CT scan, and MRI [154]. The management of anorectal malformations involves surgery. Surgical treatment involves posterior sagittal anorectoplasty (PSARP), and primary perineal repair [151, 152]. Unfortunately, after surgery, soiling can still occur, which is mostly due to associated anomalies [151].

6.2.5 *Nutrients Required for Embryonic and Fetal Growth and Development*

Development of the embryo and fetus requires adequate signaling of nutrients. One of such substances known to be actively involved in differentiation of cells of the embryo is glycogen. However, to ensure adequate growth and development, groups of nutrients that perform both plastic and energy functions must be continuously consumed to ensure continuity of life. The nutrients include amino acids, monosaccharides, fatty acids, and dietary elements. Following birth, the newborn must have constant flow of nutrients to maintain body functions which are necessary for survival. Adequate nutrient intake is inevitable for growth of the individual from birth to young adult stage and beyond [155–158].

6.3 Digestive Functions of a Newborn in the First Few Hours of Life

The first functional evidence of a normal gut formation is the passage of the meconium by the neonate within the 24–48 h after birth. Meconium is the first stool of a mammalian infant; it is viscous, sticky, and tarry-like in nature. The color of meconium is dark olive green. It is almost odorless. Unlike later feces, meconium is composed of materials ingested by the fetus from the amniotic fluid during the last trimester. The fetus may swallow as much as 750 ml per day during this period. Major constituents of the fluid include protein, free amino acids, and a variety of growth factors derived from amniotic fluid, lung and nasal secretions. The ingested fluid may provide about 10–15% of fetal nitrogen requirements per day [159–161]. In general, the swallowed fluid plays a huge role in the maturation of the fetal gut as well as regulation of release of humoral factors (such as endogenous cortisol) [160]. Thus, meconium is a digested product of swallowed amniotic fluid components and other substances *in utero*.

Passage of meconium in neonate is a developmentally programmed event that may be related to stress *in utero*, during the period of birth, or on exposure to the environment. However, meconium may be passed *in utero* by the fetus. In preterm and term neonates, intrauterine meconium passage may be due to fetomaternal stress or infection, whereas in postterm neonates it is usually associated with maturation of the GI tract [159, 161]. *In utero* passage of meconium may lead to respiratory distress—a condition called meconium aspiration syndrome and may result in death of the fetus or neonate after birth. On the other way around, if a child is unable to pass meconium after birth, it is probably due to a developmental block in one or more regions of the gut or motor dysfunction of the gut. Failure to pass meconium is seen in Hirschsprung's disease and mucoviscidosis (cystic fibrosis). Investigation of the chemical composition of meconium can provide a clue to future developmental pattern of the neonate and the maternal behavior during the period of pregnancy [162–165].

Meconium was previously thought to be sterile; however, emerging studies have challenged the sterility of meconium. It is now believed that meconium is made up of numerous enteric bacteria such as *Escherichia coli*, *Staphylococcus*, and *Lactobacilli*. The discovery of numerous microbes in meconium has challenged the etiopathogenesis of meconium aspiration syndrome. A couple of researchers believe that some severe cases of this syndrome may not be caused by the aspirated meconium per se, but the microbial content of the meconium [166–168]. It is believed that certain microbes swallowed by the fetus during intrauterine life may cause premature delivery [161, 169]. For instance, the presence of *Serratia marcescens* (a bacteria that is supposed to be present in the fecal mass of the growing neonate in first few days of life) in meconium is strongly associated with immaturity [169]. The type of microbes present in meconium together with perinatal factors such as mode of delivery, diet, and genetics all contributes to influence microbial colonization of the baby after birth [168].

Research has shown that while the weight of a mammalian fetus at term increases by 79% g kg^{-1} body weight, the activity of some digestive enzymes increases by 1.5–10-fold. The absorption of glucose and intact proteins increases by 3–6-fold. The contribution of the esophagus to body weight is estimated at 20%, suggesting that in addition to substances secreted by the esophagus, fluid that passes through this tube plays a huge role in the growth of the fetus. The esophagus contributes to intestinal weight by 43%, aminopeptidase A activity by 24%, and glucose absorption by 27%. The fluid that passes through the esophagus stimulates the fetus's cortisol production machinery. Cortisol is known to stimulate fetal GI tract maturation. As the fetus approaches term, the secretion of glucocorticoids increases, which is known to affect the functional development of the fetal liver, lungs, and GI tract. Accordingly, administration of glucocorticoids to preterm infants (infants delivered before 37 weeks of gestational age) before and after birth reduces the risk of intestinal disease. Ligation of the esophagus, however, has shown to increase the activities of lactase, sucrase, and dipeptidylpeptidase IV by about 40–50% [160]. It is, thus, evident that fetal enteral nutrition and glucocorticoid secretion have varying influences on different ages, gut regions and organs, and functions. These are the likely events that occur in esophageal obstruction. Consistent with the age differences in maturation of enzymes, research data published by Andersson et al. [170] showed that when a mammalian neonate's diet is predominantly milk, dominant lipases include the pancreatic enzymes—pancreatic lipase-related protein 2 and bile salt-stimulated lipase, which hydrolyze triglycerides to free fatty acid and glycerol [170]. Gastric lipase in addition to the said two pancreatic lipases secreted in newborns also plays a considerable role in lipid digestion. Surprisingly, however, bile salt-stimulated lipase is also present in breast milk, thus reinforcing the digestive functions of lipid in the alimentary canal. Depending on the substrate, the effect of both enzymes may be either additive or synergistic. Fatty acids released in the small bowel are efficiently absorbed by the cells and reesterified to triglycerides [170]. These triglycerides are then transported to lymphatic vessels in the form of "chylomicron". In contrast to adults, pancreatic triglyceride lipase and bile salts are found to have the lowest lipase activity [171].

During the first hours of life, the neonate is exposed to trophic feeding. Trophic feeding can be defined as the practice of feeding minute volumes of enteral feeds in order to stimulate the development of the immature GI tract. This feeding is important to reduce the incidence of necrotizing enterocolitis (NEC) and improve neurodevelopmental outcome, reduce the incidence of vomiting and constipation, improve milk tolerance, greater postnatal growth, reduce systemic sepsis, and shorter hospital stay. Trophic feeding generally affects the physiology of the GI system and alters GI enzyme activities, hormone secretion, blood flow, motility and microbial flora. It should be noted that increase in the volume of feeds can increase the incidence of NEC [172–175].

6.4 Digestive Functions of a Neonate in the Postnatal Period During the First ~48 h of Life

The gut of a neonate continues to undergo development and maturation even after birth. The neonate gut undergoes significant changes by adapting to the environment few weeks after birth. These changes are determined, in part, by the diet of the newborn as well as environmental factors and are largely modulated by genetic, neural, and humoral factors. The changes occurring in the GI tract are due to the maturation of absorptive, secretory, and transport functions of the gut [176]. These functions are not fully developed in a neonate, and thus neonate cannot utilize all nutrients effectively. Feed components apart from providing energy and molecules for growth also stimulate the genes and signaling pathways necessary for growth and maturation of the gut, especially the epithelium. This period involves dramatic changes in the microbial ecosystem of the neonate gut [177]. The initial colonizers of the gut during the first 48 h of life are aerobic species such as *Enterobacter sp.*, *streptococci*, and *staphylococci*, which enter the gut via external sources.

6.5 Digestive Functions of a Neonate in the Postnatal Period After ~48 h of Life

After about two days of life, the gut may be colonized by Bifidobacteria, Lactobacilli, Corynebacteria, Micrococci, Propionibacteria, etc. The microbial species that began the initial colonization of the gut also continues to do so. The fecals of a healthy neonate within the first week of life mainly contains a wide range of microbes including aerobic or facultative types such as *Lactococcus lactis*, *Leuconostoc citreum*, *Streptococcus mitis*, *Enterococcus faecalis*, *Escherichia coli*, *Streptococcus mitis*, *Streptococcus salivarius*, *Citrobacter*, *Clostridium difficile*, *Enterobacter sp.*, *Enterobacter cloacae*, and gram-positive cocci of *Pseudomonas sp.* [12, 178–180].

However, neonates delivered through cesarean section may experience a delay in colonization of the gut by beneficial bacteria such as bifidobacteria. The gut microflora of the neonate is determined by the mother's vaginal microflora; hence, cesarean delivery, including environmental bacteria from equipment, air, other infants, nursing staff can have significant influence on the neonate's gut microbiota. During the first week of life, mother's bacteria from the mouth, skin, or during breast sucking have influence on the neonate gut microflora [177]. Breast milk contains an estimated number of 10^9 microbes per liter in a healthy mother, but the origin of these microorganisms is not precisely known. The bacteria may originate from the nipple and surrounding skin as well as the milk ducts in the breast. Milk contains oligosaccharides and many prebiotic compounds which are excellent stimulators of microbial growth. Moreover, milk contains proteins known to inhibit the growth of harmful bacteria. The probiotic content of milk is capable of reducing feed-induced allergy. A stable intestinal microflora is also associated with reduction in frequency of allergy in neonates. After the first 2 weeks of life, the microflora ecology of the gut appears quite stable [12, 178]. The growth factors present in breast milk positively enhance the signaling of crypt–villus axis to improve growth and development of the intestinal epithelium [178, 181]. Such growth factors in breast milk include hepatocyte growth factor (HGF), erythropoietin (Epo), glucagon-like peptide-2 (GLP-2), insulin growth factor-1 (IGF-1), epidermal growth factor (EGF), growth hormone (GH), fibroblast growth factor (FGF), transforming growth factor beta (TGFβ), and keratinocyte growth factor (KGF) [178].

Gut microflora in the newborn are also important in the prevention of osmotic diarrhea and synthesis of biologically useful substances [179]. Substances produced by the gut microbiota include certain groups of essential molecules such as vitamins and short-chain fatty acids. There is increasing research on the usefulness of short-chain fatty acids synthesized by the gut microbiota. Humans are unable to synthesize short-chain fatty acids de novo and so this group of biomolecules must enter into the organism either by exogenous sources or they are produced by the commensal microbes of the gut. The usefulness of short-chain fatty acids to the organism cannot be overemphasized. They form the building block of membrane of cells and organelles, used as sources of energy, and form the good fat of the body. Short-chain fatty acids serve as fuel for colonocytes and stimulate cell replication in the colon and small intestine. More importantly, short-chain fatty acids play significant role as xenobiotics and scavengers of free radicals and toxins in the liver and other organs and tissues of the body. In addition, short-chain fatty acids that stimulate sodium and water absorption serve to prevent diarrhea, which may result from disorder in nutrient absorption. Malabsorption of sugar in the small bowel is associated with the onset of osmotic diarrhea. This type of diarrhea may be caused by high content of unfermented sugar or even abnormally excessive fermentation which results in abnormally low luminal pH. The low luminal pH inhibits bacterial enzymes. Thus, the condition can be treated with colonic motility stimulants or administration of stimulants of gut fermentation such as probiotics [179, 180]. Malabsorption is usually more serious in preterm infants due to the increased intestinal permeability and immaturity of the gut. Thus, microbes including potentially dangerous ones may enter the gut and easily translocate to systemic organs and tissues. Thus, the infant is prone to higher rate of infections and systemic diseases. This risk is worse for preterm infants since their immune system is still not fully developed to counter such pathogenic invasion. The use of antibiotics in neonates and preterm infants may contribute to the reduction in the colonization of the gut by beneficial microbes [178, 181, 182].

The urge for food ingestion is a developmental process and genetically as well as evolutionarily determined beginning from the process of ontogenesis and is aimed at maintaining the need of the growing individual and the requirements of physiological processes. Generally, the intake of nutrients is controlled by several factors —exogenous and endogenous factors. Exogenous confounding factors such as nutrient availability may determine the intake and quality of nutrients. Endogenous determinant of nutrient intake is regulated by a complex system of neural, behavioral, and hormonal origin, which constantly aid in sensing the quantity of nutrients in the blood. The amount of blood nutrients is a changing physiological constant, which is controlled by the functional system of feeding. This system regulates the intake of nutrients to maintain a normal physiological condition for the organism [182–190].

6.6 Gastrointestinal Aging

The GI tract undergoes continuous development to attain full maturity. The adult GI tract is continuously renewed even after attaining maturation. Due to aging, some parts of the tract as well as accessory organs may gradually lose their functions. Shedding of the cell of epithelium and apoptosis of cell of the GI system generally are normally counterbalanced with their renewal. However, as the individual progresses to old age, the rate of loss exceeds the rate of renewal. Apart from aging, many conditions including disease states, nutrition, and environmental factors may interfere with the balance between cell loss and renewal [191, 192].

Normal aging is associated with many GI impairments, either loss or decrease in functions, which may be divided into functional and structural disorders. The structural disorders include mucosal variations, mucosal growth, GI tract carcinogenesis, gastric mucosal changes, and intestinal bacterial overgrowth. The functional disorders include motility changes, secretory, intraluminal digestion, absorption, changes in taste and smell [193, 194]. Functional GI disorders are associated with swallowing disorder, dysphagia, dyspepsia, maldigestion, malabsorption, postprandial hypotension (due to slow gastric emptying), fecal incontinence, constipation, aspiration pneumonitis, increased Clostridium difficile infections, gallstones, and altered drug metabolism [195–200]. Normal aging is associated with both microanatomical and functional changes of the GI tract—which form the basis of physiological anorexia of aging [201, 202]. However, the aging of the gut predisposes the elderly to GI impairments. For instance, the elderly experience more cases of gastroesophageal reflux, atrophic gastritis, gastric ulcer, colon diverticulosis, irritable bowel syndrome, diverticula due to weakening of the colonic muscular wall, gallstones, chronic hepatitis, liver cirrhosis, hepatocellular carcinoma, and chronic pancreatitis [203, 204]. Some of the functional disorders may be due to dysfunctions of oropharyngeal muscle motility which may result from decreased esophageal peristalsis and lower esophageal sphincter pressures. Because of peristaltic and sphincter disorders, the elderly people frequently come down with more cases of prolonged gastric motility and emptying, intestinal and anorectal motility [203]. Changes in GI motility in the aged may be due to selective decline in the number and volume of neurons and glial cells in the nervous system of the gut accompanied by gradual loss of ICCs [205, 206]. Gomez-Pinilla et al. [207] found that the number of ICC bodies and volume in the stomach and colon significantly decreased with age at a rate of 13% per decade [206]. The gender and regional differences in ICCs are associated with age. It should be mentioned that some elderly people may not experience these functional gut disorders [203]. The reason may be due to the conservation of the relatively large reserve capacity of the intestine, pancreas, and liver as well as nutritional status [207, 208]. The nutritional state of the individual can substantially affect the development of functional GI disorders. The effect of nutritional status can be mediated, in part, through the intestinal microfloral ecology, which in turn modulates a number of GI events and functions. It should be noted that normal aging of the gut may increase the diversity of commensal microbes, but the major ones remain bifidobacteria and lactobacilli [208–212]. However, even these bacteria may express different strain types across life span. For example, neonates and infants generally have *Bifidobacterium bifidum, B. breve*, and *B. longum*, whereas adults have *B. adolescentis*, *B. catenulatum*, and *B. longum* [212].

The secretory changes associated with aging include decreased secretory activity of different regions of the GI tract: achlorhydria, hypochlorhydria, reduced enzyme and hormone secretion. The secretions of the exocrine pancreas and gallbladder are also reduced. Decreased secretions result in impairment of the mucous-bicarbonate barrier and may lead to gastric ulcer [191, 203].

Normal aging is associated with varying changes in absorption in different regions of the GI tract. Aging may increase the absorption of lipids and large size molecules. However, impairment in absorption may be found for vitamin B-12, calcium carbonate, ferric iron as observed in cases of atrophic gastritis [207]. Impairment in absorption of calcium, zinc, and other nutrients is also observed in normal processes of aging. Malabsorption of some forms of iron, calcium, and vitamin D may be due to achlorhydria [207].

Some of the changes associated with aging are due to decreased immunity possibly caused by intestinal microbial dysbalance. In the liver, aging is associated with delayed drug metabolism. Some of these changes occur basically on account of decreased blood flow to the organ or region of the tract [213–215].

Conditions that affect or worsen the evaluation of such age-related changes in GI functions include coexistence of disease conditions and history of use of medications. Such medications include broad-spectrum antibiotics, anticholinergics (antidepressants with anticholinergic effect), opioid analgesics, and calcium antagonists. Disease conditions that may coexist with age-related changes in GI functions include diabetes, depression, hypothyroidism, and chronic renal failure. Gastric emptying is significantly reduced in long-standing diabetes mellitus in about half of the sufferers. This function of the GI tract is also impaired in chronic renal failure. Whole-gut transit time is significantly prolonged in depression and may be prolonged in hypothyroidism [193].

Surprisingly, the functional changes related to aging are increasingly been considered as early features of Parkinson's disease and frequently precede the neurological manifestations [205]. This further confirms the gut–brain functional relationship in health and disease [216–221].

The marked difference in the parameters of structural and functional integrity of the gut is usually observed in individuals of about 65 years and above compared with younger people [203, 204].

6.7 Conclusion

The development of the GI tract begins early during ontogenesis and is fully formed by the 11th week of gestation. The growth and development of the gut are predetermined by thousands of genes that control multiple signaling pathways, transcription factors, morphogens, and growth factors. Environmental factors serve to modulate these factors and may have profound effects on the fetus. Disorders in any of the stages of development of the gut are associated with congenital anomalies. The process of maturity of the gut and aging is also controlled by the signaling pathways that were necessary for the formation of the gut. However, large influence on the gut during this period is due to nutritional status and other environmental and the health conditions of the individual as a whole.

Recommended Readings

Original Articles

1. Soffers JHM, Hikspoors JPJM, Mekonen HK, Koehler SE, Lamers WH (2015) The growth pattern of the human intestine and its mesentery. BMC Dev Biol 15:31
2. Wang J, Li G, Qian G-H, Hua J-H, Wang Y-Q (2016) Expression analysis of eight amphioxus genes involved in the Wnt/β-catenin signaling pathway. Dongwuxue Yanjiu 37(3):136–143

Review Articles

1. Aiello NM, Stanger BZ (2016) Echoes of the embryo: using the developmental biology toolkit to study cancer. Dis Model Mech 9(2):105–114
2. Aurora M, Spence JR (2016) hPSC-derived lung and intestinal organoids as models of human fetal tissue. Dev Biol 420(2):230–238
3. Danowitz M, Solounias N (2016) Embryology, comparative anatomy, and congenital malformations of the gastrointestinal tract. Edorium J Anat Embryo 3:39–50
4. Dasgupta S, Arya S, Choudhary S, Jain SK (2016) Amniotic fluid: source of trophic factors for the developing intestine. World J Gastrointest Pathophysiol 7 (1):38–47
5. Lake JI, Heuckeroth RO (2013) Enteric nervous system development: migration, differentiation, and disease. Am J Physiol Gastrointest Liver Physiol 305 (1): G1–G24
6. Lévy E, Delvin E, Ménard D, Beaulieu J-F (2009) Functional development of human fetal gastrointestinal tract. Methods Mol Biol 550:205–24
7. Savin T, Kurpios NA, Shyer AE, Florescu P, Liang H, Mahadevan L, Tabin CJ (2011) On the growth and form of the gut. Nature 476(7358):57–62
8. Thum C, Cookson AL, Otter DE, McNabb WC, Hodgkinson AJ, Dyer J, Roy NC (2012) Can nutritional modulation of maternal intestinal microbiota

influence the development of the infant gastrointestinal tract? J Nutr 142 (11):1921–8
9. Vaillancourt C, Lafond J (2009) Human embryogenesis: overview. Methods Mol Biol 550:3–7

Guidelines

1. Dutta S, Singh B, Chessell L, Wilson J, Janes M, McDonald K et al (2015) Guidelines for feeding very low birth weight infants. Nutrients 7:423–442
2. de la Torre AM, de Mateo Silleras B, Pérez-García A (2001) Guidelines for nutrition support in the elderly. Public Health Nutr 4(6A):1379–84
3. Kuo AA, Inkelas M, Slusser WM, Maidenberg M, Halfon N (2011) Introduction of solid food to young infants. Matern Child Health J 15(8):1185–1194
4. Torrazza RM, Neu J (2013) Evidence-based guidelines for optimization of nutrition for the very low birth weight infant. NeoReviews 14(7):e340–e349
5. Volkert D, Berner YN, Berry E, Cederholm T, Coti Bertrand P, Milne A et al (2006) ESPEN guidelines on enteral nutrition: geriatrics. Clin Nutr 25(2):330–360
6. World Health Organization (2011) Guidelines on optimal feeding of low birth-weight infants in low- and middle-income countries. Geneva, Switzerland

Books

1. Gary SD, Jason SE, Thomas PB, eds. (2012) Embryo culture: methods and protocols. Springer, New York, USA
2. Lafond J, Vaillancourt C (eds) (2009) Human embryogenesis: methods and protocols. Humana Press, New York, USA
3. Paul S, Ivor M (eds) (2009) Molecular embryology: methods and protocols, 2nd edn. Humana Press, Totowa, New Jersey, USA
4. Sadler TW (2012) Langman's medical embryology, 12th edn. Lippincott Williams & Wilkins, Philadelphia, USA

Bibliography

1. Longo LD (2017) The rise of fetal and neonatal physiology. Springer, New York
2. Nicoletto SF, Rinaldi A (2011) In the womb's shadow. EMBO Rep 12(1):30–34
3. Saffery R, Novakovic B (2014) Epigenetics as the mediator of fetal programming of adult onset disease: what is the evidence? Acta Obstet Gynecol Scand 93(11):1090–1098
4. Padmanabhan V, Cardoso RC, Puttabyatappa M (2016) Developmental programming, a pathway to disease. Endocrinology 157(4):1328–1340
5. Gluckman PD, Hanson MA, Spencer HG, Bateson P (2005) Environmental influences during development and their later consequences for health and disease: implications for the interpretation of empirical studies. Proc Biol Sci 272(1564):671–677
6. Shehzad K (2011) Neonatal birth-weights and reference intervals in sonographically monitored normal fetuses. Int J Health Sci (Qassim) 5(2 Suppl 1):27

7. Xie YJ, Peng R, Han L, Zhou X, Xiong Z, Zhang Y et al (2016) Associations of neonatal high birth weight with maternal pre-pregnancy body mass index and gestational weight gain: a case–control study in women from Chongqing, China. BMJ Open 6(8):e010935
8. Weekes AR, Flynn MJ (1975) Engagement of the fetal head in primigravidae and its relationship to duration of gestation and time of onset of labour. Br J Obstet Gynaecol 82 (1):7–11
9. Jukic AM, Baird DD, Weinberg CR, McConnaughey DR, Wilcox AJ (2013) Length of human pregnancy and contributors to its natural variation. Hum Reprod 28(10):2848–2855
10. De Santa B (2011) Molecular embryology of the foregut. J Pediatr Gastroenterol Nutr 52 (Suppl 1):S2–S3
11. Itskovitz-Eldor J, Schuldiner M, Karsenti D, Eden A, Yanuka O, Amit M et al (2000) Differentiation of human embryonic stem cells into embryoid bodies comprising the three embryonic germ layers. Mol Med 6(2):88–95
12. Lévy E, Delvin E, Ménard D, Beaulieu J-F (2009) Functional development of human fetal gastrointestinal tract. Methods Mol Biol 550:205–224
13. Foulk RA (2001) From fertilization to implantation. Early Pregnancy 5(1):61–62
14. Tang WWC, Kobayashi T, Irie N, Dietmann S, Surani MA (2016) Specification and epigenetic programming of the human germ line. Nat Rev Genet 17:585–600
15. Bleil JD, Wassarman PM (1980) Structure and function of the zona pellucida: identification and characterization of the proteins of the mouse oocyte's zona pellucida. Dev Biol 76 (1):185–202
16. Wassarman PM, Mortillo S (1991) Structure of the mouse egg extracellular coat, the zona pellucida. Int Rev Cytol 130:85–110
17. Fazleabas AT (2007) Physiology and pathology of implantation in the human and nonhuman primate. Semin Reprod Med 25(6):405–409
18. Niakan KK, Han J, Pedersen RA, Simon C, Pera RAR (2012) Human pre-implantation embryo development. Development 139(5):829–841
19. Oestrup O, Hall V, Petkov SG, Wolf XA, Hyldig S, Hyttel P (2009) From zygote to implantation: morphological and molecular dynamics during embryo development in the pig. Reprod Domest Anim 44(Suppl 3):39–49
20. Mikawa T, Poh AM, Kelly KA, Ishii Y, Reese DE (2004) Induction and patterning of the primitive streak, an organizing center of gastrulation in the amniote. Dev Dyn 229(3):422–432
21. Chuai M, Weijer CJ (2008) The mechanisms underlying primitive streak formation in the chick embryo. Curr Top Dev Biol 81:135–156
22. Vasiev B, Balter A, Chaplain M, Glazier JA, Weijer CJ (2010) Modeling gastrulation in the chick embryo: formation of the primitive streak. PLoS ONE 5(5):e10571
23. Gilbert SF (2000) Developmental biology, 6th edn. Sinauer Associates, Sunderland
24. Hall BK (2000) The neural crest as a fourth germ layer and vertebrates as quadroblastic not triploblastic. Evol Dev 2:3–5
25. Lee S (ed) (1901) Allman, George James. Dictionary of National Biography. Smith, Elder & Co, London
26. Spence JR, Lauf R, Shroyer NF (2011) Vertebrate intestinal endoderm development. Dev Dyn 240(3):501–520
27. Zorn AM, Wells JM (2009) Vertebrate endoderm development and organ formation. Annu Rev Cell Dev Biol 25:221–251
28. Artells R, Navarro A, Diaz T, Monzó M (2011) Ultrastructural and immunohistochemical analysis of intestinal myofibroblasts during the early organogenesis of the human small intestine. Anat Rec 294(3):462–471
29. Lee DH, Chung HM (2011) Differentiation into endoderm lineage: pancreatic differentiation from embryonic stem cells. Int J Stem Cells 4(1):35–42
30. Obermayr F, Hotta R, Enomoto H, Young HM (2013) Development and developmental disorders of the enteric nervous system. Nat Rev Gastroenterol Hepatol 10(1):43–57

31. Bell L, Williams L (1982) A scanning and transmission electron microscopical study of the morphogenesis of human colonic villi. Anat Embryol 165(3):437–455
32. Balbi V, Ciarletta P (2013) Morpho-elasticity of intestinal villi. J R Soc Interface 10 (82):20130109
33. Walton KD, Kolterud Å, Czerwinski MJ, Bell MJ, Prakash A, Kushwaha J, Grosse AS, Schnell S, Gumucio DL (2012) Hedgehog-responsive mesenchymal clusters direct patterning and emergence of intestinal villi. PNAS 109(39):15817–15822
34. Johnson FP (1910) The development of the mucous membrane of the esophagus, stomach and small intestine in the human embryo. Am J Anat 10:521–561
35. Johnson FP (1913) The development of the mucous membrane of the large intestine and the vermiform process in the human embryo. Am J Anat 14:187–233
36. Okamoto E, Ueda T (1967) Embryogenesis of intramural ganglia of the gut and its relation to Hirschsprung's disease. J Pediatr Surg 2(5):437–443
37. Bates MD (2002) Development of the enteric nervous system. Clin Perinatol 29(1):97–114
38. Mills JC, Gordon JI (2001) The intestinal stem cell niche: there grows the neighborhood. Proc Natl Acad Sci U S A 98(22):12334–12336
39. Takashima S, Hartenstein V (2012) Genetic control of intestinal stem cell specification and development: a comparative view. Stem Cell Rev 8(2):597–608
40. Martini E, Schneider E, Neufert C, Neurath MF, Becker C (2016) Survivin is a guardian of the intestinal stem cell niche and its expression is regulated by TGF-β. Cell Cycle 15 (21):2875–2881
41. Sadler TW (2012) Langman's medical embryology, 12th edn. Lippincott Williams & Wilkins, Philadelphia
42. Tam PP, Kanai-Azuma M, Kanai Y (2003) Early endoderm development in vertebrates: lineage differentiation and morphogenetic function. Curr Opin Genet Dev 13(4):393–400
43. Fukuda K, Kikuchi Y (2005) Endoderm development in vertebrates: fate mapping, induction and regional specification. Dev Growth Differ 47(6):343–355
44. Kluth D, Fiegel HC, Metzger R (2011) Embryology of the hindgut. Semin Pediatr Surg 20 (3):152–160
45. Kluth D, Steding G, Seidl W (1987) The embryology of foregut malformations. J Pediatr Surg 22(5):389–393
46. Kluth D, Lambrecht W (1997) Current concepts in the embryology of anorectal malformations. Semin Pediatr Surg 6(4):180–186
47. Kluth D, Fiegel H (2003) The embryology of the foregut. Semin Pediatr Surg 12(1):3–9
48. Kluth D (2010) Embryology of anorectal malformations. Semin Pediatr Surg 19(3):201–208
49. Seely S (2000) Some new thoughts on evolution: the role of the germ layers. Med Hypotheses 54(6):1019–1022
50. Evans R (1899) The structure and metamorphosis of the larva of Spongilla lacustris. Q J Microsc Sci 43:362–477
51. Purves D, Augustine GJ, Fitzpatrick D, Katz LC, LaMantia A-S, McNamara JO, Williams SM (eds) (2001) Neuroscience, 2nd edn. Sinauer Associates, Sunderland
52. Artinger KB, Bronner-Fraser M (1992) Notochord grafts do not suppress formation of neural crest cells or commissural neurons. Development 116:877–886
53. Sasselli V, Pachnis V, Burns AJ (2012) The enteric nervous system. Dev Biol 366(1):64–73
54. Lake JI, Heuckeroth RO (2013) Enteric nervous system development: migration, differentiation, and disease. Am J Physiol Gastrointest Liver Physiol 305(1):1–G24
55. Tamura Y, Matsumura K, Sano M, Tabata H, Kimura K, Ieda M et al (2011) Neural crest–derived stem cells migrate and differentiate into cardiomyocytes after myocardial infarction. Arterioscler Thromb Vasc Biol 31:582–589
56. Loring JF, Barker DL, Erickson CA (1988) Migration and differentiation of neural crest and ventral neural tube cells in vitro: implications for in vitro and in vivo studies of the neural crest. J Neurosci 8(3):1001–1015

57. Narytnyk A, Verdon B, Loughney A, Sweeney M, Clewes O, Taggart MJ, Sieber-Blum M (2014) Differentiation of human epidermal neural crest stem cells (hEPI-NCSC) into virtually homogenous populations of dopaminergic neurons. Stem Cell Rev 10(2):316–326
58. Gershon MD, Rothman TP (1991) Enteric glia. Glia 4(2):195–204
59. Young HM, Hearn CJ, Newgreen DF (2000) Embryology and development of the enteric nervous system. Gut 47:iv12–iv14
60. Fu M, Lui VCH, Sham MH, Pachnis V, Tam PKH (2004) Sonic hedgehog regulates the proliferation, differentiation, and migration of enteric neural crest cells in gut. J Cell Biol 166 (5):673–684
61. Metzger R, Wachowiak R, Kluth D (2011) Embryology of the early foregut. Semin Pediatr Surg 20(3):136–144
62. Smith LM, LaGasse LL, Derauf C, Grant P, Shah R, Arria A et al (2006) The infant development, environment, and lifestyle study: effects of prenatal methamphetamine exposure, polydrug exposure, and poverty on intrauterine growth. Pediatrics 118(3):1149–1156
63. Thompson BL, Levitt P, Stanwood GD (2009) Prenatal exposure to drugs: effects on brain development and implications for policy and education. Nat Rev Neurosci 10:303–312
64. Chiriboga CA (2003) Fetal alcohol and drug effects. Neurologist 9(6):267–279
65. Ross EJ, Graham DL, Money KM, Stanwood GD (2015) Developmental consequences of fetal exposure to drugs: what we know and what we still must learn. Neuropsychopharmacology 40(1):61–87
66. Kluth D, Jaeschke-Melli S, Fiegel H (2003) The embryology of gut rotation. Semin Pediatr Surg 12(4):275–279
67. Eksarko P, Nazir S, Kessler E, LeBlanc P, Zeidman M, Asarian AP et al (2013) Duodenal web associated with malrotation and review of literature. J Surg Case Rep 2013(12):rjt110
68. Ingoe R, Lange P (2007) The Ladd's procedure for correction of intestinal malrotation with volvulus in children. AORN J 85:300–308
69. Williams H (2007) Green for danger! Intestinal malrotation and volvulus. Arch Dis Child Educ Pract Ed 92:ep87–91
70. Lee HC, Pickard SS, Sridhar S, Dutta S (2012) Intestinal malrotation and catastrophic volvulus in infancy. J Emerg Med 43(1):e49–e51
71. Torres AM, Ziegler MM (1993) Malrotation of the intestine. World J Surg 17(3):326–331
72. Gupta P, Debi U, Sinha SK, Prasad KK (2014) Upper gastrointestinal barium evaluation of duodenal pathology: a pictorial review. World J Radiol 6(8):613–618
73. Kim WK, Kim H, Ahn DH, Kim MH, Park HW (2003) Timetable for intestinal rotation in staged human embryos and fetuses. Birth Defects Res A Clin Mol Teratol 67(11):941–945
74. Hamidi H, Obaidy Y, Maroof S (2016) Intestinal malrotation and midgut volvulus. Radiol Case Rep 11(3):271–274
75. Pickhardt PJ, Bhalla S (2002) Intestinal malrotation in adolescents and adults: spectrum of clinical and imaging features. AJR Am J Roentgenol 179(6):1429–1435
76. Macpherson RI (1993) Gastrointestinal tract duplications: clinical, pathologic, etiologic, and radiologic considerations. Radiographics 13(5):1063–1080
77. Jehangir S, Ninan PJ, Jacob TJ, Eapen A, Mathai J, Thomas RJ, Karl S (2015) Enteric duplication in children: experience from a tertiary center in South India. J Indian Assoc Pediatr Surg 20(4):174–178
78. Rattan KN, Bansal S, Dhamija A (2017) Gastrointestinal duplication presenting as neonatal intestinal obstruction: an experience of 15 years at tertiary care centre. J Neonatal Surg 6 (1):5
79. Olajide A-RL, Yisau AA, Abdulraseed NA, Kashim IOO, Olaniyi AJ, Morohunfade AOA (2010) Gastrointestinal duplications: experience in seven children and a review of the literature. Saudi J Gastroenterol 16(2):105–109
80. Chaudhary S, Raju U, Harjai M, Gupta CM (2006) Intestinal duplication cyst. Med J Armed Forces India 62(1):87–88

81. Liu R, Adler DG (2014) Duplication cysts: diagnosis, management, and the role of endoscopic ultrasound. Endosc Ultrasound 3(3):152–160
82. Bhat NA, Agarwala S, Mitra DK, Bhatnagar V (2001) Duplications of the alimentary tract in children. Trop Gastroenterol 22(1):33–35
83. Karnak I, Ocal T, Senocak ME, Tanyel FC, Büyükpamukçu N (2000) Alimentary tract duplications in children: report of 26 years' experience. Turk J Pediatr 42(2):118–125
84. Tabari AK, Mirshemirani A, Tabari NK (2012) Complete colonic duplication in children. Caspian J Intern Med 3(2):436–439
85. Plummer JM, McFarlane ME, Torrence DE, Duncan N, Shah S, Roberts P (2009) Intestinal duplication presenting with recurrent abdominal pain. Can J Surg 52(4):E103–E104
86. Ravitch MM (1953) Hind gut duplication—doubling of colon and of the genital and lower urinary tracts. Ann Surg 137(5):588–601
87. Danowitz M, Solounias N (2016) Embryology, comparative anatomy, and congenital malformations of the gastrointestinal tract. Edorium J Anat Embryo 3:39–50
88. Duran C, Oztürk E, Uraz S, Kocakuşak A, Mutlu H, Killi R (2008) Midgut volvulus: value of multidetector computed tomography in diagnosis. Turk J Gastroenterol 19(3):189–192
89. Shah MR, Levin TL, Blumer SL, Berdon WE, Jan DM, Yousefzadeh DK (2015) Volvulus of the entire small bowel with normal bowel fixation simulating malrotation and midgut volvulus. Pediatr Radiol 45(13):1953–1956
90. Langer JC (2017) Intestinal rotation abnormalities and midgut volvulus. Surg Clin North Am 97(1):147–159
91. Kapadia MR (2017) Volvulus of the small bowel and colon. Clin Colon Rectal Surg 30 (1):40–45
92. Gingold D, Murrell Z (2012) Management of colonic volvulus. Clin Colon Rectal Surg 25 (4):236–244
93. Perrot L, Fohlen A, Alves A, Lubrano J (2016) Management of the colonic volvulus in 2016. J Visc Surg 153(3):183–192
94. Bozlar U, Ugurel MS, Ustunsoz B, Coskun U (2008) CT angiographic demonstration of a mesenteric vessel "whirlpool" in intestinal malrotation and midgut volvulus: a case report. Korean J Radio 9(5):466–469
95. Applegate KE, Goske MJ, Pierce G, Murphy D (1999) Situs revisited: imaging of the heterotaxy syndrome. RadioGraphics 19:837–852
96. Mishra S (2015) Cardiac and non-cardiac abnormalities in heterotaxy syndrome. Indian J Pediatr 82(12):1135–1146
97. Kim S-J (2011) Heterotaxy syndrome. Korean. Circ J 41(5):227–232
98. Tortajada M, Moreno M, Gracia M, Sanchis A (2010) Situs ambiguous in a schoolchild. BMJ Case Rep 2010:bcr0720092071
99. Nakamura T, Hamada H (2012) Left-right patterning: conserved and divergent mechanisms. Development 139:3257–3262
100. Lirong Z, Belmont JW, Ware SM (2005) Genetics of human heterotaxias. Eur J Hum Genet 14(1):17–25
101. Chen C, Norris D, Bhattacharya S (2010) Transcriptional control of left–right patterning in cardiac development. Pediatr Cardiol 31:371–377
102. Hamada H, Tam PPL (2014) Mechanisms of left–right asymmetry and patterning: driver, mediator and responder. F1000Prime Rep 6:110. https://doi.org/10.12703/p6-110
103. Marinella G, Giovanni FB, Giuseppe P, Bassi MT, Aylsworth AS, Penman-Splitt M et al (1997) X-linked situs abnormalities result from mutations in ZIC3. Nat Genet 17(3):305–308
104. Jörnvall H, Reissmann E, Andersson O, Mehrkash M, Ibáñez CF (2004) ALK7, a receptor for Nodal, is dispensable for embryogenesis and left–right patterning in the mouse. Mol Cell Biol 24(21):9383–9389
105. Al-Salem AH (2014) Omphalocele. An illustrated guide to pediatric surgery. Springer International Publishing, Switzerland
106. Leung AKC, Robson WLM, Wong AL (2009) Omphalocele. In: Lang F (ed) Encyclopedia of molecular mechanisms of disease. Springer, Berlin

107. Hwang P-J, Kousseff BG (2004) Omphalocele and gastroschisis: an 18-year review study. Genet Med 6:232–236
108. Kumar HR, Jester AL, Ladd AP (2008) Impact of omphalocele size on associated conditions. J Pediatr Surg 43(12):2216–2219
109. Govaerts LC, Bongers MY, Lammens MM, Tuerlings JH, van de Kaa CA (1997) Monosomy X and omphalocele. Prenat Diagn 17(3):282
110. Oudesluijs G (2014) Omphalocele, radial ray defect and diaphragmatic hernia: another case of Gershoni-Baruch syndrome? Genet Couns 25(1):77–79
111. Kluth D, Lambrecht W (1996) The pathogenesis of omphalocele and gastroschisis. Pediatr Surg Int 11(2):62–66
112. Eltayeb AA, Mostafa MM (2015) Topical treatment of major omphalocoele: Acacia nilotica versus povidone-iodine: a randomised controlled study. Afr J Paediatr Surg 12(4):241–246
113. Pandey V, Gangopadhyay AN, Gupta DK, Sharma SP, Kumar V (2014) Non-operative management of giant omphalocele with topical povidone-iodine and powdered antibiotic combination: early experience from a tertiary centre. Pediatr Surg Int 30(4):407–411
114. Bauman B, Stephens D, Gershone H, Bongiorno C, Osterholm E, Acton R et al (2016) Management of giant omphaloceles: a systematic review of methods of staged surgical versus nonoperative delayed closure. J Pediatr Surg 51(10):1725–1730
115. Desai AA, Iqbal CW (2017) Omphalocele. In: Mattei P, Nichol P, Rollins M II, Muratore C (eds) Fundamentals of pediatric surgery. Springer, Cham, Switzerland
116. Ferreira-Aparicio FE, Gutiérrez-Vega R, Gálvez-Molina Y, Ontiveros-Nevares P, Athie-Gútierrez C, Montalvo-Javé EE (2012) Diverticular disease of the small bowel. Case Rep Gastroenterol 6(3):668–676
117. Martin JP, Connor PD, Charles K (2000) Meckel's diverticulum. Am Fam Phys 61(4):1037–1042
118. Rashid F, Aber A, Iftikhar SY (2012) A review on gastric diverticulum. World J Emerg Surg 7:1
119. Sagar J, Kumar V, Shah DK (2006) Meckel's diverticulum: a systematic review. J R Soc Med 99(10):501–505
120. Klarenbeek BR, de Korte N, van der Peet DL, Cuesta MA (2012) Review of current classifications for diverticular disease and a translation into clinical practice. Int J Colorectal Dis 27(2):207–214
121. Boynton W, Floch M (2013) New strategies for the management of diverticular disease: insights for the clinician. Therap Adv Gastroenterol 6(3):205–213
122. Weizman AV, Nguyen GC (2011) Diverticular disease: epidemiology and management. Can J Gastroenterol 25(7):385–389
123. Mantas D, Kykalos S, Patsouras D, Kouraklis G (2011) Small intestine diverticula: Is there anything new? World J Gastrointest Surg 3(4):49–53
124. Friedman AI (1957) Annular pancreas: a congenital abnormality manifest in the aged. Gastroenterology 32(6):1172–1177
125. Moodley S, Hegarty M, Thomson SR (2004) Annular pancreas. S Afr Med J 94(1):28–30
126. Tadokoro H, Takase M, Nobukawa B (2011) Development and congenital anomalies of the pancreas. Anat Res Int 2011:351217
127. Huisman TA, Cappelli C (2008) Congenital abnormalities, pancreatic. In: Baert AL (ed) Encyclopedia of diagnostic imaging. Springer, Heidelberg
128. Sandrasegaran K, Patel A, Fogel EL, Zyromski NJ, Pitt HA (2009) Annular pancreas in adults. AJR Am J Roentgenol 193(2):455–460
129. Lee WC, Yang HW, Lee YJ, Jung SH, Choi GY, Go H et al (2008) Brunner's gland hyperplasia: treatment of severe diffuse nodular hyperplasia mimicking a malignancy on pancreatic-duodenal area. J Korean Med Sci 23(3):540–543
130. Chen H-D, Jiang H, Kan A, Huang L-E, Zhong Z-H, Zhang Z-C, Liu J-C (2014) Intestinal obstruction due to dual gastrointestinal atresia in infants: diagnosis and management of 3 cases. BMC Gastroenterology 14:108

131. Snyder CL, Miller KA, Sharp RJ, Murphy JP, Andrews WA, Holcomb GW 3rd, Gittes GK, Ashcraft KW (2001) Management of intestinal atresia in patients with gastroschisis. J Pediatr Surg 36(10):1542–1545
132. Choudhry MS, Rahman N, Boyd P, Lakhoo K (2009) Duodenal atresia: associated anomalies, prenatal diagnosis and outcome. Pediatr Surg Int 25(8):727–730
133. Dalla Vecchia LK, Grosfeld JL, West KW, Rescorla FJ, Scherer LR, Engum SA (1998) Intestinal atresia and stenosis: a 25-year experience with 277 cases. Arch Surg 133(5):490–497
134. Federici S, Sabatino MD, Domenichelli V, Straziuso S (2015) Worst prognosis in the "complex" jejunoileal atresia: is it real? European J Pediatr Surg Rep 3(1):7–11
135. Phillips JD, Raval MV, Redden C, Weiner TM (2008) Gastroschisis, atresia, dysmotility: surgical treatment strategies for a distinct clinical entity. J Pediatr Surg 43(12):2208–2212
136. Hemming V, Rankin J (2007) Small intestinal atresia in a defined population: occurrence, prenatal diagnosis and survival. Prenat Diagn 27(13):1205–1211
137. Yeung F, Tam YH, Wong YS, Tsui SY, Wong HY, Pang KK et al (2016) Early reoperations after primary repair of jejunoileal atresia in newborns. J Neonatal Surg 5(4):42
138. Schiller M, Aviad I, Freund H (1979) Congenital colonic atresia and stenosis. Am J Surg 138 (5):721–724
139. Harthun NL, Morse JH, Shaffer HA Jr, Minasi JS (2002) Duodenal obstruction caused by intraluminal duodenal diverticulum and annular pancreas in an adult. Gastrointest Endosc 55 (7):940–943
140. Boybeyi O, Karnak I, Ekinci S, Ciftci AO, Akçören Z, Tanyel FC, Senocak ME (2010) Late-onset hypertrophic pyloric stenosis: definition of diagnostic criteria and algorithm for the management. J Pediatr Surg 45(9):1777–1783
141. Houben C, Davenport M, Ade-Ajayi N, Flack N, Patel S (2009) Closing gastroschisis: diagnosis, management, and outcomes. J Pediatr Surg 44(2):343–347
142. de Buys Roessingh AS, Damphousse A, Ballabeni P, Dubois J, Bouchard S (2015) Predictive factors at birth of the severity of gastroschisis. World J Gastrointest Pathophysiol 6(4):228–234
143. Mirza FG, Bauer ST, Van der Veer A, Simpson LL (2015) Gastroschisis: incidence and prediction of growth restriction. J Perinat Med 43(5):605–608
144. Lee SS, Hwang ST, Jang NG, Tchah H, Choi DY, Kim HY, Ryoo E (2011) A case of congenital duodenal web causing duodenal stenosis in a down syndrome child: endoscopic resection with an insulated-tip knife. Gut Liver 5(1):105–109
145. Beeks A, Gosche J, Giles H, Nowicki M (2009) Endoscopic dilation and partial resection of a duodenal web in an infant. J Pediatr Gastroenterol Nutr 48:378–381
146. Serracino-Inglott F, Smith GHM, Anderson DN (2003) Duodenal webs—no age limit. HPB (Oxford) 5(3):186–187
147. Melek M, Edirne YE (2008) Two cases of duodenal obstruction due to a congenital web. World J Gastroenterol 14(8):1305–1307
148. Sioulas AD, Grammata P, Scotiniotis I (2017) An obstructive wind sock web deformity of the proximal duodenum. Indian J Surg 79(3):270–271
149. Wani ML, Malik AA, Malik RA, Irshad I (2011) Brunner's gland hyperplasia: an unusual cause of gastrointestinal bleeding. Turk J Gastroenterol 22(4):419–421
150. Levitt MA, Peña A (2015) Anorectal malformations. Orphanet J Rare Dis 2007(2):33
151. Gangopadhyay AN, Pandey V (2015) Anorectal malformations. J Indian Assoc Pediatr Surg 20(1):10–15
152. Rintala RJ, Mildh L, Lindahl H (1996) H-type anorectal malformations: incidence and clinical characteristics. J Pediatr Surg 31(4):559–562
153. Peña A (1995) Anorectal malformations. Semin Pediatr Surg 4(1):35–47
154. Shaul DB, Harrison EA (1997) Classification of anorectal malformations–initial approach, diagnostic tests, and colostomy. Semin Pediatr Surg 6(4):187–195
155. Luke B (1994) Nutritional influences on fetal growth. Clin Obstet Gynecol 37(3):538–549

156. Wu G, Bazer FW, Cudd TA, Meininger CJ, Spencer TE (2004) Maternal nutrition and fetal development. J Nutr 134(9):2169–2172
157. Tzanetakou IP, Mikhailidis DP, Perrea DN (2011) Nutrition during pregnancy and the effect of carbohydrates on the offspring's metabolic profile: in search of the "perfect maternal diet". Open Cardiovasc Med J 5:103–109
158. Giroud A (1973) Nutritional requirements of the embryo. World Rev Nutr Diet 18:195–262
159. Ahanya SN, Lakshmanan J, Morgan BL, Ross MG (2005) Meconium passage in utero: mechanisms, consequences, and management. Obstet Gynecol Surv 60(1):45–56; quiz 73–4
160. Sangild PT, Schmidt M, Elnif J, Björnvad CR, Weström BR, Buddington RK (2002) Prenatal development of gastrointestinal function in the pig and the effects of fetal esophageal obstruction. Pediatr Res 52:416–424
161. Ardissone AN, de la Cruz DM, Davis-Richardson AG, Rechcigl KT, Li N, Drew JC et al (2014) Meconium microbiome analysis identifies bacteria correlated with premature birth. PLoS ONE 9(3):e90784
162. Hen J Jr, Dolan TF Jr, Touloukian RJ (1980) Meconium plug syndrome associated with cystic fibrosis and Hirschsprung's disease. Pediatrics 66(3):466–468
163. Keckler SJ, St. Peter SD, Spilde TL, Tsao KJ, Ostlie DJ, Holcomb GW III, Snyder CL (2008) Current significance of meconium plug syndrome. J Pediatr Surg 43(5): 896–898
164. Fanaroff AA (2008) Meconium aspiration syndrome: historical aspects. J Perinatol 28(Suppl 3):S3–S7
165. Whitfield JM, Charsha DS, Chiruvolu A (2009) Prevention of meconium aspiration syndrome: an update and the Baylor experience. Proc (Bayl Univ Med Cent) 22(2):128–131
166. Ghidini A, Spong CY (2001) Severe meconium aspiration syndrome is not caused by aspiration of meconium. Am J Obstet Gynecol 185(4):931–938
167. Jiménez E, Marín ML, Martín R, Odriozola JM, Olivares M, Xaus J et al (2008) Is meconium from healthy newborns actually sterile? Res Microbiol 159(3):187–193
168. Rodríguez JM, Murphy K, Stanton C, Ross RP, Kober OI, Juge N et al (2015) The composition of the gut microbiota throughout life, with an emphasis on early life. Microb Ecol Health Dis 26:26050
169. Moles L, Gómez M, Heilig H, Bustos G, Fuentes S, de Vos W et al (2013) Bacterial diversity in meconium of preterm neonates and evolution of their fecal microbiota during the first month of life. PLoS ONE 8(6):e66986
170. Andersson E-L, Hernell O, Bläckberg L, Fält H, Lindquist S (2011) BSSL and PLRP2: key enzymes for lipid digestion in the newborn examined using the Caco-2 cell line. J Lipid Res 52:1949–1956
171. Lindquist S, Hernell O (2010) Lipid digestion and absorption in early life: an update. Curr Opin Clin Nutr Metab Care 13(3):314–320
172. Adamkin DH (2005) Pragmatic approach to in-hospital nutrition in high-risk neonates. J Perinatol 25:S7–S11
173. McClure RJ (2001) Trophic feeding of the preterm infant. Acta Paediatr 90(s436):19–21
174. Tyson JE, Kennedy KA (2007) Trophic feedings for parenterally fed infants. Cochrane Database Syst Rev 4:CD000504
175. National Heart, Lung, and Blood Institute Acute Respiratory Distress Syndrome (ARDS) Clinical Trials Network, Rice TW, Wheeler AP, Thompson BT, Steingrub J, Hite RD et al (2012) Initial trophic versus full enteral feeding in patients with acute lung injury: the EDEN randomized trial. JAMA 307(8):795–803
176. Xu H, Collins JF, Ghishan FK (2012) Molecular physiology of gastrointestinal function during development. In: Johnson LR, Ghishan FK, Merchand JL, Said HM, Wood JD (eds) Physiology of the gastrointestinal tract, 5th edn. Elsevier, San Diego
177. Morelli L (2008) Postnatal development of intestinal microflora as influenced by infant nutrition. J Nutr 138(9):1791S–1795S
178. Cummins AG, Thompson FM (2002) Effect of breast milk and weaning on epithelial growth of the small intestine in humans. Gut 51(5):748–754

179. Kien CL (1996) Digestion, absorption, and fermentation of carbohydrates in the newborn. Clin Perinatol 23(2):211–228
180. Mackie RI, Sghir A, Gaskins HR (1999) Developmental microbial ecology of the neonatal gastrointestinal tract. Am J Clin Nutr (5):1035s–1045s
181. Buccigrossi V, de Marco G, Bruzzese E, Ombrato L, Bracale I, Polito G, Guarino A (2007) Lactoferrin induces concentration-dependent functional modulation of intestinal proliferation and differentiation. Pediatr Res 61:410–414
182. Westerbeek EA, van den Berg A, Lafeber HN, Knol J, Fetter WP, van Elburg RM (2006) The intestinal bacterial colonisation in preterm infants: a review of the literature. Clin Nutr 25(3):361–368
183. Sudakov KV (1997) A systems process of reinforcement. Neurosci Behav Physiol 27 (4):370–380
184. Sudakov KV (1997) The theory of functional systems: general postulates and principles of dynamic organization: dedicated to the Anokhin centenary. Integr Physiol Behav Sci 32 (4):392–414
185. Oomura Y (1988) Chemical and neuronal control of feeding motivation. Physiol Behav 44 (4–5):555–560
186. Lau C, Geddes D, Mizuno K, Schaal B (2012) The development of oral feeding skills in infants. Int J Pediatr 2012:572341
187. Morton GJ, Cummings DE, Baskin DG, Barsh GS, Schwartz MW (2006) Central nervous system control of food intake and body weight. Nature 443:289–295
188. Hans-Rudolf B (2003) Neural systems controlling food intake and energy balance in the modern world. Curr Opin Clin Nutr Metab Care 6(6):615–620
189. Zheng H, Berthoud H-R (2008) Neural systems controlling the drive to eat: mind versus metabolism. Physiology 23(2):75–83
190. Kromin AA (1991) The integrating role of biological motivation in the realization of feeding and drinking behaviors. Zh Vyssh Nerv Deiat Im I P Pavlova 41(1):51–9
191. Stanley S, Wynne K, McGowan B, Bloom S (2005) Hormonal regulation of food intake. Physiol Rev 85(4):1131–1158
192. Bhutto A, Morley JE (2008) The clinical significance of gastrointestinal changes with aging. Curr Opin Clin Nutr Metab Care 11(5):651–660
193. Soenen S, Rayner CK, Jones KL, Horowitz M (2016) The ageing gastrointestinal tract. Curr Opin Clin Nutr Metab Care 19(1):12–18
194. O'Mahony D, O'Leary P, Quigley EM (2002) Aging and intestinal motility: a review of factors that affect intestinal motility in the aged. Drugs Aging 19(7):515–527
195. Salles N (2007) Basic mechanisms of the aging gastrointestinal tract. Dig Dis 25(2):112–117
196. Park K-J, Singer W, Sletten DM, Low PA, Bharucha AE (2013) Gastric emptying in postural tachycardia syndrome: a preliminary report. Clin Auton Res 23(4):163–167
197. Firth M, Prather CM (2002) Gastrointestinal motility problems in the elderly patient. Gastroenterology 122:1688–1700
198. Raghavendran K, Nemzek J, Napolitano LM, Knight PR (2011) Aspiration-induced lung injury. Crit Care Med 39(4):818–826
199. Theriot CM, Young VB (2015) Interactions between the gastrointestinal microbiome and Clostridium difficile. Annu Rev Microbiol 69:445–461
200. El-Chammas KI, Sood MR (2017) Irritable bowel syndrome and functional gi disorders in inflammatory bowel disease. In: Mamula P, Grossman A, Baldassano R, Kelsen J, Markowitz J (eds) Pediatric inflammatory bowel disease. Springer, Cham, Switzerland
201. Corazziari E, ShaVer EA, Hogan WJ, Sherman S, Toouli J (1999) Functional disorders of the biliary tract andpancreas. Gut 45(Suppl II):II48–II54
202. D'Souza AL (2007) Ageing and the gut. Postgrad Med J 83(975):44–53
203. Bhutto A, Morley JE (2008) The clinical significance of gastrointestinal changes with aging. Curr Opin Clin Nutr Metab Care 11(5):651–660

204. Grassi M, Petraccia L, Mennuni G, Fontana M, Scarno A, Sabetta S, Fraioli A (2011) Changes, functional disorders, and diseases in the gastrointestinal tract of elderly. Nutr Hosp 26(4):659–668
205. Morley JE (2007) The aging gut: physiology. Clin Geriatr Med 23(4):757–67, v–vi
206. Rayner CK, Horowitz M (2013) Physiology of the ageing gut. Curr Opin Clin Nutr Metab Care 16(1):33–38
207. Gomez-Pinilla PJ, Gibbons SJ, Sarr MG, Kendrick ML, Shen KR, Cima RR et al (2011) Changes in interstitial cells of Cajal with age in the human stomach and colon. Neurogastroenterol Motil 23(1):36–44
208. Russell RM (1992) Changes in gastrointestinal function attributed to aging. Am J Clin Nutr 55(6):1203S–1207S
209. Saunier K, Doré J (2002) Gastrointestinal tract and the elderly: functional foods, gut microflora and healthy ageing. Digest Liver Dis 34(2):S19–S24
210. Valle Gottlieb MG, Closs VE, Junges VM, Schwanke CH (2017) Impact of human aging and modern lifestyle on microbiota. Crit Rev Food Sci Nutr 13:1–8
211. Jill SM (2014) Aging of the mammalian gastrointestinal tract: a complex organ system. Age (Dordr) 36(3):9603
212. Cheddani H, Dauchet L, Fumery M, Charpentier C, Marie Bouvier A, Dupas JL et al (2016) Cancer in elderly onset inflammatory bowel disease: a population-based study. Am J Gastroenterol 111(10):1428–1436
213. Arboleya S, Watkins C, Stanton C, Ross RP (2016) Gut Bifidobacteria populations in human health and aging. Front Microbiol 7:1204
214. Bhattacharjee S, Lukiw WJ (2013) Alzheimer's disease and the microbiome. Front Cell Neurosci 7:153
215. Sharon G, Sampson TR, Geschwind DH, Mazmanian SK (2016) The central nervous system and the gut microbiome. Cell 167(4):915–932
216. Winek K, Dirnagl U, Meisel A (2016) The gut microbiome as therapeutic target in central nervous system diseases: implications for stroke. Neurotherapeutics 13(4):762–774
217. Deans E (2017) Microbiome and mental health in the modern environment. J Physiol Anthropol 36:1
218. Rieder R, Wisniewski PJ, Alderman BL, Campbell SC (2017) Microbes and mental health: a review. Brain Behav Immun 66:9–17
219. Smolinska S, Groeger D, O'Mahony L (2017) Biology of the microbiome 1: interactions with the host immune response. Gastroenterol Clin North Am 46(1):19–35
220. Dinan TG, Cryan JF (2017) The microbiome-gut-brain axis in health and disease. Gastroenterol Clin North Am 46(1):77–89
221. Grenham S, Clarke G, Cryan JF, Dinan TG (2011) Brain–gut–microbe communication in health and disease. Front Physiol 2:94

Chapter 7
Gastrointestinal Motor Function

Abstract One of the major functions of the gastrointestinal (GI) tract is to carry out some types of movement that allows the aborally directed flow of luminal contents. This function is referred to as GI motor function or motility. GI motor function is the result of the activities of muscles located in different layers of the tract. The muscles of the digestive apparatus function to ensure proper chewing in the mouth, swallowing, and movement of luminal contents through the digestive tract and removal of undigested residues from the body. The functions of muscles of the gut are to a large extent determined by the influences from the surrounding environment. However, the gut at certain locations maintains a basal level of motility, which, in part, is due to the functional pacemaker activity of the intestine. The discovery of GI motility (mechanical activity or mechanistic functions of the gut) and the basis of its functionality provided important information on future research directions, which now form the basis of our understanding of GI motility. This chapter aims to identify the course and key milestones in GI motility research. The basis and regulation of gut motility are also discussed. The chapter lays down basic concepts and principles of motility of different regions of the gut and their relationship to the maintenance of the functioning of digestion.

Keywords Gut motility · GI motility · Gut motorics · Gut motor function
Intestinal cells · Pacemaker activity · Migrating motor complex
Migrating myoelectrical activity · Mechanistic theory of gut · GI stretch receptors
Migrating motor complex · History of GI motility · Chewing · Mastication
GI housekeeping · Peristalsis · Smooth muscle contraction · Excitation
Excitation-contraction coupling · Contraction-relaxation cycling
Motor unit · Dysphagia · Achalasia · Hirschsprung's disease · Congenital colonic aganglionosis · Gastroparesis · CD34-positive · PDGFRα-positive cells
Interstitial Cajal cells · Action potential · Negative Schwankung
Electrogastrography · Ileal brake effect · Hunger Pang · Alessandro Volta
Alex Bortoff · Alexander von Humboldt · Aloysio Luigi Galvani
Bennett MR · Bülbring E · Burnstock G · Carlo Matteucci · Clifford Ladd Prosser
Crema A · Edith Bülbring · Edward Banfield · Emil Bozler · Emil Heinrich du Bois-Reymond · Ernest Starling · Ewald Georg von Kleist · Forbes Alexander

M. O. Welcome, *Gastrointestinal Physiology*,
https://doi.org/10.1007/978-3-319-91056-7_7

Gaston R. Vantrappen · Giuseppe Moruzzi · Hans Christian Oersted
Herbert Spencer Gasser · Hermann von Helmholtz · Hodgkin · Huxley
Holman ME · Hoskins RG · Edgar S. Hunter · Ivan Pavlov · James Christensen
Johann Joosten van Musschenbroek · Johann Schweigger · Johannes Peter Müller
John Walsh · Joseph Erlanger · Joseph H. Szurszewski · Julius Bernstein
Kenton M. Sander · Lars Thunberg · Leopoldo Nobili · Leyden jar
Lucille J. Mahoney · Luigi Galvani · Paul Trendelenburg · Pieter (Petrus) van Musschenbroek · Prosser · von Humboldt · Walter B. Cannon
Walter C. Alvarez · Willem Einthoven · William Bayliss · William Beaumont

Abbreviations

ADF	Actin-depolymerizing factor
Ang II	Angiotensin II
Arp2/3	Actin-related protein-2 and -3
BK	Large conductance calcium-activated potassium channel
CaMKII	Ca^{2+}/calmodulin(CaM)-dependent kinase type II
CD34	Cluster of differentiation-34
CGRP	Calcitonin gene-related peptide
CPI-17	Protein kinase C-potentiated inhibitor protein or C-kinase potentiated protein phosphatase-1 inhibitor, molecular weight 17 kDa
cpm	Cycle per minute
CREB	cAMP-responsive element-binding protein
DAG	Diacylglycerol
EDHF	Endothelial-derived hyperpolarizing factor
ER	Endoplasmic reticulum
ERK	Extracellular signal-regulated kinase type
ERK	Extracellular signal responsive kinase
FC	Fast closing
FO	Fast opening
GI	Gastrointestinal
GIP	Gastric inhibitory peptide
GPCR	G protein-coupled receptor
GRP	Gastrin-releasing peptide
ICCs	Interstitial Cajal cells
ILK	Integrin-linked kinase
iNOS	Inducible nitric oxide synthase
IP_2	Phosphatidylinositol 4,5-bisphosphate
IP_3	Inositol 1,4,5-trisphosphate
IP_3R	IP_3 receptor
MAPK	Mitogen-activated protein kinase
MEK	Mitogen-activated protein kinase kinase (also known as MAPK kinase or MAPKK or MKK)
MHC	Myosin heavy chain
MLC	Myosin light chain

MLCK	Myosin light-chain kinase
MLCP	MLC phosphatase
MMC	Migrating motor complex or migrating myoelectrical activity
MyBP-C	Myosin-binding protein C
MYPT1	Myosin-binding regulatory subunit-1 of MLC phosphatase
NO	Nitric oxide
OTT	Oral transit time
PACAP	Pituitary adenylate cyclase-activating polypeptide
PDE	Phosphodiesterase
PDGFRα	Platelet-derived growth factor receptor alpha
PGI2	Prostacyclin, prostaglandin
PI3K	Phosphoinositide 3-kinase
PKC	Protein kinase C
PLCβ	Phospholipase C isozyme β
PTT	Pharyngeal transit time
RMP	Resting membrane potential
ROCK	Rho-associated coiled-coil kinase
RyR	Ryanodine receptor
SC	Slow closing
SER	Sarcoendoplasmic reticulum
SIP	**S**mooth muscle cell, **I**CC cell, **P**DGFRα+ cell
SK	Small-conductance calcium-activated potassium channel
SO	Slow opening
SR	Sarcoplasmic reticulum
VASP	Vasodilator-stimulated phosphoprotein
VEGF	Vascular endothelial growth factor
VIP	Vasoactive intestinal peptide
WASP	Wiskott–Aldrich syndrome protein
ZIPK	Zipper interacting protein kinase

7.1 Introduction

Gastrointestinal (GI) motor functions (also called GI motility or motorics) refer to a complex series of activities of the muscles of the walls of the GI tract initiated by stretch, hormones, neurotransmitters, or paracrine factors that enable the generation of movement of the associated region of the tract. GI motility can also be described as the series of coordinated contractions of the muscle layers of the GI tract. This description of GI motor functions also includes the spontaneous slow waves taking into account that a basal level of secretion of gut peptides is necessary for the regulation of these slow waves [1–3]. GI slow waves are discussed in later part of this chapter.

Motor functions of the GI tract are carried out by the muscles of the digestive apparatus and include the processes of chewing in the mouth, swallowing, and movement of chyme through the digestive tract and removal of undigested residues from the anus. Chewing is aided by the different structures and organs located in the mouth. This masticatory (motor) function of the mouth is aided by voluntary muscles, teeth, tongue, and lubrications of minor and major salivary glands. The motor function of the esophagus is aided by both voluntary (skeletal) and involuntary (smooth) muscles [1–3]. The motor functions of the rest of the GI tract involve a coordinated series of rhythmic movement of smooth muscles regulated by a number of factors including intrinsic excitation by pacemaker cells of the GI tract, extrinsic excitation, or inhibition by the peripheral nervous system or modulated by neural secretions of the intrinsic and central nervous system or a combination of two or more of these factors. The smooth muscles are involved in many disorders of the GI tract. Thus, motor activity of the GI tract is mainly determined by the functionality of the smooth muscles of the gut, which is part of the visceral motor system composed of the smooth muscles, neural, and endocrine input, which regulate the visceral signals [4].

7.2 Historical Background: The Implication of Discovery of Animal Electricity on Future Understanding of Gastrointestinal Motor Functions

7.2.1 Emergence of the Phenomenon of Electricity in Living Systems and the Pioneer Investigations on the Motor Functions of the Gastrointestinal Tract—From Galvani to Alvarez and Beyond

Following the discovery of electricity and the device for the storage of current, many scientists who came across the information on the new research results became interested in studying the phenomenon of electricity. Research on electricity in living systems started during the early 1770s. The British scientist, John Walsh (1726–1795) who later became a Fellow of the Royal Society in 1770, was actively involved in the government activities and administration of the British government in their mission abroad. He became one of the first scientists in the early 1770s to conduct experimental investigation on electricity in living organism. Walsh studied electricity of living system using the electric fish as a model. In one of his experiment, Walsh demonstrated the production of an electric spark from a discharge of the electric ray (*Torpedo marmorata*). This discovery must have enhanced development of electricity research in living systems. Unfortunately, Walsh is not usually acknowledged probably due to the fact that he was unable to publish key findings from his observation [5–7]. Around the same period, precisely in 1769, the physician and chemist Edward Banfield (1726–1795) also reported similar findings demonstrating that the electric eel emitted electric shocks [7, 8].

But it is generally believed that investigation of the origin of electricity in animals actually began following the work of the Italian physician and physiologist, Aloysio Luigi Galvani (1737–1798). The discovery of animal electricity in the second half of the eighteenth century was a huge leap forward that would later provide opportunities for understanding of cellular electrical activity of the digestive tract and other regions/organs of the body. Galvani pioneered the field of electrophysiology [9]. He set up in his home a laboratory where he carried out series of experiment in animals (frogs, in particular) after the initial dissection. It was no surprise that Galvani's wife readily provided this specimen for experimentation as it was widely used as food by Italians. Galvani's intention was to study the effect of thunderstorm on living organisms, but surprisingly and unexpectedly, the results turned out to be something that would change some scientific and philosophical concepts forever. The discovery of animal electricity was due to chance in which the nerves of a frog were prodded by a metal while the frog was on a table. In one of his experiment, his scalpel (surgical knife used in dissection) touched the body of the frog, and he saw the muscles in the frog's leg twitch. In another experiment, Galvani connected the region of the exposed muscle of a recently dead frog to a metal wire and placed it on a support such that the frog was raised above the floor. He observed a miraculous movement in the dead animal: during thunderstorm, the leg of the frog twitched and the dead frog jumped as if it was alive [10]. This Italian physiologist became curious of the phenomenon of motion in a dead animal. But there was no substantial evidence or reason to explain what he has observed. Thus, Galvani decided to improve on the methods of his experiment. In one of the modified experimental model, he used an "electric machine," an early hand-cranked generator. If this machine was made to produce electric sparks at the same time a nerve of the frog was touched with a metal, then the muscles of the dead frog were observed to contract. Galvani noted that there were movements of the frog leg when two metals were made to touch each other while one metal was in contact with the nerve and the other was in contact with the muscle of the frog. Consequently, Galvani concluded that the electricity was inherent in the animal itself. Galvani successfully published his works; unfortunately, it was met with many criticisms. One of the critics was a renowned Italian physicist, Alessandro Volta (1745–1827). Volta argued that electricity was not present in any animal, but in the metal itself. This Italian physicist opined that two different metals produced a gradient in electrical current that causes the muscle contraction. In a move to disprove Volta, Galvani improved on his experimental model. The result of the modified model showed that upon initiation of muscle contraction, a circuit running from the nerves to the muscles is observed. Galvani concluded that this phenomenon was similar to the electric circuit which is made in a Leyden jar. Leyden jar was one of the first known devices for storing electric current. The device was invented independently by the German cleric and physicist, Ewald Georg von Kleist (1700–1748) in 1745 and by the Dutch scientist Pieter (Petrus) van Musschenbroek of Leiden (Leyden) (1692–1761) in Holland in 1745.

That is why the Leyden jar is sometimes called the Kleistian (from the name of one of the inventors, Kleist) jar. The Kleistian or Leyden jar is a cylindrical glass container made up of a dielectric (an insulator such as plastic, glass)-coated inside and outside with a layer of metal foil with a connection between the inner coating and a conducting rod passed through a stopper in the top and, therefore, was not difficult to construct. With the outside surface grounded, a charge is given to the inside surface. This provides the outside an equal but opposite charge. When the outside and inside surfaces are connected by a conductor, a spark is observed and everything returns to normal [11, 12]. von Kleist spent substantial part of his research life on the development of instrument for investigating the phenomenon of electricity. Van Musschenbroek studied in Leiden University, where his father, Johann Joosten van Musschenbroek (1660–1707) worked as a renowned maker of scientific instruments in Holland. Born into the family of instrument makers, van Musschenbroek also had keen interest in instrument making. He graduated from the university with his medical degree in 1715 and later obtained a Ph.D. in 1719. Van Musschenbroek became a professor of physics and was elected a Fellow of the Royal Society in 1734. He held university positions in both medicine and physics [13–16].

Galvani's work became so widespread at the time such that the phenomenon of animal electricity was later called galvanism to mean the contraction of a muscle that is stimulated by an electric current. The end of the eighteenth century and the early nineteenth century were marked with increasing investigations, some scientists suggested that muscle and nerve cells possess an intrinsic electrical force responsible for muscle contraction and nerve conduction in living organisms [10]. Scientists who made useful contributions to the development of animal electricity were Alexander von Humboldt (1769–1859), Forbes Alexander (1882–1965), Carlo Matteucci (1811–1868), du Bois-Reymond among others [10, 17]. The extension of Galvani's concept of animal electricity into digestive physiology or medicine was made by Walter Clement Alvarez (1884–1978) in early the 1920s [18]. He is referred to as the pioneer of electrogastrography (from Greek "electro," relating to electrical activity, "gastro"—stomach, "gram"—"to write"). It is a technique of graphical recording of the electrical signals that travel through the muscles of the stomach, i.e., gastric muscle contraction. The graphic record is called electrogastrogram or EGG. It is believed that Alvarez was inspired not by Galvani, but by the Dutch physician, physiologist, and 1924 Nobel Prize winner in medicine or physiology, Willem Einthoven (1860–1927) who invented the first electrocardiograph (also called the string galvanometer) in 1903. The galvanometer (named after Galvani) instrument was used to produce graphical indications of the functioning of the heart—electrocardiogram. A galvanometer is an instrument for detecting electric current. It is an analog electromechanical device that produces a rotary deflection of a pointer in response to electric current flowing through its coil in a magnetic field [19]. Einthoven's success was due to the previously produced galvanometer by the German scientist Johann Schweigger (1779–1857). Schweigger's success was the product of the Danish physicist Hans Christian Oersted (1777–1851) who in 1820 reported that an electric current passing through a wire deflected a nearby compass needle. In 1825, the Italian physiologist Leopoldo Nobili (1784–1835) developed the astatic galvanometer, with a better precision compared to previous ones. Two years later, Nobili reported that he had successfully measured the current in a frog using the galvanometer he had previously developed. With this report, another Italian physicist Carlo Matteucci (1811–1868) decided to acquire the Nobili's galvanometer and tried to reproduce the experiment on animal electricity previously performed by his fellow citizen, Luigi Galvani. Matteucci was able to prove beyond doubt that injured tissues generated an electric current. He was able to register the current flowing between the damaged and undamaged part of the frog muscle (now called the resting membrane potential) and realized the influence of certain agents on this potential. In essence Matteucci, though indirectly, had discovered a potential difference between the extracellular and intracellular environment of the muscle fiber. Importantly, he was also the first to document the presence of electrical signals in muscle fibers, which we now refer to as the action potential (see Reference Note 7.1) [19–24].

Reference Note 7.1

Action Potential of Gastrointestinal Smooth Muscle Cell

In an experimental condition, cellular signals, or excitability of cells or tissues can be recorded using the glass microelectrode or patch clamp (Figs. 7.1 and 7.2). Patch clamp is a relatively recent technique that allows to record cellular signal from the whole cell or different locations of the cell. Details on patch clamp are discussed in Chap. 7 of the book "Gastrointestinal Physiology: Contemporary Trends, Methods and Models."

The electrical activities occurring in a GI smooth muscle are due to the differences in the concentration of ions (Ca^{2+}, K^+, Na^+) inside and outside the cell. Because GI smooth muscle cells are joined together by special intercellular network called gap junctions, these ions can flow from one cell to another and thus electric current can passively flow between the cells. This means that GI smooth muscle cells are electrically coupled. When the cell has not received stimulus, the difference in potential between the inside and outside of the cell is about −70 mV (referred to as the resting membrane potential—RMP). But this RMP may be as low as −40 to −65 mV depending on the region of the GI tract. Since GI smooth muscle cells are coupled electrically, the membrane potential can spread from one cell to another. This results to waves of initial depolarization that travel along the GI tract in a periodic manner and they are called "slow waves" [25].

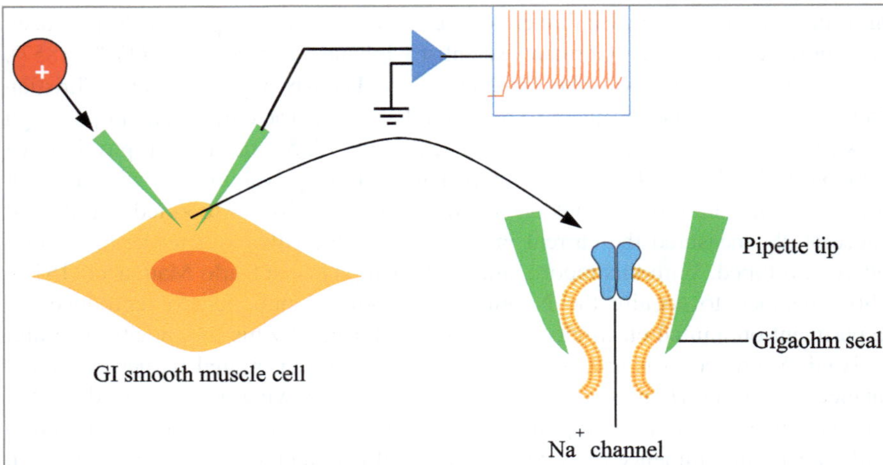

Fig. 7.1 Patch clamp intracellular recording of ion channel activity from a smooth muscle cell. Part of the membrane of the cell is clamped to form a gigaohm seal to record the activity (potential)

The slow waves can be defined as pulses of Ca^{2+} waves, which are responsible for the periodic depolarization that spread through the nexuses, which in turn provides the depolarization that activates the smooth muscle cells [25]. Though the mechanism of generation of the slow waves is complex and still being investigated, it is believed that slow waves are due to influx of Ca^{2+} and Na^{+}, which results to depolarization of the membrane potential to approximately −30 mV. At this potential value, the Na^{+} channels rapidly inactivate, whereas Ca^{2+} channel remains open, maintaining the membrane potential at a plateau. Thus, the waves are actually representing fluctuations in membrane potential by approximately 10–20 mV [26–28]. The slow waves occur as cycles of waves with a frequency or periodicity of 3 cycles per minute (cpm) in the stomach, 10–12 cpm in the duodenum, and 8–9 cpm in the ileum [29]. Slow waves are an intrinsic property of smooth muscle and are not dependent on external stimuli. So, these waves do not cause any visible contraction of the smooth muscles. These waves are important in that they regulate GI motor functions in periods of absence of meal and help to synchronize muscle contractions following the generation of action potential by stretch, chemical, or electrical stimuli [26–28].

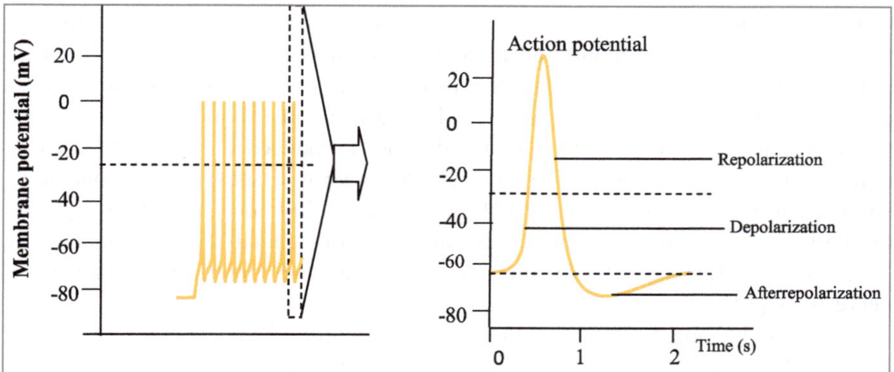

Fig. 7.2 Action potential of a GI smooth muscle cell

Stimuli such as stretch, neurotransmitters, hormones, and paracrine factors can further depolarize the membrane potential of the smooth muscle cells by increasing Ca^{2+} influx, which leads to the generation of action potential. The Ca^{2+} activates intracellular contractile proteins, which results to muscle contraction [26–28]. For the membrane potential to return to the resting phase, Ca^{2+} channels are inactivated, while K^{+} channels remain activated for a fairly prolonged period leading to repolarization of the membrane potential to the RMP [29]. GI motility is the result of coordinated contraction of smooth muscle, slow waves, and action potential [26–28]. The molecular mechanisms of this process are discussed in details in later part of this chapter.

About a century later, upon reviewing Matteucci's results, Giuseppe Moruzzi (1910–1986) noted that the phenomenon of muscle contraction initially reported by Galvani was due to an induced twitch and indicated that Matteucci's results constituted the cornerstone to modern electrophysiology. Importantly, Moruzzi highlighted that the induced twitch was due to the propagation of the signal along the nerve and muscle fibers resulting from the conduction of an electric wave capable of regenerating itself in the course of its propagation, acting as a stimulus for the next segment of the fiber [30].

The renowned German physiologist Johannes Peter Müller (1801–1858) held the view that a vital force was the cause of the previously identified animal electricity. Müller thought that electricity was not involved in the functioning of the nerve and muscle of the frog. But this view of Müller's changed immediately he read about Matteucci's latest experimental results. The German physician and physiologist, Emil Heinrich du Bois-Reymond (1818–1896) who was Müller's student, came across Matteucci's work after his teacher had introduced to him the outstanding experimental results of the Italian physiologist. So du Bois-Reymond decided to reproduce and improve on the experiment of Matteucci. In course of his experiment, du Bois-Reymond observed that contraction of the muscle was due to the stimulus that traveled from the point of stimulation along the nerve to the muscle fiber. Alas, du Bois-Reymond had discovered the nerve impulse, which was recognized as a form of transmission of information in the nervous system. He also conducted series of experiment, which indicated that some tissues did not respond to the action of a stimulus. He thus, in addition, introduced the concept of "excitation" and "excitable tissue" to distinguish between tissues that responded to stimulation and those that did not [31–33]. du Bois-Reymond measured negative oscillations or variations "negative Schwankung" now called "action potential" in 1872 as the difference between the inner and outer membrane of the muscle or nerve fiber [32].

It should be noted, however, that du Bois-Reymond and another promising scientist Hermann von Helmholtz (1821–1894) who will later become leading figures in electrophysiology were at one time mentored by the Prussian naturalist, Alexander von Humboldt (1769–1859) [10, 34]. Humboldt's philosophy and experimental findings, to some extent, must have influenced the career choice and works of du Bois-Reymond.

Initially, von Humboldt was not sure of the difference between animal electricity and metallic electricity; hence, he referred to the phenomenon as galvanic force. To investigate why the galvanic force occurred, he decided to conduct experiments on electric eel in the 1790s. In fact, he was not sure that there was anything like animal electricity. After conducting many experiments with the electric eel and painstaking study of Alessandro Volta's work, von Humboldt was convinced about the presence of intrinsic electricity in animals. In 1797, von Humboldt published key findings of his experiments on the concept of animal electricity [10, 34].

In the nineteenth century, the German physician and physicist, Hermann von Helmholtz (1821–1894) dwelling on the works of his predecessors, reported interesting experimental findings on nervous transmission. Using a scale with improved sensitivity together with attached galvanometer, Helmholtz successfully measured the speed at which a signal travels along a nerve fiber (sciatic nerve of a frog) to excite the calf muscle to which the nerve is attached. In 1849, von Helmholtz reported signal transmission speeds in the range of 24.6–38.4 m/s [35]. von Helmholtz's experiment showed that the presence of electricity in an animal was a reality.

The German organic physicist and student of du Bois-Reymond, Julius Bernstein (1839–1917) played immense role in the development of modern physiology and biophysics in the second half of the nineteenth century [36, 37]. Interestingly, at onetime Bernstein was mentored by von Helmholtz while in Heidelberg. Bernstein formulated the "membrane theory" and gave a precise description of action potential. This theory would later become the basis for a revolution in animal electricity and future molecular basis of understanding the electrical activity of cells and tissues. Bernstein's membrane theory of electrical potential in cells and tissues was published in 1902, in which he provided the first model describing the physicochemical events occurring when impulse excites a cell or tissue [37]. Bernstein, with the intention of recording the current produced when a cell or tissue is excited, first, developed an instrument called a differential rheotome, which was meant to resolve the precise time course of electric activity in a nerve and muscle fiber. With the aid of this instrument, the interval between stimulation and sampling could be varied. The instrument was also used to measure the velocity of conduction of nervous impulse. The differential rheotome of Bernstein was an improvement on the rheotome (rheo—"flow or current" + tome —"cutter," which literally means flow cutter), initially developed by du Bois-Reymond in 1849 for the measurement of current by modernizing the galvanometer, which he achieved by adding a two position switch. In the latest development of Bernstein there was a space between stimulations [31, 38]. In 1868, Bernstein reported that he had successfully measured the time course of the action potential of cells. He was the first to provide an accurate description of the action potential in a nerve and muscle cell. Bernstein is widely acknowledged for discovering the basis of bioelectricity of the cell [19, 37].

Further insight into the electrical activities of the cell was provided by Forbes Alexander, Joseph Erlanger among others. Forbes Alexander (1882–1965) was an American pioneer electrophysiologist and neurophysiologist, who was later appointed Professor of Neurophysiology at the Harvard Medical School of the Harvard University, made important contributions to animal electricity [39]. Joseph Erlanger (1874–1965), an American physiologist, who served as the 11th President of the American Physiological Society from 1926 to 1929, although best known for his contributions to the field of neuroscience, also studied digestion and absorption of food in dogs with shortened intestines. The first physiology lecture of Erlanger in the Hopkins Medical School was on digestion and metabolism. Joseph Erlanger was jointly with Herbert Spencer Gasser (1888–1963) awarded the 1944 Nobel Prize in physiology or medicine "for their discoveries relating to the highly differentiated functions of single nerve fibers" [40].

Following increasing investigations on the electrical properties of the cell beginning from the early 1900s [41], scientists began to expand the accumulating knowledge on animal electricity to other areas of physiological science. In the GI system, pioneer scientists that used different methods of research (including the use of modified galvanometer) to examine motor functions of this system made outstanding contributions that vehemently changed the traditional and philosophical views of gut activity. Pioneer scientists in this area include William Beaumont, Ivan Pavlov, Walter B. Cannon, Walter C. Alvarez, Emil Bozler, Clifford Ladd Prosser, and James Christensen among others. Beaumont was the first to report experimental findings on gastric contractions in 1823 [42–45]. Cannon was instrumental to developing a visual method of studying the physiology of the digestive tract by the invention of X-ray method of investigating peristalsis (The term peristalsis is derived from the Greek "peristallein" meaning "to wrap around," from peri —"around" and stallein—"to place"). Peristalsis can be defined as a series of symmetrical contractions and relaxations of GI muscles required to propel luminal contents through the alimentary canal aborally. The phenomenon was discovered in the nineteenth century by William Bayliss and Ernest Starling while working on the small intestines of dogs. They realized that increasing the pressure in the intestine caused the relaxation of the muscle wall below and contraction of the muscle wall above the point of stimulation) [46–49]. The peristaltic reflex in isolated small intestinal segment and their mechanism of rhythmic activity (neural basis) was studied by the German pharmacologist Paul Trendelenburg (1884–1931) around 1917 [50] and subsequently, around the middle of the last century, Feldberg and Lin [51], Bennett et al. [52] and Bülbring and Crema [53]. Edith Bülbring (1903–1990), a British scientist in the field of smooth muscle physiology, conducted further investigations to unraveling the intrinsic and neural basis of intestinal smooth muscle functioning. Alvarez who was a pioneer scientist to identify the rhythmicity of intestinal peristalsis, also contributed hugely to the field of GI tract smooth muscle physiology. Alvarez became the first to observe that the frequency of intestinal rhythmic segmenting contractions decreases from the gullet to the anus [42, 54, 55]. In the early 1920s, he observed that some types of motor activity of the gut was always present and had a basal level of activity—spontaneous slow waves [56]. In one of his experiment, which he conducted with his colleague, Lucille J. Mahoney (c. 1897–1962), the spontaneous slow waves of contractions of the wall of the stomach and small intestine propagate as a sleeve along the GI tract wall [56]. In the same year, Hoskins and Hunter (1924) showed that intestinal segment responds to different stimuli. At this point, scientists reasoned that gut functions could be regulated if the mechanisms of the slow waves are known. The German Professor Emil Bozler (1901–1995) was born in Steingebronn, Germany, and after spending his childhood period in his home country, he migrated to the USA where he conducted many groundbreaking studies in physiology including gut motility. Bozler investigated using electrical stimulation, how electrical waves were propagated in smooth muscle and showed that slow waves were always present in the intestine even in the absence of visible stimuli. In addition, he demonstrated that muscle contraction was basically accomplished by changes in the Plato of the slow waves [57–59]. But the mechanisms including the origin of the slow waves were yet to be ascertained. Clifford Ladd Prosser, James Christensen, as well as other scientists researched the origin of the slow waves (Discussed below).

7.2.2 The First Measurement of Gastrointestinal Motility Using an Electrical Device and the Origin of Spontaneous Slow Waves

The measurement of wave-like contractions of the stomach, peristalsis, using an improved galvanometer was reported in the middle of the 1920s [60]. Since the invention of electrical devices to measure digestive functions, there has been tremendous development in the area. There are computed devices (non-invasive) that measure the electrical functions of not only the stomach, but also intestines and associated organs. It is used to assess gastroenteric functions in diseases of the pancreas, gallbladder, and liver [41, 61–68].

With the improvement in experimental methods and techniques, extended researches into the origin of the slow waves of the GI tract, led to the establishment that the smooth muscle of the digestive tract can spontaneously generate, and conduct electric potentials because of the presence of pacemaker cells called interstitial Cajal cells (ICCs) and that these potentials can be recorded either from a portion of the gut or from the surface of the skin (ICC is discussed below) [69–72]. This field of digestive physiology is called electro-gastroenterography, which actually belongs to the broader field of electrophysiology. The motor activity of the gut is based on the migrating motor complex (MMC). The MMC of the gut is a spontaneous cyclic, recurring electrical, and motor activity that occurs in the stomach and small intestine during periods of fasting and is interrupted by feeding [73]. During fasting, MMC develops approximately every 90–120 min to sweep residual debris through the GI tract [74]. MMC is the pattern of GI tract contractions seen between the interdigestive periods. In contrast, the postprandial period is characterized by coordinated segmental and peristaltic contractions that facilitate digestion and absorption of nutrients, fluid, and electrolytes [75, 76]. Malfunction of MMC has been implicated in a range of diseases. The MMC is associated with gastroparesis, intestinal pseudoobstruction, and small intestinal bacterial overgrowth. Evaluation of MMC in different conditions may provide useful prognostic information on the occurrence of GI disorders in high-risk individuals. Besides it can as well serve as a prognostic measure of GI tract disorders. Thus, the MMC forms the basis of gut motor activity and peristalsis. The MMC involves four phases, which occur in their order from I through IV. The mean MMC cycle duration ranges from 105 to 114 min [77]. Mechanisms of MMC control are complex and involve both neural and non-neural mechanisms. The functions of MMC include "GI housekeeping," meaning that it preserves the normal functioning of the gut. It maintains the integrity of the gut mucosa. MMC, however, varies largely within an individual and between individuals and depends on the origin of the phase of the wave. The intraindividual variability of this duration may be as large as 90%. Secretory activities of the gut especially those of the stomach, gallbladder, and pancreas are related to the gastric (antral) component of the MMC [77]. Gastric emptying also depends on functional coupling of slow waves between the corpus and antrum (The waves are initiated in the corpus and propagate to the pyloric sphincter, thereby generating gastric peristalsis for adequate evacuation of gastric contents) [78, 79]. For further information on MMC and its history of development, review Deloose et al. [73].

Around the second half of the last century, some research groups including that of Clifford Ladd Prosser (1907–2002) and Alex Bortoff observed in their experiment that electrical slow waves originate from the longitudinal muscle cells. The longitudinal muscle layer, besides the circular muscle layer, is the layer of muscle cells that constitute the smooth muscles of the GI tract. In one of the papers published in 1965 by Alex Bortoff in the American Journal of Physiology, he wrote "*circular muscle from cat intestine exhibits spontaneous rhythmical contractions only when it is attached to longitudinal muscle. Under these conditions electrical slow waves can be recorded from circular muscle, but they disappear following complete removal of the longitudinal layer. If a small patch of longitudinal muscle remains, then slow waves can be recorded from adjacent circular muscle*" [80]. Bortoff made many contributions to the initial understanding of the motility pattern of the gut and also had a strong belief that transmission of slow waves along the gut can only occur when circular muscle is attached to the longitudinal muscle layer. Bortoff also mentioned that the amplitude of the waves decreases exponentially with distance, approaching zero at approximately 12 mm from the lateral edges and approximately 3 mm from the oral or anal side of the longitudinal muscle layer. This suggests that the waves take a circular reoccurring pattern along the gut walls. Unfortunately, a problem soon arose when over 80% of longitudinal muscle preparations isolated from the intestinal wall could not generate electrical slow waves. The remaining percentage of isolated longitudinal muscle that generated slow waves had pieces of myenteric plexus present in them. So the hypothesis of Bortoff, Prosser and colleagues came under serious scrutiny and many scientists began to question the view that slow waves originate from the longitudinal muscle. In one of Prosser's experiment, it was found that the magnitude of the slow wave is maximum at the boundary between the longitudinal and circular muscle layers. So, it became worrisome whether or not the waves originated from either the circular or longitudinal layer or both or no waves originated from the layer. Notwithstanding, however, Prosser maintained that the slow waves that originated from the longitudinal muscle layer were conducted passively through fibroblasts and interstitial cells to the circular muscle layer, where it was amplified. This explanation was quite reasonable since at the time the electrical conductivity and physiology of other cells and tissue were fairly understood. For instance, scientists were already aware that connective tissue can conduct electrical impulse between the cells of the heart. But some researchers were keen about the issue of Prosser's conflicting results and his hypothesis of the origin of slow waves in the gut [81–90]. James Christensen and coworkers successfully solved the puzzle and showed that circular smooth muscle layer was the site of origin of electrical slow waves in the large bowel. They used isolated muscle preparations from the colon and rightly pointed that spontaneously occurring electrical slow waves are dependent on the integrity of the junction between the submucosal and the innermost circular muscle layer [91–94].

In 1969, the Polish-born scientist Joseph H. Szurszewski (1940) became the first to provide a detailed description of the MMC and the site of gastric pacemaker—the origin of the spontaneous slow waves in the GI tract. Szurszewski successfully recorded intestinal MMC in a dog; he observed a cyclic pattern of muscle activity [95]. A couple of years later, precisely in 1977, Gaston R. Vantrappen (1927) and coworkers had reported similar pattern of myoelectrical activity in gut muscle of humans [96]. When electrodes are inserted into the wall of the canine gut or pressure is recorded in human stomachs by gastric manometry, periodic fluctuations of GI motility are observed. These fluctuations are accompanied by changes in both GI secretions involving the stomach and pancreas [97].

Though, the results of researches that accumulated over the years gave scientists a rough view of the origin of slow waves in the gut, nobody knew what actually was responsible for the slow waves. Scientists could not explain the phenomenon with a vital force since vatalism was not scientific and the era of using vital force to explain scientific phenomenon was dead long ago. The first functional evidence that the slow waves originate in ICCs associated with Auerbach's plexus was demonstrated by Lars Thunberg. He also showed that the slow waves were conducted through the network of the ICCs between the longitudinal and circular muscle layers of the intestine [98, 99]. ICCs are the pacemaker cells of GI tract, which are the origin of rhythmic slow waves, controlling the frequency and propagation of electrical contractile activity of the gut [99]. These cells are non-muscular and are mesenchymal in origin, located in the tunica muscularis of the GI tract [100]. Electrical slow waves originating from the network of ICCs propagate into the tunica muscularis, which initiates rhythmic contractile activity following excitation by enteric nerves [101]. It is the activity of the enteric nervous system that determines the pattern of contractile activity in GI muscles. The intrinsic properties of ICCs show rhythmic inward currents not found in typical smooth muscle cells. However, in the absence of such intrinsic slow waves, enteric cholinergic excitation evokes slow waves leading to the occurrence of rhythmic action potential. This second wave is termed an induced slow wave activity. Recent review has shown that motor neurotransmission is more complicated than simple release of transmitter from nerve terminals and binding of receptors on smooth muscle cells [100]. It is believed that the "neuroeffector" junction in the tunica muscularis is a synapse with different specialized cell types and not only the ICCs. Correspondingly, both intrinsic and induced slow wave activity are the origin of the peristaltic motor patterns of the gut. This activity is crucial for an orderly processing of food substances, absorption of nutrients and elimination of wastes from the anus.

Many scientists including Kenton M. Sander were instrumental in confirming the research results of Thunberg concerning the functional evidence that the slow waves originated in ICCs. In the mid-1980s, Sander and coworkers using intracellular recording techniques provided further evidence on the origin of spontaneous electrical slow waves. They noted that slow wave pacemaker regions were located throughout the digestive tract in regions lined with interstitial cells. Furthermore, they showed that active propagation of slow waves occurs along the borders of the muscle layer where the interstitial cells are located and beyond these regions, slow waves propagate passively through the muscle layers [90, 102–104]. Advancement in cellular research has shown that the initiation of waves in these cells is controlled mainly by the steel factor, the natural ligand for the c-Kit receptor tyrosine kinase. A knockout of this receptor annuls the generation of pacemaker potential in these cells as the network of ICCs does not develop at all [99]. Interestingly, in normal pacemaker cells, propagation of generated waves is controlled, in part, by specific ion channels and intercellular junctions located between smooth muscle cells. These intercellular junctions with low-resistance pathways allow for the propagation of electrical potential in smooth muscle of the gut by local circuit current flow [105]. The intercellular junctions can be electrical or mechanical in nature or both [106]. In whichever case, the driving force for slow wave propagation is the excitation–contraction coupling. This coupling occurs by Ca^{2+} entry via ion channels in the plasma membrane of the muscle cells and ICCs, leading to a rise in intracellular Ca^{2+} concentration, which in turn initiates the motion of contractile proteins [106].

The contractile activity of smooth muscle cells is intrinsic and maintained even after complete denervation of the enteric nerves or by blocking the nerves with pharmacological agents [105, 107]. This contractile activity that remains following complete denervation is termed myogenic. This myogenic contraction is due to the intrinsic property of the muscle to generate waves of depolarization [108–110].

The motility pattern of the gut is due to contraction cycle triggered by a long-lasting wave of depolarization. As described by Hodgkin and Huxley, initiation of activity starts with voltage-dependent activation of ion channels which produces a gradient for activation of other ion channels. The waves of depolarization called spontaneous discrete transient depolarization or unitary potentials then spreads from the Cajal cells in the circular smooth muscle to other regions [111]. For instance, proximal-to-distal slow wave frequency gradient is maintained throughout the gut. In the stomach, the contraction starts from the corpus and migrates toward the duodenum. There is electrical quiescency in the fundus. The corpus contains the dominant pacemaker frequency actively generating slow waves of about 3 cycles per minute, which in turn, activate a network of myenteric Cajal cells, which starts in the antrum and slowly conducts waves of depolarization down the stomach [105]. It was thought that the corpus and antral waves were about 3 cycles per minute; however, recent estimations indicate that cycles of waves originating from these regions of the stomach are more than 3 cycles per minute. And the wave cycles of the antral region are more than that of the corpus [112]. The waves move across the layer of circular muscle cells and activate the intramuscular Cajal cells in the circular muscle layer. Slow waves from the intramuscular Cajal cells spread radially, thereby triggering each ring of contraction [105]. Slow waves can be recorded from the respective sites using serosal electrodes or multichannel electrogastrogram [113]. Factors that affect the pattern of gastric slow waves include local influences as well as influences from higher brain centers. Wang et al. [113] showed that vasopressin infusion and sleep impair gastric slow wave coupling. Rhee et al. [112] noted that chronotropic influences (resulting from stretch, extracellular Ca^{2+}, and temperature) may affect the pattern or some characteristics of the slow waves. Cholinergic muscarinic stimulation enhances pacemaker frequency [78].

7.3 Physiologic Anatomy of the Muscles of the Gastrointestinal Tract

7.3.1 Overview of the Structural and Functional Architecture of Gastrointestinal Smooth Muscle Cells

As previously noted in Chap. 3, the muscle tissue of the gut is composed of two cell types—smooth and skeletal muscle cells (Fig. 7.3). From Chap. 3, it was mentioned that GI skeletal (striated) muscles are the masticatory muscles and the muscles located in some regions of the esophagus, in particular, the upper one-third (composed of skeletal muscles only) and the middle one-third (composed of a mixture of skeletal and smooth muscles). The lower one-third consists of only smooth muscle cells [1, 3]. The smooth muscle cells are arranged in two layers in

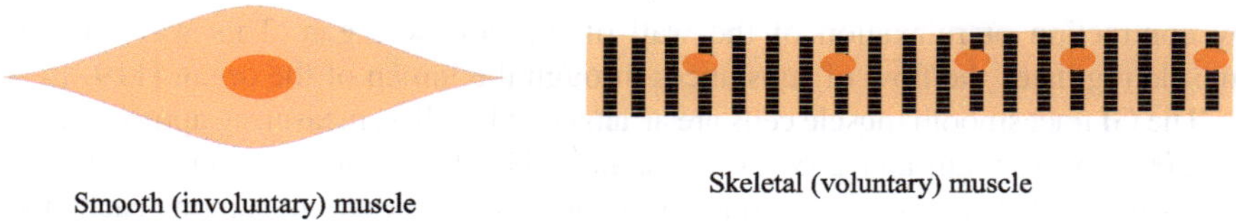

Fig. 7.3 Muscles of the GI tract. The diagrams show the two types of muscles of the GI tract: smooth (non-striated) having intrinsic excitable characteristics and skeletal (striated) muscle cells without such characteristics

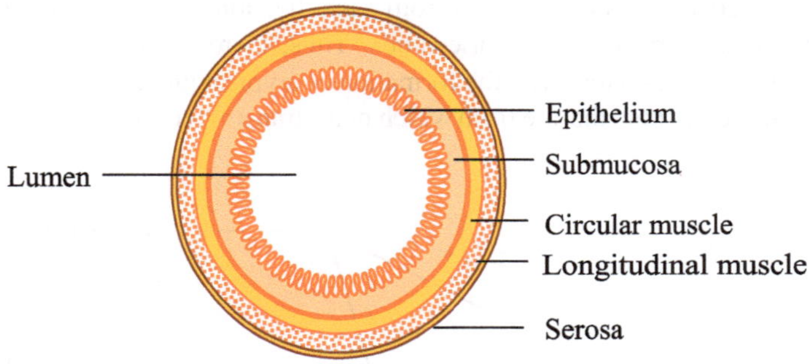

Fig. 7.4 Layers of the GI tract showing location of the smooth muscles. There are generally two muscle layers in the GI tract, except for the stomach that has three muscle layers. The first layer of muscle, known as longitudinal layer, is located between the outermost layer of GI tract (*tunica serosa*) and a myenteric or Auerbach's plexus (not shown in the diagram). The circular muscle layer (inner muscle layer) is located between the myenteric network of neurons and submucosa. Note that there is a neural or Meissner's plexus (not shown in the diagram) between the circular muscle and submucosa. These two muscle layers are called *tunica muscularis externa*, which is responsible for GI segmental contractions and peristaltic movement required for mechanical breaking up of food particles and mixing them with digestive enzymes. The motor function of the GI tract is regulated by neurons of the sympathetic and parasympathetic divisions of the autonomic nervous system as well as plexuses of the gut. The Meissner's plexus mainly provides nerve supply to the *muscularis mucosae* (singular—mucosa) of the ileum and colon. The outermost later which is the *tunica mucosa* is made up of epithelial layer of a network of cells, an underlying layer called the *lamina propria*, which contains blood vessels, mucous glands, lymph nodes, and sensory nerve endings. The *muscularis mucosae* are a layer of smooth muscle cells in most areas of the digestive tract present in the lamina propria. This muscle produces local movements of the mucosa and does not really move food through the tract [75, 114–118]

the walls of the GI tract (Fig. 7.4). Because skeletal muscle constitutes only a small part of the muscles of the GI tract, this chapter is mainly concerned with smooth muscle physiology. But striated muscle physiology will be mentioned at strategic points of the discussion.

The smooth muscle cells are not only found in the GI tract, but also line the walls of hollow organs and tubes in the body (blood vessels, uterine wall, airways, ureter, urinary bladder, uterus, and the penile and clitoral cavernosal sinuses), erector pili muscle, smooth muscle of the iris and lens [119–121]. Smooth muscles function by generating force to change the length. They maintain the activities of hollow organs by contraction or relaxation of the wall of the hollow organ. This way, smooth muscle regulates the flow of substances through the lumen of the organ [119–121].

The GI tract smooth muscle cells are arranged in bundles, measuring approximately 20–500 μm in length and 2–5 μm in diameter [122]. The smooth muscle cells are innervated by the autonomic nervous system and are interconnected among each other by communicating junctional network (Figs. 7.5 and 7.6). Smooth muscle cells of the GI tract are innervated by both the sympathetic and parasympathetic division of the autonomic nervous system. The nerve fibers from these divisions of the nervous system form synaptic contacts with the smooth muscle cells at strings of bulbs or swelling in the fiber called varicosities. Nerve endings from each division of the autonomic nervous system form their corresponding varicosities. Thus, there are parasympathetic and sympathetic varicosities. However, the formation of atypical junctions is also possible. These structures constitute the site from which neurotransmitter is released on arrival of

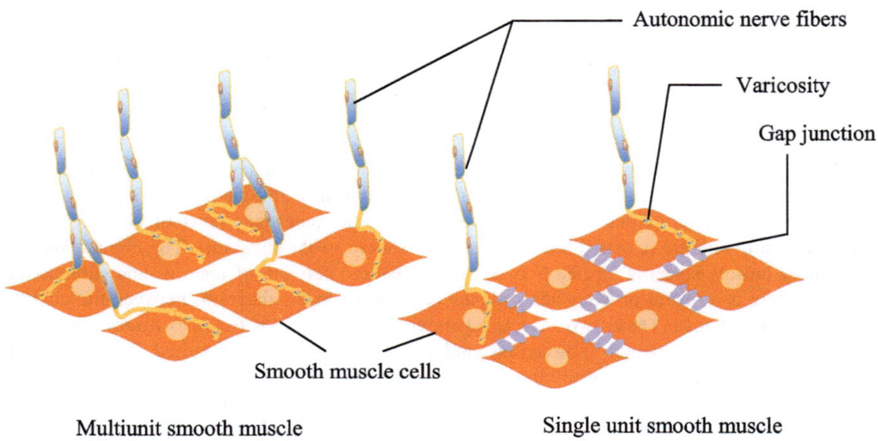

Fig. 7.5 Types of smooth muscle cells. Smooth muscle is usually described as either single unit (also called unitary) or multiunit. Single-unit muscle cells in a bundle are innervated by varicosities of autonomic nerve fiber. The bundle of single-unit muscle cells contracts as a syncytium—in a coordinated fashion. This property of smooth muscle cells is due to the presence of numerous intercellular junctions called gap junctions. Thus, when one muscle cell depolarizes, waves of calcium ions within a very short interval spread to all muscle cells of the bundle causing it to contract as one unit. However, single-unit smooth muscle can contract regularly without input from a motor neuron. Moreover, some cells within the unitary smooth muscle cells may behave as pacemaker cells, generating rhythmic action potential due to their intrinsic electrical activity. The response of single-unit visceral smooth muscle cells are thus referred to as myogenic. Single-unit muscle is found in the GI tract [123, 124]. In multiunit smooth muscle tissues, individual muscle cell is innervated by an autonomic fiber. Thus, each cell behaves (contract and relax) independently. In contrast to single-unit muscle cell, multiunit smooth muscle contraction must be initiated by a nerve fiber of the autonomic nervous system and thus it expresses a substantially lower number of gap junctions. Multiunit muscle cells are therefore referred to as neurogenic. Multiunit smooth muscle behaves like skeletal muscle as they allow for gradual responses to stimuli. Multiunit smooth muscle is present in such locations of the body as major breathing pipe leading to the lungs, large arteries, and internal eye muscles [123–126]

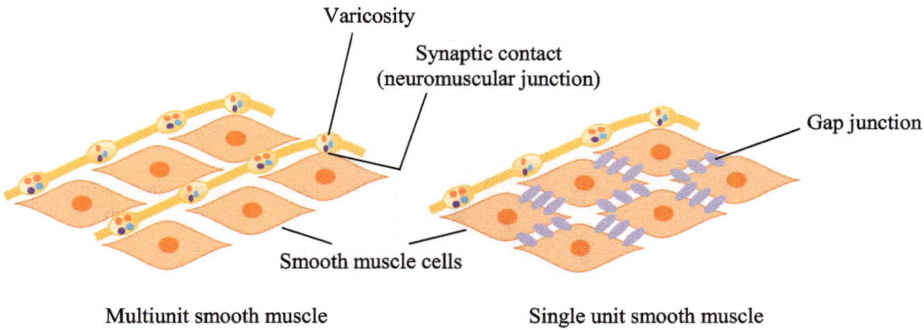

Fig. 7.6 Types of smooth muscle cells

the nerve impulse [127]. The space between varicosity and smooth muscle cells is about 50–200 nm and is called the synaptic cleft. The junction formed between varicosity and smooth muscle cell is referred to as the neuromuscular junction. Varicosity can be formed with the modified smooth muscle cell—ICC. The ICCs are in turn connected to the smooth muscle cells via gap junctions [128, 129].

The communicating junctional network allows smooth muscle to function in a group in what has been termed "effector muscle bundle." This junctional network allows for both mechanical and electrical coupling between smooth muscle cells. The mechanical coupling is thought to be mediated by intermediate junctions enabling force generation and parallel transmission to contractile proteins along the longitudinal axis of the muscle bundle [130]. For this reason, smooth muscle cells are believed to function as a mechanical syncytium. Another junctional complex believed to affect contractile activities is the gap junctions that allow the passive movement of current from one smooth muscle cell to another. Gap junctions link cells structurally and functionally by flow of ions and small molecules between the cells. Thus, smooth muscle cells form a functional syncytium [131].

The smooth muscle cells are enveloped by a semipermeable membrane called the plasma membrane (also known as sarcolemma: from Greek "sarx"—flesh, and "lemma"—sheath) (Fig. 7.7). This plasma membrane, like any other cell membrane is composed of a glycocalyx layer lying in close proximity with the lipid bilayer in which numerous integral proteins acting as enzymes, receptors, transporters, and carriers are embedded. The plasma membrane also consists of peripheral proteins. The glycocalyx is linked to the extracellular matrix (also called basement membrane), which contains numerous thin collagen fibers (primarily types I and III), elastin, glycoproteins, and proteoglycans that provide structural and mechanical support for the muscle cells and also contribute to the viscoelasticity of these tissues. The basement membrane is connected to the cytoskeleton through the plasma membrane. The sarcolemma is rich in membrane invaginations called caveolae. Caveolae are plasma membrane microdomains of lipid rafts that are rich in many proteins including caveolin, dystrophin complex (dystrophin, α-sarcoglycan, and β-dystroglycan), signaling molecules/structures (GPCRs, protein kinases, ion channels, adenylate cyclase, phospholipase C, small GTPases, etc.)

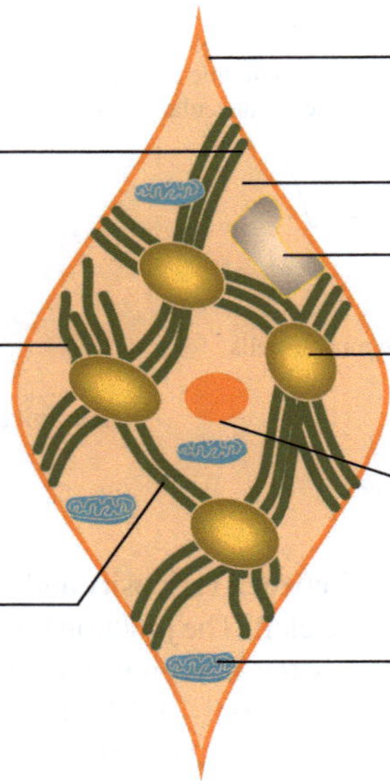

Fig. 7.7 Structure of smooth muscle cell

(Caveolae have been discussed in Chap. 3) [132]. The sarcolemma also contains many communicating junctions (e.g., gap junctions, and adherens junctions) that form sites for cellular signaling as well as anchorage to the cytoskeleton [133]. The basic role of the sarcolemma is that of a housekeeping function, enclosing the liquid portion of the smooth muscle cell. The sarcolemma is particularly important for synaptic transmission, generation, and propagation of action potential, as well as excitation–contraction coupling and contraction–relaxation cycling in the muscle fiber. The sarcoplasm is composed of numerous organelles, a liquid portion and many chemical substances [29, 134, 135]. The organelles found in smooth muscles include mitochondria, nucleus, sarcoendoplasmic reticulum (SER), and cytoskeleton (Fig. 7.7) [123, 124, 133, 136, 137]. These organelles have been discussed in Chap. 3. However, in this chapter, we shall expatiate on the smooth muscle cytoskeleton and review latest information on SER of the smooth muscle cell.

The smooth muscle SER is similar to the non-muscle ER and is adapted for protein and lipid synthesis and also serves as storage of Ca^{2+} ions [138]. These functions of the SER have been discussed in Chap. 3. Here, we shall emphasize on the mechanism by which SER sequester Ca^{2+} ions. All muscle cells contain ER, which in turn contains calsequestrin [139]. Calsequestrin is a low-affinity/high-capacity Ca^{2+} binding protein that helps to hold Ca^{2+} ions in the cisterna of the SER, thus allowing this organelle to store an enormous amount of Ca^{2+} ions. This protein can bind as much as 18–50 Ca^{2+} ions per molecule of calsequestrin. The expression of this Ca^{2+} binding protein is dependent on tissue localization. But the different isoforms of calsequestrin exhibit similar functions of sequestering Ca^{2+} ions in the SER. Other Ca^{2+} binding proteins of the ER/SER include calreticulin—a protein that is expressed not only in other cells but also present as an integral Ca^{2+} binding protein in the smooth muscle SER [140, 141].

The cytoskeleton of the smooth muscle cell is mainly composed of myosin, actin and intermediate filaments. The intermediate filaments are mainly composed of desmin and vimentin, which are the type III intermediate filament proteins. The intermediate filaments act as anchors from which actin filaments can exert force [142]. (Review Chap. 3 for further information about intermediate filaments). The myosin filament is a thick filament found in muscle cells. The smooth muscle cell myosin is the class II type. This smooth muscle thick filament protein consists of six subunits—two myosin heavy chains (MHCs) and two pairs of myosin light chains (MLCs). One pair of the MLCs associated with each of the MHC heads [125, 143, 144]. Each of the MHCs is composed of N-terminal head domain and the C-terminal tails with coiled-coil morphology. One pair of the MLCs weighs 20 kDa (MLC_{20}) and the other—17 kDa (MLC_{17}). The MLCs bind the MHCs around the head region, precisely between the tail and the head—so-called the neck region. Muscle myosins are generally classified as contractile and non-contractile types. Thus, some isoforms of myosins in smooth muscle cells may not play any role in muscle contraction [107, 125, 145]. It should be noted, however, that non-muscle cells may express some isoforms of myosins. Importantly, studies have indicated that like muscle myosin, non-muscle myosin can generate force and shortening in smooth muscle [146]. The myosin of smooth muscle cell is characterized by a side-polar arrangement of cross-bridges [147]. The myosin filaments interact with actin to generate muscle contraction and many forms of cell motility [148].

The actin (also called thin filaments) is the major contractile proteins of muscle cells weighing about 42 kDa. The thin filament primarily exists as filamentous actin (F-actin) [144, 149]. There are six isoforms of actins in vertebrate tissues; these different isoforms are produced from separate genes [149]. The isoforms of the actins that are highly expressed in smooth muscle are the α-actin and γ-actin. The α-actin and γ-actin are specific isoforms of smooth muscle cells. In smooth muscle cells, the isoforms of actin widely expressed in the cytoplasm and associated with the cytoskeleton of these types of cell are the β-actin and γ-actin [150]. These cytoplasmic isoforms are also called non-muscle actin isoforms of smooth muscle cells [151]. The interaction between these different actin isoforms is not fully understood. The α-actin is highly expressed in the smooth muscle cell of blood vessels, whereas the γ-actin is abundantly expressed in the GI tract. Like the myosin, some actin isoforms do not take part in contractile functions of the smooth muscle cell, but may serve for signaling purposes and anchorage [150–152]. Apart from the three isoforms of actin mentioned above, there are other isoforms of actin in muscle and non-muscle cells [152]. Smooth muscle actin is attached to the dense bodies (Fig. 7.7). The dense bodies are α-actinin-rich contractile structures of smooth muscle cells that connect actin, myosin, and intermediate filaments of the muscle cell together and link them up with the plasma membrane and the extracellular matrix [142]. Thus the dense bodies function as anchoring sites for contractile cytoskeletal proteins [153]. The submembranous structures to which contractile cytoskeletal proteins are attached are called costameres. These structures are found in both striated and non-striated muscles and they help to connect the contractile apparatus to the extracellular matrix via the sarcolemma in striated muscle [154, 155]. The costamere is made up of dystrophin–glycoprotein complex and the integrin–vinculin–talin complex. These protein complexes form focal adhesions (also known as dense or adhesion plaques) alternating with membrane invaginations or caveolae [151, 154–156]. The functions of the costameres are to serve as mechanical anchor, transmitting contractile forces from the sarcomere to the sarcolemma and vice versa. This transmission of force occurs via the dense plaques of the sarcolemma. From the sarcolemma, force can be transmitted to the extracellular matrix via integrin receptors (Integrin receptors have been discussed in Chap. 5) [149]. This way, the costameres protect the sarcolemma from damage due to excessive contraction [153, 154]. Because costameres also contain enzymes such as protein kinases (e.g., integrin-linked kinase, ILK), they serve as hubs for many signaling pathways initiated from inside or outside the cell [157]. The costameres were first discovered in striated muscle to play an integral role in mechanical anchorage of the sarcomere to the sarcoplasm. However, emerging studies indicate that smooth muscle also exhibits similar structural characteristics. Like in striated muscles, smooth muscle costamere–sarcolemma complex serves as a shock absorber. This mechanical behavior of smooth muscle cells is mainly due to the presence of the dense plaques [151]. Mutations to the contractile apparatus, sarcomere or costamere have been implicated in muscular dystrophies [158].

Apart from myosin, actin and intermediate filaments, smooth muscle contains substantial quantity of calmodulin, caldesmon, calponin, tropomyosin, smoothelin, actin-depolymerizing factor (ADF; also known as destrin)/cofilin, dystrophin, filamin, fascin, fimbrin, leiomodin, myosin light-chain kinase, myosin light-chain phosphatase, actinin, Arp2/3 (actin-related protein-2 and -3), VASP (vasodilator-stimulated phosphoprotein), WASP (Wiskott—Aldrich syndrome protein) proteins among others [151, 159]. These proteins are generally referred to as thin filament (actin) associated proteins [144, 160]. Of these proteins, calponin, caldesmon, and tropomyosin are the most investigated, and thus, generally observed to regulate muscle contraction [144]. In the GI tract, inflammatory diseases such as colitis and Crohn's disease have been associated with upregulation of the expression of calponin, caldesmon, and tropomyosin [144]. These proteins are also referred to as contractile proteins of the muscle cell, functioning to inhibit the ATPase activity of the myosin filaments. The ubiquitous and multifunctional protein, calmodulin is thought to be the major Ca^{2+}-binding protein in smooth muscle cells functioning through the Ca^{2+}/calmodulin(CaM)-dependent kinase type II (CaMKII) [159]. CaMKII consists of four isoforms, which include α, β, γ, and δ. The α- and β-isoforms are predominantly expressed in cells of the nervous system, whereas the γ- and δ-isoforms are ubiquitously expressed in different cells of the body. These isoforms also express different variants [149].

Tropomyosin is a dimeric (or two-chained) α-helical coiled-coil contractile protein present in all muscle cells that alternates between azimuthal positions on the surface of the F-actin in an end-to-end fashion [151, 161–164]. There are different isoforms of tropomyosin (e.g., α- and β-forms), which may vary in their expression pattern in skeletal and smooth muscle cells [165–167]. This protein spans four–seven monomers along the actin filament, depending on the length of the tropomyosin [151, 161–164]. Tropomyosin controls the spatiotemporal access of actin-binding proteins to F-actin. Thus, tropomyosin serves as a gatekeeper of F-actin [165, 166]. In all muscle cell types, tropomyosin blocks actomyosin interaction site by preventing ATP hydrolysis. Tropomyosin position on F-actin is controlled by Ca^{2+} ions [151, 161–164].

It was previously thought that smooth muscles are devoid of troponin [151], however, recent investigation revealed that smooth muscle cells of blood vessels also contain the three subunits of troponins [troponin T (tropomyosin-binding subunit), troponin I (inhibitory subunit), and troponin C (calcium binding subunit)] expressed in fast-twitch skeletal muscle [168]. This discovery widens the current understanding of smooth muscle contraction. The study by Moran et al. [168] further showed that in vascular smooth muscle cells troponin is localized with tropomyosin, suggesting that the contraction mechanism regulated by troponin-tropomyosin complex in striated muscle cells may be present in smooth muscle cells. Under resting conditions (presumably in low cytosolic Ca^{2+} concentration), in striated muscle cells, troponin I protein binds to tropomyosin in such a way that induces the movement of tropomyosin to a constrained, inhibitory position that prevents myosin cross-bridge interaction. Upon increase in intracellular Ca^{2+} ions, troponin C reversibly binds the Ca^{2+} ions resulting in change in the confirmation of the tropomyosin protein that subsequently leads to the release of the tropomyosin from blocking the actin reactive sites, thus allowing actin to interact with myosin filaments for muscle contraction to take place [169, 170].

Caldesmon is an actin- and myosin-binding protein that maintains the contractile activity of smooth muscle cells, in part, by interacting with tropomyosin dimer [171, 172]. Though it is not exactly clear how caldesmon affect tropomyosin to mediate actomyosin interaction, it is believed that in a resting state, the protein caldesmon is associated with tropomyosin in a way that prevents myosin cross-bridge interaction. Upon activation, increase in intracellular Ca^{2+} levels leads to the recruitment of CaMKII, which phosphorylates caldesmon to release its inhibitory action on actomyosin ATPase activity site. This finally allows for actin–myosin interaction. The protein kinase enzyme, ERK-1 and -2 (extracellular signal–regulated kinase type 1 and 2, a type of MAPK activated by PKC) can also phosphorylate caldesmon to release its inhibitory effect on actin by altering tropomyosin position on F-actin [149, 151, 173].

Calponin is a protein believed to function in a similar fashion as caldesmon by inhibiting the actin and myosin interaction [171, 172]. Like caldesmon, calponin phosphorylation is thought to release the inhibition of actomyosin ATPase activity by changing the position of tropomyosin on actin to allow for actomyosin interaction [151, 171–173].

Several factors and conditions including mutations, inflammation (e.g., colitis) have been shown to affect the expression of the actin-associated proteins—calponin, caldesmon, tropomyosin in the smooth muscle cells of the GI tract and other regions of the body [144]. Disorders of smooth muscle contraction–relaxation are responsible for a variety of diseases including hemorrhagic cerebral vasospasm, hypertension, pulmonary hypertension, preterm labor, asthma, muscular dystrophy, cardiomyopathy, functional bowel disease such as dysmotility [149, 174–177].

7.3.2 The Contractile Unit of Gastrointestinal Muscles

The contractile unit of striated muscle is the sarcomere, which is made up of thick filament, thin filament, and the Z-band [156]. When viewed under the light microscope, the sarcomere is a characteristic repeating pattern seen in each myofibril of a striated muscle. In smooth muscle cell, the contractile unit is made up of the thin filaments, thick filaments, and dense bodies—referred to as the "smooth muscle sarcomere" [121, 153]. Studies using electron microscope have provided substantial evidences regarding the structure of the smooth muscle cells. The "smooth muscle sarcomere" consists of thin filament, thick filament, and two dense bodies. The thick filament is situated between parallel thin filaments [178]. The thick filament extends from one dense body to the nearest dense body [178]. In contrast to striated muscle where the thick filaments are bipolar, in smooth muscle, these filaments are either side polar or row polar [121]. When the smooth muscle cell contracts (i.e., shortens), the thick filament overlaps with the thin filaments. This leads to shortening of the distance between two dense bodies that constitute the contractile unit. The degree of overlap depends on the contractile force applied. Thus, the greater the contractile force the greater the overlap [178]. So, the sarcomere, though, with a different structural architecture in smooth and striated muscles, constitute the basic structural and functional unit of all muscle cells. Like the regular and interdigitating pattern of thick and thin filaments present in striated muscles, smooth muscle filaments seem to be organized into arrays of contractile units that lie parallel to the longitudinal axis of the muscle bundle. In this arrangement, some portions of the contractile unit, in particular, the actin filaments are attached to the membrane of the nucleus, which at least, in part, allows for the even distribution of mechanical strain across the cell. The parallel arrangement of smooth muscle sarcomeres allows to sum the contractile forces generated by individual sarcomeres required to maintain a regulated contractile activity that move luminal content aborally in the GI tract [121, 130, 151, 178].

The actin filament is connected to the Z-band (in striated muscles) and also to the dense bodies (in smooth muscles). Like the Z-band, the dense bodies form the border between two contractile units (Fig. 7.8). Both the Z-band and dense bodies are connected to the costameres. These structures provide a stable but plastic architectural system for mediating mechanical strain in living systems [154, 180].

The thick filaments are linked to each other in the M-line, a structure located in the center of the thick filament, containing myosin-binding proteins. The myosin-binding proteins are thick filament–associated proteins, which include M-protein, myosin-binding protein C (MyBP-C), H-protein, M-protein, titin,

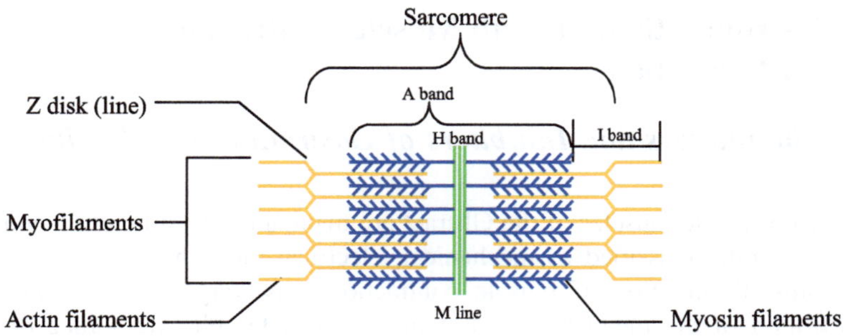

Fig. 7.8 Structure of striated muscle sarcomere. The observation of striated muscle under a light microscope reveals a series of dark-and-light bands perpendicular to the long axis—the origin of the striations seen in these muscles. These striations are due to the presence of numerous thick and light filaments, forming cylindrical bundles called myofibrils, which densely occupy the sarcoplasm and also extends to the plasma membrane to link with adjacent cells. The middle of the sarcomere consists of myosin producing a dark band—A band (this band contains a central light band—H band and a narrow dark band—M-line in the center of the H band). At either end of the sarcomere are the actin filaments. One end of the actin filaments is attached to a connecting group of proteins that form the Z-disk or line. The Z-disk is situated in the centers of the thin filaments. The sarcomere is the distance between two Z-lines. Between the A bands of two adjacent sarcomeres is the I band, which is the part of the actin filaments that do not overlap with the myosin filaments. Actin filament interacts with myosin filaments through an extension of the myosin that overlaps with a portion of the thin filament. This overlapping extension is referred to cross-bridge [179]

myomesin, skelemin, muscle isoform of creatine kinase [121, 130, 178]. Notably, however, these proteins can extend beyond the M-line. The thick filament (mainly characterized by the presence of myosin) in the M-line is connected to the third filament system (mainly characterized by the presence of titin) by myomesin [179]. The protein myomesin also functions as an elastic spring [181].

In addition to the major components of muscle contractile unit, both striated and non-striated muscles contain a third contractile protein "myofilament" called titin (also known as connectin)—an elastic giant protein (with molecular weight of approximately 3000 kDa) that develops passive tension maintaining the contractile functions of the muscle cells at rest. Titin functions as a spring for the muscle contractile unit. Each titin molecule runs through the "I" band and binds to the myosin filament. From the myosin, titin courses through the Z-line to the "M" line [182–184]. Substantial part of the titin molecule links the ends of thick filaments to the Z-line [185]. Titin is associated with another giant elastic protein called nebulin, which in striated muscle is associated with the thin filaments [154, 185].

The contractile proteins of smooth muscle are highly labile. For instance, depolymerization of the myosin filaments can be induced by mechanical strain, and their repolymerization underlies the recovery process following the action of a contractile force [186].

7.4 Gastrointestinal Smooth Muscle Contraction and Relaxation

7.4.1 Stimulators and Inhibitors of Gastrointestinal Motility

The activities of the muscles of the GI tract are maintained by different factors that may be generally classified as mechanical, electrical, and chemical. Though the mechanisms of initiation of muscle contraction may vary, these factors cause depolarization of the membrane of the smooth muscle cell. Membrane depolarization may result from the activation of L-type voltage-gated Ca^{2+} channels, ligand-gated channels may be due to passive flow of ions between two cells. The mechanical factors that initiate contraction are those that are initiated by wall stretch or strain. A stretch in the cell membrane induces the release of Ca^{2+} from internal stores, which is possibly mediated by influx of Ca^{2+} ions from the extracellular medium. Electrical stimulation or inhibition of contraction is mediated by the nervous impulses. The chemical factors that stimulate or inhibit GI motility are the GI peptides, which are classified into three broad categories: hormones [e.g., gastrin, cholecystokinin, secretin, gastric inhibitory peptide—GIP, enteroglucagon (glicentin)], neurocrines (e.g., calcitonin-gene-related peptide—CGRP, vasoactive intestinal peptide—VIP, gastrin-releasing peptide—GRP), paracrine (e.g., somatostatin), and autocrine factors. Details on GI hormones and neurotransmitters and their mechanisms of action are discussed in Chaps. 8 and 9 [75, 119, 187–190]. The factors that initiate muscle contraction or inhibit muscle activity can further be divided into pharmacomechanical and electromechanical factors. The pharmacomechanical factors are those stimuli that cause contraction by voltage-independent pathway—processes that are independent on the membrane potential. For instance, the pharmacological agent that stimulates muscarine- and nicotine-sensitive cholinergic receptors, carbachol is a pharmacomechanical factor. The electromechanical factors are agents or stimuli that affect muscle activity by membrane depolarization, with an associated influx of Ca^{2+} ions. KCl is an example of an electromechanical factor [191–193].

Stimulation of the parasympathetic division of the autonomic nervous system activates GI motility by depolarizing the muscle cell membrane via the release of acetylcholine—the most potent excitatory neurotransmitter. The cognate receptors of this neurotransmitter are the muscarinic receptor subtypes M_2 and M_3. Other excitatory neurotransmitters of the gut include substance P [2]. Stimulation of the sympathetic division impedes GI motility by hyperpolarizing the muscle membrane. Examples of a neurotransmitter that hyperpolarizes the membrane of muscle cells of the GI tract are norepinephrine, VIP, pituitary adenylate cyclase-activating polypeptide (PACAP), nitric oxide (NO), and CGRP [194–196]. The most potent inhibitory neurotransmitter is CGRP followed by NO. For details see Chap. 9.

The membrane depolarization that initiates muscle contraction can occur due to the influx of Ca^{2+} from extracellular fluid, periodic release of Ca^{2+} from intracellular stores or activities of plasma membrane oscillator. The membrane oscillator is a feature of pacemaker cells and is linked with a cytosolic oscillator [25]. These oscillators cause transient increase in Ca^{2+} waves that subsequently activate an inward current, which in turn generates pacemaker currents. This current can spread through the gap junctions to other cells providing the depolarizing signal that generate the action potential—the driving force for muscle contraction [25].

7.4.2 Pacemaker Cells of the Gut: Interstitial Cells of Cajal, CD34-Positive and PDGFRα-Positive Cells

The concept of smooth muscles as the basis of GI motility was rooted in the works of Santiago Ramón y Cajal (1852–1934) of Madrid. In 1892, y Cajal reported that he had successfully identified a novel type of cells in the intestine. He called these cells "interstitial cells" (see below how these cells of y Cajal will later become known as the pacemaker cells of the GI tract) [197]. Importantly, Ramón y Cajal observed that these cells connected the nervous system to muscles of the intestine. Following y Cajal's description of the interstitial cells, scientists began to unravel how these cells carry out their functions. In 1910–1920s, Arthur Keith (1866–1955) suggested that the y Cajal novel cells function like the electroconductive system of the heart. He suggested that the cells initially identified by y Cajal were pacemakers of the intestine. Some decades later, in 1982, Thunberg Lars confirmed the suggestions of Keith [98, 198]. During the close of the twentieth century, experimental findings identified molecular regulators and receptors of the pacemaker cells of GI tract. In 1992, Hitomi Maeda identified c-Kit as an integral regulator in the development and function of the interstitial cells of Cajal [197, 199]. c-Kit is a transmembrane glycoprotein, a tyrosine kinase receptor that occur in myenteric or muscular Cajal cells. The stimulators of these receptors are diverse and include factors produced by both neuronal and non-neuronal cells. The stimulators, generally called the kit ligands modulate the functions of the cells where the cognate receptors are found [200]. Further research on the c-Kit receptor has shown that it is required for intestinal pacemaker activity [201].

The ICCs are the best-known pacemaker cells of smooth muscles of the digestive tract, forming complex networks between smooth muscle cells and postganglionic neuronal fibers that regulate contractile functions of the digestive tract [202]. The ICCs make up about 5–9% of the total number of cells in the muscle layers of the GI tract [71, 203]. The complex network formed by these cells are believed to play a role in regulating the functions of GI cells including epithelial cells, mesenchymal cells, and immune cells as well as the glial cells and neurons [203].

It should be mentioned that the frequency of electrical rhythm of these cells varies in different part of the digestive system. For instance, the frequency is very high in the duodenum (about 11 cpm) and ileum (about 10 cpm) and very low in the stomach (about 3 cpm) and 4 cpm in the large intestine. These waves spread to smooth muscle cells and the resulting depolarization initiates calcium ion entry that generates action potential for the muscle contraction. Slow waves from several cells give rise to contractions that are organized into phases—which form the basis for peristalsis and segmentation of the GI tract. In association with the enteric nervous system, these cells play a critical role in regulating smooth muscle function and GI motility, acting to transduce neuronal and mechanical signals along the GI tract [203]. ICCs mediate neural input from enteric motor neurons. Animals lacking ICC have greatly reduced responses to the neurotransmitter acetylcholine, released from excitatory motor neurons, and to the transmitter nitric oxide, released from inhibitory motor neurons. Loss of ICC in disease, therefore, may interrupt normal neural control of GI contractions and lead to functional GI disorders, such as irritable bowel syndrome. Interestingly, ICCs also have mechanosensitive mechanisms that cause these cells to respond to stretch. Stretching GI muscles can affect the resting potentials of ICC and change the frequency of pacemaker activity. ICCs have been implicated in the pathogenesis of Hirschsprung's disease as well as other diseases including GI tumor, neuroinflammatory and neurodegenerative intestinal pathologies [204].

Besides the ICCs, CD34-positive (cluster of differentiation—a single pass transmembrane glycoprotein) and PDGFRα-positive cells (platelet-derived growth factor receptor alpha) also contribute to smooth muscle automaticity. Thus, pacemaker cells of the GI tract are those that express the c-Kit receptor, CD34, and PDGFRα. The pacemaker cells of the GI tract also express other receptors including ion channels (see below). Activation of these pacemaker receptors leads to changes in ion conductance through the membrane of the associated cells, which results to generation of waves of depolarization that are propagated across the smooth muscle cells of the intestine. The peristalsis of the digestive tract is controlled by the activities of these cells. These cells are electrically coupled to smooth muscles forming a functional syncytium that is responsible for regulated contractile functions in the GI tract (Fig. 7.9). If the network between ICC and GI smooth muscle cells is broken, then the different regions of muscle will function independently [2].

The GI pacemaker cells are generally known as telocytes. These cells are pleomorphic having an oval nucleus and long, branching, and thin cytoplasmic processes that interlace with processes of adjacent cells. They express different types of markers including CD34, vimentin, caveolin-1, inducible nitric oxide synthase (iNOS), and vascular endothelial growth factor (VEGF). These cells not only generate, but also contribute to the transmission of impulses [205]. Telocytes form a three-dimensional cellular network in the submucosa and in the interstitium between muscle layers and an almost continuous layer at the submucosal borders of *muscularis mucosae* and circular muscle layer. They also encircle muscle bundles, nerve structures, blood vessels and are found in gastric glands and intestinal crypts. There are also candidate pacemaker cells located in the myenteron (Auerbach's or myenteric plexus). Apart from the known cells, there are other cell types that express the small-conductance Ca^{2+}-activated K^+ channel type 3 (SK3) that play a role in pacemaking. Numerous channels have been shown to be involved in

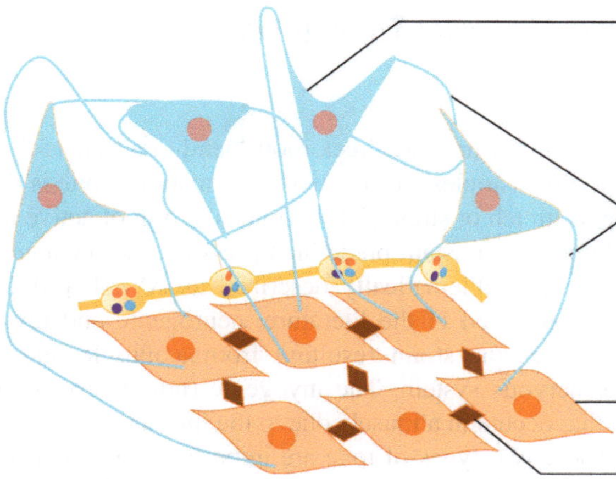

Fig. 7.9 Cellular network of pacemaker and GI smooth muscle cells. Smooth muscle cells have functional L-type calcium channels that increase conductivity after activation by pacemaker cells. It should be noted that Cajal cells are also present in other parts of the body including the brain, urinary tract, reproductive tract, and vascular system (blood vessels and lymphatics)—where they are responsible for generating pacemaker current that drive the information processing or contractile functions of these regions or organs [70]

pacemaker activity. These channels include Ca^{2+}-activated Cl^- channels that move Cl^- ion to the extracellular space, thereby, resulting in membrane depolarization; Ca^{2+}-induced Ca^{2+} release channels; membrane Ca^{2+}-ATPase; Na^+/Ca^{2+} exchanger; sarcoendoplasmic reticulum Ca^{2+}-ATPase; ryanodine receptors; transient receptor potential channels, which may be activated by influx of Ca^{2+}. These channels are important in pacemaker activity, and controls neuronal excitability in smooth muscle [25, 205, 206]. The ICCs have a high turnout of ATP and Ca^{2+}, which are major triggering factors required for effective pacemaking. The ICCs also have a high number of mitochondria, which are associated with the ER to form a cytosolic pacemaker unit [25].

Though the ICCs can generate spontaneous rhythmic electrical slow waves in the digestive tract without external influences, factors such as mechanical force or strain, neurotransmitters, and hormones can modulate the amplitude and frequency of the slow waves generated by the pacemaker cells. For instance, acetylcholine and noradrenaline can increase the activity of ICC by increasing IP3 that increases cytosolic Ca^{2+}. The neurotransmitter NO has inhibitory effect on ICC. The action of NO on ICC leads to reduction in the amplitude and frequency of slow waves, thereby reducing the pacemaking drive of these cells. The activity of NO is mediated through cyclic GMP. This second messenger activates cyclic GMP-dependent protein kinase, which phosphorylates IP_3R, thereby reducing their sensitivity to Ca^{2+} [25].

7.4.3 Neurogenic and Myogenic Tone of Gastrointestinal Muscles

The word "tone" is derived from the ancient Greek roots "tend-, tent-, tens-," which refer to stretch or strain—a dimensionless measure of deformation. Tone is defined as the resistance of muscle to stretch or strain [207]. Tone is a form of active muscle contraction defined as volume at a constant predefined pressure, or a change in the slope of the compliance curve, mathematically described as $\Delta V/\Delta P$ [207]. Two types of tone have been widely used in the literature: neurogenic and myogenic tone. Neurogenic tone indicates the strain resulting from a muscle due to the excitatory inputs from the nervous system. The myogenic (non-neurogenic) tone refers to the type of strain that occurs in a muscle due to the intrinsic property of the muscle itself [207–212]. These two types of tone are important components of GI motility [207]. For details on the classification of tone, see Gregersen and Christensen [207].

GI tract smooth muscles generate specific type rather than tetanic type of tone—a property of smooth muscle described as latching. This property of gut smooth muscles is due to several factors including the autonomic input, intrinsic gut neurons, activities of mechanoreceptors, and ICC modulatory role on the muscle cells [213]. The specific tone is characterized by a sustained contraction at a given segment of the smooth muscle cells. This type of tone is believed to be mainly myogenic. In contrast, tetanic tone, which under normal condition is not a characteristic of the gut smooth muscles, is characterized by contractions with frequencies that do not allow for the normal time interval in normal physiological situation. Tetanic tone is thought to be partly neurogenic. Tone of the gut smooth muscles affects the rate of flow and mixing of luminal contents [207].

7.4.4 Mechanisms of Gastrointestinal Smooth Muscle Contraction

Contraction is a property, characterized for all muscles. Generally, contraction can be defined as the activation of force generating sites of the muscle fibers. Muscle tissues can develop active force (contract) in response to stress or adequate stimuli. Muscle contraction occurs due to its passive physical properties and the properties due to the activation of the muscle contractile elements. Muscle contraction is characterized by both changes in length and tension [207, 214–217].

Like other smooth muscles, GI smooth muscle contraction is controlled by Ca^{2+} and Rho kinase. GI tract muscle contraction is mainly initiated by increase in intracellular Ca^{2+} ion concentration, which can result from a variety of sources—from intracellular stores or influx from the extracellular fluid [25, 218]. Influx of Ca^{2+} ions from the extracellular fluid can result from stimulation of parasympathetic fibers that innervate the smooth muscle cells, activation of mechanoreceptors of smooth muscles or structurally connected cells (e.g., ICC), and stimulation by paracrine factors or hormones. The effect of parasympathetic stimulation on gut has been extensively investigated. The major neurotransmitter of this division of the autonomic nervous system is acetylcholine—which has excitatory function on gut smooth muscle. The effect of this neurotransmitter on the gut smooth muscle is mediated through the M_3 subtype muscarinic cholinoreceptor, a type of G protein-coupled receptor. The activation of this receptor will result to dissociation of G_q, which activates membrane-associated enzymes that results to increase in intracellular Ca^{2+} and other second messengers that in turn activates other intracellular enzymes including PKC. This protein kinase activates downstream effectors that increase muscle tone and motility. This way, acetylcholine increases the digestion (mechanical) of luminal contents, providing a greater contact of the digesting food with the mucosa, which undoubtedly increases the rate of absorption [208, 218]. It should be noted that the action of acetylcholine and other mediators may vary in different layers of the GI musculature or regions of the tract [208]. For instance, in an in vitro study, acetylcholine increased the amplitude and frequency of contraction in the circular muscle layer of the rabbit small intestine, but not longitudinal muscle layer. Like acetylcholine, potassium ions can induce contraction in circular smooth muscle layer of the GI tract (and also longitudinal layer) of rabbit small intestine in vitro [219]. Several hormones (e.g., angiotensin II), growth factors and paracrine factors can stimulate increase in intracellular Ca^{2+} ions through the IP_3 pathway (Fig. 7.10).

Calcium Sensitization of Smooth Muscle Contraction

The contraction of smooth muscle occurs for a longer duration than the increase in intracellular Ca^{2+} concentration, which is often transient. This phenomenon is called Ca^{2+} sensitization of smooth muscle contraction. There are multiple mechanisms of this sensitization. The sensitization can be due to the inhibition of myosin phosphatase (i.e., MLCP) activity by Rho kinase [119]. The inhibition of myosin phosphatase increases myosin phosphorylation, thereby sustaining smooth muscle contraction, even in the absence of elevated cytosolic Ca^{2+} level. In neuron, Ca^{2+} sensitization can result to long-term depression [237]. This is currently considered the main mechanism responsible for Ca^{2+} sensitisation. The stimulatory effect of an agonist can increase the phosphorylation of CPI-17 and MYPT1, which control MLCP activity to mediate phasic or tonic contraction of GI tract smooth muscle (including sphincteric muscle) [238–241]. For instance, CPI-17 (and PKC) can inhibit MLCP to cause Ca^{2+} sensitization of smooth muscle. The inhibitory effect on this muscle phosphatase can result to tonic or phasic contractions [239]. Ca^{2+} sensitisation also involves the activation of the small G protein RhoA, which increases the activity of the Rho kinase enzyme. This protein kinase mediates the inhibition of MLCP. Thus, promoting sustained contractile state of the smooth muscle cells [119]. Ca^{2+} sensitization can be caused by phosphorylation of MLCK by ERK, or direct phosphorylation of MLC20 by other protein kinases such as ILK.

This process increases the sensitivity of MLCK to Ca^{2+} ions in the cytosol [149]. The inhibitory action of actin-binding proteins on actomyosin interaction has some effects on Ca^{2+} sensitization as intracellular enzymes that act on these inhibitory proteins regulate the availability of actin for cross-bridge cycling, which is required to promote contractile state. Cytoskeletal remodeling also contributes to this sensitization [149]. Remodeling of the cytoskeletal proteins (e.g., actin) by protein kinases results to changes in the force of contraction [119].

Sliding Filament Theory of Muscle Contraction

This theory initially developed to explain the mechanisms of contraction and relaxation in striated muscles is now well accepted as the mechanism of contraction and relaxation in smooth muscle [121]. The theory was developed in 1954 by independent researchers—the English Physiologist and Nobel Prize winner Andrew

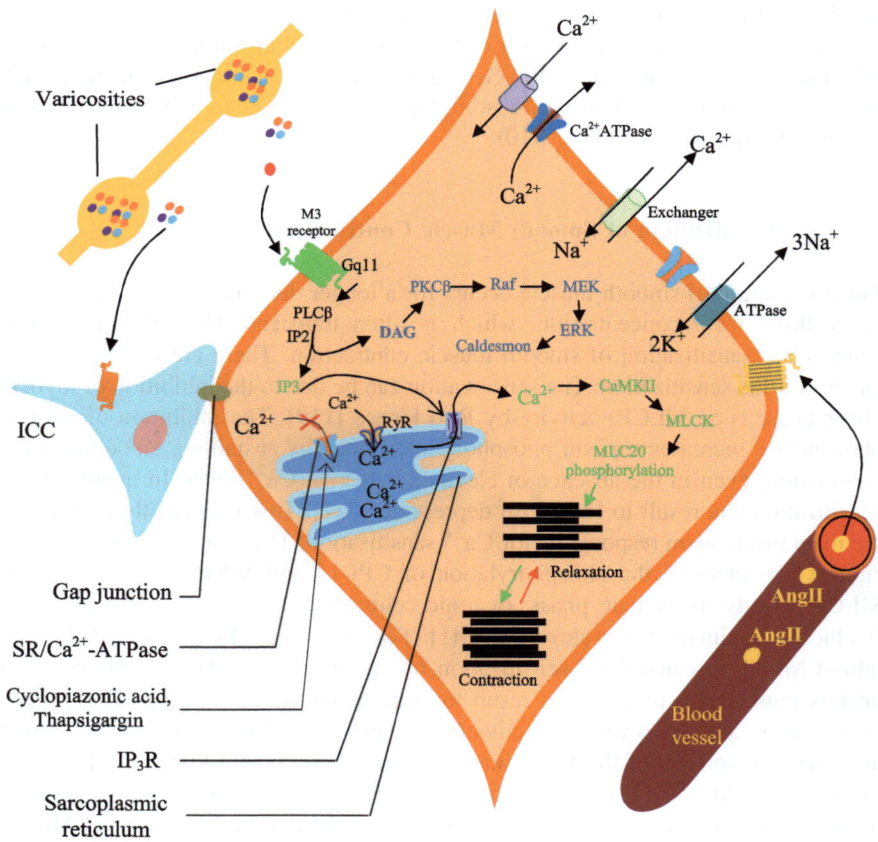

◂**Fig. 7.10** Mechanism of contraction and relaxation of smooth muscle. Muscle contraction is initiated by increase in the level of intracellular Ca^{2+} ions, which occurs via influx from extracellular medium or by the mobilization of internal stores of Ca^{2+} especially in the sarcoplasmic reticulum (SR) or sarcoendoplasmic reticulum (SER). Ca^{2+} influx from the extracellular fluid can occur through Ca^{2+} channels. Membrane Ca^{2+} channels responsible for transporting Ca^{2+} into the cell include—voltage-dependent L-type Ca^{2+} channel; Na^+/Ca^{2+} exchanger, a channel that functions in both forward and reverse modes. In the forward mode, this channel move Ca^{2+} out of the cell to cause hyperpolarization of the muscle cell membrane. In the reverse mode, the channel functions to move one Ca^{2+} ion in exchange for three Na^+ ions contributing to membrane depolarization [25, 75, 217, 220]. The number of Na^+ ions entering the cell is balanced by the activity of the Na^+–K^+-ATPase pump, which is activated following an action potential to extrude $3Na^+$ for $2K^+$, thereby contributing to the early phase of the hyperpolarization potential of the muscle cell membrane [25]. As the concentration of Ca^{2+} ions increases inside the cell, Ca^{2+} ions activate many channels from the inside. These channels include the Ca^{2+}-sensitive Cl^- channels, Ca^{2+}-activated K^+ channels. Activation of these channels contributes to the development of an action potential [25, 217]. While influx of Ca^{2+} via voltage-gated ion channels (L-type Ca^{2+} channels) can stimulate further release of Ca^{2+} from internal stores, the main activator of the release of Ca^{2+} from internal stores is believed to occur via the stimulation of metabotropic receptors that signal downstream IP_3. This can occur by the stimulation of muscarine-type cholinergic receptors M_2 or M_3 (many neurotransmitters, hormones, and receptors can also cause increase in IP_3). So, activation of GI smooth muscle M_3 receptors stimulates the GPCR, which is coupled to $G\alpha q$ or $G_{q/11}$. This G protein isoform dissociates to activate an enzyme of the inner leaflet of the plasma membrane, phospholipase C isozyme β, PLCβ [2, 75, 119]. This enzyme hydrolyzes the membrane lipid phosphatidylinositol 4,5-bisphosphate (IP_2) to produce inositol 1,4,5-trisphosphate (IP_3) and diacylglycerol (DAG). IP_3 diffuses to the SER to bind to a pocket in the IP_3R, which results to the opening of the channel. This channel transports of Ca^{2+} ions into the cytosol, producing "a cytosolic calcium spark" [220, 221]. Though IP_3 is the primary ligand, IP_3R can be stimulated by an increase in cytosolic Ca^{2+} [2]. The smooth muscle SER also contains another receptor that is responsible for release of Ca^{2+} into the cytosol. This receptor, initially discovered in skeletal muscle, is called the ryanodine receptors (RyRs) because of its high affinity to the alkaloid "ryanodine" derived from the South American *Ryania speciosa* plant [222]. Inhibition of the SER channels by this alkaloid depends on the concentration of ryanodine as well as conditions of the experiment. RyR is a mushroom-like Ca^{2+} channel with a large cytosolic head and transmembrane portion, activated by increase in cytosolic Ca^{2+} ions and caffeine. Thus, RyR is a type of Ca^{2+}-induced Ca^{2+} release channel [2, 223, 224]. RyR channels are expressed not only in GI tract smooth muscles, but also in the SR of smooth muscles of the peripheral and central blood vessels [225]. Therefore, the IP_3R and the RyR are the major Ca^{2+}-release channels in the SER [225]. The increase in intracellular Ca^{2+} mediates the activation of Ca^{2+}-dependent protein kinases such as CaMKII, which in turn activates myosin light-chain kinase (MLCK), forming a complex of CaMKII-MLCK, which phosphorylates the 20 kDa regulatory light chain of myosin II (MLC20) at position serine 19. Since MLC20 is structurally associated with the myosin filament, this phosphorylation results to a conformational change of the myosin filament, which increases the angle in the neck region of the myosin heavy chain. This region corresponds to the area of the cross-bridge cycle where the myosin head is unattached to the actin filament. The change in conformation substantially increases the myosin head ATPase activity to promote cross-bridge cycle by allowing the myosin head to bind to actin filaments and thereby develop contractile force that shortens the smooth muscle cell [125, 144, 226–228]. Thus, increase in intracellular Ca^{2+} ions and subsequent phosphorylation of the MLC20 is the main initiator of the process of smooth muscle contraction. However, phosphorylation of smooth muscle MLC20 can occur in the absence of elevated intracellular Ca^{2+} ion concentration, i.e., Ca^{2+}-independent pathway, through the activities of other protein kinases such as Rho kinase, integrin-linked kinase, and zipper-interacting protein kinase (ZIPK) [149, 229]. Though the contribution of these kinases to smooth muscle contraction is still actively been investigated, it is not completely clear how they affect contractile force and tension of smooth muscles. However, they may play a role in contributing to the phosphorylation of some proteins that are associated with the contractile filaments. These protein kinases may also be involved in cytoskeletal remodeling, which may facilitate the development and transmission of force in the muscle [149]. Recall that some actin inhibitory proteins constantly occupy the actin active site so that this contractile protein does not interact with myosin. Since myosin interaction with actin depends on actin availability, another pathway is believed to control the readiness of binding of actin filaments with myosin. For actin–myosin interaction to occur, actin-binding protein (such as caldesmon and calponin) must change their position by moving away from the active site. For instance, caldesmon having two isoforms, with the high molecular weight isoform specifically expressed in smooth muscle and is believed to play a role in stabilizing the actin filament and restricting the binding of myosin head to the actin filament, thereby reducing the generation of contractile force and tension. The heavyweight isoform has been shown to enhance intestinal peristalsis as its disruption has been associated with disorder in propulsive action of the intestine [228]. The low molecular weight caldesmon is expressed in many cells including smooth muscles, but it is believed to mediate interaction between actin and non-muscle myosin filaments [228]. For actomyosin interaction to take place, certain protein kinases (e.g., ERK, extracellular signal responsive kinase) have to phosphorylate the actin-binding protein. The activation of this protein kinase is regulated by the level of the second messengers—DAG and Ca^{2+}, which in turn activate protein kinase C, PKC. PKC signals downstream target proteins phosphorylating MEK (mitogen-activated protein kinase kinase, MAPK kinase, or MAPKK, also known as MKK), ERK, L-type Ca^{2+} channels, proteins that regulate cross-bridge cycling and other target proteins to initiate cellular response [119, 230, 231]. The actin filament associated proteins prevent cross-bridge cycling by inhibiting the actin-activated ATPase activity of myosin. Phosphorylation of caldesmon increases the availability of actin for interaction with myosin head. Interaction of the thick and thin filaments leads to cross-bridge cycling and the development of force that is evidenced by muscle shortening as well as increase in tension with dissipation of energy in the form of ATP [159, 168]. CaMKII has been shown to phosphorylate MLCK, MLC20, but also caldesmon, calponin, cAMP-responsive element-binding protein (CREB), etc. [149]. CREB is a transcription factor that signals downstream the nucleus to regulate gene expression [149]. The discovery of both smooth muscle troponin and tropomyosin has widened our knowledge on the mechanism of smooth muscle contraction. In the resting state, tropomyosin blocks the actin filament from interacting with myosin. Prior to the discovery of troponin in smooth muscle, it was thought that tropomyosin position in the actin filament was basically regulated by phosphorylation, however, the findings that tropomyosin is present in close proximity with troponin indicate that similar mechanism that regulate the position of tropomyosin in skeletal muscle may play out in smooth muscle cell. In the presence of troponin and calcium, tropomyosin azimuthally changes position on the actin filament to expose myosin-binding sites that allow myosin heads to interact with the actin filaments [163]. Under normal condition, relaxation follows immediately after contraction. This allows the muscle to return to its initial length (lengthening and reduction in tension) [159, 168]. Smooth muscle relaxation usually occurs as a result of decrease in cellular ATP and reduction in intracellular Ca^{2+}. Decrease in cytosolic Ca^{2+} is due to influx of Ca^{2+} into the SER as well as efflux to the extracellular medium [159, 168]. The main plasma membrane efflux channels of Ca^{2+} are the Na^+–Ca^{2+} exchanger and the SER Ca^{2+} ATPase. The exchanger is responsible for the acute phase of Ca^{2+} efflux. The Ca^{2+} ATPase is responsible for the sustained increase in Ca^{2+} efflux. The mitochondria have a crucial role in removal of excess Ca^{2+} from the cytosol (For details on mitochondrial role in cellular Ca^{2+} signaling, review Chap. 5) [25, 120, 221]. It should be mentioned that the recovery phase of smooth muscle is dependent, in part, on restoration of membrane potential through opening of K^+ channels [75, 218]. In particular, the small and large conductance Ca^{2+}-activated K^+ (SK and BK) channels are involved in muscle relaxation [2]. The Ca^{2+}-sensitive BK and SK channels are activated during the repolarizing phase of the action potential of the smooth muscle cells. They contribute to the hyperpolarization phase of the smooth muscle electrical activity. The decrease in intracellular Ca^{2+} allows for another sequence of depolarization of the smooth muscle [25].

The SER of smooth muscle can accumulate Ca^{2+} via the SER Ca^{2+}ATPase. The opening of this pump results to increased translocation of cytosolic Ca^{2+} into the SER [190, 232]. Pharmacological inhibition of SER Ca^{2+}-ATPase with cyclopiazonic acid and thapsigargin results to sustained muscle tone [217]. Phospholamban, a SER-associated negative regulator of SER Ca^{2+}-ATPase is abundantly expressed in the SR of cardiac muscle and slow skeletal, but in smooth muscle its expression is low and thus may not play a crucial role in smooth muscle Ca^{2+} signaling [233–236]. However, in smooth muscle cells phospholamban is phosphorylated by cGMP-dependent protein kinase [236]. In cardiac muscles, phospholamban is phosphorylated by cAMP-dependent protein kinase, cGMP-dependent protein kinase and CaM-kinase-II [149, 233, 236]. Phosphorylation of phospholamban activates the SER Ca^{2+}ATPase, thereby increasing the rate of translocation of Ca^{2+} into the SER, subsequently resulting in decrease the level of cytosolic Ca^{2+} [236]. Relaxation of smooth muscle cells can be due to dephosphorylation of MLC20 by the active MLC phosphatase (MLCP), which cleaves the high-energy phosphate bond of the MLC, thereby promoting smooth muscle relaxation [75, 144, 227]

Fielding Huxley (1917–2012), German physiologist and physician Rolf Niedergerke (1921–2011), British molecular biologist and muscle physiologist Hugh Esmor Huxley (1924–2013), and the British biophysicist and zoologist Emmeline Jean Hanson (1919–1973) [214, 215]. For further information on the history of the sliding filament theory, review the following publication: Hitchcock-DeGregori and Irving [242], Huxley [243], Maruyama [244], Squire [245], Rall [246], and Huxley [247].

The sliding filament theory was developed to describe a cycle of repetitive events that cause the myosin filament to slide along the actin filament. During this process, force and tension are developed. The theory is based on the cross-bridge mechanism [121]. The cross-bridge is used to refer to two myosin heads that interact with actin filament. Each myosin head comprises two binding sites. One site is for binding of actin, while the other is for ATP binding. The ATP-binding site of myosin has an intrinsic ATPase activity so that an ATP molecule is readily hydrolyzed upon binding to this site. The myosin head is activated when the ATP is hydrolyzed to ADP + P_i while still attached to the myosin head [130, 248, 249]. The myosin head then initiates a power stroke resulting in a conformational change in the globular portion of the myosin motor, which opens the nucleotide binding pocket and separates ADP from Pi (inorganic phosphate), releasing the Pi [250]. The power stroke generates about 1–2 piconewtons (pN) (10^{-12} N) of force. (The movement of individual myosin motor can generate up to 5–10 pN of force during a working stroke depending on the type of myosin isoform that constitutes the myosin filament.) The generated force is used to slide the myosin filament along the actin filament inwards in discrete steps, about 5–10 nm at a velocity of about 0.04–4.5 μm/s, thereby shortening the sarcomere. The force that creates this velocity of the myosin motor is inevitable in muscle contraction and transport of vesicle in the cytoplasm (The work done in joules is the product of the force in Newton and distance given in meters) [250–255]. The amount of force generated and thus the length of inward sliding depends on the degree of overlap of the filaments [121]. The detachment of the myosin from the actin filament requires MLCP activity, which brings the muscle to relaxation (discussed below). In the presence of ATP, the myosin is ready to be activated for another cycle of event to occur. The cyclic interaction of myosin cross-bridges with actin filaments in smooth muscle is responsible for force generation, motion, and muscle shortening with accompanying changes in tension [130].

In skeletal muscle, myosin filament detaches from actin filament upon binding of the ATP molecule to the myosin head (MLCP is not required here for the detachment to occur). This allows the cycle to continue in so far as ATP and Ca^{2+} are available in the cell. The Ca^{2+} in striated muscle is required to unblock the myosin-binding site on the actin filament. This is achieved by the reaction between the calcium-(C)-subunit of troponin, which results to conformational change in Tropomyosin-(T)-subunit of troponin that removes the inhibitory-(I)-subunit of troponin from the tropomyosin, causing tropomyosin to shift azimuthally from the actin filament. (Two Ca^{2+} ions bind to the C-troponin). This opens the myosin-binding site on the actin filament and the energized head attaches to this site forming a cross-bridge in which the myosin heads tilt, sliding along the actin filament over a relatively short distance, in corkscrew fashion [130, 248, 249]. Still attached to the actin filament, the myosin head then releases ADP and is ready to bind with another ATP molecule [130]. Until ATP binds to its site on the myosin head, the muscle cell continues to remain in a state of rigor [248–250].

Though there are other theories of muscle contraction, the sliding filament theory remains the most widely accepted. For instance, in the latch bridge theory [256], though not widely accepted [257], proponents believe that attached, dephosphorylated myosin cross-bridge can maintain force [258, 259]. This latch bridge theory seems to explain the relatively prolonged contraction of smooth muscle observed despite transient increase in intracellular Ca^{2+} level.

Change in Length and Tension Defines Muscle Contraction—Types of Muscle Contraction

Length and tension of muscles are key parameters that define muscle contraction [260]. The amount of force generated during contraction is related to the length of the muscle cell, in particular, the contractile unit of the muscle cell. However, muscles do not usually exceed their elastic limit, otherwise, the force of contraction of the muscle declines [260].

On the basis of the condition in which smooth and striated muscle contraction takes place, contraction can be divided into isotonic and isometric. Experimental studies have shown that both skeletal and smooth muscle cells exhibit both types of contraction. **Isometric contraction** is a type of contraction in which the muscle tension changes, but the length does not change. In **isotonic contraction**, the development of force is accompanied by changes in muscle length, but the tension remains unchanged. Force generation may lead to shortening (concentric contraction, i.e., having the same center) or lengthening (eccentric contraction, i.e., not having the same center) of muscle fiber [178, 261–264]. The shortening of muscle fiber in concentric contraction allows the tension generated to overcome the load, whereas in eccentric contraction, the lengthening of the muscle fiber occurs because the tension generated is insufficient to overcome the load on the muscle [265–270]. However, under normal conditions, both shortening and increase in tension take place during muscle contraction. Also, muscle contraction can be defined as tetanic or twitch [121]. Both tetanic and twitch contractions have been demonstrated experimentally in smooth muscle cells including those of the GI tract [209, 271]. Tetanic contraction is a sustained muscle contraction resulting from excessive and high rate generation of action potential by the motor neuron innervating the muscle. A twitch is a muscle contraction resulting from a single and adequate stimulus. When a muscle is stimulated by a single stimulus or impulse, which is adequate to cause contraction, a twitch is observed. However, when the stimulation is done more frequently, the twitches overlap and is evidenced as tetanic contraction [272–276].

7.4.5 Mechanisms of Muscle Relaxation

Smooth muscle relaxation occurs either as a result of removal of the contractile stimulus or by the direct action of a substance that stimulates inhibition of the pathways that are responsible for muscle contraction [119]. The role of decrease in Ca^{2+} and MLCP in muscle relaxation has been discussed in the previous subsections (Figs. 7.10, 7.11, and 7.12). GI smooth muscle relaxation can be mediated by mediators released by the sympathetic nervous system or by the cells surrounding the smooth muscles. Sympathetic nerves of the autonomic nervous system modulate GI tract motility indirectly through the gut's little brain—the enteric nervous system (Discussed in detail in Chap. 9). Activation of the sympathetic nerves results

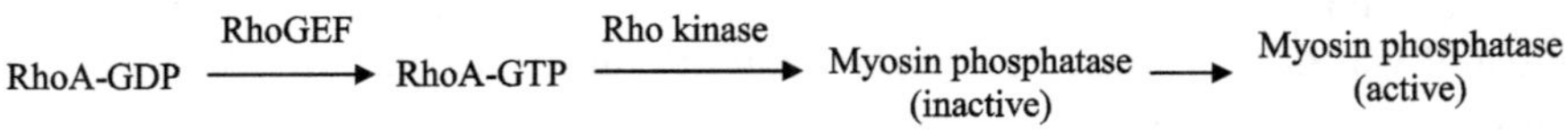

Fig. 7.11 Inactivation of myosin phosphatase. The activity of this phosphatase is controlled by Rho kinase via the stimulation of $G\alpha q_{13}$-coupled G protein, which recruits the guanine nucleotide exchange factor, RhoGEF, and the small G protein RhoA that mediate the signaling of downstream effectors, including the Rho kinase (this figure) [119]. The serine/threonine kinase, Rho kinase, phosphorylates the myosin-binding regulatory subunit-1 of MLC phosphatase (MYPT1) to render the enzyme inactive. Inactivation of MLC phosphatase is also controlled by PKC, protein kinase C-potentiated inhibitor protein, also called C-kinase potentiated protein phosphatase-1 inhibitor of molecular weight 17 kDa (CPI-17), Rho-associated coiled-coil kinase (ROCK), protein kinase N, ZIPK, and ILK [229, 237, 277, 278]. The activities of these protein kinases result to the inhibition of MLCP; maintaining increased MLC phosphorylation, and enhanced contraction. PKC can also activate other kinases by phosphorylation to mediate the inhibition of the MLC [279]. More so, the activities of these enzymes on the cytoskeleton can increase the permeability of paracellular shunt [226]

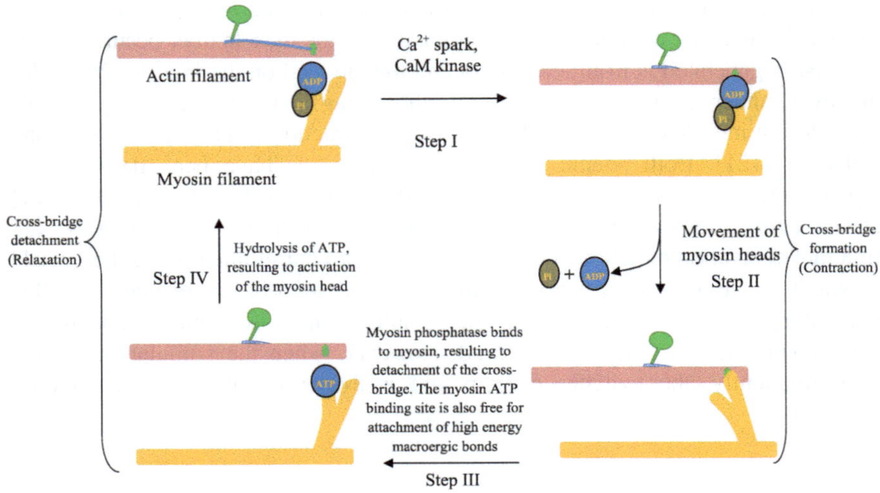

Fig. 7.12 Cross-bridge cycling. In the presence of intracellular Ca^{2+} and ATP, as well as the activity of protein kinases, the blocking sites of actin-binding proteins are freed and the myosin heads are energized for interaction with the myosin-binding site on the actin filament. When this happens, the actin filaments move in such a way that opposes the movement of the myosin filaments on either side. This causes decrease in the length of the contractile unit and also tension in the actin filament. Subsequently, the cross-bridges detach from the actin filaments. In the availability of ATP, the ATP-binding site on the myosin filament is energized again and ready to bind with actin filament. This cycle of cross-bridges continues in so far as ATP and Ca^{2+} are present in the cell [248, 249]

to release of norepinephrine, which acts on the enteric neuron via the activation of the α_2 adrenoreceptor, coupled to G_i. The activation of this G protein inhibits the activity of adenylyl cyclase so that the pathways that are responsible for increase in the second messengers cAMP and Ca^{2+} are inhibited. Norepinephrine can act on GI tract smooth muscle cells to cause relaxation through activation of β_3-adrenoceptors possibly via PKA pathway [208]. The neurotransmitter acetylcholine can mediate muscle relaxation through its action on M2 receptor coupled Gi_3. The mediator, somatostatin act via SSTR3 receptor coupled Gi_1 to inhibit muscle contraction, thereby promoting muscle relaxation. Adenosine act via adenosine A1 receptors coupled to Gi_3 [280]. The inhibitory action of these molecules, which is mediated via Gi-coupled receptors, involves successive activation of $G\beta\gamma i$, PI-3-K, Ca^{2+}-independent MLCK, ILK, and other protein kinases to cause muscle relaxation. PKA (as well as PKG) desensitizes the receptors involved in muscle contraction via upstream signaling, inhibiting adenylyl and guanylyl cyclases, and stimulating cAMP-specific phosphodiesterases (PDE) type 3 and 4 and cGMP-specific PDE-5 activity [229].

Some substances secreted by endothelial cells can diffuse to smooth muscles from the extracellular space or via endothelial-smooth muscle junctions. These substances include NO (also called endothelial relaxing factor), and endothelial-derived hyperpolarizing factors (EDHF). NO and prostaglandins (e.g., prostacyclin, PGI2) cause vasodilatation in vascular smooth muscle. But this vasodilatation remains even after the application of NO synthase and cyclooxygenase inhibitors. EDHF is responsible for this non-NO, non-prostaglandin-mediated vasodilatation. EDHF is believed to act downstream, activating K^+ channels of smooth muscle cells to allow K^+ efflux along its chemical gradient, which is the result of the membrane hyperpolarization [281, 282]. NO stimulates soluble guanylate cyclase, which catalyzes the production of cGMP. The cGMP activates the cGMP-dependent protein kinase—PKG, which in turn phosphorylates RhoA to inhibit the signaling cascade. PKG can phosphorylate MYPT1 to block the activities of the associated protein phosphatase enzyme, thus promoting muscle relaxation [119, 149]. PGI2 stimulates adenylate cyclase to increase intracellular production of cAMP, which in turn activates PKA. This protein kinase phosphorylates a couple of protein channels, receptors, and other intracellular molecules that are implicated in muscle contraction and relaxation. Depending on the isoform of the enzyme activated, the phosphorylation results to the inhibition of SER Ca^{2+} channels (such as IP_3 receptor, L-type Ca^{2+} channel), stimulation of plasma membrane and SER Ca^{2+}ATPases, decrease in MLC20 phosphorylation to increase MLCP activity, activation of Ca^{2+} sensitive K^+ channels, ATP sensitive K^+ channels, allowing K^+ efflux resulting in membrane hyperpolarization and the production of muscle relaxation [282, 283].

7.4.6 Motor Unit

Motor unit of striated muscle such as skeletal is different from the motor unit of smooth muscle. In muscle physiology, the traditional definition of motor unit is motoneuron and all of its associated muscle fibers. Motor unit can be referred to as the α-motor neuron and all the muscle fibers it innervates [284–286]. For further details on motor unit and the methods of investigating motor unit, review Kaya et al. [285]. Motor unit of GI smooth muscle is the smooth muscle cell, and GI interstitial cells which form a functional syncytium—referred to as SIP syncytium (the acronym "SIP" is derived from the first letter in each of the components of the syncytium—**s**mooth muscle cell, **I**CC cell, **P**DGFRα+ cell) [287]. The interstitial cells of the GI tract are ICC and PDGFRα+ cells that contact with smooth muscle cells by way of gap junctions [106, 287, 288]. This functional unit of the muscles of the GI tract is responsible for the spontaneous pacemaker activity, slow waves, and contractile functions of the gut. The myogenic tone is believed to be regulated by this functional unit [71].

7.5 Motor Patterns of the Gastrointestinal Tract

The GI tract motility involves both contraction and relaxation processes that give rise to different types of motor patterns. Types of GI muscle contraction include tonic, phasic, segmentation, peristaltic, antiperistaltic, pendulum-like, propulsive contraction or movement among others [212]. It should be noted that more than one pattern of GI motility may be characterized for multiple regions of the GI tract; however, some features of GI motility or motor pattern may be specific for a given region of the tract.

7.5.1 Motor Functions of the Mouth

The mouth is a cavity, which is bounded by the lips, tongue, cheeks, and palate. The space between the lips/cheeks and the teeth is called the vestibule. The gingivae (or gums), surrounding the teeth, are protective soft tissue that is bound to the underlying bone. The structures of the oral cavity help in the mechanical processing of food (mastication or chewing) and thus provide a greater surface area for the action of hydrolytic enzymes. The palate helps to maintain simultaneous mechanical processing of food and breathing process. Food is prevented from entering the nasopharynx (that would have otherwise passed into the respiratory tract) by the uvula which is a movable extension of the soft palate [289, 290].

Mastication

Mastication, also called chewing, is a series of cyclic or rhythmic movements of the jaw that is associated with mechanical food processing and bolus formation. Mastication occurs as a sequence of successive jaw openings and closings that characterizes the masticatory cycle. The cycle is defined by successive vertical jaw movement using the anterior maxillomandibular midpoint as a reference point [291–296]. Masticatory events can be studied with functional and radiological methods, which include electromyography, videofluorography, sirognathography, axiography, among others [291, 292]. The electromyography allows to investigate the electrical activity of the muscles involved in chewing (such as masseter, suprahyoid, infrahyoid muscles) by placement, implantation or insertion of surface or intramuscular electrodes [292]. Videofluorography is a radiographic technique that allows to obtain images by photographing with a camera produced during the X-ray method by fluoroscopy. Sirognathography is a technique for the registration of mandibular movements [297, 298]. Axiography is a technique used to study the temporomandibular joint movement. It is used as a measuring method for recording mandibular movements, including translations and rotations [299–301].

Each masticatory cycle consists of three phases—opening, closing, and intercuspal (occlusal) phases. The first phase, called opening, is characterized by intake of food into the mouth, a process controlled by reflex inhibition and activation of some muscles and directed movement of the condyles in the forward and downward directions. The second phase, closing phase, is characterized by contraction of specific muscles of the mouth with activities of the condyles in the backward direction and upwards. The third phase, intercuspal phase, is characterized by teeth-to-teeth contact. In each power stroke, the size of the food is reduced to smaller pieces. The contact between the cusps of the teeth as well as gliding of the teeth grinds the food particles with the aid of saliva to form a bolus [291–296, 302]. The masticatory cycle may be divided into four, instead of the three defined above. The four phases of the masticatory cycle include slow opening, fast opening, fast closing, and slow closing (Fig. 7.13) [293]. Independent on the number of phases (i.e., whether 3 or 4), each masticatory cycle lasts for about 1 s. But this duration depends on the type of food and disease conditions. There is also gender difference associated with duration of masticatory cycle. The masticatory cycle is generally longer in females compared to males [292]. The masticatory cycle has been divided into five phases when investigating chewing with a kymograph (Details are discussed in Chap. 1 of the book "Gastrointestinal Physiology: Contemporary Trends, Methods and Models").

Stages of Mastication

The cycle of mastication starts with ingestion of food to the moment just before swallowing takes place. The duration between food ingestion and the moment just before swallowing is divided into three stages or periods of mastication (for solid food). The first stage is the "**preparatory stage**" occurs with the introduction of the food into the mouth. The stage that is characterized by rhythmic chewing activity, involved in the formation of the bolus, is referred to as the "**reduction stage**." The next stage, called the "**preswallow stage**" is involved with preparation of the food for swallowing. During this period the food bolus is passed from the oral cavity into the oropharynx by the forward and upward movement of the tongue and hyoid bone

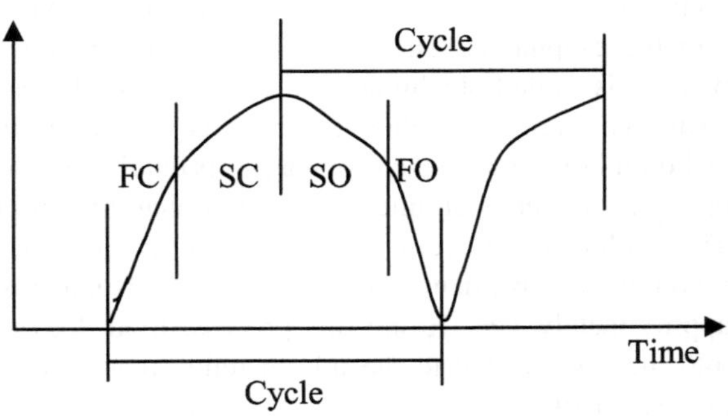

Fig. 7.13 Cycle and phases of the masticatory movement. A cycle is each successive maximum jaw closure or maximum gap. The four phases are slow opening (SO), fast opening (FO), fast closing (FC) and slow closing (SC)

(protraction), which is due to the contraction of anterior digastric and geniohyoid muscles. The mechanics of this process also involve squeezing the food by the tongue against palate, while a portion of the food bolus is passed into the pharynx [292]. Similar mechanism is involved in the passage of the food bolus into the pharynx [293]. In the majority of the cases, swallowing is initiated during the intercuspal (minimum gap) phase of the masticatory cycle. This is the stage of mastication when tongue protraction occurs [292]. Details about swallowing are discussed in Sect. 7.5.2.

Non-masticatory Functions of the Oral Cavity

The functions of oral cavity not only include mechanical processing of food, but also, lubrication of food particles with mucus and saliva, initial chemical processing of food (depending on the type of food and duration of location in the mouth), sensory analysis of food just before swallowing as well as antimicrobial activity of some components of saliva [289, 290, 303]. Details on the non-masticatory functions of the oral cavity are discussed in Chaps. 11 and 12.

7.5.2 Motor Functions of the Esophagus

During the formation of the food bolus, mechanoreceptors, and chemoreceptors of the oral cavity evaluate the bolus for consistency and if the food is adequate for swallowing, then the swallowing reflex is initiated to allow the food move into the esophagus via the pharynx [304, 305].

Swallowing

Swallowing (also known as deglutition) is a developmentally programmed process that begins during the period of ontogenesis around the 11th week of gestation, and progressively develops up to maturity after birth [306, 307]. Swallowing refers to a complex series of process from ingestion to the passage of the ingested substance into the esophagus. The scientific study of swallowing and disorders associated with it is called deglutology. The process of deglutition is controlled by the swallowing reflex. Swallowing reflex is voluntarily initiated by propulsion of food or liquid boluses into the oropharynx primarily by movements of tongue. (The oral cavity is divided anatomically into nasopharynx, oropharynx, and laryngopharynx.) This voluntary movement in the oral cavity immediately transforms into an involuntary movement in the pharynx and esophagus within some milliseconds or approximately 1 s. During this process, food does not pass through nasopharynx into the trachea due to the inborn reflex that prevents this occurrence. In normal process of swallowing the food bolus or liquid passes through the pharynx into the esophagus from where it is delivered into the stomach. According to the location of the bolus, swallowing process can be classified into oral, pharyngeal, and esophageal phases [308]. These phases are independent processes in swallowing [309].

Oral phase: Following solid food ingestion, the tongue moves the food to the postcanine region and places the food onto the occlusal surface of lower teeth. In this position, the food is ready to be processed by the masticatory action of the teeth and softened by saliva. This process continues until the consistency of the food is optimal for deglutition. The action of salivary components on the food partially breaks down the food particles in preparation for the formation of paste bolus. The mechanical action of the teeth, cyclic movement of the jaw, and movement of the tongue, cheek, soft palate, and hyoid bone are all required for the formation of food bolus [308]. The chewed and softened (i.e., triturated) food moves to the oropharynx (including the valleculae—a depression between the base of the tongue and the epiglottis) through the fauces (throat). The oropharynx is the anatomical region where the bolus is formed [308]. If liquid is ingested instead of solid food, the posterior oral cavity is sealed by tongue–palate contact during the oral preparatory stage when the bolus is held in the oral cavity [308].

The oral phase is subdivided into two subphases: oral preparatory and oral propulsive. Together, these subphases involve the mastication and breakdown of solid food into a bolus with a consistency that is ready for swallowing and propelled by the tongue into the pharynx [308–310]. In the **oral preparatory phase**, the bolus (solid or liquid) is held in the anterior part of the mouth between the tongue and hard palate (Fig. 7.14) [310]. Posteriorly the bolus is bound by the soft palate and tongue dorsum—This prevents the premature passage of the bolus into the oropharynx [308]. Within a few seconds, the tongue squeezes the bolus through the fauces into the vallecula or oropharynx where the bolus accumulates for about 0.5–5 s, depending on the type of food (**oral propulsive phase**) [308, 310]. As the food bolus accumulates in the nasopharynx, the distal tongue further presses against the hard palate, pushing the bolus backward so that the soft palate closes the nasopharynx, and food bolus is propelled into the upper pharynx [3, 308–310]. The closure of the nasopharynx is a protective mechanism that avoids nasal regurgitation—which is defined as retrograde flow of part or whole of the bolus into the nose. This mechanics of tongue movement are termed elevation and retraction and have been reported by Matsuo and Palmer [308]. It is currently believed that the mechanisms of swallowing are characterized for both liquid and solid boluses. However, liquid bolus may first pass into the laryngopharynx (hypopharynx) before it is swallowed [308].

Oral transit time (OTT) significantly varies in different studies and may depend on the volume, type of ingested material, age and gender of the participants. For liquids, the OTT may be about 0.4–1.5 s, for pasty foods—0.4–1.1, and 1.0–12.8 s for solid foods. OTT is longer in the elderly compared with younger people [312].

Pharyngeal phase: This phase of swallowing begins as food bolus is propelled into the pharynx. This phase is estimated to last for about a second [308]. During this period, the hyolaryngeal complex (hyoid bone and larynx) moves upward anteriorly by contraction of the suprahyoid and thyrohyoid muscles, sealing the

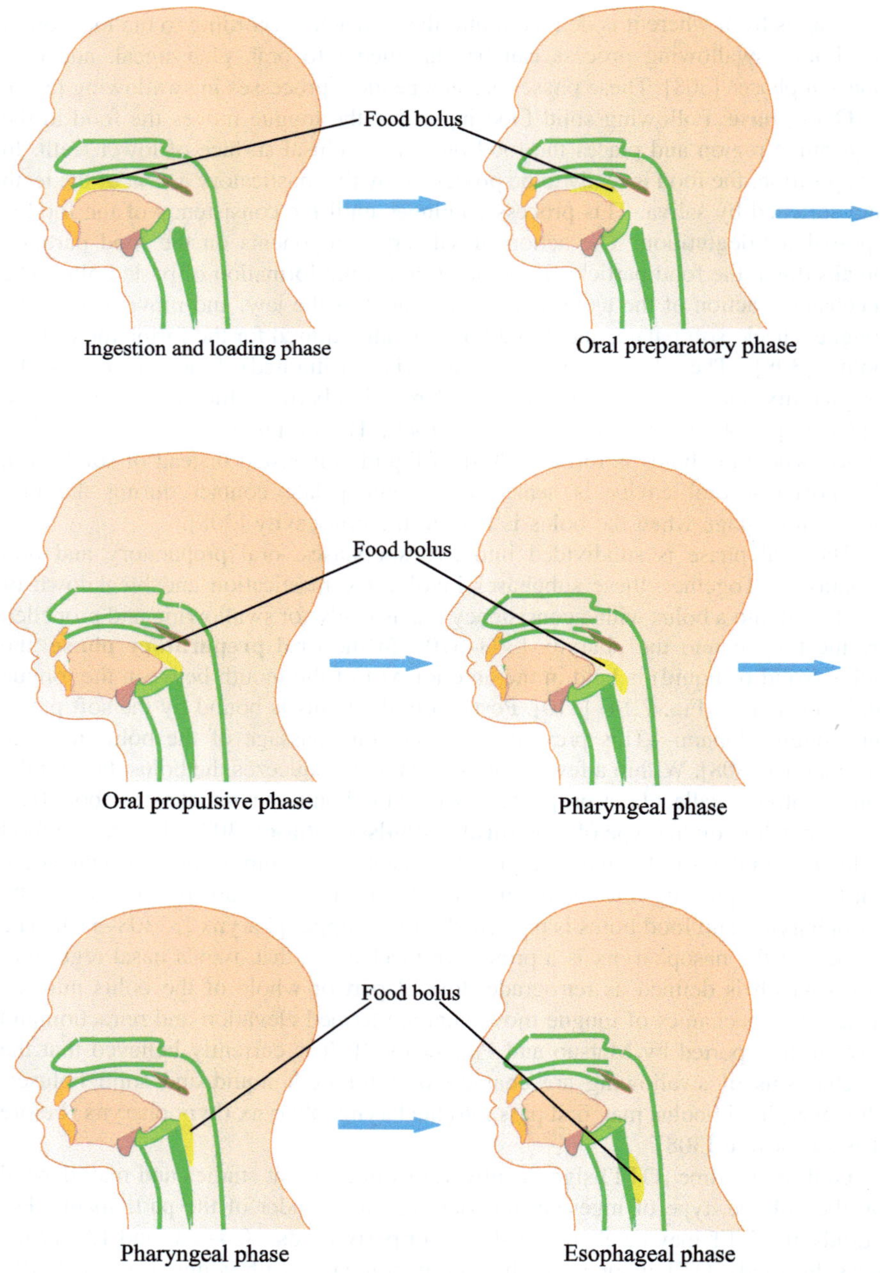
Food bolus
Ingestion and loading phase
Oral preparatory phase
Food bolus
Oral propulsive phase
Pharyngeal phase
Food bolus
Pharyngeal phase
Esophageal phase

◀**Fig. 7.14** Mechanisms and phases of swallowing. Ingestion and loading of food in the mouth initiate the oral phase of swallowing. In the first phase of oral swallowing (i.e., **oral preparatory phase**), the bolus is held between the anterior surface of the tongue and hard palate. At the same time, the anterior portion of the tongue then pushes the bolus against the hard palate just behind the upper incisors while the posterior tongue drops away from the palate and the bolus is propelled by tongue dorsum into the pharynx (**oral propulsive phase**) to contact with epiglottis resulting in its downward tilt, with subsequent hyolaryngeal excursion, and the opening of the upper esophageal sphincter. The **pharyngeal phase** is characterized by elevation of the soft palate, resulting in closure of the nasopharynx. The tongue further pushes the bolus posteriorly, squeezing the bolus into the pharynx. This results to a backward tilt of the epiglottis and the larynx is displaced upward anteriorly. Consequently, the upper esophageal sphincter opens, and the oropharyngeal swallowing is completed. The esophageal phase continues as the base of the tongue moves backward to contact the pharyngeal wall. This wall, progressively, first, contracts superiorly around the bolus and then downward toward the esophagus. The soft palate moves in a downward direction and the larynx and pharynx reopen to allow passage of air into the breathing pipe. As soon as the bolus passes, the upper esophageal sphincter returns to its closed state [308, 311]. In infant human, mechanism of swallowing is different due to the anatomical peculiarities of the structures involved in the process. For review of the mechanisms of swallowing in infants, see Matsuo and Palmer [308]

glottis by the closure of the vocal folds. Further, this causes the arytenoids to tilt forward thereby contacting the base of the epiglottis. The epiglottis moves downward posteriorly, resulting in closure of the laryngeal vestibule. These movements prevent the food bolus from entering the trachea. The presence of the food in the pharynx causes continued closure of the nasopharynx (velopharyngeal opening), at the same time, a sequential contraction of the superior, middle, and inferior constrictor muscles of the pharynx is initiated, pushing the bolus toward the esophagus [308–313]. The esophagus consists of the upper and lower esophageal sphincters that control the entrance and exit of boluses into and out of this hollow structure. In the absence of bolus in the pharynx, the upper esophageal sphincter is tonically contracted (i.e., passage is closed). Following the presence of bolus in the pharynx, the upper esophageal sphincter relaxes, resulting in opening of the passage—which is the resultant effect of anterior displacement of the hyolaryngeal complex and relaxation of the cricopharyngeus muscle. The upper esophageal sphincter comprises the inferior pharyngeal constrictor and cricopharyngeus muscles, together with proximal part of the esophagus. The esophageal swallowing begins as soon as the bolus passes into the esophagus [308, 310]. The opening of the upper esophageal sphincter and the subsequent entrance of the bolus into the esophagus promotes return of the velopharyngeal and hyolaryngeal complexes to baseline, which allow for opening of the airway for breathing to continue.

The pharyngeal transit time (PTT) may also vary depending on the factors earlier mentioned for OTT. The calculated PTT by Cassiani et al. [309] for paste bolus was 0.21–0.75 s and 0.27–0.49 s for liquid bolus.

Esophageal phase: As the bolus enters the cervical esophagus, peristaltic contraction is initiated and controlled by the activities of cranial nerve X as well as intrinsic neurons of the esophagus. This moves the bolus into the thoracic esophagus down to stomach through the lower esophageal sphincter, also called the cardiac sphincter or gastroesophageal sphincter. In a healthy person, in the absence of bolus, the lower esophageal sphincter is tonically contracted to prevent regurgitation or reflux of gastric contents. However, the sphincter relaxes as the bolus exerts pressure on it. This leads to the opening of the sphincter, allowing food or liquid bolus to pass into the stomach. The downward movement of boluses into the stomach occurs mainly by peristaltic waves of contractions and the force of gravity [308]. Peristaltic waves of this tubular anatomical structure travel at a speed of about 2.5 cm/s [310].

Esophageal Peristalsis

Peristalsis is a rhythmic and unidirectional wave of contraction and relaxation of smooth muscle that propels food and liquid boluses in an anterograde direction (distally) through the GI tract. The peristaltic wave consists of two main parts, an initial wave that accommodates the bolus, followed by a wave of contraction that propels it (Fig. 7.15) [308]. The waves of peristalsis are due to the synchronous contraction of the longitudinal and circular muscles that allows the passage of food down the GI tract [3].

Esophageal peristalsis could be initiated either by a swallow or by distension of the walls of the tubular structure. The rhythmic contraction resulting from the former process is termed primary peristalsis, whereas in the latter case—it is referred to as secondary peristalsis. Primary peristalsis is believed to occur as a result of sensory input triggered by the passage of the food or liquid bolus from the pharynx into the cervical esophagus. The rate of travel of primary peristaltic waves is independent on the rate of movement of the bolus down the esophagus. The duration of primary peristalsis is estimated to be about 9 s. The contractile strength of primary peristalsis is usually stronger and longer than contractile strength of

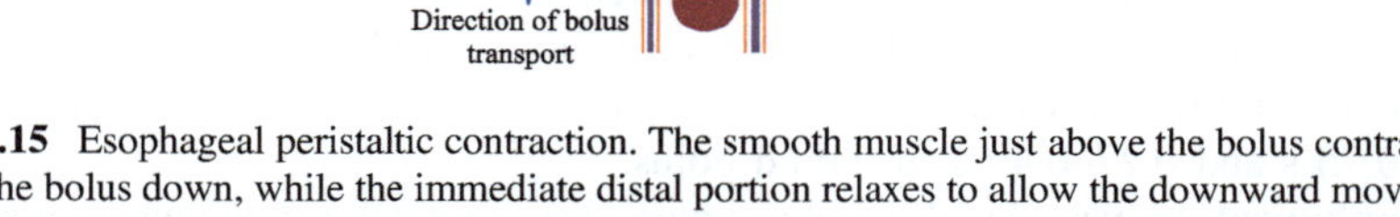

Fig. 7.15 Esophageal peristaltic contraction. The smooth muscle just above the bolus contracts to push the bolus down, while the immediate distal portion relaxes to allow the downward movement of the bolus. The coordination of these movements is due to the excitatory and inhibitory activity of the motor neurons that innervate these muscle layers. The sensory neurons that send information to the medulla synapse on both excitatory and inhibitory motor neurons. Contraction of the smooth muscle just above the bolus is due to excitation of motor neurons, while relaxation is due to the activation of inhibitory motor neurons that innervate the region just below the bolus. Pathological conditions associated with peristalsis are outlined in Clinical Correlate 7.1

secondary peristalsis of the esophagus. In secondary peristalsis, stretch receptors in the wall of the esophagus are stimulated, causing local reflex responses that push the bolus further down the esophagus [314–317]. Thus, esophageal secondary peristalsis is a reflex response to wall stretch or mechanical distension [318]. It is believed that secondary peristalsis help in acid clearance from the lumen of the esophagus. Thus, patients with gastroesophageal reflux diseases have disorders of secondary peristalsis [319]. In a manometric study, the wall tension before an evoked contraction, termed the preload tension, contractile tension, and postload tension can be studied to identify changes in the strength, amplitude, and duration of contraction in various stages of swallowing or esophageal peristalsis [317]. Esophageal peristalsis is controlled by brainstem lower motor neurons that innervate skeletal muscles of the esophagus, central, and enteric nervous systems (smooth muscles of the esophagus) and the intrinsic properties of the esophageal smooth muscle itself [320].

Clinical Correlate 7.1

Physiological and Pathological Ileus

The area of the esophageal wall just proximal to the contracting wall is said to be in physiological ileus. The word “ileus” is derived from the Greek “eileós” meaning “twisted.” In the absence of mechanical obstruction, the hypomotility area just proximal to the contracting wall above the bolus—may resume activity with subsequent movement of another liquid or food bolus. Physiological ileus can occur in the postoperative period, a complication of surgery on the GI tract. However, physiological postoperative ileus may become pathological when the temporary cessation of GI obstruction does not reverse (paralytic ileus) [321, 322]. Around the sixteenth to the eighteenth centuries, the Greeks and Romanians were the first to have defined the clinical triad that characterizes pathological ileus: abdominal pain, obstipation, and fecal vomiting [323]. Pathological ileus may be caused by mechanical obstruction possibly by twisting of part of the intestinal segment. This type of ileus due to obstruction by a region or part of the viscus is generally termed mechanical ileus [324]. Another type of pathological ileus, known as strangulation ileus, is due to encasement of a loop of the small intestine by the great omentum. Strangulation ileus may be caused by abnormal adhesion resulting in anoxic mucosal injury. Either strangulation or mechanical ileus can be caused by internal hernia or volvulus [325–327]. Volvulus is the abnormal twisting of a loop of the intestine around itself and the axis of its own mesentery, producing a mechanical bowel obstruction. Thus, the condition can result from torsion and occlusion of the vasculature of the mesentery. If the mesentery is tightly twisted such that blood flow to the affected part of the intestine is completely cut off, death of the tissues without blood flow can result—from ischemia to necrosis of the tissues. Volvulus can affect any part of the intestines [328, 329]. An internal abdominal hernia is the protrusion of a viscus through a normal or abnormal aperture within the peritoneal cavity [330]. A paraduodenal hernia is the most common type of internal abdominal hernias. Rare types include paravesical, intersigmoid, and transomental internal hernias [324]. An internal hernia may be either congenital or acquired. The reported incidence of such hernias is 1–2%. The herniation may be persistent [331]. It should be noted, however, that internal hernias are very rare causes of small bowel obstruction [331]. Following the embryonic formation of an intestinal loop, congenital defects occur in areas of the mesentery that are thin and avascular—this predisposes the individual to the development of hernia. In adults, however, most mesenteric defects are the result of previous surgery on the GI tract, abdominal trauma, or intraperitoneal inflammation [331].

Coordination of Swallowing and Breathing

Under physiological conditions, swallowing is said to be dominant to respiration. That is swallowing inhibits respiration both peripherally and centrally. Peripherally, swallowing and respiration share the same anatomical structures—mouth and pharynx. The mouth is used for breathing (when there is difficulty in breathing via the nose) and chewing as well as swallowing (oral phase of swallowing). The pharynx is separated from the lower breathing tract by the laryngeal complex and from the upper airway (nasal cavity) by velopharyngeal complex. The coordination of these complexes allows swallowing to be accomplished and also prevents aspiration. Importantly, both structures participate in speech formation [308, 332]. Centrally, neurons concerned with swallowing (swallowing center) inhibit the inspiratory neurons, which are coupled to the expiratory neurons. Though this inhibition is only transitory, it is sufficient to interrupt breathing for about 1 s (0.5–1.5 s) [308, 332–334]. The coordination of the peripheral and central controllers of swallowing gives rise to the cycle "exhalation–swallowing–exhalation" [332]. In most cases, swallowing starts immediately following expiration and usually, expiration follows thereafter. But, in case of sequential swallowing, it is possible for breathing to begin with inspiration [332].

Control of Swallowing

Swallowing is a complex and sequential activity involving coordinated action of oropharyngeal and esophageal muscles, controlled at several levels of the nervous system—brainstem, subcortex, cortex, and enteric nervous system [335]. In the brainstem, swallowing is controlled by the reticular formation and the central pattern generator. Peripheral reflexes are also involved in the control of swallowing

[309]. These central and peripheral regulators of swallowing are involved in the excitatory and inhibitory regulation of over 25 pairs of muscles of the oropharyngeal and laryngeal complexes as well as the esophagus [3, 308].

Afferent pathways: Sensory receptors located in the mucosa of the laryngeal and oropharyngeal regions comprising the soft palate, uvula, posterior tongue, epiglottal surface, faucial pillars, etc., are sensitive to the arrival of liquid, paste, or solid bolus. The first-order neurons that contact with these receptors constitute the primary afferent fibers, which travel to the brainstem via the trigeminal (cranial nerve V), glossopharyngeal (cranial nerve IX), and vagus (cranial nerve X) nerves. The first-order neuron synapse on second-order neuron located in the nucleus tractus solitarius (NTS) [333, 335, 336].

Swallowing center: The swallowing center is located in the medulla [337–339]. Samuel James Meltzer (1851–1920) was the first to postulate the presence of central component of swallowing, which he called the central pattern generator [340, 341]. Swallowing reflex is controlled by the central pattern generator of the medulla oblongata of the brainstem and peripheral reflexes. The central pattern generator is a conglomerate of neuron bodies that rhythmically function to regulate the timing of the phases of swallowing. Neurons of the central pattern generator are believed to exhibit automaticity [336, 342]. The medullary central pattern generator is divided into the dorsal and ventral regions of the swallowing centers. The dorsal region contains neurons located around the NTS and is responsible for the initiation, timing, and programming of swallowing. Primary neurons synapse in the NTS from where fibers are sent to the dorsal region. This dorsal region activates the premotor neurons of the ventral region, which corresponds to the medullary reticular formation and located above the nucleus ambiguus. Neurons of the nucleus ambiguus distribute the swallowing drive to the pools of motor neurons involved in swallowing [333, 335, 336, 343]. The different regions of the swallowing centers and nuclei have different contributions to the phases of swallowing [342]. The oral phase of swallowing is due to the activities of the trigeminal nucleus and reticular formation. The pharyngeal and esophageal phases of swallowing are due to the activities of NTS, nucleus ambiguus, and dorsal motor nucleus. Importantly, the NTS is composed of second-order sensory neurons and the central pattern generator, whereas the nucleus ambiguus and dorsal motor nucleus are composed of neurons that synapse with motor neurons. It should be mentioned that the phases of swallowing do not occur as individual entities in time but are closely related and do not occur as separate phenomenon. The beginning and completion of each phase is regulated both peripherally and centrally. For instance, the pharyngeal phase is coupled to the esophageal phase by the activities of the NTS, in particular, the ventromedial nucleus [342].

Central pattern generator: The central pattern generator is where the motor information for phase initiation, timing, sequence, and rhythm of swallowing is generated. The sensory information that is transmitted to the brainstem interneurons of pattern generator located in the dorsal medulla (NTS and the neighboring reticular formation) and ventral medulla (nucleus ambiguus and the neighboring reticular formation). Signal from the periphery is relayed from dorsal to the ventral medulla where the final motor output is delivered to the lower motor neurons that innervate different structures involved in swallowing. It is believed that the motor information is generated mainly in the ipsilateral half of the pattern generator, from where the preprocessed signal is transferred to the contralateral side of the pattern generator [335, 344, 345].

Efferent pathways: The motor nuclei involved in swallowing are the trigeminal (cranial nerve V), facial (cranial nerve VII), ambiguus (cranial nerves IX and X), and hypoglossal nuclei (cranial nerve XII). But the main nuclei involved in swallowing are the ambiguus and hypoglossal nuclei. Only a small part of trigeminal and facial nuclei are involved in normal swallowing [336]. The axons of the neurons of these nuclei course their fibers to innervate the muscles of oropharynx and larynx. For example, the nucleus ambiguus sends axons via IX and X cranial nerves to innervate the muscles of the pharynx [333].

Peripheral swallowing reflexes: The peripheral reflexes modify the output of the swallowing center via sensory feedback through pharyngeal, laryngeal, and esophageal regions [309, 336]. These reflexes have both facilitative and inhibitory role on the phases of swallowing. For example, the peripheral reflex responsible for facilitation of the pharyngeal phase of swallowing also inhibits the esophageal phase. This activity of the reflex guarantees the appropriate timing of the respective phases of swallowing [342]. Swallowing may be modified by reflexes or stimulation of receptors of the lungs and chest wall. Branches of IX and X consist of afferents that transfer sensory information to brainstem sites that control respiration, activities of stretch receptors of the lungs, baroreceptors of blood vessels, and chemoreceptors of the carotid and aortic bodies [336].

Subcortical and cortical (suprabulbar) centers modulate swallowing centers of the brainstem: The premotor interneurons that constitute part of the medullary central pattern generator are interconnected to multiple areas of the brainstem, including the pons, subcortical (e.g., basal ganglia, hypothalamus), and cortical regions of the brain [335, 336, 344, 345]. The neurons of the medullary reticular formation involved in swallowing travel to different brain regions including the pons, where they synapse on a wide range of neurons, also believed to participate in the regulation of swallowing [343]. Some of the descending neurons from the frontal cortex synapse on the cell bodies of the reticular formation and the central pattern generator of the brainstem [336]. Some of the descending neurons can modulate the swallowing center via NTS. Moreover, the NTS constitutes the first site to receive viscerosensory information from the cardiovascular, digestive, and respiratory systems. The trigger zone of the swallowing center, the dorsal region, receives inputs from subcortical and cortical centers [333, 336]. Experimental model using animals has shown that swallowing act can be elicited upon stimulation of the lateral precentral gyrus and frontal orbital gyrus [333]. It has also been shown that destruction to Brodmann's area 6 in laboratory animals resulted in persistent difficulty in making the transition between chewing and swallowing [346]. Other brain regions that control swallowing include the insular and anterior cingulate [335]. The swallowing center also interacts with other areas of the brain involved in respiration and speech. The swallowing and respiratory centers of the brainstem exert reciprocal inhibitory effects on each other to coordinate the processes of swallowing and breathing [336].

Investigation of swallowing can be done with videofluoroscopy, manometry, and electromyography while participants swallow different volumes of food or liquid bolus [347, 348].

Disorders associated with swallowing are shown in Clinical Correlates 7.2 and 7.3.

Clinical Correlate 7.2

Dysphagia

Dysphagia means difficulty with swallowing, affecting about 5–8% of persons above the age of 50. The incidence rate increases with age [310, 349, 350]. The causes of this condition include both structural and functional deficits in the oropharyngeal, laryngeal, and esophageal regions of the GI tract [308]. Functional disorders of the neuromuscular junction of these regions may contribute to a wide range of symptom complex in dysphagia. Functional deficits resulting in difficulty in swallowing may be observed in patients with Parkinson's disease, amyotrophic lateral sclerosis (ALS), or other neuromuscular disorders [310, 351, 352]. Patients with Parkinson's disease, ALS, or other neuromuscular disorders come up with dysphagia due to lower motor neuron impairment or combined impairment of the lower and upper motor neurons, which are associated with the muscles involved in the oral or pharyngeal phases of swallowing [351, 352]. The structural deficits which can either be acquired or congenital include cleft lip, cleft palate, cervical esophageal webs, esophageal stricture, enlarged thyroid gland, pharyngeal diverticula and pouches. Cleft lip, also called cheiloschisis, is a congenital defect characterized by the presence of a cleft in the middle of the upper lip. Cleft palate is a congenital fissure of the hard palate. Both cleft lip and cleft palate affect mastication, oral and pharyngeal phases of swallowing due to defects in labial manipulation, teeth malalignment, and insufficiency of closure of the velopharyngeal port, which in turn results to nasal regurgitation [308, 353]. Cervical esophageal webs are thin folds of the mucosa of the hypopharynx and cervical esophagus. Pharyngeal diverticula and pouches such as *Zenker's diverticulum* and *Killian-Jamieson diverticulum* affect swallowing and cause dysphagia [310]. *Zenker's diverticulum* [*named after the* German pathologist Friedrich Albert von Zenker (1825–1898)] is a pouch that develops in a weak spot in the hypopharyngeal muscular wall, above the cricopharyngeus muscle *in an area called the Killian's triangle or dehiscence* (This area is located between the transverse fibers of cricopharyngeus and the oblique fibers of thyropharyngeus parts of the inferior pharyngeal constrictor muscle), while *Killian–Jamieson diverticulum* [*named after the German physician Gustav Killian* (1860–1921) *and Scottish-born Australian physician James Jamieson* (*1840–1916*)] is a pouch that forms below the cricopharyngeus muscle [308, 310]. In these cases, the bolus can accumulate in the pouches, and later, in retrograde direction flow into the pharynx. In enlarged thyroid gland, the esophagus may be compressed, thus causing strain in the passage of bolus along the esophagus [310]. Radiation can result in many structural, mechanical and functional changes in the organs of the organism. Esophageal stricture constitutes one of those changes which can substantially affect swallowing. Dysphagia can result from complications of surgery to the head or neck regions. In surgical resection of tumors of the neck region, for example, damage to cranial nerve may occur, which affects swallowing if the nerve innervates the muscles involved in swallowing [310, 353]. Dysphagia can be classified according to the phases of swallowing. Thus, there are oropharyngeal and esophageal dysphagia. Oropharyngeal dysphagia is difficulty in the flow of the bolus from the oral cavity into the cervical esophagus. Thus, oropharyngeal dysphagia may compromise the functions of the airway as well as pharyngeal bolus clearance. Oropharyngeal dysphagia is usually neurologic in origin. The presenting complaints are coughing following the act of swallowing [310, 350, 354]. Esophageal dysphagia simply means difficulty in the passage of food or liquid bolus through the esophagus. The presenting complains of pharyngeal dysphagia are feeling of a lump in the throat or globus sensation, which is a sensation of food accumulation in the region of the sternal notch. The sensation occurs in dysphagia and is associated with gastroesophageal reflux and achalasia [310]. Dysphagia can be assessed with radiological techniques such as video barium esophagography and non-radiological techniques such as bedside swallowing assessment and fiberoptic endoscopy. Functional assessment of dysphagia can be done with esophageal manometry. Dysphagia may lead to such complications as dehydration, malnutrition, weight loss, airway obstruction, and aspiration pneumonia [308, 350, 354, 355].

Clinical Correlate 7.3
Achalasia

First identified in 1672 by the English physician, anatomist, and neurologist, Thomas Willis (1621–1675), the term “achalasia” simply means “failure to relax” [356]. Achalasia is characterized by absence of peristalsis (aperistalsis) and failure of the lower esophageal sphincter to relax in response to swallowing, resulting in accumulation of food and liquid boluses in the esophagus, which leads to its dilation [357–359]. The prevalence of achalasia is about 1 in every 10,000 persons of the general population, but the annual incidence rate is about 1 in every 100,000 persons. Though the disease can develop at any age, the majority of cases are identified in a person around 25–60 years of age with no gender differences [356, 359, 360]. The major symptom of achalasia is dysphagia. Other symptoms include chest pain, bland regurgitation (regurgitation of swallowed food and liquid), heartburn, hiccup, weight loss, sensation of lump in the throat or chest. The condition can occur as a primary disorder (primary achalasia) or as a condition secondary to other diseases such as Chagas disease (caused by the protozoan parasite *Trypanosoma cruzi*), esophageal adenocarcinoma (secondary achalasia or pseudoachalasia). Cases of achalasia have been reported in patients suffering from polio, Guillian-Barré, varicella, viral esophageal infections, eosinophilic esophagitis, systemic lupus, Sjogren syndrome, and scleroderma. Achalasia is one of the components that characterizes Triple A syndrome, also known as allgrove syndrome, a rare autosomal recessive disorder, which also involves a lacrimal and adrenal insufficiency. Achalasia of unknown cause is generally referred to as idiopathic. Hence, primary achalasia may be termed idiopathic achalasia [359–362]. Though the cause of the disease is not known, it is speculated that achalasia can result from pathologies that affect the vagus nuclei or intrinsic neurons of the esophagus. For instance, inflammatory response due to autoimmune diseases or invasion by pathogenic viruses or bacteria that results in the substantial destruction of the vagus nerve, vagus nuclei, or intrinsic (such as myenteric) neurons can lead to the development of the condition [356, 358, 359]. Cancer cells can also cause destruction of the myenteric neurons (especially, the NO- and VIP-secreting neurons) of the esophagus, resulting to the disease development. Though very rare, it is believed that the disease may be familial in origin. Traumatic damage to the vagus nerve can cause neuropathic disorder that may result to the development of achalasia [359, 362]. A high degree of suspicion is warranted for patients presenting with dysphagia, chest pain, and symptoms of refractory reflux, for which endoscopy does not reveal mechanical obstruction [363]. The conduction of barium esophagography may point to the presence of achalasia in the patient. The diagnosis is confirmed by esophageal manometry. There is currently no cure for achalasia. The goal of management of the condition is to relieve symptoms and improve esophageal peristalsis. Pharmacological management involves the use of smooth muscle relaxants. Pneumatic dilation and surgical myotomy have been successfully used in managing the condition [356, 363]. Achalasia can be complicated with airway obstruction and dyspnea resulting from megaesophagus. Megaesophagus is a condition characterized by abnormal enlargement of the esophagus on a background of aperistalsis. The enlarged esophagus can cause compression of the respiratory airway resulting to difficulty in breathing and other signs and symptoms associated with respiratory compromise [364].

7.5.3 Motor Functions of the Stomach

The motor functions of the stomach (gastric motility) are characterized by gastric contraction and relaxation resulting from the activity of smooth muscle cells arranged in three layers (outer longitudinal, middle circular, and inner oblique) in the five regions of the stomach: cardia, fundus, corpus, antrum, and pylorus (Fig. 7.16) [365, 366].

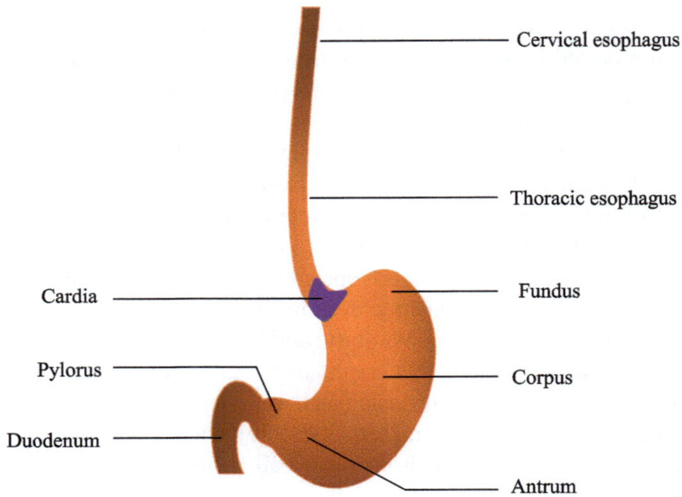

Fig. 7.16 Diagram of the esophagus and stomach

The motility of the stomach is controlled by the intrinsic and extrinsic nervous systems that are associated with this organ [365, 366]. Factors affecting gastric motility include electrical, motor, and sensory abnormalities. The electrical abnormalities that affect gastric motility are gastric arrhythmias and abnormal impulse propagation (Clinical Correlate 7.4). The motor abnormalities of the stomach include amotility, hypermotility, hypomotility, impaired fundic relaxation, antral dilation, and pyloric spasm. Finally, the sensory abnormalities are visceral hyperalgesia, impaired fundic relaxation, accommodation, gastric hypersensitivity, and hyposensitivity [367–371].

Clinical Correlate 7.4
Gastric Dysrhythmias (Arrhythmias)

Gastric dysrhythmias (also known as gastric arrhythmias) refer to slow waves of the stomach that are irregular [372, 373]. A normal gastric rhythm is usually considered as 3 cpm [373]. A gastric rhythm below this value (1.0–2.5 cpm) is referred to as bradygastria, while a more frequent gastric rhythm (3.75–10 cpm) is termed tachygastria [373]. Gastric rhythm irregularity is due to disorders in gastric conduction system. The impulse in this conduction system is generated mainly by ICC. Dysfunctions in ICC due to disruption of the network of this conduction system, cell depletion, or channelopathies can lead to the development of irregular slow waves [372]. Impairments in hormone and neurotransmitter signaling are also believed to underlie the development of irregular gastric rhythm [372]. To this end, gastric slow waves have been shown to correlate with the plasma VIP, vasopressin, epinephrine, and motilin levels. Gastric arrhythmia is an objective marker of nausea—a noxious, uncomfortable feeling in the epigastrium. Gastric arrhythmias are associated with gastroesophageal reflux disease, and other gastric functional impairments such as gastroparesis, functional dyspepsia, and idiopathic nausea [372–375]. A diagnosis of gastric arrhythmias can be made with electrogastrography (EGG) [372]. Treatment is based on the etiology, severity, and type of the disorder [376, 377]. Current treatment also involves the use of high-frequency gastric electrical stimulation and has been shown to be highly effective as a treatment option [378].

Gastric Tonic and Phasic Contractions

Tonic and phasic (rhythmic) contractions occur in response to changes in load acting on the smooth muscle cells. Tonic contraction is sustained contraction without relaxation. In tonic contraction, the smooth muscle is continuously active (Fig. 7.17) [379]. Phasic contraction is characterized by periodic contraction followed by relaxation. Phasic contraction is paced electrically by rhythmic changes in membrane potential (Fig. 7.17) [76, 119]. It should be noted, however, that these types of contractions are not peculiar to the stomach and can occur in esophagus and intestines [379]. For instance, phasic contraction of GI sphincters can result from movement of chyme or fluid along the GI tract, whereas the same sphincters can contract tonically if they need to remain closed for a considerably long period to prevent the passage of food [380–382]. The presence of both tonic and phasic contractions can be studied with the calcium channel blocker, nifedipine. Administration of this blocker to regions sensitive to it mainly produces phasic contraction which can merge to form sustained tonic contraction [382].

Gastric Peristaltic Contraction

Peristaltic contraction of the stomach frequently begins in the middle region and increases as it extends toward the pyloric sphincter. The gastric contents are moved back from the pyloric region until the food particles are suitable for emptying into the duodenum. During this period, some contents of the chyme are periodically

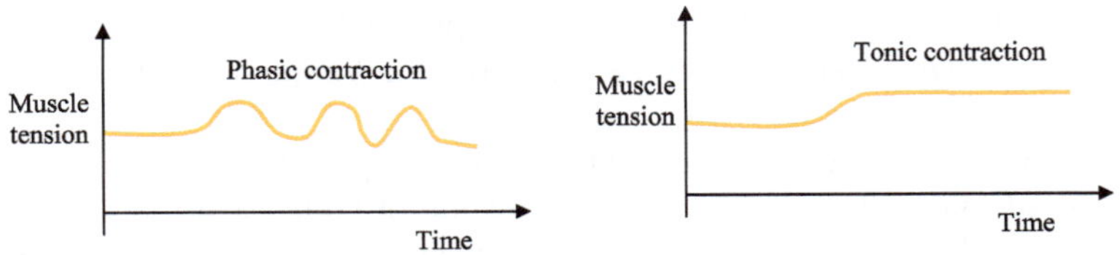

Fig. 7.17 Tonic and phasic contractions

propelled into the duodenum. Peristalsis of the stomach like other motor functions of the GI tract is generally due to the properties of the ICCs, smooth muscles, and associated functional elements. The energy for contraction is provided by mitochondria, serving as the energy powerhouse, and also responsible for the production of energy that drives gastric slow waves [383]. Though contraction of the stomach can occur in the absence of food, gastric peristaltic contraction occurs normally only in the presence of food in the stomach and is characterized by changes in gastric tone and compliance. This type of contraction, under normal condition, increases upon increase in gastric volume (food loading) [384–386].

Gastric tone, compliance, and volume characterize what is termed "gastric accommodation." **Gastric accommodation** is a neurally mediated property of the stomach that results in reduced gastric tone, increased compliance, and increased gastric volume, occurring in response to food ingestion [384–386]. During the initial period of food bolus flow into the stomach, this hollow organ accepts load without substantial increase in intragastric pressure. This period is referred to as the "receptive relaxation" and takes place within seconds of food ingestion. The insignificant rise in intragastric pressure may be due to the gastric distension that accompanies volume loading. A few seconds later, the stomach begins to respond to the properties of the ingested food by changes in tone. This second response is termed "adaptive relaxation" [385]. These responses are mediated by the central nervous system (particularly the brainstem), vagus nerve, and gastric peripheral reflexes. Disorder in gastric accommodation occurs in many disease conditions including functional dyspepsia, rumination syndrome, achalasia, gastroesophageal reflux disease, and vagus neuropathy due to diabetes (Clinical Correlate 7.5) [385, 387].

Clinical Correlate 7.5

Functional Dyspepsia, Rumination Syndrome (Merycism), Gastroesophageal Reflux Disease (GERD), and Diabetic Vagal Neuropathy

Functional dyspepsia (dyspepsia from the Greek "dys-" and "pepsis"—indigestion) is a relapsing and remitting disorder of the gastroduodenal region characterized by chronic pain or burning in the epigastrium, early satiety (inability to finish a normal-sized meal), fullness during or after a meal, or a combination of these symptoms, occurring at least weekly over a period of at least 6 months, in the absence of an organic pathology. The incidence of this condition is estimated at about 5–11% in the population. The condition is due to abnormal gastric emptying, visceral hypersensitivity, impaired gastric accommodation, and some disorders of central nervous system functions. Functional dyspepsia can be managed with prokinetics, tricyclic antidepressants, selective serotonin reuptake inhibitors, proton-pump inhibitors, and alternative therapies [388–390].

Rumination syndrome, also known as **merycism**, is a rare chronic motility disorder characterized by regurgitation of undigested food from the stomach back into the mouth, followed by either rechewing and deglutition or expulsion of the regurgitate to the exterior [390–392]. Regurgitation can be defined as the passage of refluxed gastric content into the oropharynx. The expulsion of the refluxed gastric content from the oral cavity is known as vomiting. The backward flow of gastric contents into the esophagus or mouth is termed gastroesophageal reflux [393]. Rumination is a normal process in ruminant animals, such as cow, goat, sheep, used by these animals to aid digestion of feeds, however, in humans, the condition is considered abnormal when it occurs frequently or is associated with complications. Patients with this syndrome experience regurgitation usually 1–2 h following food ingestion. Rumination syndrome is frequently seen in infants or persons who are developmentally disabled, but now generally agreed to occur in healthy persons of all ages, though less frequent [390–392]. In healthy persons including infants up to 18-months of age regurgitation that is not complicated with other conditions such as weight loss or poor weight gain, and without respiratory embarrassment, etc., is said to be normal. Normal regurgitation is more frequent in infants up to 6 months of age [393]. Infant rumination is usually associated with crying, coughing, overfeeding, and air swallowed during feeding [393]. The causes of rumination generally are multifactorial and can be grouped under the following [394, 395]:

1. Acquired and congenital anomalies of the oral cavity, larynx, trachea, and esophagus (e.g., cleft palate, esophageal stricture, tracheoesophageal fistula, enlarged thyroid gland, and pharyngeal diverticula).
2. Structural or mechanical disorders of the upper respiratory tract (e.g., choanal atresia, choanal stenosis, nasal pyriform aperture stenosis and nasal mid-line masses).
3. Neurological deficits (e.g., cerebral palsy, and autism).
4. Drug side effect (e.g., chemotherapy drugs).
5. Chronic diseases (e.g., gastroesophageal reflux disease).
6. Genetic and metabolic disorders (e.g., Down syndrome and phenylketonuria).
7. Food allergies.

The backflow of gastric contents often occurs during transient relaxation of the lower esophageal sphincter, thereby allowing the refluent enter the esophagus. However, gastric reflux can occur due to failure of the lower esophageal sphincter to compensate for a sudden increase in intraabdominal pressure or involuntary contraction of the abdominal muscles [393]. The complications of rumination syndrome include halitosis, dental erosions, weight loss, malnutrition, and disorder of electrolyte balance [390–392].

Gastroesophageal reflux disease (GERD) is a chronic, relapsing condition characterized by pyrosis (heartburn), due to backward flow (reflux) of gastric contents into the esophagus. This leads to abnormal esophageal pH usually less than 4 units and may occur more than once per hour. In the USA, the estimated lifetime prevalence of the disease is 25–35%. GERD results on account of failure of the multiple protective mechanisms of gastroesophageal reflux. The diagnosis of the disease is based on patient history and pH monitoring. Endoscopy and biopsy are required if esophagitis is suspected. Also, radiological series are required if any structural deformity is suspected. The management of GERD includes changes in lifestyle, pharmacotherapy and surgery. The latter is considered in refractory cases [393, 396, 397]. For treatment purposes, proton-pump inhibitors and H_2 receptor antagonists are recommended [393].

Diabetic vagal neuropathy is a disease of the vagus nerve resulting in decreased sensation or inadequate vagal responses to stimuli [398, 399]. Diabetic vagal neuropathy was defined by Kanatsuka et al. [400] as a heart rate variation during deep breathing of less than 9 beats per minute.

Gastric accommodation is influenced by the rate of release of gastric contents into the duodenum, referred to as gastric emptying [3, 401]. **Gastric emptying** is the process by which the stomach contents flow intermittently into the duodenum by regulated churning and peristaltic movement [402]. Although gastric emptying depends on the type of meal (energy density of meal), gastric emptying at 1 h after meal is defined as accelerated, and at 4 h after meal—as delayed emptying [401]. Gastric emptying appears to be faster in obese persons, even though gastric capacity is larger in this group of people [403–405]. This may be due to hyperkinetic and hypertonic gastric motor function [406].

Gastric compliance is the ability of the stomach to distend with increasing intragastric pressure. It can be defined as the ability of the stomach to resist recoil toward its original dimension on application of a distending force such as during food loading. It is mathematically defined as the ratio of volume change to pressure change—$\Delta V/\Delta P$ [407, 408]. Compliance is the reciprocal of elastance.

Gastric elastance is a measurement of the tendency of the stomach to recoil toward its original dimension upon removal of the distending force. The rate of gastric compliance, accommodation, and emptying is determined by the neuromuscular function of the stomach [403–405]. Neuromuscular disorders of the stomach can affect the rate of emptying, accommodation, and compliance. These disorders affect sensory and myoelectrical functions as well as contractility of the stomach [409].

Gastric compliance, accommodation, and emptying can be assessed with radiological and non-radiological methods, which include intragastric barostat, transabdominal ultrasound, magnetic resonance imaging, scintigraphy, single photon emission computed tomography, satiation drinking tests, and cutaneous electrogastrography [367, 401, 403, 406, 410–413].

Gastric Hunger Contraction

Hunger generally means a strong desire for food ingestion. It can be defined as sensation that reflects the biological need for energy intake. Hunger increases the motivation to eat food. Hunger is the opposite of satiety—the condition of being full. Both hunger and satiety are regulated by the central nervous system, which detect peripheral signals about the status of energy and metabolism and integrate this information to provide appropriate responses to ensure regulated energy homeostasis [414, 415]. The word "appetite" describes the desire to eat as well as the enjoyment of certain foods [416]. The magnitude of decline in hunger upon food ingestion depends on a range of factors, which include the properties of the food (sensory, physical, nutritive) and characteristics of the consumer (health status and cognitive state). Ingestion of meal or drink increases the satiety level, thus reducing hunger sensation. Some hours after meal, hunger sensation gradually increases until a threshold is reached and the individual continues to search for energy intake [415].

Clinical Correlate 7.6

Bulimia Nervosa and Binge Eating Disorder

Bulimia nervosa is a disease condition that describes binge eating and purging, with intermittent dietary restriction, and is considered addictive, thereby, differentiating it from other eating disorders and obesity. Diagnosis of bulimia nervosa can be made according to laid down criteria. The criteria for "bulimia" include inconspicuous eating; termination of such eating episodes by abdominal pain, sleep, social interruption, or self-induced vomiting; awareness that the eating pattern is abnormal; and depressed mood and self-deprecating thoughts following eating binges [417]. Bulimia nervosa may result in altered reward sensitivity in these individuals, particularly through the effects on the dopaminergic system [418].

Binge eating is characterized by discrete episodes of rapid and excessive food consumption not necessarily driven by hunger or metabolic need. Binge eating can also be defined as the consumption of an unusually large amount of food with a feeling of loss of control over food intake. Eating to excess (i.e., overeating) is not a disease per se, but can be a symptom of binge eating disorder. Individuals engaging in binge eating will eat until they feel uncomfortably full. Binge eating is often accompanied by feelings of loss of control and psychological distress [417]. Diagnosis of binge eating disorder can be made according to set criteria. Binge eating is associated with at least three of the following [417]:

1. Eating when not physically hungry.
2. Eating more quickly than normal.
3. Eating until uncomfortably full.

4. Eating alone because of shame.
5. Feeling disgusted with oneself, depressed, or guilty after overeating.

Hunger is closely related to the word “thirst”—defined as sensation that reflects the biological need for water intake. Thirst increases motivation to drink [415]. Thirst may be described as low, moderate, and severe. Excessive thirst is called hyperdipsia. Excessive drinking is referred to as polydipsia (from the Greek “polús,” meaning “much, many” and “dípsa,” meaning “thirst,” and altogether means “very thirsty”). Excessive drinking may characterize a disease entity in the absence of a physiologic stimulus. For instance, psychogenic polydipsia is clinical disorder characterized by excessive drinking of water. It may be associated with hyponatremia [419, 420].

Gastric hunger contractions are characteristic high-amplitude contraction occurring in the stomach during the interdigestive period—it corresponds to the Cannon and Washburn’s hunger pang, a physical sensation in the epigastrium that indicates need for food ingestion. This physical sensation in the epigastrium is associated with the phase III of the gastric and duodenal MMC. The hunger period is characterized by quiescency as well as regular cycle of “housekeeping” contraction of the gastric and ileal muscles—referred to as the MMC [73, 416, 421]. Phase III contraction is the phase of MMC, which is characterized by relatively strong contraction and propulsion of undigested food particles, microbes, debris, and secretion in the stomach and duodenum during the interdigestive period [421]. The duration of phase III contraction is estimated to be about 5–8 min. MMC usually begins at the duodenum, but it may be initiated at any region of the stomach (mostly at the antrum or mid-region) and propagates down the tract, terminating in the distal region of the ileum [416, 422]. The duodenum displays the highest number of cycles of slow waves—about 12 cycles per minute, which gradually decreases toward the terminal ileum—about 8 cycles per minute [423, 424]. The consecutive phases that comprise MMC are: quiescence, irregular spiking activity and regular spiking activity [424].

The sensation of hunger is thought to originate in the antral region of the stomach [416]. The mechanisms of MMC generation are thought to involve the enteric and central nervous systems. The phase III is controlled by the enteric nervous system and vagus nerve [416]. Phase III contractions are likely to be regulated by gut hormones such as motilin, ghrelin, somatostatin, and serotonin [416, 421, 425]. Motilin, ghrelin, and erythromycin induce phase III contraction of the MMC in the stomach. Serotonin and somatostatin induce duodenal phase III contraction [73]. However, the most potent agent that induces phase III contractions is motilin [416].

The physiological role of gastric hunger contractions is diverse. It is crucial for both mechanical and chemical cleansing of any undigested material not removed during normal gastric emptying in readiness for the next meal. The phase III is thought to play a housekeeping role [421, 426]. Phase III MMC activity is thought to help prevent bacterial overgrowth in the upper gut [416]. MMC activity is also accompanied by variations in splanchnic blood flow, gallbladder emptying, and gastric and pancreatic secretions [416]. Importantly, the slow waves coordinate motility pattern with intestinal secretory activity—referred to as the secretory component of MMC. Indeed MMC pattern corresponds to the secretory pattern of the intestine during the interdigestive period. The patterns of motility and secretion dictated by the slow waves are the main reason for the cyclic variation of pH in the GI tract [422]. The absence of phase III contraction is associated with gastroparesis (Clinical Correlate 7.7) [73].

Clinical Correlate 7.7
Gastroparesis

Gastroparesis is defined as abnormal gastric motility characterized by delayed emptying in the absence of mechanical outlet obstruction from the stomach. The incidence of the disease is estimated to be about 4% in the population [29, 367, 427].

The causes of gastroparesis are multifactorial. The most common causes are diabetes and postoperative complication of surgery on the stomach or vagus nerve. Diabetes is one of the major causes of gastroparesis—about 18% of long-term standing diabetes with poorly controlled blood sugar may have symptoms of gastroparesis which may be secondary to autonomic neuropathy [29, 367]. The postoperative cause may be considered following surgery (e.g., gastric bypass, vagotomy, gastric resection) on the upper GI tract, which may involve damage to the vagus nerve. In some cases, the etiology may be unidentified (idiopathic causes). Certain medications can cause gastroparesis. Central and peripheral diseases, including autonomic dysfunctions, enteric neuropathy, dysfunctions of ICCs, and psychosomatic disorders can result to abnormal gastric motility [367]. Specific diseases that cause gastroparesis include intestinal pseudoobstruction, liver disease, Parkinson's disease, multiple sclerosis, cerebrovascular accident, collagen vascular diseases, infectious diseases (e.g., varicella zoster, Epstein–Barr virus, Chagas disease, clostridium, botulinum), thyroid disease, and chronic renal failure [367, 427].

The symptoms of gastroparesis are non-specific and include pain in the abdomen or epigastrium, vomiting, nausea, bloating, and early signs of satiety or postprandial fullness [29, 427].

Diagnostic procedures include gastric scintigraphy, ^{13}C-octanoate breath testing, and the use of a wireless motility capsule [427]. Solid meal gastric scintigraphy is the gold standard [367, 427]. It should be noted that clinical conditions such as mechanical outlet obstruction, gastric or peptic ulcer disease, gastric cancer, or other malignancies must be excluded to make the diagnosis of gastroparesis. A severity grading scale for gastroparesis is also available [367].

Management of gastroparesis is based on the etiology. Management includes relief of symptoms, normalization of nutritional state, glycemic control, and improvement of gastric emptying. The first-line therapy involves the use of prokinetics (see Clinical Correlate 7.10). Patients with nausea and vomiting will benefit from the use of antiemetic medications. Selected patients will benefit from electrical stimulation of the stomach wall and surgery [367, 427].

Complications of gastroparesis include malnutrition, weight loss, esophagitis, Mallory-Weiss tear (resulting from chronic vomiting), acute renal failure secondary to fluid volume depletion, electrolyte dysbalance, and bezoar formation [367].

7.5.4 Motor Function of the Small Intestine

The small intestine comprises the duodenum, jejunum, and ileum in the proximal-to-distal direction and serves to further breakdown food substances to a form that is most suitable for absorption. The duodenum is the initial segment to receive the chyme that flow through the pyloric sphincter. The small intestine is the region where absorption of nutrients, water, electrolytes, and ingested drugs takes place. For effective absorption, the ingesta are propelled or moved aborally along the small intestine. There are different types of movements of the small intestine—segmentation contraction, peristalsis, peristaltic rush, and pendular movement. Peristalsis is the main type of propulsive contraction (also called propagating contraction) in the intestine [423, 428–430]. Peristalsis (including other intestinal contraction types) takes place on a background of slow waves, occurring during the interdigestive period [422].

Segmentation Contraction

Segmentation contraction is characterized by simultaneous alternating excitation (contraction) and inhibition (relaxation) of the intestinal circular muscle in both directions. This type of contraction acts to churn, mix, and locally circulate the intestinal content without pushing the content down the tract. The alternating contraction of the circular muscle divides and re-divides the luminal content, thereby providing a greater surface area for "mixing" with the intestinal juice. For this reason, this contraction type is sometimes referred to as mixing contraction. The degree of segmentation depends on the type of meal. However, both segmentation and peristalsis are required for effective digestion and absorption. The contribution of peristalsis and segmentation varies for different meals.

Under normal conditions, segmentation contraction predominates during ileal digestion of food substances with slow intestinal transit such as lipid, protein, and starchy foods. However, propulsive contraction is the primary mechanism of intestinal motility when food substances with rapid intestinal transit (such as fibers and semisolid feeds) are ingested. Segmentation contraction is primarily myogenic. But neurogenic factors also participate in the generation or modulation of this type of contraction [423, 431, 432]. Contraction that is primarily neurogenic means that motility is executed by local reflexes that are mediated by nerve plexuses of the intestinal wall. In myogenic contraction, the main contributing factors are the intrinsic properties of the muscles. The mechanism of segmentation contraction is not exactly clear, but local reflex with involvement of the enteric nervous system may play a crucial role in its initiation or regulation [423, 431, 433]. It is believed that segmentation is due to simultaneous local rhythmic muscular activity of the excitatory and the neighboring inhibitory motor neurons [434]. It should be noted however that segmentation contraction is dependent on nutrient availability, but they do not depend on the activity of intestinal slow waves [434]. Recent work indicates that segmentation contraction originates from the combined activities of the slow waves and rhythmic depolarization associated with the deep muscular plexus. In both cases, the ICC plays an integral role in the generation of waves or depolarization [431]. Like peristalsis, segmentation is affected by disorders such as inflammatory bowel diseases. In Crohn's disease, for example, segmentation is reduced, whereas propulsion is increased [423].

Small Intestinal Peristalsis

Peristalsis of the small intestine is a propulsive motor pattern type of contraction that involves contraction of the intestinal segment above the chyme and relaxation below, propelling luminal contents in the aboral direction. Peristaltic waves are characterized by the anal migration or propagation of the contractile ring, which gradually pushes the luminal content in the caudad direction (Fig. 7.18) [435–439]. The peristaltic waves progressively wane along the intestinal tract. In the fasting state, peristaltic contraction extends for an intestinal segment measuring about 1–2 cm in about 1 min, but may reach 10 min depending on several factors including the health state of the individual. The duration of contraction gradually increases beginning from the preprandial moments. In the postprandial state, peristaltic contraction extends for the same range of length but lasting just 1 s, but may reach 30–40 s depending on the type of meal [423, 440].

Peristaltic movement is predominantly neurogenic [440]. One of the mechanisms of peristaltic contraction is mediated by the peristaltic reflex, in which the neurons involved in the reflex generate action potential, relay the signal in the proximal direction, resulting in the relaxation of circular muscle and contraction of longitudinal muscle layers, and to the distal direction to cause contraction of circular muscle and relaxation of longitudinal muscle. The result is proximal-to-distal movement of intestinal luminal content. The peristaltic wave or contraction is caused by ascending

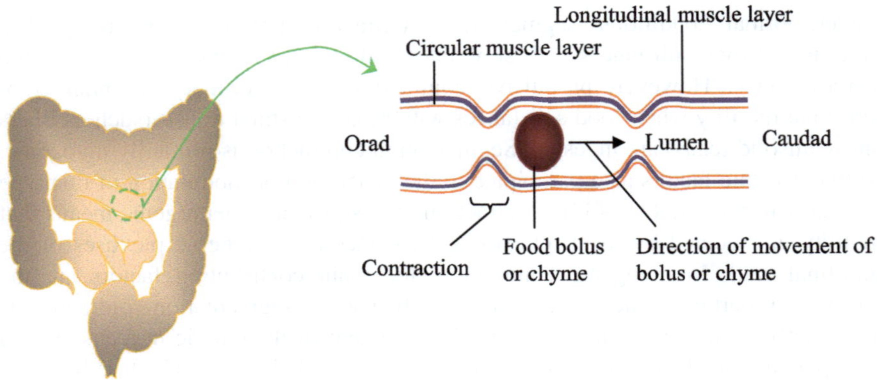

Fig. 7.18 Peristaltic pattern of contraction of the intestine

excitatory signal that evokes contraction above the bolus and descending inhibitory signal that results in relaxation below the bolus [76, 436]. The contraction above the bolus is termed ascending contraction, while the relaxation below the bolus is termed descending relaxation. Certain epithelial cells of the intestine (including enterochromaffin-like cells) can act as sensory cells that mediate numerous intestinal functions by the release of a variety of mediators, which are recognized by intrinsic primary afferent neurons of the submucosal and neural (mainly myenteric) plexuses [50, 441]. Peristaltic reflex is also regulated by intrinsic properties of the intestinal smooth muscle, and certain hormones including gastrin, CCK, motilin, secretin, and glucagon. Mediators such as gastrin, CCK, and motilin stimulate peristalsis, whereas secretin and glucagon inhibit peristalsis. The extrinsic innervation of the gut also affects peristalsis—the parasympathetic fibers increase peristalsis, while the sympathetic fibers decrease or inhibit intestinal peristalsis [50, 442]. Disorders in peristalsis occur in pregnancy with an incidence rate of 55% [443]. Abnormal peristalsis can also occur in GI tract tumor, which may be caused by abnormal synthesis and production of mediators [444]. In some instances, the abnormal peristalsis may occur in the distal-to-proximal direction (instead of the usual proximal-to-distal direction) resulting in what is termed antiperistalsis—retrograde movement of luminal content [445]. Retrograde contraction involves the emptying of the luminal content of the proximal half of ileum into the stomach. This contraction precedes vomiting, a process involving the expulsion of gastric contents by contraction of abdominal and diaphragmatic muscles [430]. Retrograde contraction occurs against the law of the intestine (or law of the gut) proposed in 1899 by William Bayliss and Ernest Starling [446, 447]. ***The Bayliss and Starling's law of the intestine states that excitation at any point of the gut causes contraction above and inhibition below*** [448, 449]. This means that contraction begins at the oral end and moves to the anal direction. In other words, luminal contents move aborally [450]. This direction of movement is due to the myenteric reflex proposed by the American physiologist Walter Bradford Cannon (1871–1945). The myenteric reflex is synonymous with the law of the intestine [451].

Peristaltic Rush

The peristaltic rush is a phenomenon that was first observed and described in 1872 by the Dutch scientist Jakob Pieter van Braam Houckgeest (1838–1889). The phenomenon was observed when small intestine of rabbits was submersed in warm saline bath [440]. Pioneers who conducted substantial investigations on peristaltic rush were the American physiologists Samuel James Meltzer (1851–1920) and his son-in-law Auer John (1875–1948). Peristaltic rush, also called "rollbewegungen," is a fast running movement extending over the whole or a large section of the small intestine [56, 440]. In contrast, normal peristalsis in the small intestine is weaker. Furthermore, the amplitude of spike potential in peristaltic rush is higher, and the aboral propagating speed is about 2 cm/s [452, 453]. The mechanism of peristaltic rush involves peripheral and central reflexes mediated by the intrinsic plexuses of the gut and the brainstem. Peristaltic rushes can be observed in pathological conditions such as gastroenteritis and diarrhea [454].

Pendular Movement

Pendular movement or contraction refers to rhythmic, swaying to-and-fro movement or "sleeve contraction" of the wall of the intestine that contribute to the mixing of luminal content of the intestine [440, 455]. Pendular movement involves contraction phase in one direction followed by a relaxation phase in the opposite direction [455]. In murine intestine, it was reported that pendular movement involves longitudinal muscle contraction, but very weak irregular circular contraction [456]. The direction of pendular movement depends on the source of pacemaker potential. Pendular contraction occurs in the anal direction if the slow waves are propagated in the oral direction and vice versa [455].

7.5.5 Motor Functions of the Large Intestine

The large intestine is divided into the cecum; ascending, transverse, descending, and sigmoid colon; the rectum and the anal canal. About 1–2 L of iso-osmotic chyme is emptied via the ileocolonic junction into the colon each day. The initial portion of the colon is predominantly concerned with absorption of salt and water from the chyme and accumulating about 200–250 ml semisolid feces in the distal portion of the colon [457, 458]. Subsequently, only about 100 ml of feces are evacuated daily through regulated defecation, which occurs 1–2 times every 24 h in healthy persons [459–461]. (See Reference Note 7.2 for composition of human feces). For feces formation to take place, the muscle wall of the colon actively contracts and relaxes in a regulated fashion that helps the aboral movement of its contents. Types of colonic contraction include mixing or segmentation, rhythmic phasic, tonic, antiperistaltic (retrograde movement), peristaltic, and mass movement. Peristaltic and mass movements are the main types of propulsive contraction in the colon that move fecal pellet to the anal canal [460, 462, 463]. However, generally, the main types of colonic movement are mass movement, mixing contraction, and antiperistalsis [463, 464]. The peristaltic contraction of the colon differs from the type of propulsion (primary propulsive contraction) observed in the intestine and generally refers to this as the mass movement or massive peristalsis [463, 464]. The tonic contraction of the colon helps rhythmic phasic contraction to move luminal content anally [465].

Reference Note 7.2
Composition of Human Solid Fecal Matter

Human fecal matter is composed of organic and inorganic constituents. The major component of human feces is water comprising about 63–85% (average of 75%) of fecal weight. However, the water content of a diarrheic stool is above 85%. Under normal condition, the remaining 25% is composed of solid matter which contains 84–93% organic materials [461, 466–468]. The organic materials in solid feces are undigested protein 2–25% (in addition 50% of bacterial biomass is protein), undigested carbohydrate 25% (including plant matter such as undigested cellulose, vegetable fibers, and pentosan), and undigested lipids 2–15%. The inorganic component of the feces is primarily undigested dietary elements [461]. The major organic material in feces is bacterial biomass consisting of 25–54% of solid feces. The bacteria in feces are mostly those that comprise the normal flora of the colon. In the colon, these bacteria synthesize different biomolecules including vitamins (vitamin B12, vitamin B1, thiamin, riboflavin, and vitamin K), gases (methane, carbon dioxide, hydrogen sulfide), indole, skatole, and mercaptans. The gases make up flatus. The action of the bacteria also plays a major role in determining the color and odor of feces [461, 469, 470]. Feces usually have a brown color, ranging from a tan hue to a darker brown color. The color is due to the action of microbes on the bilirubin contained in the feces. The odor of feces may differ among people and is influenced significantly by the foods consumed. Hydrogen sulfide is one of the most prominent odoriferous compounds responsible for the characteristic smell of the stool. The inorganic fraction is predominantly made up of calcium phosphate and iron phosphate, but also contains K, Mg, Na, Cu, Ni, Cd, Pb, Hg, and Zn. The composition of Ca in solid feces is about 5.4%. Feces also contain intestinal secretion and shredded epithelial cells. The pH values of feces are in the range 5.3–7.5 (average of 6.6) [461]. The values of the organic and inorganic constituents, color, odor, and pH depend on diet and state of health [461, 467, 468].

Clinical Correlate 7.8
Hirschsprung's Disease (Congenital Colonic Aganglionosis)

Hirschsprung's disease is named after Harald Hirschsprung (1830–1916), a Danish physician who first described the condition among infants in 1888. Hirschsprung's disease, also called colonic aganglionosis, is a congenital disease of the colon resulting in intestinal obstruction or the inability to pass stool due to the absence of ganglion cells in the distal intestine. In over three-quarter of cases, the colonic disorder is found in the rectosigmoid region of the large intestine. The disease affects about 1 in 5000 children. The cause of the disease is not exactly clear, but numerous signaling pathways and genes have been implicated in the etiopathogenesis of the disease. Mutation of the c-Kit receptor, an integral receptor of interstitial cells may cause disorders in the development of such cells and electrical arhythmicity that results in functional inability of these cells to execute their specific role in the gut [471, 472]. Research has shown that disorders of c-Kit receptor signaling in intestine are responsible for the development of Hirschsprung's disease. Certain dysfunctions that result in the migration of neuroblasts to the hindgut during the period of ontogenesis may also cause the disease. In course of fetal development, during the first 10 weeks of gestation, the incomplete migration of the cells from the neural crest to the colon to form Auerbach's and Meissner's plexuses is believed to cause the condition. Thus, the colon where these cells are malfunctioned cannot function well, and peristalsis is absent (aperistalsis), causing an obstruction [473]. The disease is diagnosed with anorectal manometry, barium enema, and rectal biopsy. The treatment for the disease involves surgical resection (removal) of the abnormal section of the colon, followed by re-anastomosis, a surgical reconnection technique required to preserve the integrity of the remaining section of the colon [472].

Mixing Contraction

The mixing contraction (also known as segmentation movement) that occurs in the colon is similar to the segmentation movement of the small intestine. The mixing activity of the colon is due to the morphology of the muscle wall—the longitudinal muscle layer form strips called teniae coli. The combination of these strips and the circular muscles results to sac-like appearance of the large intestine called haustrations. The segmentation movement results from the alternating contraction and relaxation of the circular muscle constricting the lumen and the longitudinal muscle strips [431, 463, 474]. In reality, colonic motility pattern may not occur independently of each other, but as regulated contraction involving contribution from different types of motility. Though a particular pattern of contraction may predominate in specific region of the colon, motility depends on the meal and health state of the individual. The mixing movement is largely due to phasic and tonic contraction as well as non-migrating motor activities [465, 475].

Mass Movement

Mass movement is a type of propulsive contraction characterized by forceful contractile waves that begin from the initial portion of the ascending colon propelling the chyme into the rectum. Besides the ascending, transverse, and descending colon, the rectum and sigmoid colon also exhibit mass movement. This type of contraction in these regions is generated on a background of slow waves or pacesetter potentials [464]. The frequency of mass movement in healthy persons is about 1–3 times per day each lasting for about 10–30 min [464]. Colonic mass movement is triggered by reflexes that result from gastric or duodenal distention caused by the presence of chyme—gastrocolic and duodenocolic reflexes [464, 476–478]. The duration of movement of colonic content to the anal canal in healthy subjects may be around 10–15 h or more depending on the type of meal, its fiber content as well as quantity of water intake. In constipation, colonic-to-anal transit time may increase to 23 h or more [479], whereas in diarrhea it significantly reduces [480]. (The small intestinal transit time is approximately 3 h for solids or liquids [481]). Mass movement is basically neurogenic [477].

Mass movement is produced by giant migrating contraction and is responsible for providing the force for propulsion of fecal materials and descending inhibition required to relax the internal anal sphincter during defecation [460, 465, 475]. Accumulation of the fecal matter in the rectum leads to distension of the rectal ampulla and the initiation of the sampling reflex of the anal sphincter, which relay stimuli to the brain. Since in healthy persons, defecation reflex is under control of higher brain centers, the individual decides on the socially appropriate condition, place and time to evacuate the feces [482]. Under normal condition, giant migrating contraction does not occur frequently [475]. Diarrhea observed in colonic motility disorders such as irritable bowel syndrome (IBS), inflammatory bowel disease (IBD) (e.g., ulcerative colitis, pancolitis), and diverticular disease, is caused by frequent mass movements [465]. On the contrary, decrease in the frequency of giant migrating contraction causes constipation [465, 482]. Patients with ulcerative colitis, for instance, frequently display colonic motor abnormalities, including lack of contractility, increase in propulsive contractile waves [476–478, 482]. Again, in healthy individuals, afferent signal generated by the wall of the colon upon compression by fecal matter is below nociceptive threshold; however, in patients with colonic motility disorders the nociceptive threshold exceeds a comparable value for healthy people and results to the sensation of pain [465].

Several mechanisms account for the development of mass movement, and they include neural, hormonal, chemical, mechanical, and myogenic [483]. The neural mechanism involves the enteric, autonomic, and central nervous systems. The autonomic mechanism of control of mass movement is regulated by the activities of the vagus, pelvic, lumbar colonic, hypogastric, and splanchnic nerves, which constantly evaluate the ditentability of the muscle wall of the colon and send the relevant stimuli to the central nervous system for appropriate responses. The descending input to the colon, rectum, anal canal and sphincters are required for the conscious determination of the time, place, and condition most appropriate for defecation [465, 475]. The enteric neurons, hormones, inflammatory mediators, stress mediators, and myogenic factors modulate the responses during this process [463, 465, 484]. Hormones or neurotransmitters such as epinephrine, serotonin, and glucagon inhibit colonic motility. Local distension reflex initiated by the fecal pellets due to stimulation of rectal mechanoreceptors modulate autonomic responses and aid in evacuation process [463]. These stretch-mediated responses are generated by the intrinsic mechanosensory interneurons that generate stimuli for ascending excitation and descending inhibition of the circular muscle [448, 485, 486]. Mechanical distension of the colonic wall by fecal pellets also activates enteric reflexes that mediate excitation above and inhibition below the fecal matter. The movement of the fecal matter toward the rectum is also thought to involve spontaneous slow MMC that occurs in the colon independent of the fecal contents [463, 487, 488].

The pattern of colonic motility corresponds to a circadian trend. Movement of the colon is greatest in the morning hours and after meal. Colonic movement frequently occurs in the daytime [482].

Disorders associated with motility of the intestines are briefly discussed in Clinical Correlate 7.9.

In Clinical Correlates 7.10 and 7.12 key pharmacological drugs and how they affect GI motility and secretion are discussed. Several drugs enter the market yearly and numerous drugs are currently under development/clinical trials [367, 489].

Clinical Correlate 7.9

Disorders of Intestinal Motility: Constipation, Intestinal Obstruction, Gastrointestinal Tract Disorders in Pregnancy and Thyroid Disease

Disorders of intestinal motility can lead to constipation or diarrhea, nausea and vomiting, abdominal distension and pain [490]. Constipation can be defined according to the Rome IV criteria for functional constipation or the American College of Gastroenterology definition, or the American Gastroenterological Association definition [491–493]. Constipation can be defined based on straining during defecation or difficulty passing stool, lumpy or hard stool consistency, sensation of incomplete evacuation or anorectal obstruction, feelings of abdominal cramping, manual maneuvers to facilitate defecation, and/or less than three defecations per week [494]. The estimated incidence of constipation in the general population ranges from 2 to 27%. Constipation is classified as primary (also called idiopathic) and secondary. The causes of primary constipation are not known. Secondary constipation can occur due to changes in diet, medication, or health state of the individual [491, 492, 494]. Constipation is frequently experienced by pregnant women.

Approximately, 11–38% of pregnant women experience constipation mostly during the third trimester. Constipation in pregnancy is due to many factors which include increasing level of progesterone and estrogen. These hormones inhibit the contractility of GI tract smooth muscle. Generally, constipation is mostly due to low intake of fluid and fiber in the diet [494]. The management of constipation involves dietary modification and use of bulking agents (such as soluble fiber, e.g., psyllium seed husk, insoluble fiber, e.g., wheat bran), stool softeners (such as docusate sodium, docusate calcium), stimulant laxatives (such as bisacodyl, sennosides, sodium picosulfate), osmotic laxatives (such as lactulose, polyethylene glycol, sorbitol, macrogol, milk of magnesia) and 5-HT_4 agonists (such as prucalopride) [491, 492, 494]. Bulking agents with water increase stool bulk and frequency. Stool softeners aid in softening stool by facilitating the mixing of fatty and aqueous substances. Stimulant laxatives enhance peristalsis by increasing water and electrolyte secretion into the lumen. Osmotic laxatives are hyperosmolar substances that are poorly absorbed by the intestine, thereby increasing the water content of stool. The 5-HT_4 agonists enhance peristalsis by myogenic mechanisms [491, 492, 495]. The use of probiotics has been shown to help constipated children and adults in different settings. Probiotics are defined as live microorganisms that confer immense health benefits on the host primarily by encouraging the population of gut resident microbes to synthesize short-chain fatty acids and other biologically active beneficial biomolecules. Probiotics differ in their composition of the beneficial microbes and include different strains of *Bifidobacteria (bifidum, lactis, longum), Lactobacilli (casei, plantarum, rhamnosus), and Propionibacterium freudenreichii* [494]. When used in their appropriate quantity, these bacteria increase the frequency of defecation and improve stool consistency in constipation [494]. Liquid paraffin (mineral oil) is also used in the management of constipation in both children and adults [495]. The agents used in the management of constipation have some side effects, which include diarrhea, flatulence, abdominal pain, and nausea [495].

Intestinal obstruction is the failure of normal proximal-to-distal movement of intestinal contents that results in a variety of symptoms including abdominal pain, vomiting, and constipation. Two types of intestinal obstruction can be distinguished—mechanical obstruction and paralytic ileus. The former occurs when the intestinal wall contracts against a fixed obstruction. The latter occurs in the absence of functional intestinal myoelectric activity [496, 497]. Causes of intestinal obstruction include appendicitis, internal hernia, adhesions, intestinal torsion, and colorectal carcinoma [443]. Symptoms of intestinal obstruction can also occur in pregnancy. However, it should be noted that in the first trimester of pregnancy, women usually suffer from nausea (50–90% incidence) and vomiting (30% incidence) [494]. See Clinical Correlate 7.1 for other information on mechanical obstruction and paralytic ileus.

GI tract functions are affected in thyroid disease such as hypothyroidism, in which patients suffer from constipation possibly due to delayed esophageal and gastric emptying, prolonged esophageal, and gastric transit time [490, 498]. Hypothyroidism reduces GI tract motor activity and can cause GI dysfunction [499].

Clinical Correlate 7.10

Effects of Some Pharmacological Agents on Gastrointestinal Motility. Pharmacological and Non-pharmacological Treatment of Constipation

Introduction: GI motility is controlled by hormones, neurotransmitters, drugs, and other substances. GI motility is disturbed in several conditions and diseases, including small intestinal bacterial overgrowth, IBS, functional dyspepsia, intestinal neuropathy, and aging [73]. In pathologically decreased motility, stimulation of GI tract motility is advised [208]. Numerous methods have been used to improve GI tract motility. Pharmacological and dietary agents are used to enhance GI motorics.

Dietary substances: Dietary agents are used for the promotion of gastric emptying and the movement of lumenal content in the ileum and colon. When recommending diets to improve constipation, it is important to take into consideration tolerance of solids, semisolids and liquids, fat content of diet, dietary balance, meal size, and timing of meals. To reduce the incidence of constipation, multiple small low-fat meals, but rich in fiber about four or five times each day are recommended. Research has shown, however, that the presence of fat in the terminal ileum decreases upper gut motility and inhibits ad libitum food ingestion—a phenomenon termed the ileal brake. So low-fat diet is recommended for constipation; it serves a useful means of decreasing energy ingestion. The fiber content of diet affects the rate of secretion of GI hormones. Certain fibers can reduce the levels of postprandial blood glucose and insulin, which, in addition to fat restriction, serves to limit energy intake. It is also essential to restrict consumption of carbonated liquids in other to limit gastric distention [367, 500]. The use of dietary substances is aimed at improving gastric emptying, upper small gut motility, rate of carbohydrate, fat and protein absorption in the intestine. The rhizomes of *Zingiber officinale* (ginger) have antiemetic agents, acting as weak 5-HT$_3$ receptor antagonists with gastric slow wave antidysrhythmic effects in humans. The 5-HT$_3$ antagonists present in ginger are gingerols, shogaols, and galanolactone. Preliminary clinical data suggest that administration of ginger may be effective for treatment of nausea or vomiting in a number of settings [367, 501–504].

Adrenaline (epinephrine): Epinephrine (released from the adrenal medulla) exerts an antidiarrheal effect via adrenoreceptors of smooth muscle cells and postganglionic sympathetic fibers. Therefore, the effect of application of epinephrine would be the same as sympathetic nerve activation, i.e., a reduction in smooth muscle tone and mechanical digestion and absorption [208, 505].

Morphine: This agent is usually administered to achieve pain relief in a clinical setting. However, like other opioid agonists, morphine decreases the propulsive activity of the intestine, resulting in constipation. In addition, opioid agonists (including endogenous opioid peptides) reduce both GI motility and secretion [208, 505]. The μ-, κ-, and δ-opioid receptors mediate the effects of opioids on neurons and smooth muscle cells. Opioids interact with a couple of GI neurotransmitters and also serve as neuromodulatory agents [506]. Opioid agonist-induced suppression of GI secretion and motility can be reversed by administration of lubiprostone. This drug is approved for treatment of idiopathic constipation and IBS-induced constipation in adults [507].

Low-dose tricyclic antidepressants—Tricyclic antidepressants impair GI motility through their anticholinergic activity but have been shown to relieve nausea, vomiting, and pain in functional dyspepsia [367, 508]. Examples of tricyclic antidepressants include amitriptyline, nortriptyline, and desipramine [509]. These drugs increase the level of central serotonin and norepinephrine through decrease in rate of these neurotransmitters reuptake by neurons. That is why they are termed norepinephrine and serotonin reuptake inhibitors [509, 510]. However, studies have shown that these drugs have multiple arrays of effectors inhibiting histaminic, cholinergic, and alpha1-adrenergic receptor activity. Since they are non-selective, administration of tricyclic antidepressants has been associated side effects such as constipation, dry mouth, weight gain, drowsiness, and dizziness [510–512].

Botulinum toxin: This neurotoxin is produced by *Clostridium botulinum*, a heterogeneous group of Gram-positive, rod-shaped, spore-forming, obligate anaerobic bacteria [208, 367]. This bacterium produces neurotoxins named with letters "A, B, C_1, C_2, D, E, F, and G." Of the eight toxins, five are known to cause disease in humans. Some neurotoxins have found useful application in medicine. In particular botulinum toxin A is recommended for treatment of a couple of ailments—this neurotoxin is approved by FDA (Food and Drug Administration) of the USA. The mechanism of action of botulinum neurotoxin is based on the inhibition of ACh release from presynaptic nerve endings of the neuromuscular junction, thereby causing muscle paralysis. But this action of the toxin can be applied for therapeutic purposes to weaken a muscle, in cases of excessive neural activity of central origin resulting in muscle spam, manage tension-type headache or in refractory chronic migraine, and other medical conditions [513, 514]. For further information on this toxin, review Münchau and Bhatia [515] and Nigam and Nigam [516]. For the purpose of emphasis, injection of minute amount of botulinum toxin A directly (botox injection) into the external anal sphincter is used as treatment of chronic idiopathic constipation, anal fissure and anismus in children and adults [208, 367, 517–520].

Atropine: Atropine specifically antagonizes the muscarinic type of cholinoreceptor and does not affect the nicotinic receptor. Therefore, if the response to ACh is completely absent in the presence of atropine, the response can be directly linked to the muscarinic rather than the nicotinic type cholinoreceptor [208]. Like guanethidine, atropine reduces the amplitude and tone of contraction in both layers of muscles of the GI tract. However, the frequency of contraction is reduced only in the circular muscle layer [219].

Prokinetics: These agents enhance the contractility of the GI tract, correct gastric dysrhythmias, and promote the movement of luminal contents in the antegrade direction. Prokinetics are also used to improve symptoms of nausea, vomiting, and bloating. Examples of prokinetics include erythromycin, metoclopramide, domperidone, cisapride, tegaserod, mitemcinal, bethanechol, pyridostigmine, and buspirone [367].

Erythromycin: Discovered in 1984 by Itoh and coworkers as a GI motility enhancer with comparative effects as motilin [521], this macrolide antibiotic, acting as motilin agonist, is the most potent stimulant of gastric emptying [367]. This drug facilitates ACh release from the cholinergic motor fibers of the GI tract, and also, activates other calcium-dependent pathways, and interacts with other pathways [522–524]. Similar actions have been reported for the antibiotic azithromycin [525].

Metoclopramide: This prokinetic drug possesses multiple mechanisms of action. It acts as $5HT_4$ receptor agonist, dopamine D2 receptor antagonist, and direct stimulant of GI smooth muscle contraction. This drug increases esophageal, fundic and antral contractile amplitudes, and elevates the pressure of the lower esophageal sphincter. The drug also has antiemetic effect, which is mediated through its antagonistic properties on the brainstem D2 receptor, vagal and brainstem $5\text{-}HT_3$ receptor [367].

Domperidone: This benzimidazole derivative is a peripheral dopamine D2 receptor antagonist, with a central antiemetic action [367]. Dopamine antagonists are effective in the treatment of nausea and vomiting associated with radiation, neoplasia, chemotherapeutic drugs, general anesthetics, and opioid overdose [508, 526].

Pyridostigmine, buspirone, and bethanechol: Bethanechol is an approved smooth muscle muscarinic agonist that increases lower esophageal sphincter pressure and evokes fundoantral contractions but does not induce propulsive contractions or accelerate gastric emptying [367, 527, 528]. The short-acting acetylcholinesterase inhibitor, edrophonium, when administered intravenously causes a substantial increase in the amplitude and duration of esophageal contraction, but edrophonium cannot be used for therapeutics due to its short duration of action and lack of administration per os [528]. Of the three muscarinic agonists, buspirone, bethanechol, and pyridostigmine, the latter seems to have a greater esophageal motility enhancing property [528].

Clinical Correlate 7.11

Effects of Electrical Stimulation and Acupuncture on Gastrointestinal Motility

Electrical stimulation: The application of electrical stimulation with surface electrodes, placed at different locations of the abdomen or anal region or projection points of the esophagus on the skin for given period of time (for instance, 30 min or 1 h per day or session) over a couple of weeks or months has been shown to be an effective treatment of chronic constipation [529]. The treatment method is also safe for use in pediatric patients [530].

Acupuncture: This is an ancient Chinese Traditional Medicine therapy in which acupoints on skin are manually stimulated by needles. And this type of manual stimulation is called hand-acupuncture [531]. Acupuncture has been shown to treat functional constipation in a randomized controlled trial using specific acupoints of large intestine meridians [532]. For example, Wang and Yin [531] showed that stimulation of acupoints corresponding to the large intestine reduced incidence of chronic constipation. Auricular acupuncture, which is performed at acupoints on the skin of ear, is also effective in treating various human illnesses including constipation [531].

Electroacupuncture: Electroacupuncture simply means combined electrical stimulation with acupuncture. This method involves the delivery of different intensities of electrical current to needles inserted into acupoints. The needles may be placed together with electrodes [531]. Electroacupuncture can effectively treat functional constipation. The effect may be mediated via changes in the levels of hormones/neurotransmitters such as cortisol, substance P, and VIP [533]. Apart from pharmacological treatment of constipation, acupuncture on the P6 acupoint has been shown to reduce nausea [367].

Clinical Correlate 7.12

Antiemetics

Introduction: Several drugs and dietary substances are known to serve to prevent nausea and vomiting—also known as antiemetics. Some agents have been discussed in Clinical Correlate 7.10; others will be discussed in the present Clinical Correlate. Examples of antiemetic agents include ondansetron, domperidone, metoclopramide, promethazine, dexamethasone, among others [534].

Cannabinoids: Cannabinoid drugs such as dronabinol have been studied for improvement of GI symptoms resulting from chemotherapy. These drugs are used for treatment of nausea and vomiting associated with cachexia and cytotoxic drugs. Cannabinoid drugs serve as alternatives for patients who are unresponsive to other agents. However, the administration of cannabinoids is associated with dizziness, loss of coordination and causes changes in perception [508, 526].

Benzodiazepines: This class of pharmacological agents carries out its action through activation of central ionotropic GABA type A receptors. Though benzodiazepines are primarily used as anticonvulsant, sedative-hypnotic, antiamnestic, antianxiety agents, they have been found to be effective in managing anticipatory nausea and vomiting before chemotherapy and anesthesia or as an adjunct treatment for nausea and vomiting. Examples include midazolam, lorazepam, and diazepam. The effects of these benzodiazepines can be reversed by Flumazenil [508, 526, 535].

5-HT$_3$ receptor antagonists: This class of pharmacological agents is used for prevention of postoperative, radiation-, chemotherapy-induced nausea and vomiting. The effects of 5-HT$_3$ receptor antagonists are mediated via the inhibition of central and peripheral (GI tract) 5-HT$_3$ receptors. Examples of 5-HT$_3$ receptor antagonists are alosetron, dolasetron, granisetron, ondansetron, and palonosetron. These drugs may be used in combination with corticosteroids (see below) [536–539].

Neurokinin (NK)-1 receptor antagonists: This class of drugs, in recent times, has shown promise for treating both acute and slow onset chemotherapy-induced nausea and vomiting. Examples of NK-1 receptor antagonists are aprepitant, rolapitant fosaprepitant, casopitant, ezlopitant, vestipitant, and netupitant [508, 526, 540].

Corticosteroids: Corticosteroids are used as antiemetics in the postoperative setting for patients with risk of nausea and vomiting or in the prevention of chemotherapy- and radiotherapy-induced nausea and vomiting [367, 541]. Though corticosteroids, in particular, glucocorticoids, are used primarily as antiallergic and antiinflammatory drugs, at low dose, they are found to be effective in preventing nausea and vomiting when administered alone or in combination with other antiemetics [541]. For instance, dexamethasone, a type of glucocorticoids, at low dose is effective in reducing low-level emesis. However, the antiemetic mechanism of action of this drug is not exactly clear. But recent investigations indicate that the antiemeiogenic action of dexamethasone is due to its activation of central glucocorticoid receptors in the medullary nuclei of solitary tract [542]. Glucocorticoids are also known to regulate the hypothalamic–pituitary–adrenal axis. This class of drugs can be used to reduce pain [541].

7.6 Conclusion

The motor functions of the GI tract allow the aboral flow of luminal contents. GI tract motorics or motility is due to the regulated activity of the smooth muscles of the walls of the tract. The functioning of these muscles is controlled by nutrients, myogenic, neural, and endocrine as well as paracrine factors. The muscles of the digestive apparatus serve to ensure proper chewing, swallowing, and the proximal-to-distal (anally directed) motion of luminal contents along the digestive tract and the removal of undigested residues from the anus.

Recommended Readings

Original Articles

1. Annese V, Bassotti G, Napolitano G, Usai P, Andriulli A, Vantrappen G (1997) Gastrointestinal motility disorders in patients with inactive Crohn's disease. Scand J Gastroenterol 32(11):1107–1117
2. Blonski W, Vela MF, Freeman J, Sharma N, Castell DO (2009) The effect of oral buspirone, pyridostigmine, and bethanechol on esophageal function evaluated with combined multichannel esophageal impedance-manometry in healthy volunteers. J Clin Gastroenterol 43(3):253–260
3. Bochukova E, Schoenmakers N, Agostini M, Schoenmakers E, Rajanayagam O, Keogh JM, et al (2012) A mutation in the thyroid hormone receptor alpha gene. N Engl J Med 366:243–249
4. Chen CL, Yi CH, Chou AS, Liu TT (2013) Esophageal solid bolus transit: studies using combined multichannel intraluminal impedance and manometry in healthy volunteers. Dis Esophagus 26(1):91–96
5. Chen H, Zhu W, Lu J, Fan J, Sun L, Feng X, et al (2016) The effects of auricular electroacupuncture on ameliorating the dysfunction of interstitial cells of Cajal networks and nNOSmRNA expression in antrum of STZ-induced diabetic rats. PLoS 11(12):e0166638
6. Cowles VE, Sarna SK (1990) Effect of cholera toxin on small intestinal motor activity in the fed state. Digest Dis Sci 35(3):353–359
7. Fiskerstrand T, Arshad N, Haukanes BI, Tronstad RR, Pham KD-C, Johansson S, et al (2012) Familial diarrhea syndrome caused by an activating *GUCY2C* mutation. N Engl J Med 366:1586–1595
8. Hoog CM, Lindberg G, Sjoqvist U (2007) Findings in patients with chronic intestinal dysmotility investigated by capsule endoscopy. BMC Gastroenterol 7:29
9. Lembo AJ, Lacy BE, Zuckerman MJ, Schey R, Dove LS, Andrae DA, et al (2016) Eluxadoline for irritable bowel syndrome with diarrhea. N Engl J Med 374:242–253
10. Li H, Wang YP (2013) Effect of auricular acupuncture on gastrointestinal motility and its relationship with vagal activity. Acupunct Med 31(1):57–64

11. López-Herce J, Santiago MJ, Sánchez C, Mencía S, Carrillo A, Vigil D (2008) Risk factors for gastrointestinal complications in critically ill children with transpyloric enteral nutrition. Eur J Clin Nutr 62:395–400
12. Scheerens C, Tack J, Rommel N (2015) Buspirone, a new drug for the management of patients with ineffective esophageal motility? United European Gastroenterol J 3(3):261–265
13. Shubert TR, Sitaram S, Jadcherla SR (2016) Effects of pacifier and taste on swallowing, esophageal motility, transit, and respiratory rhythm in human neonates. Neurogastroenterol Motil 28(4):532–542
14. Taguchi A, Sharma N, Saleem RM, Sessler DI, Carpenter RL, Seyedsadr M, Kurz A (2001) Selective postoperative inhibition of gastrointestinal opioid receptors. N Engl J Med 345:935–940
15. Toporowska-Kowalska E, Gębora-Kowalska B, Fendler W, Popińska K, Szlagatys-Sidorkiewicz A, Grzybowska-Chlebowczyk U, et al (2013) Diagnosis of gastroesophageal reflux and antireflux procedures among Polish children with gastrostomies: a 10-year nationwide analysis. Eur J Clin Nutr 67 (11):1169–1174
16. Törnblom H, Lang B, Clover L, Knowles CH, Vincent A, Lindberg G (2007) Autoantibodies in patients with gut motility disorders and enteric neuropathy. Scand J Gastroenterol 42(11):1289–1293
17. Tsili Z, Carla C, Samuel Z, Di Lorenzo C, Flores AF, Cocjin J, et al (2003) Gastrointestinal motility and sensory abnormalities may contribute to food refusal in medically fragile toddlers. J Pediatr Gastroenterol Nutr 37(3):287–293

Review Articles

1. Berridge MJ (2008) Smooth muscle cell calcium activation mechanisms. J Physiol 586(Pt 21):5047–5061
2. Clauss M, Schwarm A, Ortmann S, Streich WJ, Hummel J (2007) A case of non-scaling in mammalian physiology? Body size, digestive capacity, food intake, and ingesta passage in mammalian herbivores. Comp Biochem Physiol A Mol Integr Physiol 148(2):249–265
3. Fishbein TM (2009) Intestinal transplantation. N Engl J Med 361:998–1008
4. Hanson J (1952) Changes in the cross-striation of myofibrils during contraction induced by adenosine triphosphate. Nature 169:530–533
5. Hanson J, Huxley HE (1953) Structural basis of the cross-striations in muscle. Nature 172:530–532
6. Hitchcock-DeGregori SE, Barua B (2017) Tropomyosin structure, function, and interactions: a dynamic regulator. Subcell Biochem 82:253–284
7. Kaya RD, Hoffman RL, Clark BC (2014) Reliability of a modified motor unit number index (MUNIX) technique. J Electromyogr Kinesiol 24(1):18–24
8. Lehman W, Morgan KG (2012) Structure and dynamics of the actin-based smooth muscle contractile and cytoskeletal apparatus. J Muscle Res Cell Motil 33(6):461–469

9. Squire JM, Parry DAD (2017) Fibrous protein structures: hierarchy, history and heroes. Subcell Biochem 82:1–33
10. Tse G, Lai ETH, Lee APW, Yan BP, Wong SH (2016) Electrophysiological mechanisms of gastrointestinal arrhythmogenesis: lessons from the heart. Front Physiol 7:230

Books and Book Chapters

1. Bredenoord AJ, Smout A, Tack J (2016) A guide to gastrointestinal motility disorders. Springer International Publishing, Cham
2. British Medical Association (2005) Complete family health guide. Kindersley Dorling, London
3. Mittal RK (2011) Motor function of the pharynx, esophagus, and its sphincters. Morgan & Claypool Life Sciences, San Rafael
4. Palmer JB, Monahan DM, Matsuo K (2006) Rehabilitation of patients with swallowing disorders. In: Braddom R (ed) Physical medicine and rehabilitation. Elsevier, Philadelphia
5. Parkman H, McCallum R, Rao SSC (2010) GI motility testing: a laboratory and office handbook, 1st edn. Slack Incorporated, Thorofare
6. Rao SSC (ed) (1999) Gastrointestinal motility: tests and problem-oriented approach. Springer, New York
7. Rao SSC, Parkman H, McCallum R (2015) Handbook of gastrointestinal motility and functional disorders, 1st ed. Slack Incorporated, Thorofare
8. Roman C (ed) (1984) Gastrointestinal motility. MTP Press, Boston

Letters to the Editor

No Author (2006) The many facets of intestinal peristalsis. Am J Physiol Gastrointest Liver Physiol 290:G1347–G1349

Bibliography

1. Gartner LP, Hiatt JL, Strum JM (2003) Board review series. Cell biology and histology, 4th edn. Lippincott Williams & Wilkins, Philadelphia
2. Sanders KM, Don Koh S, Ward SM (2012) Organization and electrophysiology of interstitial cells of Cajal and smooth muscle cells in the gastrointestinal tract. In: Johnson L, Ghishan F, Kaunitz J, Merchant J, Said H, Wood J (eds) Physiology of the gastrointestinal tract. Elsevier, San Diego
3. Mittal RK (ed) (2011) Motor function of the pharynx, esophagus, and its sphincters. Integrated systems physiology: from molecule to function to disease. Morgan & Claypool Life Sciences, San Rafael
4. Sundler F (2004) GI tract, general anatomy (cells). In: Martini Luciano (ed) Encyclopedia of endocrine diseases. Elsevier, San Diego
5. Piccolino M, Bresadola M (2002) Drawing a spark from darkness: John Walsh and electric fish. Endeavour 26(1):19–26
6. Delbourgo J (2006) A most amazing scene of wonders: electricity and enlightenment in early America. Harvard University Press, Cambridge

7. Plumb C (2010) The ‘electric stroke’ and the ‘electric spark’: anatomists and eroticism at George Baker’s electric eel exhibition in 1776 and 1777. Endeavour 34(3):87–94
8. Carlyle EI (2004) Walsh, John (1726–1795). In: Harrison BH, Matthew HCG (eds) Oxford dictionary of national biography. Oxford University Press, New York
9. Piccolino M, Biologies CR (2006) Luigi Galvani’s path to animal electricity. CR Biol 329 (5–6):303–318
10. Kettenmann H (1997) Alexander von Humboldt and the concept of animal electricity. Trends Neurosci 20(6):239–242
11. Hopkinson J (1875) The residual charge of the leyden jar. Proc R Soc Lond 24:407–408
12. Elliott P (2008) “More subtle than the electric aura”: georgian medical electricity, the spirit of animation and the development of Erasmus Darwin’s psychophysiology. Med Hist 52(2):195–220
13. Verkhratsky A, Krishtal OA, Petersen OH (2006) From Galvani to patch clamp: the development of electrophysiology. Pflugers Arch 453(3):233–247
14. Al-Khalili J (2015) The birth of the electric machines: a commentary on Faraday (1832) ‘Experimental researches in electricity’. Philos Trans A Math Phys Eng Sci 373(2039): 20140208
15. McWhirter L, Carson A, Stone J (2015) The body electric: a long view of electrical therapy for functional neurological disorders. Brain 138(Pt 4):1113–1120
16. Steinberg H (2011) Electrotherapeutic disputes: the ‘Frankfurt Council’ of 1891. Brain 134 (4):1229–1243
17. Goldensohn ES (1998) Animal electricity from Bologna to Boston. Electroencephalogr Clin Neurophysiol 106(2):94–100
18. Alvarez WC (1922) New methods of studying gastric peristalsis. J Am Med Assoc 22:1281–1284
19. Rivera-Ruiz M, Cajavilca C, Varon J (2008) Einthoven’s string galvanometer—the first electrocardiograph. Tex Heart Inst J 35(2):174–178
20. Verkhratsky A, Parpura V (2014) History of electrophysiology and the patch clamp. Methods Mol Biol 1183:1–19
21. Moruzzi G (1996) The electrophysiological work of Carlo Matteucci. 1964. Brain Res Bull 40(2):69–91
22. Gryglewski RW (2015) The origins of electrocardiography in Poland. Kardiochir Torakochirurgia Pol 12(2):184–189
23. Ahmed OJ, Cash SS (2013) Finding synchrony in the desynchronized EEG: the history and interpretation of gamma rhythms. Front Integr Neurosci 7:58
24. Longair M (2015) ‘…a paper …I hold to be great guns’: a commentary on Maxwell (1865) ‘A dynamical theory of the electromagnetic field’. Philos Trans A Math Phys Eng Sci 373 (2039):20140473
25. Berridge MJ (2008) Smooth muscle cell calcium activation mechanisms. J Physiol 586(Pt 21):5047–5061
26. O’Grady G (2012) Gastrointestinal extracellular electrical recordings: fact or artifact? Neurogastroenterol Motil 24(1):1–6
27. Hara Y, Kubota M, Szurszewski JH (1986) Electrophysiology of smooth muscle of the small intestine of some mammals. J Physiol 372:501–520
28. Huizinga JD (1991) Action potentials in gastrointestinal smooth muscle. Can J Physiol Pharmacol 69(8):1133–1142
29. Du P, O’Grady G, Davidson JB, Cheng LK, Pullan AJ (2010) Multiscale modeling of gastrointestinal electrophysiology and experimental validation. Crit Rev Biomed Eng 38 (3):225–254
30. Piccolino M (2011) Electric fishes research in the nineteenth century, following the steps of Carlo Matteucci and Giuseppe Moruzzi. Arch Ital Biol 149(Suppl.):77–85
31. Finkelstein G (2011) Matteucci and du Bois-Reymond: a bitter rivalry. Arch Ital Biol 149 (4):29–37

32. Waclawik AJ (2014) du Bois-Reymond, Emil. In: Aminoff MJ, Daroff RB (eds) Encyclopedia of the neurological sciences. Elsevier, San Diego
33. Pearce JMS (2001) Emil Heinrich Du Bois-Reymond (1818–96). J Neurol Neurosurg Psychiatry 71:620. https://doi.org/10.1136/jnnp.71.5.620
34. Finger S, Piccolino M, Stahnisch FW (2013) Alexander von Humboldt: galvanism, animal electricity, and self-experimentation part 2: the electric eel, animal electricity, and later years. J Hist Neurosci 22(4):327–352
35. Ian G (2010) Elegance in science. Oxford University Press, Oxford
36. Marasco DD (2011) The great era of english electrophysiology: from Francis Gotch to Hodgkin and Huxley. Arch Ital Biol 149(Suppl.):77–85
37. Seyfarth E-A (2006) Julius Bernstein (1839–1917): pioneer neurobiologist and biophysicist. Biol Cybernet 94(1):2–8
38. Finkelstein G (2015) Mechanical neuroscience: Emil du Bois-Reymond's innovations in theory and practice. Front Syst Neurosci 9:133
39. Frank RG Jr, Goetzl JH (1978) The J. H. B. Archive report the Alexander Forbes papers. J Hist Biol 11(2):387–393
40. Davis H (1970) Joseph Erlanger (1874–1965). National Academy of Sciences, Washington DC
41. Kosenko PM, Vavrinchuk SA (2013) Electrogastroenterography in clinical practice. Rus Open Med J 2:0203
42. Szurszewski JH (1998) A 100-year perspective on gastrointestinal motility. Am J Physiol Gastrointest Liver Physiol 274(37):G447–G453
43. Beaumont W (1833) Experiments and observations on the gastric juice and the physiology of digestion. FP Allen, Plattsburgh
44. Mai FM (1988) Beaumont's contribution to gastric psychophysiology: a reappraisal. Can J Psychiatr 33(7):650–653
45. Dubois A, Johnson LF (1985) William Beaumont: frontier physician and founding father of gastric physiology. J Clin Gastroenterol 7(6):472–474
46. Cannon WB (1898) The movements of the stomach studied by means of the Roentgen rays. Am J Physiol 1:359–382
47. Cannon WB (1902) The movements of the intestines studied by means of the Roentgen rays. Am J Physiol 6:251–277
48. Cannon WB (1904) The passage of different food-stuffs from the stomach and through the small intestine. Am J Physiol 12:387–418
49. Cannon WB, Mosher A (1898) The movements of the food in the esophagus. Am J Physiol 1:435–444
50. Trendelenburg P (2006) Physiological and pharmacological investigations of small intestinal peristalsis. Naunyn Schmiedebergs Arch Pharmacol 373(2):101–133. Translation of the article "Trendelenburg P (1917) Physiologische und pharmakologische Versuche über die Dünndarmperistaltik. Arch Exp Pathol Pharmakol 81:55–129"
51. Feldberg W, Lin RCY (1949) The action of local anaesthetics and D-tubocurarine on the isolated intestine of the rabbit and guinea-pig. Br J Pharmacol 4:33–44
52. Bennett MR, Burnstock G, Holman ME (1966) Transmission from intramural inhibitory nerves to the smooth muscle of the guinea-pig taenia coli. J Physiol (Lond) 182:541–558
53. Bülbring E, Crema A (1958) Observations concerning the action of 5-hydroxytryptamine on the peristaltic reflex. Br J Pharmacol 13:444–457
54. Alvarez WC (1914) Functional variations in contractions of different parts of the small intestine. Am J Physiol 35:177–193
55. Alvarez WC, Mahoney LJ (1922) Action currents in stomach and intestine. Am J Physiol 58:476–493
56. Alvarez WC, Mahoney LJ (1924) Peristaltic rush as depicted in the electroenterogram. Am J Physiol 69:226–228
57. Bozler E (1938) Electric stimulation and conduction of excitation in smooth muscle. Am J Physiol 122:614–623

58. Bozler E (1942) The activity of the pacemaker previous to the discharge of a muscular impulse. Am J Physiol 136:543–552
59. Bozler E (1945) The action potentials of the stomach. Am J Physiol 144:693–700
60. Tumpeer IH, Blitzsten PW (1926) Registration of peristalsis by the Einthoven galvonometer. Am J Dis Child 21:454–455
61. Chang FY (2005) Electrogastrography: basic knowledge, recording, processing and its clinical applications. J Gastroenterol Hepatol 20(4):502–516
62. Matsuura Y, Yamamoto T, Takada M, Shiozawa T, Takada H (2011) Application of electrogastrography to public health. Nihon Eiseigaku Zasshi 66(1):54–63
63. Yin J, Chen JD (2013) Electrogastrography: methodology, validation and applications. J Neurogastroenterol Motil 19(1):5–17
64. Rebrov VG (1981) Registration of stomach and intestine potentials in clinical conditions. Ter archive 9:24–30
65. Koenig JB, Martin CE, Dobson H, Mintchev MP (2009) Use of multichannel electrogastrography for noninvasive assessment of gastric myoelectrical activity in dogs. Am J Vet Res 70(1):11–55
66. Lebedev NN, Trusov AN, Popova YuP (1991) Broadband multichannel electrogastrography and periodical motorics of gastrointestinal tract. Fiziol Chelov 17(4):54–66
67. Burduli NM, Gutnova SK (2010) Peripheral computer electrogastroentherography in the diagnostics of motor disorders in chronic pancreatitis. Rus J Gastroenterol Hepatol Coloproctol 20(5):59
68. Hamilton JW, Bellahsene BE, Reichelderfer M, Webster JG, Bass P (1986) Human electrogastrograms. Comparison of surface and mucosal recordings. Dig Dis Sci 31(1):33–39
69. Sanders KM, Don Koh S, Ward SM (2006) Interstitial cells of cajal as pacemakers in the gastrointestinal tract. Annu Rev Physiol 68:307–343
70. Sanders KM, Ward SM (2006) Interstitial cells of Cajal: a new perspective on smooth muscle function. J Physiol 576(Pt 3):721–726
71. Sanders KM, Ward SM, Don Koh S (2014) Interstitial cells: regulators of smooth muscle function. Physiol Rev 94(3):859–907
72. Sanders KM, Don Koh S, Ro S, Ward SM (2012) Regulation of gastrointestinal motility—Insights from smooth muscle biology. Nat Rev Gastroenterol Hepatol 9(11):633–645
73. Deloose E, Janssen P, Depoortere I, Tack J (2012) The migrating motor complex: control mechanisms and its role in health and disease. Nat Rev Gastroenterol Hepatol 9:271–285
74. Nucera G, Gabrielli M, Lupascu A, Lauritano EC, Santoliquido A, Cremonini F et al (2005) Abnormal breath tests to lactose, fructose and sorbitol in irritable bowel syndrome may be explained by small intestinal bacterial overgrowth. Aliment Pharmacol Ther 21(11): 1391–1395
75. Shea-Donohue T, Notari L, Sun R, Zhao A (2012) Mechansims of smooth muscle responses to inflammation. Neurogastroenterol Motil 24(9):802–811
76. Fiorenza V, Yee YS, Zfass AM (1987) Small intestinal motility: normal and abnormal function. Am J Gastroenterol 82(11):1111–1114
77. Luiking YC, van der Reijden AC, van Berge Henegouwen GP, Akkermans LMA (1998) Migrating motor complex cycle duration is determined by gastric or duodenal origin of phase III. Am J Physiol Gastrointest Liver Physiol 275(6):G1246–G1251
78. Kim TW, Koh SD, Ordög T, Ward SM, Sanders KM (2003) Muscarinic regulation of pacemaker frequency in murine gastric interstitial cells of Cajal. J Physiol 546(Pt 2):415–425
79. Forrest AS, Hennig GW, Jokela-Willis S, Park CD, Sanders KM (2009) Prostaglandin regulation of gastric slow waves and peristalsis. Am J Physiol Gastrointest Liver Physiol 296 (6):G1180–G1190
80. Bortoff A (1965) Electrical transmission of slow waves from longitudinal to circular intestinal muscle. Am J Physiol 209:1254–1260
81. Prosser CL, Burnstock G, Kahn J (1960) Conduction in smooth muscle: comparative structural properties. Am J Physiol 199:545–552

82. Prosser CL, Sperelakis N (1956) Transmission in ganglion free circular muscle from the cat intestine. Am J Physiol 187:536–545
83. Nagai T, Prosser CL (1963) Electrical parameters of smooth muscle cells. Am J Physiol 204:915–924
84. Nagai T, Prosser CL (1963) Patterns of conduction in smooth muscle. Am J Physiol 204:910–914
85. Papasova MP, Nagai T, Prosser CL (1968) Two-component slow waves in smooth muscle of cat stomach. Am J Physiol 214:695–702
86. Diamant NE, Bortoff A (1969) Nature of the intestinal slow wave frequency gradient. Am J Physiol 216:301–307
87. Diamant NE, Bortoff A (1969) Effects of transection on the intestinal slow-wave frequency gradient. Am J Physiol 216:734–743
88. Connor C, Mangel AW, Prosser CL (1979) Propagation and entrainment of slow waves in cat small intestine. Am J Physiol Cell Physiol 237(6):C237–C246
89. Connor C, Prosser CL (1974) Comparison of ionic effects on longitudinal and circular muscle of cat jejunum. Am J Physiol 226:1212–1218
90. Suzuki N, Prosser CL, Dahms V (1986) Boundary cells between longitudinal and circular layers: essential for electrical slow waves in cat intestine. Am J Physiol Gastrointest Liver Physiol 250(13):G287–G294
91. Christensen J, Caprilli R, Lund GF (1969) Electric slow waves in circular muscle of cat colon. Am J Physiol 217:771–776
92. Christensen J, Hauser RL (1971) Circumferential coupling of electric slow waves in circular muscle of cat colon. Am J Physiol 221:1033–1037
93. Christensen J, Hauser RL (1971) Longitudinal axial coupling of slow waves in proximal cat colon. Am J Physiol 221:246–250
94. Christensen J, Rasmus SC (1972) Colon slow waves: size of oscillators and rates of spread. Am J Physiol 223:1330–1333
95. Szurszewski JH (1969) A migrating electric complex of canine small intestine. Am J Physiol 217:1757–1763
96. Vantrappen G, Janssens J, Hellemans J, Ghoos Y (1977) The interdigestive motor complex of normal subjects and patients with bacterial overgrowth of the small intestine. J Clin Invest 59:1158–1166
97. Konturek PC, Konturek SJ (2003) The history of gastrointestinal hormones and the polish contribution to elucidation of their biology and relation to nervous system. J Physiol Pharmacol 54(S3):83–98
98. Thuneberg L (1982) Interstitial cells of Cajal: intestinal pacemaker cells? Adv Anat Embryol Cell Biol 71:1–130
99. Thomson L, Robinson TL, Lee JCF, Farraway LA, Hughes MJG, Andrews DW, Huizinga JD (1998) Interstitial cells of Cajal generate a rhythmic pacemaker current. Nat Med 4:848–851
100. Sanders KM, Hwang SJ, Ward SM (2010) Neuroeffector apparatus in gastrointestinal smooth muscle organs. J Physiol 588(23):4621–4639
101. Huizinga JD (1999) Gastrointestinal peristalsis: joint action of enteric nerves, smooth muscle, and interstitial cells of Cajal. Microsc Res Tech 47(4):239–247
102. Bauer AJ, Publicover NG, Sanders KM (1985) Origin and spread of slow waves in canine gastric antral circular muscle. Am J Physiol Gastrointest Liver Physiol 249(12):G800–G806
103. Smith TK, Reed JB, Sanders KM (1987) Interaction of two electrical pacemakers in the muscularis of the canine proximal colon. Am J Physiol Cell Physiol 252(21):C290–C299
104. Smith TK, Reed JB, Sanders KM (1987) Origin and propagation of electrical slow waves in circular muscle of the canine proximal colon. Am J Physiol Cell Physiol 252(21):C215–C224
105. Hirst GD, Edwards FR (2006) Electrical events underlying organized myogenic contractions of the guinea pig stomach. J Physiol 576(Pt 3):659–665

106. Sanders KM, Koh SD, Ro S, Ward SM (2012) Regulation of gastrointestinal motility-insights from smooth muscle biology. Nat Rev Gastroenterol Hepatol 9:633–645
107. Bitar KN (2003) Function of gastrointestinal smooth muscle: from signaling to contractile proteins. Am J Med 115(Suppl 3A):15S–23S
108. Suzuki H, Hirst GDS (1999) Regenerative potentials evoked in circular smooth muscle of the antral region of guinea-pig stomach. J Physiol 517(Pt 2):563–573
109. Mustafa SM, Thulesius O (2001) Cooling-induced gastrointestinal smooth muscle contractions in the rat. Fundam Clin Pharmacol 15(5):349–354
110. Suzuki H (2000) Cellular mechanisms of myogenic activity in gastric smooth muscle. Jpn J Physiol 50(3):289–301
111. Edwards FR, Hirst GD (2005) An electrical description of the generation of slow waves in the antrum of the guinea-pig. J Physiol 564(Pt 1):213–232
112. Rhee PL, Lee JY, Son HJ, Kim JJ, Rhee JC, Kim S et al (2011) Analysis of pacemaker activity in the human stomach. J Physiol 589(Pt 24):6105–6118
113. Wang ZS, Elsenbruch S, Orr WC, Chen JDZ (2003) Detection of gastric slow wave uncoupling from multi-channel electrogastrogram: validations and applications. Neurogastroenterol Motil 15(5):457–465
114. Yokoyama S, Ozaki T (1990) Contractions of the longitudinal and circular muscle of the small intestine. Prog Clin Biol Res 327:483–492
115. Kachlik D, Baca V, Stingl J (2010) The spatial arrangement of the human large intestinal wall blood circulation. J Anat 216:335–343
116. Phillips RJ, Powley TL (2007) Innervation of the gastrointestinal tract: patterns of aging. Auton Neurosci 136(1–2):1–19
117. Bitar KN, Raghavan S, Zakhem E (2014) Tissue engineering in the gut: developments in neuromusculature. Gastroenterology 146(7):1614–1624
118. Rao JN, Wang JY (2010) Regulation of gastrointestinal mucosal growth. Morgan & Claypool Life Sciences, San Rafael
119. Webb CR (2003) Smooth muscle contraction and relaxation. Adv Physiol Educ 27(4): 201–206
120. Brozovich FV, Nicholson CJ, Degen CV, Gao YZ, Aggarwal M, Morgan KG (2016) Mechanisms of vascular smooth muscle contraction and the basis for pharmacologic treatment of smooth muscle disorders. Pharmacol Rev 68(2):476–532
121. Herrera AM, McParland BE, Bienkowska A, Tait R, Paré PD, Seow CY (2005) 'Sarcomeres' of smooth muscle: functional characteristics and ultrastructural evidence. J Cell Sci 118:2381–2392
122. Koeppen B, Stanton B (2010) Berne & Levy physiology, 6th edn. Mosby, St. Louis
123. Andersson K-E, Arner A (2004) Urinary bladder contraction and relaxation: physiology and pathophysiology. Physiol Rev 84(3):935–986
124. Fisher SA (2010) Vascular smooth muscle phenotypic diversity and function. Physiol Genomics 42A(3):169–187
125. Eddinger TJ, Meer DP (2007) Myosin II isoforms in smooth muscle: heterogeneity and function. Am J Physiol Cell Physiol 293(2):C493–C508
126. Dodge JT, Bevan RD, Bevan JA (1994) Comparison of density of sympathetic varicosities and their closeness to smooth muscle cells in rabbit middle cerebral and ear arteries and their branches. Circ Res 75:916–925
127. Bennett MR (1998) Transmission at sympathetic varicosities. Physiology 13(2):79–84
128. Huizinga JD, Liu LWC, Fitzpatrick A, White E, Gill S, Wang X-Y et al (2008) Deficiency of intramuscular ICC increases fundic muscle excitability but does not impede nitrergic innervation. Am J Physiol Gastrointest Liver Physiol 294(2):G589–G594
129. Krizsan-Agbas D, Zhang R, Marzban F, Smith PG (1998) Presynaptic adrenergic facilitation of parasympathetic neurotransmission in sympathectomized rat smooth muscle. J Physiol 512(3):841–849
130. Kuo K-H, Seow CY (2004) Contractile filament architecture and force transmission in swine airway smooth muscle. J Cell Sci 117:1503–1511

131. Diep HK, Vigmond EJ, Segal SS, Welsh DG (2005) Defining electrical communication in skeletal muscle resistance arteries: a computational approach. J Physiol 568(Pt 1):267–281
132. Song KS, Scherer PE, Tang Z, Okamoto T, Li S, Chafel M et al (1996) Expression of caveolin-3 in skeletal, cardiac, and smooth muscle cells. Caveolin-3 is a component of the sarcolemma and co-fractionates with dystrophin and dystrophin-associated glycoproteins. J Biol Chem 271(25):15160–15165
133. North AJ, Galazkiewicz B, Byers TJ, Glenney JR, Small JV (1993) Complementary distributions of vinculin and dystrophin define two distinct sarcolemma domains in smooth muscle. J Cell Biol 120(5):1159–1167
134. Aickin CC (1990) Chloride transport across the sarcolemma of vertebrate smooth and skeletal muscle. In: Alvarez-Leefmans FJ, Russell JM (eds) Chloride channels and carriers in nerve, muscle, and glial cells. Springer, New York
135. Kidwai AM (1985) Biochemistry of the sarcolemma. Subcell Biochem 11:181–193
136. de Groat WC, Wickens C (2013) Organization of the neural switching circuitry underlying reflex micturition. Acta Physiol (Oxf) 207(1):66–84
137. Zhu JX, Wu XY, Owyang C, Li Y (2001) Intestinal serotonin acts as a paracrine substance to mediate vagal signal transmission evoked by luminal factors in the rat. J Physiol 530 (3):431–442
138. Liu LW, Thuneberg L, Huizinga JD (1995) Cyclopiazonic acid, inhibiting the endoplasmic reticulum calcium pump, reduces the canine colonic pacemaker frequency. J Pharmacol Exp Ther 275(2):1058–1068
139. Wuytack F, Raeymaekers L, Verbist J, Jones LR, Casteels R (1987) Smooth-muscle endoplasmic reticulum contains a cardiac-like form of calsequestrin. Biochim Biophys Acta 899(2):151–158
140. Milner RE, Baksh S, Shemanko C, Carpenter MR, Smillie L, Vance JE et al (1991) Calreticulin, and not calsequestrin, is the major calcium binding protein of smooth muscle sarcoplasmic reticulum and liver endoplasmic reticulum. J Biol Chem 266(11):7155–7165
141. Volpe P, Martini A, Furlan S, Meldolesi J (1994) Calsequestrin is a component of smooth muscles: the skeletal- and cardiac-muscle isoforms are both present, although in highly variable amounts and ratios. Biochem J 301(Pt 2):465–469
142. Tang DD (2008) Intermediate filaments in smooth muscle. Am J Physiol Cell Physiol 294 (4):C869–C878
143. Matsuoka R, Yoshida MC, Furutani Y, Imamura S, Kanda N, Yanagisawa M et al (1993) Human smooth muscle myosin heavy chain gene mapped to chromosomal region 16q12. Am J Med Genet 46(1):61–67
144. Alkahtani R, Mahavadi S, Al-Shboul O, Alsharari S, Grider JR, Murthy KS (2013) Changes in the expression of smooth muscle contractile proteins in TNBS- and DSS-induced colitis in mice. Inflammation 36(6):1304–1315
145. Perrin BJ, Ervasti JM (2010) The actin gene family: function follows isoform. Cytoskeleton (Hoboken) 67(10):630–634
146. Löfgren M, Ekblad E, Morano I, Arner A (2003) Nonmuscle myosin motor of smooth muscle. J Gen Physiol 121(4):301–310
147. Xu JQ, Harder BA, Uman P, Craig R (1996) Myosin filament structure in vertebrate smooth muscle. J Cell Biol 134(1):53–66
148. Craig R, Woodhead JL (2006) Structure and function of myosin filaments. Curr Opin Struct Biol 16(2):204–212
149. Kim HR, Appel S, Vetterkind S, Gangopadhyay SS, Morgan KG (2008) Smooth muscle signalling pathways in health and disease. J Cell Mol Med 12(6a):2165–2180
150. Drew JS, Murphy RA (1997) Actin isoform expression, cellular heterogeneity, and contractile function in smooth muscle. Can J Physiol Pharmacol 75(7):869–877
151. Lehman W, Morgan KG (2012) Structure and dynamics of the actin-based smooth muscle contractile and cytoskeletal apparatus. J Muscle Res Cell Motil 33(6):461–469

152. Miwa T, Manabe Y, Kurokawa K, Kamada S, Kanda N, Bruns G et al (1991) Structure, chromosome location, and expression of the human smooth muscle (enteric type) gamma-actin gene: evolution of six human actin genes. Mol Cell Biol 11(6):3296–3306
153. Bond M, Somlyo AV (1982) Dense bodies and actin polarity in vertebrate smooth muscle. J Cell Biol 95(2 Pt 1):403–413
154. Peter AK, Cheng H, Ross RS, Knowlton KU, Chen J (2011) The costamere bridges sarcomeres to the sarcolemma in striated muscle. Prog Pediatr Cardiol 31(2):83–88
155. Bloch RJ, Reed P, O'Neill A, Strong J, Williams M, Porter N, González-Serratos H (2004) Costameres mediate force transduction in healthy skeletal muscle and are altered in muscular dystrophies. J Muscle Res Cell Motil 25(8):590–592
156. Jaka O, Casas-Fraile L, López de Munain A, Sáenz A (2015) Costamere proteins and their involvement in myopathic processes. Expert Rev Mol Med 17:e12
157. Knöll R, Buyandelger B, Lab M (2011) The sarcomeric Z-disc and Z-discopathies. J Biomed Biotechnol 2011:569628
158. Bloch RJ, Capetanaki Y, O'Neill A, Reed P, Williams MW, Resneck WG et al (2002) Costameres: repeating structures at the sarcolemma of skeletal muscle. Clin Orthop Relat Res 403:S203–S210
159. Walsh MP (1994) Calmodulin and the regulation of smooth muscle contraction. Mol Cell Biochem 135(1):21–41
160. Bernstein BW, Bamburg JR (2010) ADF/cofilin: a functional node in cell biology. Trends Cell Biol 20(4):187–195
161. Li X(E), Tobacman LS, Mun JY, Craig R, Fischer S, Lehman W (2011) Tropomyosin position on F-actin revealed by EM reconstruction and computational chemistry. Biophys J 100(4):1005–1013
162. Bing W, Razzaq A, Sparrow J, Marston S (1998) Tropomyosin and troponin regulation of wild type and E93K mutant actin filaments from drosophila flight muscle: charge reversal on actin changes actin-tropomyosin from on to off state. J Biol Chem 273(24):15016–15021
163. Barua B (2013) Periodicities designed in the tropomyosin sequence and structure define its functions. Bioarchitecture 3(3):51–56
164. Landis CA, Bobkova A, Homsher E, Tobacman LS (1997) The active state of the thin filament is destabilized by an internal deletion in tropomyosin. J Biol Chem 272:14051–14056
165. Rao JN, Rivera-Santiago R, Li XE, Lehman W, Dominguez R (2012) Structural analysis of smooth muscle tropomyosin α and β isoforms. J Biol Chem 287(5):3165–3174
166. Schmidt WM, Lehman W, Moore JR (2015) Direct observation of tropomyosin binding to actin filaments. Cytoskeleton (Hoboken) 72(6):292–303
167. Fischer S, Rynkiewicz MJ, Moore JR, Lehman W (2016) Tropomyosin diffusion over actin subunits facilitates thin filament assembly. Struct Dyn 3(1):012002
168. Moran CM, Garriock RJ, Miller MK, Heimark RL, Mudry RE, Gregorio CC, Krieg PA (2008) Expression of the fast twitch troponin complex, fTnT, fTnI and fTnC, in vascular smooth muscle. Cell Motil Cytoskelet 65(8):652–661
169. Gomes AV, Potter JD, Szczesna-Cordary D (2002) The role of troponins in muscle contraction. IUBMB Life 54:323–333
170. Galińska-Rakoczy A, Engel P, Xu C, Jung HS, Craig R, Tobacman LS, Lehman W (2008) Structural basis for the regulation of muscle contraction by troponin and tropomyosin. J Mol Biol 379(5):929–935
171. Matthew JD, Khromov AS, McDuffie MJ, Somlyo AV, Somlyo AP, Taniguchi S, Takahashi K (2000) Contractile properties and proteins of smooth muscles of a calponin knockout mouse. J Physiol 529(Pt 3):811–824
172. Chen M, Li AY (1994) Caldesmon, calponin and contractile regulation of the vertebrate smooth muscle. Sheng Li Ke Xue Jin Zhan 25(4):314–318
173. Gerthoffer WT, Pohl J (1994) Caldesmon and calponin phosphorylation in regulation of smooth muscle contraction. Can J Physiol Pharmacol 72(11):1410–1414

174. Ulrich C, Quillici DR, Schegg K, Woolsey R, Nordmeier A, Buxton ILO (2012) Uterine smooth muscle *S*-nitrosylproteome in pregnancy. Mol Pharmacol 81(2):143–153
175. Chokhavatia S, Anuras S (1991) Neuromuscular disease of the gastrointestinal tract. Am J Med Sci 301(3):201–214
176. Duan D (2006) Challenges and opportunities in dystrophin-deficient cardiomyopathy gene therapy. Hum Mol Genet 15(2):R253–R261
177. Harvey PA, Leinwand LA (2011) Cellular mechanisms of cardiomyopathy. J Cell Biol 194 (3):355–365
178. Ali F, Paré PD, Seow CY (2005) Models of contractile units and their assembly in smooth muscle. Can J Physiol Pharmacol 83(10):825–831
179. Auerbach D, Bantle S, Keller S, Hinderling V, Leu M, Ehler E, Perriard J-C (1999) Different domains of the M-band protein myomesin are involved in myosin binding and M-band targeting. Mol Biol Cell 10(5):1297–1308
180. Kargacin GJ, Cooke PH, Abramson SB, Fay FS (1989) Periodic organization of the contractile apparatus in smooth muscle revealed by the motion of dense bodies in single cells. J Cell Biol 108:1465–1475
181. Agarkova I, Perriard JC (2005) The M-band: an elastic web that crosslinks thick filaments in the center of the sarcomere. Trends Cell Biol 15(9):477–485
182. Maruyama K (1994) Connectin, an elastic protein of striated muscle. Biophys Chem 50 (1–2):73–85
183. Tskhovrebova L, Trinick J (2017) Titin and nebulin in thick and thin filament length regulation. Subcell Biochem 82:285–318
184. Granzier H, Labeit S (2007) Structure-function relations of the giant elastic protein titin in striated and smooth muscle cells. Muscle Nerve 36(6):740–755
185. Trinick J (1992) Understanding the functions of titin and nebulin. FEBS Lett 307(1):44–48
186. Kuo K-H, Wang L, Paré PD, Ford LE, Seow CY (2001) Myosin thick filament lability induced by mechanical strain in airway smooth muscle. J Appl Physiol 90(5):1811–1816
187. Heasley LE (2001) Autocrine and paracrine signaling through neuropeptide receptors in human cancer. Oncogene 20(13):1563–1569
188. Plaisancié P, Barcelo A, Moro F, Claustre J, Chayvialle J-A, Cuber J-C (1998) Effects of neurotransmitters, gut hormones, and inflammatory mediators on mucus discharge in rat colon. Am J Physiol Gastrointest Liver Physiol 275(5):G1073–G1084
189. Tognini P (2017) Gut microbiota: a potential regulator of neurodevelopment. Front Cell Neurosci 11:25
190. Kotlikoff ML, Wang YX, Xin HB, Ji G (2002) Calcium release by ryanodine receptors in smooth muscle. Novartis Found Symp 246(108–24):221–227
191. Somlyo AP, Somlyo AV (1994) Signal transduction and regulation in smooth muscle. Nature 372:231–236
192. Borysova L, Shabir S, Walsh MP, Burdyga T (2011) The importance of Rho-associated kinase-induced Ca^{2+} sensitization as a component of electromechanical and pharmacomechanical coupling in rat ureteric smooth muscle. Cell Calcium 50(4):393–405
193. Bonnevier J, Malmqvist U, Sonntag D, Schroeter M, Nilsson H, Pfitzer G, Arner A (2002) Sustained norepinephrine contraction in the rat portal vein is lost when Ca^{2+} is replaced with Sr^{2+}. Am J Physiol Cell Physiol 282(4):C845–C852
194. McCorry LK (2007) Physiology of the autonomic nervous system. Am J Pharm Educ 71(4):78
195. Browning KN, Travagli RA (2014) Central nervous system control of gastrointestinal motility and secretion and modulation of gastrointestinal functions. Compr Physiol 4 (4):1339–1368
196. Gershon MD (1967) Inhibition of gastrointestinal movement by sympathetic nerve stimulation: the site of action. J Physiol 189(2):317–327
197. Modlin IM, Champaneria MC, Bornschein J, Kidd M (2006) Evolution of the diffuse neuroendocrine system—clear cells and cloudy origins. Neuroendocrinology 84(2):69–82
198. Keith A (1915) A new theory of the causation of enterostasis. Lancet 2(4799):371–375

199. Maeda H, Yamagata A, Nishikawa S, Yoshinaga K, Kobayashi S, Nishi K, Nishikawa S (1992) Requirement of c-Kit for development of intestinal pacemaker system. Development 116:369–375
200. Rolle U, Piotrowska AP, Nemeth L, Puri P (2002) Altered distribution of interstitial cells of Cajal in Hirschsprung disease. Arch Pathol Lab Med 126:928–933
201. Huizinga JD, Thuneberg L, Klüppel M, Malysz J, Mikkelsen HB, Bernstein A (1995) W/kit gene required for interstitial cells of Cajal and for intestinal pacemaker activity. Nature 373:347–349
202. Chen Y, Shamu T, Chen H, Besmer P, Sawyers CL, Chi P (2011) Visualization of the interstitial cells of Cajal (ICC) network in mice. J Vis Exp 53:e2802
203. Al-Sajee D, Huizinga JD (2012) Interstitial cells of Cajal: pathology, injury and repair. Sultan Qaboos Univ Med J 12(4):411–421
204. Rühl A (2009) Enteric nervous system: glial cells and interstitial cells of Cajal. In: Squire LR (ed) Encyclopedia of neuroscience. Academic Press, Oxford
205. Vannucchi M-G, Traini C, Manetti M, Ibba-Manneschi L, Faussone-Pellegrini M-S (2013) Telocytes express PDGFRα in the human gastrointestinal tract. J Cell Mol Med 17(9):1099–1108
206. Aoyama M, Yamada A, Wang J, Ohya S, Furuzono S, Goto T et al (2004) Requirement of ryanodine receptors for pacemaker Ca^{2+} activity in ICC and HEK293 cells. J Cell Sci 117:2813–2825
207. Gregersen H, Christensen J (2000) Gastrointestinal tone. Neurogastroenterol Mot 12:501–508
208. Montgomery LEA, Tansey EA, Johnson CD, Roe SM, Quinn JG (2016) Autonomic modification of intestinal smooth muscle contractility. Adv Physiol Educ 40(1):104–109
209. Donnerer J, Liebmann I, Holzer-Petsche U (2014) Hexamethonium-induced augmentation of the electrical twitch response in the guinea-pig ileum longitudinal muscle-myenteric plexus strip. Neurosci Lett 577:34–37
210. Every-Palmer S, Lentle RG, Reynolds G, Hulls C, Chambers P, Dunn H, Ellis PM (2017) Spatiotemporal mapping techniques show clozapine impairs neurogenic and myogenic patterns of activity in the colon of the rabbit in a dose-dependent manner. Front Pharmacol 8:209
211. Costa M, Dodds KN, Wiklendt L, Spencer NJ, Brookes SJ, Dinning PG (2013) Neurogenic and myogenic motor activity in the colon of the guinea pig, mouse, rabbit, and rat. Am J Physiol Gastrointest Liver Physiol 305(10):G749–G759
212. Dinning PG, Costa M, Brookes SJ, Spencer NJ (2012) Neurogenic and myogenic motor patterns of rabbit proximal, mid, and distal colon. Am J Physiol Gastrointest Liver Physiol 303(1):G83–G92
213. Lentle RG, Reynolds GW, Janssen PW (2013) Gastrointestinal tone; its genesis and contribution to the physical processes of digestion. Neurogastroenterol Motil 25(12):931–942
214. Huxley H, Hanson J (1954) Changes in the cross-striations of muscle during contraction and stretch and their structural interpretation. Nature 173(4412):973–976
215. Huxley AF, Niedergerke R (1954) Structural changes in muscle during contraction: interference microscopy of living muscle fibres. Nature 173(4412):971–973
216. Horowits R, Podolsky RJ (1987) The positional stability of thick filaments in activated skeletal muscle depends on sarcomere length: evidence for the role of titin filaments. J Cell Biol 105(5):2217–2223
217. Petkov GV, Boev KK (1996) The role of sarcoplasmic reticulum and sarcoplasmic reticulum Ca^{2+}-ATPase in the smooth muscle tone of the cat gastric fundus. Pflugers Arch 431(6):928–935
218. Wray S, Burdyga T, Noble K (2005) Calcium signalling in smooth muscle. Cell Calcium 38 (3–4):397–407
219. Grasa L, Rebollar E, Arruebo MP, Plaza MA, Murillo MD (2004) The role of Ca^{2+} in the contractility of rabbit small intestine in vitro. J Physiol Pharmacol 55:639–650

220. Bolton TB (2006) Calcium events in smooth muscles and their interstitial cells; physiological roles of sparks. J Physiol 570(Pt 1):5–11
221. Smith JB (1996) Calcium homeostasis in smooth muscle cells. New Horiz 4(1):2–18
222. Guerrero-Hernández A, Gómez-Viquez L, Guerrero-Serna G, Rueda A (2002) Ryanodine receptors in smooth muscle. Front Biosci 7:d1676–d1688
223. Van Petegem F (2012) Ryanodine receptors: structure and function. J Biol Chem 287 (38):31624–31632
224. Meissner G (1986) Ryanodine activation and inhibition of the Ca^{2+} release channel of sarcoplasmic reticulum. J Biol Chem 261(14):6300–6306
225. Lesh RE, Nixon GF, Fleischer S, Airey JA, Somlyo AP, Somlyo AV (1998) Localization of ryanodine receptors in smooth muscle. Circ Res 82:175–185
226. Ugalde-Silva P, Gonzalez-Lugo O, Navarro-Garcia F (2016) Tight junction disruption induced by type 3 secretion system effectors injected by enteropathogenic and enterohemorrhagic *Escherichia coli*. Front Cell Infect Microbiol 6:87
227. He WQ, Qiao YN, Peng YJ, Zha JM, Zhang CH, Chen C, et al (2013) Altered contractile phenotypes of intestinal smooth muscle in mice deficient in myosin phosphatase target subunit 1. Gastroenterology 144(7):1456–1465
228. Abrams J, Davuluri G, Seiler C, Pack M (2012) Smooth muscle caldesmon modulates peristalsis in the wild type and non-innervated zebrafish intestine. Neurogastroenterol Motil 24(3):288–299
229. Murthy KS (2006) Signaling for contraction and relaxation in smooth muscle of the gut. Annu Rev Physiol 68:345–374
230. Jaaro H, Rubinfeld H, Hanoch T, Seger R (1997) Nuclear translocation of mitogen-activated protein kinase kinase (MEK1) in response to mitogenic stimulation. Proc Natl Acad Sci U S A 94(8):3742–3747
231. Schaeffer HJ, Weber MJ (1999) Mitogen-activated protein kinases: specific messages from ubiquitous messengers. Mol Cell Biol 19(4):2435–2444
232. Magnier C, Papp B, Corvazier E, Bredoux R, Wuytack F, Eggermont J et al (1992) Regulation of sarco-endoplasmic reticulum Ca(2+)-ATPases during platelet-derived growth factor-induced smooth muscle cell proliferation. J Biol Chem 267(22):15808–15815
233. Raeymaekers L, Jones LR (1986) Evidence for the presence of phospholamban in the endoplasmic reticulum of smooth muscle. Biochim Biophys Acta 882(2):258–265
234. Ferguson DG, Young EF, Raeymaekers L, Kranias EG (1988) Localization of phospholamban in smooth muscle using immunogold electron microscopy. J Cell Biol 107(2):555–562
235. Paul RJ (1998) The role of phospholamban and SERCA3 in regulation of smooth muscle-endothelial cell signalling mechanisms: evidence from gene-ablated mice. Acta Physiol Scand 164(4):589–597
236. Chen W, Lah M, Robinson PJ, Kemp BE (1994) Phosphorylation of phospholamban in aortic smooth muscle cells and heart by calcium/calmodulin-dependent protein kinase II. Cell Signal 6(6):617–630
237. Eto M, Ohmori T, Suzuki M, Furuya K, Morita F (1995) A novel protein phosphatase-1 inhibitory protein potentiated by protein kinase C. Isolation from porcine aorta media and characterization. J Biochem 118(6):1104–1107
238. Perrino BA (2016) Calcium sensitization mechanisms in gastrointestinal smooth muscles. J Neurogastroenterol Motil 22(2):213–225
239. Zhang Y, Hermanson ME, Eddinger TJ (2013) Tonic and phasic smooth muscle contraction is not regulated by the PKCα—CPI-17 pathway in swine stomach antrum and fundus. PLoS ONE 8(9):e74608
240. Mizuno Y, Isotani E, Huang J, Ding H, Stull JT, Kamm KE (2008) Myosin light chain kinase activation and calcium sensitization in smooth muscle in vivo. Am J Physiol Cell Physiol 295(2):C358–C364

241. Uehata M, Ishizaki T, Satoh H, Ono T, Kawahara T, Morishita T et al (1997) Calcium sensitization of smooth muscle mediated by a Rho-associated protein kinase in hypertension. Nature 389(6654):990–994
242. Hitchcock-DeGregori SE, Irving TC (2014) Hugh E. Huxley: the compleat biophysicist. Biophys J 107(7):1493–1501
243. Huxley HE (2008) Memories of early work on muscle contraction and regulation in the 1950's and 1960's. Biochem Biophys Res Commun 369:34–42
244. Maruyama K (1995) Birth of the sliding filament concept in muscle contraction. J Biochem 117(1):1–6
245. Squire JM (2016) Muscle contraction: sliding filament history, sarcomere dynamics and the two Huxleys. Global Cardiol Sci Pract 11:1–23
246. Rall JA (2014) Mechanism of muscular contraction, perspectives in physiology. Birth of the sliding filament model of muscular contraction: proposal. Springer, New York
247. Huxley HE (2004) Fifty years of muscle and the sliding filament hypothesis. Eur J Biochem 271:1403–1415
248. Haeberle JR (1994) Calponin decreases the rate of cross-bridge cycling and increases maximum force production by smooth muscle myosin in an in vitro motility assay. J Biol Chem 269(17):12424–12431
249. Arner A, Malmqvist U (1998) Cross-bridge cycling in smooth muscle: a short review. Acta Physiol Scand 164(4):363–372
250. Preller M, Holmes KC (2013) The myosin start-of-power stroke state and how actin binding drives the power stroke. Cytoskeleton (Hoboken) 70(10):651–660
251. Lodish H, Berk A, Zipursky SL, Matsudaira P, Baltimore D, Darnell J (2000) Molecular cell biology, 4th edn. WH Freeman, New York
252. Marcucci L, Truskinovsky L (2010) Mechanics of the power stroke in myosin II. Phys Rev E Stat Nonlin Soft Matter Phys 81(5 Pt 1):051915
253. Chen B, Gao H (2011) Motor force homeostasis in skeletal muscle contraction. Biophys J 101(2):396–403
254. Tyska MJ, Dupuis DE, Guilford WH, Patlak JB, Waller GS, Trybus KM et al (1999) Two heads of myosin are better than one for generating force and motion. Proc Natl Acad Sci U S A 96(8):4402–4407
255. Greenberg MJ, Lin T, Shuman H, Ostap EM (2015) Mechanochemical tuning of myosin-I by the N-terminal region. PNAS 112(26):E3337–E3344
256. Hai C-M, Kim HR (2005) An expanded latch-bridge model of protein kinase C-mediated smooth muscle contraction. J Appl Physiol 98:1356–1365
257. Han S, Speich JE, Eddinger TJ, Berg KM, Miner AS, Call C, Ratz PH (2006) Evidence for absence of latch-bridge formation in muscular saphenous arteries. Am J Physiol Heart Circ Physiol 291(1):H138–H146
258. Stull JT, Gallagher PJ, Herring BP, Kamm KE (1991) Vascular smooth muscle contractile elements. Hypertension 17(6 Pt 1):723–732
259. Murphy RA, Rembold CM (2005) The latch-bridge hypothesis of smooth muscle contraction. Can J Physiol Pharmacol 83(10):857–864
260. Harris DE, Warshaw DM (1991) Length vs. active force relationship in single isolated smooth muscle cells. Am J Physiol 260(5 Pt 1):C1104–C1112
261. Stephens NL, Van Niekerk W (1977) Isometric and isotonic contractions in airway smooth muscle. Can J Physiol Pharmacol 55(4):833–838
262. Oishi K, Itoh Y, Isshiki Y, Kai C, Takeda Y, Yamaura K et al (2000) Agonist-induced isometric contraction of smooth muscle cell-populated collagen gel fiber. Am J Physiol Cell Physiol 279(5):C1432–C1442
263. Faulkner JA (1985) Terminology for contractions of muscles during shortening, while isometric, and during lengthening. J Appl Physiol 95(2):455–459
264. Faulkner JA (2003) Terminology for contractions of muscles during shortening, while isometric, and during lengthening. J Appl Physiol 95(2):455–459

265. Colliander EB, Tesch PA (1990) Effects of eccentric and concentric muscle actions in resistance training. Acta Physiol Scand 140(1):31–39
266. Alfredson H, Pietilä T, Jonsson P, Lorentzon R (1998) Heavy-load eccentric calf muscle training for the treatment of chronic Achilles tendinosis. Am J Sports Med 26(3):360–366
267. Satyendra L, Byl N (2006) Effectiveness of physical therapy for Achilles tendinopathy: an evidence based review of eccentric exercises. Isokinet Exerc Sci 14(1):71–80
268. Jonathan MN, Nathan C, Chris H, Van Dusseldorp T, Kravitz L, Kerksick CM (2017) The effects of eccentric contraction duration on muscle strength, power production, vertical jump, and soreness. J Strength Cond Res 31(3):773–786
269. Behrens M, Mau-Moeller A, Weippert M, Fuhrmann J, Wegner K, Skripitz R et al (2015) Caffeine-induced increase in voluntary activation and strength of the quadriceps muscle during isometric, concentric and eccentric contractions. Sci Rep 5:10209
270. Ulrich G, Parstorfer M (2016) Effects of plyometric vs. concentric and eccentric conditioning contractions on upper body postactivation potentiation. Int J Sports Physiol Perform 13:1–21
271. Duthie HL, Kirk D (1978) Electrical activity of human colonic smooth muscle in vitro. J Physiol 283:319–330
272. Yensen C, Matar W, Renaud J-M (2002) K^+-induced twitch potentiation is not due to longer action potential. Am J Physiol Cell Physiol 283(1):C169–C177
273. Petit J, Filippi GM, Gioux M, Hunt CC, Laporte Y (1990) Effects of tetanic contraction of motor units of similar type on the initial stiffness to ramp stretch of the cat peroneus longus muscle. J Neurophysiol 64(6):1724–1732
274. Renaud JM, Comtois A (1994) The effect of K^+ on the recovery of the twitch and tetanic force following fatigue in the sartorius muscle of the frog, Rana pipiens. J Muscle Res Cell Motil 15(4):420–431
275. Alipour-Haghighi F, Titze IR, Perlman AL (1989) Tetanic contraction in vocal fold muscle. J Speech Hear Res 32(2):226–231
276. Hood DA, Gorski J, Terjung RL (1986) Oxygen cost of twitch and tetanic isometric contractions of rat skeletal muscle. Am J Physiol 250(4 Pt 1):E449–E456
277. Zemlickova E, Johannes F-J, Aitken A, Dubois T (2004) Association of CPI-17 with protein kinase C and casein kinase I. Biochem Biophys Res Commun 316(1):39–47
278. Rattan S (2017) Ca^{2+}/calmodulin/MLCK pathway initiates, and RhoA/ROCK maintains, the internal anal sphincter smooth muscle tone. Am J Physiol Gastrointest Liver Physiol 312(1): G63–G66
279. Ringvold HC, Khalil RA (2017) Protein kinase C as regulator of vascular smooth muscle function and potential target in vascular disorders. Adv Pharmacol 78:203–301
280. Huang J, Mahavadi S, Sriwai W, Hu W, Murthy KS (2006) Gi-coupled receptors mediate phosphorylation of CPI-17 and MLC20 via preferential activation of the PI3K/ILK pathway. Biochem J 396(1):193–200
281. Luksha L, Agewall S, Kublickiene K (2009) Endothelium-derived hyperpolarizing factor in vascular physiology and cardiovascular disease. Atherosclerosis 202(2):330–344
282. Ozkor MA, Quyyumi AA (2011) Endothelium-derived hyperpolarizing factor and vascular function. Cardiol Res Pract 2011:156146
283. Williams SP, Dorn GW 2nd, Rapoport RM (1994) Prostaglandin I2 mediates contraction and relaxation of vascular smooth muscle. Am J Physiol 267(2 Pt 2):H796–H803
284. Roger EM, Keir PG (2013) The motor unit and muscle action. In: Kandel ER, Schwartz JH, Jessell TM, Siegelbaum SA, Hudspeth AJ (eds) Principles of neural science, 5th ed. McGraw-Hill Medical, New York
285. Kaya RD, Hoffman RL, Clark BC (2014) Reliability of a Modified Motor Unit Number Index (MUNIX) technique. J Electromyogr Kinesiol 24(1):18–24
286. Monti RJ, Roy RR, Edgerton VR (2001) Role of motor unit structure in defining function. Muscle Nerve 24(7):848–866
287. Song NN, Xu WX (2016) Physiological and pathophysiological meanings of gastrointestinal smooth muscle motor unit SIP syncytium. Sheng Li Xue Bao 68(5):621–627

288. Koh SD, Ward SM, Sanders KM (2012) Ionic conductances regulating the excitability of colonic smooth muscles. Neurogastroenterol Motil 24:705–718
289. Bogaerts M, Deggoujf N, Huart C, Hupin C, Laureyns G, Lemkens P et al (2012) Physiology of the mouth and pharynx, Waldeyer's ring, taste and smell. B-ENT 8(19):13–20
290. Donner MW, Bosnia JF, Robertson DL (1985) Anatomy and physiology of the pharynx. Gastrointest Radiol 10(1):197–212
291. Salsench J, Martínez-Gomis J, Torrent J, Bizar J, Samsó J, Peraire M (2005) Relationship between duration of unilateral masticatory cycles and the type of lateral dental guidance: a preliminary study. Int J Prosthodont 18(4):339–346
292. Palmer JB, Rudin NJ, Lara G, Crompton AW (1992) Coordination of mastication and swallowing. Dysphagia 7(4):187–200
293. Gerstner G, Madhavan S, Crane E (2011) Mammalian oral rhythms and motor control. In: Klika V (ed) Biomechanics in applications. InTech, Croatia
294. Hedegård B, Lundberg M, Wictorin L (1970) Masticatory function—a cineradiographic study. IV. Duration of the masticatory cycle. Acta Odontol Scand 28(6):859–865
295. Bates JF, Stafford GD, Harrison A (1975) Masticatory function—a review of the literature. 1. The form of the masticatory cycle. J Oral Rehabil 2(3):281–301
296. Langenbach GEJ, Zhang F, Herring SW, Hannam AG (2002) Modelling the masticatory biomechanics of a pig. J Anat 201(5):383–393
297. Rowlands JA (1984) Videofluorography: the role of temporal averaging. Med Phys 11 (2):129–136
298. Martin-Harris B, Jones B (2008) The videofluorographic swallowing study. Phys Med Rehabil Clin N Am 19(4):769–785
299. Shigezumi M, Shibata T, Arisue M, Hattori M, Ishijima T, Hirai T (2000) A case of habitual temporomandibular dislocation treated with eminoplasty using T-shaped titanium miniplate and its evaluation of mandibular movement. J Jpn Soc TMJ 12(1):62–67
300. Gedrange T, Walter B, Tetzlaff I, Kasper M, Schubert H, Harzer W, Bauer R (2003) Regional alterations in fiber type distribution, capillary density, and blood flow after lower jaw sagittal advancement in pig masticatory muscles. J Dent Res 82(7):570–574
301. Landes CA, Sterz M (2003) Evaluation of condylar translation by sonography versus axiography in orthognathic surgery patients. J Oral Maxillofac Surg 61(12):1410–1417
302. Minami I, Oogai K, Nemoto T, Nakamura T, Igarashi Y, Wakabayashi N (2010) Measurement of jerk-cost using a triaxial piezoelectric accelerometer for the evaluation of jaw movement smoothness. J Oral Rehabil 37(8):590–595
303. Chambers ES, Bridge MW, Jones DA (2009) Carbohydrate sensing in the human mouth: effects on exercise performance and brain activity. J Physiol 587(8):1779–1794
304. Steele CM, Miller AJ (2010) Sensory input pathways and mechanisms in swallowing: a review. Dysphagia 25(4):323–333
305. Rudenga K, Green B, Nachtigal D, Small DM (2010) Evidence for an integrated oral sensory module in the human anterior ventral insula. Chem Senses 35(8):693–703
306. Bu'Lock F, Woolridge MW, Bairn JD (1990) Development of coordination of sucking, swallowing, and breathing: ultrasound study of term and preterm infants. Dev Med Child Neurol 32:669–678
307. Gewolb IH, Vice FL, Schweitzer-Kenney EL, Taciak VL, Bosma JF (2001) Developmental patterns of rhythmic suck and swallow in preterm infants. Dev Med Child Neurol 43:22–27
308. Matsuo K, Palmer JB (2008) Anatomy and physiology of feeding and swallowing—normal and abnormal. Phys Med Rehabil Clin N Am 19(4):691–707
309. Cassiani RA, Santos CM, Parreira LC, Dantas RO (2011) The relationship between the oral and pharyngeal phases of swallowing. Clinics (Sao Paulo) 66(8):1385–1388
310. Jaffer NM, Edmund NG, Au FW-F, Steele CM (2015) Fluoroscopic evaluation of oro-pharyngeal dysphagia: anatomy, technique, and common etiologies. AJR Am J Roentgenol 204(1):49–58

311. Robbins JA, Bridges AD, Taylor A (2006) Oral, pharyngeal and esophageal motor function in aging. PART 1 Oral cavity, pharynx and esophagus. GI Motil. Online https://doi.org/10.1038/gimo39
312. Soares TJ, Moraes DP, de Medeiros GC, Sassi FC, Zilberstein B, de Andrade CR (2015) Oral transit time: a critical review of the literature. Arq Bras Cir Dig 28(2):144–147
313. Thompson TZ, Obeidin F, Davidoff AA, Hightower CL, Johnson CZ, Rice SL, Sokolove R-L, Taylor BK, Tuck JM, Pearson WG Jr (2014) Coordinate mapping of hyolaryngeal mechanics in swallowing. J Vis Exp 87:51476
314. Hendrix TR (1993) Coordination of peristalsis in pharynx and esophagus. Dysphagia 8 (2):74–78
315. Bredenoord AJ (2015) Minor disorders of esophageal peristalsis: highly prevalent, minimally relevant? Clin Gastroenterol Hepatol 13(8):1424–1425
316. Fleshler B, Hendrix TR, Kramer P, Ingelfinger FJ (1959) The characteristics and similarity of primary and secondary peristalsis in the esophagus. J Clin Invest 38(1 Pt 1-2):110–116
317. Liao D, Krarup AL, Lundager FH, Drewes AM, Gregersen H (2014) Quantitative differences between primary and secondary peristaltic contractions of the esophagus. Dig Dis Sci 59 (8):1810–1816
318. Christensen J (1997) Mechanisms of secondary esophageal peristalsis. Am J Med 103 (5):44S–46S
319. Schoeman MN, Holloway RH (1995) Integrity and characteristics of secondary oesophageal peristalsis in patients with gastro-oesophageal reflux disease. Gut 36(4):499–504
320. Bardan E, Xie P, Aslam M, Kern M, Shaker R (2000) Disruption of primary and secondary esophageal peristalsis by afferent stimulation. Am J Physiol Gastrointest Liver Physiol 279 (2):G255–G261
321. Horgan PG (2005) Supportive and treatment measures for patients with postoperative ileus. Semin Colon Rectal Surg 16(4):211–214
322. Hübner M, Scott M, Champagne B (2015) Postoperative ileus: prevention and treatment. In: Feldman LS, Delaney CP, Ljungqvist O, Carli F (eds) The SAGES/ERAS® society manual of enhanced recovery programs for gastrointestinal surgery. Springer International Publishing, Cham
323. Ballantyne GH (1984) The meaning of ileus. Its changing definition over three millennia. Am J Surg 148(2):252–256
324. Choi PW (2017) Strangulated small bowel obstruction caused by broad ligament hernia: report of a case and review of literature. Am J Med Case Rep 5(2):38–40
325. Takahashi R, Akagi Y, Tanaka T, Kaibara A, Kajiwara S, Shima I et al (2014) Clinicopathological evaluation of anoxic mucosal injury in strangulation ileus. BMC Surg 14:79
326. Kimura H, Earashi M, Konishi K, Tsuneda A, Tazawa K, Inoue T et al (1996) Strangulation ileus resulting from encasement of a loop of the small intestine by the great omentum, caused by abnormal adhesion. J Gastroenterol 31(5):714–716
327. Athanasakis E, Kafetzakis A, Hatzakis G, Daskalogiannaki M, Chalkiadakis G, Georgopoulos D (2002) Internal herniation and strangulation of the ileus through a post-traumatic defect of the mesentery in an intensive care unit patient. A case report. Intensive Care Med 28(4):523
328. Katis PG, Dias SM (2004) Volvulus: a rare twist on small-bowel obstruction. CMAJ 171 (7):728
329. Atamanalp SS (2013) Sigmoid volvulus: diagnosis in 938 patients over 45.5 years. Tech Coloproctol 17(4):419–424
330. Meyers MA (1988) Internal abdominal hernias. In: Dynamic radiology of the abdomen. Springer, New York
331. Ueda J, Yoshida H, Makino H, Yokoyama T, Maruyama H, Hirakata A et al (2012) Transmesocolic hernia of the ascending colon with intestinal obstruction. Case Rep Gastroenterol 6(2):344–349

332. Matsuo K, Palmer JB (2009) Coordination of mastication, swallowing and breathing. Jpn Dent Sci Rev 45(1):31–40
333. Martin-Harris B (2006) Coordination of respiration and swallowing. GI Motil. Online https://doi.org/10.1038/gimo10
334. Wilson SL, Thach BT, Brouillette RT, Abu-Osba YK (1981) Coordination of breathing and swallowing in human infants. J Appl Physiol Respir Environ Exerc Physiol 50(4):851–858
335. Erasmus CE, van Hulst K, Rotteveel JJ, Willemsen MAAP, Jongerius PH (2012) Clinical practice—swallowing problems in cerebral palsy. Eur J Pediatr 171(3):409–414
336. Nishino T (2012) The swallowing reflex and its significance as an airway defensive reflex. Front Physiol 3:489
337. Ertekin C, Aydogdu I (2003) Neurophysiology of swallowing. Clin Neurophysiol 114 (12):2226–2244
338. Roman C (1986) Neural control of deglutition and esophageal motility in mammals. J Physiol (Paris) 81(2):118–131
339. Ludlow CL (2015) Central nervous system control of voice and swallowing. J Clin Neurophysiol 32(4):294–303
340. Meltzer SJ (1899) On the causes of the orderly progress of the peristaltic movement in the esophagus. Am J Physiol 2:266–272
341. Meltzer SJ (1907) Deglutition through an esophagus partly deprived of its muscularis, with demonstration. Proc Soc Exp Biol Med 4:40–43
342. Lang IM (2009) Brain stem control of the phases of swallowing. Dysphagia 24(3):333–348
343. Amri M, Car A, Jean A (1984) Medullary control of the pontine swallowing neurones in sheep. Exp Brain Res 55(1):105–110
344. Jean A (2001) Brain stem control of swallowing: neuronal network and cellular mechanisms. Physiol Rev 81(2):929–969
345. Car A, Jean A, Roman C (1998) Deglutition: physiologic and neurophysiologic aspects. Rev Laryngol Otol Rhinol (Bord) 119(4):219–225
346. Enomoto S, Schwartz G, Lund JP (1987) The effects of cortical ablation on mastication in the rabbit. Neurosci Lett 82(2):162–166
347. Dantas RO, Kern MK, Massey BT, Dodds WJ, Kahrilas PJ, Brasseur JG et al (1990) Effect of swallowed bolus variables on oral and pharyngeal phases of swallowing. Am J Physiol 258(5 Pt 1):G675–G681
348. Conklin JL (2013) Evaluation of esophageal motor function with high-resolution manometry. J Neurogastroenterol Motil 19(3):281–294
349. Cook IJ (2008) Diagnostic evaluation of dysphagia. Nat Clin Pract Gastroenterol Hepatol 5:393–403
350. Khan A, Carmona R, Traube M (2014) Dysphagia in the elderly. Clin Geriatr Med 30(1):43–53
351. Baijens LWJ, Speyer R (2009) Effects of therapy for dysphagia in Parkinson's disease: systematic review. Dysphagia 24(1):91–102
352. Restivo DA, Casabona A, Nicotra A, Zappia M, Elia M, Romano MC et al (2013) ALS dysphagia pathophysiology—differential botulinum toxin response. Neurology 80(7):616–620
353. King SN, Dunlap NE, Tennant PA, Pitts T (2016) Pathophysiology of radiation-induced dysphagia in head and neck cancer. Dysphagia 31(3):339–351
354. Marshall JB (1985) Dysphagia. Pathophysiology, causes, and evaluation. Postgrad Med 77 (5):58–63, 67–68
355. Kuo P, Holloway RH, Nguyen NQ (2012) Current and future techniques in the evaluation of dysphagia. J Gastroenterol Hepatol 27(5):873–881
356. Richter JE (2010) Achalasia—an update. J Neurogastroenterol Motil 16(3):232–242
357. Zendehdel K, Nyrén O, Edberg A, Ye W (2011) Risk of esophageal adenocarcinoma in achalasia patients, a retrospective cohort study in Sweden. Am J Gastroenterol 106:57–61
358. Perkin GD, Murray-Lyon I (1998) Neurology and the gastrointestinal system. J Neurol Neurosurg Psychiatry 65:291–300

359. Farrokhi F, Vaezi MF (2007) Idiopathic (primary) achalasia. Orphanet J Rare Dis 2:38
360. Agrusa A, Romano G, Frazzetta G, De Vita G, Chianetta D, Di Buono G et al (2014) Achalasia secondary to submucosal invasion by poorly differentiated adenocarcinoma of the cardia, Siewert II: consideration on preoperative workup. Case Rep Surg 2014:654917
361. Stavropoulos SN, Friedel D, Modayil R, Parkman HP (2016) Diagnosis and management of esophageal achalasia. BMJ 354:i2785
362. Swaney JM, Smith YM, Sachai W (2016) Primary achalasia: practice implications. JNP 12 (7):473–478
363. Pandolfino JE, Gawron AJ (2015) Achalasia—a systematic review. JAMA 313(18):1841–1852
364. Kaths MJ, Foltys DB, Scheuermann U, Strempel M, Niebisch S, Ebert M et al (2015) Achalasia with megaesophagus and tracheal compression in a young patient: a case report. Int J Surg Case Rep 14:16–18
365. Soybel DI (2005) Anatomy and physiology of the stomach. Surg Clin 85(5):875–894
366. Daniels IR, Allum WH (2005) The anatomy and physiology of the stomach. In: Fielding JWL, Hallissey MT (eds) Upper gastrointestinal surgery. Springer, London
367. Waseem S, Moshiree B, Draganov PV (2009) Gastroparesis: current diagnostic challenges and management considerations. World J Gastroenterol 15(1):25–37
368. Wilmer A, Van Cutsem E, Andrioli A, Tack J, Coremans G, Janssens J (1998) Ambulatory gastrojejunal manometry in severe motility-like dyspepsia: lack of correlation between dysmotility, symptoms, and gastric emptying. Gut 42:235–242
369. Thumshirn M (2002) Pathophysiology of functional dyspepsia. Gut 51(Suppl I):i63–i66
370. Shoji T, Fukudo S, Nomura T, Satake M, Endo Y, Karahashi K et al (2001) Gastric hyposensitivity and large gastric capacity in patients with bulimia nervosa, but not in patients with anorexia nervosa. Gastroenterology 120(5):A462
371. Ziessman HA, O'Malley JP, Thrall JH (2014) Gastrointestinal system. In: Thrall J (ed) Nuclear medicine: the requisites, 4th edn. Saunders, Philadelphia
372. O'Grady G, Wang TH-H, Du P, Angeli T, Lammers WJEP, Cheng LK (2014) Recent progress in gastric arrhythmia: pathophysiology, clinical significance and future horizons. Exp Pharmacol Physiol 41(10):854–862
373. Koch KL (2014) Gastric dysrhythmias: a potential objective measure of nausea. Exp Brain Res 232(8):2553–2561
374. Mcnearney TA, Sallam HS, Hunnicutt SE, Doshi D, Wollaston DE, Mayes MD, Chen JDZ (2009) Gastric slow waves, gastrointestinal symptoms and peptides in systemic sclerosis patients. Neurogastroenterol Motil 21(12):1269
375. Ray K (2012) Motility: mapping gastric dysrhythmias in gastroparesis—a slow wave of electrical activity. Nat Rev Gastroenterol Hepatol 9:363
376. Koch KL (2003) Diagnosis and treatment of neuromuscular disorders of the stomach. Curr Gastroenterol Rep 5(4):323–330
377. Angeli TR, Du P, Midgley D, Paskaranandavadivel N, Sathar S, Lahr C et al (2016) Acute slow wave responses to high-frequency gastric electrical stimulation in patients with gastroparesis defined by high-resolution mapping. Neuromodulation 19(8):864–871
378. Scorza K, Williams A, Phillips JD, Shaw J (2007) Evaluation of nausea and vomiting. Am Fam Physician 76(1):76–84
379. Zhao J-B, Chen P-M, Gregersen H (2013) Changes of phasic and tonic smooth muscle function of jejunum in type 2 diabetic Goto-Kakizaki rats. World J Diabetes 4(6):339–348
380. Sanders KM, Berry RG (1985) Effects of ethyl alcohol on phasic and tonic contractions of the proximal stomach. J Pharmacol Exp Ther 235(3):858–863
381. Thornbury KD (1999) Tonic and phasic activity in smooth muscle. Ir J Med Sci 168(3):201–207
382. Golenhofen K, Mandrek K (1991) Phasic and tonic contraction processes in the gastrointestinal tract. Dig Dis 9(6):341–346
383. Sanjeevi A (2007) Gastric motility. Curr Opin Gastroenterol 23(6):625–630

384. Bredenoord AJ, Chial HJ, Camilleri M, Mullan BP, Murray JA (2003) Gastric accommodation and emptying in evaluation of patients with upper gastrointestinal symptoms. Clin Gastroenterol Hepatol 1(4):264–272
385. Schwizer W, Steingötter A, Fox M, Zur T, Thumshirn M, Bösiger P, Fried M (2002) Non-invasive measurement of gastric accommodation in humans. Gut 51(Suppl I):i59–i62
386. Xiao A, Wang H, Lu X, Zhu J, Huang D, Xu T et al (2015) H_2S, a novel gasotransmitter, involves in gastric accommodation. Sci Rep 5:16086
387. Talley NJ (2011) Rumination syndrome. Gastroenterol Hepatol (NY) 7(2):117–118
388. Brun R, Kuo B (2010) Functional dyspepsia. Therap Adv Gastroenterol 3(3):145–164
389. Lacy BE, Everhart K, Crowell MD (2011) Functional dyspepsia is associated with sleep disorders. Clin Gastroenterol Hepatol 9(5):410–414
390. Talley NJ, Ford AC (2015) Functional dyspepsia. N Engl J Med 373:1853–1863
391. Papadopoulos V, Mimidis K (2007) The rumination syndrome in adults: a review of the pathophysiology, diagnosis and treatment. J Postgrad Med 53:203–206
392. Cunningham CL, Banez GA (2006) Rumination and cyclic vomiting syndrome - biopsychosocial assessment and treatment. Pediatric gastrointestinal disorders. Springer, Boston
393. Indrio F, Riezzo G, Raimondi F, Cavallo L, Francavilla R (2009) Regurgitation in healthy and non healthy infants. Ital J Pediatr 35:39
394. Arts-Rodas D, Benoit D (1998) Feeding problems in infancy and early childhood: identification and management. Paediatr Child Health 3(1):21–27
395. Başal Y, Akcan AB, Polat YD, Günel C, Eryilmaz A, Başak S (2016) Rarely seen nasal congenital problems causing neonatal upper respiratory obstruction: a case series. Pediatr Rep 8(1):6456
396. Scott M, Gelhot AR (1999) Gastroesophageal reflux disease: diagnosis. Am Fam Physician 59(5):1161–1169
397. Bredenoord AJ, Pandolfino JE, Smout AJPM (2013) Gastro-oesophageal reflux disease. Lancet 381(9881):1933–1942
398. Tesfaye S, Boulton AJM, Dyck PJ, Freeman R, Horowitz M, Kempler P et al (2010) Diabetic neuropathies: update on definitions, diagnostic criteria, estimation of severity, and treatments. Diabetes Care 33:2285–2293
399. Vinik AI (1999) Diabetic neuropathy: pathogenesis and therapy. Am J Med 107(2):17–26
400. Kanatsuka A, Osegawa M, An T, Suzuki T, Hashimoto N, Makino H (1984) Augmented gastrin responses in diabetic patients with vagal neuropathy. Diabetologia 26(6):449–452
401. Szarka LA, Camilleri M (2009) Methods for measurement of gastric motility. Am J Physiol Gastrointest Liver Physiol 296(3):G461–G475
402. Yokrattanasak J, De Gaetano A, Panunzi S, Satiracoo P, Lawton WM, Lenbury Y (2016) A simple, realistic stochastic model of gastric emptying. PLoS ONE 11(4):e0153297
403. Park MI, Camilleri M (2005) Gastric motor and sensory functions in obesity. Obes Res 13 (3):491–500
404. Thumshirn M, Camilleri M, Choi MG, Zinsmeister AR (1999) Modulation of gastric sensory and motor functions by nitrergic and α2-adrenergic agents in humans. Gastroenterology 116 (3):573–585
405. Wisén O, Hellström PM (1995) Gastrointestinal motility in obesity. J Intern Med 237 (4):411–418
406. Maev IV, Samsonov AA, Vorob'ev LP, Salova LM, Zandanov AO, Dronova OB (2000) Motor and secretion function of stomach and duodenum, duodenogastric reflux in patients with duodenal ulcer. Klin Med (Mosk) 78(6):39–42
407. Drummond JC (1995) Elastance versus compliance. Anesthesiology 82(5):1309–1310
408. Römer M, Painsipp E, Schwetz I, Holzer P (2005) Facilitation of gastric compliance and cardiovascular reaction by repeated isobaric distension of the rat stomach. Neurogastroenterol Motil 17(3):399–409
409. Parkman HP, Jones MP (2009) Tests of gastric neuromuscular function. Gastroenterology 136(5):1526–1543

410. De Schepper HU, Cremonini F, Chitkara D, Camilleri M (2004) Assessment of gastric accommodation: overview and evaluation of current methods. Neurogastroenterol Motil 16 (3):275–285
411. Penagini R, Hebbard G, Horowitz M, Dent J, Bermingham H, Jones K, Holloway RH (1998) Motor function of the proximal stomach and visceral perception in gastro-oesophageal reflux disease. Gut 42(2):251–257
412. Tack J (2009) Gastric motor and sensory function. Curr Opin Gastroenterol 25(6):557–565
413. Tack J, Janssen P (2010) Gastroduodenal motility. Curr Opin Gastroenterol 26(6):647–655
414. Smith PM, Ferguson AV (2008) Neurophysiology of hunger and satiety. Dev Disabil Res Rev 14(2):96–104
415. Mattes RD (2010) Hunger and thirst: issues in measurement and prediction of eating and drinking. Physiol Behav 100(1):22–32
416. Sanger GJ, Hellström PM, Näslund E (2010) The hungry stomach: physiology, disease, and drug development opportunities. Front Pharmacol 1:145
417. Mathes WF, Brownley KA, Mo X, Bulik CM (2009) The biology of binge eating. Appetite 52(3):545–553
418. Umberg EN, Shader RI, Hsu LK, Greenblatt DJ (2012) From disordered eating to addiction: the "food drug" in bulimia nervosa. J Clin Psychopharmacol 32(3):376–389
419. Cronin RE (1987) Psychogenic polydipsia with hyponatremia: report of eleven cases. Am J Kidney Dis 9(5):410–416
420. McKiernan F, Hollis JH, McCabe GP, Mattes RD (2009) Thirst-drinking, hunger-eating; tight coupling? J Am Diet Assoc 109(3):486–490
421. Ciampolini M, Lovell-Smith HD, Kenealy T, Bianchi R (2013) Hunger can be taught: hunger recognition regulates eating and improves energy balance. Int J Gen Med 6:465–478
422. Laplace JP (1984) Motility of the small intestine: organization, regulation and functions. 15 years' research on migrating complexes. Reprod Nutr Dev 24(5B):707–765
423. Gwynne RM, Thomas EA, Goh SM, Sjovall H, Bornstein JC (2004) Segmentation induced by intraluminal fatty acid in isolated guinea-pig duodenum and jejunum. J Physiol 556 (2):557–569
424. Connell AM (1961) The motility of the small intestine. Postgrad Med J 37(434):703–716
425. Deloose E, Tack J (2016) Redefining the functional roles of the gastrointestinal migrating motor complex and motilin in small bacterial overgrowth and hunger signaling. Am J Physiol Gastrointest Liver Physiol 310(4):G228–G233
426. Takahashi T (2012) Mechanism of interdigestive migrating motor complex. J Neurogastroenterol Motil 18(3):246–257
427. von Arnim U (2015) Gastroparesis: definition, diagnostics, and therapy. Internist (Berl) 56 (6):625–630
428. Siegle ML, Bühner S, Schemann M, Schmid HR, Ehrlein HJ (1990) Propagation velocities and frequencies of contractions along canine small intestine. Am J Physiol 258(5 Pt 1): G738–G744
429. Lenz H, Blömer A, Düx A (1971) Analysis of the propulsive movements of the small intestine. Am J Digest Dis 16(12):1107–1115
430. Sarna SK, Otterson MF (1989) Small intestinal physiology and pathophysiology. Gastroenterol Clin North Am 18(2):375–404
431. Huizinga JD, Chen J-H, Zhu YF, Pawelka A, McGinn RJ, Bardakjian BL et al (2014) The origin of segmentation motor activity in the intestine. Nat Commun 5:3326
432. Hennig GW, Gregory S, Brookes SJ, Costa M (2010) Non-peristaltic patterns of motor activity in the guinea-pig proximal colon. Neurogastroenterol Motil 22(6):e207–e217
433. Chambers JD, Bornstein JC, Thomas EA (2008) Insights into mechanisms of intestinal segmentation in guinea pigs: a combined computational modeling and in vitro study. Am J Physiol Gastrointest Liver Physiol 295(3):G534–G541
434. Gwynne RM, Bornstein JC (2007) Mechanisms underlying nutrient-induced segmentation in isolated guinea pig small intestine. Am J Physiol Gastrointest Liver Physiol 292(4):G1162–G1172

435. Huizinga JD, Parsons SP, Chen J-H, Pawelka A, Pistilli M, Li Chunpei et al (2015) Motor patterns of the small intestine explained by phase-amplitude coupling of two pacemaker activities: the critical importance of propagation velocity. Am J Physiol Cell Physiol 309(6): C403–C414
436. Huizinga JD, Lammers WJEP (2008) Gut peristalsis is governed by a multitude of cooperating mechanisms. Am J Physiol Gastrointest Liver Physiol 296(1):G1–G8
437. Huizinga JD, McKay CM, White EJ (2006) The many facets of intestinal peristalsis. Am J Physiol Gastrointest Liver Physiol 290(6):G1347–G1349
438. Quigley EMM (1996) Gastric and small intestinal motility in health and disease. Gastroenterol Clin 25(1):113–145
439. Sternini C, Anselmi L, Guerrini S, Cervio E, Pham T, Balestra B et al (2004) Role of galanin receptor 1 in peristaltic activity in the guinea pig ileum. Neuroscience 125(1):103–112
440. Meltzer SJ, Auer J (1907) Peristaltic rush. Am J Physiol 20:259–281
441. Gershon MD (2004) Serotonin receptors and transporters—roles in normal and abnormal gastrointestinal motility. Aliment Pharmacol Ther 20(Suppl 7):3–14
442. Nezami BG, Srinivasan S (2010) Enteric nervous system in the small intestine: pathophysiology and clinical implications. Curr Gastroenterol Rep 12(5):358–365
443. Stukan M, Kruszewski JW, Dudziak M, Kopiejć A, Preis K (2013) Intestinal obstruction during pregnancy. Ginekol Pol 84(2):137–141
444. von der Ohe MR, Camilleri M, Kvols LK, Thomforde GM (1993) Motor dysfunction of the small bowel and colon in patients with the carcinoid syndrome and diarrhea. N Engl J Med 329(15):1073–1078
445. Basha ME, Duke GE (1999) Effect of fasting on small intestinal antiperistalsis in the Nicholas turkey (Meleagris gallopavo). J Exp Zool 283(4–5):469–477
446. Bayliss WM, Starling EH (1899) The movements and innervation of the small intestine. J Physiol 24:99–143
447. Bayliss WM, Starling EH (1900) The movements and innervation of the large intestine. J Physiol 26:107–118
448. Spencer NJ, Hennig GW, Smith TK (2002) A rhythmic motor pattern activated by circumferential stretch in guinea-pig distal colon. J Physiol 545(Pt 2):629–648
449. Alvarez WC (1915) The motor functions of the intestine from a new point of view. JAMA LXV(5):388–394
450. Alvarez WC (1924) Bayliss and Starling's law of the intestine or the myenteric reflex. Am J Physiol 69:229–248
451. Cannon WB (1912) Peristalsis, segmentation, and the myenteric reflex. Am J Physiol 30:114–128
452. Fleckenstein P (1978) Migrating electrical spike activity in the fasting human small intestine. Am J Dig Dis 23(9):769–775
453. Rayner V, Wenham G (1986) Small intestinal motility and transit by electromyography and radiology in the fasted and fed pig. J Physiol 379:245–256
454. Kruis W, Azpiroz F, Phillips SF (1985) Contractile patterns and transit of fluid in canine terminal ileum. Am J Physiol 249(2 Pt 1):G264–G270
455. Lammers WJEP (2005) Spatial and temporal coupling between slow waves and pendular contractions. Am J Physiol Gastrointest Liver Physiol 289(5):G898–G903
456. Seerden TC, Lammers WJEP, De Winter BY, De Man JG, Pelckmans PA (2005) Spatiotemporal electrical and motility mapping of distension-induced propagating oscillations in the murine small intestine. Am J Physiol Gastrointest Liver Physiol 289(6):G1043–G1051
457. Campbell I (2012) The mouth, stomach and intestines. Anaesth Inten Care Med 13(2):56–58
458. Kerlin P, Phillips S (1983) Differential transit of liquids and solid residue through the human ileum. Am J Physiol 245:G38–G43
459. Seow-Choen F (2009) The physiology of colonic hydrotherapy. Colorectal Dis 11(7):686–688

460. Maurer AH (2015) Gastrointestinal motility, part 2: small-bowel and colon transit. J Nucl Med 56(9):1395–1400
461. Rose C, Parker A, Jefferson B, Cartmell E (2015) The characterization of feces and urine: a review of the literature to inform advanced treatment technology. Crit Rev Environ Sci Technol 45(17):1827–1879
462. Smith TK, Oliver GR, Hennig GW, O'Shea DM, Vanden Berghe P, Kang SH, Spencer NJ (2003) A smooth muscle tone-dependent stretch-activated migrating motor pattern in isolated guinea-pig distal colon. J Physiol 551(Pt 3):955–969
463. D'Antona G, Hennig GW, Costa M, Humphreys CM, Brookes SJ (2001) Analysis of motor patterns in the isolated guinea-pig large intestine by spatio-temporal maps. Neurogastroenterol Motil 13(5):483–492
464. Shafik A (1997) Electrosigmoidogram, electrorectogram and their relation. Front Biosci 2: b12–b16
465. Sarna SK (2010) Colonic motility: from bench side to bedside. Integrated systems physiology: from molecule to function to disease. Morgan & Claypool Life Sciences, San Rafael
466. Fischer B, Hoh S, Wehler M, Hahn EG, Schneider HT (2001) Faecal elastase-1: lyophilization of stool samples prevents false low results in diarrhoea. Scand J Gastroenterol 36(7):771–774
467. Schouw NL, Danteravanich S, Mosbaek H, Tjell JC (2002) Composition of human excreta—a case study from Southern Thailand. Sci Total Environ 286(1–3):155–166
468. Tare V, Yadav KD (2009) Fate of physico-chemical parameters during decomposition of human feces. Glob J Environ Res 3(1):18–21
469. Santiago A, Panda S, Mengels G, Martinez X, Azpiroz F, Dore J et al (2014) Processing faecal samples: a step forward for standards in microbial community analysis. BMC Microbiol 14:112
470. Misawa H, Hoshi T, Kitame F, Homma M, Nakamura K (1986) Isolation of an antigenically unique methanogen from human feces. Appl Environ Microbiol 51(2):429–431
471. Ward SM, Burns AJ, Torihashi S, Sanders KM (1994) Mutation of the proto-oncogene c-Kit blocks development of interstitial cells and electrical rhythmicity in murine intestine. J Physiol 480:91–97
472. Fukuzawa M (2014) Progress in the treatment of and research on Hirschsprung's disease. Nihon Geka Gakkai Zasshi 115(6):312–316
473. Yamataka A, Kato Y, Tibboel D, Murata Y, Sueyoshi N, Fujimoto T et al (1995) A lack of intestinal pacemaker (c-Kit) in aganglionic bowel of patients with Hirschsprung's disease. J Pediatr Surg 30:441–444
474. Vanasin B, Ustach TJ, Schuster MM (1974) Electrical and motor activity of human and dog colon in vitro. Johns Hopkins Med J 134(4):201–210
475. Sarna SK (1991) Physiology and pathophysiology of colonic motor activity (2). Dig Dis Sci 36(7):998–1018
476. Gonella J, Bouvier M, Blanquet F (1987) Extrinsic nervous control of motility of small and large intestines and related sphincters. Physiol Rev 67(3):902–961
477. Holdstock DJ, Misiewicz JJ, Smith T, Rowlands EN (1970) Propulsion (mass movements) in the human colon and its relationship to meals and somatic activity. Gut 11(2):91–99
478. Bassotti G, Crowell MD, Whitehead WE (1993) Contractile activity of the human colon: lessons from 24 hour studies. Gut 34(1):129–133
479. Bhate PA, Patel JA, Parikh P, Ingle MA, Phadke A, Sawant PD (2015) Total and segmental colon transit time study in functional constipation: comparison with healthy subjects. Gastroenterol Res 8(1):157–159
480. Read NW, Miles CA, Fisher D, Holgate AM, Kime ND, Mitchell MA et al (1980) Transit of a meal through the stomach, small intestine, and colon in normal subjects and its role in the pathogenesis of diarrhea. Gastroenterology 79(6):1276–1282
481. Camilleri M (2004) Chronic diarrhea: a review on pathophysiology and management for the clinical gastroenterologist. Clin Gastroenterol Hepatol 2(3):198–206

482. Bassotti G, Antonelli E, Villanacci V, Baldoni M, Dore MP (2014) Colonic motility in ulcerative colitis. United European Gastroenterol J 2(6):457–462
483. Sarna SK (1993) Colonic motor activity. Surg Clin North Am 73(6):1201–1223
484. Misiewicz JJ (1984) Human colonic motility. Scand J Gastroenterol 93:43–51
485. Suzuki N, Gomi Y (1992) Evidence that stretch receptors for peristalsis are located in the mucosal layer of the guinea pig ileum. Jpn J Pharmacol 60(3):261–267
486. Tsuji S, Anglade P, Ozaki T, Sazi T, Yokoyama S (1992) Peristaltic movement evoked in intestinal tube devoid of mucosa and submucosa. Jpn J Physiol 42(3):363–375
487. Hennig GW, Costa M, Chen BN, Brookes SJH (1999) Quantitative analysis of peristalsis in the guinea-pig small intestine using spatio-temporal maps. J Physiol 517(Pt 2):575–590
488. Smith TK, Robertson WJ (1998) Synchronous movements of the longitudinal and circular muscle during peristalsis in the isolated guinea-pig distal colon. J Physiol 506(Pt 2):563–577
489. Lee KJ (2015) Pharmacologic agents for chronic diarrhea. Intest Res 13(4):306–312
490. Quigley EM (1987) Small intestinal motor activity—its role in gut homeostasis and disease. Q J Med 65(246):799–810
491. Canadian Agency for Drugs and Technologies in Health (2014) Dioctyl sulfosuccinate or docusate (calcium or sodium) for the prevention or management of constipation: a review of the clinical effectiveness. CADTH Rapid Response Reports, Ottawa, Canada
492. Canadian Agency for Drugs and Technologies in Health (2014) Treatments for constipation: a review of systematic reviews. CADTH Rapid Response Reports, Ottawa, Canada
493. Schmulson MJ, Drossman DA (2017) What is new in Rome IV. J Neurogastroenterol Motil 23(2):151–163
494. de Milliano I, Tabbers MM, van der Post JA, Benninga MA (2012) Is a multispecies probiotic mixture effective in constipation during pregnancy? 'A pilot study'. Nutr J 11:80
495. Gordon M, Naidoo K, Akobeng AK, Thomas AG (2013) Cochrane review: osmotic and stimulant laxatives for the management of childhood constipation (review). Evid Based Child Health 8(1):57–109
496. Kortz WJ (1985) Intestinal obstruction and pregnancy. In: Gleicher N (ed) Principles of medical therapy in pregnancy. Springer, New York
497. Vanagunas A (2008) Gastrointestinal complications in pregnancy. Glob Libr Women's Med. https://doi.org/10.3843/GLOWM.10172
498. İlhan M, Arabaci E, Turgut S, Karaman O, Danalioglu A, Tasan E (2014) Esophagus motility in overt hypothyroidism. J Endocrinol Invest 37(7):639–644
499. Yaylali O, Kirac S, Yilmaz M, Akin F, Yuksel D, Demirkan N, Akdag B (2009) Does hypothyroidism affect gastrointestinal motility? Gastroenterol Res Pract 2009:529802
500. Leeds AR (1987) Dietary fibre: mechanisms of action. Int J Obes 11(1):3–7
501. Abdel-Aziz H, Windeck T, Ploch M, Verspohl EJ (2006) Mode of action of gingerols and shogaols on 5-HT3 receptors: binding studies, cation uptake by the receptor channel and contraction of isolated guinea-pig ileum. Eur J Pharmacol 530(1–2):136–143
502. Huang Q, Iwamoto Y, Aoki S, Tanaka N, Tajima K, Yamahara J et al (1991) Anti-5-hydroxytryptamine3 effect of galanolactone, diterpenoid isolated from ginger. Chem Pharm Bull 39(2):397–399
503. Marx WM, Teleni L, McCarthy AL, Vitetta L, McKavanagh D, Thomson D, Isenring E (2013) Ginger (Zingiber officinale) and chemotherapy-induced nausea and vomiting: a systematic literature review. Nutr Rev 71(4):245–254
504. Lete I, Allué J (2016) The effectiveness of ginger in the prevention of nausea and vomiting during pregnancy and chemotherapy. Integr Med Insights 11:11–17
505. Ahrens FA, Zhu BL (1982) Effects of epinephrine, clonidine, L-phenylephrine, and morphine on intestinal secretion mediated by Escherichia coli heat-stable enterotoxin in pig jejunum. Can J Physiol Pharmacol 60(12):1680–1685
506. Kromer W (1993) Gastrointestinal effects of opioids. In: Herz A, Akil H, Simon EJ (eds) Opioids II, vol 104. Springer, Heidelberg

507. Sun X, Wang X, Wang GD, Xia Y, Liu S, Qu M et al (2011) Lubiprostone reverses the inhibitory action of morphine on mucosal secretion in human small intestine. Dig Dis Sci 56(2):330–338
508. Glare P, Pereira G, Kristjanson LJ, Stockler M, Tattersall M (2004) Systematic review of the efficacy of antiemetics in the treatment of nausea in patients with far-advanced cancer. Support Care Cancer 12(6):432–440
509. Gillman PK (2007) Tricyclic antidepressant pharmacology and therapeutic drug interactions updated. Br J Pharmacol 151(6):737–748
510. Artigas F, Nutt DJ, Shelton R (2002) Mechanism of action of antidepressants. Psychopharmacol Bull 36(2):123–132
511. Feighner JP (1999) Mechanism of action of antidepressant medications. J Clin Psychiatry 60(4):4–13
512. Schwartz TL, Goradia V (2013) Managing insomnia: an overview of insomnia and pharmacologic treatment strategies in use and on the horizon. Drugs Context 2013:212257
513. Menezes C, Rodrigues B, Magalhães E, Melo A (2007) Botulinum toxin type A in refractory chronic migraine: an open-label trial. Arq Neuropsiquiatr 65(3A):596–598
514. Wieckiewicz M, Grychowska N, Zietek M, Wieckiewicz G, Smardz J (2017) Evidence to use botulinum toxin injections in tension-type headache management: a systematic review. Toxins (Basel) 9(11):E370
515. Münchau A, Bhatia KP (2000) Uses of botulinum toxin injection in medicine today. BMJ 320(7228):161–165
516. Nigam PK, Nigam A (2010) Botulinum toxin. Indian J Dermatol 55(1):8–14
517. Ahmadi J, Azary S, Ashjaei B, Paragomi P, Khalifeh-Soltani A (2013) Intrasphincteric botulinum toxin injection in treatment of chronic idiopathic constipation in children. Iran J Pediatr 23(5):574–578
518. Arbizu RA, Rodriguez L (2015) Use of Clostridium botulinum toxin in gastrointestinal motility disorders in children. World J Gastrointest Endosc 7(5):433–437
519. Keshtgar AS, Ward HC, Clayden GS (2009) Transcutaneous needle-free injection of botulinum toxin: a novel treatment of childhood constipation and anal fissure. J Pediatr Surg 44(9):1791–1798
520. Emile SH, Elfeki HA, Elbanna HG, Youssef M, Thabet W, Abd El-Hamed TM et al (2016) Efficacy and safety of botulinum toxin in treatment of anismus: a systematic review. World J Gastrointest Pharmacol Ther 7(3):453–462
521. Itoh Z, Nakaya M, Suzuki T, Arai H, Wakabayashi K (1984) Erythromycin mimics exogenous motilin in gastrointestinal contractile activity in the dog. Am J Physiol 247: G688–G694
522. Armstrong DN, Ballantyne GH, Modlin IM (1992) Erythromycin stimulates ileal motility by activation of dihydropyridine-sensitive calcium channels. J Surg Res 52(2):140–146
523. Broad J, Mukherjee S, Samadi M, Martin JE, Dukes GE, Sanger GJ (2012) Regional- and agonist-dependent facilitation of human neurogastrointestinal functions by motilin receptor agonists. Br J Pharmacol 167(4):763–774
524. Deloose E, Vos R, Janssen P, Van den Bergh O, Van Oudenhove L, Depoortere I, Tack J (2016) The motilin receptor agonist erythromycin stimulates hunger and food intake through a cholinergic pathway. Am J Clin Nutr 103(3):730–737
525. Broad J, Sanger GJ (2013) The antibiotic azithromycin is a motilin receptor agonist in human stomach: comparison with erythromycin. Br J Pharmacol 168(8):1859–1867
526. Basch E, Prestrud AA, Hesketh PJ, Kris MG, Feyer PC, Somerfield MR et al (2011) Antiemetics: American Society of Clinical Oncology clinical practice guideline update. J Clin Oncol 29(31):4189–4198
527. Agrawal A, Hila A, Tutuian R, Mainie I, Castell DO (2007) Bethanechol improves smooth muscle function in patients with severe ineffective esophageal motility. J Clin Gastroenterol 41(4):366–370
528. Blonski W, Vela MF, Freeman J, Sharma N, Castell DO (2009) The effect of oral buspirone, pyridostigmine, and bethanechol on esophageal function evaluated with combined

multichannel esophageal impedance-manometry in healthy volunteers. J Clin Gastroenterol 43(3):253–260
529. Vitton V, Benezech A, Honoré S, Sudour P, Lesavre N, Auquier P, Baumstarck K (2015) CON-COUR study: interferential therapy in the treatment of chronic constipation in adults: study protocol for a randomized controlled trial. Trials 16:234
530. Lu ML, He J, Lu S (2015) Electrical stimulation therapy for slow transit constipation in children: a systematic review. Int J Colorectal Dis 30(5):697–702
531. Wang X, Yin J (2015) Complementary and alternative therapies for chronic constipation. Evid Based Complement Alternat Med 2015:396396
532. Li Y, Zheng H, Zeng F, Zhou S-Y, Zhong F, Zheng H-B et al (2012) Use acupuncture to treat functional constipation: study protocol for a randomized controlled trial. Trials 13:104
533. Xiong F, Wang Y, Li S, Tian M, Zheng C-H, Huang G-Y (2014) Clinical study of electro-acupuncture treatment with different intensities for functional constipation patients. J Huazhong Univ Sci Technol (Med Sci) 34(5):775–781
534. DeCamp LR, Byerley JS, Doshi N, Steiner MJ (2008) Use of antiemetic agents in acute gastroenteritis—a systematic review and meta-analysis. Arch Pediatr Adolesc Med 162(9): 858–865
535. Olkkola KT, Ahonen J (2008) Midazolam and other benzodiazepines. Handb Exp Pharmacol 182:335–360
536. Enrico P, Giuseppe V, Gabriele F, Walter M, Salvatore G, Leonardo G et al (1997) Ondansetron versus metoclopramide in the treatment of postoperative nausea and vomiting. Anesth Analg 85(2):395–399
537. Savant K, Khandeparker RV, Berwal V, Khandeparker PV, Jain H (2016) Comparison of ondansetron and granisetron for antiemetic prophylaxis in maxillofacial surgery patients receiving general anesthesia: a prospective, randomised, and double blind study. J Korean Assoc Oral Maxillofac Surg 42(2):84–89
538. Gupta K, Singh I, Gupta PK, Chauhan H, Jain M, Rastogi B (2014) Palonosetron, ondansetron, and granisetron for antiemetic prophylaxis of postoperative nausea and vomiting—a comparative evaluation. Anesth Essays Res 8(2):197–201
539. Board T, Board R (2006) The role of 5-HT3 receptor antagonists in preventing postoperative nausea and vomiting. AORN J 83(1):209–216, 219–220
540. Liu M, Zhang H, Du B-X, Xu F-Y, Zou Z, Sui B, Shi X-Y (2015) Neurokinin-1 receptor antagonists in preventing postoperative nausea and vomiting. A systematic review and meta-analysis. Medicine (Baltimore) 94(19):e762
541. Chu C-C, Hsing C-H, Shieh J-P, Chien C-C, Ho C-M, Wang J-J (2014) The cellular mechanisms of the antiemetic action of dexamethasone and related glucocorticoids against vomiting. Eur J Pharmacol 722:48–54
542. Ho CM, Ho ST, Wang JJ, Tsai SK, Chai CY (2004) Dexamethasone has a central antiemetic mechanism in decerebrated cats. Anesth Analg 99(3):734–739

Chapter 8
Gastrointestinal Hormones

Abstract Gastrointestinal (GI) hormones are internal or endocrine secretions of the gut released from special cells of the GI system that participate in modulating the functions of the gut or extragut tissues and organs. Although proposed to have evolved many millions of years ago, internal secretions of the gut were discovered relatively recently—around the nineteenth century. Founding fathers of internal secretions include Claude Bernard (1813–1878), Ernest Henry Starling (1866–1927), William Maddock Bayliss (1860–1924), Konstantinovich Kulchitsky (1856–1925), Sir Edward Albert Sharpey-Schafer (1850–1935), and Paul Langerhans (1847–1888). The concept of endocrine regulation was borne out from the 1855 pioneering work of Claude Bernard, who observed a novel role of the liver in the regulation of blood sugar. Bernard concluded from his observation on how the liver was able to regulate blood sugar to maintain a constant range of values by the release of internal secretions from ductless glands. This concept of ductless glands gradually gained recognition following discoveries of intestinal regulation of some secretions. Currently, over 60 types of hormones have been identified in the gut alone. This chapter presents data on the path of discoveries of internal secretions of the gut and trends in contemporary science and also describes the mechanisms of functioning of the internal secretions of the gut. The chapter is a key reference source on gut hormones, including recently discovered hormones, their structural and functional characteristics. Importance of the hormones to clinical medicine and diagnostics is outlined at specific points of the discussion.

Keywords 5-lipoxygenase · Adrenocorticotropic hormone · ACTH-like peptide Adenes · Adipo-fibrokines · Adipokines · Adrenomedullin · Alpha- and gamma-endorphins · Alpha-gustducin · Amylin · Islet amyloid polypeptide APUD · Arachidonic acid · Beta-endorphin · Bombesin related peptides Bombesin · Calcitonin gene-related peptide · Calcitonin receptor and calcitonin-like receptor · Celecoxib · Chemokines Cholecystokinin-pancreozymin · Cholecystokinin · Chromogranin A Cortistatin · Coxibs · Calcitonin receptor-like receptor · Crypts of lieberkühn Cyclooxygenase · Cystic fibrosis transmembrane conductance regulator D cells · DNES · Ductless glands · Duodenal glands · Eicosanoids

M. O. Welcome, *Gastrointestinal Physiology*,
https://doi.org/10.1007/978-3-319-91056-7_8

Leptin · Endocrine · Endoperoxides · Enterochromaffin · Enteroglucagon Enterokines · Etoricoxib · Fibrokines · Follicle-stimulating hormone–releasing hormone · G cells · G1 cells · GAL receptors · Galanin · Galenism Gamma-aminobutyric acid · Gastrin · Ghrelin · Stomatostatin · Gastric inhibitory polypeptide · Glicentin · Glicentin-related pancreatic polypeptide Glucagon · Glucagon-like peptide 1 · Glucagon-like peptide-2 Glucose-dependent insulinotropic polypeptide · Gonadotropin-releasing hormone Gonadoliberin · Luliberin · GPR119 · GPR120 · GPR40 (FFA1) GPRC6A · Growth hormone–releasing hormone · Growth hormone secretagogue-type 1A receptor · Histamine · Hormone database Hormones · Hormono · Incretins · Inflammatory bowel diseases Intermedin · Adrenomedullin-2 · Internal secretion · Intestinal (mucous) glands Intestinal L cell · K cells · Kulchitsky cells · Leukotriene A4 · Leukotriene B4 Leukotriene C4 · LPAR5 · Lipoprotein lipase · Lumaricoxib · Lumencrine (exocrine) · Luteinizing hormone · Lutropin · Lutrophin · Luteinizing hormone–releasing hormone · Lymphokines · Major proglucagon fragment Met-encephalin · Misoprostol · Motilin · Myokines · Neuracrines Neuronostatin · Neurotenin- related peptide · Non-steroidal anti-inflammatory drugs · Organic cation transporter · Oxyntomodulin · Oxytocin Pancreastatin · Paraneuron concept · Peptide YY · Phospholipase A2 PP (F) cells · Proglucagon · Prohormone convertase 2 · Prostacyclins Prostaglandins · Prostanoids · Receptor activity modifying proteins Receptor fatty acid translocase · CD36 · Rofecoxib · Secretin · Selective COX-2 inhibitors · Serotonin · SGLT1 · Somatostatin receptors · Somatostatin Substance P · Sucralfate/antacids · Thromboxane A2 · Thyrotropin-releasing hormone · Thyrotropin-releasing factor · Thyroliberin · Thyrotropin-releasing hormone · Valdecoxib · Valosin · Vasoactive intestinal peptide Vasopressin · Vesicular monoamine transporter subtype 2 · Vesiglandins Vital spirits · Xenin · Xenopsin · A cells · B cells · Δ-1 cell · Δ cells E cells · Epidermal growth factor · Endothelins · Insulin-like growth factor Nitric oxide · Neuropeptide Y · Neurotensin · Peptide YY · Vascular endothelial growth factor · Alister J. moody · Andrew Conway Ivy · Anthony Guy Everson Pearse · Bengt Ingemar Samuelsson · Benjamin Moore · Charles Best Christian de Duve · Claude Bernard · David M. Kipnis · Earl Wilbur Sutherland jr Edgar Zunz · Edward Albert Sharpey-Schafer · Edward S. Edie Efendic S · Emil Theodor Kocher · Erick Oldberg · Ernest Henry Starling Frederick Grant Banting · Friedrich Feyrter · George Barger · George Oliver Gerd Hamscher · Gerhard E. Feurle · Gustave-Édouard Laguesse Hartmut Kratzin · Hippocrates · J. Michael Conlon · Jean La Barre Jerzy Kaulbersz · Johann Conrad Brunner · Johann Nathanael Lieberkühn John Hill Abram · John Jacob Abel · John James Rickard Macleod John Sydney Edkins · Jorg W. Metzgers · Julian Walawski · Kazuhiko Tatemoto Lars Thim · Luis de Lecea · M.C. Ciacco · Mats Carlquist · Michael J. Perley Nikolai Konstantinovich Kulchitsky · Paul Langerhans · Roderic Alfred Gregory Roger C. L. Guillemin · Rosalyn Sussman Yalow · Rudolf Peter Heidenhain

Sir Henry Hallett Dale • Sir John Robert Vane • Sir William Maddock Bayliss
Solomon Aaron Berson • Sune K. Bergström • Tatemoto K • Thomas Addison
Ulf Svante von Euler • Viktor Mutt • Vittorio Erspamer • W. Hardy
Werner Creutzfeldt • Willis K. Samson • Wolfgang E. Schmidt

8.1 Introduction

Gastrointestinal (GI) hormones, also known as internal secretions of the gut, are the incretory (endocrine) substances of the GI system, directly released into the bloodstream in response to specific stimulus and they possess a relatively long duration of action. The cells that secrete these substances have no duct; hence, they are called ductless organs or glands. In distinction, exocrine organs (glands with duct) secrete their substances via a duct system into the environment or lumen [1–8].

GI hormones are secreted into the interstitium and blood from where they travel to the target to perform predetermined functions (Fig. 8.1). These hormones are released from some glands in the oral cavity, mucosa of the esophagus, stomach, and intestine by nervous activity, distension, and chemical stimulation and may occur simultaneously with food ingestion or with other stimuli. From the circulatory system, the hormones may be transported via portal circulation into the liver, then to the heart, and back to the digestive system to regulate its motility, secretory and growth of the stomach, small intestine and pancreas or other regions of the tract or even extragut tissues. Numerous hormones are released from the glands of the oral cavity, esophagus, stomach, pancreas and liver, gallbladder, and intestines to regulate local, peripheral and central physiological functioning. It should be remembered, however, that hormones are also released into intercellular space and synaptic clefts. Contrary to the traditional definition of hormones, accumulating evidences now indicate that hormones are also released into lumen similar to the mechanism of release of substances by exocrine glands. Moreover, some hormones can directly stimulate nerve endings [9–12].

Secreted hormones may act via plasma membrane receptors or via intracellular receptors (cytoplasmic or nuclear) (For details see Chap. 5). While some endocrine hormones that act on the gut may not be synthesized in the GI system, these hormones can still exert their activities on the GI because of the presence of their functional receptors. These hormones may be peptide or steroid signaling molecules. The intracellular target of these hormones may be different for different signals and different cells. While for some cells, signals may regulate the synthesis and exocytosis of metabolic enzymes, in other cells the signals may regulate the expression of certain proteins or may have a gene regulatory function or even regulate the activity of motor proteins [13–17].

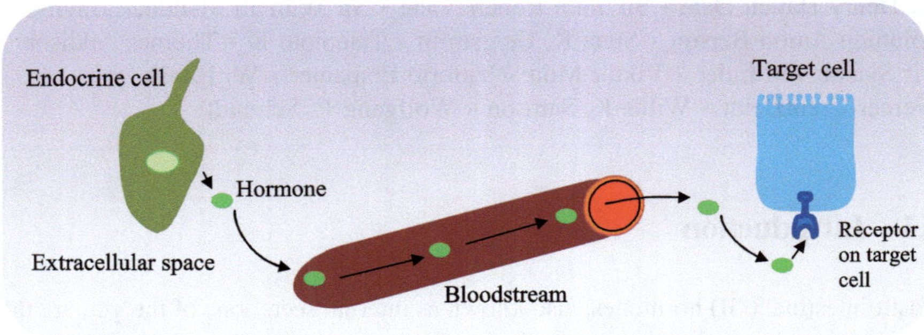

Fig. 8.1 Endocrine signaling of peptide hormones. The endocrine cell secretes a peptide hormone that enters the bloodstream and travels to the target cell where it localizes to the surface plasma membrane or intracellular receptor to initiate cellular activity

8.2 Discovery of Internal Secretion of the Gut: Origin of the Endocrine Concept of Regulation of Functions

The path to the discovery of GI epithelial (neuroendocrine and enteroendocrine) cells and their functions (secretions) was an indirect one. (The cells and their chemical secretions were discovered separately.) For example, it was previously thought that each epithelial cell of the GI tract is specialized in secreting a particular type of molecule; however, it is now becoming clearer that one and same epithelial cell could release molecules (biological acid, enzyme, hormone, neuropeptide neurotransmitter) that will act locally and in distant locations (described in detail in Table 8.1).

The endocrine system evolved about 800 million years ago. Of the endocrine organs of humans, the endocrine gut must have been the first to evolve, considering the importance of the gut in the body, its evolution, and the colonization of the gut by the beneficial microbial world. The identification of endocrine glands generally and their role in health and disease were noted during the ancient times. For instance, the ancient Chinese identified the role of thyroid gland in the pathophysiology of goiter. The ancient Egyptians observed that the ovaries play a role in reproduction, the endocrine glands of the central nervous system—in some forms of psychics. The ancient Egyptians were also keen in hypothesizing the role of the thyroid gland in human health. Some endocrine diseases were identified in Egypt as early as 5000 BC. However, no substantial progress on the endocrine glands and their role in disease were made until the reign of the Greek philosophers. The widely acclaimed father of medicine, Hippocrates (460–370 BC) was responsible for coining the term "adenes," meaning glands. While it was believed that glands were important in regulating the body humors, a vast number of Greek and Roman philosophers centered their arguments on the psychic and spiritual roles of glands in health and disease. No substantial advancement was made during this period until the second century BC. Recall that during the reign of Galenism, spirituality could

Table 8.1 Description of course of discovery of digestive hormones and their structural–functional characteristics

Hormone	Description	Discoverer	Year of discovery
Secretin	A 27 amino acid hormone secreted by enterochromaffin cell (S cell), located on the wall of the upper part of the small intestine—duodenum and the crypts of Lieberkühn. Secretin is a member of the secretin–VIP–glucagon family [44]. This hormone was isolation and synthesized in vitro in the 1960s. The gene that directs the production of secretin, located on chromosome 11p15.5, was first cloned in the 1990s [44] The cytoplasm of mucosal S cells is abundant in secretory granules containing secretin. This peptide is also produced in the CNS (pituitary, pineal glands, hypothalamus, thalamus, and olfactory lobe), where it is believed to play a role in neuromodulation and water regulation. Functional secretin receptors have been identified in the brain. Though the mechanism of action is not completely understood, it is believed that secretin may act via central angiotensin-II, in addition to stimulating its cognate secretin receptors. Secretin is a dipsogenic hormone that controls water intake behavior and can also stimulate the dipsogenic effects of angiotensin-II. Secretin may stimulate the expression of several receptor systems including vasopressin secretion and its receptor expression. Central secretin acts as an antidiuretic hormone, as such it stimulates the process of water reabsorption in the kidney by a mechanism independent of vasopressin. The stimulus for release of secretin from neural cells of certain regions of the brain (e.g., neurohypophysis) into circulation includes water deprivation, electrical stimulation of the paraventricular nuclei, and stimulation by some neurochemical agents. Central secretin is also involved in metabolic functions which may be mediated via the estrogen-related receptor α transcription factor. This transcription factor upregulates the secretin gene resulting to increased secretin expression and thus greater stimulation of central angiotensin-II and vasopressin expression and release [129, 130]. The relationship between the central and peripheral systems of secretin action has not been completely explored and so the influence of peripheral secretin on central secretin and vice versa is not precisely understood	Ernest Henry Starling (1866–1927) and Sir William Maddock Bayliss (1860–1924). The same authors introduced the concept of hormones in 1905 [30]	1902

(continued)

Table 8.1 (continued)

Hormone	Description	Discoverer	Year of discovery
	Peripheral secretin release is stimulated by decrease in duodenal pH (~ 4.5, the relatively high acidity is due to the passage of HCl-rich chyme from the stomach via the pyloric sphincter), chyme and the products of protein digestion. Upon activation, secretin is released into circulation from where it is transported to the site of action. But studies have indicated that secretin is also released into the intestinal lumen suggesting that this hormone may be produced by exocrine glands of intestinal mucosa [131, 132]. Any substance that reduces the pH in the GI tract is likely to inhibit secretin secretion. Thus, H_2 antagonists can inhibit secretin release. This is because secretin release, in addition to other factors, is regulated by the duodenal pH level. If the pH is above 4.5, secretin release is significantly inhibited [131, 132] Human secretin receptor is a G protein-coupled receptor which binds secretin. The gene for the secretin receptor is localized on chromosome 2p14.1 [44]. *The secretin receptor* is a prototypic member of the class B GPCR subfamily [133]. The ligand-binding site (amino-terminal domain) of this receptor assumes a conformation that fits the helix of the carboxyl-terminal region of secretin [134]. The amino-terminal region of secretin determines its biological activity; this region is responsible for activating the stimulating subunit of G protein-coupled receptor or secretin receptor. Secretin exerts its biological effects via its receptors expressed in the basolateral domain of several cells. Secretin receptors are expressed in the basolateral epithelial cells of the hepatopancreatobiliary system and GI tract to regulating secretion, trophic activity, and gene expression [135] More recently a novel receptor was discovered for the hormone secretin. Sundaresan et al. (2013) reported that fatty acids stimulated secretin release via the receptor fatty acid translocase (FAT) (also referred to as Cluster of Differentiation, CD36) [132]. This receptor is widely expressed in the enterocyte of the proximal small intestine [132]. The contribution of this receptor to secretin or CCK release may constitute about 50–60% [132]. The effect of activation of this receptor is the increase in intracellular Ca^{2+} or cAMP levels, which in turn activates the enzyme PKA. This enzyme then phosphorylates other peptides located in the cytosol and nucleus to increase the expression and secretion of gut peptides [132]. Activated PKA stimulates a range of proteins including ion channel, cystic fibrosis transmembrane conductance regulator (CFTR), which in turn induces activation of the Cl^-/HCO_3^- anion exchanger [135]		

(continued)

Table 8.1 (continued)

Hormone	Description	Discoverer	Year of discovery
	Though the mechanism of secretin action is cAMP-dependent, cross-signaling may up-regulate or down-regulate cAMP, thus affecting the activity of secretin on the target cells. Up- or down-regulation of cAMP may occur via the action of other hormones such as somatostatin, gastrin, endothelin-1, and phenylephrine—an α1-adrenergic agonist. Up- or down-regulation may occur via Ca^{2+}-dependent PKC pathways to affect secretion such as the flow of bile from the biliary duct (a process termed choleresis) [135] Secretin stimulates the secretion of bicarbonate-rich intestinal, pancreatic, and biliary fluids into the gut lumen [44]. The bicarbonate is required to neutralize the acid, which is needed for a favorable pH for the action of intestinal enzymes and to prevent the traumatic effects of acid on the mucosa [131, 132]. Thus, secretin plays a role in pH regulation of duodenal content [135]. Secretin has been termed "nature's antacid" because it inhibits the secretion of gastric HCl. Secretin possesses a growth stimulatory effect on many tissues and organs including the exocrine pancreas. The hormone secretin has antagonistic effect to the release of gastrin. Secretin also stimulates insulin secretion from the endocrine pancreas. Secretin has been termed a neuroendocrine hormone due to its pleiotropic signaling effects on many organs including heart, kidney, lung, and brain [135]. Further information about secretin can be found at [136, 137]		
Gastrin	A 14-, 17-, and 34-amino acid hormone secreted by the wall of the stomach (at the pyloric end). The gastrin gene is located on the long arm of the chromosome 17q21 [138–141] Gastrin stimulates secretion of gastric acid by the parietal cells of the stomach and aids in gastric motility. This hormone also enhances insulin secretion from the endocrine pancreas. Gastrin can stimulate growth of the mucosa of the intestines [141]	The British physiologist John Sydney Edkins (1863–1940) [29]. In 1905, Edkins proposed a hypothesis indicating that the mucosa of the gastric antrum produces a substance that might be responsible for HCl secretion. This substance, named gastrin, was isolated in 1964 by the British scientist, Roderic Alfred Gregory (1913–1990) at the University of Liverpool. The structure of gastrin was determined in 1964 [142, 143]. Though the second hormone on the discovery chain, gastrin was the first GI hormone for which the structure was identified	1905; 1964

(continued)

Table 8.1 (continued)

Hormone	Description	Discoverer	Year of discovery
	Gastrin is released by the neuroendocrine ("G") cells located mostly in the antrum of the stomach, but is also produced in the duodenum, and the pancreas. Gastrin release can be stimulated by chemical, electrical, and mechanical signals. Distension of the gastric antrum can cause the release of gastrin via stretch receptors. Chemical stimulators of gastrin release include **alcohol, caffeine, and protein substances** [136, 137, 141]. **The CaSR calcium-sensing receptor, GPRC6A, and LPAR5 lysophosphatidic acid receptor on the surface of G cells can recognize these molecules to initiate the release of the hormone** [136, 137]. The hormone acts through gastrin receptors—CCK-2 or CCK-B receptors, which are G protein-coupled receptors [144, 145]. Activated CCK-B receptor mediates the stimulation of several signaling pathways via Gq (PLCbeta or PLCgamma) and PI3 K. Activation of PLCbeta mediates the formation of DAG and Ca^{2+}, which in turn activate PKC. CCK-1 receptor in addition to mediating the above signaling pathway also activates Gs subunit of the G protein to activate adenylate cyclase [138–141, 145]. Upon binding of gastrin to CCK-B receptors, histamine secretion is stimulated in enterochromaffin-like cells, and, also inducing the insertion of K^+/H^+-ATPase pump into the apical membrane of parietal cells (which in turn increases H^+ release into the gastric pit) [141]. Gastrin can also stimulate the gastric chief cells to release pepsinogen [141] Some hormones such as bombesin or GRP can stimulate gastrin release via neural stimulation through the vagus nerve [141]		
	Gastrin release is inhibited mainly by negative feedback loop. Increase in acid secretion above the threshold causes inhibitory effect on gastrin release. Gastrin release can be inhibited by somatostatin. Somatostatin also mediates the inhibition of secretion of other hormones including secretin, histamine, GIP, VIP, glucagon, calcitonin. In addition to somatostatin, the activity of gastrin is inhibited by secretin, GIP, NT, and low pH [136, 137, 141]		
	Gastrin is highly homologous to CCK (see CCK below). The receptors and ligands of CCK/gastrin have a high degree of structural and functional similarity in the animal kingdom. CCK/gastrin plays an important role in the regulation of feeding behavior and energy homeostasis in animals, including humans. The CCK-gastrin signaling system is a highly conserved system in the animal kingdom. This system is believed to have evolved since the past 500 million years ago [146]. The gastrin-CCK system also plays a crucial role in vertebrates such as fishes, amphibians, reptiles, and birds [147]		

(continued)

Table 8.1 (continued)

Hormone	Description	Discoverer	Year of discovery
	Continued research in physiology of the pancreas and other parts of the GI tract in the twentieth century resulted to further understanding of the mechanisms of functioning of gastrin in health and disease. In 1955, the American professors of surgery, Robert Milton Zollinger (1903–1992) and Edwin Homer Ellison (1918–1970) of Columbus, Ohio, discovered that certain tumors (precisely called gastrinoma) excessively produced the hormone gastrin. The secreted gastrin leads to hypersecretion of gastric acid, subsequently resulting to diarrhea, gastritis, severe gastroesophageal reflux disease, and peptic ulcer disease. The ulcers are usually severe and recurrent affecting the esophagus, stomach, duodenum, and jejunum [148, 149]		
	Gastrinoma usually occurs in the duodenum and sometimes in the pancreas. Zollinger and Ellison were the first to have made a detailed description of the classic triad—peptic ulcer disease, gastric acid hypersecretion, and islet cell tumors of the pancreas that constitute the condition. This clinical syndrome was later termed Zollinger–Ellison syndrome and occurs secondary to a gastrinoma. The condition is characterized by elevated fasting circulating gastrin levels (>200 pg/mL) and gastric acid hypersecretion (basal acid output >15 mEq/h with an intact stomach or >5 mEq/h after ulcer surgery). To confirm the diagnosis of Zollinger–Ellison syndrome, a provocative test by secretin stimulation is carried out. Serum level of gastrin is measured at intervals (2, 5, 10, 15, and 20 min) following intravenous administration of secretin 0.4 mcg/kg over 1 min (2 units/kg bolus). If the patient is suffering from gastrinoma, it is expected that the serum gastrin level will increase. An increase up to 200 pg/mL or more in 15 min interval is strongly suggestive of a gastrinoma [149, 150]. It should be mentioned that in gastrinoma hypochlorhydria or achlorhydria may be observed instead of increased acid secretion. This is referred to as a false-positive result, which may occur in about 15% of cases [151]. Hypochlorhydria is the reduced production of stomach acid, while achlorhydria refers to state of the absent production of hydrochloric acid in gastric secretions of the stomach [151]. The observation of hypochlorhydria and achlorhydria in gastrinoma following secretin stimulation gastric atrophy and drug intake (e.g., use of PPIs). The false-positive result in this test can be confirmed by structural and functional imaging studies [151, 152]		

(continued)

Table 8.1 (continued)

Hormone	Description	Discoverer	Year of discovery
	Gastrin has functional homology with a molecule that is useful in diagnostics. A synthetic polypeptide with gastrin-like effects, L-Phenylalaninamide, N-((1,1-dimethylethoxy)carbonyl)-beta-alanyl-L-tryptophyl-L-methionyl-L-alpha-aspartyl also called pentagastrin is administered intravenously to study the secretory activity of gastric acid, pepsin, intrinsic factor and also used as diagnostic tool in pentagastrin-stimulated calcitonin test. In radiological diagnostics, pentagastrin is used to stimulate ectopic gastric mucosa for the detection of Meckel's diverticulum and to detect medullary thyroid carcinoma following thyroidectomy by stimulation of calcitonin release from residual parafollicular C cells [153]. Pentagastrin-stimulated calcitonin test is a pentagastrin stimulation test using a calcitonin immunoradiometric assay to investigate medullary thyroid carcinoma and individual with familial medullary thyroid carcinoma. In other conditions such as C cell hyperplasia, pheochromocytoma carriers, hyperparathyroidism, and thyroid nodules, calcitonin level has been shown to be above normal (more than 10 ng/L). A calcitonin level of 30 ng/L and above is an indication of medullary thyroid carcinoma. The abnormal calcitonin response to pentagastrin is used as criterion for surgical treatment. Individual with a peak calcitonin response above 100 ng/L in the serum is recommended for surgery irrespective of the rational for the test [154]. However, the diagnostic value (sensitivity and specificity) increases at higher cutoff for serum calcitonin concentration [155] Since B-type CCK receptors are located in the brain [147], pentagastrin has been shown to activate these receptors via PLC pathway. It should be noted that intravenous administration of pentagastrin may cause panic attacks. Hence caution must be taken during intravenous administration of this hormone		
Histamine	Histamine is an organic nitrogenous compound acting as a hormone, neurotransmitter, and immunomodulator, involved in local immune responses as well as regulating physiological function in central and peripheral tissues [156–159]	The British chemist, George Barger (1878–1939) and the English physiologist, Sir Henry Hallett Dale (1875–1968) were the first to have isolated histamine from the plant fungus ergot in 1910. A year later, they succeeded in isolating the same substance from animal tissues [157, 159].	1910, 1911

(continued)

Table 8.1 (continued)

Hormone	Description	Discoverer	Year of discovery
	Histamine is produced through the decarboxylation of histidine, and the reaction is catalyzed by histidine decarboxylase [158]. In the presence of adequate stimuli, histamine is released to the extracellular space by vesicular exocytosis or via histamine transporters [160, 161]. Histamine may be released from enterochromaffin-like cells by exocytosis of pools of secretory vesicles, and the strongest stimulus for this exocytosis is believed to be gastrin [162]. The vesicles maybe released by proton-histamine countertransport mechanism via the vesicular monoamine transporter subtype 2 (VMAT-2) [162–164]. Over the past decades other transporters have been identified, which include organic cation transporters (OCT)-2 and -3. The release of the vesicles is initiated by the activation of the appropriate ion channels on the membrane of enterochromaffin-like cells. Traditional ion channels localized to the membrane include voltage-gated K^+ and Ca^{2+} channels [165]. The exocytosis process is catalyzed by Ca^{2+} cellular sensors involved in vesicle mechanics such as synaptobrevin and synaptosomal-associated protein of 25 kDa [162]. The secreted histamine localizes to its receptor, and excess histamine is degraded by oxidative deamination to imidazole acetic acid. However, only a small proportion of histamine is degraded via this pathway. The majority of histamine is inactivated via the enzymatic pathway which involves methylation of the imidazole ring by imidazole-*N*-methyltransferase [165, 166] Histamine is produced in many cells of the organism: neurons, non-neuronal cells such as endocrine, epithelial, mast, and dendritic cells as well as lymphocytes [164, 166, 168]. Histamine is highly expressed in the GI tract (enteric neurons and endocrine cells including enterochromaffin-like cells), pituitary, and adrenal gland as well as central and peripheral neurons [169] The effect of histamine is mediated by postsynaptic H1-, H2-, H3-, H4-histamine receptors [170]. These receptors are distributed in different proportions in the different cells and tissues of the body. The recently discovered H4-receptor is known to have a greater affinity for histamine compared to the conventional H1- and H2-receptors by over 10,000-fold. Interestingly, H4-receptor has been implicated in many autoimmune diseases. Thus, in conditions where H1- and H2-receptors-antagonists are non-effective, a better treatment option is to use H4-receptor antagonist [168]	Dale also contributed enormously to the study of acetylcholine in nerve transmission for which he shared the Nobel Prize in Physiology or Medicine with Otto Loewi in 1936 [156, 167].	

(continued)

Table 8.1 (continued)

Hormone	Description	Discoverer	Year of discovery
	The intracellular mechanism leading to the activation histamine release is coupled to adenylate cyclase. This pathway is also believed to be the acid stimulatory pathway in parietal cells. But histamine receptor signaling is also coupled to other signaling pathways that mediate numerous cellular processes including growth and differentiation [158] Histamine is involved in the stimulation of gastric acid secretion; contraction of smooth muscle tissues (e.g., of stomach and lungs). Histamine dilates blood vessel and increases its permeability, thus lowering the blood pressure [157, 159]. Histamine also mediates inflammatory response and pruritus. The molecule is involved in neuroendocrine regulation of the secretion of glands including the pituitary. In the central nervous system, histamine modulates pituitary hormone secretion via indirect activity of histaminergic neurons of the hypothalamus. For instance, histamine inhibits release of growth hormone and thyroid-stimulating hormone (TSH, also known as thyrotropin) [170] However, histamine stimulates the secretion of luteinizing hormone (LH, also known as lutropin or lutrophin) in females (mediated by GnRH—gonadotropin-releasing hormone, also known as follicle-stimulating hormone–releasing hormone, FSH-RH; luteinizing hormone–releasing hormone, LHRH; gonadoliberin; and luliberin), vasopressin (AVP), oxytocin, adrenocorticotropic hormone (ACTH), beta-endorphin, and prolactin [170].		
Bombesin and bombesin related peptides	A 14 amino acid peptide first discovered in the skin of the European fire-bellied toads *Bombina bombina* and *Bombina variegata. The fact that this hormone was first identified in the skin of these animals indicate some sorts of connections in the hormone function and the major points of actions (brain, gut, skin) to be functionally related* [171, 172]. *This connectivity was suggested by the discoverer* [173]	Vittorio Erspamer and colleagues (1909–1999) [171]	1970

(continued)

Table 8.1 (continued)

Hormone	Description	Discoverer	Year of discovery
	The release of bombesin is regulated by different stimuli ranging from mechanical, electrical to chemical and involves changes in intracellular Ca^{2+} concentration [174]. The hormone bombesin stimulates gastrin release from G cells. It also induces the release of CCK in the GI tract. However, histamine is known to reduce bombesin stimulated release of these hormones [175]. Bombesin also induces the release of somatostatin and is mediated via cholinergic pathway as both tetrodotoxin and atropine inhibit bombesin-induced somatostatin release [176]. Bombesin also stimulates the release of neurotensin; however, it is independent on cholinergic pathway. Bombesin stimulated release of hormone or neurotransmitter may occur either by neural or non-neural pathways [176, 177]		
	Bombesin is the ligand for the G protein-coupled receptors, Bombesin receptor-1, -2, and -3. These receptors are located in the GI tract, brain, and are abundant in tumor cells. Bombesin and CCK are hormones that provide negative feedback signals necessary for regulating eating behavior. These hormones are responsible for terminating the feeling of hunger [178–180]. It is a gastroprotective agent [181]. Bombesin can stimulate mitogen-activated protein and increase phosphorylation of extracellular regulated kinase and transactivate epidermal growth factor receptor. These actions of bombesin on non-classical receptors occur in pathological conditions such as tumors [182]		
	At least two types of bombesin-related peptide hormones are known: neuromedin B and gastrin-related peptide. Neuromedin B is an endocrine hormone, as well as a neurotransmitter, present in GI tissues and CNS. Its functions include regulation of cellular growth, glycemia, blood pressure, exocrine and endocrine secretions, smooth muscle contraction, feeding, and body temperature [183]. The activities of neuromedin B are mediated by the binding to the neuromedin B receptor (NMB-R), a type of G protein-coupled receptor. Activation of this receptor leads to downstream signaling cascades involving activation of phospholipase, PKC, and mobilization of calcium stores. The activation of these pathways also leads to expression of several genes and DNA synthesis. Altogether, the results of these cellular events are evident as secretion and replenishment of the secretory vesicular pools, motility, cellular growth, and other related physiological functions [180, 183, 184]		

(continued)

Table 8.1 (continued)

Hormone	Description	Discoverer	Year of discovery
Prostaglandins and vesiglandins	The name "prostaglandin" is derived from the seminal fluid of the prostate gland. The hormone was discovered in semen by Ulf von Euler (see third column), who thought that they were secreted by the prostate gland; hence, he named the molecule prostaglandin [185–188]. Prostaglandins are physiologically active lipid compounds with hormone-like effects (lipid mediators or autacoids). They exert their functions via autocrine and paracrine mechanisms [189, 190] Prostaglandins are made up of unsaturated fatty acids that contain a cyclopentane hydrocarbon ring. These lipid mediators belong to the class of biomolecules called prostanoids, a subclass of eicosanoids, and are defined as cyclic lipid mediators formed from enzymatic cyclooxygenation of linear polyunsaturated fatty acids (e.g., arachidonic acid). In addition to prostaglandins (PGD2, PGE2, PGF2alpha, PGI2), thromboxane A2 and prostacyclins are all classified as prostanoids. Being the main component of phospholipids, arachidonic acid together with other biomolecules form the plasma membrane of the cell. Stimuli (chemical, e.g., hormones; physical; mechanical) impinging on the cell membrane may stimulate phospholipases (e.g., phospholipase A_2), which catalyzes the release of arachidonic acid from phospholipids [191–193] Upon release, arachidonic acid is converted to leukotriene A4 by the activity of the enzyme 5-lipoxygenase. Hydrolysis of leukotriene A4 produces leukotriene B4. The incorporation of glutathione results in the formation of leukotriene C4. Lipoxygenases such as 12- and 15-lipoxygenases can catalyze the conversion of arachidonic acid to lipoxins A and B. Leukotrienes mediate the inflammatory process in various cells. The leukotrienes were discovered in 1938 as a smooth muscle-contracting factor in lung perfusates in leucocytes by Feldberg and Kellaway [195–199] In the second pathway, catalysis by lipoxygenase yields various hydroperoxy acids. In platelets, for instance, 12-hydroperoxyeicosatetraenoic acid is the predominant product. In polymorphonuclear leukocytes, 5-hydroperoxyeicosatetraenoic acid is produced. These products are primarily reduced to 12-hydroxyeicosatetraenoic acid and 5-hydroxyeicosatetraenoic acid in platelets and polymorphonuclear leukocytes, respectively. However, 5-hydroperoxyeicosatetraenoic acid may be dehydrated to leukotriene A4. Enzymatic hydrolysis of leukotriene A4 yields leukotriene B4, a potent mediator of leukocyte function and chemotaxis stimulator [195–197]	The Swedish physiologist and pharmacologist Ulf Svante von Euler of Karolinska Institutet in Stockholm, Sweden. Following his prostaglandin pioneering investigation, syntheses of several prostaglandins have been carried out by other scientists. One of the major works on prostaglandin was carried out by the Swedish biochemists Sune K. Bergström and Bengt Ingemar Samuelsson and the British biochemist Sir John Robert Vane (1927–2004). They were awarded the 1982 Nobel Prize for Physiology or Medicine for the isolation, identification, and analysis of numerous prostaglandins [185–188, 194] Bergström, Samuelsson and Vane demonstrated beyond possible doubt that prostaglandins are involved in diverse functions and processes in the living cells. Their research results opened new window for therapeutic applications in thrombosis, inflammation and allergy [185–188]	1935

(continued)

Table 8.1 (continued)

Hormone	Description	Discoverer	Year of discovery
	Leukotriene A4 may undergo reaction to form leukotriene C4 (a natural component of slow-reacting substance of anaphylaxis), leukotriene D4, leukotriene E4, and the 11 trans-isomers. Leukotrienes C4, D4, and E4 play a crucial role in inflammatory reaction by increasing the permeability of blood vessels. Leukotrienes are implicated in bronchoconstriction that is observed in asthma. In addition, these substances also mediate cellular adhesion, chemotaxis of white blood cells, and degranulation reaction [198] The pathways mentioned above occur not only in leukocytes, but also platelets and other cells [197] Arachidonic acid may as well be converted to endoperoxides (e.g., endoperoxides G2 and H2) by the action of the enzyme—cyclooxygenase. Endoperoxides, via chemical transformation, can form prostaglandins, prostacyclin, and thromboxane A2. Endoperoxides and their transformed products may induce rapid irreversible aggregation of human platelets and smooth muscle contraction [197, 200] From the above-mentioned facts, it can be deduced that the fate of arachidonic acid is dependent on the type of signal impinging on the cell and the activated enzyme Prostaglandins have a variety of roles to play in health and disease. They modulate the functioning of the CNS, cardiovascular, GI, genitourinary, endocrine, respiratory, and immune systems. This group of bioactive molecules mediates inflammatory and anaphylactic reactions. Prostaglandins regulate reproductive processes (ovulation, implantation, menstruation). Prostaglandin signaling is involved in angiogenesis, cell adhesion, morphology, motility, invasion, and metastases [201]. Prostaglandins are important mediators of smooth muscle relaxation and contraction [202]. Prostacyclins mediate inflammatory reactions. Both prostaglandin E2 and prostacyclin act as vasodilators. Thromboxanes mediate vasoconstriction reaction. Prostaglandins, thromboxanes, and some hydroxyeicosatetraenoic acid exert chemotactic effects on polymorphonuclear leukocytes [198, 203]		

(continued)

Table 8.1 (continued)

Hormone	Description	Discoverer	Year of discovery
	The activities of prostanoids are mediated via the prostanoid receptors, which are G protein-coupled rhodopsin-type receptors [203]. The prostanoid receptor subfamily consists of nine members, which include DP, EP1-4, FP, IP, TP, and CRTH2—chemoattractant receptor homologous molecule expressed on Th2 cells [191] There are at least ten prostaglandin receptors on various cell types also belonging to the family of G protein-coupled receptors. Prostaglandins bind to these receptors to activate intracellular signaling that regulates a host of cellular processes and gene transcription [201] In the GI tract, prostaglandins (particularly prostaglandin E2 and prostacyclin) which are synthesized and degraded in intestinal mucosa by a 15-prostaglandin dehydrogenase are involved in cytoprotection of the GI epithelium and also stimulate motility. Prostaglandins reduce absorption and stimulate secretion of electrolytes and water in the intestines through increase in cyclic AMP formation. Prostaglandin E2 is responsible for contraction of the gallbladder, relaxation of the sphincter of Oddi and inhibition of isosmotic fluid transport. Prostaglandins also inhibit pancreatic secretion (including insulin secretion). Prostaglandin signaling dysfunction plays a role in inflammatory bowel diseases (IBD) and GI neoplasia. Prostaglandins prevent gastric and duodenal ulcers and accelerate the rate of healing of duodenal ulcers in humans [204]. The cytoprotective effect of prostaglandins appears to result from their ability to stimulate mucosal mucus and bicarbonate secretion, to increase mucosal blood flow and, particularly in the stomach, to limit back diffusion of acid into the epithelium [205, 206] Some pharmacological agents interfere with prostaglandin functions. For example, non-steroidal anti-inflammatory drugs (NSAIDs) such as aspirin and indomethacin are associated with gastric erosions and ulcers: they inhibit prostaglandin synthesis. NSAIDs inhibit cyclooxygenase. Steroids (corticosteroids), which are sometimes used to treat inflammatory reactions inhibit phospholipases to prevent release of arachidonic acid [198]		

(continued)

Table 8.1 (continued)

Hormone	Description	Discoverer	Year of discovery
	In clinical setting, NSAIDs are generally prescribed to ameliorate symptoms associated with acute pain and chronic inflammatory diseases such as arthritis. More importantly, studies have shown that NSAIDs and cyclooxygenase (COX)-2 selective inhibitors are associated with a reduced risk of certain malignancies, especially GI cancer [207]. Unfortunately, NSAIDs cause damage to the gastric mucosa leading to NSAID gastropathy, a clinical condition characterized by dyspepsia, ulceration, and upper GI bleeding [208]. Although the mechanisms of NSAID-induced GI injury are not fully understood, it is believed that disruption of the epithelial barrier allows back diffusion of acid into the mucosa, which further worsens the prognosis of ulcer and gastric bleeding [208, 2010] Inhibition of cyclooxygenases (e.g., by NSAIDs) block the formation of proinflammatory and gastroprotective prostaglandins. Thus, the functions of prostaglandins to maintaining gastric mucosal blood flow, increase protective mucus, bicarbonate production are interrupted. It was later discovered that different isoforms of COX possess different functions. The gastroprotective role is played by COX-1 for instance, whereas COX-2, at present, has no known positive role in gastroprotection. Stimulation of COX-2 leads to inflammation and pain. So, non-selective COX inhibition enhances gastric injury, while selective COX-2 inhibition maintains the integrity of the gut mucosa. Because of the need for NSAIDs therapy, and the resulting effects, strategies have been taken to ensure reduction of gut mucosa injury due to NSAIDs. Such strategies include coprescription of gastroprotective agents, use of selective COX-2 inhibitors, and eradication of H. pylori. Considerable effort is underway to develop NSAIDs that fail to inhibit mucosal prostaglandin synthesis, but will still exhibit antinociceptive action [207–211]		

(continued)

Table 8.1 (continued)

Hormone	Description	Discoverer	Year of discovery
	Gastroprotective drugs include oral misoprostol[a] (synthetic prostaglandin E_1 analogue); sucralfate/antacids, which diminish acid exposure to the damaged epithelium by forming a protective gel (sucralfate) or by neutralization of gastric acid (antacids); inhibitors of acid secretion (e.g. H_2-receptor antagonists and proton-pump inhibitors); selective COX-2 inhibitors/Coxibs. The selective COX-2 inhibitors confer GI tract protection. Examples include Rofecoxib and celecoxib. The coxibs of the second generation such as valdecoxib, etoricoxib, lumaricoxib, and the water-soluble parecoxib possess a several-fold higher selectivity for COX-2 [207–212]. Unfortunately, two coxibs (rofecoxib and valdecoxib) were withdrawn from the market due to their adverse effects on the cardiovascular system [213]. Gastroprotection is also enhanced via use of antioxidants/vitamin C and E, which play a role in preserving gastric mucosal integrity [207, 208]		
Incretin hormones: glucose-dependent insulinotropic polypeptide (GIP, formerly known as gastric inhibitory polypeptide) and glucagon-like peptide 1 (GLP-1)	The term "incretin" was introduced in the 1930s by the Belgian Professor of Pharmacology and Medical Toxicology at the University of Brussel, Jean La Barre on account that hypoglycemia results from the administration of mucosal extracts of the upper gut. However, this extract did not stimulate exocrine pancreatic secretion [215]. Frankly, the emergence of incretin concept has a long history and dates as far back as 1906 when Benjamin Moore (1867–1922), Edward S. Edie (1865–1935), and John Hill Abram (1863–1933) of Liverpool, UK, proposed that the duodenum produced a substance that increased pancreatic secretion and showed that mucosa extracts of the gut may increase endocrine pancreatic secretion [216]. The criteria for considering a substance as an "incretin" were defined by Werner. Based on the criteria, a substance is considered as an incretin, if it fulfills the following. Firstly, the substance must be released by nutrients, particularly carbohydrates. Second, the substance must stimulate insulin secretion at physiological levels of the substance and in the presence of elevated blood glucose [220, 221] A few years later, the Belgian physiologists Edgar Zunz (1874–1939) and Jean La Barre showed that intestinal mucosal extracts containing duodenal extracts (secretin) produced hypoglycemia, which was mediated by the pancreas. This compound also influenced the secretion of the exocrine pancreas [227]. It was found with contradictory results from different laboratories that administration of duodenal extracts in normal or diabetic mammals including humans leads to hypoglycemia [220]	GIP was discovered in 1973 on the basis of its ability to inhibit gastric acid secretion. Different groups of scientists who contributed to this discovery include Brown and coworkers [217–219] GLP-1, the second major incretin was discovered in 1987. The scientists who contributed to this discovery include [222–226]	1973 and 1987

(continued)

Table 8.1 (continued)

Hormone	Description	Discoverer	Year of discovery
	It was not until in 1964 when an independent group of investigators led by McIntyre Neil [228] and Harold Elrick [229] simultaneously showed that orally administered glucose evokes a greater insulin response than intravenously administered glucose. From the late 60s of that century, scientists began to confirm the results of the greater response resulting from oral glucose intake compared to intravenous administration. One of the earliest confirmations was made by Michael Perley and David M. Kipnis (1927–2014) in 1967. Furthermore, Perley and Kipnis also calculated the gut contribution to this phenomenon to be approximately 50% [230] GIP and GLP-1 are the incretins that are released in the gut in response to food intake. These two hormones have almost the same percentage of contribution to increase in insulin secretion. This incretin effect accounts for 50–70% of the total insulin secreted following oral glucose challenge [231–233] The first incretin to be identified, GIP, is a peptide of 42 amino acids produced predominantly in duodenal K cell (an open-type enteroendocrine cell) and in reduced expression in small intestinal mucosa. The precursor is a 153-amino acid molecule. GIP was initially thought to be gastric inhibitory polypeptide (enterogastrone) because it inhibited gastric acid secretion in dogs [234]. However, it was in 1973 that its insulinotropic effect was identified. Since it exhibited weak gastric inhibitory effects, hence the molecule was renamed "glucose-dependent insulinotropic polypeptide" [235] GIP binds to its receptor to initiate cellular response. The GIP receptor is a type II G protein-coupled receptor, the ligands of which (with a few exceptions) constitute members of the secretin–VIP–glucagon family of peptides. The GIP receptors are expressed in the pancreatic islets, gut, adipose tissue, heart, pituitary, adrenal cortex, and brain. The ligands of these receptors include ingested carbohydrates and lipids [221, 236–239]. GIP secretion also depends on health state and age as well as meal [239]. The secretion of this hormones is dependent on the availability of fat, carbohydrates, and certain amino acids—which activate SGLT1, GPR40 (FFA1), GPR119, and GPR120, resulting to downstream signaling mediated by alpha-gustducin [136, 137]		

(continued)

Table 8.1 (continued)

Hormone	Description	Discoverer	Year of discovery
	The functions of GIP include inhibition of acid secretion and insulinotropic activity. It is the main incretin factor of the enteroinsular axis. Its effects on increased insulin secretion ultimately lead to enhanced insulin-stimulated incorporation of fatty acids into triglycerides. This facilitates synthesis and deposition of fatty acids in tissues. Consequently, antagonism of GIP receptor can reduce obesity incidence. In addition, it increases insulin receptor affinity, and sensitivity of insulin-stimulated glucose transport. GIP activates lipoprotein lipase in adipose tissue. It has metabolic functions in liver, muscle, GI tract, brain, heart, adrenal cortex, pancreas, and adipose tissue—These tissues, organs, and systems have a high expression of GIP receptors [136, 137, 240] GLP-1 (7–36 amino acids), the second incretin hormone is believed to have a higher insulinopotentiation effect. GLP-1 is rapidly degraded by dipeptide peptidase-4 into GLP-1 (9–36 amino acids) or GLP-1 (9–37 amino acids) amide following its release from gut L (ultrastructurally "L" for large granules) cells of the distal ileum, which are the inactive forms of the native GLP-1 [231]. Only a very small proportion of GLP-1 passes into circulation. The half-life of this hormone is estimated to be about 2 min [136, 137]. Incretins are rapidly cleared from the body by the kidney within 1–2 min of their release [240]. The precursor, proglucagon is produced in the pancreas and "L" cells of the distal ileum and colon. Pancreatic proglucagon is cleaved to glucagon, glicentin-related pancreatic peptide and a fragment of proglucagon. However, in intestinal L cells, proglucagon is cleaved to produce GLP-1, GLP-2, and glicentin. All three molecules are released during food digestion. The production of these molecules from proglucagon depends on the enzyme type undergoing the cleavage process. For instance, the prohormone convertase PC2 is responsible for the processing of proglucagon to glucagon, whereas a related endoprotease PC3 (or prohormone convertase PC3) is responsible for the formation of GLP-1. Apart from GLP-1, GLP-2, and glicentin, cleavage of proglucagon also produces miniglucagon and oxyntomodulin. Oxyntomodulin (proglucagon 33–69) and glicentin (proglucagon 1–69) are important regulators of secretions of gastric acid and inorganic substances including water in the intestines. They reduce digestive secretions and decrease gastric emptying, at least in part, by reducing the duration of postprandial myoelectrical activity. Thus, they reduce gastric motility pattern in the fed state. In addition, oxyntomodulin is involved in		

(continued)

Table 8.1 (continued)

Hormone	Description	Discoverer	Year of discovery
	control of food intake and energy expenditure [220, 241–243]. At physiologic concentration, oxyntomodulin inhibits pancreatic secretion through a vagal mechanism [244]. Glicentin inhibits the secretion of glucagon but induces insulin release [245]. Both glicentin and oxyntomodulin receptors are coupled via specific type of G protein. The activation of these receptors leads to changes in cytosolic Ca^{2+} and cAMP. These hormones reduce forskolin-activated cAMP level by 30–40% and the effect is sensitive to the action of pertussis toxin [246]. Miniglucagon is an undecapeptide that inhibits glucose-, glucagon- and GLP-1-stimulated insulin release at subpicomolar concentrations. Surprisingly, miniglucagon lowers insulin response to glucose with no change in glycemia. This is probably due to its exhibition of insulin-like properties at the cellular level through some of insulin pathways [247]. The functions of GLP-2 include the regulation of growth of intestinal cells [248]. Although not fully understood, it is possible that each of the cleavage products of proglucagon might have their specific receptors in smooth muscle cells or other cell types. However, some molecules may signal through common pathways [220] The secretion of GLP-1 is dependent on the availability of monosaccharides, medium- and long-chain fatty acids. Importantly, physical exercise can also induce the secretion of this incretin. These factors are known to stimulate the release of this incretin by the activation of the taste receptors—T2Rs, TIR2-TIR3, GPR120, and LPAR5. The released GLP-1 stimulates GLP-1 receptors located on different targets including afferent endings of the vagus nerve, afferent neurons of specific tissues, and organs of the body [136, 137]. GLP-1 receptors are plasma membrane receptors belonging to the 7-transmembrane domain receptor family coupled to G proteins. They are expressed in the GI tract, endocrine pancreas, lung, kidneys, heart, and brain (hypothalamus, nucleus of the solitary tract, area postrema) [249–252] The functions of GLP-1 include stimulation of insulin secretion, increase in beta cell mass. The hormone also inhibits glucagon secretion, inhibits gastric emptying, and reduces food intake. Thus, GLP-1 plays an important role in glucose homoeostasis. By inhibiting food intake, it prevents body weight increase. GLP-1 has specific effects on pancreatic alpha cells, hypothalamus, GI, and cardiovascular systems [219]. GLP-1 also mediates decrease in gastric acid secretion, cell growth, and proliferation [136, 137]		

(continued)

Table 8.1 (continued)

Hormone	Description	Discoverer	Year of discovery
	The properties of GLP-1 have enhanced the application of this hormone into clinical practice. Thus GLP-1 receptor agonists are effective agents in the treatment of diabetes type 2 [253]. But since this incretin, including GIP are rapidly degraded by the enzyme dipeptidyl peptidase-4, its action wanes very quickly. This has led to the development of dipeptidyl peptidase-4 inhibitors, which have shown promise for the treatment of diabetes type 2. Both GLP-1 receptor agonists and dipeptidyl peptidase-4 inhibitors are known to effectively decrease blood glucose levels [254]		
Enterogastrones	The name of the hormone is derived from “entero-” (substance of the intestine), “-one” (that inhibits), “gastro-” (the stomach). Enterogastrones simply refer to hormones that decrease the secretory activity of the gut—the substances produced in the mucosa of the duodenum in response to dietary lipids and inhibit the stomach secretion and large bowel movement of chyme [29]. Under conditions of elevated blood sugar, the hormones stimulate insulin release. Examples of enterogastrones include secretin, CCK, and GIP [239, 240, 255, 256]	Julian Walawski (1898–1979) and Jerzy Kaulbersz (1891–1986)	1920s
Endocrine GI leptin	A 16-kDa satiety protein hormone, encoded by the obesity gene, initially discovered as adipose tissue hormone in 1994. The hormone is secreted by mature white adipocytes. But leptin is now known to be secreted by other tissues. The hormone acts locally in peripheral and central tissues. In the peripheral tissues, it mainly regulates growth, plasticity, motility, inflammation, metabolism, and absorption of nutrients. In the central tissues (hypothalamus), it regulates appetite, food intake, energy expenditure, body composition, and activity of the sympathetic nervous system. The exocrine and endocrine cells located in the gastric mucosa actively secrete leptin through a regulated mechanism involving the rough endoplasmic reticulum–Golgi-granule secretory pathway of chief and specific endocrine cells [259–261] Leptin receptors are present on the luminal and basolateral membranes and cytoplasm as well as plasma membranes of gastric, intestinal (duodenal, jejunal, ileal) enterocytes and have been shown to regulate several intestinal activities [262, 263]	Zhang and colleagues discovered leptin in 1994 [257]. But the gastric origin of leptin was first reported in 1998 by Bado et al. [258]	1994

(continued)

Table 8.1 (continued)

Hormone	Description	Discoverer	Year of discovery
	Leptin is not affected by acidic conditions of the stomach as it binds to a protein that gives it a protective covering. The protease activator furin and proprotein convertase 7 found in chief cell granules probably confer a protective role on leptin in the stomach and duodenum where acid concentration is high [258, 264, 265]. The hormone is also present in the salivary glands [266]. Exocrine leptin activates CCK (see CCK functions below) Endocrine secretion of leptin increases during fasting. In this physiological state, the concentration of leptin increases in both the serum and gastric mucosa [267]. Although the mechanisms are not clearly understood, during refeeding gastric endocrine release of leptin is increased. This, in part, may be due to the tandem pattern of leptin-pepsinogen secretion from the chief cells [263, 268]		
Motilin	A 22-amino acid GI peptide, secreted by duodenal mucosal endocrine M cells. Motilin was discovered on the basis of the contractile effect of extracts of duodenal mucosa from pigs on canine stomach. The peptide is released upon neural or mechanical stimulation. The pattern of release is usually cyclically occurring every 90 min postprandially. The hormone regulates the motility of the digestive tract. Motilin increases contractility (motility) of the small intestine [269] Motilin and other related intestinal peptide hormones such as villikinine, enterocrinin, and duocrinin increase the contractile activity of the villi and microvilli of the intestine and associated structures. Villikinine stimulates the motility of the villi. Enterocrinin promotes the production of large amounts of alkaline mucus by the submucosal glands of the small intestine. Duocrinin stimulates the secretory activity of the duodenal glands (Brunner's glands) [270–272] The secretion of motilin is stimulated by gastric distension, lipids, bile acids, low pH in the duodenum. The release of motilin can be induced by neural impulses. The cognate receptors of these nutrients (e.g., bile acid receptors) stimulate motilin release [136, 137]. The action of motilin is mediated via motilin receptors present in specific areas of the GI tract [275, 276]	The hypothesis on the presence of a substance in duodenal extracts that stimulated gastric motor activity was initially made by Brown in 1967 [273]. The substance "motilin" was isolated from a side fraction produced during the purification of secretin by Brown, Mutt, and Dryburgh in 1971 [218]. The amino acid sequence of motilin was determined in 1973 [274]	1971

(continued)

Table 8.1 (continued)

Hormone	Description	Discoverer	Year of discovery
Gastric ghrelin	A hunger peptide hormone, containing 28 amino acids produced by ghrelin cells in the GI tract and functioning as a neuropeptide in the CNS. The name "ghrelin" is formed from the old time European language root word "ghre," meaning to grow. Therefore, the hormone name, in other words, is "growth hormone release-inducing or growth hormone releasing peptide" [269, 277–279] In humans, ghrelin is produced by P/D_1-like cells. This cell type also produces other related peptides such as acyl ghrelin, des-acyl ghrelin, obestatin, and nesfatin-1. In other members of the animal kingdom, for instance, in rodents, ghrelin producing cells are designated X/A-like cells. About 20% of ghrelin-producing cells are located in the oxyntic glands, while glands in the pyloric region of the stomach and small intestine produce the remaining 80%. But ghrelin-producing cells have low expression in the intestines. However, ghrelin may be produced by other tissues and organs including the exocrine pancreas [136, 137, 269, 277–279] The secretion of the hormone gradually increases following stomach emptying (fasting). Only the basal secretion is maintained when the stomach is filled with food substances. Thus, ghrelin-producing cells are activated on fasting. Food substances like carbohydrates and long-chain fatty acids inhibit the activities of these cells [136, 137, 284, 285] Ghrelin and the closely related peptides are involved in body weight regulation. In particular, ghrelin stimulates food ingestion. However, the ghrelin-related peptides may have varying and independent roles in the regulation of body weight and food ingestion [284–286] The mechanism of regulation of food ingestion by ghrelin is based on its central signaling effects. The hormone acts on the hypothalamus by increasing hunger, gastric secretion, and GI motility. Both ghrelin and leptin receptors are found on the same cells of the hypothalamus. Besides the regulation of feeding, ghrelin plays a role in modulating the activity of neurons of the ventral tegmental area and nucleus accumbens of the brain reward, perception and reinforcement systems. Therefore, disorders in	The receptor (called growth hormone secretagogue-type 1A) of this hormone was first discovered and cloned before the hormone itself was discovered. Ghrelin was discovered in the stomach in 1999 by Kojima et al. [280–282]. In 2000, ghrelin was rediscovered as motilin-related peptide. Motilin-related peptide is a hormone produced in gut and acts to stimulate smooth muscle motility patterns in the GI tract. It enhances MMC [283]	1996; 1999

(continued)

Table 8.1 (continued)

Hormone	Description	Discoverer	Year of discovery
	feeding behavior could be related to the dysfunctions of some neural pathways in these brain regions [269, 277–279]		
Cholecystokinin-Pancreozymin (CCK-PZ)—shortened to "CCK" (Greek "chole" for bile, "cysto" for sac, "kinin" for movement. So, it altogether means move the bile-sac—cholecyst, which is the gallbladder) CCK and PZ are the same hormone [see details on third column]	A satiety peptide hormone synthesized by the triangular-shaped endocrine I cells in the mucosal epithelium of the proximal small intestine, responsible for stimulating the digestion of fat and protein. This hormone and neuropeptide belong to the CCK and gastrin families of peptides. CCK is highly homologous to gastrin. There are different lengths of CCK, which include CCK-4, CCK-8, CCK-27, and CCK-33 [141] CCK-PZ receptors belong to the CCK/gastrin family of receptors, which are GPCRs. There are two classes of CCK receptors to which either CCK or gastrin can bind to: CCK1R (CCK-AR) and CCK2R (CCK-BR). However, the affinity of CCK for CCK-AR is higher compared to CCK-BR. The activation of these receptors leads to stimulation of PLC and adenylyl cyclase [289, 290]. The duodenal I cells express a range of receptors which include alpha-gustducin, T2R, FFA1, LPAR5, PepT1, CaSR, and FAT (fatty acid transporter). Activation of these receptors by lipids (mostly long-chain fatty acids) as well as amino acids activates the pool of basally located granules containing CCK and other peptides, which results to their subsequent secretion. For instance, FAT receptor is activated by fatty acids in the duodenum to increase intracellular calcium signaling which is mediated via cAMP pathway independent of PKA to increase both the expression and release of CCK. Calcium released by FAT activity may be mediated via CaM-KII activation [132] CCK is produced in both peripheral and central tissues and cells of the body [290]. Central CCK takes part in the regulation of learning, memory, thermoregulation, satiety, anxiety, analgesia, and dopamine-mediated behavior [290, 292, 293] CCK facilitates nutrient digestion by inhibiting gastric emptying, reducing gastric acid secretion, stimulating gallbladder contraction and pancreatic enzyme secretion [136, 137]. The hormone is involved in GI motility and secretion [289]. Importantly, CCK enhances the release of enzymes from the pancreas and bile from the gallbladder [141].	CCK was discovered by Andrew Conway Ivy (1893–1978) and Erick Oldberg [287]. PZ was discovered by Harper and Raper [288]. Following the discovery of PZ in 1943, researchers realized that either PZ or CCK could stimulate the contraction and evacuation of both the gallbladder and pancreatic enzyme secretion. The successful purification of CCK in 1966, enabled scientists to have a glimpse of the structure of this hormone [56, 291]. The structural identity of CCK to PZ led scientists to rename the hormone CCK-PZ	1928; 1943

(continued)

Table 8.1 (continued)

Hormone	Description	Discoverer	Year of discovery
	Increasing investigations over the past decades have shown that the hormone also acts as a hunger suppressant by promoting satiety, delaying gastric emptying. Thus, it could have useful role in pathological conditions such as obesity and bulimia (binge eating syndrome). CCK also potentiates insulin secretion and regulates inflammatory reaction in the gut. The secretion of CCK is associated with significant reduction in intestinal inflammation. Substantial ingestion of dietary fats can induce CCK elevation. Hence, in condition when fats cannot adequately elevate CCK, the organism becomes prone to the development of inflammation in the gut [293, 294] CCK can potentiate the effect of secretin. It stimulates glucagon and insulin release. It also has a growth stimulatory effect on the exocrine pancreas [295–297]. CCK-2 receptor antagonists such as spiroglumide and itriglumide possess antisecretory and antiulcer effects, which is achieved by reducing acid production. These CCK antagonists have high affinity for CCK [298]. A selective cholecystokinin CCK-A receptor antagonist, 2-naphthalenesulphonyl l-aspartyl-(2-phenethyl) amide (2-NAP) was found to have at least 300-fold greater affinity for CCK-A receptors than gastrin and CCK-BR [299]		
GI somatostatin	Somatostatin is an endocrine, local (paracrine) regulator, and neurotransmitter tetradeca- or octacosapeptide (14 or 28 amino acids) produced by the D- (delta) cells, responsible chiefly for inhibiting the release of lumencrine (exocrine), endocrine, and neural secretions [300, 301]. But a somatostatin analogue with eight amino acids (octapeptide) is available [302] Somatostatin is found in both central and peripheral tissues. In the peripheral tissues, the GI tract (especially the duodenum) and pancreas are most abundant sites of production of somatostatin. However, the peptide is expressed throughout the gut in luminal (mucosal) D cells, enteric neurons and gut-associated lymphoid tissue [305, 306]. Somatostatin-producing D cells in different regions of the GI tract differ in their morphofunctional properties. Relatively more recently, however, the cells that produce this hormone have been found in virtually every organ of the body [300, 307]	Somatostatin was isolated from sheep hypothalamic extract by Roger Guillemin (1924–) and coworkers [303, 304] The hormone was named for inhibiting secretion of growth hormone from the pituitary gland. Growth hormone secretion is controlled by somatostatin and growth hormone–releasing hormone (GHRH). GHRH and somatostatin are secreted by the hypothalamus. Following the initial discovery, somatostatin was found to be secreted by a broad range of tissues, including pancreas, intestinal tract, and regions of the central nervous system outside the hypothalamus [303, 304, 313]	1972

(continued)

Table 8.1 (continued)

Hormone	Description	Discoverer	Year of discovery
	There are two forms of somatostatin with varying physiological activities: SS-14 and SS-28. The cytoplasm of D cells contains secretory granules of both SS-14 and SS-28. Either form of somatostatin is generated by the proteolytic cleavage of prosomatostatin. This propolypeptide is derived from preprosomatostatin. The secretory pool of vesicles containing somatostatin in the cytoplasm is made up of different proportions of SS-14 and SS-28. The proportion of these two molecules depends on the tissue where they are synthesized. In central tissues, SS-14 is the predominant form, whereas both forms occur in varying proportions in peripheral tissues depending on the tissues involved [308–311] The pancreas produces mostly if not only-14, whereas the gut produces mostly SS-28. The SS-14 is more potent in inhibiting glucagon release than the SS-28 [312]. However, the SS-28 is about tenfold more potent in inhibiting growth hormone secretion compared to the SS-14 [308–311] The release of somatostatin is regulated chemically via GPRC6A (G protein-coupled receptor family C group 6 member A—receptor–sensor of L-amino acids, calcium, and steroids), CaSR, and LPAR5 receptors. It was shown that gastric calcitonin gene-related peptide (CGRP) as well as exogenous CGRP stimulates gastric somatostatin release [136, 137, 314–317]. Somatostatin release can be modulated by neural and mechanical stimulation. Somatostatin exerts its activities via the five somatostatin receptors (SST_{1-5}), which belong to the G protein-coupled receptor superfamily. However, the main mediators of GI activities are SST_2 and SST_5. Somatostatin and its receptors are ubiquitously expressed with high expression levels in the GI tract, pancreas, and brain. All five receptors are known to inhibit adenylyl cyclase. However, each of the five receptors activates distinct signaling mechanisms [136, 137] Somatostatin family of peptides produced by the somatostatin gene product, prosomatostatin includes two other peptides—cortistatin and the more recently identified neuronostatin [318]. Cortistatin and neuronanostatin are briefly discussed in the column below		

(continued)

Table 8.1 (continued)

Hormone	Description	Discoverer	Year of discovery
	Somatostatin exerts an inhibitory action on numerous physiological functions [300]. In central tissues, somatostatin is involved in neuromodulation, neural transmission, and regulation of growth. It may be associated with arousal and decreased sleep, impairment of certain aspects of motor responses. In the peripheral tissues via paracrine mechanism, somatostatin inhibits the secretion of GI hormones, including gastrin, VIP, glucagon, insulin, CCK, secretin. Thus, somatostatin may inhibit pancreatic exocrine secretions via CCK/secretin-dependent pathway. This way the peptide reduces smooth muscle contraction and blood flow within the intestine. In the stomach, somatostatin suppresses secretion of gastric acid and pepsin, thus lowering the rate of gastric emptying and delaying gastric transit time. Collectively, these activities seem to have an overall effect of decreasing the rate of nutrient absorption [300, 319] In the intestine, somatostatin may be involved in regulating water and electrolyte absorption and secretion. The peptide affects both the epithelial transport and intestinal motility. This secretory and motility function of this peptide may be, at least, in part due to its effect on CCK and secretin or other peptides [305, 319] Dysfunction in somatostatin production or overexpression of somatostatin receptors has been associated with certain pathological conditions, particularly neuroendocrine tumors of the GI tract [300, 306]. The majority of carcinoids and islet cell carcinomas (including their metastases) have a high density of somatostatin receptor expression. However, only a very few of this receptor is expressed in colorectal carcinomas [306]. Somatostatin reduces stool output in chronic diarrhea due to endocrine tumor [305]		
Neuronostatin	Neuronostatin is a somatostatin-like peptide of 13 amino acid residues first identified from porcine tissues [320]. The peptide is encoded by the preprosomatostatin gene [322] Neuronostatin is produced in central (anterior pituitary, cerebellum, and hippocampus) and peripheral (GI) tissues [320, 321]. In the pancreas, for instance, neuronostatin is produced δ-cells of the endocrine pancreas [323] The neuropeptide function via stimulation of somatostatin receptors [321]. The peptide also acts via receptor—GPCR107 [323]	The peptide was discovered by Willis K. Samson and coworkers [320]	2008

(continued)

Table 8.1 (continued)

Hormone	Description	Discoverer	Year of discovery
	Neuronostatin in the central tissues directly depolarizes paraventricular neurons of the hypothalamus. This action of the peptide regulates the production of hormones by the anterior pituitary. The neuropeptide is found in other brain regions and may modulate learning, memory, and growth. Generally, this somatostatin-like neuropeptide increases blood pressure, suppresses food intake and water drinking, and plays a key role in glucose homeostasis. The action of neuronostatin on glucose homeostasis is mainly modulated via α-cells. Neuronostatin promotes glucagon secretion at low glucose concentration in pancreatic α-cells but seems not to have any direct effect on insulin release from the β-cells. Surprisingly, neuronostatin inhibits glucose-stimulated insulin secretion from pancreatic islet cells. Neuronostatin maintains glucose homeostasis by reducing insulin secretion. This suggests that the action of neuronostatin on insulin may be indirect and mediated via substances released from α-cells [320, 321, 323] Neuronostatin may play a role in glucose dysallostasis, GI cancers and metastasis		
Cortistatin (CORT or CST)	CORT is a somatostatin-like neuropeptide with 14-amino acid produced by cleavage of 105-amino acid residue precortistatin, which is produced from the 112-amino acid precursor, preprocortistatin. The precursor protein of cortistatin is produced from the CORT gene located on chromosome locus 1p36.22 [324]. CORT belongs to the somatostatin neuropeptide family, having strong structural similarities with somatostatin. The cleavage of precortistatin produces different forms of the neuropeptide: CORT-14, CORT-17, CORT-29. CORT-17 is one most active peptide expressed in inhibitory neurons of the cerebral cortex [324, 325]. The neuropeptide is named after its predominantly cortical expression and ability to depress cortical activity [326] Cortistatin binds to all known somatostatin receptors [326] Cortistatin is produced in both central and peripheral tissues, and its receptors are widely distributed in these tissues. In the central nervous system, the neuropeptide is found in the cortex, hippocampus, and amygdala. In peripheral tissue, it is produced in GI tract, nociceptive neurons and immune system cells including macrophages, lymphocytes, and T helper cells [324–329]	Luis de Lecea and coworkers discovered CST as cortical neuropeptide with neuronal depressant and sleep-modulating properties [327]. The subsequent year, they succeeded in cloning the precursor peptide, preprocortistatin [324].	1996

(continued)

Table 8.1 (continued)

Hormone	Description	Discoverer	Year of discovery
	CORT enhances slow-wave sleep, reduces locomotor activity, depresses neuronal activity, deactivates inflammatory responses, and plays a role in nociception [324, 326] CORT is involved in the pathogenesis of inflammatory disorders such as Crohn's disease, colitis, endotoxemia, sepsis, septic shock, arthritis, and multiple sclerosis [328]. Administration of CORT downregulates inflammatory and T helper cell-mediated autoimmune responses and regulates inflammatory mediators. This way, the neuropeptide, in the GI tract for instance, is able to deactivate intestinal inflammatory response and restores mucosal immune tolerance [329]		
Thyrotropin-releasing hormone (TRH)	Thyrotropin-releasing hormone (TRH) [also known as thyrotropin-releasing factor (TRF) or thyroliberin] is a releasing hormone, produced primarily by the hypothalamus, to stimulate the release of thyrotropin (thyroid-stimulating hormone or TSH) and prolactin from the anterior pituitary [330]. TRH is a tripeptide pyroglutamyl–histidyl–prolineamide (pyro-Glu–His–Pro-amide) neuropeptide, hypothalamic releasing factor involved in controlling the secretion of all (anterior) pituitary hormones [331] TRH is produced not only in the brain, but also in GI tract, and pancreas [332]. Not only is the GI tract origin of the hormone known to affect the GI functions but also the central TRH, which influence smooth muscle and secretory activity in the GI tract [333]. The cerebrogastrointestinal pathways send efferent viscerotropic projections to the gut and influence GI motility and secretion [334]. TRH influences the activity of some GI hormones. For instance, it inhibits gastrin-stimulated gastric acid secretion. However, TRH do not influence the activity of secretin [335]. The effects of TRH are opposite to those of bombesin, endorphin, and neurotensin [334]. Bombesin and endorphin are discussed in the next chapter TRH may play a role in gastric ulcers and other GI pathologies [334]	Around the 1950–60s, the French neuroendocrinologist, Roger C. L. Guillemin (1924–), and a Polish scientist, Andrew Schally (1926–) were the first researchers to show that hypothalamic extract control the release of hormones from the pituitary and thyroid glands as well as gonads. For further details see Thorner (1999) [335] and Bauer et al. (1999) [336] They subsequently isolated and characterized the tripeptide during the turn of the 70s of the last century [337–339]. Other contributors to the discovery of the peptide include but are not limited to Paul Brazeau and Wylie Vale [337, 340]	1968
Galanin (GAL)	GAL is a 29-amino acid peptide hormone, neurotransmitter, neuromodulator, and trophic factor named after its N-terminal glycine and C-terminal alanine. The peptide was identified by C-terminal amidation in porcine intestinal extract. GAL is produced from the gene product, preprogalanin [341, 342]	The peptide was discovered by Kazuhiko Tatemoto, Viktor Mutt, and colleagues at Karolinska Institute, Stockholm [342]	1983

(continued)

Table 8.1 (continued)

Hormone	Description	Discoverer	Year of discovery
	GAL is produced in the nervous system (central—hypothalamus, arcuate nucleus, septum, neurointermediate lobe of the pituitary, and the spinal cord, and peripheral nervous tissues), GI tract, as well as in enteroendocrine, neuroendocrine cells, pancreas, and tumors [342, 343]. Galanin is also expressed in keratinocytes, eccrine sweat glands, blood vessels, and macrophages [344] GAL functions by activation of its receptors: GALR1, GALR2, and GALR3. These receptors belong to the G protein-coupled receptor (GPCR) superfamily [345, 346] Apart from GAL-29, GAL with amino acids (GAL 2–11), (GAL 1–11) lengths has been shown to act on GAL receptors [346]. Although discovered as a 29 peptide, the human galanin is 30 amino acids in length. About 50% of the structure of galanin is conserved among various species [344] The galanin family of peptides consists of galanin, galanin-like peptide, 60-amino acid galanin-message associated peptide, and alarin. These peptides are produced from the galanin prepropeptide [344]. The galanins play a variety of roles (physiological and pathological) in health and disease in both central and peripheral tissues [344]. GAL is involved in a number of physiological processes such as hormone and neurotransmitter secretion, cardiovascular activity, feeding, and cognition. The neurotransmitter may be involved in modulating the secretion of several hormones and neurotransmitters co-secreted together with the neuropeptide, including adrenaline, noradrenaline, galanin, including acetylcholine, serotonin, glutamate, GABA, noradrenalin, dopamine, enkephalin, NPY, substance P, vasopressin, calcitonin gene-regulated peptide, gonadotropin-releasing hormone [136, 137, 344]. Galanin stimulates the secretion of growth hormone and prolactin. Galanin inhibits the release of somatostatin, insulin, pancreatic peptide, and dopamine. Galanin stimulates food intake [136, 137]. Galanin can cause the contraction of smooth muscle of the GI and genitourinary tracts. The neuropeptide possesses hyperglycemic effect which may be mediated via multiple pathways including inhibition of insulin release [136, 137, 342]		

(continued)

Table 8.1 (continued)

Hormone	Description	Discoverer	Year of discovery
	The galanins are involved in a host of pathological conditions. The peptide is upregulated in Alzheimer's disease, nerve injury involving sensory and motor neural systems. However, a depletion of galanin was observed in epilepsy [344]. Down- or up-regulation of this peptide may be due to the increased expression of certain factors. Up-regulation of galanin expression may occur via the action of leukemia inhibitory factor, whereas down-regulation takes place through the activity of nerve growth factor [344]. This peptide may play a role in nociception, depression, addiction, tumorigenesis, and metastasis Several analogues of GAL have been synthesized. For review see [347, 348]		
Amylin (AMY) or islet amyloid polypeptide (IAPP)	IAPP is a 37 amino acid peptide hormone that was originally identified as the chief constituent of amyloid in insulinoma and islets of patients with non-insulin-dependent diabetes mellitus [349, 350]. IAPP is structurally similar to calcitonin gene-related peptide (CGRP), adrenomedullin, and intermedin [351, 352]. Consequently, IAPP is considered a member of the calcitonin family of peptides, which include calcitonin, CGRP, adrenomedullin, and intermedin. Wimalawansa (1997) suggested that the calcitonin family peptides be referred to as insulin superfamily of peptides due to the similarity of B-chain of insulin to the peptides [353]. They also have similar biological activity such as growth factor-like effects. These facts indicate that the peptides may have diverged from a common ancestral gene during the course of evolution [353] Amylin is produced in mucosa of the GI tract (especially stomach and duodenum), Beta- and D cells of the islets of Langerhans [352, 354]. Amylin colocalizes with somatostatin in endocrine cells of the gastric fundus [352]. The peptide is co-stored and co-secreted with insulin in pancreatic B cells [355] The secretion of amylin is stimulated by both receptor-dependent and receptor independent mechanisms. Receptor-dependent stimulation of either of the Ca^{2+}/PKC and adenylate cyclase/cAMP pathways leads to secretion of amylin [355]. The inhibition of amylin release involves the activation of muscarinic receptors and auto-regulation by somatostatin [352]	IAPP was identified as the major component of diabetes-associated islet amyloid deposits, by two independent groups led by Cooper and Westermark [349, 350]	1987

(continued)

Table 8.1 (continued)

Hormone	Description	Discoverer	Year of discovery
	Activation of PKC increases the release of amylin. Activation of adenylate cyclase with forskolin also leads to increase in amylin secretion [355]. Other hormones and neurotransmitters can initiate the secretion of amylin. CCK, GLP-1, and epinephrine activate amylin secretion [355]. Amylin secretion can be inhibited with carbachol and octreotide. In physiological and pathological conditions, amylin is co-secreted with other hormones and neurotransmitters [355] The action of amylin is mediated via AMY_1, AMY_2, and AMY_3 receptors—heterodimers formed by the complexation of receptor activity modifying proteins (RAMPs1—3) with calcitonin receptor. RAMPs, calcitonin receptor and calcitonin-like receptor comprise the calcitonin family of peptide receptors [356–359]. RAMP is a single-domain protein and not necessarily a receptor. The calcitonin receptor is a type of G protein-coupled receptor [360]. The AMY receptors have several isoforms [361] Amylin possesses variety of physiological responses on activation. Amylin from the gastric fundus stimulates somatostatin and thus inhibits histamine and acid secretion. In gastric fundus, the released of amylin from somatostatin cells interacts with distinct amylin receptors to enhance somatostatin secretion via an autocrine pathway that leads to inhibition of histamine and acid secretion [351, 352]. Thus, amylin decreases acid secretion [352]. The peptide possesses diabetogenic effects by virtue of its ability to inhibit pancreatic insulin release and to interfere with insulin action [352]. Thus, amylin is involved in the regulation of glucose homeostasis [355]. Amylin inhibits food intake (anorectant effect) and gastric emptying [362]. The peptide is also involved in the regulation of gastric motor and secretory function as well as gastroprotection [355]. Intravenous amylin is a potent inhibitor of basal, pentagastrin, and 2-deoxy-D-glucose stimulated gastric acid secretion in the rat [355]. Amylin inhibits basal and insulin-stimulated glycogen synthesis, it inhibits insulin, somatostatin, and glucagon secretion from pancreatic islets, and it reduces postprandial glucagon and insulin secretion. Amylin may also contribute to glycemic control by slowing gastric emptying [355] Pramlintide, an amylin analogue, delays gastric emptying, possibly by a centrally mediated mechanism [363].		

(continued)

Table 8.1 (continued)

Hormone	Description	Discoverer	Year of discovery
Adrenomedullin (AM)	AM is a vasodilator peptide hormone and neurotransmitter of the calcitonin peptide family, consisting of 52 amino acids, produced from a 185-amino acid precursor, preproadrenomedullin—the adrenomedullin gene product [364]. The processing of this precursor forms not only AM, but also proadrenomedullin N-terminal 20 peptide (PAMP) [365, 366] The hormone is expressed in peripheral and central tissues. In the GI tract, the peptide is expressed in stomach, duodenum, jejunum, ileum, colon, liver, and pancreas [366, 367]. The peptide is produced by endocrine cells of the GI tract, subpopulation of the enterochromaffin (serotonin-containing) cells [368]. Immunochemistry analysis conducted by Mulder et al. (1996) showed that AM is not expressed in the pancreas, however, the peptide was found to stimulate insulin release from isolated rat islets in the presence of glucose [368]. Therefore, AM possesses insulinotropic action [368]. The mechanism of action of the peptide on insulin secretion is not exactly known, but it may be related to its stimulation of incretins AM actions are mediated via CGRP type 1 (CGRP-1) receptors and by specific AM receptors. CGRP-1 receptors are represented by CRLR (calcitonin receptor-like receptor) and RAMP1. AM receptors (AM_1 and AM_2) are formed from a complex of the **GPCR superfamily RAMP2 or RAMP3 with RAMP 2 or RAMP 3**—CRLR/RAMP2 or CRLR/RAMP3 [360, 369, 370]. CRLR, like other type II family G protein-coupled receptors, signals via stimulation of G(s) and adenylate cyclase. The activation of AM_1 and AM_2 leads to the stimulation of Gs subunit of the GPCR, which leads to activation of adenylate cyclase [366] AM is important in the regulation of nervous system, cardiovascular, and respiratory functions [371]. AM circulates in the plasma; thus, it exerts a wide range of physiological effects. The hormone was found to elicit a long-lasting hypotensive effect [364, 366]. The hormone increases mucosal blood flow, thus enhancing angiogenesis, proliferation, and restitution of epithelial cell, which all leads to promotion of gastroprotection [366]. This gastroprotective effect might be due to, in part, by	Adrenomedullin was first isolated by Kitamura K and coworkers [364] as a hypotensive peptide from human pheochromocytoma, a tumor of the adrenal medulla, hence the name	1993

(continued)

Table 8.1 (continued)

Hormone	Description	Discoverer	Year of discovery
	inhibition of histamine and acid secretion via stimulation of somatostatin secretion. The inhibition of gastric emptying by AM might play a role in weight control and obesity. This leads to reduced gastric transit time which may be due to the inhibition of smooth muscle activity, mediated via cAMP-dependent pathway. AM has the highest potency among the calcitonin peptide family but equal potency with CGRP in inhibiting CCK-8-induced contraction of smooth muscle cells [372] The hormone has been found to inhibit pancreatic secretion of amylase. AM also inhibits sodium absorption in the colon [372] Emerging studies indicate a possible role of the peptide in human pregnancy complications [373]. The peptide might be involved a variety of pathologies involving the cardiovascular system [374, 375], infertility, impotence, varicocele [376], tumor angiogenesis [377]		
Intermedin also called adrenomedullin-2 (IMD or AM2)	IMD is a peptide hormone, neuromediator, and neurotransmitter belonging to the calcitonin peptide family peptide [371] produced from the 148 (146–150 amino acid) prepropeptide encoded by a gene located on the distal arm of human chromosome 22q3. Cleavage of the precursor at the N-terminus produces different lengths of peptides including the 47 amino acid form of the peptide, IMD_{1-47}, and a 40 amino acid peptide IMD_{8-47} [379, 380]. However, proteolytic processing of the precursor may yield a third biologically active C-terminal fragment, IMD_{1-53} [379, 380] IMD is distributed throughout the human body, but the peptide is less extensively expressed in mammals compared with AM. IMD is found in hypothalamus, pituitary, heart, kidney, GI tract, plasma, pancreas, lung, spleen, thymus, skin, submaxillary gland, pancreas, lung, spleen, thymus, and ovary, but not in testis or adrenal gland. The highest distribution of the peptide is found in kidney, pituitary, hypothalamus, and stomach [379, 380]. In the hypothalamus, IMD is colocalized with arginine vasopressin, and in intermediate and anterior lobes of pituitary [379]. In the GI tract IMD is especially found in the muscularis mucosa of stomach and in jejunum [371]. The cardiovascular distribution of IMD has some peculiarities as it is present in neonatal heart and blood vessels at considerable levels, but sparsely distributed or absent in adult heart and blood vessels [379]	Two groups of researchers discovered the peptide: [371, 378]. The peptide was named IMD due to its abundant expression in the intermediate lobe of the anterior pituitary	2004

(continued)

Table 8.1 (continued)

Hormone	Description	Discoverer	Year of discovery
	IMD acts through the CGRP receptors and CRLR/RAMP complexes [371]. Recall that the CGRP receptor is formed from heterodimerization of CRLR/RAMP1. Compared to CGRP, IMD is less potent at activating $CGRP_1$ receptors but more potent at AM_1 receptors and AM_2 receptors. Compared to AM, IMD is more potent at stimulating $CGRP_1$ receptors but less potent at AM_1 and AM_2 receptors [379–381]		
	IMD possesses a wide range of physiological actions including local and systemic effects in central and peripheral tissues. IMD reduces blood pressure in both normal and hypertension [371, 380]. The peptide functions as a systemic and pulmonary vasodilator. However, in experimental conditions, the hypotensive effect of the peptide is observed only when given peripherally. Central injection of the peptide into the CNS causes sympathetic activation and increases blood pressure [380]. IMD augments cardiac contractile activity and protects myocardium from the deleterious effects of oxidative stress associated with ischemia-reperfusion injury. It has an antigrowth effect on cardiomyocytes, a process necessary to oppose the influence of hypertrophic stimuli [379–382]. In the GI tract, IMD suppresses gastric emptying and inhibits food intake—which are important mechanisms in weight control and obesity management [371]		
	IMD also functions as an antidiuretic and a natriuretic hormone [380]. The localization of IMD in hypothalamus, pituitary, and kidney points to its role in central and peripheral regulation of water and electrolyte homeostasis [379]. The functions of IMD described above suggest that the hormone acts as an auto-paracrine regulator in the hypothalamo–pituitary–adrenal axis. The peptide and its receptors are expressed in the paraventricular and supraoptic nuclei of the hypothalamus, cells of the anterior pituitary and adrenal gland, especially the medulla. Hypothalamic IMD colocalizes with arginine vasopressin and stimulates the secretion of ACTH, prolactin, and oxytocin. Hypothalamic IMD suppresses the secretion of GH [383, 384]		
	IMD and its receptors are abundantly expressed in adrenocortical tumors (e.g., aldosterone-secreting adenomas, and pheochromocytomas) [383]		

(continued)

Table 8.1 (continued)

Hormone	Description	Discoverer	Year of discovery
Enteroglucagon	Enteroglucagon is the gut glucagon derived from preproglucagon precursor synthesized primarily by the gut L-type endocrine cells of the terminal ileum and colon [385]. Enteroglucagon is a distinct peptide with proglucagon activities derived from the product of the glucagon gene, preproglucagon. The cleavage of this precursor produces glucagon and GLP-1, (from proglucagon$_{78\text{-}107}$), GLP-2 (from proglucagon$_{126\text{-}158}$), oxyntomodulin (from proglucagon$_{33\text{-}69}$), glicentin (from proglucagon$_{1\text{-}69}$), glicentin-related pancreatic polypeptide (from GRPP, proglucagon$_{1\text{-}30}$), the major proglucagon fragment, MPGF [386], intervening peptide-1 & 2 (IP-1 & 2, from proglucagon$_{111\text{-}122}$) [386–388]	The word "glucagon" came into existence when Kimball and Murlin first extracted a pancreatic substance known to have opposing effect to the then discovered insulin [389]. The substance was later called glucagon, which means glucose-giving. It is a hormone with 29 amino acids and a molecular weight of 3485. During that time, it was known that glucagon is produced only by the pancreas [390]	
	Apart from the pancreas and gut, the peptide is produced in neurons of the nucleus of the solitary tract in the hindbrain. But proglucagon is processed in a tissue-specific manner [253]. In the pancreas, for instance, proglucagon processing by the prohormone convertase 2 (PC2) generates predominantly glucagon. In intestinal L cells of the jejunum, ileum, and colon, the prohormone convertase enzyme is PC 1/3, which predominantly produces glicentin, OXM, GLP-1, and GLP-2. The major proglucagon products produced in the brain are GLP-1 and GLP-2 [253] Enteroglucagon decreases intestinal motility to allow sufficient time for nutrients to be absorbed [385] Enteroglucagon is released when fats and glucose are present in the small intestine; However, this peptide may also be produced by the cells of the stomach [385]	The concept of gut glucagon was first proposed by Christian de Duve (1917–2013) and Earl Wilbur Sutherland, Jr. (1915–1974) among others around the middle of the last century, precisely in 1948, when they described the glucagon-like substance in the intestinal mucosa [391, 392]. The invention of biochemical technique of peptide analysis, precisely, the radioimmunoassay (RIA), made it possible to confirm the glucagon-like immunoreactivity in intestinal extracts [393, 394], and the substance was named enteroglucagon. However, there were observed peaks of several lengths in the chromatographic technique used in the study. Based on speculation that the extracted substance may contain several other substances, the name enteroglucagon was given the name glucagon-like immunoreactivity. The original name enteroglucagon is still valid and may refer to several hormones produced from the proteolytic cleavage of preproglucagon precursor [395, 396]. More often, the peptides are referred to the proglucagon derived peptides [397]	

(continued)

Table 8.1 (continued)

Hormone	Description	Discoverer	Year of discovery
		Sutherland was an American pharmacologist and biochemist who won the 1971 Nobel Prize in Physiology or Medicine for his discoveries concerning the mechanisms of the action of hormones (epinephrine) via cyclic AMP. de Duve was Belgian cytologist and biochemist who discovered two cell organelles: peroxisome and lysosome. He shared the 1974 Nobel Prize in Physiology or Medicine with Albert Claude and George E. Palade for their discoveries concerning the structural and functional organization of the cell. de Duve also invented the names "autophagy, endocytosis, and exocytosis" [398, 399]	
Oxyntomodulin (OXM)	A proglucagon (PG)-derived gut peptide with 37 amino acids produced from preproglucagon precursor, secreted by L cells localized in the distal small intestine (especially the jejunum and ileum) and large intestine, in response to specific food substances [136, 137, 253]. The processing of the precursor produces different lengths of OXM. The peptide is composed of the glucagon sequence extended by a C-terminal octapeptide [400] OXM is widely distributed in the mucosa of the GI tract [400] The OXM-producing cells are activated by lipids via stimulation of GPR119 receptor [136, 137, 253] OXM exerts its actions by stimulation of receptors including the GLP-1 receptor (GLP-1R) and the glucagon receptor (GCGR) [253]. Thus, OXM is a dual agonist acting on both GLP-1R and GCGR, making the hormone potentially more effective in the treatment of obesity than GLP-1R agonists [253]	OXM was first identified as component in gut glucagon-like immunoreactivity (GLI) in 1968 by two independent groups of researchers [401, 402]. After some years of research, GLI C-terminal octapeptide extension (intervening peptide-1) was named OXM for its ability to inhibit gastric acid secretion in gastric oxyntic glands [403–405]. This enteroglucagon was also named glucagon-37. It inhibited pentagastrin-stimulated acid secretion in rats [406]	1968, 1981

(continued)

Table 8.1 (continued)

Hormone	Description	Discoverer	Year of discovery
	OXM inhibit histamine-stimulated gastric acid secretion [400]. OXM promotes weight loss and energy expenditure, possibly by reducing food intake and appetite in humans. The plasma level of OXM is known to increase postprandially [253]. OXM administration improved glucose tolerance in diet-induced obese mice. This may be due to the fact that OXM promotes insulin secretion [253]. Therefore, OXM plays an important role in glucose homeostasis. Chronic treatment with OXM results in superior weight-lowering and comparable antihyperglycemic effect to a GLP1R-selective agonist [253]. OXM decreases gastric acid and pancreatic exocrine secretion and increases intestinal glucose uptake. OXM, via neural mechanism, also stimulates the secretion of somatostatin and glucagon. Glucagon and GLP-1 have positive inotropic and chronotropic effects on the heart. OXM administration was shown to increase heart rate in wild-type mice [253]		
Glicentin	A proglucagon (PG)-derived peptides from L-type gut cells, consisting of 69 amino acid residues. It is the largest molecule processed from proglucagon precursor. Glicentin is co-secreted with GLP-1, GLP-2, and OXM [136, 137] Glicentin is produced in the GI tract and endocrine pancreas [408, 409]. The secretion of this peptide is stimulated by protein and lipids (especially long-chain fatty acids). However, this nutrient-mediated peptide secretion is greater for lipids compared to proteins. The food substances stimulate glicentin secretion via membrane receptors coupled to alpha-gustducin [136, 137] Glicentin is involved in a variety of physiological functions. It promotes insulin secretion and thus represents a key molecule in glucose homeostasis. Glicentin also inhibits gastric acid secretion and delays gastric emptying. It is one of the main growth hormones of the gut, stimulating mucosal epithelial cell proliferation [408, 409] The role of glicentin in pathological conditions has been documented [413]. Dysfunctional expression of glicentin-producing cells has been implicated in tumors of the APUD system, including intestinal metaplasia, adenoma, and carcinoma of the stomach, as well as rectal tumors [412, 413]	Following the work of [402, 407] which suggested that enteroglucagon comprised of, at least, two peptides, the GLI's C-terminal extension plus an N-terminal extension of 30 amino acids was named glicentin [410, 411]. Sundby Jacobsen and Moody [410] first isolated the hormone from porcine pancreas and later from the intestinal mucosa. The structure of the peptide revealed that C-terminal portion of glicentin consists of the sequence of glucagon extended at its C terminus by an octapeptide, and differs slightly from the sequence of a proposed fragment of proglucagon [408, 411, 414]	1976

(continued)

Table 8.1 (continued)

Hormone	Description	Discoverer	Year of discovery
Glicentin-related pancreatic polypeptide (GRPP)	GRPP is a proglucagon (PG)-/glucagon gene-derived peptide from the gut and contains 30 amino acid residues, and it is identical to glicentin (1–30) [415, 416]. The mechanism of action of this peptide is still under speculation. Recent investigation suggests that the hormone functions as a potent glucoregulatory peptide by inhibition of glucose-stimulated insulin secretion from the rat pancreas [415]	Lars Thim and Alister J. Moody discovered the peptide as a proglucagon fragment from porcine pancreas [416]	1982
Xenin (xenopsin-, neurotenin- related peptide)	A 25-amino acid neurotensin (NT)-related peptide hormone, and xenopsin-related peptide, secreted by a subpopulation of chromogranin A(CGA)-positive endocrine cells in the mucous membrane of the duodenum and jejunum (K cells) [417]. The peptide is produced from the 35-amino acid precursor proxenin [418]. Xenin is a member of the xenopsin/neurotensin/xenin peptide family [418]. Xenin represents the N-terminus of a cytosolic coat protein (α-COP, coatomer protein complex subunit α) from which xenin can be cleaved by aspartic proteinases such as pepsin and cathepsin E [418]. α-COP is the precursor of the hormone [419] Xenin is secreted from intestinal K cells in the proximal intestine in response to carbohydrates, fatty acids, proteins via gut receptor–sensors such as alpha-gustducin (taste-specific G protein alpha subunit), FFA1, GRP120, and GRP119 [136, 137, 417] Xenin is also produced in the pancreas [417] and brain [420] The xenin-producing cells designated K cells are distinct from enterochromaffin, somatostatin, neurotensin, motilin, secretin, cholecystokinin-producing cells. To reiterate, xenin-producing cells are comprised of certain proportion of CGA-positive cells in the duodenum [420] In the gut, the action of xenin is mediated via interactions with the neurotensin receptors [418] The actions of xenin are similar to those of the members of the xenopsin/neurotensin/xenin peptide family. Xenin enhances insulin secretion, inhibits gastric acid secretion and gastric emptying, and stimulates the activity of LPL (lipoprotein lipase) in adipose tissue [136, 137]. The hormone also stimulates exocrine pancreatic secretion and inhibits the gastrin-stimulated secretion of gastric acid [418, 419, 421]	Gerhard E. Feurle and coworkers [417] were the first to isolate xenin from human gastric mucosa in the course of searching for a counterpart to the amphibian octapeptide xenopsin [418]	1992

(continued)

Table 8.1 (continued)

Hormone	Description	Discoverer	Year of discovery
	Xenin is an anorexigenic gut hormone. The plasma level increases after a meal [419]. Xenin reduces food intake, in part, by acting via hypothalamic pathways that are leptin- or melanocortin-independent [419, 420]. Postprandial xenin increases both frequency and the percentage of aborally propagated smooth muscle contraction [419, 422]. Xenin is involved in glucose homeostasis. Administration of xenin increases the insulinotropic response to GIP [136] Xenin may be overexpressed in certain tumors of the gut. The hormone is produced in neuroendocrine tumors of the duodenal mucosa. Thus, xenin may serve as a marker of duodenal neuroendocrine tumors [423]		
Valosin	A 25 amino acid residue gut peptide hormone with N-terminal valine and C-terminal tyrosine [424, 425] Valosin is expressed in GI tract [425, 426]. It functions to modulate gastric and exocrine pancreatic secretion. It also modulates the intestinal myoelectric activity. Valosin promotes pepsin release, pentagastrin-stimulated gastric secretion, pancreatic bicarbonate secretion, and protein secretion [425]	Wolfgang E. Schmidt, Viktor Mutt, Mats Carlquist, Hartmut Kratzin, J.Michael Conlon, and Werner Creutzfeldt [424] isolated and characterized a novel peptide from porcine intestine. The peptide was initially called peptide VQY [426] and was later renamed valosin by the same group of researchers [424]. The name valosin is derived from N-terminal valine and C-terminal tyrosine	1985
Pancreastatin	Pancreastatin is a 49 amino acid residue peptide hormone produced [427, 428] from the prohormonal precursor, chromogranin A (CGA). CGA is a glycoprotein present in neuroendocrine cells, including the endocrine pancreas and adrenal gland [429]. The processing of CGA is tissue specific, with the pancreatic islet and antral gastric endocrine cells being the major source of fully processed pancreastatin. Most of the circulating CGA is secreted by chromaffin tissue. Therefore, peripheral processing of CGA is probably the major indirect source of pancreastatin [429]. After cleavage from CGA, pancreastatin requires carboxyl-terminal amidation by the peptide α-amidating monooxygenase for activation [430]. Interestingly pancreastatin shows structural homology with CGA [431]. In fact, human pancreastatin is $CGA_{250-301}$. Different variants of pancreastatin are produced depending on the point of cleavage and type of proteolytic enzyme involved. The cleavage of the prohormone by proteolytic enzymes yield not only pancreastatin but also other fragments including vasostatin (human CGA_{1-76}), a vasodilator peptide; catestatin (human $CGA_{352-372}$), an inhibitor of catecholamine release [430]	Tatemoto, Efendic, Mutt, Makk, Feistner, and Barchas discovered the peptide as a pancreatic secretion (insulin) inhibitor [432]; hence, the name "pancreastatin." Pancreastatin was discovered in porcine pancreatic extracts by the presence of its unique C-terminal structure, a glycine amide [432]. Pancreastatin was subsequently isolated from human gut in 1988 and 1989 [427, 428]	1986

(continued)

Table 8.1 (continued)

Hormone	Description	Discoverer	Year of discovery
	Pancreastatin is widely expressed in endocrine cells of the GI tract, pancreas, thyroid, anterior pituitary, and adrenal glands [428, 431, 433] Generally, pancreastatin regulates the function of tissues near to its point of secretion via autocrine and paracrine mechanisms. Pancreatic pancreatin colocalizes with insulin, somatostatin secreting cell of the islet. Pancreastatin inhibits the function of a variety of cells including islet cells, parietal cells, parathyroid cells. The peptide inhibits glucose-stimulated insulin release from pancreatic islet β-cells and also inhibits glucose uptake by adipocytes and hepatocytes [430]. It is a peptide with metabolic function counter-regulatory to insulin action [429]. Research has indicated that pancreastatin can decrease glucose uptake by approximately 48–50% [430]. Pancreastatin regulates not only glucose level but also free fatty acid metabolism; however, the peptide has not been shown to affect amino acid metabolism [430]. The peptide also regulates the functions of the exocrine pancreas and gastric secretions [433] CGA and its cleavage product, pancreastatin are overexpressed in human diseases such as neuroendocrine neoplasias and diabetes [430]. Pancreastatin is overexpressed in the majority of carcinoid tumors of ileum, rectum, ovary, and lungs [431]. CGA (210–301), which contains 92 amino acid residues, is the major form of pancreastatin in neuroenteroendocrine tumors and associated metastasis as well as in human gut [434]. The normal level of pancreastatin in tissues is less than 1.4 pmol g^{-1} [435]. The plasma level of pancreastatin is about 7.4–22.4 pmol l^{-1}. Pancreastatin level in fasting increases to 4–37.5 pmol l^{-1}. The plasma level increases in individuals with GI metastases and carcinoid tumors [435]		

Note The liver produces specific hormones including insulin-like growth factors I and II. Accessory cells of the GI tract such as adipocytes, fibroblasts, macrophages, leucocytes, endothelial cells produce cytokines (interleukins, interferons, tumor necrosis factors, colony stimulating factors), adipokines, which are all considered as hormones. The hormones discussed here are not the complete list of hormones produced in the GI tract. Some of the GI tract hormones (which also act as neurotransmitters) are discussed in Chap. 9

[a]Misoprostol is widely used in the practice of obstetrics and gynecology. But exposure in early pregnancy causes congenital defects; hence, misoprostol is considered a teratogen [214]

not be separated from real-world events. Galen suggested that "vital spirits" in the blood regulated human bodily functions [18–25]. Galen introduced the concept of vital spirits in the studies of endocrine glands, in particular, the pituitary gland. Spiritism of bodily functions continued for several centuries, not until, research evidences began showing that spiritual phenomena could be reproduced and observed as physical phenomena in the laboratory [6, 26–28]. However, Galenic school of endocrine mechanisms of regulation of physiological functions continued to flourish for several centuries and even after the rebirth. Progressive development in knowledge of endocrine regulation of functions began during the eighteenth and nineteenth centuries. For instance, Johann Conrad Brunner (1653–1727) discovered duodenal glands, while intestinal (mucous) glands which were discovered by Johann Nathanael Lieberkühn (1711–1756) in the mid-eighteenth century [29]. These glands, though discovered as exocrine, also contain a vast number of endocrine cells. The discoveries and investigations on endocrine glands marked the beginning of modern knowledge in hormonal regulation of physiological functions at the cellular level. Key pioneers that made breakthrough studies in identifying cellular mechanisms of functioning of the GI tract and also laid the basis of humoral (endocrine) regulation include Claude Bernard (1813–1878), William Maddock Bayliss (1860–1924), Frederick Grant Banting (1891–1941), Ernest Henry Starling (1866–1927), Sir Edward Albert Sharpey-Schafer (1850–1935), Gustave-Édouard Laguesse (1861–1927), Paul Langerhans (1847–1888), M.C. Ciacco (1877–1956), Rudolf Peter Heidenhain (1834–1897), Nikolai Konstantinovich Kulchitsky (1856–1925) [6, 24, 25, 30–32]. The contributions of these pioneer scientists will be highlighted, while other key contributions will be discussed in later part of this chapter.

The concept of endocrine regulation of the gut originated from the 1855 pioneering work of Claude Bernard, who observed a novel role of the liver in the regulation of blood sugar. Bernard concluded from his observation on how the liver was able to regulate blood sugar to maintain a constant range of values by the release of "internal secretions" from "ductless glands." During this time, the investigations by Claude Bernard on the concept of internal secretions and milieu intérieur created a better understanding of how hormones could carry out their functions in regulating physiological activities. The investigations on other areas of endocrine organs and regulation that were ongoing at the same time with Bernard's works on internal secretions provided an impetus to the progress of endocrinology as a whole. The classical and pioneering clinical works of the English physician and scientist Thomas Addison (1793–1860) and the Swiss physician, neurosurgeon and physiologist, Emil Theodor Kocher (1841–1917) provided additional information on modeling of endocrine functions in animals. In 1855, Addison published a monograph, describing a degenerative disease of the suprarenal capsules "adrenal glands," which was later named after him—Addison's disease or adrenal insufficiency. He also discovered pernicious anemia, latter known as Addisonian anemia, a disorder of vitamin B12 absorption from the upper intestine [24, 25, 33, 34]. Kocher made contributions to the understanding of the surgery and functions of the thyroid gland for which he was awarded the 1909 Nobel Prize in Physiology or Medicine [34].

Around the middle of the nineteenth century, Heidenhain devised a novel method of studying the mechanisms of digestion. He produced an artificial stomach, which was later called the "Heidenhain pouch," and was used for studying gastric secretions. In 1868–70, Heidenhain reported a novel class of clear cells in the gastric mucosa of the rabbit and dog. Heidenhain noted the cells stained yellow in the presence of chromic acid—they were the cells that would be later called enterochromaffin cells—a group of endocrine cells of the gut responsible for secreting serotonin and other hormones. In the following year, Langerhans while still a medical student identified pancreatic islets as clusters of cells in the pancreas. The English physiologist, Sir Edward Albert Sharpey-Schafer (1850–1935) and French pathologist and histologist, Gustave-Édouard Laguesse (1861–1927) provided further descriptions of the functions of the endocrine cells of the gut and pancreas. Laguesse was a French pathologist and histologist who began actively conducting scientific research after receiving his medical doctorate in Paris in 1885. Laguesse conducted numerous works on the pancreas, and in 1893, he named the small cellular clusters of the pancreas the "Islets of Langerhans" in honor of Paul Langerhans (1847–1888)—the first to have studied this group of cells. Even though the functions of the islet of Langerhans were not known at the time, the anatomical architecture was fairly understood. Importantly, it was Laguesse who suggested that the islets of Langerhans were responsible for secretions that played a role in digestion. We now know that some secretions of islets of Langerhans (generally called hormones) modulate the functions of the exocrine pancreas that play an important role in intestinal digestion. Laguesse's proposal represented one of the key steps in the identification of insulin—the first pancreatic hormone to be discovered. In 1894 together with George Oliver (1841–1915), he discovered and demonstrated the existence of a substance that constricted blood vessels—a molecule that was later isolated and named epinephrine (adrenaline) in 1897 by John Jacob Abel (1857–1938). It was Laguesse who coined the term "endocrine" for the secretions of the ductless glands. Schafer coined the word "insulin" and proposed that this substance secreted by the pancreas was responsible for diabetes mellitus [6, 18, 35–39].

Toward the end of the nineteenth century, precisely in 1897, Kulchitsky described certain cells that stained yellow with Ehrlich-Biondi mixture after a day of observation using the intestinal mucosa of cats and dogs. Further investigation by Kulchitsky showed that these cells were similar in their staining characteristics with the cells previously described by Heidenhain. The name "enterochromaffin" was coined later in 1906 by M.C. Ciacco (1877–1956) [39–41]. Kulchitsky also described epithelial leucocytes of the GI system. Enterochromaffin cells or Kulchitsky (or K) cells are widely distributed in the gut and currently known to play a huge role in health and disease. Further advancement to the study of hormones and how they regulate physiological functions was made by Bayliss and Starling who suggested that a bodily chemical could regulate the physiological process by stimulating or inhibiting organismal functions through the secretion of such chemical. Further, they hypothesized that the duodenal content can stimulate the release of pancreatic juice. In one of their observations, in the early 1900s, Bayliss and Starling were studying the effects of an extract of small intestine in pancreatic secretions of a dog and observed pancreatic juice being released from the pancreatic ducts. Starling named the chemical "secretin." In 1902, such chemical agent that accelerates reaction was named hormone (Greek "hormao" meaning "to stimulate") by Starling and Bayliss, having read about W. Hardy's term "hormono" (Greek, "I excite") [18, 29]. Bayliss and Starling theorized that the chemical was released from the intestinal lining into the blood, from where it travelled to the pancreas to stimulate pancreatic secretion into the duodenum. The direct release of this chemical into the blood to stimulate another tissue was the second observation made following Claude Bernard's initial proposal—the second birth of ductless (endocrine) organs. This was the birth of the gut as an endocrine organ. These scientists made a colossal shift in concepts from Pavlov's inspired conception of nervism to the chemical basis of transmission of information in the body. The discovery of the first peptide hormone, secretin ignited further investigations by scientists who had read about the new experimental results of Starling and Bayliss. Secretin is produced by the "S cells" of the duodenum, located precisely in the crypts of Lieberkühn. In humans, the secretin peptide is encoded by the SCT gene on chromosome locus 11p15.5. This peptide hormone plays a variety of roles in the human body. Secretin can regulate the pH of the duodenum by inhibiting the secretion of gastric acid from the parietal cells of the stomach, and stimulating the production of bicarbonate from the centroacinar cells and intercalated ducts of the pancreas. Secretin is a peptide hormone that regulates water homeostasis throughout the body and influences the environment of the duodenum by regulating secretions in the stomach and pancreas [42–44]. In 2007, secretin was discovered to play a role in osmoregulation by acting on the hypothalamus, pituitary, and kidney [45, 46]. It was thought that secretin deficiency is involved in autistic syndrome and that the hormone may have neuroendocrine functions; however, systemic reviews and meta-analysis on multiple randomized controlled trials did not find any benefit on the administration of secretin to children with autism [44, 47]. It should be noted however that many neuropsychiatric disorders usually involve the dysfunctions of multiple hormones and neurotransmitters.

A breakthrough in the concept of hormone regulation of physiological functions was the result of the exceptional work of Sir Frederick Grant Banting (1891–1941), Charles Best (1899–1978), and John James Rickard Macleod (1876–1935) in the early 1920s. They discovered and subsequently isolated insulin from canine (dog) and bovine (cow) pancreas. In 1923, the Nobel Prize in Physiology or Medicine was awarded to Banting and McLeod alone for their contribution to science. This is one of the cases of Nobel Prize awards that have raised criticisms [48, 49].

Further progress on the endocrine system of the gut was based on the results of earlier investigations on the endocrine cell identified ab initio by Heidenhain, Kulchitsky, and Ciacco that would later become the basis of a diffuse neuroendocrine system (DNES) located in the gut and other areas of the body [39, 40]. In the early 1900s, scientists began actively working on the enterochromaffin cells and related cells in the gut. In 1914, Pierre Masson (1880–1959) proposed that scattered endocrine (Kulchitsky or enterochromaffin) cells within the gut formed a diffuse endocrine organ. After obtaining his medical degree from the University of Paris in 1909, he continued his studies at the Pasteur Institute in Paris between 1909 and 1914. After the World War, around 1923–8, Masson observed that the cells which he had termed "endocrine" were also neural in origin, which now forms the basis of the concept of "neurocrine secretion" [50]. The concept of neurocrine secretion is believed to be a key mechanism regulating gut activity. The endocrine and neurocrine mechanisms of gut activity are studied by a branch of science called gastroneuroendocrinology. One of the pioneer gastroneuroendocrinologists, Friedrich Feyrter (1895–1973), an Australian scientist, accurately described the DNES. He made many contributions to the regulation of GI functions that allowed for the understanding of a syncytial regulatory system comprising of endocrine and neural origin. In 1938, Feyrter used Masson's staining techniques to identify "Helle Zellen" (meaning, "clear cells") within the pancreatic ductal system and the GI epithelium. Another researcher of enduring memory, Anthony Guy Everson Pearse (1916–2003) was creditworthy for grouping the various cells belonging to the system outlined by Feyrter under a unifying system—"Amine Precursor Uptake Decarboxylase" (APUD) system (Amine—for high amine content; Precursor Uptake—for high uptake of amine precursors; Decarboxylase—for high content of the enzyme amino acid decarboxylase), which converts the precursors to amines. This system is concerned with the regulation of the activities of the gut by endocrine, paracrine, and neurocrine mechanisms. Over the years, the concept of APUD system has encompassed the gut, to involve certain cells of the respiratory tract and other tracts of the body with epithelial lining [18, 51].

Further advancement in this area of GI physiology (i.e., GI neuroendocrinology) was predicted by progress in the peptide biochemistry of protein analysis [52–56] and the invention of the radioimmunoassay in 1959 by the American physician and scientist Solomon Aaron Berson (1918–1972) and the American medical physicist Rosalyn Sussman Yalow (1921–2011) [57] that enabled the measurement of minute quantities of hormone [58, 59]. By the turn of the last century, there had been a recorded development in sequencing techniques used to identify a range of molecules. Current methods of novel peptide identification use the techniques of genomic sequencing, a process in which the genome is screened to determine the order of nucleotides within a DNA molecule. The method is used to identify novel genes and peptide molecules [60, 61]. More endocrine and neuronal molecules have been discovered using state-of-the-art sequencing technologies and hybrid devices. There are over 60 types of hormones in the gut alone. An extensive list of hormones of the gut is given in Table 8.1.

8.3 Gut as the Largest Neuroendocrine Organ in the Human Body: An Integral Part of the Diffuse Neuroendocrine System (DNES)/Amine Precursor Uptake Decarboxylase (APUD) System

Endocrine functions of the GI tract are due to the presence of neuroendocrine and enteroendocrine cells, which are distributed within the epithelium throughout the tract. The neuroendocrine and enteroendocrine cells are specialized endocrine cells of the tract and pancreas. These cells respond to varying physiological stimuli by the production of hormones and neurotransmitters, which function to regulate the activities of the cells of the enteric nervous system and other cells of the digestive system as well as other tissues and organs of the body. The gut endocrine cells are actively involved in sensing the taste of food. Thus, they help to scan through the bad and good substances in food and mobilize the necessary physiological response [62–69].

Endocrine cells of the gut are diffusively located throughout the tract and pancreas, but are more concentrated in the stomach, intestine, and pancreas. These cells altogether belong to the gastroenteropancreatic diffuse endocrine system. Endocrine cells of this system are located on the base of the epithelium of the GI tract. The cytoplasm of these cells always contains secretory vesicles that are released into the lamina propria upon stimulation, but they also confer substantial effect on digestive activity [64–71].

In the oral cavity, endocrine cells secrete various types of hormones, peptides, and immunomodulatory peptides involved in a variety of physiological processes. In the esophagus, the endocrine cells include the enterochromaffin cells and other types that secrete substances known to modulate gut activities. In the stomach, the endocrine cells include the enterochromaffin cells, which secrete a couple of hormones including histamine. The gastric hormone, gastrin, stimulates the release of gastric histamine. Enterochromaffin cells also secrete serotonin [72]. Gastric endocrine cells produce numerous hormones (vasoactive intestinal peptide, somatostatin, substance P, alpha- and gamma-endorphins, etc.) in varying quantities [63–65, 73]. In the intestine, enteroendocrine cells include enterochromaffin cells, K cells—secrete gastric inhibitory peptide (GIP—an incretin hormone), L cells—secrete glucagon-like peptide-1 (also an incretin), glucagon-like peptide-2. Other hormones secreted in the intestine are cholecystokinin-pancreozymin, gastrin, secretin, motilin, somatostatin, cholecystokinin, neurotensin, vasoactive intestinal peptide, enteroglucagon, etc. The level of secretions of these hormones varies from one region of the GI tract to another. In the GI tract, endocrine cells are primarily found in the stomach and proximal small intestine (mostly gastrin, ghrelin, GIP, secretin, cholecystokinin). In the ileum and colon, the predominant hormones are considered peptide YY, GLP-1, GLP-2, and neurotensin. The large intestine has but a reduced number of enteroendocrine cells. Hormones that are found throughout the tract include somatostatin, serotonin, histamine, and substance P [63–65, 74]. In the pancreas, enteroendocrine cells are located in the islets of Langerhans and they secrete most importantly the hormones insulin (produced by β-cells) and glucagon (produced by α-cells). The α-cells also secrete thyrotropin-releasing hormone, glicentin, GLP-1, cholecystokinin, peptide YY, and pancreastatin [75]. Apart from insulin, β-cells also secrete gamma-aminobutyric acid (GABA), amylin, thyrotropin-releasing hormone, calcitonin gene-related peptide (CGRP), gastrin, and pancreastatin. Other types of endocrine cells present in the pancreas include δ-cells, δ-1, PP (also called F) cells, enterochromaffin cells, G1 cells, and ε cells. The pancreatic δ-cells secrete somatostatin, met-encephalin, CGRP, and pancreastatin. The PP cells secrete pancreatic polypeptide, met-encephalin, and peptide YY. The pancreatic δ-1 cells, which are believed to be less than 1% of the endocrine cells of the pancreas, secrete vasoactive intestinal polypeptide. The enterochromaffin cells, also less than 1% of the endocrine pancreatic cells, secrete substance P, serotonin among others. The pancreatic G1 cells, which are also less than 1%, secrete gastrin and ACTH-like peptide. It is not exactly clear whether pancreatic ghrelin is secreted only from ε (epsilon)-cell or other endocrine cells of the pancreas. Notice that pancreastatin may be secreted by more than one endocrine pancreatic cell [75–77]. The liver, salivary glands, and gallbladder also secrete endocrine substances involved in the regulation gut functions [78–82]. The secretions of these cells are regulated by nutrient availability, the nervous system (enteric and autonomic nervous systems), and hormones. For instance, the parasympathetic division of the autonomic nervous system which stimulates secretion (e.g., insulin) and inhibits secretion (e.g., glucagon) and sympathetic stimulation having opposite effect [83–86].

The endocrine cells belonging to the gastroenteropancreatic diffuse endocrine system described above form part of the DNES. DNES is comprised of endocrine cells diffusively located in various parts of tracts of hollow organs including the respiratory and reproductive tracts. Several immunohistochemistry and microscopy methods have been used to study and distinguish different types of endocrine cells [8]. Sometimes the cells of this system are grouped under the APUD system. The cells of this system share a common function of secreting low molecular weight polypeptide hormones [87]. But they also have some differences. The enteroendocrine cells can be differentiated through their morphological, histochemical, and functional characteristics [88].

The DNES of the GI tract produces numerous hormones, making the gut a relatively large endocrine organ. Correspondingly, [89] considered the gut as the largest endocrine organ in the body [89]. The hormones produced by the gut are discussed in detail in Table 8.1 [88, 90].

It should be borne in mind, however, that one and the same hormone maybe produced by different regions of the digestive system. The hormones produced in one region of the digestive system may function not only locally but also in other regions of the digestive tract as well as in extragut tissues. For instance, some hormones produced in the pancreas can modulate several functions of the GI tract, adipose tissue, heart, brain, among others.

8.4 The Changing Views on the Origin of Humoral and Neurohumoral Secretions of the Gut: The Origin of Enteroendocrine Cells

Accumulating evidences suggest need to revise the traditional concept of the origin of APUD cells. Emerging studies indicate that traditionally defined non-enteroendocrine cells also possess endocrine functions, and the GI cells originally identified as enteroendocrine are now playing both roles. The cells that were initially thought to be solely responsible for secretions of acids or enzymes are now known to be a secretory vessel of hormones. Moreover, numerous hormones and/or neurotransmitters that were initially identified in other parts of the body are now found to be produced by the gut (see Table 8.1). It was previously believed that enteroendocrine cells of the gastroenteropancreatic system are derived from the neural crest during embryonic development [91]. However, progress achieved in cytochemical techniques investigating the fate and differentiation of these cells over the past decades indicates that many cells of this system did not originate from the neural crest [92].

Clinical Correlate 8.1

Neuroendocrine Neoplasms of the Gastroenteropancreatic System

Neuroendocrine neoplasms of the gastroenteropancreatic system are epithelial neoplasms of the gut with predominant neuroendocrine differentiation. A neoplasm is a tumor that results from an abnormal growth or progressive multiplication of the cells that constitute a tissue or organ. This type of tumors originates from the neuroendocrine cells of the GI system [93]. Some authors term the tumors of the gastroenteropancreatic system apudomas (from "APUD") [94]. Since DNES is found in the respiratory tract, this tumor sometimes may originate or spread to this region. The prevalence of this tumor is estimated to be less than 1% of GI cancers [93]. Neoplasms of the gastroenteropancreatic system include carcinoid tumors, small-, intermediate- and large-cell carcinomas [93]. For the purpose of diagnosis, histology, immunohistochemistry, biochemistry, and genetic techniques are used [94–97]. Biochemical analysis of Ki-67 nuclear antigen (index of cell proliferation) is a key parameter of diagnosis of neuroendocrine neoplasms of the gastroenteropancreatic system [98]. The treatment options include surgery, chemotherapy (with cisplatin, carboplatin, paclitaxel), and radiotherapy [93, 99].

8.5 The Paraneuron Concept: Is an Enteroendocrine Cell a Type of Neuron?

The traditional way of classifying body systems in recent times has been seriously challenged. Classical examples of systems with contention are the endocrine and nervous systems. The relationship between these two systems has been known several decades ago. However, increasing evidences based on the results of morphological, physiological, biochemical immunocytochemical investigation on the nervous and endocrine systems have continuously indicated that numerous cells that were traditionally classified as endocrine alone are now believed to function as a neuron, particularly with regard to their secretory functions and types of hormones or neurotransmitters produced. In fact some neurotransmitters are now classified as hormones vice versa [100, 101]. First proposed by Shigeru Kobayashi, Tsuneo Fujita, and Tanemichi Chiba, the paraneuron concept, which is based on the idea that neurons are not as specialized a cell type as traditionally believed, may be in essence true. According to this concept, some neurons are more like endocrine cells, whereas functional characteristics of others may be closer to the traditionally classified neuronal cell type [100–105]. In reality, though, these transitional forms of neurons or endocrine cells coexist with both cells. Although with varying differences, at the molecular level, both endocrine cells and neural cells display memory functions that were initially thought to be peculiar to neurons alone. Therefore, none of these cells can be completely classified as neuronal or endocrine in origin, rather they bear some endocrine as well as neuronal characteristics. These transitional cells, called paraneurons are receptor-secretory cells that produce neuronal substances. They are similar to neurons in that their secretory contents are stored in membrane-bound granules and vesicles which contain peptides, amines, messengers, adenine nucleotides, and acidic carrier proteins including chromogranins [105].

8.6 Classification of Hormones

Hormones can be classically defined as endogenous chemical substances that are released by ductless glands into circulation, localize, and transduce signal via specific receptors on the membrane, cytosol, or nucleus of the target cell, generate a second messenger, and cause a concrete response in the target cell at a very small concentration $\sim$10.6 to 0.12 mmol/l [106].

Hormones can be classified according to the origin of synthesis and mechanism of action. Thus, hormones may be classified as oral, esophageal, gastric (antral, corporal, pyloric, postpyloric, etc.), duodenal, jejunal, ileal, colonic, intestinal, ileocolonic, pancreatic, salivary [108]. However, this regional classification of hormones may not have overall clinical usefulness.

Hormones can also be classified according to their chemical structure—peptides (e.g., neurotensin, somatostatin, insulin, glucagon); derivatives of amino acids (e.g., serotonin, adrenaline, melatonin); steroids—derivatives of cholesterol (e.g., gonadal hormones); eicosanoids—derivatives of arachidonic acids (e.g., prostaglandins and thromboxanes) [106, 107].

Functionally, hormones can be classified as effectors, which act on the target cells. The effectors include trophic hypophyseal hormones, which regulate the secretion of peripheral glands (including endocrine glands of the gut) and hypophyseal controlling (hypothalamic) hormones that regulate the secretion of pituitary hormones [109–112].

Hormones can be classified according to their cell sources: myokines (hormones secreted by muscle cells) [113, 114], adipokines (hormones secreted by adipocytes), enterokines (hormones secreted by enterocytes) [115], fibrokines or adipo-fibrokines (hormones secreted by fibroblasts or adipocytes) [116], neurocrines (secreted by enteral neurons) [117, 118], lymphokines (secreted by lymphocytes) [119], chemokines (secreted by monocyte, neutrophil, tumor cells) [120–124]. Chemokines can be secreted by other cell types such as CD34 + cells, myeloblasts, erythroblasts, and megakaryoblasts. Some secreted substances by these cells act as growth factors [125]. It should be noted, however, that the hormones may be classified according to their mechanisms of action. Thus, GI hormones may be paracrine secretions (somatostatin), endocrine secretions (gastrin, CCK), neurocrine secretions (VIP, substance P) [117, 118, 126, 127]. These hormones can be searched in the hormone database "Hmrbase" (http://crdd.osdd.net/raghava/hmrbase/), which contains about 2000 hormones, including information regarding their function, structures, source, and receptors [128].

8.7 Gastrointestinal Hormones: Timeline on History of Discovery, Their Structural and Functional Characteristics, as well as Clinical Application

The table shown below is a list of all known hormones of the gut, their characteristic architecture, and functional properties. The course of discovery of these molecules as well as the personalities responsible for their discovery and the year of discovery are provided in the Table.

8.8 Conclusion

GI hormones represent a crucial aspect of the regulation of gut structure and functions. These mediators not only exert their actions on the gut, but also have numerous roles to play in extragut tissues. Though the functions of hormones were speculated already during the ancient times, substantial efforts on this area were made only beginning from the nineteenth century. Currently, over 60 hormones have been discovered to be produced in the gut. The GI-derived hormones are inevitably useful in the maintenance of health, and their dysfunctions can cause a range of diseases.

Recommended Readings

Original Articles

1. Mazzawi T, Hausken T, Gundersen D, El-Salhy M (2016) Dietary guidance normalizes large intestinal endocrine cell densities in patients with irritable bowel syndrome. Eur J Clin Nutr 70:175–181
2. Fjällskog ML, Sundin A, Westlin JE, Öberg K, Janson ET, Eriksson B (2002) Treatment of malignant endocrine pancreatic tumors with a combination of α-interferon and somatostatin analogs. Med Oncol 19(1):35–42
3. Pellegrini S, Sordi V, Bolla AM, Saita D, Ferrarese R, Canducci F, et al (2017) Show more. Duodenal mucosa of patients with type 1 diabetes shows distinctive inflammatory profile and microbiota. J Clin Endocrinol Metab 102(5):1468–1477
4. Qin J, Li Y, Cai Z, Li S, Zhu J, Zhang F, et al (2012) A metagenome-wide association study of gut microbiota in type 2 diabetes. Nature 490(7418):55–60
5. Pietroletti R, Bishop AE, Carlei F, Bonamico M, Lloyd RV, Wilson BS, et al (1986) Gut endocrine cell population in coeliac disease estimated by immunocytochemistry using a monoclonal antibody to chromogranin. Gut 27(7):838–43
6. Hays MT (1988) Thyroid hormone and the gut. Endocr Res 14(2–3):203–224

Review Articles

1. Deane A, Chapman MJ, Fraser RJL, Horowitz M (2010) Bench-to-bedside review: The gut as an endocrine organ in the critically ill. Crit Care 14(5): 228
2. Furness JB (2012) The enteric nervous system and neurogastroenterology. Nat Rev Gastroenterol Hepatol 9:286–294
3. Merlino DJ, Blomain ES, Aing AS, Scott A (2014) Waldman. Gut-brain endocrine axes in weight regulation and obesity pharmacotherapy. J Clin Med 3:763–794
4. Rindi G, Bordi C (2003) Highlights of the biology of endocrine tumours of the gut and pancreas. Endocrine-Related Cancer 10:427–436
5. Rindi G, Wiedenmann B (2012) Neuroendocrine neoplasms of the gut and pancreas: new insights. Nat Rev Endocrinol 8:54–64
6. Schonhoff SE, Giel-Moloney M, Leiter AB (2004) Minireview: development and differentiation of gut endocrine cells. Endocrinology 145(6): 2639–2644
7. Wozniak SE, Gee LL, Wachtel MS, Frezza EE (2009) Adipose tissue: the new endocrine organ? Dig Dis Sci 54(9):1847–1856
8. Zhang Y, Ning G, Handelsman Y, Bloomgarden ZT (2010) Gut hormones and the brain. J Diabetes 2(3):138–45

Books

1. Brierley S, Costa M (eds) (2016) The enteric nervous system. Springer International Publishing, Cham, Switzerland
2. George G (2010) Gastrointestinal endocrinology. Humana Press, Totowa, New Jersey, United States
3. Thompson J (ed) (1990) Gastrointestinal endocrinology: receptors and post-receptor mechanisms. Academic Press, San Diego, CA, USA

Bibliography

1. Lechago J, Bencosme SA (1973) The endocrine elements of the digestive system. Int Rev Exp Pathol 12:119–201
2. Bencosme SA, Lechago J (1973) Staining procedures for the endocrine cells of the upper gastrointestinal mucosa: light-electron microscopic correlation for the gastrin-producing cell. J Clin Pathol 26(6):427–434
3. Eknoyan G (2004) Emergence of the concept of endocrine function and endocrinology. Adv Chronic Kidney Dis 11(4):371–376
4. Buchanan KD (1976) The gut as an endocrine organ. Ir J Med Sci 145(4):103–115
5. Lloyd RV (1990) Endocrine pathology. Springer-Verlag, New York, USA
6. Liu EH, Oberg K (2010) The history and development of the gastroenteropancreatic endocrine axis. Endocrinol Metab Clin 39(4):697–711
7. McConalogue K, Furness JB (1994) Gastrointestinal neurotransmitters—the gut as an endocrine organ. Baillière's Clin Endocrinol Metab 8(1):51–76
8. Sundler F (2004) GI tract, general anatomy (cells). In: Martini L (ed) Encyclopedia of endocrine diseases. Elsevier, San Diego, CA, USA
9. Arimura A, Fishback JB (1981) Somatostatin: regulation of secretion. Neuroendocrinol 33(4):246–256
10. Chandra R, Liddle RA (2009) Neural and hormonal regulation of pancreatic secretion. Curr Opin Gastroenterol 25(5):441–446
11. Bloom SR, Polak JM (1980) Physiology of gastrointestinal hormones. Biochem Soc Trans 8 (1):15–17
12. Rehfeld JF (2014) Gastrointestinal hormones and their targets. In: Lyte M, Cryan JF (eds) Microbial endocrinology: the microbiota-gut-brain axis in health and disease. Springer, New York, USA
13. Pacheco-Pantoja EL, Ranganath LR, Gallagher JA, Wilson PJM, Fraser WD (2011) Receptors and effects of gut hormones in three osteoblastic cell lines. BMC Physiology 11:12
14. Wren AM, Bloom SR (2007) Gut hormones and appetite control. Gastroenterology 132(6): 2116–2130
15. Côté CD, Zadeh-Tahmasebi M, Rasmussen BA, Duca FA, Lam TK (2014) Hormonal signaling in the gut. J Biol Chem 289(17):11642–11649
16. Perry B, Wang Y (2012) Appetite regulation and weight control: the role of gut hormones. Nutr Diabetes 2:e26
17. Pressley L, Funder JW (1975) Glucocorticoid and mineralocorticoid receptors in gut mucosa. Endocrinol 97(3):588–596
18. Modlin IM, Champaneria MC, Bornschein J, Kidd M (2006) Evolution of the diffuse neuroendocrine system–clear cells and cloudy origins. Neuroendocrinol 84(2):69–82

19. Medvei VC (1983) A history of endocrinology. Kluwer Academic Publishers Group, Dordrecht, Netherlands
20. David AR (1997) Disease in Egyptian mummies: the contribution of new technologies. Lancet 349(9067):1760–1763
21. Tassinari J, Sisti A (2016) Endocrinology and art. First dwarf represented: ancient Egypt 2520 B.C. J Endocrinoll Invest 39(4):489
22. Emerson CH (1987) Thyroid function and disease in the female. In: Gold JJ, Josimovich JB (eds) Gynecologic endocrinology. Springer, New York, USA
23. Kleine B, Rossmanith WG (2016) History of hormones and endocrinology. In: Hormones and the endocrine system. Springer International Publishing, Cham, Switzerland
24. Encyclopædia Britannica (2003) Encyclopædia Britannica, Inc, Chicago, IL
25. Koertge N (ed) (2007) New dictionary of scientific biography, 1st edn. Charles Scribners & Sons, Detroit, Michigan
26. Stoka AM (1999) Phylogeny and evolution of chemical communication: an endocrine approach. J Mol Endocrinol 22:207–225
27. Trinkaus JP (1984) The forces that shape the embryo. Prentice-Hall Inc, Englewood, NJ, USA
28. Alexander JS, Ganta VC, Jordan PA, Witte MH (2010) Gastrointestinal lymphatics in health and disease. Pathophysiol 17(4):315–335
29. SRódka A (2003) The short history of gastroenterology. J Physiol Pharmacol 54(S3):9–21
30. Goet JP (1953) Gustave Edouard Laguesse; his demonstration of the significance of the islands of Langerhans. Diabetes 2(4):322–324
31. Grossman MI (1950) Gastrointestinal hormones. Physiol Rev 30(1):33–90
32. de Herder WW (2014) Heroes in endocrinology: nobel prizes. Endocr Connect 3(3):R94–R104
33. Smans LCCJ, Zelissen PMJ (2012) Thomas addison and the adrenal gland. Ned Tijdschr Geneeskd 156(32):A4788
34. Gautschi OP, Hildebrandt G (2009) Emil Theodor Kocher (25/8/1841-27/7/1917)—a swiss (neuro-)surgeon and Nobel Prize winner. Br J Neurosurg 23(3):234–236
35. Hill L (1935) Sir Edward Albert Sharpey-Schafer. 1850–1935. Obit Not Fell R Soc 1(4): 400
36. Sneader W (2005) Drug discovery: a history. John Wiley and Sons Ltd, Chichester, UK
37. Power HJ (2007) Sharpey-Schafer, Edward Albert (1850–1935). In: Considine GD (ed) Van Nostrand's scientific encyclopedia. John Wiley & Sons Inc, New York, United States
38. Hoet JP (1953) Gustave Edouard Laguesse: his demonstration of the significance of the islands of Langerhans. Diab 2:322–324
39. Drozdov I, Modlin IM, Kidd M, Goloubinov VV (2009) From Leningrad to London: the saga of Kulchitsky and the legacy of the enterochromaffin cell. Neuroendocrinol 89(1):1–12
40. Drozdov I, Modlin IM, Kidd M, Goloubinov VV (2009) Nikolai Konstantinovich Kulchitsky (1856–1925). J Med Biogr 17(1):47–54
41. Marincola P, Liu EH (2015) History of neuroendocrine tumors. In: Yalcin S, Öberg K (eds) Neuroendocrine tumours. Springer, Berlin, Germany
42. Häcki WH (1980) Secretin. Clin Gastroenterol 9(3):609–632
43. Kopin AS, Wheeler MB, Leiter AB (1990) Secretin: structure of the precursor and tissue distribution of the mRNA. Proc Natl Acad Sci USA 87(6):2299–2303
44. Whitmore TE, Holloway JL, Lofton-Day CE, Maurer MF, Chen L, Quinton TJ et al (2000) Human secretin (SCT): gene structure, chromosome location, and distribution of mRNA. Cytogenet Cell Genet 90(1–2):47–52
45. Chu JY, Chung SC, Lam AK, Tam S, Chung SK, Chow BK (2007) Phenotypes developed in secretin receptor-null mice indicated a role for secretin in regulating renal water reabsorption. Mol Cell Biol 27(7):2499–2511
46. Chu JY, Lee LT, Lai CH, Vaudry H, Chan YS, Yung WH, Chow BK (2009) Secretin as a neurohypophysial factor regulating body water homeostasis. Proc Natl Acad Sci USA 106(37):15961–15966

47. Krishnaswami S, McPheeters ML, Veenstra-Vanderweele J (2011) A systematic review of secretin for children with autism spectrum disorders. Pediatr 127(5):e1322–e1325
48. Simoni RD, Hill RL, Vaughan M (2002) The discovery of insulin: the work of Frederick Banting and Charles Best. J Biol Chem 277:e15
49. Majumdar SK (2001) Glimpses of the history of insulin. Bull Indian Inst Hist Med Hyderabad 31(1):57–70
50. Masson P (1928) Carcinoids (argentaffin-cell tumors) and nerve hyperplasia of the appendicular mucosa. Am J Pathol 4:181–212
51. Champaneria MC, Modlin IM, Kidd M, Eick GN (2006) Friedrich Feyrter: a precise intellect in a diffuse system. Neuroendocrinol 83(5–6):394–404
52. Gregory H, Hardy PM, Jones DS, Kenner GW, Sheppard RC (1964) The antral hormone gastrin. Nat 204:931–933
53. Gregory RA, Tracy HJ (1964) The constitution and properties of two gastrins extracted from hog antral mucosa. Gut 5:103–117
54. Gregory RA (1979) A review of some recent developments in the chemistry of the gastrins. Bioorg Chem 8(4):497–511
55. Mutt V, Magnusson S, Jorpes JE, Dahl E (1965) Structure of porcine secretin. Biochem 4:2358–2362
56. Jorpes E, Mutt V (1966) Cholecystokinin and pancreozymin—a single hormone? Acta Physiol Scand 66:196–202
57. Yalow RS, Berson SA (1960) Immunoassay of endogenous plasma insulin in man. J Clin Invest 39:1157–1175
58. Rehfeld JF (1978) Immunochemical studies on cholecystokinin. I. Development of sequencespecific radioimmunoassays against porcine triacontatriapeptide cholecystokinin. J Biol Chem 253:4016–4021
59. Konturek PC, Konturek SJ (2003) The history of gastrointestinal hormones and the polish contribution to elucidation of their biology and relation to nervous system. J Physiol Pharmacol 54(S3):83–98
60. Yeom J, Kabir MH, Lim B, Ahn H-S, Kim S-Y, Lee C (2016) A proteogenomic approach for protein-level evidence of genomic variants in cancer cells. Sci Rep 6:35305
61. Johnston PR, Dobson AJ, Rolff J (2016) Genomic signatures of experimental adaptation to antimicrobial peptides in Staphylococcus aureus. G3 Genes Genom Genet 6(6):1535–1539
62. Rehfeld JF (1998) The new biology of gastrointestinal hormones. Physiol Rev 78:1087–1108
63. Rawdon BB, Kramer B, Andrew A (1984) The distribution of endocrine cell progenitors in the gut of chick embryos. J Embryol Exp Morphol 82:131–145
64. Janssen S, Depoortere I (2013) Nutrient sensing in the gut: new roads to therapeutics? Trends Endocrinol Metab 24(2):92–100
65. Sternini C (2007) Taste receptors in the gastrointestinal tract. IV. Functional implications of bitter taste receptors in gastrointestinal chemosensing. Am Jf Physiol Gastrointest Liver Physiol 292(2):G457–G461
66. Sundaresan S, Abumrad NA (2015) Dietary lipids inform the gut and brain about meal arrival via CD36-mediated signal transduction. J Nutr 145(10):2195–2200
67. Depoortere I (2014) Taste receptors of the gut: emerging roles in health and disease. Gut 63 (1):179–190
68. Rawal S, Huedo-Medina TB, Hoffman HJ, Swede H, Duffy VB (2017) Structural equation modeling of associations among taste-related risk factors, taste functioning, and adiposity. Obesity (Silver Spring) 25(4):781–787
69. Ano Y, Dohata A, Taniguchi Y, Hoshi A, Uchida K, Takashima A, Nakayama H (2017) Iso-α-acids, bitter components of beer, prevent inflammation and cognitive decline induced in a mouse model of alzheimer's disease. J Biol Chem 292(9):3720–3728
70. Klinger A, Orzekowsky-Schroeder R, von Smolinski D, Blessenohl M, Schueth A, Koop N et al (2012) Complex morphology and functional dynamics of vital murine intestinal mucosa revealed by autofluorescence 2-photon microscopy. Histochem Cell Biol 137:269–278

71. Solcia E, Capella C, Buffa R, Usellini L, Fiocca R, Frigerio B et al (1981) The diffuse endocrine-paracrine system of the gut in health and disease: ultrastructural features. Scand J Gastroenterol 70:25–36
72. Kiba T (2004) Relationships between the autonomic nervous system and the pancreas including regulation of regeneration and apoptosis: recent developments. Pancreas 29(2): e51–e58
73. Mantyh PW, Catton MD, Allen CJ, Labenski ME, Maggio JE, Vigna SR (1992) Receptor binding sites for cholecystokinin, galanin, somatostatin, substance P and vasoactive intestinal polypeptide in sympathetic ganglia. Neurosci 46(3):739–754
74. Schonhoff SE, Giel-Moloney M, Leiter AB (2004) Minireview: development and differentiation of gut endocrine cells. Endocrinol 145(6):2639–2644
75. Škrha J (2006) Pancreatic hormones and hormonal regulation of insulin secretion. Čas Lék Čes 145(8):599–605
76. Petersen MBK, Azad A, Ingvorsen C, Hess K, Hansson M, Grapin-Botton A, Honoré C (2017) Single-cell gene expression analysis of a human esc model of pancreatic endocrine development reveals different paths to β-cell differentiation. Stem Cell Rep 9(4):1246–1261
77. Andralojc KM, Mercalli A, Nowak KW, Albarello L, Calcagno R, Luzi L et al (2009) Ghrelin-producing epsilon cells in the developing and adult human pancreas. Diabetologia 52(3):486–493
78. Amano O, Mizobe K, Bando Y, Sakiyama K (2012) Anatomy and histology of rodent and human major salivary glands: -overview of the Japan salivary gland society-sponsored workshop. Acta Histochem Cytochem 45(5):241–550
79. Tierney S, Pitt HA, Lillemoe KD (1993) Physiology and pathophysiology of gallbladder motility. Surg Clin North Am 73(6):1267–1290
80. Rybakova MG (1978) Endocrine function of the salivary glands. Arkh Patol 40(2):85–91
81. Houten SM, Watanabe M, Auwerx J (2006) Endocrine functions of bile acids. EMBO J 25(7):1419–1425
82. Pietrangelo A (2007) Hemochromatosis: an endocrine liver disease. Hepatol 46:1291–1301
83. Solomon TE (1984) Regulation of pancreatic secretion. Clin Gastroenterol 13(3):657–678
84. Rossi J, Santamäki P, Airaksinen MS, Herzig K-H (2005) Parasympathetic innervation and function of endocrine pancreas requires the glial cell line–derived factor family receptor $\alpha 2$ (GFR$\alpha 2$). Diab 54(5):1324–1330
85. Barber DL (1993) Regulation of peptide secretion from gastroenteric endocrine cells. In: Brown DR (ed) Gastrointestinal regulatory peptides. Springer, Berlin, Germany
86. Morris KE, St. Laurent CD, Hoeve RS, Forsythe P, Suresh MR, Mathison RD, Befus AD (2009) Autonomic nervous system regulates secretion of anti-inflammatory prohormone SMR1 from rat salivary glands. Am J Physiol Cell Physiol 296(3):C514-C524
87. Pearse AG (1969) The cytochemistry and ultrastructure of polypeptide hormone-producing cells of the APUD series and the embryologic, physiologic and pathologic implications of the concept. J Histochem Cytochem 17(5):303–313
88. Montuenga LM, Guembe L, Burrell MA, Bodegas ME, Calvo A, Sola JJ et al (2003) The diffuse endocrine system: from embryogenesis to carcinogenesis. Prog Histochem Cytochem 38(2):153–272
89. Ahlman H, Nilsson O (2001) The gut as the largest endocrine organ in the body. Ann Oncol 12(Suppl 2):S63–S68
90. Gunawardene AR, Corfe BM, Staton CA (2011) Classification and functions of enteroendocrine cells of the lower gastrointestinal tract. Int J Exp Pathol 92(4):219–231
91. Pearse AGE, Polak JM (1971) Neural crest origin of the endocrine polypeptide (APUD) cells of the gastrointestinal tract and pancreas. Gut 12:783–788
92. Luca IC (1998) Endocrine diffuse system. Histological and functional aspects interrelated with tumoral pathology. Rom J Morphol Embryol 44(1–4):17–22
93. Ilett EE, Langer SW, Holst IO, Federspiel B, Kjær A, Knigge U (2015) Neuroendocrine carcinomas of the gastroenteropancreatic system: a comprehensive review. Diagn 5(2): 119–176

94. Welbourn RB (1977) Current status of the apudomas. Ann Surg 185(1):1–12
95. Rindi G, Wiedenmann B (2012) Neuroendocrine neoplasms of the gut and pancreas: new insights. Nat Rev Endocrinol 8:54–64
96. Moran GW, Leslie FC, Levison SE, McLaughlin JT (2008) Review: enteroendocrine cells: neglected players in gastrointestinal disorders? Therap Adv Gastroenterol 1(1):51–60
97. Bosman FT, Louwerens J-WK (1981) APUD cells in teratomas. Am J Pathol 104:174–180
98. Delektorskaya VV, Kozlov NA, Chemeris GY (2013) Clinico-morphological analysis of the neuroendocrine neoplasms of the gastroenteropancreatic system. Klin Lab Diagn 10: 48–50, 10–3
99. Fazio N, Spada F (2016) Management of neuroendocrine tumors. In: Ambrosini V, Fanti S (eds) PET/CT in neuroendocrine tumors. Part of the series clinicians' guides to radionuclide hybrid imaging. Springer, Cham, Switzerland
100. Fujita T, Kanno T, Kobayashi S (2011) The paraneuron. Springer Verlag, Tokyo, Japan
101. Fujita T, Kobayashi S (1979) Current views on the paraneurone concept. Trend Neurosci 2:27–30
102. Kobayashi S, Chiba T (1977) Paraneurons: new concepts on neuro-endocrine relatives. In: Proceedings of the symposium on the paraneurons, Morioka, Japan
103. Fujita T (1976) Gastro-enteric endocrine cell and its paraneuronic nature. In: Coupland RE, Fujita T (eds) Chromaffin, enterochromaffin and related cells. Elsevier, Amsterdam, Netherlands
104. Kobayashi S, Segi M (1984) Gut paraneurons and Segi's cap. In: Motta PM (ed) Ultrastructure of endocrine cells and tissues. Electron microscopy in biology and medicine (current topics in ultrastructural research), vol 1. Springer, Boston, MA
105. Fujita T (1989) Present status of paraneuron concept. Arch Histol Cytol 52:1–8
106. Kubarko AI, Semenovich AA, Pereverzev VA (eds) (2013) A textbook of human physiology. Part I & 2. Visheishaya Shkola, Minsk, Belarus
107. Nikitina OS, Welcome MO, Pereverzev VA (2016) Human anatomy and physiology. In: 2 parts. Part 1 & 2. Belarusian State Medical University Press, Minsk, Belarus
108. Gomez GA, Englander EW, Greeley GH Jr (2012) Postpyloric gastrointestinal peptides. In: Physiology of the gastrointestinal tract. Academic Press, San Diego, CA, USA
109. Xiong X, Zhong A, Xu H (2014) Effect of cyanotoxins on the hypothalamic–pituitary–gonadal axis in male adult mouse. PLoS ONE 9(11):e106585
110. Labrie F, Borgeat P, Godbout M, Barden N, Beaulieu M, Lagacé L et al (1980) Mechanism of action of hypothalamic hormones and interactions with sex steroids in the anterior pituitary gland. In: Jutisz M, McKerns KW (eds) Synthesis and release of adenohypophyseal hormones. Springer, Boston, MA
111. Parker HE, Gribble FM, Reimann F (2014) The role of gut endocrine cells in control of metabolism and appetite. Exp Physiol 99(9):1116–1120
112. Hill JW, Elmquist JK, Elias CF (2008) Hypothalamic pathways linking energy balance and reproduction. Am J Physiol Endocrinol Metab 294:E827–E832
113. Pedersen BK (2009) The diseasome of physical inactivity—and the role of myokines in muscle–fat cross talk. J Physiol 587(23):5559–5568
114. Pedersen BK, Åkerström TCA, Nielsen AR, Fischer CP (2007) Role of myokines in exercise and metabolism. J Appl Physiol 103(3):1093–1098
115. Gosman GG, Katcher HI, Legro RS (2006) Obesity and the role of gut and adipose hormones in female reproduction. Hum Reprod Update 12(5):585–601
116. Venteclef N, Guglielmi V, Balse E, Gaborit B, Cotillard A, Atassi F et al (2015) Human epicardial adipose tissue induces fibrosis of the atrial myocardium through the secretion of adipo-fibrokines. Eur Heart J 36(13):795–805
117. Larsson LI (1980) On the possible existence of multiple endocrine, paracrine and neurocrine messengers in secretory cell systems. Invest Cell Pathol 3(1):73–85
118. Larsson LI (1980) Gastrointestinal cells producing endocrine, neurocrine and paracrine messengers. Clin Gastroenterol 9(3):485–516

119. O'Byrne P (1988) Lymphokine activated killer cell activity in patients with GI cancer. Gut 29(5):693–694
120. Lehmann FS, Terracciano L, Carena I, Baeriswyl C, Drewe J, Tornillo L et al (2002) In situ correlation of cytokine secretion and apoptosis in Helicobacter pylori-associated gastritis. Am J Physiol Gastrointest Liver Physiol 283(2):G481–G488
121. Lin HY, Chang KT, Hung CC, Kuo CH, Hwang SJ, Chen HC et al (2014) Effects of the mTOR inhibitor rapamycin on monocyte-secreted chemokines. BMC Immunol 26(15):37
122. López-Lago MA, Posner S, Thodima VJ, Molina AM, Motzer RJ, Chaganti RSK (2013) Neutrophil chemokines secreted by tumor cells mount a lung antimetastatic response during renal cell carcinoma progression. Oncogene 32(14):1752–1760
123. Lazennec G, Richmond A (2010) Chemokines and chemokine receptors: new insights into cancer-related inflammation. Trends Mol Med 16(3):133–144
124. Olsnes C, Stavang H, Brokstad K, Olofsson J, Aarstad HJ (2009) Chemokines are secreted by monocytes following OK-432 (lyophilized Streptococcus pyogenes) stimulation. BMC Immunol 10:6
125. Majka M, Janowska-Wieczorek A, Ratajczak J, Ehrenman K, Pietrzkowski Z, Kowalska MA et al (1999) Numerous growth factors, cytokines, and chemokines are secreted by human CD34 + cells, myeloblasts, erythroblasts, and megakaryoblasts and regulate normal hematopoiesis in an autocrine/paracrine manner. Blood 94(suppl 1):465a
126. Arnold R (1979) Pathophysiology and clinical significance of gut hormones. Acta Med Austriaca 6(3):77–90
127. Zhao CM, Martinez V, Piqueras L, Wang L, Taché Y, Chen D (2008) Control of gastric acid secretion in somatostatin receptor 2 deficient mice: shift from endocrine/paracrine to neurocrine pathways. Endocrinol 149(2):498–505
128. Rashid M, Singla D, Sharma A, Kumar M, Raghava GPS (2009) Hmrbase: a database of hormones and their receptors. BMC Genom 10:307
129. Henriksen JH, de Muckadell OB (2000) Secretin, its discovery, and the introduction of the hormone concept. Scand J Clin Lab Invest 60(6):463–471
130. Lee VHY, Lam IPY, Choi H-S, Chow BKC, Lee LTO (2012) The estrogen-related receptor alpha upregulates secretin expressions in response to hypertonicity and angiotensin II stimulation. PLoS ONE 7(6):e39913
131. Pelletier MJ, Chayvialle JA, Minaire Y (1979) Uneven and transient secretin release after a liquid test meal. Gastroenterol 75(6):1124–1132
132. Sundaresan S, Shahid R, Riehl TE, Chandra R, Nassir F, Stenson WF et al (2013) CD36-dependent signaling mediates fatty acid-induced gut release of secretin and cholecystokinin. FASEB J 27(3):1191–1202
133. Dong M, Miller LJ (2002) Molecular pharmacology of the secretin receptor. Recept Channels 8(3–4):189–200
134. Miller LJ, Dong M, Harikumar KG (2012) Ligand binding and activation of the secretin receptor, a prototypic family B G protein-coupled receptor. Br J Pharmacol 166(1):18–26
135. Afroze S, Meng F, Jensen K, McDaniel K, Rahal K, Onori P et al (2013) The physiological roles of secretin and its receptor. Ann Transl Med 1(3):29
136. Posovszky C, Wabitsch M (2015) Regulation of appetite, satiation, and body weight by enteroendocrine cells. Part 1: characteristics of enteroendocrine cells and their capability of weight regulation. Horm Res Paediatr 83(1):1–10
137. Posovszky C, Wabitsch M (2015) Regulation of appetite, satiation, and body weight by enteroendocrine cells. part 2: therapeutic potential of enteroendocrine cells in the treatment of obesity. Horm Res Paediatr 83(1):11–18
138. Bassalyk LS (1970) Gastrin (biochemistry, physiology and clinical importance). Ter Arkh 42(12):4–9
139. Peggion E, Foffani MT, Mammi S (1991) Structure-function relationship of gastrin hormones: a rational approach to drug design. Bioorganic Chemistry in Healthcare and Technology. Vol 207 of the series NATO ASI Series. Plenum Press, New York, USA

140. Wiborg O, Berglund L, Boel E, Norris F, Norris K, Rehfeld JF et al (1984) Biochemistry structure of a human gastrin gene. Proc Natl Acad Sci USA 81:1067–1069
141. Beinfeld MC (2009) CCK/gastrin and receptors. In: Squire LR (ed) Encyclopedia of neuroscience. Elsevier, San Diego, CA, USA
142. Modlin IM, Kidd M, Marks IN, Tang LH (1997) The pivotal role of John S. Edkins in the discovery of gastrin. World J Surg 21(2):226–234
143. Gregory RA, Tracy HJ (1964) The constitution and properties of two gastrins extracted from hog antral mucosa: Part I the isolation of two gastrins from hog antral mucosa. Gut 5(2):103
144. Dufresne M, Seva C, Fourmy D (2006) Cholecystokinin and gastrin receptors. Physiol Rev 86(3):805–847
145. Janssen T, Meelkop E, Lindemans M, Verstraelen K, Husson SJ, Temmerman L et al (2008) Discovery of a cholecystokinin-gastrin-like signaling system in nematodes. Endocrinol 149 (6):2826–2839
146. Johnsen AH (1998) Phylogeny of the cholecystokinin/gastrin family. Front Neuroendocrinol 19(2):73–99
147. McRoberts JW (1986) Cholecystokinin and pain. Anesth Prog 33(2):87–90
148. Lewis L, Van Brimmer B (2003) A biography of Robert Milton Zollinger, M.D. Am J Surg 186(3):205–207
149. Strosberg JR (2014) Systemic treatment of gastroenteropancreatic neuroendocrine tumors (GEP-NETS): current approaches and future options. Endocr Pract 20(2):167–175
150. Libutti SK, Alexander HR Jr (2006) Gastrinoma: sporadic and familial disease. Surg Oncol Clin 15(3):479–496
151. Shah P, Singh MH, Yang Y-X, Metz DC (2013) Hypo-and achlorhydria are associated with false-positive secretin stimulation testing for Zollinger-Ellison syndrome. Pancreas 42 (6):932–936
152. Goldman JA, Blanton WP, Hay DW, Wolfe MM (2009) False-positive secretin stimulation test for gastrinoma associated with the use of proton pump inhibitor therapy. Clin Gastroenterol Hepatol 7(5):600–602
153. Braganza JM, Herman K, Hine P, Kay G (1979) The effect of pentagastrin on peptic secretion in man. J Physiol 289:9–16
154. Barbot N, Calmettes C, Schuffenecker I, Saint-André JP, Franc B, Rohmer V et al (1994) Pentagastrin stimulation test and early diagnosis of medullary thyroid carcinoma using an immunoradiometric assay of calcitonin: comparison with genetic screening in hereditary medullary thyroid carcinoma. J Clin Endocrinol Metab 78(1):114–120
155. Milone F, Ramundo V, Chiofalo MG, Severino R, Paciolla I, Pezzullo L et al (2010) Predictive value of pentagastrin test for preoperative differential diagnosis between C-cell hyperplasia and medullary thyroid carcinoma in patients with moderately elevated basal calcitonin levels. Clin Endocrinol (Oxf) 73(1):85–88
156. Riley JF (1965) Histamine and Sir Henry Dale. Br Med J 1(5448):1488–1490
157. Dale HH, Laidlaw PP (1910) The physiological action of beta-iminazolylethylamine. J Physiol 41(5):318–344
158. Beales ILP (2002) Easy as 1, 2, 3? Histamine receptors and gastric acid. Gut 50(6):747–748
159. Barger G, Dale HH (1911) beta-Iminazolylethylamine a depressor constituent of intestinal mucosa. J Physiol 41(6):499–503
160. Dvorak AM, MacGlashan DW Jr, Morgan ES, Lichtenstein LM (1996) Vesicular transport of histamine in stimulated human basophils. Blood 88(11):4090–4101
161. Xu Y, An F, Borycz JA, Borycz J, Meinertzhagen IA, Wang T (2015) Histamine recycling is mediated by CarT, a carcinine transporter in Drosophila photoreceptors. PLoS Genet 11(12): e1005764
162. Prinz C, Zanner R, Gerhard M, Mahr S, Neumayer N, Höhne-Zell B, Gratzl M (1999) The mechanism of histamine secretion from gastric enterochromaffin-like cells. Am J Physiol Cell Physiol 277(5):C845–C855

163. Travis ER, Wang Y-M, Michael DJ, Caron MG, Wightman RM (1999) Differential quantal release of histamine and 5-hydroxytryptamine from mast cells of vesicular monoamine transporter 2 knockout mice. PNAS 97(1):162–167
164. Merickel A, Edwards RH (1995) Transport of histamine by vesicular monoamine transporter-2. Neuropharmacol 34(11):1543–1547
165. Ogasawara M, Yamauchi K, Satoh Y, Yamaji R, Inui K, Jonker JW et al (2006) Recent advances in molecular pharmacology of the histamine systems: organic cation transporters as a histamine transporter and histamine metabolism. J Pharmacol Sci 101(1):24–30
166. Stegaev V, Nies AT, Porola P, Mieliauskaite D, Sánchez-Jiménez F, Urdiales JL et al (2013) Histamine transport and metabolism are deranged in salivary glands in Sjogren's syndrome. Rheumatol (Oxford) 52(9):1599–1608
167. Feldberg WS (1970) Henry Hallett Dale. 1875–1968. Biogr Mem Fellows R Soc 16:77–174
168. Stegaev V, Sillat T, Porola P, Hänninen A, Falus A, Mieliauskaite D et al (2012) Brief report: first identification of H_4 histamine receptor in healthy salivary glands and in focal sialadenitis in Sjögren's syndrome. Arthritis Rheum 64(8):2663–2668
169. Panula P, Kaartinen M, Mäcklin M, Costa E (1985) Histamine-containing peripheral neuronal and endocrine systems. J Histochem Cytochem 33(9):933–941
170. Knigge U, Warberg J (1991) The role of histamine in the neuroendocrine regulation of pituitary hormone secretion. Acta Endocrinol (Copenh) 124(6):609–719
171. Erspamer V, Erspamer GF, Inselvini M (1970) Some pharmacological actions of alytesin and bombesin. J Pharm Pharmacol 22:875–876
172. Orloff MS, Reeve JR Jr, Ben-Avram CM, Shively JE, Walsh JH (1984) Isolation and sequence analysis of human bombesin-like peptides. Peptides 5(5):865–870
173. Erspamer V, Melchiorri P, Broccardo M, Erspamer GF, Falaschi P, Improta G et al (1981) The brain-gut-skin triangle: new peptides. Peptides 2(2):7–16
174. Guo YS, Thompson JC, Singh P (1988) Role of Ca2 + in bombesin -stimulated release of gastrin and somatostatin from isolated perfused rat stomach. Am J Physiol 255(5 Pt 1): G627–G632
175. Kleibeuker JH, Kauffman GL Jr, Walsh JH (1985) Intravenous histamine reduces bombesin-stimulated gastrin release in dogs. Regul Pept 11(3):209–215
176. Guo Y-S, Thompson JC, Singh P (1990) Role of gastrin in bombesin-stimulated somatostatin release. Gastroenterol 99(5):1297–1302
177. Fletcher DR, Shulkes A, Bladin PHD, Hardy KJ (1983) The effect of atropine on bombesin and gastrin releasing peptide stimulated gastrin, pancreatic polypeptide and neurotensin release in man. Regul Pept 7(1):31–40
178. Gonzalez N, Moody TW, Igarashi H, Ito T, Jensen RT (2008) Bombesin-related peptides and their receptors: recent advances in their role in physiology and disease states. Curr Opin Endocrinol Diabetes Obes 15(1):58–64
179. Ohlsson B, Fredäng N, Axelson J (1999) The effect of bombesin, cholecystokinin, gastrin, and their antagonists on proliferation of pancreatic cancer cell lines. Scand J Gastroenterol 34(12):1224–1229
180. Jensen RT, Battey JF, Spindel ER, Benya RV (2008) International Union of Pharmacology. LXVIII. Mammalian bombesin receptors: nomenclature, distribution, pharmacology, signaling, and functions in normal and disease states. Pharmacol Rev 60(1):1–42
181. West SD, Mercer DW (2005) Bombesin-induced gastroprotection. Ann Surg 241(2):227–231
182. Xiao D, Qu X, Weber HC (2003) Activation of extracellular signal-regulated kinase mediates bombesin-induced mitogenic responses in prostate cancer cells. Cell Sig 15 (10):945–953
183. Ohki-Hamazaki H (2000) Neuromedin B. Prog Neurobiol 62(3):297–312

184. Krane IM, Naylor SL, Helin-Davis D, Chin WW, Spindel ER (1988) Molecular cloning of cDNAs encoding the human bombesin-like peptide neuromedin B. Chromosomal localization and comparison to cDNAs encoding its amphibian homolog ranatensin. J Biol Chem 263(26):13317–13323
185. von Euler US (1935) Uber die spezifische blutdrucksenkende substanz des menschlichen prostate- und samenblasensekrets. Klin Wochenschr 14:1182–1183
186. Kresge N, Simoni RD, Hill RL (2006) The prostaglandins, Sune Bergström and Bengt Samuelsson. J Biol Chem 281:e9–e11
187. Goldblatt MW (1935) Properties of human seminal plasma. J Physiol 84:208–218
188. Stanley D, Kim Y (2014) Eicosanoid signaling in insects: from discovery to plant protection. Crit Rev Plant Sci 33(1):20–63
189. Shimizu N, Nakamura T (1985) Prostaglandins as hormones. Dig Dis Sci 30(11 Suppl): 109S–113S
190. Lupulescu A (1996) Prostaglandins, their inhibitors and cancer. Prostaglandins Leukot Essent Fatty Acids 54(2):83–94
191. Hata AN, Breyer RM (2004) Pharmacology and signaling of prostaglandin receptors: multiple roles in inflammation and immune modulation. Pharmacol Ther 103(2):147–166
192. Brash AR (2001) Arachidonic acid as a bioactive molecule. J Clin Invest 107(11):1339–1345
193. Kuksis A, Myher JJ (1995) Application of tandem mass spectrometry for the analysis of long-chain carboxylic acids. J Chromatogr B Biomed Sci Appl 671(1–2):35–70
194. Flower RJ (2006) Prostaglandins, bioassay and inflammation. Br J Pharmacol 147(1):S182–S192
195. Hammarström S (1983) Leukotrienes. Annu Rev Biochem 52:355–377
196. Feldberg W, Kellaway CH (1938) Liberation of histamine and formation of lysocithin-like substances by cobra venom. J Physiol 94:187–226
197. Malmsten CL (1986) Prostaglandins, thromboxanes, and leukotrienes in inflammation. Am J Med 80(4B):11–17
198. Samuelsson B (1987) An elucidation of the arachidonic acid cascade. Discovery of prostaglandins, thromboxane and leukotrienes. Drugs 33(1):2–9
199. Weinberger B, Hirsch D, Yin K, Spur BW (2015) Lipid mediators and lung function. In: Parent RA (ed) Comparative biology of the normal lung. Elsevier, San Diego, CA, USA
200. Davies P, Bailey PJ, Goldenberg MM, Ford-Hutchinson AW (1984) The role of arachidonic acid oxygenation products in pain and inflammation. Ann Rev Immunol 2:335–357
201. Jabbour HN, Sales KJ (2004) Prostaglandin receptor signalling and function in human endometrial pathology. Trends Endocrinol Metab 15(8):398–404
202. Leonhardt A, Glaser A, Wegmann M, Hackenberg R, Nüsing RM (2003) Expression of prostanoid receptors in human lower segment pregnant myometrium. Prostaglandins Leukot Essent Fatty Acids 69(5):307–313
203. Narumiya S, Sugimoto Y, Ushikubi F (1999) Prostanoid receptors: structures, properties, and functions. Physiol Rev 79(4):1193–1226
204. Robert A (1984) Prostaglandins: effects on the gastrointestinal tract. Clin Physiol Biochem 2 (2–3):61–69
205. Dey I, Lejeune M, Chadee K (2006) Prostaglandin E2 receptor distribution and function in the gastrointestinal tract. Br J Pharmacol 149(6):611–623
206. Occhipinti M (1978) Prostaglandins and gastrointestinal function. Adv Pediatr 25:205–221
207. Wang D, Mann JR, Dubois RN (2005) The role of prostaglandins and other eicosanoids in the gastrointestinal tract. Gastroenterol 128(5):1445–1461
208. Becker JC, Domschke W, Pohle T (2004) Current approaches to prevent NSAID-induced gastropathy—COX selectivity and beyond. Br J Clin Pharmacol 58(6):587–600
209. Scarpignato C, Hunt RH (2010) Nonsteroidal antiinflammatory drug-related injury to the gastrointestinal tract: clinical picture, pathogenesis, and prevention. Gastroenterol Clin 39 (3):433–464

210. Johnell K, Fastbom J (2008) Concomitant use of gastroprotective drugs among elderly NSAID/COX-2 selective inhibitor users—a nationwide register-based study. Clin Drug Invest 28(11):687–695
211. Wolfe F, Anderson J, Burke TA, Arguelles LM, Pettitt D (2002) Gastroprotective therapy and risk of gastrointestinal ulcers: risk reduction by COX-2 therapy. J Rheumatol 29(3):467–473
212. Pasina L, Nobili A, Tettamanti M, Riva E, Lucca U, Piccinelli R et al (2010) Co-prescription of gastroprotective agents in patients taking non-selective NSAIDs or COX-2 selective inhibitors: analysis of prescriptions. Int J Clin Pharmacol Ther 48(11):735–743
213. Sun SX, Lee KY, Bertram CT, Goldstein JL (2007) Withdrawal of COX-2 selective inhibitors rofecoxib and valdecoxib: impact on NSAID and gastroprotective drug prescribing and utilization. Curr Med Res Opin 23(8):1859–1866
214. Allen R, O'Brien BM (2009) Uses of misoprostol in obstetrics and gynecology. Rev Obstet Gynecol 2(3):159–168
215. La Barre J (1932) Sur les possibilite´s d'un traitement du diabète par l'incrétine. Bull Acad R Med Belg 12:620–634
216. Moore N, Edie ES, Abram TH (1906) On the treatment of diabetes mellitus by acid extract of duodenal mucous membrane. Biochem J 1:28–38
217. Brown JC, Pederson RA (1970) Cleavage of a gastric inhibitory polypeptide with cyanogen bromide and the physiological action of the C-terminal fragment. J Physiol 210:52–53
218. Brown JC (1971) A gastric inhibitory polypeptide. I. The amino acid composition and the tryptic peptides. Can J Biochem 49:255–261
219. Dupre J, Ross SA, Watson D, Brown JC (1973) Stimulation of insulin secretion by gastric inhibitory polypeptide in man. J Clin Endocrinol Metab 37:826–828
220. Creutzfeldt W (2005) The [pre-] history of the incretin concept. Regul Pept 128(2):87–91
221. Kieffer TJ, Habener JF (1999) The glucagon-like peptides. Endocr Rev 20(6):876–913
222. Orskov C, Nielsen JH (1988) Truncated glucagon-like peptide-1(proglucagon 78–107 amide), an intestinal insulin-releasing peptide, has specific receptors on rat insulinoma cells (RIN 5AH). FEBS Lett 229:175–178
223. Orskov C, Holst JJ, Poulsen SS, Kirkegaard P (1987) Pancreatic and intestinal processing of proglucagon in man. Diabetologia 30:874–881
224. Orskov C, Holst JJ, Nielsen OV (1988) Effect of truncated glucagon-like peptide-1 [proglucagon-(78–107)amide] on endocrine secretion from pig pancreas, antrum, and nonantral stomach. Endocrinol 123:209–213
225. Kreymann B, Williams G, Ghatei MA, Bloom SR (1987) Glucagon-like peptide-1 7-36: A physiological incretin in man. Lancet 2:1300–1304
226. Mojsov S, Weir GC, Habener JF (1987) Insulinotropin: glucagon-like peptide I (7–37) co-encoded in the glucagon gene is a potent stimulator of insulin release in the perfused rat pancreas. J Clin Invest 79:616–619
227. Zunz E, LaBarre J (1928) Hyperinsulinémie consécutive a l'injection de solution de secrétine non hypotensive. C R Soc Biol (Paris):1435–1438
228. McIntyre N, Holdsworth CD, Turner DS (1964) New interpretation of oral glucose tolerance. Lancet 2:20–21
229. Elrick H, Stimmler L, Hlad CJ, Arai Y (1964) Plasma insulin responses to oral and intravenous glucose administration. J Clin Endocrinol Metab 24:1076–1082
230. Perley MJ, Kipnis DM (1967) Plasma insulin responses to oral and intravenous studies in normal and diabetic subjects. J Clin Invest 46:1954–1962
231. Brubaker PL, Drucker DJ (2002) Structure-function of the glucagon receptor family of G protein–coupled receptors: the glucagon, GIP, GLP-1, and GLP-2 receptors. Recept Channels 8:179–188
232. Girard J (2008) The incretins: From the concept to their use in the treatment of type 2 diabetes. Part A: Incretins: Concept and physiological functions. Diabetes Metab 34:550–559

233. Takeda J, Seino Y, Tanaka K, Fukumoto H, Kayano T, Takahashi H et al (1987) Sequence of an intestinal cDNA encoding human gastric inhibitory polypeptide precursor. Proc Natl Acad Sci USA 84:7005–7008
234. Brown JC, Mutt V, Pederson RA (1970) Further purification of a polypeptide demonstrating enterogastrone activity. J Physiol (Lond) 209:57–64
235. Lauritsen KB, Moody AJ, Christensen KC, Jensen SL (1980) Gastric-inhibitory polypeptide (GIP) and insulin release after small bowel resection in man. Scand J Gastroenterol 15:833–840
236. Mojsov S, Heinrich G, Wilson IB, Ravazzola M, Orci L, Habener JF (1986) Preproglucagon gene expression in pancreas and intestine diversifies at the level of post-translational processing. J Biol Chem 261:11880–11889
237. Holst JJ, Gromada J (2004) Role of incretin hormones in the regulation of insulin secretion in diabetic and nondiabetic humans. Am J Physiol Endocrinol Metab 287:E199–E206
238. Orskov C, Holst JJ, Knuhtsen S, Baldissera FG, Poulsen SS, Nielsen OV (1986) Glucagon-like peptides GLP-1 and GLP-2, predicted products of the glucagon gene are secreted separately from pig small intestine but not pancreas. Endocrinol 119:1467–1475
239. Yip RG, Wolfe MM (2000) GIP biology and fat metabolism. Life Sci 66(2):91–103
240. Beck B (1989) Gastric inhibitory polypeptide: a gut hormone with anabolic functions. J Mol Endocrinol 2(3):169–174
241. Pellissier S, Sasaki K, Le-Nguyen D, Bataille D, Jarrousse C (2004) Oxyntomodulin and glicentin are potent inhibitors of the fed motility pattern in small intestine. Neurogastroenterol Motil 16(4):455–463
242. Steiner DF, Albrechtsen NJW, Rehfeld JF, Holst JJ (2017) Glucagon processing. In: Stein J, Bennett D, Coen C, Dunbar R, Goodwin G, Husain M et al (eds) Reference module in neuroscience and biobehavioral psychology. Elsevier, San Diego, CA, USA
243. Mineo I, Matsumura T, Shingu R, Namba M, Kuwajima M, Matsuzawa Y (1995) The role of prohormone convertases PC1 (PC3) and PC2 in the cell-specific processing of proglucagon. Biochem Biophys Res Commun 207(2):646–651
244. Anini Y, Jarrousse C, Chariot J, Nagain C, Yanaihara N, Sasaki K et al (2000) Oxyntomodulin inhibits pancreatic secretion through the nervous system in rats. Pancreas 20 (4):348–360
245. Ohneda A, Ohneda K, Nagasaki T, Sasaki K (1995) Insulinotropic action of human glicentin in dogs. Metab 44(1):47–51
246. Rodier G, Magous R, Mochizuki T, Le Nguyen D, Martinez J, Bali JP et al (1999) Glicentin and oxyntomodulin modulate both the phosphoinositide and cyclic adenosine monophosphate signaling pathways in gastric myocytes. Endocrinol 140(1):22–28
247. Bataille D, Dalle S (2014) The forgotten members of the glucagon family. Diabetes Res Clin Pract 106(1):1–10
248. Vilsbøll T (2004) On the role of the incretin hormones GIP and GLP-1 in the pathogenesis of Type 2 diabetes mellitus. Dan Med Bull 51(4):364–370
249. Sørensen G, Reddy IA, Weikop P, Graham DL, Stanwood GD, Wortwein G et al (2015) The glucagon-like peptide 1 (GLP-1) receptor agonist exendin-4 reduces cocaine self-administration in mice. Physiol Behav 149:262–268
250. Seufert J, Gallwitz B (2014) The extra-pancreatic effects of GLP-1 receptor agonists:a focus on the cardiovascular, gastrointestinal and centralnervous systems. Diabetes Obes Metab 16:673–688
251. Körner M, Stöckli M, Waser B, Reubi JC (2007) GLP-1 receptor expression in human tumors and human normal tissues: potential for in vivo targeting. J Nucl Med 48(5):736–743
252. Runge S, Wul BS, Madsen K, BraÈ uner-Osborne H, Knudsen LB (2003) Different domains of the glucagon and glucagon-like peptide-1 receptors provide the critical determinants of ligand selectivity. Br J Pharmacol 138:787–794
253. Pocai A (2012) Unraveling oxyntomodulin, GLP1's enigmatic brother. J Endocrinol 215 (3):335–346

254. Baggio LL, Drucker DJ (2007) Biology of incretins: GLP-1 and GIP. Gastroenterol 132 (6):2131–2157
255. You C, Chey W (1987) Secretin is an enterogastrone in humans. Dig Dis Sci 32(5):466–471
256. Lloyd K, Maxwell V, Chuang C, Wong H, Soll A, Walsh J (1994) Somatostatin is released in response to cholecystokinin by activation of type A CCK receptors. Peptides 15(2):223–227
257. Zhang Y, Proenca R, Maffei M, Barone M, Leopold L, Friedman JM (1994) Positional cloning of the mouse obese gene and its human homolog. Nat 372:425–432
258. Bado A, Levasseur S, Attoub S, Kermorgant S, Laigneau JP, Bortoluzzi MN et al (1998) The stomach is a source of leptin. Nat 394:790–793
259. Cammisotto PG, Renaud C, Gingras D, Delvin E, Levy E, Bendayan M (2005) Endocrine and exocrine secretion of leptin by the gastric mucosa. J Histochem Cytochem 53:851–860
260. Yarandi SS, Hebbar G, Sauer CG, Cole CR, Ziegler TR (2011) Diverse roles of leptin in the gastrointestinal tract: Modulation of motility, absorption, growth, and inflammation. Nutr 27 (3):269–275
261. Guilmeau S, Buyse M, Bado A (2004) Gastric leptin: a new manager of gastrointestinal function. Curr Opin Pharmacol 4(6):561–566
262. Barrenetxe J, Villaro AC, Guembe L, Pascual I, Munoz-Navas M, Barber A, Lostao MP (2008) Distribution of the long leptin receptor isoform in brush border, basolateral membrane, and cytoplasm of enterocytes. Gut 50:797–809
263. Cammisotto P, Bendayan M (2012) A review on gastric leptin: the exocrine secretion of a gastric hormone. Anat Cell Biol 45(1):1–16
264. Havel PJ (2000) Role of adipose tissue in body-weight regulation: mechanisms regulating leptin production and energy balance. Proc Nutr Soc 59:359–371
265. Guilmeau S, Buyse M, Tsocas A, Laigneau JP, Bado A (2003) Duodenal leptin stimulates cholecystokinin secretion: evidence of a positive leptin-cholecystokinin feedback loop. Diab 52:1664–1672
266. Bohlender J, Rauh M, Zenk J, Groschl M (2003) Differential distribution and expression of leptin and the functional leptin receptor in major salivary glands of humans. J Endocrinol 178:217–223
267. Cinti S, de Matteis R, Ceresi E, Picó C, Oliver J, Oliver P et al (2001) Leptin in the human stomach. Gut 49:155–159
268. Boelen A, Kwakkel J, Vos XG, Wiersinga WM, Fliers E (2006) Differential effects of leptin and refeeding on the fasting-induced decrease of pituitary type 2 deiodinase and thyroid hormone receptor beta2 mRNA expression in mice. J Endocrinol 190(2):537–544
269. Casanueva FF, Dieguez C (1999) Growth hormone secretagogues: physiological role and clinical utility. Trends Endocrinol Metab 10:30–38
270. Mojsilović M, Katić V, Ilić I, Radovanović B, Petrović B, Lazić M, Krstić M (2015) A thirty-seven-year follow-up of Peutz-Jeghers syndrome across three generations. Acta Fac Med Naiss 32(3):221–226
271. Modlin IM, Sank A, Albert D (1981) Current aspects of gut hormones. J Surg Res 30 (6):602–618
272. Kokas E, Ludány G (1938) Relation between the "villikinine" and the absorption of glucose from the intestine. Q J Exp Physiol 28(1):15–22
273. Brown JC (1967) Presence of a gastric motor-stimulating property in duodenal extracts. Gastroenterol 52(2):225–229
274. Brown JC, Cook MA, Dryburgh JR (1973) Motilin, a gastric motor activity stimulating polypeptide: the complete amino acid sequence. Can J Biochem 51(5):533–537
275. Sakai T, Satoh M, Sonobe K, Nakajima M, Shiba Y, Itoh Z (1994) Autoradiographic study of motilin binding sites in the rabbit gastrointestinal tract. Regul Peptides 53(3):249–257
276. Pasricha P (1999) Identifying the motilin receptor: Science, serendipity, or simply business? Gastroenterol 117:1502–1504
277. Bowers CY (1998) Growth hormone-releasing peptide (GHRP). Cell Mol Life Sci 54:1316–1329

278. Simonsson M, Eriksson S, Håkanson R, Lind T, Lönroth H, Lundell L et al (1988) Endocrine cells in the human oxyntic mucosa: a histochemical study. Scand J Gastroenterol 23(9):1089–1099
279. Perello M, Scott MM, Sakata I, Lee CE, Chuang JC, Osborne-Lawrence S et al (2012) Functional implications of limited leptin receptor and ghrelin receptor coexpression in the brain. J Comp Neurol 520(2):281–294
280. Kojima M, Hosoda H, Date Y, Nakazato M, Matsuo H, Kangawa K (1999) Ghrelin is a growth-hormone-releasing acylated peptide from stomach. Nat 402:656–660
281. Sato T, Nakamura Y, Shiimura Y, Ohgusu H, Kangawa K, Kojima M (2012) Structure, regulation and function of ghrelin. J Biochem 151(2): 119–128
282. Sakata I, Sakai T (2010) Ghrelin cells in the gastrointestinal tract. Int J Peptides 2010:945056
283. Tomasetto C, Karam SM, Ribieras S, Masson R, Lefèbvre O, Staub A et al (2000) Identification and characterization of a novel gastric peptide hormone: the motilin-related peptide. Gastroenterol 119(2):395–405
284. Broom DR, Miyashita M, Wasse LK, Pulsford R, King JA, Thackray AE, Stensel DJ (2017) Acute effect of exercise intensity and duration on acylated ghrelin and hunger in men. J Endocrinol 232(3):411–422
285. Kojima M, Kangawa K (2005) Ghrelin: structure and function. Physiol Rev 85:495–522
286. St-Pierre DH, Wang L, Taché Y (2003) Ghrelin: a novel player in the gut-brain regulation of growth hormone and energy balance. News Physiol Sci 18:242–246
287. Ivy AC, Oldberg E (1928) Hormone mechanism for gallbladder contraction and evacuation. Am J Physiol 65:599–613
288. Harper AA, Raper HS (1943) Pancreozymin, a stimulant of secretion of pancreatic enzymes in extracts of the small intestine. J Physiol 102(1):115–125
289. Wank SA (1998) G protein-coupled receptors in gastrointestinal physiology. I. CCK receptors: an exemplary family. Am J Physiol 274(4 Pt 1):G607–13
290. Wank SA (1995) Cholecystokinin receptors. Am J Physiol 269(5 Pt 1):G628–G646
291. Mutt V, Jorpes J (1968) Structure of porcine cholecystokinin-pancreozymin. Eur J Biochem 6:156–162
292. Liddle RA (2003) CCK (Cholecystokinin). In: Martini L (ed) Encyclopedia of endocrine diseases. Elsevier, San Diego, CA, USA
293. Rashmi C, Liddle RA (2007) Cholecystokinin. Curr Opin Endocrinol Diabetes Obes 14 (1):63–67
294. López H, Molina-Ruiz R, Molón L, Díaz-Marsá M (2011) Cholecystokinin in the pathogenesis of bulimia nervosa. Int J Clin Med 2(4):509–515
295. Williams RH, Champagne J (1979) Effects of cholecystokinin, secretin, and pancreatic polypeptide on secretion of gastric inhibitory polypeptide, insulin, and glucagon. Life Sci 25 (11):947–956
296. Verspohl EJ, Ammon HPT (1987) Cholecystokinin (CCK8) regulates glucagon, insulin, and somatostatin secretion from isolated rat pancreatic islets: interaction with glucose. Pflugers Arch 410(3):284–287
297. MacIntosh CG, Morley JE, Wishart J, Morris H, Jansen JB, Horowitz M, Chapman IM (2001) Effect of exogenous cholecystokinin (CCK)-8 on food intake and plasma CCK, leptin, and insulin concentrations in older and young adults: evidence for increased CCK activity as a cause of the anorexia of aging. J Clin Endocrinol Metab 86(12):5830–5837
298. Lehmann F, Hildebrand P, Beglinger C (2003) New molecular targets for treatment of peptic ulcer disease. Drugs 63(17):1785–1797
299. Hull RAD, Shankley NP, Harper EA, Gerskowitch VP, Black JW (1993) 2-Naphthalenesulphonyl l-aspartyl-(2-phenethyl)amide (2-NAP)-a selective cholecystokinin CCKA-receptor antagonist. Br J Pharmacol 108(3):734–740
300. Martinez V (2013) Somatostatin. In: Kastin A (ed) Handbook of biologically active peptides. Academic Press, San Diego, CA, USA

301. O'Shaughnessy DJ, Long RG, Adrian TE, Christofides ND, Ghatei MA, Sarson DL, Bloom SR (1985) Somatostatin-14 modulates postprandial glucose levels and release of gastrointestinal and pancreatic hormones. Dig 31(4):234–242
302. Barnes AJ, Long RG, Adrian TE, Vale W, Brown MR, Rivier JE et al (1981) Effect of a long-acting octapeptide analogue of somatostatin on growth hormone and pancreatic and gastrointestinal hormones in man. Clin Sci (Lond) 61(5):653–656
303. Vale W, Brazeau P, Grant G, Nussey A, Burgus R, Rivier J et al (1972) Premières observations sur le mode d'action de la somatostatine, un facteur hypothalamique qui inhibe la sécrétion de l'hormone de croissance. C R Acad Sci Hebd Seances Acad Sci D 275 (25):2913–2916
304. Brazeau P, Vale W, Burgus R, Ling N, Butcher M, Rivier J, Guillemin R (1973) Hypothalamic polypeptide that inhibits the secretion of immunoreactive pituitary growth hormone. Sci 179(4068):77–79
305. Dharmsathaphorn K (1985) Intestinal somatostatin function. Adv Exp Med Biol 188:463–473
306. Reubi JC (1992) Somatostatin receptors in the gastrointestinal tract in health and disease. Yale J Biol Med 65(5):493–536
307. Dubois MP (1975) Presence of immunoreactive somatostatin in discrete cells of the endocrine pancreas. Proc Natl Acad Sci (Wash) 72:1340–1343
308. Patel YC, Murthy KK, Escher EE, Banville D, Spiess J, Srikant CB (1990) Mechanism of action of somatostatin: an overview of receptor function and studies of the molecular characterization and purification of somatostatin receptor proteins. Metab 39(9 Suppl 2):63–69
309. Patel YC, Galanopoulou AS, Rabbani SN, Liu J-L, Ravazzola M, Amherdt M (1997) Somatostatin-14, somatostatin-28, and prosomatostatin[1–10] are independently and efficiently processed from prosomatostatin in the constitutive secretory pathway in islet somatostatin tumor cells (1027B2). Mol Cell Endocrinol 131(2):183–194
310. Schreff M, Schulz S, Händel M, Keilhoff G, Braun H, Pereira G et al (2000) Distribution, targeting, and internalization of the sst4 somatostatin receptor in rat brain. J Neurosci 20 (10):3785–3797
311. Fodor M, Slama A, Guillaume V, Videau C, Csaba Z, Oliver C, Epelbaum J (1997) Distribution and pharmacological characterization of somatostatin receptor binding sites in the sheep brain. J Chem Neuroanat 12(3):175–182
312. Seno M, Seino Y, Taminato T, Tsuda K, Kitano N, Takeda J et al (1988) Comparative effect of somatostatin-14 and somatostatin-28 on glucagon-induced glycogenolysis from the perfused rat liver. Metab, Clin Exp 37(6):587–590
313. Brazeau P, Vale W, Burgus R, Ling N, Rivier J, Guillemin R (1972) Hypothalamic polypeptide that inhibits the secretion of immunoreactive pituitary growth hormone. Sci 129:77–79
314. Inui T, Kinoshita Y, Yamaguchi A, Yamatani T, Chiba T (1991) Linkage between capsaicin-stimulated calcitonin gene-related peptide and somatostatin release in rat stomach. Am J Physiol Gastrointest Liver Physiol 261(5):G770–G774
315. Wellendorph P, Hansen KB, Balsgaard A, Greenwood JR, Egebjerg J, Bräuner-Osborne H (2005) Deorphanization of GPRC6A: a promiscuous L-alpha-amino acid receptor with preference for basic amino acids. Mol Pharmacol 67(3):589–597
316. Pi M, Faber P, Ekema G, Jackson PD, Ting A, Wang N et al (2005) Identification of a novel extracellular cation-sensing G-protein-coupled receptor. J Biol Chem 280(48):40201–40209
317. Pi M, Parrill AL, Quarles LD (2010) GPRC6A mediates the non-genomic effects of steroids. J Biol Chem 285(51):39953–39964
318. Epelbaum J (2003) Peptides and non peptides of oncologic and neuroendocrine relevance. In: Müller EE (ed) Physiology and pharmacology of somatostatin and its receptors. Springer, Milan, Italy
319. Tannenbaum GS, Epelbaum J (1999) Somatostatin. In: Kostyo JL, Goodman HM (eds) Handbook of physiology. Oxford University Press, Oxford, New York

320. Samson WK, Zhang JV, Avsian-Kretchmer O, Cui K, Yosten GLC, Klein C et al (2008) Neuronostatin encoded by the somatostatin gene regulates neuronal, cardiovascular, and metabolic functions. J Biol Chem 283(46):31949–31959
321. Salvatori AS, Elrick MM, Samson WK, Corbett JA, Yosten GLC (2014) Neuronostatin inhibits glucose-stimulated insulin secretion via direct action on the pancreatic α-cell. Am J Physiol Endocrinol Metab 306(11):E1257–E1263
322. Mo J, Yang A, Chen Z, Shao T, Zhang Y, Chen Q (2013) Neuronostatin ameliorates sodium taurocholate-induced acute pancreatitis in rats. Digest Dis Sci 58(10):2903–2907
323. Elrick MM, Samson WK, Corbett JA, Salvatori AS, Stein LM, Kolar GR et al (2016) Neuronostatin acts via GPR107 to increase cAMP-independent PKA phosphorylation and proglucagon mRNA accumulation in pancreatic α-cells. Am J Physiol Regul Integr Comp Physiol 310(2):R143–R155
324. de Lecea L, Ruiz-Lozano P, Danielson PE, Peelle-Kirley J, Foye PE, Frankel WN, Sutcliffe JG (1997) Cloning, mRNA expression, and chromosomal mapping of mouse and human preprocortistatin. Genomics 42:499–506
325. de Lecea L, Castaño JP (2006) Cortistatin: not just another somatostatin analog. Nat Rev Endocrinol 2:356–357
326. Spier AD, de Lecea L (2000) Cortistatin: a member of the somatostatin neuropeptide family with distinct physiological functions. Brain Res Rev 33(2–3):228–241
327. de Lecea L, Criado JR, Prospero-Garcia O, Gautvik KM, Schweitzer P, Danielson PE et al (1996) A cortical neuropeptide with neuronal depressant and sleep-modulating properties. Nat 381:242–245
328. Leavy O (2006) Inflammatory disorders: cortistatin to the rescue. Nat Rev Drug Discov 5:376
329. Gonzalez-Rey E, Pedreño M, Delgado-Maroto V, Souza-Moreira L, Delgado M (2015) Lulling immunity, pain, and stress to sleep with cortistatin. Ann N Y Acad Sci 1351:89–98
330. Fliers E, Bauer K, Visser T (2004) Thyrotropin-releasing hormone (TRH). In: Martini L (ed) Encyclopedia of endocrine diseases. Elsevier, Amsterdam, Netherlands
331. Morley JE, Garvin TJ, Pekary AE, Hershman JM (1977) Thyrotropin-releasing hormone in the gastrointestinal tract. Biochem Biophys Res Commun 79(1):314–318
332. LaHann TR, Horita A (1977) Thyrotropin releasing hormone and the gastrointestinal tract: the effect of central administration on colonic smooth muscle activity. Proc West Pharmacol Soc 20:305–306
333. Hernandez DE (1989) Neurobiology of brain—gut interactions. Digest Dis Sci 34(12):1809–1816
334. Morley JE, Steinbach JH, Feldman EJ, Solomon TE (1979) The effects of thyrotropin releasing hormone (TRH) on the gastrointestinal tract. Life Sci 24(12):1059–1065
335. Thorner MO (1999) The discovery of growth hormone-releasing hormone. J Clin Endocrinol Metab 84(12):4671–4676
336. Bauer K, Schomburg L, Heuer H, Schäfer MK (1999) Thyrotropin releasing hormone (TRH), the TRH-receptor and the TRH-degrading ectoenzyme; three elements of a peptidergic signalling system. Results Probl Cell Differ 26:13–42
337. Guillemin R (2008) Somatostatin. The beginnings, 1972. Mol Cell Endocrinol 286(1–2):3–4
338. Wade N (1978) Guillemin and Schally: the Years in the wilderness. Sci 200(4339):279–282
339. Bowers CY, Schally AV, Hawley WD, Gual C, Parlow A (1968) Effect of thyrotropin-releasing factor in man. J Clin Endocr 28:978
340. Montminy M, Lee K-F, Rivier JE, Rivier C, Reichlin S (2012) Wylie vale: neuroendocrine master. Proc Natl Acad Sci U S A 109(10):3604–3605
341. Evans H, Baumgartner M, Shine J, Herzog H (1993) Genomic organization and localization of the gene encoding human preprogalanin. Genomics 18(3):473–477
342. Tatemoto K, Rökaeus Å, Jornvall H, McDonald TJ, Mutt V (1983) Galanin—a novel biologically active peptide from porcine intestine. FEBS Lett 164:124–128

343. Mechenthaler I (2008) Galanin and the neuroendocrine axes. Cell Mol Life Sci 65(12):1826–1835
344. Webling KEB, Runesson J, Bartfai T, Langel Ü (2012) Galanin receptors and ligands. Front Endocrinol (Lausanne) 3:146
345. Mitsukawa K, Lu X, Bartfai T (2008) Galanin, galanin receptors and drug targets. Cell Mol Life Sci 65(12):1796–1805
346. Lu X, Lundström L, Bartfai T (2005) Galanin (2-11) binds to GalR3 in transfected cell lines: limitations for pharmacological definition of receptor subtypes. Neuropeptides 39(3):165–167
347. Yanaihara N, Mochizuki T, Takatsuka N, Iguchi K, Sato K, Kakuyama H et al (1993) Galanin analogues: agonist and antagonist. Regul Pept 46(1–2):93–101
348. Ruczyński J, Konstański Z, Korolkiewicz R, Petrusewicz J, Rekowski P (2002) Galanin and its analogues: A structure-activity relationship studies in rat isolated gastric smooth muscles. Lett Pept Sci 9(2–3):91–99
349. Cooper GJ, Willis AC, Clark A, Turner RC, Sim RB, Reid KB et al (1987) Purification and characterization of a peptide from amyloid-rich pancreases of type 2 diabetic patients. Proc Natl Acad Sci USA 84(23):8628–8632
350. Westermark P, Wernstedt C, Wilander E, Hayden DW, O'Brien TD, Johnson KH et al (1987) Amyloid fibrils in human insulinoma and islets of Langerhans of the diabetic cat are derived from a neuropeptide-like protein also present in normal islet cells. Proc Natl Acad Sci USA 84(11):3881–3885
351. Tingstedt J-E, Edlund H, Madsen OD, Larsson L-I (1999) Gastric amylin expression: cellular identity and lack of requirement for the homeobox protein PDX-1. A study in normal and PDX-1-deficient animals with a cautionary note on antiserum evaluation. J Histochem Cytochem 47(8):973–980
352. Zaki M, Koduru S, McCuen R, Vuyyuru L, Schubert ML (2002) Amylin, released from the gastric fundus, stimulates somatostatin and thus inhibits histamine and acid secretion in mice. Gastroenterol 123(1):247–255
353. Wimalawansa SJ (1997) Amylin, calcitonin gene-related peptide, calcitonin, and adrenomedullin: a peptide superfamily. Crit Rev Neurobiol 11(2–3):167–239
354. Li Z, Shi A, Karlsson FA (2002) Transient increase of rat gastric amylin in the neonatal period and in experimental ulcers. J Gastroenterol 37(3):172–176
355. Beales ILP, Calam J (2003) Regulation of amylin release from cultured rabbit gastric fundic mucosal cells. BMC Physiol 3:13
356. Hay DL, Garelja ML, Poyner DR, Walker CS (2018) Update on the pharmacology of calcitonin/CGRP family of peptides: IUPHAR Review. Br J Pharmacol 175(1):3–17
357. Lee SM, Hay DL, Pioszak AA (2016) Calcitonin and amylin receptor peptide interaction mechanisms: insights into peptide-binding modes and allosteric modulation of the calcitonin receptor by receptor activity-modifying proteins. J Biol Chem 291(16):8686–8700
358. Hay DL, Christopoulos G, Christopoulos A, Poyner DR, Sexton PM (2005) Pharmacological discrimination of calcitonin receptor: receptor activity-modifying protein complexes. Mol Pharmacol 67(5):1655–1665
359. Sexton PM, Poyner DR, Simms J, Christopoulos A, Hay DL (2012) RAMPs as drug targets. In: RAMPs. Advances in experimental medicine and biology, vol 744. Spielman WS, Parameswaran N eds. Springer, New York, USA
360. Fu W, Patel A, Jhamandas JH (2013) Amylin receptor: a common pathophysiological target in Alzheimer's disease and diabetes mellitus. Front Aging Neurosci 5:42
361. Adeghate E, Kalász H (2011) Amylin analogues in the treatment of diabetes mellitus: medicinal chemistry and structural basis of its function. Open Med Chem J 5:78–81
362. El-Rasheidy OF, Amin DA, Ahmed HA, El Masry H, Montaser ZM (2012) Amylin level and gastric emptying in obese children: before and after weight loss. J Egypt Soc Parasitol 42(2):431–442

363. Vella A, Lee JS, Camilleri M, Szarka LA, Burton DD, Zinsmeister AR et al (2002) Effects of pramlintide, an amylin analogue, on gastric emptying in type 1 and 2 diabetes mellitus. Neurogastroenterol Motil 14(2):123–131
364. Kitamura K, Kangawa K, Kawamoto M, Ichiki Y, Nakamura S, Matsuo H, Eto T (1993) Adrenomedullin: a novel hypotensive peptide isolated from human pheochromocytoma. Biochem Biophys Res Commun 192(2):553–560
365. Kitamura K, Kangawa K, Eto T (2002) Adrenomedullin and PAMP: discovery, structures, and cardiovascular functions. Microsc Res Tech 57(1):3–13
366. Schubert ML (2006) Adrenomedullin in gastrointestinal function. In: Kastin AJ (ed) Handbook of biologically active peptides. Elesvier, London, UK
367. Sakata J, Asada Y, Shimokubo T, Kitani M, Inatsu H, Kitamura K et al (1998) Adrenomedullin in the gastrointestinal tract. Distribution and gene expression in rat and augmented gastric adrenomedullin after fasting. J Gastroenterol 33(6):828–834
368. Mulder H, Ahrén B, Karlsson S, Sundler F (1996) Adrenomedullin: localization in the gastrointestinal tract and effects on insulin secretion. Regul Peptides 62(2–3):107–112
369. Turner AJ (2007) Adrenomedullin receptor. In: Enna SJ, Bylund DB (eds) xPharm: The comprehensive pharmacology reference. Elsevier, Amsterdam, Netherlands
370. Smith DM, Coppock HA, Withers DJ, Owji AA, Hay DL, Choksi TP et al (2002) Adrenomedullin: receptor and signal transduction. Biochem Soc Trans 30(4):432–437
371. Roh J, Chang CL, Bhalla A, Klein C, Hsu SY (2004) Intermedin is a calcitonin/calcitonin gene-related peptide family peptide acting through the calcitonin receptor-like receptor/ receptor activity-modifying protein receptor complexes. J Biol Chem 279(8):7264–7274
372. Ochiai T, Chijiiwa Y, Motomura Y, Yasuda O, Harada N, Nawata H (2001) Direct inhibitory effect of adrenomedullin, calcitonin gene-related peptide, calcitonin, and amylin on cholecystokinin-induced contraction of guinea-pig isolated caecal circular smooth muscle cells. Peptides 22(6):909–914
373. Lenhart PM, Caron KM (2012) Adrenomedullin and pregnancy: perspectives from animal models to humans. Trends Endocrinol Metab 23(10):524–532
374. Shimosawa T, Shibagaki Y, Ishibashi K, Kitamura K, Kangawa K, Kato S et al (2002) Adrenomedullin, an endogenous peptide, counteracts cardiovascular damage. Circ 105 (1):106–111
375. Ando K, Shimosawa T, Fujita T (2004) Adrenomedullin in vascular diseases. Curr Hypertens Rep 6(1):55–59
376. Ozbek E, Yurekli M, Soylu A, Davarci M, Balbay MD (2000) The role of adrenomedullin in varicocele and impotence. BJU Int 86(6):694–698
377. Nikitenko LL, Fox SB, Kehoe S, Rees MCP, Bicknell R (2006) Adrenomedullin and tumour angiogenesis. Br J Cancer 94:1–7
378. Takei Y, Hyodo S, Katafuchi T, Minamino N (2004) Novel fish-derived adrenomedullin in mammals: structure and possible function. Peptides 25:1643–1656
379. Bell D, McDermott BJ (2008) Intermedin (adrenomedullin-2): a novel counter-regulatory peptide in the cardiovascular and renal systems. Br J Pharmacol 153(1):S247–S262
380. Hong Y, Hay DL, Quirion R, Poyner DR (2012) The pharmacology of adrenomedullin 2/ intermedin. Br J Pharmacol 166(1):110–120
381. Smith RS Jr, Gao L, Bledsoe G, Chao L, Chao J (2009) Intermedin is a new angiogenic growth factor. Am J Physiol Heart Circ Physiol 297:H1040–H1047
382. Li H, Bian Y, Zhang N, Guo J, Wang C, Lau WB, Xiao C (2013) Intermedin protects against myocardial ischemia-reperfusion injury in diabetic rats. Cardiovasc Diabetol 12:91
383. Takahashi K, Morimoto R, Hirose T, Satoh F, Totsune K (2011) Adrenomedullin 2/ intermedin in the hypothalamo–pituitary–adrenal axis. J Mol Neurosci 43(2):182–192
384. Gan X-B, Sun H-J, Chen D, Zhang L-L, Zhou H, Chen L-Y, Zhou Y-B (2014) Intermedin in the paraventricular nucleus attenuates cardiac sympathetic afferent reflex in chronic heart failure rats. PLoS ONE 9(4):e94234
385. Holst JJ (1988) Enteroglucagon. Adv Metab Dis 11:393–419
386. Holst JJ (1997) Enteroglucagon. Annu Rev Physiol 59:257–271

387. Bloom SR (1987) Gut hormones in adaptation. Gut 28:31–35
388. Drucker DJ, Ehrlich P, Asa SL, Brubaker PL (1996) Induction of intestinal epithelial proliferation by glucagon-like peptide 2. Proc Natl Acad Sci USA 93(15):7911–7916
389. Kimball CP, Murlin JR (1923) Aqueous extracts of pancreas: III. Some precipitation reactions of insulin. J Biol Chem 58:337–346
390. Moody AJ, Holst JJ, Thim L, Jensen SL (1981) Relationship of glicentin to proglucagon and glucagon in the porcine pancreas. Nat 289(5797):514–516
391. Sutherland EW, De Duve C (1948) Origin and distribution of the hyperglycemic-glycogenolytic factor of the pancreas. J Biol Chem 175:663–674
392. Sutherland EW, Cori CF, Haynes R, Olsen NS (1949) Purification of hyperglycemic-glycogenolytic factor from insulin and from gastric mucosa. J Biol Chem 180:825
393. Unger RH, Eisentraut AM, McCall MS, Keller S, Lanz HC, Madison LL (1959) Glucagon antibodies and their use for immunoassay for glucagon. Proc Soc Exp Biol Med 102:621–623
394. Unger RH, Eisentraut AM, McCall MS, Madison LL (1961) Glucagon antibodies and an immunoassay for glucagon. J Clin Invest 40:1280–1289
395. Baldissera FGA, Holst JJ (1984) Glucagon-related peptides in the human gastrointestinal mucosa. Diabetologia 26(3):223–228
396. Gutman RA, Fink G, Voyles N, Selawry H, Penhos JC, Lepp A, Recant L (1973) Specific biologic effects of intestinal glucagon-like materials. J Clin Invest 52:1165–1175
397. Holst JJ (1997) Enteroglucagon. Ann Rev Physiol 59:257–271
398. Günter B (2013) Christian de Duve (1917–2013) Biologist who won a Nobel prize for insights into cell structure. Nat 498(7454):300
399. Klionsky DJ (2008) Autophagy revisited: a conversation with Christian de Duve. Autophagy 4(6):740–743
400. Caries-Bonnet C, Jarrousse C, Niel H, Martinez J, Bataille D (1991) Oxyntomodulin and its (19–37) and (30–37) fragments inhibit histamine-stimulated gastric acid secretion in the conscious rat. Eur J Pharmacol 203(2):245–252
401. Unger RH, Eisentraut AM (1969) Entero-insular axis. Arch Intern Med 123:261–266
402. Valverde I, Rigopoulou D, Exton J, Ohneda A, Eisentraut A, Unger RH (1968) Demonstration and characterization of a second fraction of glucagon-like immunoreactivity in jejunal extracts. Am J Med Sci 255:415–420
403. Bataille D, Gespach C, Coudray AM, Rosselin G (1981) 'Enteroglucagon': a specific effect on gastric glands isolated from the rat fundus. Evidence for an 'oxyntomodulin' action. Biosci Rep 1:151–155
404. Bataille D, Gespach C, Tatemoto K, Marie JC, Coudray AM, Rosselin G, Mutt V (1981) Bioactive enteroglucagon (oxyntomodulin): present knowledge on its chemical structure and its biological activities. Peptides 2(2):41–44
405. Bataille D, Tatemoto K, Gespach C, Jornvall H, Rosselin G, Mutt V (1982) Isolation of glucagon-37 (bioactive enteroglucagon/oxyntomodulin) from porcine jejuno-ileum characterization of the peptide. FEBS Lett 146:79–86
406. Dubrasquet M, Bataille D, Gespach C (1982) Oxyntomodulin (glucagon-37 or bioactive enteroglucagon): a potent inhibitor of pentagastrin-stimulated acid secretion in rats. Biosci Rep 2:391–395
407. Unger RH, Ohneda A, Valverde I, Eisentraut AM, Exton J (1968) Characterization of the responses of circulating glucagon-like immunoreactivity to intraduodenal and intravenous administration of glucose. J Clin Invest 47:48–65
408. Moody AJ, Hoist JJ, Thim L, Jensen SL (1981) Relationship of glicentin to pro-glucagon and glucagon in the porcine pancreas. Nat 289:514–518
409. Sasaki M, Fitzgerald AJ, Mandir N, Sasaki K, Wright NA, Goodlad RA (2001) Glicentin, an active enteroglucagon, has a significant trophic role on the small intestine but not on the colon in the rat. Aliment Pharmacol Ther 15(10):1681–1686
410. Sundby F, Jacobsen H, Moody AJ (1976) Purification and characterization of a protein from porcine gut with glucagon-like immunoreactivity. Horm Metab Res 8:366–371

411. Thim L, Moody AJ (1981) The amino acid sequence of porcine glicentin. Peptides 2(2):37–39
412. Fiocca R, Rindi G, Capella C, Grimelius L, Polak JM, Schwartz TW et al (1987) Glucagon, glicentin, proglucagon, PYY, PP and proPP-icosapeptide immunoreactivities of rectal carcinoid tumors and related non-tumor cells. Regul Peptides 17(1):9–29
413. Ito H, Yokozaki H, Hata J, Mandai K, Tahara E (1984) Glicentin-containing cells in intestinal metaplasia, adenoma and carcinoma of the stomach. Virchows Archiv A 404(1): 17–29
414. Holst JJ (1982) Evidence that enteroglucagon (II) is identical with the C-terminal sequence (residues 33–69) of glicentin. Biochem J 207:381–388
415. Whiting L, Stewart KW, Hay DL, Harris PW, Choong YS, Phillips AR, et al (2015) Glicentin-related pancreatic polypeptide inhibits glucose-stimulated insulin secretion from the isolated pancreas of adult male rats. Physiol Rep 3(12). pii: e12638
416. Thim L, Moody AJ (1982) Purification and chemical characterization of a glicentin-related pancreatic peptide (proglucagon fragment) from porcine pancreas. BBA Protein Struct Mol Enzymol 703(2):134–141
417. Feurle GE, Hamscher G, Kusiek R, Meyer HE, Metzger JW (1992) Identification of xenin, a xenopsin-related peptide, in the human gastric mucosa and its effect on exocrine pancreatic secretion. J Biol Chem 267(31):22305–22309
418. Feurle GE (1998) Xenin—a review. Peptides 19(3):609–615
419. Leckstrom A, Kim ER, Wong D, Mizuno TM (2009) Xenin, a gastrointestinal peptide, regulates feeding independent of the melanocortin signaling pathway. Diab 58(1):87–94
420. Alexiou C, Zimmermann JP, Schick RR, Schusdziarra V (1998) Xenin-a novel suppressor of food intake in rats. Brain Res 800(2):294–299
421. Feurle GE, Pfeiffer A, Schmidt T, Dominguez-Munoz E, Malfertheiner P, Hamscher G (2001) Phase III of the migrating motor complex: associated with endogenous xenin plasma peaks and induced by exogenous xenin. Neurogastroenterol Motil 13(3):237–246
422. Anlauf M, Weihe E, Hartschuh W, Hamscher G, Feurle GE (2000) Localization of xenin-immunoreactive cells in the duodenal mucosa of humans and various mammals. J Histochem Cytochem 48(12):1617–1626
423. Feurle GE, Anlauf M, Hamscher G, Arnold R, Klöppel G, Weihe E (2002) Xenin-immunoreactive cells and extractable xenin in neuroendocrine tumors of duodenal origin. Gastroenterol 123(5):1616–1626
424. Schmidt WE, Mutt V, Carlquist M, Kratzin H, Conlon JM, Creutzfeldt W (1985) Valosin: isolation and characterization of a novel peptide from porcine intestine. FEBS Lett 191(2): 264–268
425. Konturek SJ, Schmidt WE, Mutt V, Konturek JW, Creutzfeldt W (1987) Valosin stimulates gastric and exocrine pancreatic secretion and inhibits fasting small intestinal myoelectric activity in the dog. Gastroenterol 92(5):1181–1186
426. Schmidt WE, Mutt V, Konturek SJ, Creutzfeldt W (1985) Peptide VQY: isolation and characterization of a new biologically active gastrointestinal peptide. Dig Dis Sci 29:755
427. Funakoshi S, Tamamura H, Ohta M, Yoshizawa K, Funakoshi A, Miyasaka K et al (1989) Isolation and characterization of a tumor-derived human pancreastatin-related protein. Biochem Biophys Res Commun 164(1):141–148
428. Schmidt WE, Siegel EG, Kratzin H, Creutzfeldt W (1988) Isolation and primary structure of tumor-derived peptides related to human pancreastatin and chromogranin A. Proc Natl Acad Sci USA 85:8231–8235
429. Sánchez-Margalet V, Lucas M, Goberna R (1996) Pancreastatin: further evidence for its consideration as a regulatory peptide. J Mol Endocrinol 16:1–8
430. O'Connor DT, Cadman PE, Smiley C, Salem RM, Rao F, Smith J et al (2005) Pancreastatin: multiple actions on human intermediary metabolism in vivo, variation in disease, and naturally occurring functional genetic polymorphism. J Clin Endocrinol Metab 90:5414–5425

431. Bishop AE, Bretherton-Watt D, Hamid QA, Fahey M, Shepherd N, Valentino K et al (1988) The occurrence of pancreastatin in tumours of the diffuse neuroendocrine system. Mol Cell Probes 2(3):225–235
432. Tatemoto K, Efendic S, Mutt V, Makk G, Feistner GJ, Barchas JD (1986) Pancreastatin, a novel pancreatic peptide that inhibits insulin secretion. Nat 324:476–478
433. Tatemoto K (1991) Novel peptides and islet function. In: Vranic M, Efendic S, Hollenberg CH (eds) Fuel homeostasis and the nervous system. Springer, NY, USA
434. Woussen-Colle M-C, Lingier P, Vertongen P, Vandermeers-Piret M-C, Vandermeers A, Robberecht P (1994) Chromogranin A (210–301) is the major form of pancreastatin-like material in human gut extracts and endocrine tumors. Peptides 15(5):869–874
435. McGrath-Linden SJ, Johnston CF, O'Connor DT, Shaw C, Buchanan KD (1991) Pancreastatin-like immunoreactivity in human carcinoid disease. Regul Pept 33(1):55–70

Chapter 9
Neural Secretions and Regulation of Gut Functions

Abstract Neural secretions of the gut are substances released from the neurons or nerve terminals of the gastrointestinal (GI) tract to regulate the functioning of the GI system as well as extragut tissues and organs. The neural secretions may be mediated and released centrally or peripherally to act on the GI tract and extragut tissues/organs. The central neural secretions that act on the gut are mediated by the brain and spinal cord, whereas peripherally mediated neural secretions are regulated and released from the intrinsic nervous system of the gut (enteric nervous system) and extrinsic nervous system. Progressive advancement of knowledge on neural secretions of the gut was dictated mainly by technological progress. This area of study (regarding neural secretions and regulation of gut) has increasingly been recognized as a new field of science called neurogastroenterology. The science of neural secretions and regulation of GI functioning began around the eighteenth to nineteenth centuries. The founding fathers of this new science were William Maddock Bayliss (1860–1924), Ernest Henry Starling (1866–1927), Georg Meissner (1829–1905), Leopold Auerbach (1828–1897), Santiago Ramón y Cajal (1852–1934), Alexander Stanislavovich Dogiel (1852–1922), John Newport Langley (1852–1925), Paul Trendelenburg (1884–1931), and Harold Hirschsprung (1830–1916) among others. The second half of the twentieth century was marked by substantial advancement in neurogastroenterology, and generally in science. Development of the broader field "gastroenteroneuro endocrinology," which comprises neurogastroenterology and gut endocrinology, was enhanced by progress in genetics and molecular biology pioneered by James Watson (1928–), Francis Crick (1916–2004), and Maurice Hugh Frederick Wilkins (1916–2004); peptide biochemistry, and the invention of the radioimmunoassay, which enabled the measurement of minute quantities of neuropeptides. The gut produces over 50 types of neural secretions (neurotransmitters, neuromodulators), and tens of different types of neurons have been identified in the gut alone. This makes the gut one of the most important organs that can mediate neural signals at the peripheral level and in extragut tissues and organs such as the brain and adipose tissues. With extensively growing data on neurotransmitters of the gut and their functions, it is necessary to provide state-of-the-art information on neural secretions of the gut and their mechanisms of regulation. This chapter provides contemporary information on

M. O. Welcome, *Gastrointestinal Physiology*,
https://doi.org/10.1007/978-3-319-91056-7_9

fundamental aspects of the neural secretions of the gut, gives an account of the course of discovery of these secretions (neurotransmitters) of the gut, and provides the mechanisms of neural regulation of GI functions. The clinical importance of the neurotransmitters is systematically described.

Keywords Neural secretions · Mediator · Neurotransmitter · Hormone
Neuromodulator · Gastrin-releasing peptide (human bombesin)
The amphibian · Acetylcholine · ATP · Serotonin · 5-hydroxytryptamine
Histamine · 2-(1*H*-imidazol-4-yl) ethanamine · Γ-aminobutyric acid
Tachykinins · Substance P · Neurokinin A · B-tachykinin (neurokinin B or neuropeptide beta) · Γ-tachykinin (also called gamma-neuropeptide)
Norepinephrine (noradrenaline or noradrenalin) · Epinephrine (adrenalin or adrenaline) · Neurotensin · Glutamate · Neuropeptide Y or neuropeptide tyrosine
Peptide YY · Vasoactive intestinal polypeptide · Peptide histidine isoleucine or peptide histidine isoleucinamide · Peptide histidine methionine or peptide MI
Peptide histidine valine or peptide HV · Pituitary adenylate cyclase-activating peptide · Calcitonin gene-related peptide · Nitric oxide · Carbon monoxide
Hydrogen sulfide · Enkephalins · B-endorphin · Dynorphin · Anandamide
2-arachidonoylglycerol · Virodhamine (O-arachidonoyl-ethanolamine)
Noladin (2-arachidonyl-glyceryl ether) · And *N*-arachidonoyl dopamine
Ajay Verma · Alan N. Taylor · Alfred Nobel · Anne Marie Staub
Candace Beebe Pert · Cyrus Hartwell Fiske · Daniel Bovet · Ferid Murad
George Barger · Hans Walter Kosterlitz · Henry Hallett Dale · John Hughes
John Jacob Abel · Julius Axelrod · Karl Lohmann · Kazuhiko Tatemoto
Kikunae Ikeda · Lev Popielskij (Leon Popielski) · Louis J. Ignarro
Mats Carlquist · Otto Loewi · Raphael Mechoulam · Robert E. Carraway
Robert Francis Furchgott · Robert H. Wasserman · Sami I. Said
Sir Bernard Katz · Sir John Henry Gaddum · Sir Patrick Playfair Laidlaw
Solomon Halbert Snyder · Susan E. Leeman · Thomas J Mcdonald
Ulf Svante Von Euler · Viktor Mutt · Vittorio Erspamer · Yechiel Gaoni
Yellagaprada SubbaRow

Abbreviations

5-HT	5-hydroxytryptamine (serotonin)
ACE	Angiotensin-converting-enzyme
Acetyl-CoA	Acetyl co-enzyme A
Ach	Acetylcholine
AChR	Acetylcholine receptor
ADP	Adenosine diphosphate
Ang	Angiotensin
ARB	Angiotensin receptor blocker
ATP	Adenosine triphosphate
BB-3 or BRS-3	Bombesin receptor subtype 3
Ca^{2+}	Calcium ion

cAMP	Cyclic adenosine monophosphate
CAT	Choline acetyltransferase
CGRP	Calcitonin gene-related peptide
CO	Carbon monoxide
DAG	Diacylglycerol
DYN	Dynorphin
ENK	Encephalin
GABA	Γ-aminobutyric acid
Gq/11-type G proteins	Gq type G protein
GRP	Gastrin-releasing peptide
GRPR	GRP receptors
H_2S	Hydrogen sulfide
Hz	Hertz
IBS	Irritable bowel syndrome
$InsP_3$	Inositol triphosphate
K^+	Potassium ion
K_{ir}	Inward-rectifier potassium ion channel or GIRK (G protein inwardly rectifying potassium channel)
L-DOPA	L-3,4-dihydroxyphenylalanine
M_4, M_3	Subtypes of muscarine-sensitive acetylcholine receptors
mAChR	Muscarinic acetylcholine receptor
Na^+	Sodium ion
nAChR	Nicotinic acetylcholine receptor
NADA	*N*-arachidonoyl dopamine
NO	Nitric oxide
NPY	Neuropeptide Y or neuropeptide tyrosine
PACAP	Pituitary adenylate cyclase-activating peptide
Peptide HI or PHI	Peptide histidine isoleucin or peptide histidine isoleucinamide
PHM	Peptide histidine methionine or peptide MI
PHV	Peptide histidine valine or peptide HV
PLC	Phospholipase C
PTX	Pertussis toxin
PYY	Peptide YY
RAAS	Renin–angiotensin–aldosterone system
SERT	Serotonin reuptake transporter
TK	Tachykinins
TpH1, TpH2	Tryptophan hydroxylases
VAChT	Vesicular ACh transporter
VDCCs	Voltage-dependent Ca^{2+} channels
VIP	Vasoactive intestinal polypeptide

9.1 Introduction

Neural secretions are molecules that are released from the neurons of the central or peripheral nervous system. Many of such secretions are known to affect the functioning of the gut [1–3]. Neural secretions of the gut are secretions that originate from the neurons and glial cells of the gut as well as extrinsic nerves that innervate the gut [4, 5].

The nervous system of the GI tract comprises the extrinsic nervous system and the intrinsic nervous system. The former comprises part of the autonomic nervous system (ANS). The extrinsic nervous system is further divided into parasympathetic and sympathetic (Fig. 9.1). The extrinsic and intrinsic nervous systems form part of the peripheral nervous system (PNS). In addition to the somatic nervous system which takes part in mediating voluntary movement, the autonomic and intrinsic nervous systems represent major divisions of the PNS [6–9]. The PNS is composed of nerves which are enclosed in bundles of axons and cells' bodies, sometimes forming ganglia in certain locations which connect the central nervous system (CNS) to other parts of the body. (The CNS, which comprises the brain and spinal cord, plays a pivotal role in GI tract functioning.) The PNS neurons receive information from the target organs, tissues, and cells of the body and transmit the information upstream to the central analyzer via relay centers. Neurons that transmit signals from body organs or tissues to the CNS are called afferent or sensory neurons. In contrast, efferent (motor) neurons transmit information from the CNS to the body organs, tissues, or cells. It should be mentioned, however, that majority of the nerves are both afferent and efferent—known as mixed nerves. The neural centers receive numerous information, and also function as sites of integration and can decode (interpret) the content of the information and send back the processed information (signal) through another pathway to the target. The complex control of GI functions, including food intake and body weight, involves CNS integration of information from different sources, including the peripheral nervous system and humoral signals from the gut, liver, pancreas, and adipocytes [10–13]. Integration of information from the various sources in the CNS helps to accurately monitor the quality and quantity of luminal contents and determine the plasma level of metabolites (e.g., level of glucose, lipids, ketone bodies) at any given time. The processing, integration, and analysis of this information are useful in controlling feeding behavior [10, 13].

The enteric nervous system (ENS) is composed of the intrinsic neurons and glial cells of the gut, which regulate GI and extragut activities. The ENS is located in the walls of the GI tract (from esophagus to the anus) and associated glands/organs—salivary glands, pancreas, and gallbladder. The ENS is sometimes regarded as part of the autonomic nervous system because it functions without control of our will. In a living human, however, the activities of the enteric neurons are constantly been modulated by the activities of the extrinsic nervous system. But, ENS is able to function independently without external influence from the extrinsic neurons of the

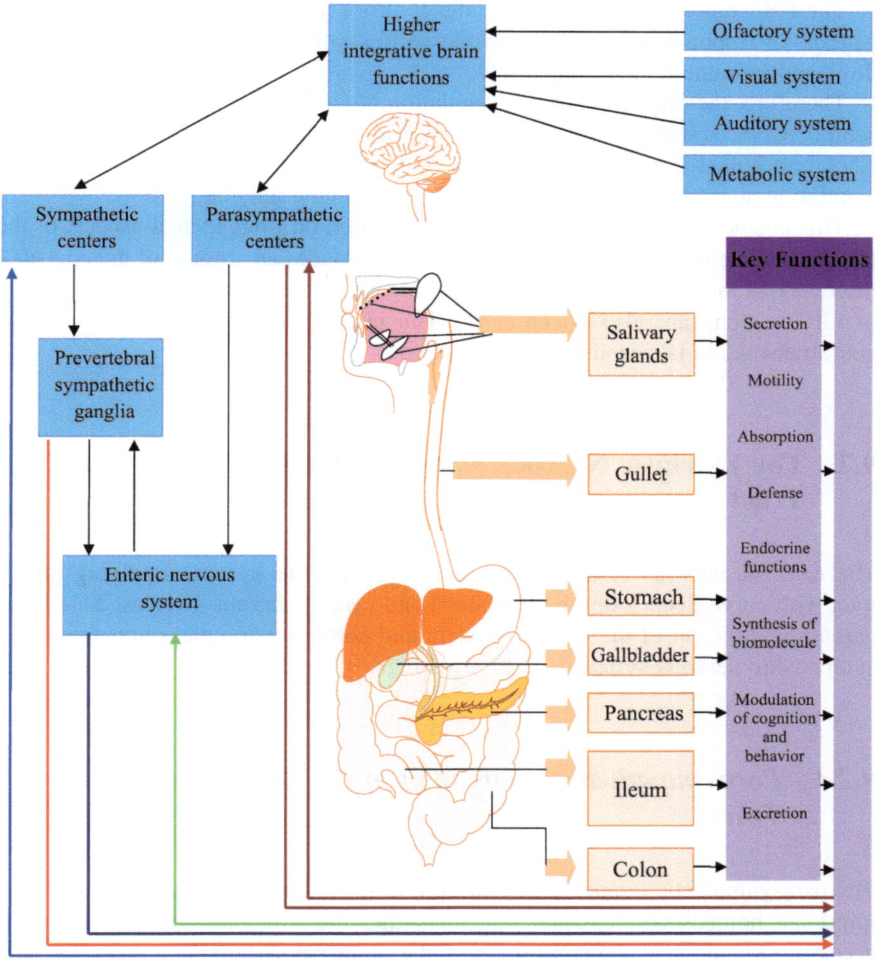

Fig. 9.1 General schematic layout of influence of nervous system on GI functioning. The parasympathetic centers are located in the brainstem and sacral region. The nerves that arise from the centers are preganglionic efferents that project to the GI tract. Examples of such nerves include vagus and pelvic nerves. Both nerves form synapses with neurons of the enteric nervous system. The vagus nerves form synapses with intrinsic neurons of the esophagus, stomach, small intestine, colon, gallbladder, and pancreas. The pelvic nerves form synapses with the intrinsic neurons of the colon. These connections (synapses) provide media for interaction between the extrinsic and intrinsic nervous system to modulate secretion, motility, absorption in the respective region of the GI tract. The centers of the sympathetic nervous system are located in the thoracic and lumbar regions of the spinal cord (preganglionic ganglia)

autonomic nervous system. It should be noted that link between the CNS and the ENS is bidirectional. Thus, not only the brain and spinal cord influence the activities of the ENS, but the ENS also influences the functions of the CNS [14, 15].

The ENS functions by integrating signals from different sources into organized patterns of response through neural reflexes to regulate motility, secretions, blood flow, and immune responses [15, 16].

Therefore, the different levels of regulation of GI functions may be classified as central (CNS, spinal cord) and peripheral. The central regulatory centers are the CNS and spinal cord. The peripheral centers are the prevertebral sympathetic ganglia, paravertebral sympathetic ganglia, and enteric ganglia [15].

This chapter provides contemporary information on fundamental aspects of the GI neural secretions (neurotransmitters). It also provides an account of the course of discovery of the neural secretions of the gut. The mechanisms of neural regulation of GI functions are also documented. The clinical importance of the GI tract neurotransmitters is systematically described.

9.2 The Extrinsic Nervous System of the Gastrointestinal Tract

The extrinsic nervous system is the part of the nervous system supplying the GI tract with nerve fibers from the peripheral and central nervous systems. This is the autonomic division of the nervous system and comprises the parasympathetic and sympathetic nervous system [14, 17, 18] (Figs. 9.2 and 9.3).

9.2.1 Parasympathetic Innervation of the Gastrointestinal Tract

The parasympathetic (craniosacral) division of the autonomic nervous system is an important homeostasis regulator that innervates the structures of the oral cavity, accessory glands, and the entire tract of the digestive system. The nuclei or integrating centers of this division is located in some brainstem nuclei and cell bodies of the sacral spinal cords (S2–S4). The preganglionic fibers emerging from these nuclei synapse with postganglionic nerve fibers, which then innervate the associated region [14, 17].

The parasympathetic fibers of S2–S4 innervate (pelvic afferents) the transverse colon, distal colon, and rectum. But the postganglionic fibers arising from S2 to S4 also innervate the bladder and reproductive organs [14, 25, 26]. The pelvic afferent fibers extend into the serosa, muscle, and mucosa of the tracts where they sense, transmit signal (noxious, mechanical distension due to passage of stool, urgency, desire to defecate, and other sensations) to the regulating centers [14].

Three types of parasympathetic nerves innervate the digestive system—facial (VII), glossopharyngeal (IX), and vagus (X) nerves. The salivary glands are innervated by the postganglionic parasympathetic nerves that arise from the otic

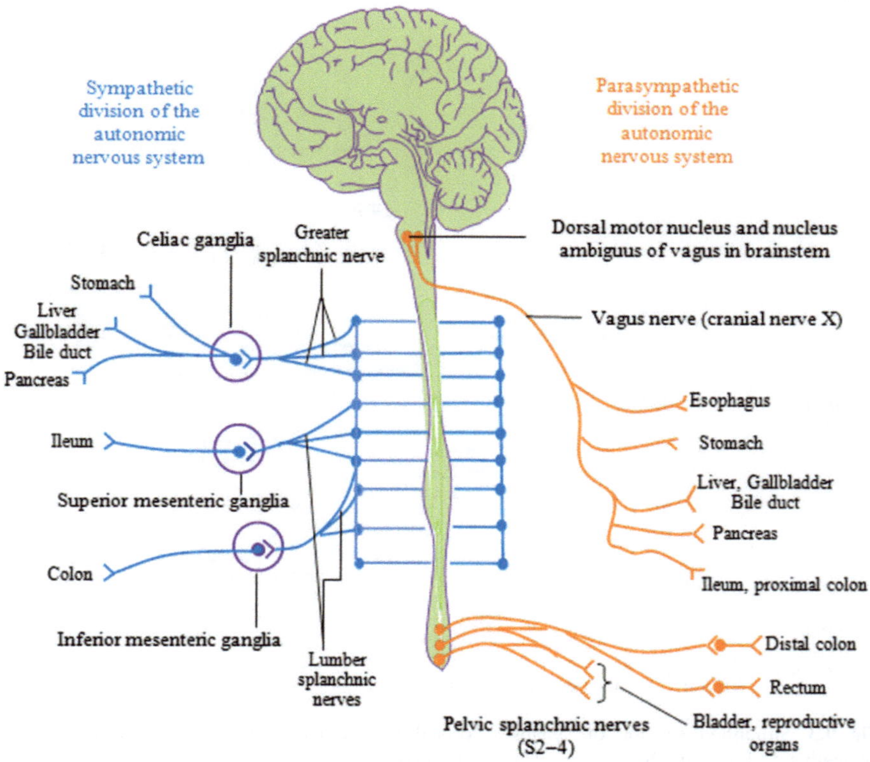

Fig. 9.2 Simplified schematic representation of sympathetic and parasympathetic innervation of the ENS. The activity of the gut is regulated by both the sympathetic and the parasympathetic divisions of the visceral system. The dorsal motor nucleus of vagus nerve in the brainstem and the intermediate gray zone in the sacral spinal cord segments are the major points from where preganglionic neurons of the parasympathetic system that innervate the gut originate. The vagus innervates the esophagus, stomach, liver, pancreas, gallbladder, small intestine, and proximal colon, but not the transverse and distal colon and rectum. The transverse colon is innervated by the cruciform parasympathetic nucleus of the spinal cord. This area of the spinal cord also innervates the ileocecal sphincter and ascending colon [19]. The distal rectum or colorectum is innervated by extrinsic nerves organized into plexuses in the pelvis. This plexus also innervates the bladder and reproductive organs. The preganglionic neurons of the sympathetic system that modulates the action of the gut plexuses are derived from the thoracolumbar spinal cord, which gives rise to the celiac, superior, and inferior mesenteric ganglia [14, 20–22]

ganglion and submandibular ganglion. Specifically, the parotid gland is innervated by the postganglionic nerves arising from the otic ganglion, while the submandibular and sublingual are innervated from the postganglionic fibers that arise from the submandibular ganglion (cranial nerves VII or IX). The minor salivary glands of the mouth also receive innervation from the postganglionic parasympathetic fibers of the VII nerve [17, 27–31]. It should be mentioned that some structures of the mouth are innervated by lingual nerve, which is a branch of the

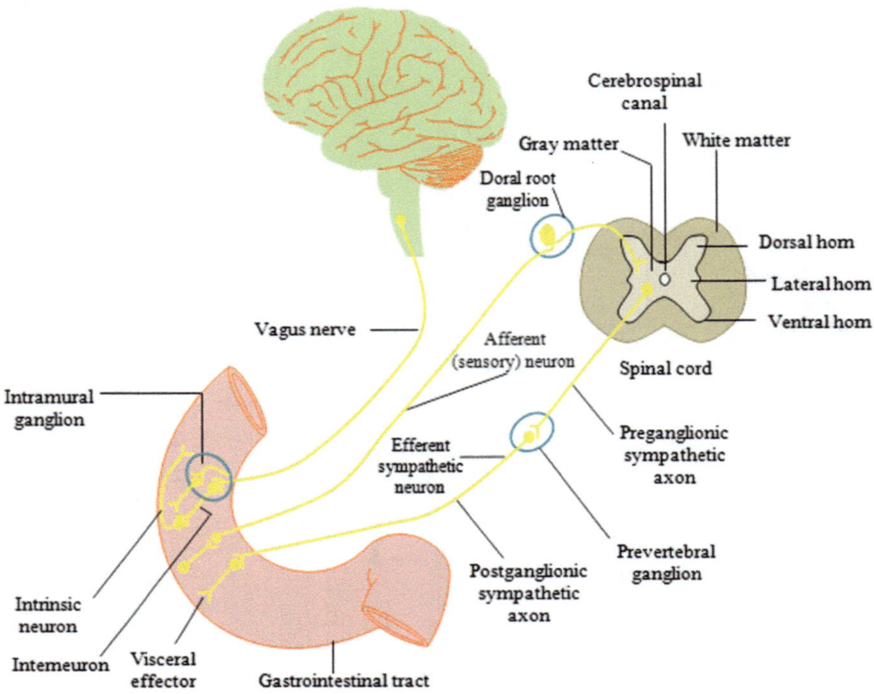

Fig. 9.3 Simplified scheme of synapses formed by the parasympathetic (craniosacral) and sympathetic (thoracolumbar) nerves. The vagus nerve exits the brainstem to form ganglion in the GI tract. The human vagus nerve is believed to contain about 100,000 nerves fibers [23, 24]. About 80% of the fibers are afferents, while the remaining 20% are efferent fibers. Mechanoreceptors are found on the serosal and mucosal. Intrafugal neuron exits the prevertebral ganglion extending to the myenteric plexus with its receptors in the circular muscle. Sensory neurons from the epithelium or other sites transfer information about the gut via the prevertebral ganglion or dorsal root ganglion to the spinal cord and then to the brainstem. However, primary afferent neurons may synapse with the myenteric or submucosal neurons [17, 18]

mandibular division of the trigeminal nerve (cranial nerve IV). The sensory information from the tongue is transmitted to the brain via this nerve. The lingual nerve also gathers taste information from the anterior two-thirds of the tongue, returning the information to the chorda tympani—a branch of facial nerve [32–34].

Except the vagus, the rest of the parasympathetic nerve ganglia are located in the head, where preganglionic neurons synapse onto the postganglionic neurons. The nerve VII originates from the pons and form synapse at the submandibular ganglia, which give rise to postganglionic nerve fibers that innervate the parotid gland. The submandibular ganglion is situated laterally to the hyoglossus muscle, below the lingual nerve. The IX nerve has preganglionic nerve that originates from the medulla oblongata, projecting to form the otic ganglia which send postganglionic nerve fibers to innervate the submandibular and sublingual salivary glands. The otic ganglion is located medially to the mandibular nerve, just beneath the oval foramen. The term "vagus" is derived from the Latin "vag" meaning vagrant or wandering due to its wide distribution in the body. The vagus nerve forms preganglionic neuron that originates from the medulla and travels through the neck to the thoracic cavity and through the esophageal opening in the diaphragm to the abdominal cavity. The vagus nerve sends preganglionic fibers' branches along its path to innervate visceral structures. The vagus nerve forms terminal ganglia near or inside the organ. In addition to the cranial nerve X, IX and X nerves contain both sensory and motor neurons and are called mixed nerves [17, 27–31]. The descending colon and the rectum receive their innervation from the pelvic splanchnic nerves that originate from the sacral spinal cords at S2–S4. These parasympathetic nerves convey different mechano- and chemosensory information from the colon to the spinal cord [35–40].

Autonomic malfunction of the pelvic nerves can result in impairment in colorectal motility, including fecal incontinence and constipation. These disorders may occur spontaneously in individuals who recently underwent hysterectomy and colorectal surgery. Disorders of colorectal motility may also occur after childbirth [40].

Parasympathetic Afferent and Efferent Pathways, Regulating Gastrointestinal Functioning

The vagus nerve is responsible for majority of the parasympathetic innervation of the gut. The vagus nerve carries signals in both directions between the brain and the gut. Thus, it contains both afferent and efferent nerves. The afferent nerve fibers exist in close association with different types of sensory receptors including stretch (mechano-), osmo-, and chemoreceptors, which constitute the initial points of signal transduction in the gut. These receptors are localized on the epithelial cells, smooth muscle cells, and other cells of the GI tract. These receptors sense, code, and transfer information about the milieu of the GI tract to the dorsal vagal complex (located in the medulla oblongata) through the afferent nerve fibers of the vagus. GI tract information is also transferred by sensory neurons to the CNS via the lumbosacral region of the spinal cord [41]. Such information may include the presence or absence of food in the GI tract, volume of food, chemical composition of food, sensations of fullness, bloating, nausea, urgency, discomfort, and pain. For instance, mechanoreceptors of the smooth muscles respond to mechanical distortion in response to food entry into the GI tract and activate vagal afferents, which transmit the information to the CNS. The brain interprets the signal and relays a motor command for execution. The mechanical distorting force also activates tension receptors especially at a higher level of distension. Tension receptors are active especially in normal peristaltic contractions. Thus, afferent vagal nerves comprise an integral structure for bottom-up transfer of information from the GI tract to the brain [14, 41–43]. The nerve fibers of the gut transmit information about the signaling of mechanosensation resulting in nociception in the gut or discomfort (i.e., mechanonociception). The mechanoreceptors are low-threshold nociceptors, activated by cell membrane distension, and can generate tonic muscle contraction. But, if distension is maintained, these receptors respond by generating peristaltic activity. The high-threshold afferents are thought to be the sensory transmitters of pain. Pathophysiological information resulting from extreme distension, excessive or prolonged contraction, injury or inflammation is primarily encoded by the afferent fibers that run through the spinal cord. The afferent fibers of the vagus nerve primarily encode information within the physiological range [41].

Information Coding by Sensory Receptors

Stimulation of Sensory Receptors

Receptors are activated by different stimuli. The sensory (afferent) nerves of the vagus located in different parts of the GI tract are triggered by the release of peptides. The type of peptide released depends on the type and content of food intake or the presence of other activating stimuli [42]. In response to chemical (food substances such as carbohydrates, proteins, and lipids), mechanical, or electrical stimuli, substances such as hormones and neuropeptides are released from GI primary sensors (enteroendocrine cells) through the activation of plasma membrane receptors including transporters. The hormones and neuropeptides may diffuse to activate vagal or spinal afferent neurons, which harbor a variety of receptors for different hormones and neuropeptides. The terminals of intrinsic and extrinsic afferent neurons are located in close proximity to the GI mucosa [44]. For instance, in response to food (fatty) intake, cholecystokinin (CCK) is released from neuroendocrine cells. The secreted CCK localizes to its receptors on the target cell which include the vagal afferent neuron. Activation of the corresponding receptors on the vagal afferent fiber of the GI tract can evoke fast, nicotinic synaptic responses, or action potentials. Activation of this type of receptor has been implicated in vago-vagal reflex controlling GI function and in regulation of food intake. The secreted hormones or neuropeptides may act on local environ containing epithelial, immune and other surrounding cells via autocrine, paracrine pathways. The secreted substances can be transported by the circulatory system to act on exocrine glands (e.g., mucosal and submucosal glands as well as exocrine pancreas) [44, 45]. Food bolus or gastric load (meal size) can mechanically and chemically activate vagal afferent via the release of gut peptide and can integrate multiple modalities to regulate feeding. In the presence of chemical and mechanical stimuli, such as in consumption of increasing quantity of oily food, CCK, and gastric load synergistically suppress food intake by activating afferent nerve fiber. The signal is transmitted through the vagal fiber usually directly to the nucleus tractus solitarius (NTS) [42, 46]. The neural cells of the NTS relay the information to the nucleus ambiguus and dorsomotor nucleus of the vagus nerve (but some vagal fibers directly synapse with neurons of the nucleus ambiguus and dorsomotor nucleus of the vagus nerve from where the signal is relayed via interneurons to the NTS) [15, 47]. The vagus nucleus also synapses with neurons of the hypothalamus and other regions of the brain [48]. Thus, increase in meal size during feeding will gradually reduce satiety by inhibiting the central hunger centers through the neural connections outlined above, thereby regulating ingestive behavior. The molecular mechanism of these processes involves activation of neuromediators and signaling cascades that also involve some transcription factors including the c-Fos within caudal brainstem, where gut vagal afferents terminate. This feedback mechanism is required to adequately regulate the quantity or size of ingested meal [46, 49]. Also note that, genes regulating the production of brain-derived neurotrophic factor (BDNF) and neurotrophin-3 (NT-3), which are essential for vagal afferent development, have been shown to play a significant role in the functioning of sensory receptors [50]. Detail mechanisms of these processes are discussed in Table 9.1.

Table 9.1 Effects of autonomic stimulation of the GI tract [17, 145, 146]

Effector organs and regions of GI tract	Effects of sympathetic stimulation. Postganglionic stimulation (mediated via α or β adrenoreceptor; preganglionic stimulation via N or M receptors)	Effects of parasympathetic stimulation (mediated via N- or M-cholinoreceptor)
Esophagus	↓ Motility and tone esophageal sphincter Secretion	↑ Motility and tone esophageal sphincter secretion
Stomach	↓ Motility and tone (β-receptor); sphincter muscle contracts (α-adrenoreceptor); secretion is inhibited	↑ Motility and tone, sphincter relaxes, secretion is stimulated
Intestine	↓ Motility and tone (α and β-receptors), sphincter muscle contracts (α-adrenoreceptor); secretion is inhibited	↑ Motility and tone sphincter relaxes, secretion is stimulated
Mucous glands	Inhibition of secretion	↑ Secretion
Gallbladder and ducts	Relaxation	Contraction
Salivary glands	Thick or viscous, sparse secretion	Profuse, watery low protein secretion, serous saliva
Liver	Stimulation of glycogen hydrolysis ↑ Glycogenolysis (β) ↑ Gluconeogenesis ↑ Glycolysis	Glycogenesis is activated
Endocrine pancreas	Inhibition of insulin and glucagon secretion (α-adrenoreceptor); activation of insulin and glucagon secretion (β-adrenoreceptor)	Insulin and glucagon secretion
Exocrine pancreas	↓ Secretion	↑ Secretion

Note ↑—Increase; ↓—decrease

Mechanism of Information Coding by Sensory Receptors

The sensory receptors are the first anatomical structures that sense the immediate environment to trigger adequate response, required for the maintenance of homeostasis. Information or events sensed by the receptors are coded by as amplitude and frequency. There are at least two mechanisms by which receptors in the GI tract convert stimulus into nerve impulse. The stimulus acts directly on the membrane receptors located on the sensory neuron to initiate a direct (primary) response (Fig. 9.4). This interaction leads to changes in membrane permeability to sodium ions (or calcium), which results in passive receptor potential (electrotonic receptor potential) spread along the fibers. The amplification of graded potential changes into much larger ones leads to the generation of action potential. The propagation of action potential along the nerve or muscle fiber occurs as a consequence of the axonal cable structure (for review, see the cable theory in Bressloff [51] and Bell [52]) [53]. This action potential spreads along the nerve fiber in the orthodromic direction. Thus, transformation of the energy of the stimulus and nervous impulse takes place in the sensory neuron itself. In primary receptor cell, the receptor potential is called generator potential. The local responses or potential that occurs in the cell upon action of an electrical or chemical signal, produced by the movement of positively charged ions (e.g., sodium or calcium ions) into the cell through membrane ion channels, is referred to as the receptor potential. The stimulus is transduced, or transformed, into an electrical response. If the response is high enough (amplitude above the threshold), then it generates the nerve impulse [54–56].

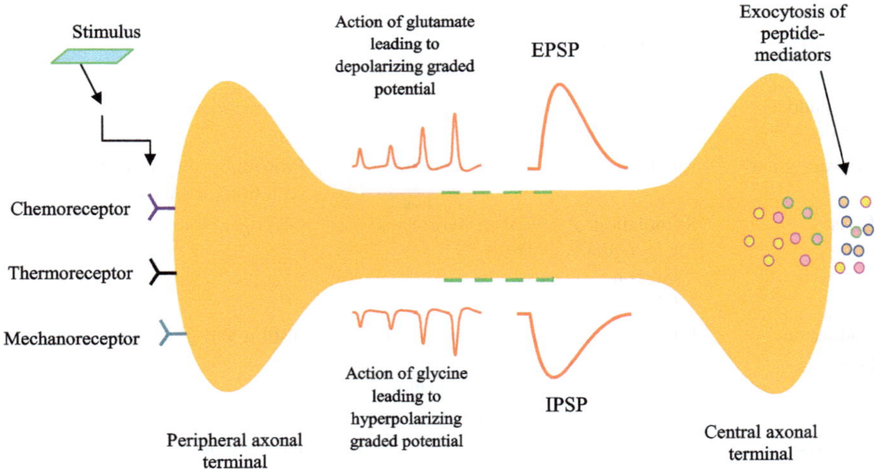

Fig. 9.4 Schematic representation of generator potential. The afferent nerve ending has receptors or it may function like a receptor (modified receptor), which responds to stimulus such as electrical or chemical signal. This stimulus activates voltage-gated channel leading to the opening of sodium or calcium channel. The potential resulting from this activation is called generator potential

A receptor potential is a graded response to a stimulus; the transmembrane potential difference produced by activation of a sensory receptor, which transduces the stimulus into an electrical signal. Depending on the type of ion channel activated, receptor potential may be depolarizing or hyperpolarizing. Receptor potential has a threshold in stimulus amplitude that must be reached before a response is generated. It is generally a depolarizing event resulting from inward current flow. An example of such process is the one involved in taste receptor signaling in the GI tract. When a taste receptor is stimulated by a tasty substance, series of chemical reactions trigger the release of neurotransmitter through exocytosis of synaptic vesicles from the presynaptic or basolateral membrane. The neurotransmitter molecules diffuse across the synaptic cleft to the postsynaptic membrane of the primary sensory neuron, where they elicit an action potential that travels along the nerve fiber [13, 57].

Secondary sensory receptors are those receptors that have additional receptor cell between the stimulus and sensory neuron. The receptor cells are connected by way of synapses with the sensory neurons (Fig. 9.5). The impinging stimulus acts on the membrane receptor of the receptor cell to initiate a response leading to changes in membrane permeability to sodium or calcium ions. The potential of the receptor cell can initiate the release of a neurotransmitter in the presynaptic membrane of the

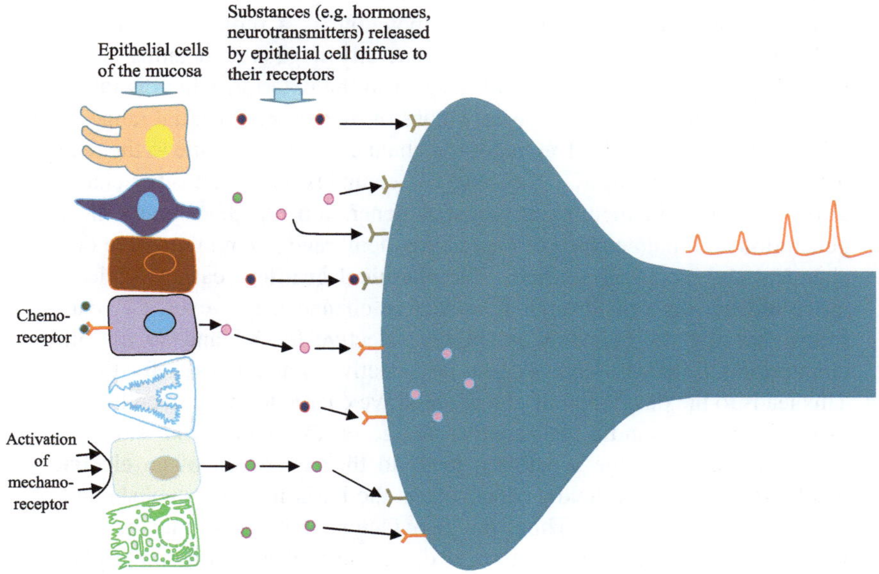

Fig. 9.5 Simple schematic representation of receptor potential formation. The activation of receptor in nerve terminal by substances released by a separate cell leads to opening of sodium channels in the nerve terminal. This type of signaling is mediated by ligand-gated channels. The resulting effect (electric signal) is the summation of the effect of activation of the different ion channels

receptor cell. The released neurotransmitter localizes to its receptor on the postsynaptic membrane of the nerve fiber, mediating cellular signaling pathways that result in depolarization of the postsynaptic membrane of the nerve fiber, which produces a local response referred to the postsynaptic potential, which in turn results in the formation of the generator potential. Thus, the sensory neuron is indirectly excited by the potential of the receptor cell. The generator potential spreads electrotonically on the nerve fiber. Summation of several generator potentials passing through the nerve fiber produces the action potential at the electrogenic zones of the nerve fiber [58–62]. The formation of action potential by summation of individual potential traveling toward the electrogenic zones is a graded electrical response of the receptor cell, which simply indicates the proportionality of the amplitude of the receptor potential to the stimulus size. Graded potential arises from the summation of the individual actions of ligand-gated ion channel proteins. The activities of voltage-gated ion channel proteins basically do not contribute to this potential [62–64].

Reference Note 9.1

Resting Membrane and Action Potentials

Membrane potential (transmembrane potential or membrane voltage, V_m) is a measure of the potential difference across the membrane that separates the two compartments of the cell—intracellular (*i*) and extracellular (*e*) fluid compartments. The resting membrane potential is a property of quiescent cells. The numerical value of V_m is generally negative, meaning that the interior of the cell is negative with respect to the exterior, which is taken as the reference or zero value. Typical values of membrane potential range from −50 to −95 mV [65, 66]. The major ion channels that contribute to the resting membrane potential are K^+, Na^+, and Cl^- channels; however, other channels also play a role in membrane potential generation of the resting cell. The generation and maintenance of the resting membrane potential involve both a chemical and electrical gradient. The **chemical gradient** causes K^+ ions to move outside the cell through K^+-selective channels. However, Cl^- cannot follow because the channels are highly selective for K^+ and do not allow counterion Cl^- to pass and because no selective channels exist for this ion. This leads to the development of a net positive charge to the extracellular side of the membrane, and a net negative charge on the intracellular side of the membrane. This charge separation leads to the generation of an **electrical gradient** (i.e., electric field), which forms the basis for the establishment of the membrane potential. Thus, the two factors that determine the resting membrane potential are asymmetrical distribution of ions across the plasma membrane and selectivity of ion channels in the plasma membrane of the receptor cell, nerve fiber, or neuron [67–71].

Recall that in the previous scenario, more K^+ ions diffuse out of the cell, thus the size of the electrical gradient increases. Ultimately, a point is reached when diffusion of K^+ ions down its concentration gradient from *i* to *e* side is opposed by the excess positive charge on the extracellular side of the membrane. At this point, the size of the electrical gradient is large enough to exactly balance and stop the net movement of K^+ from *i* to *e* side down its concentration gradient. This is the K^+ electrochemical equilibrium: the electrical gradient, which is the voltage potential difference, across the plasma membrane, balances the K^+ chemical gradient. The membrane potential established is the equilibrium potential ($V_{eq.}$) for the ion. This equilibrium potential is referred to as the thermodynamic equilibrium potential ($V_{eq.}$)—This is the potential at which the chemical and the electrical forces acting on K^+ ions exactly balance each other in the cell. At equilibrium, there is no net movement of K^+ between the two compartments. Because K^+ is the ion under consideration, this potential is referred to as the K^+ equilibrium potential and is designated V_K. To calculate the equilibrium potential for K^+ and other ions, the Nernst equation is applied. The Cl^-, Na^+, and Ca^{2+} only have a little contribution to the resting membrane potential. The contribution of these minor ions is more pronounced, though, still little, at low extracellular K^+ levels. Little amount of Na^+ and Ca^{2+} ions constantly enter the cell even at the resting phase. The depolarizing force resulting from the influx of these positive ions contributes to the expulsion of K^+ ions by active transport mechanism mediated by Na^+/K^+-ATPase and $3Na^+/Ca^{2+}$ exchanger (the functions and mechanisms of the Na^+/K^+-ATPase and $3Na^+/Ca^{2+}$ exchanger have been discussed in Chap. 5). Note that apart from Cl^- ions, other negatively charged molecules such as phosphates and anionic proteins are present in high concentration in the cytosol. The phosphates and anionic proteins can only leave the cytosol by passive transport and not by active transport mechanism [this phenomenon is referred to as Donnan–Gibbs distribution, named after the American physicist Josiah Willard Gibbs (1839–1903) and the British chemist Frederick G. Donnan (1870–1956)] [67–71]. It was thought that charged proteins may not contribute to the development of the resting potential [72], but recent reports suggest that anionic proteins can interact with molecules (including proteins) in the inner leaflet of the plasma membrane and form particles that can aid in the formation of charged membrane domains and participate in the development of transmembrane potential [73–75].

The resting membrane potential of enteric neurons is normally less negative than in the CNS (about −40 to −70 mV) and is largely determined by the activity of potassium channels. A stimulus impinging on the cell can either excite or inhibit the cell by depolarizing or hyperpolarizing the cell membrane. The phenomenon of depolarization and hyperpolarization leads to excitation and inhibition, respectively. Excitation or inhibition depends, at least in part, on the activating neurotransmitter. If the released neurotransmitter activates downstream targets resulting in the opening of Na^+ and K^+ channels, then Na^+ ions diffuse into the cell while K^+ ions diffuse outward. However, the inward Na^+ current is normally greater than the outward K^+ current, causing the postsynaptic membrane to depolarize, which subsequently results in the generation of action potential and excitatory potential referred to as the excitatory postsynaptic potential (EPSP). Such synapses are referred to as excitatory synapses. The opening of Ca^{2+} channel can also cause the generation of excitatory potential. In fact, Ca^{2+} channel activity is the primary basis for the generation of excitatory potential in intrinsic neurons of the gut. (The predominant cause of excitatory potential in extrinsic neurons is the activity of Na^+ channels.) The effect of inhibitory synapse on signal transmission is opposite to that of the excitatory synapse. In inhibitory synapse, the release of neurotransmitter leads to opening of K^+ or Cl^- ion channels that hyperpolarize the postsynaptic membrane. The K^+ diffuses outward, whereas Cl^- diffuses into the postsynaptic membrane. The result is hyperpolarization of the membrane such that potential becomes more negative than the resting membrane potential, which produces an inhibition called inhibitory postsynaptic potential (IPSP). It should be noted, however, that excitation or inhibition occurs in groups of neurons and thus it will be expected that EPSP or IPSP occurring in different neuronal synapses is summed up simultaneously over the dendrites and soma [76–83]. If the sum of the local EPSPs produces an electrotonic current that is able to depolarize the axon hillock, and if the depolarizing current reaches the excitation threshold, an action potential is generated and propagated along the nerve fiber. This represents a crucial way by which neuronal dendrites contribute to the production of an action potential on the hillock, and the phenomenon is termed spatial summation. However, a single neuron may fire repeatedly at a higher rate in such a way that the EPSPs sum up to produce a greater degree of depolarization compared to the one that would result from a single firing. If the depolarizing current could reach the excitation threshold, then an action potential is generated in the axon hillock and the signal is transmitted along the nerve fiber. This phenomenon involves the summation of EPSPs over time and is called temporal summation. However, signal or impulse generated may decay over time; thus, excitation may arise as a step-like progression. Another phenomenon observed to occur in neuronal transmission either in extrinsic or intrinsic nervous system of the gut is called synaptic integration—a process by which the neuron computes (integrate) input signals to determine and produce outputs in the form of nerve impulses. A motor neuron that receives input from neuronal sources (for instance, from both excitatory and inhibitory synapses) compute the summed value of the IPSPs and EPSPs at the axon hillock to determine the output signal. If the integrated potential is greater than the excitation threshold, then the action potential is generated and the impulse is transmitted along the nerve fiber. Thus, transmission of signal across the axon depends on the excitation threshold that is determined by the number, type, and firing rate of input synapses. It should be mentioned that the potential that builds up at the axon hillock is not a simple algebraic sum of the number of excitatory and inhibitory synapses firing at any given time, but may involve higher level order spatiotemporal computing of signal frequency and amplitude [84–89]. The formation of action potential at the axon hillock initiates transmission of impulses toward the axonal terminal—referred to as orthodromic transmission of action potential. Orthodromic transmission of impulse results in the release of neurotransmitter with subsequent excitation or inhibition of the postsynaptic membrane. The backward transmission of impulses toward the soma or dendrite is referred to as antidromic transmission. Pathological conditions (e.g., diseases of neuronal and glial cell) and physiological state (e.g., circumstances that initiate negative or positive emotions) can affect transmission of impulse in the central and peripheral (enteric) nervous systems [90–93] (Fig. 9.6).

The phases of an action potential: The action potential (also known as impulse, excitation, or spike potential) is a universal, high-amplitude, fast-spreading signal on the membrane of excitable cell (neural cell, astrocyte, muscle cell, some secretory cells in glands) that enables the transfer of information in the nervous system. It is characterized by a rapid change in membrane potential (accompanied by corresponding change in the cell polarity) in response to a stimulus, resulting from the opening of voltage-dependent ion channel, which participates in the generation of the transmembrane ion current. The action potential curve is made up of the following phases: depolarization (rising phase of the curve), fast repolarization (falling phase of the curve), slow repolarization (negative trace potential), hyperpolarization (positive trace potential) (Fig. 9.6) [69, 107, 108].

At the resting phase, ion channels are inactivated. When a stimulus (electrical) impinges on the cell, ion channel receptors sensitive to voltage changes become activated (voltage-gated ion channel) and open to allow influx of ions (Na^+ or Ca^{2+}). The rapid influx of Na^+ ions causes the polarity of the plasma membrane to reverse, and the ion channels then rapidly inactivate. During this period, the K^+ gate is closed. As the membrane potential increases, more Na^+ channels open to further depolarize the cell. At this stage, some Na^+ and the K^+ gates are closed. When the depolarization of the cell membrane increases up to the critical level (threshold), there is an avalanche voltage-dependent opening of Na^+ channels. The threshold is characterized by the difference between the membrane resting potential and the critical level of depolarization. The level of excitability of the cell is defined by the threshold potential. When the cell reaches the threshold, greater number of Na^+ ions enters the cell along a concentration gradient referred to as Na^+ current. This leads to a rapid decrease in membrane potential to 0 and reversal of the membrane potential to a positive value. The period during which the membrane potential has a positive value is referred to as overshoot. The reversal of the potential to a positive value leads to opening of the K^+ channels.

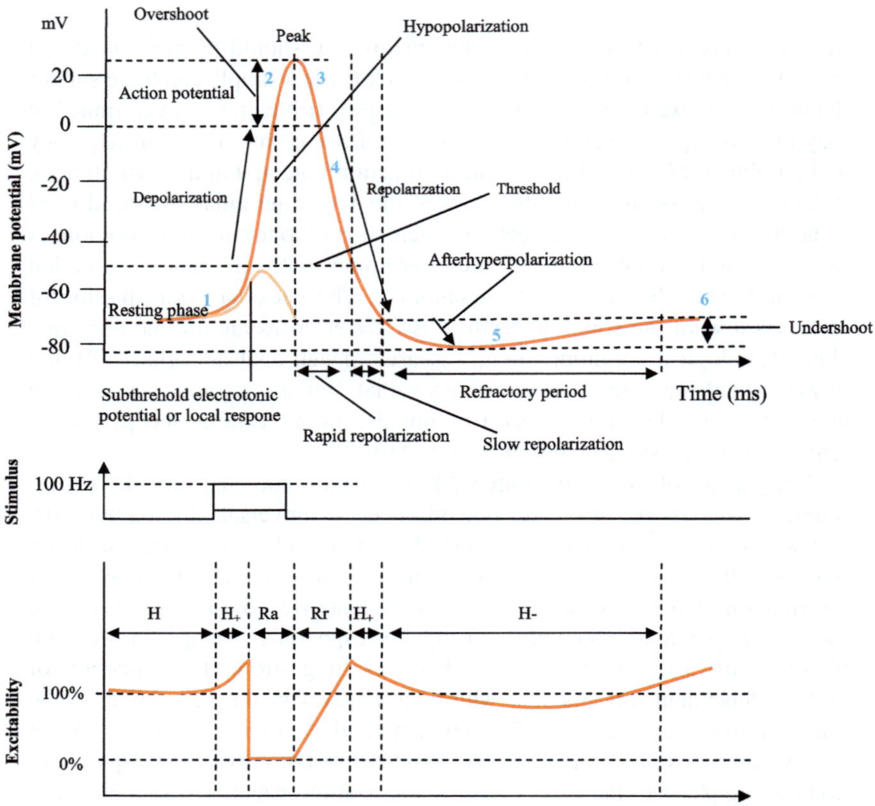

Fig. 9.6 Schematic representation of action potential. Phase "Ra" of excitability corresponds to depolarization phase of the action potential. "Ra" is the absolute refractory period. "Rr" is the relative refractory period. "H" is the period of normal excitability. "H⁻" of excitability corresponds to hyperpolarization phase. "H⁻" is the period of subnormal excitability (secondary refractory period). H^+ is the period of supernormal excitability. The resting phase is the property of all quiescent excitable cells. During the depolarization phase, the cell is non-excitable, because all the Na^+ channels are open. Absolute refraction is the state in which the excitability of the cell falls to zero. At this point, even the strongest stimulus cannot cause another excitation of the cell. Relative refraction of the cell is a condition in which the excitability of the cell is significantly lower than normal, but very strong stimuli can cause excitation of cell. During repolarization phase, the ion channels return to the closed state as the excitability of the cell is gradually restored [94–97]. Subnormal excitability is characterized by a slight decrease in excitability of the cells below the normal level. This decrease is due to the increase in the excitability threshold potential during hyperpolarization phase. Supernormal excitability (exaltation) is the state of the cell in which the excitability is above the normal level. Supernormal excitability is observed during the initial depolarization and during the phase of slow repolarization. Increased excitability of the cells in these phases is caused by the reduction of the excitation threshold compared to normal [98–106]

The activation of K^+ channels leads to an outward current of K^+ ions, returning the potential of the cell to the value of the resting membrane potential. The decrease in potential could be divided into fast and slow repolarization phases depending on the type of excitable cell involved. In the fast phase of repolarization, K^+ moves rapidly out of the cell, and subsequently the rate of transport of this ion reduces giving rise to slow phase of repolarization. The movement of K^+ out of the cell by density gradient (K^+ current) leads to the restoration of the resting membrane potential. In some situations, the membrane potential may further decrease below the resting value, producing a transient negative shift, called the afterhyperpolarization (AHP). This phase is also called hyperpolarization. AHP can be defined as a change in the membrane potential that makes it more negative relative to the resting potential. It is a component of some action potential referred to as the undershoot phase. AHP inhibits the transmission of impulse by increasing the stimulus required to move the membrane potential to the critical level of depolarization. AHP prevents an action potential from traveling the way it was generated. AHP phase develops due to residual potassium current as well as the direct electrogenic effect of activated Na^+/K^+ pump. This undershoot phase may result from efflux of K^+ or influx of Cl^- or inhibition of Na^+ or Ca^{2+} transport in and out of the cell. The hyperpolarization phase may last for about 2–4 ms. The resting membrane potential is restored through the activity of Na^+/K^+-ATPase or pump, which redistributes K^+ and Na^+ ions bringing the resting potential to about −70 mV [109–115].

The AHP can be classified as fast, medium, and slow; each of which has distinct ionic mechanisms. Fast and medium AHP can be produced by single action potential. During single action potential, transient depolarization of the membrane opens more voltage-gated K^+ channels than are open in the resting state. Many of these channels do not close immediately when the membrane returns to its normal resting voltage. The Ca^{2+}-activated K^+ channels that open in response to the influx of Ca^{2+} during the action potential carry much of the K^+ current as the membrane potential becomes more negative. AHP-type Ca^{2+}-activated K^+ channels contribute to a more depolarized membrane potential. Hyperpolarization persists until the membrane K^+ permeability returns to its usual value [109–115]. Other channels or receptors including the hyperpolarization-activated (Ih or HCN) channels are involved in the formation of AHP current. The Ih potential varies among cells and may be around −20 to 30 mV [116, 117]. Other types of channels involved in AHP current formation are the L- and R-type Ca^{2+} channels. However, T-type Ca^{2+} channels also play a role in AHP current [117]. It is believed that AHP is a useful phenomenon that ensures propagation of action potential in the naturally occurring direction—orthodromic conduction of action potential (Greek "orthos"—correct; "dromos"—run) [118]. Orthodromic action potential runs in anterograde direction along the axon, away from the soma. In contrast, antidromic action potential is the conduction of impulse along the axon, away from the axon terminal, toward the soma. In antidromic conduction, the action potential produced in the dendrites, soma, or axons may move toward the soma or move along the axon [92]. Antidromic and orthodromic conductions are the two possible way in which action potential is propagated along the cell surface [53, 93]. Many activities of the cell are believed to be mediated through AHP. AHP is involved in pacemaker potential formation [117]. AHP is also involved in synaptic plasticity and learning [119]. A special type of plasticity called metaplasticity characterizes the previous history of neuronal activity which determines the induction threshold, and the magnitude and direction of changes in synaptic plasticity, by integrating and coordinating synaptic activity [119].

9.2.2 Sympathetic Innervation of the Gastrointestinal Tract

The sympathetic nerves that innervate the GI tract arise from the celiac, superior, and inferior mesenteric ganglia as well as the pelvic splanchnic nerves (Fig. 9.2). The celiac, superior, and inferior mesenteric ganglia are generally called the prevertebral ganglia. The sympathetic nerves of the salivary glands arise from the superior cervical ganglion [17, 120, 121]. Nerves forming the prevertebral ganglia or splanchnic network arise from the intermediolateral cell column or lateral horn of the spinal cord at T1–L2. The afferent nerve fibers contained in the splanchnic nerves project to the thoracolumbar segments T10–L2 of the spinal cord via the thoracolumbar dorsal root ganglion. (Figure 9.7 shows pairs of ganglia and nerves arising from the spinal segments.) The cell bodies of these nerves are located in the dorsal root ganglion. The neurons of the dorsal root ganglion are predominantly pseudounipolar, but it also contains different population of neurons and peripheral glial cells. It should be noted that some of the afferent nerve fibers course through all three prevertebral ganglia [14]. But afferent nerves from the esophagus project to the lateral horns of the cervical and thoracic cords at C1–T9 via the corresponding dorsal root ganglia [18, 122]. The splanchnic nerves form varicose branching afferent axons near submucosal blood vessels [14]. Afferent signals, for instance, from the salivary glands (submandibular), are transferred via the preganglionic fibers that synapse with the postganglionic fibers at the submandibular ganglion, within the chorda tympani nerve, which subsequently merges with the facial nerve at the level where the facial nerve exits the skull. The fibers of the facial nerve transfer information to the pons where information is interpreted and relayed to higher centers, and the efferent copy is transmitted to the effectors [123–125]. The afferent fibers are involved in the transmission of various signals including the sensation of pain (nociception), discomfort, fullness, bloating, nausea, urgency, etc. [14, 41].

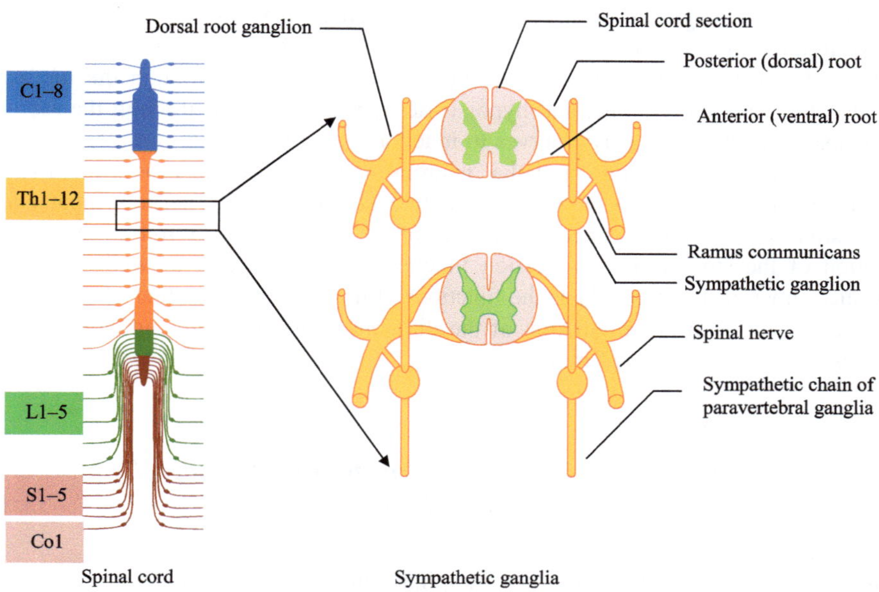

Fig. 9.7 Spinal cord showing the 31 pairs of ganglia and nerves arising from the cervical (C1–C8), thoracic (Th1–Th12), lumbal (L1–L5), sacral (S1–S5), coccygeal (1 spinal cord segment) regions and section of the spinal cord showing sympathetic (thoracolumbal) ganglia. The preganglionic fibers of the sympathetic division exit the spinal cord from the ventral roots at position T1–L2. The preganglionic neurons synapse with postganglionic neurons within a double row of sympathetic ganglia, called paravertebral ganglia. These ganglia are located on both sides of the spinal cord and are interconnected in a row to form a sympathetic chain of paravertebral ganglia. However, some of the fibers of the preganglionic neurons, especially those below the diaphragm, do not synapse with paravertebral ganglia; they diverge from the spinal nerves within the white rami communicantes (ramus communicans), passing through the sympathetic chain of the paravertebral ganglia to form splanchnic nerves, which synapse in ganglia located proximal to the target (visceral) organs, referred to as prevertebral ganglia. There are three types of prevertebral ganglia: celiac, superior mesenteric, and inferior mesenteric ganglia. In the prevertebral ganglia, each nerve fiber forms synapse with their postganglionic nerve, which innervates the target organs. The axons of the postganglionic sympathetic fibers form the gray rami communicantes as they return to the spinal nerves and travel as part of the spinal nerves to their effectors. Sensory fibers also pass through the rami communicantes into the spinal cord. Postganglionic fibers that arise from the collateral ganglia innervate organs of the digestive, urinary, and reproductive systems. Preganglionic nerves are mostly myelinated (ensheathed with myelin, about 80–90% of the fibers), whereas the postganglionic nerves are predominantly unmyelinated (do not contain myelin sheath). Majority of the preganglionic neurons are cholinergic, though they also contain other neuropeptides and transmitter molecules [132, 133, 138–144], whereas the postganglionic neurons are noradrenergic but also transmit other peptides [134–137]

The splanchnic nerves terminate in various region of the GI tract including serosa, mucosa, submucosa, myentericus [14]. Remember that, part of the network of nerve fibers of the lumbosacral splanchnic nerves innervates the rectum as well as the bladder, prostate, and external genitalia [126–131]. The sympathetic nerves that innervate the gut are mainly composed of efferent fibers (~90%). The efferents exit through the ventral root of the spinal cord to their effector organs [14, 18, 132, 133]. The sympathetic output of the lumbar region projects through the lumbar splanchnic, lumbar colonic, and hypogastric nerves, as well as through the sacral sympathetic chain and pelvic nerves to innervate vascular smooth muscle, visceral smooth muscles, secretory epithelia, neurons in the enteric nervous system (including vasoconstrictor neuron) and the pelvic ganglia [134–137]. The sympathetic division of the autonomic nervous system of the gut plays a key role in the regulation of the storage and evacuation functions of the hindgut, as well as lower urinary tract, and also regulates the functions of the reproductive organs [134]. The effects resulting from stimulation of the different divisions of the autonomic nervous system are presented in Table 9.1.

9.2.3 Central Transmission and Processing of Visceral Signals of the Gastrointestinal Tract

Afferent signal from the gut reaches the spinal cord via the dorsal root ganglion (see diagram of spinal cord section in Fig. 9.8). The neurons of the dorsal root ganglion project to the spinal cord via dorsal posterior roots and travel via the anterolateral pathway or dorsal column medial lemniscal pathway to the brain. Afferent signal also travels directly to the brain via the sensory fibers of the cranial nerves VII, IX, and X [14].

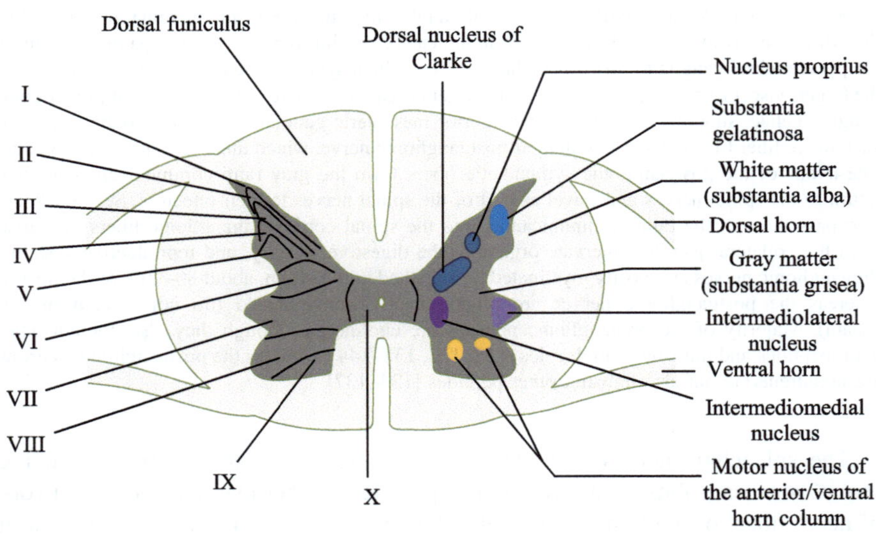

◂**Fig. 9.8** Cross section of the spinal cord showing the white matter and nomenclature of the gray matter and its associated nuclei. White matter is found in the CNS (brain and superficial part of the spinal cord) and is composed of glial cells and myelinated axons that transmit signals from one region of the CNS to another. The color of this area of the CNS is mainly due to the large content of lipid tissue (that make up myelin) and capillaries. The white matter underlies the gray matter. In many central nervous diseases, the degree of myelination is reduced leading to impairments in neuronal functions. The gray matter is the H-shaped structure and comprises neuronal cell bodies, neuropil (neuronal dendrites, spines, and unmyelinated axons), glial cells, synapses, as well as capillaries. However, gray matter also contains myelinated axons, although very few. Importantly, the large population of cell bodies found in gray matter is a key distinguishing feature between gray and white matter. The gray matter of the spinal cord can be divided into different layers, called Rexed laminae. Each layer constitutes a distinct topography of specific group of cells that execute a particular function. The location of the cells is identified according to a widely naming system. Rexed laminae nomenclature was introduced by the Swedish physician, neuroscientist and professor at Uppsala University, Bror Rexed (1914–2002) [154–159]. Rexed laminae represent a functional and structural zone in the spinal gray matter. It is used as a mark in defining the precise location of neurons and synaptic connection in physiological studies. The Rexed laminae comprise a system of ten layers of gray matter (laminae I–X) that corresponds to the spinal nuclei or synaptic connections and serves as a key landmark for some neurosurgical procedures. Rexed laminae may be divided into dorsal (laminae I–VI), intermediate (laminae VII and part of X), and ventral (laminae VIII–IX). Rexed lamina I and part of II contain the marginal or posteromarginal nucleus. Rexed lamina II comprises substantia gelatinosa of Rolando (named after the Italian anatomist Luigi Rolando (1773–1831). Laminae III and IV consist of the nucleus proprius. Short relay neurons (interneurons) are located in Rexed lamina II and III. Rexed lamina VI contains the dorsal nucleus. However, the nucleus dorsalis also extends into lamina VII. Rexed lamina VII contains the intermediomedial, intermediolateral and dorsal nuclei. The dorsal nucleus of Clarke is found in the thoracic and upper lumbar region. The lamina VII also contains the inhibitory interneurons called Renshaw cells. The laminae are also used as landmarks for spinal tracts. Rexed lamina VII, for instance, contains the spinocerebellar tract. Lamina VIII is composed of commissural nucleus and efferent interneurons. Lamina IX is composed of lateral and medial motor neurons; phrenic and spinal accessory nuclei at cervical spinal cords; and Onuf's nucleus, first described by the Russian-American neuropsychiatrist and neuropathologist, Onufrowicz (Onuf) Bronislav (1863–1928) in 1989–1900, is a group of motoneurons in anterior horn S2 sacral spinal cord, involved in innervation of striated muscles (external anal sphincter muscle and external urethral sphincter muscles). Onuf's nucleus is believed to play a role in erection and ejaculation. Lamina X is the central lamina located in the gray matter around the central canal [160–167]. The integrity of the laminae is sequel to the maintenance of myelination, which is important in preserving the electrical activity of the CNS neurons by regulation of the velocity and synchrony of impulse conduction between different regions [168]. The determination of the integrity of neural electrical conductivity could be made with gray–white matter ratio in cranial computed tomography. A gray–white matter ratio less than 1.16 is a good predictor of electrical conductivity. The ratio is obtained by calculating the ratio of Hounsfield units in gray and white matter [169]

Transmission of Gut Signal to the Brain

Part of the information received in the dorsal vagal complex may be transmitted to autonomic centers in the medulla (nucleus of the solitary tract or NTS), hypothalamus, amygdala, and thalamus where different aspects of the signal are processed, interpreted, stored in accordance with the neural program that ensures adequate maintenance of GI functions [42] (see diagrams of medulla and limbic structures in Figs. 9.9 and 9.10). Though most of the afferent signal from the gut travels to the brain via the cranial nerves, GI tract information is also transferred by sensory neurons to the CNS via the thoracolumbar and lumbosacral spine [41].

The NTS, located in the dorsomedial medulla, is one of the sites of visceral sensory modulation and autonomic regulation of a variety of homeostatic functions including heart rate, respiration, blood pressure, blood flow, and GI motility (Fig. 9.9). The NTS represents a critical brainstem region that participates in analysis and integration of visceral signal. The primary input of neural fibers to the NTS comes from vagus nerves, which is the largest sensory pathway in the human body (the facial and glossopharyngeal nerves also have afferents that converge at the NTS). The vagus nerve comprises about 80% of afferent fibers, while the remaining 20% are efferent fibers. The afferent nerves of the vagus travel to the NTS where they synapse with another neuron [14]. Some neurons of the NTS relay the partly processed signal in the NTS to other regions of the brain to regulate homeostatic functions and behavioral responses [147]. Behavioral responses to feeding are evident following food intake, which is associated with satiety and sedation [42]. In addition, vagal afferent signaling has been implicated modulating mood and affect, including distinct forms of anxiety and fear. Klarer et al. (2014) reported that subdiaphragmatic vagal deafferentation in laboratory animals

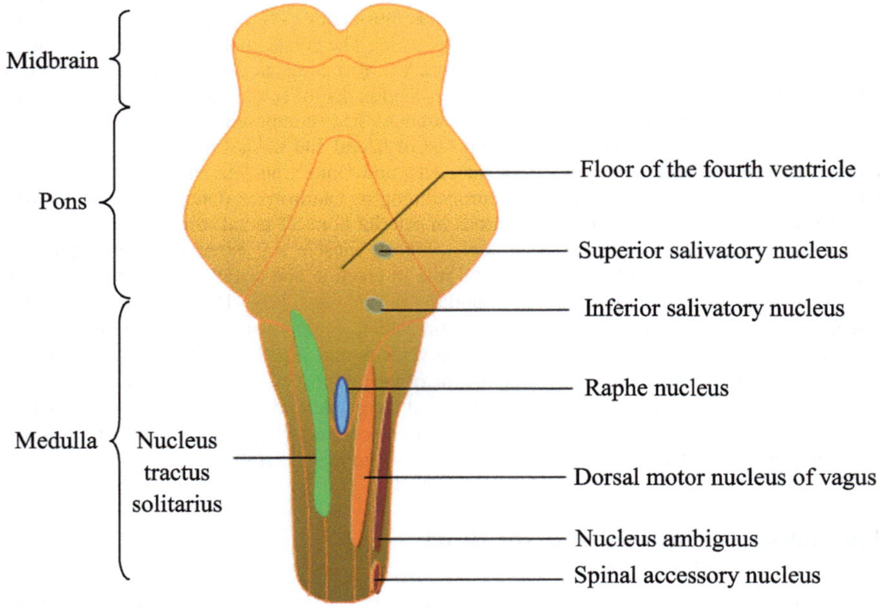

Fig. 9.9 Structure of the brainstem and the location of some nuclei involved in the regulation of GI activities. The brainstem extends from the medulla through the pons to the midbrain. The raphe nuclei of the reticular formation are largely implicated in modulation of GI functions. The raphe nuclei is a region of serotonin synthesis, a neurotransmitter involved in many functions of the central and peripheral nervous system, especially in regulation of mood [20, 170, 171]. The role of other nuclei depicted in the diagram is discussed in the text

leads to reduced innate anxiety-like behavior in elevated plus maze test, open field test, and food neophobia test [148]. These behavioral changes were associated with region-dependent changes in the level of noradrenaline and GABA (γ-aminobutyric acid) in the limbic system. (See limbic system in Fig. 9.10.) However, such alterations were not observed in the hypothalamus-pituitary-adrenal axis, suggesting that innate anxiety is subject to visceral modulation through GI vagal afferents, and that this modulation is achieved via regulation of neurotransmitters of the limbic system [148]. The vagus nerve modulates higher brain activity (emotion, anxiety), control feeding behavior, hunger, appetite, fullness, satiety, bloating, nausea, inflammation, and pain [14, 149–152]. The vagus nerve controls feeding behavior, at least in part, by GI reflexes that regulate gastric compliance, gastric accommodation, gastric emptying, gastric and pancreatic secretion, and emesis [14, 151]. The role of the vagus on inflammation appears to be indirect and is modulated via its action on the tone of cholinergic neurons, which in turn regulates the dampening of the activity of immune response cells such as macrophages and dendricytes. The vagus nerve (afferents) is also involved in chemonociception [14, 153].

The role of the vagus in digestion and modulation of higher brain functions can be studied with total subdiaphragmatic vagotomy, selective subdiaphragmatic vagal deafferentation, selective subdiaphragmatic vagal deefferentation, or sham surgery. The effect of the hormones or neuropeptides can be studied by intravenous,

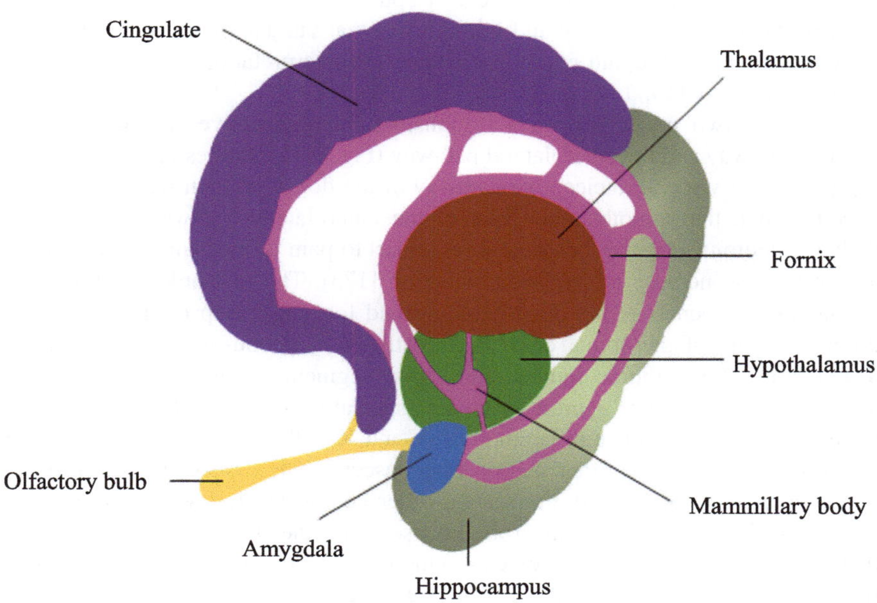

Fig. 9.10 Limbic system showing structures to which GI information is transmitted. The limbic system is located on either side of the thalamus and under the cortex. It plays important role in motivation, emotion, behavior, olfaction, and memory

intraperitoneal, or intracerebral injection in any of the above methods when different types of food substances are passed into the gut of the animal under study [46, 49, 148].

Central Processing of Gastrointestinal Nociception

Pain arising from the gut is a characteristic of many diseases of the gut. The nociceptive (pain) signal of the gut is usually sensed by mechanoreceptors located next to the sensory neuron—the first-order neuron. The cell body of this neuron is located in dorsal root ganglion. The pain signal from nociceptors and thermoreceptors is usually transmitted via the anterolateral pathway. But part of the signal is also transmitted via the dorsal column pathway (Fig. 9.11) [172, 173]. In the presence of inflammatory reactions of the dorsal column, the dorsal column medial lemniscal pathway compromises transmission of visceral pain leading to increased, decreased, or even absence of sensation. Thus, symptoms of visceral allodynia and hyperalgesia may not be well represented in the somatosensory cortex [172, 174]. Visceral allodynia is the inability of the visceral organs to respond to painful stimulation, hence pain is not perceived. An increase in pain perception to stimuli that are normally perceived as painful is referred to as hyperalgesia. These disorders of pain-signaling pathway are found in GI disorders such as inflammatory bowel diseases [14, 175, 176]. The mechanisms of visceral signal disorders include sensitization of peripheral visceral afferent or spinal cord dorsal horn neurons; alteration in descending excitatory and inhibitory signal via the spinal cord neurons to the effectors; cognitive and emotional bias in the interpretation of innocuous sensation as noxious [14, 177].

There are two major pathways responsible for nociceptive signaling—dorsal column pathway and the anterolateral pathway (Fig. 9.11). Interestingly, it has been suggested that visceral nociceptive pathway of the dorsal column pathway is more effective than the spinothalamic tract of the anterolateral pathway in activating thalamic neurons, eliciting behavioral responses to pain or mechanical stimulation, and triggering increase in cerebral circulation [173]. The first-order neurons send collaterals that penetrate the dorsal column and forming synapse with neurons in different layers of the spinal cord, but mostly with second-order neurons in the dorsal horn of the thoracolumbar and sacral segments of the spinal cord. The terminals of the sensory neuron terminate especially in laminae I and II and to a lower extent also terminate in laminae III–V and X of the gray matter (Fig. 9.8). In the spinal cord, the first-order neuron fibers ascend. The first-order neuron of the dorsal column synapses with the second-order neuron at the level of the medulla oblongata. The second-order neuron synapses with the third-order neuron in the thalamus from where the axons extend to the primary somatosensory cortex, where final processing and interpretation of the signal takes place. The processing of the signal in the cortex results in conscious feeling of the signal as pain [14, 178–181].

The anterolateral pathway transmits noxious (chemosensory) and nociceptive signals to the brain centers responsible for processing of GI signals.

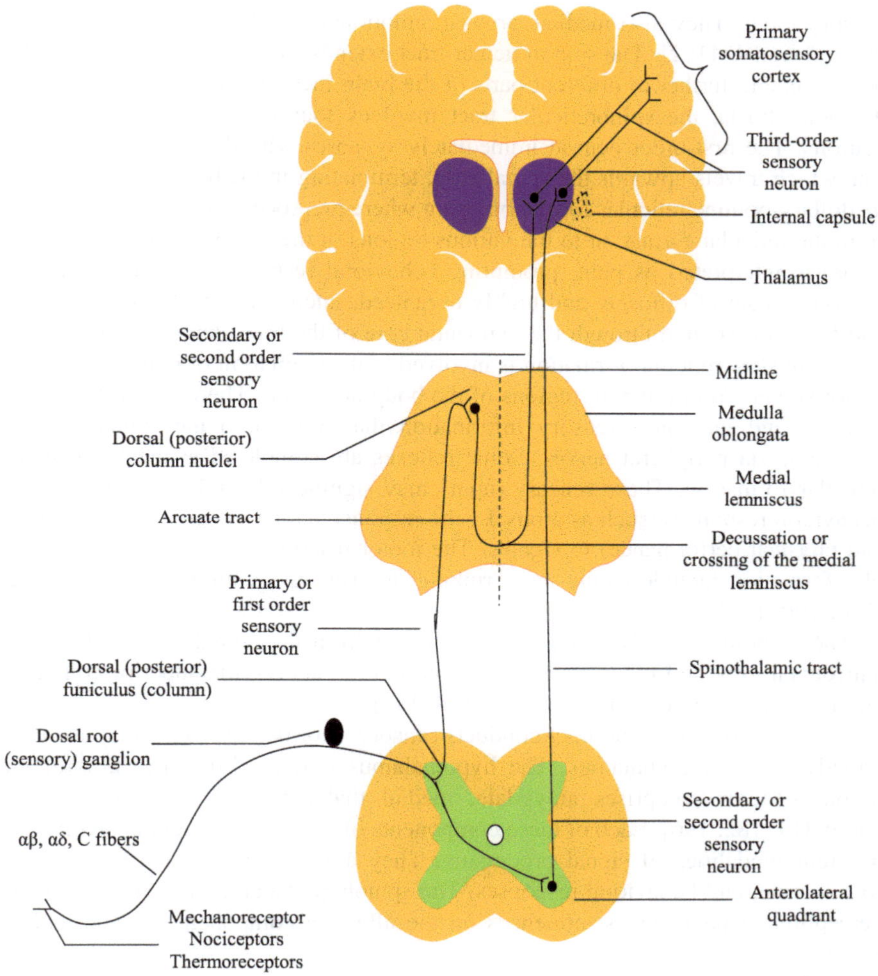

Fig. 9.11 Central branches of the visceral afferents innervating the gut travel via the dorsal root ganglia and project into second-order neurons in the spinal gray matter. Ascending pathways arise from the dorsal column (dorsal column pathway) and from the anterolateral quadrant projecting to the thalamus, (spinothalamic tract), reticular formation (spinoreticular tract), cerebellum (spinomesencephalic tract), hypothalamus (spinohypothalamic tract). Sensory neurons in the gut via first-order extrinsic afferent neuron convey stimuli to second-order neurons. Second-order neurons cross the white commissure and ascend in the anterolateral column and in the ipsilateral dorsal column to higher brain centers, first reaching the brainstem and then to other brain region such as hypothalamus, thalamus, amygdala. Each of these centers analyzes, interprets, and sends the appropriate response to the effector organ [14, 20, 175–177]

The anterolateral pathway is made up of spinothalamic, spinoreticular, spinomesencephalic, and spinohypothalamic tracts. The spinoreticular, spinomesencephalic, and spinohypothalamic tracts are responsible for unconscious responses to visceral sensory input. They also mediate arousal, emotional, and behavioral responses to visceral signals [182]. The spinoreticular tract extends from spinal cord to reticular formation and further to different parts of the brain including the thalamus. Unlike the other tracts, the spinoreticular tract involves four levels of organization of neurons: The first-order neuron immediately synapses with the second-order neuron, which travels upwards the spinal cord, terminating in the brainstem within the medullary-pontine reticular formation, from where preprocessed information is sent from the reticular formation to the various regions of the cerebral cortex where the signal is interpreted as pain, prompting behavioral responses. The reticular formation is a set of diffusely and highly organized, interconnected brainstem nuclei and fibers distributed through the tegmental core of the brainstem [170, 183–189]. This brainstem reticular formation is involved in the regulation of both sensory and motor signals from different regions of the body and brain. The reticular formation receives and integrates sensory information that flows into the spinal cord or brainstem via peripheral nerves. Some reflexes are completed at the level of the reticular formation. These sensory inputs may significantly influence the level of behavioral responses (such as arousal—the responsive tone of conscious awareness and physical performance) to signals. The motor responses regulated by the reticular formation include eating and drinking, as well as breathing movement, etc. [170, 190–193].

The spinomesencephalic tract carries signal from the spinal cord to the periaqueductal gray and other areas of the midbrain including the superior colliculus and midbrain reticular formation [14, 194–196].

The spinohypothalamic tract conducts sensory information from the spinal cord directly to the hypothalamus. The hypothalamus is one of the structures of the limbic system—comprises amygdala, medial thalamus, and anterior cingulate cortex [14, 182, 196]. Each of these components of the limbic system is involved in different modalities of signal processing. They are involved in autonomic, neuroendocrine, and behavioral responses. The spinohypothalamic tract modulates the perception of pain via its influences on the affective components of nociception [196].

In the spinothalamic tract, the second-order neuron forms synapse with a third-order neuron of the ventral posterior lateral, medial dorsal, and ventral medial posterior nuclei of the thalamus. The spinothalamic tract mediates pain, cold, heat, and touch sensation. The thalamus is a major relay station in the brain that allows the convergence of multiple somatic and visceral signals. Information in the thalamus is partly processed, analyzed, and interpreted before it is relayed to the cortex via the tertiary neuron. This tract is responsible for the conscious awareness of signal such as pain [14, 170, 182, 186]. Multiple areas of the cortex receive afferent signal from thalamic nuclei and send back information to different thalamic nuclei. Thalamic fibers are organized in a contralateral pattern; its neurons relay information to the somatosensory cortex (SI and SII), pregenual anterior cingulate cortex, mid-cingulate cortex, and the insula. Notably, visceral sensation is primarily represented in the secondary somatosensory cortex. The SI and SII cortices are referred to as lateral pain system. This system encodes the intensity and localization of visceral stimuli. The cingulate cortex is referred to as the medial pain system. It is predominantly responsible for the anticipatory, unpleasantness, anxiety-related characteristics of pain. Though not considered as one of the regions of processing of pain, the insula (generally called the interoceptive cortex) receives sensory signal about the viscera, including the digestive tract, and integrates the signal with emotional aspects of pain processing. A set of neurons of the insula projects to periaqueductal gray, amygdala, hypothalamus, modulating the descending responses of pain. The higher brain centers such as the prefrontal cortex have cognitive influences on pain perception. Pain sensation is perceived only when the cortical areas have processed, integrated, and interpreted it as painful [14, 182, 197–199]. The corticofugal neurons (neurons that send information downstream the cerebral cortex) serve as a major filter and selection point of information in the brain, providing positive feedback to the "correct" input, while at the same time, suppressing irrelevant information. It should be, however, noted that two pathways may have some role in common or even intersect in functionality. For instance, the ascending nociceptive pathways and temperature fibers of the spinothalamic tract send information to the periaqueductal gray via the spinomesencephalic tract [199].

The somatosensory system is a complex sensory system that gathers, transmits, analyzes, and interprets incoming information from different receptor types, including thermoreceptors, photoreceptors, mechanoreceptors, proprioceptors, and chemoreceptors. These sensory receptors are located in different parts of the body and receive information about the surrounding environment of not only the GI tract, but also musculoskeletal, respiratory, cardiovascular, and other systems of the body. The somatosensory cortex is usually divided into primary and secondary cortices. The primary somatosensory (SI) cortex (Brodmann areas 1, 2 and 3) is the location of the lateral postcentral gyrus in the parietal lobe of the human brain. The primary somatosensory cortex is adjacent to the secondary somatosensory cortex (SII) (Brodmann areas 5 and 40) [20, 171]. The Brodmann areas are regions of the cerebral cortex defined by its histocytology and organization of cells. The concept was introduced by the German neurologist, Korbinian Brodmann (1868–1918). On the basis of the histocytological characteristics, Brodmann divided the cerebral cortex into 52 regions, now called Brodmann areas 1–52. Brodmann areas 1, 2, 3, 5, and 40 are the location of the somatosensory cortex. The somatosensory cortex contains a map of sensory space, called the sensory homunculus (meaning little man). Sensory homunculus diagram is a somatotopy in sensory representation which was introduced in 1950 by the American neurologist and neurosurgeon, Wilder Penfield (1891–1976), and the Canadian neurosurgeon, neurologist, and neuropathologist, Theodore Brown Rasmussen (1910–2002) [200–203].

Visceral Signals of the Gastrointestinal Tract are Also Processed in Peripheral Neural Centers—Ganglia

Neurons in the ganglia receive continuous synaptic input from peripheral (including GI tract sensory mechano- or other receptors) and from central preganglionic neurons [204]. Information received from central preganglionic neurons may be due to endogenous oscillator activity located in the spinal cord or within the ganglion or input signal from higher centers. The endogenous activity of the neurons of the prevertebral ganglia operates as a neurogenic tone capable of bidirectional transmission of neural signals [204–207].

The ganglia occur in different locations of the peripheral nervous system. Ganglia may be intramural or located outside the organ. It may be located close to the vertebrae or (e.g., prevertebral ganglia) or further away from the vertebrae. The key functions of ganglia include integration, temporal, and spatial summation of signals, analysis, interpretation, and relay of signal to the appropriate quarters. The information processing at the ganglia also involves mechanisms of divergence and convergence. Divergence of neural signals occurs when preganglionic fiber synapses with numerous postganglionic neurons. The phenomenon of divergence involves the transmission of signal from a single afferent neuron out into multiple neurons in a neural center. In contrast, convergence involves the integration of multiple afferent signals out into a single neuron in a neural center. It occurs when a postganglionic neuron receives synaptic input from a large number of preganglionic fibers [204, 208–1012]. The ganglia function to continuously integrate convergent synaptic input originating from the peripheral and central nervous systems to alter the pattern of contraction of GI smooth muscle and other activities of the viscera [45]. Convergence and divergence are integral processes that constitute the computing power of a given neural center. The integrative power of ganglia, in part, is due to the gating of continuous central and peripheral inputs onto the ganglionic neurons. These neurons are capable of adjusting its firing potential in proportionality to amount of synaptic input. This property of ganglionic neurons is due to the pacemaker (automaticity) activity of a set of neurons [144, 204, 209].

Neural integration can be referred to as a higher order computing or summation of input excitatory or inhibitory postsynaptic potentials, which govern the output potential for firing in the postsynaptic neuron. Neural integration occurs in both time (temporal) and space (spatial). In temporal integration, the postganglionic neuron receives input signals from a single preganglionic neuron within a short period. In spatial integration, the postganglionic neuron receives multiple inputs from many preganglionic neurons [212, 213]. The ganglia constantly modulate and adjust motor functions to achieve set goals. Pattern formation of motor activity of the gut is achieved by reflex activity within the enteric plexuses, whose functions are in part modulated by the activities of peripheral and central neural centers [45].

9.2.4 Descending Neural Pathways Regulating Gastrointestinal Tract Activities

The pathways that send the command to the effectors from the CNS are called descending pathways [14]. Signals from several brain regions through interconnected pathways influence visceromotor or nociceptive activities of the GI tract

[214]. The limbic system generally can influence GI functions via the spinohypothalamic tract and limbic descending pathways. GI tract descending pathways (inhibitory and excitatory) arise from the periventricular gray of the hypothalamus, midbrain periaqueductal gray, nucleus raphe, locus coeruleus, anterior cingulate cortex, and somatosensory cortex. Other brain regions that influence descending pathway include amygdala, dorsal striatum, substantia nigra, ventral tegmental area. The descending pathways send their axons down the spinal cord to synapse with neurons at the anterior column of the spinal cord [14, 196, 214–217].

The tracts involved in transmission of motor command include reticulospinal, lateral and anterior corticospinal, medullary reticulospinal, pontoreticulospinal, etc. The corticospinal and corticobulbar descending tracts are classified as pyramidal. The reticulospinal (lateral and medial), vestibulospinal, olivospinal, rubrospinal, tectospinal, and vestibulospinal tracts are classified as extrapyramidal. The extrapyramidal tract is a neural network that forms part of the motor system that controls involuntary motor activities. The extrapyramidal tracts reach their targets without passing through pyramids of medulla by indirect regulation or modulation of the neurons of the anterior horn [196, 216]. The medullary pyramids are paired white matter structures of the medulla oblongata. The pyramidal tracts reach their target directly by synapsing with motor neurons of cranial nerve nuclei of the brainstem or anterior horn of the spinal cord. The extrapyramidal system is generally responsible for gross, synergic movements which require the activity of large groups of muscles, while the pyramidal system is responsible for fine, isolated, precise, and specific motor activity [218–220].

It is important to note that some tracts, which extend from the cortex to subcortical structures or diencephalon, play an important modulatory role in the activities of the descending pathways. For instance, the corticolimbic tracts play modulatory role on the affective behavioral dimensions of GI visceral signals. Also, the amygdala, an important brain structure that is important in affective behavior, is connected with anterior paralimbic cortex including ventral anterior cingulate, orbitofrontal, insular and temporopolar cortices [221].

The anterior cingulate cortex is one of the most important sources of descending modulatory (inhibitory or excitatory) pathways, projecting to the amygdala and the periaqueductal gray. Example of a descending inhibitory pathway is the periaqueductal gray–rostral ventromedial medulla–dorsal horn circuitry. This pathway is also involved in nociception. The periaqueductal gray is a key control center for descending nociceptive pathway. The gray matter of the periaqueductal gray is located around the cerebral aqueduct in the tegmental core of the midbrain. The periaqueductal gray neurons project to the nucleus raphe and send descending autonomic tracts that influence effector responses to stimuli. The perigenual anterior cingulate cortex (pACC)-pontomedullary pathway is another inhibitory pathway that influences visceral signaling [14, 214]. Also, the pACC is functionally connected to the amygdala. Both the amygdala and the periaqueductal gray are involved in the integration of behavioral and physiological responses to innate and learned threats [222, 223].

The reticular formation pathways are immensely implicated in modulating the activity of descending neural pathways. The pathways originating from the reticular formation are under constant inhibition by higher brain centers. For instance, the reticulospinal tract is under constant influence of cortical projections. This tract projects to the ventral horn of the spinal cord to affect output of signal to the effector [170, 190, 193].

Tsuruoka et al. (2005) reported that stimulation of the nucleus locus coeruleus modulates motor functions of the viscera, as evidenced by colorectal distention [214]. Stimulation of this nucleus leads to inhibitory action on the visceral motor system. Thus, visceromotor activities of the GI tract are controlled by centrifugal (descending) pathways originating from the locus coeruleus. The locus coeruleus is a nucleus in the posterior area of the rostral pons in the lateral floor of the fourth ventricle, serving as the principal brain site for synthesis of norepinephrine (noradrenaline) [214]. The locus coeruleus is involved in stress and panic responses, sleep, and arousal. The neurons of the locus coeruleus send fibers to the spinal cord, brainstem, cerebellum, hypothalamus, thalamic nuclei, amygdala, basal ganglia, and cortex [217, 224–226]. Their diffusely arborizing terminal network makes them so useful in a wide range of physiological processes. The projections of the neurons of the locus coeruleus serve a dual effect due to the activation of the predominant receptor types. For instance, excitatory locus coeruleus–cerebral cortical pathway (via α_1-adrenoceptors) is involved in promoting wakefulness, whereas inhibitory locus coeruleus–cerebral cortical pathway (via α_2-adrenoceptors) is involved in promoting sleep. The medullary reticular nuclei project to preganglionic sympathetic and parasympathetic neurons. These nuclei are involved in other responses including cognition, behavior, but also mediate visceral responses associated with the GI tract [215].

The preganglionic motor fibers from the dorsal vagal nucleus and the visceral efferents from the nucleus ambiguus, which are also under influence of descending pathways, descend and travel to the gut as part of the nerve fibers that constitute the vagus nerve [199, 227]. The efferent vagal nerves allow top-down transmission of processed information from the brain to regulate secretion, motility, and neuroendocrine functions of GI effectors [42]. The effectors or effector organs of the autonomic nervous system of the GI tract are the blood vessels, smooth muscles, enteric neurons, and gut-associated lymphoid tissues [228].

It is important to note that not only neurons but also glial cells (astrocytes) of the brain regions (NTS, hypothalamus, amygdala, thalamus, and cortex) participate in regulating GI functions. Over the past decades, there has been increasing evidence pointing to the useful role of astrocytes in regulating and modulating synaptic activities and integrating signal activities via a special type of synapse called tripartite synapse [229, 230]. This synapse is formed by the pre- and postsynaptic terminal of neurons and the glial cell, characterized by bidirectional communication between astrocytes and neurons, and mainly involved in the regulation of the concentration of intracellular Ca^{2+} concentrations. The activity in this synapse is coordinated, in part, by gliotransmitters (e.g., glutamate, D-serine, GABA, and ATP), which also modulate neuronal activity. These neuroactive substances diffuse through membrane channels, translocate via membrane transporters or by vesicular exocytosis, and reach neurons to modulate activity [147, 231, 232]. This modern view of information transmission is advancement on the traditional bipartite synapse information transmission model between the pre- and postsynaptic neurons [233]. Receptors that regulate the activities of the tripartite synapse are discussed in later part of this chapter [147, 231, 232]. Consequently, dysfunctions in astrocyte physiology may lead to disorders in neuronal network regulating feeding behavior and other GI activities.

9.2.5 Gastrointestinal Reflexes—Automatic Responses to Stimuli, Regulating Gastrointestinal Functioning

GI reflexes are automatic response of GI cells (smooth muscle cells, secretory cells, etc.) evoked by stimulation of sensory receptors in a particular region of the GI tract. The responses may be elicited by physical, electrical, chemical, or mechanical stimuli [234]. The signal is transmitted by the sensory receptor via the afferent nerve to the neural center where it is analyzed and relayed back via the efferent nerve to the effector. Majority of the neural centers of such reflexes are located in the spinal cord, which contains groups of neurons capable of integration, interpretation of neural signals, and these neurons also possess automaticity. However, some GI reflexes may be completed at the level of the brain. GI reflexes are comprised of a sensory receptor, afferent neuron, central neuron or relay neuron, efferent neuron, and the effector. In case of the spinal cord, afferent signals of the smooth muscle of the intestines are transferred by intestinofugal neurons to the spinal cord via sympathetic prevertebral ganglia. The signal is processed and the command relayed back to the effector for appropriate response. The neural pathway involved in such reflexes is called a reflex arc. The reflexes are known to play an important role in the onset or pathophysiology of various diseases of the GI tract [235–239].

GI reflexes represent a major way by which information is processed or coordinated in the GI tract. Such reflexes include upper airway reflexes (e.g., swallowing reflex), esophago-salivary reflex, esophageal reflex, enteroenteric reflex (e.g., colo-colonic reflex), enterospinal reflex, vago-vagal reflex, gastroileal reflex, gastrocolic reflex, and cologastric inhibitory reflex. GI reflexes may be intrinsic (i.e., all components of the reflex are located in the wall of the GI tract), enteroenteric reflexes (in which the reflexes arise in one part of the digestive system and affects a different region), or centrally coordinated (in which the reflex pathway passes through or originates in the CNS). These reflexes are involved in the regulation of GI activities, and they represent an important way through which the functions of the esophagus, stomach, intestine, intestinal sphincters, pancreas, and biliary system are regulated [118]. Although these reflexes have their specific roles, they are all involved in the control of motility, peristalsis, and local blood flow in the GI tract. The reflexes also function as nociceptors that initiate tissue-protective propulsive and secretory reflexes to rid the gut of pathogens [118, 235, 240]. The mechanism for the initiation of these reflexes involves release of neuropeptides acting as mediators—serotonin, neurotensin, cholecystokinin (CCK), gastrin, tachykinins, nitric oxide (NO), vasoactive intestinal peptide (VIP), pituitary adenylyl cyclase-activating peptide (PACAP), ATP [118, 235, 240] as well as multiple ion channels and ion channel regulators. The strength of enteric reflexes depends on regulation of excitability of intrinsic primary afferent neurons [235].

A few of the GI tract reflexes will be considered here. So, in cologastric inhibitory reflex, passage of feces into the colon inhibits the motor activity of the stomach. This reflex may be triggered by inappropriate colonic emptying [241].

In enteroenteric inhibitory reflexes, the afferent signal is transferred from intestinofugal neurons through the prevertebral ganglion to the spinal cord [235].

In vago-vagal reflex, gut stimuli such as mechanical distension caused by food ingestion can initiate coordinated responses by afferent and efferent nerves of the vagus via the dorsal motor nucleus and nucleus ambiguus. The vago-vagal reflex is responsible for controlling gastric smooth muscle contraction in response to food intake which is necessary for controlling intragastric pressure. The vago-vagal reflex is active during the receptive relaxation of the stomach in response to food swallowing when the stretch or mechanoreceptor are activated resulting in stimulation of the afferent fibers of the vagus nerve. The activated afferent fibers of the vagus send information to the dorsal vagus complex. This vagal complex carries out information processing and relays the command via the efferent fibers of the vagus nerve to postganglionic cholinergic nerves innervating the motor endplates leading to relaxation of the stomach muscle and secretory effect. The preganglionic neurons from the dorsal motor nucleus or nucleus ambiguus implement parasympathetic motor and secretory functions. The preganglionic cholinergic neurons form parallel inhibitory and excitatory vagal pathways to smooth muscle viscera and stimulate postganglionic neurons via nicotinic and muscarinic receptors. Excitation of the postganglionic inhibitory neurons leads to the release of adenosine triphosphate, vasoactive intestinal polypeptide, and nitric oxide—which are all inhibitory neurotransmitters. The stimulation of excitatory neurons leads to the release of such excitatory neurotransmitters as acetylcholine (Ach) and substance P [242–247]. The associated effects of vagus motor nerve stimulation depend on the interaction resulting from the afferents and efferents of the vagus as well as smooth muscle activity. The motor effects include esophageal peristalsis, gastric motility, lower esophageal sphincter, and pyloric sphincter, whereas the secretory effects through the vagal secretory pathway involve release of substances into the lumen of the GI tract, secretion of hormones or neuropeptides. Associated effects of this stimulation include release of gastric HCl, release of histamine from enterochromaffin cells, release of gastrin-releasing peptide from peptidergic neurons, reduced somatostatin secretion from inhibition of delta cells [242, 245].

The gastroileal reflex involves an increase in motility of the colon in response to stretch in the stomach and by-products of digestion in the small intestine. The reflex is important in peristalsis, ileocecal sphincter relaxation, initiated by gastric emptying and secretion. This reflex is responsible for the urge to defecate following a meal. It is coordinated with gastrocolic and duodenocolic reflexes to stimulate defecation. Immediately after a meal, pressure begins to build-up in the rectum, which initiates a reflex action that stimulates defecation. The reflex also provides opportunity for motility of the content of the stomach and helps create space for more food [248].

Dysfunction of GI reflexes has been implicated in a row of gut disorders including irritable bowel syndrome, heightened visceral sensitivity, abdominal pain, diarrhea, and constipation [240].

9.2.6 Gastrointestinal Motility Responses to Stimuli

The two types of motility responses (phasic and tonic) vary across the GI tract. A tonic response is associated with repeated firing as long as the stimulus persists. Tonic sensory input adapts slowly to a stimulus and continues to produce action potentials over the duration of the stimulus (Fig. 9.12). In this way, it conveys information about the duration of the stimulus. Thus, tonic response is sustained over time. The type of response produced by Ruffini corpuscle (a slowly adapting mechanoreceptor of the skin) is a classic example of a tonic response [20, 248]. However, GI smooth muscle cells and some neurons also produce tonic response—characterized by rapid adaptability to stimulus in which the response fades away quickly. Phasic response on the other hand is transient (Fig. 9.12). It conveys information on changes in stimulus intensity and rate. Example of phasic response is the type of response produced by mechanoreceptors [248]. Both phasic and tonic responses are characteristic responses in GI smooth muscle and neurons. Neural centers in the CNS or ganglia interpret the combinations of tonic and phasic signals. Both types of responses vary across different regions of the GI tract and types of cells and density of associated receptors. In gastrocolic reflex, for instance, phasic response is greater in the sigmoid colon compared to other parts of the colon, while the tonic response may vary across the entire colon [118, 249–252].

9.2.7 Mechanosensitive Responses of Enteric Neurons to Stimulation

On the basis of speed of adaptation, enteric neuron responses to stimulation can be classified as—rapidly adapting, slowly adapting, ultraslowly adapting, and probably non-adapting (Fig. 9.12). These responses are exhibited by different populations

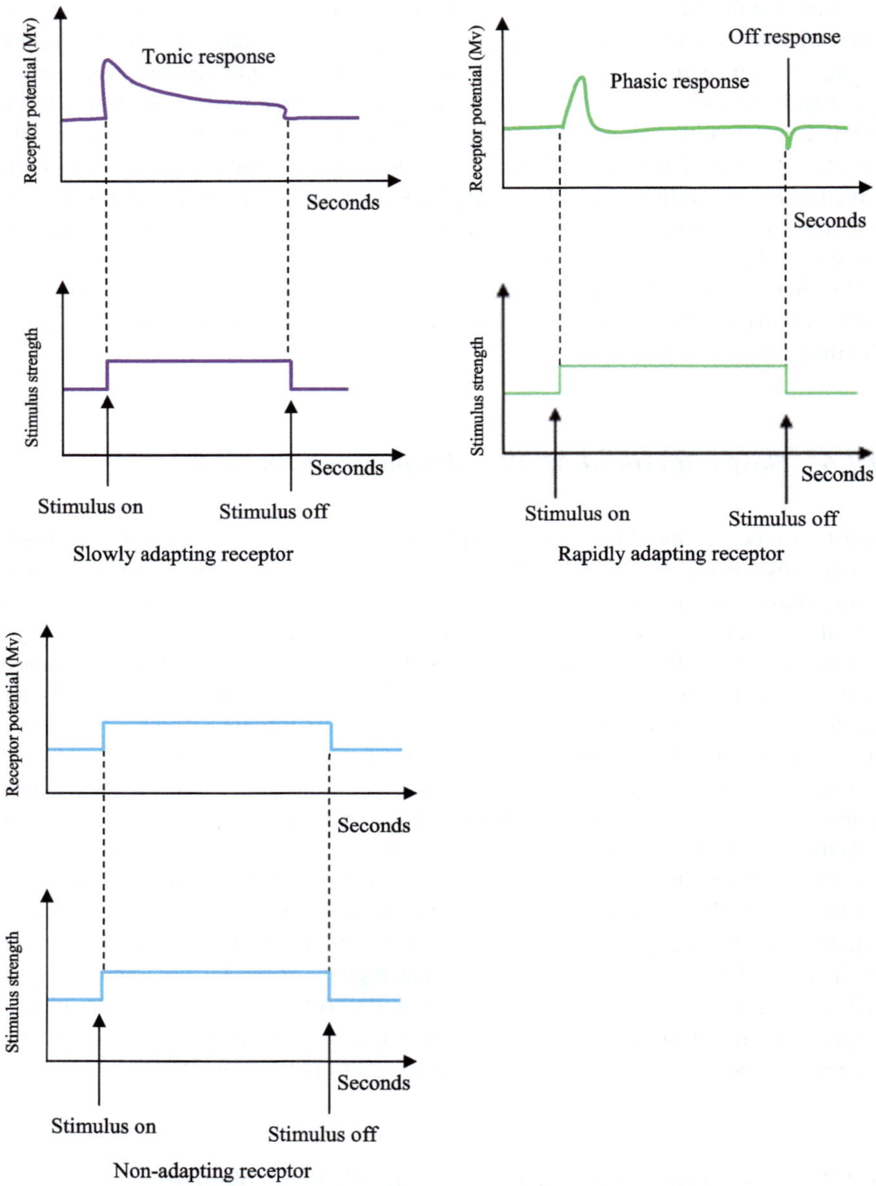

Fig. 9.12 Schematic representation of tonic and phasic and other types of receptor responses to the action of a stimulus

of mechanosensitive neurons of the gut. These neurons can trigger reflex action upon mechanical deformation (stretch) of the gut wall. This unique property of these neurons is due to the presence in their plasma membrane mechanoreceptors or channels. Approximately 60% of the entire population of enteric neurons is sensitive to stretch. Some mechanosensitive neurons of the gut can function as interneurons. These neurons participate in sensing and controlling muscle activity and also serve as feedback regulators of muscle activity [253–255]. Further details are discussed in Sect. 9.3.4.2.

It should be mentioned that the receptors of these cells exhibit differential sensitivity (i.e., threshold of stimulation) to stimulation. On the basis of sensitivity, the receptors of these cells can be classified as low-threshold (e.g., mechanoreceptors), and high-threshold (e.g., nociceptors) [253–255].

9.3 Intrinsic (Enteric) Nervous System

9.3.1 Enteric Nervous System and its Anatomo-functional Characteristics

The intrinsic nervous system of the gut (or ENS) refers to the entire system of enteric glial cells and enteric nerve cell bodies of sensory, interneurons, and motor neurons grouped into ganglia and interconnected by bundles of nerve processes forming plexuses, controlling the GI system activities, and is capable of functioning without influence from the CNS (brain or spinal cord) and does not depend on innervation from the autonomic nervous system through the vagus nerve or prevertebral ganglia. The enteric neurons play a major role in regulating motility and secretions of the gut. Like the CNS, the neurons of the ENS communicate with each other via the secretions of neuromediators. The branch of science that is concerned with this aspect of gut functioning is a fairly new area called neurogastroenterology and was born in the year 1902 on the basis of the groundbreaking work of Bayliss and Starling when they observed movement of an isolated intestinal segment of a dog. Interestingly, when this segment of the intestine (the nerves connecting the gut and the CNS) was deinnervated, the scientists still observed the movement of the intestinal segment on stimulation. Bayliss and Starling attributed this phenomenon to the presence of a local nervous mechanism. The phenomenon was referred to as "the law of the intestine." This law of the intestine or gut is now called the peristaltic reflex [15, 256].

The ENS is widely distributed on the walls of the esophagus, stomach, small intestine, and colon [14]. The estimated number of neurons in these regions of the human gut is estimated to be about 5×10^8 [257]. Thus, the neurons of the gut wall are more than 100 million. The ENS controls motility, exocrine (mucosal and submucosal secretion) and endocrine secretion, absorption, growth, local blood flow (microcirculation), GI sensation, immune and inflammatory processes in the gut [14, 15, 258, 259].

The brain and ENS system seem to have numerous roles in common. Like the CNS, the ENS consists of ganglia, primary interganglionic fiber tracts, and secondary and tertiary fibers projecting to the effector systems (muscle cells, glands, blood vessels, and immune cells). The ENS comprises many types of functional neurons, which include sensory neurons, interneurons, muscle motor neurons, and secretomotor neurons. Like the CNS, the ENS contains pacemaker cells, which are capable of generating impulses without external input. That is the primary reason why the gut system of neurons is able to function independently without external influence. Hence, the ENS is sometimes referred to as the little brain of the gut or minibrain or second brain or gut–brain [15, 260, 261].

The enteric nervous system is important not only in physiological regulation of GI activities, but also in development of a couple of neuropathologies including Hirschsprung's disease, Chagas (acquired), diabetic gastroparesis, drug-induced (opioid bowel syndrome), postoperative ileus, Alzheimer's disease, and Parkinson's disease. More importantly, emerging studies have indicated that numerous diseases involving the CNS show cellular changes in the enteric neurons earlier before their appearance in CNS. Thus, the ENS represents a crucial division of the nervous system, mediating multiple regulatory functions at the peripheral and central levels [256, 258, 259, 262, 263].

9.3.2 Types of Cells in the Enteric Nervous System

Neurons and glial cells are the major types of cells that make up the ENS. There are different types of glial cells and neurons of the ENS distributed mainly in two intramuscular plexuses that extend along the entire length of the gut [14, 256].

Enteric Glial Cells

The glial cells are key components of the ENS and are found in the intrinsic ganglia of the intestinal plexuses and the network of neural fibers connecting the ganglia of the digestive system [264–266]. The enteric glial cells are also found outside the plexuses, within the circular muscle and in the lamina propria of the mucosa. These glial cells of the ENS outnumber the enteric neurons by about fourfold (in the CNS, the glial cells outnumber neurons by about tenfold) [264]. Although they have similar functions, their locations, at least in part, determine their functions. Thus, glial cells of the enteric ganglia are mainly responsible for neurogliatransmission [267]. The glial cells of the neural fibers connecting the ganglia not only support but also feed and protect the surrounding neurons [261, 265]. The neuroprotective role of glial cells is, in part, due to the release of glutathione and other functionally related substances [268]. The enteric glial cells can inhibit the proliferation of intestinal epithelial cells and reduce intestinal epithelial paracellular permeability. Interestingly, ablation of enteric glial cells has been shown not only to increase paracellular permeability, but also intestinal epithelial cell proliferation and reduce gastric emptying as well as intestinal transit [268]. Specifically, subepithelial glia cells play a trophic and supportive role in the intestinal epithelial cells and also maintain the integrity of the intestinal epithelial barrier. More interestingly, enteric glia can form new neurons. However, this phenomenon may be a rare one [265]. It is believed that enteric gliogenesis is more pronounced in conditions of gut injury, inflammation, or other GI disorders, whereas enteric neurogenesis especially in the adult may be minimal. The enteric glial cells form sheaths around the cells bodies and axons of enteric neurons. Both cell types play a role in gut homeostasis, coordinated smooth muscle activity, mucosal secretion, immune response and inflammatory processes of the GI tract [264, 269]. They are also involved in regulation of blood flow, uptake of nutrients, and secretion in the gut [264]. Dysfunction of ENS has been implicated in the etiopathogenesis of necrotizing enterocolitis, inflammatory bowel disease (e.g., Crohn's disease), slow transit constipation, postoperative ileus, idiopathic megacolon, colonic diverticular disease, and other diseases such as Parkinson's disease, and obesity. The reduction in the number of enteric glia is one of the main causes of these diseases. A major mechanism for the involvement of enteric glial cells in some of these pathological conditions is their ability to secrete cytokines (interleukins, chemokines, etc) in response to inflammatory reactions. As a result, enteric glial cells can attract immune cells to the ENS and functionally interact with lymphocytes to mediate a range of physiological processes. Thus, enteric glial cells may function as antigen-presenting cells [260, 261, 270, 271]. Whether or not different types or subtypes of enteric glial cells exist is not exactly clear. The enteric glial cells are similar to astroglia (Fig. 9.13) [260, 261]. The corresponding cell types in other regions of the PNS are called Schwann cells [260, 261].

Fig. 9.13 Glial cell (astroglia). It is believed that the glial cells of the ENS are similar to the astroglia of the CNS. Apart from the enteric glial cells, known glial cells of the PNS include Schwann cells and satellite (or capsule) cells. Schwann cells form myelin in the PNS. Satellite cells surround cell bodies of neurons in sensory and autonomic ganglia and regulate the exchange of materials between somata of neurons and interstitial fluid. In contrast to the glial cells of other regions of the PNS, enteric glial cells do not form basal laminae and they ensheathe axons of enteric neurons in groups [272–280]. In general, glial cells are known to regulate neuronal survival and differentiation and maintain ion and neurotransmitter homeostasis [281, 282]

Enteric Neurons

The complexity in the arrangement of enteric (or intrinsic) neurons, in contrast to other peripheral organs, confers the GI tract a unique ability to mediate intrinsic reflexes [15, 260, 261]. Figure 9.14 is a schematic representation of a typical enteric neuron.

The neurons of the ENS can be classified on the basis of a number of criteria, which include functional, morphological, biophysical, immunological, neurochemical [15, 286–289]. Structurally, neurons may be referred to as unipolar, bipolar, multipolar, and pseudounipolar. A unipolar neuron consists of one continuous process that branches off the cell body and includes both the axon and dendrite. A bipolar neuron consists of two separate processes that branch off the cell body, the dendrite and the axon. The multipolar neuron has many processes that branch off the cell body. A pseudounipolar neuron is a type of sensory neuron in the PNS, comprising of a single axon that divides into two branches. One of the branches of the axon runs to the periphery and the other to the spinal cord. The soma of the pseudounipolar neuron is located in the dorsal root ganglion [290–293].

Morphological Classification of Enteric Neurons

Morphologically, enteric neurons are classified into Dogiel type I–VII and the giant neurons. The majority of enteric neurons are types I–III. This morphological classification of enteric neurons is based on staining techniques developed by A. S. Dogiel [15, 294].

Electrophysiological Classification of Enteric Neurons

Electrophysiological studies have indicated that the intrinsic neurons of the GI tract differentially respond to excitation and inhibition. Thus, neurons can be classified as inhibitory or excitatory. Electrophysiological experiments also indicate that enteric

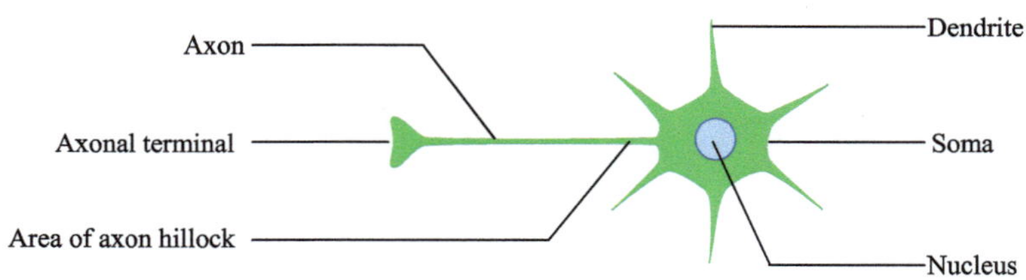

Fig. 9.14 Schematic representation of a typical enteric neuron. Cell body or soma contains the nucleus and organelles that are responsible for the production machinery of the peptides and other substances. The dendrites and the spiny parts receive impulses and other substances from adjacent cells and also carry them to the cell body. The axons carry impulses away from cell body. The axon hillock is the initial region of axon that braches off the cell body and is referred to as the "action potential trigger zone." It is believed that the neurons of the ENS are similar to the ones located in the CNS. They are responsible for transmission of impulse and processing of signals [283–285]

neurons can be classified as type S ("S" for synaptic) and type AHP ("AHP" for afterhyperpolarisation, or abbreviated "AH") neurons. The synapse-type neurons are those that form connections with neighboring neurons or other adjacent cells. The types of connections (synapses) formed include but are not limited to axoaxonal, axosomatic, and axodendritic synapses [15, 240, 295]. In a study of synapses of the myenteric ganglia, Pompolo and Furness (1990) noted that about 40–50% of synapses were dendritic, 20–25% were somatic, and 30–35% were formed on the axon hillock or first 50–70 μm of the axon [296]. The events occurring in the synapse determine the physiological outcome of the activities related to the cells that constitute the synaptic connections. The electrophysiological events occurring at these synapses include fast excitatory postsynaptic potential (fast EPSP), slow excitatory postsynaptic potential, inhibitory postsynaptic potential (IPSP), presynaptic inhibition, and presynaptic facilitation, etc. [15, 296–298].

The S-type neurons exhibit EPSP. Examples of S-type neurons are some interneurons and motor neurons. The AHP-type neurons are intrinsic sensory (or primary afferent) neurons which exhibit action potential that is usually followed by a prolonged hyperpolarization [15, 240, 295, 299]. The development of hyperpolarization may result from activation of inhibitory motor neurons to the GI muscles layer by electrical stimulation or physiological reflexes. This results in smooth muscle relaxation. The potential that leads to this hyperpolarization is called inhibitory junctional potential. The inhibitory junctional potential is the hyperpolarization of smooth muscle produced by activation of inhibitory neurons [15, 240, 300–303]. The inhibitory or excitatory processes are due to the type of neurotransmitter acting on the cognate receptor at the postsynaptic terminal. Inhibitory postsynaptic potential may result from the action of adenosine, ATP, serotonin or 5-hydroxytryptamine (5-HT), norepinephrine, opioids, and somatostatin. The excitatory postsynaptic potential may be due to the activities of ACh, CCK, calcitonin gene-related peptide (CGRP), 5-HT, histamine, PACAP. Notice that more than one mediator may mediate synaptic inhibitory or excitatory processes. This is due to the activation of different receptor subtypes by the same mediator [15, 240].

Functional Classification of Enteric Neurons

The ENS comprises numerous types of neurons categorized as sensory neurons (intrinsic primary afferent neurons), motor neurons, and interneurons. Sensory neurons are the largest population of neurons in the ENS. Interneurons and motor neurons occur in equal proportion [15, 304, 305].

Sensory neurons: These are specialized neurons that transform stimulus into signals that are coded as action potentials [15]. There are different types of sensory neurons. Brookes et al. (2013) [41] grouped visceral sensory neurons into five types on the basis of their sensory endings in the gut wall: intraganglionic sensory neuron, mucosal sensory neuron, muscular-mucosal (muscularis mucosae) sensory neuron, intramuscular (smooth muscle) sensory neuron, and vascular sensory neuron. Brookes et al. (2013) [41] also identified a new group of neurons known as "silent sensory neuron," which is non-excitable under normal condition, but becomes awaken or activated by inflammatory mediators (e.g., interleukins) in conditions of inflammation [41]. A widely recognized classification is based on the divisions of the nervous system regulating the activities of the gut. So, the sensory neurons of the gut include the extrinsic (vagal and spinal afferents, which have their cell bodies located outside the gut wall) and intrinsic primary afferent neurons. The intrinsic primary afferent neurons have their cell bodies located within the gut wall [306–311]. The intrinsic primary afferent neurons receive thermal, chemical, osmotic, and mechanical information from the sensory receptors in the mucosa, muscle, vascular wall. However, these neurons can detect these stimuli themselves. Therefore, the neurons may be further divided into neurons that serve as thermo-, chemo-, osmo-, mechanoreceptors. These receptors are present in high density and respond to stimuli that activate assemblies of hundreds or thousands of intrinsic primary afferent neurons. The receptors can also detect muscle tension and stretch of the GI wall and deliver the stimulus energy to the sensory neurons. Thus, these neurons play essential roles in regulating digestion, absorption, and transport. The intrinsic primary afferent neurons make a neural circuitry by connecting with motor neurons and with ascending and descending interneurons. The extrinsic afferents that connect with the intrinsic sensory neurons notify the brain about processes that are relevant to energy and fluid homeostasis, sensation of discomfort and pain, and other physiological and pathological conditions [15, 309, 312–314].

Interneurons: These neurons mediate information transmission between two neurons, thus forming interconnecting chains in the neural circuit. The interneurons, also known as relay neurons, are responsible for processing or modulating sensory or motor information and regulate the behavior of efferent neurons that control the effector organs. The interneurons primarily convey signals from the intrinsic primary afferent neurons to the motor neurons and also synapse with other neuron types. The interneurons can be classified on the basis of the direction in which they mediate signaling: descending-type interneurons and ascending-type interneurons. The majority of interneurons are the descending type. The activity of the interneurons is determined by the type of neurotransmitter released into the synaptic cleft. The ascending interneurons are mainly cholinergic. The descending interneurons transmit several mediators including ACh, nitric oxide, vasoactive intestinal polypeptide, 5-HT, and somatostatin. The neuromediators ACh and 5-HT are involved in local secretomotor reflexes, whereas the rest are involved in local motility reflexes. The interneurons may be orally directed or aborally directed. In electrophysiological terms, interneuron may be of the afterhyperpolarization type or synaptic type. Morphologically, interneurons are usually Dogiel type II neurons [15, 240]. An interneuron may be excitatory or inhibitory, determined by the predominant type of neurotransmitter that is secreted (Fig. 9.15). An interneuron can be a bipolar, unipolar, or multipolar neuron [315–318].

Motor Neurons: These neurons are responsible for the responses (e.g., motility, secretion, and absorption) of the GI tract. There are three main types of motor neurons present in the gut: muscle motor neurons, secretomotor neurons, and vasodilator neurons [240]. The **muscle motor neurons** innervate the smooth muscle

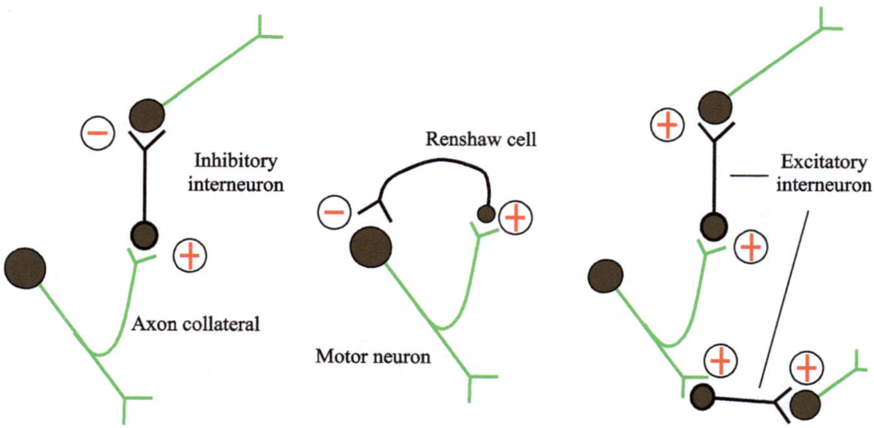

Fig. 9.15 Signaling mediated by inhibitory interneuron. The inhibitory neurons (including Renshaw cell) through the action of glycine, the most abundant inhibitory neurotransmitter in the central nervous system, inhibit the neural firing of motor neurons of the central and peripheral as well as enteric nervous systems. GABA and many other transmitters inhibit neural transmission of information. Renshaw cell named after Birdsey Renshaw (1911–1948), an American neurologist who conducted extensive investigation on this nerve cell of the spinal cord. Renshaw cells are inhibitory interneurons found in the gray matter of the spinal cord and are associated in two ways with an alpha motor neuron [319–322]

layers including the muscularis mucosae of the GI tract. These neurons may be excitatory or inhibitory—secrete mediators that either cause muscle contraction or relaxation, respectively. The excitatory neurons are predominantly muscarinic cholinergic and tachykinergic, whereas the inhibitory neurons are primarily nitriergic (nitric oxide), VIPergic (vasoactive intestinal polypeptide), peptidergic (ATP), GABAergic (gamma-aminobutyric acid), NPYergic (neuropeptide Y), COergic (carbon monoxide), and possibly PACAPergic (pituitary adenylate cyclase-activating polypeptide). The **secretomotor neurons** regulate the secretory activity of the GI tract. The secretomotor neurons are located in different regions of the gut. They are primarily found just beneath the mucosal epithelium. Others are found in the submucosa and muscularis mucosae. It should be noted that some secretomotor neurons actually project to the mucosa, especially the crypts to stimulate secretion. Secretomotor neurons innervate the mucosa. These neurons can be classified according to different criteria. On the basis of their neurochemical properties, secretomotor neurons can be classified as cholinergic and non-cholinergic. ACh released from a cholinergic secretomotor neuron acts on muscarinic receptors, to mediate secretory activity of the gut. The non-cholinergic (e.g., VIP) neurons mediate most of the local reflexes that regulate gut secretion [20, 240]. The **vasomotor (vasodilator) neurons** have their cell bodies in the submucous layer and control blood flow. The most potent vasodilators released by vasomotor neurons include NO, CO, etc. [323, 324]. However, CGRP is currently known to be most potent vasodilator. It is by far more active vasodilator neurotransmitter compared to NO and CO (see Table 9.2 for details).

Table 9.2 Structural-functional characteristics and course of discovery of GI neurotransmitters (neuromediators)

Neural secretions of the gut	Description	Discoverer	Year of discovery
Gastrin-releasing peptide, GRP, the mammalian homologue of bombesin from the amphibian Bombina bombina. The name "GRP" was derived from observation that the first known activity of the peptide-induced gastrin secretion in porcine gastric tissue [471]	GRP is a heptacosapeptide (27 amino acid neuropeptide) that is currently known to play a considerable role in normal physiological and pathophysiological processes [471] The production of GRP is regulated by the GRP gene. The human GRP gene is located on chromosome 18 [472] GRP is produced from the postganglionic fibers of the vagus nerve that innervate gastric G (gastrin-secreting) cells. The amino acid sequence of this peptide from porcine non-antral gastric and intestinal tissue was identified as Ala-Pro-Val-Ser-Val-Gly-Gly-Gly-Thr-Val-Leu-Ala-Lys-Met-Tyr-Pro-Arg-Gly-Asn-His-Trp-Ala-Val-Gly-His-Leu-Met-NH2. The peptide is homologous to the C-terminal region as seen in bombesin. Thus, the two peptides have the same biological activities [473]. All forms of GRP are derived from the precursor proGRP1-125. Proteolytic cleavage of this precursor produces N-terminal-derived GRP1-27 and GRP18-27. GRP is the human form of bombesin [475] The actions of GRP are mediated by the GRP receptors (GRPR), which include GRPR, neuromedin B receptor, and bombesin receptor subtype 3 (BB-3 or BRS-3). Like GRPR and neuromedin B receptor, BRS3 is a G protein-coupled receptor that signals via a Gq protein to increase intracellular calcium. Although BRS-3 has low affinity for bombesin, many synthetic high-affinity ligands have been created for this receptor. The selective agonist for BRS-3 includes Bag-1, Bag-2. An example of BRS-3 antagonist is Bantag-1 [476]. The precise endogenous (native) ligands for the receptor have not been completely identified. The orphan G protein-coupled receptor, BRS-3, has long been implicated in the regulation of energy homeostasis [477]. In 2011, Guan et al. (2011) successfully synthesized a BRS-3 agonist, MK-5046: (2S)-1,1,1-trifluoro-2-[4-(1H-pyrazol-1-yl)phenyl]-3-(4-{[1-(trifluoromethyl)cyclopropyl]methyl}-1H-imidazol-2-yl)propan-2-ol that significantly inhibited food intake in mice and led to persistent weight loss suggesting that the compound was a promising agent for the treatment of obesity [477]. BRS-3 regulates not only body weight, but also glucose homeostasis via islet-independent pathway [478] GRP stimulates gastrin secretion from the G cells of the GI tract and thus regulates gastric acid secretion and enteric motor function. It also regulates pancreatic enzyme output, gallbladder contraction, and plasma hormone release. These functions of GRP are equipotent to bombesin's functions [479]. GRP is also involved in brain functioning, maintenance of circadian rhythm, in spinal transmission of the itch sensation, inflammation and wound repair [475]. GRP increases plasma levels of gastrin, pancreatic polypeptide, glucagon, gastric inhibitory peptide, and insulin [480]	Thomas J. Mcdonald and co-researchers discovered the peptide in gastric tissue [473, 474]	1978–9
Acetylcholine (Ach)	Ach is the primary transmitter of parasympathetic preganglionic neurons, enteric interneurons, certain secretomotor neurons in the intestine and motor neurons controlling gastric acid secretion. Ach is an organic molecule, an ester of acetic acid and choline, with chemical formula CH3COO(CH2)2N$^+$(CH3)3 or C7H16NO^{2+} (systematic name: 2-acetoxy-*N*,*N*,*N*-trimethylethanaminium) [481–484] The functions of Ach are achieved through its action on the cognate receptors. There are two main classes of acetylcholine receptors (AChR)—nicotinic Ach receptors (nAChR) and muscarinic Ach receptors (mAChR). These receptors are named after the ligands,	Henry Hallett Dale (1875–1968) and Otto Loewi (1873–1961). They won the 1936 Nobel Prize in Physiology or Medicine "for their discoveries relating to	1913

(continued)

Table 9.2 (continued)

Neural secretions of the gut	Description	Discoverer	Year of discovery
	for which they are sensitive to. The nAChR are ionotropic receptors classically activated by nicotine. There are two types of nAChR: muscle type and neuronal type. The former can be selectively blocked by curare and the latter by hexamethonium. The nAChRs have four different subunits (α1-7, α9-10, β1-4, γ-, and δ-subunits, as well as additional ε subunit), which combine in various compositions (depending on the cell type) to form a functional homopentamer or heteropentamer receptor [485–492]. These receptors are expressed where there is need for fast excitatory transmission of signal [240] In the GI system, Ach is a common mediator, produced in the ENS, to induce GI smooth muscle contraction. On the surface of smooth muscle cells of the gut are muscarinic Ach receptors [494]. In animal model, it was shown that both nicotinic and muscarinic Ach receptors are expressed in the GI tract, including the intestinal mucosa [495, 496] nAChRs are usually expressed on muscle endplate, in autonomic ganglia, enteric nervous system, and CNS. nAChRs are also localized in adrenal medulla where they stimulate release of catecholamine and other transmitters. The most abundant type expressed in neurons is α4β2, followed by α7 nAChR [497] Several receptor subunits and homomeric heteromeric (α3, α5, α7, β2, and β4), similar to those found in the PNS, have been detected in the enteric nervous system. The PNS neurons mainly express α3(α5)β4, followed by α3β2 nAChRs [499] All receptor types of nAChR are linked to Na^+ and K^+ channels, but neurons have an additional channel for the translocation of Ca^{2+} ions. Thus is nAChR is a non-selective cation channel. For further review, see Carignano et al. (2016) and McKay et al. (2007). In skeletal muscle, for instance, activation of nACh receptor (α1β1γδ) by Ach leads to the opening of ligand-gated sodium channels of the plasma membrane. The result is increased influx of sodium into the muscle cell, which leads to muscle contraction. In neurons, however, activation of nAChR leads to the inhibition of K^+ channel activity. The result is increase in excitability [485–492] Five different subtypes of muscarinic receptor subtypes ($M_{1–5}$) have been identified and characterized. All muscarinic receptors are G protein-coupled (metabotropic) receptors [502] The mAChR ($M_{1–5}$) are expressed in both the CNS and PNS of the heart, lungs, upper GI tract, and sweat glands and are activated by muscarine and acetylcholine as their classical ligands. The name "muscarinic Ach receptor" is derived from the fact that these receptors were first discovered to be activated by a toxin from the mushroom Amanita muscaria, called muscarine, and inhibited by a toxin from Atropa belladonna, called atropine. These toxins were discovered on the basis of their influences on postganglionic parasympathetic nervous system functions. The receptors $M_{1–3}$ subtypes are activated by muscarine, methacholine, oxotremorine, bethanechol, and pilocarpine. Pharmacological agents that act both on muscular and neuronal type Ach receptors and $M_{1–3}$ include carbachol, acetylcholinesterase inhibitors (physostigmine, galantamine, neostigmine, pyridostigmine, donepezil, edrophonium, echothiophate and isofluorphate) [485, 503, 504]	chemical transmission of nerve impulses." Both met in 1902 when Loewi paid a short visit to Starling's laboratory at the University College, London [493] Dale discovered naturally occurring Ach in 1913, and showed its role as a neurotransmitter at autonomic ganglia, postganglionic parasympathetic nerve terminals and the neuromuscular junction [493, 498] From 1905 to 1921, Loewi actively investigated the role of naturally occurring chemicals in nerve transmission. In one of his reports, he described a novel parasympathetic substance, which is named "Vagusstoff" (acetylcholine). He realized that this substance was closely related to adrenaline and played a corresponding	

(continued)

Table 9.2 (continued)

Neural secretions of the gut	Description	Discoverer	Year of discovery
	Activation of M_1 receptor-linked Gq-type G protein alpha subunits induces calcium-release from intracellular stores, which then activate a calcium-activated potassium channel that results in the inhibition of the firing rate of associated neurons. In the autonomic nervous system, Ach is released from pre- and postganglionic parasympathetic neurons as well as from preganglionic sympathetic neurons. The preganglionic sympathetic neuron releases Ach, which in turn inhibits the adrenal medulla (comprises modified postganglionic neurons) release of norepinephrine [505, 511] Of the five receptor subtypes, only M_2 and M_3 have been shown to be expressed in gut smooth muscles, in a ratio of about 3–5:1. This means that gut muscarinic Ach signaling is largely mediated by M_2, whereas M_3 has little role to play. Research has shown that intramuscular and myenteric interstitial cells of Cajal (the pacemaker cells) also express M_2 receptor [494]. M_2 receptor via pertussis toxin (PTX)-sensitive mechanism induces intestinal muscle contraction that depends on voltage-dependent Ca^{2+} entry into the cell [494, 512]	role at the sympathetic nerve endings (Sourkes 2009). This experiment was the result of the dream he had the previous night, so he decided to reproduce the experiment (Raju 1999; Lembeck 1973)	
	Reference Note 9.3 **Pertussis Toxin (PTX) and Cell Signaling** PTX is a protein exotoxin produced by the gram-negative bacterium Bordetella pertussis, which causes whooping cough or pertussis. Apart from PTX, Bordetella pertussis also produces adenylate cyclase toxin (ACT), another important virulence factor of the bacteria that cause the respiratory disease. PTX catalyzes the ADP-ribosylation of the αi-subunit of the heterotrimeric G protein. This prevents the G protein from interacting with G protein-coupled receptors—a necessary step in transmission of cellular signals. The inhibition of adenylate cyclase is lost due to the maintained inactive state of the G protein in which Gi subunit remains locked and bound to GDP. The result is increase in cellular concentration of cAMP, which in turn has several cellular influences. ACT induces the formation of transient cation-selective pores in cell membrane, which causes hemolysis in erythrocytes. The effect of ACT on the cell occurs by many mechanisms. ACT can bind to a specific cell-surface receptor (such as αMβ2 integrin or CR3, which is also called Mac-1 or CD11b/CD18) by interacting with glycosyl residues on the integrin receptor. One key mechanism of intracellular activity of ACT is the binding of the adenylate cyclase domain to the cytosolic protein, calmodulin, which catalyzes cAMP synthesis from cellular ATP to raise the level of the second messenger. Excessive increase in cAMP disrupts cellular signaling. PTX and ACT provide media for studying cellular communication mechanisms involving the G protein-coupled receptor in normal and pathology [513–516] M_2 receptor-coupled PTX-sensitive G proteins (Gi/o) mediate the inhibition of adenylyl cyclase. Go-mediated activation of smooth muscle leads to activation of cationic channels, which open upon activation. Binding of M_2 receptor agonist activates the Gi/o family of G protein-coupled receptor, resulting in dissociation of the constituent subunits. In particular, the dissociated Gβγ subunit then activates the G protein-gated potassium channels (inwardly rectifying K^+ channels), which leads to hyperpolarization of the plasma membrane. Activation of the M_2 receptor-Gi/Go-cationic channels reverses the action of adenylyl cyclase activators such as isoprenaline or isoproterenol (stimulates β1 and β2 adrenergic receptors that are structurally similar to adrenaline) or forskolin (a diterpenoid obtained from the plant Coleus forskohlii and stimulates adenylyl cyclase resulting in elevation of cAMP level in the cytosol to increase cellular activity). The less-studied M_4 has similar mechanism of action with the M_2 receptor [516–523]		

(continued)

Table 9.2 (continued)

Neural secretions of the gut	Description	Discoverer	Year of discovery
	M_3 receptor is coupled preferentially to Gq/11-type G protein. Upon binding of the appropriate ligand–agonist, the Gq protein is stimulated resulting in the activation of phospholipase C (PLC) and the formation of inositol trisphosphate (IP3 or InsP3) and diacylglycerol (DAG). InsP3 and DAG through many cellular mechanisms increase the contractile activity of smooth muscle cells. DAG activates protein kinase C, which phosphorylates cellular proteins and can directly activate non-selective cationic channels. InsP3 causes Ca^{2+} release from intracellular stores and can also mobilize Ca^{2+} release via Ca^{2+}-sensitive or Ca^{2+}-store-dependent mechanisms [494]. Intracellular Ca^{2+} increases and several channels are activated, whereas others are inhibited. Inhibited channel includes the voltage-dependent calcium channel. Activated channels such as calcium-activated potassium channels open to cause depolarization of the membrane and so increase action potential. Membrane depolarization causes the opening of certain ion channels that are permeable to Na^+ and K^+ but not to Cl^-. The result is contraction of smooth muscles [512, 524]. The signaling mechanism of M_3 is similar to M_1 and M_5 receptors. M_3 signals through Gq proteins to induce intestinal smooth muscle contraction. This muscarinic-sensitive receptor can function by activating both potential-dependent and -independent Ca^{2+} stores [525] M_4 receptor is coupled to Gi/o, so activation of this receptor leads to the opening of inwardly rectifying potassium channel. The result is inhibition of signal transmission [517, 518, 526] Like the M_1 and M_3, M_5 receptor is primarily coupled to Gq/11 proteins. Thus, the activation of this receptor will signal to activate the IP3 pathway [517, 518, 527]. Importantly, the tissue expression of M_5 receptor is different from the other receptor types. The M_5 receptor is abundantly expressed in the cortex, hippocampus, basal ganglia, nucleus accumbens, and olfactory tubercle. This receptor is involved in modulating the activities of dopaminergic neurons and thus plays a crucial role in reward, punishment, and memory [528] Activation of muscarinic AChRs is relatively slow (milliseconds to seconds) compared with ionotropic cationic nicotinic receptor channel (nAChR), which occurs very rapidly [485] The release of Ach from the cytosol into the synaptic cleft or extracellular space leads to activation of the respective receptor subtypes depending on the tissue. Used Ach are hydrolyzed or degraded by the enzyme, acetylcholinesterase, also known as acetylhydrolase. The enzyme is found primarily at neuronal synapses and neuromuscular junctions, where its activity serves to terminate synaptic transmission. One enzyme can degrade as many as 25,000 molecules of ACh per second [529] Synthesis of Ach occurs by the enzyme choline acetyltransferase (CAT). CAT is produced in the cholinergic neuron cell body and transported down the axon to the nerve endings where it catalyzes the production of Ach from choline and acetyl-CoA. The latter is produced from the mitochondria. Choline is acquired via uptake from the synaptic cleft. Dietary choline increases Ach neurotransmission. Choline can be produced via biochemical breakdown of phospholipid or phosphatidylcholine. Synthesized Ach is stored in nerve endings in vesicles, measuring about 100 μm diameter. These vesicles are protected from premature degradation by cytosolic enzymes. Uptake and reuptake of Ach occurs through ion pumps that acidifies the vesicle. Acidified vesicle uses a vesicular ACh transporter (VAChT) to exchange 2 protons for one ACh molecule. Vesicle-bound ACh is not accessible to degradation by acetylcholinesterase. So, it is not prematurely degraded before exocytosis [529–534]		

(continued)

Table 9.2 (continued)

Neural secretions of the gut	Description	Discoverer	Year of discovery
Adenosine triphosphate, nucleoside triphosphate, adenosine 5′-triphosphate (ATP)	ATP is a highly priced energy currency of the cell. Following the initial discovery of this molecule (see third column), the Russian biochemist, Academician of the USSR Academy of Sciences and Academy of Medical Sciences and Laureate of the Russian State Prize, Vladimir Aleksandrovich Engelhardt (1894–1984) noted in 1935 that muscle contraction requires ATP. On the background of this initial observation, Fritz Albert Lipmann (1899–1986), around 1935–1941, discovered a novel role of ATP as the energy currency of a living cell. Lipman shared the 1953 Nobel Prize in Physiology or Medicine with Hans Adolf Krebs: one half of the prize was given to Krebs "for his discovery of the citric acid cycle" and the other half to Lipmann "for his discovery of co-enzyme A and its importance for intermediary metabolism" [535–538] In 1937, it was established by Herman Moritz Kalckar (1908–1991) that ATP synthase (the enzyme that provides energy for the cell by catalyzing the formation of ATP from ADP and inorganic phosphate, Pi) is linked with cellular respiration. The first artificial synthesis of ATP was carried out by Alexander Todd in 1948 [535, 543] ATP was not known to be a neurotransmitter even though it was widely recognized as the energy currency of living cells. In 1963, the British pharmacologist and Professor at Oxford University, William Drummond Macdonald Paton ((D.M. William (Bill) Paton, 1917–1993)), and Sir John Robert Vane (1927–2004) became the first to identify muscle relaxation, not mediated by cholinergic neurotransmission, occurring in response to vagal nerve stimulation and opined that the phenomenon was due to vagal adrenergic neurotransmission [544, 545]. Vane was an English pharmacologist who became the first to describe how aspirin administration resulted in relief of pain and inflammation. Vane made numerous contributions that led to new treatments for heart and blood vessel diseases and the introduction of angiotensin-converting-enzyme inhibitors. It was Vane who founded the Research Institute "William Harvey Research Institute" named after the English physician and physiologist, William Harvey (1578–1657) [546]. This research institute was founded to recognize the contributions of his countryman to knowledge in systemic circulation and blood physiology. Harvey building on the works of other writers such as the French physician Jacques Dubois (1478–1555), who is sometimes called Jacobus Sylvius and widely known as the first to describe venous valves, provided a detailed description of the functions of valve. Dubois also studied medicine in France under the guidance of Jean Tagault (1478–1555), who was a famous French physician from Paris [547–549] Vane together with the Swedish biochemist and a former Director of Nobel Foundation Board in Stockholm, Sweden, Sune Karl Bergström (1916–2004), and the Swedish biochemist Bengt Ingemar Samuelsson (1934–) were awarded the 1982 Nobel Prize in Physiology or Medicine "for their discoveries concerning prostaglandins and related biologically active substances" [546, 550–552] Nobody knew that ATP could turn out to be a key neurotransmitter regulating bodily functions. Fortuitously, around the 1970s scientists began documenting that ATP was, in fact, not only an energy currency of living cells, but also, a neurotransmitter. A pioneer scientist in this move was Geoffrey Burnstock (1929–). Burnstock had previously worked on the physiology of smooth muscle in the laboratory of Edith Bülbring in Oxford, which probably enabled him to make outstanding discoveries.	Karl Lohmann (1898–1978), Cyrus Hartwell Fiske (1890–1978), and Yellagaprada SubbaRow (1896–1948) [539–542]	1929

(continued)

Table 9.2 (continued)

Neural secretions of the gut	Description	Discoverer	Year of discovery
	Edith Bülbring (1903–1990) was a British scientist and one of the most influential smooth muscle physiologists during her time. She was one of the first women accepted to the Royal Society as a Fellow. She was a professor of pharmacology at the University of Oxford [553–555] While working in Oxford, Geoffrey Burnstock gained experiences that would later spur him to attain greater heights in the scientific world. He trained in King's College, University of London, where he earned a Bachelor of Science degree in 1953, majoring in mathematics and physics. Four years later, he proceeded to obtain his Ph.D. in the field of zoology (precisely in gut motility in fish) at King's College and University College London, University of London. During the period of his dissertation, he developed a "sucrose gap technique" for recording smooth muscle activity. Using this technique, Burnstock studied the electrical changes accompanying the mechanical activities in smooth muscle cells. Following several years of scientific works and positions in different parts of the world, he settled for a long-term appointment at the Department of Zoology, University of Melbourne (1959), where he was later promoted from senior lecturer to reader to professor and chair of the department in 1964. While in Melbourne, Burnstock made key discoveries that put several scientists all around the world to skepticism. The energy currency of cells, ATP, will now be a neurotransmitter—this was too uneasy with scientists throughout the globe and many did not believe [553, 554, 556–558]. With modifications to the earlier technique he had developed for studying smooth muscle physiology, Burnstock observed that a non-adrenergic, non-cholinergic neurotransmitter was co-released with acetylcholine or noradrenaline in the gut and mediated via the activities of intrinsic enteric neurons and extrinsic parasympathetic nerves. The neurotransmitter was identified as ATP [559]. This purine neurotransmitter was later shown to be a co-transmitter present in all neuronal and non-neuronal cells of the PNS and CNS. It was Burnstock who coined the term purinergic signaling to indicate the transmission of nerve impulse mediated by ATP. In the mid-70 s of the last century, Burnstock returned home to continue research at the University College London Medical School in London, UK. Burnstock is widely acknowledged as the single-most cited researcher in the field of pharmacology [560–563] It should be noted, however, that the research results of Burnstock were unexpected and very hard to accept [557]. It took decades before ATP was generally accepted as a neurotransmitter. The receptors for this molecule were cloned and characterized in the 1990s [561–563]. It is now known that ATP transmission in cells takes place by channel- or receptor-mediated mechanisms. The receptors for ATP are ionotropic P2X receptors (ligand-gated cation channels) and metabotropic P2Y receptors (G protein-coupled receptors). The P2Y family has subclasses P2Y1, 2, 4, 6, 11, 12, 12, 14, and the P2X has P2X1-7 [561, 564, 565] ATP is an excitatory co-transmitter in the autonomic nervous system (sympathetic and parasympathetic divisions). ATP acts as a neurotransmitter in the CNS. In the nervous system, ATP signaling occurs in a bidirectional manner between neurons and neurons; neurons and glia cell or glia cell and glial cell. In peripheral tissues, such as smooth muscle, on stimulation, co-stored ATP within synaptic vesicles and released with noradrenaline in postganglionic sympathetic nerves innervating smooth muscle. ATP is also co-released with acetylcholine from postganglionic parasympathetic nerves such as those innervating the urinary bladder. Purinergic signaling is made possible through the presence of ATP receptors. In case of peripheral tissue, for instance,		

(continued)

Table 9.2 (continued)

Neural secretions of the gut	Description	Discoverer	Year of discovery
	the stimulated receptor is functional postjunctional P2X1 receptor. This receptor may induce the signaling of several cellular events and molecules including Ca^{2+} signaling and membrane currents. ATP evokes depolarization, leading to influx of Ca^{2+}, Ca^{2+} sensitization and elicits smooth muscle contraction [566, 567] ATP activities are regulated by several cellular events including dephosphorylation/phosphorylation. ATP actions on the cell are terminated by dephosphorylation by extracellular, membrane-bound enzymes and soluble nucleotidases released from postganglionic nerves. ATP may be involved in control of blood pressure. Although the mechanism is not clear, it may involve pressure changes associated with ATP-evoked calcium influx through L-type Ca^{2+} channels in vascular smooth muscle [567]. Increase in arterial blood pressure leads to increase in neurovascular transmission due to increased corelease of ATP and noradrenaline [568] To study the actions of ATP, noradrenaline and sympathetic neurotransmission in mesenteric arteries, Rummery et al. (2007) used pressure myography to induce artificial increase in arterial pressure to varying levels [568]. Increase in arterial pressure to 30–90 mmHg resulted in increase in the frequency of stimulation of sympathetic axons leading to vasoconstriction which was dependent on both frequency (0.5–10 Hz) and amplitude of the signal. The P2-receptor antagonist, suramin, abolishes pressure-induced increase in vasoconstriction. However, prazosin, an α1-adrenoceptor antagonist, does not have any effect on the phenomenon. But this prazosin-resistant vasoconstriction was abolished by the administration of nifedipine, an L-type Ca^{2+} channel antagonist. This suggests that ATP might be an integral molecule regulating blood pressure in the mesentery of mammals [568] In the CNS, purinergic signaling is involved in neuroprotection, central control of autonomic functions, neural–glial interactions, control of vascular tone and angiogenesis, pain and mechanosensory transduction and the physiology of the special senses. Many of these functions of ATP are conserved in peripheral tissues [561]		
Serotonin, 5-hydroxytryptamine (5-HT)	5-HT is a polyfunctional signaling molecule, acting as a neurotransmitter, hormone and growth factor. In the GI tract, it is secreted by enterochromaffin cells and by paracrine mechanism exerts its influence on the target cells. The molecule is also produced in both sensory and enteric neurons. In enterochromaffin cells, serotonin synthesis is regulated by tryptophan hydroxylase type I, TpH1, whereas in neurons it is TpH2. Following release of the hormone into the synaptic cleft, it activates the corresponding receptors and is subsequently inactivated—a crucial regulatory mechanism that controls the action of this agent. The inactivation of serotonin occurs through serotonin reuptake transporter (SERT)-mediated uptake into enterocytes or neurons [569–572] The functions of 5-HT are modulated by its receptors and transporters [569]. Serotonin receptor subtypes have different signaling mechanisms with varying functions on the target cells. In the gut, they differentially modulate motility and peristalsis, secretion, and sensation. This neurotransmitter also regulates the growth of neurons and interstitial cells of Cajal, and also plays a crucial role in inflammation [569, 573, 574]. The intrinsic reflexes due to serotonin, in part, leads to propulsive and segmentation motility patterns, epithelial secretion, and vasodilation in the GI tract. The extrinsic reflexes due to stimulation of	Vittorio Erspamer (1909–1999) [569]	1937

(continued)

Table 9.2 (continued)

Neural secretions of the gut	Description	Discoverer	Year of discovery
	vagal and spinal afferent fibers lead to slowed gastric emptying, pancreatic secretion, satiation, pain, discomfort, nausea, and vomiting [575]. The intrinsic primary afferent neurons activated by serotonin initiate peristalsis, whereas activated extrinsic primary afferent neurons initiate transmission of information to the CNS. The long descending myenteric interneurons use serotonin as a neurotransmitter to convey signal to the gut [569, 573] Research data have shown that bulk of 5-HT in the body is found in enterochromaffin cells, and to a lesser extent is contained in the ENS [573, 574]. Over 90% of the body's serotonin is produced in the gut [570, 576] Serotonin was identified by Erspamer as an amine in enteric tissues, and thus he called it "enteramine." Further research into this molecule, led to reconciliation of different sources of data by the American biochemist Maurice Rapport (1919–2011) [577–579]. The first well-defined proof on the role of serotonin in the gut can be attributed to Edith Bülbring and colleagues [580–582]. Edith Bülbring (1903–1990) was a smooth muscle physiologist and professor of pharmacology at the University of Oxford. Before the work of Bülbring and colleagues it was widely known that serotonin was abundant molecule in GI tract, however, its functions were not exactly known. It was initially thought that serotonin may activate peristalsis since pathological condition (e.g., carcinoid syndrome), characterized by excessive production of 5-HT from mucosal enterochromaffin cells, is accompanied by increased intestinal activity and associated with diarrhea [583]. So, it was somewhat moderate to reason that normal concentration of serotonin may activate intestinal activity. However, a problem soon arose. The addition of serotonin to isolated intestinal segment inhibited peristalsis [573]. It was interesting to note the differences in behavior of peristalsis if 5-HT is applied to the serosal or mucosal sides. Bülbring and colleagues were concerned with these differences associated with administration of serotonin to different regions of the layers of the GI tract. So, mucosal application of 5-HT mimicked endogenous release of the molecule and activated receptors on the mucosal side of the gut which resulted in concentration-dependent inhibition of peristalsis, but peristalsis was not completely abolished. The researchers also realized that increase in luminal pressure of the gut may cause distension of the walls thereby leading to overflow of 5-HT into the lumen. But this will rather contribute to peristalsis. The increase was proportional to the degree of distension but declined rapidly over time. Amine oxidase inhibitor, 1-isonicotinyl-2-isopropylhydrazine slows down the breakdown of 5-HT, and thus the decline in output over time was also slowed [573]. When 5-HTP, a precursor of 5-HT was added to the medium, peristalsis was augmented. Administration of 5-HTP in decreased level of serotonin has been shown to ameliorate endogenous level of the neurohormone such as in phenylketonuria. 5-HTP has been successfully used to treat pathological conditions involving the CNS [584, 585] (Fig. 9.29) Some of the differences in 5-HT action are now believed to be due to its differential expression of the receptor subtypes and polymorphisms. 5-HT receptor subtypes have differential effect on peristalsis. For instance, phenyldiguanide, a 5-HT3/4 receptor agonist, is known to stimulate peristalsis on application to the lumen [588–591], whereas 5-HT7 receptor is known to inhibit peristalsis [592]		

(continued)

Table 9.2 (continued)

Neural secretions of the gut	Description	Discoverer	Year of discovery
	5-HT signaling in the gut is altered in many diseases of the gut. Disorders in serotoninergic signaling pathways are implicated in functional gut disturbances. To this end, serotoninergic agents have been successfully used to treat functional bowel diseases [593, 594]. It is known that radiotherapy and pharmacological agents used in the treatment of cancer leads to GI disturbances such as nausea and vomiting. The mechanisms involve activation of enterochromaffin cell and the release of 5-HT, which acts on vagal mucosal afferents. In other GI disturbances and pathologies such as constipation, diarrhea, irritable bowel syndrome, it has been found that ligands of 5-HT are effective in treating these conditions [573, 595–598] 5-HT_3 and 5-HT_4 receptor antagonists have been used to correct functional disorders of diarrhea and constipation, respectively [575, 599]. Thus, serotonergic agents that might not necessarily be absorbed into the epithelial cells, but only differentially activates intrinsic or extrinsic reflexes are excellent agents in addressing some gut disorders A representative of 5-HT_4 partial agonist, tegaserod is currently used to treat irritable bowel syndrome (IBS) with constipation or chronic diarrhea [569]. The pharmacological agent ondansetron widely used for treating nausea associated with cancer chemotherapy is a representative member of 5-HT_3 antagonists. Other examples of 5-HT_3 antagonists include granisetron, alosetron. The former is used for postcancer chemotherapy, while the latter is used for treating IBS and diarrhea. Other pharmacological agents for functional gut disturbances include tricyclic antidepressants and serotonin selective reuptake inhibitors, and 5-HT_1 agonists [569, 573]		
Histamine, 2-(1H-imidazol-4-yl) ethanamine	A neurotransmitter, organic nitrogenous compound, biogenic amine, involved in local immune responses. It is produced by epithelial enterochromaffin cells of the intestine, and by basophils and mast cells found in nearby connective tissues and in the CNS, and uterine smooth muscles. It was actually discovered as β-iminazolylethylamine [600–603]. In 1916, further works on histamine were conducted by one of Pavlov's student, Lev Popielski and also found that the amine is the most potent and direct stimulant of gastric acid secretion [604]. It was later known that iminazolylethylamine is histamine [605]. Lev Popielskij (also known as Leon Popielski) (1868–1920) also contributed to the discovery of pancreatic secretion and the mechanism of secretory functioning [606] Histamine executes its functions by binding to histamine receptors. There are 4 types of histamine receptors (H1-4R), which are GPCRs. H1R was the first receptor to be discovered and cloned [601–603, 610, 611]. The receptor is associated with Gq protein that activates phosphatidylinositol (PIP2) and the phospholipase C to mediate cellular response by increase in intracellular calcium level, which affects cAMP level. The activation of this signaling pathway is related to increase MAPK activity and transcription factors that regulate expression of inflammatory genes. The H1R signaling pathway is mainly responsible for allergic reactions such as rhinitis. Thus, histamine antagonists (antihistamines) act on this receptor to inhibit its activity. Antihistamines are used as antiallergic drugs [612–616]	Sir Henry Hallett Dale (1875–1968) worked with George Barger (1878–1939) and Sir Patrick Playfair Laidlaw (1881–1940) discovered histamine [607–609]. Subsequently in 1911 Dale and Barger showed that histamine is a natural substance in the intestinal mucosa. Dale and Laidlaw showed that histamine stimulates the	1910

(continued)

Table 9.2 (continued)

Neural secretions of the gut	Description	Discoverer	Year of discovery
	H_2R is coupled to the Gs subunit of the GPCR. Binding of an agonist to this receptor leads to the activation of adenylate cyclase and increase in cAMP production, which results in the activation of protein kinase A. This protein kinase phosphorylates several intracellular proteins to mediate cellular activity/response. Functional H_2R receptors are found in the GI tract, and immune cells (e.g., T cells). In the GI tract, for instance, activation of these receptors regulates GI secretion and motility. The discovery of H2R around the 1970–80s revolutionized the treatment of gastric and peptic ulcer disease [613, 616, 617]. However, there is a continuous search for better therapy because H_2R antagonist therapy may be associated with relapse in some group of patients [618]. Recent studies have shown that the susceptibility of diabetic patients to infection compared to healthy individuals may be related to the higher signaling activity of the H_2R [619] H3R mediates cellular signaling in both peripheral and central tissues and cells of the organism by activating the Gi subunit of the GPCR, which leads to inhibition of cAMP production. In neuronal cells, they act as presynaptic autoreceptors in histaminergic neurons. The H_3R receptors are responsible for presynaptically inhibiting activities of neurotransmitters such as GABA, dopamine, ACh, noradrenaline, tachykinins, and serotonin as well as controlling histamine synthesis and release [614, 618, 621, 622]. H_3R is present in the GI tract where it may influence gastric secretion, and may be responsible for allergic reactions, encephalomyelitis, and also associated with the severity of cerebral malaria [621]. Signaling pathways of this receptor may be involved in the pathophysiology of multiple sclerosis, memory impairment and other central nervous pathologies, as well as GI secretory disorders [618] H_4R is currently the last member of the histamine receptors discovered in 1999. This receptor is mainly found in neurons of the peripheral and central nervous systems [610, 616, 621]. In the central nervous system, the receptor is expressed in the posterior hypothalamic neurons and other brain regions including the amygdala, and cerebrum. In peripheral cells, the receptor is found predominantly in hematopoietic cells [616, 621] The H_4R is highly expressed in cells of hematopoietic origin including mast cells, eosinophils, monocytes, dendritic cells, T cells, leukocytes and natural killer cells. The receptor is also expressed in spleen, thymus, colon, and bone marrow [621, 623, 624]. The H_4R activation is mediated via $G\alpha i/o$ proteins and increases in intracellular Ca^{2+} level [623, 625] The receptor mediates the chemotaxis of leukocytes and mast cells to sites of inflammation. Thus, the receptor may play useful role in the pathophysiology of asthma, allergy (e.g., allergic rhinitis), chronic pruritus as well as autoimmune disorders [613, 616, 624] Thus, the signaling pathway of the histamine receptors is involved in the pathophysiology of asthma and inflammation and certain brain diseases [610, 626, 627]. Emerging data suggest that histaminergic neurotransmission represents an integral aspect of signal transduction that mediates some domains of cognitive functioning [628]	smooth muscle of the intestine and many other tissues [609] The first effects of the antihistamine, thymoxidiethylamine, were studied by Daniel Bovet (1907–1992) and Anne Marie Staub (1914–2012) in 1937, which led to synthesis of many antihistamines [620] In 1936 H Dale was awarded the Nobel Prize in Physiology or Medicine, which he shared with Otto Loewi "for their discoveries relating to chemical transmission of nerve impulses"	

(continued)

Table 9.2 (continued)

Neural secretions of the gut	Description	Discoverer	Year of discovery
γ-aminobutyric acid, GABA	GABA is the most important inhibitory neurotransmitter in the mammalian CNS, but it is also present in many tissues of mammals including the gut. This neurotransmitter was first discovered as an amino acid in 1863. A few decades later, precisely in 1910, GABA was identified in living tissues. It was identified in the brain in 1950 by independent laboratories led by Eugene Roberts (1920–) and Sam Frankel [629] and Awapara et al. [630] and was recognized as an inhibitory neurotransmitter in 1967 [631, 632] GABA was proposed to be a neurotransmitter in the enteric nervous system by Kristjan R. Jessen and co-workers in 1979. Thereafter, a number of research papers were published by Kristjan R. Jessen and colleagues in the early to mid-1980s identifying GABA as a key neurotransmitter in the gut tissues. Functional receptors of GABA were also identified [633–639]. GABA was later identified as an autonomic neurotransmitter found in the intrinsic GABAergic neurons of the myenteric plexus of the gut [640] GABA is produced in cells via enzymatic reactions that involve molecules obtained from the tricarboxylic acid cycle (Fig. 9.30) Apart from the CNS, GABA is abundantly expressed in the ENS. This neurotransmitter is secreted by neuronal and endocrine cells of the gut [643, 644]. Surprisingly, the gut microbiota has been found to enhance the production and functions of GABA [645]. In a relatively recent investigation, Bravo et al. (2011) observed that ingestion of probiotic containing Lactobacillus was associated with regulation of emotional behavior and central GABA functions mediated via the vagus nerve [645] Secreted GABA localizes its cognate receptors on the target cells to mediate a range of physiological processes. GABA receptors are classified under three classes: GABAA (bicuculline-sensitive) and GABAB (baclofen-sensitive), GABAA-ρ (bicuculline/baclofen-insensitive, picrotoxin-sensitive). The third class was previously called GABAC until recently when the IUPHAR renamed it GABAA-ρ (GABAArho). These three classes of receptors also have their subtypes. The class GABAA receptors are ionotropic—binding of GABA molecule triggers the opening of a selective chloride ion pore leading to influx of Cl^- ion, which reverses the cell potential approximately to −70 mV and inhibits action potential. The class GABAB receptors are metabotropic—they are GPCRs. The binding of the ligand to the metabotropic GABA receptor leads to the opening of G protein inwardly rectifying potassium channels. The activation of this channel decreases Ca^{2+} current in the cell. Consequently, the production of cAMP is inhibited—resulting in the inhibition of cellular activity [646–648]. The class GABAA-ρ receptors are ionotropic—they activate Cl^- channels [646, 648]. The class GABAA-ρ receptors are insensitive to drugs that modulate GABAA or GABAB receptors and are activated selectively by cis-4-aminocrotonic acid [649, 650] GABA is involved in a variety of roles in the gut. It modulates intestinal contractility and regulates peristalsis [651–653]. It also regulates the neuroimmune and secretory activities of the GI tract [643]	See description	1863; 1950; 1979

(continued)

Table 9.2 (continued)

Neural secretions of the gut	Description	Discoverer	Year of discovery
Tachykinins (TK)	TKs are a family of neuropeptides that rapidly induce contraction of gut tissue, hence the name. The tachykinin family is characterized by a common C-terminal sequence, Phe-X-Gly-Leu-Met-NH2, where X is either an aromatic or an aliphatic amino acid. There are three TK genes (TAC 1, 3 and 4) that produce different types of the neuropeptides. Products of TAC1 gene include substance P, neurokinin A (previously known as substance K, neurokinin α, neuromedin L), neuropeptide K and neuropeptide γ (a form of neurokinin A), neuropeptide-kappa. TAC3 gene products include neurokinin B (previously known as neurokinin β, neuromedin K). The products of TAC4 gene include hemokinin (HK-1), endokinin A, B, C, and D). However, it should be noted that each of these genes produces multiple mRNA isoforms, i.e., two or more gene may produce substance P [654, 655]. These TKs are produced as preprotachykinins which are then cleaved in the cytoplasm by peptidases [656, 657] Tachykinins are produced in peripheral and central tissues and cells of the organism. In the central nervous system, certain excitatory neurons are known to produce tachykinins. Though the role of central tachykinins is not exactly clear, they may be involved in neuronal excitation and modulation of several neurotransmitters [656, 658]. In the peripheral tissues, tachykinins are produced by neuronal and non-neuronal cells [658]. Tachykinins are produced in both central and peripheral nerve varicosities [659] The most abundant tachykinins in the GI tract are beta-tachykinin (neurokinin B), gamma-tachykinin as well as substance P and neurokinin A. The last two TKs are mostly produced in excitatory extrinsic and intrinsic neurons (including secretomotor neurons, and interneurons), which innervate smooth muscles of the gut. The products of other TAC genes may be synthesized in the gut. These tachykinins are also produced by the ascending neurons of the circular and longitudinal muscle. These neurons have their cell bodies in the dorsal root ganglion [660, 661] There are three known mammalian tachykinin receptors—NK1R, NK2R, and NK3R, which are members of the 7 TM GPCRs. Tachykinin receptors induce the activation of phospholipase C, producing IP3 via G protein signaling. Though neurokinin receptors are not specific to any tachykinin, substance P, neurokinin A, and neurokinin B, selectively bind to the known NK1R, NK2R, and NK3R, respectively [654, 664–666] Smooth muscle cells express NK1R, NK2R, and NK3R. Absorptive enterocytes express NK1R and NK2R. Secretomotor neurons express NK1R and NK3R. Interstitial cells of Cajal and blood vessels express type 1 receptor. Neurons of the intestinal plexuses, vasomotor neurons express NK3R [654, 667, 668] Tachykinins are involved in GI motility and modulate release of several excitatory neurotransmitters [667, 669]. Secreted tachykinins localize their receptors on the target cells by initiating G protein signaling to evoke action potential with subsequent cellular responses which may include vasodilation (blood vessels), smooth muscle contraction [654, 667]. Importantly, tachykinins are co-released with other excitatory neurotransmitters such as ACh in motor neurons. However, compared to ACh, tachykinins play lesser or relatively transient role in motor neuron excitability. Tachykinins are implicated in a range of diseases involving the CNS and gut [666, 669, 670]	Vittorio Erspamer (1909–1999) discovered, characterized and sequenced gut molecules with substance P-like biological activities and were thus named "tachykinins" meaning substances that cause rapid relaxation of vascular smooth muscle [662, 663] In the early 1980s, three independent groups discovered novel mammalian TKs which mammalian TKs were identified as hemokinin-1 were named neurokinin A and B, neuropeptide-kappa and -gamma. Novel (HK-1) and endokinin A and B. For review, see Lecci et al. [654]	

(continued)

Table 9.2 (continued)

Neural secretions of the gut	Description	Discoverer	Year of discovery
Substance P	An undecapeptide (an 11 amino acid) neurotransmitter molecule that belongs to the tachykinin neuropeptide family. The peptide was identified in extracts of equine intestine with characteristic rapid stimulatory actions on vascular and non-vascular smooth muscle cells. The letter "P" in the name "substance P" was derived from the first letter in the word "powder," which is the physical form of the extract [654]. The substance was purified and structure of this molecule was unraveled by the same group of scientists around the 1970–1971 [671, 672] Substance P receptors are the neurokinin (tachykinin) receptor-1, -2, -3. These receptors are associated with G proteins that activate a phosphatidylinositol-calcium second messenger system [675] Substance P and its receptors are highly expressed in the CNS (hypothalamus) and GI tract. Substance P plays a variety of roles in the organism. It has a powerful spasmogenic activity on smooth muscle [675]. Substance P also exerts an antiapoptotic action on neuronal network. However, substance P may be produced in high quantity in tumors of the nervous system from where the neuropeptide may be transported to other parts of the body via the circulatory system. Interestingly, NK-1 receptors are overexpressed in cancer cells [676]. Tachykinin receptors can be modeled with tracers to study their anatomo-physiological properties with the help of modern radiological instrument such as the positron-emission tomography or hybrid devices [677] Tachykinin receptor seems to have a special role in chemotherapy-induced nausea and vomiting. (See Clinical Correlate 9.1). Of particular interest is the tachykinin receptor, neurokinin-1 receptor, which has been proposed to play a significant role in chemotherapy-induced nausea and vomiting [677]. Substance P and other tachykinins are ligands to this receptor subtype. The antagonist for this receptor may play a beneficial role in ameliorating chemotherapy-induced nausea and vomiting. Thus, antagonist of substance P and other neurokinins may serve as prophylactic therapy for such group of patients. However, the signaling activities of this receptor are probably modulated by factors such as age, sex and behavioral contingencies such as smoking and alcohol consumption [677] *Clinical Correlate 9.1* **Vomiting (Emetic) Reflex and Tachykinins** Vomiting is the expulsion of the content of GI tract to the outside. In 1954, Gaddum and co-workers suggested that substance P was abundantly produced in brain regions known to be involved in emesis [679]. Consequently, the depletion of substance P using resiniferatoxin prevented emesis [679] Emesis is the result of complex interactions of neurotransmitters (neurokinin, serotonin, dopamine, acetylcholine, and histamine) and their receptor subtypes within GI and central pathways. The neurotransmitters are released peripherally or centrally to modulate signaling in the emetic centers of the brain. Vomiting reflex starts by the initiation of release of 5-HT which is sensed by the afferents of the vagus nerve. Both chemo- and mechanoreceptors can initiate the vomiting reflex. In the GI tract, release of emetic neurotransmitters stimulates vagal afferents, which send signal to the NTS. Neurons that signal emetic information extend to the medulla through the NTS. The medulla contains a chemoreceptor trigger zone located at the area	Ulf von Euler (1905–1983) and Sir John Henry Gaddum (1900–1965) [673, 674]	1931

(continued)

Table 9.2 (continued)

Neural secretions of the gut	Description	Discoverer	Year of discovery
	postrema at base of the fourth ventricle and a vomiting center (Fig. 9.31). The chemoreceptor trigger zone evaluates and integrates chemical signals (such as toxins) from the blood with neuronal inputs from the GI tract [680]. The vomiting center contains NK-1, Dopamine, and 5-HT3 receptors to which neurotransmitters bind. However, neural signals other than the chemoreceptor signals are also directed to the emetic center to initiate vomiting [680]. The vomiting center also integrates somatic signals, information from vestibular apparatus (labyrinth signal), postchemotherapy or radiotherapy conditions (that produce cytotoxics), higher centers such as hypothalamus, limbic system, and cortical areas (of smell, odor, pain, sight) [681–684]. Neurons conveying emesis-related signal from higher centers synapse on neurons located in other brain regions before the signal is finally transmitted to the vomiting center. The labyrinth signal, for example, (by activating H_1, M_1 receptors) passes through the cerebellum to the vomiting center or via the chemoreceptor trigger zone. However, some emesis-related signals may directly reach the vomiting center without multiple connections. Pharmacological agents such as apomorphine (dopamine D2 agonist), opioid drugs (morphine, hydromorphone, fentanyl, meperidine, methadone, buprenorphine, butorphanol, and nalbuphine), ipecac syrup (gastric irritant) can also induce the vomiting reflex [684–688]. Veratrine and copper sulfate can also serve as emetic agents [689]. Studies have shown that syrup of ipecac (ipecacuanha) can be used to safely induce vomiting in conditions of poison ingestion [690, 691]. Vomiting signal can originate from the larynx, pharynx, esophagus, or any part of the GI tract. The vomiting center plays a role in the control of peristalsis, tachy- or bradycardia, salivation, contraction of GI (including pharyngeal), respiratory and abdominal muscles. Thus, the vomiting center has both somatic motor neurons and autonomic neurons that innervate the structures involved in the vomiting reflex (Fig. 9.31). The neurons of the vomiting center form synapses with the dorsal vagal complex from where motor neurons extend to innervate the stomach. The vomiting center also sends neural fibers through the cervical and lumbar segments. Part of the fibers forms the phrenic nerve that project to the diaphragm (crural diaphragm). The respiratory muscles contract during vomiting. The abdominal muscles (external oblique muscle) that take part in vomiting by contraction to eject abdominal content are innervated by the motor neurons originating from the medullary reticular fibers/lumbar segments. The autonomic division of the vomiting reflex controls retrograde peristalsis, salivation, tachy- and bradycardia. Contraction of the muscles of the pylorus, stomach wall, increases the intra-abdominal pressure causing relaxation of the lower esophageal sphincter. During vomiting, as the content of the food or liquid moves back from the stomach toward the esophagus, the lower esophageal sphincter relaxes, the soft palate closes the air passage and the glottis also closes, and the content of the stomach is ejected to the outside [681–684, 692]		
Neurokinin A	Neurokinin A is a decapeptide of the TK family, discovered in porcine spinal cord. The substance was subsequently found to stimulate intestinal contraction [695, 696]. This peptide is produced in both central and peripheral tissues. The neurons of the CNS are responsible for producing neurokinin A. It is one of the most abundant tachykinins produced in the GI tract. The peptide is produced in peripheral neurons of adrenal glands and other sites of the body [673, 697] The activities of neurokinin are mediated via the activation of neurokinin receptor particularly, NK-2 receptor. Apart from the decapeptide neurokinin A, other fragments of neurokinins such as heptapeptide neurokinin A (4–10) also activates NK-2 receptor [698, 699]	Kangawa et al. [695] and Kimura et al. [696]	1983

(continued)

Table 9.2 (continued)

Neural secretions of the gut	Description	Discoverer	Year of discovery
	Neurokinin A participates in a range of body functioning in both peripheral and central systems. In peripheral tissues, it stimulates smooth muscles in different regions of the body. In the cardiovascular system, the peptide induces vasodilation and contributes to bradycardia. However, in the lungs, the peptide is responsible for bronchoconstriction and induces mucus production, a condition experienced by asthmatics [700]. Serum level of neurokinin A is an independent prognostic factor in certain cancers of neuroendocrine system [701, 702]. Patients with neurokinin A plasma concentrations above 50 pmol/l are likely to have a poorer survival rate compared with patients with lower level of plasma neurokinin A [703, 704] The central secretion of this peptide is important in many roles including neuromodulation, neuroprotection. It is one of the peptides of the hypothalamic–pituitary–adrenal axis that regulate among many other things, output of reproductive hormones [705–707] Dysfunction in neurokinin signaling is implicated in a host of diseases and malfunctions of central origin including stress-induced neurological disorders (e.g., depression, anxiety), fluctuation in mood, thought, emotion, and behavior. Some tachykinins (such as neurokinin A, substance P) may be involved in the pathogenesis of epilepsy. Neurokinin A is involved in the pathophysiology of migraine which may possibly be mediated through modulation of vascular response and primary nociception [708]. The discovery of neurokinin antagonist has made the future brighter for some pathologies that have recorded a surge in the past few decades despite improved treatment modalities [709]		
β-tachykinin (neurokinin B or neuropeptide beta)	This molecule is a decapeptide with amino acid sequence H-Asp-Met-His-Asp-Phe-Phe-Val-Gly-Leu-Met-NH2. The peptide is produced in both central and peripheral tissues [710–712] In the CNS, β-tachykinin is expressed together with other peptides namely, kisspeptin and dynorphin A in the arcuate nucleus of the hypothalamus. This group of neurons producing kisspeptin, neuropeptide β, and dynorphin is collectively referred to as KNDy neuronal subpopulation. This subpopulation forms a network that regulates the pulsatile neural generator of gonadotropin-releasing hormone (GnRH) and its feedback pathway integrating steroid signals with GnRH production [710, 712]. GnRH (follicle-stimulating hormone–releasing hormone, luteinizing hormone–releasing hormone, gonadoliberin, luliberin, gonadorelin) is a tropic peptide hormone produced in the hypothalamus and is responsible for the release of follicle-stimulating hormone and luteinizing hormone from the anterior pituitary [713–717] Neurokinin B is involved in the secretion, control and feedback regulation of gonadotropin-releasing hormone [710, 712]. Neurokinin B and kisspeptin, another GPCR (specifically called GPR54 or Kiss-1 receptor) ligand are involved reproductive function. The regulation of GnRH secretion and luteinizing hormone is believed to be controlled by a feedback mechanism which directly links neurokinin B with the KNDy subpopulation. Interestingly, neurokinin B agonist, senktide, acting via NK3R stimulates the secretion of luteinizing hormone. Both kisspeptin and neuropeptide beta are involved in sexual maturation and puberty. Thus, impaired sexual maturation, puberty, or infertility could be due to malfunction signaling of kisspeptin, neurokinin B or their receptors [710]	Kangawa et al. [695] and Kimura et al. [696]	1983

(continued)

Table 9.2 (continued)

Neural secretions of the gut	Description	Discoverer	Year of discovery
	The KNDy neuronal subpopulation may play a role in thermoregulation. Rance et al. (2013) found that activation of NK3R in the median preoptic nucleus, which constitutes part of the heat-defense pathway, reduces body temperature [718]. This thermoregulatory effect may be due to the regulation of estrogen signal by KNDy neurons—also regulates flushes [718] The neuropeptide-β is highly expressed in the GI tract, and placenta [711]. The expression of this peptide in the placenta suggests it may play a role in some forms of maternal pathologies during pregnancies such as preeclampsia [711]. Its endogenous receptor, NK3R is found in many locations of the human body suggesting that the peptide may be produced in other regions apart from the widely recognized sites of expression [711]		
γ-tachykinin (also called gamma-neuropeptide)	A 21 amino acid peptide initially isolated from rabbit intestine with amino acid sequence Asp-Ala-Gly-His-Gly-Gln-Ile-Ser-His-Lys-Arg-His-Lys-Thr-Asp-Ser-Phe-Val-Gly-Leu-Met.NH2. The peptide is synthesized from the precursor, gamma-preprotachykinin, and belongs to the TK family [719]. The neuropeptide is expressed in the CNS and peripheral tissues [720] In peripheral tissues, the neuropeptide is produced in lungs, GI and urogenital tracts [721, 722]. In the respiratory tract, for instance, the release of the peptide causes bronchoconstriction acting mainly via the NK-2 receptor. It is also a vasodepressor [721, 723] Centrally the peptide is produced in the spinal cord and brain regions including the hypothalamus. The neuropeptide gamma of central origin among other functions can initiate the release of catecholamine from the adrenal medulla via NK1 receptors [722] The neuropeptide plays a useful role in salivation, hypotension, and smooth muscle contraction of urinary, respiratory, and GI tract [720]. The release of the substance into circulation can cause increase in mean arterial blood pressure, heart rate, face-washing, and head-scratching [724]. The effect of neuropeptide gamma is mediated via NK1R, NK2R, and NK3R receptors and possibly by the putative neuropeptide gamma receptor or a subtype of one of the three known receptors of the tachykinin family of neuropeptides [724] Neuropeptide gamma is involved in many pathogenic conditions including asthma [721, 723]	Kage et al. [719]	1988
Norepinephrine (noradrenaline or noradrenalin)	This biomolecule is both a hormone and a neuromediator found in many parts of the body. It belongs to the group of biologically active agents produced in vivo, called catecholamines. Apart from noradrenaline, catecholamines include epinephrine (adrenaline) and dopamine. These biomolecules are produced in many cells and tissues of the body from the aromatic amino acid (L-tyrosine) taken in as food. L-tyrosine can be produced from L-phenylalanine via hydroxylation by the enzyme biopterin-dependent aromatic amino acid hydroxylase [725]. The catecholamines are abundantly synthesized and released in the adrenal medulla of the adrenal glands. The cells that synthesize catecholamines through several steps and enzymes convert tyrosine to L-DOPA (L-3,4-dihydroxyphenylalanine) and then to dopamine. L-DOPA, first isolated from seedlings of Vicia faba (broad bean, used as a staple food) by the German von Marcus Guggenheim (1885–1970) in 1913,	George Barger and Henry Dale discovered norepinephrine (noradrenaline) in 1911, but the molecule may have been discovered in 1907. Noradrenaline was identified as both a	1911

(continued)

Table 9.2 (continued)

Neural secretions of the gut	Description	Discoverer	Year of discovery
	initially thought to be inactive substance, however, it was later shown to be an active agent in rabbit during the late 1920s [726]. This important finding was advanced by the British physiologist, Henry Stanley Raper (1882–1951) and Walter Louis Dulière through a series of experimental evidences published between 1926 and 1930 [727–729] In 1938, Peter Holz (1902–1970) found that L-DOPA decarboxylizes in mammalian tissue to produce dopamine. He discovered the enzyme L-dopa decarboxylase, which was responsible for this reaction. Dopamine was the first biologically active amine in the biosynthetic pathway of catecholamines to be discovered. Dopamine is used as a cellular messenger or neurotransmitter. However, depending on the cell type and activity of enzymes, dopamine can be further converted to norepinephrine. The later may then be converted to epinephrine [731] Among other functions, noradrenaline is known to increase vascular tone (i.e., tension of vascular smooth muscle) via the activation of α-adrenergic receptor [732]. The three types of α-adrenergic receptors through which adrenaline relays its signal downstream the cell are α(1-3) adrenoreceptors. The catecholamines also act via a second receptor type: β1-3 adrenoreceptors [733–735]. The adrenergic receptors belong to the class of G protein-coupled receptors. These receptors have been shown to exhibit polymorphism which affects the functionality of catecholamines [736]. The drug tamsulosin and terazosin have been shown to selectively bind functional α1-adrenoceptors and competitively inhibit the contractile responses of smooth muscle to noradrenaline. On the basis of the tissue type, these drugs may activate α1D-, α1B- or α1L-subtype to initiate cellular response [736]. Adrenoreceptors (particularly β1-adrenoceptors) are present in a wide range of cells of the gut, including intestinal Cajal cells, which upon stimulation by noradrenaline inhibit pacemaker currents. These currents are the driving force behind motor patterns of the GI tract. Noradrenaline activates its cognate GPCR independent on adenylate cyclase, guanylate cyclase or ATP-sensitive-K^+-channel-dependent pathway as apamin, a Ca^{2+}-dependent K^+ channel blocker or glibenclamide, a K^+-channel blocker or tetraethylammonium, a voltage-dependent K^+ channel blocker, does not affect the action of noradrenaline administration. The catecholamines have been shown to signal through different types of protein kinases (such as protein kinases type A, C depending on the cell type) [736, 737]	biosynthetic precursor of epinephrine, and also aneurotransmitter [730] Somewhat detailzed information about this transmitter only became available after some years of research. In 1946, Ulf Svante von Euler made the fundamental discovery that noradrenaline was produced and stored in intracellular vesicles at nerve terminals, and from here it was released upon stimulation [738]. Together with Julius Axelrod (1912–2004) and Sir Bernard Katz (1911–2003), von Euler was awarded the 1970 Nobel Prize in Physiology or Medicine "for their discoveries concerning (neurotransmitters) and the mechanism for their storage, release, and inactivation" [493].	
Epinephrine (adrenalin or adrenaline)	Like noradrenaline, adrenaline is both a hormone and a neurotransmitter produced in various cell types. Epinephrine is a active sympathomimetic hormone synthesized mainly in the adrenal medulla. It stimulates both the alpha- and beta-adrenoreceptors [739, 740] Adrenaline is a drug that is widely used in emergency medicine for cardiopulmonary resuscitation (CPR) to reverse cardiac arrest and in anaphylaxis. Administration of the drug causes systemic vasoconstriction and stimulates the heart, and dilates bronchi and blood vessels. It is used in asthma and heart failure. In anesthesiology, it is used to delay absorption of local	Adrenaline was first identified and isolated in 1898 by John Jacob Abel (1857–1938), an American physiologist and biochemist, who received his PhD in 1883 from the	1898

(continued)

Table 9.2 (continued)

Neural secretions of the gut	Description	Discoverer	Year of discovery
	anesthetics. Like noradrenaline, the hormone is involved in many activities including fight-or-flight response. Adrenaline causes GI relaxation. These effects result from the action of the hormone on the different adrenoreceptors differentially located in various tissues [741–745] Adrenaline research marked tremendous achievements during the last decade of the 19th century. Before Abel's work, other scientists made useful contribution to the discovery of adrenaline. Though it was not clear whether or not the substance reported was really adrenaline or it was other previously unknown (or known) substance, the American physician, William Horatio Bates (1860–1931), is sometimes believed by some to be the first to discover adrenaline. He reported the discovery of a substance produced by the adrenal gland in 1886 [745] Initial contributions in the field were set by the Polish physiologist Napoleon Cybulski (1854–1919), a pioneer in the field of endocrinology and electroencephalography. Cybulski first obtained extracts of the adrenal gland as early as 1895, which was thought to contain catecholamines including adrenaline. However due to level of technology at the time, obtaining pure adrenaline remained a huge challenge. Other researchers who had interest in studying the extract encountered similar challenges. However, in addition to obtaining the adrenal extracts in 1899, the German physician from Berlin, Max Lewandowsky (1876–1916) observed that the actions of adrenal extracts on visual smooth muscle of cats were similar to the effects observed on activation of sympathetic nerve fibers [748–752] In 1901, the Japanese biochemist, Jokichi Takamine (1854–1922) successfully isolated and purified the hormone from the adrenal glands of sheep and oxen, which he named "adrenalin" [753]. Adrenaline was first synthesized in the laboratory in 1904 independently by the German chemist Friedrich Stolz (1860–1936) and the English chemist, Henry Drysdale Dakin (1880–1952) [754] The British physiologist who later became a physician, Thomas Renton Elliott (1877–1961) documented the result of his investigation conducted in Cambridge that confirmed the previous results of Max Lewandowsky. While in Langley's laboratory, in 1906 Elliott reported that the active substance discovered by the British physiologist was "adrenaline," released from sympathetic nerve terminals to act on smooth muscle cells. The actual name "adrenalin" was coined by the Japanese scientist Jōkichi Takamine (1854–1922) [746, 755, 756]. In recognition of his research achievements, he was elected to the Fellowship of the Royal Society in 1913 [757] The American physiologist, Cannon W.B. around 1920–1930s reported the sympathetic activity of this substance in tissue extract. Consequently, he called the substance sympathin E (the letter "E" for excitatory). In addition, Cannon noted that another substance, opposite in action was also present in the tissue extract, and called it sympathin I (the letter "I" for inhibitory) [758]. Cannon's sympathin E and I were actually adrenaline and noradrenaline, respectively [759]	University of Michigan. Abel, who later became a professor of pharmacology (1893–1932) at Johns Hopkins University, worked particularly in the area of endocrine physiology which resulted in his discovery of the mediator [746, 747]	

(continued)

Table 9.2 (continued)

Neural secretions of the gut	Description	Discoverer	Year of discovery
	A clear distinction about these two substances present in the extract was made in 1946 when von Euler showed that demethylated adrenaline (noradrenaline) was the sympathetic transmitter [760, 761]		
	In 1948 the American pharmacologist, Raymond Perry Ahlquist (1914–1983), who had developed interest in the study of adrenoactivity [762, 763], made outstanding observations which he published in the Journal of Physiology. It was Ahlquist who advanced our understanding of receptor and mediator, in particular, as regards adrenaline. His work showed that adrenoceptors are divided into α- and β-adrenoceptor subtypes and that these receptors are all activated by adrenaline. Worthy of note, he mentioned that adrenergic receptors (which he named adrenotropic receptors at the time) were hypothetical structures affected by epinephrine. Ahlquist investigated the activity (including contraction and relaxation of vascular and smooth muscles) on catecholamine in the intact animals and isolated it from tissue preparation. His conclusion was that regardless of the excitatory or inhibitory adrenotropic effects, catecholamine activities are mediated by two types of receptors, which were called alpha- and beta-adrenotropic receptors [762, 763]		
	However, it was not recognized that this chemical messenger could act at nerve terminals, precisely on presynaptic terminals to excite the receptor. In 1971 Daryl D Christ and Syogoro Nishi showed that this was a possibility [764]		
	Activation of postsynaptic adrenoreceptors leads to downstream signaling mediated particularly by the GPCR [765]. According to the nomenclature outlined by the International Union of Basic and Clinical Pharmacology, adrenoreceptors are divided into α and β types. The α-type receptor is further subdivided into α1 and α2. The α1 has 4 subtypes: α1A-D-adrenoceptors, while α2 has 3 subtypes: α2A-C-adrenoceptors. The β type has three subtypes: β1-3-adrenoceptors. The receptors belong to the GPCR family and are activated by norepinephrine (noradrenaline) and epinephrine (adrenaline) to varying degree of potencies [766]. There are also species variations in the expression of these receptors. Details about selective agonists and antagonists for these receptors and their potencies could be found in [766]. Activation of α1-adrenoceptors by the agonists (adrenaline and noradrenaline) leads to the activation of the Gq subunit of the GPCR. Subsequently, the membrane enzyme phospholipase C is activated, which results in increased IP3 production with the formation of DAG. The result is increase in intracellular calcium level. These molecular events are observed in the cell as smooth muscle contraction (e.g., smooth muscles of the GI tract, sphincter, and blood vessels). Both adrenaline and noradrenaline have equal potency in the activation of the α1 receptor. The agonists stimulate the α2 receptor subtype with relative potency leading to the recruitment of Gi subunit of the GPCR, inactivating adenylate cyclase, with subsequent decrease in cAMP level. The result is smooth muscle relaxation [766–768]		
	Both adrenaline and noradrenaline activate β-adrenoceptors. However, the synthetic adrenoreceptor agonist, isoprenaline is selective for β-adrenoreceptors relative to α1- and α2-adrenoceptors. The drug propranolol used for treating blood pressure is relatively selective antagonist for β1 and β2 adrenoreceptors. Noradrenaline is a selective agonist for β1-adrenoceptors. Binding of agonist to β1 leads to activation of Gs subunit of the GPCR. The membrane-bound enzyme adenylate cyclase is thus activated, resulting in increase in cAMP level. The resultant effect is a positive chronotropic, dromotropic and inotropic effects.		

(continued)

Table 9.2 (continued)

Neural secretions of the gut	Description	Discoverer	Year of discovery
	In the process, secretory cells release their contents to the lumen, e.g., enzyme secretion. Binding of agonist to β2 and β3 activate the Gs subunit of the cognate GPCR [769, 770]. The secretion of ghrelin may be stimulated by β1-adrenoreceptor activation [771] In blood vessels, both the alpha1- and the alpha2-adrenoreceptor subtypes are present postsynaptically. However, there is predominance in expression of alpha1-adrenoreceptor in vascular smooth muscle where they mediate vasoconstriction. Interestingly, released transmitter on noradrenergic nerve terminals were found to modulate neurotransmission via presynaptically localized activity of alpha-adrenoreceptors [772]. The stimulation of presynaptic inhibitory alpha2-adrenoreceptors on noradrenergic nerve terminals, leads to the reduction of transmitter secretion due to membrane depolarization [772]		
Glutamate	An amino acid and the primary excitatory transmitter in the brain. This substance was first isolated as glutamic acid from acid hydrolysate of wheat gluten in 1866 by Karl Heinrich Leopold Ritthausen (1826–1912) and thus named as "glutamic acid or L-glutamic acid." In 1911, the German chemist Hermann Emil Louis Fischer (1852–1919) and recipient of the 1902 Nobel Prize in Chemistry "in recognition of his work on sugar and purine syntheses," reported that glutamic acid has a non-palatable taste, tasted sour initially in the oral cavity, and later developed an insipid taste. Glutamate was identified by Ikeda when working with kelp—he identified the taste of this substance as the salt of glutamic acid and coined the term "umami" to describe this taste. In addition to umami, humans taste receptors can identify at least five basic taste types: sweet, sour, salty, and bitter tastes. These tastes have their respective receptors or receptor dimers that sense them to initiate downstream signaling that modulate several cellular processes [773–775]. The receptors for these tastes were only recently identified and characterized. All types of taste receptors are present in the GI tract and also found in other cells and tissues of the body. Glutamate receptors are present in the digestive tract and elsewhere in the body [230] Apart from the taste receptors, glutamate receptors are generally classified as metabotropic (mGluRs) and ionotropic glutamate receptors (iGluRs) [776]. The iGluRs are voltage-gated ion channel that allows influx of ions following the binding of glutamate to its active site on the receptor. The iGluRs are responsible for fast excitatory transmission of signal across the synapse. Examples of iGluR include N-methyl-D-aspartate (NMDA), α-amino-3-hydroxy-5-methyl-4-isoxazolepropionic acid (AMPA), and kainate receptors. The receptors are named for the chemical agonist that selectively binds to them [776–778] The mGluRs are 7TMS GPCRs that initiate signaling cascades or influx of cations upon binding of a glutamate molecule. The mGluRs are divided into group I, II, and III subfamilies. This division is based on the amino acid sequence identity and pharmacology of the receptors [776, 777]. These receptors upon activation usually show excitatory effects. Receptor types 1 and 5 mGluR of the group I are coupled to Gq subunit of GPCR signaling, which involves phospholipase C, IP3, and DAG. These molecules stimulate downstream effectors which are coupled to the activation of Ca^{2+} , K^{+}, and Na^{+} channels, mitogen-activated protein kinase, protein kinase B, Src kinase signaling pathways [776]	Kikunae Ikeda (1864–1936)	1908

(continued)

Table 9.2 (continued)

Neural secretions of the gut	Description	Discoverer	Year of discovery
	The activation of the groups II and III mGluRs is coupled to the inhibitory subunit Gi of GPCR. The stimulation of this receptor results in decrease in the activity of adenylate cyclase, with subsequent reduction in cAMP levels. Examples of group II receptors include mGluR2 and 3. The mGluR4, 6, 7, and 8 belong to group III [776] The mGluRs mainly modulate secretory processes in cells; regulate synaptic functions, and neuronal excitability [777] Both mGluRs and iGluRs are implicated in cognitive functions, rhythmogenesis—a critical factor of automaticity and signal generation. For instance, in the CNS, inspiratory neurons of the preBötzinger complex (preBötC) have been known to generate rhythm by producing local excitation of about 10–30 mV transient depolarization, referred to as inspiratory drive potentials [779]. These neurons also utilize glutamate as their neurotransmitter. To this end, Pace et al. (2007) reported that group I mGluRs contribute to the generation of inspiratory drive potentials. The mGluR1 probably regulates a K^+ channel or inwardly rectifying potassium channel to augment the generation of the drive potential. The mGluR5 augments drive potential generation by IP3 receptor-dependent mechanism [779] Dysfunction in glutamate signaling has been associated with pancreatic diseases including diabetes. Glutamate is involved in brain development, memory and learning, neuroprotection, and the pathophysiology of hypoxic injury, epilepsy, Parkinson's disease, Alzheimer's disease, pain, anxiety, schizophrenia, and drug addiction [777]		
Neurotensin	An endogenous tridecapeptide (13 amino acids), one of the satiation hormones, neurotransmitters and neuromodulators produced in both the central (brain) and peripheral (gut's minibrain) nervous systems. It was initially discovered as a vasoactive peptide that regulates blood flow mainly by dilating the endothelial and smooth muscle cells leading to hypotension. Hence the name "neurotensin." The amino acid composition of neurotensin was successfully sequenced in 1975. Neurotensin is made of the following amino acid sequence: pGlu-Leu-Tyr-Glu-Asn-Lys-Pro-Arg-Arg-Pro-Try-Ile-Leu-OH [780, 781] About 10% of the neurons that secrete this neuropeptide are found in brain (especially in hypothalamus and pituitary gland). However, the majority (80%) of the neuropeptide is secreted in the GI tract. The remaining percentage exists in other tissues including the adrenal glands [782–784]. In the GI tract, neurotensin is secreted by the neuro- , enteroendocrine cells called N-(neurotensin) cells which are scattered mainly in the mucosa of the jejunum and scantly in the mucosa of the esophagus, gastric fundus, corpus, antrum, duodenum, ileum, colon, and pancreas. The neuropeptide is also secreted by the intrinsic neurons of the GI tract [783–785] Apart from neural signal, certain food substances such as lipids (oleate) are known to adequately activate the release of the neuropeptide. Other food substances such as glucose, amino acids have little effect on neurotensin release [783–785]	Robert E. Carraway, an American physiologist, and Susan E. Leeman (1930–), an American neuroendocrinologist, discovered the peptide in bovine hypothalamic extracts [786]. The peptide was later identified by the same group of authors in ileum and stomach of rats by radioimmunoassay (RIA) [787], and later found in human intestines [788].	1973, 1976

(continued)

Table 9.2 (continued)

Neural secretions of the gut	Description	Discoverer	Year of discovery
	The signaling molecule (lipid) activates the corresponding receptors on the N cells. The release of neurotensin is activated by the N cell receptors of fatty acid type 1, 2, and 3 (FAR1-3), which sense intraluminal fatty acids. The activation of these receptors leads to increase in intracellular calcium level [784]. The calcium ions recruit vesicles (containing neurotensin) to the plasma membrane. In the GI tract, the cytoplasm of N cells is packed with neurotensin granules which are recruited to the membrane by docking and fusion on arrival of adequate signal and subsequently exocytosis of the vesicles releasing the neurotensin molecules into the surrounding. The base of the N cells is located in close proximity to capillaries so that contents of neurotensin granule released diffuse not only to nearby cells, but also into the surrounding capillaries. Molecules of the neuropeptide in the capillaries empty into the hepatic-portal circulation from where the peptide is transported to different tissues including other regions of the GI tract to regulate cellular activity. However, neurotensin (amino acids 1–13) has a very short half-life ($t1/2 = 0.55$ min), and thus it is quickly degraded into neurotensin fragments. The major proportion of neurotensin that is initially released in the GI tract is degraded, producing neurotensin (1–8) and neurotensin (1–11). The half-life of these fragments of neurotensin is 5 min and 9 min, respectively. The fragments—neurotensin (1–8) and neurotensin (1–11)—are the amino acid terminal fragments of neurotensin (1–13). The carboxyl fragment is neurotensin (8–13) [784] The secreted neurotensin localizes its receptors to regulate or modulate cellular activity through several mechanisms. There are presently three types of neurotensin receptors—neurotensin receptor 1, 2, and 3 (high affinity of nanomolar range receptor—NTSR1, low-affinity receptors—NTSR2 and NTSR3). NTSR1 and -2 are G protein-coupled receptors. The stimulation of NTSR1 is coupled to IP3 signaling pathway. NTSR2 may have multiple signaling pathways which include activation of extracellular kinases in addition to stimulation of distinct GPCR. NTSR3 is also called sortilin-1 (SORT1) or gp95, a sorting protein comprising one transmembrane domain. NTSR3 is a non-G protein-coupled receptor [790–794]. The receptor is mainly expressed in organelles including the Golgi apparatus [794]. The stimulation of neurotensin receptors is also associated with differential expression of transcription factors (e.g., c-Fos) that regulate gene expression. These receptors are differentially expressed depending on the tissue involved. The three receptors have different roles and activity levels. Besides, the effects of neurotensin may be mediated via stimulation of release of histamine, catecholamines and prostaglandins [795] The peptide neurotensin has a range of roles to play in an organism, and it is referred to as a peptide with multitasking effect. Over thirty biological effects of neurotensin have been documented. These effects are mediated via neurocrine, paracrine, and neuromodulatory mechanisms [796]. The functions of neurotensin include regulation of GI motility and pancreatic and biliary secretion. Peripheral administration of the neuropeptide leads to hyperglycemia, decreased gastric acid secretion, decreased gut motility, stimulation of pancreatic secretion, enhances fatty acid transport, acts as a growth factor [782, 788, 790, 797, 798]. Neurotensin regulates the release of gut hormones such as glucagon, insulin, somatostatin, histamine, catecholamines, prostaglandins and pancreatic polypeptide [785] Surprisingly, not only injection of neurotensin intraperitoneally, but also intracerebroventricular results in acute reduction in food intake. This effect is believed to be mediated via NTSR1 [782, 790]	These authors are widely acknowledged as some of the founders of the field of gastroneuroendocrinology [789]	

(continued)

Table 9.2 (continued)

Neural secretions of the gut	Description	Discoverer	Year of discovery
	Neurotensin has an analgesic effect which is believed to be mediated via NTSR1 and NTSR2 [799]. The peptide also exhibits hypothermic effect, which is mediated via NTSR1 [791]. Neurotensin has a hypotensive effect mediated via NTSR1 [791] Apart from the functions mentioned above, neurotensin also regulates neurodegeneration, metabolism, pituitary hormone secretion and also possesses antipsychotic properties [783, 784, 791] Neurotensin has a huge role in cardiovascular disease outcome. All three receptors of neurotensin have been discovered in the myocardium. Neurotensin secreting neurons are found in close proximity with the atrial and ventricular myocytes. These neurons are found in intracardiac ganglia, and also on coronary vessels. Thus, the neuropeptide regulates heart rate, myocardial contractility, blood pressure, vascular and smooth muscle tone, regional blood flow in GI tract and other tissues. Investigations have shown that neurons secreting neurotensin in cardiovascular system have reduced density in cardiac disease [795]. Increased secretion of the peptide has been implicated in pathogenesis of cardiovascular disease and is associated with cardiovascular mortality [790, 794, 795] The elevated secretion of this molecule is also related to increased risk of diabetes mellitus, and breast cancer. Neurotensin exerts a wide range of physiological effects and it has been found to play a critical role in a number of neurobehavioral diseases, such as schizophrenia, Parkinson's disease and drug addiction, and obesity [790, 794, 795] Thus, goal-driven research aimed at the defining the architectural-functional relationship of neurotensin and its receptors may provide important cues to the treatment of some diseases that have recorded a surge in recent times. Recently researchers have created neurotensin analog that selectively binds to NTSR2. One of such molecule is called NT79. This novel molecule reduces pain sensation, psychotic effects of amphetamine, hyperactivity, blood pressure, and temperature [791]		
Neuropeptide Y or neuropeptide tyrosine (NPY)	NPY is a neuronal peptide consisting of 36 amino acids, functioning as a neurotransmitter, hormone, and neuromediator and belongs to a family of structurally related amidated peptides, which includes peptide YY, and pancreatic polypeptide [800]. NPY is generally referred to as a stress-resilience neuromediator. NPY, peptide YY, and pancreatic polypeptide all belong to the NPY family [801] or pancreatic polypeptide family of peptides [811]. NPY exhibits a 70% homology with peptide YY and a 50% homology with pancreatic polypeptide. All three peptides are processed from 94–95 amino acid prepro-polypeptide translated from the corresponding mRNA [801]. NPY is amidated at the COOH terminus [802] NPY was initially discovered in brain tissues of pig. The peptide is found in a variety of species including humans. It is one of the most conserved neuroendocrine peptides throughout evolution. Even for separated with large evolutionary distance (hundreds of years), over 90% amino acid sequence identity for NPY has been observed. The peptide was characterized in 1984 by Minth et al. [804]. The name is derived from the single-letter database code (Y) for the amino acid tyrosine, since it contains five tyrosine residues including an amidated C-terminal tyrosine [805–807]	Kazuhiko Tatemoto, Mats Carlquist and Viktor Mutt [803]	1982

(continued)

Table 9.2 (continued)

Neural secretions of the gut	Description	Discoverer	Year of discovery
	The NPY gene is located on chromosome 7 (7p15.1). The NPY gene is about 8 kilobases in length with four small exons (each having less than 200 base pairs [bp]) separated by three introns of about 965, 4300, and 2300 bp. The NPY gene synthesizes the prepro-NPY, a precursor peptide containing 97 amino acid residues, which is cleaved, resulting in the pro-NPY, a 69-amino acid peptide. The pro-NPY undergoes cleavage by the proconverting enzymes (PC)1/3 and/or PC2 at a single dibasic site (Lys38-Arg39), releasing a C-terminal peptide (with 30 amino acids) and NPY1-39, which is further processed by a carboxypeptidase-like enzyme. The resulting NPY1-37 is finally amidated at its C-terminal end by the peptidylglycine-amidating monooxygenase in a process that removes a glycine molecule resulting in the production of the active peptide NPY1-36. This peptide can further undergo processing by the enzyme dipeptidyl peptidase-4 (DPP-4) to produce NPY3-36 and another enzyme aminopeptidase P to produce NPY2-36 [800, 802, 804]. Immediately after synthesis, the peptide folds to avoid degradation producing a three-dimensional structure called PP-fold. The neuropeptide is then packaged and stored into secretory vesicles in the cytoplasm within the cell body and nerve terminals. It is secreted upon activation by specific stimuli [801, 802, 807] NPY is one of the most abundant peptides expressed in the brain (hypothalamus, amygdala, hippocampus, and cerebrum), spinal cord, and peripheral organs, tissues, and cells (liver, spleen, heart, adrenal glands, kidney, GI tract, and endothelial cells of blood vessels). The peptide shows activity at pre- and postsynaptic neural sites. The peptide is also found in substantial level in glial cells [801, 802, 808]. Even though the neuropeptide was first discovered in brain, the adrenals are the organ believed to have the highest level of NPY expression followed by the GI tract. In the GI tract, the peptide is produced by certain neuroendocrine epithelial cells, and the submucous plexus [807, 809] The neuropeptide family has a variety of actions which are mediated through a family of Gi protein-coupled receptors called Y receptors. Currently, there are at least 5 subtypes: Y1-R, Y2-R, Y4-R, Y5-R, and Y6-R. Y6-R is a pseudogene in primates, with no known activity as yet. Y3-R has not yet been cloned. Y4-R subtype is activated specifically by pancreatic polypeptide. Thus, Y1R, Y2R, and Y5R are the three major subtypes of NPY receptors that mediate the biological effects of NPY in humans and animals. All three receptors are localized on human chromosome 4q31-q32. The receptors are also receptive for other members of the protein family such as peptide YY, pancreatic polypeptide. In the GI tract, the NPY receptors also include the GPCR—Gpr119 [801, 802, 810] Activation of NPY receptors results in inactivation of adenylyl cyclase by the alpha-G subunit which in turn leads to decreased cAMP production, depressed Ca^{2+} channel activity, and enhanced G protein-coupled inwardly rectifying potassium channel (GIRK) currents [801]. The beta–gamma subunit of the G protein activates a range of kinases. These signaling pathways activate downstream molecules resulting in a range of physiological responses. The effects of downstream signaling also include initiation of gene transcription, stimulation or inhibition of hormone or neurotransmitter secretion [801, 802]		

(continued)

Table 9.2 (continued)

Neural secretions of the gut	Description	Discoverer	Year of discovery
	NPY is involved in the regulation of a wide range of physiological processes in both central and peripheral cells and tissues. NPY is involved in the regulation of cardiovascular activity, higher brain functions (e.g., learning and memory), circadian rhythm, lung function, anxiogenesis, release of hypothalamic and pituitary hormones, reproductive processes, growth hormone secretion, and is also implicated in cancer, and analgesia. NPY is also involved in modulation of sweet and umami taste sensation. The neuropeptide also exerts sedative and anticonvulsant effects [801, 802, 811, 812]. NPY modulates other neurotransmitters including catecholamines (e.g., norepinephrine) and adenosine triphosphate (ATP) [813]. It is co-secreted with norepinephrine and ATP in the sympathetic nervous system [801, 802] NPY regulates hunger and body weight homeostasis. NPY stimulates food ingestion. Thus, the neuropeptide has a role to play in the pathophysiology of obesity. In affirmation of this hypothesis, it has been shown that increased expression of NPY in the dorsomedial hypothalamus increases food intake. In contrast, decreasing the expression of NPY by ablation in this hypothalamic area reduces hyperphagia and the incidence of obesity. Thus, inhibition of NPY expression or it signaling pathways results in reduction in food intake with subsequent effect on weight loss. This effect may be related to the role of the peptide in mediating the activity of intestinal GPR119 and the control of glucose tolerance [814, 815] In the cardiovascular system NPY functions as a vasoconstrictor and increase vascular motility. In heart failure, septic shock, serum level of NPY has been found to be elevated. The neuropeptide is currently been evaluated as a possible marker of cardiovascular disease and may be additional means for enhancing therapy in endotoxic and septic shock. Interestingly, NPY enhances diuresis and natriuresis through direct effects renal tubules independent of its hemodynamic properties making it a promising agent for controlling edema. Moreover, NPY also acts as a vasopressor [814, 816] Disorder in the expression of NPY may be related to endocrine cancers, hyperlipidemia, atherosclerosis, coronary heart disease, diabetes, obesity, higher birth weight, alcohol dependence, alcoholism, depression, schizophrenia, and stress-related psychiatric illnesses [801, 802, 817–820]		
Peptide YY (PYY)	PYY (also known as peptide tyrosine-tyrosine or pancreatic peptide YY) belongs to the pancreatic polypeptide family and comprises a peptide chain of 36-amino acids, produced by central and peripheral cells, discovered and characterized in porcine intestinal extracts using a novel method of identifying peptides that was invented some years previously [821]. Kazuhiko Tatemoto and Viktor Mutt made a breakthrough in the 1970s when they developed a novel method by for identifying carboxyamidated peptides in tissue extracts. Interestingly, a vast number of regulatory peptides are carboxyamidated, so it was no surprise that the scientists had a great success in identifying many peptides [822] PYY was subsequently found to have structural similarities with other regulatory peptides including pancreatic polypeptide and NPY [803]. In the brain, PYY is produced by certain group of cells in the medulla oblongata, hypothalamus, amygdala, hippocampus, thalamus, and hindbrain [823]. In the peripheral tissues, particularly in the GI tract, PYY is produced by the epithelial L cells of the jejunum, ileum, and colon in response to food (mostly protein) intake. Though scantly, PYY is also produced in the esophagus, stomach, duodenum, and rectum. Thus, serum concentration of the peptide increases following food	Kazuhiko Tatemoto [821]	1982

(continued)

Table 9.2 (continued)

Neural secretions of the gut	Description	Discoverer	Year of discovery
	intake, but decreases on fasting [824–826]. Significant increase in blood PYY level is observed following oleate meal, whereas local increase is observed following carbohydrate meal [825]. PYY is also produced by the cells of the ENS [827] PYY is produced by the islets of Langerhans. PYY is co-secreted with pancreatic polypeptide or glucagon in the gut and pancreas [824, 828, 829] The peptide PYY exists in two forms: PYY (1–36) and PYY (3–36) [830, 831]. The signaling pathways of these peptides are mediated predominantly by NPY receptors—Y1, Y2, and Y4 receptors [832] PYY plays a range of physiological roles. Research has shown that intravenous and intracerebral injection of PYY or introduction into the gut reduces appetite. Thus, the peptide is anorexigenic. The peptide is also known to reduce gastric secretion [824]. PYY also inhibits gastric motility, and thus it slows gastric emptying and intestinal transit time. It also increases exocrine pancreatic secretion, water, and electrolyte absorption in the colon. It may also have a role in endocrine pancreatic secretion. PYY possesses hypoglycemic effect, which may be due to multiple signaling pathways that may involve insulin secretion or action [833, 834]		
	Clinical Correlate 9.2 **Mechanisms of Regulation of Appetite and Satiety—Background Information for Obesity Therapy and Translational Physiology** Emerging evidences indicate that the molecule PYY plays a crucial role in the pathophysiology of obesity through multiple mechanisms that involve regulation of energy homeostasis in the gut–brain axis [835]. The mechanism of action of PYY on CNS involves the modulation of the activity of anorexigenic (appetite suppression) and oxerigenic (appetite promotion) neurons. For example, insulin and leptin are anorexigenic substances that suppress appetite; whereas ghrelin is oxerigenic as it stimulates appetite [836]. Interestingly, apart from ghrelin, the ghrelin gene produces other anorexigenic peptides such as des-acyl ghrelin and obestatin [837]. Ghrelin is released during hunger to signals specific group of neurons located in various brain regions. Ghrelin activates these neurons via the growth hormone secretagogue (GHS) receptors. In addition, this oxerigenic hormone (ghrelin) stimulates the production of other oxerigenic peptides of arcuate nucleus such as NPY and agouti-related protein (AgRP) and modulates anorexigenic pro-opiomelanocortin neurons. One of these brain regions which play a significant role in integration of hormonal and metabolic signals to maintain energy homeostasis is arcuate nucleus and the paraventricular nucleus of the hypothalamus as well as the brainstem [837]. The major neurons in the arcuate nucleus controlling food intake are POMC and CART (anorexigenic) referred to as POMC/CART neurons (control appetite), and AgRP, NPY (orexigenic) referred to as NPY/AgRP/GABA neurons—control satiety signals [838]. All these neurons project to the paraventricular, lateral and dorsomedial hypothalamus which contain high proportion of leptin and POMC neurons. The population of NPY/AgRP/GABA neurons of arcuate nucleus projects to the paraventricular nucleus and send neuronal fibers from GABAergic neurons to the arcuate nucleus POMC neurons. GABA blocks the anorexigenic effect of α-MSH [839–841]		

(continued)

Table 9.2 (continued)

Neural secretions of the gut	Description	Discoverer	Year of discovery
	Several neurons form synapses with alpha-MSH neurons located in the arcuate nucleus which inhibits food intake. The alpha-MSH neurons produce POMC, a precursor of melanocortins which include melanin-stimulating hormone, adrenocorticotrophic hormone (ACTH), and β-endorphin. Melanin-stimulating hormone is a strong orexigenic molecule. The receptors for this hormone are widely distributed in the hypothalamus, brainstem and include the G protein-coupled receptor family, melanocortin receptors specifically type 3 and 4 (MC3R and MC4R). There are altogether five receptor subtypes: MCR1-5. The paraventricular nucleus of the hypothalamus has high expression of melanocortin receptors. In addition, this hypothalamic region also produces anorexigenic peptides including thyrotropin-releasing hormone (TRH), corticotrophin releasing hormone (CRH), and oxytocin. The lateral hypothalamus and perifornical area produce the orexigenic substances orexin A and melanin-concentrating hormone [839, 842] The arcuate neurons also extend to the brainstem, nucleus accumbens, and amygdala. These brain regions also contain MC4R, which is required for regulating food signal [843] In the brainstem, areas involved in appetite and satiety regulation include dorsal vagal complex (dorsomotor nucleus), area postrema (AP), NTS, parabrachial, hypoglossal, trigeminal, lateral reticular and cochlear nuclei, locus coeruleus and inferior olive. These areas contain leptin receptors to which the hormone leptin localizes to suppress feeding. The main areas believed to control feeding are the NTS, DMN, AP, which have high expression of MC4R. Neurons secreting POMC from the arcuate nucleus extend to the NTS. These neurons can be activated by CCK. Neurons of the dorsal vagal complex extend to the NTS. Moreover, dorsal motor complex of the vagus also receive fibers from the POMC neurons of the arcuate nucleus. Neurons of the dorsal motor complex reach the nucleus tractus solitarius. Activation of the sensory receptors extending to the vagus in the GI tract allows the transmission of satiety signals to the vagus motor complex and then to the NTS. The afferent neurons of the NTS project to the arcuate nucleus of the hypothalamus. This is where satiety signals are integrated with appetite signals. The axons of the neurons of the hypothalamus also reach the nucleus accumbens [839, 842] Nitric oxide is another anorexigenic neuromodulator and neurotransmitter that inhibits the activities of ghrelin [844]. NO is produced in the CNS and in peripheral tissues including the GI tract. NO is synthesized by nitric oxide synthase (NOS) and induces the formation of cyclic guanosine monophosphate (cGMP) via stimulation of guanylate cyclase. In experimental condition, sodium nitroprusside, an artificial donor of NO is used to generate the neurotransmitter [844]. Apart from the signal of energy balance mentioned above, other anorexigenic hormones and molecules which are also called satiety signals include leptin, amylin, adiponectin, insulin, glucagon-like peptide 1 (GLP-1), glucose, fatty acids, lactate, pyruvate, amino acids, histamine, melanocyte stimulating hormone, α-lipoic acid, PYY, pancreatic polypeptide, cholecystokinin, oxyntomodulin, cocaine- and amphetamine-regulated transcript (CART) and pro-opiomelanocortin (POMC), ciliary neurotrophic factor (member of the cytokine family), serotonin, nesfatin-1 [842, 845–849], and oxerigenic hormones are ghrelin, cannabinoids, agouti-related peptide (AgRP), NPY [847, 848]. Central serotonin is produced in the dorsal raphe nucleus [849]. Nesfatin-1 is a novel satiety molecule found in hypothalamus (PVN) and GI tract. The level of nesfatin-1 reduces with fasting		

(continued)

Table 9.2 (continued)

Neural secretions of the gut	Description	Discoverer	Year of discovery
	NPY is an orexigenic molecule expressed in arcuate neurons that are activated by food deprivation and inhibited by feeding in a nutrient-dependent manner. PYY and leptin also prevent fasting-induced activation of the arcuate neurons. The anorexigenic hormone, amylin, secreted mainly by the pancreas and stomach, acts via the area postrema to control energy intake [845]. The mechanism for anorexigenic and orexigenic regulation may be related to AMP-activated protein kinase (AMPK) activity and mammalian target of rapamycin complex-1 (mTORC1). Anorexigenic peptides activate mTORC1 in a phosphoinositide-3-kinase (PI3 K)- and protein kinase B (PKB)-dependent manner, whereas orexigenic peptides activate AMPK [836, 847]. Interestingly, increased AMPK activity can block the activity of mTORC1		
	The secretion of PYY increases postprandially (immediately following food intake) and activates vagus afferent nerve fibers or via the circulatory system activate NPY neurons of the arcuate nucleus, which also contains receptors for orexigenic molecules. The expression of c-Fos increases in arcuate neurons on fasting. The administration of PYY has been shown to reverse this increase [837]. The inhibitions of NPY by PYY signals are similar to the mechanism of inhibition of POMC neurons by PYY. PYY inhibits POMC cells by reducing action potential, hyperpolarizing the membrane potential, decreasing input resistance and inward calcium currents, increasing G protein-gated inwardly rectifying K^+ channel currents and presynaptically inhibiting release of glutamate [850]. However, PYY may activate anorexigenic pathways by reducing the synaptic inhibition mediated by Y2 receptor. Thus, it is possible that PYY signaling in the CNS can occur via multiple pathways. Importantly, researchers have shown a dual role of PYY in satiety regulation [851]. While peripheral administration of PYY results in inhibition of food intake, intracerebral administration leads to stimulation of food intake [851]		
	In addition to the peptide released by the central neurons, GI PYY travels to the CNS (hypothalamus) via the circulatory system. In the hypothalamus, PYY acts as an agonist in the neurons of the arcuate nucleus by activating Y2 receptor, a presynaptic inhibitory receptor that blocks the orexigenic effects of the NPY/AgRP neurons. One of the effects of the inhibition of hypothalamic neurons (arcuate POMC neurons) by PYY is the release of the inhibitory effect of the GABA on these groups of neurons, which results in the removal of the inhibitory effect on POMC neurons. So, POMC neurons now release alpha-MSH, a powerful inhibitor of appetite. Therefore, appetite is inhibited [842, 845–849, 851]		
	PYY reduces appetite by a distinct mechanism other than acting on MC4R like ghrelin and leptin. However, PYY may act on other subtypes of melanocortin receptors. PYY inhibits ghrelin neurons of the arcuate nucleus [846]		
	Endogenous Molecules with both Orexigenic and Anorexigenic Properties: Some molecules serve both as orexigenic and anorexigenic signals. These molecules exhibit dual properties of regulating both appetite and satiety. The peptide molecules include noradrenaline, dopamine and the orexins. This dual property of regulating both appetite and satiety is due to their expression of different receptor types in different regions, which signal by different pathways. The secreted noradrenaline in the paraventricular nucleus of the hypothalamus activates either α1- or α2-adrenoreceptors. The former has anorexigenic effect, whereas the latter is orexigenic [849]. In addition to α1, β2, and β3 also exhibit anorexigenic effect. However, the level of		

(continued)

Table 9.2 (continued)

Neural secretions of the gut	Description	Discoverer	Year of discovery
	noradrenaline in the body exhibits dark-and-light cycle, which may also influence feeding. Noradrenaline is synthesized in the dorsal vagal complex and the locus coeruleus [849]. Dopamine possesses anorexigenic effect in the arcuate nucleus and lateral hypothalamus, whereas in the ventromedial hypothalamus, it possesses orexigenic effect. The anorexigenic effect of dopamine may be mediated via D1, D2 or D1/D2 receptor, whereas the orexigenic effect may be mediated via D5 receptor [849] The orexins (also known as hypocretins) are both anorexigenic and orexigenic in nature, derived from the precursor protein preproorexin. The orexins were discovered in 1998. Examples of orexins include orexin A, orexin B, and OXA with differential expression in regions of the brain. OXA is expressed in neurons of the PFA, LHA perifornical area, lateral hypothalamic area and dorsomedial nucleus. The neurons of these brain regions send projection fibers to neighboring hypothalamic nuclei and extrahypothalamic areas including the NTS. The orexins generally mediate signaling events via stimulation of the GPCR—OX1 (HCRTR1) and OX2 (HCRTR2) receptors [849, 852–854]. The orexins regulate not only feeding, but also sleep, and thus remain a key therapeutic target for disorders in sleep-wake cycle [852]. Emerging studies indicate that certain genetic variants or polymorphisms of orexins may be associated with excessive daytime sleepiness and narcolepsy. Though mutations in the orexin gene are rare, they may occur and may result in other disorders including polydipsia-hypernatremia observed in schizophrenia and affective disorders [853]		
Vasoactive intestinal polypeptide (VIP)	VIP is a basic octacosapeptide (28 amino acid residue peptide) hormone and non-cholinergic, non-adrenergic neurotransmitter with vasoactive properties and belongs to the secretin-glucagon peptide family [855]. The peptide was discovered unexpectedly during the purification process of secretin [856]. The peptide was initially reported in 1969-70 by the same group of researchers when they discovered that a novel peptide substance from normal lung tissue caused vasodilation and relaxed smooth muscle [855, 857]. This led the scientists to name the peptide vasoinhibitory peptide. About 2 years later, the same group of researchers isolated and characterized a peptide from porcine duodenum. The duodenal peptide was found to be identical to the one they had previously reported. However, the characterized gut peptide was found to have a greater vasodilatory effect (vasopressor effect) compared to the one identified in lung tissue. The vasopressor effect of the intestinal peptide was not any worse than that of bradykinin. Hence the peptide was renamed vasoactive intestinal peptide [858, 859] VIP is a ubiquitous peptide produced in both central and peripheral tissues [859, 860]. Centrally, the peptide is produced in the brain and spinal cord. In the peripheral tissues, it is produced in neurons innervating smooth muscles such as those of the blood vessels, genitourinary and GI tracts. Neurons secreting VIP which innervate smooth muscles, blood vessels and the conduction system of the heart are found in the peripheral and intrinsic nervous systems of the GI tract, tracheobronchial tree, genitourinary tract, pancreas, liver, and heart [861–868] When an adequate stimulus is directed toward the cell, cytoplasmic granules containing VIP are released. VIP may diffuse to the surrounding cells or diffuse into circulation from where it is transported to the target to localize its receptors. At least two types of VIP receptors have been identified: VIRP1 and VIRP2 (also known as VPAC1 and VPAC2, respectively). They all belong to the GPCR family [869]	Sami I. Said (1928–2013) and Viktor Mutt (1923–1998), both physicians who at one time worked collaboratively at Karolinska Institute, Stockholm, Sweden. Their laboratory identified a host of gut peptides which included various lengths of secretins (secretin-27, -28, -29, -71); of cholecystokinins (cholecystokinin-33, -39, –58) as well as many other gut peptides [855–857]	1969-70

(continued)

Table 9.2 (continued)

Neural secretions of the gut	Description	Discoverer	Year of discovery
	Interestingly, VPAC1 and VPAC2 bind not only VIP, but also PACAP, which is also produced in GI tract [see below]. Type one and two receptors are widely distributed in the body: GI tract (including stomach, intestine), liver, pancreas, lung, heart, muscle, kidney, adipose tissue, and internal genital organs. In the CNS, the receptor is differentially expressed in various regions [870, 871]. The binding of VIP to their respective receptors are associated with several signaling cascades. Type 2 VIP receptor activation in turn leads to the activation of the membrane enzyme, adenylate cyclase. This reaction leads to the activation of several downstream signaling pathways. However, the signaling of VIP depends on its concentration as low-level VIP has been associated with vasodilation which is mediated via the two types of VIP receptors activating downstream adenylate cyclase and potassium channels. In contrast, higher concentration of VIP has been associated with increase in contractility of the ventricle, heart rate, decrease refractory period of both the atrium and ventricles and decrease in atrioventricular conduction time [861]. These effects on the heart are mediated via cyclic AMP [872]. Thus, VIP is a potent chronotropic and inotropic agent [872] Apart from the ligands activating these receptors, other peptides produced in the GI tract can bind to the receptor subtypes. The peptides include peptide histidine isoleucineamide (also known as peptide histidine isoleucine, PHI), peptide histidine methionineamide (also known as peptide histidine methionine, PHM) and peptide histidine valine (PHV). Compared with other ligands, PACAP-27 and PACAP-38 exhibit over a 100-fold potency for VIP receptors [873]. These peptides will be briefly discussed below PACAP receptor pharmacology and agonist bias: analysis in primary neurons and glia from the trigeminal ganglia and transfected cells The VIP peptide exhibit a wide range of biological activities including smooth muscle relaxation and systemic vasodilation leading to hypotension, influence of secretory process of glands [860, 874]. VIP signaling is known to stimulate respiration and also cause hyperglycemia [875] VIP regulates circadian clock by integrating multiple signals from the suprachiasmatic nucleus, growth, energy expenditure, and immunomodulation [870, 876] Excess VIP is metabolized in the liver and lungs by neutral endopeptidase [861]. Thus, excessive metabolic functions of this enzyme may predispose an individual to diseases associated with decrease in the concentration of VIP. VIP plays a role in anti-inflammatory response. The density of VIP receptor and peptide concentration decrease with hypertension, obesity, diabetes, and hypothyroidism indicating that normal receptor density and signaling by the VIP are important in the cells and tissues involved in these diseases [861, 877–879]		

(continued)

Table 9.2 (continued)

Neural secretions of the gut	Description	Discoverer	Year of discovery
Peptide histidine isoleucin or peptide histidine isoleucinamide (Peptide HI or PHI)	PHI is a neuropeptide consisting of 27 amino acid residues and belongs to glucagon-secretin family, having similar amino acid sequence homology with VIP, secretin, glucagon, and gastric inhibitory polypeptide. Indeed, some of the biological activities of PHI are similar with the effects exerted by the secretin-glucagon family of peptides. The peptide was discovered, isolated and characterized from porcine upper intestinal extract [822, 880]. Characterization of the peptide showed the presence of COOH-terminal isoleucine amide structure. The amino acid histidine is the first amino acid in the sequence, hence the name "peptide histidine isoleucine" [880] Central PHI interacts with substances produced in the brain particularly those in the hypophysial portal system and maybe prolactin-releasing hormone. PHI and VIP are coded by the same gene in rats and humans. However, in humans, instead of isoleucine, the COOH-terminal has a methionine; hence the peptide in humans is called peptide histidine methionine (PHM) [881] Peptide HI is found in central and peripheral tissues. The peptide is secreted in the intestine and many organs including the liver, pancreas, and gallbladder. It has a wide range of biological actions, affecting the cardiovascular, GI, respiratory, and central nervous systems [882–884] PHI is involved in GI transport, secretion of fluid and electrolytes [884]. In the CNS, the peptide may modulate or control the release or functions of other neurotransmitters. Like VIP and PACAP, PHI increases the release of prolactin, insulin, and glucagon [881, 885] Peptide HI is a ligand to VIP or PACAP receptors activating downstream adenylate cyclase [883, 886]. While there are presently two known types of VIP receptors, there are three PACAP receptors (PAC1-R, VPAC1-R, and VPAC2-R) [887]	Kazuhiko Tatemoto and Viktor Mutt (1923–1998) [822, 880]	1981-1980
Peptide histidine methionine (PHM) or Peptide MI	PHM (also known as peptide histidine methioninamide) is a human peptide neurotransmitter containing 27 amino acid residues and belongs to the glucagon-secretin family of peptides. It is the human counterpart of the PHI neuropeptide. PHM is produced from the same prohormone as VIP. The COOH-terminal in the neuropeptide has a methionine, while the other end is composed of histidine [881, 885] The peptide is expressed in central and peripheral tissues. PHM prohormone is produced by neurons in the central and peripheral nervous systems as well as in GI tissues. PHM is found in high concentration in the GI and urogenital tracts and nasal mucosa. The peptide is co-secreted with ACh and a range of neuropeptides [881, 885] PHM binds to two VIP receptors, VPAC1-R, and VPAC2-R. However, PHM possesses a lower binding affinity compared with VIP or PACAP [886, 887] It has a wide range of biological actions, affecting the cardiovascular, GI, respiratory, and central nervous systems. PHM increases the release of prolactin, insulin, and glucagon. The neuropeptide causes vasodilation. PHM may reinforce and prolong the effects of some neurotransmitters. The transmitter stimulates GI secretion [881, 885]	Tatemoto and Mutt (1923–1998) [880]	1981

(continued)

Table 9.2 (continued)

Neural secretions of the gut	Description	Discoverer	Year of discovery
	PHM might be a key regulator of prolactin-releasing hormone. This is evident in prolactinomas in which a low level of the neuropeptide is observed [881]. Importantly, the neuropeptide is highly expressed in hypothalamus. This suggests that PHM may participate in the regulation of anterior pituitary hormone secretion [888–890] High plasma level of PHM is a key diagnostic marker of peptide-producing tumor (VIPoma) that causes watery diarrhea-hypokalemia-achlorhydria syndrome (also known as Verner-Morrison syndrome or watery diarrhea syndrome or pancreatic cholera)—WDHA syndrome [891–893]. VIPoma is a rare neuroendocrine neoplasm characterized by excessive and autonomous production of VIP from certain tumor cells. The tumor cells are usually cells of the pancreatic islets [894–896], but adrenal pheochromocytoma has been implicated in this disease [896] The annual incidence of the disease is estimated at 1 per 10 million individuals [894, 896, 897] VIPoma is characterized by high plasma VIP level (989 pg/mL), hypokalemia, hypercalcemia, and hyperglycemia [894, 896, 897]. A relatively recent report indicated that VIPoma also secrete dopamine [897] In WDHA syndrome, chronic or refractory watery diarrhea is usually observed in 100% cases, hypokalemia in 100% (which leads to dehydration). However, it should be noted that instead of achlorhydria, hypochlorhydria may be experienced in addition to acidosis due to bicarbonate wasting, hence the acronym WDHHA [896]. VIPoma may be associated with pulmonary thromboembolism [897]		
Peptide histidine valine (PHV) or Peptide HV	Like PHI, PHM, and VIP, PHV is synthesized from the same 170-amnio acid residue precursor regulatory peptide "VIP" (prepro-VIP) and has identical structural features and biological functions with other members of the VIP-related peptides [887] PHI/PHM/PHV/VIP gene is expressed in many cells and tissues including intestine, brain, and gallbladder. However, PHI/VIP gene is expressed in the central (brain) and peripheral (intestine, gallbladder) organs. The PHV receptor gene is primarily expressed in the pituitary and to a lesser extends in the intestine and gallbladder, suggesting that PHV may play a role in the regulation of pituitary function [887] PHV has a greater potency than peptide histidine methionine in relaxing smooth muscle (of the GI tract, uterus, and trachea), which is in turn associated with reduction of blood pressure [898]		
Pituitary adenylate cyclase-activating peptide (PACAP)	PACAP is a pleiotropic neuropeptide that functions as a neurotransmitter, neuromodulator, vasodilator, neurotrophic factor, neuroprotective factor, hypothalamic hormone, neuroendocrine hormone, and belongs to the secretin/glucagon/growth hormone-releasing factor/VIP family. The peptide exists in two major forms. The 38-amino acid residue peptide is more abundant compared to the 27-amino acid form. Relatively recently, however, a PACAP-related peptide has been isolated, though its functions are not well determined [901, 902]. PACAP was actually discovered as a hypothalamic hypophysiotrophic hormone which stimulated adenylate cyclase in perfused pituitary gland (adenohypophysis), specifically cells of the anterior pituitary [899]. Like the classical hypothalamic hypophysiotrophic hormones (thyrotropin-releasing hormone, luteinizing	Miyata et al. [899, 900]	1989, 1990

(continued)

Table 9.2 (continued)

Neural secretions of the gut	Description	Discoverer	Year of discovery
	hormone-releasing hormone, growth hormone-releasing hormone, corticotropin-releasing hormone, somatostatin, and dopamine, prolactin) which were already discovered by the 1980s, were known to stimulate adenylate cyclase [903–905]. PACAP is a highly conserved peptide during the evolution from protochordate to mammals [902] PACAP is widely distributed in the brain (hypothalamus, anterior adenohypophysis and cerebellum) and peripheral organs. PACAP is found in peripheral organs such as endocrine pancreas, adrenal medulla, gonads, GI, respiratory, and urogenital organs and tracts [902, 906, 907] PACAP signals through three known receptors which had been cloned: VPAC1, VPAC2, and PAC1. These receptors are activated by VIP, PACAP-38 (also activates ADCYAP1R1), PACAP-27 (also activates ADCYAP1R2), PHI, PHM, and PHV [904, 905]. All three receptors are GPCRs of class B. The specificities of these receptors for the noted substrates differ. They also differ in their pattern of expression in various tissues [908]. PAC1 receptor is PACAP-specific and is coupled to several transduction pathways. PAC1 receptors are particularly abundant in the brain and pituitary and adrenal glands. The VPAC1 and VPAC2 receptors signal primarily via adenylyl cyclase to increase intracellular cAMP with subsequent increase in intracellular calcium concentration and activation of protein kinases. Activated protein kinases cause phosphorylation of several intracellular proteins and activation of genes. VPAC receptors are expressed mainly in the lung, liver, and testis [902, 909]. Importantly, these receptors exist in isoforms which are differentially expressed in central and peripheral tissues [910, 911] The site of secretion of PACAP has a significant implication on their physiological functions. PACAP regulates both hypophysiotrophic and neurohypophysial hormones. Whereas dopamine inhibits prolactin release, PACAP stimulates the release of hypophysiotrophic hormones that specifically stimulate the release of follicle-stimulating hormone or prolactin, as well as a hormone that regulates the function of non-glandular pituitary cells such as the folliculostellate cells [909]. For instance, PACAP stimulates the secretion of the hormone interleukin-6 from the folliculostellate cells of the pituitary [901]. PACAP's influence on hormonal and peptide synthesis is believed to be mediated via paracrine or autocrine mechanism. PACAP stimulates the release of arginine vasopressin, a naturally occurring antidiuretic hormone regulating water balance in the body [901]. PACAP also stimulate the secretion of oxytocin [912] PACAP aids the growth and survival of cortical neuronal progenitors, cortical neurons, dorsal root ganglion cells, cerebellar granule cells, and peripheral sympathetic neurons [913–916] PACAP regulates the secretory activity of GI exocrine and endocrine cells. The neuropeptide play a role in relaxation of gastric smooth muscles to decrease of gastric motility [912]. PACAP stimulates electrogenic ion secretion in the GI tract [916]. PACAP in GI tract has been found in both neurons and glial cells. Surprisingly, excitatory responses in over 95% of afterhyperpolarization in the small intestine were reported to involve PACAP. The peptide plays a role in membrane depolarization, and suppression of hyperpolarizing afterpotential. Excitatory PACAP receptors are linked to adenylate cyclase activity in certain groups of myenteric neurons [917]		

(continued)

Table 9.2 (continued)

Neural secretions of the gut	Description	Discoverer	Year of discovery
	PACAP-specific receptor PAC1 is widely distributed in the cardiovascular system. The hypotensive effect of PACAP is mainly due to its role in regulating smooth muscle contractility [872] The peptide is known to be useful in maintaining numerous cerebral and cognitive functions including memory, learning, psychomotor behavior [912]. Investigation has shown that loss of function or gain of function of the PACAP–PAC1–VPAC2 signaling pathway may be associated with psychiatric illness [918]. For instance, the gene encoding the VPAC2 receptor is located on chromosome 7q36.3. Disorder in gene expression associated with this gene locus has been associated with schizophrenia and posttraumatic stress disorder. PACAP also regulates the neurotransmitter synthesis and signaling in glial cells, and thus the hormone is involved in metabolism, survival, growth, and secretion of neurons and glial cells [919]. The peptide also plays a role in the dysfunction of lipid and carbohydrate metabolism and may be associated with death [912]. PACAP also modulates glutamate signaling in neurons—a pathway implicated in metabolism [919] The peptide has a possible role in stimulating the release of insulin from the endocrine pancreas [901]. More recently, this fact has provided the basis of researching better therapies for diabetic mellitus. The major complications of dysglycaemia associated with long-standing diabetes are endothelial and pancreatic islets dysfunction, retinopathy, and nephropathy [920]. It is currently hypothesized that the prolonged hyperglycemic state in long-standing diabetes may be related to PACAP role in modulating the functions of the endocrine pancreas or adrenal medulla. PACAP functions as a secretagogue for adrenaline from the adrenal medulla through activation of thyrothropin [901]. Apart from the modulation of ACTH to regulate catecholamines, the sympathoadrenal neurotransmitter, PACAP is also present in the adrenomedullary synapse, where it is thought to be required for survival during prolonged hypoglycemia [921]. Even though the mechanism is not exactly clear, it is proposed that the absence of PACAP inhibits the activation of tyrosine hydroxylase, the rate-limiting enzyme in catecholamine synthesis, which is necessary to maintain normoglycemia through gluconeogenesis. Thus, PACAP is involved in glucose homeostasis [922]		
Calcitonin gene-related peptide (CGRP)	CGRP is a 37-amino acid peptide neurotransmitter, belonging to the calcitonin peptide family [923, 924]. The calcitonin family of peptides consists of at least 15 members, which include calcitonin, amylin, adrenomedullin, calcitonin gene-related peptides (αCGRP and βCGRP—was previously known as CGRPI & II, respectively), and intermedin (also called adrenomedullin-2). The chain length of these peptides ranges from 32 to 52 amino acids [925–928] The two forms of CGRP (α and β) are formed from different genes on chromosome 11 [924, 929] The receptor to which the peptide localizes is called CGRP receptor. This functional receptor is formed from the GPCR calcitonin receptor-like receptor (CALCRL or CLR) and a receptor activity-modifying protein (RAMP). RAMP is an accessory protein. The major transduction pathway involves the stimulating Gs subunit of the G protein [924]. CLR itself binds no known endogenous ligand, however, in the presence of RAMPs; functional receptors of calcitonin gene products (CGRP, adrenomedullin, and adrenomedullin-2/intermedin) are formed. The two subunits of the CGRP receptors belong to the family of secretin-like (or family B) GPCRs or calcitonin/CGRP family of peptides [925]. There is currently no detailed mechanism of functional CGRP activities and thus warrants further investigations [925]	Amara and co-workers [930] for the first time successfully cloned the calcitonin gene. The cloning resulted in the production of different products related to the gene, but, the neuropeptide CGRP was only identified the following year by Amara SG, Rosenfeld MG and colleagues [931]	1982, 1983

(continued)

Table 9.2 (continued)

Neural secretions of the gut	Description	Discoverer	Year of discovery
	CGRP receptors are expressed in enteric neurons, smooth muscle, endocrine cells, and vascular structures. The receptor is a key player in maintaining homeostasis in GI tract. The stimulation of this receptor leads to inhibition of gastric acid secretion and gut motility and causes vasodilation by increased gastric mucosal blood flow and mucosal resistance to injury, and modulation of visceral nociception [923, 932] CGRP is produced in both peripheral and central neurons and may be synthesized in certain neuroendocrine cells. In the central nervous system, the peptide is produced in many regions of the brain, cortex, neural cell bodies of the trigeminal ganglion, cerebrovascular system, meninges, brainstem nuclei, and also, but particularly in the dorsal horn of the spinal cord. In the peripheral tissues, the peptide is produced in, endocrine and immune-responsive organs, urogenital, respiratory, GI, and cardiovascular, smooth muscle, skeletal muscle, and cochlear systems [923, 925, 929]. The neuropeptide is distributed in the thyroid gland [925, 933]. In the GI tract, the peptide is produced by both extrinsic and intrinsic neurons, as well as enteroendocrine cells [923, 924, 934] CGRP is the most potent endogenous vasodilatory peptide discovered so far. Its vasodilatory effect by far exceeds that of NO [928] CGRP has been implicated in many pathophysiological dysfunctions in the peripheral and central tissues. For instance, some disorders of the CGRP in the trigeminovascular system have been associated with migraine and subarachnoid hemorrhage [925]. Subarachnoid hemorrhage is now thought to involve diminished level of CGRP or reduced functionality of its receptor resulting in vasospasm [932]. CGRP receptor antagonists (e.g., olcegepant, telcagepant, and ubrogepant) have shown promise for treating a range of disorders associated with CGRP receptor signaling such as migraine [933, 935–937]		
Nitric oxide (NO)	NO is a free radical gas acting as inhibitory messenger molecule that is produced in central and peripheral neurons as well as vascular endothelial cells. In central neurons, NO is produced presynaptically or postsynaptically by enzymatic reactions in response to adequate stimuli, which may be triggered by excitatory amino acid [938] In 1992, "NO" was named the molecule of the year by "Science"—one of the world-leading scientific journals. The molecule is widely distributed in the body and is highly significant to human life [939] NO activate soluble guanylate cyclase and so raise cGMP levels in target cells. Soluble guanylate cyclase is the receptor for NO [938] NO transmission (nitrergic neurotransmission) is widespread in nervous system (peripheral and central) as well as in the GI tract. NO is produced in vascular wall, macrophages, and neurons of nervous system [938, 940]. It functions in the GI, and respiratory tracts to mediate relaxation of the smooth muscles. It also functions in the genital organ (causes penile erection). Importantly noradrenergic and cholinergic transmission is controlled by release of NO, which is necessary to counteract and dominate the response to the noradrenergic or cholinergic stimulus [941, 942]	The discovery of nitric oxide as a biological agent and neurotransmitter was made by the American scientists—Louis J. Ignarro (1941–), Robert Francis Furchgott (1916–2009) and Ferid Murad (1936–). The three scientists won the 1998 Nobel Prize in Physiology or Medicine for discovering the signaling properties of nitric oxide [493] Furchgott and	1980

(continued)

Table 9.2 (continued)

Neural secretions of the gut	Description	Discoverer	Year of discovery
	In the vascular system, NO plays a role not only in vasodilation, but also inhibition of platelet aggregation (Figs. 9.32 and Fig. 9.33). Thus, increased production of NO may lead to hypotension resulting from both vasodilation, and vascular leakage. Reduced production of NO may result in elevated blood pressure due to vasoconstriction and platelet aggregation [943–945] In the nervous system, NO is associated with long-term depression in the cerebellum, long-term potentiation in the CA1 region of the hippocampus. These phenomena are important in synaptic plasticity and memory [946]. Research has shown that stimulation of *N*-methyl-*D*-aspartate (NMDA) receptor leads to the production of NO, which in turn stimulates the release of other neurotransmitters [947]. The NO-evoked neurotransmitter release is mediated by a calcium-dependent pathway. The inhibitory action involving NO is mediated via sodium-dependent pathway [948]. Increased production of NO in neurons is associated with neurotoxicity that may result in stroke and neurodegenerative diseases. The pathophysiology involves the excessive activation of NMDA receptors leading to the overproduction of NO, which becomes cytotoxic to the neurons and glial cells [948]	Zawadzki are biochemists who became the first to identify NO as a biologicalagent in 1980. Furchgott was investigating the effect of acetylcholine on blood vessels, when he realized that the neurotransmitter acetylcholine relaxed blood vessels only whentheendothelium was intact. Subsequently, this phenomenon was observed for other neurotransmitters including bradykinin, 5-HT, and histamine. Furchgott and Zawadzki reasoned that endothelial cells produced a substance, yet unknown, that caused blood vessels to relax. This substance was termed endothelium-derived relaxing factor (EDRF) [949, 950] Before the 1980s, scientists were already aware that certain substances, including drugs released NO which relaxed smooth muscles. For instance, in 1977, Murad and co-researchers, while investigating the mechanism of functioning	

(continued)

Table 9.2 (continued)

Neural secretions of the gut	Description	Discoverer	Year of discovery
		of nitroglycerin observed that this substance released NO which caused smooth muscle relaxation that resulted in vasodilation [951] It was surprising, however, that a colorless, odorless gas (but toxic), NO could be released from nitroglycerin, a major substance used in making dynamite, which was invented by the Swedish Alfred Nobel [952] Until 1985 nobody knew what EDRF was, even though NO and this factor both caused relaxation of blood vessels. The trend and believe at this time was that EDRF is produced from the endothelial cells in vivo, whereas, NO was produced from others sources that donate NO when added into the experimental milieu. Azuma et al. (1986) was one of the first to have showed that NO and EDRF had similar potency in inhibiting platelet aggregation [958]. In the following year, Furchgott and Kahn reported evidences indicating that EDRF and NO were the	

(continued)

Table 9.2 (continued)

Neural secretions of the gut	Description	Discoverer	Year of discovery
		same molecules [959]. Similar report was simultaneously made by Ignarro et al. [960]. Apart from vasodilation, the molecules similarly are inhibited by myoglobin and hemoglobin; their activities are enhanced by superoxide dismutase. Moreover they had similar half-life. Many other scientists also contributed to delineating the similarities between the two molecules. Several aspects of the pharmacology of EDRF and NO are similar [961] Soon after it was realized that the mechanism of action of NO is similar to that of EDRF. Beginning from the 1970s when Murad F and co-workers were studying NO-producing compounds and drugs, they found that these compounds activated guanylate cyclase [951]. The mechanism of action of EDRF was also known to involve activation of guanylate cyclase to accumulate cGMP. As early as 1984, Ignarro (1941–) and co-workers had shown that the Ach-induced blood	

(continued)

Table 9.2 (continued)

Neural secretions of the gut	Description	Discoverer	Year of discovery
		vessel relaxation was associated with accumulation of cGMP [962]. These amazing pharmacological similarities between EDRF and NO were necessary to relinquish any doubt that the two molecules were different In the late 1980s it was undoubtedly showed that NO was synthesized in vivo by endothelial cells from L-arginine [963, 964]. Alongside with delineation of the NO pathway, scientists observed that not only endothelial cells produced NO, but also macrophages, various neurons of the central and peripheral nervous system [938, 940, 946]	
Carbon monoxide (CO)	Like NO, CO is a polar gas that functions as auto-paracrine signaling molecule via canonical pathways in many cells of the body including peripheral and central neurons and neuroendocrine cells. Hydrogen sulfide (H2S) is another gasotransmitter that is synthesized in vivo and functions like NO and CO. It was previously thought that these gases do not have plasma membrane receptor, and thus they signal by diffusing into any cell along a concentration gradient [965, 966]. However, more recently, studies have suggested the presence of membrane or cytoplasmic receptors for the gasotransmitters currently identified (see Chap. 5 for details, and also see below). CO is a tasteless, odorless, colorless, and non-irritating gas, first considered a neurotransmitter in the early 1990s. Like NO, CO is a poisonous molecule, however, in low concentration, it is needed by the body to avert many pathological conditions. In toxic doses these molecules can cause cerebral and peripheral tissue injury [965, 966]. The normal concentration of CO in humans is less than 1%. The concentration increases in chronic smokers (9–12%). Increase in the level of CO can be due to inhalation of CO produced by incomplete combustion of gasoline or wood, exhaust of motor vehicles, heaters, and cooking equipment, or during fuel combustion. CO readily binds with hemoglobin to form carboxyhemoglobin (COHb). Its affinity for hemoglobin is 230 times that of oxygen. Thus, CO readily displaces oxygen in binding hemoglobin, which accounts for its high toxicity. Clinical symptoms of CO inhalation or poisoning appear at a	Ajay Verma and colleagues were the first to suggest that CO is a neurotransmitter on account of its physiological similarities with NO and other gasotransmitters, particularly H2S. CO, an activator of guanylyl cyclase, is formed by the action of the enzyme heme oxygenase [968]	1993

(continued)

Table 9.2 (continued)

Neural secretions of the gut	Description	Discoverer	Year of discovery
	carboxyhemoglobin level of 20–30%. However, even a concentration as low as 5% may be associated with reduction in visual acuity or perception, cognitive and driving performance and attention. Initial symptoms of CO poisoning include headache, vertigo, dizziness, weakness, nausea, vomiting, disorientation, confusion. Continued inhalation or inhalation of high dose may lead to collapse, coma, convulsion, and respiratory arrest. Death may result in severe cases [965–967]		
	CO is produced by the enzyme heme oxygenase of soluble guanylyl cyclase, activation of which produces cyclic GMP. There are two forms of the heme oxygenase: type 1 and 2. These enzymes are the CO receptor. It is stimulated by many factors particularly neurotransmitters [965, 966]. Like NO, for CO biosynthesis, molecular oxygen must be present. Thus like NO, CO may be regarded as a second messenger [969]. In the biosynthetic pathway, the enzyme converts heme substrate to CO, ferrous iron, and biliverdin, while oxidizing 1 mol of NADPH. CO is unevenly distributed in the nervous system and in every cell of the body including cells of the ENS [966]. There are two forms of heme oxygenase (HO-1 and HO-2). HO-1 is mainly found in the liver and spleen whereas HO-2 is mostly localized in neurons, glial cells, and cerebral vasculature. HO-1 activity is rapidly upregulated by seizures, hypoxia, and hypotension. HO is found in such brain regions as olfactory bulb, cortex, hippocampus, hypothalamus, cerebellum, brainstem [966]		
	CO plays diverse roles in the body. The neurotransmitter is involved in GI motility [970, 971]. CO-synthesizing enzymes are located in glomus cells of the carotid body, a very important structure in the regulation of blood pressure. CO is an inhibitory neurotransmitter of carotid sensory neurons that mediate its activity via regulation of intracellular Ca^{2+} level in the glomus cells [969, 970]. CO may be involved in memory formation, which may be mediated through long-term potentiation. Long-term potentiation is a phenomenon characterized by persistent depolarization following tetanic stimulation or induction phase of stimulation [966]		
	CO plays a role in immunity and release of hypothalamic factors [972]. CO is involved in vasodilation, and it is mediated via interactions with soluble guanylyl cyclase, which subsequently mediates activity of various ion channels including Ca^{2+} -channels of vascular smooth muscles. The vasomotor effect of CO is lower than that of NO. The CO-gasotransmitter is possibly involved in the control of circadian rhythms [966]. Like NO, CO has been proposed to stimulate GnRH secretion, which may be mediated via CO or NO action on glutamate. Glutamate controls GnRH secretion. Heme-containing gases may inhibit hypothalamic-pituitary-adrenal (HPA) axis, vasopressin, and oxytocin secretion [966, 973]		
	Increased cGMP produced on activation of the enzyme-soluble guanylate cyclase by CO leads to activation of such ion channels as smooth muscle large conductance Ca^{2+} -activated K^+ or big K^+ ion channel, voltage-activated K^+ , L-type Ca^{2+} channel, P2X receptor, and the epithelial Na^+ channel. Though mechanisms for the activation of the ion channels vary, majority of the ion channels are activated by changes in intracellular calcium. For instance, in CO-mediated vasodilation, increased intracellular calcium may activate K^+-channels which allow the flow of K^+ ions down an electrochemical gradient to cause hyperpolarization of the cell. Another mechanism for K^+-channel activation may be related to the activation of one of the subunits of the big K^+ channel by CO, which increases intracellular Ca^{2+}-concentration leading to the activation of K^+-channel [974, 975]		

(continued)

Table 9.2 (continued)

Neural secretions of the gut	Description	Discoverer	Year of discovery
	The protective role of CO is mediated through the MAPK pathway. Thus, CO is a useful molecule in both peripheral (some GI disorders) and neurological diseases (e.g., ischemic and hemorrhagic stroke, neuropathic pain, multiple sclerosis and other autoimmune inflammatory conditions) [966, 971]		
Gut endocannabinoids	Endocannabinoids are pleiotropic endogenous signaling lipid mediators localized in certain tissues of the body. They mimic the action of THC and are responsible for modulating pain perception and other physiological processes [976–979]. These lipid mediators include amides, esters, and ethers of long-chain polyunsaturated fatty acids. The best-known endocannabinoids are anandamide (arachidonoyl ethanolamide) and 2-arachidonoylglycerol. Anandamide, a lipid-derived neurotransmitter, is named after the word “ananda,” which means “delight” or “bliss” [980–982]. In the first decade of the twenty-first century, more molecules were added to the list. They include virodhamine (O-arachidonoyl-ethanolamine), noladin (2-arachidonyl-glyceryl ether), and N-arachidonoyl dopamine (NADA). Other lipid mediators include homo-linolenylethanolamide, docosate-traenylethanolamide, palmitoylethanolamide, oleoylethanolamide [982]. The exogenous cannabinoid, THC, and endogenous cannabinoids, anandamide and 2-arachidonoylglycerol are altogether referred to as lipid neurotransmitters [983–985] The endocannabinoid system is localized in the brain, spinal cord, endocrine glands, connective tissue and immune cells. They are also found in other organs and systems including the GI system [987, 988] Cannabinoids act on cannabinoid receptors (CB1 and CB2), transient receptor potential vanilloid 1 receptors (TRPV1), peroxisome proliferator-activated receptor alpha receptors (PPARs) and the orphan G protein-coupled receptors, GPR55 (a proposed CB3), and GPR119 [982, 989, 990]. CB1 and CB2 also belong to the family of G protein-coupled receptors. CB1 receptors are expressed mainly in neurons of the CNS and PNS, as well as ENS [989]. CB2 is highly expressed in peripheral tissues and cells, particularly in immune cells [983, 989] Activation of CB1 and CB2 is coupled to the stimulation of Gi/o protein, resulting in decrease in cellular adenylyl cyclase and subsequently decrease in cAMP/protein kinase A, increase in potassium channel permeability, decrease in calcium channel activity and transport. TRPV1 activation increases intracellular calcium concentration and intracellular proteins. PPARs activation increases tyrosine kinase activity [982, 994] Following their biosynthesis in the intracellular milieu, endocannabinoids are released into the extracellular space via endocannabinoid membrane transporter [982, 994]. These molecules in the extracellular milieu may be transported into the cell by carrier-mediated uptake or inactivated by enzymes. Endocannabinoids are hydrolyzed by fatty acid amide hydrolase. The half-life of endocannabinoids is about 5 min [989]. These lipid mediators can diffuse into the cell via Na^+ -independent facilitated transport (anandamide membrane transporter), a selective, saturable, temperature-dependent process involving the transport of the lipophilic molecules (e.g., AEA and 2-AG) through the plasma membrane. This diffusion process occurs if the concentration of extracellular endocannabinoids is higher than their concentration on the intracellular environ [994]	The Israeli organic chemist, Raphael Mechoulam (1930 –), together with Yechiel Gaoni, carried out the isolation, structure elucidation and total synthesis of Δ9-tetrahydrocannabinol (THC), the primary active agent of cannabis [986] After three decades following the initial discovery of cannabinoids, the endogenous cannabinoids anandamide (arachidonoylethanolamine or AEA) and 2-arachidonoyl glycerol (2-AG) were isolated around the late 80s to early 90s of the last century [991–993] In the 1960–1980s, it was hard to believe that lipid molecules could serve as neurotransmitters. Moreover, it was widely known that neurotransmitters could only be amino acids, peptides, or amines. However, following	1964

(continued)

Table 9.2 (continued)

Neural secretions of the gut	Description	Discoverer	Year of discovery
	Unlike other neurotransmitters that are released in the presynaptic neuron to excite the postsynaptic neuron, endocannabinoids are released from postsynaptic neuron and travel backward (retrograde transmission) across the synapse, to stimulate endocannabinoid receptors on adjacent nerve terminals. This leads to inhibition of the release of both excitatory and inhibitory neurotransmitters. This way, postsynaptic neuron can influence its own functioning or incoming synaptic signal [983] The endocannabinoids are generally involved in the regulation of appetite, mood, memory, and other physiological functions [979, 990]. In the gut, vagal afferent CB1 receptors are involved in the control of satiety, motor activity, emesis, etc. In addition to endocannabinoids, cholecystokinin, and ghrelin may also activate these receptors [979]. The brainstem control of gut functions may occur by endocannabinoid activity on CB1, CB2, and TRPV-1 receptors. Recall that the major endocannabinoid receptor in the gut is CB1 and is largely expressed on enteric cholinergic nerves where the receptors when activated inhibit ACh release in neurons of the myenteric plexus. Thus, endocannabinoids and their receptors can mediate inhibition of GI activities such as gastric emptying, intestinal peristalsis, and secretion [989, 997–999]. The lipid neurotransmitters are also involved GI inflammation, pain sensation, nausea, and emesis [979, 990]. Thus, the gut endocannabinoids serve as protective molecules in pathological conditions and aid physiological functioning of the gut and extragut tissues	decades of search for the receptors and behavior of endocannabinoids, a breakthrough was made in 1990 and 1993 when CB1 and CB2, respectively, were isolated and cloned [995, 996]. The discovery of these receptors relinquished doubts about the function of lipid molecules as neurotransmitters	
Gut opioids	Opioids belong to the pleiotropic signaling system localized in certain tissue and cells of the body and are primarily responsible for modulating pain perception and other physiological processes [976, 977]. Enkephalins (ENK), dynorphin (DYN), endorphins (END) are three well-characterized opioid peptides and together with their receptors belong to the gut opioid system. Other endogenous opioids are endomorphins and nociceptin/orphanin FQ [978, 979] Opioids are produced in certain neuroendocrine cells (e.g., tuft cells—these cell also produce prostanoids) and neurons of the gut plexuses and central nervous system [1000]. For instance, met-encephalin, leu-enkephalin, β-endorphin, and dynorphin are produced in enteric neurons and mucosal endocrine cells [978]. Their tissue distribution also varies. The concentration of met-enkephalin in the gut is by far higher than that of leu-enkephalin by about 10 times. Met-enkephalin (amino acid sequence: Tyr-Gly-Gly-Phe-Met) and leu-enkephalin (amino acid sequence: Tyr-Gly-Gly-Phe-Leu) are the two major types of endogenous enkephalins [978, 1001]. Most of the circulating met-enkephalin originates in the gut, sympathetic nervous system and adrenal medulla [1002, 1003] Opioids exert their influence via the opioid receptors—a group of metabotropic 7TMS inhibitory G protein-coupled receptors. These receptors are coupled to the stimulation of the Gi/Go subtypes of G proteins, which subsequently leads to inhibition of physiological processes [1004–1006]. The receptors for enkephalin are the delta (δ) opioid receptors. Beta-endorphins bind to mu (μ) opioid receptors. Dynorphins bind to kappa (κ) receptors. Nociceptin binds to the nociceptin receptor (NOR) [984, 985, 1007]	Hans Walter Kosterlitz (1903–1996), a German-born British biologist, and John Hughes, identified leucine-enkephalin and methionine-enkephalin as the first endogenous opioids in 1975 (enkephalin means "in the head"). The receptors of these molecules were earlier identified by the American neuroscientists Candace Beebe Pert (1946–2013) and Solomon Halbert Snyder (1938–). They first discovered the opiate receptor in 1972 [1008]	1975

(continued)

Table 9.2 (continued)

Neural secretions of the gut	Description	Discoverer	Year of discovery
	Opioid receptors are distributed widely in the brain and are also found in the spinal cord and GI tract [1004–1006]. They have differential expression in various organs, tissues, and cells of the body. NOR is expressed in the CNS; whereas, μ-, κ-, and δ-subtypes are highly expressed in peripheral cells, such as in cells of the GI tract. The μ-, κ-, and δ-receptors are highly expressed in enteric neurons and mucosal endocrine cells [978]. Opioids receptors are also expressed in epithelial and immune cells [990] Opioid receptors are activated by opioid peptides on the membrane of the respective cells. The activated receptor relays the signal downstream activating intracellular acceptors that mediate a range of cellular processes. The receptor then undergoes endocytosis [978]. In the GI tract, the activation of opioid receptor in the enteric plexuses may lead to inhibition of calcium channels, opening of potassium channels, and decrease in the level of cyclic AMP. The resultant effect is hyperpolarization of the cell membrane [978] Gut opioids function both as hormones and neurotransmitters as well as neuromodulators [1009–1011]. Opioidergic neurotransmitters are involved in modulation of a host of other peptide neurotransmitters in various parts of the CNS and PNS. In the central tissues, met-enkephalin stimulates delta-receptors, located predominantly in the basal ganglia and limbic systems, which leads to tonic inhibition of hormone and neurotransmitter release. For instance, enkephalins are involved in inhibiting corticotropin-releasing hormone. In both central and peripheral tissues, enkephalins may participate in modulating the response of CCK to various physiological stimuli by inhibiting CCK-induced contraction. Enkephalins can associate with receptors for CCK [1002, 1003, 1009–1011]. Apart from modulation of pain perception, beta-endorphin is involved in the stimulation of release of prolactin, growth hormone and thyroid-stimulating hormone [1002]. The activation of kappa-receptors by dynorphin leads to inhibition of the release of many neurotransmitters and hormones such as vasopressin and gonadotropin-releasing hormone [1002] The gut opioid system interacts with the gut endocannabinoid system. The cross talks between opioid and endocannabinoids involve multiple signaling pathways and interaction between their receptors [998]. Opioids and endocannabinoids are distinct class of molecules but with overlapping physiological responses [983, 999]. Receptors of opioids and endocannabinoids may dimerise or the peptides may activate the receptors which are not traditionally activated in the presence of the peptides [983] Opioids may be synthesized by cancer cells. The secretion of these molecules has been reported to increase in cases of tumor and other diseases. For instance, elevations in circulating met-enkephalin may occur in cardiovascular and psychiatric diseases [1002, 1012]		

Note The discovery date for some peptides is given as two separate years. The first date represents the year of discovery of the peptide in extragut tissues and the second date when the peptide was found in the GI tract. Otherwise, reasons for the separate dates are discussed in the text

Neurochemical Classification of Enteric Neurons

Neurons carry out their functions by expression of a variety of neurotransmitters and peptides. These molecules aid in information coding in any given neuron. This process, which is generally termed neurochemical coding by neurons, represents a useful marker of neural signaling in the ENS. It was previously thought that each neurotransmitter is synthesized and transmitted by one neuron, a concept of chemical coding known as Henry Hallett Dale's principle. However, contemporary understanding about the mechanism of neurotransmission indicates that many transmitter molecules are released from a single neuronal terminal to mediate signal transmission (see Table 9.2 for details).

The coding molecules include gastrin-releasing peptide or human bombesin (GRP), ACh, ATP, 5-HT, histamine ((2-(1*H*-imidazol-4-yl) ethanamine)), GABA, tachykinins, substance P, neurokinin A, β-tachykinin (neurokinin B or neuropeptide beta), γ-tachykinin (also called gamma-neuropeptide), norepinephrine (noradrenaline or noradrenalin), epinephrine (adrenalin or adrenaline), neurotensin, glutamate, neuropeptide Y or neuropeptide tyrosine (NPY), Peptide YY (PYY), vasoactive intestinal polypeptide (VIP), peptide histidine isoleucin or peptide histidine isoleucinamide (Peptide HI or PHI), peptide histidine methionine (PHM) or peptide MI, peptide histidine valine (peptide HV or PHV), PACAP, CGRP, NO, CO, hydrogen sulfide (H_2S), enkephalins (ENK), β-endorphin, dynorphin (DYN), anandamide, 2-arachidonoylglycerol, virodhamine (O-arachidonoyl-ethanolamine), noladin (2-arachidonyl-glyceryl ether), and *N*-arachidonoyl dopamine (NADA) (see Table 9.2).

Generally, the transmitters may function as neuromodulators, neurotransmitters, and neurohormones. Some of the molecules actually function as a neurotransmitter and neurohormone, whereas others may function as neuromodulator or neurotransmitter. These transmitter molecules can be grouped as catecholamines or biogenic amines (monoaminergic neurotransmitters), amino acid derivatives (amino acid neurotransmitters), lipid neurotransmitters, gasotransmitters, peptides, purines, cholinergic transmitters (see Table 9.2). Catecholamines are monoamines, organic compounds that have a catechol (benzene ring with two hydroxyl side groups) and a side-chain amine. The catecholamines are derived from the amino acid tyrosine. These monoaminergic neurotransmitters include serotonin, dopamine, norepinephrine (noradrenaline), epinephrine (adrenaline), and histamine. However, there are also trace amines which include phenethylamine, *N*-methylphenethylamine, tyramine, 3-iodothyronamine, octopamine, tryptamine, etc. The amino acid neurotransmitters are glutamate, aspartate, D-serine, GABA, and glycine. The neurotransmitters that occur as gases (gasotransmitters) are NO, CO, and H_2S. The peptides neurotransmitters are somatostatin, substance P, cocaine- and amphetamine-regulated transcript, opioid peptides, and endorphins. Purine neurotransmitter includes ATP and β-nicotinamide (Table 9.2; Hwang et al. 2011). The lipid neurotransmitter includes anandamide, 2-arachidonoylglycerol, virodhamine, noladin, and NADA (see Table 9.2). The lipid-derived neurotransmitters are produced from the non-oxidative metabolism of arachidonic acid, an essential ω-6 polyunsaturated fatty acid [325–329] (see Table 9.2). The neurotransmitters enkephalins, β-endorphin, and dynorphin are called opioidergic neurotransmitters. They are produced in central and peripheral neurons and endocrine cells, including cells of the GI tract, and are involved in a variety of physiological functioning, notably, nociception, motility, secretion. The molecules may function as hormones or neuromodulators (see Table 9.2).

Neurotransmitters may be classified as excitatory or inhibitory. Glutamate is an example of excitatory transmitter. GABA, ATP, VIP, and NO are inhibitory transmitters of neurons of ENS [329]. They are also abundantly found in the CNS. In the gut, activation of excitatory motor neurons by ACh and substance P evokes smooth muscle contraction that can be observed as characteristic motility pattern of the gut. The motor pattern may be associated with peristaltic movement (muscle cell) or secretion (secretomotor neuron). The release of the neurotransmitter in the presynaptic terminal of the neuron, followed by its diffusion toward the effector (glandular or muscle cell) activate their corresponding receptors in the secretory cell resulting in the release of mucin, electrolytes, water, and other molecules. It should be noted that excessive secretion (secretory diarrhea) in pathological conditions may be neurogenic, resulting from activation of secretomotor neurons of intestinal or submucosal secretory glands that release histamine (see Table 9.2). Some neurotransmitters or neuromodulators may function as an autocrine, paracrine, or juxtacrine agent [330–332].

Neurotransmitters can also be classified according to their site, tissue, or organ of secretion. Gut sources of neurotransmitters include enteric neurons, enteric glial cells, neuroendocrine cells, non-neural or non-glial cell lineage. However, some neurotransmitters found in the gut may originate from any other tissue apart from the gut (see Table 9.2). They are also classified on the basis of the division of the nervous system from where they are synthesized. Thus, there are parasympathetic and sympathetic neurotransmitters; enteric neurotransmitters. The major neurotransmitter in parasympathetic nerves is ACh, whereas the predominant transmitter in sympathetic nerves is noradrenaline. Ach is a cholinergic transmitter. Non-cholinergic transmitters include nitric oxide, substance P, CGRP, tachykinins, etc. (see Table 9.2).

9.3.3 Synthesis of Neural Secretions (Neurotransmitters, Neuromodulators, Neurohormones) and Packaging for Export, Exocytosis, and Recycling

Details on mechanisms of synthesis of molecules in the cell have been discussed in Chap. 3. The signal for synthesis of neurotransmitters is regulated by multiple mechanisms/pathways including concentration of neurotransmitter vesicles or granules in the cytoplasm, available starting products for the synthesis, signal type, and intensity impinging on the cell [333–335]. Materials destined for secretion pass through the endoplasmic reticulum and Golgi complex to the site of secretion (Fig. 9.16).

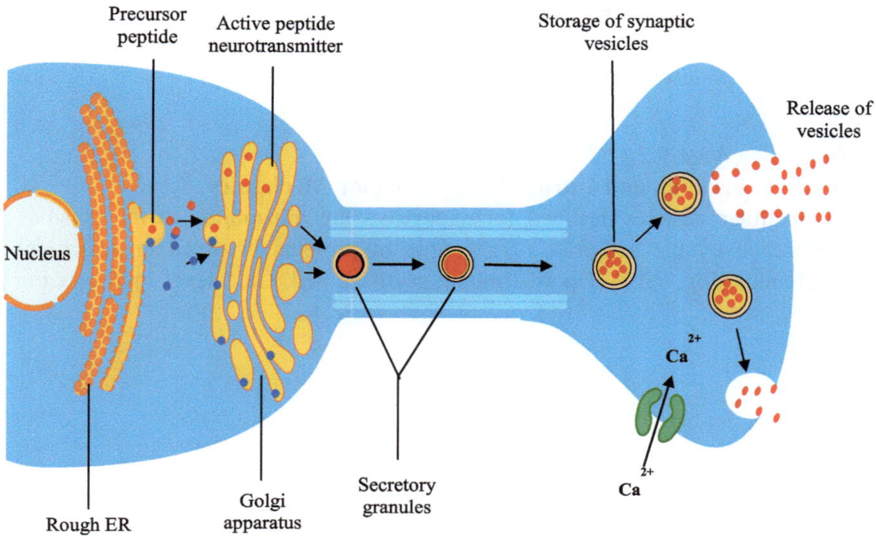

Fig. 9.16 Synthesis, storage, and exocytosis of neurotransmitter molecules. The cytoplasmic components in dendrites and axons of neurons comprise scattered granular vesicles, mitochondria, Golgi apparatus and endoplasmic reticulum and other organelles. The nucleus is located in the cell body. These organelles constitute the secretory apparatus in the cells of the ENS. For detail, see Chap. 3 [296, 325, 333–335]. The mechanisms of synthesis of neurotransmitters are similar in almost all cells of the nervous and endocrine systems [336–339]

Secretion of neurotransmitter occurs in many cells of the ENS—enteric neurons, glial cells, neuroendocrine cells. There are two major pathways of secretion—the constitutive secretory and regulated secretory pathways. The regulated secretory pathway involves a process of exocytosis in which soluble proteins and other substances are initially stored in secretory vesicles for later release. This secretory pathway is found mainly in cells that are specialized for secreting products such as hormones, neurotransmitters, or digestive enzymes rapidly on demand. The cells that use this secretory pathway include neurons, endocrine and exocrine cells [340–342]. In constitutive "non-regulated" secretory pathway, which is present in all cells of the body, substances are secreted continuously, regardless of external factors or stimuli. The non-regulated secretory pathway is also used by all eukaryotic cells to deliver newly synthesized proteins to the cell surface and to the extracellular milieu maintain the cell's plasma membrane [343–345]. The materials secreted may range from neuropeptides to gliotransmitters and neurotransmitters. To emphasize, unlike the release mechanism of exocytosis characterized for neurons and neuroendocrine cells, in enteric glial cells, the main release mechanisms of gliotransmitters is thought to occur via gap junctions (e.g., connexin 43), hemichannels, and by exocytosis mediated by Ca^{2+}-dependent pathways [267]. Materials to be secreted can easily diffuse through gap junctions or hemichannels if the size and charge of the molecules are small enough to be transported or the charge of the gap junction proteins does not repel the molecule. The mechanism of exocytosis involves the activities of numerous intracellular and membrane proteins and ions (particularly calcium) (see *Reference Note 9.2*).

Reference Note 9.2

SNARE Proteins and Intracellular Vesicular Mechanics

Intracellular vesicles destined for exocytosis are transported to the plasma membrane by the activities of a couple of proteins involved in vesicular mechanics (Fig. 9.17). Upon stimulation of the cell, Ca^{2+} level increases in

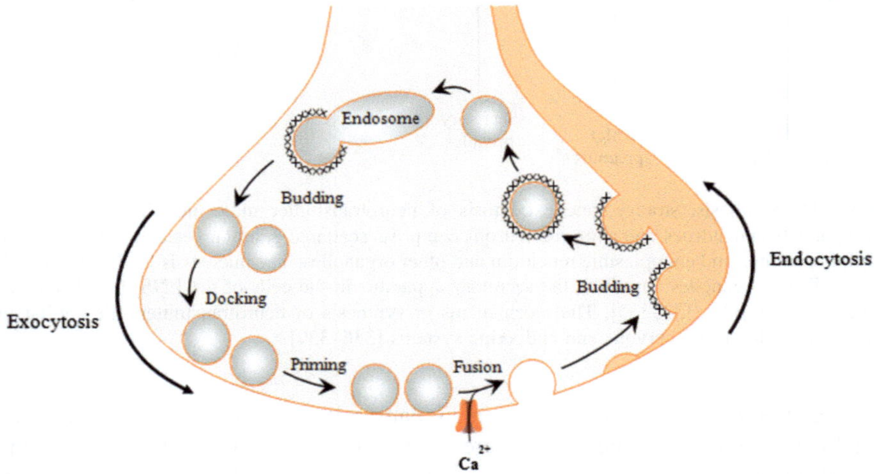

Fig. 9.17 Exocytosis–endocytosis cycling of synaptic vesicles. Following fusion of synaptic vesicles, the remaining vesicle contents are reinternalization and recycled for reuse by endocytosis. Subsequently, vesicles accumulate in the endosome until the signal for release of the vesicles is received for export of the endosomal vesicles to the plasma membrane. Vesicles budding off the endosome dock to the plasma membrane. It is believed that docking occurs at specific region of the plasma membrane (reviewed in Chap. 3). Importantly, certain plasma membrane structures called porosome appears to play a crucial role in the docking of vesicles to the plasma membrane, immediately before priming, followed by fusion. Porosomes are universal secretory cup-shaped portal in cells, discovered about 20 years ago [352, 353]. These secretory structures of plasma membrane of eukaryotic cells are where secretory vesicles transiently dock in the process of vesicle fusion and secretion. Porosomes are composed of different ion channels, motor proteins, SNARE, lipids, etc. [354, 355]. The fusion of the vesicles takes place at the base of the porosome via SNARE proteins, resulting in increase in intravesicular pressure with the subsequent formation of a fusion pore, required for the release of the vesicle contents into the exterior. The size of the fusion pore of the porosome varies from cell to cell. In acinar cells of the exocrine pancreas, for instance, the size of the fusion pore is estimated to be about 150 nm in diameter. In neurons, the fusion pore size is estimated to be about 12 nm in diameter [352]. The fusion pore formed at the base of the porosome is sealed following completion of secretion [354]. Vesicles that did not empty all their contents into the exterior are reinternalized for further use and the cycle continues [348, 356, 357]

the cytosol and is sensed by synaptogamin, which in turn recruits a number of associated proteins that aid in vesicle translocation to the plasma membrane. A major class of proteins implicated in vesicular mechanics is the SNARE proteins (SNAP Receptor; SNAP—Soluble NSF Attachment Protein; NSF—*N*-ethylmaleimide sensitive factor). The SNAREs are a large protein superfamily forming integral membrane proteins that mediate vesicle fusion with plasma membrane or with organelles such as a lysosome. The SNAREs mediate docking and fusion of synaptic vesicles with the plasma membrane, but they remain the target of the microbial neurotoxins including those that cause botulism and tetanus. The SNARE protein mechanics is the molecular machinery driving vesicle fusion during secretion of neuromediator and other substances in eukaryotic cells [346–351].

The mechanisms involved in the release of the contents of secretory vesicles to the exterior include: kiss and run, cavipunture, and full fusion of vesicles with the plasma membrane. In the kiss and run mechanism, the vesicle merges with the plasma membrane, and within a short period, small quantity of molecules escape into the synaptic cleft or exterior, while the remaining (larger) amount of the vesicular molecules is returned to the cytoplasm. Release of secretory vesicles in the kiss and run mechanism is faster compared with the other two. The cavipunture allows greater quantity of the contents of the vesicles to pass into the extracellular space compared with the kiss and run [349, 358, 359].

9.3.4 *Types of Neural Signaling*

The main types of neural signaling include chemical, electrical, and mechanical [360–362].

Electrical Signaling

Electrical signaling is a type of cell-to-cell communication in which changes in the electrical conductivity of the plasma membrane of one cell initiate the movement of ions via gap junctions to the neighboring cell, triggering changes in membrane potential (Fig. 9.18). This type of signaling is due to the close proximity of the cells, which are separated by a very small distance, but connected by communicating junctions (gap junctions, pannexons). Coupling of one cell to the other via electrical signals are crucial to the functioning of the gut and the organism as a whole [360, 361, 363].

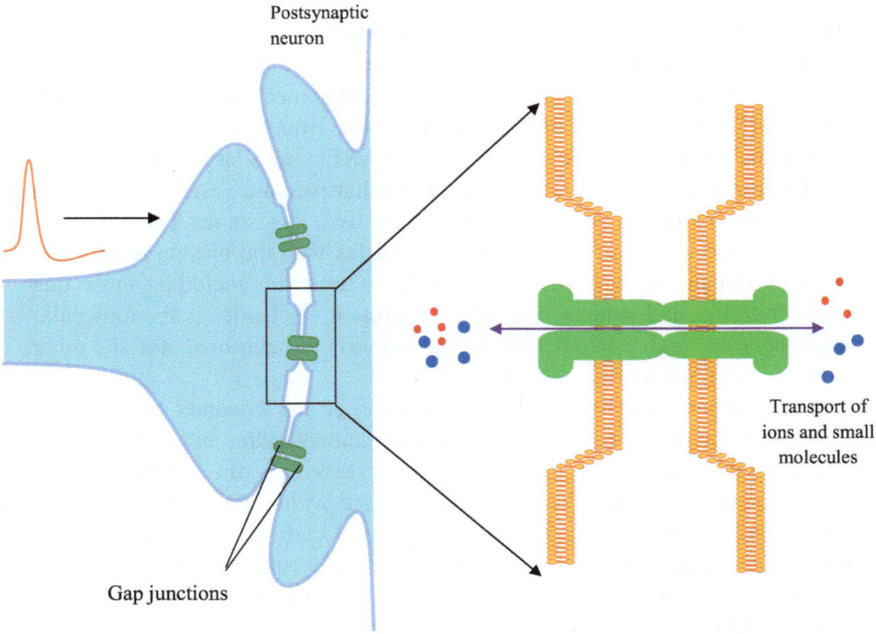

Fig. 9.18 Two neurons electrically coupled by gap junction (connexons). These communicating junctions are highly expressed in neurons and glial cells. The gap junction types found in enteric neurons include connexons (Cx) Cx43, Cx36, and Cx30, as well as the pannexons. Gap junctions allow the bidirectional transport of ions and small molecules such as ATP, glucose, inositol-1,4,5-trisphosphate, cyclic AMP, calcium ions, amino acids (glutamate, aspartate, taurine), nucleotides (ADP, ATP, NAD), metabolites (glucose, glucose-6-phosphate, lactate), small peptides (glutathione) across the plasma membrane of cells. Large molecules such as nucleic acids, proteins, and lipids may not be transmitted. The hemichannels of connexons and pannexons also allow transmission of biomolecules between the cell and extracellular space [364, 365]

Mechanical Signaling

When mechanosensitive neurons or stretch receptors on the surface of the plasma membrane of responsive cells are activated by tension or pressure, a cascade of signaling event is triggered (Fig. 9.19). The receptors and cells that are sensitive to mechanical signal are located in different region of the GI tract—mucosa, muscle layers, and in the neural plexuses. The mechanosensitive neurons of the gut include the enteric viscerofugal (intestinofugal) neurons. About 60% of the entire population of ENS cells can function as mechanosensitive neurons depending on the region. The intestinofugal neurons represent about 0.25% of myenteric neurons. These neurons also receive synaptic inputs from other neurons. The viscerofugal neurons serve as interneurons. The cell bodies of enteric viscerofugal neurons are located in the myenteric plexus. These neurons sense and receive information about distension of the intestine and transmit this information centrally to postganglionic

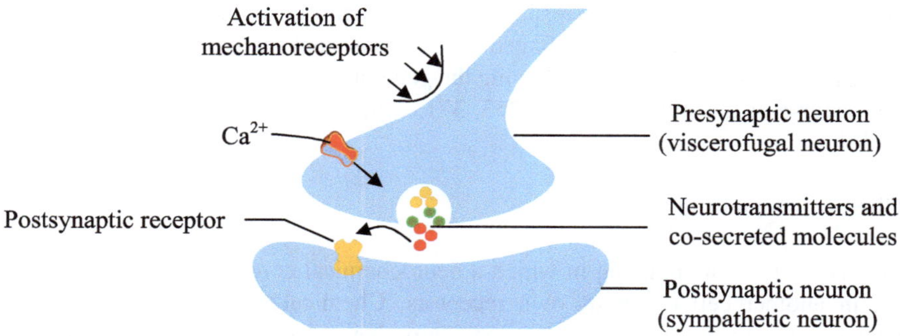

Fig. 9.19 Mechanosensitive neuron (intestinofugal or viscerofugal neuron) activated by mechanical distension. If the mechanical signal is strong enough to cause depolarization up to the critical level via the release of transmitter molecules at the presynaptic membrane, then the transmitter molecules diffuse toward the postsynaptic membrane to trigger the generation of action potential that travels along the nerve to regulate specific physiological process [372]

sympathetic neurons in the prevertebral ganglia. The viscerofugal neurons are uniaxonal neurons with multiple dendrites and lamellar expansions. They are cholinergic and their stimulation elicits excitatory postsynaptic potentials [253, 360–362, 366–371].

The intestinofugal neurons together with sympathetic prevertebral ganglia neurons form the basis of intestino-intestinal reflexes that coordinate motor activity of the GI tract even without the involvement of the CNS. The intestinofugal neurons form the afferent limbs of enteroenteric reflexes. The axons of intestinofugal neurons (usually one axon per neuron) project to prevertebral ganglia where they provide excitatory synaptic drive to sympathetic neurons. This link with sympathetic nervous system provides mechanosensitive neurons the ability to control gut motility and secretion [367, 368, 373–376]. These neurons are significantly reduced in inflammatory insult. Because intestinofugal neurons are a major driver of sympathetic output to the gut, the loss of intestinofugal neurons may have a profound pathophysiological significance—any decrease in this population would be expected to have significant effects on the coordination of motor activities in the lower and upper regions of the GI tract [367].

There are different types of mechanosensitive intestinofugal neurons. They can be distinguished on the basis of their location in the GI tract region or layers. These neurons are more abundant in the distal colon (colonofugal neurons) than other regions of the gut [372, 373, 377, 378]. The mechanosensitive intestinofugal neurons (mechanoreceptors) can be classified as fast-adapting, slowly adapting, and non-adapting neurons or receptors. Upon mechanical stimulation, the rapidly (or fast)-adapting mechanosensitive neurons discharge spike potential, which ceases after a period of sustained stimulus of constant intensity. Upon stimulation, the slowly adapting mechanosensitive neurons maintain their discharge at frequencies proportional to the intensity of the stimulus. A subset of the slowly adapting neurons is termed ultraslowly adapting neurons encoding tensile forces. The non-adapting neurons, also called the tonic response-type mechanosensitive neurons, upon constant stimulation, generate a prolonged train of consistent spike potential. Increase or decrease in stimulus intensity following the generation of spikes does not change the time course of the spike potential [253, 371, 379, 380].

Chemical Signaling

This is the type of signaling in which a neurochemical is released from one cell to act in another cell or on its own receptors. Chemical signaling may occur by neuracrine, neuroendocrine, juxtacrine, autocrine, or paracrine mechanism. The mechanisms in chemical signaling involve the activation of plasma membrane or intracellular receptors such as ion channel receptors, G protein-coupled receptors (metabotropic or ionotropic), enzyme-linked receptors (e.g., tyrosine kinase receptors). In real-world neural-environment, cells interact via multiple signaling systems with varying contributions from each signaling type and multiple interactions of different receptor types. Enteric neurons interact with each other via mixed signaling with contribution from electrical and chemical signaling types [360, 381, 382]. Accumulating evidences indicate that another type of neural signaling termed "ephaptic" may have a crucial role to play in neural functions [383].

9.3.5 *Modes of Neural Signaling*

Neurons communicate with each other and with neighboring cells through various mechanisms, which may be designated as modes of neural signaling. The chemical released by a given neuron diffuses across a narrow distance to activate its cognate receptor in a neighboring cell or neuron, thus influencing its function. However, substances released by the neuron itself can influence its own functioning by activating the receptor on its membrane to trigger downstream signaling cascades. The modes of signaling in the gut ENS include neuracrine, paracrine, autocrine, and juxtacrine signaling [384, 385].

Neuracrine Signaling

Neuracrine is a mode of cell signaling that involves the release of neurohormone by a given neuron or nerve cell to influence signaling processes in a neighboring neuron or cell of the gut. The neurohormone produced in the gut cells can also act on extraenteric tissues at distant sites. The neuracrine molecules are transported to their target cells by diffusion or via the bloodstream. Such neuracrine molecules include somatostatin, neuropeptide Y, peptide YY, pancreatic polypeptide, enteroglucagon, neurotensin. Unlike classical endocrine hormones that exert its effect over long distances (telecrine), neuracrine substances signal by synaptic or

Fig. 9.20 Schematic representation of a type of neuracrine signaling. In this type of neural signaling, a neurochemical is released from a neighboring neuron and diffuses across a relatively short distance to act on the target neuron

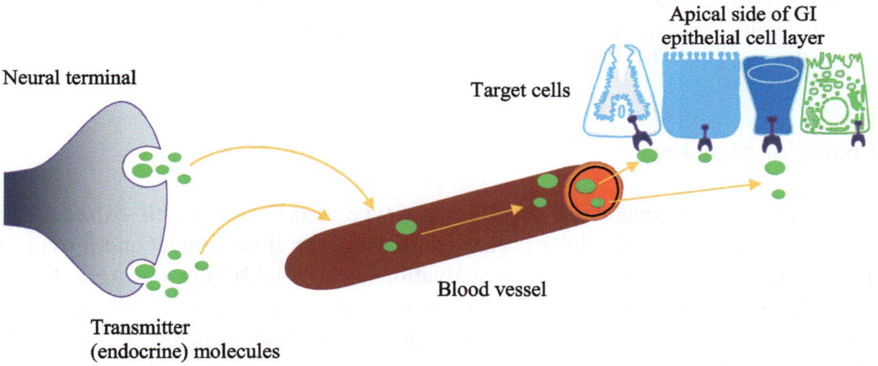

Fig. 9.21 Neuroendocrine signaling. Neuroendocrine molecules are released from the neuron into the bloodstream from where they are transported to the target cell to exert their actions [391, 392]

neuroendocrine means within the gut cells. However, it should be borne in mind that some neuracrine molecules may act on distant tissues and cells (Figs. 9.20 and 9.21) [386–392].

Autocrine Signaling

Autocrine signaling is a form of cell signaling in which a cell secretes a chemical messenger (called the autocrine factor) that binds to autocrine receptors on that same cell, leading to changes in the cell functions. An example of an autocrine agent is the cytokine—interleukin-1, secreted by monocytes. When interleukin-1 is produced in response to external stimuli, it binds to cell-surface receptors on the same cell that produced it [393–397].

Paracrine Signaling

Paracrine signaling is a form of cell-to-cell communication in which a cell produces a signaling molecule to induce changes in nearby cells, thereby altering the behavior or differentiation of the cells [395, 397, 398].

Paracrine factors (or signaling molecules) released from a neuroendocrine cell or neuron diffuse through the extracellular space to their target cells, which harbors the cognate receptor for the diffusing molecule (Fig. 9.23). The signaling molecules, also known as paracrine factors, diffuse over a relatively short distance (local action), as opposed to classical endocrine factors (hormones), which travel considerably longer distances via the bloodstream. Cells that produce paracrine factors secrete them into the immediate extracellular environment. These factors then travel to nearby cells in which the factor gradient of the diffusing molecules determines the outcome. However, the exact distance that paracrine factors can travel is not certain. Although paracrine signaling elicits a diverse array of responses in the activated cells, most paracrine factors utilize a relatively streamlined set of receptors and pathways [397–401] (Fig. 9.22).

Juxtacrine Signaling

Juxtacrine (contact-dependent) signaling is a type of cell–cell or cell–extracellular matrix signaling in multicellular organisms that require close contact and provide a tight spatiotemporal control of cell activation by a neighboring one (Fig. 9.24). Thus, there is no diffusion of the signaling molecule across the extracellular space. The term "juxtacrine" was initially introduced by Anklesaria et al. [402] and Massagué [403] to describe a possible way of signal transduction between transforming growth factor alpha (TGF-α) and epidermal growth factor receptor (EGFR) [397, 402, 403]. However, this receptor not only functions by juxtacrine, but also

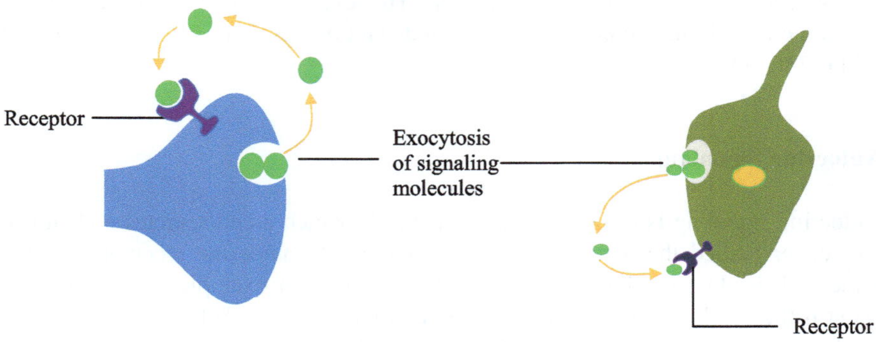

Fig. 9.22 Autocrine signaling. The signaling peptide is released from the cell to act on the same cell. In the gut, some peptides may be released from the basolateral side of the cell to act on the same cell

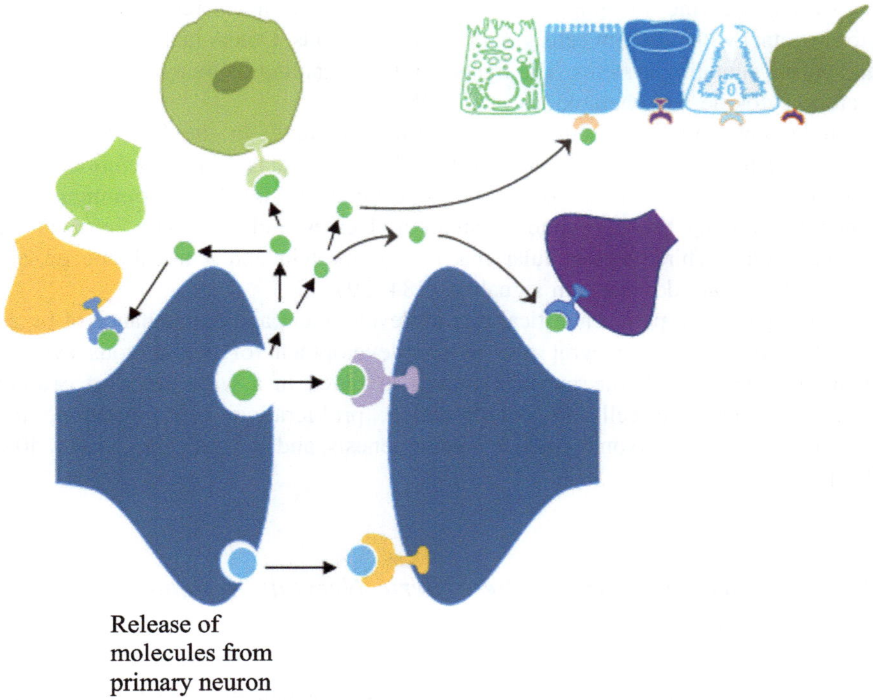

Fig. 9.23 Paracrine signaling. Molecules exocytosed from the primary neuron diffuse to neighboring neurons to localize its cognate receptors, where they exert their effects

Fig. 9.24 Contact-dependent signaling. The ligand on neuron B directly localizes the receptor on neuron A, without any need for diffusion

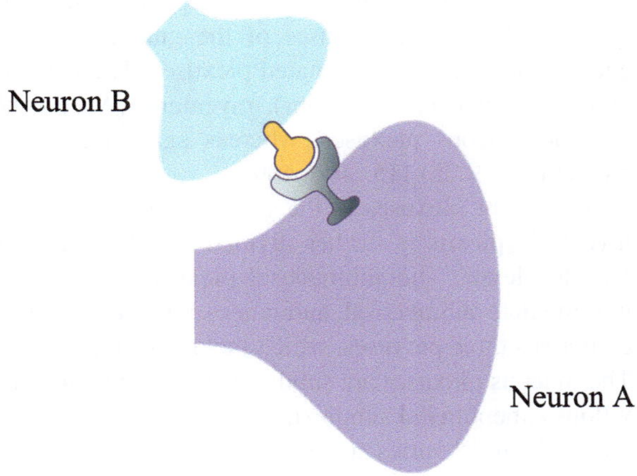

autocrine, paracrine, or endocrine means. For instance, notch receptors signal by close contact with adjacent cells via morphogen ligands. Others include semaphorin and ephrin signaling systems. The ligands of this signaling pathway include growth factors, cytokine and chemokines [402–405].

Juxtacrine signaling occurs in three different ways. In the first instance, a membrane ligand (oligosaccharide, protein, lipid) interacts with a membrane protein of an adjacent cell. Second, two adjacent cells signal via communicating junction allowing the exchange of small molecules and ions with intracellular compartments. Third, extracellular matrix glycoprotein and a membrane protein interact to initiate downstream signaling [384, 397].

This signaling type has a critical role in development and maintenance of tissue integrity. It plays an integral role in the development of the nervous system, including the ENS. Juxtacrine signaling is involved in a wide range of cellular processes, including cell fate determination, proliferation, cell migration, and apoptosis. It controls axonal guidance, neurogenesis, and synaptogenesis [384, 404, 405].

9.3.6 Neural Network of the Enteric Nervous System—The Plexuses

The neurons and the complex enteric plexuses in which they are found (plexus means "network," plural. plexi or plexuses) operate more or less independently according to their own rules of reflexive behaviors. As a result, many gut functions continue perfectly well without sympathetic or parasympathetic supervision. For instance, peristalsis can occur in isolated gut segments in vitro. Thus, most investigators classify the ENS as a separate component of the neurovisceral motor system. The neural plexuses of the gut are divided into two broad categories: ganglionated and aganglionated plexuses. The ganglionated plexuses consist of the outer muscular/serosal (Stöhr), myenteric (Auerbach's), submucous (Meissner's), and the mucous plexuses, whereas aganglionated plexuses consist of subserous plexus (Fig. 9.25) [15, 406–409].

The major plexuses of the gut are the myenteric and submucous plexuses. The myenteric plexus is further divided into longitudinal muscle plexus and circular muscle plexus. The submucosal plexuses can be divided into outer submucosal, intermediate submucosal, and inner submucosal subplexuses. The outer submucosal comprises three plexuses, which include Henle's, Schabadasch's, external plexuses. The mucous plexuses are subdivided into subglandular, periglandular, vascular, and villous subepithelial subplexuses. It should be pointed that the density of neurons in these plexuses varies across different regions of the GI tract. For instance, the myenteric plexus has higher density of neurons compared to the submucosal plexus [15, 262, 421]. Though the above classification of the enteric plexuses gives a detailed overview of the neural network in the GI tract, a somewhat simpler

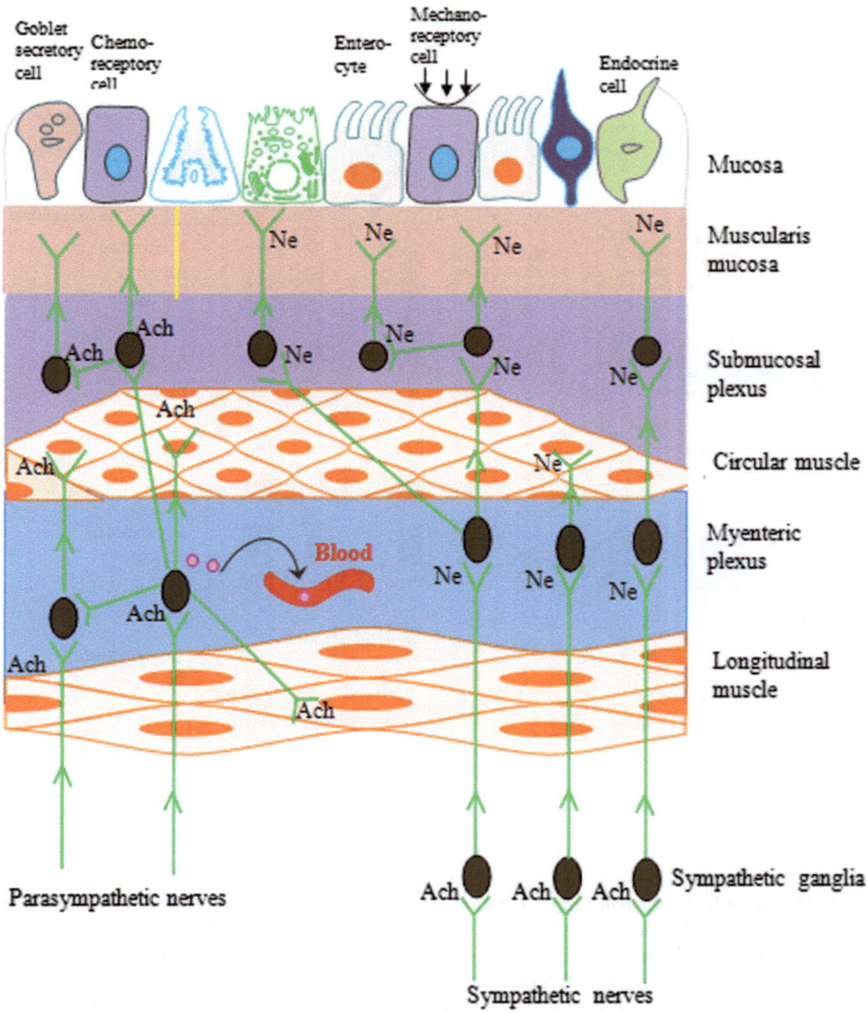

Fig. 9.25 Oversimplified scheme of the connections between the extrinsic nervous system and the ENS. Ne—noradrenaline. Ach—acetylcholine

classification can be used. The gut plexuses can be simply divided into three comprising the myenteric, submucosal, and mucosal plexuses (Fig. 9.25) [262]. The mucosal and submucosal plexuses are located within the mucosa and submucosa, respectively, and are largely responsible for the control of fluid absorption and secretion in gut as well as detection of stimuli. The myenteric plexus is mainly involved in motility [20, 288].

The enteric plexuses and intramuscular neurons contain different kinds of neurons and receptors—sensory receptors (chemoreceptors, nociceptors, and

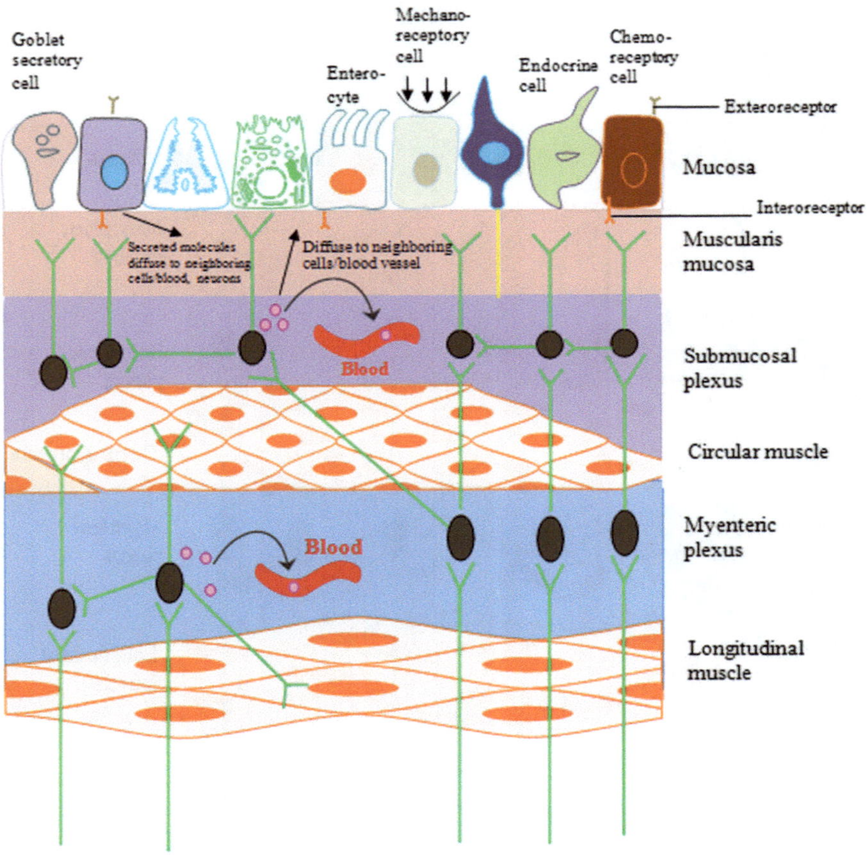

Fig. 9.26 Simplified schematic representation of GI wall showing the intrinsic nervous system and associated cells. Different receptor types which sense the environment of the lumen as well as nutrients trigger responses that mediate a wide range of physiological processes including absorption and transport: exteroreceptors such as receptors that sense peptones (CaR), bile (TGR5), sugars (GLUT2, TIR3, SGLT1, SGLT3), fats (FFAR1 (GPR40), FFAR2 (GPR43), FFAR 3 (GPR41), GPR109A, GPR119, GPR120, FATP4 (Fatty Acid Transport Protein). Free fatty acid sensors are nuclear receptors peroxisome proliferator-activated receptors (PPARs) and fatty acid-binding proteins (FABPs) [410–420]

mechanoreceptors) (Fig. 9.26). These receptors provide signal input to the interneurons of the ENS. The interneurons then relay the signal to the CNS, where the information is processed for appropriate response [422].

Besides the neurons of the submucosal and myenteric plexuses, there are solitary or groups of neurons in other locations including the mucosa of the gut. The intramucosal neurons may lie close to the muscularis mucosae and extend into the muscularis mucosae. These neurons have processes that project into the submucosa [422].

The neurons of the plexuses may be excitatory or inhibitory. Thus, they may positively or negatively, depending on the mediator release, affect the rhythm of contraction, muscle tone, rhythm rate, and velocity of conduction of excitatory waves. These functions of the ENS are under constant modulation by external influence (e.g., parasympathetic nerves). The ENS participates in the control of gut motility, fluid exchange, local blood flow, secretion, defense reactions, and many reflexes arising from the gut. Specifically, the movement of food or liquid boluses from the mouth to the aboral direction is controlled by the ENS together with coordinated activities of the sympathetic and parasympathetic nervous systems. The movement of water between the gut lumen and tissue fluid compartments is regulated by the ENS through the activity of secretomotor neurons. These neurons innervate the GI secretory glands such as the goblet cells, Brunner's glands, small intestinal crypts of Lieberkühn, and colonic crypts as well as glands of the submucosa. The somata of these neurons are located in the submucosal ganglia. Stimulation of secretomotor neurons triggers the release of ACh, VIP, and other neuropeptides, which mediate the secretory processes in the intestinal glands. The secretomotor neurons also innervate glands in the esophagus and stomach [286, 294, 423, 424].

Many pathological conditions of the gut involve the dysfunction of the enteric neurons. In inflammatory bowel disease, for example, functions of the ENS are negatively affected, resulting in malfunction not only in the motor activity but also secretory activity of the gut. In patients with small intestinal or colonic ulceration or inflammation, there is observed hyperplasia of the intestinal plexuses. However, not only neurons of the gut are affected but also glial cells as in neurogliopathies, a group of diseases that are characterized by deficiency in enteric neurons and glial cells resulting in abnormal GI motility and secretion [262, 425, 426].

The anatomo-physiological properties of the ENS are studied with many methods and techniques, which include but are not limited to electrophysiological (e.g., patch clamp), immunohistochemistry, immunofluorescence, real-time imaging, and polymerase chain reaction (PCR) [287, 295, 427, 428]. In immunofluorescence technique, for instance, fluorescence materials are used (e.g., carboxyfluorescein) to assess the cytoarchitecture and functionality of the neurons of the intestinal plexuses [428].

Submucosal (Meissner's) Plexus

Meissner's plexus is a network comprised of intrinsic neurons and their processes, located just beneath the mucosa of the gut. Meissner's plexus runs from Auerbach's plexus to the muscularis mucosae of the GI wall. The plexus is named after the German anatomist and physiologist, Georg Meissner (1829–1905), who was the first in 1857 to describe this network of intrinsic neurons and processes of the submucosal layer of the alimentary canal. He made other contribution on the physiochemical aspects of digestion including chemical of proteins in the gut.

Meissner's plexus provides a pathway for the innervation of the mucosal layer of the GI tract wall [429, 430].

Majority of authors have considered Meissner's plexus as a single entity; however, it is sometimes divided into two separate plexuses: the plexus lying adjacent to the muscularis mucosae (Meissner's plexus), plexus lying close to the circular muscle (Henle's plexus). This classification system and terminology was first introduced by Schabadasch in the first half of the twentieth century. When using this classification system, Meissner's plexus, also called plexus submucosus internus (internal submucosal plexus), is considered to be located in the innermost side or at the abluminal (serosal) side of the muscularis mucosae. The internal submucosal plexus is multilayered and forms irregular mesh. The Henle plexus—plexus submucosus externus (external submucosal plexus)—is located in the outermost side, adjacent to the luminal side of the circular muscle layer. The outermost plexus of the submucosa is also called Schabadasch plexus. In the human intestine, there is an intermediate plexus that exists between Meissner's plexus proper and Henle's plexus. These plexuses can be differentiated by the ganglia lying deep connecting those lying at higher layer [15, 431, 432]. It should be noted that the connections between the plexuses may occur in the absence of ganglia.

The submucous plexus plays numerous physiological functions, which include control of secretion, motility of the muscularis mucosae and smooth muscle of the villi, coordination of intestinal motility, sensory and secretory activity, functional integration of GI activities [431]. It should be noted that the major function of the neurons of the submucous plexus is the control of GI tract secretion [20, 433]. The secretion of digestive enzymes, water, mucus, electrolyte, ion, stomach acid, and bile as well as local blood flow in the gut is controlled by the neurons of Meissner's plexus. The plexus monitors the chemical content of lumen and regulates glandular secretory activity accordingly [288, 433].

Myenteric (Auerbach's) Plexus

Auerbach's plexus is a collection of unmyelinated fibers and postganglionic autonomic cell bodies that lie between the circular and longitudinal layers of the muscularis externa of the GI tract, extending from the esophagus to the rectum. This plexus is named after Leopold Auerbach (1828–1897), a German scientist, who solved many anatomical and neuropathological problems of the human body. After becoming a medical doctor in 1849, Auerbach successfully defended his habilitation in 1863. The habilitation (Habil.) is the highest academic degree awarded in some countries including Germany. The degree is equivalent to the Doctor of Science (D.Sc.) degree in other countries. The D.Sc. and Habil. are the highest academic degrees after the Ph.D. (Doctor of Philosophy). The D.Sc. and Habil. are awarded, in partial fulfillment, on the basis of high-level research achievements in a given field. Auerbach, a specialist in staining techniques of tissues and cells, discovered myenteric plexus that now bears his name "plexus myentericus Auerbachi (Auerbach's plexus)." Because of its location, Auerbach's plexus is sometimes called intermuscular plexus of Auerbach. The neurons of this layer provide motor inputs to both layers of the muscularis externa and also ensure parasympathetic/sympathetic input and output to and from the gut. The intermuscular plexus of Auerbach is a layer of ganglionated neurons and their processes, primarily concerned with the regulation of the motor functions of the GI tract, including peristalsis [20, 434, 435].

9.3.7 *Afferent and Efferent Nerve Fibers Connecting the Enteric Nervous System*

The neurons of the ENS exist in close connection with the parasympathetic and sympathetic divisions of the autonomic nervous system. Both input and output pathways pass through the prevertebral ganglia or the basal regions of the brain to the gut. The neural plexuses of the gut receive CNS information from the parasympathetic and sympathetic divisions. The nerve fibers of the parasympathetic and sympathetic systems synapse with neurons of the ENS [436–439].

Afferent signals from the gut are sent to the CNS via sensory neurons whose cell bodies are located in the ENS, and their axons send information to the posterior horn of the spinal cord via the dorsal root ganglion. Afferent signals are also transferred to the CNS (brainstem) via the vagus nerve or other cranial nerves. A few afferent signals are also transferred to the spinal cord via the celiac, mesenteric, and hypogastric ganglia. Information is processed and sent back to the GI system via motor neurons that synapse with neurons in the anterior horn of the spinal cord or as a component of the vagus nerve. Motor neurons of the sympathetic and parasympathetic divisions terminate in the ENS [440–443].

9.3.8 *Enteric Neurons Synapse with Smooth Muscle and Interstitial Cells of Cajal to Mediate Exocytosis and Other Physiological Processes*

Enteric neurons, smooth muscle, and interstitial cells of Cajal form anatomo-functional unit that participates in the regulation of GI motor pattern. These cells are excitable, receiving signal (mechanical, electrical, chemical) for activation through depolarization to generate action potential that results in transmission of impulses from one cell to another or characteristic pattern of functioning (secretion, peristalsis) of the gut [264, 444–447]. Transmission of signal takes place between the enteric neurons, smooth muscle, and interstitial cells. For instance, the enteric neurons can transmit signal to smooth muscle cell to regulate its functions. The structural unit that ensures the coupling of enteric neuron to smooth muscle cell is termed neuromuscular junction or synapse (Fig. 9.27). The neuron forms the

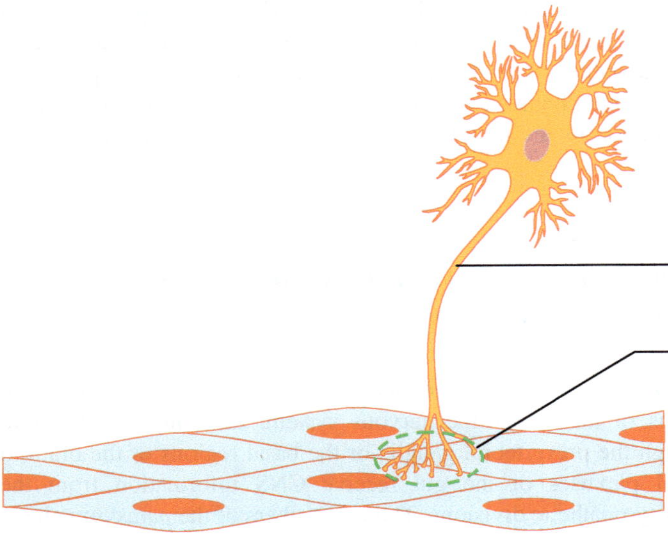

Fig. 9.27 Neuromuscular junction (neuron synapsing with smooth muscle cells). When a neuron is excited to reach the threshold potential, an action potential is formed, which travels along the nerve to excite the presynaptic machinery of secretory vesicle exocytosis. The released molecules into the neuromuscular junction (region where the nerve terminal forms synapse with muscle cells) diffuse to the presynaptic membrane to stimulate postsynaptic receptors. Depending on the type of molecule secreted, the resulting effect on the smooth muscle cell will be changed in excitability, contractility, conductivity, relaxablity, and rhythmicity [264, 444, 452–455]. Certain molecules that are secreted may not necessarily cause excitation, but modulate the architectural integrity of neurons and glial cells. Such molecules include neurotrophins (e.g., glial-derived neurotrophic factor, GDNF, neurotrphin-3) are involved in maintenance of morphofunctionality of neurons and pacemaker cells. The morphofunctional integrity of neuron-smooth muscle cell unit affects the contractility, rhythmicity, and other properties of the muscle cell [456, 457]

presynaptic membrane, while the smooth muscle cell forms the postsynaptic membrane. The arrival of action potential at the postsynaptic cell triggers the opening of voltage-dependent calcium channels, allowing the influx of calcium into the axon terminal. Increase in intracellular calcium leads to the activation of calcium-sensing proteins located around neurotransmitter-containing synaptic vesicles, which subsequently leads to the recruitment of several intracellular membrane and vesicle-associated proteins, which in turn mediates vesicle docking and fusion with the presynaptic membrane and release of neurotransmitter from the motor neuron into the synaptic cleft through exocytosis. The released neurotransmitter binds to the receptors of the postsynaptic membrane to initiate an array of cellular processes [447–451].

Depending on the type of neurotransmitter released, muscle cell either contracts or relaxes. If the neurotransmitter is ACh, the binding activates sodium channel and allows sodium ions to flow across the membrane into the muscle cell. The flow of sodium ions across the membrane into the muscle cell generates an action potential in the postsynaptic membrane, resulting in muscle contraction. Approximately 60% of enteric neurons are believed to be cholinergic and/or tachykinergic, functioning as interneurons and excitatory motor neurons, which innervate both layers of circular and longitudinal muscle cells [458–460]. If the neurotransmitter released is NO, the muscle cell relaxes. Thus, the motility of the GI tract is controlled by enteric inhibitory and excitatory motor neurons that innervate the layers of smooth muscle cells of the gut. Thus, inhibition of the enzyme that synthesizes NO (NO synthase) will result in cessation of NO production, and thus relaxation of smooth muscle cell is inhibited. But smooth muscle relaxation is not inhibited by NO synthase inhibitors alone, as neuropeptides such as VIP, PACAP, and ATP may also contribute to neuron-mediated inhibition of smooth muscle cells of the gut. The neurons that secrete these substances also serve as gut interneurons and constitute about 25% of the total population of enteric neurons [447, 458–460]. Another neurotransmitter widely distributed in the gut, and considered to be one of the most abundant inhibitory mediators in the brain is GABA. GABA-positive neurons innervate the circular and longitudinal muscle layers to mediate inhibitory processes (smooth muscle relaxation) [296]. The excitatory and inhibitory processes are responsible for gut peristaltic reflex [460].

Muscle tone of the GI tract is also maintained by neurotransmission. However, even in the absence of neurotransmitter activity, the smooth muscle cells still maintain a level of activity, referred to the basal tone [461–466].

One of the novel roles of enteric glial cell identified in the twenty-first century is its role in GI motility. Enteric glial cells function in a coordinated manner with the interstitial cells and neurons to ensure adequate functioning of the GI tract (Fig. 9.28). The malfunctions of these cells have been implicated in a range of gut pathologies. For instance, slow transit constipation reported in megacolon disorders is due to the reduction in the number of enteric neurons with concomitant reduction in the number of Cajal cells [444]. It is believed that glial cell-derived neurotrophic factor, neutrophins, ciliary neurotrophic factor, and leukemia inhibitory factor play a crucial role in the regulation of the enteral microenvironment. These factors regulate the differentiation, development, growth of neuronal and glial phenotypes, as well as cell apoptosis [264, 444, 447, 467]. Some of the pathologies affecting the ENS may be due to genetic disorders as well as other causes that result in decrease in the number of neurons, glial cells, and interstitial cells [468–470].

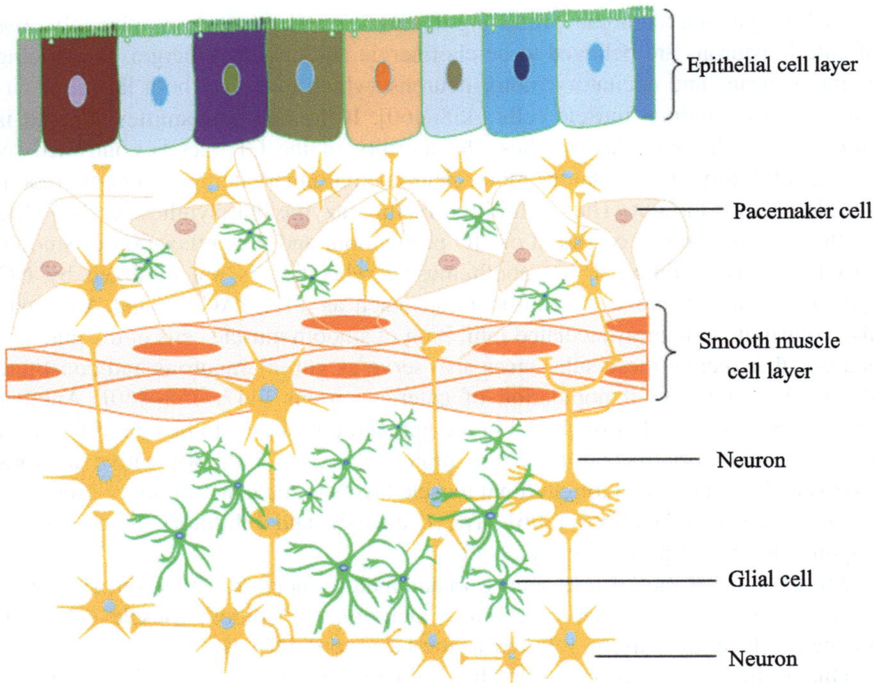

Fig. 9.28 Cellular network of pacemaker and related cells of smooth muscle. Smooth muscle cells have functional L-type calcium channels that increase conductivity after activation by pacemaker cells

9.4 Gastrointestinal Neurotransmitters: Course of Discovery, Their Structural–Functional Properties, Mechanisms of Action, and Clinical Application

Table 9.2 is a list of neurotransmitters of the gut (including recently discovered ones), with their structural and functional properties. The course of discovery of these molecules, the discoverers, and the year of discovery are provided.

$$\text{L-tryptophan} \xrightarrow{E_1} \text{5-HTP} \xrightarrow{E_2} \text{Serotonin}$$

Fig. 9.29 Synthesis of serotonin from the amino acid tryptophan. The enzyme controlling the rate-limiting step in serotonin synthesis is tryptophan hydroxylase (E1). The enzyme responsible for the production of serotonin is aromatic amino acid decarboxylase (E2) [586]. It should be noted, however, that serotonin biosynthesis involves a series of biochemical steps. For further review, see Nakamura and Hasegawa [587], Del Pino et al. [586], and Frick et al. [572]

$$\alpha\text{-ketoglutarate} \rightarrow \text{Glutamate} \rightarrow \text{GABA}$$

Fig. 9.30 Synthesis of GABA. The starting product for GABA synthesis is derived from the Krebs cycle. Alpha-ketoglutarate (may be produced from glucose) is converted to glutamate (a neurotransmitter) under the action of GABA α-oxoglutarate transaminase (GABA-T). Glutamate under the action of the enzyme, glutamic acid decarboxylase (glutamate decarboxylase), is converted to GABA. GABA maybe recycled to succinic semialdehyde by GABA-T. Succinic semialdehyde under the action of succinic semialdehyde dehydrogenase is converted to succinate, which enters the Krebs cycle. GABA and glutamate are released from neurons and taken up by astrocytes via Na^+-dependent glutamate transporters GLT-1 and GLAST (in humans—$EAAT_{1–4}$). The glutamate that passes into astrocytes can be converted by glutamine synthetase to glutamine, which is then released by the astrocytes and moved into neurons for neurotransmitter synthesis. Hence the cycle is called glutamate/GABA/glutamine cycle [641, 642]. This cycle plays an integral role in the pathogenesis of numerous neurodegenerative diseases including Alzheimer's disease, Parkinson's disease, amyotrophic lateral sclerosis, Huntington's disease, as well as other neurobehavioral disorders such as psychosis which are characterized by inefficiency of the glial and neuronal clearance of excess glutamine/glutamate [642]

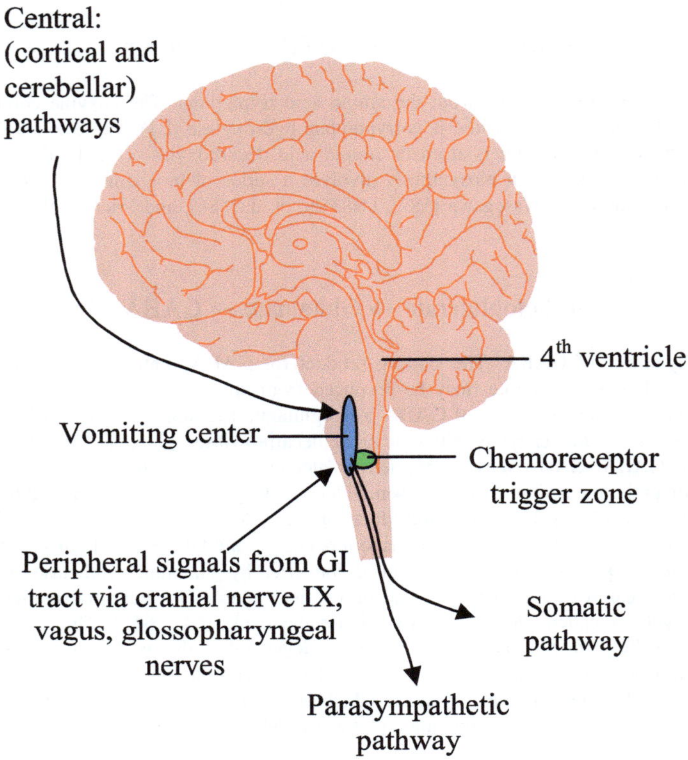

Fig. 9.31 Vomiting pathway. The afferent signals reach the emetogenic center through the nucleus tractus solitarius. The nucleus tractus solitarius also receive signal from higher brain regions [677, 693]. The mechanism of emetic or nausea signaling is inevitable in treatment of patients with various genesis including cancer patients [694]

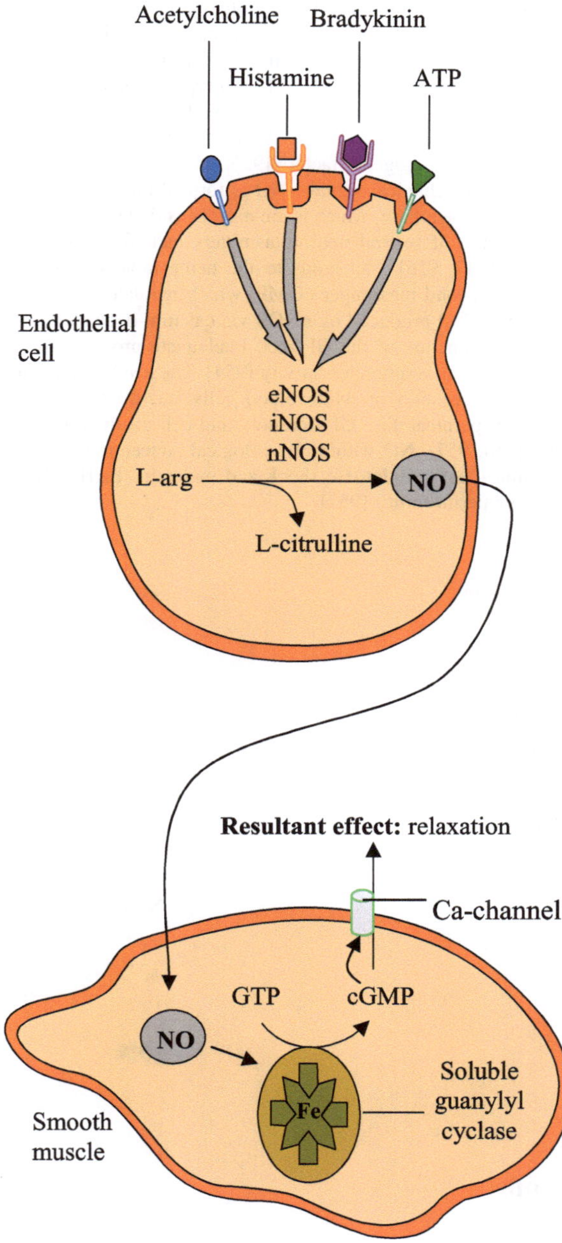
Acetylcholine
Bradykinin
Histamine
ATP
Endothelial cell
eNOS
iNOS
nNOS
L-arg
NO
L-citrulline
Resultant effect: relaxation
Ca-channel
GTP
cGMP
NO
Fe
Soluble guanylyl cyclase
Smooth muscle

◂**Fig. 9.32** Smooth muscle relaxation mediated by endothelial factors. nNOS, iNOS, and eNOS are neuronal, inducible, and endothelial nitric oxide synthase, respectively. The neuronal type is mostly located in neurons. The inducible type is mostly located in immune cells. The endothelial type is mostly located in endothelial cells [953, 954]. All types of NOS are present in the GI tract. Several types of stimuli can activate these enzymes to synthesize NO. Signals that activate NOS include neurotransmitters, metabolic stress (via AMPK), inflammatory cytokines (IFN, TNFα which activate NFκB, HIF1α), angiogenic factors [955, 956]. The production of NO can also be stimulated by electrical and mechanical signals. NO is formed through oxidative deamination of L-arginine by the enzyme NO synthase, which is the most sensitive physiological receptor for NO. NO can stimulate the release of several neurotransmitters including ACh, catecholamines, and neuroactive amino acids [947, 956]. NO binds to the heme moiety of the cyclase induces its capacity to synthesize the second messenger cGMP, which regulates intracellular calcium level. Nanomolar concentration of NO produced by nNOS via calcium-dependent pathway is important in peristalsis and sphincter function of the GI tract. Inadequate production of NO may result in decreased peristalsis or obstructive sphincter activity [954]. The eNOS maintains GI mucosa blood flow by preventing the aggregation of white blood cells and dilating the vessels. The iNOS pathway maintains the gut permeability, GI secretion, and cell death and proliferation as well as inflammatory reaction [954, 957]. NO within physiological concentration protects the GI against ischemia/reperfusion injury, early endotoxic shock and maintains barrier function, cytoskeleton, tight junctions, and normopermeability [957]

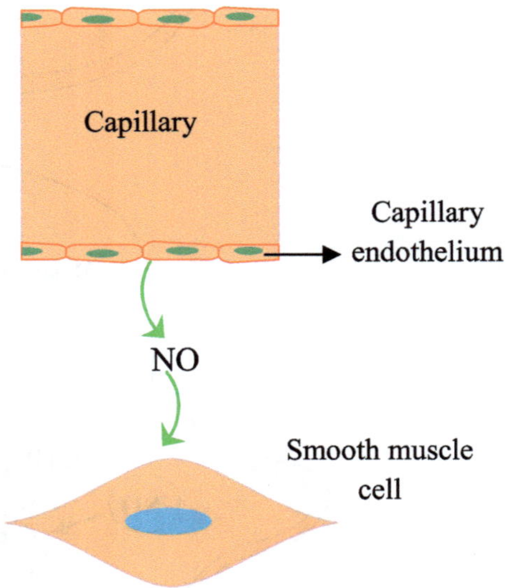

Fig. 9.33 Endothelial cell in the intima of capillary produces NO which diffuses into smooth muscle located in the tunica media of blood vessels to activate soluble guanylate cyclase. The reaction produces cGMP which act on calcium channels. The resultant effect on actin and myosin relaxes smooth muscle

9.5 Conclusion

The neural secretions of the gut are molecules released from the glial cells, neurons, nerve terminals, or neuroendocrine cells of the GI tract to regulate the functioning of the GI tract and the entire organs and systems of the body. The gut produces over

60 types of neural secretions that mediate a host of signaling cascades that are implicated in the functioning of the body in normal and disease states.

Recommended Readings

Original Articles

1. Brehmer A, Schrödl F, Neuhuber W, Hens J, Timmermans J-P (1999) Comparison of enteric neuronal morphology as demonstrated by DiI-tracing under different tissue-handling conditions. Anat Embryol 199(1):57–62
2. da Silveira ABM, D'Avila Reis D, de Oliveira EC, Neto SG, Luquetti AO, Poole D, et al (2007) Neurochemical coding of the enteric nervous system in chagasic patients with megacolon. Dig Dis Sci 52(10):2877–2883
3. Fu M, Tam PKH, Sham MH, Lui VCH (2004) Embryonic development of the ganglion plexuses and the concentric layer structure of human gut: a topographical study. Anat Embryol 208(1):33–41
4. Jiang M, Pu J-R, Yang D-H, Zhang X, Tang S-T (2016) Neuregulin 1 is required for the development of enteric neurons and vagus in zebrafish. Int J Clin Exp Pathol 9(2):1416–1423
5. Tsai ML, Rothstein JD, Gershon MD, Kirchgessner AL (1997) Glutamatergic enteric neurons. J Neurosci 17(12):4764–4784

Review Articles

1. Araque A, Parpura V, Sanzgiri RP, Haydon PG (1999) Tripartite synapses: glia, the unacknowledged partner. Trends Neurosci 22(5):208–215
2. Benarroch EE (2007) Enteric nervous system—Functional organization and neurologic implications. Neurology 69(20):1953–1957
3. Drossman D, Hasler WL (2016) Rome IV—functional GI disorders: disorders of gut-brain interaction. Gastroenterology 150(6):1257–1261
4. Enck P, Martens U, Klosterhalfen S (2007) The psyche and the gut. World J Gastroenterol 13(25):3405–3408
5. Lake JI, Heuckeroth RO (2013) Enteric nervous system development: migration, differentiation, and disease. Am J Physiol Gastrointest Liver Physiol 305 (1):G1–G24
6. Mayer EA, Labus JS, Tillisch K, Cole SW, Baldi P (2015) Towards a systems view of IBS. Nat Rev Gastroenterol Hepatol 12:592–605
7. Million M, Larauche M (2016) Stress, sex, and the enteric nervous system. Neurogastroenterol Motil 28(9):1283–9
8. Rajagopalan M, Kurian G, John JK (1996) Psychological aspects of irritable bowel syndrome. Indian J Psychiatry 38(4):217–224
9. Rhee SH, Pothoulakis C, Mayer EA (2009) Principles and clinical implications of the brain–gut–enteric microbiota axis. Nat Rev Gastroenterol Hepatol 6:306–314
10. Ringel Y, Drossman DA (2001) Psychosocial aspects of Crohn's disease. Surg Clin 81(1):231–252

11. Söderholm JD, Perdue MH II (2001) Stress and intestinal barrier function. Am J Physiol Gastrointest Liver Physiol 280(1):G7–G13
12. Sood R, Ford AC (2016) Diagnosis: Rome IV criteria for FGIDs—an improvement or more of the same? Nat Rev Gastroenterol Hepatol 13:501–502
13. Tougas G (2000) The autonomic nervous system in functional bowel disorders. Gut 47:iv78–iv80
14. Vanner SJ, Meerveld BGV, Mawe GM, Grundy D (2016) Fundamentals of neurogastroenterology: basic science. Gastroenterology 150(6):1280–1291
15. Vermeulen W, De Man JG, Pelckmans PA, De Winter BY (2014) Neuroanatomy of lower gastrointestinal pain disorders. World J Gastroenterol 20(4):1005–1020
16. Wood J (2016) Enteric nervous system: neuropathic gastrointestinal motility. Digestive Diseases and Sciences 61(7):1803–1816
17. Wood JD (2004) Enteric neuroimmunophysiology and pathophysiology. Gastroenterology 127(2):635–657

Methods and Protocol

1. Berghe VP (2004) Fluorescent molecules as tools to study Ca^{2+} signaling, mitochondrial dynamics and synaptic function in enteric neurons. Verh K Acad Geneeskd Belg 66(5–6):407–25
2. Berghe VP, Bisschops R, Tack J (2001) Imaging of neuronal activity in the gut. Curr Opin Pharmacol 1(6):563–567
3. Grundmann D, Klotz M, Rabe H, Glanemann M, Schäfer K-H (2015) Isolation of high-purity myenteric plexus from adult human and mouse gastrointestinal tract. Sci Rep 5:9226
4. Metzger M, Bareiss PM, Danker T, Wagner S, Hennenlotter J, Guenther E, Obermayr F, Stenzl A, Koenigsrainer A, Skutella T, Just L (2009) Expansion and differentiation of neural progenitors derived from the human adult enteric nervous system. Gastroenterology 137(6):2063–2073
5. Middlemiss PJ, Jiang S, Wang J, Rathbone MP (2002) A method for purifying enteric glia from rat myenteric plexus. In Vitro Cell Dev Biol Anim 38(4):188–90
6. Rauch U, Hänsgen A, Hagl C, Holland-Cunz S, Schäfer KH (2006) Isolation and cultivation of neuronal precursor cells from the developing human enteric nervous system as a tool for cell therapy in dysganglionosis. Int J Colorectal Dis 21(6):554–559
7. Schäfer KH, Hagl CI, Wink E, Holland-Cunz S, Klotz M, Rauch U, Waag KL (2003) How to approach the ENS: various ways to analyse motility disorders in situ and in vitro. Eur J Pediatr Surg 13(3):163–169
8. Smith TH, Ngwainmbi J, Grider JR, Dewey WL, Akbarali HI (2013) An in vitro preparation of isolated enteric neurons and glia from the myenteric plexus of the adult mouse. J Vis Exp (78):50688
9. Tack J, Smith TK (2004) Calcium imaging of gut activity. Neurogastroenterol Motil 16(1):86–95

Books

1. Corazziari E (ed) (1997) Neurogastroenterology. Walter De Gruyter Inc, Berlin, Germany
2. Denning PW, Maheshwari A (2013) Necrotizing enterocolitis, an issue of clinics in perinatology. Elsevier, San Diego, CA, USA
3. Faure C, Di Lorenzo C, Thapar N (eds) (2013) Pediatric neurogastroenterology—gastrointestinal motility and functional disorders in children. Springer, New York
4. Krammer H-J, Singer MV (eds) (2000) Neurogastroenterology—from the basics to the clinics (Falk symposium). Springer, Heidelberg, Germany
5. Langnas AN, Goulet O, Quigley EMM, Tappenden KA (2009) Intestinal failure: diagnosis, management and transplantation. Blackwell Publishing Ltd., Oxford, UK
6. Matarese LE, Steiger E, Seidner DL (eds) (2004) Intestinal failure and rehabilitation: a clinical guide. CRC Press, New York
7. Quigley EMM (ed) (2002) Gastroenterology highlights 2001–2002. Health Press, Oxford, UK
8. Quigley EMM, Pfeiffer R (eds) (2004) Neuro-gastroenterology. Butterworth-Heinemann, Philadelphia, Pennsylvania, USA
9. Rintala RJ, Pakarinen M, Wester T (eds) (2016) Current concepts of intestinal failure. Springer, Cham, Switzerland
10. Siegel GJ, Agranoff BW, Albers RW, Fisher SK, Uhler MD (eds) (1999) Basic Neurochemistry: molecular, cellular and medical aspects. Lippincott-Raven, Philadelphia, Pennsylvania, USA

Website for Review

Rome Foundation http://romeonline.org (Functional GI Disorders; Diagnostic Algorithms; Research Tools and Clinical Practice Materials).

Bibliography

1. McCorry LK (2007) Physiology of the autonomic nervous system. Am J Pharm Educ 71(4):78
2. Browning KN, Travagli RA (2014) Central nervous system control of gastrointestinal motility and secretion and modulation of gastrointestinal functions. Compr Physiol 4(4): 1339–1368
3. Altaf MA, Sood MR (2008) The nervous system and gastrointestinal function. Dev Disabil Res Rev 14(2):87–95
4. McConalogue K, Furness JB (1994) Gastrointestinal neurotransmitters. Baillieres Clin Endocrinol Metab 8(1):51–76
5. Goyal RK, Hirano I (1996) The enteric nervous system. N Engl J Med 334:1106–1115
6. Constantinescu M (2016) The enteric nervous system. In: Constantinescu C, Arsenescu R, Arsenescu V (eds) Neuro-immuno-gastroenterology. Springer International Publishing, Cham, Switzerland

7. Lundgren O (2000) Sympathetic input into the enteric nervous system. Gut 47:iv33–iv35
8. Rao M, Gershon MD (2016) The bowel and beyond: the enteric nervous system in neurological disorders. Nat Rev Gastroenterol Hepatol 13:517–528
9. Gershon MD (1981) The enteric nervous system. Annu Rev Neurosci 4:227–272
10. Crowell MD, Wessinger SB (2007) 5-HT and the brain-gut axis: opportunities for pharmacologic intervention. Expert Opin Investig Drugs 16(6):761–765
11. Nikitina OS, Welcome MO, Pereverzev VA (2016) Human anatomy and physiology. In: 2 parts. Part 1. Belarusian State Medical University Press, Minsk
12. Nikitina OS, Welcome MO, Pereverzev VA (2016) Human anatomy and physiology. In: 2 parts. Part 2. Belarusian State Medical University Press, Minsk, Belarus
13. Kubarko AI, Semenovich AA, Pereverzev VA, Lobanok LM, Kharlamova AN, Aleksandrov DA (2014) Normal physiology. In 2 parts. Part 2. Vysheishaya shkola, Minsk, Belarus
14. Vermeulen W, De Man JG, Pelckmans PA, De Winter BY (2014) Neuroanatomy of lower gastrointestinal pain disorders. World J Gastroenterol 20(4):1005–1020
15. Hansen MB (2003) The enteric nervous system I: organisation and classification. Pharmacol Toxicol 92(3):105–113
16. Hansen MB (2003) The enteric nervous system II: gastrointestinal functions. Pharmacol Toxicol 92(6):249–257
17. Segal K, Lisnyansky I, Nageris B, Feinmesser R (1996) Parasympathetic innervation of the salivary glands. Operat Tech Otolaryngol Head Neck Surg 7(4):333–338
18. Beyak M, Bulmer D, Jiang W, Keating W, Grundy D (2006) Extrinsic sensory afferent nerves innervating the gastrointestinal tract. In: JoHnson LR (ed) Physiology of the GI tract. Academic Press, San Diego, CA, USA
19. Dorofeeva AA, Panteleev SS, Makarov FN (2008) Parasympathetic innervation of the initial segments of the large intestine in cats. Neurosci Behav Physiol 38(9):923–927
20. Purves D, Augustine GJ, Fitzpatrick D, Katz LC, LaMantia A-S, McNamara JO, Williams SM (eds) (2001) Neuroscience, 2nd edn. Sinauer Associates, Sunderland, Massachusetts, United States
21. Rohner-Jeanrenaud F, Ionescu E, Jeanrenaud B (1983) The origins and role of efferent vagal nuclei in hyperinsulinemia in hypothalamic and genetically obese rodents. J Auton Nerv Syst 9(1):173–184
22. De Groat WC, Booth AM, Milne RJ, Roppolo JR (1982) Parasympathetic preganglionic neurons in the sacral spinal cord. J Auton Nerv Syst 5(1):23–43
23. Fox D (2017) The shock tactics set to shake up immunology: an experimental procedure is exposing the links between the nervous and immune systems. Could it be the start of a revolution? Nature 545:20–22
24. Howland RH (2014) Vagus nerve stimulation. Curr Behav Neurosci Rep 1(2):64–73
25. Phillips RJ, Powley TL (2007) Innervation of the gastrointestinal tract: patterns of aging. Auton Neurosci 136(1–2):1–19
26. Uesaka T, Young HM, Pachnis V, Enomoto H (2016) Development of the intrinsic and extrinsic innervation of the gut. Dev Biol 417(2):158–167
27. Dumont L (1955) Parasympathetic innervation of the salivary glands. C R Hebd Seances Acad Sci 240(2):240–242
28. Aminoff MJ, Daroff RB (eds) (2014) Encyclopedia of the neurological sciences, 2nd edn. Elsevier, San Diego
29. Siéssere S, Vitti M, Sousa LG, Semprini M, Iyomasa MM, Regalo SC (2008) Anatomic variation of cranial parasympathetic ganglia. Braz Oral Res 22(2):101–105
30. Senger M, Stoffels HJ, Angelov DN (2014) Topography, syntopy and morphology of the human otic ganglion: a cadaver study. Ann Anat 196(5):327–335
31. Khosravani N, Sandberg M, Ekström J (2006) The otic ganglion in rats and its parotid connection: cholinergic pathways, reflex secretion and a secretory role for the facial nerve. Exp Physiol 91(1):239–247

32. Piagkou M, Demesticha T, Piagkos G, Ioannis C, Skandalakis P, Johnson EO (2012) The mandibular nerve: the anatomy of nerve injury and entrapment, maxillofacial surgery. In: Assael L (ed) InTech, Rijeka, Croatia
33. Vranish JR, Bailey EF (2015) A comprehensive assessment of genioglossus electromyographic activity in healthy adults. J Neurophysiol 113(7):2692–2699
34. Zaidi FN, Meadows P, Jacobowitz O, Davidson TM (2013) Tongue anatomy and physiology, the scientific basis for a novel targeted neurostimulation system designed for the treatment of obstructive sleep apnea. Neuromodulation 16(4):376–386
35. Brierley SM, Jones RC 3rd, Xu L, Gebhart GF, Blackshaw LA (2005) Activation of splanchnic and pelvic colonic afferents by bradykinin in mice. Neurogastroenterol Motil 17(6):854–862
36. Brierley SM, Jones RC 3rd, Gebhart GF, Blackshaw LA (2004) Splanchnic and pelvic mechanosensory afferents signal different qualities of colonic stimuli in mice. Gastroenterology 127(1):166–178
37. Dorofeeva AA, Panteleev SS, Makarov FN (2008) Participation of sacral parasympathetic nucleus in the innervation of the descending colon and the rectum in cats. Morfologiia 133 (1):38–41
38. Dujovny N, Quiros RM, Saclarides TJ (2004) Anorectal anatomy and embryology. Surg Oncol Clin 13(2):277–293
39. Birnbaum E (2015) Surgical anatomy of the colon, rectum and anus. In: Ratto C, Parrello A, Dionisi L, Litta F (eds) Coloproctology. Springer International Publishing, Cham, Switzerland
40. Tong WD, Ridolfi TJ, Kosinski L, Ludwig K, Takahashi T (2010) Effects of autonomic nerve stimulation on colorectal motility in rats. Neurogastroenterol Motil 22(6):688–693
41. Brookes SJH, Spencer NJ, Costa M, Zagorodnyuk VP (2013) Extrinsic primary afferent signalling in the gut. Nat Rev Gastroenterol Hepatol 10:286–296
42. Uvnäs-Moberg K (1994) Role of efferent and afferent vagal nerve activity during reproduction: integrating function of oxytocin on metabolism and behaviour. Psychoneuroendocrinology 19(5–7):687–695
43. Contreras RJ, Beckstead RM, Norgren R (1982) The central projections of the trigeminal, facial, glossopharyngeal and vagus nerves: an autoradiographic study in the rat. J Auton Nerv Syst 6(3):303–322
44. de Lartigue G, de La Serre CB, Raybould HE (2011) Vagal afferent neurons in high fat diet-induced obesity; intestinal microflora, gut inflammation and cholecystokinin. Physiol Behav 105(1):100–105
45. Stapelfeldt WH, Szurszewski JH (1989) Central neurotensin nerves modulate colo-colonic reflex activity in the guinea-pig inferior mesenteric ganglion. J Physiol 411:347–365
46. Moran TH, Baldessarini AR, Salorio CF, Lowery T, Schwartz GJ (1997) Vagal afferent and efferent contributions to the inhibition of food intake by cholecystokinin. Am J Physiol 272(4 Pt 2):R1245–R1251
47. Chang HY, Mashimo H, Goyal RK (2003) IV. Current concepts of vagal efferent projections to the gut. Am J Physiol Gastrointest Liver Physiol 284(3):G357–G366
48. Schwartz GJ, McHugh PR, Moran TH (1993) Gastric loads and cholecystokinin synergistically stimulate rat gastric vagal afferents. Am J Physiol 265(4 Pt 2):R872–R876
49. Schwartz GJ (2000) The role of gastrointestinal vagal afferents in the control of food intake: current prospects. Nutrition 16(10):866–873
50. Fox EA (2006) A genetic approach for investigating vagal sensory roles in regulation of gastrointestinal function and food intake. Auton Neurosci 126–127:9–29
51. Bressloff P (2009) Cable theory of protein receptor trafficking in a dendritic tree. Physical Rev E 79(4 Pt 1):041904
52. Bell J (2009) Cable theory. In: Binder MD, Hirokawa N, Windhorst U (eds) Encyclopedia of neuroscience. Springer-Verlag, Berlin, Germany
53. Windhorst U, Lalley PM (2009) Action potential propagation. In: Binder MD, Hirokawa N, Windhorst U (eds) Encyclopedia of neuroscience. Springer, Heidelberg

54. Nakajima S, Onodera K (1969) Adaptation of the generator potential in the crayfish stretch receptors under constant length and constant tension. J Physiol 200(1):187–204
55. Belmonte C (2013) Nociceptor generator potential. In: Gebhart GF, Schmidt RF (eds) Encyclopedia of pain. Springer, Heidelberg
56. Loewenstein WR, Altamirano-Orrego R (1958) The refractory state of the generator and propagated potentials in a pacinian corpuscle. J Gen Physiol 41(4):805–824
57. Gees M, Alpizar YA, Boonen B, Sanchez A, Everaerts W, Segal A et al (2013) Mechanisms of transient receptor potential vanilloid 1 activation and sensitization by allyl isothiocyanate. Mol Pharmacol 84(3):325–334
58. Mazzone SB (2005) An overview of the sensory receptors regulating cough. Cough 1:2
59. Kennedy WR, deF Webster H, Yoon KS (1975) Human muscle spindles: fine structure of the primary sensory ending. J Neurocytol 4(6):675–695
60. Dimitriou M, Edin BB (2008) Discharges in human muscle receptor afferents during block grasping. J Neurosci 28(48):12632–12642
61. Iglesias PA (2012) A systems biology view of adaptation in sensory mechanisms. Adv Exp Med Biol 736:499–516
62. Frings S, Bradley J (eds) (2005) Transduction channels in sensory cells. Wiley-VCH Verlag GmbH, Weinheim
63. Terre V, Ashmore JF, Lamb TD, Menini A (1995) Transduction and adaptation in sensory receptor cells. J Neurosci 75(12):7757–7768
64. Møller AR (2011) Anatomy and physiology of sensory systems. In: Intraoperative neurophysiological monitoring. Springer, New York
65. Kääb S, Nuss HB, Chiamvimonvat N, O'Rourke B, Pak PH, Kass DA et al (1996) Ionic mechanism of action potential prolongation in ventricular myocytes from dogs with pacing-induced heart failure. Circ Res 78(2):262–273
66. Yakovlev AV, Koroleva KS, Valiullina FF, Khazipov RN (2013) Resting membrane potential of rat ventroposteriomedial thalamic neurons during postnatal development. Biochem (Mosc) Suppl Ser A Membr Cell Biol 7(3):207–212
67. Lodish H, Berk A, Zipursky SL, Matsudaira P, Baltimore D, Darnell J (2000) Molecular cell biology, 4th edn. WH Freeman, New York
68. Forsythe ID, Redman SJ (1988) The dependence of motoneurone membrane potential on extracellular ion concentrations studied in isolated rat spinal cord. J Physiol 404:83–99
69. Bean BP (2007) The action potential in mammalian central neurons. Nat Rev Neurosci 8:451–465
70. Clay JR, Forger DB, Paydarfar D (2012) Ionic mechanism underlying optimal stimuli for neuronal excitation: role of Na^+ channel inactivation. PLoS ONE 7(9):e45983
71. Andreoli TE, Hoffman JF, Fanestil DD, Schultz SG (eds) (1986) Physiology of membrane disorders. Springer, New York
72. Kurbel S (2011) Donnan effect on chloride ion distribution as a determinant of body fluid composition that allows action potentials to spread via fast sodium channels. Theor Biol Med Model 8:16
73. Shin EH, Li Y, Kumar U, Sureka HV, Zhang X, Payne CK (2013) Membrane potential mediates the cellular binding of nanoparticles. Nanoscale 5(13):5879–5886
74. Platre MP, Jaillais Y (2017) Anionic lipids and the maintenance of membrane electrostatics in eukaryotes. Plant Signal Behav 12(2):e1282022
75. Ma Y, Poole K, Goyette J, Gaus K (2017) Introducing membrane charge and membrane potential to T cell signaling. Front Immunol 8:1513
76. Osorio N, Delmas P (2011) Patch clamp recording from enteric neurons in situ. Nat Protoc 6:15–27
77. Carbone SE, Jovanovska V, Nurgali K, Brookes SJH (2014) Human enteric neurons: morphological, electrophysiological, and neurochemical identification. Neurogastroenterol Motil 26(12):1812–1816

78. Hawkins EG, Dewey WL, Anitha M, Srinivasan S, Grider JR, Akbarali HI (2013) Electrophysiological characteristics of enteric neurons isolated from the Immortomouse. Dig Dis Sci 58(6):1516–1527
79. Galligan JJ, Akbarali HI (2014) Molecular physiology of enteric opioid receptors. Am J Gastroenterol Suppl 2:17–21
80. Strichartz GR, Ritchie JM (1987) The action of local anesthetics on ion channels of excitable tissues. In: Strichartz GR (ed) Local anesthetics. Handbook of experimental pharmacology, vol 81. Springer, Berlin
81. North RA (1982) Electrophysiology of the enteric neurons. Mediators and drugs in gastrointestinal motility I. In: Bertaccini G (ed) Handbook of experimental pharmacology. Springer, Berlin
82. Roth BJ (1994) Mechanisms for electrical stimulation of excitable tissue. Crit Rev Biomed Eng 22(3–4):253–305
83. Franconi F, Huxtable RJ (1987) Introduction: pharmacology of excitable tissues. In: Huxtable RJ, Franconi F, Giotti A (eds) The biology of taurine. Advances in experimental medicine and biology. Springer, New York
84. Hao J, Wang X-D, Dan Y, Poo M-M, Zhang X-H (2009) An arithmetic rule for spatial summation of excitatory and inhibitory inputs in pyramidal neurons. PNAS 106 (51):21906–21911
85. Britten KH, Heuer HW (1999) Spatial summation in the receptive fields of MT neurons. J Neurosci 19(12):5074–5084
86. Oertel D (1985) Use of brain slices in the study of the auditory system: spatial and temporal summation of synaptic inputs in cells in the anteroventral cochlear nucleus of the mouse. J Acoust Soc Am 78(1 Pt 2):328–333
87. Zhang Y, Shao S, Zhang J, Wang L, Wang K, Svensson P (2017) Temporal summation and motor function modulation during repeated jaw movements in patients with temporomandibular disorder pain and healthy controls. Pain 158(7):1272–1279
88. Magee Jeffrey C (2000) Dendritic integration of excitatory synaptic input. Nat Rev Neurosci 1:181–190
89. Hiroyoshi M (2009) Synaptic integration. In: Binder MD, Hirokawa N, Windhorst U (eds) Encyclopedia of neuroscience. Springer, Berlin, Germany
90. Stewart M (1997) Antidromic and orthodromic responses by subicular neurons in rat brain slices. Brain Res 769(1):71–85
91. Grill WM, Cantrell MB, Robertson MS (2008) Antidromic propagation of action potentials in branched axons: implications for the mechanisms of action of deep brain stimulation. J Comput Neurosci 24(1):81–93
92. Bakkum DJ, Frey U, Radivojevic M, Russell TL, Müller J, Fiscella M et al (2013) Tracking axonal action potential propagation on a high-density microelectrode array across hundreds of sites. Nat Commun 4:2181
93. Gogan P, Gueritaud JP, Tyc-Dumont S (1983) Comparison of antidromic and orthodromic action potentials of identified motor axons in the cat's brain stem. J Physiol 335:205–220
94. Bielajew C, Lapointe M, Kiss I, Shizgal P (1982) Absolute and relative refractory periods of the substrates for lateral hypothalamic and ventral midbrain self-stimulation. Physiol Behav 28(1):125–132
95. Fouriezos G, Walker S, Rick J, Bielajew C (1987) Refractoriness of neurons mediating intracranial self-stimulation in the anterior basal forebrain. Behav Brain Res 24(1):73–80
96. Lowitzsch K, Hopf HC, Galland J (1977) Changes of sensory conduction velocity and refractory periods with decreasing tissue temperature in man. J Neurol 216(3):181–188
97. Yeomans JS (1979) The absolute refractory periods of self-stimulation neurons. Physiol Behav 22(5):911–919
98. Gregory JE, Harvey RJ, Proske U (1977) A 'late supernormal period' in the recovery of excitability following an action potential in muscle spindle and tendon organ receptors. J Physiol 271(2):449–472

99. Joung B, Park H-W, Maruyama M, Tang L, Song J, Han S et al (2011) Intracellular calcium and the mechanism of anodal supernormal excitability in Langendorff perfused rabbit ventricles. Circ J 75(4):834–843
100. Parkin PJ, Le Quesne PM (1982) Effect of a synthetic pyrethroid deltamethrin on excitability changes following a nerve impulse. J Neurol Neurosurg Psychiatry 45(4):337–342
101. Stöhr M, Gilliatt RW, Willison RG (1981) Supernormal excitability of human sensory fibers after ischemia. Muscle Nerve 4(1):73–75
102. Stöhr M (1981) Activity-dependent variations in threshold and conduction velocity of human sensory fibers. J Neurol Sci 49(1):47–54
103. McIntyre CC, Richardson AG, Grill WM (2002) Modeling the excitability of mammalian nerve fibers: influence of afterpotentials on the recovery cycle. J Neurophysiol 87(2):995–1006
104. Barth DS, Di S (1991) Laminar excitability cycles in neocortex. J Neurophysiol 65(4):891–898
105. Stockbridge N (1988) Etiology of the supernormal period. Biophys J 54(5):777–780
106. Ng A, Burke D, Al-Shehab A (1987) Hyperexcitability of cutaneous afferents during the supernormal period. Relevance to paraesthesiae. Brain 110(Pt 4):1015–1031
107. Trombin F, Gnatkovsky V, de Curtis M (2011) Changes in action potential features during focal seizure discharges in the entorhinal cortex of the in vitro isolated guinea pig brain. J Neurophysiol 106(3):1411–1423
108. Xie C, Lin Z, Hanson L, Cui Y, Cui B (2012) Intracellular recording of action potentials by nanopillar electroporation. Nat Nanotechnol 7:185–190
109. Oswald MJ, Oorschot DE, Schulz JM, Lipski J, Reynolds JNJ (2009) I_H current generates the afterhyperpolarisation following activation of subthreshold cortical synaptic inputs to striatal cholinergic interneurons. J Physiol 587(Pt 24):5879–5897
110. Lancaster B, Hu H, Ramakers GMJ, Storm JF (2001) Interaction between synaptic excitation and slow afterhyperpolarization current in rat hippocampal pyramidal cells. J Physiol 536(Pt 3):809–823
111. Shah M, Haylett DG (2000) Ca2 + channels involved in the generation of the slow afterhyperpolarization in cultured rat hippocampal pyramidal neurons. J Neurophysiol 83:2554–2561
112. Gu N, Vervaeke K, Hu H, Storm JF (2005) Kv7/KCNQ/M and HCN/h, but not KCa2/SK channels, contribute to the somatic medium afterhyperpolarization and excitability control in CA1 hippocampal pyramidal cells. J Physiol 566:689–715
113. Andrade R, Foehring RC, Tzingounis AV (2012) Essential role for phosphatidylinositol 4,5-bisphosphate in the expression, regulation, and gating of the slow afterhyperpolarization current in the cerebral cortex. Front Cell Neurosci 6:47
114. Gulledge AT, Dasari S, Onoue K, Stephens EK, Hasse JM, Avesar D (2013) A sodium-pump-mediated afterhyperpolarization in pyramidal neurons. J Neurosci 33:13025–13041
115. Wu WW, Chan CS, Disterhoft JF (2004) Slow afterhyperpolarization governs the development of NMDA receptor-dependent afterdepolarization in CA1 pyramidal neurons during synaptic stimulation. J Neurophysiol 92(4):2346–2356
116. Aponte Y, Lien CC, Reisinger E, Jonas P (2006) Hyperpolarization-activated cation channels in fast-spiking interneurons of rat hippocampus. J Physiol 574(Pt 1):229–243
117. Cloues RK, Sather WA (2003) Afterhyperpolarization regulates firing rate in neurons of the suprachiasmatic nucleus. J Neurosci 23(5):1593–1604
118. Binder MD, Hirokawa N, Windhorst U (eds) (2009) Encyclopedia of neuroscience. Springer, Heidelberg
119. Le Ray D, De Sevilla DF, Porto AB, Fuenzalida M, Buño W (2004) Heterosynaptic metaplastic regulation of synaptic efficacy in CA1 pyramidal neurons of rat hippocampus. Hippocampus 14(8):1011–1025

120. Emmelin N (1981) Nervous control of mammalian salivary glands. Philos Trans R Soc Lond B Biol Sci 296(1080):27–35
121. Proctor GB, Carpenter G (2007) Regulation of salivary gland function by autonomic nerves. Auton Neurosci 133(1):3–18
122. Mittal RK (2011) Motor function of the pharynx, esophagus, and its sphincters. Morgan & Claypool Life Sciences, San Diego
123. Liu L, Arnold R, Robinson M (2012) Dissection and exposure of the whole course of deep nerves in human head specimens after decalcification. Int J Otolaryngol 2012:418650
124. McManus LJ, Dawes PJD, Stringer MD (2012) Surgical anatomy of the chorda tympani: a micro-CT study. Surg Radiol Anat 34(6):513–518
125. Laage-Hellman JE, Stromblad BC (1960) Secretion from human submaxillary gland after section of the chorda tympani. J Appl Physiol 15:295–297
126. Possover M, Chiantera V, Baekelandt J (2007) Anatomy of the sacral roots and the pelvic splanchnic nerves in women using the LANN technique. Surg Laparosc Endosc Percutan Tech 17(6):508–510
127. Burnett AL, Wesselmann U (1999) History of the neurobiology of the pelvis. Urology 53 (6):1082–1089
128. Tong WD, Ridolfi TJ, Kosinski L, Ludwig K, Takahashi T (2010) Effects of autonomic nerve stimulation on colorectal motility in rats. Neurogastroenterol Motil 22(6):688–693
129. Ridolfi TJ, Tong W-D, Takahashi T, Kosinski L, Ludwig KA (2009) Sympathetic and parasympathetic regulation of rectal motility in rats. J Gastrointest Surg 13(11):2027–2033
130. Maruyama S, Okabe S, Endo M, Sato K, Iwai T (2003) The role of the rectal branches of pelvic plexus in defecation and colonic motility in a canine model. J Med Dent Sci 50 (4):275–284
131. Carlstedt A, Fasth S, Hultén L, Nordgren S (1988) The sympathetic innervation of the internal anal sphincter and rectum in the cat. Acta Physiol Scand 133(3):423–431
132. Coggeshall RE, Hancock MB, Applebaum ML (1976) Categories of axons in mammalian rami communicantes. J Comp Neurol 167(1):105–123
133. Coggeshall RE, Galbraith SL (1978) Categories of axons in mammalian rami communicantes. Part II. J Comp Neurol 181(2):349–359
134. Jänig W, McLachlan EM (1987) Organization of lumbar spinal outflow to distal colon and pelvic organs. Physiol Rev 67(4):1332–1404
135. Jänig W, McLachlan EM (1986) The sympathetic and sensory components of the caudal lumbar sympathetic trunk in the cat. J Comp Neurol 245(1):62–73
136. Kuo DC, Hisamitsu T, de Groat WC (1984) A sympathetic projection from sacral paravertebral ganglia to the pelvic nerve and to postganglionic nerves on the surface of the urinary bladder and large intestine of the cat. J Comp Neurol 226(1):76–86
137. Baron R, Jänig W, McLachlan EM (1985) The afferent and sympathetic components of the lumbar spinal outflow to the colon and pelvic organs in the cat. I. The hypogastric nerve. J Comp Neurol 238(2):135–146
138. Appenzeller O, Ogin G (1974) Myelinated fibres in human paravertebral sympathetic chain: white rami communicantes in alcoholic and diabetic patients. J Neurol Neurosurg Psychiatry 37(10):1155–1161
139. Aminoff MJ, Daroff RB (eds) (2014) Encyclopedia of the Neurological Sciences, 2nd edn. Academic Press, San Diego
140. Nógrádi A, Vrbová G (2006) Anatomy and physiology of the spinal cord. In: Antal N (ed) Transplantation of neural tissue into the spinal cord part of the series neuroscience intelligence unit. Springer, New York, US
141. Song ZF, Sun MM, Wu ZY, Xia CL (2010) Anatomical study and clinical significance of the rami communicantes between cervicothoracic ganglion and brachial plexus. Clin Anat 23(7):811–814
142. Barson AJ, Sands J (1977) Regional and segmental characteristics of the human adult spinal cord. J Anat 123(Pt 3):797–803

143. Ko H-Y, Park JH, Shin YB, Baek SY (2004) Gross quantitative measurements of spinal cord segments in human. Spinal Cord 42:35–40
144. Miolan JP, Niel JP (1996) The mammalian sympathetic prevertebral ganglia: integrative properties and role in the nervous control of digestive tract motility. J Auton Nerv Syst 58 (3):125–138
145. Gilman AG, Goodman LS, Rail TW, Murad F (eds) (1985) The pharmacological basis of therapeutics, 7th edn. Macmillan Publishing Company, New York
146. Brunton L, Knollman B, Hilal-Dandan R (2017) Goodman and Gilman's pharmacological basis of therapeutics, 13th edn. McGraw-Hill, New York
147. McDougal DH, Hermann GE, Rogers RC (2011) Vagal afferent stimulation activates astrocytes in the nucleus of the solitary tract via AMPA receptors: evidence of an atypical neural–glial interaction in the brainstem. J Neurosci 31(39):14037–14045
148. Klarer M, Arnold M, Gunther L, Winter C, Langhans W, Meyer U (2014) Gut vagal afferents differentially modulate innate anxiety and learned fear. J Neurosci 34(21):7067–7077
149. Bodenlos JS, Kose S, Borckardt JJ, Nahas Z, Darlene S, O'Neil PM et al (2007) Vagus nerve stimulation and emotional responses to food among depressed patients. J Diabetes Sci Technol 1(5):771–779
150. Park MC, Goldman MA, Carpenter LL, Price LH, Friehs GM (2007) Vagus nerve stimulation for depression: rationale, anatomical and physiological basis of efficacy and future prospects. Acta Neurochir Suppl 97(Pt 2):407–416
151. Berthoud H-R (2008) The vagus nerve, food intake and obesity. Regul Pept 149(1–3):15–25
152. Jin H, Guo J, Liu J, Lyu B, Foreman RD, Yin J et al (2017) Anti-inflammatory effects and mechanisms of vagal nerve stimulation combined with electroacupuncture in a rodent model of TNBS-induced colitis. Am J Physiol Gastrointest Liver Physiol 313(3):G192–G202
153. Browning KN, Verheijden S, Boeckxstaens GE (2017) The vagus nerve in appetite regulation, mood, and intestinal inflammation. Gastroenterology 152(4):730–744
154. Rexed B (1952) The cytoarchitectonic organization of the spinal cord in the cat. J Comp Neurol 96(3):414–495
155. Rexed B (1954) A cytoarchitectonic atlas of the spinal cord in the cat. J Comp Neurol 100 (2):297–379
156. Room R, Carstairs C (2003) Bror Rexed: an architect of drug control but far from a fundamentalist. Addiction 98(1):127–128
157. Sarikcioglu L, Ozsoy U (2008) Bror Rexed (1914–2002). J Neurol 255(12):1988–1989
158. Pearce JMS (2008) The development of spinal cord anatomy. Eur Neurol 59:286–291
159. O'Brien JS, Sampson EL (1965) Lipid composition of the normal human brain: gray matter, white matter, and myelin. J Lipid Res 6:537–544
160. Caputi F, Spaziante R, de Divitiis E, Nashold BS Jr (1995) Luigi Rolando and his pioneering efforts to relate structure to function in the nervous system. J Neurosurg 83 (5):933–937
161. Petras JM (1968) The substantia gelatinosa of Rolando. Experientia 24(10):1045–1047
162. Willis WD (1985) Nociceptive pathways: anatomy and physiology of nociceptive ascending pathways. Philos Trans R Soc Lond B Biol Sci 308(1136):253–270
163. Gebhart GF, Schmidt RF (eds) (2013) Encyclopedia of pain. Springer-Verlag, Heidelberg, Germany
164. Ristanović D, Milosević NT (2007) A confirmation of Rexed's laminar hypothesis using the Sholl linear method complemented by nonparametric statistics. Neurosci Lett 414 (3):286–290
165. Mannen T, Iwata M, Toyokura Y, Nagashima K (1982) The Onuf's nucleus and the external anal sphincter muscles in amyotrophic lateral sclerosis and Shy-Drager syndrome. Acta Neuropathol 58(4):255–260

166. Onufrowicz B (1899) Notes on the arrangement and function of the cell groups in the sacral region of the spinal cord. J Nerv Ment Dis 26:498–504
167. Onufrowicz B (1900) On the arrangement and function of the cell groups of the sacral region of the spinal cord in man. Arch Neurol Psychopathol 3:387–411
168. Fields RD (2008) White matter in learning, cognition and psychiatric disorders. Trends Neurosci 31(7):361–370
169. Scheel M, Storm C, Gentsch A, Nee J, Luckenbach F, Ploner CJ, Leithner C (2013) The prognostic value of gray-white-matter ratio in cardiac arrest patients treated with hypothermia. Scand J Trauma Resusc Emerg Med 21:23
170. Strominger NL, Demarest RJ, Laemle LB (2012) The reticular formation and the limbic system. In: Strominger NL, Demarest RJ, Laemle LB (eds) Noback's human nervous system: structure and function, 7th edn. Springer, NY
171. Willis WD, Westlund KH (2013) Spinal dorsal horn pathways, dorsal column (visceral). In: Gebhart GF, Schmidt RF (eds) Encyclopedia of pain. Springer, Berlin
172. Palecek J, Willis WD (2003) The dorsal column pathway facilitates visceromotor responses to colorectal distention after colon inflammation in rats. Pain 104(3):501–507
173. Willis WD Jr, Westlund KN (2001) The role of the dorsal column pathway in visceral nociception. Curr Pain Headache Rep 5(1):20–26
174. Palecek J (2004) The role of dorsal columns pathway in visceral pain. Physiol Res 53: S125–S130
175. Wood JD (2011) Visceral pain: spinal afferents, enteric mast cells, enteric nervous system and stress. Curr Pharm Des 17(16):1573–1575
176. Elsenbruch S (2011) Abdominal pain in Irritable Bowel Syndrome: a review of putative psychological, neural and neuro-immune mechanisms. Brain Behav Immun 25(3):386–394
177. Curatolo M (2011) Pathophysiology of abdominal pain. Ther Umsch 68(8):415–419
178. Todd AJ (2017) Identifying functional populations among the interneurons in laminae I–III of the spinal dorsal horn. Mol Pain 13:1744806917693003
179. Wood JD (2007) Neuropathophysiology of functional gastrointestinal disorders. World J Gastroenterol 13(9):1313–1332
180. Holmes GM (2012) Upper gastrointestinal dysmotility after spinal cord injury: is diminished vagal sensory processing one culprit? Front Physiol 3:277
181. Boyle KA, Gutierrez-Mecinas M, Polgár E, Mooney N, O'Connor E, Furuta T et al (2017) A quantitative study of neurochemically defined populations of inhibitory interneurons in the superficial dorsal horn of the mouse spinal cord. Neuroscience 363:120–133
182. Van Oudenhove L, Coen SJ, Aziz Q (2007) Functional brain imaging of gastrointestinal sensation in health and disease. World J Gastroenterol 13(25):3438–3445
183. Tran L, Wiskur B, Meerveld BG-V (2012) The role of the anterolateral bed nucleus of the stria terminalis in stress-induced nociception. Am J Physiol Gastrointest Liver Physiol 302 (11):G1301–G1309
184. Dubin Adrienne E, Patapoutian Ardem (2010) Nociceptors: the sensors of the pain pathway. J Clin Invest 120(11):3760–3772
185. Smith ES, Lewin GR (2009) Nociceptors: a phylogenetic view. J Comp Physiol A Neuroethol Sens Neural Behav Physiol. 195(12):1089–1106
186. Elie DAC (2009) Neuroanatomy of pain and pain pathways. In: Biobehavioral approaches to pain, pp 17–43. Springer, New York
187. Boadas-Vaello P, Castany S, Homs J, Álvarez-Pérez B, Deulofeu M, Verdú E (2016) Neuroplasticity of ascending and descending pathways after somatosensory system injury: reviewing knowledge to identify neuropathic pain therapeutic targets. Spinal Cord 54 (5):330–340
188. Sharma A, Lelic D, Brock C, Paine P, Aziz Q (2009) New technologies to investigate the brain-gut axis. World J Gastroenterol 15(2):182–191
189. De Giorgio R, Barbara G, Blennerhassett P, Wang L, Stanghellini V, Corinaldesi R et al (2001) Intestinal inflammation and activation of sensory nerve pathways: a functional and morphological study in the nematode infected rat. Gut 49:822–827

190. Guska NI (1966) Role of the reticular formation in regulation of the interoceptive reflex reactions of the stomach and small intestine. Biull Eksp Biol Med 62(8):14–18
191. Davison JS (1992) Central organization of gastrointestinal vagal reflexes. In: Thomson ABR, Shaffer E (eds) Modern concepts in gastroenterology part of the series topics in gastroenterology. Springer, New York
192. Camilleri M, Saslow SB, Bharucha AE (1996) Gastrointestinal sensation: mechanisms and relation to functional gastrointestinal disorders. Gastroenterol Clin 25(1):247–258
193. McC Brooks C, Koizumi K, Siebens AA (1956) Inhibitory action of bulbar and suprabulbar reticular formation on the spinal reflex pathway. Am J Physiol 184:497–504
194. Yezierski RP (2013) Spinomesencephalic tract. In: Gebhart GF, Schmidt RF (eds) Encyclopedia of pain. Springer, Berlin
195. Blomqvist A, Craig AD (1991) Organization of spinal and trigeminal input to the PAG. In: Depaulis A, Bandler R (eds) The midbrain periaqueductal gray matter volume 213 of the series NATO ASI series. Springer, New York, USA
196. Zhang X (2013) Spinohypothalamic tract, anatomical organization, and response properties. In: Gebhart GF, Schmidt RF (eds) Encyclopedia of pain. Springer, Berlin
197. Donald D, Price G (2002) Nicholas Verne. Does the spinothalamic tract to ventroposterior lateral thalamus and somatosensory cortex have roles in both pain sensation and pain-related emotions? J Pain 3(2):105–108
198. Todd AJ (2013) Spinothalamic tract neurons: neurotransmitter inputs and receptors. In: Gebhart GF, Schmidt RF (eds) Encyclopedia of pain. Springer, Berlin
199. Herrero MT, Barcia C, Navarro JM (2002) Functional anatomy of thalamus and basal ganglia. Childs Nerv Syst 18(8):386–404
200. Stahnisch FW, Nakashima AS (2013) Theodore Brown Rasmussen (1910–2002). J Neurol 260(10):2694–2696
201. Penfield W, Rasmussen T (1950) The cerebral cortex of man: a study of localization of function. Macmillan, New York
202. Kell CA, von Kriegstein K, Rösler A, Kleinschmidt A, Laufs H (2005) The sensory cortical representation of the human penis: revisiting somatotopy in the male homunculus. J Neurosci 25(25):5984–5987
203. Seriès P, Stocker AA, Simoncelli EP (2009) Is the homunculus “aware” of sensory adaptation? Neural Comput 21(12):3271–3304
204. Szurszewski JH (1981) Physiology of mammalian prevertebral ganglia. Annu Rev Physiol 43:53–68
205. Travagli RA, Hermann GE, Browning KN, Rogers RC (2006) Brainstem circuits regulating gastric function. Annu Rev Physiol 68:279–305
206. Williams EK, Chang RB, Strochlic DE, Umans BD, Lowell BB, Liberles SD (2016) Sensory neurons that detect stretch and nutrients in the digestive system. Cell 166(1):209–221
207. Wood JD, Alpers DH, Andrews PLR (1999) Fundamentals of neurogastroenterology. Gut 45(Suppl II):II6–II16
208. King BF, Szurszewski JH (1989) Peripheral reflex pathways involving abdominal viscera: transmission of impulses through prevertebral ganglia. Am J Physiol 256(3 Pt 1):G581–G588
209. Hartman DA, Krier J (1984) Synaptic and antidromic potentials of visceral neurones in ganglia of the lumbar sympathetic chain of the cat. J Physiol 350:413–428
210. Meyer K, Damasio A (2009) Convergence and divergence in a neural architecture for recognition and memory. Trend Neurosci 32(7):376–382
211. Man K, Kaplan J, Damasio H, Damasio A (2013) Neural convergence and divergence in the mammalian cerebral cortex: From experimental neuroanatomy to functional neuroimaging. J Comp Neurol 521(18):4097–4111
212. Satinoff E (1974) Neural integration of thermoregulatory responses. In: Dicara LV (ed) Limbic and autonomic nervous systems research. Plenum, New York, USA

213. Jackson S, Blake R (2010) Neural integration of information specifying human structure from form, motion, and depth. J Neurosci 30(3):838
214. Tsuruoka M, Maeda M, Inoue T (2005) Stimulation of the nucleus locus coeruleus/subcoeruleus suppresses visceromotor responses to colorectal distention in the rat. Neurosci Lett 381(1–2):97–101
215. Szabadi E (2013) Functional neuroanatomy of the central noradrenergic system. J Psychopharmacol 27(8):659–693
216. van Wassenaer-van Hall HN, van den Heuvel AG, Jansen GH, Hoogenraad TU, Mali WP (1995) Cranial MR in Wilson disease: abnormal white matter in extrapyramidal and pyramidal tracts. AJNR Am J Neuroradiol 16(10):2021–2027
217. Ennis M, Aston-Jones G (1988) Activation of locus coeruleus from nucleus paragigantocellularis: a new excitatory amino acid pathway in brain. J Neurosci 8(10):3644–3657
218. de Oliveira-Souza R (2012) The human extrapyramidal system. Med Hypotheses 79 (6):843–852
219. Nielson JL, Sears-Kraxberger I, Strong MK, Wong JK, Willenberg R, Steward O (2010) Unexpected survival of neurons of origin of the pyramidal tract after spinal cord injury. J Neurosci 30(34):11516–11528
220. Rea P (2015) Essential clinical anatomy of the nervous system. Academic Press, London
221. Wang F, Bobrow L, Liu J, Spencer L, Blumberg HP (2012) Corticolimbic functional connectivity in adolescents with bipolar disorder. PLoS ONE 7(11):e50177
222. Graeff FG, Cristina MLS, Nogueira RL, Audi EA, Oliveira RMW (1993) Role of the amygdala and periaqueductal gray in anxiety and panic. Behav Brain Res 58(1–2):123–131
223. Wang F, Kalmar JH, He Y, Jackowski M, Chepenik LG, Edmiston EE et al (2009) Functional and structural connectivity between the perigenual anterior cingulate and amygdala in bipolar disorder. Biol Psychiatry 66(5):516–521
224. Ferrucci M, Giorgi FS, Bartalucci A, Busceti CL, Fornai F (2013) The effects of locus coeruleus and norepinephrine in methamphetamine toxicity. Curr Neuropharmacol 11 (1):80–94
225. Mark MF, Purpura DP (2009) Autism, fever, epigenetics and the locus coeruleus. Brain Res Rev 59(2):388–392
226. Mouton PR, Pakkenberg B, Gundersen HJ, Price DL (1994) Absolute number and size of pigmented locus coeruleus neurons in young and aged individuals. J Chem Neuroanat 7 (3):185–190
227. Goncharuk VD (1994) Changes in ganglion nodosum neurons associated with stress-related cardiac deficiency. Exp Toxicol Pathol 46(6):457–464
228. Jänig W (1988) Integration of gut function by sympathetic reflexes. Baillieres Clin Gastroenterol 2(1):45–62
229. Welcome MO, Pereverzev VA (2013) A mini-review of the mechanisms of glucose memory enhancement. Int J Med Pharm Sci 4(1):17–30
230. Welcome MO, Mastorakis NE, Pereverzev VA (2015) Sweet taste receptor signaling network: possible implication for cognitive functioning. Neurol Res Int (Hindawi Publishing Corporation) 2015:606479
231. Halassa MM, Fellin T, Haydon PG (2007) The tripartite synapse: roles for gliotransmission in health and disease. Trends Mol Med 13(2):54–63
232. Zorec R, Verkhratsky A, Rodríguez JJ, Parpura V (2016) Astrocytic vesicles and gliotransmitters: slowness of vesicular release and synaptobrevin2-laden vesicle nanoarchitecture. Neuroscience 323:67–75
233. Perea G, Navarrete M, Araque A (2009) Tripartite synapses: astrocytes process and control synaptic information. Trends Neurosci 32(8):421–431
234. Uher E-M, Swash M (1998) Sacral reflexes: physiology and clinical application. Dis Colon Rectum 41(9):1165–1177
235. Furness JB (2006) Novel gut afferents: intrinsic afferent neurons and intestinofugal neurons. Auton Neurosci 125(1–2):81–85

236. Lang IM, Medda BK, Shaker R (2001) Mechanisms of reflexes induced by esophageal distension. Am J Physiol Gastrointest Liver Physiol 281(5):G1246–G1263
237. Zehr EP, Stein RB (1999) What functions do reflexes serve during human locomotion? Prog Neurobiol 58(2):185–205
238. Nishino T (2000) Physiological and pathophysiological implications of upper airway reflexes in humans. Jpn J Physiol 50(1):3–14
239. Widdicombe J (1998) Upper airway reflexes. Curr Opin Pulm Med 4(6):376–382
240. Bornstein JC, Costa M, Grider JR (2004) Enteric motor and interneuronal circuits. Neurogastroenterol Motil 16(Suppl 1):34–8
241. Enck P, Greving I, Klosterhalfen S, Wietek B (2006) Upper and lower gastrointestinal motor and sensory dysfunction after human spinal cord injury. Prog Brain Res 152:373–384
242. Chang HY, Mashimo H, Goyal RK (2003) Musings on the wanderer: what's new in our understanding of vago-vagal reflex? IV. Current concepts of vagal efferent projections to the gut. Am J Physiol Gastrointest Liver Physiol 284(3):G357–G366
243. Hsieh JH, Chen RF, Wu JJ, Yen CT, Chai CY (1998) Vagal innervation of the gastrointestinal tract arises from dorsal motor nucleus while that of the heart largely from nucleus ambiguus in the cat. J Auton Nerv Syst 70(1–2):38–50
244. Harper AA, Kidd C, Scratcherd T (1959) Vago-vagal reflex effects on gastric and pancreatic secretion and gastrointestinal motility. J Physiol (Lond) 148417–36
245. Thor PJ, Sobocki J (2000) Neuromodulation of vago-vagal reflex. Przegl Lek 57(5):79–81
246. Grabauskas G, Wu X, Song I, Zhou SY, Lanigan T, Owyang C (2016) Increased activation of the TRESK K^+ mediates vago-vagal reflex malfunction in diabetic rats. Gastroenterology 151(5):910–922
247. Bonaz B, Sinniger V, Pellissier S (2017) The vagus nerve in the neuro-immune axis: implications in the pathology of the gastrointestinal tract. Front Immunol 8:1452
248. Zhang Y, Hermanson ME, Eddinger TJ (2013) Tonic and phasic smooth muscle contraction is not regulated by the PKCα—CPI-17 pathway in swine stomach antrum and fundus. PLoS ONE 8(9):e74608
249. Condon RE, Cowles V, Schulte WJ, Frantzides C, Matsumoto T (1986) The effect of whole-gut lavage on colon motility and gastrocolic response in a monkey. Surgery 99 (5):531–536
250. Niv Y (2007) Cost, benefit, tonic, phasic: what do response rates tell us about dopamine and motivation? Ann N Y Acad Sci 1104:357–376
251. Niv Y, Daw ND, Joel D, Dayan P (2007) Tonic dopamine: opportunity costs and the control of response vigor. Psychopharmacology 191(3):507–520
252. Ostry DJ, Gribble PL, Levin MF, Feldman AG (1997) Phasic and tonic stretch reflexes in muscles with few muscle spindles: human jaw-opener muscles. Exp Brain Res 116(2):299–308
253. Mazzuoli-Weber G, Schemann M (2015) Mechanosensitivity in the enteric nervous system. Front Cell Neurosci 9:408
254. Schemann M, Mazzuoli G (2010) Multifunctional mechanosensitive neurons in the enteric nervous system. Auton Neurosci 153(1–2):21–25
255. Alcaino C, Farrugia G, Beyder A (2017) Mechanosensitive piezo channels in the gastrointestinal tract. Curr Top Membr 79:219–244
256. Ochoa-Cortes F, Turco F, Linan-Rico A, Soghomonyan S, Whitaker E, Wehner S et al (2016) Enteric glial cells: a new frontier in neurogastroenterology and clinical target for inflammatory bowel diseases. Inflamm Bowel Dis 22(2):433–449
257. Lake JI, Heuckeroth RO (2013) Enteric nervous system development: migration, differentiation, and disease. Am J Physiol Gastrointest Liver Physiol 305(1):G1–G24
258. Heuckeroth RO, Schäfer KH (2016) Gene-environment interactions and the enteric nervous system: Neural plasticity and Hirschsprung disease prevention. Dev Biol 417(2):188–197

259. de Souza SR, de Miranda Neto MH, Martins Perles JV, Vieira Frez FC, Zignani I et al (2017) Antioxidant effects of the quercetin in the jejunal myenteric innervation of diabetic rats. Front Med (Lausanne) 4:8
260. Bush TG (2002) Enteric glial cells. An upstream target for induction of necrotizing enterocolitis and Crohn's disease? BioEssays 24(2):130–140
261. Neunlist M, Rolli-Derkinderen M, Latorre R, Van Landeghem L, Coron E, Derkinderen P, De Giorgio R (2014) Enteric glial cells: recent developments and future directions. Gastroenterology 147(6):1230–1237
262. Smith TH, Ngwainmbi J, Grider JR, Dewey WL, Akbarali HI (2013) An in-vitro preparation of isolated enteric neurons and glia from the myenteric plexus of the adult mouse. J Vis Exp 78:50688
263. Vieira-Frez FC, Martins-Colombo-Perles JV, Robert-Linden D, Gibbons SJ, Amilcar-Martins H, Almeida-Brito-Romualdo D et al (2017) Restoration of density of interstitial cells of Cajal in the jejunum of diabetic rats after quercetin supplementation. Rev Esp Enferm Dig 109(3):190–195
264. Bassotti G, Villanacci V, Antonelli E, Morelli A, Salerni B (2007) Enteric glial cells: new players in gastrointestinal motility? Lab Invest 87(7):628–632
265. Gulbransen BD, Sharkey KA (2012) Novel functional roles for enteric glia in the gastrointestinal tract. Nat Rev Gastroenterol Hepatol 9(11):625–632
266. Sasselli V, Pachnis V, Burns AJ (2012) The enteric nervous system. Dev Biol 366(1):64–73
267. Grubišić V, Parpura V (2017) Two modes of enteric gliotransmission differentially affect gut physiology. Glia 65(5):699–711
268. Abdo H, Derkinderen P, Gomes P, Chevalier J, Aubert P, Masson D et al (2010) Enteric glial cells protect neurons from oxidative stress in part *via* reduced glutathione. FASEB J 24(4):1082–1094
269. Joseph NM, He S, Quintana E, Kim YG, Núñez G, Morrison SJ (2011) Enteric glia are multipotent in culture but primarily form glia in the adult rodent gut. J Clin Invest 121 (9):3398–3411
270. Liñán-Rico A, Turco F, Ochoa-Cortes F, Harzman A, Needleman BJ, Arsenescu R et al (2016) Molecular signaling and dysfunction of the human reactive enteric glial cell phenotype: implications for GI infection, IBD, POI, neurological, motility, and GI disorders. Inflamm Bowel Dis 22(8):1812–1834
271. Chow AK, Gulbransen BD (2017) Potential roles of enteric glia in bridging neuroimmune communication in the gut. Am J Physiol Gastrointest Liver Physiol 312(2):G145–G152
272. Gershon MD, Rothman TP (1991) Enteric glia. Glia 4(2):195–204
273. Jessen KR, Mirsky R (2005) The origin and development of glial cells in peripheral nerves. Nat Rev Neurosci 6:671–682
274. Cameron-Curry P (1995) Glial lineage of the peripheral nervous system. C R Seances Soc Biol Fil 189(2):253–261
275. Zhang S-C (2001) Defining glial cells during CNS development. Nat Rev Neurosci 2:840–843
276. Zhang Y, Sloan SA, Clarke LE, Caneda C, Plaza CA, Blumenthal PD et al (2016) Purification and characterization of progenitor and mature human astrocytes reveals transcriptional and functional differences with mouse. Neuron 89(1):37–53
277. Yokoyama A, Yang L, Itoh S, Mori K, Tanaka J (2004) Microglia, a potential source of neurons, astrocytes, and oligodendrocytes. Glia 45(1):96–104
278. Rusznák Z, Henskens W, Schofield E, Kim WS, Fu Y (2016) Adult neurogenesis and gliogenesis: possible mechanisms for neurorestoration. Exp Neurobiol 25(3):103–112
279. Faria TP (2015) Glial cells. In: Parnham M (ed) Encyclopedia of inflammatory diseases. Springer, Basel, Switzerland
280. Echevarría D, Martínez S (2004) The myelin glial cell of the peripheral nervous system: the schwann cell. In: Herdegen T, Delgado-García J (eds) Brain damage and repair. Springer, Dordrecht

281. Jessen KR (2004) Glial cells. Int J Biochem Cell Biol 36(10):1861–1867
282. Jessen KR, Mirsky R (1980) Glial cells in the enteric nervous system contain glial fibrillary acidic protein. Nature 286(5774):736–737
283. Costa M, Brookes SJH, Hennig GW (2000) Anatomy and physiology of the enteric nervous system. Gut 47(5):iv15–iv19
284. Llinás R, Jahnsen H (1982) Electrophysiology of mammalian thalamic neurones in vitro. Nature 297:406–408
285. Janiszewski L (1972) Electrophysiology of the neuron. Acta Physiol Pol 5:941–953
286. Furness JB (2000) Types of neurons in the enteric nervous system. J Auton Nerv Syst 81 (1–3):87–96
287. Krammer H-J, Kühnel W (1992) Immunohistochemistry for intermediate filaments in the enteric nervous system of the porcine small intestine. Ann Anat 174(4):275–278
288. Foong JP, Tough IR, Cox HM, Bornstein JC (2014) Properties of cholinergic and non-cholinergic submucosal neurons along the mouse colon. J Physiol 592(Pt 4):777–793
289. Dong H, Jiang Y, Srinivasan S, Mittal RK (2013) Morphological, immunocytochemical, and functional characterization of esophageal enteric neurons in primary culture. Am J Physiol Gastrointest Liver Physiol 305(2):G129–G138
290. Walberg F (1984) On the morphology of the mesencephalic trigeminal cells. New data based on tracer studies. Brain Res 322(1):119–123
291. Ueck M, Sato T, Ohba S, Wake K, Kobayashi H (1989) The neural organization of the pineal complex in the frog: stratification and regional differences. Arch Histol Cytol 52:459–467
292. Howard D, Bull L, De Lacy BC (2015) Evolving unipolar memristor spiking neural networks. Connection Sci 27(4):397–416
293. Wirsig CR, Leonard CM (1986) Acetylcholinesterase and luteinizing hormone-releasing hormone distinguish separate populations of terminal nerve neurons. Neuroscience 19 (3):719–740
294. Wood JD (2012) Cellular neurophysiology of enteric neurons. In: Wood J, Johnson L (eds) Physiology of the gastrointestinal tract, 5th edn. Elsevier, San Diego
295. Carbone SE, Jovanovska V, Nurgali K, Brookes SJH (2014) Human enteric neurons: morphological, electrophysiological, and neurochemical identification. Neurogastroenterol Motil 26(12):1812–1816
296. Pompolo S, Furness JB (1990) Ultrastructure and synaptology of neurons immunoreactive for gamma-aminobutyric acid in the myenteric plexus of the guinea pig small intestine. J Neurocytol 19(4):539–549
297. Hu H-Z, Gao N, Zhu MX, Liu S, Ren J, Gao C et al (2003) Slow excitatory synaptic transmission mediated by P2Y1 receptors in the guinea-pig enteric nervous system. J Physiol 550(Pt 2):493–504
298. Bertrand PP, Thomas EA (2004) Multiple levels of sensory integration in the intrinsic sensory neurons of the enteric nervous system. Clin Exp Pharmacol Physiol 31(11):745–755
299. Takaki M, Nakayama S (1990) Electrical behavior of myenteric neurons induced by mesenteric nerve stimulation in the guinea pig ileum. Acta Med Okayama 44(5):257–261
300. Ohno N, Xue L, Yamamoto Y, Suzuki H (1996) Properties of the inhibitory junction potential in smooth muscle of the guinea-pig gastric fundus. Br J Pharmacol 117(5):974–978
301. Komori K, Suzuki H (1986) Distribution and properties of excitatory and inhibitory junction potentials in circular muscle of the guinea-pig stomach. J Physiol 370:339–355
302. Beckett EAH, Sanders KM, Ward SM (2017) Inhibitory responses mediated by vagal nerve stimulation are diminished in stomachs of mice with reduced intramuscular interstitial cells of Cajal. Sci Rep 7:44759
303. Baker SA, Mutafova-Yambolieva V, Monaghan K, Horowitz B, Sanders KM, Koh SD (2003) Mechanism of active repolarization of inhibitory junction potential in murine colon. Am J Physiol Gastrointest Liver Physiol 285(5):G813–G821

304. Dong H, Jiang Y, Dong J, Mittal RK (2015) Inhibitory motor neurons of the esophageal myenteric plexus are mechanosensitive. Am J Physiol Cell Physiol 308(5):C405–C413
305. Hotta R, Stamp LA, Foong JPP, McConnell SN, Bergner AJ, Anderson RB et al (2013) Transplanted progenitors generate functional enteric neurons in the postnatal colon. J Clin Invest 123(3):1182–1191
306. Mueller MH, Xue B, Glatzle J, Hahn J, Grundy D, Kreis ME (2009) Extrinsic afferent nerve sensitivity and enteric neurotransmission in murine jejunum in vitro. Am J Physiol Gastrointest Liver Physiol 297(4):G655–G662
307. Berthoud HR, Blackshaw LA, Brookes SJH, Grundy D (2004) Neuroanatomy of extrinsic afferents supplying the gastrointestinal tract. Neurogastroenterol Motil 16(1):28–33
308. Song X, Chen BN, Zagorodnyuk VP, Lynn PA, Blackshaw LA, Grundy D et al (2009) Identification of medium/high threshold extrinsic mechanosensitive afferent nerves to the gastrointestinal tract. Gastroenterology 137(1):274–284
309. Chiocchetti R, Grandis A, Bombardi C, Lucchi ML, Dal Lago DT, Bortolami R, Furness JB (2006) Extrinsic and intrinsic sources of calcitonin gene-related peptide immunoreactivity in the lamb ileum: a morphometric and neurochemical investigation. Cell Tissue Res 323(2):183–196
310. Kugler EM, Michel K, Zeller F, Demir IE, Ceyhan GO, Schemann M, Mazzuoli-Weber G (2015) Mechanical stress activates neurites and somata of myenteric neurons. Front Cell Neurosci 9:342
311. Browning KN, Verheijden S, Boeckxstaens GE (2017) The vagus nerve in appetite regulation, mood, and intestinal inflammation. Gastroenterology 152(4):730–744
312. Kunze WA, Furness JB (1999) The enteric nervous system and regulation of intestinal motility. Annu Rev Physiol 61:117–142
313. Clerc N, Furness JB (2004) Intrinsic primary afferent neurones of the digestive tract. Neurogastroenterol Motil 16(1):24–27
314. Furness JB, Jones C, Nurgali K, Clerc N (2004) Intrinsic primary afferent neurons and nerve circuits within the intestine. Prog Neurobiol 72(2):143–164
315. Porter JT, Johnson CK, Agmon A (2001) Diverse types of interneurons generate thalamus-evoked feedforward inhibition in the mouse barrel cortex. J Neurosci 21 (8):2699–2710
316. Zaitsev AV (2013) Classification and function of GABAergic interneurons of the mammalian cerebral cortex. Biochem (Mosc) Suppl Ser A Membr Cell Biol 7(4):245–259
317. Kepecs A, Fishell G (2014) Interneuron cell types are fit to function. Nature 505:318–326
318. Tsugorka TM, Dovgan' OV, Stepanyuk AR, Cherkas VP (2007) Variety of types of cortical interneurons. Neurophysiology 39(3):227–236
319. Bui TV, Cushing S, Dewey D, Fyffe RE, Rose PK (2003) Comparison of the morphological and electrotonic properties of Renshaw cells, Ia inhibitory interneurons, and motoneurons in the cat. J Neurophysiol 90(5):2900–2918
320. Koehler W, Windhorst U (1985) Responses of the spinal alpha-motoneurone-Renshaw cell system to various differentially distributed segmental afferent and descending inputs. Biol Cybern 51(6):417–426
321. Maltenfort MG, Heckman CJ, Rymer WZ (1998) Decorrelating actions of Renshaw interneurons on the firing of spinal motoneurons within a motor nucleus: a simulation study. J Neurophysiol 80(1):309–323
322. Fyffe RE (1990) Evidence for separate morphological classes of Renshaw cells in the cat's spinal cord. Brain Res 536(1–2):301–304
323. Marshall I, Al-Kazwini SJ, Holman JJ, Craig RK (1988) Human alpha-calcitonin gene-related peptide (CGRP) is a potent vasodilator in human mesenteric vasculature. Br J Clin Pharmacol 26(6):691–695
324. Cooke JP, Dzau VJ (1997) Nitric oxide synthase: role in the genesis of vascular disease. Annu Rev Med 48:489–509
325. Hwang SJ, Durnin L, Dwyer L, Rhee P-L, Ward SM, Koh SD, Sanders KM, Mutafova-Yambolieva VN (2011) ß-nicotinamide adenine dinucleotide is an enteric

inhibitory neurotransmitter in human and non-human primate colons. Gastroenterology 140 (2):608–617
326. Adinolfi B, Romanini A, Vanni A, Martinotti E, Chicca A, Fogli S, Nieri P (2013) Anticancer activity of anandamide in human cutaneous melanoma cells. Eur J Pharmacol 718(1–3):154–159
327. Clapper JR, Moreno-Sanz G, Russo R, Guijarro A, Vacondio F, Duranti A et al (2010) Anandamide suppresses pain initiation through a peripheral endocannabinoid mechanism. Nat Neurosci 13:1265–1270
328. Bishay P, Häussler A, Lim HY, Oertel B, Galve-Roperh I, Ferreirós N, Tegeder I (2013) Anandamide deficiency and heightened neuropathic pain in aged mice. Neuropharmacology 71:204–215
329. Nichols K, Staines W, Wu J-Y, Krantis A (1995) Immunopositive GABAergic neural sites display nitric oxide synthase-related NADPH diaphorase activity in the human colon. J Auton Nerv Syst 50(3):253–262
330. Denef C (2008) Paracrinicity: the story of 30 years of cellular pituitary. J Neuroendocrinol 20(1):1–70
331. Singh AB, Harris RC (2005) Autocrine, paracrine and juxtacrine signaling by EGFR ligands. Cell Signal 17(10):1183–1193
332. Clinton JM, Davis CJ, Zielinski MR, Jewett KA, Krueger JM (2011) Biochemical regulation of sleep and sleep bio-markers. J Clin Sleep Med 7(5):S38–S42
333. Siegel GJ, Agranoff BW, Albers RW, Fisher SK, Uhler MD (eds) (1999) Basic neurochemistry: molecular, cellular and medical aspects, 6th edn. Lippincott-Raven, Philadelphia
334. Schousboe A, Bak LK, Waagepetersen HS (2013) Astrocytic control of biosynthesis and turnover of the neurotransmitters glutamate and GABA. Front Endocrinol (Lausanne) 4:102
335. Takeda K, Ueda T (2017) Effective mechanism for synthesis of neurotransmitter glutamate and its loading into synaptic vesicles. Neurochem Res 42(1):64–76
336. Behar KL (2009) GABA synthesis and metabolism. In: Squire LR (ed) Encyclopedia of neuroscience. Academic Press, San Diego
337. Rama Sastry BV, Henderson GI (1976) Mechanisms of acetylcholine synthesis by rat brain choline acetyltransferase and its regulation by products and inhibitors. In: Bradley PB, Dhawan BN (eds) Drugs and central synaptic transmission. Macmillan, London, UK
338. Kumar GK (2011) Hypoxia and neurotransmitter synthesis. Am J Physiol Cell Physiol 300 (4):C743–C751
339. Welford RWD, Vercauteren M, Trébaul A, Cattaneo C, Eckert D, Garzotti M et al (2016) Serotonin biosynthesis as a predictive marker of serotonin pharmacodynamics and disease-induced dysregulation. Sci Rep 6:30059
340. Bark SJ, Hook V (2007) The future of proteomic analysis in biological systems and molecular medicine. Mol BioSyst 3:14
341. Kulkarni S, Zou B, Hanson J, Micci M-A, Tiwari G, Becker L et al (2011) Gut-derived factors promote neurogenesis of CNS-neural stem cells and nudge their differentiation to an enteric-like neuronal phenotype. Am J Physiol Gastrointest Liver Physiol 301(4):G644–G655
342. Berghe PV (2008) Electrochemical detection of neurotransmitters in the gut wall. Neurogastroenterol Motil 20(11):1185–1188
343. Chavez RA, Miller SG, Moore HP (1996) A biosynthetic regulated secretory pathway in constitutive secretory cells. J Cell Biol 133(6):1177–1191
344. Morvan J, Tooze SA (2008) Discovery and progress in our understanding of the regulated secretory pathway in neuroendocrine cells. Histochem Cell Biol 129(3):243–252
345. Ponnambalam S, Baldwin SA (2003) Constitutive protein secretion from the trans-Golgi network to the plasma membrane. Mol Membr Biol 20(2):129–139
346. Ungar D, Hughson FM (2003) SNARE protein structure and function. Annu Rev Cell Dev Biol 19:493–517

347. Hay JC, Scheller RH (1997) SNAREs and NSF in targeted membrane fusion. Curr Opin Cell Biol 9(4):505–512
348. De Camilli P (1995) The eighth Datta Lecture. Molecular mechanisms in synaptic vesicle recycling. FEBS Lett 369(1):3–12
349. Yu I-M, Hughson FM (2010) Tethering factors as organizers of intracellular vesicular traffic. Annu Rev Cell Dev Biol 26:137–156
350. Ungar D, Hughson FM (2003) SNARE protein structure and function. Annu Rev Cell Dev Biol 19:493–517
351. Chen YA, Scheller RH (2001) SNARE-mediated membrane fusion. Nat Rev Mol Cell Biol 2:98–106
352. Jena BP (2009) Porosome: the secretory portal in cells. Biochemistry 48(19):4009–4018
353. Neher E (1993) Secretion without full fusion. Nature 363:497–498
354. Jena BP (2015) 'Porosome' discovered nearly 20 years ago provides molecular insights into the kiss-and-run mechanism of cell secretion. J Cell Mol Med 19(7):1427–1440
355. Jena BP (2012) Porosome: the secretory portal. Exp Biol Med 237(7):748–757
356. Bryant AS, Li B, Beenhakker MP, Huguenard JR (2009) Maintenance of thalamic epileptiform activity depends on the astrocytic glutamate-glutamine cycle. J Neurophysiol 102(5):2880–2888
357. Cremona O, De Camilli P (1997) Synaptic vesicle endocytosis. Curr Opin Neurobiol 7 (3):323–330
358. Scheller RH (2013) In search of the molecular mechanism of intracellular membrane fusion and neurotransmitter release. Nat Med 19:1232–1235
359. MacDonald PE, Braun M, Galvanovskis J, Rorsman P (2006) Release of small transmitters through kiss-and-run fusion pores in rat pancreatic β cells. Cell Metab 4(4):283–290
360. Bhalla US (2004) Models of cell signaling pathways. Curr Opin Genet Dev 14(4):375–381
361. Canales A, Jia X, Froriep UP, Koppes RA, Tringides CM, Selvidge J et al (2015) Multifunctional fibers for simultaneous optical, electrical and chemical interrogation of neural circuits in vivo. Nat Biotechnol 33:277–284
362. Sánchez D, Anand U, Gorelik J, Benham CD, Bountra C, Lab M et al (2007) Localized and non-contact mechanical stimulation of dorsal root ganglion sensory neurons using scanning ion conductance microscopy. J Neurosci Methods 159(1):26–34
363. Tang M, Song Q, Li N, Jiang Z, Huang R, Cheng G (2013) Enhancement of electrical signaling in neural networks on graphene films. Biomaterials 34(27):6402–6411
364. Connors BW (2009) Electrical signaling with neuronal gap junctions. In: Harris AL, Locke D (eds) Connexins. Humana Press, New York, USA
365. Fields D (2011) Nonsynaptic and nonvesicular ATP release from neurons and relevance to neuron–glia signaling. Semin Cell Dev Biol 22(2):214–219
366. Sharkey KA, Lomax AE, Bertrand PP, Furness JB (1998) Electrophysiology, shape, and chemistry of neurons that project from guinea pig colon to inferior mesenteric ganglia. Gastroenterology 115(4):909–918
367. Linden DR (2012) Colitis is associated with a loss of intestinofugal neurons. Am J Physiol Gastrointest Liver Physiol 303(10):G1096–G1104
368. Hibberd TJ, Zagorodnyuk VP, Spencer NJ, Brookes SJ (2012) Identification and mechanosensitivity of viscerofugal neurons. Neuroscience 225:118–129
369. Franze K, Janmey PA, Guck J (2013) Mechanics in neuronal development and repair. Annu Rev Biomed Eng 15:227–251
370. Arulmoli J, Pathak MM, McDonnell LP, Nourse JL, Tombola F, Earthman JC, Flanagan LA (2015) Static stretch affects neural stem cell differentiation in an extracellular matrix-dependent manner. Sci Rep 5:8499
371. Mayer CJ, Wood JD (1975) Properties of mechanosensitive neurons within Auerbach's plexus of the small intestine of the cat. Pflugers Arch 357(1–2):35–49
372. Mazzuoli G, Schemann M (2012) Mechanosensitive enteric neurons in the myenteric plexus of the mouse intestine. PLoS ONE 7(7):e39887

373. Furness JB, Koopmans HS, Robbins HL, Lin HC (2000) Identification of intestinofugal neurons projecting to the coeliac and superior mesenteric ganglia in the rat. Auton Neurosci 83(1–2):81–85
374. Sha L, Linden DR, Farrugia G, Szurszewski JH (2013) Hydrogen sulfide selectively potentiates central preganglionic fast nicotinic synaptic input in mouse superior mesenteric ganglion. J Neurosci 33(31):12638–12646
375. Sharkey KA, Lomax AEG, Bertrand PP, Furness JB (1998) Electrophysiology, shape, and chemistry of neurons that project from guinea pig colon to inferior mesenteric ganglia. Gastroenterology 115(4):909–918
376. Szurszewski J, Ermilov L, Miller S (2002) Prevertebral ganglia and intestinofugal afferent neurons. Gut 51(1):i6–i10
377. Davison JS (1972) The electrophysiology of gastrointestinal chemoreceptors. Digestion 7 (5):312–317
378. Furness JB (2003) Intestinofugal neurons and sympathetic reflexes that bypass the central nervous system. J Comp Neurol 455(3):281–284
379. Purali N, Rydqvist B (1998) Action potential and sodium current in the slowly and rapidly adapting stretch receptor neurons of the crayfish (Astacus astacus). J Neurophysiol 80:2121–2132
380. Sur M, Wall JT, Kaas JH (1984) Modular distribution of neurons with slowly adapting and rapidly adapting responses in area 3b of somatosensory cortex in monkeys. J Neurophysiol 51(4):724–744
381. Bourinet E, Altier C, Hildebrand ME, Trang T, Salter MW, Zamponi GW (2014) Calcium-permeable ion channels in pain signaling. Physiol Rev 94(1):81–140
382. Ennes HS, Young SH, Goliger JA, Mayer EA (1999) Chemical signaling from colonic smooth muscle cells to DRG neurons in culture. Am J Physiol Cell Physiol 276(3):C602–C610
383. Anastassiou CA, Perin R, Markram H, Koch C (2011) Ephaptic coupling of cortical neurons. Nat Neurosci 14:217–223
384. Singh AB, Harris RC (2005) Autocrine, paracrine and juxtacrine signaling by EGFR ligands. Cell Signal 17(10):1183–1193
385. Chalazonitis A, Gershon MD, Greene LA (2012) Cell death and the developing enteric nervous system. Neurochem Int 61(6):839–847
386. Mayer EA, Tillisch K, Gupta A (2015) Gut/brain axis and the microbiota. J Clin Invest 125 (3):926–938
387. Rosell S, Al-Saffar A, Thor K (1984) The role of neurotensin in gut motility. Scand J Gastroenterol 96:69–75
388. Bueno L, Fioramonti J, Fargeas MJ, Primi MP (1985) Neurotensin: a central neuromodulator of gastrointestinal motility in the dog. Am J Physiol 248(1 Pt 1):G15–G19
389. Mayer EA, Knight R, Mazmanian SK, Cryan JF, Tillisch K (2014) Gut microbes and the brain: paradigm shift in neuroscience. J Neurosci 34(46):15490–15496
390. Chen P, Hu B, Tan Q, Liu L, Li D, Jiang C et al (2010) Role of neurocrine somatostatin on sphincter of Oddi contractility and intestinal ischemia reperfusion-induced acute pancreatitis in macaques. Neurogasteroenterol Motil 22(8):935–e240
391. Stengel A, Taché Y (2009) Neuroendocrine control of the gut during stress: corticotropin-releasing factor signaling pathways in the spotlight. Annu Rev Physiol 71:219–239
392. Flemström G, Sjöblom M (2005) Epithelial cells and their neighbors. II. New perspectives on efferent signaling between brain, neuroendocrine cells, and gut epithelial cells. Am J Physiol Gastrointest Liver Physiol 289(3):G377–G380
393. Osman D, Buchon N, Chakrabarti S, Huang Y-T, Su W-C, Poidevin M et al (2012) Autocrine and paracrine unpaired signaling regulate intestinal stem cell maintenance and division. J Cell Sci 125:5944–5949

394. Lee B-C, Kim H-S, Shin T-H, Kang I, Lee JY, Kim J-J et al (2016) PGE_2 maintains self-renewal of human adult stem cells via EP2-mediated autocrine signaling and its production is regulated by cell-to-cell contact. Sci Rep 6:26298
395. Kolterud Å, Grosse AS, Zacharias WJ, Walton KD, Kretovich KE, Madison B et al (2009) Paracrine Hedgehog signaling in stomach and intestine: new roles for Hedgehog in gastrointestinal patterning. Gastroenterology 137(2):618–628
396. Liñán-Rico A, Wunderlich JE, Grants IS, Frankel WL, Xue J, Williams KC et al (2013) Purinergic autocrine regulation of mechanosensitivity and serotonin release in a human EC model: ATP-gated P2X3 channels in EC are downregulated in ulcerative colitis. Inflamm Bowel Dis 19(11):2366–2379
397. Grumolato L, Aaronson SA (2013) Autocrine/paracrine. In: Maloy S, Hughes K (eds) Brenner's encyclopedia of genetics, 2nd edn. Elsevier, San Diego, CA, USA
398. Osman D, Buchon N, Chakrabarti S, Huang YT, Su WC, Poidevin M et al (2012) Autocrine and paracrine unpaired signaling regulate intestinal stem cell maintenance and division. J Cell Sci 125(Pt 24):5944–5949
399. Galligan JJ (2008) Purinergic signaling in the gastrointestinal tract. Purinergic Signal 4 (3):195–196
400. Roy S, Kornberg TB (2015) Paracrine signaling mediated at cell–cell contacts. BioEssays 37(1):25–33
401. Krantis A (2000) GABA in the mammalian enteric nervous system. Physiology 15(6):284–290
402. Anklesaria P, Teixidó J, Laiho M, Pierce JH, Greenberger JS, Massagué J (1990) Cell-cell adhesion mediated by binding of membrane-anchored transforming growth factor a to epidermal growth factor receptors promotes cell proliferation. PNAS 87:3289–3293
403. Massagué J (1990) Transforming growth factor-A. A model for membrane-anchored growth factors. J Biol Chem 265:21393–21396
404. Yaron A, Sprinzak D (2012) The *cis* side of juxtacrine signaling: a new role in the development of the nervous system. Trends Neurosci 35(4):230–239
405. Ashton RS, Conway A, Pangarkar C, Bergen J, Lim KI, Shah P et al (2012) Astrocytes regulate adult hippocampal neurogenesis through ephrin-B signaling. Nat Neurosci 15 (10):1399–1406
406. Lynch AC, Antony A, Dobbs BR, Frizelle FA (2001) Bowel dysfunction following spinal cord injury. Spinal Cord 39:193–203
407. Sasselli V, Pachnis V, Burns AJ (2012) The enteric nervous system. Dev Biol 366(1):64–73
408. Braak H, de Vos RA, Bohl J, Del Tredici K (2006) Gastric alpha-synuclein immunoreactive inclusions in Meissner's and Auerbach's plexuses in cases staged for Parkinson's disease-related brain pathology. Neurosci Lett 396(1):67–72
409. Burns AJ, Le Douarin NM (1998) The sacral neural crest contributes neurons and glia to the post-umbilical gut: spatiotemporal analysis of the development of the enteric nervous system. Development 125:4335–4347
410. Hara T, Hirasawa A, Ichimura A, Kimura I, Tsujimoto G (2011) Free fatty acid receptors FFAR1 and GPR120 as novel therapeutic targets for metabolic disorders. J Pharm Sci 100 (9):3594–3601
411. Ichimura A, Hirasawa A, Hara T, Tsujimoto G (2009) Free fatty acid receptors act as nutrient sensors to regulate energy homeostasis. Prostaglands Other Lipid Mediat 89(3–4):82–88
412. Mohamed HA, Khaled DM, Mohamed HA (2014) Homology modeling and explicit membrane molecular dynamics simulation to delineate the mode of binding of thiazolidinediones into FFAR1 and the mechanism of receptor activation. Bioorg Med Chem Lett 24(22):5330–5336
413. Thorens B, Mueckler M (2010) Glucose transporters in the 21st century. Am J Physiol Endocrinol Metab 298(2):E141–E145

414. Mace OJ, Lister N, Morgan E, Shepherd E, Affleck J, Helliwell P et al (2009) An energy supply network of nutrient absorption coordinated by calcium and T1R taste receptors in rat small intestine. J Physiol 587:195–210
415. Li T, Holmstrom SR, Kir S, Umetani M, Schmidt DR, Kliewer SA, Mangelsdorf DJ (2011) The G protein-coupled bile acid receptor, TGR5, stimulates gallbladder filling. Mol Endocrinol 25(6):1066–1071
416. Peng Y, Gillis-Smith S, Jin H, Tränkner D, Ryba NJP, Zuker CS (2015) Sweet and bitter taste in the brain of awake behaving animals. Nature 527:512–515
417. Oka Y, Butnaru M, von Buchholtz L, Ryba NJP, Zuker CS (2013) High salt recruits aversive taste pathways. Nature 494:472–475
418. Li F (2013) Taste perception: from the tongue to the testis. Mol Hum Reprod 19(6):349–360
419. Greenfield JR, Chisholm DJ (2013) How sweet it is: intestinal sweet taste receptors in type 2 diabetes. Diabetes 62(10):3336–3337
420. Shaik FA, Singh N, Arakawa M, Duan K, Bhullar RP, Chelikani P (2016) Bitter taste receptors: extraoral roles in pathophysiology. Int J Biochem Cell Biol 77:197–204
421. Christensen J, Rick GA (1987) Intrinsic nerves in the mammalian colon: confirmation of a plexus at the circular muscle-submucosal interface. J Auton Nerv Syst 21(2–3):223–231
422. Fang S, Wu R, Christensen J (1993) Intramucosal nerve cells in human small intestine. J Auton Nerv Syst 44(2–3):129–136
423. Furness JB (2012) The enteric nervous system and neurogastroenterology. Nat Rev Gastroenterol Hepatol 9:286–294
424. Reed DE, Vanner S (2007) Mucosal stimulation activates secretomotor neurons via long myenteric pathways in guinea pig ileum. Am J Physiol Gastrointest Liver Physiol 292(2): G608–G614
425. Nadorra R, Landing BH, Wells TR (1986) Intestinal plexuses in Crohn's disease and ulcerative colitis in children: pathologic and microdissection studies. Pediatr Pathol 6(2–3):267–287
426. Siemers PT, Dobbins WO 3rd (1974) The Meissner plexus in Crohn's disease of the colon. Surg Gynecol Obstet 138(1):39–42
427. Santer RM, Baker DM (1988) Enteric neuron numbers and sizes in Auerbach's plexus in the small and large intestine of adult and aged rats. J Auton Nerv Syst 25(1):59–67
428. Eaker EY (1997) Neurofilament and intermediate filament immunoreactivity in human intestinal myenteric neurons. Digest Dis Sci 42(9):1926–1932
429. Brehmer A, Rupprecht H, Neuhuber W (2010) Two submucosal nerve plexus in human intestines. Histochem Cell Biol 133(2):149–161
430. Howard ER (2013) Joseph Lister: his contributions to early experimental physiology. Notes Rec R Soc Lond 67(3):191–198
431. Hoyle CHV, Burnstock G (1989) Neuronal populations in the submucous plexus of the human colon. J Anat 166:7–22
432. Saffrey MJ, Hassall CJ, Allen TG, Burnstock G (1992) Ganglia within the gut, heart, urinary bladder, and airways: studies in tissue culture. Int Rev Cytol 136:93–144
433. Goerg KJ, Roux M, Wanitschke R, Meyer zum Büschenfelde KH (1986) Significance of Meissner's plexus for the secretory function of the colonic mucosa. Z Gastroenterol 24 (3):70–8
434. Csendes A, Smok G, Braghetto I, González P, Henríquez A, Csendes P, Pizurno D (1992) Histological studies of Auerbach's plexuses of the oesophagus, stomach, jejunum, and colon in patients with achalasia of the oesophagus: correlation with gastric acid secretion, presence of parietal cells and gastric emptying of solids. Gut 33(2):150–154
435. Dahm R (2010) From discovering to understanding. EMBO Rep 11(3):153–160
436. Inamura N, Kimura T, Tada S, Kurahashi T, Yanagida M, Yanagawa Y et al (2012) Intrinsic and extrinsic mechanisms control the termination of cortical interneuron migration. J Neurosci 32(17):6032–6042

437. Costes LM, Boeckxstaens GE, de Jonge WJ, Cailotto C (2013) Neural networks in intestinal immunoregulation. Organogenesis 9(3):216–223
438. Mueller MH, Xue B, Glatzle J, Hahn J, Grundy D, Kreis ME (2009) Extrinsic afferent nerve sensitivity and enteric neurotransmission in murine jejunum in vitro. Am J Physiol Gastrointest Liver Physiol 297(4):G655–G662
439. Uesaka T, Young HM, Pachnis V, Enomoto H (2016) Development of the intrinsic and extrinsic innervation of the gut. Dev Biol 417(2):158–167
440. Benarroch EE (2014) Sympathetic system—overview. In: Aminoff MJ, Daroff RB (eds) Encyclopedia of the neurological sciences, 2nd edn. Elsevier, San Diego
441. Costa M, Brookes SH, Zagorodnyuk V (2004) How many kinds of visceral afferents? Gut 53(2):ii1–4
442. Aldskogius H, Elfvin LG, Forsman CA (1986) Primary sensory afferents in the inferior mesenteric ganglion and related nerves of the guinea pig. An experimental study with anterogradely transported wheat germ agglutinin-horseradish peroxidase conjugate. J Auton Nerv Syst 15(2):179–190
443. Sengupta JN (2009) Visceral pain: the neurophysiological mechanism. Handb Exp Pharmacol 194:31–74
444. Bassotti G, Villanacci V, Fisogni S, Rossi E, Baronio P, Clerici C et al (2007) Enteric glial cells and their role in gastrointestinal motor abnormalities: Introducing the neuro-gliopathies. World J Gastroenterol 13(30):4035–4041
445. Mei F, Guo S, He YT, Zhu J, Zhou DS, Niu JQ et al (2009) Apoptosis of interstitial cells of Cajal, smooth muscle cells, and enteric neurons induced by intestinal ischemia and reperfusion injury in adult guinea pigs. Virchows Arch 454(4):401–409
446. Wang XY, Sanders KM, Ward SM (1999) Intimate relationship between interstitial cells of cajal and enteric nerves in the guinea-pig small intestine. Cell Tissue Res 295(2):247–256
447. Klein S, Seidler B, Kettenberger A, Sibaev A, Rohn M, Feil R et al (2013) Interstitial cells of Cajal integrate excitatory and inhibitory neurotransmission with intestinal slow-wave activity. Nat Commun 4:1630
448. Messerli B, Robinson L (2010) Neuromuscular junction disease or neuropathic disorder? Am J Phys Med Rehabil 89(10):862
449. Wallace AS, Burns AJ (2005) Development of the enteric nervous system, smooth muscle and interstitial cells of Cajal in the human gastrointestinal tract. Cell Tissue Res 319 (3):367–382
450. Ward SM, Sanders KM (2006) Involvement of intramuscular interstitial cells of Cajal in neuroeffector transmission in the gastrointestinal tract. J Physiol 576(Pt 3):675–682
451. Guo T, Li J, Li J, Kong D, Bi C, He Z et al (2017) Association between hyperpolarization-activated channel in interstitial cells of Cajal and gastrointestinal dysmotility induced by malignant ascites. Oncol Lett 13(3):1601–1608
452. Thiriet M (2014) Physiology and pathophysiology of arterial flow. In: Lanzer P (ed) Panvascular medicine. Springer, Heidelberg, Germany
453. Krishnamurthy S, Schuffler MD (1987) Pathology of neuromuscular disorders of the small intestine and colon. Gastroenterology 93(3):610–639
454. Gil V, Gallego D, Grasa L, Martín MT, Jiménez M (2010) Purinergic and nitrergic neuromuscular transmission mediates spontaneous neuronal activity in the rat colon. Am J Physiol Gastrointest Liver Physiol 299(1):G158–G169
455. Wali FA (1985) Actions of polypeptides at the neuromuscular junction. Agents Actions 16 (6):558–566
456. Hansebout CR, Su C, Reddy K, Zhang D, Jiang C, Rathbone MP, Jiang S (2012) Enteric glia mediate neuronal outgrowth through release of neurotrophic factors. Neural Regen Res 7(28):2165–2175
457. Korsak K, Dolatshad NF, Silva AT, Saffrey MJ (2012) Ageing of enteric neurons: oxidative stress, neurotrophic factors and antioxidant enzymes. Chem Cent J 6:80
458. Brookes SJ (1993) Neuronal nitric oxide in the gut. J Gastroenterol Hepatol 8(6):590–603

459. Olsson C, Holmgren S (2000) PACAP and nitric oxide inhibit contractions in the proximal intestine of the atlantic cod, Gadus morhua. J Exp Biol 203(Pt 3):575–583
460. Grider JR, Murthy KS, Makhlouf GM (2004) Peptide neurotransmitters and smooth muscle in the gut. In: Martini L (ed) Encyclopedia of endocrine diseases. Academic Press, San Diego
461. Wang YY, Park SY, Oh KH, Min Y, Lee Y-J, Lee S-Y, Sohn UD (2011) Characteristics of 5-hydroxytryptamine receptors involved in contraction of Feline Ileal longitudinal smooth muscle. Korean J Physiol Pharmacol 15(5):267–272
462. Gregersen H, Christensen J (2000) Gastrointestinal tone. Neurogastroenterol Motil 12 (6):501–508
463. Zhang CH, Wang P, Liu DH, Chen CP, Zhao W, Chen X et al (2016) The molecular basis of the genesis of basal tone in internal anal sphincter. Nat Commun 7:11358
464. Xue L, Camilleri M, Locke GR III, Schuurkes JAJ, Meulemans A, Coulie BJ et al (2006) Serotonergic modulation of murine fundic tone. Am J Physiol Gastrointest Liver Physiol 291(6):G1180–G1186
465. Jimenez M (2017) Is the muscular tone of the internal anal sphincter a property of the syncytium? J Physiol 595(6):1853–1854
466. Gibbons SJ, Farrugia G (2004) The role of carbon monoxide in the gastrointestinal tract. J Physiol 556(Pt 2):325–336
467. Ward S (2000) Interstitial cells of Cajal in enteric neurotransmission. Gut 47(4):iv40–iv43
468. Sanders KM, Salter AK, Hennig GW, Koh SD, Perrino BA, Ward SM, Baker SA (2014) Responses to enteric motor neurons in the gastric fundus of mice with reduced intramuscular interstitial cells of cajal. J Neurogastroenterol Motil 20(2):171–184
469. Park KS, Cho KB, Hwang IS, Park JH, Jang BI, Kim KO et al (2016) Characterization of smooth muscle, enteric nerve, interstitial cells of Cajal, and fibroblast-like cells in the gastric musculature of patients with diabetes mellitus. World J Gastroenterol 22 (46):10131–10139
470. Beckett EA, Sanders KM, Ward SM (2017) Inhibitory responses mediated by vagal nerve stimulation are diminished in stomachs of mice with reduced intramuscular interstitial cells of Cajal. Sci Rep 7:44759
471. Song G, Satterfield MC, Kim J, Bazer FW, Spencer TE (2008) Gastrin-releasing peptide (GRP) in the ovine uterus: regulation by interferon tau and progesterone. Biol Reprod 79:376–386
472. Merali Z, McIntosh J, Anisman H (1999) Role of bombesin-related peptides in the control of food intake. Neuropeptides 33(5):376–386
473. McDonald TJ, Jornvall H, Nilsson G, Vagne M, Ghatei M, Bloom SR, Mutt V (1979) Characterization of a gastrin releasing peptide from porcine non-antral gastric tissue. Biochem Biophys Res Commun 90:227–233
474. McDonald TJ, Nilsson G, Vagne M, Ghatei M, Bloom SR, Mutt V (1978) A gastrin releasing peptide from the porcine nonantral gastric tissue. Gut 19:767–774
475. Ischia J, Patel O, Shulkes A, Baldwin GS (2009) Gastrin-releasing peptide: different forms, different functions. BioFactors 35(1):69–75
476. Guan XM, Chen H, Dobbelaar PH, Dong Y, Fong TM, Gagen K et al (2010) Regulation of energy homeostasis by bombesin receptor subtype-3: selective receptor agonists for the treatment of obesity. Cell Metab 11(2):101–112
477. Guan XM, Metzger JM, Yang L, Raustad KA, Wang SP, Spann SK et al (2011) Antiobesity effect of MK-5046, a novel bombesin receptor subtype-3 agonist. J Pharmacol Exp Ther 336(2):356–364
478. Majumdar ID, Weber HC (2012) Biology and pharmacology of bombesin receptor subtype-3. Curr Opin Endocrinol Diabetes Obes 19(1):3–7
479. Hildebrand P, Werth B, Beglinger C, Delco F, Jansen JBMJ, Lamers CBHW, Gyr K (1991) Human gastrin-releasing peptide: biological potency in humans. Regul Pept 36 (3):423–433

480. Naylor SL, Sakaguchi AY, Spindel E, Chin WW (1987) Human gastrin-releasing peptide gene is located on chromosome 18. Somat Cell Mol Genet 13(1):87–91
481. Kandel ER, Schwartz JH, Jessell TM (2000) Principles of neural science, 4th edn. McGraw-Hill, New York
482. McConalogue K, Furness JB (1994) Gastrointestinal neurotransmitters. Baillières Clin Endocrinol Metab 8(1):51–76
483. Furness JB, Callaghan BP, Rivera LR, Cho H-J (2014) Microbial endocrinology: the microbiota-gut-brain axis in health and disease. In: Lyte M, Cryan JF (eds) The enteric nervous system and gastrointestinal innervation: integrated local and central control. Springer, New York
484. Corthay J, Dunant Y, Eder L, Loctin F (1985) Incorporation of acetate into acetylcholine, acetylcarnitine, and amino acids in the Torpedo electric organ. J Neurochem 45(6):1809–1819
485. Albuquerque EX, Pereira EFR, Alkondon M, Rogers SW (2009) Mammalian nicotinic acetylcholine receptors: from structure to function. Physiol Rev 89(1):73–120
486. Lindstrom JM (2008) Acetylcholine receptor structure. In: Myasthenia gravis and related disorders, part of the series current clinical neurology. Humana Press, New York
487. Lindstrom J (1997) Nicotinic acetylcholine receptors in health and disease. Mol Neurobiol 15(2):193–222
488. Fambrough DM (1979) Control of acetylcholine receptors in skeletal muscle. Physiol Rev 59(1):165–227
489. Zwart R, Vijverberg HP (1998) Four pharmacologically distinct subtypes of alpha4beta2 nicotinic acetylcholine receptor expressed in Xenopus laevis oocytes. Mol Pharmacol 54 (6):1124–1131
490. Zoli M, Léna C, Picciotto MR, Changeux JP (1998) Identification of four classes of brain nicotinic receptors using beta2 mutant mice. J Neurosci 18(12):4461–4472
491. Itier V, Bertrand D (2001) Neuronal nicotinic receptors: from protein structure to function. FEBS Lett 504(3):118–125
492. Giniatullin R, Nistri A, Yakel JL (2005) In muscle, the acetylcholine ligand binds to two regions, one region is between the alpha and delta subunits and the other, between the alpha and gamma subunits. Desensitisation of nicotinic ACh receptors: shaping cholinergic signaling. Trends Neurosci 28(7):371–378
493. Nobel Lectures (1965) Physiology or medicine 1922–1941, Elsevier Publishing Company, Amsterdam
494. Iino S, Nojyo Y (2006) Muscarinic M(2) acetylcholine receptor distribution in the guinea-pig gastrointestinal tract. Neuroscience 138(2):549–559
495. Schneider DA, Perrone M, Galligan JJ (2000) Nicotinic acetylcholine receptors at sites of neurotransmitter release to the guinea pig intestinal circular muscle. J Pharmacol Exp Ther 294(1):363–369
496. Unno T, Matsuyama H, Sakamoto T, Uchiyama M, Izumi Y, Okamoto H, Yamada M, Wess J, Komori S (2005) M_2 and M_3 muscarinic receptor-mediated contractions in longitudinal smooth muscle of the ileum studied with receptor knockout mice. Br J Pharmacol 146(1):98–108
497. Feduccia AA, Chatterjee S, Bartlett SE (2012) Neuronal nicotinic acetylcholine receptors: neuroplastic changes underlying alcohol and nicotine addictions. Front Mol Neurosci 5:83
498. Tansey EM (2006) Henry Dale and the discovery of acetylcholine. C R Biol 329(5–6):419–425
499. Garza A, Huang LZ, Son J-H, Winzer-Serhan UH (2009) Expression of nicotinic acetylcholine receptors and subunit messenger RNAs in the enteric nervous system of the neonatal rat. Neuroscience 158(4):1521–1529
500. Carignano C, Pablo EB, Spitzmaul G (2016) Analysis of neuronal nicotinic acetylcholine receptor $\alpha4\beta2$ activation at the single-channel level. BBA Biomembranes 1858(9):1964–1973

501. McKay BE, Placzek AN, Dani JA (2007) Regulation of synaptic transmission and plasticity by neuronal nicotinic acetylcholine receptors. Biochem Pharmacol 74(8):1120–1133
502. Ishii M, Kurachi Y (2006) Muscarinic acetylcholine receptors. Curr Pharm Des 12(28): 3573–3581
503. Heilbronn E, Bartfai T (1978) Muscarinic acetylcholine receptor. Prog Neurobiol 11(3–4): 171–188
504. Alexander SP, Kelly E, Marrion NV, Peters JA, Faccenda E, Harding SD et al (2017) The concise guide to pharmacology 2017/18: overview. Br J Pharmacol 174(1):S1–S16
505. Noronha-Blob L, Lowe VC, Hanson RC, U'Prichard DC (1987) Heterogeneity of muscarinic receptors coupled to phosphoinositide breakdown in guinea pig brain and peripheral tissues. Life Sci 41(8):967–975
506. Brown DA (2013) The other (muscarinic) acetylcholine receptors in sympathetic ganglia: actions and mechanisms. Neurophysiology 45(1):60–66
507. Sorimachi M, Nishimura S (1982) Mechanism of choline-induced secretion of catecholamines from the cat adrenal medulla: involvement of nicotinic receptor. Jpn J Physiol 32(4):541–551
508. Sage HJ, Smith WJ, Kirshner N (1967) Mechanism of secretion from the adrenal medulla. I. A microquantitative immunologic assay for bovine adrenal catecholamine storage vesicle protein and its application to studies of the secretory process. Mol Pharmacol 3(1):81–89
509. Soll AH, Walsh JH (1979) Regulation of gastric acid secretion. Annu Rev Physiol 41: 35–53
510. Malomouzh AI, Nikol'skii EE (2007) Non-quantal secretion of acetylcholine from motor nerve endings: Molecular mechanisms, physiological role, and regulation. Neurophysiology 39(4):307–317
511. Bogatcheva PO, Ezhova EV, Balezina OP (2010) Mechanism of inhibition of acetylcholine secretion in newly formed mouse synapses involving Ca^{2+}-dependent kinases and voltage-gated K^+-channels. Bull Exp Biol Med 149(2):170–173
512. Eglen RM, Reddy H, Watson N, Challiss RA (1994) Muscarinic acetylcholine receptor subtypes in smooth muscle. Trends Pharmacol Sci 15(4):114–119
513. Nürnberg B (2000) Pertussis toxin as a pharmacological tool. bacterial protein toxins. Handb Exp Pharmacol 145:187–206
514. Carbonetti NH (2010) Pertussis toxin and adenylate cyclase toxin: key virulence factors of *Bordetella pertussis* and cell biology tools. Future Microbiol 5:455–469
515. Lee HY, Kim SD, Shim JW, Kim HJ, Yun J, Baek S-H et al (2010) A pertussis toxin sensitive G-protein-independent pathway is involved in serum amyloid A-induced formyl peptide receptor 2-mediated CCL2 production. Exp Mol Med 42:302–309
516. Leaney JL, Tinker A (2000) The role of members of the pertussis toxin-sensitive family of G proteins in coupling receptors to the activation of the G protein-gated inwardly rectifying potassium channel. Proc Natl Acad Sci U S A 97(10):5651–5656
517. Berizzi A, Gentry PR, Rueda P, den Hoedt S, Sexton PM, Langmead CJ, Christopoulos A (2016) Molecular mechanisms of action of m5 muscarinic acetylcholine receptor allosteric modulators. Mol Pharmacol 116:104182
518. Wang SZ, Lee SY, Zhu SZ, el-Fakahany EE (1996) Differential coupling of m1, m3, and m5 muscarinic receptors to activation of neuronal nitric oxide synthase. Pharmacology 53 (5):271–80
519. Lai J, Waite SL, Bloom JW, Yamamura HI, Roeske WR (1991) The m2 muscarinic acetylcholine receptors are coupled to multiple signaling pathways via pertussis toxin-sensitive guanine nucleotide regulatory proteins. J Pharmacol Exp Ther 258(3): 938–944
520. Lechleiter J, Peralta E, Clapham D (1989) Diverse functions of muscarinic acetylcholine receptor subtypes. Trends Pharmacol Sci Suppl 34–38
521. Vorobiov D, Kanti AB, Keren-Raifman T, Barzilai R, Dascal N (2000) Coupling of the muscarinic m2 receptor to g protein-activated K^+ channels via $G\alpha z$ and a receptor-$G\alpha z$

fusion protein: fusion between the receptor and Gα2 eliminates catalytic (collision) coupling. J Biol Chem 275:4166–4170
522. Ostrom RS, Ehlert FJ (1998) M_2 muscarinic receptors inhibit forskolin—but not isoproterenol-mediated relaxation in bovine tracheal smooth muscle. J Pharmacol Exp Ther 286(1):234–242
523. Ostrom RS, Ehlert FJ (1999) Comparison of functional antagonism between isoproterenol and M2 muscarinic receptors in guinea pig ileum and trachea. J Pharmacol Exp Ther 288 (3):969–976
524. Bolton TB, Lim SP (1991) Action of acetylcholine on smooth muscle. Z Kardiol 80(7):73–77
525. Morel JL, Macrez N, Mironneau J (1997) Specific Gq protein involvement in muscarinic M3 receptor-induced phosphatidylinositol hydrolysis and Ca^{2+} release in mouse duodenal myocytes. Br J Pharmacol 121:451–458
526. Guo M-L, Mao L-M, Wang JQ (2010) Modulation of M4 muscarinic acetylcholine receptors by interacting proteins. Neurosci Bull 26(6):469–473
527. Liao CF, Schilling WP, Birnbaumer M, Birnbaumer L (1990) Cellular responses to stimulation of the M5 muscarinic acetylcholine receptor as seen in murine L cells. J Biol Chem 265:11273–11284
528. Reever CM, Ferrari-DiLeo G, Flynn DD (1997) The M5 (m5) receptor subtype: fact or fiction? Life Sci 60(13–14):1105–1112
529. Branch A, Rafacz J, Boudreaux L (2011) Prolonged neuromuscular block in a 74-year-old patient. AANA J 79(4):317–321
530. Devi KL, Yellamma K (2013) The stimulatory effect of zinc, pyridoxine and methoprene on cholinergic system in the silkworm, *Bombyx mori* (L). J Entomol Zool Stud 1(6):74–84
531. Cash F, Vernon SW, Phelan P, Goodchild J, Baines RA (2016) Central cholinergic synaptic vesicle loading obeys the set-point model in Drosophila. J Neurophysiol 115 (2):843–850
532. Liu Y, Edwards RH (1997) Differential localization of vesicular acetylcholine and monoamine transporters in PC12 cells but not CHO cells. J Cell Biol 139(4):907–916
533. Eiden LE, Schäfer MKH, Weihe E, Schütz B (2004) The vesicular amine transporter family (SLC18): amine/proton antiporters required for vesicular accumulation and regulated exocytotic secretion of monoamines and acetylcholine. Pflugers Arch 447(5):636–640
534. Prado VF, Roy A, Kolisnyk B, Gros R, Prado MAM (2013) Regulation of cholinergic activity by the vesicular acetylcholine transporter. Biochem J 450(2):265–274
535. Langen P, Hucho F (2008) Karl Lohmann and the discovery of ATP. Angew Chem Int Ed Engl 47(10):1824–1827
536. Prebble JN (2010) The discovery of oxidative phosphorylation: a conceptual off-shoot from the study of glycolysis. Stud Hist Philos Biol Biomed Sci 41(3):253–262
537. Lipmann F (1981) The ATP-phosphate cycle. Curr Top Cell Regul 18:301–311
538. Szent-Györgyi AG (2004) The early history of the biochemistry of muscle contraction. J Gen Physiol 123(6):631–641
539. Lohmann K (1929) Über die pyrophosphatfraktion im muskel. Naturwissenschaften 17 (31):624–625
540. Lipmann F (1941) Metabolic generation and utilization of phosphate bond energy. Adv Enzymol 1:99–162
541. North RA, Verkhratsky A (2006) Purinergic transmission in the central nervous system. Pflügers Archiv 452(5):479–485
542. Fiske CH, SubbaRow Y (1929) Phosphorous compounds of muscle and liver. Science 70:381–382
543. Meier T, Faraldo-Gomez J, Börsch B (2011) ATP synthase—a paradigmatic molecular machine. In: Frank J (ed) Molecular machines in biology. Cambridge University Press, Cambridge
544. Paton WDM (1963) Cholinergic transmission and acetylcholine output. Can J Biochem Physiol 41:2637–2653

545. Paton WDM, Vane JR (1963) An analysis of the responses of the isolated stomach to electrical stimulation and to drugs. J Physiol (Lond) 165:10–46
546. Keates F (2013) Papers of Sir John Robert Vane (1927–2004). Notes Rec R Soc Lond 67 (2):179–183
547. Tubbs RS, Linganna S, Loukas M (2007) Jacobus Sylvius (1478-1555): physician, teacher, and anatomist. Clin Anat 20(8):868–870
548. Bylebyl JJ (1982) Boyle and Harvey on the valves in the veins. Bull Hist Med 56(3):351–367
549. Aird WC (2011) Discovery of the cardiovascular system: from Galen to William Harvey. J Thromb Haemost 9(1):118–129
550. Kresge N, Simoni RD, Hill RL (2006) The prostaglandins, Sune Bergström and Bengt Samuelsson. J Biol Chem 281:e9
551. Harding A (2004) Sir John Robert Vane. Lancet 364(9451):2090
552. Oates JA (1982) The 1982 nobel prize in physiology or medicine. Science 218(4574):765–768
553. Brading AF (2006) Smooth muscle research: from Edith Bülbring onwards. Trends Pharmacol Sci 27(3):158–165
554. Bolton TB, Brading AF (1992) Edith Bülbring. 27 December 1903—5 July 1990. Biogr Mem Fellows R Soc 38:68–95
555. Born GV (1985) Edith Bülbring. Experientia 41(7):839–840
556. King BF (2001) Reflections on the purinergic hypothesis: the Burnstock Festschrift in the millennial year. Auton Neurosci 87(2–3):173–177
557. Burnstock G (2011) Purinergic signaling in the gastrointestinal tract. World J Gastrointest Pathophysiol 2(2):31–34
558. Burnstock G (2009) Purinergic cotransmission. Experiment Physiol 94(1):20–24
559. Burnstock G (2014) Purinergic signalling: from discovery to current developments. Exp Physiol 99(1):16–34
560. Burnstock G (2004) Geoffrey Burnstock: most highly cited scientist. Mol Interv 4(4):192–195
561. Burnstock G (1996) P2 purinoceptors: historical perspective and classification. Ciba Found Symp 198:1–34
562. Edwards FA, Gibb AJ (1993) ATP - a fast neurotransmitter. FEBS Lett 325(1–2):86–89
563. Zimmermann H (2008) ATP and acetylcholine, equal brethren. Neurochem Int 52(4–5):634–648
564. Oury C, Lecut C, Hego A, Wéra O, Delierneux C (2015) Purinergic control of inflammation and thrombosis: role of P2X1 receptors. Comput Struct Biotechnol J 13:106–110
565. Burnstock G (2007) Purine and pyrimidine receptors. Cell Mol Life Sci 64:1471–1483
566. Verkhratsky A, Krishtal O (2009) Adenosine triphosphate (ATP) as a neurotransmitter. In: Squire LR (ed) Encyclopedia of neuroscience. Academic Press, San Diego, CA, USA
567. Kennedy C (2015) ATP as a cotransmitter in the autonomic nervous system. Auton Neurosci 191:2–15
568. Rummery NM, Brock JA, Pakdeechote P, Ralevic V, Dunn WR (2007) ATP is the predominant sympathetic neurotransmitter in rat mesenteric arteries at high pressure. J Physiol 582(2):745–754
569. Gershon MD, Tack J (2007) The serotonin signaling system: from basic understanding to drug development for functional gi disorders. Gastroenterology 132(1):397–414
570. Kim D-Y, Camilleri M (2000) Serotonin: a mediator of the brain–gut connection. Am J Gastroenterol 95:2698–2709
571. Li Z, Chalazonitis A, Huang Y-Y, Mann JJ, Gross KM, Yang QM et al (2011) Essential roles of enteric neuronal serotonin in gastrointestinal motility and the development/survival of enteric dopaminergic neurons. J Neurosci 31(24):8998–9009

572. Frick A, Åhs F, Engman J, Jonasson M, Alaie I, Björkstrand J et al (2015) Serotonin synthesis and reuptake in social anxiety disorder: a positron emission tomography study. JAMA Psychiatry 72(8):794–802
573. Gershon MD (2013) 5-Hydroxytryptamine (serotonin) in the gastrointestinal tract. Curr Opin Endocrinol Diabetes Obes 20(1):14–21
574. Gershon MD (2012) Serotonin is a sword and a shield of the bowel: serotonin plays offense and defense. Trans Am Clin Climatol Assoc 123:268–280
575. Mawe GM, Hoffman JM (2013) Serotonin signalling in the gut—functions, dysfunctions and therapeutic targets. Nat Rev Gastroenterol Hepatol 10:473–486
576. Ruddell RG, Mann DA, Ramm GA (2008) The function of serotonin within the liver. J Hepatol 48(4):666–675
577. Erspamer V (1937) Experimental research on the biological significance of enterochromaffin cells. Arch Fisiol 37:156–169
578. Erspamer V (1953) Concerning the 5-hydroxytryptamine (enteramine) content of the gastrointestinal tract lining. Naturwissenschaften 40:318–319
579. Page IH (1976) The discovery of serotonin. Perspect Biol Med 20:1–8
580. Bülbring E, Lin RCY, Schofield G (1958) An investigation of the peristaltic reflex in relation to anatomical observations. Q J Exp Physiol 43:26–37
581. Bülbring E, Lin RCY (1958) The effect of intraluminal application of 5-hydroxytryptamine and 5-hydroxytryptophan on peristalsis, the local production of 5-hydroxytryptamine and its release in relation to intraluminal pressure and propulsive activity. J Physiol (Lond) 140:381–407
582. Bülbring E, Crema A (1958) Observations concerning the action of 5-hydroxytryptamine on the peristaltic reflex. Br J Pharmacol 13:444–457
583. Grundy D (2006) Serotonin and sensory signalling from the gastrointestinal lumen. J Physiol 575(1):1–2
584. Jennifer H, Mariana P, Karyn B (2017) Serotonin syndrome from 5-hydroxytryptophan supplement ingestion in a 9-month-old labrador retriever. J Med Toxicol 13(2):183–186
585. Turner EH, Loftis JM, Blackwell AD (2006) Serotonin a la carte: supplementation with the serotonin precursor 5-hydroxytryptophan. Pharmacol Ther 109(3):325–338
586. Del Pino J, Moyano P, Ruiz M, Anadón MJ, Díaz MJ, García JM, Labajo-González E, Frejo MT (2017) Amitraz changes NE, DA and 5-HT biosynthesis and metabolism mediated by alterations in estradiol content in CNS of male rats. Chemosphere 181:518–529
587. Nakamura K, Hasegawa H (2009) Roles of 5-HTP, a precursor of serotonin. Int J Tryptophan Res 2:37–43
588. Kim HS (2009) 5-Hydroxytryptamine4 receptor agonists and colonic motility. J Smooth Muscle Res 45(1):25–29
589. Tuladhar BR, Kaisar M, Naylor RJ (1997) Evidence for a 5-HT3 receptor involvement in the facilitation of peristalsis on mucosal application of 5-HT in the guinea pig isolated ileum. Br J Pharmacol 122(6):1174–1178
590. Tuladhar BR, Costall B, Naylor RJ (1996) 5-HT3 and 5-HT4 receptor-mediated facilitation of the emptying phase of the peristaltic reflex in the marmoset isolated ileum. Br J Pharmacol 117(8):1679–1684
591. Spencer NJ (2015) Constitutively active 5-HT receptors: an explanation of how 5-HT antagonists inhibit gut motility in species where 5-HT is not an enteric neurotransmitter? Front Cell Neurosci 9:487
592. Tuladhar BR, Ge L, Naylor RJ (2003) 5-HT7 receptors mediate the inhibitory effect of 5-HT on peristalsis in the isolated guinea-pig ileum. Br J Pharmacol 138:1210–1214
593. Kindt S, Tack JB (2007) Mechanisms of serotonergic agents for treatment of gastrointestinal motility and functional bowel disorders. Neurogastroenterol Motil 19(2):32–39
594. Gaman A, Bucur MC, Kuo B (2009) Therapeutic advances in functional gastrointestinal disease: irritable bowel syndrome. Therap Adv Gastroenterol 2(3):169–181

595. Manocha M, Khan WI (2012) Serotonin and GI disorders: an update on clinical and experimental studies. Clin Transl Gastroenterol 3(4):e13
596. Karantanos T, Markoutsaki T, Gazouli M, Anagnou NP, Karamanolis DG (2010) Current insights into the pathophysiology of Irritable Bowel Syndrome. Gut Pathogens 2:3
597. Jones MP, Dilley JB, Drossman D, Crowell MD (2006) Brain–gut connections in functional GI disorders: anatomic and physiologic relationships. Neurogastroenterol Motil 18(2):91–103
598. Böttner M, Barrenschee M, Hellwig I, Harde J, Egberts JH, Becker T et al (2013) The enteric serotonergic system is altered in patients with diverticular disease. Gut 62 (12):1753–1762
599. Dong Y, Yang C, Wang Z, Qin Z, Cao J, Chen Y (2016) The injury of serotonin on intestinal epithelium cell renewal of weaned diarrhoea mice. Eur J Histochem 60(4):2689
600. Willets JM, Taylor AH, Shaw H, Konje JC, Challiss RA (2008) Selective regulation of H1 histamine receptor signaling by G protein-coupled receptor kinase 2 in uterine smooth muscle cells. Mol Endocrinol 22(8):1893–1907
601. Tsunada S, Fujimoto K, Gotoh Y, Sakai T, Kang M, Sakata T et al (1994) Role of histamine receptors in intestinal repair after ischemia-reperfusion in rats. Gastroenterology 107(5):1297–1304
602. Smolinska S, Jutel M, Crameri R, O'Mahony L (2014) Histamine and gut mucosal immune regulation. Allergy 69(3):273–281
603. Schwittlick U, Junginger J, Hahn K, Habierski A, Hewicker-Trautwein M (2017) Histamine receptor expression in the gastrointestinal tract of dogs. Anat Histol Embryol 46 (1):33–42
604. Konturek SJ (2003) Gastric secretion—from Pavlov's *nervism* to Popielski's histamine as direct secretagogue of oxyntic glands. J Physiol Pharmacol 54(S3):43–68
605. Grandi D, Shenton FC, Chazot PL, Morini G (2008) Immunolocalization of histamine H3 receptors on endocrine cells in the rat gastrointestinal tract. Histol Histopathol 23(7):789–798
606. Henriksen JH, de Muckadell OB (2000) Secretin, its discovery, and the introduction of the hormone concept. Scand J Clin Lab Invest 60(6):463–472
607. Dale HH, Laidlaw PP (1910) The physiological action of β-iminazolylethylamine. J Physiol (Lond) 41(5):318–344
608. Masini E, Mannaioni PF (1994) Effect of histamine on smooth muscle. Handb Exp Pharmacol 111:193–225
609. Barger G, Dale HH (1911) Beta-Iminazolylethylamine a depressor constituent of intestinal mucosa. J Physiol 41(6):499–503
610. Parsons ME, Ganellin CR (2006) Histamine and its receptors. Br J Pharmacol 147(1): S127–S135
611. Criado PR, Criado RF, Maruta CW, Machado Filho Cd (2010) Histamine, histamine receptors and antihistamines: new concepts. An Bras Dermatol 85(2):195–210
612. Beermann S, Bernhardt G, Seifert R, Buschauer A, Neumann D (2015) Histamine H_1- and H_4-receptor signaling cooperatively regulate MAPK activation. Biochem Pharmacol 98 (3):432–439
613. Lim HD, van Rijn RM, Ling P, Bakker RA, Thurmond RL, Leurs R (2005) Evaluation of histamine H1-, H2-, and H3-receptor ligands at the human histamine H4 receptor: identification of 4-methylhistamine as the first potent and selective H4 receptor agonist. J Pharmacol Exp Ther 314(3):1310–1321
614. Hegyesi H, Darvas Z, Thurmond RL, Falus A (2005) Histamine receptors and signaling. In: Damjanovich S (ed) Biophysical aspects of transmembrane signaling. Springer, Heidelberg
615. Hill SJ (1990) Distribution, properties, and functional characteristics of three classes of histamine receptor. Pharmacol Rev 42(1):45–83

616. Panula P, Chazot PL, Cowart M, Gutzmer R, Leurs R, Liu WL et al (2015) International union of basic and clinical pharmacology. XCVIII. Histamine receptors. Pharmacol Rev 67 (3):601–655
617. Delvalle J, Wang L, Gantz I, Yamada T (1992) Characterization of H2 histamine receptor: linkage to both adenylate cyclase and [Ca^{2+}]i signaling systems. Am J Physiol Gastrointest Liver Physiol 263(6):G967–G972
618. Lehmann F, Hildebrand P, Beglinger C (2003) New molecular targets for treatment of peptic ulcer disease. Drugs 63(17):1785–1797
619. Carlos D, Spiller F, Souto FO, Trevelin SC, Borges VF, de Freitas A et al (2013) Histamine H_2 receptor signaling in the pathogenesis of sepsis: studies in a murine diabetes model. J Immunol 191(3):1373–1382
620. Church DS, Church MK (2011) Pharmacology of antihistamines. World Allergy Organ J 4 (3):S22–S27
621. Beghdadi W, Porcherie A, Schneider BS, Morisset S, Dubayle D, Peronet R et al (2009) Histamine H(3) receptor-mediated signaling protects mice from cerebral malaria. PLoS ONE 4(6):e6004
622. Leurs R, Smit MJ, Timmerman H (1995) Molecular pharmacological aspects of histamine receptors. Pharmacol Ther 66(3):413–463
623. Zampeli E, Tiligada E (2009) The role of histamine H_4 receptor in immune and inflammatory disorders. Br J Pharmacol 157(1):24–33
624. Walter M, Kottke T, Stark H (2011) The histamine H_4 receptor: targeting inflammatory disorders. Eur J Pharmacol 668(1–2):1–5
625. Nakamura T, Itadani H, Hidaka Y, Ohta M, Tanaka K (2000) Molecular cloning and characterization of a new human histamine receptor, HH4R. Biochem Biophys Res Commun 279(2):615–620
626. Barcik W, Pugin B, Westermann P, Perez NR, Ferstl R, Wawrzyniak M et al (2016) Histamine-secreting microbes are increased in the gut of adult asthma patients. J Allergy Clin Immunol 138(5):1491–1494
627. Di Giovanni G, Svob Strac D, Sole M, Unzeta M, Tipton KF, Mück-Šeler D et al (2016) Monoaminergic and histaminergic strategies and treatments in brain diseases. Front Neurosci 10:541
628. Provensi G, Fabbri R, Munari L, Costa A, Baldi E, Bucherelli C et al (2017) Histaminergic neurotransmission as a gateway for the cognitive effect of oleoylethanolamide in contextual fear conditioning. Int J Neuropsychopharmacol 20(5):392–399
629. Roberts E, Frankel S (1950) Gamma-aminobutyric acid in brain: its formation from glutamic acid. J Biol Chem 187:55–63
630. Awapara J, Landua A, Fuerst R, Seale B (1950) Free gamma-aminobutyric acid in brain. J Biol Chem 187:35–39
631. Bowery NG, Smart TG (2006) GABA and glycine as neurotransmitters: a brief history. Br J Pharmacol 147(1):S109–S119
632. Froestl W (2011) An historical perspective on GABAergic drugs. Future Med Chem 3 (2):163–175
633. Jessen KR, Hills JM, Dennison ME, Mirsky R (1983) Aminobutyrate as an autonomic neurotransmitter: release and uptake of [3H]$\gamma\gamma$-aminobutyrate in guinea-pig large intestine and cultured enteric neurons using physiological methods and electron microscopic autoradiography. Neuroscience 10:1427–1442
634. Jessen KR, Hills JM, Saffrey MJ (1986) Immunohistochemical demonstration of GABAergic neurons in the enteric nervous system. J Neurosci 6:1628–1634
635. Jessen KR (1981) GABA and the enteric nervous system. Dev Mol Cell Biochem 1:69–76
636. Kerr DIB, Ong J (1984) GABA and GABA-receptors in the enteric nervous system. Neuropharmacology 23(7):835–836
637. Baetge G, Gershon MD (1986) GABA in the PNS: demonstration in enteric neurons. Brain Res Bull 16(3):421–424

638. Jessen KR, Hills JM, Limbrick AR (1988) GABA immunoreactivity and ^{3}H-GABA uptake in mucosal epithelial cells of the rat stomach. Gut 29:1549–1556
639. Krantis A, Kerr DIB, Dennis BJ (1986) Autoradiographic study of the distribution of [^{3}H]γ-aminobutyrate-accumulating neural elements in guinea-pig intestine: evidence for a transmitter function of γ-aminobutyrate. Neuroscience 17:1243–1255
640. Jessen KR, Mirsky R, Hills JM (1987) GABA as an autonomic neurotransmitter: studies on intrinsic GABAergic neurons in the myenteric plexus of the gut. Trends Neurosci 10:255–262
641. Bak LK, Schousboe A, Waagepetersen HS (2006) The glutamate/GABA-glutamine cycle: aspects of transport, neurotransmitter homeostasis and ammonia transfer. J Neurochem 98 (3):641–653
642. Figiel M, Engele J (2000) Pituitary adenylate cyclase-activating polypeptide (PACAP), a neuron-derived peptide regulating glial glutamate transport and metabolism. J Neurosci 20 (10):3596–3605
643. Auteri M, Zizzo MG, Serio R (2015) GABA and GABA receptors in the gastrointestinal tract: from motility to inflammation. Pharmacol Res 93:11–21
644. Geiman EJ, Zheng W, Fritschy JM, Alvarez FJ (2002) Glycine and GABA(A) receptor subunits on Renshaw cells: relationship with presynaptic neurotransmitters and postsynaptic gephyrin clusters. J Comp Neurol 444(3):275–289
645. Bravo JA, Forsythe P, Chew MV, Escaravage E, Savignac HM, Dinan TG, Bienenstock J, Cryan JF (2011) Ingestion of Lactobacillus strain regulates emotional behavior and central GABA receptor expression in a mouse via the vagus nerve. PNAS 108(38):16050–16055
646. Bormann J, Feigenspan A (2001) $GABA_C$ receptors: structure, function and pharmacology. In: Möhler H (ed) Pharmacology of GABA and glycine neurotransmission. Springer, Heidelberg
647. Johnston GAR (1996) $GABA_C$ receptors: relatively simple transmitter-gated ion channels? Trends Pharmacol Sci 17(9):319–323
648. Cherubini E, Strata F (1997) GABAC receptors: a novel receptor family with unusual pharmacology. News Physiol Sci 12(3):136–141
649. Sigel E, Steinmann ME (2012) Structure, function, and modulation of $GABA_A$ receptors. J Biol Chem 287:40224–40231
650. Bormann J (2000) The 'ABC' of GABA receptors. Trends Pharmacol Sci 21(1):16–19
651. Seifi M, Brown JF, Mills J, Bhandari P, Belelli D, Lambert JJ et al (2014) Molecular and functional diversity of GABA-a receptors in the enteric nervous system of the mouse Colon. J Neurosci 34(31):10361–10378
652. Tonini M, De Petris G, Onori L, Manzo L, Rizzi CA, Crema A (1989) The role of GABAA receptor function in peristaltic activity of the guinea-pig ileum: a comparative study with bicuculline, SR 95531 and picrotoxinin. Br J Pharmacol 97:556–562
653. Tonini M, Crema A, Frigo GM, Rizzi CA, Manzo L, Candura SM, Onori L (1989) An in vitro study of the relationship between GABA receptor function and propulsive motility in the distal colon of the rabbit. Br J Pharmacol 98:1109–1118
654. Lecci A, Altamura M, Capriati A, Maggi CA (2008) Tachykinin receptors and gastrointestinal motility: focus on humans. Eur Rev Med Pharmacol Sci 12(1):69–80
655. Maggio JE (1988) Tachykinins. Annu Rev Neurosci 11:13–28
656. Sternini C, Anderson K, Frantz G, Krause JE, Brecha N (1989) Expression of substance P/neurokinin A-encoding preprotachykinin messenger ribonucleic acids in the rat enteric nervous system. Gastroenterology 97(2):348–356
657. Carter MS, Krause JE (1990) Structure, expression, and some regulatory mechanisms of the rat preprotachykinin gene encoding substance P, neurokinin A, neuropeptide K, and neuropeptide gamma. J Neurosci 10(7):2203–2214
658. Pennefather JN, Lecci A, Candenas ML, Patak E, Pinto FM, Maggi CA (2004) Tachykinins and tachykinin receptors: a growing family. Life Sci 74(12):1445–1463

659. Christofi FL, McDonald TJ, Cook MA (1990) Neuropeptide release from isolated myenteric nerve endings derived from the guinea pig myenteric plexus. Regul Pept 30 (2):165–177
660. Arai H, Emson PC (1986) Regional distribution of neuropeptide K and other tachykinins (neurokinin A, neurokinin B and substance P) in rat central nervous system. Brain Res 399 (2):240–249
661. Patacchini R, Maggi CA, Holzer P (2000) Tachykinin autoreceptors in the gut. Trends Pharmacol Sci 21(5):166
662. Erspamer V (1981) The tachykinin peptide family. Trends Neurosci 4:267–269
663. Erspamer V, Ronzoni G, Falconieri Erspamer G (1981) Effects of active peptides on the isolated muscle of the human urinary bladder. Invest Urol 18(4):302–304
664. Pinto FM, Ravina CG, Subiran N, Cejudo-Román A, Fernández-Sánchez M, Irazusta J et al (2010) Autocrine regulation of human sperm motility by tachykinins. Reprod Biol Endocrinol 8:104
665. Patacchini R, Maggi CA (2004) The nomenclature of tachykinin receptors. In: Holzer P (ed) Tachykinins volume 164 of the series handbook of experimental pharmacology. Springer, Heidelberg, Germany
666. Steinhoff MS, von Mentzer B, Geppetti P, Pothoulakis C, Bunnett NW (2014) Tachykinins and their receptors: contributions to physiological control and the mechanisms of disease. Physiol Rev 94(1):265–301
667. Holzer-Petsche U (1995) Tachykinin receptors in gastrointestinal motility. Regul Pept 57 (1):19–42
668. Torrens Y, Saffroy M, Glowinski J, Beaujouan JC (1997) Substance P(6–11) and natural tachykinins interact with septide-sensitive tachykinin receptors coupled to a phospholipase C in the rat urinary bladder. Neuropeptides 31(3):243–251
669. Grider JR, Langdon LE (2003) Physiological role of neuropeptide Y in the regulation of the ascending phase of the peristaltic reflex. Am J Physiol Gastrointest Liver Physiol 285(6): G1139–G1146
670. Montaño LM, Carbajal V, Arreola JL, Barajas-López C, Flores-Soto E, Vargas MH (2003) Acetylcholine and tachykinins involvement in the caffeine-induced biphasic change in intracellular Ca^{2+} in bovine airway smooth muscle. Br J Pharmacol 139:1203–1211
671. Chang MM, Leeman SE (1970) Isolation of a sialogogic peptide from bovine hypothalamic tissue and its characterization as substance P. J Biol Chem 245(18):4784–4790
672. Chang MM, Leeman SE, Niall HD (1971) Amino-acid sequence of substance P. Nature New Biol 232:86–87
673. Harrison S, Geppetti P (2001) Substance P. Int J Biochem Cell Biol 33(6):555–576
674. von Euler US, Gaddum JH (1931) An unidentified depressor substance in certain tissue extracts. J Physiol 72(1):74–87
675. Hanley MR, Iversen LL (1980) Substance P receptors. In: Enna SJ, Yamamura HI (eds) Neurotransmitter receptors: part 1 amino acids, peptides and benzodiazepines. Chapman and Hall, New York
676. Muñoz M, Coveñas R (2010) Neurokinin-1 receptor: a new promising target in the treatment of cancer. Discov Med 10(53):305–313
677. Aziz F (2012) Neurokinin-1 receptor antagonists for chemotherapy-induced nausea and vomiting. Ann Palliat Med 1(2):130–136
678. Andrews PLR, Okada F, Woods AJ, Hagiwara H, Kakaimoto S, Toyoda M, Matsuki N (2000) The emetic and anti-emetic effects of the capsaicin analogue resiniferatoxin in Suncus murinus, the house musk shrew. Br J Pharmacol 130(6):1247–1254
679. Amin AH, Crawford TBB, Gaddum JH (1954) The distribution of substance P and 5-hydroxytryptamine in the central nervous system of the dog. J Physiol 126(3):596–618
680. Aapro MS (2007) Palonosetron as an anti-emetic and anti-nausea agent in oncology. Ther Clin Risk Manag 3(6):1009–1020
681. Becker DE (2010) Nausea, vomiting, and hiccups: a review of mechanisms and treatment. Anesth Prog 57(4):150–157

682. Koga T, Fukuda H (1992) Neurons in the nucleus of the solitary tract mediating inputs from emetic vagal afferents and the area postrema to the pattern generator for the emetic act in dogs. Neurosci Res 14(3):166–179
683. Horn CC (2016) The physiology of vomiting. In: Koch KL, Hasler WL (eds) Nausea and vomiting. Springer International Publishing, Cham
684. Pleuvry BJ (2012) Physiology and pharmacology of nausea and vomiting. Anaesth Intens Care Med 13(12):598–602
685. McKeen MJ, Quraishi SA (2013) Clinical review of intravenous opioids in acute care. J Anesthesiol Clin Sci 2:1
686. Schechter MD (1980) Different dopaminergic mechanisms for amfonelic acid, amphetamine and apomorphine. Pharmacol Biochem Behav 13(4):497–500
687. Smith HS, Laufer A (2014) Opioid induced nausea and vomiting. Eur J Pharmacol 722:67–78
688. Mercadante S, Sapio M, Serretta R (1998) Ondansetron in nausea and vomiting induced by spinal morphine. JPSM 16(4):259–262
689. Horn CC, Kimball BA, Wang H, Kaus J, Dienel S, Nagy A et al (2013) Why can't rodents vomit? A comparative behavioral, anatomical, and physiological study. PLoS ONE 8(4): e60537
690. Manoguerra AS, Cobaugh DJ (2005) Guidelines for the management of poisoning consensus panel. Guideline on the use of ipecac syrup in the out-of-hospital management of ingested poisons. Clin Toxicol (Phila) 43(1):1–10
691. Minton N, Swift R, Lawlor C, Mant T, Henry J (1993) Ipecacuanha-induced emesis: a human model for testing antiemetic drug activity. Clin Pharmacol Ther 54(1):53–57
692. Andrews PLR, Horn CC (2006) Signals for nausea and emesis: implications for models of upper gastrointestinal diseases. Auton Neurosci 125(1–2):100–115
693. Herrstedt J (2008) Antiemetics: an update and the MASCC guidelines applied in clinical practice. Nat Clin Pract Oncol 5:32–43
694. Hogan CM, Grant M (1997) Physiologic mechanisms of nausea and vomiting in patients with cancer. Oncol Nurs Forum 24(Suppl 7):8–12
695. Kangawa H, Kangawa K, Minamino N, Fukuda A, Matsuo H (1983) Neuromedin K: a novel mammalian tachykinin identified in porcine spinal cord. Biochem Biophys Res Commun 114:533–540
696. Kimura S, Okado M, Sugita Y, Kanazawa I, Munekata E (1983) Novel neuropeptides, neurokinin A and B isolated from porcine spinal cord. Proc Jpn Acad 59B:101–104
697. Wang JM, De Ridder EFMM, De Potter WP, Weyns ALM (1994) Localization of neurokinin A and chromogranin A immunoreactivity in the developing porcine adrenal medulla. Histochem J 26(5):431–436
698. Regoli D, Rhaleb NE, Dion S, Tousignant C, Rouissi N, Jukic D, Drapeau G (1990) Neurokinin A. A pharmacological study. Pharmacol Res 22(1):1–14
699. Rovero P, Pestellini V, Rhaleb N-E, Dion S, Rouissi N, Tousignant C et al (1989) Structure-activity studies of neurokinin A. Neuropeptides 13(4):263–270
700. Joos GF, Vincken W, Louis R, Schelfhout VJ, Wang JH, Shaw MJ et al (2004) Dual tachykinin NK1/NK2 antagonist DNK333 inhibits neurokinin A-induced bronchoconstriction in asthma patients. ERJ 23(1):76–81
701. Eriksson B, Oberg K, Stridsberg M (2000) Tumor markers in neuroendocrine tumors. Digestion 62(1):33–38
702. Halfdanarson TR, Howe JR, Haraldsdottir S, O'Dorisio TM (2015) Circulating tumor markers in patients with neuroendocrine tumors—a clinical perspective. Int J Endocr Oncol 2(1):89–99
703. Turner GB, Johnston BT, Mccance DR, McGinty A, Watson RGP, Patterson CC, Ardill JES (2006) Circulating markers of prognosis and response to treatment in patients with midgut carcinoid tumours. Gut 55(11):1586–1591

704. Diebold AE, Boudreaux JP, Wang YZ, Anthony LB, Uhlhorn AP, Ryan P et al (2012) Neurokinin A levels predict survival in patients with stage IV well differentiated small bowel neuroendocrine neoplasms. Surgery 152(6):1172–1176
705. Malendowicz LK, Warchol JB, Nussdorfer GG, Nowak M, Filipiak K (1995) Effects of neurokinin-A on the rat hypothalamo-pituitary-adrenal axis. Endocr Res 21(4):757–767
706. Malendowicz LK, Warchol JB, Nussdorfer GG, Nowak M, Filipiak K (1995) Effects of neurokinin-A on the rat hypothalamo-pituitary-adrenal axis. Endocr Res 21(4):757–767
707. De Laurentiis A, Candolfi M, Pisera D, Seilicovich A (2003) Effects of lipopolysaccharide on neurokinin A content and release in the hypothalamic-pituitary axis. Regul Pept 111(1–3):91–95
708. Tuxen C, Jensen K, Kjeldsen M, Edvinsson L, Jansen I, Olesen J (1989) Effect of neurokinin A on human temporal muscle blood flow. Peptides 10(5):921–924
709. Takaya T (1996) Discovery of neurokinin antagonists. Pure Appl Chem 68(4):875–880
710. Navarro VM (2013) Interactions between kisspeptins and neurokinin B. Adv Exp Med Biol 784:325–347
711. Page NM, Woods RJ, Lowry PJ (2001) A regulatory role for neurokinin B in placental physiology and pre-eclampsia. Regul Pept 98(3):97–104
712. Sakamoto R, Osada H, Litsuka Y, Masuda K, Kaku K, Seki K, Sekiya S (2003) Profile of neurokinin B concentrations in maternal and cord blood in normal pregnancy. Clin Endocrinol 58(5):597–600
713. Hofmann HA (2006) Gonadotropin-releasing hormone signaling in behavioral plasticity. Curr Opin Neurobiol 16(3):343–350
714. Bliss SP, Navratil AM, Xie J, Roberson MS (2010) GnRH signaling, the gonadotrope and endocrine control of fertility. Front Neuroendocrinol 31(3):322–340
715. Santoro N, Filicori M, Spratt D, Crowley WF Jr (1986) Gonadotropin-releasing hormone (GnRH) physiology in men and women. Acta Med Hung 43(2):201–221
716. Kumasaka T, Masaoka K, Watanabe H, Mori T, Ohya H, Tadokoro N (1986) Neuroendocrine regulation of LHRH and gonadotropin. Adv Contracept Deliv Syst 2(2–3):176–181
717. Millar RP, King JA (1988) Evolution of gonadotropin-releasing hormone: multiple usage of a peptide. Physiology 3(2):49–53
718. Rance NE, Dacks PA, Mittelman-Smith MA, Romanovsky AA, Krajewski-Hall SJ (2013) Modulation of body temperature and LH secretion by hypothalamic KNDy (kisspeptin, neurokinin B and dynorphin) neurons: a novel hypothesis on the mechanism of hot flushes. Front Neuroendocrinol 34(3):211–227
719. Kage R, McGregor GP, Thim L, Conlon JM (1988) Neuropeptide-gamma: a peptide isolated from rabbit intestine that is derived from gamma-preprotachykinin. J Neurochem 50(5):1412–1417
720. Lee K, Lee S, Kim Y, Park NG (2003) Structures of neuropeptide γ from goldfish and mammalian neuropeptide γ, as determined by ^{1}H NMR spectroscopy. J Pept Res 61 (5):274–285
721. Burcher E, Alouan LA, Johnson PRA, Black JL (1991) Neuropeptide gamma, the most potent contractile tachykinin in human isolated bronchus, acts via a 'non-classical' NK2 receptor. Neuropeptides 20(2):79–82
722. Poulat P, de Champlain J, Couture R (1996) Cardiovascular responses to intrathecal neuropeptide gamma in conscious rats: receptor characterization and mechanism of action. Br J Pharmacol 117(2):250–257
723. Chandrashekar IR, Dike A, Cowsik SM (2004) Membrane-induced structure of the mammalian tachykinin Neuropeptide gamma. J Struct Biol 148(3):315–325
724. Picard P, Couture R (1996) Intracerebroventricular responses to neuropeptide gamma in the conscious rat: characterization of its receptor with selective antagonists. Br J Pharmacol 117(2):241–249
725. Fernstrom JD, Fernstrom MH (2007) Tyrosine, phenylalanine, and catecholamine synthesis and function in the brain. J Nutr 137(6):1539S–1547S

726. Hornykiewicz O (2002) L-DOPA: from a biologically inactive amino acid to a successful therapeutic agent. Amino Acids 23(1–3):65–70
727. Raper HS (1927) The tyrosinase-tyrosine reaction: production from tyrosine of 5: 6-dihydroxyindole and 5: 6-dihydroxyindole-2-carboxylic acid–the precursors of melanin. Biochem J 21(1):89–96
728. Raper HS (1926) The tyrosinase-tyrosine reaction: production of l-3.4-dihydroxyphenylalanine from tyrosine. Biochem J 20(4):735–742
729. Dulière WL, Raper HS (1930) The tyrosinase-tyrosine reaction: the action of tyrosinase on certain substances related to tyrosine. Biochem J 24(2):239–249
730. Sneader W (2001) The discovery and synthesis of epinephrine. Drug News Perspect 14 (8):491–494
731. Hornykiewicz O (2010) A brief history of levodopa. J Neurol 257(2):249–252
732. Lanska DJ (2014) Kühne, Wilhelm (Willy) Friedrich. In: Daroff R, Aminoff MJ (eds) Encyclopedia of the neurological sciences, 2nd edn. Academic Press, San Diego
733. Dogrul A, Coskun I, Uzbay T (2006) The contribution of alpha-1 and alpha-2 adrenoceptors in peripheral imidazoline and adrenoceptor agonist-induced nociception. Anesth Analg 103(2):471–477
734. Galitzky J, Carpéné C, Bousquet-Mélou A, Berlan M, Lafontan M (1995) Differential activation of beta 1-, beta 2- and beta 3-adrenoceptors by catecholamines in white and brown adipocytes. Fundam Clin Pharmacol 9(4):324–331
735. Martin DJ (1999) Preclinical pharmacology of alpha1-adrenoceptor antagonists. Eur Urol 36(1):35–41; discussion 65
736. Muramatsu I, Taniguchi T, Okada K (1998) Tamsulosin: alpha1-adrenoceptor subtype-selectivity and comparison with terazosin. Jpn J Pharmacol 78(3):331–335
737. Jun JY, Choi S, Yeum CH, Chang IY, Park CK, Kim MY et al (2004) Noradrenaline inhibits pacemaker currents through stimulation of beta 1-adrenoceptors in cultured interstitial cells of Cajal from murine small intestine. Br J Pharmacol 141(4):670–677
738. von Euler US (1946) A specific sympathomimetic ergone in adrenergic nerve fibres (sympathin) and its relations to adrenaline and nor-adrenaline. Acta Physiol Scand 12:73–97
739. Furst J, Kurra S (2015) Adrenal physiology. In: Pasieka JL, Lee JA (eds) Surgical endocrinopathies. Springer International Publishing, Cham, Switzerland
740. Luckensmeyer GB, Keast JR (1998) Activation of α- and β-adrenoceptors by sympathetic nerve stimulation in the large intestine of the rat. J Physiol 510 (Pt 2)(2):549–561
741. Angelos MG, Butke RL, Panchal AR, Torres CA, Blumberg A, Schneider JE, Aune SE (2008) Cardiovascular response to epinephrine varies with increasing duration of cardiac arrest. Resuscitation 77(1):101–110
742. Callaway CW (2012) Questioning the use of epinephrine to treat cardiac arrest. JAMA 307 (11):1198–1200
743. McLean-Tooke APC, Bethune CA, Fay AC, Spickett GP (2003) Adrenaline in the treatment of anaphylaxis: what is the evidence? BMJ 327(7427):1332–1335
744. Burrage WS, Irwin JW, Gallemore JI, Wang DMK (1954) Effects of histamine and epinephrine on the small pulmonary blood vessels of living rabbits. J Allergy Clin Immunol 25(4):293–301
745. William BH (1896) The use of extract of suprarenal capsule in the eye. N Y M J 647–650
746. Bennett MR (1999) One hundred years of adrenaline: the discovery of autoreceptors. Clin Auton Res 9(3):145–159
747. Abel JJ (1899) Ueber den blutdruckerregendes Bestandtheil des Nebenniere, das Epinephrin. Hoppe-Seylers Z Physiol Chem 28:318–362
748. Koehler PJ, Stahnisch FW (2014) Three twentieth-century multiauthored neurological handbooks—a historical analysis and bibliometric comparison. J Hist Neurosci Basic Clin Perspect 23(1):1–30
749. Widacki J (2015) Professor Ivan Tarchanoff. Przegl Lek 72(1):45–48
750. Smith JE (1920) Max Lewandowsky. J Nerv Ment Dis 51(4):404

751. Bilski R, Kaulbersz J (1987) Napoleon Cybulski, 1854–1919. Acta Physiol Pol 38(2):74–90
752. Grzybowski A, Pietrzak K (2013) Napoleon Cybulski (1854–1919). J Neurol 260 (11):2942–2943
753. Takamine J (1901) The isolation of the active principle of the suprarenal gland. J Physiol (Lond) 27:xxix–xxx
754. Yamashima T (2003) Jokichi Takamine (1854–1922), the samurai chemist, and his work on adrenalin. J Med Biogr 11(2):95–102
755. Elliott TR (1905) The action of adrenalin. J Physiol 32(5–6):401–467
756. Elliott TR, Herbert DE (1906) On subcutaneous injections of adrenalin. J Physiol 34 (6):490–498
757. Dale HH (1961) Thomas Renton Elliott. 1877–1961. Biogr Mem Fellows R Soc 7:52–74
758. Cannon WB, Rosenbueth A (1933) Studies on the conditions of activity in endocrine organs: XXIX. Sympathin E and sympathin I. Am J Physiol 104:557–574
759. Bacq ZM (1979) Early work on the adrenergic mediator: how to go wrong. Trends Pharmacol Sci 1(1):141–143
760. von Euler US (2008) A specific sympathomimetic ergone in adrenergic nerve fibres (sympathin) and its relations to adrenaline and nor-adrenaline. Acta Physiol Scand 12 (1):73–97
761. von Euler US (1946) A specific sympathomimetic ergone in adrenergic nerve fibres (sympathin) and its relations to adrenaline and nor-adrenaline. Acta Physiol Scand 12:73–96
762. Ahlquist RP (1948) A study of adrenotropic receptors. Am J Physiol 153:586–600
763. Bylund DB (2007) Alpha- and beta-adrenergic receptors: Ahlquist's landmark hypothesis of a single mediator with two receptors. Am J Physiol Endocrinol Metab 293(6):E1479–E1481
764. Christ DD, Nishi S (1971) Effects of adrenaline on nerve terminals in the superior cervical ganglion of the rabbit. Br J Pharmacol 41(2):331–338
765. Wenger NK, Greenbaum LM (1984) From adrenoceptor mechanisms to clinical therapeutics: Raymond Ahlquist, Ph.D., 1914–1983. J Am Coll Cardiol 3 (2 Pt 1):419–421
766. Alexander SPH, Benson HE, Faccenda E, Pawson AJ, Sharman JL, Spedding M et al (2013) The concise guide to pharmacology 2013/14: G protein-coupled receptors. Br J Pharmacol 170:1459–1581
767. García-Sáinz JA, Vázquez-Prado J, Villalobos-Molina R (1999) Alpha 1-adrenoceptors: subtypes, signaling, and roles in health and disease. Arch Med Res 30(6):449–458
768. Koshimizu TA, Tanoue A, Hirasawa A, Yamauchi J, Tsujimoto G (2003) Recent advances in alpha1-adrenoceptor pharmacology. Pharmacol Ther 98(2):235–244
769. Chen-Izu Y, Xiao RP, Izu LT, Cheng H, Kuschel M, Spurgeon H, Lakatta EG (2000) G(i)-dependent localization of beta(2)-adrenergic receptor signaling to L-type Ca^{2+} channels. Biophys J 79(5):2547–2556
770. Nisoli E, Tonello C, Landi M, Carruba MO (1996) Functional studies of the first selective beta 3-adrenergic receptor antagonist SR 59230A in rat brown adipocytes. Mol Pharmacol 49(1):7–14
771. Zhao TJ, Sakata I, Li RL, Liang G, Richardson JA, Brown MS et al (2010) Ghrelin secretion stimulated by {beta}1-adrenergic receptors in cultured ghrelinoma cells and in fasted mice. PNAS 107(36):15868–15873
772. Langer SZ, Hicks PE (1984) Alpha-adrenoreceptor subtypes in blood vessels: physiology and pharmacology. J Cardiovasc Pharmacol 6(4):S547–S558
773. Jinap S, Hajeb P (2010) Glutamate. Its applications in food and contribution to health. Appetite 55(1):1–10
774. Kurihara K (2009) Glutamate: from discovery as a food flavor to role as a basic taste (Umami). Am J Clin Nutr 90(3):719S–722S
775. Young VR, Ajami AM (2000) Glutamate: an amino acid of particular distinction. J Nutr 130(4S):892S–900S

776. Willard SS, Koochekpour S (2013) Glutamate, glutamate receptors, and downstream signaling pathways. Int J Biol Sci 9(9):948–959
777. Benarroch EE (2008) Metabotropic glutamate receptors: synaptic modulators and therapeutic targets for neurologic disease. Neurology 70(12):964–968
778. Greget R, Pernot F, Bouteiller JM, Ghaderi V, Allam S, Keller AF et al (2011) Simulation of postsynaptic glutamate receptors reveals critical features of glutamatergic transmission. PLoS ONE 6(12):e28380
779. Pace RW, Mackay DD, Feldman JL, Del Negro CA (2007) Inspiratory bursts in the preBötzinger complex depend on a calcium-activated non-specific cation current linked to glutamate receptors in neonatal mice. J Physiol 582(Pt 1):113–125
780. Carraway RE, Leeman SE (1975) The amino acid sequence of a hypothalamic peptide, neurotensin. J Biol Chem 250:1907–1911
781. Tyler-McMahon BM, Boules M, Richelson E (2000) Neurotensin: peptide for the next millennium. Regul Pept 93(1–3):125–136
782. Doyle H, Greeley GH, Mate L, Sakamoto T, Townsend CM, Thompson JC (1985) Distribution of neurotensin in the canine gastrointestinal tract. Surgery 97(3):337–341
783. Atoji Y, Hirasawa Y, Yamamoto Y, Suzuki Y (1995) Distribution of neurotensin-immunoreactive neurons in the digestive tract of the chicken. J Auton Nerv Syst 53(2–3):185–194
784. Ganten D, Pfaff DW (2011) Neuroendocrinology of mood. Springer, Heidelberg
785. Eysselein VE (1984) Neurotensin-what is known about its role as a hormone in the gastrointestinal tract? Klin Wochenschr 62(11):523–530
786. Carraway R, Leeman SE (1973) The isolation of a new hypotensive peptide, neurotensin, from bovine hypothalami. J Biol Chem 248(19):6854–6861
787. Carraway R, Leeman SE (1976) Characterization of radioimmunoassayable neurotensin in the rat; its differential distribution in the central nervous system, small intestine, and stomach. J Biol Chem 251:7045–7052
788. Mogard MH, Maxwell V, Sytnik B, Walsh JH (1987) Regulation of gastric acid secretion by neurotensin in man evidence against a hormonal role. J Clin Invest 80:1064–1067
789. Holmgren S, Jönsson A-C (1995) Neurotensin in the gut of the amphibian necturus maculosus and the effects of neurotensin on stomach motility. Physiol Zool 68(2):177–191
790. Melander O, Maisel AS, Almgren P, Manjer J, Belting M, Hedblad B et al (2012) Plasma proneurotensin and incidence of diabetes, cardiovascular disease, breast cancer, and mortality. JAMA 308(14):1469–1475
791. Boules M, Liang Y, Briody S, Miura T, Fauq I, Oliveros A et al (2010) NT79: A novel neurotensin analog with selective behavioral effects. Brain Res 1308:35–46
792. Mazella J (2001) Sortilin/neurotensin receptor-3: a new tool to investigate neurotensin signaling and cellular trafficking? Cell Signal 13(1):1–6
793. Di Fruscia P, He Y, Koenig M, Tabrizifard S, Nieto A, McDonald PH, Kamenecka TM (2014) The discovery of indole full agonists of the neurotensin receptor 1 (NTSR1). Bioorg Med Chem Lett 24(16):3974–3978
794. Gendron L, Perron A, Payet MD, Gallo-Payet N, Sarret P, Beaudet A (2004) Low-affinity neurotensin receptor (NTS2) signaling: internalization-dependent activation of extracellular signal-regulated kinases 1/2. Mol Pharmacol 66(6):1421–1430
795. Osadchii OE (2015) Emerging role of neurotensin in regulation of the cardiovascular system. Eur J Pharmacol 762:184–192
796. Shulkes A (2004) Neurotensin. In: Martini L (ed) Encyclopedia of endocrine diseases. Academic Press, San Diego, CA, USA
797. Carraway R, Demers L, Leeman SE (1976) Hyperglycemic effect of neurotensin, a hypothalamic peptide. Endocrinol 99:1452–1462
798. Posovszky C, Wabitsch M (2015) Regulation of appetite, satiation, and body weight by enteroendocrine cells. Part 1: characteristics of enteroendocrine cells and their capability of weight regulation. Horm Res Paediatr 83:1–10

799. Feng YP, Wang J, Dong YL, Wang YY, Li YQ (2015) The roles of neurotensin and its analogues in pain. Curr Pharm Des 21(7):840–848
800. Larhammar D (2004) Neuropeptide Y, evolution of. In: Martini L (ed) Encyclopedia of endocrine diseases. Academic Press, San Diego
801. Brothers SP, Wahlestedt C (2010) Therapeutic potential of neuropeptide Y (NPY) receptor ligands. EMBO Mol Med 2(11):429–439
802. Sheriff S, Ali M, Yahya A, Haider KH, Balasubramaniam A, Amlal H (2010) Neuropeptide Y Y5 receptor promotes cell growth through extracellular signal-regulated kinase signaling and cyclic AMP inhibition in a human breast cancer cell line. Mol Cancer Res 8(4):604–614
803. Tatemoto K, Carlquist M, Mutt V (1982) Neuropeptide Y-a novel brain peptide with structural similarities to peptide YY and pancreatic polypeptide. Nature 296:659–660
804. Minth CD, Bloom SR, Polak JM, Dixon JE (1984) Cloning, characterization, and DNA sequence of a human cDNA encoding neuropeptide tyrosine. Proc Natl Acad Sci U S A 81 (14):4577–4581
805. Larhammar D (1996) Evolution of neuropeptide Y, peptide YY and pancreatic polypeptide. Regul Pept 62(1):1–11
806. Xu B, Kalra PS, Farmerie WG, Kalra SP (1999) Daily changes in hypothalamic gene expression of neuropeptide Y, galanin, proopiomelanocortin, and adipocyte leptin gene expression and secretion: effects of food restriction. Endocrinology 140(6):2868–2875
807. Cerda-Reverter JM, Larhammar D (2000) Neuropeptide Y family of peptides: structure, anatomical expression, function, and molecular evolution. Biochem Cell Biol 78(3):371–392
808. Ubink R, Calza L, Hokfelt T (2003) Neuro-peptides in glia: focus on NPY and galanin. Trends Neurosci 26(11):604–609
809. Cavadas C, Ribeiro CA, Cotrim M, Mosimann F, Brunner HR, Grouzmann E (2002) Catecholamine and neuropeptide Y secretion from human adrenal chromaffin cells: effect of nicotine and KCl. Ann N Y Acad Sci 971:332–334
810. Dumont Y, Martel JC, Fournier A, St-Pierre S, Quirion R (1992) Neuropeptide Y and neuropeptide Y receptor subtypes in brain and peripheral tissues. Prog Neurobiol 38 (2):125–167
811. Memminger M, Keller M, Lopuch M, Pop N, Bernhardt G, von Angerer E, Buschauer A (2012) The neuropeptide Y Y_1 receptor: a diagnostic marker? Expression in MCF-7 breast cancer cells is down-regulated by antiestrogens in vitro and in xenografts. PLoS ONE 7 (12):e51032
812. Amlal H, Faroqui S, Balasubramaniam A, Sheriff S (2006) Estrogen up-regulates neuropeptide Y Y1 receptor expression in a human breast cancer cell line. Cancer Res 66 (7):3706–3714
813. Dimitrijević M, Stanojević S (2012) Neuropeptide Y: the story, the players, the outcomes. In: Levite M (ed) Nerve-Driven Immunity. Springer, Vienna
814. Nguyen AD, Herzog H, Sainsbury A (2011) Neuropeptide Y and peptide YY: important regulators of energy metabolism. Curr Opin Endocrinol Diabetes Obes 18(1):56–60
815. Chao P-T, Yang L, Aja S, Moran TH, Bi S (2011) Knockdown of NPY expression in the dorsomedial hypothalamus promotes development of brown adipocytes and prevents diet-induced obesity. Cell Metab 13(5):573–583
816. Pedrazzini T, Pralong F, Grouzmann E (2003) Neuropeptide Y: the universal soldier. Cell Mol Life Sci 60(2):350–377
817. Grouzmann E, Werffeli-George P, Fathi M, Burnier M, Waeber B, Waeber G (1994) Angiotensin-II mediates norepinephrine and neuropeptide-Y secretion in a human pheochromocytoma. J Clin Endocrinol Metab 79(6):1852–1856
818. Pettersson-Fernholm K, Karvonen MK, Kallio J, Forsblom CM, Koulu M, Pesonen U et al (2004) Leucine 7 to proline 7 polymorphism in the preproneuropeptide Y is associated with proteinuria, coronary heart disease, and glycemic control in type 1 diabetic patients. Diabetes Care 27(2):503–509

819. van Rossum CT, Pijl H, Adan RA, Hoebee B, Seidell JC (2006) Polymorphisms in the NPY and AGRP genes and body fatness in Dutch adults. Int J Obes (Lond) 30(10):1522–1528
820. Enman NM, Sabban EL, McGonigle P, Van Bockstaele EJ (2015) Targeting the neuropeptide Y system in stress-related psychiatric disorders. Neurobiol Stress 1:33–43
821. Tatemoto K (1982) Isolation and characterization of peptide YY (PYY), a candidate gut hormone that inhibits pancreatic exocrine secretion. Proc Natl Acad Sci USA 79:2514–2518
822. Tatemoto K, Mutt V (1980) Isolation of two novel candidate hormones using a chemical method for finding naturally occurring polypeptides. Nature 285:417–418
823. Blevins JE, Chelikani PK, Haver AC, Reidelberger RD (2008) PYY (3–36) induces Fos in the arcuate nucleus and in both catecholaminergic and non-catecholaminergic neurons in the nucleus tractus solitarius of rats. Peptides 29(1):112–119
824. Murphy KG, Bloom SR (2006) Gut hormones and the regulation of energy homeostasis. Nature 444(7121):854–859
825. McFadden DW, Rudnicki M, Nussbaum MS, Balasubramaniam A, Fischer JE (1989) Independent release of peptide YY (PYY) into the circulation and ileal lumen of the awake dog. J Surg Res 46(4):380–385
826. Adrian TE, Ferri G-L, Bacarese-Hamilton AJ, Füeßl HS, Polak JM, Bloom SR (1985) Human distribution and release of a putative new gut hormone, peptide YY. Gastroenterology 89:1070–1077
827. El-Salhy M, Grimelius L, Emson PC, Falkmer S (1987) Polypeptide YY- and neuropeptide Y-immunoreactive cells and nerves in the endocrine and exocrine pancreas of some vertebrates: an onto-and phylogenetic study. Histochem J 19:111–117
828. Böttcher G, Sjöberg J, Ekman R, Hâkanson R, Sundler F (1993) Peptide YY in the mammalian pancreas: immunocytochemical localization and immunochemical characterization. Regul Pept 43:115–130
829. Ali-Rachedi A, Varndell IM, Adrian TE, Gapp DA, Van Noorden S, Bloom SR, Polak JM (1984) Peptide YY (PYY) immunoreactivity is co-stored with glucagon-related immunoreactants in endocrine cells of the gut and pancreas. Histochemistry 80:487–491
830. Gomez G, Udupi V, Greeley GH Jr (1998) Peptide YY. In: Greeley GH (ed) Gastrointestinal endocrinology. Humana Press, Totowa
831. Eberlein GA, Eysselein VE, Schaeffer M, Layer P, Grandt D, Goebell H et al (1989) A new molecular form of PYY: structural characterization of human PYY (3–36) and PYY (1–36). Peptides 10:797–803
832. Ballantyne GH (2006) Peptide YY(1–36) and peptide YY(3–36): Part I. Distribution, release and actions. Obes Surg 16(5):651–658
833. Füeßl HS, Adrian TE, Uttenthal LO, Bloom SR (1988) Peptide YY in diabetics treated chronically with an intestinal glucosidase inhibitor. Klin Wochenschr 66(19):985–989
834. Adrian TE, Sagor GR, Savage AP, Allen JM, Bacarese-Hamilton AJ, Tatemoto K et al (1983) Low dose PYY inhibits gastric secretion in man. Gut 24:479A
835. Hanusch-Enserer U, Roden M (2005) News in gut-brain communication: a role of peptide YY (PYY) in human obesity and following bariatric surgery? Eur J Clin Invest 35(7):425–430
836. Watterson KR, Bestow D, Gallagher J, Hamilton DL, Ashford FB, Meakin PJ, Ashford ML (2013) Anorexigenic and orexigenic hormone modulation of mammalian target of rapamycin complex 1 activity and the regulation of hypothalamic agouti-related protein mRNA expression. Neurosignals 21(1–2):28–41
837. Gil-Campos M, Aguilera CM, Cañete R, Gil A (2006) Ghrelin: a hormone regulating food intake and energy homeostasis. Br J Nutr 96(2):201–226
838. Druce M, Bloom SR (2006) The regulation of appetite. Arch Dis Child 91(2):183–187
839. Mimee A, Kuksis M, Ferguson AV (2014) α-MSH exerts direct postsynaptic excitatory effects on NTS neurons and enhances GABAergic signaling in the NTS. Neuroscience 262:70–82

840. Sohn J-W, Elmquist JK, Williams KW (2013) Neuronal circuits that regulate feeding behavior and metabolism. Trends Neurosci 36(9):504–512
841. Cowley MA, Pronchuk N, Fan W, Dinulescu DM, Colmers WF, Cone RD (1999) Integration of NPY, AGRP, and melanocortin signals in the hypothalamic paraventricular nucleus: evidence of a cellular basis for the adipostat. Neuron 24(1):155–163
842. Valassi E, Scacchi M, Cavagnini F (2008) Neuroendocrine control of food intake. Nutr Metab Cardiovasc Dis 18:158–168
843. Millington GWM (2007) The role of proopiomelanocortin (POMC) neurones in feeding behaviour. Nutr Metab 4:18
844. Riediger T, Giannini P, Erguven E, Lutz T (2006) Nitric oxide directly inhibits ghrelin-activated neurons of the arcuate nucleus. Brain Res 1125(1):37–45
845. Riediger T (2012) The receptive function of hypothalamic and brainstem centres to hormonal and nutrient signals affecting energy balance. Proc Nutr Soc 71(4):463–477
846. Riediger T, Bothe C, Becskei C, Lutz TA (2004) Peptide YY directly inhibits ghrelin-activated neurons of the arcuate nucleus and reverses fasting-induced c-Fos expression. Neuroendocrinology 79(6):317–326
847. Folmes CD, Lopaschuk GD (2007) Role of malonyl-CoA in heart disease and the hypothalamic control of obesity. Cardiovasc Res 73(2):278–287
848. Cui H, Cai F, Belsham DD (2005) Anorexigenic hormones leptin, insulin, and alpha-melanocyte-stimulating hormone directly induce neurotensin (NT) gene expression in novel NT-expressing cell models. J Neurosci 25(41):9497–9506
849. Wellman PJ (2000) Norepinephrine and the control of food intake. Nutrition 16(10):837–842
850. Acuna-Goycolea C, van den Pol AN (2005) Peptide YY_{3-36} inhibits both anorexigenic proopiomelanocortin and orexigenic neuropeptide Y neurons: implications for hypothalamic regulation of energy homeostasis. J Neurosci 25(45):10510–10519
851. Ghamari-Langroudi M, Colmers WF, Cone RD (2005) PYY3–36 inhibits the action potential firing activity of POMC neurons of arcuate nucleus through postsynaptic Y2 receptors. Cell Metab 2(3):191–199
852. Yin J, Babaoglu K, Brautigam CA, Clark L, Shao Z, Scheuermann TH et al (2016) Structure and ligand-binding mechanism of the human OX_1 and OX_2 orexin receptors. Nat Struct Mol Biol 23:293–299
853. Thompson MD, Xhaard H, Sakurai T, Rainero I, Kukkonen JP (2014) OX_1 and OX_2 orexin/hypocretin receptor pharmacogenetics. Front Neurosci 8:57
854. Lu XY, Bagnol D, Burke S, Akil H, Watson SJ (2000) Differential distribution and regulation of OX_1 and OX_2 orexin/hypocretin receptor messenger RNA in the brain upon fasting. Horm Behav 37(4):335–344
855. Said SI, Mutt V (1970) Potent peripheral and splanchnic vasodilator peptide from normal gut. Nature 225:863–864
856. Fahrenkrug J (1980) Vasoactive intestinal polypeptide. Trends Neurosci 3(1):1–2
857. Said SI, Mutt V (1969) Long acting vasodilator peptide from lung tissue. Nature (Lond) 224:699–700
858. Said SI, Mutt V (1972) Isolation from porcine intestinal wall of a vasoactive octacosapeptide related to secretin and to glucagon. Eur J Biochem (Tokyo) 28:199–204
859. Dimaline R (1989) Vasoactive intestinal peptide. In: Holmgren S (ed) The comparative physiology of regulatory peptides. Chapman and Hall, London
860. Jünemann KP, Lue TF, Melchior H, Tanagho EA (1987) VIP—A peripheral neurotransmitter in penile erection. Invest Urol 2:163–166
861. Sawmiller DR, Henning RJ (2006) Vasoactive intestinal peptide. In: Kastin AJ (ed) Handbook of biologically active peptides. Academic Press, San Diego
862. Larsen JJ, Ottesen B, Fahrenkrug J, Fahrenkrug L (1981) VIP in the male genitourinary tract. Concentration and motor effect. Invest Urol 19(3):211–213

863. Strauss GI, Edvinsson L, Larsen FS, Møller K, Knudsen GM (2001) Circulating levels of neuropeptides (CGRP, VIP, NPY) in patients with fulminant hepatic failure. Neuropeptides 35(3–4):174–180
864. Robberecht P, Gourlet P, Cauvin A, Buscail L, De Neef P, Arimura A, Christophe J (1991) PACAP and VIP receptors in rat liver membranes. Am J Physiol Gastrointest Liver Physiol 260(1):G97–G102
865. Altiere RJ, Diamond L (1984) Relaxation of cat tracheobronchial and pulmonary arterial smooth muscle by vasoactive intestinal peptide: lack of influence by peptidase inhibitors. Br J Pharmacol 82(2):321–328
866. Arimura A (1992) Receptors for pituitary adenylate cyclase-activating polypeptide comparison with vasoactive intestinal peptide receptors. Trends Endocrinol Metab 3 (8):288–294
867. Widdicombe JG (1989) Nervous receptors in the tracheobronchial tree: airway smooth muscle reflexes. In: Coburn RF (ed) Airway smooth muscle in health and disease. Plenum Press, New York
868. Rodríguez-Henche N, Rodríguez-Pena MS, Guijarro LG, Prieto JC (1994) Characterization of vasoactive intestinal peptide receptors in human liver. BBA Mol Cell Res 1221(2):193–198
869. Delgado M, Ganea D (2012) Vasoactive intestinal peptide: immune mediator and potential therapeutic agent. In: Levite M (ed) Nerve-driven immunity. Springer, Vienna
870. Laburthe M, Couvineau A, Marie JC (2002) VPAC receptors for VIP and PACAP. Receptors Channels 8(3–4):137–153
871. Laburthe M, Couvineau A (2002) Molecular pharmacology and structure of VPAC Receptors for VIP and PACAP. Regul Pept 108(2–3):165–173
872. Miyata A (2003) Function of PACAP in the cardiovascular system. In: Vaudry H (ed) Pituitary adenylate cyclase-activating polypeptide. Kluwer Academic Publishers, Norwell
873. Wei M-X, Naruse S, Ozaki T, Hu P, Wray V, Nokihara K (2009) Differences in action of PACAP-27 and PACAP-38 on guinea pig gallbladder smooth muscle using synthetic C-terminally modified PACAP peptides. Int J Pept Res Ther 15(3):227–232
874. Ottesen B, Ulrichsen H, Wagner G, Fahrenkrug J (1979) Vasoactive intestinal polypeptide (VIP) inhibits oxytocin induced activity of the rabbit myometrium. Acta Physiol Scand 107 (3):285–287
875. Said SI, Mutt V (1970) Polypeptide with broad biological activity: isolation from small intestine. Science 169(3951):1217–1218
876. Reubi JC, Laderach U, Waser B, Gebbers JO, Robberecht P, Laissue JA (2000) Vasoactive intestinal peptide/pituitary adenylate cyclase-activating peptide receptor subtypes in human tumors and their tissues of origin. Cancer Res 60(11):3105–3112
877. Walker CS, Sundrum T, Hay DL (2014) PACAP receptor pharmacology and agonistbias: analysis in primaryneurons and glia from the trigeminal ganglia and transfected cells. Br J Pharmacol 171:1521–1533
878. Kodali S, Ding W, Huang J, Seiffert K, Wagner JA, Granstein RD (2004) Vasoactive intestinal peptide modulates langerhans cell immune function. J Immunol 173(10):6082–6088
879. Dvoráková MC (2005) Cardioprotective role of the VIP signaling system. Timely Top Med Cardiovasc Dis 9:E33
880. Tatemoto K, Mutt V (1981) Isolation and characterization of the intestinal peptide porcine PHI (PHI-27), a new member of the glucagon-secretin family. Proc Natl Acad Sci U S A 78 (11):6603–6607
881. Sasaki A, Sato S, Shinkawa O, Go M, Otsuka T, Sugawara A et al (1988) Peptide histidine methionine may be a prolactin-releasing hormone in humans. J Clin Endocrinol Metab 66 (6):1202–1207
882. Tatemoto K, Carlquist M, Mutt V (1982) Neuropeptide Y—a novel brain peptide with structural similarities to peptide YY and pancreatic polypeptide. Nature 296:659–660

883. Buscail L, Cauvin A, Gourlet P, Gossen D, De Neef P, Rathé J et al (1038) Purification and amino acid sequence of vasoactive intestinal peptide, peptide histidine isoleucinamide (1–27) and secretin from the small intestine of guinea pig. BBA Protein Struct Mol Enzymol 3:355–359
884. Anagnostides AA, Manolas K, Christofides ND, Yiangou Y, Welbourn RB, Bloom SR, Chadwick VS (1983) Peptide histidine isoleucine (PHI). A secretagogue in porcine intestine. Digest Dis Sci 28(10):893–896
885. Watanabe J (2016) Peptide histidine isoleucine/methionine. In: Takei Y, Ando H, Tsutsui K (eds) Handbook of hormones: comparative endocrinology for basic and clinical research. Academic Press, San Diego
886. Robberecht P, Coy DH, De Neef P, Camus J-C, Cauvin A, Waelbroeck M, Christophe J (1987) [d-Phe^4]Peptide histidine-isoleucinamide ([d-Phe^4]PHI), a highly selective vasoactive-intestinal-peptide (VIP) agonist, discriminates VIP-preferring from secretin-preferring receptors in rat pancreatic membranes. Eur J Biochem 165(2):243–249
887. Tse DL-Y, Pang RTK, Wong AO-L, Chan S-M, Vaudry H, Chow BKC (2002) Identification of a potential receptor for both peptide histidine isoleucine and peptide histidine valine. Endocrinology 143(4):1327–1336
888. Watanobe H, Tamura T (1994) Stimulation by peptide histidine methionine (PHM) of adrenocorticotropin secretion in patients with Cushing's disease: a comparison with the effect of vasoactive intestinal peptide (VIP) and a study on the effect of combined administration of corticotropin-releasing hormone with PHM or VIP. J Clin Endocrinol Metab 78(6):1372–1377
889. Palle C, Bredkjaer HE, Ottesen B, Fahrenkrug J (1990) Peptide histidine methionine (PHM) increases vaginal blood flow in normal women. Peptides 11(3):401–404
890. Bennett V (2007) Peptide histidine methionine. In: Enna SJ, Bylund DB (eds) xPharm—the comprehensive pharmacology reference. Elsevier, New York
891. Priest WM, Alexander MK (1957) Islet cell tumour of the pancreas with peptic ulceration, diarrhoea, and hypokalaemia. Lancet 273(7006):1145–1147
892. Verner JV, Morrison AB (1958) Islet cell tumor and a syndrome of refractory watery diarrhea and hypokalemia. Am J Med 25(3):374–380
893. Matsumoto KK, Peter JB, Raymond G (1966) Watery diarrhea and hypokalemia associated with pancreatic islet cell adenoma. Gastroenterology 50:231–242
894. Abu-Zaid A, Azzam A, Abudan Z, Algouhi A, Almana H, Amin T (2014) Sporadic pancreatic vasoactive intestinal peptide-producing tumor (VIPoma) in a 47-year-old male. Hematol/Oncol Stem Cell Ther 7(3):109–115
895. Ikuta S, Yasui C, Kawanaka M, Aihara T, Yoshie H, Yanagi H et al (2007) Watery diarrhea, hypokalemia and achlorhydria syndrome due to an adrenal pheochromocytoma. World J Gastroenterol 13(34):4649–4652
896. Vinik A (2013) Vasoactive intestinal peptide tumor (VIPoma). In: Chrousos G, Dungan K, Feingold KR, Grossman A, Hershman JM (eds) De Groot LJ. MDText.com Inc, South Dartmouth
897. Nilubol N, Freedman EM, Quezado MM, Patel D, Kebebew E (2016) Pancreatic neuroendocrine tumor secreting vasoactive intestinal peptide and dopamine with pulmonary emboli: a case report. J Clin Endocrinol Metab 101(10):3564–3567
898. Yiangou Y, Di Marzo V, Spokes RA, Panico M, Morris HR, Bloom SR (1987) Isolation, characterization, and pharmacological actions of peptide histidine valine 42, a novel prepro-vasoactive intestinal peptide-derived peptide. J Biol Chem 262:14010–14013
899. Miyata A, Arimura A, Dahl RR, Minamino N, Uehara A, Jiang L et al (1989) Isolation of a novel 38 residue-hypothalamic polypeptide which stimulates adenylate cyclase in pituitary cells. Biochem Biophys Res Commun 164:567–574
900. Miyata A, Jiang L, Dahl RD, Kitada C, Kubo K, Fujino M et al (1990) Isolation of a neuropeptide corresponding to the N-terminal 27 residues of the pituitary adenylate cyclase activating polypeptide with 38 residues (PACAP38). Biochem Biophys Res Commun 170:643–648

901. Arimura A (1998) Perspectives on pituitary adenylate cyclase activating polypeptide (PACAP) in the neuroendocrine, endocrine, and nervous systems. Jpn J Physiol 48(5):301–331
902. Vaudry D, Gonzalez BJ, Basille M, Yon L, Fournier A, Vaudry H (2000) Pituitary adenylate cyclase-activating polypeptide and its receptors: from structure to functions. Pharmacol Rev 52(2):269–324
903. Culler MD, Kenjo T, Obara N, Arimura A (1984) Stimulation of pituitary cAMP accumulation by human pancreatic GH-releasing factor (1–44). Am J Physiol 247:E609–E615
904. Labrie F, Borgeat P, Drouin J, Beaulieu M, Lagace L, Ferland L, Raymond V (1979) Mechanism of action of hypothalamic hormones in the adenohypophysis. Annu Rev Physiol 41:555–569
905. Labrie F, Veilleux R, Lefevre G, Coy DH, Sueiras-Diaz J, Schally AV (1982) Corticotropin-releasing factor stimulates accumulation of adenosine 3′,5′-monophosphate in rat pituitary corticotrohs. Science 216:1007–1008
906. Basille M, Cartier D, Vaudry D, Lihrmann I, Fournier A, Freger P et al (2006) Localization and characterization of pituitary adenylate cyclase-activating polypeptide receptors in the human cerebellum during development. J Comp Neurol 496(4):468–478
907. Arimura A (1992) Pituitary adenylate cyclase activating polypeptide (PACAP): discovery and current status of research. Regul Pept 37(3):287–303
908. Reubi JC (2000) In vitro evaluation of VIP/PACAP receptors in healthy and diseased human tissues: clinical implications. Ann N Y Acad Sci 921(1):1–25
909. Arimura A (2003) PACAP: discovery, gene, receptors. In: Vaudry H, Arimura A (eds) Pituitary adenylate cyclase-activating polypeptide. Kluwer Academic Publishers, Norwell
910. Vaudry D, Falluel-Morel A, Bourgault S, Basille M, Burel D, Wurtz O et al (2009) Pituitary adenylate cyclase-activating polypeptide and its receptors: 20 years after the discovery. Pharmacol Rev 61(3):283–357
911. Bokaei PB, Ma X-Z, Byczynski B, Keller J, Sakac D, Fahim S, Branch DR (2006) Identification and characterization of five-transmembrane isoforms of human vasoactive intestinal peptide and pituitary adenylate cyclase-activating polypeptide receptors. Genomics 88(6):791–800
912. Freson K, Hashimoto H, Thys C, Wittevrongel C, Danloy S, Morita Y et al (2004) The pituitary adenylate cyclase–activating polypeptide is a physiological inhibitor of platelet activation. J Clin Invest 113(6):905–912
913. Scharf E, May V, Braas KM, Shutz KC, Mao-Draayer Y (2008) Pituitary adenylate cyclase-activating polypeptide (PACAP) and vasoactive intestinal peptide (VIP) regulate murine neural progenitor cell survival, proliferation, and differentiation. J Mol Neurosci 36 (1–3):79–88
914. Waschek JA (2013) VIP and PACAP: neuropeptide modulatorsof CNS inflammation, injury, and repair. Br J Pharmacol 169:512–523
915. Nielsen KM, Chaverra M, Hapner SJ, Nelson BR, Todd V, Zigmond RE, Lefcort F (2004) PACAP promotes sensory neuron differentiation: blockade by neurotrophic factors. Mol Cell Neurosci 25(4):629–641
916. Cox HM (1992) Pituitary adenylate cyclase activating polypeptides, PACAP-27 and PACAP-38: stimulators of electrogenic ion secretion in the rat small intestine. Br J Pharmacol 106(2):498–502
917. Christofi FL, Wood JD (1993) Effects of PACAP on morphologically identified myenteric neurons in guinea pig small bowel. Am J Physiol Gastrointest Liver Physiol 264(3):G414–G421
918. Shen S, Gehlert DR, Collier DA (2013) PACAP and PAC1 receptor in brain development and behavior. Neuropeptides 47(6):421–430

919. Figiel M, Engele J (2000) Pituitary adenylate cyclase-activating polypeptide (PACAP), a neuron-derived peptide regulating glial glutamate transport and metabolism. J Neurosci 20 (10):3596–3605
920. Marzagalli R, Scuderi S, Drago F, Waschek JA, Castorina A (2015) Emerging role of PACAP as a new potential therapeutic target in major diabetes complications. Int J Endocrinol 2015:160928
921. Stroth N, Holighaus Y, Ait-Ali D, Eiden LE (2011) PACAP: a master regulator of neuroendocrine stress circuits and the cellular stress response. Ann N Y Acad Sci 1220 (1):49–59
922. Hamelink C, Tjurmina O, Damadzic R, Young WS, Weihe E, Lee H-W, Eiden LE (2002) Pituitary adenylate cyclase activating polypeptide is a sympathoadrenal neurotransmitter involved in catecholamine regulation and glucohomeostasis. Proc Natl Acad Sci USA 99 (1):461–466
923. Martinez V, Taché Y (2006) Calcitonin gene-related peptide and gastrointestinal function. In: Kastin AJ (ed) Handbook of biologically active peptides. Academic Press, San Diego
924. Ma H (2004) Calcitonin gene-related peptide (CGRP). Nat Sci 2(3):41–47
925. Russo AF, Dickerson IM (2006) CGRP: a multifunctional neuropeptide. In: Lajtha A, Lim R (eds) Handbook of neurochemistry and molecular neurobiology. Springer, New York
926. Barwell J, Gingell JJ, Watkins HA, Archbold JK, Poyner DR, Hay DL (2012) Calcitonin and calcitonin receptor-like receptors: common themes with family B GPCRs? Br J Pharmacol 166(1):51–65
927. Poyner D, Sexton PM, Marshall I, Smith DM, Quirion R, Born W et al (2002) The mammalian calcitonin gene-related peptides, adrenomedullin, amylin, and calcitonin receptors. Pharmacol Rev 54(2):233–246
928. Wimalawansa SJ (1997) Amylin, calcitonin gene-related peptide, calcitonin, and adrenomedullin: a peptide superfamily. Crit Rev Neurobiol 11(2–3):167–239
929. Kimura S, Sugita Y, Kanazawa I, Saito A, Golo K (1987) Isolation and amino acid sequence of calcitonin gene related peptide from porcine spinal cord. Neuropeptides 9 (1):75–82
930. Amara SG, Jonas V, Rosenfeld MG, Ong ES, Evans RM (1982) Alternative RNA processing in calcitonin gene expression generates mRNAs encoding different polypeptide products. Nature 298:240–244
931. Rosenfeld MG, Mermod JJ, Amaraa SG, Swanson LW, Swchenko PE, Rivier J et al (1983) Production of a novel neuropeptide encode by the calcitonin gene via tissue-specific RNA processing. Nature 304:129–135
932. Brain SD (2004) Calcitonin gene-related peptide (CGRP) antagonists: blockers of neuronal transmission in migraine. Br J Pharmacol 142(7):1053–1054
933. Tepper SJ, Stillman MJ (2008) Clinical and preclinical rationale for CGRP-receptor antagonists in the treatment of migraine. Headache 48(8):1259–1268
934. Chen LJ, Zhang FG, Li J, Song HX, Zhou LB, Yao BC et al (2010) Expression of calcitonin gene-related peptide in anterior and posterior horns of the spinal cord after brachial plexus injury. J Clin Neurosci 17(1):87–91
935. Tfelt-Hansen P, Olesen J (2011) Possible site of action of CGRP antagonists in migraine. Cephalalgia 31(6):748–750
936. de Prado BM, Russo AF (2006) CGRP receptor antagonists: a new frontier of anti-migraine medications. Drug Discov Today Ther Strateg 3(4):593–597
937. Doods H, Arndt K, Rudolf K, Just S (2007) CGRP antagonists: unravelling the role of CGRP in migraine. Trends Pharmacol Sci 28(11):580–587
938. Garthwaite J (1991) Glutamate, nitric oxide and cell-cell signalling in the nervous system. Trends Neurosci 14(2):60–67
939. SoRelle R (1998) Nobel Prize awarded to scientists for nitric oxide discoveries. Circulation 98:2365–2366

940. Toda N, Okamura T (2011) Modulation of renal blood flow and vascular tone by neuronal nitric oxide synthase-derived nitric oxide. J Vasc Res 48:1–10
941. Cellek S, Moncada S (1997) Nitrergic modulation of cholinergic responses in the opossum lower oesophageal sphincter. Br J Pharmacol 122:1043–1046
942. Cellek S, Moncada S (1997) Nitrergic control of peripheral sympathetic responses in the human corpus cavernosum: A comparison with other species. PNAS USA 94:8226–8231
943. Rajendran P, Rengarajan T, Thangavel J, Nishigaki Y, Sakthisekaran D, Sethi G, Nishigaki I (2013) The vascular endothelium and human diseases. Int J Biol Sci 9 (10):1057–1069
944. Avontuur JA, Biewenga M, Buijk SL, Kanhai KJ, Bruining HA (1998) Pulmonary hypertension and reduced cardiac output during inhibition of nitric oxide synthesis in human septic shock. Shock 9(6):451–454
945. Avontuur JAM, Boomsma F, van den Meiracker AH, de Jong FH, Bruining HA (1999) Endothelin-1 and blood pressure after inhibition of nitric oxide synthesis in human septic shock. Circulation 99:271–275
946. Shibuki K, Okada D (1991) Endogenous nitric oxide release required for long-term synaptic depression in the cerebellum. Nature 349:326–328
947. Kuriyama K, Ohkuma S (1995) Role of nitric oxide in central synaptic transmission: effects on neurotransmitter release. Jpn J Pharmacol 69(1):1–8
948. Snyder SH (1992) Nitric oxide and neurons. Curr Opin Neurobiol 2(3):323–327
949. Furchgott RF, Zawadzki JV (1980) The obligatory role of endothelial cells in the relaxation of arterial smooth muscle by acetylcholine. Nature 288:373–376
950. Furchgott RF, Cherry PD, Zawadzki JV, Jothianandan D (1984) Endothelial cells as mediators of vasodilation of arteries. J Cardiovasc Pharmacol 6:S336–S343
951. Murad F, Mittal CK, Arnold WP, Katsuki S, Kimura H (1978) Guanylate cyclase activation by azide, nitro compounds, nitric oxide, and hydroxyl radical and inhibition by hemoglobin and myoglobin. Adv Cyclic Nucleotide Res 9:145–158
952. Kantha SS (1997) Could nitroglycerine poisoning be the cause of Alfred Nobel's anginal pains and premature death? Med Hypotheses 49(4):303–306
953. Ermert M, Ruppert C, Günther A, Duncker HR, Seeger W, Ermert L (2002) Cell-specific nitric oxide synthase-isoenzyme expression and regulation in response to endotoxin in intact rat lungs. Lab Invest 82(4):425–441
954. Dijkstra G, van Goor H, Jansen PL, Moshage H (2004) Targeting nitric oxide in the gastrointestinal tract. Curr Opin Investig Drugs 5(5):529–536
955. Schuman EM, Madison DV (1994) Nitric oxide and synaptic function. Annu Rev Neurosci 17:153–183
956. Martin E, Berka V, Tsai AL, Murad F (2005) Soluble guanylyl cyclase: the nitric oxide receptor. Methods Enzymol 396:478–492
957. Salzman AL (1995) Nitric oxide in the gut. New Horiz 3(2):352–364
958. Azuma H, Ishikawa M, Sekizaki S (1986) Endothelium-dependent inhibition of platelet aggregation. Br J Pharmacol 88:411–415
959. Kahn MT, Furchgott RF (1987) Additional evidence that endothelium-derived relaxing factor is nitric oxide. In: Rand MJ, Raper C (eds) Pharmacology. Elsevier, Amsterdam
960. Ignarro LJ, Buga GM, Wood KS, Byrns RE, Chaudhuri G (1987) Endothelium-derived relaxing factor produced and released from artery and vein is nitric oxide. PNAS USA 84:9265–9269
961. Palmer RMJ, Ferrige AG, Moncada S (1987) Nitric oxide release accounts for the biological activity of endothelium-derived relaxing factor. Nature 327:524–526
962. Ignarro LJ, Burke TM, Wood KS, Wolin MS, Kadowitz PJ (1984) Association between cyclic GMP accumulation and ACh-elicited relaxation of bovine intrapulmonary artery. J Pharmacol Exp Ther 228:682–690
963. Hibbs JB Jr, Taintor RR, Vavrin Z (1987) Macrophage cytotoxicity: role for L-arginine deiminase and imino nitrogen oxidation to nitrite. Science 235:473–476

964. Palmer RMJ, Ashton DS, Moncada S (1988) Vascular endothelial cells synthesize nitric oxide from L-arginine. Nature 333:664–666
965. Shi ZQ, Sunico CR, McKercher SR, Cui J, Feng GS, Nakamura T, Lipton SA (2013) S-nitrosylated SHP-2 contributes to NMDA receptor-mediated excitotoxicity in acute ischemic stroke. PNAS U S A 110(8):3137–3142
966. Hanafy KA, Oh J, Otterbein LE (2013) Carbon monoxide and the brain: time to rethink the dogma. Curr Pharm Des 19(15):2771–2775
967. Raub JA, Benignus VA (2002) Carbon monoxide and the nervous system. Neurosci Biobehav Rev 26(8):925–940
968. Verma A, Hirsch DJ, Glatt CE, Ronnett GV, Snyder SH (1993) Carbon monoxide: a putative neural messenger. Science 259(5093):381–384
969. Prabhakar NR (1999) NO and CO as second messengers in oxygen sensing in the carotid body. Respir Physiol 115(2):161–168
970. Snyder SH, Jaffrey SR, Zakhary R (1998) Nitric oxide and carbon monoxide: parallel roles as neural messengers. Brain Res Brain Res Rev 26(2–3):167–175
971. Johnson RA, Johnson FK (2000) The effects of carbon monoxide as a neurotransmitter. Curr Opin Neurol 13(6):709–713
972. Herman ZS (1997) Carbon monoxide: a novel neural messenger or putative neurotransmitter? Pol J Pharmacol 49(1):1–4
973. Brann DW, Bhat GK, Lamar CA, Mahesh VB (1997) Gaseous transmitters and neuroendocrine regulation. Neuroendocrinology 65(6):385–395
974. Wilkinson WJ, Kemp PJ (2011) Carbon monoxide: an emerging regulator of ion channels. J Physiol 589(Pt 13):3055–3062
975. Jaggar JH, Li A, Parfenova H, Liu J, Umstot ES, Dopico AM, Leffler CW (2005) Heme is a carbon monoxide receptor for large-conductance Ca^{2+}-activated K^{+} channels. Circ Res 97 (8):805–812
976. Di Marzo V (2009) The endocannabinoid system: its general strategy of action, tools for its pharmacological manipulation and potential therapeutic exploitation. Pharmacol Res 60 (2):77–84
977. De Petrocellis L, Di Marzo V (2009) An introduction to the endocannabinoid system: from the early to the latest concepts. Best Pract Res Clin Endocrinol Metab 23(1):1–15
978. Holzer P (2009) Opioid receptors in the gastrointestinal tract. Regul Pept 155(1–3):11–17
979. Storr MA, Sharkey KA (2007) The endocannabinoid system and gut-brain signalling. Curr Opin Pharmacol 7(6):575–582
980. Mechoulam R (2007) Conversation with Raphael Mechoulam. Addiction 102(6):887–893
981. Mechoulam R (1986) Interview with Prof. Raphael Mechoulam, codiscoverer of THC. Interview by Stanley Einstein. Int J Addict 21(4–5):579–587
982. Battista N, Di Tommaso M, Bari M, Maccarrone M (2012) The endocannabinoid system: an overview. Front Behav Neurosci 6:9
983. Christie McD J (2006) Opioid and cannabinoid receptors: friends with benefits or just close friends? Br J Pharmacol 148(4):385–386
984. Corbett AD, Henderson G, McKnight AT, Paterson SJ (2006) 75 years of opioid research: the exciting but vain quest for the Holy Grail. Br J Pharmacol 147(1):S153–S162
985. Stein C, Schäfer M, Machelska H (2003) Attacking pain at its source: new perspectives on opioids. Nat Med 9(8):1003–1008
986. Gaoni Y, Mechoulam R (1964) Isolation, structure and partial synthesis of an active constituent of Hashish. J Am Chem Soc 86:1646–1647
987. Di Iorio G, Lupi M, Sarchione F, Matarazzo I, Santacroce R, Petruccelli F et al (2013) The endocannabinoid system: a putative role in neurodegenerative diseases. Int J High Risk Behav Addict 2(3):100–106
988. Pacher P, Bátkai S, Kunos G (2006) The endocannabinoid system as an emerging target of pharmacotherapy. Pharmacol Rev 58(3):389–462
989. Vigna SR (2003) Cannabinoids and the gut. Gastroenterology 125(3):973–975

990. Izzo AA, Sharkey KA (2010) Cannabinoids and the gut: new developments and emerging concepts. Pharmacol Ther 126(1):21–38
991. Sugiura T, Kondo S, Sukagawa A, Nakane S, Shinoda A, Itoh K et al (1995) 2-Arachidonoylglycerol: a possible endogenous cannabinoid receptor ligand in brain. Biochem Biophys Res Commun 215:89–97
992. Devane WA, Dysarz FA 3rd, Johnson MR, Melvin LS, Howlett AC (1988) Determination and characterization of a cannabinoid receptor in rat brain. Mol Pharmacol 34:605–613
993. Devane WA, Hanus L, Breuer A, Pertwee RG, Stevenson LA, Griffin G et al (1992) Isolation and structure of a brain constituent that binds to the cannabinoid receptor. Science 258:1946–1949
994. De Petrocellis L, Cascio MG, Di Marzo V (2004) The endocannabinoid system: a general view and latest additions. Br J Pharmacol 141:765–774
995. Matsuda LA, Lolait SJ, Brownstein MJ, Young AC, Bonner TI (1990) Structure of a cannabinoid receptor and functional expression of the cloned cDNA. Nature 346:561–564
996. Munro S, Thomas KL, Abu-Shaar M (1993) Molecular characterization of a peripheral receptor for cannabinoids. Nature 365:61–65
997. Bisogno T, Ligresti A, Di Marzo V (2005) The endocannabinoid signalling system: biochemical aspects. Pharmacol Biochem Behav 81(2):224–238
998. Tallett AJ, Blundell JE, Rodgers RJ (2008) Endogenous opioids and cannabinoids: system interactions in the regulation of appetite, grooming and scratching. Physiol Behav 94 (3):422–431
999. Desroches J, Beaulieu P (2010) Opioids and cannabinoids interactions: involvement in pain management. Curr Drug Targets 11(4):462–473
1000. Linnoila RI, DiAugustine RP, Miller RJ, Chang KJ, Cuatrecasas P (1978) An immunohistochemical and radioimmunological study of the distribution of [Met^5]- and [Leu^5]-enkephalin in the gastrointestinal tract. Neuroscience 3(12):1187–1196
1001. Herz A (ed) (1993) Opioids I. Springer, Heidelberg
1002. Grossman A, Clement-Jones V (1983) Opiate receptors: enkephalins and endorphins. Clin Endocrinol Metab 12(1):31–56
1003. Murphy RB, Pincus MR, Beinfeld M, Dykes DC, Chen JM, Schneider LH et al (1992) Enkephalin is a competitive antagonist of cholecystokinin in the gastrointestinal tract, as predicted from prior conformational analysis. J Protein Chem 11(6):723–729
1004. Dhawan BN, Cesselin F, Raghubir R, Reisine T, Bradley PB, Portoghese PS, Hamon M (1996) International union of pharmacology. XII. Classification of opioid receptors. Pharmacol Rev 48(4):567–592
1005. Janecka A, Fichna J, Janecki T (2004) Opioid receptors and their ligands. Curr Top Med Chem 4(1):1–17
1006. Waldhoer M, Bartlett SE, Whistler JL (2004) Opioid receptors. Annu Rev Biochem 73:953–990
1007. Porreca F, Filla A, Burks TF (1983) Studies in vivo with dynorphin-(1–9): Analgesia but not gastrointestinal effects following intrathecal administration to mice. Eur J Pharmacol 91 (2–3):291–294
1008. Pert CB, Snyder SH (1973) Opiate receptor: demonstration in nervous tissue. Science 179 (4077):1011–1014
1009. Koneru A, Satyanarayana S, Rizwan S (2009) Endogenous opioids: their physiological role and receptors. Global J Pharmacol 3(3):149–153
1010. Jhamandas KH (1984) Opioid-neurotransmitter interactions: significance in analgesia, tolerance and dependence. Prog Neuro-Psychopharmacol Biol Psychiatry 8(4–6):565–570
1011. Froehlich JC (1997) Opioid peptides. Alcohol Health Res World 21(2):132–136
1012. Bénard A, Cavaillès P, Boué J, Chapey E, Bayry J, Blanpied C et al (2010) μ-Opioid receptor is induced by IL-13 within lymph nodes from patients with Sézary syndrome. J Invest Dermatol 130(5):1337–1344

Chapter 10
Immunomodulatory Functions of the Gastrointestinal Tract

Abstract The gastrointestinal (GI) tract executes a range of functions. GI response to invaders constituting the immunomodulatory function of the gut is crucial to life. This function is mainly carried out by the gut-associated lymphoid tissue (GALT), the largest component of the mucosa-associated lymphoid tissue. This function of the gut is of immense importance not only to the GI tract, but also to the entire organism. The GI tract lymphoid tissues are influenced by several environmental signals which may predispose the organism to the development of diseases including cancers, lymphoma, infections, chronic inflammation, and autoimmune diseases. This chapter is concerned with structure–function relationship of the lymphoid tissues of the GI tract and associated pathological conditions. The molecular mechanisms of functioning of GALT in normal and diseases are also discussed.

Keywords Adalimumab · Adamantiades-Behçet's disease · Adaptive immunity Adenoids · Antibody class isotype switching · Antigen · Antikörper Antimicrobial peptides · Antisomatogen + Immunkörperbildner Aphthae · Aphthous stomatitis · Aschoff's lympho-reticuloendothelial system Atypical antigen-presenting cells · Basophils · Behçet's disease Bronchus-associated lymphoid tissue · Canker sores or aphthae Celiac disease · Cell-mediated immunity · Central lymphoid organs Chemokine receptors · Chemokines · Chemotaxins · Chemotaxis Classical pathway · Clinical trial · Collectins · Complement activation Complement components · C-reactive protein · Crohn's disease Custocyte system · Custocytes · Cytokines · Damage-associated molecular patterns · Defensins · Dendritic cell · Duct-associated lymphoid tissue Eosinophils · Epithelial restitution · Fc region · Fibroblasts · Ficolins Fucose · Gastrokines · Glycoproteins · Granulocytes · Granzymes Gut immune system · Gut-associated lymphoid tissue · Helicobacter-induced gastritis · Hematopoietic stem cell · Histiocytes · Histocompatibility tests Homing · Human leukocyte antigen · Immunity · Immunogenic substances Immunoglobulins · Immunostimulants · Infection · Inflammation Inflammatory bowel disease · Infliximab · Innate immunity · Integrins

M. O. Welcome, *Gastrointestinal Physiology*,
https://doi.org/10.1007/978-3-319-91056-7_10

Interferon · Interleukins · Intestinal crypt · Intestinal trefoil factor Intrapancreatic lymphoid tissue · Isolated lymphoid tissues · Kallikrein Killer immunoglobulin-like receptor · Killer inhibitory receptor Kupffer cell · Lamina propria · Larynx-associated lymphoid tissue Lectins · Leukocyte · Lipopolysaccharide · Liver immunity · Lymph node Lympho-histiocytic reticular · Lympho-histiocytic · Lymphoid aggregates Lymphoid tissue inducer · Lymphoid tissue initiator · Lymphoid tissue organizer Lymphokines lymphopoietin receptor · Lympho-reticulo-histiocytic Lymphotoxin receptor · Lysis · Macrophage · Mast cells · MHC Microsurveillance · Monoclonal antibodies · Monocytes · Mucin Mucosa-associated lymphoid tissue · Multiorgan failure · Multipotent stem cell Myofibroblasts · *N*-acetylglucosamine · *N*-acetylneuraminic acid Nasopharynx-associated lymphoid tissue · Natural killer cells · Neuraminic acid Neutrophils · Nonself antigens · Opsonin · Opsonization · Oral mucosal diseases Pancreas-associated lymphoid tissue · Pathogen-associated immunostimulants Pathogen-associated molecular patterns · Pattern recognition receptors Pentraxin · Peptidoglycan · Perforins · Peripancreatic lymphoid tissue Peyer's patch · Phagocytosis · Pharyngeal ring · Plasma cells · Primary lymphoid organs · Proinflammatory cytokines · Proteasome · Quality of life Receptor-mediated endocytosis · Resident microbes · Reticular cells Reticuloendothelial system · Reticulo-histiocytic · Salivary gland-associated lymphoid tissue · Secondary lymphoid organs · Secretory diarrhea Self-antigens · Sentinel system · Sepsis · Sialic acids · Sialidases Stellate cells · Surfactant proteins · Systemic inflammatory response syndrome Teichoic acids · Tertiary lymphoid tissue · Thymopoietic growth factors TLR signaling · Tolerance · Toll-like receptor · Toll-interacting protein Tonsil · Trefoil factors · Urinary tract infection · Waldeyer's tonsillar Zinc finger protein · αβ T cells · γδ T cells · Arne Wilhelm Kaurin Tiselius Baruj Benacerraf · Bruce A. Beutler · César Milstein · Charles Alderson Janeway Jr. · Christiane Nüsslein-Volhard · Edward Butts Lewis · Elie Metchnikoff Elvin Abraham Kabat · Emil von Behring · Eric Francis Wieschaus George Davis Snell · Georges Jean Franz Köhler · Gerald Maurice Edelman Gunnar Blix · Heinrich Wilhelm Gottfried von Waldeyer-Hartz Henry Bence Jones · Ruslan M. Medzhitov · Jean Dausset · Jules A. Hoffmann Karl Albert Ludwig Aschoff · László Detre · Niels Kai Jerne · Paul Ehrlich Peter Alfred Gorer · Ralph Marvin Steinman · Rodney Robert Porter Shibasaburo Kitasato · Sir Gregory Paul Winter · Zanvil Alexander Cohn

Abbreviations

Ab	Antibody
AIF-1	Allograft inflammatory factor 1
AP1	Activator protein 1
APC	Antigen-presenting cell
BALT	Bronchus-associated lymphoid tissue

CCR, CXC	Chemokine receptor
CREB	Cyclic AMP-responsive element-binding protein
CRP	C-reactive protein
CXCL12	C-X-C motif chemokine 12
DALT	Duct-associated lymphoid tissue
DAMPs	Damage-associated molecular patterns
DAP10	DNAX-activating protein of molecular mass 10 kD
DLL4	Delta-like 4
DNAM-1	DNAX accessory molecule 1
ELISA	Enzyme-linked immunosorbent assay
Fab	Fragment, antigen-binding
Fc	Fragment, crystallizable
GALT	Gut-associated lymphoid tissue
GI	Gastrointestinal
GM-CSF	Granulocyte-macrophage colony-stimulating factor
HLA	Human leukocyte antigen
IBD	Inflammatory bowel disease
ICAM-1	Intercellular adhesion molecule-1
IFN	Interferon
Ig	Immunoglobulin
IL	Interleukin
IPLT	Intrapancreatic lymphoid tissue
IRAKs	IL-1R-associated kinases
J chain	Joining chain
KARAP	Killer cell-activating receptor-associated protein
KIR	Killer inhibitory or immunoglobulin-like receptors
KLRG-1	Killer cell lectin-like receptor subfamily G member 1
LALT	Larynx-associated lymphoid tissue
LFA-1	Lymphocyte function-associated antigen 1
LPS	Lipopolysaccharide
Ltin	Lymphoid tissue initiator
Ltind	Lymphoid tissue inducer
Lto	Lymphoid tissue organizer
LTβR	Lymphotoxin (LT) receptor
LY6-G	Lymphocyte antigen 6, subtype G
M cell	Membrane or microfold cell
mAbs	Monoclonal antibodies
MAC	Membrane attack complex
MALT	Mucosa-associated lymphoid tissue
MAp	MBL-associated protein
MASP	MBL-associated serine protease
MBL	Mannan-binding lectin
MHC	Major histocompatibility complex
MIP	Macrophage inflammatory protein

MYD88	Myeloid differentiation primary response protein 88
NALT	Nasopharynx-associated lymphoid tissue
NANA	*N*-acetylneuraminic acid
NKG2D	Natural killer group 2, member D
NOD-2	Oligomerization domain-like receptor
PALT	Pancreas-associated lymphoid tissue
PAMPs	Pathogen-associated molecular patterns
Pax1	Paired box protein-1
PCR	Polymerase chain reaction
PPAR-γ	Peroxisome proliferator-activated receptor gamma
PPLT	Peripancreatic lymphoid tissue
PRRs	Pattern recognition receptors
PTX	Pentraxin
RLRs	RIG-1-like receptors
SALT	Salivary gland-associated lymphoid tissue
SAP	Serum amyloid P
SDF1	Stromal cell-derived factor 1
SIGIRR	Single Ig IL-1-related receptor
SP	Surfactant proteins
TAP	Transporter associated with antigen processing
TCR	T cell receptor
TFF	Intestinal trefoil factor
TFF1	Trefoil factors
Th cell	Helper T cell
TLR	Toll-like receptor
Tollip	Toll-interacting protein
TRAF	TNF receptor-associated factor
TRAM	TRIF-related adaptor molecule
TRIF	TIR-domain-containing adapter-inducing interferon
TYROBP	Tyrosine kinase-binding protein
UBX	Ubiquitin-associated domain-containing ubiquitin regulatory X
UTI	Urinary tract infection
VCAM-1	Vascular cell adhesion molecule

10.1 Introduction

Many exogenous materials continuously enter the gastrointestinal (GI) tract. Some of these materials are microorganisms, which may be pathogenic. To prevent infection or ensure uninterrupted activity caused by the pathogenic microorganisms, the host in the course of evolution has developed some levels of protection and measures against microbial aggression or invasion. The epithelium of the GI tract serves as an anatomical barrier that prevents invasion by pathogenic microbes. In addition, the GI tract harbors a host of immune cells that mediate several responses not only against foreign invaders, but also tumor cells. These protective measures constitute the GI tract immune system, often referred to as gut-associated lymphoid tissue (GALT). The GALT is made up of several types of lymphoid tissues that harbor immune cells such as the T cells, B lymphocytes, antigen-presenting cells, including dendritic cells, and macrophages [1–5]. Thus, the immune system of the gut ensures adequate functioning of the GI system by maintaining the structural integrity of the gut epithelium and the associated lymphoid tissues. Both the gut architectural integrity and GALT are affected by a range of diseases that affect the physiological functioning and hence the quality of life of the individual (see Clinical Correlates 10.1 and 10.2). This chapter is concerned with structural constituents and functions of the lymphoid tissues of the GI tract. The molecular mechanisms of functioning of GALT in normal and disease states are also discussed.

Clinical Correlate 10.1

Oral Mucosal Diseases: Behçet's Disease, Aphthae, and Associated Diseases.

Behçet's Disease (Also Known as Adamantiades-Behçet's Disease). Malfunction of gut-associated immune cells has been implicated in number of disorders of the GI and other systems. Such disorder in the functions of the gut occurs in Adamantiades-Behçet's disease. Behçet's disease is a multi-system disease with recurrent ulceration affecting genital mucosa, eye, skin, joints, GI tract (including oral mucosa), vascular, respiratory, and central nervous systems. The disease is characterized by the onset of oral aphthous ulceration (oral aphthosis) and gradually developing into systemic disorders [6]. The etiopathogenesis of Behçet's disease is currently not clear. However, it is believed that Behçet's disease is mediated via T cell immune response. There is evidence that pathogenic oral microbes may play chief role in the etiology of the disease and is mediated through interleukins of the cytokine family. Research data indicate that the disease is characterized by increased proportion of $CD3^{+}CD4^{-}CD8^{-}$ T cells in peripheral blood [7]. There is considerable success in the use of CO_2 laser for the management of Behçet's disease and recurrent aphthous stomatitis. Research has shown that the use of CO_2 ablative laser as a monotherapy relieves the pain experienced by patients suffering from the disease. In the majority of cases, however, anesthetic topical treatment, topical or systemic steroids, or antibiotics are used [8].

Aphthous Stomatitis: Aphthous stomatitis (also known as recurrent aphthous stomatitis, canker sores, or aphthae) is a chronic inflammatory, ulcerative condition of the oral mucosa, characterized by multiple recurrent small, round, or ovoid ulcers with circumscribed margins, erythematous haloes, and yellow or gray floors [9, 10]. It is the most common oral mucosal disease, with the non-keratinized oral mucosa being the prime site of lesions

[10]. It usually presents first in children or adolescents. The disorder can cause considerable pain and may interfere with eating, speaking, and swallowing [11].

Classification of Aphthous Stomatitis: Recurrent aphthous stomatitis is classified on the basis of the size and number of ulcers: minor, major forms, and herpetiform ulcers. The most common presentation is minor recurrent aphthous stomatitis: recurrent, round, clearly defined, small, painful ulcers that heal in 10–14 days without scarring. In major types, lesions are larger, usually greater than 5 mm, and can last for 6 weeks or longer and frequently appear as scar. The third type is herpetiform ulcer, which presents as multiple small clusters of pinpoint lesions that can coalesce to form large irregular ulcers and last about 7–10 days [12, 13].

Epidemiology of Aphthous Stomatitis: Recurrent aphthous stomatitis affects about 5–25% people depending on the population and group of individuals studied [14]. However, prevalence of the disease may be as high as 60% or as low as 0.5% [15]. It is estimated that at least one in five individuals has once had aphthous ulcers [11, 13]. The disease may reoccur every 3 months with a probability of 50% [11].

Etiopathogenesis of Aphthous Stomatitis: The cause of the disease is believed to be multifactorial. In the majority of cases, the cause of aphthae is idiopathic—the principal cause of the disease has not been identified. There is possibility for an autoimmune or hypersensitivity mechanism in the development of the disease [16]. Certain trigger factors including mechanical injury, stress, or microbial antigens may stimulate immune system response that results in the development of the disease. People with certain genetic composition are more likely to develop the disease. For instance, the disease has been observed to run more frequently in some families. Certain gene polymorphisms encoding proinflammatory cytokines and other molecules of immune response, which play a role in aphthae, may be predisposing factors. A genetic predisposition for the disease has a strong association with genotypes of IL-1beta and IL-6. Also, hematinic deficiency is found in up to 20% of patients [9, 10, 13, 14, 17, 18]. Hematinics are the minerals and vitamins essential for normal erythropoiesis, which include iron, cobalt, copper, vitamins A, B_{12}, B_6, C, E, folate, riboflavin, and nicotinic acid. Their deficiency can lead to many diseases including disorders characterized by defect in gut epithelium [19, 20]. Emerging data indicate that serum levels of immunoglobulins may play a role in the pathogenesis of the disease. More recently, research data point to the role of salivary immunoglobulins in the pathogenesis of the disease [21]. While Sistig et al. [21] did not find any differences in immunoglobulin subclasses between major and minor acute aphthae, in the remission period, IgG1 and IgG4 returned to normal values while IgG2, IgG3, and IgA2 remained increased. In HIV-positive patients, research indicates that salivary IgA2 was significantly reduced when compared to the healthy controls, but no differences in salivary IgA1 level between HIV patients and healthy controls were found [21]. Compared to the healthy individuals, HIV patients had increased salivary IgG1, IgG2, IgG3, and IgG4 subclasses [22]. It is believed that stress is a crucial precipitating factor for the condition. Systemic diseases, nutritional deficiencies, food allergies, immune disorders, use of certain medications may also predispose an individual to developing the disease [11].

Risk factors of Aphthous Stomatitis: Risk factors include local trauma, stress, food intake, drugs, hormonal changes, and vitamin and trace element deficiencies [13]. Emotional and environmental factors may contribute to the development of the disease [18].

Diagnosis of Aphthous Stomatitis: The diagnosis of the disease is based on history and clinical criteria. Evidences pointing to the use of laboratory procedures for aiding diagnosis are only emerging. It should be noted that the disease may be a marker of an underlying systemic illness such as celiac disease, or may present as one of the features of Behcet's disease. Even when there is only recurrent aphthous stomatitis and no additional body systems are affected, patients may appear fit and well. In addition to the description of the type of ulcers, important clinical features that are considered if a patient has this disease include pain, recurrence, and destruction of the oral epithelium [13, 17, 23].

In minor form of aphthae, levels of salivary IgA are significantly increased in acute phase (106.22–136.31 μg/mL) in comparison with the levels in healthy individuals (76.85–93.91 μg/mL) [24]. Mohammad et al. [24] did not find any difference in gender, disease phases, and salivary flow rate between the sufferers and healthy individuals. Other researchers have also found significant elevation of salivary IgA in patients with minor aphthae, compared to the control. But the serum IgG and IgM levels did not show any differences between individuals with either minor or major aphthae and the healthy control [25].

Low serum levels of IgG2 may play a role in the pathogenesis of aphthae, just as in other recurrent infectious diseases. However, serum IgG and total IgA levels may transiently change depending on the different periods of activity and quiescence of the disease [26].

Differential diagnoses include conditions capable of producing erosive and ulcerative oral mucosal lesions [17]. It should be noted that lesions in aphthae may be similar to other diseases. Similar ulcers may be found in HIV disease and other immune disorders, and negative reactions to the use of drugs such as non-steroidal anti-inflammatory drugs [9].

Management of Aphthous Stomatitis: There is presently no specific treatment for aphthae. The present treatment is symptomatic aimed at reducing pain, and ulcer, and also restores normal oral function. Nevertheless, certain medications can be used to reduce or eradicate the lesions, but recurrence may not be prevented. It is important to establish whether or not aphthae is a sign of systemic disease, so that the correct medications can be initiated. It is also necessary to determine possible causes contributing to the disease so that the factors can be corrected. Different therapy approaches can be used depending on the nature of the disease. Topical medications, such as antimicrobial chlorhexidine mouthwashes, tetracycline oral rinses, thalidomide, fluocinonide, colchicines, and topical corticosteroids (e.g., dexamethasone, triamcinolone, fluocinonide, hydrocortisone, clobetasol), can achieve the primary goal of reducing pain and to improve healing time. Systemic medications can be tried if topical therapy is ineffective [9, 16, 27]. However, these medications do not improve recurrence or remission rates. Evidences indicate that the antihelminthic drug, levamisole, is effective in providing long-term benefits for patients with aphthae [28]. However, it has been suggested that the use of levamisole should be reserved for severe cases of major aphthae that do not respond to topical agents [11]. The agent, thalidomide, is effective in major aphthae; unfortunately, its toxicity limits its usage in practice [29–31].

Clinical investigations have shown effectiveness of irsogladine (2–4 mg/day) administered orally to patients with Behçet's disease [32]. Although not completely clear, the mechanism of action of irsogladine might be associated with enhancement of gap junctional intercellular signaling [32].

In a placebo-controlled trial, Femiano et al. [33] compared the therapeutic effectiveness and adverse effects of systemic prednisone and systemic montelukast in patients who are unresponsive to topical therapy. Results showed that although negative reactions with drug are more common with prednisone than montelukast (10%) or placebo/cellulose (10%), prednisone had higher effectiveness in pain cessation and in accelerating ulcer healing [33].

Laser therapy appears to be effective against aphthae. Treatment with a 940-nm diode laser has found positive effects for the relief of pain and ulcer healing [27].

Immune boosters such as levamisole, vitamin therapy, topical interferon α-2a, and probiotics could have significant positive outcomes [27].

Other Diseases of the Oral Mucosa: Apart from the diseases mentioned above, there are other oral mucosal diseases that result in substantial reduction in quality of life of the sufferer. These diseases may occur predominately in the oral mucosa or may be associated with other systems [34]. **Erythema multiforme** is an acute mucocutaneous hypersensitivity reaction characterized by a skin eruption, which may occur with or without oral or other mucous membrane lesions. The disease manifests as toxic epidermal necrolysis. **Stevens–Johnson syndrome** is extensive skin disorder with a mortality rate ranging from 5 to 15%, which may involve internal organs. The disorder may be due to reactions against medications, infections caused by herpes zoster, varicella zoster. In these disorders, tissue damage is mediated by soluble factors such as the cell death receptor Fas (also known as Apo-1 or CD95, a member of the tumor necrosis factor, TNF, receptor), Fas ligand (FasL) [34, 35]. Many investigators have documented association between skin diseases with oral mucosal lesions [36]. Oral mucosal lesions in patients with skin disease are usually vesiculobullous reaction pattern (72%), lichenoid reaction pattern (61%), infectious lesion (57%), psoriasiform reaction pattern (57%), and spongiotic reaction pattern (47%). **Lichen planus** is a common dermatologic disease that manifests in the oral mucosa as white striations and patches associated with mucosal atrophy [37]. The fact that oral mucosal diseases are more prevalent in older age groups suggests some immune associations of oral mucosal functions with skin integrity [36, 38, 39].

Even though the etiopathogenesis of these diseases is not clearly known, emerging evidences suggest the involvement of nitric oxide (NO) as a mediator in mucosal diseases of the oral cavity and other regions of the gut [37, 40, 41]. NO is a free radical gas present in the body and plays a huge role in several functions of the body. NO is an endothelial-derived relaxant of vascular smooth muscle, an inhibitor of platelet aggregation and adhesion as well as a neurotransmitter. As a cytotoxic molecule produced by macrophages, it is involved in phagocytosis. NO can significantly cause damage to DNA, cellular proteins, lipids, and, thus, affect organ integrity and function. Oral mucosal diseases are associated with increase in the concentration of NO in saliva [37, 42, 43].

Clinical Correlate 10.2

Quality of Life in Oral Mucosal Diseases

The functions of the oral cavity, including food intake, speech, social contact, are very important not only for physiological processing of food and respiration, but also for social contact. The importance of these functions is vividly noticed when any of the functions is impaired due to disorders in oral cavity. These disorders affect the quality of life of the individual. In addition, disorders of the mouth also affect the ability to eat; thus, the nutritional status of the individual becomes poor. Masticatory functions may be also impaired in oral diseases [15, 44–46].

Oral mucosal diseases are associated with considerable level of reduction in quality of life of the sufferers [15]. To assess the impact of oral mucosal disease on various aspects of health of the individual, specific methodological approaches are used. The use of special questionnaires has found to be effective in studying the relationship of oral health with other systemic manifestations and outcomes such as interpersonal relationships. Widely used instruments (generic) for assessing the health-related quality of life include the Medical Outcomes Study 36-Item Short Form (SF-36) health survey; Nottingham Health Profile (NHP); Sickness Impact Profile (SIP); Dartmouth Primary Care Cooperative Information Project (COOP) charts; Quality of Well-Being (QWB) Scale; Health Utilities Index (HUI); EuroQol Instrument

(EQ-5D) [15, 47]. For the assessment of health-related quality of life in oral mucosal diseases, SF-36 and the short form of Oral Health Impact Profile (OHIP-14) have proven very effective. The reliability, validity, population differences vary with various instruments [47, 48]. Therefore, it is recommended to use two or more instruments for investigating health-related quality of life.

The standard SF-36 contains eight domains: physical functioning, role limitation physical, bodily pain, general medical health, vitality, social functioning, role limitation emotional, and mental health. The instrument is designed for people aged 14 years and above and evaluates the health quality of life during the 4 weeks prior to the interview. Higher scores indicate better health [15]. The OHIP-14 contains seven domains: functional limitation items, physical pain items, psychological discomfort items, physical disability items, psychological disability items, social disability items, and handicap items. It evaluates individual's quality of life in the preceding 12 months on a 5-point Likert-scale coded: 4 = very often, 3 = fairly often, 2 = occasionally, 1 = hardly ever, and 0 = never. Higher scores indicate worse oral health. The instrument can be evaluated by summing the total number of scores or finding the number of times respondents noted their answer as "very often," "often," etc. [15].

Oral mucosal disease patients have significantly higher scores compared with healthy individuals. Assessment of quality of life is useful not only in improving the quality of medical services, but also helps to evaluate the effectiveness of therapy [15].

10.2 Gut-Associated Lymphoid Tissue

The GALT forms the largest part of the mucosa-associated lymphoid tissue (MALT) comprising about 70% of the immune cells of the entire organism [2–5, 49]. MALT also includes smaller divisions of lymphoid aggregates such as salivary gland-associated lymphoid tissue (SALT), pancreas-associated lymphoid tissue (PALT), duct-associated lymphoid tissue (DALT), larynx-associated lymphoid tissue (LALT), bronchus-associated lymphoid tissue (BALT) [4, 50]. Each of these subdivisions of MALT can be further divided into smaller lymphoid unit. For instance, PALT can be further divided into intrapancreatic lymphoid tissue (IPLT) and peripancreatic lymphoid tissue (PPLT) [51]. Likewise, MALT of the gut can be divided into esophageal mucosa-associated lymphoid tissue, gastric mucosa-associated lymphoid tissue, intestine-associated lymphoid tissue, etc. [52–54]. However, some divisions of MALT can be merged to form larger units. Therefore, PALT and SALT belong to DALT [55]. These subdivisions though not widely used and not officially recommended occur in the literature. The lymphoid tissues transfer and present soluble and particulate antigens (e.g., pathogenic microbes) from the lumen of the GI tract directly from the mucosal surfaces via specialized epithelial cells called M (M for "Membrane" or "Microfold") cells, which are especially present in follicle-associated epithelium [4].

Of the component parts of MALT, the GI tract is believed to have the largest mass of lymphoid tissue in the human body [2, 3]. The lymphoid tissue of the gut (GALT) consists of tonsils (Waldeyer's tonsillar or pharyngeal ring), adenoids (also known as unpaired nasopharyngeal tonsil), Peyer's patches, lymphoid aggregates in the esophagus, colon, and appendix as well as diffusely located lymphoid and plasma cells of the GI tract lamina propria [2, 3, 5]. The GALT also includes isolated lymphoid follicles of the intestine [4, 49]. Waldeyer's tonsillar or pharyngeal ring, adenoids, paired palatine tonsils are referred to as nasopharynx-associated lymphoid tissue (NALT) in the mouse, rat, and rabbit. Waldeyer's tonsillar or pharyngeal ring is named after Heinrich Wilhelm Gottfried von Waldeyer-Hartz (1836–1921), a German anatomist and neuroscientist, who described the structure in 1884 [4, 5, 56].

The GALT belongs to the secondary lymphoid tissue as defined by the Terminologia Anatomica (International Anatomical Terminology). In addition, secondary lymphoid tissues or organs also consist of spleen, lymph nodes, and skin [57]. The secondary lymphoid tissues or organs (also known as peripheral lymphoid organs) are responsible for generating immune responses and tolerance especially in response to antigenic challenge, maintaining mature and naïve lymphocytes, initiating adaptive immune response, serving as site for lymphocyte activation, discriminating between dangerous foreign (nonself) antigens and the non-dangerous self-antigens [56, 58–60]. The secondary lymphoid organs are the sites where pathogenic microbes and nonpathogenic microbes (e.g., resident bacteria) are distinguished [49]. In secondary lymphoid tissues, T and B cells encounter antigen to generate effector cells or tolerance [56]. The development of secondary lymphoid tissues depends on expression of lymphoid cytokines (e.g., chemokines, interleukins), which are actively secreted by immune cells in these tissues. The developmental processes of secondary lymphoid tissues involve the activities of lymphoid tissue initiator (LTin), lymphoid tissue inducer (LTind), and lymphoid tissue organizer (LTo) cells [56, 61–65]. Several transcription factors and genes including Tlx1 and Pbx1 are involved in the regulation of secondary lymphoid tissue. The secondary lymphoid tissues are genetically preprogrammed and begin to develop during embryonic life through a tightly regulated process that involves interactions between hematopoietic and mesenchymal stem cells [63, 66–68].

The primary lymphoid organs "also known as central lymphoid organs" serve to generate diverse populations of functionally mature, but naïve lymphocytes from immature progenitor cells in the absence of foreign antigens. The cells produced here then migrate to the secondary lymphoid tissues where they can respond to pathogenic invaders. Mature T cells are derived from T cell progenitors, which originate in the hematopoietic tissues (bone marrow) and, through series of developmental changes, move to the thymus, where they further develop to form functional T cells. The B cells develop in bone marrow. The primary lymphoid organs are the bone marrow, thymus, and fetal liver. These developmental stages involve the activities of transcription factors (e.g., Pax1—paired box protein 1), receptors (e.g., lymphopoietin receptor), thymopoietic growth factors (e.g., granulocyte-macrophage colony-stimulating factor, GM-CSF; interleukin-7, IL-7; stromal cell-derived factor 1, SDF1, also known as C-X-C motif chemokine 12, CXCL12; thymus and activation-regulated chemokine, CCL17/TARC; delta-like 4, DLL4; thymus-expressed chemokine, TECK), which mediate the differentiation, proliferation, selection, survival, maturation, and homing of the T cells [56, 69–77]. The primary lymphoid organs develop during embryogenesis [72].

The third type of lymphoid tissues termed tertiary lymphoid tissues develop during unfavorable environmental conditions. The tertiary lymphoid tissues can be defined as ectopic aggregates or accumulations of lymphoid cells that arise in non-lymphoid organs of adult in response to environmental influences from chronic inflammation, microbial infection, graft rejection, cancers, autoimmune disease, and lymphatic malformations [3, 68, 78–81]. These tertiary lymphoid tissues resemble normal secondary lymphoid organs (e.g., lymph nodes) with regard to their cellular composition and functions. Like the secondary lymphoid organs, tertiary lymphoid tissues contain chemokines, dendritic cells, B cell, T cell, high endothelial venules, and lymphatic vessels [56, 61, 82–87]. However, tertiary lymphoid tissues have lesser number of lymphocytes and their lymph nodes are un-encapsulated and embedded in non-lymphoid organs at the sites of chronic inflammation, infection, graft rejection, cancers, autoimmune disease, and lymphatic malformations [84]. It should be noted, however, that unfavorable environmental conditions (e.g., exposure to pathogenic microbes) also modulate the structure, composition, and functions of secondary lymphoid organs [88].

In the GI tract, major components of GALT include Peyer's patches, lymph nodes, and tertiary lymphoid organs [4]. But the isolated lymphoid follicles also play a crucial role in GI immunity. It is estimated that the human gut has approximately 30,000 isolated lymphoid follicles. The density of these follicles tends to increase in the oral-to-anal direction. In the healthy human, the jejunum is believed to harbor on the average approximately 1 follicle per 269 villi increasing to 1 per 28 villi in the ileum, and progressively increasing toward the descending colon [4, 89].

10.2.1 Peyer's Patch—Structural and Functional Aspects

Peyer's patches are clusters of 5–200 lymphoid nodules located primarily in the lamina propria of the ileum. The presence of lymphoid clusters differentiates the duodenum and jejunum from the rest of the ileum. Each cluster of lymphoid nodules is positioned on the side of the intestine away from the mesentery and forms a bulge that may protrude into the lumen as well as into the submucosa [4, 49, 90]. It is named after the seventeenth-century Swiss anatomist Johann Conrad Peyer, who carried out a detailed description of the structure in 1677 [49]. But the

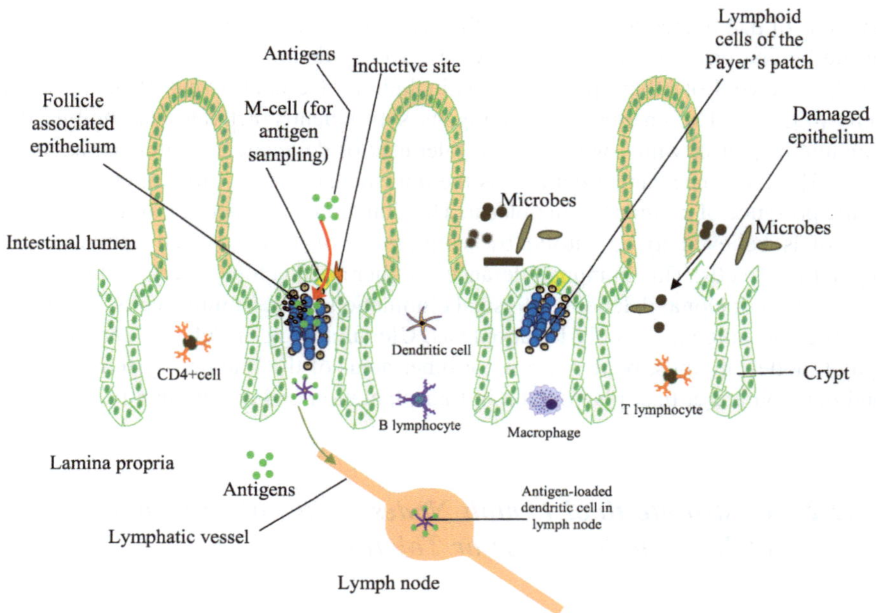

Fig. 10.1 Payer's patch. The human Peyer's patch serves as a site for antigen sampling and induction. Antigens from the lumen of the gut are channeled from the mucosal surface of the intestinal Peyer's patch to a region called the subepithelial dome via the specialized epithelial M cells. The subepithelial dome is also composed of resident dendritic cells, B lymphocytes, T lymphocytes, macrophages, and other migrating cells. (Reticular cells are mainly located in the outer surface of the subepithelial dome.) These immune cells are contained in each follicle of Peyer's patch [4, 49, 92–101]. These immune cells recognize, phagocytose microbes, or load antigens on their surface to initiate the appropriate immune responses required to curb microbial invasion. For instance, the resident dendritic cells of the subepithelial dome are required for antigen uptake. When loaded with antigens, dendritic cells translocate with the help of integrin into mesenteric lymph nodes, where further processing of the antigen takes place. But some antigens may escape into the systemic circulation via lymphatic vessels [89, 99, 102]. Details on the migration of the immune cells (also known as immunocytes) are discussed in later part of this chapter. These immune cells have differing roles, which are outlined in later part of this chapter [93]. Peyer's patches also represent a crucial site responsible for the production of immunoglobulin A [103]. The functionality of Peyer's patch and its interaction with antigens and microbes are regulated by pattern recognition receptors, which will be discussed in later part of this chapter. The components of Peyer's patch are also involved in other functions including control of paracellular permeability and also play a considerable role in the pathophysiology of inflammatory bowel diseases (e.g., Crohn's disease), cell/tissue/organ transplantation rejection reactions, etc. [49]

Italian Marco Aurelio Severino is thought to have made an earlier description of these lymphoid structures in 1645 [91, 92] (Fig. 10.1).

The organized aggregated lymphoid nodules or Peyer's patches are surrounded by a special epithelium called the follicle-associated epithelium. The epithelium connects the lumen of the GI tract to the GALT. This epithelium is mainly composed of the epithelial gatekeeper cells "M cells" that are specialized in channeling antigens of pathogenic microbes from the lumen of the gut to adjacent immune cells of the lamina propria and ileal enterocytes. The epithelium in this region (Peyer's patch) is devoid of other epithelial cells including enteroendocrine, Paneth, goblet cells [49, 93]. The channeling of antigens subsequently activates or inhibits the immune response, which will result in tolerance or downstream immune signaling [49, 93]. (Tolerance can be defined as the unresponsiveness of the immune system to the presence of antigens or neoplasia. This functional unreactivity of the immune system is believed to be caused by suppression of T and B cell activation by regulatory T cells. This immune tolerance is either due to clonal anergy—functional inactivation or clonal deletion—physical elimination of the immune cells specific to a particular antigen, possibly by apoptosis. Clonal anergy or deletion can occur in lymph nodes, thymus, bone marrow, or other areas of the immune system. Anergy and deletion are critical phenomena in transplantation and autoimmunity [94–97].)

10.2.2 Gastrointestinal Lymph Nodes—Sites of Induction of Immune Response or Tolerance

Lymph nodes are secondary lymphoid organs constituting a crucial site for immune responses to antigens of pathogenic microbes or neoplasia [97, 104]. There are many lymph nodes in the GI tract, which include myenteric lymph nodes (for details, see Chap. 2) [105]. The lymph nodes consist of parenchyma and spaces lined with epithelial cells. Each lymph node also contains venules and arterioles. The smooth muscles of lymph node capsules are involved in the movement of lymphocytes and lymph [106]. Lymph nodes have high abundance of macrophages, T cells, B cells, dendritic cells, hematopoietic cells, and resident stromal cells. These cells secrete a range of cytokines that regulate the activities of the immune system [104, 105, 107–109]. The lymph nodes are specialized in initiating immune response to antigens from lymph drainage regions (in the form of antigen-loaded dendritic cells) or circulating lymphocytes from blood or tolerance [97, 106]. Of particular note, the antigen-presenting dendritic cells and fibroblast dictate to the T cells how they should respond to the microbial aggression [104, 105, 107–109]. During an invasion, dendritic cells present pathogenic peptides (antigens) to recirculating T cells in order to activate the T cells with the receptors specific for that particular antigen. This process occurs within secondary lymphoid organs in zones of T cell abundance [110].

In the lymph node, the T cell-rich zone is the paracortex (Fig. 10.2) [97, 104]. The germinal centers contain dendritic cells [112]. The antigen filtration ability of draining lymph makes the node to adequately localize potential invasion by pathogenic microbes. When an antigen is spotted in the lymph node, two responses are initiated depending on the type of antigen and functionality of the responsible immune cells. Under normal circumstances, recognition of pathogenic antigens initiates immune response. However, if the antigen is not harmful (e.g.,

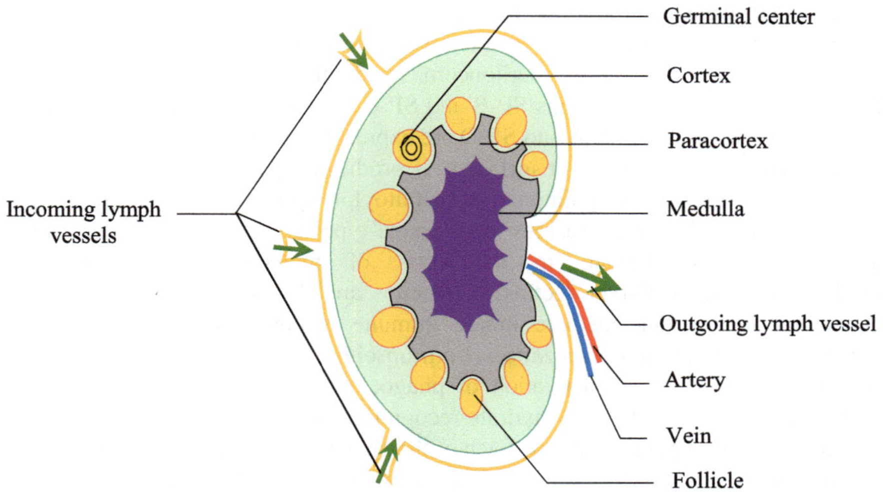

Fig. 10.2 Lymph node. Basic microanatomy of the lymph node showing afferent, efferent vessels, structures of the cortex and medulla [109, 111]

self-antigen), then tolerance takes place. Antigen-presenting cells such as dendritic cells, macrophages carrying antigens from afferent lymphatics, enter the paracortex to present the antigens to T lymphocytes. (Lymph node-resident macrophages are found in medulla [97, 104].) The recognition of the antigen by T cell leads to its activation with its subsequent differentiation and proliferation. A type of T cell found in the paracortex, referred to as T helper lymphocytes, upon this activation translocates to the cortex to help B cells [97]. The B cells are mostly found in the germinal centers [112]. The activated B cells differentiate into plasma cells—which produce the antibody. The effector cells activated upon antigen challenge (plasma cells, natural killer, cytotoxic T cells) translocate to the medulla. From here, they leave the lymph node via efferent lymphatics and travel to the specific area of inflammation, microbial invasion, or neoplasia [97, 104].

10.3 The Gastrointestinal Tract as an Anatomical Barrier to Potential Pathogenic Invaders—First Line of Defense

Like the skin, the GI tract forms a physical barrier that prevents the invasion by microbes—this forms the first line of defense. The mechanisms of this first-line defense include peristalsis, secretion of acids, enzymes, mucus, saliva, antimicrobial peptides (e.g., defensins). The gut resident microbes also play a crucial role in this defense [113–116]. It was previously thought that phospholipid-rich surfactant proteins (SP) were specific to the lungs; however, accumulating evidences indicate the presence of surfactants in the GI tract (stomach and intestine) and other regions of the body including pleura, peritoneum, synovial fluid [117]. The surfactants designated hydrophobic proteins SP-B and SP-C are specific to the lungs, whereas the hydrophilic proteins SP-A and SP-D are expressed in the GI tract and other sites of the body [117]. The GI tract surfactant is produced by the epithelial cells of the mucosa. In contrast to the pulmonary dipalmitoylphosphatidylcholines, gastric and intestinal epithelial cells secrete unsaturated-type phosphatidylcholines [118]. These surfactants belong to the class of molecules called collectins (collagen-containing C-type lectins). Collectins are oligomeric collagenous Ca^{2+}-dependent lectins that mediate innate immune response through aggregation, complement activation, opsonization (from opsonin, Greek *opsōneîn*, meaning "to prepare for eating"), and activation of phagocytosis [119]. The subunits of the oligomeric proteins with carbohydrate-recognition domains "collectins" are composed of three polypeptide chains. Each of the chains contains a lectin domain and a collagenous region [119]. The lectin domain binds carbohydrates on microbes, while the collagenous regions serve as ligands for the collectin receptor on phagocytes recruiting them to destroy the invading microbes [120]. (Lectins are carbohydrate-binding proteins that are specific for sugar moieties and play a crucial role in recognition of microbes.) The collectins form soluble pattern recognition receptors (PRRs) that bind pathogen-associated molecular patterns (PAMPs) of oligosaccharide unit or lipid origin on the membrane of microbes. Examples of collectins are SP-A, SP-D, mannan-binding lectin, conglutinin, and collectin-43. The later three proteins are produced by the liver and delivered into blood [121]. The collectins play a crucial role in innate immunity [122].

10.3.1 Innate Immunity

This first line of defense described above constitutes the **innate immunity**. The **innate immunity**, also known as the non-specific or antigen-independent immunity, is a rapid response, occurring within minutes or hours, following the invasion by pathogenic microbes—which is particularly required for the recruitment of immune cells to the site of microbial aggression. This first line of immune response does not have immunologic memory; hence, it cannot recognize a second invasion by the same pathogenic microbe [123, 124].

Innate immunity is ensured, first, by the epithelial barrier through the presence of glycoproteins (e.g., mucins), tight junctions, and antimicrobial proteins (e.g., defensins). These agents are discussed in their respective chapters (mucins and glycoproteins are discussed in Chap. 11; tight junctions are discussed in Chap. 4; defensins are discussed in Chap. 11). It will be briefly pointed that the antimicrobial proteins, defensins, are represented by short-chain polypeptide of about 12–50 amino acids, have hydrophobic or amphipathic domains in their structure, are a major first-line defense system against a wide spectrum of microbes such as gram-positive and gram-negative bacteria, fungi, parasites (e.g., protozoa), and can envelope viruses such as the human immunodeficiency virus, thus reducing possibilities for infections. These proteins are the most abundant on the mucus membrane and are also present in some leukocytes [125–128]. Additional defense is ensured by the specific cells of the immune system by distinguishing self from nonself. Immune cells do this by recognizing molecules present on the surface of microbes. These molecules are called pathogen-associated immunostimulants. These immunostimulants stimulate phagocytosis and inflammation. Examples of immunostimulants include short-repeat DNA of microbes, membrane lipopolysaccharide of gram-negative bacteria, and teichoic acids on gram-positive bacteria, peptidoglycan cell wall and flagella of bacteria, fungal cell wall molecules such as zymosan, glucan, chitin, and the membrane glycosylphosphatidylinositol of parasites such as plasmodium. The pathogen-associated immunostimulants are recognized by high-precision sentinel or surveillance molecules of specialized host cells. The high-precision surveillance molecules are referred to as pattern recognition receptors. The recognition receptors include membrane receptors of the Toll-like receptor family and components of the complement system [129–136]. The pattern recognition receptors and the complement system form integral parts of the innate immunity [137].

The Complement System

The molecules of the complement system consist of over 35 soluble and insoluble (membrane-bound) proteins (receptors) in the serum and extracellular fluid, including the serosal fluid of the GI tract; their major production machinery is in hepatocytes and phagocytes [138]. They amplify and complement the functions of B cells (antibody production) and other phagocytic cells by enhancing phagocytosis (opsonization), stimulating homing of phagocytes to sites of inflammation (chemotaxis), cluster pathogens together (agglutination), and lysis of the cell. The complement proteins participate in local inflammatory response and killing of microbial pathogens. It is also involved in physiological waste disposal by removal of pathogenic, tumorigenic, and other cellular debris [137, 139–142].

The complement proteins are constantly found in circulation, but in their inactive forms. The complement molecules are usually activated by microbial antigen; however, they may be activated in a non-classic pattern without the presence of microbial antigen. Both involve similar functional pathways that regulate their behavior and performance. The complement system that is activated first is termed the early complement components [137, 139, 140]. The signaling pathways responsible for their activation are the classical pathway, alternative pathway, and the mannan-binding lectin pathway. Thereafter, a late activation occurs in the late complement system. The activation of signaling pathways of the complement systems leads to the recruitment of different proteins that constitute a cascade of events, which result in lysis, phagocytosis of the microbe, and other functions [137]. The classical pathway, which is the first pathway to be discovered in the complement system, is activated by the binding of an antibody to an antigen (e.g., bacterial cell-surface molecule) with resultant lysis of the microbe. The complement proteins of the classical pathway are designated C1–C9 (The letter "C" denotes "complement"). This naming system is based on the order of discovery of the complement proteins [137]. Upon reaction initiation, C1 is successively followed by C4, C2, C3, C5 and then proceeds sequentially through C9 (Fig. 10.3) [137]. The classical pathway begins with the activation of component C1, a calcium-dependent complex proteinase. This first component of the classical pathway comprises subunits C1q, C1r, C1s, which play different roles in the activation process [143]. C1r and C1s are serine protease enzymes that initiate the activation of this pathway [144, 145]. C1r and C1s exist as protease zymogen tetramer $C1r_2s_2$ or C1s-C1r-C1r-C1s or C1r2-C1s2 [145]. In this tetrameric form (proenzyme form), the serine protease C1r is stable [145]. The tetramer readily forms complexes with the pattern recognition molecule C1q, a non-enzymatic protein subunit of C1 [144, 145]. This complex binds to a variety of activators such as antibody–antigen complexes containing IgG or IgM, resulting in autocatalytic autoactivation of C1r and the subsequent transactivation of C1s by C1r-mediated cross-proteolysis. The C1s then sequentially cleaves C4 and C2, producing C4a, C4b, C2a, and C2b (C2a and C2b are the large and small fragments, respectively). C4b and C2a bind to form C4bC2a complex—a classical pathway C3 convertase, which promotes cleavage of C3 into C3a and C3b [146–150]. The activation of C1 is controlled by C1 inhibitor, which is serpin peptidase inhibitor, clade G [149]. It should be noted, however, that C3 is unstable in aqueous medium as it is easily hydrolyzed spontaneously through the cleavage of the internal thioester bond. To prevent the spontaneous cleavage of this complement, the organism has development defense mechanisms that make the host cell produce substances that regulate the activities of complement proteins [142, 151–155]. The third complement molecule, C3, is a scaffold protein, constituting the central component of the complement system as all signaling pathways are related to this molecule [137, 139]. The C3 component is cleaved by an active protease C3 convertase into C3b and C3a, which, respectively, represents the large and small cleavage fragments. C3b is an opsonin (i.e., molecule that enhances phagocytosis, causing the phagocyte to savor cells marked for destruction) that is deposited on the cell. If C3b is not deactivated, it proceeds to activate the next step. C3b joins with C4b2a (the C3 convertase) to form C5 convertase (C4b2a3b complex). The activation of 5C by the C5 convertase leads to the formation of 5Ca and 5Cb. Both C3a and C5a diffuse away serving as a chemotactic and inflammatory paracrine factor, functioning in similar way as inflammatory cytokines. (The influences of C3b and C5b can be manifested in any tissue or cell that express the receptors for these complement proteins—including cells and tissues of the peripheral immune system and central nervous system [156].) C3 and C5 are also involved in the generation of anaphylatoxin. C5b interacts with C6, C7, C8, and C9 to form the membrane attack complex (MAC). Thus, the result of MAC formation is the assembly of the late complement components. The formation of this complex is preceded by local inflammatory response and recruitment of macrophages to the sites of aggression.

MAC formation is followed by tissue damage—MAC assembles around the surface of the pathogen, forming pores through the bilayer membrane, and thus makes the membrane leaky, which can lead to lysis of the microbial cell. This mechanism of lysis of microbes is similar to the mechanism used by defensins and other lytic proteins to destroy pathogens. Components of the complement system such as C1 and C4 can lyse microbes (including viruses) via antibody-mediated immune mechanism. These complement proteins neutralize microbes by phagocytosis and opsonization. The components are also involved in anaphylaxis (Figs. 10.3 and 10.4) [137, 139, 157–161].

Many factors can initiate the activation of the classical pathway by binding to C1q (Fig. 10.4). Of particular interest is pentraxin (PTX), a calcium-dependent ligand-binding pattern recognition protein molecule, involved in acute inflammatory response to tissue injury [163–165]. PTX3 interacts with several ligands, including growth factors, extracellular matrix component, and selected pathogens,

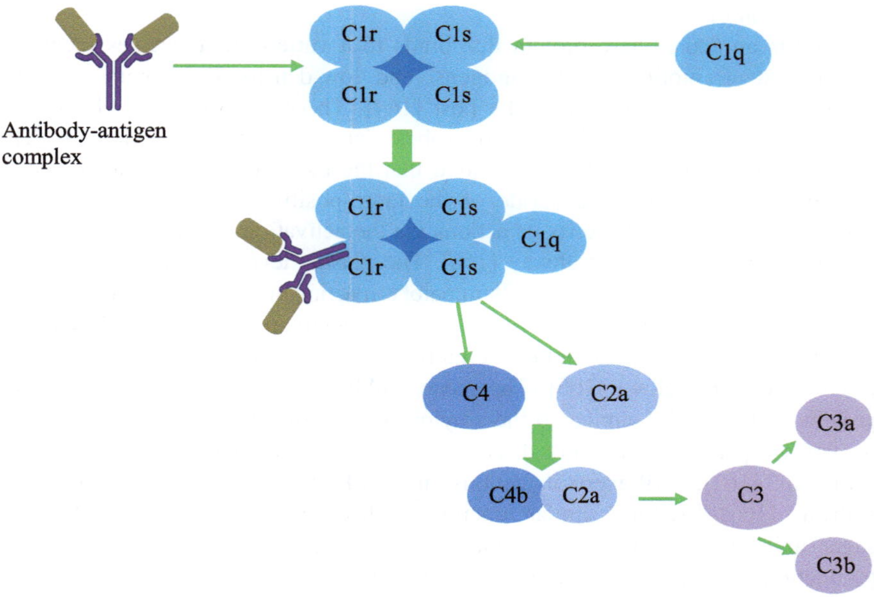

Fig. 10.3 Classical pathway of the complement system. The pathway is activated when antibody–antigen complex interacts with C1q and the tetramer C1r2C1s2. This activation results in autocatalysis of C1r, which transactivates C1s, which in turn cleaves C4 and C2 to form C4bC2a. Under the action of this convertase, C3a and C3b are formed from C3. The C3b proceeds to activate C5 to form C5a and C5b. The latter combines with C6, C7, C8, and C9 to form the membrane attack complex, responsible for mediating several activities of the classical pathway of the complementary system. The C3 is a central component liking all three activation pathways. The complement components (C5–C9) are the terminal or late-phase molecules of the complement activation system [138, 162]. Apart from immune complexes (e.g., antigen–antibody complex), the classical pathway can be activated by non-immune substances (e.g., DNA and small RNA molecules, beta-amyloid protein, and prions) [138]

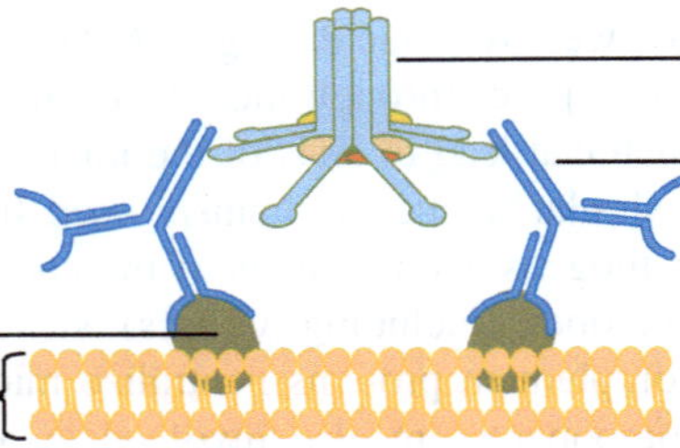

Fig. 10.4 C1-antigen-bound IgG complex

playing a role in complement activation and facilitating pathogen recognition by phagocytes [163]. PTXs are divided into short and long types. Examples of short PTXs are C-reactive protein (CRP) and serum amyloid P component (SAP). Examples of long PTXs are neuronal pentraxins and PTX-3 [164]. The long PTXs are predominantly synthesized in the liver especially during periods of acute inflammation or infection [164].

CRP is an acute-phase protein that binds to a variety of molecules including bacterial sugar motifs, nuclear antigens, and low-density lipoproteins, thereby facilitating their clearance from the host [164]. These molecules to which CRP binds are expressed on pathogenic microbes, damaged, and neoplastic cells. The name "CRP" evolved on the background that the molecule was first identified to interact with the C-polysaccharide of the gram-positive bacteria, *Streptococcus pneumoniae* [166]. CRP exerts its actions via the activation of Fc receptors (FcRs) and through the classical pathway [166]. The plasma level of CRP is used as an indication of acute infection or inflammatory reaction in the body [166]. CRP is also associated with the alternative pathway (see alternative pathway below). CRP confers an inhibitory effect on the alternative pathway through recruitment of factor H (see factor H below) [166]. The name "SAP" derived from the fact that the protein is structurally similar to the amyloid P component (AP), which is a small pentameric peptide, is initially discovered in pathological deposits referred to as "amyloid" [167]. SAP exerts its actions through FcRs, macrophages, and classical pathway [166]. SAP has similar functions with CRP. It binds to amyloid fibrils, sugar units, nuclear materials of pathogenic microbes, and diseased cells. The production of SAP is aimed at resolving certain pathologies such as amyloidosis, infectious, and autoimmune diseases [163, 166].

In the second pathway (i.e., alternative pathway), activation occurs in the absence of an antibody and is mediated through the activities of C3 and complement protein factors on the pathogen or neoplastic cell [139, 168, 169]. The proteins of this pathway are called factors and designated with letter "F" [137]. Examples include fB, fD, fH [161, 168, 170–172]. fB is a serine protease with single-chain polypeptide present in the plasma. It is composed of a non-catalytic subunit Ba and catalytic subunit Bb. The alternative pathway is activated by spontaneous hydrolysis of the internal thioester bond in C3, thereby initiating association with fB, which is then cleaved by fD producing Ba and Bb. The serine protease activity of fB is exhibited by the Bb subunit—it associates with C3 and cleaves it to form C3a and C3b. The large subunit C3b is deposited on target surfaces with the formation of C3 convertase, C3bBb—the protein that amplifies the alternative pathway [146, 170, 172]. fH is a glycoprotein present in plasma—it plays a crucial role in the regulation of the complement system. This factor is responsible for deactivating the complement system [161]. Upon activation, this factor binds to the host surfaces protecting the host cells against complement activation, which occurs, at least in part, by disassociating factor B from C3b on host surfaces. The fH is a major inhibitor of the alternative complement pathway. It ensures that the host cell is not mistaken for pathogenic microbes [161, 172].

In the lectin pathway (also called mannose- or mannan-binding lectin, MBL, pathway), the C-type lectin (e.g., MBL and ficolins) associates with MBL-associated serine protease-2 (MASP-2), forming MBL–MASP-2 complex, which recognizes and subsequently binds to mannose groups of bacterial cell wall, resulting in the cleavage of both C4 and C2 to form the C3 convertase, C4bC2a, that amplifies the pathway [139, 146, 162, 173–177]. The recognition of microbial sugars is mediated by MBL unit of the complex, whereas MASP-2 mediates the cleavage reaction exposing the microbe to the reactive thioester [173, 175]. MBL is a Ca^{2+}-dependent lectin belonging to the collectin (col + lectin) family of proteins, which is characterized by collagen-like domain and a carbohydrate-recognition domain that binds to D-mannose, *N*-acetylglucosamine, and fucose groups on microbial cell wall [162, 177, 178]. The collectins include MBL, collectin 43, conglutinin, SP-A, and SP-D [178]. The ficolins (fi + col + lin) are a group of oligomeric lec**tins** consisting of an N-terminal collagen (Col)-like domain and a C-terminal globular **fi**brinogen-like domain. Ficolins are pattern recognition molecules of the innate immune systems which recognize carbohydrate molecules on pathogens, apoptotic, and necrotic cells. Ficolins specifically recognize *N*-acetyl compounds such as *N*-acetylglucosamine, components of bacterial and fungal cell walls. Three ficolins have been identified in humans: L-ficolin, H-ficolin, and M-ficolin also known as ficolin-2, -3, and -1, respectively [179–181]. Like MBL, ficolins are opsonins that exist in complexes with MASPs. MBL oligomers and ficolins can form complexes with at least four MBL-associated serine proteases (MASP-1, MASP-2, MASP-3, and MASP-4) and with two non-enzymatic MBL-associated proteins-19 and -44 (MAp-19 and MAp-44) [146]. The order of activation of these enzymes and non-enzymatic components of MBL pathway is not fully understood. However, preliminary evidences suggest that MASP-1 aids the activity of MASP-2, but MASP-2 can autoactivate the cleavage reaction without MASP-1 [146, 175]. MASP-1 and MASP-3 are involved in activation of profactor D and factor B [146]. The MASPs are controlled by C1 inhibitor and α2-macroglobulin [162, 182].

It should be noted however that in the lectin pathway, other mechanisms of MBL-dependent activation of C3 may bypass C2 activation, mediated by direct cleavage of native C3 by mannan-binding lectin-associated serine protease-2 [162].

Late-Phase Complement Activation Independent on C3: Previously, it was mentioned that C3 serves as the central component in the complement system.

However, accumulating evidences suggest that plasmin, factor XIIa (a serum serine protease enzyme of the coagulation cascade formed from the zymogen factor XII or Hageman factor), kallikrein, and other proteases released by neutrophils and macrophages can activate complement system downstream 3C [183]. The activation of C5 can occur through the stimulation of Fc receptors or thrombin to generate C5a and C5b with subsequent amplification of the pathway to form MAC [184–186].

Biosynthesis of Complement Proteins: The complement soluble and membrane-bound proteins are produced by the liver and cells of the immune system [140, 187–189]. The immune cells found in peripheral blood and those resident in tissues (e.g., leukocytes, such as neutrophils, macrophages, monocytes, and dendritic cells) synthesize both soluble and insoluble complement proteins [138, 187–189]. Following their synthesis, the complement proteins circulate throughout the body fluids including bloodstream, serosal fluid, synovial fluid, saliva, intestinal fluid [138, 190–192]. The liver hepatocytes produce the bulk of the soluble proteins [140].

Liver Immunology—The Liver as an Innate Immune Organ: The liver is a crucial innate immune organ. In addition to its biosynthetic role of complement proteins, the liver plays a crucial role in non-specific phagocytosis of particles, pinocytosis of molecules, and cell killing [169]. The hepatic cells play an important role in the removal of waste molecules and elimination of microbes from the body [193]. Fetal liver produces hematopoietic cells, and this organ continues to produce cells of the leukocyte lineage from resident hepatic hematopoietic stem cells even after birth. Thus, the liver is a key hematopoietic organ. The liver has a high number of innate immune cells such as macrophages, natural killer, natural killer, and other T cell subtypes—see Table 10.1 for details. The non-hematopoietic cells of the liver such as hepatic endothelial cells, parenchymal cells, and stellate cells of the subendothelial space are also believed to play some roles in liver orchestrated immunity [194]. The Kupffer cells of the liver can initiate reactions that result in cytotoxicity of neoplastic and infected cells. The hepatic Kupffer cells produce signaling molecules (e.g., interleukin-12, -18) in response to bacterial lipopolysaccharide and superantigens, activating hepatic natural killer cells to produce interferon (INF)-γ, which in turn mediate the killing of neoplastic and infected cells [195]. The liver is also involved in adaptive immunity by deletion of activated T cells, induction of tolerance to ingested and self-antigens, extrathymic proliferation of T cells. The liver is a major site of extrathymic T cell development [169]. In liver disease such as cirrhosis, there is considerable dysfunction in the immune system that results in alterations in both innate and acquired immunities. Such immune dysfunctions of the liver are associated with decrease in GI motility, increase in intestinal permeability, altered balance or composition of resident microbes of the gut. The result is increase in the population of intestinal pathogenic microbes. Thus, higher number of microbes is translocated from the gut to other tissues and organs of the body. Consequently, the bacterial titer in bloodstream increases and predisposes an individual to the development of systemic inflammatory response syndrome and sepsis. In conditions when the body cannot cope

Table 10.1 Lymphocytes: types, functions, and their proportion in peripheral blood

Types	Functions	Examples/surface receptors[a]	Proportion (%)
Natural killer (NK) cells	Lyse tumor cells and cells infected with virus (cytotoxic function). NK cells possess the ability to kill transformed or virally infected cells without prior sensitization. NK cells are found in lymph nodes and tonsils and have a helper role in the production of interferon in response to IL-12, IL-15, and IL-18 stimulation. NK cells also enable maternal adaptation to pregnancy. The destructive functions of these cells, at least in part, are due to their ability to secrete perforins and granzymes	$CD3^{-}CD16^{+}$ $CD56^{+}$ expressing lymphocytes. The expression of surface receptors depends on their tissue location	5–20
Helper T cells	Regulate other cells of the immune system through the secretion of cytokines and growth factors	$CD4^{+}$ lymphocytes	~30.8–49.6
Cytotoxic T cells	Lyse tumor cells, allografts, cells infected with virus	$CD8^{+}$ lymphocytes	~20.1–42.5
γδ T cells	Their functions have not been fully resolved, but they seem to release substances that regulate the immune system functions. They are probably the first line of defense. These cells adequately recognize antigens of lipids, proteins, and phosphoantigens, without requiring human leukocyte antigen (HLA) presentation. They respond to the presence of mycobacteria, *Listeria monocytogenes, Epstein–Barr virus, Cytomegalovirus*, and tumor cells. They also play a role in cytotoxicity. Besides, they may	A type of antigen-specific T cell receptor (TCR). Over 70% of these cells express CD3 receptor and are largely found in the mucosa, and are referred to as $CD3^{+}CD4^{-}$ $CD8^{-}$ cells	0.5–10

(continued)

Table 10.1 (continued)

Types	Functions	Examples/surface receptors[a]	Proportion (%)
	participate in binding to stress-induced molecules. $CD3^+CD4^+$ γδ T cells positively correlate with serum level of IgE		
αβ T cells	Provide help to B cells through the secretion of cytokines, such as interleukin (IL)-4 or γ-interferon (γIFN). These secretions promote synthesis of immunoglobulin (Ig) G by B cells. αβ T cells can also bind to secreted molecules induced by stress	A type of antigen-specific TCR, expressing $CD4^-$ $CD8^-$, $CD4^+$, $CD8^+$, $CD3^+$, $CD56^+$, HSA + TCRαβ+	0.1–10
B cells	Secrete antibodies that bind to their specific antigen. This binding also recruits various white blood cells and complement system. Leukocytes and activated complement proteins work in concert with each other. B cells also function as antigen-presenting cells and develop into memory B cells. B cells can also release cytokines	Cells expressing $CD27^+$, IgD^-, etc.	26.6–36

References to Table: [123, 387–398]

[a]The expression of certain surface receptors significantly changes during activation of these cells. They might have interchangeability of functions in certain situations, or their expression is tissue-dependent. The proportion of some of the cells is affected by method of receptor analysis. In this regard, it is known that the most frequent subset of γδ T cells in the peripheral blood is the γδ $CD3^+CD4^-CD8^-$ type lymphocytes. In one analysis, however, it was shown that, of the isolated $CD3^+$ T cells, 75% were αβ and 25% were γδ T cells. The proportion of $NK1.1^+$αβ T cells was higher [399]. The proportion of the cells is also affected by age and season of the year as well as ethnicity. It is important to consult with an expert for current and most appropriate reference values for these cell subtypes as the values may change in different circumstances. Moreover, the values depend on many factors including gender, state of health. [399, 400]. For instance, γδ T cells tend to decrease at about 70–79 years. It should be noted, however, that these cells have different subsets some of which may not be affected by the confounding factors. Notwithstanding, however, due to immunesenescence, aging remains a major factor that causes changes in immune cell activity that might necessarily affect the prognosis of infections and propensity to the development of tumors [399]. As regards the health state, a recent research report indicates that obesity may negatively influence γδ T and αβ T cell signaling and antiviral (e.g., anti-influenza virus) functions. These cells aid in minimizing viral load through their production of IFN-γ and their cytotoxic actions on cells infested with microbes [401]. In tuberculosis, αβ and γδ T cells lacking

both CD4 and CD8 (i.e., $CD4^-CD8^-$ $\alpha\beta$ and $\gamma\delta$ T cells) exercise both inflammatory and regulatory response to the tuberculosis toxins [402]. In patients who are severely infected with *M. tuberculosis*, there is a reported high proportion of $CD4^-CD8^-$ $\alpha\beta$ T cells and decreased level of $\gamma\delta$ $CD4^-CD8$ T cells. Cells lacking these clusters of differentiation markers have been shown to express CD69, which is an early activation marker. Contrary to severe cases of tuberculosis, non-severe cases may be characterized by display of proinflammatory cytokines including IFN-γ. These T cells have been implicated in gut malfunctions such as celiac disease—an intestinal autoimmune disease due to dietary gluten [403]. In certain groups of individuals, gluten in diet triggers activation of $CD4^+$ and $CD8^+$ $\alpha\beta$ T cells and $\gamma\delta$ T cell receptors. These groups of T cell actively home to the gut from peripheral blood, resulting in an increase in intestinal intraepithelial lymphocytes containing the subtype mentioned above. It is these lymphocytes that are responsible for the negative effects of gluten in celiac disease. The negative effects of this condition include tissue damage, development of lymphoma. The homing of T cells to the site of inflammation or microbial embarrassment is a process by which naïve T cells travel to T cell regions of secondary lymphoid organs in search of antigen presented by dendritic cells [271]. Upon encounter of antigens on the surface of dendritic cells, the T cells become activated. This leads to differentiation, proliferation, and the generation of effector cells that can migrate to B cell regions of lymphoid organs or the areas of inflammation. Some T cells remain as circulating memory cells—which are immune cells that that can confer protection and enhanced response upon secondary challenge by similar antigen [271, 404]. There are two main types of memory cells—central memory and effector memory cells. The latter expresses CCR7 receptors (i.e., CCR7+memory cell) and thus home to lymph nodes to stimulate dendritic cells and differentiate into effector memory cells following a subsequent exposure to the same antigen. The effector memory cells are T cells that lack CCR7 receptors, but express receptors required for migration to inflamed tissues [271, 404, 405]. Memory cells have been discovered for all types of lymphocytes given in Table 10.1. $CD27^+$ (memory) B cells; CD27 has been found to be expressed on somatically mutated B cells and is thus a positive marker for memory B cells in peripheral blood [387, 406]

with the microbial insult and treatment is ineffective or not commenced timely, the result is multiorgan failure, and subsequently, death [196]. The complement system of the liver or its production machinery is influenced by pathogens such as hepatitis C virus and the malaria parasite. These pathogens can strive, at least in part, due to the negative effects they exert on the immune functions of the liver [194]. For details on liver immunobiology, review Kmieć [197] and Parker and Picut [169].

Pathology of the Complement Proteins: The deficiency of the complement molecules can cause increased susceptibility to infections and immune diseases including inflammatory bowel disease (IBD), Helicobacter-induced gastritis, GI manifestations of HIV infection (e.g., chronic diarrhea), rheumatoid arthritis, some CNS disorders (multiple sclerosis, Alzheimer's disease), some heart and vascular diseases, asthma, rhinitis, glomerulonephritis, hereditary angioedema, lupus erythematosus, paroxysmal nocturnal hemoglobinuria, and hemolytic uremic syndrome (hemolytic anemia, acute kidney failure, thrombocytopenia) [137, 183, 198–205]. Genetic abnormalities or mutations of genes responsible for the production of the complement molecules can lead to serious immune pathologies that predispose the individual to the development of infections. Such abnormalities can prolong time of recovery if the sufferer contracts any illness [137, 206].

Pattern Recognition Receptors

The first hypothesis that host cells express certain receptors that sense pathogenic microbes as foreign by recognizing structures that pathogens express by a mechanism described as "pattern recognition" was made in 1989 by the renowned immunologist Charles Alderson Janeway, Jr. (1943–2003) [207, 208]. Further, he reasoned that the interaction between the host cell and pathogen permitted lymphocytes to mount defense against the invading pathogen [207]. Unfortunately, knowledge about signaling pathways or cellular or subcellular structures that were associated with control of host cell–microbial interactions was only emerging and so Janeway possibly had not come across evidences that could point to the structural or physiological substrate of pattern recognition. Prior to this hypothesis of pattern recognition, the gene that would turn out to be the genetic material that harbors the signal for the production of the first discovered receptor of pattern recognition was discovered; unfortunately, its functions were not fully known [209].

In 1980, the German developmental biologist Christiane Nüsslein-Volhard (1942–) and the American fellow scientist Eric Francis Wieschaus (1947–) reported that certain mutations of Drosophila melanogaster resulted in a disorder of segmentation and polarity—this made the flies looked bizarre [210]. Nüsslein-Volhard and Wieschaus were so excited that they exclaimed in German "Das ist ja toll" ("That's great")—indicating that their results were "*Toll!*" ("Great!") [209]. Nüsslein-Volhard, Wieschaus together with Edward B. Lewis won the 1995 Nobel Prize in Physiology or Medicine "for their discoveries concerning the genetic control of early embryonic development." Edward Butts Lewis (1918–2004), an American geneticist, introduced several methods of modeling drosophila development and developmental mutations associated with this fly [211]. By the 1990s, several mutations associated with segmentation, polarity, and development of this fly had been discovered [212–216].

In 1996 the French biologist, Jules A. Hoffmann (1941–), and coworkers reported that mutations in the Toll gene of *drosophila* resulted in inadequate response of this fly to fungal infections and further suggested that interleukin signaling may be similar to the pathway initiated by Toll [216]. In the following year, Janeway and Ruslan M. Medzhitov (1966–) [217] reported that they had identified a human homolog of Toll—which was the type 4 Toll-like receptor (TLR4)—one of the products of the Toll gene. In 1998, Bruce Beutler and colleagues reported that TLR4 mutant laboratory animals (mice) did not respond to the bacterial lipopolysaccharide suggesting that this toxin may be the ligand for the Toll receptor [218]. Bruce Beutler, Jules Hoffmann, and Ralph Steinman were together awarded the 2011 Nobel Prize in Physiology or Medicine. Ralph Steinman discovered the dendritic cell in spleen and lymph nodes of mice in 1973 and suggested their role in activation of the immune system [219, 220].

Toll-like receptor (TLR) is an example of pattern recognition receptors (PRRs). PRRs are the anatomical units of host cell plasma or endosomal membrane, specialized for sensing certain structural motifs associated with pathogenic microbes or cellular stress (called pathogen-associated molecular patterns, PAMPs), and signaling molecules released from damaged cell (called damage-associated molecular patterns, DAMPs) (Fig. 10.5) [221]. However, it should be pointed that PRR ligands, which activate TLRs, are produced not only by pathogenic microbes, damaged cells, but also by the commensal flora of the gut. This suggests that PRRs evolved, at least in part, to mediate the bidirectional cross-talk between symbionts and their hosts [222]. Thus, PRR probably evolved before components of the adaptive immunity [221]. PRRs are responsible for maintaining the delicate balance of host–microbial interactions in the GI tract, composition of resident microbes, mucosal permeability, and commensal tolerance. PRRs are localized on GI epithelial cells such as Paneth cells [223–225]. Examples of PAMPs are peptidoglycan, lipopolysaccharide, lipoteichoic acid, teichoic acid, lipoarabinomannan, arabinogalactan, lipopeptides, flagellin, and foreign nuclear materials such as bacterial DNA and viral RNA [226–229]. Examples of DAMPs are certain extracellular matrix components released during cell damage (e.g., fibrinogen, heparan sulfate, laminin, elastin and collagen-derived peptides, fibronectin, matrix

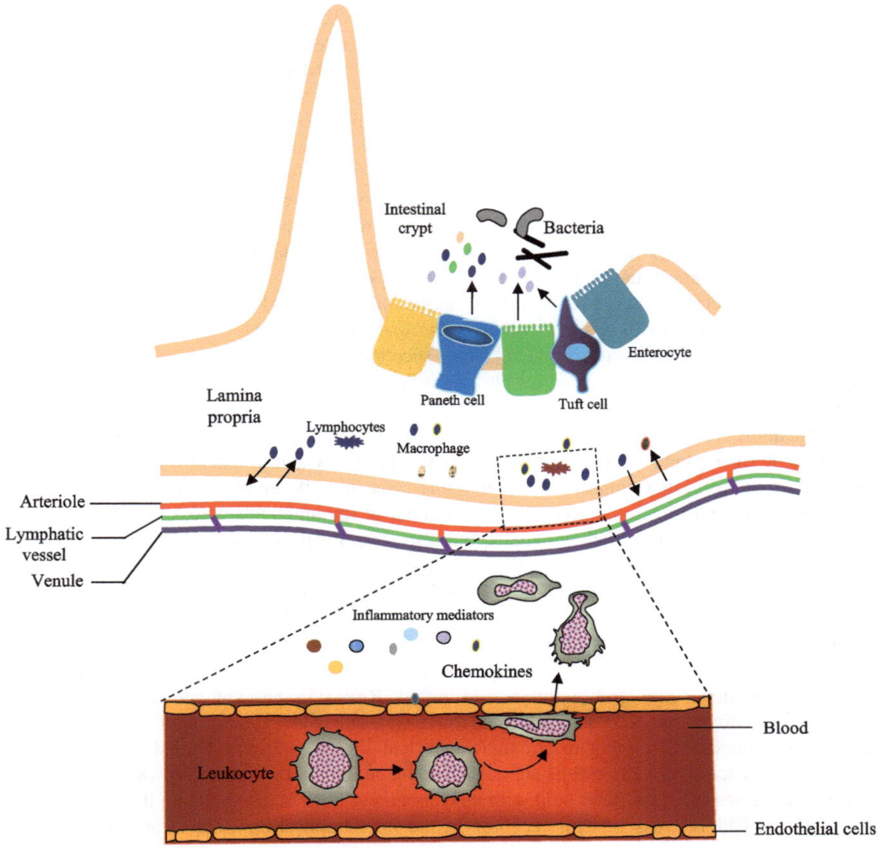

◀**Fig. 10.5** Intestinal crypt is a site of active defense against microbial aggression. The PRRs of gut epithelial cells sense the presence of pathogenic microbes in the intestine and initiate signaling events that culminate in the secretion of different molecules by the epithelial cells. The substances released into the gut by the epithelial cells include various types of antimicrobial peptides, proinflammatory cytokines such as chemokines (lymphokines, monokines, etc.), interleukins (e.g., IL-6), and mucosal immunoglobulin A [234–244]. A recently discovered group of gut peptides referred to as trefoil factors (TFF) appear to play a critical role in modulating cytokine signaling and thus innate immunity. TFF enhances healing of the gut mucosa through a process called restitution. These factors are secretory products of mucin-producing cells [245]. There are currently three classes of TFF: gastric peptides (TFF1), spasmolytic peptide (TFF2), and intestinal trefoil factor (TFF3) [245, 246]. TFF1 is known to function as heterodimer with gastrokine-2 (GKN-2). GKNs are a new family of proteins, secreted by gastric mucus-producing cells, and are known to modulate gastric mucosal inflammation [247]. Disordered GKN expression has been implicated in a range of GI tract diseases including cancers [248]. There are three types of GKNs —GKN-1, -2, -3. Emerging studies suggest that apart from GKN-2, other types may form dimers with TFFs to mediate inflammatory responses [249]. TFF3, which is abundantly present in human breast milk, has been shown to downregulate IL8 and IL6 and promote the production of defensins in epithelial cells of the intestine [245, 246]. Some secretions of epithelial cells are produced by immune response cells of the intestine. For instance, upon activation, tuft cells secrete IL-25 to mediate immune response [250]. Intestinal macrophages and dendritic cells present in the lamina propria secrete IL-6, TNF; tissue fibroblasts of the gut secrete IL-6, -10. Th cells of the lamina propria also secrete a range of cytokines including IL-6, -22, IFNγ, TNF. Intestinal neutrophils release IL-1 among other factors. Intestinal eosinophils secrete IL-1α, IL-2–6, IL-8, IL-10, IL-13, TGF-α, TGF-β, GM-CSF, TNF-α, MIP-1α [244, 251–255]. Mast cells also produce proinflammatory cytokines [244]. The antimicrobial peptides (e.g., defensins) produced by the epithelial cells (as well as certain immune cells) act to kill the microbes. The secreted cytokines by the epithelial cells induce the recruitment of innate immune cells such as macrophages, neutrophils, lymphocytes (e.g., Th2 cells, B cells), monocytes to the sites of microbial aggression by a process termed transepithelial migration (the movement is termed chemotaxis—which is defined as spatiotemporal directed movement of immune cells dictated by the gradient of chemoattractants or chemotaxins [256, 257]) [258–261]. Of these immune cells, macrophages and mast cells are thought to be the main initiators of innate immune responses [229]. In addition, inflamed tissues can also produce chemotaxins such as interferon-inducible protein 10, monocyte chemoattractant protein-1 [262]. These cytokines are critical inflammatory factors that play an important role in preventing excessive inflammation [261]. The secreted cytokines play a crucial role in the development of Th cells, responsible for curbing inflammatory responses in the GI tract and other regions of the body. Thus, these secreted molecules protect the epithelium against damage. The loss of the ability of these cells to secrete some of these molecules predisposes the individual to the development of a range of inflammatory diseases [263]. A fairly recent study suggests that intestinal myofibroblasts (nonprofessional immune cells) may be involved in peripheral immune regulation of the gut functions [264]. The recruitment of immune cells across the gut epithelium involves not only diffusion of chemotaxins, but also several cell adhesion molecules and several steps of immune cell movement. Cell adhesion molecules involved in such chemotaxis include β_2 integrin, Mac-1 (CD11b/CD18) [260, 264–266]. Integrins and other extracellular matrix receptors are discussed in Chap. 4. Mac-1 (CD11b/CD18) is a macrophage integrin or macrophage-1 antigen. Mac-1 is a complement receptor 3 that binds to C3b and C4b. This complement receptor is also required for the binding of leukocytes to factor H [267–269]. Other adhesion molecules required for immune cell migration into tissues include P-selectin, vascular cell adhesion molecule (VCAM-1), intercellular adhesion molecule-1 (ICAM-1) [262]. The immune cells (including leukocytes, monocytes) homing in circulation are usually non-adhesive to endothelial surfaces. However, upon activation, these cells are quickly tethered to endothelium of postcapillary venules, rolling under force of blood flow, bind with its receptors to ligands displayed on the surface of the endothelium of blood vessels. This receptor–ligand interaction is believed to activate integrins, which enhance adhesion by a process known as diapedesis (or extravasation). The formation of receptor–ligand association allows the extravasation (migration) of the activated immune cells out of the blood vessel into tissues (such as lamina propria, a region of the mucosa where majority of intestinal immune cells especially intestinal macrophages are found) from where they migrate to the site of aggression. This site of aggression may be lymphoid (e.g., Peyer's patches) and non-lymphoid (villus) regions of the intestinal mucosa [1, 237, 257, 262, 270]. Similar mechanism is involved in the migration (homing) of peripheral blood naive T cells to lymph nodes [271]. During migration, the cytokines can induce differentiation and maturation of certain blood cells. For instance, monocytes actively migrate into the lamina propria and intestinal crypt where they differentiate and mature to form resident intestinal macrophages [262, 272]. The macrophages can reside in the crypt and other regions of the GI tract throughout life [273–276]. The migrating immune cells (e.g., monocytes, macrophages, neutrophils) in turn secrete a couple of cytokines, present antigens, phagocytize foreign bodies, secrete proteolytic enzymes and oxygen radicals required to curb the microbial aggression. The secretion of these molecules are initially aimed at resolving the injury, but the activities of the vast number of molecules secreted by these immune cells may injure epithelial cells and may also induce secretory diarrhea. Excessive transepithelial migration of these cells substantially affects the permeability of epithelial barrier [265, 277]. The immune cells permeating the intestinal epithelium play a crucial role in the initiation of inflammatory reactions of the gut mucosa [270]. The epithelial cells express a number of cytokine receptors that mediate the activities of the secreted cytokines by the migrating immune cells. The cytokine (chemokine) receptors including CCR1-8, CXCR3, and CXCR4 are abundantly but differentially expressed on the apical and basolateral sides of epithelial cells. For instance, the apical side has high expression of CXCR4 and CCR5. The macrophage inflammatory protein (MIP)-1α is a ligand for the epithelial cell chemokine receptor CCR5 and CXCR4. The CXCR3 receptor resists parasitic invasion in gut [278, 279]. The membrane-bound lymphotoxin (LT) receptor (LTβR) is primarily expressed on epithelial cells. LT belongs to the TNF family of cytokines [280, 281]. It appears that a subset of immune cells can themselves recognize microbial cell wall component. To this end, a relatively recent study suggests that monocytes express microbial lipopolysaccharide receptor CD14. These cells are designated CD14^{+}CD16^{+} and contribute to production of proinflammatory cytokines [270]. Corticosteroids are effective in reducing inflammation, at least in part, due to their ability to decrease the number of CD14^{+}CD16^{+} monocytes [282]. The cytokines mainly function via signaling pathways (JAK/STAT, NF-κB, p38 pathways, etc.) in the gut and exhibit cross-talks with other signaling pathways such as the phosphatidylinositol-3-kinase (PI3 K), Akt, Wnt pathways [283]

metalloproteinase-3 and -13, proteoglycans, such as hyaluronan, versican, and biglycan [230, 231]) and cytoplasmic proteins such as heat shock proteins, S100 proteins (e.g., S100B), RNA and mitochondrial DNA, nuclear DNA, IL-1, high-mobility group box 1 protein (an intranuclear protein that plays a crucial role as lethal mediator of sepsis and disseminated intravascular coagulation), histones, adenosine triphosphate, neutrophil-derived alarmins such as α-defensins, lactoferrin, and cathelicidin [226, 231–234]. These DAMPs are generally referred to as endogenous damage signals [233]. For details on DAMPs and their receptor types and subtypes, review Rosin and Okusa (2011) [231].

The process by which the host cell receptors detect the pathogenic foreign or damage signals is called pattern recognition, and it is responsible for induction and bridging the two lines of defense in mammals—innate and adaptive immunities [229, 235]. TLRs recognize microbes by binding to PAMPs or DAMPs, with resultant activation of transcription factors such as nuclear factor kappa of B cell

(NF-κB) including the interferon regulatory factors (IRFs), leading to the synthesis of cytokines, interferons (IFNs). These products of transcription activation orchestrate adaptive immune response required for effective immunity [226, 229, 236, 237]. NF-κB is a protein complex that controls transcription of DNA, cytokine production [238, 239]. IRFs are proteins that control the transcription of interferons [240, 241]. The cytokines mainly use the JAK/STAT pathway to exert their activities on the cell [241].

TLRs: TLRs are single transmembrane non-catalytic receptors usually expressed in immune cells that sense structurally conserved motifs of microbes as well as molecules released from host cells during damage reactions. There are at least ten different TLRs (TLR1–10) in humans. These receptors sense a broad range of microbes in different compartments of the cell. TLR-1, -2, -5, and -6 are predominantly found in the plasma membrane. These receptors recognize components of bacterial cell wall [229, 284, 285]. TLR-3, -7, -8, and -9 are found in endosomal and lysosomal membranes. These organellar TLRs sense microbial nucleic acids [229, 284]. TLR4 is found in both the plasma and the endosomal membranes [284, 286].

TLRs are predominantly expressed on antigen-presenting cells, such as macrophages, dendritic cells. TLRs are expressed in mast cells. They are also found in considerable number on epithelial cells of the GI tract. TLR signaling activates antigen-presenting cells to initiate innate immunity and is required for effective adaptive immunity [287, 288].

Signaling Pathways of TLRs: In the presence of pathogenic microbes such as bacteria, TLRs sense the endotoxin lipopolysaccharide and other components of the bacterial cell wall [289]. TLRs bind to this endotoxin to initiate downstream signaling cascades. If the bacterial toxin is released into the bloodstream, it binds to LPS-binding protein (LBP) in blood plasma and subsequently delivered to the cell-surface receptor CD14. Next, LPS is transferred to the Toll-like receptor 4 (TLR4). The downstream signaling involves the recruitment of NF-κB, MAPK, etc. [286, 290–293]. The result of activation of these signaling pathways is the production of different cytokines (e.g., IL-1 and IL-18) and chemokines, antimicrobial peptides, initiation of the differentiation, and maturation of antigen-presenting cells that mediate inflammatory reactions [223, 236, 292, 294]. Research evidences indicate that TLR signaling mediates downstream effect via the cytoplasmic adaptor protein myeloid differentiation primary response protein 88 (MyD88) as well as other unrelated proteins [295, 296]. The signaling pathway of the former is MyD88-dependent, which controls the synthesis of proinflammatory cytokines such as IL-1β, IL-2, IL-6, IL-8, IL-12, IL-15, IL-21, INF-γ, and TNF-α [297–299], whereas the latter is MyD88-independent and controls the IFNβ activity, differentiation, and maturation of dendritic cells [295, 296].

In the MyD88-dependent pathway, the activation of TLRs results in their dimerization, stimulating association with adaptor proteins. For instance, bacterial lipopeptide can induce the dimerization of TLR-2 with TLR-1 or TLR-6 to form TLR1–TLR2 or TLR2–TLR6, respectively (Fig. 10.6) [300, 301]. TLR4 signaling is also MyD88-dependent. The association with the adaptor proteins such as MyD88 leads to activation of downstream effectors kinase enzymes that catalyze

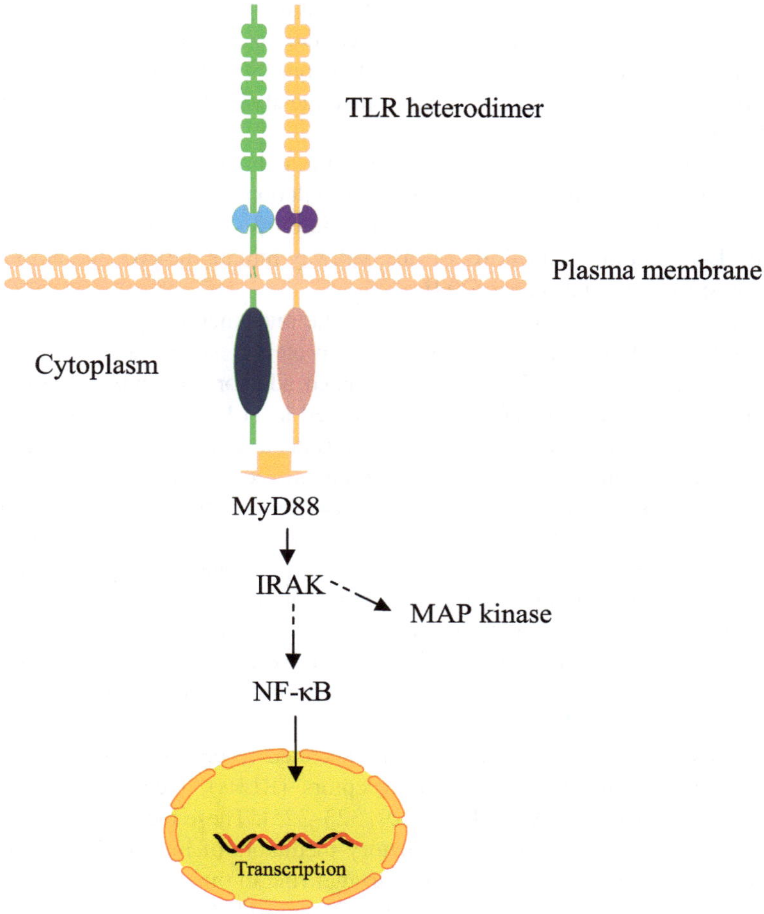

Fig. 10.6 TLR signaling

several downstream pathways associated with transcription factors that regulate gene expression. One of such activated enzymes includes IL-1R-associated kinases (IRAKs), which through several downstream enzymes activates mitogen-activated protein kinases, c-Jun N-terminal kinase, p38, etc. These proteins in turn mediate the activation of transcription factors such as NF-κB, IRFs, activator protein 1 (AP1), cyclic AMP-responsive element-binding protein (CREB), which control the production of specific biological molecules/cytokines by the respective (activated) gene (Fig. 10.6) [229, 284, 290, 302, 303]. TLR signaling is negatively regulated by molecules such as Toll-interacting protein (TOLLIP), nucleotide-binding, and oligomerization domain-like receptor (NOD-2), single Ig IL-1-related receptor (SIGIRR or Toll–interleukin-1 receptor), A20 (also known as TNF alpha-induced protein 3, TNFAIP3, zinc finger protein), A20-binding inhibitor of NF-κB

(ABIN-1, -2, -3), soluble CD83, peroxisome proliferator-activated receptor gamma (PPAR-γ), ubiquitin-associated domain-containing ubiquitin regulatory X (UBX) protein (UBXN-1), and F-box protein only 7 (FBXO7) [236, 294, 304–312]. Many of these cytokine regulators mediate their influence by inhibition of NF-κB activation [286].

The MyD88-independent pathway occurs through the activation of TLR3 and TLR4, and it is mediated by the stimulation of the adaptor protein TIR-domain-containing adapter-inducing interferon-β (TRIF). (TIR is an acronym for Toll/IL-1 receptor.) The TLRs of this pathway initiate the production of IFN-β in response to microbial nuclei [299].

A couple of adaptor molecules other than the ones mentioned above have been implicated in TLR signaling. Adaptor proteins such as TIRAP, TNF receptor-associated factor (TRAF), TRIF-related adaptor molecule (TRAM), TNF receptor-associated factors (TRAFs) are involved in TLR signaling [284, 290, 299, 313, 314]. For further review, see Akira and Takeda [298], Lim and Staudt [315].

Emerging evidences indicate that certain molecules acting as agonists or antagonists of TLRs can substantially influence the occurrence or development of diseases such as infectious diseases, immune disorders, allergy, cancers, sepsis, atherosclerosis, kidney failure, liver disease, pulmonary disease, myocardial ischemia/reperfusion injury, traumatic brain injury. These molecules show promising therapeutic potential for use in the clinic [229, 287, 316–318]. For example, the TLR4 antagonists, Eritoran and Resatorvid, are pharmacological agents that have shown promise for treating the diseases mentioned above [318–322]. For review, see [317, 319, 320].

PRRs also include NOD-like receptors, secreted receptors (collectins, ficolins, and pentaxins), cytosolic RIG-1-like receptors (RLRs), receptor for advanced glycation end products (RAGE) [225, 229, 323–325]. These receptors, in particular TLR and NOD, represent key inflammatory mediators of intestinal epithelial and mucosal immune cells and play essential trigger role in the production of cytokines and antimicrobial factors [225, 323, 324]. NOD and TLR collaborate to strengthen the immune system. Dysfunctions in the receptor signaling can predispose an individual to the development of IBD (Crohn's disease and ulcerative colitis), a group of diseases, characterized by chronic recurring inflammation of the intestinal mucosa [270, 324]. TLRs mediate inflammatory reactions in other regions of the body including the urinary tract. For example, urinary tract infections (UTIs) can result from TLR (TLRs 2, 4, and 5) dysfunctions [326]. The activation of NF-κB plays a central role in the establishment of the stimulatory effects of several PRRs including NOD-2 in human intestinal epithelial cells [327].

Recognition Mechanisms of Self and Nonself

The mechanisms by which epithelial cells discriminate between pathogenic and commensal microorganisms are based on pattern recognition described in the previous section. Following recognition by the cells of the host, foreign bodies routed for destruction via several pathways including lysosomal and proteosomal pathways. The lysosome and proteasome are major organelles involved in intracellular digestion. The functions of lysosomes are discussed in Chap. 3. Proteasomes are intracellular organelles that maintain functional proteome of the cell, regulating proteins involved in cell cycle, growth, and apoptosis. This organelle degrades peptides of pathogenic microbes and damaged proteins, tagged for degradation. The organelle harbors a variety of hydrolases (including enzymes with trypsin-like activities). This enzyme complex exhibits its functions by several chemical reactions including carbonylation, glutathionylation, glycoxidation, and other modification reactions [328, 329].

Clinical Correlate 10.3
Pathogenic Microbes, Residents, and NANA

NANA is the predominant sialic acid (neuraminic acid) and the chief component of glycoconjugates such as glycolipids, glycoproteins, and proteoglycans (sialoglycoproteins) where it confers selective binding property to the glycosylated component. The name "sialic acid" comes from the Greek word sialon (saliva). Sialic acid was first discovered in bovine submaxillary mucin in 1936 by Gunnar Blix (1894–1981), a German biochemist and son of the famous physiologist, Magnus Blix, and in 1941 found in brain matter (as neuroamine) by the German biochemist Klenk Ernst (1896–1971) [330–333]. NANA is negatively charged residue found in complex glycans on mucins and glycoproteins at the membrane of cells including neurons. Sialic acid contains over 40 naturally occurring nine-carbon keto sugars acids derived from the parent compound 2-keto-3-deoxy-5-acetamido-D-*glycero*-D-*galacto*-nonulosonic acid (*N*-acetylneuraminic acid—NANA). The sialic acids and related nonulosonates are unique in nature, representing the only nine-carbon sugars found in prokaryotes. NANA is useful for stabilizing glycoconjugates and cell membranes due to (i) charge–charge repulsion, (ii) mediation of cell–cell regulation, acting as chemical messenger, (iii) regulation of transmembrane receptor function, (iv) its effect on membrane transport, (v) its control of the half-life of circulating glycoproteins and viable cells, and (vi) contribution to the permselectivity of the glomerular endothelium and slit diaphragm. Sialic acid is not only a major component of animal cell membrane, but also microbes including those that are responsible for some human maladies. Sialic acid is located in the interface between the host and pathogenic microorganisms or the residents (commensals). An important function of host sialic acid is to regulate innate immunity. But some pathogenic microbes have evolved various strategies for subverting this process by their surface expression of sialylated oligosaccharides that mimic those of the host. These subversive strategies include a de novo synthetic pathway and at least two truncated pathways that depend on scavenging host-derived intermediates. A fourth strategy involves modification of sialidases so that instead of transferring sialic acid to water (hydrolysis), a second active site is created for binding alternative acceptors. Sialic acids also are excellent sources of carbon, nitrogen, energy, and precursors of cell wall biosynthesis. The catabolic strategies for exploiting host sialic acids as nutritional sources are as diverse as the biosynthetic mechanisms, including examples of horizontal gene transfer and multiple transport systems [332, 334]. Human NANA acts as a receptor for influenza viruses, allowing attachment to mucous membrane via hemagglutinin. Sialic acid may be used by pathogenic bacteria either as a nutrient, providing both carbon and nitrogen to the bacterium, or can be deposited on the cell surface. Bacteria have evolved transporters for Neu5Ac to enable them capture it from their environment. A couple of such transporters have been characterized including the NanT protein from Escherichia coli, SiaPQM TRAP transporter from Haemophilus influenza, and the SatABCD ABC transporter from Haemophilus ducreyi. The molecule can be used as a signaling molecule to accomplish many cellular functions of both animal and microbial cells [335–338].

10.4 The Reticuloendothelial System—Cells of the Innate Immune System

The reticuloendothelial system (also known as lympho-histiocytic reticular, lympho-histiocytic, reticulo-histiocytic, lympho-reticulo-histiocytic, or custocyte system) comprises a system of immune cells such as monocytes, macrophages, dendritic cells, lymphocytes that participate in functional maintenance of the immune system (Fig. 10.7). An integral characteristic of the cells of the custocyte system is phagocytosis of foreign bodies [339–341].

The phagocytic mononuclear cells have a long history of changes in classification names. The first classification was advanced by Elie Metchnikoff (1845–1916) and named the "macrophage system" in 1887. But his description of digestion of foreign bodies by wondering cells was made earlier in 1883. Metchnikoff classified phagocytes (Greek "phago" for devour and "cytes" for cells) as "macrophages" (large eaters) and "microphages" (a smaller type of phagocytic cell, the polymorphonuclear leukocyte, now known as granulocytes). He maintained that both types of phagocytes played an important role in protecting the host against infections. Metchnikoff and Paul Ehrlich were awarded the 1908 Nobel Prize for their work on phagocytosis [340].

A few decades later, precisely in 1924, Karl Albert Ludwig Aschoff (1866–1942) introduced the lympho-reticuloendothelial system, which later bears his name "Aschoff's lympho-reticuloendothelial system." Thereafter, Volterra in 1927 proposed the "reticulo-histiocyte system," as the researcher noted that most endothelial

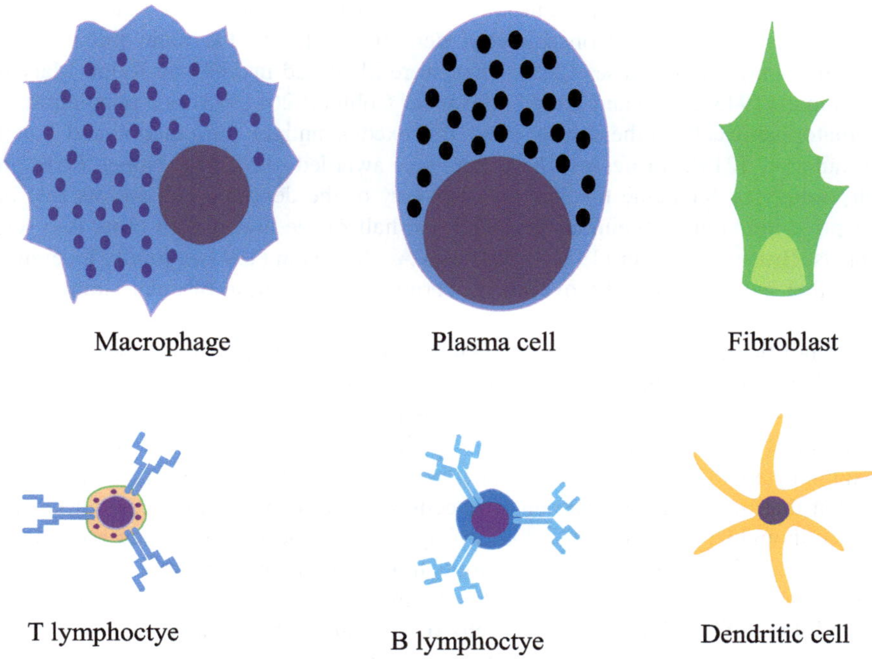

Fig. 10.7 Immune response cells of the GI tract

cells were not phagocytic, rather to a substantial extent comprise cells found in tissues [341].

Further, in 1969, a group of scientists in a Conference on Mononuclear Phagocytes, held in Leiden (Netherlands), on September 2–5, 1969, proposed a new classification "mononuclear phagocyte system." This system includes the promonocytes and their precursors in the bone marrow, the monocytes in the peripheral blood, and the macrophages in the tissues. This new classification was based on the similarities in the morphology, function, origin, and kinetics of the phagocytes. The histiocytes (also called tissue macrophages) are included in this system. However, some cells of the phagocytic system did not fall into this classification: reticular cells, dendritic cells, endothelial cells, fibroblasts [342, 343]. Reticular cells have cytoplasmic projections extending out to the plasma membrane. Such cells include fibroblasts, and they are found in connective tissues, spleen, and lymphatic system and direct B and T cells to specific places for immune reinforcement [344–346].

Dendritic (stellate-like) cells are antigen-presenting cells (accessory cells) and subpopulation of monocytes having special markers or receptors—CD33, CD14, CD11c on their membrane (Fig. 10.7) [339, 347]. The dendritic cells capture, phagocytose, process antigen, and present it on their cell surfaces to the T cell. Hence, they mediate response between innate and adaptive immune systems. Thus, dendritic cells act as important messengers between the innate and adaptive immunities [107, 123]. Consequently, dendritic cells are the main initiators of adaptive immune responses [229]. They were identified in 1973 by Ralph Marvin Steinman (1943–2011) and Zanvil Alexander Cohn (1926–1993) as a population of hematopoietic cells in the mouse spleen that excel at antigen presentation and T cell stimulation [219]. Professor Steinman was awarded the 2011 Nobel Prize in Physiology or Medicine for "for his discovery of the dendritic cell and its role in adaptive immunity." Steinman received one half of the award with other two scientists, Bruce A. Beutler (1957–) and Jules A. Hoffmann (1941–), jointly receiving the other half "for their discoveries concerning the activation of innate immunity" [348].

Dendritic cells are present in the skin (Langerhans cell previously described by Paul Langerhans), inner lining of the nose, lungs, stomach, and intestines. In the lymph node, dendritic cells interact with B and T lymphocytes. In fact, the functions of B and T cells are controlled by the dendritic cells via the secretion of lymphocyte stimulatory molecules such as cytokines [107]. In contrast to previous notion that macrophages and dendritic cells were resident immune cells, it is currently thought that monocytes, macrophages, and dendritic cells may recirculate following a first pass in tissue or margination in the lungs and spleen since these cells have been localized in peripheral circulation. Morphologically, the term "dendritic cell" is shared by other cells of the human body including fibroblasts, myofibroblasts, oligodendrocytes, dermal dendrocytes [107, 108, 339, 349].

In 1996, Goerdt et al. [339] proposed another term for both the dendritic and mononuclear phagocyte system—custocyte system (Latin "custos"—sentinel, guard). This proposal was based on evidences suggesting that monocytes/macrophages and dendritic cells have single progenitor, share a sentinel, receptor, and effector, as well as mode of presentation of antigen. Goerdt et al. further divided cells of this system into type I, type II, and type III custocytes. Type I custocytes are those cells of the system that exhibit good presenter functions [339]. Type II custocytes are the effectors that activate specific immune reactions. Type III custocytes are bipolar cells, which during their development and in period of inflammatory responses pass through types I and II phases of development [339].

Although mentioned by Metchnikoff as the principal combatant, neutrophils were later in course of reclassification of the phagocyte system, left out for reasons not exactly understood. However, it is possible that the macrophago-centric view of phagocytosis could have been responsible for this. In 2010, Silva [350] proposed a new name for the phagocyte system that accounted for the involvement of the neutrophils, one of the professional phagocytes, besides inflammatory monocytes, macrophages, and immature dendritic cells [351], make up the phagocyte system that was called myeloid phagocyte system, since these cells originated from a common myeloid progenitor via the granulocyte/macrophage lineages. It should be noted that apart from neutrophils, other granulocytes are also phagocytic. However, only the neutrophils exhibit avid phagocytosis [351]. The distinguishing receptors (markers) of human neutrophils are the lymphocyte antigen 6, subtype G (LY6-G); [351, 352]. Leukocytes generally fall into two groups—granulocytes and agranulocytes. In granulocytes, staining shows a granular cytoplasm. Types of granulocytes are neutrophils, eosinophils, and basophils [353–357]. In agranulocytes, no granularity of the cytoplasm is observed upon staining. Examples of agranulocytes are lymphocytes and monocytes [356, 358, 359]. The monocytes that participate in inflammatory reaction (known as inflammatory monocytes) express the receptors CD11b and LY6C [351, 352]. The tissue monocytes express CD33 receptor [351, 352]. Accumulating studies have indicated that mast cells, basophils, and eosinophils play a substantial role in innate immunity [123]. First described by Paul Ehrlich in 1878, mast cells are granulocytes, containing granules with high content of heparin and histamine [123]. These cells also synthesize considerable amount of proinflammatory cytokines [360]. The mast cells rise from committed mast cell precursors in the bone marrow. The committed cells exit the bone marrow and complete their maturation process in the peripheral tissues [361, 362]. When they are fully differentiated, mast cells home from the peripheral circulation to intestine and other regions of the body including the skin, lungs, and uterus [363]. In these organs and tissues, the immune cells are mainly localized in connective tissue surrounding blood vessels [123]. In these areas of the body, mast cells are primarily involved in allergy and anaphylaxis. Recently, however, evidences indicate that they play substantial role in inflammation by detecting and responding to pathogenic microbial embarrassment via the secretion of cytokines [364–366]. Mast cell dysfunctions have been implicated not only in allergic disorders, but also inflammatory, autoimmune disorders as well as cancer [364]. These dysfunctions in mast cells may occur as excessive production of these cells (a condition referred to as mastocytosis) or inadequate production of these cells—referred to as mastopenia [366–369].

First described in 1879 by Paul Ehrlich, basophils are multifunctional effector cells of the immune system with progenitors originating from multipotent hematopoietic stem cells [123]. The differentiation and maturation of basophils are completed in the bone marrow. Like mast cells, the cytoplasm of basophils contains granules with high content of histamine and can also produce large amounts of the regulatory cytokines (e.g., IL-4, IL-13) [370]. Both the mast cells and basophils express the immunoglobulin E (IgE) receptor (FcεRI). Activation of this receptor with the corresponding antigen (allergen) leads to degranulation and the release of inflammatory mediators [361, 371]. The basophilic leukocytes are found predominantly in the peripheral circulation, with a total proportion less than 1% [123]. These cells are implicated in a couple of diseases that involves allergic and immune response dysfunctions [361, 370, 371].

The eosinophilic leukocytes (or eosinophilic granulocytes, eosinophils) are multifunctional granulocytes that contribute to initiation and modulation of inflammatory reactions. These cells carry out their functions through the release of cytotoxic mediators and cytokines [372–374]. Eosinophils can phagocytose foreign bodies and destroy parasites. Like the mast cells and basophils, eosinophils are also involved in inflammatory diseases, allergy, asthma, and a host of autoimmune diseases. In the GI tract, dysfunctional signaling of eosinophils has been implicated in the pathogenesis of inflammatory bowel diseases and primary biliary cirrhosis [123, 375].

A very important property of these phagocytic cells that makes them exceptionally active is "homing" and recirculation, ensuring timely movement and drift to the site where they are needed according to the body needs [376]. This property is particularly characterized for lymphocyte. Lymphocyte is a type of white blood cell classified as agranulocytes: natural killer cells, T cells (responsible for cell-mediated, cytotoxic adaptive immunity), and B cell (responsible for humoral, antibody-driven adaptive immunity) (Table 10.1). These cells are present in circulation and in large numbers in the lymphatic system (e.g., lymph nodes), intestinal crypt, and lamina propria, where they exert their functions and cooperate with other cells of the epithelia and cells of the immune system, and thus regulate the microarchitecture of the mucosa [377–383]. The highest proportion of T lymphocytes in peripheral circulation is the T lymphocytes, which comprise about 70% of all lymphocytes in circulation. Maturation of T lymphocytes takes place in the thymus. The B cells comprise about 15% of circulating lymphocytes. Maturation of B lymphocytes starts from the red bone marrow and ends in the peripheral lymphoid organs. About 15% of circulating lymphocytes are natural killer (NK) cells, also called large granular lymphocytes [123, 384]. A subset of circulating lymphocytes observed to be without receptors as at the time were referred to as zero cells. These zero lymphocytes were thought to be the precursors of all other types of lymphocytes [385, 386]. Several researches conducted around the globe have provided definite evidences about the types and subsets of lymphocytes and their proportion in peripheral blood. Table 10.1 shows the current list of types and proportion of lymphocytes present in peripheral blood. Figures 10.8, 10.9, 10.10, and 10.11 are schematic representation of the functions of the major lymphocytes.

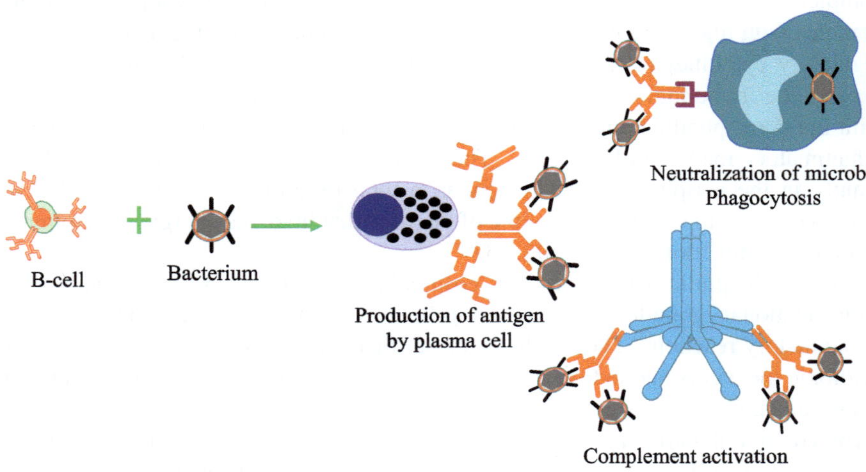

Fig. 10.8 Functions of B lymphocytes

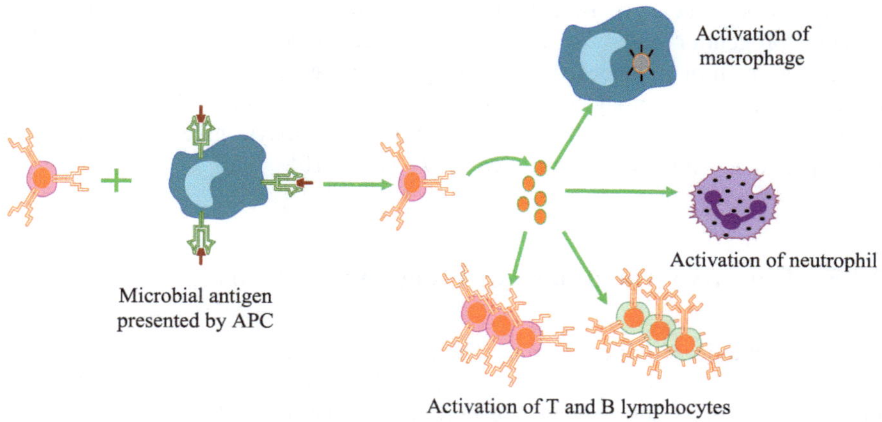

Fig. 10.9 Functions of helper T lymphocyte

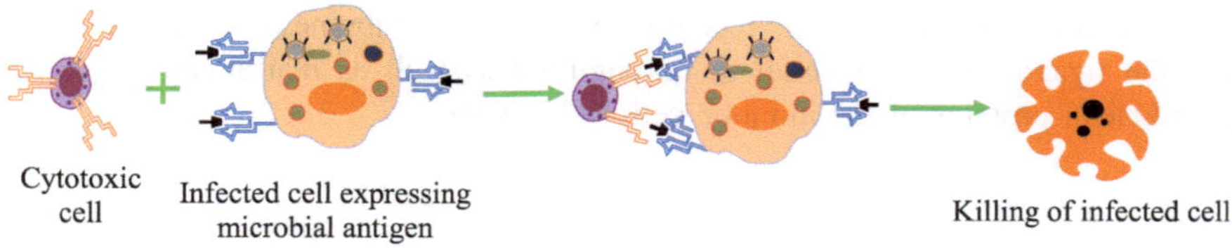

Fig. 10.10 Functions of cytotoxic T lymphocyte

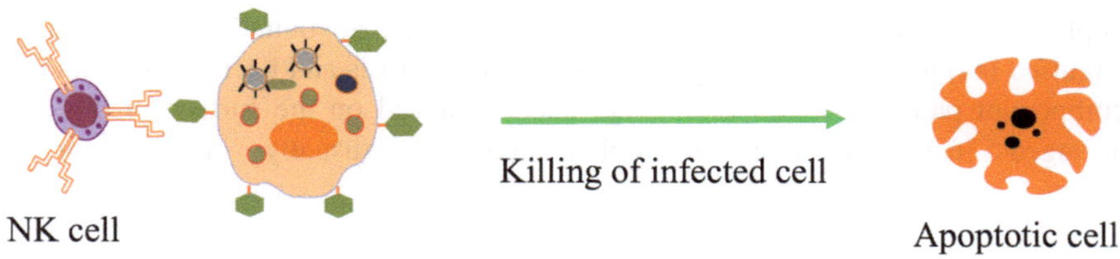

Fig. 10.11 Functions of NK T lymphocyte

10.4.1 Origin of Immune Response Cells

Immune cells are produced from a common hematopoietic stem cell (multipotent stem cell or hemocytoblast) through a series of cell divisions involving interactions with cytokines and their corresponding receptors. The differentiation of hemocytoblasts gives rise to a common myeloid and lymphoid progenitor cells. The lymphoid progenitor produces lymphoblast (which divides to produce B lymphocyte, T lymphocyte, NK cell, etc.) and lymphoid dendritic cell. Upon activation (in the presence of the required cytokines), B lymphocyte further differentiates to produce plasma cells (plasmacyte), which synthesize specific antibodies. The myeloid progenitor gives rise to myeloblast, red blood cells, and platelets. The myeloblast in turn produces the granulocytes through the differentiation of their corresponding promyelocytes and monoblast. The monoblast differentiates and matures to form monocyte. The differentiation of monocyte gives rise to resident macrophage and myeloid dendritic cell [277, 407–410].

10.5 Digestive Machinery of Phagocytotic Cells

The phagocytotic activities of innate immune cells involve the internalization of foreign bodies (including pathogenic microbes) into the cytoplasm forming phagosome with its subsequent digestion or degradation [123]. Many cells of the immune system can perform phagocytotic functions, but the main phagocytes are the neutrophils and macrophages [123] (Fig. 10.12). Dendritic cells also represent an integral member of the immune phagocytes [411]. Phagocytosis is mediated by the interaction between specific receptors of the cell membrane of the immune cell and the ligands of the material to be engulfed as well as rearrangement of the cytoskeletal proteins of the immune cells [412–414].

10.5.1 Phagocytosis and Antigen Presentation

The receptors of B cell can bind antigens such as diphtheria toxoid of the Diphtheria, Tetanus, and Pertussis (DTP) vaccine. The bound antigen is then engulfed into the B cell by receptor-mediated endocytosis and later digested into its component parts. The components are later displayed on the surface of the cell, arranged inside a histocompatibility molecule (usually class I or II

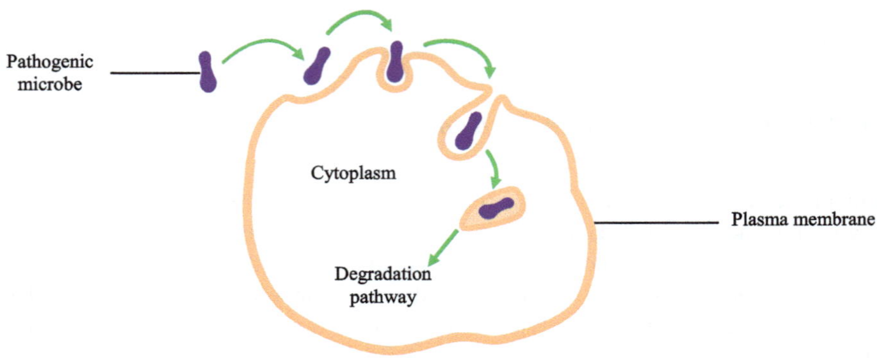

Fig. 10.12 Phagocytosis

histocompatibility complex, MHC). Like B cells, macrophages and dendritic cells can perform similar functions by engulfing foreign bodies including pathogenic microbes with degradation of the internalized contents and subsequent generation of antigenic peptides for presentation by MHC molecules classes localized on the surface of the macrophages and dendritic cells. The MHC molecule class I usually presents the processed antigenic peptides to $CD8^{+}$ T cells, whereas the class II MHC molecules present to $CD4^{+}$ T cells [411]. For this reason, the B cells, macrophages, and dendritic cells are referred to as phagocytic antigen-presenting cells [415–419]. The epithelial cells of the thymus also function as professional antigen-presenting cells [419]. A type of T cell—called Helper T cell (Th cell)—having a complementary T cell receptor can bind to the antigen-presenting (B lymphocyte) cell. This stimulates the production of lymphokines from the T cells, which in turn stimulate B cell to enter the cell cycle and develop by mitosis to form a clone of cells with identical B cell receptors. A subset of the B cell can differentiate into plasma cells that secrete soluble B cell receptors referred to as antibodies [415–419].

10.5.2 Major Histocompatibility Complex: A Key Component of the Digestive Machinery of Phagocytotic Cells

Peter Alfred Gorer (1907–1961) became the first in 1936 to describe major histocompatibility complex (MHC) [420]. However, it was George Davis Snell (1903–1996) who identified the gene locus of MHC. He shared the 1980 Nobel Prize in Physiology or Medicine with Baruj Benacerraf (1920–2011) and Jean Dausset (1916–2009) for "their discoveries concerning genetically determined structures on the cell surface that regulate immunological reactions" [421–424].

The MHC is a set of cell-surface molecules encoded by a group of genes that controls a major part of the immune system functions of mammals. The genes are essential to both the adaptive and innate immune systems [425]. MHCs bind to major peptide fragments derived from pathogens and display them on the cell surface for recognition by the appropriate T cells. Another name for MHC in humans is human leukocyte antigen (HLA). The HLA that the body's cells carry on their cell membranes are determined by the genes, located on chromosome 6. Some of the genes encoding MHC molecules have been conserved throughout the period of evolution, spanning over 300 million years ago, and predating the divergence of the amphibian lineage [425–428].

The major functions of MHC molecules include mediation of the interactions among leukocytes and with other cells of the body. Compatibility of donors for organ transplantation is determined by the MHC molecules. The MHC also determines the susceptibility of an organism to autoimmune disease. MHC displays only a part of the molecular registry of the host's immune system fraction. The part that is displayed is the epitope. Thus, antigen can be recognized either as self or nonself. MHC is known to be involved in a number of other functions including plastic functions, for instance, in the central nervous system [154, 429].

Classes of MHC

Figure 10.13 is a diagrammatic representation of MHC class I molecule. There are four classes of MHC. These classes also have their subtypes. Class I MHC encodes the human leukocyte antigen (HLA)-A, -B, and -C that all nucleated cells and platelets in the body carry to identify them as self [430, 431]. They present epitopes (also called antigenic determinants, which are the region on the surface of the antigen molecule or pathogen-derived peptides to which an antibody or T cell attaches itself) to cytotoxic T cells or natural killer cells, thus informing the immune system about the presence of pathogenic microbes, so that necessary steps are taken to curb their aggression. This way, MHC class I molecules act as ligands to the cytotoxic T and NK cells. Cell killing is carried out by a specific killer cell by

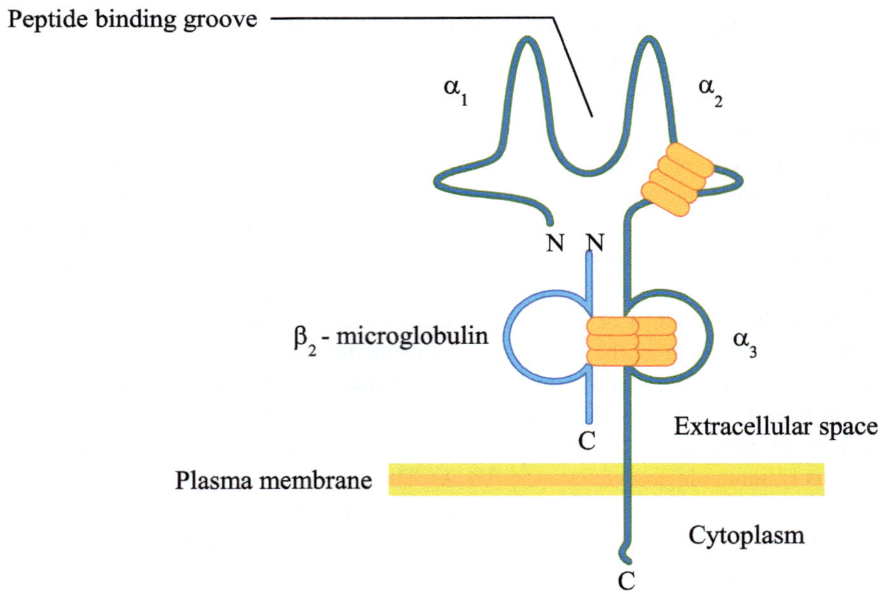

Fig. 10.13 Structure of MHC class I molecule. The class I MHC is made up of an α chain or subunit, which comprises three domains—α1, α2, α3. Domain α1 is located on β2 microglobulin molecule. The latter is not an MHC molecule; it is encoded on chromosome 15. The MHC class I molecule is attached to plasma membrane by domain α3. The different domains are covalently linked to each other. The antigenic peptide to be presented is carried in the peptide-binding cleft or groove. The specificity of the interaction between the MHC molecule and the presented antigenic peptide depends on the sequence of amino acids in the groove [445–447]

locating an MHC class I molecule to initiate the killing process of cells infected by pathogenic microbes [123, 425, 432–436].

The mechanisms of processing of MHC are complex and involve numerous biomolecular reaction steps. In a nutshell, the factory for peptide generation is the proteosome consisting of 28 domains or subunits. About 50% of these domains affect proteolytic activity. Proteolysed peptide residues are released into the cytosol and translocate through the transporter associated with antigen processing (TAP) into the endoplasmic reticulum (ER) to dock with MHC class I molecule in the ER lumen. TAP, a heterodimeric multimembrane-spanning polypeptide belongs to the family of ABC transporters. TAP is made up of two subunits that bind the peptide as well as with ATP. The process of MHC assembly in the ER involves peptide loading and dissociation of molecules. When the MHC is completely loaded with peptides, it is transported to the cell membrane through the secretory pathway. Before docking with the plasma membrane for externalization, loaded MHC protein molecule is modified by several cellular proteins in the Golgi apparatus and cytosol (Figs. 10.14 and 10.15). Some of the complexes may be recycled back to the ER for reuse [437–444]. Figure 10.16 shows how the loaded complex is recognized by T cell.

Figure 10.17 is a diagrammatic representation of class II MHC. MHC class II (also known as HLA-DP, -DQ, and -DR) is expressed in all professional antigen-presenting cell types such as macrophages, B cells, dendritic cells, group 3 innate lymphoid cells (e.g., plasmacytoid dendritic cells). But MHC class II may sometimes be present in other cells types induced by interferon. (The professional

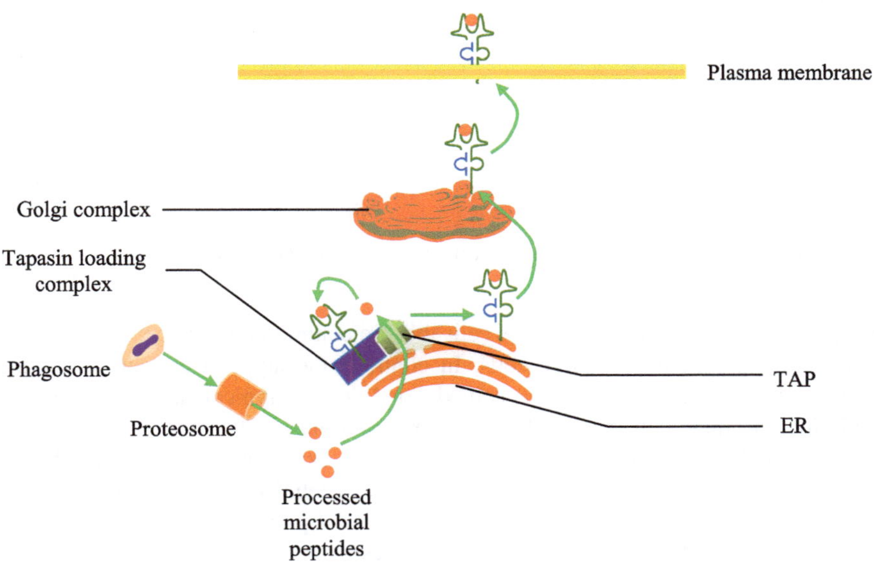

Fig. 10.14 Antigen processing and loading of MHC and trafficking to the plasma membrane (presentation) for recognition by T cell

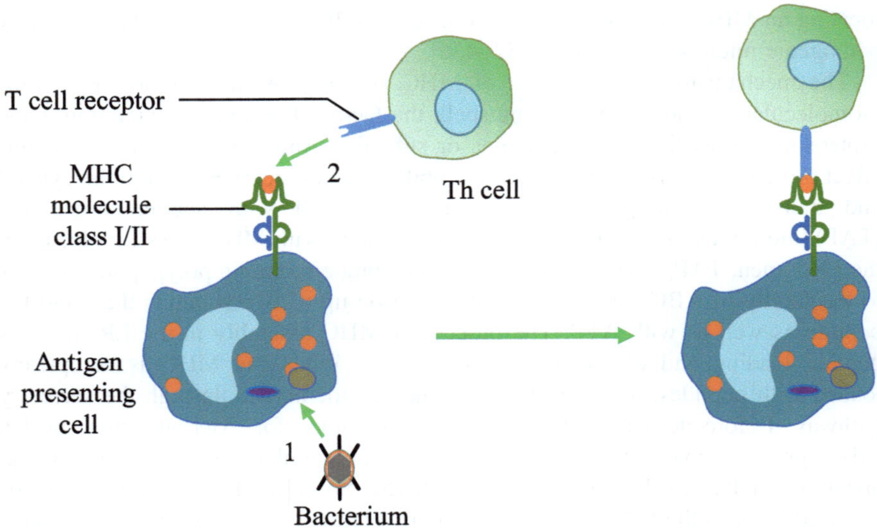

Fig. 10.15 Recognition of MHC molecule/antigen by Th cell

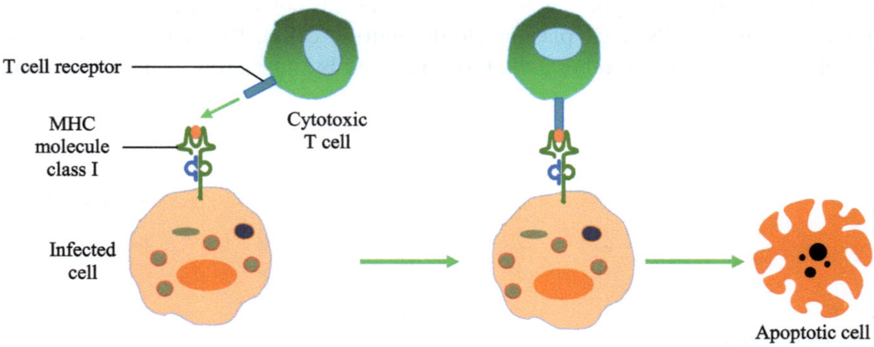

Fig. 10.16 Apoptotic activity of cytotoxic cell recognition of MHC molecule/antigen

antigen-presenting cells are immune cells that are committed to antigen presentation —an integral function of these cells. In contrast, the atypical antigen-presenting cells present antigens only on specified conditions. Atypical antigen-presenting cells include mast cell, basophil, eosinophil, fibroblasts.) In addition to the cells mentioned above, MHC class II molecules can be expressed by some endothelial, epithelial (e.g., intestinal, pulmonary, thymic epithelial cells), and stromal (e.g., lymph node) cells of [123, 448, 449]. MHC class II plays a useful role mediating antibody-mediated immunity, graft and host response in disease, and rejection of organ/tissue transplant [449–452].

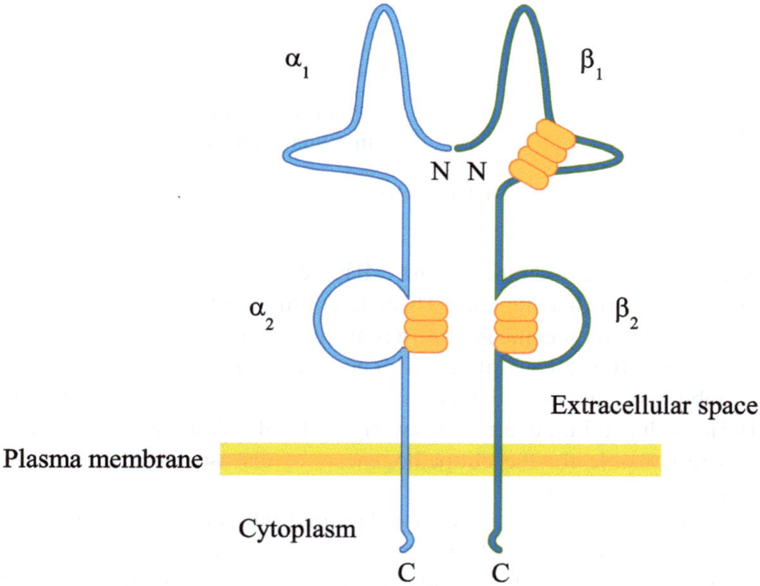

Fig. 10.17 Structure of MHC class II molecule. MHC-II comprises two α (α1, α2) and two β (β1, β2) chains. Both chains are anchored to the plasma membrane. The peptide-binding cleft is formed by both α1 and β1 molecules. The proteins of class II MHC are synthesized in the endoplasmic reticulum and are combined to form complexes. The complexed molecule is transported into the Golgi apparatus and passes through several modifications before it is presented to the surface of the cell [462–464]

Similar to the MHC I–peptide complex, formation of peptide necessary for complex formation with MHC-II molecule mediates antigen presentation and displays the epitope on the cell surface. Another similarity is that both classes I and II regions contain genes of related function and encode molecules that are responsible for antigen presentation to T cells [425]. However, in contrast to class I, antigens presented by class II peptides are derived from extracellular (exogenous) proteins, not cytosolic or intracellular (endogenous) [123, 453]. The phagocytes of MHC class II take up particles by phagocytosis into phagosomes. But some phagocytes may internalize particles by endocytosis into vesicles called endosomes, which subsequently fuse with lysosomes. The enzymes of lysosomes cleave the particles to form different peptides. The peptide fragments are then loaded into MHC class II molecule and are transported to the cell surface [453]. The helper T cell receptors identify MHC class II molecule on antigen-presenting cells and docks to the molecule. Antigen-presenting cells secrete cytokines that may polarize naive helper T cell (Th0) into a memory Th cell or an effector Th cell with different phenotypes —type 1 (Th1), type 2 (Th2), type 17 (Th17), or regulatory/suppressor (Treg) (see Fig. 10.18 for Treg cell function). Thus, the class II MHC molecules interact mainly with the T helper cell ($CD4^+$ cell). The MHC class II is important in

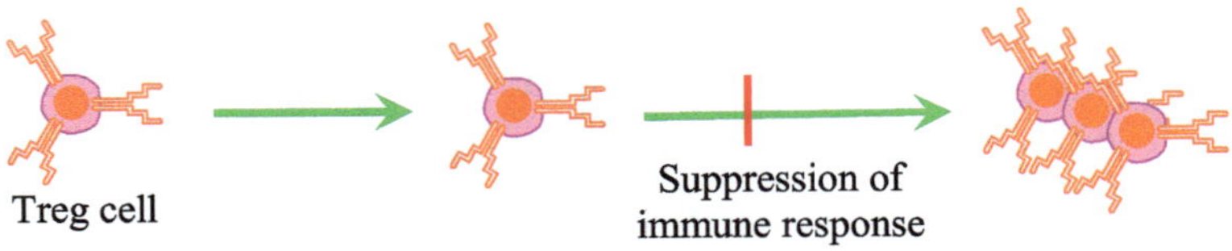

Fig. 10.18 Functions of regulatory T lymphocyte

mediation of immune tolerance. This MHC class is responsible for skewing immune response so that memory Th cells coordinate well when their memory recall is triggered upon secondary exposure to similar antigens. For instance, B cells with MHC class II present antigen to Th0. Activated B cells produce soluble immunoglobulins (antibody) which mediates humoral immune response. The MHC-II/Th-mediated immune responses trigger local inflammation, which, at least in part, is responsible for the etiopathogenesis of diseases such as inflammatory bowel diseases [454–461].

Deficiencies, defects, or genetic disorders of MHC class I or II result in rare primary immune deficiencies, which are inherited in an autosomal recessive pattern. Disorders in MHC class I or II and their signaling pathways may result from defects in the regulatory factors [465].

MHC class III molecules are encoded between classes I and II in the short arm of chromosome 6. The MHC class III region contains 57–61 structural genes spanning 654–759 kb of genomic DNA [425, 453, 466]. Proteins of class III molecules include complement components C2, C4a, C4b, factor B, cytokines (TNF-α; lymphotoxin A, LTA; lymphotoxin B, LTB), heat shock proteins. However, some genes encoded in this region have not been linked to immune response. Since these molecules are not membrane proteins, they do not have a role in antigen presentation. Class III MHC encodes the immunoglobulins [421, 467–469].

MHC class IV is encoded in the telomeric end of the class III region. Molecules encoded in this region are involved in both global and specific inflammatory responses. Hence, class IV region is sometimes termed "inflammatory region." The molecules include TNF family, allograft inflammatory factor 1, and heat shock protein 70 [425, 453, 470]. Deakin et al. [425] showed that genes involved in inflammation such as those encoding TNF, LTA, apolipoprotein M, HLA-B-associated transcripts (BAT_{2-5}) have been conserved for over 450 million years. The sequences, transcriptional regulation, and functionality are conserved within the inflammatory region in the different living things [425]. The inflammatory region seems to be associated with many diseases. For instance, a region between NF-κB inhibitor-like protein 1 and MHC class I polypeptide sequence may control susceptibility to inflammatory diseases [425]. Increasing evidences have implicated MHC genes in many inflammatory responses in the GI tract [471–474]. There is likelihood that disorders of MHC classes resulting in inflammatory diseases may run within members of a given family [475, 476]—this can be traced by such techniques as histocompatibility testing (see Reference Note 10.2).

Clinical Correlate 10.4
Applications of MHC—The Histocompatibility Tests

The histocompatibility tests are a set of investigations that analyzes the HLA alleles of an individual. Histocompatibility testing can be used to trace family lineage. It is useful in anthropology. In paternity testing, it is believed that HLA haplotype is shared by a man and his child, so if they do not have similar HLA haplotype, then the man is not the father of the child. Histocompatibility testing is also very useful generally in transplantation. Organ or tissue donor can be traced and matched by HLA typing. The histocompatibility test has also found important application in forensics. There are currently numerous methods used in histocompatibility testing [477–481]. **Serologic methods of histocompatibility testing**: This method uses sera containing specific antibodies to HLA antigens [482]. **Microcytotoxicity assay**: This assay involves the isolation of lymphocytes, possibly by centrifugation. The isolated lymphocytes are placed in wells of known antisera HLA. Complement is then added to each well and incubated—this leads to activation of the complement molecules with lysis of the lymphocyte, upon binding of the antibodies bind to the lymphocytes. The damaged cells allow uptake of vital stains such as eosin Y, trypan blue, or fluorescent stains such as ethidium bromide. Cells that are not damaged in the process do not take up the stain [479, 480, 483]. **Mixed leukocyte reaction**: In this reaction, the T cells of one individual interact with allogenic class II MHC antigen-bearing cells (e.g., B cells) of unrelated individual. When lymphocytes from individuals of different class II haplotypes are cultured together, blast cell transformation and mitosis occur. These changes are recorded by the addition of radioactive (tritiated, 3H) thymidine into the culture, and its incorporation into DNA is monitored [484]. **Antibody screening**: The method uses solid-phase ELISA to detect serum antibodies. This method is useful in organ transplantation. It is used to detect the presence of HLA antibodies in potential transplant recipients [479, 480]. **Molecular technique**: This method uses sequence-specific PCR, restriction fragment length polymorphism, and sequence-specific oligonucleotide probe to detect the genes that code for antigens [479, 480, 485–487].

10.5.3 Killer Receptor Signaling: To Die or Not To Die?

The killing of microbes is mediated by receptors generally referred to as killer receptors, expressed on the surface of certain immune cells especially the cytotoxic T and NK cells. The cytotoxic T cells, in addition to their widely recognized receptors, express killer receptors such as natural killer group 2, member D (NKG2D) receptor (also known as CD94), killer inhibitory receptors (KIR), p58.1, p58.2, and p70. These receptors of cytotoxic cells recognize their molecules produced from microbes, infected cells, tumor cells, stress molecules. The killer cells are activated on binding to these ligands, thereby exerting their effector functions, proliferating actively, and producing proinflammatory cytokines such as IFN-γ, TNF-α, or releasing lytic enzymes to destroy the microbe, infected cells, tumor cells, or neutralize the stress molecules [488–491]. It should be noted that lytic functions can also be carried out by other cells of the immune system, in particular the lymphocytes. For example, hepatic $CD3^+$ lymphocytes could be induced to lyse NK-sensitive K562 target cells [492]. Memory $\alpha\beta$ T cells and $\gamma\delta$ T cells may also express killer receptors such as lectin-like receptor [493].

Killer receptor cells form highly effective microsurveillance system of the body through a tightly regulated activating and inhibiting signaling network. The resulting action of killer receptors is determined by the integration of both inhibiting and activating signals. For instance, the functional outcome of NK cells is determined by integrating both activating and inhibitory signals resulting in a highly controlled response, which mediates cytotoxicity against transformed cells and, in addition, release a couple of cytokines critical to the immune response [494]. The number of such killer receptors exceeds 20. Examples of activating receptors in human NK cells and a subset of lymphocytes include CD16, the short-tail members of KIRs, CD94/NKG2C, NKG2D, Fc receptor CD16, 2B4, NKp30, NKp44, and NKp46. Activating receptors signal through association with either DNAX-activating protein of molecular mass 10 kD (DAP10; also known as tyrosine kinase-binding protein, TYROBP, or killer cell-activating receptor-associated protein, KARAP) which allows PI-3-kinase activation. Adhesion molecules, such as lymphocyte function-associated antigen 1 (LFA-1, an integrin) and DNAX accessory molecule 1 (DNAM-1), are also important for the lytic function of natural killer cells, but they do not associate directly with the known immunotyrosine-based activation motifs-containing molecules. The inhibitory receptors in natural killer cells include the long-tail members of the KIR family, CD94/NKG2A, leukocyte-associated immunoglobulin-like receptor-1, and killer cell lectin-like receptor subfamily G member 1 (KLRG-1) which signal via cytoplasmic immunotyrosine-based inhibitory motifs. The inhibitory receptors found on NK cells include the KIR receptors killer cell immunoglobulin-like receptor, two Ig domains and long cytoplasmic tail 1 (KIR2DL1) and KIR2DL2, which usually occur in complex with peptide-bound MHC class I molecules, as well as CD94 and NKG2A receptors, which bind to peptide–HLA-E complex. The C-type lectin receptor KLRG-1 binds to cadherin, and the NK cell receptor LAIR-1 molecule binds to collagen [494–498].

KIRs are surface inhibitory receptors, which are specific for allelic forms of molecules of MHC class I. The binding of KIR to MHC class I leads to inhibition of NK cell activation. In normal situation, the activating receptors of NK cells predominate. As consequence, cells not expressing MHC class I are killed by NK cells.

The NK-mediated killing of these cells represents an important defense mechanism that prevents the spread of pathogens and tumors in the body [497–500].

The interaction of KIRs of NK cells with self-MHC class I molecules is the basis of tolerance of NK cells to self, at the same time maintaining functional competence to foreign particles [501–503].

Inhibiting and activating receptors are regulated by various kinases and phosphatases [501].

10.5.4 Antibodies: Origin, Structure, Functions, and Signaling Mechanisms

An antibody (abbr. Ab), also called immunoglobulin (abbr. Ig), is a heterodimeric Y-shaped protein produced by activated B lymphocytes (called plasma cells) that is used by the immune system to identify and neutralize pathogens such as bacteria, protozoa, and viruses. Immunoglobulins are globular plasma proteins (measuring about 150 kDa in weight), containing both amino and sugar residues. Hence, they are also called glycoproteins [504–506]. Immunoglobulin recognizes a special molecule of the invader, cellular debris or abnormal cell (antigen) in the body. Immunoglobulins are composed of two heavy (H) and two light (L) chains. They can be separated functionally into variable (V) domains that bind the antigens and the constant (C) domains that specify effector functions such as activation of complement or binding to Fc receptors [504].

Both antibody and antigen have a long history. The first immunoglobulin fraction, Bence Jones protein, was discovered in 1845 [507]. Bence Jones protein is named after the discoverer, Henry Bence Jones (1813–1873), an English physician who successfully described a globulin protein or immunoglobulin light chain, weighing 22–24 kDa in urine [508]. In 1890, the Japanese physician and bacteriologist, Shibasaburo Kitasato (1853–1931), and the German physiologist, Emil von Behring (1854–1917), using experimental rabbits as a model, reported the existence of an agent in the serum that could neutralize tetanus antitoxin. Von Behring is credited for discovering diphtheria antitoxin. Kitasato and von Behring discovered passive immunity: the acquisition of resistance to pathogens through the transference of that property from an immunized donor. The 1901 Nobel Prize in Physiology or Medicine was awarded to von Behring "for his work on serum therapy, especially its application against diphtheria." In 1891, Paul Ehrlich introduced the term "antikörper" (antikörper is the German word for antibodies) to mean an agent in the serum that is able to discriminate between two immune substances. His experiment on animals using two different toxins (rycin and abrin) showed that two different antikörpers were generated. Thus, he had discovered the mechanisms of defense by adaptive immune system, i.e., the humoral immune response mediators that began the era of modern immunology [509–511].

In 1899, another renowned physician László Detre (Ladislas Deutsch or Ladislaus Deutsch) (1874–1939) from Hungary observe that certain substances from microbes induced the production of immunoglobulins. He named the substances "substances immunogenes ou antigens" (a French phrase meaning "antigenic or immunogenic substances"). Detre maintained that these substances were precursors of antibodies. After several works and collaboration with the renowned Russian biologist Élie Metchnikoff, in 1903, the duo published a paper showing that antigen induced the production of antibodies. They opined that antigen was a contraction of "antisomatogen = immunkörperbildner" [520, 521]. "Antisomatogen + immunkörperbildner" is the substance that induced the production of antibodies. Thus, an antibody and its antigen are repetitive terms that make no sense—in essence, it is a tautology [504, 522].

In the beginning of the late twentieth century, Gerald Maurice Edelman (1929–2014) and Rodney Robert Porter (1917–1985) working on the structure of antibodies successfully described the "Y" shape structure of antibodies. They subsequently won the 1972 Nobel Prize in Physiology or Medicine "for their discoveries concerning the chemical structure of antibodies." So, by this time, both the structure and functions of antibodies were well known [513, 523–525]. The structure of a typical antibody is shown in Fig. 10.19.

An antigen is an antibody generator produced from various sources. Antigens are processed in the body, but they either enter the organism from the outside or are produced in vivo. Antigen-presenting cells identify, take up, and process antigens into fragment via specific mechanisms such as endocytosis or phagocytosis. Using MHC class I/II, antigen-presenting cells present fragments of the processed antigens to T helper cells. T helper cells become activated thereby secreting cytokines, which then activate other immune cells including cytotoxic T lymphocytes, B cells, and macrophages. Antigen is produced in the body de novo in course of biochemical processes, or its production may occur as a result of microbial infection. Fragments of the antigen are presented on the cell surface in association with MHC class I/II molecules. T cells recognize these molecules and secrete substances that lyse the infected cells. Some antigens are formed from the body's proteins that become recognized by the body's immune system as foreign. These antigens are called autoantigens, which may induce the development of autoimmune diseases. Tumor or cancer cell-associated autoantigen can give rise to the development of tumor or cancer in the body [504, 526, 527].

Types of Immunoglobulins: Structure and Functions

In 1939, the Swedish biochemist, Arne Wilhelm Kaurin Tiselius (1902–1971), and the American immunochemist, Elvin Abraham Kabat (1914–2000), using electrophoresis technique, separated immunized serum into albumin, alpha globulin, beta globulin, and gamma globulin fractions. Absorption of the serum against the antigen depleted the gamma globulin fraction, yielding gamma globulin, immunoglobulin (Ig), and IgG. Further separation was made with columns of

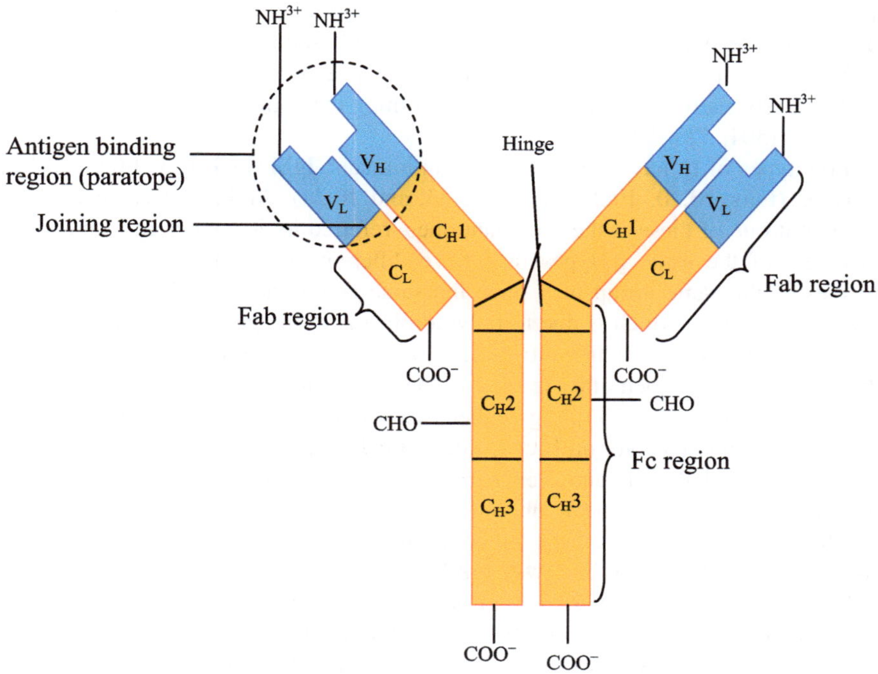

Fig. 10.19 Structure of Ig monomer. The Ig monomer polypeptide chain comprises two heavy chains and two light chains connected by disulfide bonds. Each chain is composed of structural domains called immunoglobulin domains, which are further classified as variable (V_H and V_L) or constant (C_H1, C_H2, and C_H3 and sometimes with C_H4 plus C_L). Each part of the chain is linked together by disulfide bridges. In addition, the heavy chain contains a hinge region. The heavy chains have different classes, which include alpha (α), gamma (γ), delta (δ), epsilon (ε), and mu (μ), and they define the antibody's isotypes IgA, G, D, E, and M, respectively. All Ig isotypes have identical C region. The V region varies among Ig isotypes. Individual immunoglobulin antigenic determinant is called *idiotype* and is contained within V domains. Each V or C domain consists of approximately 110–130 amino acids, with a molecular weight of about 12,000–13,000 kDa [504, 512–515]. The antigen-binding region of the constant chain is called fragment, antigen-binding (Fab) and determines the specificity of the antigen. The paratope is the terminal end of the Ig containing the heavy- and light-chain variable domains—it is where the antigen binds to. The variable domain (F_V region) is the most important region for binding to antigens. Precisely, variable loops of β-strands, three each on V_L and V_H chains, are responsible for binding to the antigen. These loops are called complementarity determining regions. The site on the antigen that is bound to the paratope is called the *epitope*. Forces involved in the interaction between epitope and Fab include electrostatic forces, hydrogen bonds, hydrophobic interactions, and van der Waals forces. The interaction between the Fab and epitope is relatively weak and non-specific; hence, the antigen–antibody complex is relative and reversible. There is also possibility that an Ig may interact with different antigens [504, 512–519]. The fragment, crystallizable (Fc) region modulates immune activity and determines the Ig class effect. Class effects of antibodies include: opsonization, agglutination, hemolysis, complement activation, and degranulation of mast cells, basophils, and eosinophils, as well as neutralization. Hence, the Fc-mediated effects are directed at effector cells or molecules. The Fc region ensures that each antibody generates an appropriate immune response for a given antigen, through its binding to a specific class of Fc receptors, and complement proteins [504, 514, 515, 519]

different sizes (size-exclusion chromatography) to yield immunoglobulins "heavy" (IgM) and "regular" (IgA, IgE, IgD, IgG). Arne Tiselius later received the 1948 Nobel Prize in Chemistry "for his research on electrophoresis and adsorption analysis, especially for his discoveries concerning the complex nature of the serum proteins" [504, 528, 529].

Five types of Ig have been described in humans: IgM, IgG, IgA, IgD, and IgE isotypes (Figs. 10.20, 10.21, 10.22, 10.23, and 10.24). Antigens have many epitopes, but one of these is the dominant one—referred to as the determinant. Some members of the same isotypes may have different determinants, whereas other species of the same group may have the same determinant. Such determinant is called an allotype, which is due to polymorphisms of the gene product. Mutations can occur in any part of the antibody. Also, different parts of Ig may interact in different ways with each other leading to the formation of different Ig isotypes and subclasses [504, 534–536]. Ig either occurs in its soluble form or as a membrane-bound receptor on B cell, called B cell receptor. This receptor facilitates the activation of B cells and their subsequent differentiation into plasma cells or memory B cells (see the next subsection for details) [512, 513, 537, 538].

The isotypes also have their different subclasses. For instance, IgG has four subclasses, IgG1–4; IgA has two, IgA1–2. Some of these subclasses still have different isoforms [531, 539, 540]. Changes in the level of these antibodies are observed in infection, immunization against a particular organism. Serum and salivary level of IgA subclasses in HIV disease and after immunization have been reported to change disproportionally [541, 542].

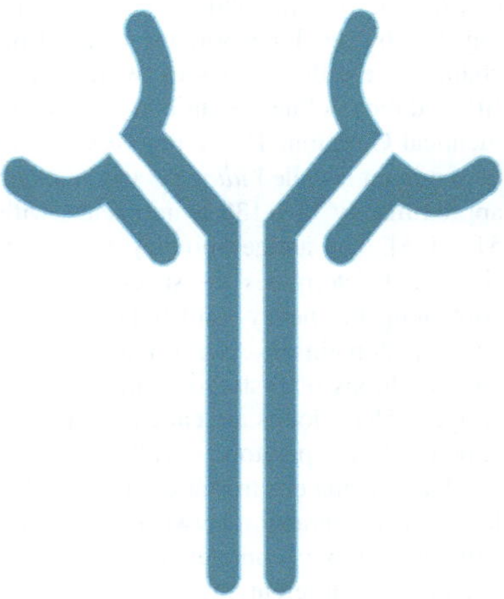

Fig. 10.20 Schematic representation of IgG. This Ig is the major antibody present in the blood, but it is able to enter tissue spaces to recognize and bind to antigens. IgG can cross the placenta to provide passive immunity to the fetus [530–533]

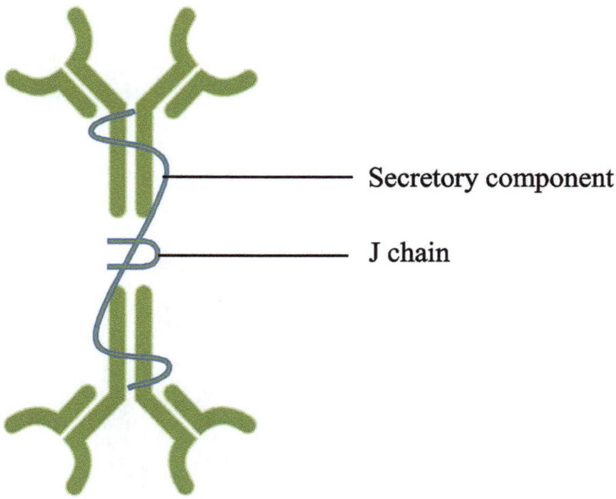

Fig. 10.21 Schematic representation of IgA. This Ig occurs as a dimer secreted to mucosal areas, such as the GI (especially saliva), respiratory, and urogenital tract. The Ig is also found in tears and breast milk. IgA is released in body fluids to protect the body canal and prevents colonization by pathogens [519, 543–547]

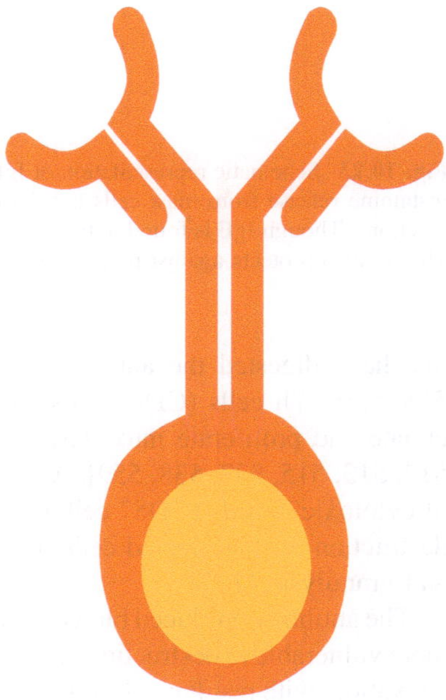

Fig. 10.22 Schematic representation of IgD. This Ig occurs as membrane-bound antibody. The primary function of this Ig is to act as an antigen receptor on B cells that have not been exposed to antigens. Ig can activate mast cells and basophils to synthesize antimicrobial factors [548–550]

Production of Antibodies and Their Roles in Immune Response

The B cells are the antibody-producing cells in the body. B cells are activated by the presence of an antigen. The antigen binds to the B cell antigen receptor and is internalized and processed into peptides that activate helper T cells, which have been previously produced from T cells, upon activation by antigen-presenting cells

Fig. 10.23 Schematic representation of IgE. This Ig is a dimer that binds to allergens and triggers histamine release from mast cells and basophils. The activity of IgE forms the basis of allergic reactions. Though IgE is found in trace amounts in the blood, it is able to trigger allergic reactions. This Ig also protects against parasitic worm invasion [551–554]

that have digested the antigen, and displaying the MHC–antigen complex [154, 515, 559]. Th cells ($CD4^+$ cells) secrete cytokines that assist the B cell to differentiate and proliferate into plasma cells that produce specific antibody [123, 154, 512, 513, 515, 537, 538, 559]. Antigen-presenting cells also stimulate the formation of cytotoxic T cells ($CD8^+$ cells)—responsible for the killing of infected cells. The destruction of the infected cells may result from the release of granzyme, perforin, and granulysin [123].

The antibody produced binds to antigens on the surface of pathogens, thus marking them vulnerable to destruction. This can occur either by neutralization or complement activation, with resultant elimination of the foreign body [123]. Neutralization occurs when the antibodies (e.g., IgE, IgG) bind to the corresponding antigen, making the pathogen more susceptible to attachment by the phagocyte—a process known as opsonization. The antibody Fab region binds with epitopes of the antigen, while the Fc region binds to macrophages or neutrophils, initiating a series of reactions that culminate in the killing of the pathogen. In addition, antibodies can initiate the activation of the complement proteins, in which C3b or C4b binds the antigen thus making the pathogen susceptible to phagocytosis [123, 154, 515, 559–562].

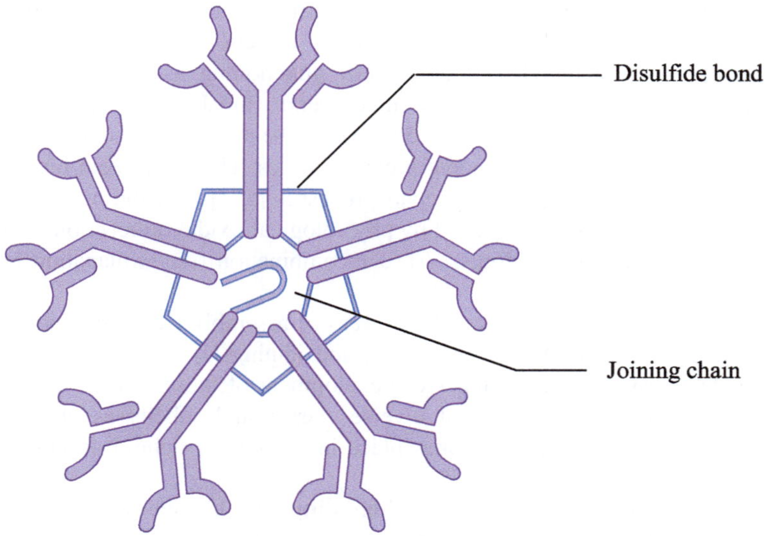

Fig. 10.24 Schematic representation of IgM. This Ig is the largest antibody, occurring in two different forms—on the surface of B cells (as monomer) and in a secreted form (as pentamer). This Ig is responsible for eliminating pathogenic microbes in the early stages of B cell-mediated (humoral) immunity—this ensures ongoing immune response before sufficient IgG is produced. This Ig tends to remain in the blood, which is required for adequate response and killing of pathogenic bacteria [555–558]

This branch of acquired immune system that is mediated by B cell antibody production is referred to as antibody-mediated immunity. This type of immunity is active during the acute phase of infection [123, 154]. Cell-mediated immunity involves the activation of antigen-specific cytotoxic T cells, natural killers, and macrophages that leads to destruction of infected or damaged cells. Cell-mediated immunity is active in circumstances of chronic infections [123].

Antibody Class Isotype Switching: Mature, unactivated B cells, referred to as naïve B lymphocytes, homing in circulation, have both IgM and D on their cell surface. However, not all B cells express these two Ig. So, the appearance of these two antibodies makes the cell ready to respond to the presence of antigens. A subset of the differentiating cells of the activated B lymphocytes pass through a process referred to as antibody class isotype switching [563–566]. Isotype switching is a mechanism of DNA recombination in which genes that code for antibody diversify effector functions of Ig [567]. During this process, the production and the expression of antibodies on lymphocytes can change from either IgM or IgD to IgE, IgA, or IgG [563]. The mechanisms of isotype switching are not fully understood, but emerging evidences indicate that Ku and ICOS play a crucial role [567–570]. ICOS is the acronym for inducible costimulator protein—which is functionally related to CD28, an integral costimulatory receptor expressed on resting T lymphocytes [569, 570]. Ku is named after the two letters of the name of the initial patient with connective tissue disease whose serum was used as a prototype [571]. Ku is a high molecule weight acidic nuclear protein antigen, composed of a complex of two proteins (Ku70 and Ku80), functioning as a heterodimer, and responsible for binding DNA double-strand breaks via activation of DNA-dependent protein kinases [567, 568, 572]. The protein also contains catalytic subunit of 470 kDa, DNA-dependent protein kinases. The Ku protein binds to the ends of DNA molecules, initiating repair processes [573]. Discontinuity or breaks in DNA double strand can be caused by physiological oxidation reactions, ionizing radiation, chemotherapeutic drugs, V(D)J recombination (see explanation below), and intranuclear processes [573, 574].

V(D)J Recombination: This is the process of assembly (recombination) of the gene segments variable (V), diversity (D), and joining (J), occurring in B and T lymphocytes during the early stages of development of these cells. V(D)J recombination leads to the generation of diverse classes and subclasses of functional antibodies. This process usually takes place in primary lymphoid organs such as bone marrow and thymus [575–578].

Antibody Fc Region–Fc Receptor Signaling: For efficient antibody activity, the Fc region of the antibody must be recognized by its receptors (Fc receptors, FcR) present on immune cells, in particular T lymphocytes. Antibodies communicate with cells of the immune system via their Fc region. The association of Fc region of an antibody with FcR of immune cells leads to activation of the effector function of the cell. If the cell is a macrophage, it will phagocytose foreign bodies. If the immune cell is a natural killer, it will trigger the production of destructive peptides and cytokines that will lead to the killing of the invader. If the cell is a mast cell, basophil, or neutrophil, then the result is degranulation [579–584]. The deletion or loss of FcR of immune cells can have implications on the loss of tolerance for the FcR-specific antigen or play some roles in the pathophysiology of autoimmune diseases [585].

Monoclonal Antibodies

In 1975, the German biologist Georges Jean Franz Köhler (1946–1995) and the Argentinian biochemist César Milstein (1927–2002) discovered monoclonal antibodies (mAbs) produced by "hybridoma technology." They made fusions of B cells and myeloma cells to produce hybridomas, which produced antibodies to known antigens [586, 587]. Georges J. F. Köhler and César Milstein together with Niels K. Jerne were awarded the 1984 Nobel Prize in Physiology or Medicine "for theories concerning the specificity in development and control of the immune system and the discovery of the principle for production of monoclonal antibodies" [588, 589]. The Danish immunologist Niels Kai Jerne (1911–1994) also contributed immensely to the contemporary understanding of the immune system functions. Among other things, he opined that the body's immune system produces specific antibodies to fight pathogenic microbes. He also proposed that cells of the immune system learn to recognize self—a phenomenon that takes place in the thymus [589].

Table 10.2 List of clinical trial databases

Name	Website
ClinicalTrials.gov	http://clinicaltrials.gov/
EU Clinical Trials Register	http://www.clinicaltrialsregister.eu/
NHLBI Trials	http://www.nhlbi.nih.gov/studies/nhlbi-trials/browse-category
International Clinical Trials Registry Platform (ICTRP)	http://www.who.int/ictrp/en/
CenterWatch	http://www.centerwatch.com/
ClinicalTrials.com	http://www.clinicaltrials.com/
National Institute of Mental Health Clinical Trials	https://www.nimh.nih.gov/health/trials/index.shtml
Clinical Connections	https://www.clinicalconnection.com/default

The works of these Nobelists led to the production of antibodies from laboratory animals (mice) for use in humans. However, the application of mice-derived antibodies was difficult because of the negative reaction observed due to injections of mice-derived antibodies in humans. In 1988, the British biochemist, Sir Gregory Paul Winter (1951–) and coworkers pioneered the techniques to humanize mAbs—which allowed to get rid of the reactions caused by the use of mAbs in some patients [590–592].

Monoclonal Abs are monospecific antibodies made by identical immune cells that are all clones of a unique parent cell. Monoclonal Abs can only bind to the same epitope. Due to its clinical significance, mAbs over the past decades have gained worldwide application in the treatment of various diseases related to the immune system, inflammation, allergy, and cancer. Numerous medications have been produced using the technology of mAbs. The names of these medications have the suffix—"mab" [593, 594]. Examples of clinically important monoclonal antibodies with anti-inflammatory actions are infliximab and adalimumab. Both of these medications function via inhibition of TNF-α and have been successfully used in IBD (Crohn's disease and ulcerative colitis). Other monoclonal antibodies have been used for other inflammatory diseases, allergic diseases, asthma, cancer, and transplant rejection [595–601]. Newer mAbs are currently under development or are in their various phases of clinical trials [602–604] (Table 10.2).

10.6 Initiation of Antibody Production Is Cooperatively Linked to the Induction of Adaptive Immunity

The adaptive immune system is the second arm (line of defense) of the immune system. Adaptive immunity also known as the acquired immune system or specific immune system can be defined as the presence of cells (e.g., T and B lymphocytes in higher vertebrates) displaying the clonal expression of a colossal repertoire of receptors (i.e., T cell and B cell antigen receptors), the diversity of which results from somatic DNA rearrangements [123, 605, 606]. The adaptive immune system creates immunological memory after an initial response to a specific pathogen and leads to an enhanced response to subsequent encounters with that pathogen. This process of acquired immunity is the basis of vaccination [124].

The main functions of the adaptive immune response are the recognition of nonself antigens, generation of specific effector responses that culminate in elimination of pathogens, and the development of an immunologic memory [123, 416, 607–609].

It should be noted that innate and adaptive immune responses are inseparable entities, rather are complementary [123]. Both lines of immune defense comprise humoral and cellular components of host cell defense [154].

10.7 Conclusions

The GI tract immune subsystem represents a crucial division of the body's immune system that responds to invasion by pathogenic or other pathological signals. The gut immune system is the largest immune system in the body and plays a critical role in the pathogenesis of many diseases including GI cancers, lymphoma, infections, chronic inflammation, and autoimmune diseases.

Recommended Readings
Original Articles

1. Akhmatova NK, Egorova NB, Kurbatova EA, Akhmatov EA (2014) Activation of innate immunity by bacterial ligands of Toll-like receptors. Front Immunol 5:89
2. Barone F, Patel P, Sanderson JD, Spencer J (2009) Gut-associated lymphoid tissue contains the molecular machinery to support T-cell-dependent and T-cell-independent class switch recombination. Mucosal Immunol 2(6):495–503
3. DeWitt RC, Wu Y, Renegar KB, Kudsk KA (2000) Bombesin recovers gut-associated lymphoid tissue and preserves immunity to bacterial pneumonia in mice receiving total parenteral nutrition. Ann Surg 231(1):1–8
4. Severson KM, Mallozzi M, Driks A, Knight KL (2010) B cell development in GALT: role of bacterial superantigen-like molecules. J Immunol 184(12):6782–6789
5. Stanisavljević S, Lukić J, Soković S, Mihajlovic S, Stojković MM, Miljković D, Golić N (2016) Correlation of gut microbiota composition with resistance to experimental autoimmune encephalomyelitis in rats. Front Microbiol 7:2005

Review Articles

1. Adamczyk-Sowa M, Medrek A, Madej P, Michlicka W, Dobrakowski P (2017) Does the gut microbiota influence immunity and inflammation in multiple sclerosis pathophysiology? J Immunol Res 2017:7904821
2. Hashi H, Yoshida H, Honda K, Fraser S, Kubo H, Awane M et al (2001) Compartmentalization of Peyer's patch anlagen before lymphocyte entry. J Immunol 166(6):3702–3709
3. Jung C, Hugot JP, Barreau F (2010) Peyer's patches: the immune sensors of the intestine. Int J Inflam 2010:823710
4. McGhee JR, Fujihashi K (2012) Inside the mucosal immune system. PLoS Biol 10(9):e1001397
5. Nayak A, Ferluga J, Tsolaki AG, Kishore U (2010) The non-classical functions of the classical complement pathway recognition subcomponent C1q. Immunol Lett 131(2):139–150
6. Nesargikar PN, Spiller B, Chavez R (2012) The complement system: history, pathways, cascade and inhibitors. Eur J Microbiol Immunol (Bp) 2(2):103–111
7. Paul WE (2011) Bridging innate and adaptive immunity. Cell 147(6):1212–1215
8. Ruth MR, Field CJ (2013) The immune modifying effects of amino acids on gut-associated lymphoid tissue. J Anim Sci Biotechnol 4:27
9. Sanderson IR (2007) Dietary modulation of GALT. J Nutr 137(11):2557S–2562S
10. Susskind B (2007) Methods for histocompatibility testing in the early 21st century. Curr Opin Organ Transplant 12(4):393–401
11. Watzl B, Girrbach S, Roller M (2005) Inulin, oligofructose and immunomodulation. Br J Nutr 93(1):S49–55

Books

1. Beutler B, Wagner H (eds) (2002) Toll-like receptor family members and their ligands. Springer, Germany
2. Gadjeva M (ed) (2014) The complement system—methods and protocols. Springer, New York
3. Murphy K, Travers P, Walport M (2007) Janeway's immunobiology, 7th edn. Garland Science, Connecticut, United States
4. Spickett G (ed) (2013) Oxford handbook of clinical immunology and allergy, 3rd edn. Oxford University Press, Oxford

Nobel Lectures

1. Bruce Beutler (2011) How mammals sense infection: from endotoxin to the Toll-like receptors. In physiology or medicine. http://www.nobelprize.org/nobel_prizes/medicine/laureates/2011/beutler-lecture_slides.pdf. Accessed 5 Sept 2017

2. Hoffmann J (2011) The host defense of insects: a paradigm for innate immunity. Nobel Lecture in physiology or medicine. https://www.nobelprize.org/nobel_prizes/medicine/laureates/2011/hoffman-lecture.pdf. Accessed 20 Oct 2017
3. Lewis EB (1995) The bithorax complex: the first fifty years. Nobel Lecture in physiology or medicine. https://www.nobelprize.org/nobel_prizes/medicine/laureates/1995/lewis-lecture.pdf. Accessed 27 Oct 2017
4. Nussenzweig MC (2011) Ralph Steinman and the discovery of dendritic cells. Nobel Lecture in physiology or medicine. https://www.nobelprize.org/nobel_prizes/medicine/laureates/2011/steinman_lecture.pdf. Accessed 20 Oct 2017
5. Nüsslein-Volhard C (1995) The identification of genes controlling development in flies and fishes. Nobel Lecture in physiology or medicine. https://www.nobelprize.org/nobel_prizes/medicine/laureates/1995/nusslein-volhard-lecture.pdf. Accessed 20 Oct 2017
6. Piro A, Tagarelli A, Tagarelli G, Lagonia P, Quattrone A (2008) Paul Ehrlich: the Nobel Prize in physiology or medicine 1908. Int Rev Immunol 27(1–2):1–17
7. Schmalstieg FC Jr, Goldman AS (2008) Ilya Ilich Metchnikoff (1845–1915) and Paul Ehrlich (1854–1915): the centennial of the 1908 Nobel Prize in physiology or medicine. J Med Biogr 16(2):96–103
8. Wieschaus E (1995) From molecular patterns to morphogenesis—the lessons from drosophila. Nobel Lecture in physiology or medicine. https://www.nobelprize.org/nobel_prizes/medicine/laureates/1995/wieschaus-lecture.pdf. Accessed 20 Oct 2017

Bibliography

1. Mowat AM, Agace WW (2014) Regional specialization within the intestinal immune system. Nat Rev Immunol 14:667–685
2. Verbrugghe P, Kujala P, Waelput W, Peters PJ, Cuvelier CA (2008) Clusterin in human gut-associated lymphoid tissue, tonsils, and adenoids: localization to M cells and follicular dendritic cells. Histochem Cell Biol 129(3):311–320
3. Ree HJ, Rege VB, Knisley RE, Thayer WR, D'Amico RP, Song JY, Crowley JP (1980) Malignant lymphoma of Waldeyer's ring following gastrointestinal lymphoma. Cancer 46 (7):1528–1535
4. Brandtzaeg P, Kiyono H, Pabst R, Russell MW (2008) Terminology: nomenclature of mucosa-associated lymphoid tissue. Mucosal Immunol 1:31–37
5. Plank L (2016) Waldeyer's ring. In: Volavšek M (ed) Head and neck pathology. Springer International Publishing, Cham
6. Kaneko F, Togashi A, Saito S, Sakuma H, Oyama N, Nakamura K et al (2011) Behçet's disease (Adamantiades-Behçet's disease). Clin Dev Immunol 2011:681956
7. Ling E, Shubinsky G, Press J (2007) Increased proportion of $CD3^{+}CD4^{-}CD8^{-}$ double-negative T cells in peripheral blood of children with Behcet's disease. Autoimmun Rev 6(4):237–240

8. Demetriades N, Hanford H, Laskarides C (2009) General manifestations of Behçet's syndrome and the success of CO_2-laser as treatment for oral lesions: a review of the literature and case presentation. J Mass Dent Soc 58(3):24–27
9. Jurge S, Kuffer R, Scully C, Porter SR (2006) Mucosal disease series. VI. Recurrent aphthous stomatitis. Oral Dis 12(1):1–21
10. Preeti L, Magesh K, Rajkumar K, Karthik R (2011) Recurrent aphthous stomatitis. J Oral Maxillofac Pathol 15(3):252–256
11. Barrons RW (2001) Treatment strategies for recurrent oral aphthous ulcers. Am J Health Syst Pharm 58(1):41–50; quiz 51–3
12. Ship JA (1996) Recurrent aphthous stomatitis. An update. Oral Surg Oral Med Oral Pathol Oral Radiol Endod 81(2):141–147
13. Natah SS, Konttinen YT, Enattah NS, Ashammakhi N, Sharkey KA, Häyrinen-Immonen R (2004) Recurrent aphthous ulcers today: a review of the growing knowledge. Int J Oral Maxillofac Surg 33(3):221–234
14. Slebioda Z, Szponar E, Kowalska A (2013) Recurrent aphthous stomatitis: genetic aspects of etiology. Postepy Dermatol Alergol 30(2):96–102
15. Liu LJ, Xiao W, He QB, Jiang WW (2012) Generic and oral quality of life is affected by oral mucosal diseases. BMC Oral Health 12:2
16. Femiano F, Lanza A, Buonaiuto C, Gombos F, Nunziata M, Piccolo S, Cirillo N (2007) Guidelines for diagnosis and management of aphthous stomatitis. Pediatr Infect Dis J 26(8): 728–732
17. Rogers RS 3rd (1977) Recurrent aphthous stomatitis: clinical characteristics and evidence for an immunopathogenesis. J Invest Dermatol 69(6):499–509
18. Getz II, Bader HI (1967) Recurrent aphthous stomatitis: a family case report. Oral Surg Oral Med Oral Pathol 24(2):186–190
19. Pawson R, Mehta A (1998) The diagnosis and treatment of haematinic deficiency in gastrointestinal disease. Aliment Pharmacol Ther 12:687–698
20. Lin HP, Wang YP, Chen HM, Kuo YS, Lang MJ, Sun A (2013) Significant association of hematinic deficiencies and high blood homocysteine levels with burning mouth syndrome. J Formos Med Assoc 112(6):319–325
21. Sistig S, Vucićević-Boras V, Lukac J, Kusić Z (2002) Salivary IgA and IgG subclasses in oral mucosal diseases. Oral Dis 8(6):282–286
22. Sistig S, Lukac J, Vucicevic-Boras V, Delic D, Kusic Z (2003) Salivary immunoglobulin A and G subclasses in HIV positive patients. Eur J Med Res 8(12):543–548
23. Chavan M, Jain H, Diwan N, Khedkar S, Shete A, Durkar S (2012) Recurrent aphthous stomatitis: a review. J Oral Pathol Med 41:577–583
24. Mohammad R, Halboub E, Mashlah A, Abou-Hamed H (2013) Levels of salivary IgA in patients with minor recurrent aphthous stomatitis: a matched case-control study. Clin Oral Investig 17(3):975–980
25. Brozović S, Vucićević-Boras V, Buković D (2001) Serum IgA, IgG, IgM and salivary IgA in recurrent aphthous ulceration. Coll Antropol 25(2):633–637
26. Vicente M, Soria A, Mosquera A, Pérez J, Lamas A, Castellano T, Ramos A (1996) Immunoglobulin G subclass measurements in recurrent aphthous stomatitis. J Oral Pathol Med 25(10):538–540
27. Misra N, Maiti D, Misra P, Singh AK (2013) 940 nm diode laser therapy in management of recurrent apthous ulcer. BMJ Case Rep pii:bcr2012008489
28. Sharma S, Ali FM, Saraf K, Mudhol A (2014) Anti-helminthic drugs in recurrent apthous stomatitis: a short review. J Pharm Bioallied Sci 6(2):65–68
29. Popovsky JL, Camisa C (2000) New and emerging therapies for diseases of the oral cavity. Dermatol Clin 18(1):113–125
30. Radomsky CL, Levine N (2001) Thalidomide. Dermatol Clin 19(1):87–103
31. Hassan I, Dorjay K, Anwar P (2015) Thalidomide in dermatology: revisited. Indian J Dermatol 60:213

32. Nanke Y, Kamatani N, Okamoto T, Ogiuchi H, Kotake S (2008) Irsogladine is effective for recurrent oral ulcers in patients with Behçet's disease: an open-label, single-centre study. Drugs R D 9(6):455–459
33. Femiano F, Buonaiuto C, Gombos F, Lanza A, Cirillo N (2010) Pilot study on recurrent aphthous stomatitis (RAS): a randomized placebo-controlled trial for the comparative therapeutic effects of systemic prednisone and systemic montelukast in subjects unresponsive to topical therapy. Oral Surg Oral Med Oral Pathol Oral Radiol Endod 109(3):402–407
34. Farthing P, Bagan JV, Scully C (2005) Mucosal disease series. Number IV. Erythema multiforme. Oral Dis 11(5):261–267
35. Kishore BN, Ankadavar NS, Kamath GH, Martis J (2014) Varicella zoster with erythema multiforme in a young girl: a rare association. Indian J Dermatol 59(3):299–301
36. Suliman NM, Åstrøm AN, Ali RW, Salman H, Johannessen AC (2011) Oral mucosal lesions in skin diseased patients attending a dermatologic clinic: a cross- sectional study in Sudan. BMC Oral Health 11:24
37. Sunitha M, Shanmugam S (2006) Evaluation of salivary nitric oxide levels in oral mucosal diseases: a controlled clinical trial. Indian J Dent Res 17:117
38. Collins BM (2006) Oral mucosal pathology in the geriatric population. In: Calhoun KH, Eibling DE (eds) Geriatric otolaryngology. CRC Press, New York
39. Gonsalves WC, Wrightson AS, Henry RG (2008) Common oral conditions in older persons. Am Fam Physician 78(7):845–852
40. Ratajczak-Wrona W, Jablonska E, Antonowicz B, Dziemianczyk D, Grabowska SZ (2013) Levels of biological markers of nitric oxide in serum of patients with squamous cell carcinoma of the oral cavity. Int J Oral Sci 5:141–145
41. Kimura H, Miura S, Shigematsu T, Ohkubo N, Tsuzuki Y, Kurose I et al (1997) Increased nitric oxide production and inducible nitric oxide synthase activity in colonic mucosa of patients with active ulcerative colitis and Crohn's disease. Dig Dis Sci 42(5):1047–1054
42. Keklikoglu N, Koray M, Kocaelli H, Akinci S (2008) iNOS expression in oral and gastrointestinal tract mucosa. Dig Dis Sci 53(6):1437–1442
43. Marletta MA (1989) Nitric oxide: biosynthesis and biological significance. Trends Biochem Sci 14(12):488–492
44. Tadakamadla J, Kumar S, Lalloo R, Johnson NW (2017) Qualitative analysis of the impact of oral potentially malignant disorders on daily life activities. PLoS One 12(4):e0175531
45. Ni Riordain R, Meaney S, McCreary C (2011) Impact of chronic oral mucosal disease on daily life: preliminary observations from a qualitative study. Oral Dis 17(3):265–269
46. Stoopler ET, Sollecito TP (2014) Oral mucosal diseases: evaluation and management. Med Clin North Am 98(6):1323–1352
47. Coons SJ, Rao S, Keininger DL, Hays RD (2000) A comparative review of generic quality-of-life instruments. Pharmacoeconomics 17(1):13–35
48. López-Jornet P, Camacho-Alonso F, Lucero Berdugo M (2009) Measuring the impact of oral mucosa disease on quality of life. Eur J Dermatol 19(6):603–606
49. Jung C, Hugot JP, Barreau F (2010) Peyer's patches: the immune sensors of the intestine. Int J Inflam 2010:823710
50. Halle S, Dujardin HC, Bakocevic N, Fleige H, Danzer H, Willenzon S et al (2009) Induced bronchus-associated lymphoid tissue serves as a general priming site for T cells and is maintained by dendritic cells. JEM 206(12):2593–2601
51. Korpershoek E, Leenen PJ, Drexhage HA, De Krijger RR (2004) Cellular composition of pancreas-associated lymphoid tissue during human fetal pancreatic development. Histopathology 45(3):291–297
52. Yano S, Usui N, Dobashi N, Yahagi Y, Takahara S, Sugiyama K et al (2009) A case of primary esophageal mucosa-associated lymphoid tissue lymphoma with a numerical abnormality of 18q21 detected by fluorescence in situ hybridization. Ann Hematol 88 (7):703–704

53. Morgner A, Bayerdörffer E, Neubauer A, Stolte M (2000) Malignant tumors of the stomach. Gastric mucosa-associated lymphoid tissue lymphoma and Helicobacter pylori. Gastroenterol Clin North Am 29(3):593–607
54. Carter JW, Lancaster HK, Hardman WE, Cameron IL (1994) Distribution of intestine-associated lymphoid tissue, aberrant crypt foci, and tumors in the large bowel of 1,2-dimethylhydrazine-treated mice. Cancer Res 54(16):4304–4307
55. Nakamura T, Ito T, Izutsu K, Gibo J, Itaba S, Kawabe K et al (2008) Primary pancreatic low-grade mucosa-associated lymphoid tissue lymphoma presenting with multiple masses. Clin J Gastroenterol 1(4):168–173
56. Ruddle NH, Akirav EM (2009) Secondary lymphoid organs: responding to genetic and environmental cues in ontogeny and the immune response. J Immunol 183(4):2205–2212
57. Drake RL (2011) Terminologia anatomica: international anatomical terminology, 2nd edn. Thieme, Stuttgart
58. Baptista A, Olivier B, Goverse G, Greuter M, Knippenberg M, Kusser K et al (2012) Colonic patch and colonic SILT development are independent and differentially-regulated events. Mucosal Immunol 6(3):511–521
59. Matloubian M, Lo CG, Cinamon G, Lesneski MJ, Xu Y, Brinkmann V et al (2004) Lymphocyte egress from thymus and peripheral lymphoid organs is dependent on S1P receptor 1. Nature 427:355–360
60. Konenknov VI, Borodin YI, Dergacheva TI, Shurlygina AV, Mel'nikova EV, Tenditnik MV et al (2016) Cell composition of central and peripheral lymphoid organs of wistar rats treated with a biomedical cell product. Bull Exp Biol Med 161(1):150–154
61. Butler JA, Cosgrove J, Alden K, Timmis J, Coles MC (2017) Model-driven experimentation: a new approach to understand mechanisms of tertiary lymphoid tissue formation, function, and therapeutic resolution. Front Immunol 7:658
62. Coles M, Kioussis D, Veiga-Fernandes H (2010) Cellular and molecular requirements in lymph node and Peyer's patch development. Prog Mol Biol Transl Sci 92:177–205
63. Coles M, Veiga-Fernandes H (2013) Insight into lymphoid tissue morphogenesis. Immunol Lett 156(1–2):46–53
64. Ager A (2017) High endothelial venules and other blood vessels: critical regulators of lymphoid organ development and function. Front Immunol 8:45
65. Genovese L, Brendolan A (2016) Lymphoid tissue mesenchymal stromal cells in development and tissue remodeling. Stem Cells Int 2016:8419104
66. Brendolan A, Rosado MM, Carsetti R, Selleri L, Dear TN (2007) Development and function of the mammalian spleen. BioEssays 29(2):166–177
67. Lenti E, Farinello D, Yokoyama KK, Penkov D, Castagnaro L, Lavorgna G et al (2016) Transcription factor TLX1 controls retinoic acid signaling to ensure spleen development. J Clin Invest 126(7):2452–2464
68. Drayton DL, Liao S, Mounzer RH, Ruddle NH (2006) Lymphoid organ development: from ontogeny to neogenesis. Nat Immunol 7:344–353
69. Boehm T, Bleul CC (2007) The evolutionary history of lymphoid organs. Nat Immunol 8:131–135
70. Boehm T, Hess I, Swann JB (2012) Evolution of lymphoid tissues. Trends Immunol 33 (6):315–321
71. Quaglino D, Accorsi A, Boraldi F, Ottaviani E (2014) Thymic maturation and programmed cell death. In: Malagoli D, Ottaviani E (eds) Eco-immunology. Springer, Dordrecht
72. Koch U, Radtke F (2011) Mechanisms of T cell development and transformation. Annual Rev Cell Dev Biol 27:539–562
73. Savchenko AS, Hasegawa G, Naito M (2006) Development and maturation of thymic dendritic cells during human ontogeny. Cell Tissue Res 325(3):455–460
74. Wallin J, Eibel H, Neubuser A, Wilting J, Koseki H, Balling R (1996) Pax1 is expressed during development of the thymus epithelium and is required for normal T-cell maturation. Development 122:23–30

75. Lee HW, Na YJ, Jung PK, Kim MN, Kim SM, Chung JS et al (2008) Nerve growth factor stimulates proliferation, adhesion and thymopoietic cytokine expression in mouse thymic epithelial cells in vitro. Regul Pept 147(1–3):72–81
76. Alves NL, Huntington ND, Rodewald HR, Di Santo JP (2009) Thymic epithelial cells: the multi-tasking framework of the T cell "cradle". Trends Immunol 30(10):468–474
77. Vicari AP, Figueroa DJ, Hedrick JA, Foster JS, Singh KP, Menon S et al (1997) TECK: a novel CC chemokine specifically expressed by thymic dendritic cells and potentially involved in T cell development. Immunity 7(2):291–301
78. Lochner M (2011) Tertiary lymphoid tissues in the colon: friend and foe. Gut Microbes 2 (3):193–197
79. Germain C, Gnjatic S, Dieu-Nosjean MC (2015) Tertiary lymphoid structure-associated B cells are key players in anti-tumor immunity. Front Immunol 6:67
80. Bordon Y (2014) Lymphoid tissues: tertiary education. Nat Rev Immunol 14:284–285
81. Bergomas F, Grizzi F, Doni A, Pesce S, Laghi L, Allavena P, Mantovani A, Marchesi F (2012) Tertiary intratumor lymphoid tissue in colo-rectal cancer. Cancers (Basel) 4(1):1–10
82. Chinn IK, Blackburn CC, Manley NR, Sempowski GD (2012) Changes in primary lymphoid organs with aging. Semin Immunol 24(5):309–320
83. Neyt K, Perros F, GeurtsvanKessel CH, Hammad H, Lambrecht BN (2012) Tertiary lymphoid organs in infection and autoimmunity. Trends Immunol 33(6):297–305
84. Kirsh AL, Cushing SL, Chen EY, Schwartz SM, Perkins JA (2011) Tertiary lymphoid organs in lymphatic malformations. Lymphat Res Biol 9(2):85–92
85. Tang H, Zhu M, Qiao J, Fu YX (2017) Lymphotoxin signalling in tertiary lymphoid structures and immunotherapy. Cell Mol Immunol 14(10):809–818
86. Ruddle NH (2014) Lymphatic vessels and tertiary lymphoid organs. J Clin Invest 124(3): 953–959
87. Dieu-Nosjean MC, Goc J, Giraldo NA, Sautès-Fridman C, Fridman WH (2014) Tertiary lymphoid structures in cancer and beyond. Cell 35(11):571–580
88. Martino A, Sacchi A, Volpe E, Agrati C, De Santis R, Pucillo LP et al (2005) Non-pathogenic Mycobacterium smegmatis induces the differentiation of human monocytes directly into fully mature dendritic cells. J Clin Immunol 25(4):365–375
89. Dunay IR, DaMatta RA, Fux B, Presti R, Greco S, Colonna M, Sibley (2008) LD $Gr1^+$ ($Ly6C^+$) inflammatory monocytes are required for mucosal resistance to the pathogen *Toxoplasma gondii*. Immunity 29(2):306–317
90. Pakurar AS, Bigbee JW (2004) Digital histology: an interactive CD atlas with review text. Wiley, Hoboken
91. Loukas M, Bellary SS, Kuklinski M, Ferrauiola J, Yadav A, Shoja MM et al (2011) The lymphatic system: a historical perspective. Clin Anat 24(7):807–816
92. Reynolds JD (1985) Evidence of differences between Peyer's patches and germinal centers. In: Klaus GGB (ed) Microenvironments in the lymphoid system. Plenum Press, New York
93. de Lau W, Kujala P, Schneeberger K, Middendorp S, Li VSW, Barker N et al (2012) Peyer's patch M cells derived from $Lgr5^+$ stem cells require SpiB and are induced by RankL in cultured "miniguts". Mol Cell Biol 32(18):3639–3647
94. Ramsdell F, Fowlkes BJ (1990) Clonal deletion versus clonal anergy: the role of the thymus in inducing self tolerance. Science 248(4961):1342–1348
95. Anderson DE, Hafler DA (2001) Immune tolerance and the nervous system. Adv Exp Med Biol 490:79–98
96. Kamradt T, Mitchison NA (2001) Tolerance and autoimmunity. N Engl J Med 344(9):655–664
97. Buettner M, Bode U (2012) Lymph node dissection—understanding the immunological function of lymph nodes. Clin Exp Immunol 169(3):205–212
98. Lelouard H, Fallet M, de Bovis B, Méresse S, Gorvel JP (2011) Peyer's patch dendritic cells sample antigens by extending dendrites through M cell-specific transcellular pores. Gastroenterology 142(3):592–601

99. Guy-Grand D, Griscelli C, Vassali P (1974) The gut-associated lymphoid system: nature and properties of the large dividing cells. Eur J Immunol 4(6):435–443
100. Kagnoff MF (1975) Functional characteristics of Peyer's patch cells. III. Carrier priming of T cells by antigen feeding. J Exp Med 142(6):1425–1435
101. Kagnoff MF, Campbell S (1974) Functional characteristics of Peyer's patch lymphoid cells. I. Induction of humoral antibody and cell-mediated allograft reactions. J Exp Med 139 (2):398–406
102. Inoue T, Tsuzuki Y, Matsuzaki K, Matsunaga H, Miyazaki J, Hokari R et al (2005) Blockade of PSGL-1 attenuates $CD14^+$ monocytic cell recruitment in intestinal mucosa and ameliorates ileitis in SAMP1/Yit mice. J Leukoc Biol 77(3):287–295
103. Hirota K, Turner JE, Villa M, Duarte JH, Demengeot J, Steinmetz OM, Stockinger B (2013) Plasticity of TH17 cells in Peyer's patches is responsible for the induction of T cell–dependent IgA responses. Nat Immunol 14:372–379
104. Yang CY, Vogt TK, Favre S, Scarpellino L, Huang HY, Tacchini-Cottier F, Luther SA (2014) Trapping of naive lymphocytes triggers rapid growth and remodeling of the fibroblast network in reactive murine lymph nodes. Proc Natl Acad Sci U S A 111(1):E109–E118
105. Alexander JS, Ganta VC, Jordan PA, Witte MH (2010) Gastrointestinal lymphatics in health and disease. Pathophysiology 17(4):315–335
106. Ohtani O, Ohtani Y (2012) Recent developments in morphology of lymphatic vessels and lymph nodes. Ann Vasc Dis 5(2):145–150
107. Banchereau J, Steinman RM (1998) Dendritic cells and the control of immunity. Nature 392 (6673):245–252
108. Sallusto F, Lanzavecchia A (2002) The instructive role of dendritic cells on T-cell responses. Arthritis Res 4(3):S127–S132
109. Chyou S, Ekland EH, Carpenter AC, Tzeng TC, Tian S, Michaud M, Madri JA, Lu TT (2008) Fibroblast-type reticular stromal cells regulate the lymph node vasculature. J Immunol 181(6):3887–3896
110. Luther SA, Vogt TK, Siegert S (2011) Guiding blind T cells and dendritic cells: a closer look at fibroblastic reticular cells found within lymph node T zones. Immunol Lett 138(1):9–11
111. Miller MJ, Newberry RD (2010) Microanatomy of the intestinal lymphatic system. Ann N Y Acad Sci 1207(1):E21–E28
112. Yoshida T, Takaya K (1989) Follicular dendritic reticular cells in the germinal center of the rat lymph node as studied by immuno-electron microscopy. Arch Histol Cytol 52(4):327–335
113. Bengmark S, Jeppsson B (1995) Gastrointestinal surface protection and mucosa reconditioning. JPEN J Parenter Enteral Nutr 19(5):410–415
114. Isolauri E, Sütas Y, Kankaanpää P, Arvilommi H, Salminen S (2001) Probiotics: effects on immunity. Am J Clin Nutr 73(2):444s–450s
115. Aiello SE, Moses MA (2016) The merck veterinary manual, 11th edn. Wiley, Hoboken
116. Yandrapu H, Sarosiek J (2015) Protective factors of the gastric and duodenal mucosa: an overview. Curr Gastroenterol Rep 17:24
117. Bourbon JR, Chailley-Heu B (2001) Surfactant proteins in the digestive tract, mesentery, and other organs: evolutionary significance. Comp Biochem Physiol A: Mol Integr Physiol 129 (1):151–161
118. Bernhard W, Postle AD, Rau GA, Freihorst J (2001) Pulmonary and gastric surfactants. A comparison of the effect of surface requirements on function and phospholipid composition. Comp Biochem Physiol A: Mol Integr Physiol 129(1):173–182
119. Håkansson K, Reid KB (2000) Collectin structure: a review. Protein Sci 9(9):1607–1617
120. Holmskov U, Malhotra R, Sim RB, Jensenius JC (1994) Collectins: collagenous C-type lectins of the innate immune defense system. Immunol Today 15(2):67–74
121. Holmskov UL (2000) Collectins and collectin receptors in innate immunity. APMIS Suppl 100:1–59
122. Sim RB, Clark H, Hajela K, Mayilyan KR (2006) Collectins and host defence. Novartis Found Symp 279(170–86):216–219

123. Warrington R, Watson W, Kim HL, Antonetti FR (2011) An introduction to immunology and immunopathology. Allergy Asthma Clin Immunol 7(1):S1
124. Chaplin DD (2010) Overview of the immune response. J Allergy Clin Immunol 125(2):S3–S23
125. Ganz T (2003) Defensins: antimicrobial peptides of innate immunity. Nat Rev Immunol 3(9):710–720
126. Sankaran-Walters S, Hart R, Dills C (2017) Guardians of the gut: enteric defensins. Front Microbiol 8:647
127. Touchette MH, Seeliger JC (2017) Transport of outer membrane lipids in mycobacteria. Biochim Biophys Acta 1862(11):1340–1354
128. Deplancke B, Gaskins HR (2001) Microbial modulation of innate defense: goblet cells and the intestinal mucus layer. Am J Clin Nutr 73(6):1131S–1141S
129. Butowski N (2010) Immunostimulants for malignant gliomas. Neurosurg Clin 21(1):53–65
130. Selsted ME, Ouellette AJ (2005) Mammalian defensins in the antimicrobial immune response. Nat Immunol 6(6):551–557
131. Yang D, Biragyn A, Hoover DM, Lubkowski J, Oppenheim JJ (2004) Multiple roles of antimicrobial defensins, cathelicidins, and eosinophil-derived neurotoxin in host defense. Annu Rev Immunol 22:181–215
132. Junqueira-Kipnis AP, Neto LMM, Kipnis A (2014) Role of fused *Mycobacterium tuberculosis* immunogens and adjuvants in modern tuberculosis vaccines. Front Immunol 5:188
133. Gordon S (2002) Pattern recognition receptors—doubling up for the innate immune response. Cell 111(7):927–930
134. Oh JE, Lee HK (2014) Pattern recognition receptors and autophagy. Front Immunol 5:300
135. Paramo T, Tomasio SM, Irvine KL, Bryant CE, Bond PJ (2015) Energetics of endotoxin recognition in the toll-like receptor 4 innate immune response. Sci Rep 5:17997
136. Hajishengallis G, Lambri JD (2010) Crosstalk pathways between toll-like receptors and the complement system. Trends Immunol 31(4):154–163
137. Walport MJ (2001) Complement. N Engl J Med 344:1058–1066
138. Galvan M (2014) The complement system. In: Wells RD, Bond JS, Klinman J, Masters BSS, Bell E (eds) Molecular life sciences. Springer, New York
139. Taylor P, Botto M, Walport M (1998) The complement system. Curr Biol 8(8):R259–R261
140. Lubbers R, van Essen MF, van Kooten C, Trouw LA (2017) Production of complement components by cells of the immune system. Clin Exp Immunol 188(2):183–194
141. Afshar-Kharghan V (2017) The role of the complement system in cancer. J Clin Invest 127 (3):780–789
142. Merle NS, Noe R, Halbwachs-Mecarelli L, Fremeaux-Bacchi V, Roumenina LT (2015) Complement system part II: role in immunity. Front Immunol 6:257
143. Gigli I, Porter RR, Sim RB (1976) The unactivated form of the first component of human complement, C1. Biochem J 157(3):541–548
144. Gál P, Dobó J, Závodszky P, Sim RBM (2009) Early complement proteases: C1r, C1s and MASPs. A structural insight into activation and functions. Mol Immunol 46(14):2745–2752
145. Thielens NM, Illy C, Bally IM, Arlaud GJ (1994) Activation of human complement serine-proteinase C1r is down-regulated by a Ca(2+)-dependent intramolecular control that is released in the C1 complex through a signal transmitted by C1q. Biochem J 301(Pt 2):509–516
146. Degn SE, Jensen L, Hansen AG, Duman D, Tekin M, Jensenius JC, Thiel S (2012) Mannan-binding lectin-associated serine protease (MASP)-1 is crucial for lectin pathway activation in human serum, whereas neither MASP-1 nor MASP-3 is required for alternative pathway function. J Immunol 189(8):3957–3969
147. Mortensen SA, Sander B, Jensen RK, Pedersen JS, Golas MM, Jensenius JC et al (2017) Structure and activation of C1, the complex initiating the classical pathway of the complement cascade. Proc Natl Acad Sci U S A 114(5):986–991

148. Arlaud GJ, Thielens NM, Illy C (1993) Assembly of the C1 complex. Behring Inst Mitt 93:189–195
149. Colomb MG, Arlaud GJ, Villiers CL (1984) Activation of C1. Philos Trans R Soc Lond B Biol Sci 306(1129):283–292
150. Dodds AW, Sim RB, Porter RR, Kerr MA (1978) Activation of the first component of human complement (C1) by antibody-antigen aggregates. Biochem J 175(2):383–390
151. Rawal N, Pangburn MK (2001) Structure/function of C5 convertases of complement. Int Immunopharmacol 1(3):415–422
152. Rawal N, Pangburn MK (1998) C5 convertase of the alternative pathway of complement. Kinetic analysis of the free and surface-bound forms of the enzyme. J Biol Chem 273(27): 16828–16835
153. Rawal N, Pangburn MK (2003) Formation of high affinity C5 convertase of the classical pathway of complement. J Biol Chem 278(40):38476–38483
154. Janeway CA Jr, Travers P, Walport M, Shlomchik MJ (2001) Immunobiology: the immune system in health and disease, 5th edn. Garland Science, New York
155. Noris M, Remuzzi G (2013) Overview of complement activation and regulation. Semin Nephrol 33(6):479–492
156. Ohinata K, Yoshikawa M (2008) Food intake regulation by central complement system. Adv Exp Med Biol 632:35–46
157. Toapanta FR, Ross TM (2006) Complement-mediated activation of the adaptive immune responses. Immunol Res 36(1):197–210
158. Mathern DR, Heeger PS (2015) Molecules great and small: the complement system. Clin J Am Soc Nephrol 10(9):1636–1650
159. Zipfel PF, Skerka C (2009) Complement regulators and inhibitory proteins. Nat Rev Immunol 9:729–740
160. Regal JF (2014) Complement system. Encyclopedia of immunotoxicology. Vohr H-W, Springer
161. Clark SJ, Bishop PN (2015) Role of factor H and related proteins in regulating complement activation in the macula, and relevance to age-related macular degeneration. J Clin Med 4 (1):18–31
162. Selander B, Mårtensson U, Weintraub A, Holmström E, Matsushita M, Thiel S et al (2006) Mannan-binding lectin activates C3 and the alternative complement pathway without involvement of C2. J Clin Invest 116(5):1425–1434
163. Agrawal A, Singh PP, Bottazzi B, Garlanda C, Mantovani A (2009) Pattern recognition by pentraxins. Adv Exp Med Biol 653:98–116
164. de la Torre MY, Fabbri M, Jaillon S, Bastone A, Nebuloni M, Vecchi A et al (2010) Evolution of the pentraxin family: the new entry PTX4. J Immunol 184(9):5055–5064
165. Pepys MB, Baltz ML (1983) Acute phase proteins with special reference to C-reactive protein and related proteins (pentaxins) and serum amyloid A protein. Adv Immunol 34:141–212
166. Du Clos TW (2013) Pentraxins: structure, function, and role in inflammation. ISRN Inflamm 2013:379040
167. Cathcart ES, Shirahama T, Cohen AS (1967) Isolation and identification of a plasma component of amyloid. Biochim Biophys Acta 147:392–393
168. Pangburn MK, Müller-Eberhard HJ (1984) The alternative pathway of complement. Springer Semin Immunopathol 7(2):163–192
169. Parker GA, Picut CA (2005) Liver immunobiology. Toxicol Pathol 33(1):52–62
170. Ambrus JL, Peters MG, Fauci AS, Brown EJ (1990) The Ba fragment of complement factor B inhibits human B lymphocyte proliferation. J Immunol 144(5):1549–1553
171. Maibaum J, Liao SM, Vulpetti A, Ostermann N, Randl S, Rüdisser S et al (2016) Small-molecule factor D inhibitors targeting the alternative complement pathway. Nat Chem Biol 12:1105–1110

172. Loyet KM, DeForge LE, Katschke KJ Jr, Diehl L, Graham RR, Pao L et al (2012) Activation of the alternative complement pathway in vitreous is controlled by genetics in age-related macular degeneration. Invest Ophthalmol Vis Sci 53:6628–6637
173. Wallis R, Dodds AW, Mitchell DA, Sim RB, Reid KB, Schwaeble WJ (2007) Molecular interactions between MASP-2, C4, and C2 and their activation fragments leading to complement activation via the lectin pathway. J Biol Chem 282(11):7844–7851
174. Yaseen S, Demopulos G, Dudler T, Yabuki M, Wood CL, Cummings WJ et al (2017) Lectin pathway effector enzyme mannan-binding lectin-associated serine protease-2 can activate native complement C3 in absence of C4 and/or C2. FASEB J 31(5):2210–2219
175. Vorup-Jensen T, Jensenius JC, Thiel S (1998) MASP-2, the C3 convertase generating protease of the MBLectin complement activating pathway. Immunobiology 199(2):348–357
176. Pangburn MK, Müller-Eberhard HJ (1986) The C3 convertase of the alternative pathway of human complement. Enzymic properties of the bimolecular proteinase. Biochem J 235 (3):723–730
177. Petersen SV, Thiel S, Jensenius JC (2001) The mannan-binding lectin pathway of complement activation: biology and disease association. Mol Immunol 38(2–3):133–149
178. Kase T, Suzuki Y, Kawai T, Sakamoto T, Ohtani K, Eda S et al (1999) Human mannan-binding lectin inhibits the infection of influenza A virus without complement. Immunology 97(3):385–392
179. Endo Y, Matsushita M, Fujita T (2011) The role of ficolins in the lectin pathway of innate immunity. Int J Biochem Cell Biol 43(5):705–712
180. Gout E, Garlatti V, Smith DF, Lacroix M, Dumestre-Pérard C, Lunardi T et al (2010) Carbohydrate recognition properties of human ficolins. J Biol Chem 285(9):6612–6622
181. Zhang XL, Ali MAM (2008) Ficolins: structure, function and associated diseases. In: Lambris JD (ed) Current topics in complement II volume 632 of the series advances in experimental medicine and biology. Springer, New York
182. Terai I, Kobayashi K, Matsushita M, Fujita T, Matsuno K (1995) alpha 2-Macroglobulin binds to and inhibits mannose-binding protein-associated serine protease. Int Immunol 7(10):1579–1584
183. Sarma JV, Ward PA (2011) The complement system. Cell Tissue Res 343(1):227–235
184. Auger JL, Haasken S, Binstadt BA (2012) Autoantibody-mediated arthritis in the absence of C3 and activating Fcγ receptors: C5 is activated by the coagulation cascade. Arthritis Res Ther 14(6):R269
185. Tsao PY, Arora V, Ji MQ, Wright AC, Eisenberg RA (2011) KRN/I-A^{g7} mouse arthritis is independent of complement C3. J Clin Immunol 31(5):857–863
186. Foley JH, Walton BL, Aleman MM, O'Byrne AM, Lei V, Harrasser M et al (2016) Complement activation in arterial and venous thrombosis is mediated by plasmin. EBioMedicine 5:175–182
187. Bentley C, Fries W, Brade V (1978) Synthesis of factors D, B and P of the alternative pathway of complement activation, as well as of C3, by guinea-pig peritoneal macrophages in vitro. Immunology 35(6):971–980
188. McPhaden AR, Whaley K (1993) Complement biosynthesis by mononuclear phagocytes. Immunol Res 12(3):213–232
189. Colten HR, Ooi YM, Edelson PJ (1979) Synthesis and secretion of complement proteins by macrophages. Ann N Y Acad Sci 332:482–490
190. Barros VC, Assumpção JG, Cadete AM, Santos VC, Cavalcante RR, Araújo RN et al (2009) The role of salivary and intestinal complement system inhibitors in the midgut protection of triatomines and mosquitoes. PLoS One 4(6):e6047
191. Ferreira VP, Vale VF, Pangburn MK, Abdeladhim M, Mendes-Sousa AF, Coutinho-Abreu IV et al (2016) SALO, a novel classical pathway complement inhibitor from saliva of the sand fly *Lutzomyia longipalpi*. Sci Rep 6:19300
192. Edgar WM, Bowen WH, Cole MF (1982) Protein components in saliva and plaque fluid from irradiated primates. J Oral Pathol 11(3):252–259

193. Gao B, Jeong WI, Tian Z (2008) Liver: an organ with predominant innate immunity. Hepatology 47(2):729–736
194. Crispe IN (2009) The liver as a lymphoid organ. Annu Rev Immunol 27:147–163
195. Seki S, Habu Y, Kawamura T, Takeda K, Dobashi H, Ohkawa T, Hiraide H (2000) The liver as a crucial organ in the first line of host defense: the roles of Kupffer cells, natural killer (NK) cells and NK1.1 Ag^+ T cells in T helper 1 immune responses. Immunol Rev 174: 35–46
196. Noor MT, Manoria P (2017) Immune dysfunction in cirrhosis. J Clin Transl Hepatol 5(1): 50–58
197. Kmieć Z (2001) Cooperation of liver cells in health and disease. Adv Anat Embryol Cell Biol 161:III–XIII, 1–151
198. Gösswein T, Kocot A, Emmert G, Kreuz W, Martinez-Saguer I, Aygören-Pürsün E et al (2008) Mutational spectrum of the C1INH (SERPING1) gene in patients with hereditary angioedema. Cytogenet Genome Res 121(3–4):181–188
199. Ismail HF, Zhang J, Lynch RG, Wang Y, Berg DJ (2003) Role for complement in development of *Helicobacter*-induced gastritis in interleukin-10-deficient mice. Infect Immun 71(12):7140–7148
200. Turnberg D, Cook HT (2005) Complement and glomerulonephritis: new insights. Curr Opin Nephrol Hypertens 14(3):223–228
201. Crum-Cianflone NF (2010) HIV and the gastrointestinal tract. Infect Dis Clin Pract (Baltim Md) 18(5):283–285
202. Vecchio AL, Giannattasio A, Guarino A (2015) HIV and the intestine. In: Guandalini S, Dhawan A, Branski D (eds) Textbook of pediatric gastroenterology, hepatology and nutrition. Springer International Publishing, Cham
203. Gerard NP, Gerard C (2002) Complement in allergy and asthma. Curr Opin Immunol 14(6):705–708
204. Jain U, Otley AR, Van Limbergen J, Stadnyk AW (2014) The complement system in inflammatory bowel disease. Inflamm Bowel Dis 20(9):1628–1637
205. Okroj M, Heinegård D, Holmdahl R, Blom AM (2007) Rheumatoid arthritis and the complement system. Ann Med 39(7):517–530
206. Ingels C, Vanhorebeek I, Steffensen R, Derese I, Jensen L, Wouters PJ et al (2014) Lectin pathway of complement activation and relation with clinical complications in critically ill children. Pediatr Res 75:99–108
207. Janeway CA (1989) Approaching the asymptote? Evolution and revolution in immunology. Cold Spring Harb Symp 54:1–13
208. Gayed PM (2011) Toward a modern synthesis of immunity: Charles A. Janeway Jr. and the immunologist's dirty little secret. Yale J Biol Med 84(2):131–138
209. Pisetsky DS (2013) The effectors of innate immunity: DAMPs, DAMEs, or DIMEs? Arthritis Res Ther 15:123
210. Nüsslein-Volhard C, Wieschaus E (1980) Mutations affecting segment number and polarity in Drosophila. Nature 287(5785):795–801
211. Lipshitz HD (2005) From fruit flies to fallout: ed lewis and his science. Dev Dynam 232:529–546
212. Weiner AJ, Scott MP, Kaufman TC (1984) A molecular analysis of fushi tarazu, a gene in Drosophila melanogaster that encodes a product affecting embryonic segment number and cell fate. Cell 37(3):843–851
213. Kuroiwa A, Hafen E, Gehring WJ (1984) Cloning and transcriptional analysis of the segmentation gene fushi tarazu of Drosophila. Cell 37(3):825–831
214. Martinez-Arias A, Ingham PW (1985) The origin of pattern duplications in segment polarity mutants of Drosophila melanogaster. J Embryol Exp Morphol 87:129–135
215. Frohnhöfer HG, Nüsslein-Volhard C (1986) Organization of anterior pattern in the Drosophila embryo by the maternal gene bicoid. Nature 324:120–125

216. Lemaitre B, Kromer-Metzger E, Michaut L, Nicolas E, Meister M, Georgel P et al (1995) A recessive mutation, immune deficiency (imd), defines two distinct control pathways in the Drosophila host defense. Proc Natl Acad Sci U S A 92(21):9465–9469
217. Medzhitov R, Preston-Hurlburt P, Janeway CA Jr (1997) A human homologue of the Drosophila Toll protein signals activation of adaptive immunity. Nature 388(6640):394–397
218. Poltorak A, He XL, Smirnova I, Liu MY, VanHuffel C, Du X et al (1998) Defective LPS signaling in C3H/Hej and C57BL/10ScCr mice: mutations in TLR4 gene. Science 282:2085–2088
219. Steinman RM, Cohn ZA (1973) Identification of a novel cell type in peripheral lymphoid organs of mice. I. Morphology, quantitation, tissue distribution. J Exp Med 137:1142–1162
220. Steinman RM, Cohn ZA (1974) Identification of a novel cell type in peripheral lymphoid organs of mice. II. Functional properties in vitro. J Exp Med 139:380–397
221. Pott J, Hornef M (2012) Innate immune signaling at the intestinal epithelium in homeostasis and disease. EMBO Rep 13(8):684–698
222. Chu H, Mazmanian SK (2013) Innate immune recognition of the microbiota promotes host-microbial symbiosis. Nat Immunol 14:668–675
223. Fukata M, Arditi M (2013) The role of pattern recognition receptors in intestinal inflammation. Mucosal Immunol 6(3):451–463
224. Cario E, Podolsky DK (2005) Intestinal epithelial tollerance versus intollerance of commensals. Mol Immunol 42(8):887–893
225. Wells JM, Loonen LMP, Karczewski JM (2010) The role of innate signaling in the homeostasis of tolerance and immunity in the intestine. Int J Med Microbiol 300(1):41–48
226. Lien E, Ingalls RR (2002) Toll-like receptors. Crit Care Med 30(1):S1–S11
227. Brown L, Wolf JM, Prados-Rosales R, Casadevall A (2015) Through the wall: extracellular vesicles in Gram-positive bacteria, mycobacteria and fungi. Nat Rev Microbiol 13:620–630
228. Ginsberg C, Brown S, Walker S (2008) Bacterial cell wall components. In: Fraser-Reid BO, Tatsuta K, Thiem J (eds) Glycoscience. Springer, Heidelberg
229. Christmas P (2010) Toll-like receptors: sensors that detect infection. Nat Educ 3(9):85
230. Spiesz EM, Thorpe CT, Chaudhry S, Riley GP, Birch HL, Clegg PD, Screen HRC (2015) Tendon extracellular matrix damage, degradation and inflammation in response to in vitro overload exercise. J Orthop Res 33:889–897
231. Rosin DL, Okusa MD (2011) Dangers within: DAMP responses to damage and cell death in kidney disease. JASN 22(3):416–425
232. Sharma SK, Naidu G (2016) The role of danger-associated molecular patterns (DAMPs) in trauma and infections. J Thorac Dis 8(7):1406–1409
233. Kang R, Lotze MT, Zeh HJ, Billiar TR, Tang D (2014) Cell death and DAMPs in acute pancreatitis. Mol Med 20:466–477
234. Ito T, Kawahara K, Nakamura T, Yamada S, Nakamura T, Abeyama K et al (2007) High-mobility group box 1 protein promotes development of microvascular thrombosis in rats. J Thromb Haemost 5(1):109–116
235. Sandor F, Buc M (2005) Toll-like receptors. I. Structure, function and their ligands. Folia Biol (Praha) 51(5):148–157
236. Melmed G, Thomas LS, Lee N, Tesfay SY, Lukasek K, Michelsen KS et al (2003) Human intestinal epithelial cells are broadly unresponsive to Toll-like receptor 2-dependent bacterial ligands: implications for host-microbial interactions in the gut. J Immunol 170(3):1406–1415
237. Arpaia N, Rudensky AY (2014) Microbial metabolites control gut inflammatory responses. PNAS 111(6):2058–2059
238. Oeckinghaus A, Ghosh S (2009) The NF-κB family of transcription factors and its regulation. Cold Spring Harb Perspect Biol 1(4):a000034
239. Guinn Z, Lampe AT, Brown DM, Petro TM (2016) Significant role for IRF3 in both T cell and APC effector functions during T cell responses. Cell Immunol 310:141–149

240. Driggers PH, Ennist DL, Gleason SL, Mak WH, Marks MS, Levi BZ et al (1990) An interferon gamma-regulated protein that binds the interferon-inducible enhancer element of major histocompatibility complex class I genes. Proc Natl Acad Sci USA 87(10):3743–3747
241. Majoros A, Platanitis E, Kernbauer-Hölzl E, Rosebrock F, Müller M, Decker T (2017) Canonical and non-canonical aspects of JAK-STAT signaling: lessons from interferons for cytokine responses. Front Immunol 8:29
242. Fukata Masayuki, Abreu Maria T (2012) Mucosal bacterial recognition and signaling systems in the intestine. In: Johnson LR, Ghishan FK, Merchand JL, Said HM, Wood JD (eds) Physiology of the gastrointestinal tract, 5th edn. Elsevier, San Diego
243. Reinecker HC, Podolsky DK (1995) Human intestinal epithelial cells express functional cytokine receptors sharing the common gamma c chain of the interleukin 2 receptor. Proc Natl Acad Sci U S A 92(18):8353–8357
244. Lampinen M, Carlson M, Hakansson LD, Venge P (2004) Cytokine-regulated accumulation of eosinophils in inflammatory disease. Allergy 59:793–805
245. Barrera GJ, Sanchez G, Gonzalez JE (2012) Trefoil factor 3 isolated from human breast milk downregulates cytokines (IL8 and IL6) and promotes human beta defensin (hBD2 and hBD4) expression in intestinal epithelial cells HT-29. Bosn J Basic Med Sci 12(4):256–264
246. Barrera GJ, Tortolero GS (2016) Trefoil factor 3 (TFF3) from human breast milk activates PAR-2 receptors, of the intestinal epithelial cells HT-29, regulating cytokines and defensins. Bratisl Lek Listy 117(6):332–339
247. Xiao P, Ling H, Lan G, Liu J, Hu H, Yang R (2015) Trefoil factors: Gastrointestinal-specific proteins associated with gastric cancer. Clin Chim Acta 450:127–134
248. Menheniott TR, Kurklu B, Giraud AS (2013) Gastrokines: stomach-specific proteins with putative homeostatic and tumor suppressor roles. Am J Physiol Gastrointest Liver Physiol 304(2):G109–G121
249. Menheniott TR, Peterson AJ, O'Connor L, Lee KS, Kalantzis A, Kondova I et al (2010) A novel gastrokine, GKN3, marks gastric atrophy and shows evidence of adaptive gene loss in humans. Gastroenterology 138(5):1823–1835
250. von Moltke J, Ji M, Liang HE, Locksley RM (2016) Tuft-cell-derived IL-25 regulates an intestinal ILC2–epithelial response circuit. Nature 529(7585):221–225
251. Neurath MF (2014) Cytokines in inflammatory bowel disease. Nat Rev Immunol 14:329–342
252. Akhtar S, Li X, Chaudry IH, Choudhry MA (2009) Neutrophil chemokines and their role in IL-18-mediated increase in neutrophil O_2^- production and intestinal edema following alcohol intoxication and burn injury. Am J Physiol Gastrointest Liver Physiol 297(2):G340–G347
253. Owens BMJ, Simmons A (2013) Intestinal stromal cells in mucosal immunity and homeostasis. Mucosal Immunol 6:224–234
254. Lacy P, Stow JL (2011) Cytokine release from innate immune cells: association with diverse membrane trafficking pathways. Blood 118:9–18
255. Dignass AU, Podolsky DK (1993) Cytokine modulation of intestinal epithelial cell restitution: Central role of transforming growth factor β. Gastroenterology 105(5):1323–1332
256. Nuzzi PA, Lokuta MA, Huttenlocher A (2007) Analysis of neutrophil chemotaxis. In: Coutts AS (ed) Adhesion protein protocols volume 370 of the series methods in molecular biology™. Humana Press, Totowa
257. Rot A (1992) The role of leukocyte chemotaxis in inflammation. In: Whicher JT, Evans SW (eds) Biochemistry of inflammation, volume 18 of the series immunology and medicine series. Springer, Dordrecht
258. Fukata M, Vamadevan AS, Abreu MT (2009) Toll-like receptors (TLRs) and Nod-like receptors (NLRs) in inflammatory disorders. Semin Immunol 21(4):242–253
259. Watanabe T, Kitani A, Murray PJ, Wakatsuki Y, Fuss IJ, Strober W (2006) Nucleotide binding oligomerization domain 2 deficiency leads to dysregulated TLR2 signaling and induction of antigen-specific colitis. Immunity 25(3):473–485

260. Parkos CA (1997) Molecular events in neutrophil transepithelial migration. BioEssays 19 (10):865–873
261. Song F, Ito K, Denning TL, Kuninger D, Papaconstantinou J, Gourley W et al (1999) Expression of the neutrophil chemokine KC in the colon of mice with enterocolitis and by intestinal epithelial cell lines: effects of flora and proinflammatory cytokines. J Immunol 162(4):2275–2280
262. Ishii N, Tsuzuki Y, Matsuzaki K, Miyazaki J, Okada Y, Hokari R et al (2004) Endotoxin stimulates monocyte–endothelial cell interactions in mouse intestinal Peyer's patches and villus mucosa. Clin Exp Immunol 135(2):226–232
263. Hruz P, Dann SM, Eckmann L (2010) STAT3 and its activators in intestinal defense and mucosal homeostasis. Curr Opin Gastroenterol 26(2):109–115
264. Powell DW, Pinchuk IV, Saada JI, Chen X, Mifflin RC (2011) Mesenchymal cells of the intestinal lamina propria. Annu Rev Physiol 73:213–237
265. Blake KM, Carrigan SO, Issekutz AC, Stadnyk AW (2004) Neutrophils migrate across intestinal epithelium using β_2 integrin (CD11b/CD18)-independent mechanisms. Clin Exp Immunol 136(2):262–268
266. Carrigan SO, Pink DB, Stadnyk AW (2007) Neutrophil transepithelial migration in response to the chemoattractant fMLP but not C5a is phospholipase D-dependent and related to the use of CD11b/CD18. J Leukoc Biol 82(6):1575–1584
267. DiScipio RG, Daffern PJ, Schraufstätter IU, Sriramarao P (1998) Human polymorphonuclear leukocytes adhere to complement factor H through an interaction that involves alphaMbeta2 (CD11b/CD18). J Immunol 160(8):4057–4066
268. Vetvicka V, Thornton BP, Ross GD (1996) Soluble beta-glucan polysaccharide binding to the lectin site of neutrophil or natural killer cell complement receptor type 3 (CD11b/CD18) generates a primed state of the receptor capable of mediating cytotoxicity of iC3b-opsonized target cells. J Clin Invest 98(1):50–61
269. Springer T, Galfré G, Secher DS, Milstein C (1979) Mac-1: a macrophage differentiation antigen identified by monoclonal antibody. Eur J Immunol 9(4):301–306
270. Koch S, Kucharzik T, Heidemann J, Nusrat A, Luegering A (2010) Investigating the role of proinflammatory $CD16^+$ monocytes in the pathogenesis of inflammatory bowel disease. Clin Exp Immunol 161(2):332–341
271. Sallusto F, Lenig D, Förster R, Lipp M, Lanzavecchia A (1999) Two subsets of memory T lymphocytes with distinct homing potentials and effector functions. Nature 401(6754):708–712
272. Tyrer PC, Bean EG, Foxwell AR, Pavli P (2011) Effects of bacterial products on enterocyte–macrophage interactions in vitro. Biochem Biophys Res Commun 413(2):336–341
273. Epelman S, Lavine KJ, Randolph GJ (2014) Origin and functions of tissue macrophages. Immunity 41(1):21–35
274. Pollard JW (2009) Trophic macrophages in development and disease. Nature Rev Immunol 9:259–270
275. Wood W, Martin P (2017) Macrophage functions in tissue patterning and disease: new insights from the fly. Dev Cell 40(3):221–233
276. Wynn TA, Chawla A, Pollard JW (2013) Macrophage biology in development, homeostasis and disease. Nature 496:445–455
277. Imhof BA, Aurrand-Lions M (2004) Adhesion mechanisms regulating the migration of monocytes. Nat Rev Immunol 4:432–444
278. Cohen SB, Maurer KJ, Egan CE, Oghumu S, Satoskar AR, Denkers EY (2013) CXCR3-dependent $CD4^+$ T cells are required to activate inflammatory monocytes for defense against intestinal infection. PLoS Pathog 9(10):e1003706
279. Dwinell MB, Eckmann L, Leopard JD, Varki NM, Kagnoff MF (1999) Chemokine receptor expression by human intestinal epithelial cells. Gastroenterology 117(2):359–367
280. Wang Y, Koroleva EP, Kruglov AA, Kuprash DV, Nedospasov SA, Fu YX, Tumanov AV (2010) Lymphotoxin beta receptor signaling in intestinal epithelial cells orchestrates innate immune responses against mucosal bacterial infection. Immunity 32(3):403–413

281. Upadhyay V, Fu YX (2014) Lymphotoxin organizes contributions to host defense and metabolic illness from innate lymphoid cells. Cytokine Growth Factor Rev 25(2):227–233
282. Hanai H, Iida T, Takeuchi K, Watanabe F, Yamada M, Kikuyama M et al (2008) Adsorptive depletion of elevated proinflammatory CD14 $^{+}$CD16^{+}DR^{++} monocytes in patients with inflammatory bowel disease. Am J Gastroenterol 103(5):1210–1216
283. Brown JB, Cheresh P, Goretsky T, Managlia E, Grimm GR, Ryu H et al (2011) Epithelial phosphatidylinositol-3-kinase signaling is required for β-catenin activation and host defense against Citrobacter rodentium infection. Infect Immun 79(5):1863–1872
284. O'Neill LAJ, Golenbock D, Bowie AG (2013) The history of Toll-like receptors—redefining innate immunity. Nat Rev Immunol 13:453–460
285. Takeda K, Kaisho T, Akira S (2003) Toll-like receptors. Annu Rev Immunol 21:335–376
286. Bulut Y, Faure E, Thomas L, Equils O, Arditi M (2001) Cooperation of Toll-like receptor 2 and 6 for cellular activation by soluble tuberculosis factor and Borrelia burgdorferi outer surface protein A lipoprotein: role of Toll-interacting protein and IL-1 receptor signaling molecules in Toll-like receptor 2 signaling. J Immunol 167(2):987–994
287. Kaisho T, Akira S (2006) Toll-like receptor function and signaling. J Allergy Clin Immunol 117(5):979–987
288. Bauer S, Müller T, Hamm S (2009) Pattern recognition by Toll-like receptors. Adv Exp Med Biol 653:15–34
289. Cario E, Rosenberg IM, Brandwein SL, Beck PL, Reinecker HC, Podolsky DK (2000) Lipopolysaccharide activates distinct signaling pathways in intestinal epithelial cell lines expressing Toll-like receptors. J Immunol 164(2):966–972
290. Guha M, Mackman N (2001) LPS induction of gene expression in human monocytes. Cell Signal 13(2):85–94
291. Abreu MT, Vora P, Faure E, Thomas LS, Arnold ET, Arditi M (2001) Decreased expression of Toll-like receptor-4 and MD-2 correlates with intestinal epithelial cell protection against dysregulated proinflammatory gene expression in response to bacterial lipopolysaccharide. J Immunol 167(3):1609–1616
292. Ishihara S, Rumi MA, Ortega-Cava CF, Kazumori H, Kadowaki Y, Ishimura N, Kinoshita Y (2006) Therapeutic targeting of toll-like receptors in gastrointestinal inflammation. Curr Pharm Des 12(32):4215–4228
293. Takeuchi O, Akira S (2001) Toll-like receptors; their physiological role and signal transduction system. Int Immunopharmacol 1(4):625–635
294. Shibolet O, Podolsky DK (2007) TLRs in the Gut. IV. Negative regulation of Toll-like receptors and intestinal homeostasis: addition by subtraction. Am J Physiol Gastrointest Liver Physiol 292(6):G1469–G1473
295. Premkumar V, Dey M, Dorn R, Raskin I (2010) MyD88-dependent and independent pathways of Toll-Like Receptors are engaged in biological activity of Triptolide in ligand-stimulated macrophages. BMC Chem Biol 10:3
296. Kawai T, Takeuchi O, Fujita T, Inoue J, Mühlradt PF, Sato S et al (2001) Lipopolysaccharide stimulates the MyD88-independent pathway and results in activation of IFN-regulatory factor 3 and the expression of a subset of lipopolysaccharide-inducible genes. J Immunol 167(10):5887–5894
297. Sakai J, Cammarota E, Wright JA, Cicuta P, Gottschalk RA, Li N et al (2017) Lipopolysaccharide-induced NF-κB nuclear translocation is primarily dependent on MyD88, but TNFα expression requires TRIF and MyD88. Sci Rep 7(1):1428
298. Akira S, Takeda K (2004) Toll-like receptor signaling. Nat Rev Immunol 4:499–511
299. Yamamoto M, Sato S, Hemmi H, Uematsu S, Hoshino K, Kaisho T et al (2003) TRAM is specifically involved in the Toll-like receptor 4-mediated MyD88-independent signaling pathway. Nat Immunol 4(11):1144–1150
300. Jin MS, Kim SE, Heo JY, Lee ME, Kim HM, Paik SG et al (2007) Crystal structure of the TLR1-TLR2 heterodimer induced by binding of a tri-acylated lipopeptide. Cell 130 (6):1071–1082

301. Farhat K, Riekenberg S, Heine H, Debarry J, Lang R, Mages J et al (2008) Heterodimerization of TLR2 with TLR1 or TLR6 expands the ligand spectrum but does not lead to differential signaling. J Leukoc Biol 83(3):692–701
302. Keating SE, Maloney GM, Moran EM, Bowie AG (2007) IRAK-2 participates in multiple Toll-like receptor signaling pathways to NFκB via activation of TRAF6 ubiquitination. J Biol Chem 282:33435–33443
303. Sun L, Deng L, Ea CK, Xia ZP, Chen ZJ (2004) The TRAF6 ubiquitin ligase and TAK1 kinase mediate IKK activation by BCL10 and MALT1 in T lymphocytes. Mol Cell 14 (3):289–301
304. Xiao H, Gulen MF, Qin J, Yao J, Bulek K, Kish D et al (2007) The Toll-interleukin-1 receptor member SIGIRR regulates colonic epithelial homeostasis, inflammation, and tumorigenesis. Immunity 26(4):461–475
305. Fernandes P, MacSharry J, Darby T, Brint E (2016) Differential expression of key regulators of Toll-like Receptors in Ulcerative Colitis and Crohn's disease: A role for Tollip and PPARγ? Clin Exp Immunol 183(3):358–368
306. Zhou J, Wu R, High AA, Slaughter CA, Finkelstein D, Rehg JE et al (2011) A20-binding inhibitor of NF-κB (ABIN1) controls Toll-like receptor-mediated CCAAT/enhancer-binding protein β activation and protects from inflammatory disease. PNAS 108(44):E998–E1006
307. Horvatinovich JM, Grogan EW, Norris M, Steinkasserer A, Lemos H, Mellor AL et al (2017) Soluble CD83 inhibits t cell activation by binding to the TLR4/MD-2 complex on $CD14^+$ monocytes. J Immunol 198(6):2286–2301
308. Gon Y, Asai Y, Hashimoto S, Mizumura K, Jibiki I, Machino T et al (2004) A20 inhibits toll-like receptor 2- and 4-mediated interleukin-8 synthesis in airway epithelial cells. Am J Respir Cell Mol Biol 31(3):330–336
309. Verstrepen L, Carpentier I, Verhelst K, Beyaert R (2009) ABINs: A20 binding inhibitors of NF-kappa B and apoptosis signaling. Biochem Pharmacol 78(2):105–114
310. Wang YB, Tan B, Mu R, Chang Y, Wu M, Tu HQ et al (2015) Ubiquitin-associated domain-containing ubiquitin regulatory X (UBX) protein UBXN1 is a negative regulator of nuclear factor κB (NF-κB) signaling. J Biol Chem 290(16):10395–10405
311. Garlanda C, Riva F, Bonavita E, Mantovani A (2013) Negative regulatory receptors of the IL-1 family. Semin Immunol 25(6):408–415
312. Kuiken HJ, Egan DA, Laman H, Bernards R, Beijersbergen RL, Dirac AM (2012) Identification of F-box only protein 7 as a negative regulator of NF-kappaB signalling. J Cell Mol Med 16(9):2140–2149
313. Yamamoto M, Sato S, Mori K, Hoshino K, Takeuchi O, Takeda K, Akira S (2002) Cutting edge: a novel Toll/IL-1 receptor domain-containing adaptor that preferentially activates the IFN-beta promoter in the Toll-like receptor signaling. J Immunol 169(12):6668–6672
314. Horng T, Barton GM, Flavell RA, Medzhitov R (2002) The adaptor molecule TIRAP provides signalling specificity for Toll-like receptors. Nature 420(6913):329–333
315. Lim KH, Staudt LM (2013) Toll-like receptor signaling. Cold Spring Harb Perspect Biol 5: a011247
316. Rakoff-Nahoum S, Medzhitov R (2009) Toll-like receptors and cancer. Nat Revs Cancer 9:57–63
317. Leon CG, Tory R, Jia J, Sivak O, Wasan KM (2008) Discovery and development of toll-like receptor 4 (TLR4) antagonists: a new paradigm for treating sepsis and other diseases. Pharm Res 25(8):1751–1761
318. Feng Y, Gao J, Cui Y, Li M, Li R, Cui C, Cui J (2017) Neuroprotective effects of resatorvid against traumatic brain injury in rat: involvement of neuronal autophagy and TLR4 signaling pathway. Cell Mol Neurobiol 37(1):155–168
319. Shirey KA, Lai W, Scott AJ, Lipsky M, Mistry P, Pletneva LM et al (2013) The TLR4 antagonist Eritoran protects mice from lethal influenza infection. Nature 497:498–502
320. Bhardwaj N, Gnjatic S, Sawhney NB (2010) TLR agonists: are they good adjuvants? Cancer J 16(4):382–391

321. Hua F, Tang H, Wang J, Prunty MC, Hua X, Sayeed I, Stein DG (2015) TAK-242, an antagonist for Toll-like receptor 4, protects against acute cerebral ischemia/reperfusion injury in mice. J Cereb Blood Flow Metab 35(4):536–542
322. Feng Y, Cui Y, Gao JL, Li MH, Li R, Jiang XH et al (2016) Resveratrol attenuates neuronal autophagy and inflammatory injury by inhibiting the TLR4/NF-κB signaling pathway in experimental traumatic brain injury. Int J Mol Med 37(4):921–930
323. Ferrero RL, Philpott DJ (2005) Nod1 signaling to extracellular mucosal bacteria in epithelial cells. The innate immune system: strategies for disease control. Int Congr Ser 1285:68–77
324. Mueller T, Podolsky DK (2005) Nucleotide-binding-oligomerization domain proteins and toll-like receptors: sensors of the inflammatory bowel diseases' microbial environment. Curr Opin Gastroenterol 21(4):419–425
325. Kato J, Svensson CI (2015) Role of extracellular damage-associated molecular pattern molecules (DAMPs) as mediators of persistent pain. Prog Mol Biol Transl Sci 131:251–279
326. Behzadi E, Behzadi P (2016) The role of toll-like receptors (TLRs) in urinary tract infections (UTIs). Cent European J Urol 69(4):404–410
327. Kim JG, Lee SJ, Kagnoff MF (2004) Nod1 is an essential signal transducer in intestinal epithelial cells infected with bacteria that avoid recognition by toll-like receptors. Infect Immun 72(3):1487–1495
328. Höhn TJ, Grune T (2014) The proteasome and the degradation of oxidized proteins: Part III—redox regulation of the proteasomal system. Redox Biol 2:388–394
329. Adams J (2003) The proteasome: structure, function, and role in the cell. Cancer Treat Rev 29(1):3–9
330. Lundblad A (2015) Gunnar Blix and his discovery of sialic acids. Fascinating molecules in glycobiology. Ups J Med Sci 120(2):104–112
331. Hopkins AP, Hawkhead JA, Thomas GH (2013) Transport and catabolism of the sialic acids N-glycolylneuraminic acid and 3-keto-3-deoxy-D-glycero-D-galactonononic acid by Escherichia coli K-12. FEMS Microbiol Lett 347(1):14–22
332. Corfield AP, Schauer R (1982) Occurrence of sialic acids. In: Schauer R (ed) Sialic acids: chemistry, metabolism, and function. Springer, Vienna
333. Faillard H (1989) The early history of sialic acids. Trends Biochem Sci 14(6):237–241
334. Vimr ER, Kalivoda KA, Deszo EL, Steenbergen SM (2004) Diversity of microbial sialic acid metabolism. Microbiol Mol Biol Rev 68(1):132–153
335. Severi E, Hood DW, Thomas GH (2007) Sialic acid utilization by bacterial pathogens. Microbiology 153:2817–2822
336. Severi E, Randle G, Kivlin P, Whitfield K, Young R, Moxon R et al (2005) Sialic acid transport in Haemophilus influenzae is essential for lipopolysaccharide sialylation and serum resistance and is dependent on a novel tripartite ATP-independent periplasmic transporter. Mol Microbiol 58:1173–1185
337. Severi E, Hosie AH, Hawkhead JA, Thomas GH (2010) Characterization of a novel sialic acid transporter of the sodium solute symporter (SSS) family and in vivo comparison with known bacterial sialic acid transporters. FEMS Microbiol Lett 304:47–54
338. Vimr ER (2013) Unified theory of bacterial sialometabolism. How and why bacteria metabolize host sialic acids. ISRN Microbiol 2013:816713
339. Goerdt S, Kodelja V, Schmuth M, Orfanos CE, Sorg C (1996) The mononuclear phagocyte-dendritic cell dichotomy: myths, facts, and a revised concept. Clin Exp Immunol 105(1):1–9
340. Hashimoto D, Miller J, Merad M (2011) Dendritic cell and macrophage heterogeneity in vivo. Immunity 35(3):323–335
341. Volterra M (1927) Ricerche sul sistema reticulo-istiociterio. Lo Sperimentale (Researches on the reticulo-histiocyte system. Experimental). Arch Biol Pathol 81:319
342. van Furth R, Cohn ZA, Hirsch JG, Humphrey JH, Spector WG, Langevoort HL (1972) The mononuclear phagocyte system: a new classification of macrophages, monocytes, and their precursor cells. Bull World Health Organ 46(6):845–852
343. Hume DA (2006) The mononuclear phagocyte system. Curr Opin Immunol 18(1):49–53

344. Fletcher AL, Acton SE, Knoblich K (2015) Lymph node fibroblastic reticular cells in health and disease. Nat Rev Immunol 15:350–361
345. Zhao L, Liu L, Gao J, Yang Y, Hu C, Guo B, Zhu B (2014) T lymphocytes maintain structure and function of fibroblastic reticular cells via lymphotoxin (LT)-B. BMC Immunol 15:33
346. Nagasawa T, Omatsu Y, Sugiyama T (2011) Control of hematopoietic stem cells by the bone marrow stromal niche: the role of reticular cells. Trends Immunol 32(7):315–320
347. Steinman RM, Hawiger D, Nussenzweig MC (2003) Tolerogenic dendritic cells. Annu Rev Immunol 21:685–711
348. Volchenkov R, Sprater F, Vogelsang P, Appel S (2012) The 2011 nobel prize in physiology or medicine. Scand J Immunol 75(1):1–4
349. Lane PJ, Brocker T (1999) Developmental regulation of dendritic cell function. Curr Opin Immunol 11(3):308–313
350. Silva MT (2010) When two is better than one: macrophages and neutrophils work in concert in innate immunity as complementary and cooperative partners of a myeloid phagocyte system. J Leukoc Biol 87:93–106
351. Silva MT, Correia-Neves M (2012) Neutrophils and macrophages: the main partners of phagocyte cell systems. Front Immunol 3:174
352. Lee PY, Wang JX, Parisini E, Dascher CC, Nigrovic PA (2013) Ly6 family proteins in neutrophil biology. J Leukoc Biol 94:585–594
353. Bosch X, Ramos-Casals M (2014) Granulocytes: neutrophils, basophils, eosinophils. In: Mackay IR, Rose NR (eds) The autoimmune diseases, 5th edn. London
354. Evans K, Amenhotep Z, Dawson D, Waters H, Ardern J (2013) Hematology. In: Wild D (ed) The immunoassay handbook—theory and applications of ligand binding, ELISA and related techniques. Elsevier, Amsterdam
355. Parmley RT, Ogawa M, Spicer SS, Wright NJ (1976) Ultrastructure and cytochemistry of bone marrow granulocytes in culture. Exp Hematol 4(2):75–89
356. Shousha S, Mitchell TR (1983) Orcein staining of leukocytes. Histochem J 15(6):563–569
357. Burtin P, Flandrin G, Fondaneche MC (1980) Presence of NCA (nonspecific cross-reacting antigen) in the cells of the human granulocyte series. Blood Cells 6(2):263–278
358. Gemmell MA, Anderson NL (1982) Lymphocyte, monocyte, and granulocyte proteins compared by use of two-dimensional electrophoresis. Clin Chem 28(4 Pt 2):1062–1066
359. Barton JC, Parmley RT, Butler TW, Williamson SE, Lilly MB, Gualtieri RJ, Heck LW Jr (1988) Differential staining of neutrophils and monocytes: surface and cytoplasmic iron-binding proteins. Histochem J 20(3):147–155
360. Taylor ML, Metcalfe DD (2001) Mast cells in allergy and host defense. Allergy Asthma Proc 22(3):115–119
361. Valent P (1995) Cytokines involved in growth and differentiation of human basophils and mast cells. Exp Dermatol 4(4 Pt 2):255–259
362. Jamur MC, Oliver C (2011) Origin, maturation and recruitment of mast cell precursors. Front Biosci (Schol Ed) 3:1390–1406
363. Kumar V, Sharma A (2010) Mast cells: emerging sentinel innate immune cells with diverse role in immunity. Mol Immunol 48(1–3):14–25
364. da Silva EZ, Jamur MC, Oliver C (2014) Mast cell function: a new vision of an old cell. J Histochem Cytochem 62(10):698–738
365. Abraham SN, St John AL (2010) Mast cell-orchestrated immunity to pathogens. Nat Rev Immunol 10(6):440–452
366. Metcalfe DD, Peavy RD, Gilfillan AM (2009) Mechanisms of mast cell signaling in anaphylaxis. J Allergy Clin Immunol 124(4):639–46; quiz 647–8
367. Valent P, Horny HP, Escribano L, Longley BJ, Li CY, Schwartz LB et al (2001) Diagnostic criteria and classification of mastocytosis: a consensus proposal. Leuk Res 25(7):603–625
368. Hatch EW, Geeze MB, Martin C, Salama ME, Hartmann K, Eisenwort G et al (2018) Variability of PD-L1 expression in mastocytosis. Blood Adv 2(3):189–199

369. Theoharides TC, Valent P, Akin C (2015) Mast cells, mastocytosis, and related disorders. N Engl J Med 373(2):163–172
370. Falcone FH, Haas H, Gibbs BF (2000) The human basophil: a new appreciation of its role in immune responses. Blood 96:4028–4038
371. Akin C (2007) What does a basophil do? Blood 110:790–791
372. Legrand F, Woerly G, Driss V, Capron M (2008) Innate immune function of eosinophils—from antiparasite to antitumor cells. In: Ewbank J, Vivier E (eds) Innate immunity. Humana Press, New Jersey
373. Bandeira-Melo C, Weller PF (2005) Mechanisms of eosinophil cytokine release. Mem Inst Oswaldo Cruz 100(1):73–81
374. Spencer LA, Weller PF (2010) Eosinophils and Th2 immunity: contemporary insights. Immunol Cell Biol 88:250–256
375. Diny NL, Rose NR, Čiháková D (2017) Eosinophils in autoimmune diseases. Front Immunol 8:484
376. Picker LJ (1992) Mechanisms of lymphocyte homing. Curr Opin Immunol 4(3):277–286
377. MacDonald TT, Bajaj-Elliott M, Pender SLF (1999) T cells orchestrate intestinal mucosal shape and integrity. Trends Immunol 20(11):505–510
378. Elson CO, Machelski E, Weiserbs DB (1985) T cell-B cell regulation in the intestinal lamina propria in Crohn's disease. Gastroenterology 89(2):321–327
379. Panja A, Barone A, Mayer L (1994) Stimulation of lamina propria lymphocytes by intestinal epithelial cells: evidence for recognition of nonclassical restriction elements. J Exp Med 179 (3):943–950
380. Bailey M, Plunkett F, Clarke A, Sturgess D, Haverson K, Stokes C (1998) Activation of T cells from the intestinal lamina propria of the pig. Scand J Immunol 48(2):177–182
381. James SP (1991) Mucosal T-cell function. Gastroenterol Clin North Am 20(3):597–612
382. Pabst O (2013) Trafficking of regulatory T cells in the intestinal immune system. Int Immunol 25(3):139–143
383. Jones JC, Knox SJ (2013) Lymph nodes, thymus, spleen, and lymphatics. In: Rubin P, Constine LS, Marks LB (eds) ALERT—adverse late effects of cancer treatment, part of the series medical. Springer, Heidelberg
384. Israels LG, Israels ED (1999) Lymphocytes. Oncologist 4(2):129–137
385. Ivanova LA, Vasil'eva EV, Sokolov VV (1979) Identification of B-, T- and "zero" lymphocytes by staining for acid phosphatase. Lab Delo 10:593–596
386. Pluzańska A (1977) Comparative studies of the proportions of T, B and zero lymphocytes and the degrees of blastic transformation of lymphocytes after stimulation by PHA and CON A in patients with Hodgkin's disease. Haematol Blood Transfus 20:203–207
387. Bohnhorst JØ, Thoen JE, Natvig JB, Thompson KM (2001) Significantly depressed percentage of $CD27^+$ (memory) B cells among peripheral blood B cells in patients with primary Sjögren's syndrome. Scand J Immunol 54(4):421–427
388. Pampena MB, Levy EM (2015) Natural killer cells as helper cells in dendritic cell cancer vaccines. Front Immunol 6:13
389. Shahabuddin S, Al-Ayed I, Gad El-Rab MO, Qureshi MI (1998) Age-related changes in blood lymphocyte subsets of saudi arabian healthy children. Clin Vaccine Immunol 5(5): 632–635
390. Uppal SS, Verma S, Dhot PS (2003) Normal values of CD4 and CD8 lymphocyte subsets in healthy indian adults and the effects of sex, age, ethnicity, and smoking. Cytometry B Clin Cytom 52B(1):32–36
391. Bukowska-Straková K, Kowalczyk D, Baran J, Siedlar M, Kobylarz K, Zembala M (2009) The B-cell compartment in the peripheral blood of children with different types of primary humoral immunodeficiency. Pediatr Res 66:28–34
392. Campbell AC, Wiernik G, Wood J, Hersey P, Waller CA, Maclennan ICM (1976) Characteristics of the lymphopenia induced by radiotherapy. Clin Exp Immunol 23(2):200–208

393. Pao W, Wen L, Smith AL, Gulbranson-Judge A, Zheng B, Kelsoe G et al (1996) γδ T cell help of B cells is induced by repeated parasitic infection, in the absence of other T cells. Curr Biol 6(10):1317–1325
394. Déchanet J, Merville P, Lim A, Retière C, Pitard V, Lafarge X et al (1999) Implication of γδ T cells in the human immune response to cytomegalovirus. J Clin Invest 103(10):1437–1449
395. Palmer C, Diehn M, Alizadeh AA, Brown PO (2006) Cell-type specific gene expression profiles of leukocytes in human peripheral blood. BMC Genom 7:115
396. Bruno L, Fehling HJ, von Boehmer H (1996) The αβ T cell receptor can replace the γδ receptor in the development of γδ lineage cells. Immunity 5(4):343–352
397. Coico R, Sunshine G (2009) Immunobiology: a short course, 6th edn. Wiley, Hoboken
398. Mixter PF, Russell JQ, Morrissette GJ, Charland C, Aleman-Hoey D, Budd RC (1999) A model for the origin of TCR-$\alpha\beta^+$ $CD4^-CD8^-$ $B220^+$ cells based on high affinity TCR signals. J Immunol 162(10):5747–5756
399. Saitoh–Inagawa W, Hiroi T, Yanagita M, Iijima H, Uchio E, Ohno S et al (2000) Unique characteristics of lacrimal glands as a part of mucosal immune network: high frequency of IgA-committed B-1 cells and $NK1.1^+$ αβ T cells. Invest Ophthalmol Vis Sci. 41(1):138–144
400. Andreu-Ballester JC, García-Ballesteros C, Benet-Campos C, Amigó V, Almela-Quilis A, Mayans J, Ballester F (2012) Values for αβ and γδ T-lymphocytes and $CD4^+$, $CD8^+$, and $CD56^+$ subsets in healthy adult subjects: assessment by age and gender. Cytometry B Clin Cytom 82B:238–244
401. Costanzo AE, Taylor KR, Dutt S, Han PP, Fujioka K, Jameson JM (2015) Obesity impairs γδ T Cell homeostasis and antiviral function in humans. PLoS One 10(3):e0120918
402. Pinheiro MB, Antonelli LR, Sathler-Avelar R, Vitelli-Avelar DM, Spindola-de-Miranda S, Guimarães TM et al (2012) $CD4^-CD8^-$ αβ and γδ T cells display inflammatory and regulatory potentials during human tuberculosis. PLoS One 7(12):e50923
403. Han A, Newell EW, Glanville J, Fernandez-Becker N, Khosla C, Chien YH, Davis MM (2013) Dietary gluten triggers concomitant activation of $CD4^+$ and $CD8^+$ alphabeta T cells and gammadelta T cells in celiac disease. Proc Natl Acad Sci 110:13073–13078
404. Sallusto F, Geginat J, Lanzavecchia A (2004) Central memory and effector memory T cell subsets: function, generation, and maintenance. Annu Rev Immunol 22:745–763
405. Maggioli MF, Palmer MV, Thacker TC, Vordermeier HM, Waters WR (2015) Characterization of effector and memory T cell subsets in the immune response to bovine tuberculosis in cattle. PLoS One 10(4):e0122571
406. Dieli F, Poccia F, Lipp M, Sireci G, Caccamo N, Di Sano C, Salerno A (2003) Differentiation of effector/memory Vδ2 T cells and migratory routes in lymph nodes or inflammatory sites. J Exp Med 198(3):391–397
407. Doulatov S, Notta F, Laurenti E, Dick JE (2012) Hematopoiesis: a human perspective. Stem Cell 10(2):120–136
408. Takizawa H, Boettcher S, Manz MG (2012) Demand-adapted regulation of early hematopoiesis in infection and inflammation. Blood 119:2991–3002
409. McCabe A, MacNamara KC (2016) Macrophages: key regulators of steady-state and demand-adapted hematopoiesis. Exp Hematol 44(4):213–222
410. Liang HCY, Zúñiga-Pflücker JC (2015) Hematopoiesis: from start to immune reconstitution potential. Stem Cell Res Ther 6:52
411. Ramachandra L, Simmons D, Harding CV (2009) MHC molecules and microbial antigen processing in phagosomes. Curr Opin Immunol 21(1):98–104
412. Richards DM, Endres RG (2014) The mechanism of phagocytosis: two stages of engulfment. Biophys J 107(7):1542–1553
413. Kovari DT, Wei W, Chang P, Toro JS, Beach RF, Chambers D et al (2016) Frustrated phagocytic spreading of J774A-1 macrophages ends in myosin II-dependent contraction. Biophys J 111(12):2698–2710
414. Lavoie PM, Levy O (2017) Mononuclear phagocyte system. In: Polin RA, Abman SH, Rowitch D, Benitz WE (eds) Fetal and neonatal physiology, 5th edn. Philadelphia

415. Nutt SL, Hodgkin PD, Tarlinton DM, Corcoran LM (2015) The generation of antibody-secreting plasma cells. Nat Rev Immunol 15:160–171
416. Alberts B, Johnson A, Lewis J, Raff M, Roberts K, Walter P (eds) (2002) Molecular biology of the cell, 4th edn. Garland Science, New York
417. Knight AM, Lucocq JM, Prescott AR, Ponnambalam S, Watts C (1997) Antigen endocytosis and presentation mediated by human membrane IgG1 in the absence of the Igα/Igβ dimer. EMBO J 16:3842–3850
418. Sevigny LM, Booth BJ, Rowley KJ, Leav BA, Cheslock PS, Garrity KA et al (2013) Identification of a human monoclonal antibody to replace equine diphtheria antitoxin for treatment of diphtheria intoxication. Infect Immun 81(11):3992–4000
419. Roche PA, Furuta K (2015) The ins and outs of MHC class II-mediated antigen processing and presentation. Nat Rev Immunol 15:203–216
420. Medawar PB (1961) Peter Alfred Gorer. 1907–1961. Biogr Mem Fellows R Soc 7:95–109
421. Relle M, Schwarting A (2012) Role of MHC-linked susceptibility genes in the pathogenesis of human and murine lupus. Clin Dev Immunol 2012:584374
422. Charron DJ (2009) Obituary: Jean Dausset (1916–2009). Nature 460:338
423. Penn DJ, Ilmonen P (2005) Major histocompatibility complex (MHC). In: Encyclopedia of life sciences. Wiley, Chichester
424. Adiko AC, Babdor J, Gutiérrez-Martínez E, Guermonprez P, Saveanu L (2015) Intracellular transport routes for MHC I and their relevance for antigen cross-presentation. Front Immunol 6:335
425. Deakin JE, Papenfuss AT, Belov K, Cross JGR, Coggill P, Palmer S et al (2006) Evolution and comparative analysis of the MHC Class III inflammatory region. BMC Genom 7:281
426. Trowsdale J, Knight JC (2013) Major histocompatibility complex genomics and human disease. Annu Rev Genomics Hum Genet 14:301–323
427. Meyer D, C Aguiar VR, Bitarello BD, C Brandt DY, Nunes K (2018) A genomic perspective on HLA evolution. Immunogenetics 70(1):5–27
428. Viļuma A, Mikko S, Hahn D, Skow L, Andersson G, Bergström TF (2017) Genomic structure of the horse major histocompatibility complex class II region resolved using PacBio long-read sequencing technology. Sci Rep 7:45518
429. Candon S, Margulies DH (2004) MHC: structure and function. In: Wilkes DS, Burlingham WJ (eds) Immunobiology of organ transplantation. Springer, New York
430. Marsh SG, WHO Nomenclature Committee for Factors of the HLA System (2010) Nomenclature for factors of the HLA system, update April 2010. Tissue Antigens 76 (6):501–508
431. Redman CW, McMichael AJ, Stirrat GM, Sunderland CA, Ting A (1984) Class 1 major histocompatibility complex antigens on human extra-villous trophoblast. Immunology 52 (3):457–468
432. Nepom GT (2012) MHC class II tetramers. J Immunol 188(6):2477–2482
433. Sparer TE, Gooding LR (1998) Suppression of MHC class I antigen presentation by human adenoviruses. In: Whitton JL (ed) Antigen presentation volume 232 of the series current topics in microbiology and immunology. Springer, Heidelberg
434. Bubeník J, Šímová J (2009) MHC class I deficiency. In: Lang F (ed) Encyclopedia of molecular mechanisms of disease. Springer, Heidelberg
435. Robbins GR, Truax AD, Davis BK, Zhang L, Brickey WJ, Ting JP-Y (2012) Regulation of class I major histocompatibility complex (MHC) by nucleotide-binding domain, leucine-rich repeat-containing (NLR) proteins. J Biol Chem 287(29):24294–24303
436. Guiliano DB, Antoniou AN (2013) Measuring synthesis and degradation of MHC class I molecules. In: van Endert P (ed) Antigen processing volume 960 of the series methods in molecular biology. Springer, New York
437. Vukmanović S, Lilić M, Santori FR, Demaria S, Kulig K (2001) Peptide loading of nascent MHC class I molecules. Arch Immunol Ther Exp (Warsz) 49(3):195–201
438. Simone LC, Wang X, Solheim JC (2009) A transmembrane tail: interaction of tapasin with TAP and the MHC class I molecule. Mol Immunol 46(10):2147–2150

439. Rock KL, Goldberg AL (1999) Degradation of cell proteins and the generation of MHC class I-presented peptides. Annu Rev Immunol 17:739–779
440. Monaco JJ (1992) A molecular model of MHC class-I-restricted antigen processing. Trends Immunol 13(5):173–179
441. Day PM, Yewdell JW, Porgador A, Germain RN, Bennink JR (1997) Direct delivery of exogenous MHC class I molecule-binding oligopeptides to the endoplasmic reticulum of viable cells. PNAS 94(15):8064–8069
442. Schumacher TN, Heemels MT, Neefjes JJ, Kast WM, Melief CJ, Ploegh HL (1990) Direct binding of peptide to empty MHC class I molecules on intact cells and in vitro. Cell 62 (3):563–567
443. Kloetzel PM, Ossendorp F (2004) Proteasome and peptidase function in MHC-class-I-mediated antigen presentation. Curr Opin Immunol 16(1):76–81
444. Kloetzel PM (2001) Antigen processing by the proteasome. Nat Rev Mol Cell Biol 2:179–188
445. Li XC, Raghavan M (2010) Structure and function of major histocompatibility complex (MHC) class I antigens. Curr Opin Organ Transplant 15(4):499–504
446. Clark MS, Pontarotti P, Gilles A, Kelly A, Elgar G (2000) Identification and characterization of a β proteasome subunit cluster in the Japanese pufferfish (*Fugu rubripes*). J Immunol 165 (8):4446–4452
447. Moreland LW (ed) (2004) Rheumatology and immunology therapy. Springer, Heidelberg
448. Baudeau C, Delarue F, Hé CJ, Nguyen G, Adida C, Peraldi MN et al (1994) Induction of MHC class II molecules HLA-DR, -DP and -DQ and ICAM 1 in human podocytes by gamma-interferon. Exp Nephrol 2(5):306–312
449. Kambayashi T, Laufer TM (2014) Atypical MHC class II-expressing antigen-presenting cells: can anything replace a dendritic cell? Nat Rev Immunol 14:719–730
450. Singh N, Agrawal S, Rastogi AK (1997) Infectious diseases and immunity: special reference to major histocompatibility complex. Emerg Infect Dis 3(1):41–49
451. Guo X, Fu YX (2015) The tragic fate of group 3 innate lymphoid cells during HIV-1 infection. J Clin Invest 125(9):3430–3432
452. Hepworth MR, Fung TC, Masur SH, Kelsen JR, McConnell FM, Dubrot J et al (2015) Group 3 innate lymphoid cells mediate intestinal selection of commensal bacteria–specific $CD4^+$ T cells. Science 348(6238):1031–1035
453. Gruen JR, Weissman SM (2001) Human MHC class III and IV genes and disease associations. Front Biosci 6:D960–D972
454. Kokuho T, Nariuchi H, Gyotoku Y, Kakiuchi T (1995) Role of newly synthesized MHC class II molecules in antigen-specific antigen presentation by B cells. Immunobiology 193 (1):42–58
455. Kämpgen E, Koch N, Koch F, Stöger P, Heufler C, Schuler G, Romani N (1991) Class II major histocompatibility complex molecules of murine dendritic cells: synthesis, sialylation of invariant chain, and antigen processing capacity are down-regulated upon culture. Proc Natl Acad Sci USA 88(8):3014–3018
456. Morel A-S, Coulton G, Londei M (2002) Regulation of major histocompatibility complex class II synthesis by interleukin-10. Immunology 106:229–236
457. Hou T, Rinderknecht CH, Hadjinicolaou AV, Busch R, Mellins E (2013) Pulse-chase analysis for studies of MHC class II biosynthesis, maturation, and peptide loading. Methods Mol Biol 960:411–432
458. Young LJ, Wilson NS, Schnorrer P, Proietto A, ten Broeke T, Matsuki Y et al (2008) Differential MHC class II synthesis and ubiquitination confers distinct antigen-presenting properties on conventional and plasmacytoid dendritic cells. Nat Immunol 9:1244–1252
459. Naldi A, Carneiro J, Chaouiya C, Thieffry D (2010) Diversity and plasticity of Th cell types predicted from regulatory network modelling. PLoS Comput Biol 6(9):e1000912
460. Leung S, Liu X, Fang L, Chen X, Guo T, Zhang J (2010) The cytokine milieu in the interplay of pathogenic Th1/Th17 cells and regulatory T cells in autoimmune disease. Cell Mol Immunol 7:182–189

461. Eisenstein EM, Williams CB (2009) The Treg/Th17 cell balance: a new paradigm for autoimmunity. Pediatr Res 65:26R–31R
462. Stern LJ, Potolicchio I, Santambrogio L (2006) MHC class II compartment subtypes: structure and function. Curr Opin Immunol 18(1):64–69
463. Chen S, Li Y, Depontieu FR, McMiller TL, English AM, Shabanowitz J et al (2013) Structure-based design of altered MHC II-restricted peptide ligands with heterogeneous immunogenicity. J Immunol 191(10):5097–5106
464. Jasanoff A, Wagner G, Wiley DC (1998) Structure of a trimeric domain of the MHC class II-associated chaperonin and targeting protein Ii. EMBO J 17(23):6812–6818
465. Hanna S, Etzioni A (2014) MHC class I and II deficiencies. J Allergy Clin Immunol 134 (2):269–275
466. Yu CY, Yang Z, Blanchong CA, Miller W (2000) The human and mouse MHC class III region: a parade of 21 genes at the centromeric segment. Immunol Today 21(7):320–328
467. Torres AR, Sweeten TL, Johnson RC, Odell D, Westover JB, Bray-Ward P et al (2016) Common genetic variants found in HLA and KIR immune genes in autism spectrum disorder. Front Neurosci 10:463
468. Fernando MMA, Vyse TJ (2011) Major histocompatibility complex class II. In: Tsokos G, Buyon JP, Koike T, Lahita RG (eds) Systemic lupus erythematosus, 5th edn. Academic Press, Cambrigde
469. Milner CM, Campbell RD (2001) Genetic organization of the human MHC class III region. Front Biosci 6:D914–D926
470. Schneiders A, Thiel S, Winkler J, Möller P, Koch N (2005) Antibodies generated by a novel DNA vaccination identify the MHC class III encoded BAT2 polypeptide. Vaccine 23 (19):2540–2550
471. Beswick EJ, Reyes VE (2009) CD74 in antigen presentation, inflammation, and cancers of the gastrointestinal tract. World J Gastroenterol 15(23):2855–2861
472. Sun M, He C, Cong Y, Liu Z (2015) Regulatory immune cells in regulation of intestinal inflammatory response to microbiota. Mucosal Immunol 8:969–978
473. Danese S (2011) Immune and nonimmune components orchestrate the pathogenesis of inflammatory bowel disease. Am J Physiol Gastrointest Liver Physiol 300(5):G716–G722
474. Harney SMJ, Vilariño-GüellI C, Adamopoulos E, Sims AM, Lawrence RW, Cardon LR et al (2008) Fine mapping of the MHC Class III region demonstrates association of AIF1 and rheumatoid arthritis. Rheumatol (Oxford) 47(12):1761–1767
475. Matzaraki V, Kumar V, Wijmenga C, Zhernakova A (2017) The MHC locus and genetic susceptibility to autoimmune and infectious diseases. Genome Biol 18:76
476. Ting JP-Y, Kastner DL, Hoffman HM (2006) CATERPILLERs, pyrin and hereditary immunological disorders. Nat Rev Immunol 6:183–195
477. Ayala García MA, González Yebra B, López Flores AL, Guaní Guerra E (2012) The major histocompatibility complex in transplantation. J Transplant 2012:842141
478. Iannone R, Davies SM (2005) Tissue typing for hematopoietic cell transplantation: newer techniques and newer antigens for which cross-matching is helpful. Pediatr Transplant 9(7):76–80
479. Howell WM, Carter V, Clark B (2010) The HLA system: immunobiology, HLA typing, antibody screening and crossmatching techniques. J Clin Pathol 63(5):387–390
480. Tinckam KJ (2011) Basic histocompatibility testing methods. In: Chandraker A, Sayegh MH, Singh AK (eds) Core concepts in renal transplantation. Springer, New York
481. Rani R (2013) Major histocompatibility complex (MHC), applications. In: Dubitzky W, Wolkenhauer O, Cho K-H, Yokota H (eds) Encyclopedia of systems biology. Springer, New York
482. Mittal KK, Kaur D (1981) Serological testing of the HLA antigens—problems and approaches to standardization. In: Reisfeld RA, Ferrone S (eds) Current trends in histocompatibility. Springer, New York

483. Ward FE, MacQueen JM, Amos DB, Delmas-Marsalet Y, Johnson A (1975) The antiglobulin microcytotoxicity assay in HL-A genotyped families. Transplantation 19(4):286–290
484. Osoba D, Falk J (1974) The mixed-leukocyte reaction in man: effect of pools of stimulating cells selected on the basis of crossreacting HL-A specificities. Cell Immunol 10(1):117–135
485. Abraham LJ, Marley JV, Nedospasov SA, Cambon-Thomsen A, Crouau-Roy B, Dawkins RL, Giphart MJ (1993) Microsatellite, restriction fragment-length polymorphism, and sequence-specific oligonucleotide typing of the tumor necrosis factor region comparisons of the 4AOHW cell panel. Hum Immunol 38(1):17–23
486. Moribe T, Kaneshige T, Inoko H (1997) Complete HLA-A DNA typing using the PCR-RFLP method combined with allele group- and sequence-specific amplification. Tissue Antigens 50(5):535–545
487. Tinckam K (2009) Histocompatibility methods. Transplant Rev 23(2):80–93
488. Poggi A, Zocchi MR (2014) γδT lymphocytes as a first line of immune defense: old and new ways of antigen recognition and implications for cancer immunotherapy. Front Immunol 5:575
489. Campbell KS, Colonna M (2001) Human natural killer cell receptors and signal transduction. Int Rev Immunol 20(3–4):333–370
490. Vély F, Vivier E (2005) Natural killer cell receptor signaling pathway. Sci STKE 2005(292): cm6
491. Treanor B, Lanigan PMP, Kumar S, Dunsby C, Munro I, Auksorius E et al (2006) Microclusters of inhibitory killer immunoglobulin–like receptor signaling at natural killer cell immunological synapses. J Cell Biol 174(1):153–161
492. Norris S, Doherty DG, Collins C, McEntee G, Traynor O, Hegarty JE, O'Farrelly C (1999) Natural T cells in the human liver: cytotoxic lymphocytes with dual T cell and natural killer cell phenotype and function are phenotypically heterogenous and include Vα24-JαQ and γδ T cell receptor bearing cells. Hum Immunol 60(1):20–31
493. Eberl M, Engel R, Aberle S, Fisch P, Jomaa H, Pircher H (2005) Human Vγ9/Vδ2 effector memory T cells express the killer cell lectin-like receptor G1 (KLRG1). J Leukoc Biol 77(1): 67–70
494. Joyce MG, Sun PD (2011) The structural basis of ligand recognition by natural killer cell receptors. J Biomed Biotechnol 2011:203628
495. Enqvist M, Ask EH, Forslund E, Carlsten M, Abrahamsen G, Béziat V et al (2015) Coordinated expression of DNAM-1 and LFA-1 in educated NK cells. J Immunol 194(9): 4518–4527
496. Pegram HJ, Andrews DM, Smyth MJ, Darcy PK, Kershaw MH (2011) Activating and inhibitory receptors of natural killer cells. Immunol Cell Biol 89:216–224
497. Parham P (2005) Immunogenetics of killer cell immunoglobulin-like receptors. Mol Immunol 42(4):459–462
498. Parham P (2004) Killer cell immunoglobulin-like receptor diversity: balancing signals in the natural killer cell response. Immunol Lett 92(1–2):11–13
499. Moretta L, Moretta A (2004) Killer immunoglobulin-like receptors. Curr Opin Immunol 16 (5):626–633
500. Augusto DG (2016) The impact of KIR polymorphism on the risk of developing cancer: not as strong as imagined? Front Genet 7:121
501. Yu J, Heller G, Chewning J, Kim S, Yokoyama WM, Hsu KC (2007) Hierarchy of the human natural killer cell response is determined by class and quantity of inhibitory receptors for self-HLA-B and HLA-C ligands. J Immunol 179(9):5977–5989
502. Tiegs G, Lohse AW (2010) Immune tolerance: what is unique about the liver. J Autoimmun 34(1):1–6
503. Orr MT, Lanier LL (2010) Natural killer cell education and tolerance. Cell 142(6):847–856
504. Schroeder HW Jr, Cavacini L (2010) Structure and function of immunoglobulins. J Allergy Clin Immunol 125(2):S41–S52

505. El-Loly MM (2007) Bovine milk immunoglobulins in relation to human health. Int J Dairy Sci 2:183–195
506. Litman GW, Rast JP, Shamblott MJ, Haire RN, Hulst M, Roess W et al (1993) Phylogenetic diversification of immunoglobulin genes and the antibody repertoire. Mol Biol Evol 10 (1):60–72
507. Black CA (1997) A brief history of the discovery of the immunoglobulins and the origin of the modern immunoglobulin nomenclature. Immunol Cell Biol 75:65–68
508. Jones HB (1848) On a new substance occurring in the urine of a patient with mollities ossium. Philos Trans R Soc 138:55–62
509. Ribatti D (2015) Edelman's view on the discovery of antibodies. Immunol Lett 164(2): 72–75
510. Kantha SS (1991) A centennial review; the 1890 tetanus antitoxin paper of von Behring and Kitasato and the related developments. Keio J Med 40(1):35–39
511. Ramos-Bello D, Llorente L (2009) Fiftieth anniversary of the description of the chemical structure of antibodies. Reumatol Clin 5(6):280–284
512. del Guercio P (1987) Regulatory structures on the V and C regions of immunoglobulin molecules. Trends Immunol 8(10):304–308
513. DeFranco AL (1993) Structure and function of the B cell antigen receptor. Annu Rev Cell Biol 9:377–410
514. Mestecky J (1972) Structure of antibodies. J Oral Pathol 1(6):288–300
515. Roux K (1999) Immunoglobulin structure and function as revealed by electron microscopy. Int Arch Allergy Immunol 120(2):85–99
516. Vernersson M, Belov K, Aveskogh M, Hellman L (2010) Cloning and structural analysis of two highly divergent IgA isotypes, IgA1 and IgA2 from the duck billed platypus, *Ornithorhynchus anatinus*. Mol Immunol 47(4):785–791
517. Vernersson M, Aveskogh M, Munday B, Hellman L (2002) Evidence for an early appearance of modern post-switch immunoglobulin isotypes in mammalian evolution (II); cloning of IgE, IgG1 and IgG2 from a monotreme, the duck-billed platypus, *Ornithorhynchus anatinus*. Eur J Immunol 32(8):2145–2155
518. Aveskogh M, Hellman L (1998) Evidence for an early appearance of modern post-switch isotypes in mammalian evolution; cloning of IgE, IgG and IgA from the marsupial *Monodelphis domestica*. Eur J Immunol 28(9):2738–2750
519. Putnam FW, Liu YS, Low TL (1979) Primary structure of a human IgA1 immunoglobulin. IV. Streptococcal IgA1 protease, digestion, Fab and Fc fragments, and the complete amino acid sequence of the alpha 1 heavy chain. J Biol Chem 254(8):2865–2874
520. Gergely J (1992) How immunology was won in Hungary. Immunol Today 13(8):A1–A3
521. Jean L (1984) Origins of the terms 'antibody' and 'antigen'. Scand J Immunol 19(4):281–285
522. Strohl WR, Strohl LM (2012) Introduction to biologics and monoclonal antibodies. In: Therapeutic antibody engineering. Current and future advances driving the strongest growth area in the pharmaceutical industry. Woodhead Publishing, Philadelphia
523. Rutishauser U (2014) Gerald Edelman (1929–2014). Nature 510(7506):474
524. Raju TN (1999) The Nobel chronicles. 1972: Gerald M Edelman (b 1929) and Rodney R Porter (1917–85). Lancet 354(9183):1040
525. Tononi G (2014) Gerald M. Edelman (1929–2014). Science 344(6191):1457
526. Murphy K, Travers P, Walport M (2007) Janeway's immunobiology, 7th edn. Garland Science, USA
527. Spickett G (ed) (2013) Oxford handbook of clinical immunology and allergy, 3rd edn. Oxford University Press, Oxford
528. Willis MS (2009) Arne Tiselius: clinical chemistry. Lab Med 40(10):627–628
529. Marcus DM, Schlossman SF (2001) In Memoriam Elvin Abraham Kabat September 1, 1914–June 16, 2000. J Immunol 166(6):3635–3636
530. Gorevic PD, Prelli FC, Frangione B (1985) Immunoglobulin G (IgG). Methods Enzymol 116:3–25

531. Ligaarden SC, Lydersen S, Farup PG (2012) IgG and IgG4 antibodies in subjects with irritable bowel syndrome: a case control study in the general population. BMC Gastroenterology 12:166
532. Yan M, Zhang X, Pu Q, Huang T, Xie Q, Wang Y et al (2016) Immunoglobulin G expression in human sperm and possible functional significance. Sci Rep 6:20166
533. Braun J, Saxon A, Wall R, Morrison SL (1992) The second century of the antibody. Molecular perspectives in regulation, pathophysiology, and therapeutic applications. West J Med 157(2):158–168
534. Jackson KJ, Wang Y, Collins AM (2014) Human immunoglobulin classes and subclasses show variability in VDJ gene mutation levels. Immunol Cell Biol 92(8):729–733
535. Spiegelberg HL (1989) Biological role of different antibody classes. Int Arch Allergy Appl Immunol 90(1):22–27
536. Wingren C, Michaelsen TE, Magnusson CG, Hansson UB (1996) Comparison of surface properties of human IgA, IgE, IgG and IgM antibodies with identical and different specificities. Scand J Immunol 44(5):430–436
537. Wingren C, Hansson UB, Magnusson CG, Ohlin M (1995) Antigen-binding sites dominate the surface properties of IgG antibodies. Mol Immunol 32(11):819–827
538. Xu Y, Xu L, Zhao M, Xu CG, Fan Y, Pierce SK, Liu W (2014) No receptor stands alone: IgG B-cell receptor intrinsic and extrinsic mechanisms contribute to antibody memory. Cell Res 24:651–664
539. Caturegli P, Kuppers RC, Mariotti S, Burek CL, Pinchera A, Ladenson PW, Rose NR (1994) IgG subclass distribution of thyroglobulin antibodies in patients with thyroid disease. Clin Exp Immunol 98(3):464–469
540. Bazin H, Platteau B, Beckers A, Pauwels R (1978) Differential effect of neonatal injections of Anti-μ or Anti-δ antibodies on the synthesis of IgM, IgD, IgE, IgA, IgG1, IgG2a, IgG2b, and IgG2c immunoglobulin classes. J Immunol 121(5):2083–2087
541. Sweet SP, Rahman D, Challacombe SJ (1995) IgA subclasses in HIV disease: dichotomy between raised levels in serum and decreased secretion rates in saliva. Immunology 86 (4):556–559
542. Engström PE, Gustafson R, Granberg M, Engström GN (2002) Specific IgA subclass responses in serum and saliva: a 12-month follow-up study after parenteral booster immunization with tetanus toxoid. Acta Odontol Scand 60(4):198–202
543. Andreu-Ballester JC, Pérez-Griera J, Ballester F, Colomer-Rubio E, Ortiz-Tarín I, Peñarroja Otero C (2007) Secretory immunoglobulin A (sIgA) deficiency in serum of patients with GALTectomy (appendectomy and tonsillectomy). Clin Immunol 123(3):289–297
544. Cebra JJ, Gearhart PJ, Kamat R, Robertson SM, Tseng J (1977) The secretory immunoglobulin A response in the gut. Biochem Soc Trans 5(5):1565–1569
545. Engstrom JF, Arvanitakis C, Sagawa A, Abdou NI (1978) Secretory immunoglobulin deficiency in a family with inflammatory bowel disease. Gastroenterology 74(4):747–751
546. Woof JM (2007) The structure of IgA. In: Kaetzel CS (ed) Mucosal immune defense: immunoglobulin A. Springer, Boston
547. Underdown B, Schiff J (1986) Immunoglobulin A: strategic defense initiative at the mucosal surface. Annu Rev Immunol 4(1):389–417
548. Vladutiu AO (2000) Immunoglobulin D: properties, measurement, and clinical relevance. Clin Diagn Lab Immunol 7(2):131–140
549. Garcia-Pardo A (1978) Structure of immunoglobulin D: evidence for the absence of an extra disulfide bridge in the Fc fragment. Z Immunitatsforsch Immunobiol 154(3):218–225
550. Spiegelberg HL (1977) The structure and biology of human IgD. Immunol Rev 37:3–24
551. Pate MB, Smith JK, Chi DS, Krishnaswamy G (2010) Regulation and dysregulation of immunoglobulin E: a molecular and clinical perspective. Clin Mol Allergy 8:3
552. Bennich H, Johansson SG (1971) Structure and function of human immunoglobulin E. Adv Immunol 13:1–55
553. Buckley RH, Becker WG (1978) Abnormalities in the regulation of human IgE synthesis. Immunol Rev 41:288–314

554. Bennich H, Johansson SGO, von Bahr-Lindström H, Karlsson T (1976) Function and structure of immunoglobulin E (IgE). In: Johansson SGO, Strandberg K, Uvnäs B (eds) Molecular and biological aspects of the acute allergic reaction. Plenum Press, New York
555. Kishimoto T, Onoue K, Yamamura Y (1968) Structure of human immunoglobulin M. 3. Pepsin fragmentation of IgM. J Immunol 100(5):1032–1040
556. Hickman S, Kornfeld R, Osterland CK, Kornfeld S (1972) The structure of the glycopeptides of a human M-immunoglobulin. J Biol Chem 247(7):2156–2163
557. Onoue K, Kishimoto T, Yamamura Y (1968) Structure of human immunoglobulin MII. Isolation of a high molecular Q eight Fc fragment of IgM composed of several Fc subunits. J Immunol 100(2):238–244
558. Casali P, Schettino EW (1996) Structure and function of natural antibodies. Curr Top Microbiol Immunol 210:167–179
559. Forthal DN (2014) Functions of antibodies. Microbiol Spectr 2(4):1–17
560. Tarasenko T, Dean JA, Bolland S (2007) FcgammaRIIB as a modulator of autoimmune disease susceptibility. Autoimmunity 40(6):409–417
561. Sarantis H, Grinstein S (2012) Subversion of phagocytosis for pathogen survival. Cell Host Microbe 12(4):419–431
562. van Kessel KPM, Bestebroer J, van Strijp JAG (2014) Neutrophil-mediated phagocytosis of *Staphylococcus aureus*. Front Immunol 5:467
563. Capolunghi F, Rosado MM, Sinibaldi M, Aranburu A, Carsetti R (2013) Why do we need IgM memory B cells? Immunol Lett 152(2):114–120
564. Lanasa MC, Weinberg JB (2011) Immunoglobulin class switch recombination in chronic lymphocytic leukemia. Leuk Lymphoma 52(7):1398–1400
565. Geisberger R, Lamers M, Achatz G (2006) The riddle of the dual expression of IgM and IgD. Immunology 118(4):889–898
566. Murphy MJ, Hay JB, Morris B, Bessis MC (1972) Ultrastructural analysis of antibody synthesis in cells from lymph and lymph nodes. Am J Pathol 66(1):25–42
567. Casellas R, Nussenzweig A, Wuerffel R, Pelanda R, Reichlin A, Suh H et al (1998) Ku80 is required for immunoglobulin isotype switching. EMBO J 17(8):2404–2411
568. Ouyang H, Nussenzweig A, Kurimasa A, Soares VC, Li X, Cordon-Cardo C et al (1997) Ku70 is required for DNA repair but not for T cell antigen receptor gene recombination in vivo. J Exp Med 186(6):921–929
569. Loke P, Zang X, Hsuan L, Waitz R, Locksley RM, Allen JE, Allison JP (2005) Inducible costimulator is required for type 2 antibody isotype switching but not T helper cell type 2 responses in chronic nematode infection. Proc Natl Acad Sci U S A 102(28):9872–9877
570. Kopf M, Coyle AJ, Schmitz N, Barner M, Oxenius A, Gallimore A et al (2000) Inducible costimulator protein (ICOS) controls T helper cell subset polarization after virus and parasite infection. J Exp Med 192(1):53–61
571. Mimori T, Akizuki M, Yamagata H, Inada S, Yoshida S, Homma M (1981) Characterization of a high molecular weight acidic nuclear protein recognized by autoantibodies in sera from patients with polymyositis-scleroderma overlap. J Clin Invest 68(3):611–620
572. Dynan WS, Yoo S (1998) Interaction of Ku protein and DNA-dependent protein kinase catalytic subunit with nucleic acids. Nucleic Acids Res 26(7):1551–1559
573. Featherstone C, Jackson SP (1999) Ku, a DNA repair protein with multiple cellular functions? Mutat Res 434(1):3–15
574. Zhu C, Lee V, Finn A, Senger K, Zarrin AA, Du Pasquier L, Hsu E (2012) Origin of immunoglobulin isotype switching. Curr Biol 22(10):872–880
575. Ramsden DA, Weed BD, Reddy YVR (2010) V(D)J recombination: born to be wild. Semin Cancer Biol 20(4):254–260
576. Lewis SM, Wu GE (1997) The origins of V(D)J recombination. Cell 88(2):159–162
577. Market E, Papavasiliou FN (2003) V(D)J recombination and the evolution of the adaptive immune system. PLoS Biol 1(1):E16

578. Dudley DD, Chaudhuri J, Bassing CH, Alt FW (2005) Mechanism and control of V(D)J recombination versus class switch recombination: similarities and differences. Adv Immunol 86:43–112
579. Forthal DN, Landucci G, Phan TB, Becerra J (2005) Interactions between natural killer cells and antibody Fc result in enhanced antibody neutralization of human immunodeficiency virus type 1. J Virol 79(4):2042–2049
580. Woof JM, Burton DR (2004) Human antibody–Fc receptor interactions illuminated by crystal structures. Nat Rev Immunol 4:89–99
581. Beningo KA, Wang Y (2002) Fc-receptor-mediated phagocytosis is regulated by mechanical properties of the target. J Cell Sci 115:849–856
582. Forthal DN, Moog C (2009) Fc receptor-mediated antiviral antibodies. Curr Opin HIV AIDS 4(5):388–393
583. Takai T (2002) Roles of Fc receptors in autoimmunity. Nat Rev Immunol 2:580–592
584. Addadi L, Rubin N, Scheffer L, Ziblat R (2008) Two and three-dimensional pattern recognition of organized surfaces by specific antibodies. Acc Chem Res 41(2):254–264
585. Nimmerjahn F, Ravetch JV (2007) Fc-receptors as regulators of immunity. Adv Immunol 96:179–204
586. Köhler G, Milstein C (1975) Continuous cultures of fused cells secreting antibody of predefined specificity. Nature 256(5517):495–497
587. Alkan SS (2004) Monoclonal antibodies: the story of a discovery that revolutionized science and medicine. Nat Rev Immunol 4:153–156
588. Freysd'ottir J (2000) Production of monoclonal antibodies. Methods Mol Med 40:267–279
589. Raju TN (2000) The Nobel chronicles. 1984: Niels Kai Jerne, (1911–94); César Milstein (b 1926); and Georges Jean Franz Köhler (1946–95). Lancet 355(9197):75
590. Winter G, Harris WJ (1993) Humanized antibodies. Trends Immunol 14(6):243–246
591. Winter G, Griffiths AD, Hawkins RE, Hoogenboom HR (1994) Making antibodies by phage display technology. Annu Rev Immunol 12:433–455
592. McCafferty J, Griffiths A, Winter G, Chiswell D (1990) Phage antibodies: filamentous phage displaying antibody variable domains. Nature 348(6301):552–554
593. Sukupolvi-Petty S, Austin SK, Engle M, Brien JD, Dowd KA, Williams KL et al (2010) Structure and function analysis of therapeutic monoclonal antibodies against dengue virus type 2. J Virol 84(18):9227–9239
594. Melero I, Hervas-Stubbs S, Glennie M, Pardoll DM, Chen L (2007) Immunostimulatory monoclonal antibodies for cancer therapy. Nat Rev Cancer 7(2):95–106
595. Tansey EM, Catterall PP (1994) Monoclonal antibodies: a witness seminar in contemporary medical history. Med Hist 38(3):322–327
596. Beck A, Wurch T, Bailly C, Corvaia N (2010) Strategies and challenges for the next generation of therapeutic antibodies. Nat Rev Immunol 10(5):345–352
597. Khawli LA, Goswami S, Hutchinson R, Kwong ZW, Yang J, Wang X et al (2010) Charge variants in IgG1: isolation, characterization, in vitro binding properties and pharmacokinetics in rats. MAbs 2(6):613–624
598. Siegel DL (2002) Recombinant monoclonal antibody technology. Transfus Clin Biol 9(1): 15–22
599. Chadd HE, Chamow SM (2001) Therapeutic antibody expression technology. Curr Opin Biotechnol 12(2):188–194
600. Robak T, Robak E (2011) New anti-CD20 monoclonal antibodies for the treatment of B-cell lymphoid malignancies. BioDrugs 25(1):13–25
601. Weiner GJ (2010) Rituximab: mechanism of action. Semin Hematol 47(2):115–123
602. Fanale M, Assouline S, Kuruvilla J, Solal-Céligny P, Heo DS, Verhoef G et al (2014) Phase IA/II, multicentre, open-label study of the CD40 antagonistic monoclonal antibody lucatumumab in adult patients with advanced non-Hodgkin or Hodgkin lymphoma. Br J Haematol 164(2):258–265
603. Donato F, Gay F, Bringhen S, Troia R, Palumbo A (2014) Monoclonal antibodies currently in Phase II and III trials for multiple myeloma. Exp Opin Biol Ther 14(8):1127–1144

604. Gold M (2017) Phase II clinical trials of anti–β-amyloid antibodies: when is enough, enough? Alzheimers Dement 3(3):402–409
605. Iwasaki A, Medzhitov R (2015) Control of adaptive immunity by the innate immune system. Nat Immunol 16(4):343–353
606. Spiering MJ (2015) Primer on the immune system. Alcohol Res 37(2):171–175
607. Gadjeva M, Verschoor A, Carroll MC (2003) The role of complement in innate and adaptive immunity. In: Ezekowitz RAB, Hoffmann JA (eds) Innate immunity, part of the series infectious disease. Humana Press Inc., Totowa
608. Liu Y-M, Luo J, Bennett C (2010) Adaptive immunity—based on the dual recognition responses of αβT cells. Self Nonself 1(1):62–66
609. Lebedev KA, Ponyakina ID (2009) Physiology of adaptive immunity regulation via signaling pattern-recognition receptors. Hum Physiol 35(1):108–116

Chapter 11
Gastrointestinal Exocrine (Lumencrine) Secretions. The Reception Theory as the Basis for Developing the First Antisecretory Pharmacotherapy Drugs

Abstract Secretions of the gastrointestinal (GI) tract were identified since antiquity. However, the role of these secretions in the process of digestion was not recognized after a couple of centuries. Modern knowledge about the secretory activity and regulation of the gut began with the work of Camillo Golgi (1843–1926), Jan Evangelista Purkyně (1787–1869), William Beaumont (1785–1853), Rudolph Heidenhain (1834–1897), Ivan Petrovich Pavlov (1849–1936). GI secretions include oral secretions (saliva), gastric juice, pancreatic juice, intestinal juice, bile, and other co-released components produced by the glandular cells of the digestive tract. Enzymes and their cofactors represent major components of these digestive juices that help to break down proteins, fats, and carbohydrates into simple absorbable substances. The GI juices also contain water, ions, mineral salts, and other endogenous proteins. In its broader sense, however, GI secretions include neuromediators and hormones. Of special attention is the secretion of the cells of the stomach (known as gastric juice), which has immense clinical implications in gastric pathology. Though it was widely accepted that hydrochloric acid is a major component of stomach juice, the clinical importance of gastric acid was not appreciated until the cellular and molecular mechanisms of gastric secretion were unraveled. It was known that gut secretions are controlled by a complex network involving the nervous and humoral systems as well as components of ingested food. However, nothing was known about the molecular processes regulating the secretory activity of the stomach. Compelling evidences on the mechanisms of regulation of gastric secretion came in the second half of the twentieth century following the groundbreaking investigations led by Sir James Whyte Black (1924–2010), which was rooted on chemical reception theory proposed around the beginning of the twentieth century by John Newport Langley (1852–1925) and Paul Ehrlich (1854–1915). The theory of chemical reception posits that chemical receptive substances (receptors), which are plasma membrane proteins, are required for receiving incoming chemical messengers (hormones and neurotransmitters) that initiate cellular response. The discovery of the hormone gastrin by John Sydney Edkins (1863–1940) in 1906 and of the GI source and functions of the hormone histamine by a student of Pavlov, Popielski Leon Bernardovich (Popielski Lev Bernardovich) (1866–1920) in 1916, coupled with the discovery of proton pumps

M. O. Welcome, *Gastrointestinal Physiology*,
https://doi.org/10.1007/978-3-319-91056-7_11

in 1973 by Allen L. Ganser (1942–), and John Gaetano Forte (1934–2012), made it possible for the pioneer investigator Dr. George Sachs to extensively study proton pump inhibitors (PPI) and histamine (H) type 2 receptor blockers, which provided a good and superior alternative to gastric surgery that was initially the mainstay of treatment of gastric ulcer. This chapter not only gives a historic account on discoveries and the clinical importance of gastric secretions, but also discusses the mechanisms and secretory functions of the various regions of the GI tract.

Keywords Reception theory · Chemical reception · Chemical signaling Ion transport · Gastrointestinal secretions · Saliva · Exocrine glands Duacrine glands · Gastric juice · Pancreatic juice · Intestinal juice Bile · Proton pumps · Proton pump inhibitors · Histamine type 2 receptor blockers Gastric pharmacology · Gastric history · John Newport Langley Paul Ehrlich · Sir James Whyte Black · John Sydney Edkins · Popielski Leon Bernardovich (Popielski Lev Bernardovich) · Allen L. Ganser · John Gaetano Forte George Sachs

Abbreviations

HCl	Hydrochloric acid
M_1, M_2, M_3, M_4, and M_5	Muscarinic acetylcholine receptor types
μg/kg	Microgram per kilogram
μg/kg/h	Microgram per kilogram per hour
PPIs	Proton pump inhibitors
GABA	Gamma-aminobutyric acid
NO	Nitric oxide
CCK	Cholecystokinin
CNS	Central nervous system
5-HT(2A)	Serotonin 2A (5-HT(2A)) receptor
LD_{50} or LC_{50}	Lethal dose or concentration 50%
nmol/kg	Nanomole per kilogram
microM	Micromole
4-DAMP	4-Diphenyl-acetoxy-*N*-methyl-piperidine
ATP4A	Adenosine triphosphate type 4A
TM4, TM5, TM6, and TM8	Transmembrane segments
Kir4.1	ATP-dependent inwardly rectifying potassium
KCNQ1	Voltage-gated potassium channel, KQT-like subfamily Q, member 1
KCNE2	Member 2 of the potassium voltage-gated channel subfamily E also known as MinK-related peptide 1 (MiRP1)
CFTR	Cystic fibrosis transmembrane conductance regulator
CLIC-6	Chloride intracellular channel protein 6
Cl^-	Chloride ion

K^+	Potassium ion
KCC4	K^+-Cl^- cotransporter type 4
Å	Armstrong
GERD	Gastroesophageal reflux disease
SLC26A9	Solute carrier family 26 (anion exchanger), member 9

11.1 Introduction

Gastrointestinal (GI) secretions are the various substances that are released by the classical exocrine glandular cells of the digestive tract. The secretions include saliva, gastric juice, pancreatic juice, intestinal juice, bile, and other components from other cells of the GI tract. The fluids contain enzymes that break down proteins, fats, and carbohydrates into simple chemical compounds. The components of the fluids also include water, ions, mineral salts, proteins. These are the components of secretions of the GI tract. These secretions together with those of associated organs enable the breakdown of food into small molecules that can be easily absorbed in the gut. GI secretions are controlled by a complex network of regulatory systems involving hormones, neurotransmitters, regulatory peptides as well as the components of the food nutrients [1–5]. In this chapter, the secretions of the various regions of the GI tract are discussed. The chapter also gives a concise account of historic developments of GI lumencrine secretions. The clinical implications of GI secretions are strategically discussed.

11.2 Components of Gastrointestinal Lumencrine Secretions

11.2.1 Water

Water constitutes a major component of GI secretions. The total volume of water in the GI tract alone is estimated to be about 20% of the total body water. Approximately, 8.5–9 L of fluid are presented to the GI tract per day [6]. But this volume of fluid may reach 10 L per day in certain conditions [7]. (During fasting, the rate of fluid secretion is lower—at ~2.5 mL/min, compared to the fed state—5–6.3 mL/min [6].) Of the roughly 8.5–9 L of fluid, 1.5–2 L are ingested and ~7 L are derived from endogenous secretions of the salivary glands, stomach, gallbladder, pancreas, and the intestines. The salivary glands produce about 1.5 L of saliva daily (normal volume of saliva production in some individuals may be as low as 0.5 L per day [8]). The stomach secretes about 2 L of gastric juice per day. The gallbladder has a daily production of 0.5 L of bile. The pancreas secretes approximately 1.5 L of pancreatic juice per day. The intestines secrete approximately 1.5–2.5 L of fluid per day (Fig. 11.1). The intestinal fluid flux occurs via the permeable junctions of the intestinal mucosa along a gradient of osmosis. Some

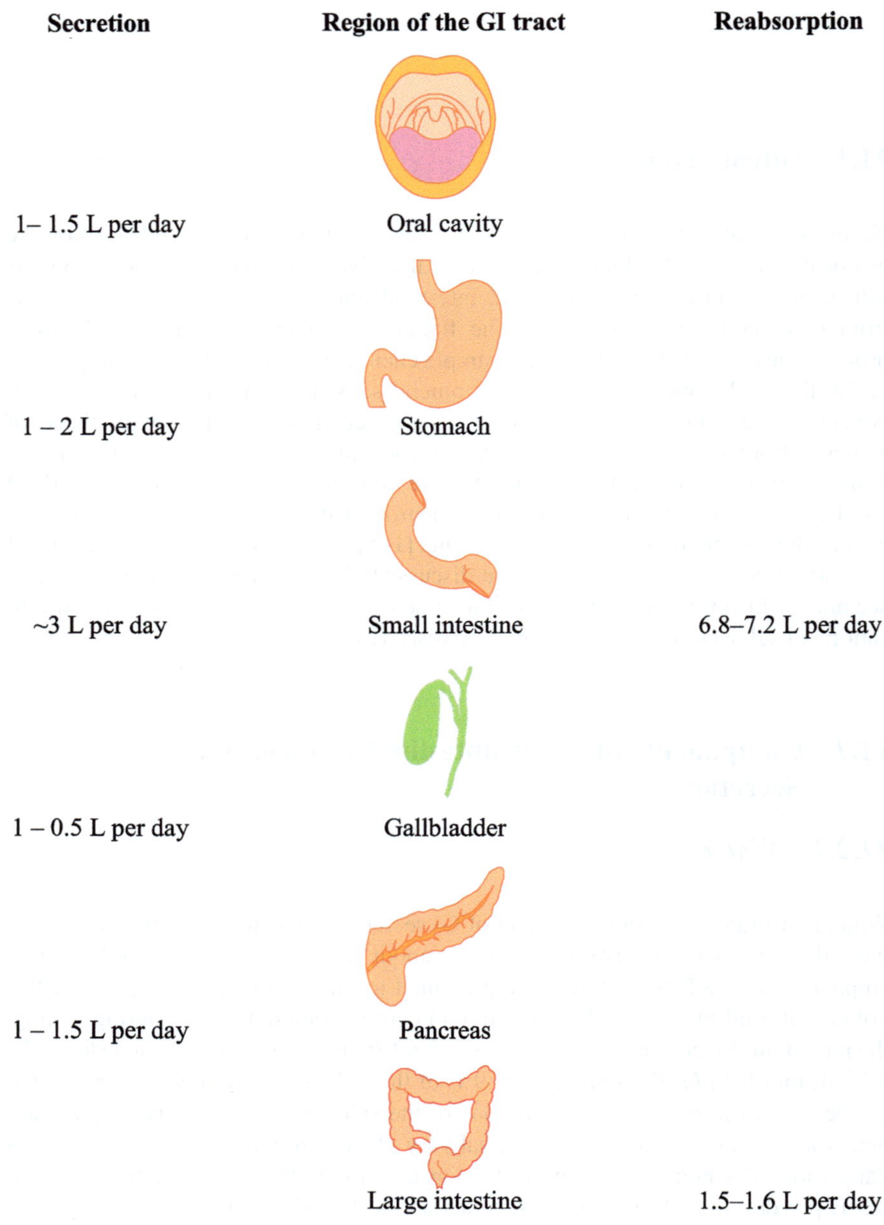

Fig. 11.1 Gastrointestinal fluid fluxes

quantities of the fluid that drain into the lumen represent temporary losses from the intravascular space of the gut [6, 7, 9].

The volume of fluid secreted is adequately balanced with the amount of fluid reabsorbed in the gut. Of the total volume of fluid available to the GI tract per day, the ileum reabsorbs 80% (~6.8–7.2 L of fluid per day), the colon reabsorbs 18% (~1.5–1.6 L of fluid per day), and the remaining 2% is excreted in stool (~100–200 mL of fluid per day) (Fig. 11.1) [7, 9]. It should be mentioned that the reserve of the absorptive capacity of the colon can reach 3 L per day depending on the functional state of the individual [6]. The rate of absorption of fluid by the colon determines colonic transit time and stool consistency [6].

Water reabsorbed from the intestine escapes to the basolateral side into interstitium; from here the water flows into blood and lymph circulation. The water is used for different activities of the body, and excess fluid is removed by excretory organs [9]. It should be noted, however, that numerous factors influence both the secretory capacity and the reabsorptive capacity of the gut. These factors include state of health, environmental factors, nutritional, age. Thus, the volume of gut secretions mentioned above can significantly vary depending on prevailing factors or group of factors, counteracting the secretory activities of the gut [10–12]. The fluid fluxes in the GI tract are to a large extent due to the epithelial transport of electrolytes primarily Na^+, Cl^-, and $HCO^-{}_3$ ions [7].

11.2.2 Ions, Mucus, Enzymes, and Other Biologically Active Molecules

Apart from water, the GI tract continuously secretes numerous ions, enzymes, and other biologically active molecules, which are all required for the normal functioning of the gut and extragut tissues or organs [9]. The main ions secreted by the gut include bicarbonate ion ($HCO^-{}_3$), chloride ion (Cl^-), sodium ion (Na^+), hydroxyl ion (OH^-), proton (H^+), potassium ion (K^+) [6]. In a normal situation, approximately 800 mmol of Na^+, 100 mmol of K^+, and 700 mmol of Cl^- are transported across the gut per day [7]. Na^+ is the predominant cation secreted [6]. Cl^- in addition to $HCO^-{}_3$ ions are the predominantly secreted anions [6, 13]. The high concentration of $HCO^-{}_3$ in the GI mucosa is required, at least in part, to neutralize the acidity of the secreted protons. The $HCO^-{}_3$ ions are mostly produced by the cystic fibrosis transmembrane conductance regulator (CFTR)—a type of ion channel localized on the plasma membrane of epithelial cells of the GI tract [13]. Majority of the secreted electrolytes are reabsorbed in the intestines [6].

The ion fluxes in the GI tract are controlled by coupled transport and by the bulk transport of water along osmotic gradient. The transport of water is concomitantly associated with the transport of electrolytes across the intestinal mucosa—a phenomenon referred to as solvent drag [6, 9]. Both ion and fluid fluxes are mainly controlled by the transport proteins of the apical brush border and basolateral aspects of the intestinal epithelium in addition to the intercellular linkages [7].

GI enzymes are secreted in various regions of the gut (mouth, stomach, intestines) to break down the ingested food particles to the size that are absorbable by the enterocytes. Mucin comprises a crucial component of GI secretions in all regions of the tract. Other molecules including gastric intrinsic factor, growth factors, secretions of resident microbes play an integral role in regulation of gut activities [14–17]. The ion and water fluxes are influenced by nutrition, health, and physiological states of the individual [6].

11.3 Synthesis of Gastrointestinal Secretory Molecules

The secretory functions of the GI cells were poorly understood some decades ago; however, continuous progress in the technique of cellular visualization enabled extensive study of cellular secretory dynamics. The advancement of techniques of cell imaging made it possible for investigating signaling pathways involved in GI tract secretion in the cellular region of interest or whole in real time and at a high spatial resolution and nanometer scale. Nowadays, modern techniques of cell imaging allow for high throughput analysis of receptor–ligand interactions and mechanics of intracellular vesicular cargo transport—these processes are integral in secretory of substances in the GI tract [18].

For initiation of synthesis, the signal arriving at the cell membrane is relayed to the cytoplasm and then to the nucleus to begin the process of synthesis of molecules required for the formation of the secretory cargo (Fig. 11.2). Further details about the mechanisms of secretion are discussed in Chaps. 3, 8, and 9.

11.3.1 Cellular Signaling Pathways Regulating Gastrointestinal Secretory Activity

Cellular signaling pathways in secretion include—but are not limited to G protein-coupled receptors, enzyme-linked receptor (tyrosine kinase)—mitogen-activated protein kinases, epidermal growth factor, JAK/STAT, and NFκB pathways, as well as ion channels, transporters, steroid receptor pathways. For instance, pathways of G protein-coupled receptors and tyrosine kinases are involved in regulating mucin secretion [18, 23–28]. Furthermore, salivary secretion occurs through the activities of several ion channels, and G protein-coupled receptors [25]. Many of the pathways exhibit extensive cross-talks that regulate a vast number of physiological processes. The signaling pathways are discussed in Chap. 5.

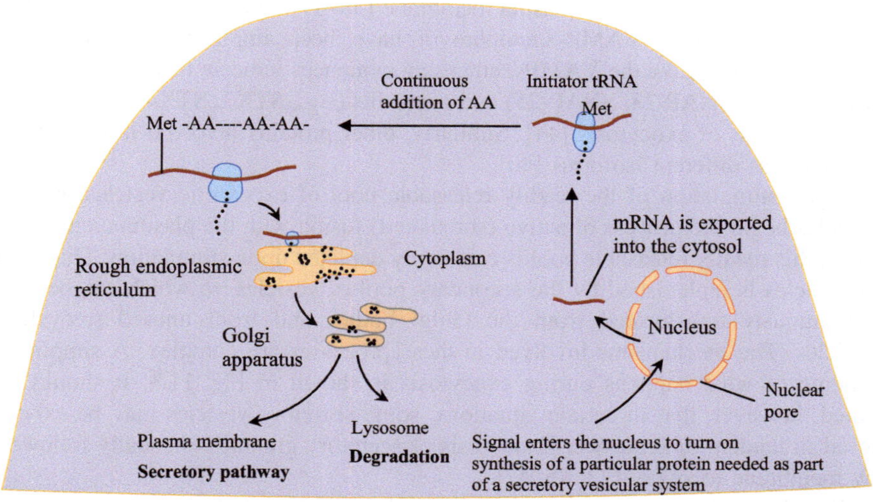

Fig. 11.2 Production of secretory protein for export. Signal from outside the cell initiates a cascade of cellular events leading to the activation of transcription factors and protein kinases. These transcription factors or protein kinases are responsible for activating gene transcription process which forms mRNA that is exported through the nuclear pore into the cytoplasm. The information in this mRNA is then transcribed into a sequence of amino acids that are further processed in the endoplasmic reticulum and Golgi apparatus to pass through the initial packaging stage. The packaged protein is trafficked away from the site of packaging where it undergoes further modifications awaiting to receive a signal for translocation toward the plasma membrane or it is transported into lysosomes for degradation [19–22]

11.3.2 Exocytic Machinery of Secretory Vesicles of the Gastrointestinal Cells

The exocytosis of secretory granules occurs in exocrine, endocrine, and neurosecretory cells, as well as neurons through the activities of several cytoskeleton-associated proteins and enzymes. The mechanism of the release of exocytotic vesicles in these cells was proposed about 40 years ago [29–32]. Over the past decades, accumulating evidences have broadened our understanding of the multi-step processes involved in vesicle exocytosis [33–35].

Following synthesis and packaging of proteins and other molecules destined for exocytosis, the cargoes (vesicles) containing substances for secretion are exported toward the plasma membrane. The movement of secretory cargoes to the plasma membrane for release of the content (exocytosis) is controlled by many cytoplasmic and membrane-bound proteins, cytoskeletal components, including small protein (e.g., GTPases), SNARE (soluble *N*-ethylmaleimide-sensitive factor attachment protein receptor), SNAP-25 (synaptosomal-associated protein of 25 kDa), syntaxins, synaptophysin, and synaptobrevin/vesicle-associated membrane protein (VAMP2). These proteins play a crucial role in translocation, tethering, docking, and fusion of vesicles to the plasma membrane [38–41]. Different types of VAMPs including VAMP-7, VAMP-8/endobrevin have been implicated in exocytotic process [42, 43]. Like the VAMPs, emerging evidences indicate that many types of SNAPs (e.g., SNAP-23, SNAP-25) and syntaxins (e.g., STX3, STX4) are involved in the process of exocytosis [44]. Similarly, other proteins involved in exocytosis also express different isoforms [40].

Upon stimulation of the readily releasable pool of exocytotic vesicles, nearby vesicles begin the process of active (stimulated) fusion with the plasma membrane. Thus, this readily releasable pool is constantly depleted upon stimulation. This pool of vesicles is replenished by the secondary pool of vesicles, to which cargoes are continuously translocated from the Golgi bodies and from unused (recycled) vesicles. The mechanisms involved in these processes are complex. A simplified scheme of what happens during exocytosis is shown in Fig. 11.3. It should be noted, however, that in certain situations, some secretory vesicles may be exocytosed in tandem or sequence. Exocytosis of secretory granules is usually followed by membrane recovery [22, 45–47].

Details on the molecular mechanisms and mechanics of the steps involved in translocation of secretory vesicles to the plasma membrane and exocytosis can be assessed in modern textbooks of molecular biology.

11.4 Regional Gastrointestinal Secretions

11.4.1 Salivary (Buccal) Secretions

Secretions of the oral cavity (also referred to as buccal secretions) are produced by the minor and major salivary glands. In addition to salivary secretions, buccal secretions also include gingival crevicular fluid. These secretions are sometimes called whole saliva or oral fluid and also contain oral microbes and food debris. The fluid provides the necessary milieu for dental health. The minor salivary glands are located throughout the oral cavity and include buccal, labial, and palatine glands. The major salivary glands are the parotid, sublingual, and submandibular (Fig. 11.4) [48–51]. Secretions from the different glands and sites differ in many aspects. The main differences are in the composition of saliva produced by the minor and major salivary glands. Compared to the major glands, the minor salivary glands have higher sodium, low phosphate, and bicarbonate ion concentrations. Minor salivary glands have low concentration of amylase, but high protein concentration. Furthermore, the concentration of IgA and certain blood group molecules are higher in minor salivary secretions [49, 52–54]. The secretions of the salivary glands pass through ducts into the oral cavity. Each day approximately 1–2 L of saliva are secreted by the salivary glands. Saliva is a mucin-rich alkaline (or slightly acidic) fluid containing substantial quantity of α-amylase (ptyalin) and other enzymes, ions, as well as other biomolecules (discussed below) [1, 49, 55, 56].

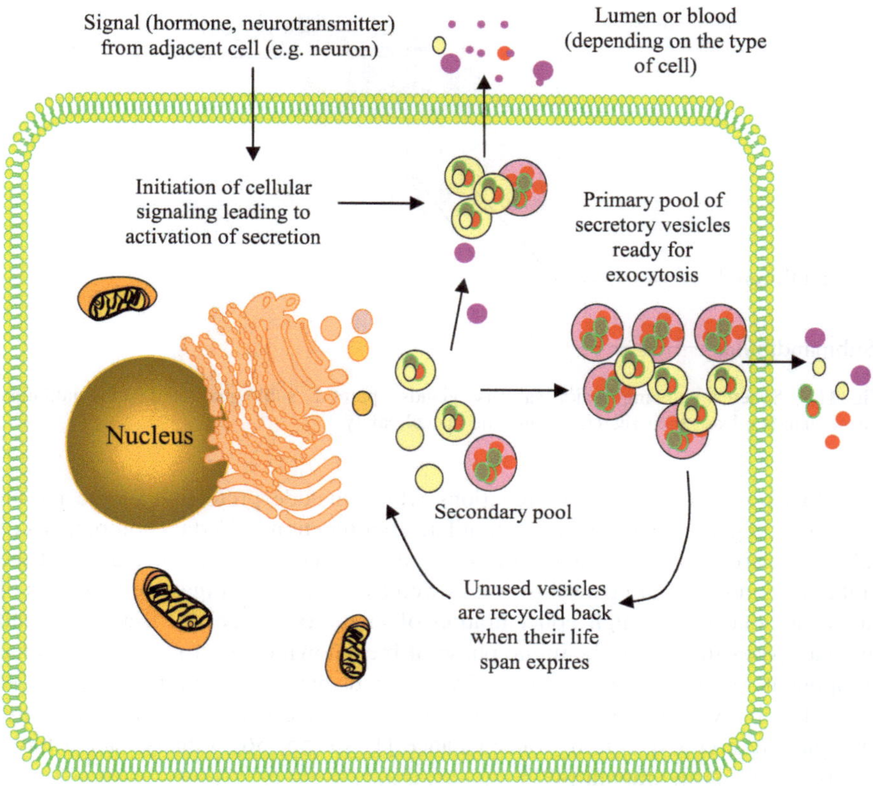

Fig. 11.3 Recruitment of pools of secretory vesicles during exocytosis. The mechanisms of secretion in the GI tract vary depending on the cell type and the signal impinging on the cell. The diagram shows a general view of the processes involved in the secretion of substances into the lumen or circulation of the GI system. Available evidences show that hormones, neurotransmitters, enzymes, and other products are secreted according to this model. The contents of vesicles may be secreted directly into the lumen, lymphatic, or blood vessels. In some cases, cellular secretions may act as ligands for the receptors of another cell by switching on or off (activation or inhibition) a specific activity of the cell [32]. The readily available pool of secretory vesicles is refilled from the secondary pool. The secondary pool is made up of de novo synthesized or recycled vesicles. The density of these pools of secretory vesicles is controlled by cellular signaling pathways. Transport of secretory vesicles is carried out through an orderly series of events that includes tethering, docking, membrane apposition, fusion with the plasma membrane, and subsequent exocytosis of the vesicle cargo [18, 36, 37]. The signal that determines the translocation of these vesicles are chemical (hormone, neurotransmitter, ions, etc.), mechanical (strain, stretch), among others. For instance, signaling pathways involving the increase in cytosolic calcium waves can initiate the exocytotic events. The signal may be initiated by the central nervous system, chemical components in food or physical agents. These signals activate cellular receptors, leading to downstream signaling, which activate the recruitment of vesicles for exocytosis [18, 36, 37]

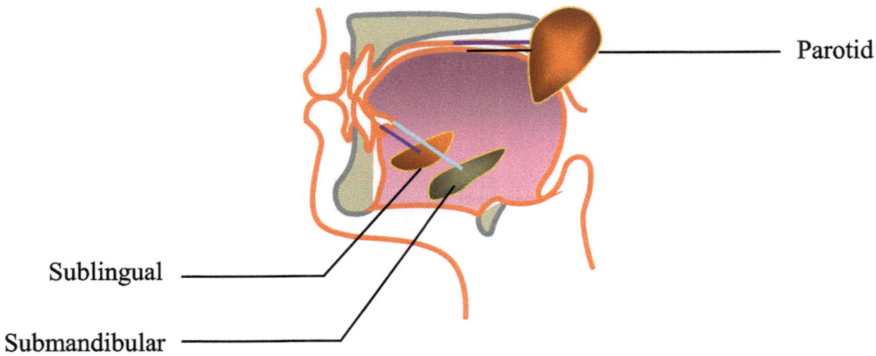

Fig. 11.4 Schema showing major salivary glands—parotid, sublingual, and submandibular. These glands release their secretions into the buccal cavity

Saliva possesses a range of functions which include protection of the mouth against pathogenic microorganisms (antibacterial functions). Certain components of saliva, in particular, mucin acts as a lubricant, protecting against ulceration of the oral cavity and facilitating swallowing as well as speech. The digestive functions of saliva are due to the high concentration of enzymes, especially α-amylase and lingual lipases that begin the initial phase of breakdown of macromolecules. Some components of the secretions of the salivary glands have a buffering function. Over 99% of salivary secretions, which comprises water, are reabsorbed into the body, thus playing a crucial role in water balance [1, 49, 55, 56]. Finally, saliva has a cleansing role in the mouth [49].

A couple of dysfunctions in salivary secretions occur in a range of diseases. For instance, reduction in the volume of salivary secretions (known as hyposalivation) resulting in dry mouth represents a key symptom in xerostomia [57]. Evidently, the rate of salivary flow represents a function of oral dryness [58]. The condition, hyposalivation, can result from Sjögren's syndrome, use of certain pharmacological drugs (e.g., in chemotherapy), radiotherapy, presence of sialolithiasis or salivary calculi (salivary stones), neoplasia of the salivary glands [57, 59, 60]. However, the commonest disorders of the salivary glands are usually associated with acute inflammatory diseases due to pathogenic bacterial and viral invasion [59]. Decrease in salivary secretions can predispose an individual to the development of oral mucosal disorders such as dental caries [57]. Hypersalivation on the other hand, referred to as sialorrhea (also known as drooling or ptyalis), is a symptom, characterized by excess saliva in the mouth beyond the lip margin [61]. The condition is normal for children aged 15–36 months, but also represents a key symptom occurring as a side effect of psychotropic medications (e.g., clozapine) and in neurodegenerative disorders such as cerebral palsy, amyotrophic lateral sclerosis, Parkinson's disease, and other diseases [61–64]. Pathological drooling is treated with botulinum toxin. In particular, abobotulinumtoxin A and rimabotulinumtoxin B have shown considerable success for use in addressing cases of sialorrhea [65, 66].

Major Salivary Glands

The major salivary glands—parotid, sublingual, and submandibular (submaxillary) glands—occur as pairs on the right and left sides. They are responsible for about 90% of saliva that is secreted in the mouth [67]. The names of these glands were derived following the discovery of their ducts during the seventeenth century. The discovery of the ducts of these glands formed the basis of the concept of exocrine secretion [49, 57, 68]. The nomenclature of the salivary glands is based not only on the location of the opening of the duct, but also on the histology as well as the autonomic innervation of the glands [57, 59]. The major salivary glands have secretory endpiece composed of acinar cells and arborization of the ductal tree, which opens into the mouth [69]. Salivary glands are surrounded by a capsule, and they have a basal lamina that surrounds the epithelium [69]. The epithelium of the salivary glands is composed of three main cell types—acinar, ductal, and myoepithelial cells [59, 68]. Some of the epithelial cells secrete mucin. The epithelial cells express microvilli or cilia [56]. The salivary glands are composed of many other cell and tissue types, which include endothelial cells, immune cells, myofibroblasts, stromal cells, nerve fibers [69].

The acinar cells are responsible for primary secretion of saliva (primary saliva) that drains into the ductal epithelium. The ductal epithelium comprises ductal cells that produce the secondary from the primary saliva [59, 68]. The myoepithelial cells are elongated cells that surround both the acinar and ductal cells. Those surrounding the acinar cells are stellate myoepithelial cells (for this reason, they are also known as basket cells), while those surrounding the ductal cells are myoepithelial cells [68, 69]. The myoepithelial cells play many roles in maintaining the structural and functional integrity of the salivary glands. These cells produce certain factors that regulate the growth of acinar and ductal cells. The contraction of these cells facilitates secretion of salivary components. It can act as sensory cell, playing an integral role in chemoreception. The chemoreceptive functions of the myoepithelial cells are due to their expression of cilia, which projecting into the invaginations of neighboring salivary secretory cells [70, 71]. The cells of the salivary glands are linked laterally by communicating junctions such as desmosomes, tight and adherent junctions [56].

Salivary glands are composed of several divisions termed lobes. Each salivary lobe can be further divided into smaller portions termed lobules. Each lobule is separated by a septum of connective tissue [68]. The smaller ducts pass within lobules (intralobular ducts) to drain into interlobular ducts and then to the interlobar ducts. The intralobular ducts comprise two parts—intercalated ducts and striated ducts. The former connects the acinus to the later. Both the intercalated and striated ducts comprise the same type of cells—simple columnar cells (Fig. 11.5). The striated ducts drain into the interlobar ducts, also known as proper or main excretory ducts, from where saliva flows into the oral cavity, which is the initial region of the GI tract with underlining mucous membrane [57, 68]. The ducts are lined with simple columnar cells. A population of columnar cells, called the brush (tuft) cells, possesses microvilli at their luminal side and is mostly involved in secretion and

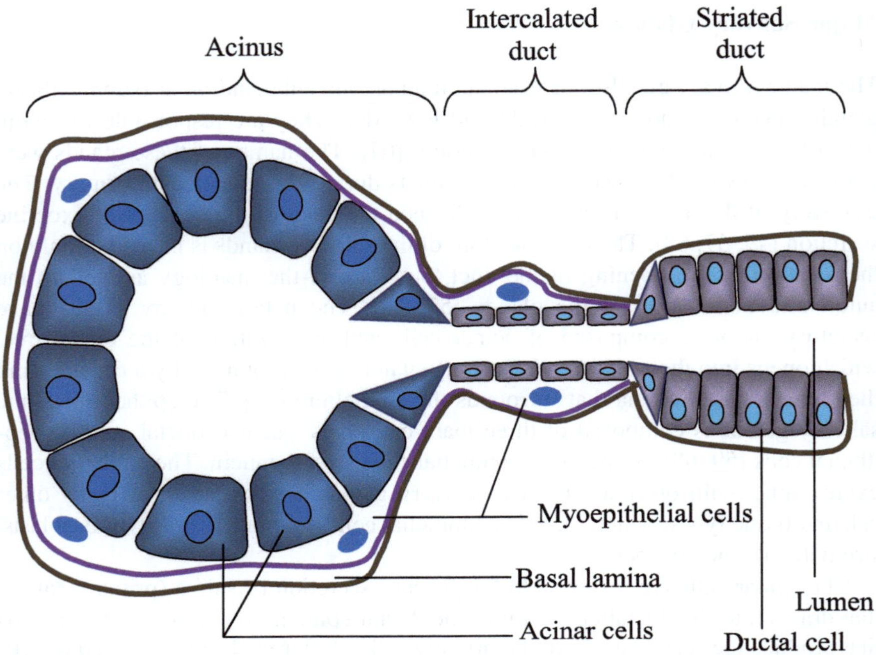

Fig. 11.5 Simplified schematic representation of salivon (basic secretory unit of salivary gland), comprising mostly of acinar and ductal cells. The acinar cells in the soma of the glands produce saliva, which flows sequentially through the intercalated ducts, striated ducts, and excretory ducts into the oral cavity [57, 72]

absorption of ions from primary saliva [57]. But tuft cells also act as chemosensory cells, which is due to their expression of the alpha-gustducin and receptive [73, 74]. Other cells of the salivary glands including the acinar cells, ductal epithelial goblet cells are believed to express microvilli [56, 75].

The secretory cells of the salivary glands are the acinar cells (Fig. 11.5). The acinar cells of the salivary glands produce three types of saliva—serous, mucous, or mixed (seromucous) saliva [57]. The serous acinar cells secrete a watery fluid that is devoid of mucus. The mucous acinar cells secrete mucus-rich fluid. The seromucous cells are believed to secrete both fluid types. The three types of acinar cells can be distinguished on the basis of their morphological properties. The serous acinar cells are comprised of small pyramidal cells with spherical nuclei. The mucous acinar cells contain large columnar cells with flattened nuclei located near the basal membrane. The seromucous acinar cells contain large mucous portion and a semilunar serous portion referred to as serous demilune [68, 69, 76].

Parotid glands: The parotid glands are the largest of the three pairs of glands that drain into the oral cavity. The gland is located opposite the first/second molars of the upper jaw. The parotid provides about 25% of the daily salivary output. The secretions of the parotid are serous [51, 57, 59, 77, 78]. Salivary secretions of the parotid are delivered into the mouth through the parotid duct, discovered in 1661 by Nicholas Stenson (Niels Stensen) (1638–1686), a Danish anatomist and geologist who later became a Catholic bishop [57, 79]. This duct is sometimes named after the discoverer—Stensen's duct [49]. The parotid duct empties into the mouth at the level of the second upper molar tooth [50]. The parotid is involved in a range of diseases as earlier outlined in the previous section. In addition, surgery on the parotid or development of tumor can damage the facial nerve that transverses the gland [50, 80, 81]. Thus, the topography and course of the facial nerve are essential for safe excision of tumors of the parotid glands [82].

Submandibular glands: The submandibular glands are located in the floor of the mouth and lie inferior to the body of the mandible [49]. These glands produce about 70% of the total daily salivary output. The submandibular gland secretion is a combination of serous and mucous. However, the latter is slightly higher than the former [83]. The submandibular gland secretions flow through the submandibular duct into the floor of the mouth on either side of the lingual frenulum, at the point where it is crossed by the lingual nerve [49, 50, 57, 59, 84]. The duct is also known as Wharton's duct, named after Thomas Wharton (1614–1673), an English physician and anatomist who first described the anatomical structure in his book published in 1656 [57, 79]. The submandibular glands are affected by a range of diseases including sialolithiasis [49].

Sublingual glands: The sublingual glands are located submucosally in the floor of the mouth on either side of the tongue. These glands are the smallest of all the major salivary glands. The glands have numerous short ducts that drain into the oral cavity [49, 50, 57, 85]. The acinar cells of the sublingual glands produce about 5% of the daily salivary output. The secretions of these glands are mostly mucous in nature [57, 86]. The sublingual glands have several small ducts called ducts of Rivinus, named after the German physician Augustus Quirinus Rivinus (1652–1723). There are about 20 of such ducts in the sublingual glands [85]. The ducts drain into the floor of mouth along sublingual papillae and folds. Some of these ducts may unite to form Bartholin's (major sublingual) duct—the main excretory duct of the sublingual gland, named after Caspar Bartholin (1655–1738), a Danish anatomist who was the first to describe the drainage system of this anatomical structure in 1685 [57, 59, 79, 85]. Bartholin's duct drains into Wharton's duct at the sublingual caruncula [69, 85].

Minor Salivary Glands

The minor salivary glands are minute auxiliary glands scattered beneath the mucosa of the oral cavity [50]. The term "minor," in contrast to "major," just refers to smaller anatomic size of these oral cavity glands [51]. Their size may vary but is usually in the range $\sim$10–25 mm in length and $\sim$8 mm in width. The minor salivary glands can be divided into posterior and anterior groups [87–89]. These glands can be further divided on the basis of their anatomical position or depth of localization—lateral, middle, upper, lower, etc. [90]. The minor salivary glands are also classified on the basis of their predominant localization in the region of the oral cavity—palatine, lingual, labial, pharyngeal, buccal glands correspondingly are found in the palate, tongue, lower lip, pharynx, and cheeks [51, 89].

The anterior lingual salivary glands, also known as Blandin-Nuhn's glands named after the French surgeon Philippe-Frédéric Blandin (1798–1849) and the German anatomist Anton Nuhn (1814–1889), are deep seromucous glands that are located near the tip of the tongue on either side of the lingual frenulum. Some of the anterior lingual glands may be distinctly mucous in nature. The glands drain by ~3–4 ducts beneath the tip of the tongue [91, 92]. The morphology of the minor salivary glands is similar to that of the major salivary glands [87].

The posterior lingual glands consist of two subgroups of minor salivary glands—the posterior superficial (Weber) and deep (von Ebner) lingual glands [88, 89]. Von Ebner's glands, named after the Austrian histologist, Anton Gilbert Victor von Ebner (1842–1925), are a group of serous glands located beneath the lingual circumvallate and foliate papillae. The ducts of these glands drain into the oral cavity at the base of the papillae [88, 89]. The Weber glands, named after German anatomist Moritz Ignaz Weber (1795–1875), are mucin-producing cells located in the peritonsillar space, which is an area of the root of the tongue on the lateral sides of the foliate papillae, just behind the circumvallate papillae. The ducts of Weber glands drain into the crypts of lingual tonsils [88–90].

Secretions of minor salivary glands are useful for maintaining the fluid composition of the buccal cavity. The minor salivary glands contribute approximately 7–10% of the total volume of saliva. The rate of secretion of saliva generally varies according to the anatomical location and size of glands as well as genetic specifications [52–54, 93]. The flow rate of a single minor salivary gland is estimated at 0.1 μL/min. The flow rate for entire minor salivary population in the oral cavity is estimated at approximately 0.05–4.8 $\mu L/cm^2/min$, which corresponds to the flow rate of human labial salivary glands [52]. The minor salivary glands are useful source of mucins, immunoglobulins, lipases and play a crucial role in the maintenance of the physiology of the oral cavity [93]. The mucus secreted by these glands participates in cleansing the lingual/tonsilar crypts and also facilitates swallowing act [88–90]. The secretions of the minor glands (especially those at the posterior region) modulate oral chemoreception and thus influence perception of taste in the mouth. Minor salivary secretions also contain some growth factors that affect the integrity of the oral microenvironment. The lingual glands, in particular von Ebner's glands, secrete acid-resistant lingual lipase that begins the initial phase fat (triacylglycerol) hydrolysis in the mouth, which is continued in the acidic gastric lumen [88].

Mechanism of Secretion of Saliva

Secretion of fluid takes place in acinar cells of the salivary glands. The salivary ducts are composed of absorptive epithelium. One of the main functions of the ductal epithelium is the absorption of NaCl from primary saliva (Fig. 11.6). The

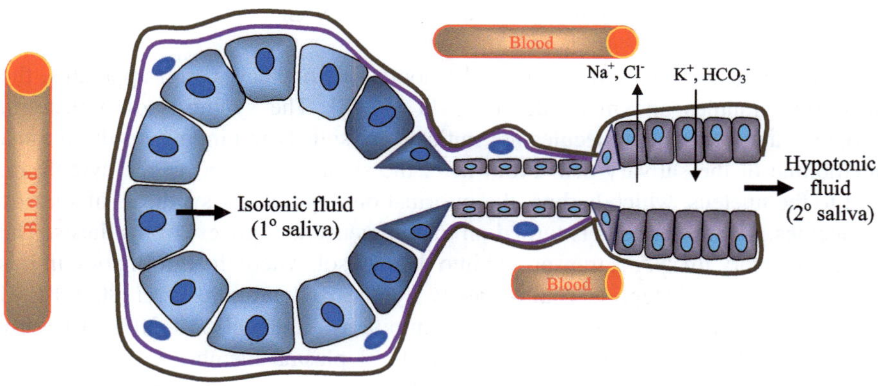

Fig. 11.6 Simplified schematic representation of salivary secretion. The two major cell types of the salivary glands are—acinar and ductal cells. The acinar cells secrete large amount of isotonic fluid consisting of mainly NaCl. This fluid is the primary saliva. The major functions of salivary glands are to secrete water and electrolytes. (But keep in mind that saliva also contains lubricants, antimicrobial substances, enzymes, and growth factors.) The secretion of water and electrolytes by the salivary glands is an energy-consuming process that occurs in two stages. The first stage involves the secretion of isotonic primary saliva into the lumen of the glands. The isotonic saliva is then modified by electrolyte reabsorption in the ductal region of the salivon to form hypotonic or isotonic saliva (depending on the flow rate of the fluid) [55]. The salivary excretory duct also consists of columnar-shaped goblet cells, which are principal cells of the epithelium that secrete mucin [56]. The principal cells mitochondria, central nucleus with a cytoplasm-filled granules and vesicles [95]. It is important to note that some authors consider the star-shaped or basket cells as part of the secretory unit of the human salivary gland. These basket cells are located between the secretory cells (acinar and ductal cells) and basal lamina [96, 97]. Because these cells are similar to epithelial and smooth muscle cells, they are called myoepithelial cells [70]. The myoepithelial cells play essential role in salivon contractility, thereby promoting the expulsion of saliva from acinar lobules. The cells are also involved in other functions including transport of metabolites, propagation of neural stimuli, synthesis of basement membrane proteins, and suppression of tumor development [96, 97]. Consequently, the myoepithelial cells may be better referred to as supportive and protective cells of the secretory unit of the human salivary gland

ability of the ductal cells to absorb NaCl can reduce the NaCl concentration by six times at an initial concentration of approximately 150 mEq/L. However, this absorptive capacity of the ductal epithelium depends on flow rate of primary saliva. Thus, hypotonic saliva is formed at low flow rate, while isotonic saliva is formed at higher flow rate (tonicity of saliva is used in reference to plasma) [8]. But the ductal cells of the salivary glands also secrete about 15% of proteins found in saliva [1, 94]. Flow of saliva in the acinar-to-ductal cell direction is due to the driving force of movement of chloride ions. The chloride channels of the basolateral membrane mediating the influx of chloride ions ensure this directed flow. These channels are activated by numerous signaling pathways including those involving Ca^{2+} [28]. The functions of acinar cells and their mechanisms of secretion of salivary components are discussed in Fig. 11.7.

Production of Saliva by Acinar Cells of the Salivary Glands

Acinar cells have extensive rough and smooth ER, nucleus, Golgi apparatus, the secretory granules and molecules (Fig. 11.7) [98]. The synthesis of molecules, proteins, and other biomolecules are initiated by signals impinging on the plasma membrane of the salivary cell. From here, the signal is relayed to the cytoplasm, and to the nucleus, which harbors the information required for synthesis of specific molecules. The target gene is turned on to commence the process of synthesis. The complementary mRNA is transported into the cytosol, where the information in it is translated into polypeptide chain—newly synthesized protein. Over 80–90% of freshly synthesized proteins are secreted into saliva and only 10–20% of freshly synthesized proteins are used as components of plasma membrane. Freshly synthesized proteins are transported into the ER through a signal recognition complex. Further modification of the proteins takes place in the Golgi network. Following posttranslational modifications in the Golgi complex, proteins are coated to form granules, which are further condensed to form secretory vesicles ready for transport to the plasma membrane for exocytosis [99–102]. The information for the synthesis of a particular or group of biomolecules varies depending on the activated gene loci

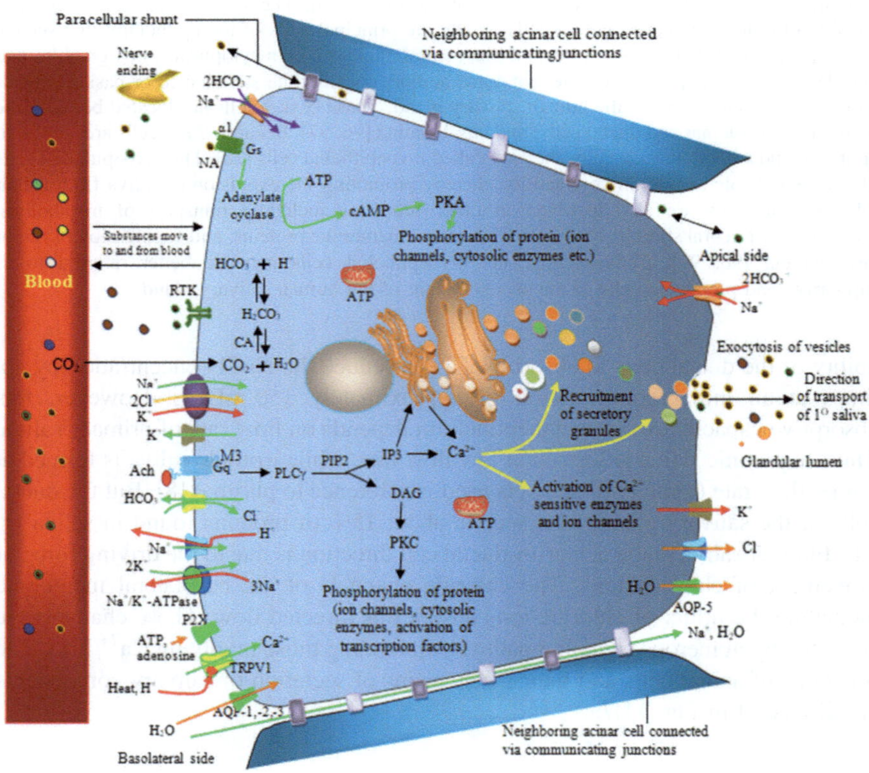

◀**Fig. 11.7** Production and secretion of salivary components by acinar cells. The acinar cell nucleus is located near the basal pole. It has abundance of Golgi apparatus, endoplasmic reticulum (ER), and secretory granules (In contrast, the ductal cells are poor in Golgi apparatus, rough ER, and secretory granules [56]). The secretory granules (of varying diameters of approximately 1 pm) of the acinar cells are packed at the apical pole. The apical pole constitutes about 10% of the cell surface of acinar cell, and the secretory pathway is usually oriented toward the apical surface. Intercellular communicating junction links one acinar cell to another and also participates in the regulation of flow of biomolecules. Acinar cells have a high expression of G protein-coupled receptors, ion channels, and other receptors such as glucose transporters, fatty acid transporters, amino acid transporters [56, 99, 125]. These receptors are located on the apical and basolateral sides of the acinar cell. Upon stimulation, the nerves (parasympathetic and sympathetic nerves) that innervate the salivary glands release neurotransmitters (acetylcholine, noradrenaline, etc.) to activate the corresponding plasma membrane receptors. Parasympathetic stimulation leads to high secretion of saliva due to activation of M3 muscarinic receptors. Sympathetic stimulation on the other hand activates alpha1-adrenergic receptors and also leads to a high secretion of saliva [126]. Hormones, physical factors (stretch, heat, etc.), or other transmitter molecules (e.g., peptidergic signals) can also stimulate or inhibit salivary secretion [127–129]. The salivary glands express a high level of histamine H1 receptor, which upon stimulation inhibits increase in cytosolic Ca^{2+} level [130]. For further details on peptidergic signaling, see mechanisms of regulation of salivary secretion in Sections "Mechanism of Hydrochloric Acid (HCl) Secretion by the Parietal Cells", "Reception Theory and Gastric Exocrine Secretions", and "The Reception Theory as the Basis for Developing the FirstAntisecretory Pharmacotherapy Drugs". Stimulation of adrenergic receptors results in the formation cAMP, which can activate ion channels, transcription factors, or activates PKA. The enzyme PKA mediates phosphorylation of numerous cytosolic and membrane-associated proteins including ion channels and other enzymes. Of the neural influences, the most important is muscarinic signaling from parasympathetic nerve fibers. The released acetylcholine activates M3 muscarinic receptors, which is coupled to $G_{q/11}$ subunit. This G protein subunit activates the phospholipase C γ (PLCγ), which splits phospho-inositol bisphosphate (PIP2) to diacylglycerol (DAG) and inositol 1,4,5-trisphosphate (IP3). The IP3 activates Ca^{2+} intracellular stores to increase cytosolic level of Ca^{2+}. This increase in cytosolic Ca^{2+} is believed to be one of the most integral signals required for the secretion of saliva [130–134]. The lipid messenger DAG activates protein kinase C (PKC), which in turn phosphorylates several intracellular and membrane protein acceptors such as ion channels, enzymes resulting in the mobilization of secretory vesicles toward the plasma membrane for exocytosis. In addition, ion channels and other receptors are translocated to the plasma membrane at a higher rate [28, 98, 125]. Also, the activated protein kinases (e.g., PKC, p38/MAPK) also influence the activity of transcription factors and early response genes, which subsequently affects gene expression and synthesis of salivary components [28, 98, 125]. The exocytosed salivary granules and fluid as well as ion secretion by the endpiece secretory apparatus into the lumen of the secretory unit determine the composition of primary saliva. Electrical signal arriving at the acinar cell can also initiate secretory process. The resting membrane potential of human acinar cells is approximately −50 to −60 mV [28, 98]. Following electrical stimulation of the salivary cell, the membrane potential increases, and this leads to progressive activation of different ion channels. The Ca^{2+} waves in acinar cells are usually initiated at the apical membrane and propagated to the basolateral membrane [122]. In the activated cells, the efflux of K^+ is stimulated leading to increase in membrane conductance [28, 98, 135]. There are different types of K^+ channels, which include Ca^{2+}-dependent K^+ channel, voltage-dependent K^+ channel, ATP-dependent K^+ channel, large conductance K^+, and intermediate conductance K^+ channel [129, 135–138]. In addition, the H^+/K^+-ATPase also mediate K^+ transport in the acinar cell [139]. As the cell returns to the initial membrane potential, influx of K^+ ions into the cytosol is observed [138]. The Ca^{2+} waves also activate other ion channels, notably Cl^- channel at the apical membrane. The apical Ca^{2+}-activated Cl^- channel is the major means of apical Cl^- efflux in secretory endpiece cells [56, 67]. But these cells also express apical cAMP-activated Cl^- channels [56, 139, 140]. Apical K^+ channels enhance

Cl^- efflux by hyperpolarizing the apical membrane [136]. Cl^- and HCO^-_3 exit involves apical CFTR [139]. Cl^- is imported at the basolateral membrane primarily by Na^+ cotransport ($Na^+/K^+/2Cl^-$ cotransporter). This transporter is activated upon decrease in cell volume and is required to restore the volume of the cell [28, 67, 98, 136, 139]. The $Na^+/K^+/2Cl^-$ transporter is responsible for about 70% of the total Cl^- influx that is released at the apical membrane of the acinar cells of not only the salivary glands, but also pancreas. Some diuretics such as furosemide is known to inhibit $Na^+/K^+/2Cl^-$ activity, thereby preventing water retention by the cells [56, 125, 126]. Apart from decrease in cell volume that activate the $Na^+/K^+/2Cl^-$ transporter uptake of Cl^-, the inwardly directed Na^+ gradient generated by the basolateral Na^+/K^+ pump is also known to facilitate Cl^- influx. Moreover, Cl^- transport can occur through K^+ channel (Cl^- leaky channel) [125, 136]. Excess Cl^- ions can be removed by apical Cl^- channels—which are facilitated by the basolateral Ca^{2+} activated K^+ channel. The latter helps to build up negative potential that pulls Na^+ via paracellular shunt pathway. This increases the potential that pulls water into the acinar lumen, thereby creating an isoosmotic medium, known as isotonic fluid [124]. The efflux of H_2O occurs via the AQP5 channel resulting to reduction in cell volume. This secretion of water at the luminal membrane is replenished by the basolateral water channels—AQP-1,2,3 [56, 67, 126]. The activities of the basolateral Na^+/H^+ exchangers regulate the level of intracellular Na^+ and the acidity of the cytosol at a level required for physiological activity [56, 136]. The Na^+ ions are also transported via acinar cell gap junctions [28, 139]. The acinar cell is believed to function as a syncytium as the gap junctions enable synchronization of waves of ions throughout the cell. The acinar cell expresses many junctional proteins including connexins Cx26 and Cx32 [28, 98]

[103–106]. For synthesis of α-amylase, for instance, the amylase gene must be activated. In humans, the gene for amylase synthesis is located on chromosome 1p21. Salivary amylase genes are AMY1A, AMY1B, and AMY1C. (Genes responsible for pancreatic amylase production are AMY2A, AMY2B.) These genes and their signaling pathways direct the synthesis of the enzymes. The gene products and their functions are related to salivary gland morphogenesis and differentiation of cellular components of the glands [99, 105–110]. The manufacturing process of proteins in the cytoplasm of salivary acinar cells is controlled by a variety of factors regulating gene expression. When the gene for synthesis of a particular isoform of the enzyme or protein is tuned on, the mRNA is transcribed from the complementary DNA strand in the nucleus. This mRNA is transported into the cytoplasm to produce the polypeptide chain of the molecule. Further processing of the polypeptide takes place in the ER and Golgi complex. The proteins are coated in vesicles, which are special compartment to safeguard the polypeptide from unnecessary degradation. Vesicles of salivary proteins are constantly produced, packaged, and exported at varying rates depending on the requirements and physiological state of the organism. These vesicles are released into the extracellular space by multiple pathways of which exocytosis is the major one. Acinar cells also ensure a regulated transport of biomolecules, fluid, water, and electrolytes into the lumen—which are all important components in saliva [99, 103]. About 85% of the protein found in saliva is secreted by acinar cell, and only about 15% is produced by the ductal cells [94]. However, the various salivary glands have different contribution to the composition of saliva. For instance, acinar cells of the parotid gland produce enzymes including amylase, DNase, RNase, lysozyme, peroxidases as well as elongate polypeptides such as proline-rich proteins, histatins. The acinar cells of the sublingual gland produce mucins as its major component.

The submandibular acinar cells produce mucins, enzymes (proteases), elongate polypeptides, and endocrine molecules, including growth factors [28, 98, 100, 111]. Nevertheless, certain substances found in saliva such as antimicrobial peptides (cathelicidin, trefoil factor family 3, salivary chaperokine heat shock proteins-70 kDa, bactericidal/permeability-increasing proteins, palate lung, and nasal epithelial clone proteins or PLUNC proteins), growth factor (such as vascular endothelial growth factor) are contribution from both the major and minor salivary glands [28, 98, 100, 111–118].

Calcium-Dependent Signaling Pathway Regulating Salivary Gland Functions: ER Ca^{2+} stores are the most important factors in salivary secretion [119]. ER is one of the most important internal calcium stores in the cell (for review, see Chap. 5) [120–122]. Increase in intracellular ion calcium activates a number of ion channels and calcium-dependent enzymes (e.g., some protein kinases) resulting in the modulation of several signaling pathways involved in synthesis and secretion of salivary components (Fig. 11.7) [122–125]. The increase in cytosolic Ca^{2+}, first, activates the apical Cl^- channel Ca^{2+}-dependent Cl^- channel, and the exocytotic machinery followed by subsequent opening of K^+ channels of the basolateral membrane [124]. Numerous cellular factors affect the secretion of salivary components. These factors include protein–protein interactions, lipid–protein interactions, phosphorylation mediated by protein kinases, expression of different isoforms of the signaling molecules involved in salivary synthesis and secretion [124].

Calcium-Independent Signaling Pathways Regulating Salivary Gland Functions: A Case of Synergism or Cross-talks between Calcium and Calcium-Independent Molecules? The Ca^{2+}-independent pathways involved in salivary synthesis and secretion are the cAMP/PKA and the purinergic signaling pathways. The cAMP/PKA signaling pathway is activated by Gs-coupled receptors as in acinar and ductal cells. Receptor activation leads to increase in cAMP and subsequent activation of PKA that mediate downstream signaling of several intracellular effectors. PKA phosphorylates many proteins including ion channels, enzymes. For instance, phosphorylation of IP3 receptor by PKA can increase probability of channel opening. The activation of PKA can also lead to stimulation of apical cystic fibrosis transmembrane conductance regulator (CFTR), basolateral Na^+-$HCO^-{}_3$ cotransporter, among others. Purinergic receptors are also stimulated [122, 124].

Over the past decades, accumulating research has shown that there is a relationship between calcium and cAMP signaling pathways. A synergistic relationship between these two pathways appears to modulate the secretory activity of the salivary glands. Moreover, Ca^{2+} may activate CFTR to regulate Cl^-/$HCO^-{}_3$ transporter activity. Increase in cytosolic Ca^{2+} is believed to activate IRBIT (IP_3 receptor-binding protein released with inositol 1,4,5-trisphosphate). Traditionally, at the resting state, IRBIT binds to the ER Ca^{2+} channel IP3 receptors to inhibit the functions of this receptor. However, IRBIT also activates Na^+-$HCO^-{}_3$ cotransporter, Cl^-/$HCO^-{}_3$ exchanger, Cl^- channel CFTR, and Na^+/H^+ exchanger. Increase in IP3 may result in dissociation of IRBIT from the IP3 receptors, which increases its binding to these ion channels, activating them to increase ion and fluid secretion [56, 122, 140–146].

Protein Sorting: Localizing Proteins Destined for Export

Only a given subset of synthesized or recycled proteins is transported to the plasma membrane for exocytosis. This is because certain proteins that are destined for export localizes to the corresponding molecule needed to complete their route. For instance, if a given protein is to be transported into the plasma membrane or needed for the formation of membrane of intracellular organelles, such protein is not exported. So the signal needed to localize to the plasma membrane or organelle is predetermined already during the protein synthesis. This is a useful mechanism regulating secretion that has been termed "sorting" [99, 147–152]. The cleavage of certain amino acids off a protein can delete the sorting signal, and the protein can be degraded or used by the cell intracellularly. The cell, through the sorting signal, can decide which proteins are needed to be packaged for storage into secretory vesicles prior to secretion, degraded or used by the cell as part of its component. The association between proteins and secretory vesicles is regulated by sorting receptors or enzymes (e.g., carboxypeptidase E, endothelin-converting enzyme-1, sortilin), which function to link secretory proteins to secretory granule membranes [99, 147, 153–156]. These sorting pathways are required to produce a mature vesicle for export [99].

Pools of Secretory Vesicles—Main Sources for Vesicle Exocytosis

Experimental studies have shown that release of secretory granules takes place in two phases. Pulse chase experiment revealed that exocytosis of granules is relatively rapid with a sharp peak during secretion in the initial "first" phase. The first phase contains mostly proteins with high molecular weight. The pool is rapidly replenished by a closely located pool of secretory vesicles—known as the secondary pool. The second phase of secretion of secretory granules follows immediately after the initial one. The second phase is usually broad when number of exocytosed vesicles is plotted against time. Thus, the secondary phase of exocytosis lasts longer compared to the initial phase of secretion [157–159].

Basal (Resting) and Stimulated Salivary Secretion

The two major types of salivary secretions are the basal (also known as resting) and stimulated secretions. The basal secretion is responsible for saliva secretion in unstimulated state. Thus, this type of salivary secretion primarily maintains the anatomo-physiological integrity of the oral cavity [157]. The basal secretion is mainly carried out by the submandibular and sublingual glands [56, 157]. Estimates show that these two major glands of the oral cavity contribute about 73% of salivary secretions in the basal state [51, 61]. More specifically, the submandibular contributes 65% and the sublingual contributes about 8% to unstimulated flow. The remaining percentages are from the parotid glands (20%) and minor salivary glands (about 7%) [51]. The flow rate for unstimulated saliva in a healthy adult (above 15 years) is not less than 0.1 mL/min (average of 0.3–0.4 mL/min). The normal range depends on the reference values. However, it is generally accepted that any flow rate below 0.1 mL/min indicates salivary gland hypofunction [8, 51]. In a healthy human, during sleep, salivary flow rate is approximately zero ($\sim$0.1 mL/min) [8, 51]. This basal secretion is sometimes termed "constitutive" (i.e., stimulus-independent) and is responsible for about 10% of secreted saliva [99, 160].

The stimulated secretion is the major means of salivary secretion in the oral cavity. This major secretion is mainly responsible for food digestion [38, 157]. The parotid gland is the main organ of the oral cavity that contributes major proportion of stimulated saliva [56, 61, 157]. The parotid glands are believed to contribute over 50% of total stimulated salivary secretions [51]. In a healthy person, the flow rate of stimulated saliva is usually around 0.2–7 mL/min and usually occurs during eating, chewing, or other stimulating activities [8, 51]. Stimulated saliva accounts for about 80–90% of the average daily salivary output [51]. The stimulation of the salivary gland can release as much as 70–80% of the stored secretory vesicles [99]. The major pathway involved in stimulated salivary secretion is the regulated secretory exocytosis of mature granules [38]. This regulated secretory pathway of saliva involves a complex network of cellular and subcellular components, proteins, and other biomolecules [161]. In the major pathway, mature granules upon neural or hormonal stimulation are translocated to the apical plasma membrane by the activities of some proteins (such as synaptotagmin, SNARE) involved in vesicular mechanics and nucleation of membrane protein that ensure correct translocation of the vesicles to the site of exocytosis [102, 160].

A third type of saliva secretion also exists—the minor pathway, which, like the basal secretion, is responsible for secretion of small quantities of newly synthesized secretory components of saliva [38, 160]. The minor secretion may be classified as part of the basal saliva secretion. Thus, the basal secretion is the sum of secreted saliva from the minor pathway and constitutive pathway. In both pathways of secretion, the immature granules are the main sources of secretion [38, 162–164].

Fluid and Electrolyte Secretion by the Salivary Glands

Electrolyte and fluid secretion is one of the main functions of the salivary gland [72, 165]. Electrolyte and fluid secretion is primarily stimulated by parasympathetic nervous system, whereas protein secretion is preferentially stimulated by sympathetic division of the autonomic nervous system, but there are some levels of cross-talks and synergism between the two neural systems [157]. Upon activation, secretory vesicles empty their contents into the lumen of the glands. The fluid formed is a NaCl-rich plasma-like isotonic solution, also called primary saliva [129, 166]. The fluid is transported to the ductal system, where it is modified to form hypotonic saliva. This model of saliva formation is termed the two-stage hypothesis [167], which is the currently accepted model of formation of saliva [168]. The cells of the salivary duct absorb Cl^- and Na^+, but secrete $HCO^-{}_3$ and K^+ into the fluid, thus forming a hypotonic solution or final saliva. Several ion channels are involved in the transport of ions across the salivary ductal cells. The epithelial Na^+ channel (ENaC, also known as amiloride-sensitive Na^+ channel) is the main ion channel responsible for Na^+ influx in the ductal cells [56]. Other apical transporters include Na^+/H^+ antiporter (imports Na^+ in exchange for H^+ that is secreted into the lumen), Na^+/HCO^-_3 symporter (transports both ions into the cytosol of the ductal cells), CFTR (transports HCO^-_3 into the lumen for Cl^- that is imported into the cytosol—an associated Cl^- channel that can be activated by Ca^{2+} (CaCC, calcium-activated chloride channel) resulting in Cl^- efflux in the ductal cell), K^+ channels (responsible for K^+ efflux into the lumen. The K^+ channel can be activated by cAMP, Ca^{2+}, ATP) [56, 136, 169]. Recall that Na^+ and water can be released into the lumen via paracellular shunt pathway [56, 136, 169]. But transport of molecules across the paracellular pathway is highly selected because the acinar and ductal cell communicating junctions (e.g., tight junctions) allow selective paracellular solute transport in the salivary glands [170].

The main basolateral transporters are Na^+/K^+ pump (transports three Na^+ into the ductal cytosol in exchange for $2K^+$ that are expelled), Na^+/HCO^-_3 symporter (see apical transporter described above), Na^+/H^+ antiporter (see apical transporter described above), basolateral Cl^-/HCO^-_3 exchangers (responsible for HCO^-_3 efflux and Cl^- influx) (Fig. 11.8). Ductal cell cytoplasm also exhibits considerable activity of carbonic anhydrase that regulate the concentration of H^+, HCO^-_3, CO_2, H_2O, H_2CO_3 [56, 136, 169]. The luminal membrane of the principal cells of the ductal system possesses microvilli, which increase the capacity for absorption. The activation of K^+ channels (e.g., Ca^{2+} activated K^+ channels, large and small channel

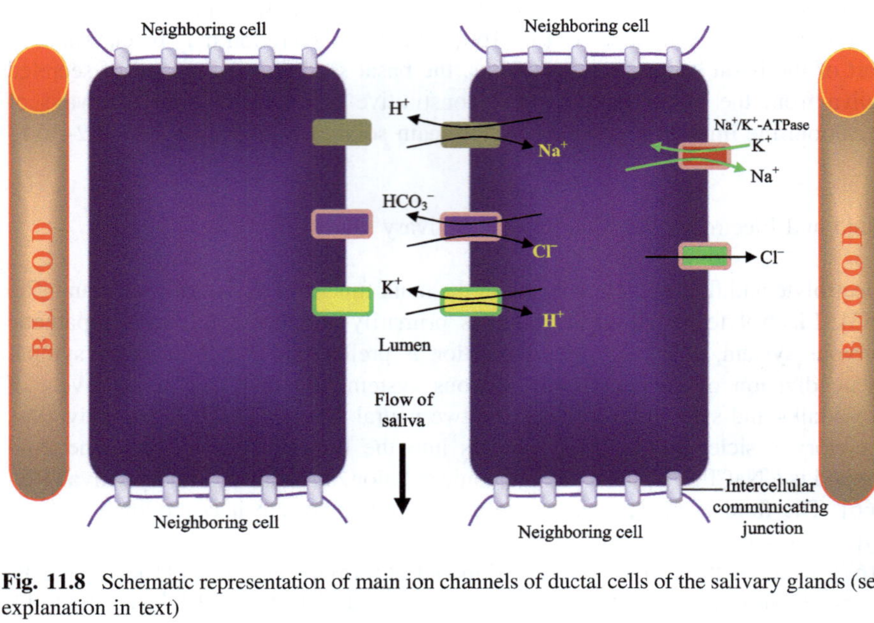

Fig. 11.8 Schematic representation of main ion channels of ductal cells of the salivary glands (see explanation in text)

conductance) allows the secretion of K^+ into the ductal fluid. The anion exchanger (Cl^-/HCO^-_3 exchanger) is highly expressed in intercalated ducts and transports HCO^-_3 into the fluid, while the Cl^- is moved in the opposite direction into the basolateral membrane [56, 125].

Another transporter or channel widely expressed in salivary gland cells is aquaporin. Aquaporins are classical "specific" water channels expressed in many tissues and cells of the body. The aquaporin family consists of 13 members [56, 68, 126, 171–173]. Aquaporin types 1, 2, 3, and 5 provide transcellular flow of water in both the acinar and ductal cells [56, 131, 136, 169, 174, 175]. Aquaporins-1,-2,-3 are widely expressed in the basolateral membrane of the acinar cells [131]. Aquaporin-5 is expressed in the luminal membrane and responsible for at least 60% of water transport. Aquaporin-5 is also expressed in the apical membrane of ductal cell and mediates the secretion of water into the lumen. Aquaporins-1 and 3 are expressed in the basolateral acinar and ductal cells. The human salivary glands also have high expression of aquaporin-4 [56, 126, 136, 169, 174]. Besides water, some aquaporin family members transport other molecules like glycerol, urea, ions, and CO_2. They also take part in cell adhesion, proliferation, migration, and survival. The origin of water transport is ion secretion which is accompanied by obligatory osmotic water flow. Water is transported by numerous pathways which besides aquaporins include certain transporters of the membrane bilayer [56, 68, 171, 172]. Certain autoimmune diseases of the salivary glands (such as Sjögren's syndrome) that are associated with xerostomia and hyposalivation are believed to be, at least in part, due to disordered expression of acinar cell aquaporins including the aquaporin-5 [176].

Cell Volume Regulation in Salivary Glands: The volume of salivary gland cells is important for their functioning. Increase in cell volume will subsequently lead to decrease in cell volume. This cell volume is regulated by cytosolic Ca^{2+} signal through IP3-induced Ca^{2+} efflux into the cytosol and the store-operated Ca^{2+} influx from the exoplasm (for details on Ca^{2+} signaling, review Chap. 5). In addition, Cl^- and K^+ channels also play a role in volume regulation. The inhibition of Cl^- and K^+ channels (which can occur via reduction in cytosolic Ca^{2+} levels) can reduce both electrolyte and fluid secretion. The volume-sensitive channels (Cl^-/HCO^-_3 exchanger, Na^+/H^+ exchanger, and Na/K/Cl cotransporter) also play a crucial role in cell volume regulation. The volume-sensitive channels are activated upon decrease in cell volume [28, 56, 98].

Composition of Saliva

Saliva is a clear, slightly acidic or alkaline (pH varies 5–7.8) viscoelastic muco-serous exocrine secretion of the oral cavity that act as a lubricant, protective agent, allowing for normal speech, facilitating taste, mastication, formation of food bolus, swallowing, and mechanochemical breakdown of the bolus in the GI tract [8, 51, 177, 178]. During the period of low flow, the salivary pH can be as low as 5, which increases with increase in flow rate, reaching 7.8 at peak salivary flow. The major constituent of saliva is water, comprising ~99% of saliva. The remaining 1% is composed of both organic and inorganic substances. The inorganic substances include sodium, potassium, chloride, phosphate, and bicarbonate ions. The organic components are the enzymes, antimicrobial proteins, mucus, blood group factors, growth factors, etc. Salivary mucus is a sticky substance that keeps the food in a mass (i.e., bolus) and serves as a lubricant in the GI tract [51, 72, 177, 179]. For further review, see de Almeida Pdel et al. [180] and Damle et al. [181].

Endocrine Secretions of the Salivary Glands

Salivary Glands as Exocrine Glands Possessing Endocrine Secretory Functions—Duacrine Glands

Salivary glands are generally classified as exocrine glands as they release their secretions into the lumen of the digestive tract. In this classical view of salivary gland functions, salivary enzymes, fluid, and other substances are secreted only into the lumen of the GI tract [182–184]. However, accumulating evidences over the years indicate that salivary glands under normal conditions also function as endocrine organs [94]. This later role of the salivary glands was previously thought to occur only in pathological condition in which salivary enzymes are released into bloodstream. Trace amount of salivary proteins in the nanogram per milliliter range can leak into the bloodstream and may not be of substantial physiological usefulness. Under different physiological stimuli, endocrine secretion of digestive enzymes by exocrine glands is a normal process that can occur in considerable amounts in healthy individuals. However, increase in blood level of salivary proteins is a crucial diagnostic parameter in inflammation of GI tract accessory glands such as the salivary glands and pancreas [182–186].

Therefore, salivary glands function as duacrine glands as they are able to secrete their products into the lumen of the organ and directly into bloodstream [182, 187, 188]. Examples of salivary endocrine factors are shown in Table 11.1.

Digestive enzymes released into the bloodstream are regulated by numerous mechanisms such as those that regulate their production, secretion, inhibition of enzyme action, and hormone metabolism [182]. There are different schools of thought on the mechanisms by which digestive enzymes pass into blood. One school of thought believes that salivary components pass (diffuse) in blood as result of increase in flow rate [8, 196, 197]. However, basolateral exocytosis in the acinar cells represents a crucial means by which salivary components can enter the bloodstream [198–200]. Proponents of paracellular transport argue that increase in digestive enzymes in serum is regulated by the pressure gradient of the intra- and extracellular fluid of the salivary gland cells and that even if there is no increase in flow rate, there should be increased values of the enzyme in serum [182, 188].

Table 11.1 Classification of hormones of the salivary glands [57, 182, 188–190]

Class	Examples	Comments
Peptide hormones	Growth factors (e.g., epidermal growth factor, transforming growth factors, vascular endothelial growth factors, parotin), leptin, ghrelin, adrenomedullin, etc.	The salivary glands produce a couple of growth hormones that control the structural and functional integrity of both the salivary glands, oral mucosa, and also exert some systemic influences [161]. Salivary leptin plays an essential role in mucosa growth and regeneration [191]. Salivary ghrelin is thought to function via ghrelin receptor located in oral keratinocytes. Like salivary leptin, salivary ghrelin enhances cell proliferation and survival. To study the proliferative effect of these hormones on salivary glands or mucosal cells, 5-bromo-2′-deoxyuridine may be used as a tracer. As the cell replicates, the tracer is incorporated into the molecules of the newly differentiated cells. Degree of incorporation of 5-bromo-2′-deoxyuridine determines the measure of cell proliferation [192]. The hormone adrenomedullin produced in salivary glands is responsible for the maintenance of health of the oral cavity by stimulating oral cell proliferation and also involved in antibacterial defense [193]
Amines	Melatonin, serotonin, dopamine, adrenaline, noradrenaline	Different types of amines are known to be produced by the cells of the salivary glands, and they stimulate their cognate receptors in salivary glands cell and may also function in some distant locations [194]
Steroids	Androgen, estrogens	

Note These salivary hormones function via the neuroimmunoregulatory network of the salivary gland and regulate many aspects of salivary functions including regulate protein synthesis. However, other hormones (e.g., steroid hormones and cytokines—interleukin-1, tumor necrosis factor, interferon gamma) delivered into the salivary gland from the systemic circulation can influence salivary protein secretion [195]

The concentration and function of the hormones in salivary gland is regulated by the hypothalamic salivary gland endocrine axis of which parotid, submandibular, and sublingual glands are the major ones connecting the hypothalamus [187, 201–204]. In the hypothalamic parotid gland endocrine axis, for instance, a parotid hormone that stimulates dentinal fluid transport is centrally controlled by the hypothalamus and the level of the hormone is believed to be regulated by its activities in the endocrine parotid gland [201, 203, 204]. Experimental investigation indicates that tissue extract of the hypothalamo-thalamic region stimulates dentinal fluid transport. Interestingly, the response is blunted when the parotid gland is completely removed by surgical means (i.e., parotidectomy), which consequently leads to substantial increase in the serum level of immunoreactive parotid hormone [202].

Other Hormones Found in Saliva: The Adipokines

Adipocytes, particularly the white adipose tissue, are widely distributed in salivary glands, and they secrete a range of hormones, which include cytokines, and more specifically, adiponectin, chemerin, leptin, omentin, resistin, vaspin, and visfatin, retinol-binding protein 4, interleukin-6, tumor necrosis factor alpha, and monocyte chemoattractant protein-1. These molecules secreted by adipocytes or other components of the adipose tissue are collectively called adipokines. However, adipose tissues in other parts of the body also secrete these hormones. So the hormones either enter the salivary gland by active transport or are secreted by the glands. Adipokines pleiotropically signal to organs such as brain, liver, skeletal muscle, and the immune system to regulate hemostasis, blood pressure, lipid and glucose metabolism, inflammation, and atherosclerosis. Adipokines also contribute to insulin sensitization. Thus, adipokines may play some role in obesity, insulin resistance, energy imbalance, and stress response [205–209]. Apart from salivary gland adipocytes, the salivary glands harbor specific group of macrophages, which can also release adipokines [208].

The salivary concentration of some adipokines has been shown to correlate with serum levels of the hormone. In addition, both the serum and salivary levels of adipokines (e.g., leptin, resistin, visfatin, and adiponectin) have been observed to change in disorders such as cardiovascular and metabolic diseases. In disease conditions, some adipokines may show reduced value whereas others may have increased values [207, 209]. The inflammatory marker of periodontitis, visfatin, which is highly expressed in this disease, has been shown to decrease after treatment [210, 211]. The salivary visfatin level is positively related to indices of dental health such as the plaque and gingival indices [211].

Proteome of the Salivary Gland

Proteome of the salivary gland is the total composition of proteins making up the salivary glands. Salivary gland proteome can be divided according to the type and size of salivary glands. The protein secreted into saliva is part of this proteome [93, 212]. At least 2290 proteins are present in whole saliva. Of these, 27% similar proteins have been identified in blood plasma [213].

The proteome of minor salivary gland secretion contains at least 56 different proteins [93]. Minor salivary gland proteome includes mucins, immunoglobulins which are known to protect the oral mucosa from desiccation, mechanical injury, and microbial aggression [93]. But these proteins are also secreted by the major salivary glands. The salivary glands of humans secrete two major mucins—MUC5B (also known as major glycoprotein, MG1) and MUC7 (also known as MG2). MUC5B is an oligomeric salivary mucin with high molecular mass above 1 MDa—this mucin is predominantly produced by mucous cells. MUC7 is a monomeric mucin with low molecular mass of 200–250 kDa secreted primarily by submandibular/sublingual glands. MUC7 is primarily produced by serous cells

[214–218]. Mucins perform many functions of which agglutination of microbes is crucial to host defense against invasion [217].

Some salivary proteins form complexes with each other (homotypic salivary complexes) or with other molecules (heterotypic salivary complexes). For instance, mucins can form complexes with salivary IgA, lactoferrin, lysozyme, and histatins. These complexes confer saliva additional functions of protecting its components (e.g., enzymes and other proteins) from degradation and also enhance the functions of salivary components by acting as chaperones to target dental surfaces [212, 219–222].

Salivary protein family includes histatins, statherins, cystatins, and proline-rich proteins [223]. These salivary gland proteins mitigate the negative effects of inflammatory processes. They play a useful role in innate oral host defense system and effectively kill the pathogenic *Candida albicans* and other pathogenic microbes and exhibit growth-inhibitory activity against microbial population of the GI tract [93].

Statherin

Statherin is a 43-residue tyrosine-rich acidic salivary peptide with molecular weight of 5380 Da, responsible for inhibiting undesirable precipitation of calcium phosphate salts in the salivary glands and oral cavity. Thus, the protein maintains high calcium concentration in saliva available for remineralization of enamel and high phosphate levels for buffering. Hence, human salivary secretions are supersaturated with respect to the calcium phosphate salts which form dental enamel [224, 225]. Like statherins, the closely related families of salivary peptides such as histatins and cystatins also inhibit calcium phosphate precipitation on dental enamel [226].

The concentration of statherin in saliva is estimated to be around 3.0–27.3 μM, with an average value of 7.3–18.3 μM [227]. Reduction of the level of this protein has been associated with increased incidences of oral diseases and diabetes. Diabetes reduces the ability of acinar cell to secrete adequate level of statherin and similar molecules and predisposes diabetic individuals to oral diseases due to the reduced statherin secretion [228–230]. To this end, changes in the expression of salivary proteins in diabetes mellitus have been identified in salivary glands [229, 231].

Like other glycopeptides, statherin serves as a receptor for adhesion of microorganisms to the host. This binding represents a crucial step in biofilm formation. While this is a normal phenomenon, some pathogenic microbes can as well bind to statherin and possibly trigger the onset of disease process [232].

Histatins

Histatins are low molecular weight histidine-rich, small, cationic multifunctional proteins of the salivary gland, having potential therapeutic inhibitory influence against fungi such as *Candida albicans* and *Cryptococcus neoformans*, and also possess bactericidal effect against *Streptococcus mutans* [233, 234]. Thus, histatins can provide first-line defense against oral candidiasis [234, 235]. All human major salivary glands are involved in the secretion of histatins. However, relatively recently, Piludu et al. [234] discovered that histatins are also produced from the serous-secreting cells of minor salivary glands, in particular von Ebner's glands.

Proline-Rich Proteins

Proline-rich proteins are a group of low molecular weight proteins, produced in the salivary glands, specifically in acinar cell, containing several repeats of proline-rich sequence. This family of salivary proteins comprises approximately 70% of the total proteins found in saliva. They aid in the regulation of mineral homeostasis and maintenance of tooth enamel integrity and pellicle dynamics [93]. Proline-rich proteins can be divided into acidic, basic, and glycosylated proteins. The acidic proline-rich proteins bind calcium to maintain the salivary concentration of calcium. This subset of proline-rich proteins can prevent the formation of crystals of hydroxyapatite on the surface of the teeth [236]. Increased secretion of proline-rich proteins can be induced by dietary tannins and other substances. These proteins function to neutralize the detrimental effects of dietary tannins and other polyphenols [237] and also initiate a reflex due to the bitter or sour taste leading to contraction of the mucosal structures and subsequent reduction in salivary flow (astringency reflex) [238]. This reflex is caused by complex formation of molecules through the interaction between tannins and proline-rich proteins. Therefore, salivary proline-rich proteins constitute part of the defense mechanism against tannins and other substances such as polyhydroxylated phenols ingested with food or liquid [239].

Cystatins

Cystatins are low molecular weight proteins belonging to the family of cysteine protease inhibitors [240]. Members of the cystatin protein family in humans include cystatin A, B, C, D, S, SA, SN, human kininogen, human α2HS-glycoprotein, human histidine-rich glycoprotein [240, 241].

Cystatins can be classified on the basis of sequence similarity, the presence or lack of disulfide bonds, and physiological localization as stefins (family I), cystatins (family II), kininogens (family III), and cystatin-related proteins (family IV). The kininogens and cystatins are extracellular protease, whereas stefins are intracellular [240, 242, 243].

The main function of cystatins is to inhibit the enzyme, peptidase (protease) belonging to peptidase families C1 (papain family) and C13 (legumain family) [244].

Cystatins may misfold to form amyloid deposits and are implicated in several disorders such as neurodegenerative and cardiovascular diseases, osteoporosis, arthritis, and cancer. (see Clinical Correlate 11.1 for amyloid deposits and salivary diseases.) In these diseases, the activities of such peptidases are increased. However, misfolding of this protein is usually prevented by heat shock proteins. Deposits of denatured proteins can be cleared by intracellular digestive apparati of the cell. The diseases mentioned above can result when these protective measures of the cell are not effective enough [245].

Cystatin-Related Proteins

Cystatin-related proteins are abundant androgen-regulated secretory glycoproteins that are specifically synthesized in the ventral prostate and lachrymal gland [246]. An example of such proteins includes cystatin 3, cystatin 11, cystatin-related epididymal spermatogenic proteins, or cystatin 8 [246–248]. Cystatin-related proteins have not been extensively studied compared to other proteins of the cystatin superfamily, and their expression in salivary glands is not clearly understood. Some members of this salivary protein family exhibit antibacterial activity (e.g., cystatin 8) [248].

Kininogens

Kininogens are multifunctional glycosylated proteins related to cystatins. Kininogens are cysteine protease inhibitors. Kininogens are important in angiostasis and pro-apoptosis, proliferation of endothelial cells, and are also involved in angiogenesis [249, 250]. Kininogens stimulate fibrinolysis, contribute to the constitutive anticoagulant nature of the intravascular compartment, and block calpain's role (a family of cytosolic Ca^{2+}-dependent cysteine proteinases) in forming the heterodimeric complex of platelet integrin alpha IIbβ3 that mediates cell migration, block thrombin from binding to the thrombin receptors on platelets [251–253].

Kininogens play substantial role as a cofactor in coagulation and inflammatory processes [254]. Kininogens are the precursors of kinins—endogenous peptide hormones with vasodilatory properties, forming an integral component of the kallikrein–kinin system. The enzyme responsible for this conversion is called kallikrein. This enzyme is formed from prekallikrein through the action of trypsin, Hageman factor, and preformed kallikrein. The enzyme is expressed in the salivary glands, and other tissues of the body including the intestines, kidneys, pancreas [255–258].

Salivary Renin–Angiotensin System

Renin was discovered in 1898 by the Finnish-born physician and physiologist Robert Tigerstedt (1853–1923), together with his student Per Gunnar Bergman at the Karolinska Institute, Stockholm. Tigerstedt and Bergman reported the pressor effect of renal extracts and named the substance responsible for this pressor effect—renin, on the basis of its origin "renal" [259]. Using a clamping technique of the renal artery invented in 1934 by Harry Goldblatt, different laboratories around the world began to document renal secretion having the pressor effect earlier reported in the late nineteenth century. Subsequent extraction of this pressor agent revealed that it is the same substance reported by Tigerstedt and Bergman. However, this pressor agent did not directly act to increase blood pressure, but via the enzymatic cleavage of a plasma protein to form the agent that is responsible for the pressor effect. This cleavage product was initially called angiotonin. But the name of this substance was reconciled in 1958 by Eduardo Braun Menéndez and Irving H. Page, who named it angiotensin [260]. For further review of the history of renin and associated proteins, see Ian Phillips and Schmidt-Ott [261], Basso and Terragno [260], Van Epps [262]. The evolutionary path of this system is documented elsewhere [263].

The renin–angiotensin system plays a crucial role in regulating water-mineral homeostasis, blood pressure, tissue perfusion, cell growth and proliferation. The renin–angiotensin system is implicated in the pathophysiology of hypertension, myocardial ischemia, heart failure, and diabetes mellitus [259, 264].

Apart from the kidneys, renin is produced in other tissues including the salivary glands [264, 265]. Although its role is not exactly clear, it may be involved in regulation of local microcirculation, water-mineral homeostasis, cell growth and proliferation.

Clinical Correlate 11.1

Amyloid Deposits and Salivary Diseases

Amyloid deposits are due to the misfolding of certain groups of proteins in tissues and cells of the body [266]. These deposits cause damage to cells and tissues that lead to irreversible pathological changes. The condition may be due to genetic disorders, posttranslational disorders, activities of infectious agents, and autoimmune reactivity [266–268].

In certain salivary diseases, biopsy study has shown development of amyloid deposits, which may sometimes be associated with tumor cells [269]. Biopsy specimens are taken from suspected sites for analysis of the presence of amyloid deposits. Common sites for tissue specimens of biopsy include gingiva, salivary gland (e.g., labial salivary gland). But biopsy may be taken from visceral organ if necessary [270]. Amyloid deposits generally are used for the diagnosis or serve as biomarker of some diseases of human such as Alzheimer's disease, certain cancers, and neoplastic changes of salivary gland or other tissues/organs. Salivary amyloid protein $A\beta_{42}$ is considered a potential marker of salivary gland disorders [266, 271, 272].

Salivary Components as Potential Biomarkers in Health and Disease

Clinical and diagnostic usefulness of saliva has been known for more than five decades. Of the total number of proteins produced by the salivary glands, about 40% is thought to serve as markers for diseases [213, 273]. Salivary biomolecules such as proteins, steroids, and amines have considerable promise in diagnostics and are increasingly been considered for application in various disciplines including psychiatry, stress research, sports medicine, veterinary medicine, psychology, clinical endocrinology, fertility, and behavioral research. Therefore, components of saliva are sources of noninvasive or alternative means for identification of health state and diseases [273, 274]. For instance, certain groups of salivary proteins have been identified as markers of type 2 diabetes mellitus [212]. Tens of salivary proteins have shown significant differences in concentration between healthy people and those suffering from diabetes type 2. These proteins are components of immune response and metabolic pathways. Further research into these proteins could provide useful information into the early identification and markers in prediabetes, diagnosis of diabetes, and monitoring of the condition [212]. Some of the salivary markers of diabetes are transported into the salivary gland by different mechanisms [274]. Though it is not exactly clear, it is thought that peptide hormones such as insulin are taken into saliva by active transport. Immediately after glucose load, salivary concentration of insulin has been shown to equal that of serum concentration [183, 274–276]. Although some of these salivary markers of diseases including diabetes are produced in the salivary glands, others are transported into the salivary glands or salivary fluid via the circulatory system. For instance, cytokines (e.g., epidermal growth factor, tumor necrosis factor), leptin, and other growth factors and hormones are produced and secreted by the salivary glands themselves [183, 274, 277]. Steroids can enter the salivary glands by diffusion. Steroid hormones conjugated with transport proteins in the blood passively diffuse into the salivary glands, where they might undergo various reactions including enzymatic conversion as in the case of cortisol and cortisone [274, 278–280]. High rate of conversion of cortisol to cortisone in the salivary glands by 11β-hydroxysteroid dehydrogenase II causes a shift from cortisol to the inactive ketoform. This conversion is often not accounted for when salivary cortisol is measured, and this is one of the reasons for some discrepancies in the cortisol concentrations in experimental findings. It is believed that salivary concentrations of molecules are better criteria and predictors of health state compared to urine when similar molecules are considered for analysis. The kidney excretes molecules which are usually water-soluble, whereas salivary glands may secrete molecules even if they are not metabolized to more water-soluble or polar molecules [274, 279, 280].

Molecules in Saliva May not Reflect Serum Concentration: A Disadvantage to the Application of Saliva in Diagnostics?

Concentration of molecules taken into salivary glands is determined by specific transport mechanisms. When molecules pass from the blood into the salivary glands by endocytosis, their concentration may not reflect the serum concentration of these molecules. This can occur for steroids. While some salivary molecules may have higher or similar concentration in saliva (epidermal growth factor), compared to the serum concentration, others show lower level (tumor necrosis factor and leptin) [274, 281–283].

Not only the type of transport mechanism, but also the speed of transport determined by the composition of plasma membrane and type of transported molecule affects the salivary concentration of salivary markers of health and disease. Transport of molecules from the blood into the salivary glands is determined by the layers of the capillaries and epithelial cells of the glands. Thus, lipophilic molecules (such as steroids) pass into salivary glands more quickly than hydrophilic molecules (such as peptides) [197, 274, 284]. Lipid-soluble molecules such as steroids and amines enter the salivary gland through passive diffusion along a concentration gradient. Thus, salivary concentration of lipid-soluble, unconjugated steroids such as cortisol reflect the approximately 10% unbound plasma concentration, whereas lipid-insoluble, conjugated steroids such as dehydroepiandrosterone sulfate make up less than 1% of the unbound plasma concentration [274].

Regulation of Salivary Secretion

Salivary secretion is controlled by the autonomic nervous system [285–287]. Both the parasympathetic (muscarinic cholinergic) and sympathetic (adrenergic) divisions of the autonomic nervous system stimulate salivary secretion, but they have some differences in terms of the quantity of protein, fluid, and electrolyte secretion

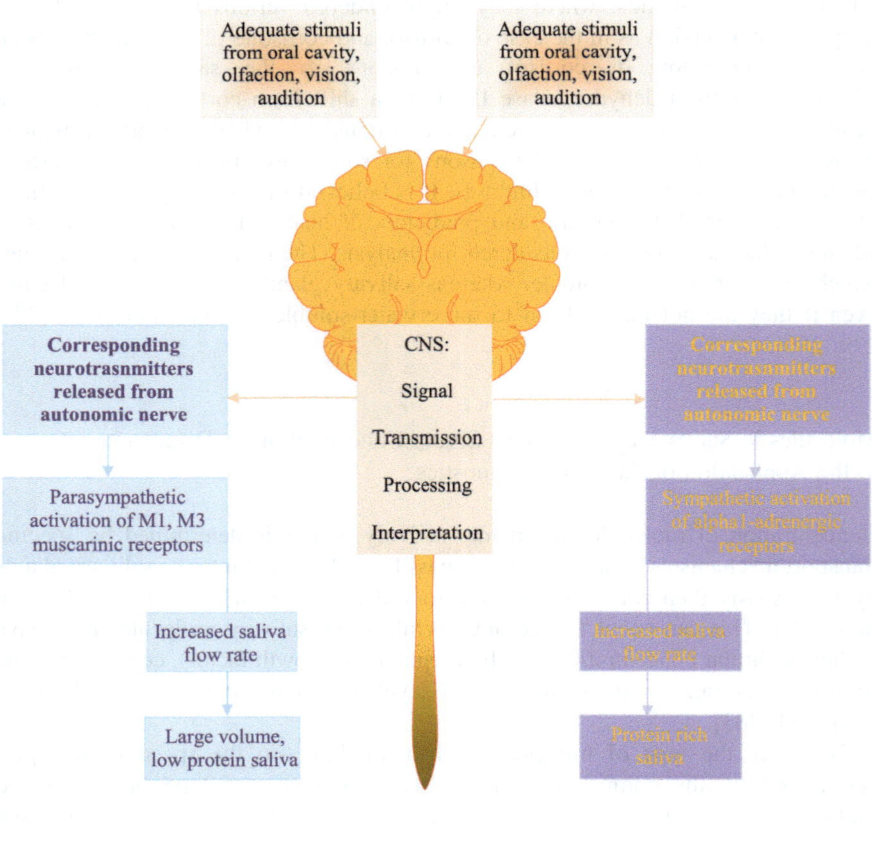

◀**Fig. 11.9** Schematic representation of neural stimulation of the salivary cells. Cells of salivary gland are controlled by many signals from different sources, and they include endocrine, paracrine, neural. Emerging research has indicated that many hormones produced in or outside the salivary gland participate in the regulation of salivary functions. Certain substances produced by the salivary cells are involved in regulating neighboring salivary cells due to the presence of intercellular communication network. Neural mechanism is possibly the most widely studied mechanism regulating salivary functions. The neural regulation is spearheaded by parasympathetic and sympathetic divisions of the autonomic nervous system [56]. The sum total of signals regulating secretory activity of the salivary gland is the salivary gland reflex, which occurs at resting "basal" rate. However, during periods of food intake, there is about ten times increase in salivary secretion [175, 296]. Afferent signals arise from chewing, taste, olfaction, vision, and audition and are transmitted, processed, and interpreted in the central nervous system before efferent signals are delivered to salivary glands [175]. Salivary glands are innervated with cholinergic parasympathetic nerves which release acetylcholine, the ligand to M3 and M1 muscarinic receptors [56, 175]. The interaction between ligand and receptor initiates signaling downstream the cell, leading to increased secretion of saliva by acinar cells in the endpieces of the salivary gland ductal tree. It should be noted, however, that activation of M3 muscarinic receptors by its ligand is far greater than activation of M1 muscarinic receptors. The neuromodulator, acetylcholine, activates M1 and M3 receptors in both acinar and duct cells to increase cytosolic Ca^{2+}. The result is increase in fluid secretion to greater extent and only a little protein secretion [56, 175]. Thus, parasympathetic stimulation evokes output of a large volume and a low protein concentration of saliva [99]. Apart from acetylcholine, substance P can also be released from the postganglionic parasympathetic nerve endings following parasympathetic stimulation. Interactions between various cells of the salivary gland are also possible. Thus, contraction of myoepithelial cells, mechanically promoting secretion, is due to the interaction between motor nerves and secretory nerves [99, 296]. Salivary glands are also innervated with sympathetic nerves which release noradrenaline upon stimulation. This neurotransmitter causes greater secretion of proteins from both acinar and ductal cells [175]. But acinar cell protein secretion is far greater than protein secretion from the ductal cells. Moreover, acinar and ductal cells receive nerves from both parasympathetic and sympathetic regions of the autonomic nervous system and there are possible interactions between the signaling pathways of the different nerves [175, 296]. Secretion of fluid by the salivary glands is mainly dependent on cholinergic parasympathetic nerve activation, while secretion of protein is dependent on, in addition to the former, activation by neuropeptides and by noradrenergic sympathetic nerve activation in the major salivary glands [297]. Sympathetic β-adrenergic stimulation has a role to play in both acinar and duct cells. Activation of β-adrenergic receptors of these salivary cells results in increase production of cAMP. In ductal cells, increase in cAMP mediates secretion of small volume of saliva, but contains high concentration of $HCO^-{}_3$ and low concentration of Cl^- [56]. However, in acinar cells, β-adrenergic stimulation represents the major pathway for protein (e.g., enzyme) secretion. Therefore, sympathetic stimulation evokes release of saliva that has a relatively small volume and high protein concentration [99]. Isoproterenol is a known beta-adrenergic agonist that stimulates alpha-amylase secretion via increase in intracellular cAMP concentration. But this isoproterenol-induced stimulation of amylase and cAMP is blocked by propranolol, a beta-adrenergic antagonist, suggesting a role of specific pathway dedicated to beta-adrenergic receptor-mediated secretion of salivary components [298]. Salivary gland acinar cells also contain alpha-adrenergic receptors that are activated by the sympathetically released norepinephrine, leading to increase in intracellular Ca^{2+} concentration [99]. Fluid secretion in alpha-adrenergic receptor stimulation is usually higher compared to stimulation by acetylcholine. In addition to the receptors mentioned above, salivary cells possess other receptor, including ionotropic and metabotropic P2 receptors, which are located in the luminal and basal poles. The non-adrenergic, non-cholinergic neuropeptides released from autonomic nerves stimulate salivary gland secretion. Purinergic signaling through P2Y receptors is important in fluid and electrolyte secretion. The mechanisms involved in salivary purinergic signaling are associated with increase in cytosolic Ca^{2+}. Vasointestinal peptide released from parasympathetic nerve terminals acts on nitric oxide in endothelial cell of the salivary glands, leading to vasodilatation reflex that is associated with salivary secretion. Also, nitric oxide released from nerve endings may play a role in protein secretion in salivary glands [56, 175]. Some forms of synergism or cross-talks occur between sympathetic and parasympathetic nerves of the salivary glands [296]. Some diseases of the salivary gland may result from disorders in salivary flow, hyposalivation (Clinical Correlate 11.2)

released upon stimulation (Fig. 11.9) [38, 195, 285]. The effects of both parasympathetic and sympathetic stimulation on salivary secretion are synergistic [160].

The major salivary glands are mainly innervated by parasympathetic nerve fibers; thus, parasympathetic stimulation provides a much greater stimulation compared to sympathetic nerve stimulation [285]. The parasympathetic innervation of both the submandibular and sublingual glands is derived from the chorda tympani branch of the facial nerve (cranial nerve VII). The integrating center of this nerve is located in the superior salivatory nucleus. The nerve exits the brainstem via the lower portion of the pons. As the preganglionic fibers course through the chorda tympani nerve, they join the lingual nerve (a branch of the marginal mandibular division of the trigeminal nerve) and extend to the submandibular ganglion where the fibers synapse with postganglionic fibers. The postganglionic fibers pass from this ganglion to innervate the submandibular and sublingual glands [59, 84, 288–291]. The submandibular gland is also closely associated with the hypoglossal nerves (cranial nerve XII) [82]. The hypoglossal (cranial nerve XII) together with glossopharyngeal (cranial nerve IX) nerves innervate the minor glands [88].

The salivary center of the parotid is located in the inferior salivatory nucleus in the upper medulla of the brainstem. Parasympathetic preganglionic fibers of this nucleus exit the brainstem as part of the glossopharyngeal nerve (cranial nerve IX), synapsing with postganglionic fibers in the otic ganglion. The postganglionic fibers course through the auriculotemporal nerve to provide parasympathetic secretory innervation to the parotid gland [175, 288–291]. The fibers of the facial nerves (cranial nerve VII) also innervate the parotid gland [57].

Salivary secretion is regulated by reflexes. Upon adequate stimulation by the action of chewing (mechanical factors), sweet tasting substances in the mouth (gustatory factors), sight of palatable food (visual factors), and smell of palatable food (olfactory factors) (Fig. 11.9) [51, 57], mechanoreceptors in the periodontal ligament, chemoreceptors in taste buds, or the corresponding transducing-associated receptors of the visual or olfactory pathway are activated to relay impulse to the salivary nuclei of the medulla (salivation center of the medulla oblongata) via the afferent fibers of cranial nerves VII, IX, and XII [61]. Factors such as the sight, taste, and smell of palatable food induce salivary secretion by conditioned reflex. The processed information in the medulla is relayed back to the salivary glands via the efferent pathway. Parasympathetic efferent signals reach the submandibular, sublingual, and other minor glands through cranial nerve VII. The parotid glands receive efferent signals via cranial nerve IX [61]. Parasympathetic stimulation of the salivary glands results in exocytosis of secretory vesicles containing acetylcholine to activate M3 (to a greater extent) and M1 (to a lesser extent) muscarinic receptors [175, 286]. Cholinergic muscarinic stimulation of the salivary glands leads to secretion of a large quantity of saliva (dilute saliva) with poor protein content (e.g., low amylase concentration) and high electrolyte (high K^+, and $HCO^-{}_3$) levels [160, 182, 195]. This cholinergic stimulation is mainly dependent on calcium signaling [126]. Reduction in the volume of salivary secretions can be observed during sleep, periods of anxiety, fear, and conditions of dehydration [178, 292–294].

The salivary glands are innervated by sympathetic nerves with the salivary centers originating from the upper thoracic segments of the spinal cord (intermediolateral column T1–T4). The preganglionic fibers course through thoracic ganglion of the paravertebral sympathetic trunk and travel to the upper (superior) cervical ganglion, where the fibers synapse with postganglionic fibers. The postganglionic fiber extends from this ganglion coursing through the external carotid artery plexus to reach and innervate the major salivary glands [175, 195, 285, 288–291].

Upon stimulation, these nerves release noradrenaline from secretory vesicles near the plasma membrane of the nerve terminal to stimulate alpha- and beta-adrenergic receptors on the salivary cells. The released neurotransmitter causes greater release of proteins, compared to parasympathetic stimulation. The acinar cells account for the major part of proteins (such as amylase) secreted, but ductal cells also secrete certain quantity of proteins (Fig. 11.9) [160, 175, 182, 195, 285, 286]. The secreted fluid also contains variable amounts of fluid and electrolytes [160]. The secretion of proteins (including zymogens) is mainly controlled by the cAMP-dependent/protein kinase signaling [295]. It should be mentioned that, apart from protein, fluid and electrolyte secretion, stimulation of autonomic nerves of the salivary glands also exert synthetic and trophic effects [286].

A third pathway that controls salivary gland secretion is known as non-cholinergic, non-adrenergic pathway (or peptidergic pathway). Some peptides released from autonomic nerves or from blood can affect salivary secretion [175]. The peptidergic signals include substance P, vasoactive intestinal peptide, dopamine, serotonin, enkephalin, neuropeptide Y, neurokinin A, pituitary adenylate cyclase-activating peptide, calcitonin gene-related peptide, and nitric oxide. Adenosine triphosphate can also trigger peptidergic signaling [69, 195, 287, 288]. For instance, the purinergic ionotropic receptors P2X4 and P2X7 are activated by extracellular ATP that evoke Ca^{2+} waves [287]. The neurotransmitter dopamine activates D1-receptor of salivary gland cells that leads to hyperpolarization of the plasma membrane with increased volume of salivary secretion but with poor protein level [299–301]. The effect of DI receptor stimulation may be due to its selective effect on ductal cell [302, 303]. In contrast, serotonin stimulates the secretion of saliva rich in protein content, but poor in water content. This suggests that serotonin selectively activates its receptors on acinar cells [302, 303].

Some factors are known to affect the rate of salivary secretion. These factors include nutritional, health state (local or systemic diseases affecting the glands themselves), psychological (such as anxiety, pain), pharmacological (certain medications) [51].

Clinical Correlate 11.2

Xerostomia and Sjögren's Syndrome

Sjögren's syndrome is a chronic or progressive systemic, multiorgan disease that primarily affects the salivary and lacrimal glands. The disease is named after the Swedish ophthalmologist Henrik Sjögren (1899–1986) who first described the condition in 1933 [304, 305]. Sjögren's syndrome is sometimes called sicca complex syndrome or Mikulicz's disease. Sicca is a Latin word meaning dryness, while Mikulicz's disease is named after Johann von Mikulicz–Radecki (1850–1905), a Polish–Austrian scientist who in 1888 first described the clinical features of enlargement of the parotid, submandibular, and lacrimal glands [304, 305]. The cause of the disease is not known, but immune disorders are believed to play a crucial role [305, 306]. For further information on epidemiology, etiology, pathogenesis, diagnosis, and treatment of Sjögren's syndrome, review the following publications.

1. Ferro F, Marcucci E, Orlandi M, Baldini C, Bartoloni-Bocci E (2017) One year in review 2017: primary Sjögren's syndrome. Clin Exp Rheumatol 35:179–191
2. Katsiougiannis S, Wong DT (2016) The proteomics of saliva in Sjögren's syndrome. Rheum Dis Clin North Am 42(3):449–456
3. Rusthen S, Young A, Herlofson BB, Aqrawi LA, Rykke M, Hove LH, Palm Ø, Jensen JL, Singh PB (2017) Oral disorders, saliva secretion, and oral health-related quality of life in patients with primary Sjögren's syndrome. Eur J Oral Sci (June 23)
4. Stefanski A, Tomiak C, Pleyer U, Dietrich T, Burmester GR, Dörner T (2017) The diagnosis and treatment of Sjögren's syndrome. Dtsch Arztebl Int 114(20):354–361
5. Zhang Q, Wang X, Chen H, Shen B (2017) Sjögren's syndrome is associated with negatively variable impacts on domains of health-related quality of life: evidence from short form 36 questionnaire and a meta-analysis. Patient Prefer Adherence 11:905–911

Xerostomia (derived from the Greek "*xeros*" meaning "dry" and "*stoma*" meaning "mouth") can be defined as dryness of the mouth resulting from salivary gland hypofunction. The condition is characterized by substantial reduction in daily salivary secretions. The cause of the disease is believed to be multifactorial, but immune disorders are considered crucial causative factors. Other causes include, but are not limited to, use of medications, irradiation, cancers, diabetes mellitus, pernicious anemia. It should be noted, however, that a considerable proportion of etiological factors of the condition is yet to be identified—generally termed idiopathic factors [28, 98]. Xerostomia can increase the risk of other oral diseases such as periodontal disease, dental caries, oral candidiasis, esophageal and gastric ulcers [28]. For further information on epidemiology, etiology, pathogenesis, diagnosis, and treatment of xerostomia, review the following publications.

1. Gabryś HS, Buettner F, Sterzing F, Hauswald H, Bangert M (2017) Parotid gland mean dose as a xerostomia predictor in low-dose domains. Acta Oncol 56(9):1197–1203
2. Gil-Montoya JA, Silvestre FJ, Barrios R, Silvestre-Rangil J (2016) Treatment of xerostomia and hyposalivation in the elderly: a systematic review. Med Oral Patol Oral Cir Bucal 21(3):e355–e366
3. Kaae JK, Stenfeldt L, Eriksen JG (2016) Xerostomia after radiotherapy for oral and oropharyngeal cancer: increasing salivary flow with tasteless sugar-free chewing gum. Front Oncol 6:111
4. Okamoto A, Miyachi H, Tanaka K, Chikazu D, Miyaoka H (2016) Relationship between xerostomia and psychotropic drugs in patients with schizophrenia: evaluation using an oral moisture meter. J Clin Pharm Ther 41(6):684–688
5. Ouanounou A (2016) Xerostomia in the geriatric patient: causes, oral manifestations, and treatment. Compend Contin Educ Dent 37(5):306–311; quiz312
6. López-Pintor RM, Casañas E, González-Serrano J, Serrano J, Ramírez L, de Arriba L, Hernández G (2016) Xerostomia, hyposalivation, and salivary flow in diabetes patients. J Diab Res 2016:4372852
7. Veerabhadrappa SK, Chandrappa PR, Patil S, Roodmal SY, Kumarswamy A, Chappi MK (2016) Evaluation of xerostomia in different psychological disorders: an observational study. J Clin Diagn Res 10(9): ZC24–ZC27
8. Tanasiewicz M, Hildebrandt T, Obersztyn I (2016) Xerostomia of various etiologies: a review of the literature. Adv Clin Exp Med 25(1):199–206

11.4.2 Esophageal Secretions

Along the walls of the GI tract, different types of cells to varying percentages produce their secretions into the lumen. In the esophagus, glands present in the lamina propria and submucosa secrete mainly mucus, bicarbonate, and other products into the lumen (Fig. 11.10). The submucosal gland architecture is similar to that of the salivary glands as discussed earlier. The esophageal submucosal glands consist of mostly mucin-producing cells, demilunar cells, serous cells, acinar cells, and ductal cells [307, 308]. The secretory cells (e.g., acinar cells) can secrete substances directly into the blood [309]. In the esophageal glands, luminal secretions are transported from the site of exocytosis/synthesis (e.g., secretory endpieces) into the ductal system, where the primary secretion is modified by the ductal cells and subsequently released into the lumen [307, 308].

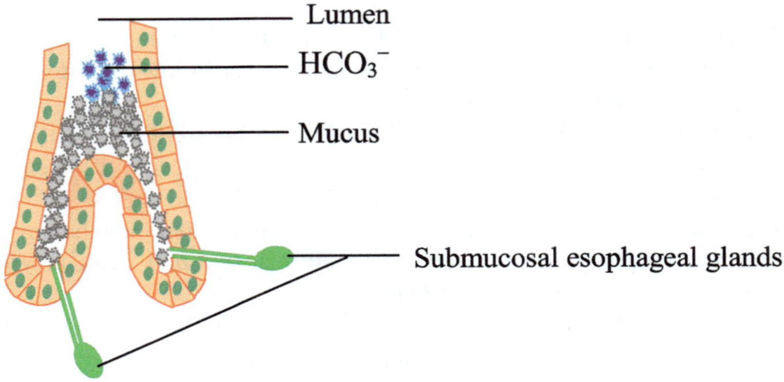

Fig. 11.10 Schematic representation of the esophageal wall, associated glands, and their secretions. The esophageal wall is coated with mucus and large concentration of bicarbonate ions which are constantly secreted together with mucus into the lumen. The mucus covers the epithelial cell layer preventing it from injury or "wounding." The ducts of submucosal gland pass through the mucosa into the lumen where their secretions regulate esophageal functions. The epithelial cell layer is made up of various cell populations but mainly include mucus-producing goblet cell. There are also enteroendocrine cells lining the epithelia of the organ [310, 311]. In disease states, any cell population or tissue of the esophagus can be affected. For instance, specific forms of cancer are known to affect esophageal mucus-producing cells, enteroendocrine cells, stem cells as well as the ductal epithelium [310]

The wall of the esophagus is abundant in mucus-secreting glands, other glands, and cells that secrete different products into the lumen of the organ. The esophageal mucus has a crucial protective role and represents one of the three protective measures, where this hollow structure possesses. The esophagus has three tiers of defense that act to limit the degree of gastroesophageal reflux (see Clinical Correlate 11.3). These tiers of defense serve as an anti-reflux barrier, which aids in minimizing the risk of mucosal injury due to reflux of acidic contents of the stomach into the esophagus [312]. The first tier of defense is believed to be the lower esophageal sphincter and the "pinchcock" effect of the right crus of the diaphragm termed the "diaphragmatic pinchcock" [312]. The "diaphragmatic pinchcock" comprises the combination of a flap valve and a circular anatomic structure of striated muscle of the diaphragmatic crural fibers functioning as a sphincter at the lower esophagus [313]. The diaphragmatic pinchcock forms the esophageal hiatus in the diaphragm [314]. This defense initiated by the lower esophageal sphincter is thought to be reflexive in nature and is caused by distension of the stomach due to accumulation of food or increase in the intra-abdominal pressure. The reflex is called gastroesophageal reflex, also known as straining-crural reflex, and it ensures crural contraction and increase in the lower esophageal pressure that tightens the lower esophageal sphincter, thus reducing the quantity if refluxed gastric contents [313]. This defense mechanism involves reduction of the frequency and volume of refluxed contents of the stomach into the gullet. Failure of this defense mechanism substantially increases the functionality of esophageal clearance along the wall of the hollow organ. However, esophageal clearance is functional even at basal level of reflux of stomach contents [312]. The main function of this second mechanism is to reduce the duration of contact between refluxed gastric contents and the esophageal epithelium. Reduction in the contact duration is further strengthened by gravity, esophageal peristalsis, upright position, salivary secretion, and the bicarbonate secretion of the esophageal submucosal glands—which are seromucous glands [312, 315, 316]. This third tier of defense comes into play when the duration of contact of refluxed gastric content with the esophageal wall is prolonged. This defense can be substantially stimulated during sleep or disease states when esophageal clearance is not effective [312]. It should be noted, however, that these tiers of protection that maintain the mucosa and tissue integrity of the esophagus do not have clear-cut boundaries. Although as defined above, the role of any tier actually varies, depending on the involvement of the embarrassing factor; their functions are interwoven [312].

The bicarbonate ions secreted by the cells of the esophageal walls aid in the neutralization of acidity of the immediate medium, thereby helping to maintain a stable environment for the functioning of the esophagus. The bicarbonate ions are the major ions secreted by the esophageal submucosal glands. The squamous epithelium of the esophagus does not secrete HCO^{-}_{3}. The secretion of HCO^{-}_{3} is regulated by both basolateral and apical ion channels. The basolateral (Cl^{-}/HCO^{-}_{3} exchanger, luminal CFTR, Na^{+}/HCO^{-}_{3}) plays a substantial role in the secretion of HCO^{-}_{3} by the serous and ductal cells of the submucosal glands [315]. The apical Cl^{-}/HCO^{-}_{3} exchanger plays a crucial role in luminal secretion of HCO^{-}_{3} [307]. The esophageal submucosal glands express other functional ion channels, which include basolateral $Na^{+}/K^{+}/2Cl^{-}$, basolateral Na^{+}/K^{+}-ATPase, calcium channels, proton pump of mucus and ductal cells. Some ions secreted into the lumen are controlled by the paracellular shunt pathway [307, 317, 318].

Substances such as growth factors are also secreted by the cells of the esophageal glands and exert a proliferative role. These substances are synthesized in the cells of the esophageal epithelia and secreted into the surrounding tissue (or may be transported to relatively distant sites), where they act on cognate receptors to modulate functions and structural remodeling at the molecular level [315, 319, 320].

Increase in esophageal secretion is stimulated by activation of the vagus nerve and is mediated via cholinergic muscarinic M_1 receptors of the cells of the submucosal glands [307].

Clinical Correlate 11.3

Secretions of the Gastroesophageal Junction and Gastroesophageal Reflux Disease (GERD)

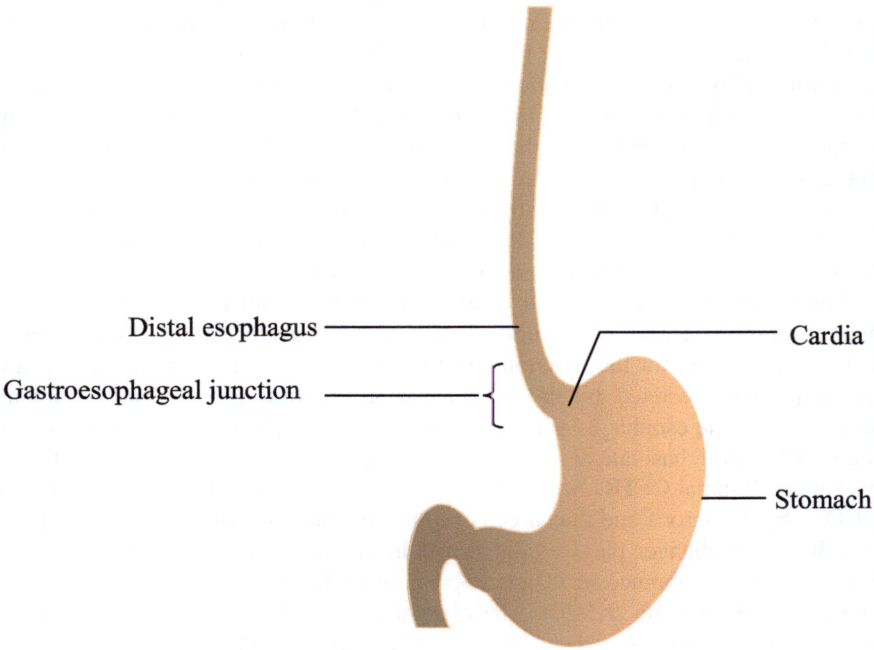

Fig. 11.11 Diagram of the esophagus and stomach showing the most vulnerable regions of gastric reflux

The wall of the transition region of the esophagus into the stomach is abundant in mucus-secreting glands—esophageal and gastric cardiac glands. Mucus secretions of these regions protect it from the high acidity [321]. Reduction of the secretions of this region can result in various diseases. The gastroesophageal junction is the short anatomic region where the distal esophagus meets the proximal stomach (known as cardia) (Fig. 11.11) [322]. The gastric cardia is a short segment measuring less than 0.4 mm long of mucosa and containing oxyntic glands [322]. The distal esophagus is mostly composed of mucous-secreting columnar epithelium, but squamous epithelium is found in other areas of the esophagus [322, 323].

The gastroesophageal junction is the region that is commonly implicated in the deleterious effects of GERD [322]. According to the Montreal Consensus, GERD is defined as a chronic condition characterized by reflux of gastric contents into the esophagus resulting in troublesome symptoms, with or without mucosal damage, and/or complications [321, 324, 325]. GERD symptoms are considered troublesome if they adversely affect the well-being of the person. These symptoms include heartburn, discomfort, chest pain, dysphagia, regurgitation of food or sour liquid (acid reflux) [321, 324, 325]. Of particular interest is heartburn, which is the most common and a classic symptom of GERD, resulting from acid reflux in an individual with erosive esophagitis, and/or activation of chemoreceptors or mechanoreceptors that possibly trigger central and peripheral sensitization [325, 326].

GERD comprises a spectrum of diseases which includes reflux esophagitis, Barrett's esophagus, and nonerosive reflux disease (NERD) [326]. These disorders may lead to inflammation and intestinal metaplasia of this anatomic region, predisposing the individual to the development of adenocarcinoma of the esophagus [327, 328].

Reflux esophagitis: This is a condition characterized by esophageal inflammation caused by acid and biliary reflux. The classic symptoms include acute chain pain and severe dysphagia. Treatment depends on the severity of the disease and involves use of medication, endoscopy, and surgery [329, 330].

Nonerosive reflux disease (NERD): NERD can be defined as the absence of visible esophageal mucosal injury during upper endoscopy, despite the presence of classic GERD symptoms such as heartburn and chest pain due to intraesophageal reflux [331, 332]. NERD is the most common form of GERD, and about 60% of patients who are known to have heartburn and/or chest pain do not have any visible evidence of esophageal mucosal injury at upper GI endoscopy [325]. NERD diagnosis can be made with 24-h pH monitoring (usually by multichannel intraluminal impedance) and the presence of heartburn and normal upper GI endoscopy [332]. However, even at normal acid reflux, there may be associated positive symptoms of acid reflux, a condition that is termed reflux hypersensitivity. In the absence of both abnormal acid reflux and symptom of acid reflux, the condition is known as functional heartburn [332].

Barrett's esophagus: The disease is named after the Australian-born British thoracic surgeon Norman Rupert Barrett (1903–1979), who first in 1953 described patients with the condition [328, 333]. Although Barrett's esophagus does not have a precise definition [323], it is known that the condition is an acquired disorder characterized by the replacement of the normal esophageal squamous epithelium with metaplastic intestinal columnar epithelium, which occurs on a background of severe esophageal mucosal injury with chronic gastroesophageal acid reflux association [328, 333]. In Barrett's esophagus, the structure of the columnar epithelium is also called specialized intestinal metaplasia because of the similarities of this epithelium to that of the intestine [323, 327, 334]. Barrett's specialized intestinal metaplasia is an adaptation to the unfavorable condition in the esophagus due to chronic gastroesophageal acid reflux. The epithelium is specialized because it now possesses a couple of functions not known to be present in the normal esophageal epithelium [328].

11.4.3 Gastric Secretions

The stomach comprises three anatomical divisions—fundus, corpus, and antrum—and two functional regions—oxyntic and pyloric glands. The oxyntic region comprises about 80% of the stomach and thus 80% of gastric glands. The pyloric gland region constitutes about 20% of total area of the stomach (Fig. 11.12) [335].

The secretions of the different regions of the stomach mostly include mucus, hydrochloric acid (HCl), pepsinogen, hormones (gastrin, leptin, histamine etc.), intrinsic factor of Castle, growth factors, immune factors [1, 17, 336–338]. These substances are produced by different gastric mucosal cells, which comprise the gastric glands, also known as gastric pits (Fig. 11.13). Gastric pits are long tubular structures of the gastric mucosa that open into the gastric lumen. They are characterized by invagination between goblet cells of the gastric epithelium [335, 339, 340].

The major types of cells of the gastric pits include goblet, chief, parietal, enterochromaffin cells, and G cells (Fig. 11.13) [1, 17].

The goblet cells produce mainly mucus, which serves to protect the wall of the stomach from the adverse effect of acid secretion and also lubricate the GI tract. The goblet cells are mainly found in gastric cardia, a region of the stomach containing high proportion of mucous glands. But the gastric cardia may also contain oxyntic glands [1, 322]. Mucus is secreted by exocytotic mechanism. Trigger factors of mucus production include high-fiber diets, distention of the stomach wall. Although the mechanisms are not clearly understood, mechanical distension of the mucosa can initiate mechanotransduction event that culminate in increased secretion of mucus [341].

The chief or zymogen cells of the gastric fundus produce mainly pepsinogen and gastric lipase [1, 339]. The zymogen is activated by gastric acid (vide infra) usually below pH 4.5 and in the presence of its substrate. The gastric secretions of humans comprise a number of isozymes of protein-hydrolyzing enzymes. The major proteinase is pepsin 3b, which constitute about 70% of the total secreted protein-hydrolyzing enzymes, followed by pepsin 5 (also known as gastricsin) constituting about 20% of the total gastric protein-hydrolyzing enzymes. Gastric secretions also contain other isozymes of pepsin such as pepsin 1 [342, 343].

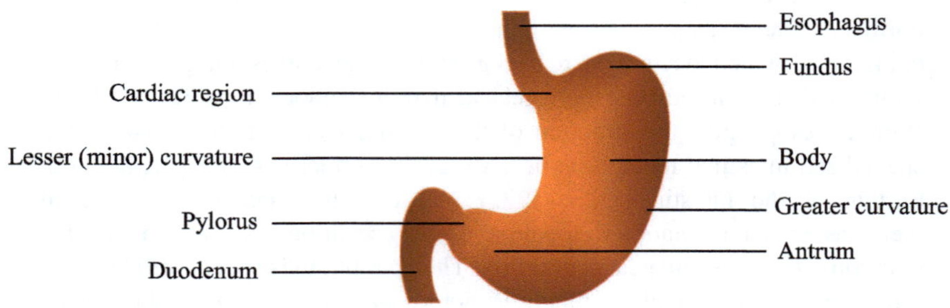

Fig. 11.12 Stomach (*gaster*) and its parts

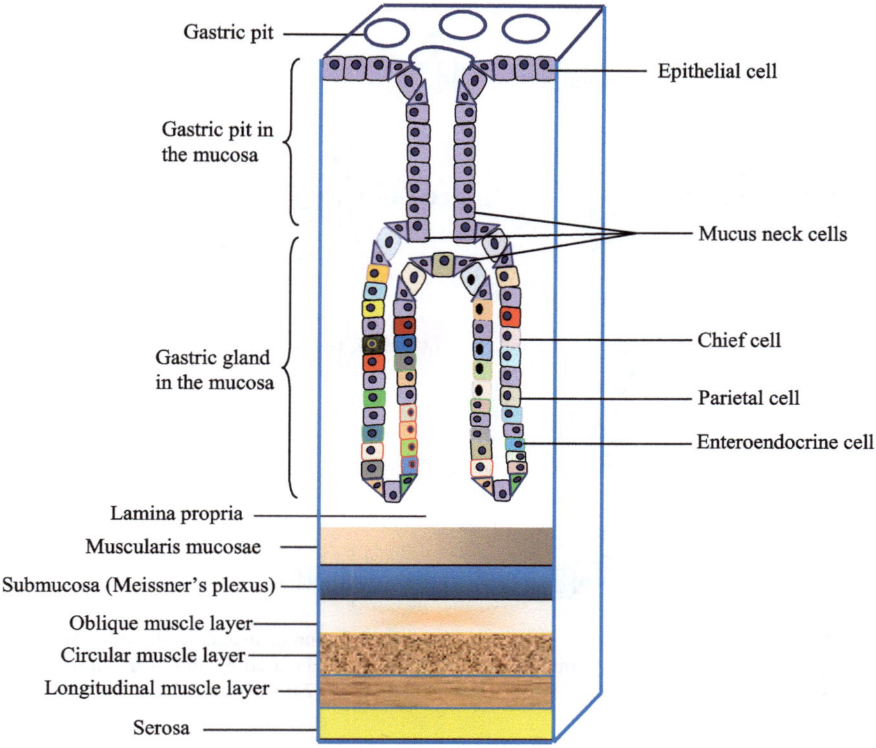

Fig. 11.13 Schematic diagram of the gastric wall. Note that the muscularis externa contains Auerbach's plexus

The parietal cells (oxyntic glands) of the fundic glands of the stomach produce mainly hydrochloric acid (HCl) and intrinsic factor. It should be noted that H^+ and Cl^- are secreted separately into the lumen of the glands (Fig. 11.14) [1, 17, 343]. The secreted acid kills microbes and converts pepsinogen into pepsin, thus initiating the process of hydrolysis of proteinous food. The enzyme pepsin activates further secretion of pepsinogen. The intrinsic factor forms complexes with vitamin B12, which is required for its absorption in the upper small intestine [1, 339].

G cells produce gastrin that stimulates gastric acid secretion. The enterochromaffin cells produce histamine that stimulates HCl secretion by the oxyntic glands. D cells secrete somatostatin, which inhibits HCl secretion by the stomach [1, 344–346].

Phases of Gastric Secretions

The secretions of the stomach occur in three phases: cephalic, gastric, and intestinal [1, 349]. The cephalic phase begins immediately before food enters the mouth and

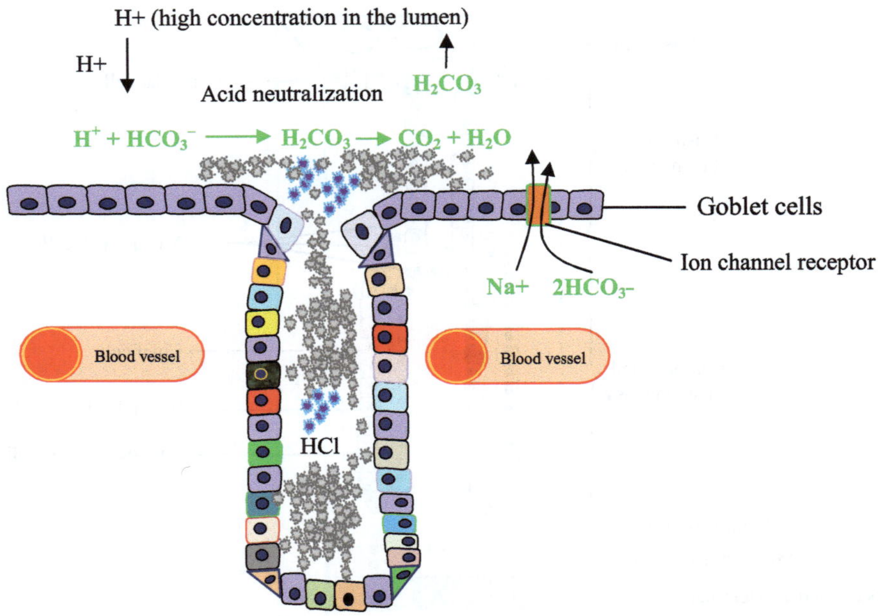

Fig. 11.14 Gastric gland. Gastric glands are interspersed among mucus neck cells. *Note* The lumen of the canaliculus pH is about 1.5–1.9 units, which is increased by almost five times in the mucus coat. The pH in the cytosol of the parietal cell is 7.4 (7.0–7.4) [347, 348]

prepares the stomach for digestion. In the cephalic phase, stimulus such as sight of food activates specific visual receptors that relay the impulse to the cerebral cortex. The cortex in turn signals downstream to the medulla oblongata, which relays the signal to the stomach via the vagus nerve. Through this pathway, negative emotion and depression can affect food intake. Other stimuli that can stimulate cephalic phase of gastric secretion include smell and taste of food. Furthermore, thought about palatable food can initiate gastric secretion. Receptors that play essential role in the initiation of gastric secretion in the cephalic phase are the M_3, H_2, and gastrin receptors—which are expressed on parietal cells. For instance, acetylcholine released by efferent ending of the vagus nerve stimulates parietal M_3 receptors to mediate the process of acid secretion [350–352]. The neurotransmitter pituitary adenylate cyclase-activating polypeptide released upon neural stimulation binds to its cognate receptor on enterochromaffin cells to mediate histamine secretion. Histamine stimulates the histamine H_2 receptor on the membrane of parietal cells [349].

In the gastric phase, products of food digestion and mechanical distension of the stomach are the main stimuli responsible for gastric secretion. Mechanical distension of the stomach due to food ingestion (and subsequently food accumulation in the stomach) activates stretch receptors that mediate vago-vagal and local reflexes (neuronal mechanism). Products of food digestion and other molecules ingested as food (e.g., peptides, caffeine) can activate gastrin secretion by stimulation of mucosal chemoreceptors of the antral G cell, which release gastrin to bind to cholecystokinin-2 (CCK-2 or CCK-B) receptors on the basolateral membrane of parietal cells, thus stimulating acid production (hormonal mechanism). Gastrin can also bind to CCK-B receptors on enterochromaffin cell to initiate histamine secretion [343, 349, 353]. Thus, increase in gastrin secretion subsequently leads to decrease in luminal pH. At a luminal pH of about 1.5, gastrin secretion is inhibited via the release of somatostatin from the D cells. Somatostatin is a paracrine factor that acts via type 2 somatostatin receptor to inhibit parietal cell secretion of HCl and also inhibit G cell secretion of gastrin and enterochromaffin secretion of histamine. Secretion of somatostatin can be inhibited by vasoactive intestinal polypeptide via a negative feedback loop [343, 349, 350, 353]. The gastric phase is responsible for about 70% of gastric secretion [350].

The intestinal phase begins when the stomach contents are emptied into the duodenum. This phase has an inhibitory action on gastric secretion and is mediated via GIP, secretion, and cholecystokinin, as well as enteropancreatic and enterogastric reflexes [350, 354–356].

Mechanism of Hydrochloric Acid (HCl) Secretion by the Parietal Cells

Secretion of HCl by the parietal cells of the stomach is controlled by acetylcholine, histamine, and gastrin. The vagus nerve is crucial in this regulation (Fig. 11.15) [1, 17]. Parietal cells express a number of ion channels and receptors (receptors for acetylcholine, histamine, and gastrin on the basolateral cell membrane) that are required for secretion of gastric acid. In addition, they express secretory canaliculi —twisted small channels containing numerous microvilli that penetrate the cytoplasm of the parietal cell. These secretory canaliculi can expand during stimulation of the cell (Fig. 11.15) [357]. Secretion of HCl by the parietal cell is stimulated by acetylcholine, gastrin, and histamine. The inhibition of the parietal cell is mediated by somatostatin (Fig. 11.15) [358].

When a stimulus impinges on the parietal cell, the acid-secreting cell undergoes several transformations that prime the cell for efficient exocytosis. This activation process drives the translocation of different membrane receptors to the apical pole. For instance, cytoplasmic tubulovesicles comprising proton pumps (H^+/K^+-ATPase) are translocated to the apical membrane of the parietal cell to fuse with the membrane of the canaliculi. The intracellular pathway for activation depends on the stimulating neurotransmitter or hormone. For example, gastrin, histamine, and PACAP stimulate acid secretion via cAMP-dependent pathway, whereas acetylcholine acts via a calcium-dependent pathway (Fig. 11.15) [349, 358]. But gastrin, histamine, and PACAP to some extent also contribute to elevation of cytosolic Ca^{2+} in parietal cells. In addition, release of histamine from enterochromaffin cells can be initiated by gastrin and acetylcholine activation [358].

Apart from neural, endocrine, and paracrine factors that initiate acid secretion upon ingestion of food, other substances present in food or liquid can also stimulate acid secretion. For example, caffeine stimulates gastric acid secretion by inhibition

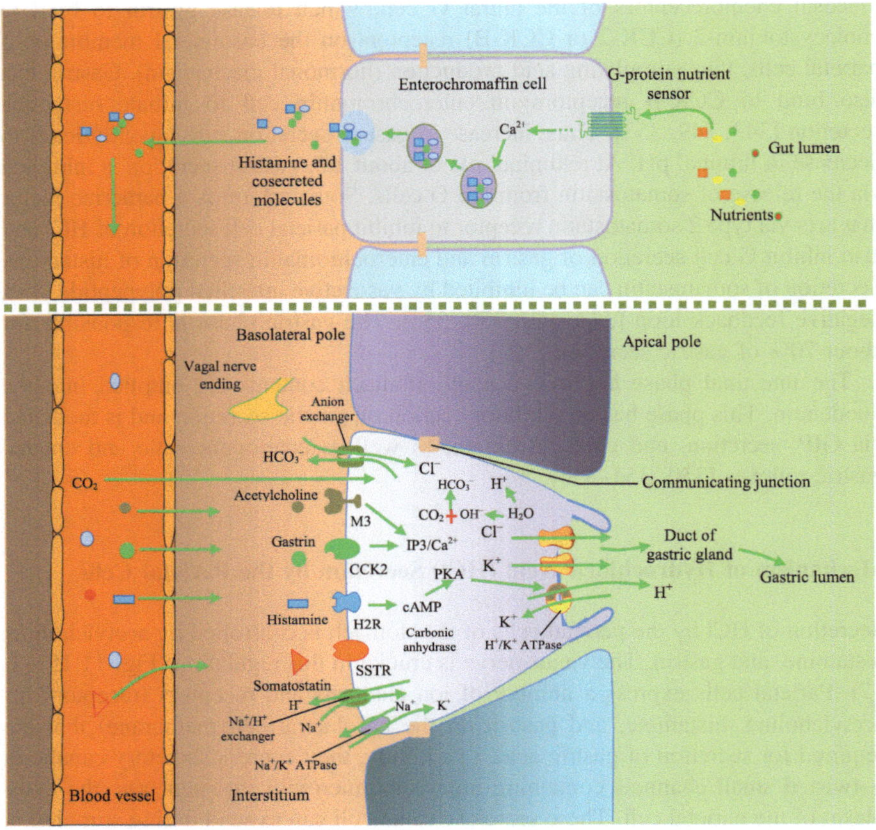

Fig. 11.15 Parietal (oxyntic) cell physiology. Acetylcholine and cholecystokinin (or gastrin) activation of M3 and CCK2 receptors, respectively, stimulate the membrane-bound lipid IP2 resulting in the production of DAG and IP3, which, respectively, in turn, stimulates PKC and Ca^{2+}. Histamine stimulates the production of cAMP/PKA via activation of H2 receptor. These products activate the translocation of secretory vesicles, phosphorylation of cytosolic and membrane proteins, and their increased translocation to the apical pole of the parietal cell. The tubulovesicle containing H^+/K^+-ATPases are increasingly translocated to the canaliculus membrane of the parietal cell. This gastric acid pump exports hydrogen ions into the lumen of the gland in exchange for potassium ions in a 1:1 ratio. An apical membrane transporter for K^+ and for Cl^- exports these ions out of the cell into the lumen. The transport of H^+ and Cl^- into the gland lumen significantly increases the acidity of the immediate environment. The parietal cytosol pH is about 7.2–7.4, whereas the pH in gastric lumen drops to about 1.0 (the blood pH is about 7.2). The ion fluxes are balanced by the apical and basolateral transport systems. One bicarbonate ion is transported out of the basolateral pole in exchange for Cl^- for every hydrogen ion that is secreted in the secretory canaliculi. The concentration of generated bicarbonate ion and H^+ is controlled by Na^+/H^+ and anion exchangers. The basolateral Na^+/K^+-ATPase also regulates the concentration of both ions in the cytosol of the parietal cell [357, 359, 360]

of cAMP-dependent phosphodiesterase, which increases the concentration of cytosolic cAMP [361]. Cyclic AMP-dependent phosphodiesterase (cAMP-dependent PDE) is an enzyme that breaks down cAMP to AMP [362, 363]. Consequently, frequent caffeine intake should be avoided by people suffering from GERD and peptic ulcer disease [361].

Reception Theory and Gastric Exocrine Secretions

A receptor is a polypeptide molecule usually located in the plasma membrane or the interior of a cell that recognizes and associates with a chemical messenger or is triggered by other factors to elicit signaling cascades that result in a biologically relevant response in the cell. The way hormones and other bodily chemicals carry out their functions would have remained only a mystery if not for the work of the scientists who conducted works on the receptor theory—chemical reception. When hormones or neurotransmitters are released, they find their appropriate cell by locating and attaching themselves to a molecule located on the plasma membrane (note that some receptors are located in the cytoplasm, nucleus, or other organelles). The initial hypothesis about chemical reception (transmitter receptor and chemical receptive substances) that formed the basis of information transmission in the body was proposed by John Newport Langley (1852–1925) and Paul Ehrlich (1854–1915). During the first half of the nineteenth century, scientists began to think about specificity in nature and wondered if this was the case with functioning of the human body. With constant progress in chemistry and the behavior of chemical extracts in the body, there was increasing chance for discovery of "receptors." The German scientist, Friedrich Sobernheim (1803–1846), noted that strychnine (a poisonous alkaloid obtained from seeds of the *nux vomica* tree and related plants of the genus *Strychnos*; and a glycine receptor antagonist) was specific in affecting the spinal cord; alcohol—the brain, mercury—the salivary glands. Although we now know that these early observations on specificity were narrow, the truth that remained was that chemical effect to the body had a property of specificity. In 1844, Claude Bernard identified the presence of an intermediate between nerve and muscle when he injected a poisonous compound (curare) into the subcutaneous tissue at the internal part of the thigh of a rabbit. After about 6 min of observation, the poisonous compound had resulted in completely paralyzing the rabbit. But the heartbeat was still present. However, there were no identified reflex responses to peripheral cutaneous stimuli. Dissection of the animal following the chemically induced death did not show any sign that could have explained the cause of death [364–367].

Langley, a British physiologist, who spent his working life at the Cambridge University, and researching mainly neural transmission, and at about 31 years of age, he became a Fellow of the Royal Society and later the vice president of this scientific society. In 1901, while working with extracts of the adrenal gland, Langley expanded his research in the area of neurotransmitters and receptors. In one of his observations, the Cambridge scientist realized that extracts of adrenal gland produced characteristic responses when put into tissues. Interestingly, the responses were similar to those induced by nerve stimulation [368, 369]. Langley introduced the concept of "receptive substance" or "receptors" in his paper published in 1905, in the journal of physiology. In the next couple of years, his research on nervous transmission at the somatic neuromuscular junction founded the idea of transmitter receptors [366, 370]. Langley, who coined the terms "autonomic nervous system" and "parasympathetic nervous system," was also the first to explicitly formulate the receptor theory of drug action [371]. In one of his experiments, Langley observed distinct effects between the application of pilocarpine (muscarinic agonist) and atropine (anticholinergic or antiparasympathetic (parasympatholytic) agent) on salivary secretion in mammals. Pilocarpine had a stimulative effect on salivary secretion, whereas atropine inhibited secretion. As part of his concepts, Langley proposed that in addition to pilocarpine and atropine, for which he was sure, had receptors, he added to the list, secretin, adrenaline, strychnine, steroid, and sex hormones as substances having their receptors in the body [366]. For details on these agent/drugs, see Clinical Correlate 11.4.

Clinical Correlate 11.4

Some Pharmacological Drugs and Their Mechanisms of Action on Gastrointestinal Tract

With continuous progress in receptor physiology and biochemistry, the mechanisms of functioning of chemical substances were becoming clearer; and their receptors were also being discovered. Pilocarpine, atropine, secretin, adrenaline, and strychnine were some of the first chemical substances identified by Langley to function through a receptor [367, 371].

Pilocarpine is a non-selective muscarinic receptor agonist that stimulates secretions in the GI tract such as salivary secretion. It is a specific stimulator of M_3 acetylcholine receptors. It is a parasympathomimetic alkaloid obtained from the leaves of tropical American shrubs from the genus *Pilocarpus*. Pilocarpine has therapeutic uses in treating dry mouth and glaucoma [372–375].

Atropine is a naturally occurring tropane alkaloid extracted from the deadly *Atropa belladonna* (nightshade) and other plants of the family *Solanaceae*. It is a competitive antagonist for the muscarinic acetylcholine receptor types M_1, M_2, M_3, M_4, and M_5. The agent acts as a parasympatholytic (anticholinergic) drug. Atropine decreases intestinal motility by inhibition of MMC-related contractions in the gut. In the heart, atropine non-selectively antagonizes the functions of the muscarinic acetylcholinergic receptor to increase firing of the sinoatrial node and conduction through the atrioventricular node of the heart, and thus heart rate increases. Atropine dilates the pupils and reduces salivation and other secretions in the human body. Atropine inhibits transient relaxation of the lower esophageal sphincter in both normal and patients with GERD. Although the mechanism for this inhibition is not exactly clear, it is thought to involve a central mechanism of atropine on the integrating mechanisms in the brainstem or altering of response of the stomach to distension at the peripheral level. Distension of the proximal stomach is a major stimulus for transient relaxation of the lower esophageal sphincter and also represents a major stimulus important in the mechanism of some GI disorders such as GERD [375–382]. The reflex for transient relaxation is mediated through the vago-vagal pathways. The signal of this pathway is initiated by tension receptors in the gastric musculature from where impulses are then sent to the central terminals in the brainstem to produce a response following evaluation with the central program. The response is sent as efferent signals to the lower esophageal sphincter via activation of the vagal motor neurons, to stimulate inhibitory enteric motor neurons, leading to smooth muscle relaxation. The efferent signal can inhibit esophageal peristalsis at the central or peripheral level [383]. Atropine thus can affect gastric compliance and both fasting and postprandial gastric tone, which are all controlled by cholinergic nerves. Intravenous injection of atropine (15 μg/kg bolus and a maintenance dose of 4 μg/kg/h) was shown to reduce the frequency of spontaneous reflux and gastric distension-induced transient relaxation of the lower esophageal sphincter [384].

Previously, it was thought that the use of pharmacological agents known to inhibit acid production in the stomach could address disorders of the stomach such as GERD. However, the use of proton pump inhibitors (PPIs) and prokinetics has some limitations. For instance, PPIs are more effective when used for patients with heartburn. However, with increase in knowledge on the mechanism of reflux, pharmacological agents to control transient relaxations of the lower esophageal sphincter were sought for. Among such pharmacological agents, baclofen, a GABA-B-receptor agonist, is known to inhibit effectively transient relaxation of the lower esophageal sphincter. The effectiveness can reach 40%. Unfortunately, baclofen has negative effects on the central nervous system functioning. Antagonists of the metabotropic glutamate type 5 receptor also inhibit the transient relaxation of the lower esophageal sphincter. The fact that these pharmacological agents have not been successful in completely addressing disorders such as GERD means that the disease mechanism is yet to be completely unraveled. Apart from dysfunction of the lower esophageal sphincter and proximal gastric distensibility, the crural diaphragm is also involved in the disease mechanism of occurrence. In normal, transient relaxation of the lower esophageal sphincter and the crural diaphragm is necessary for the passage of bolus from the esophagus down to the stomach. Importantly, this relaxation dysfunction is responsible for about 90% of acid reflux diseases. This relaxation takes place secondary to gastroesophageal reflux and allows gas venting [383, 385–388].

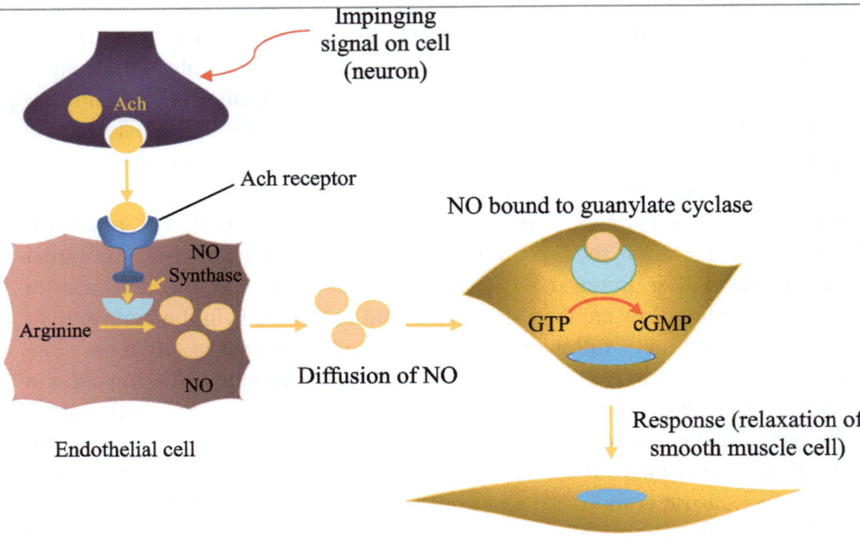

Fig. 11.16 Nitric oxide in smooth muscle relaxation

Another pharmacological agent, loxiglumide, a selective CCK1-receptor antagonist, reduces transient relaxation of the lower esophageal sphincter. Cholecystokinin (CCK) causes a reduction in lower esophageal sphincter pressure and increases the frequency of transient relaxation of the lower esophageal sphincter and a reduction in gastric emptying [389, 390].

Inhibitors of nitric oxide synthase (e.g., the amino acid, L-arginine), the enzyme that synthesizes nitric oxide in vivo, also reduce transient relaxation of the lower esophageal sphincter (Fig. 11.16) [391, 392].

The enzyme nitric oxide synthase synthesizes NO, a small uncharged molecule with a half-life of about 30 s that is freely diffusible through several structures in the cells and tissues of the organism (Fig. 11.16). However, due to the negative effects (on the GI, cardiovascular, urinary, and respiratory systems) that are yet to be fully solved, there is substantial limitation in the use of some of these agents for therapy [390].

Like atropine, antagonists of acetylcholine receptors have many negative effects including constipation and reduced acid clearance in the supine position [393, 394].

The M_1 selective antagonist, pirenzepine, reduces gastric acid secretion and reduces muscle spasm. The drug is used in the treatment of peptic ulcers. It also has a role to play in water intake and pressor responses [395]. At low doses, pirenzepine inhibits gastric acid secretion, GI motility, salivary, central nervous system, cardiovascular, ocular, and urinary functions. At doses of 100–150 mg/day, the drug promotes duodenal ulcer healing and diminishes pain. At such low doses, the effectiveness of the drug is similar to that of cimetidine 1 g/day in treating duodenal ulcers. Pirenzepine is a superior drug to the GI mucosa-enhancing agent used in GERD, gefarnate (geranyl farnesylacetate) at 300 mg/day [396, 397].

Methoctramine is an antagonist of both muscarinic M_2 and M_4 receptors and at high concentration (greater than or equal to 10 μM) may block postjunctional muscarinic M_3 receptors. The inhibitory effect of methoctramine on acetylcholine receptors is evident even at doses as low as 7–240 nmol/kg [398, 399].

The compound 4-diphenyl-acetoxy-*N*-methyl-piperidine (4-DAMP) is a selective antagonist of both M_1 and M_3 receptors. 4-DAMP covalently binds to muscarinic receptors to cause inhibition [400].

Apart from their known analgesic properties, agonists of μ(mu)-opioid receptors (e.g., morphine) have been shown to reduce the frequency of transient relaxation of the lower esophageal sphincter. The risk of addiction limits the use of μ(mu)-opioid receptor agonist in the treatment of transient relaxation of the lower esophageal sphincter. The opioid agonist antimotility drug, loperamide, is a phenylpiperidine derivative that decreases intestinal motility (decrease peristalsis) and fluid secretion, resulting in longer GI transit time and increased absorption of fluids and electrolytes from the GI tract. This effect of loperamide and other opiate receptor agonists such as diphenoxylate and haloperidol can be reversed by naloxone, which is an opioid antagonist [393, 401–403].

There is promise in using cannabinoid receptor 1 agonists (e.g., anandamide or *N*-arachidonoyl ethanolamine, *N*-arachidonoyl dopamine, 2-arachidonoylglycerol, 2-arachidonyl glyceryl ether, epigallocatechin gallate, nabilone, dronabinol, Δ^9-tetrahydrocannabinol, cannabidiol) as potential inhibitors of transient relaxation of the lower esophageal sphincter. The cannabinoid receptors belong to the G protein-coupled receptors, located in central and peripheral neurons and can be activated by cannabis and related compounds. It is quite possible that receptor coactivation (activation of an unrelated receptor by another neighboring receptor) may influence the mechanism of relaxation of the lower esophageal sphincter [404, 405]. Recent studies have shown that non-selective agonists of cannabinoid receptors can upregulate serotonin 2A (5-HT-$_{2A}$) receptor. The latter is widely distributed in the brain and peripheral system including the GI system. Disorders in 5-HT-$_{2A}$ receptor signaling are associated with several pathophysiological and neuropsychiatric disorders, such as stress response, anxiety, depression, cognitive and mood disorders, and schizophrenia, and may play some role in endocrine pancreatic functioning [406].

Strychnine is a colorless, bitter crystalline convulsant alkaloid often used as a pesticide. Thus, strychnine can be inhaled, swallowed, or absorbed through the gut and pass into the eyes. Strychnine is a naturally occurring alkaloid found in plants of the genus *Strychnos*, family *Loganiaceae*. It is a highly toxic compound with an LD_{50} of 1–2 mg/kg when ingested orally in humans. LD_{50} or LC_{50} (lethal dose or concentration, 50%) is the median lethal dose of a toxin, radiation, or pathogen and refers to the dose required to kill half the members of a tested population after a specified test duration. LD is an indication of the lethality of a given substance, pathogen, or radiation. The lower the LD_{50}, the higher the toxicity. However, LD has individual differences due to variance in resistance to drug effect. The concept of LD was created in 1927 by the British pharmacologist John William Trevan (1887–1956) [407–410]. Because strychnine is a bitter substance, it activates bitter taste receptors of the GI tract. It is rapidly absorbed from the GI tract. In the nervous system, strychnine functions as a high-affinity neurotoxin by selectively and competitively antagonizing glycine and acetylcholine receptors. Glycine is one of the most abundant inhibitory neurotransmitters in the CNS and functions through its attachment to the glycine receptor, which belongs to the class of ionotropic G protein-coupled receptor. Glycine is an agonist of glycine receptor, a ligand-gated chloride channel abundantly expressed in neurons of the spinal cord and the brain. The channel allows an influx of chlorine when the channel is activated leading to a hyperpolarization. Strychnine acts as an antagonist of glycine, so that glycine—the natural ligand of this receptor—can no longer bind. Thus, the inhibitory signals of glycine are inhibited or at least most of the receptors are not available for glycine binding. This leads to more excitation at a lower level of excitatory neurotransmitters because the inhibitory signals are prevented. The individual thus develops muscular spasms, resulting in death through asphyxia [411, 412]. Apart from strychnine and glycine, the glycine receptor can be stimulated by β-alanine and taurine. Like strychnine, caffeine is a competitive antagonist of the glycine receptor [413]. Strychnine also affects cholinergic transmission. It competitively binds to nicotine-binding sites on the nicotinic acetylcholine receptors and modifies the cooperativity of the nicotinic receptor, thus inhibiting the nicotine-induced membrane depolarization and increase in intracellular Ca^{2+} concentration. This inhibitory effect of strychnine is, however, reversible and is selective for nicotinic stimulation. In chromaffin cells of the adrenal medulla, the inhibitory effect blocks catecholamine release by antagonizing the binding site of the nicotinic acetylcholine receptor. Surprisingly, strychnine has no effect on catecholamine release in the absence of nicotine [414].

Therefore, it is obvious that if a compound or substance must exert its functions, then it must pass through a receptor. Arguably, the definition or meaning of a receptor will now determine whether or not this statement is true or false. Ehrlich, the author of "chemoreception," conducted studies on how different types of dye stained body cells and tissues. He was also interested on the interaction between bacterial toxins and antitoxins (antibodies) that were produced by the body [366, 367]. Ehrlich asserted that the "molecule" of the cell protoplasm had certain side chains that were able to attach itself to the toxins of the bacteria through chemical interaction. In this way, the engaged side chains could not carry out their normal functions in the processing of food substances and oxygen consumption, causing the affected cell to produce more side chains. He further reasoned that excess side chains were discharged into the blood as complexes of antitoxins–bacterial toxins, which formed the basis of immunity. Ehrlich's observations led him to formulate a theory of antitoxin formation "theory of side chains," which he successfully published in 1897. In 1900, Ehrlich replaced the term "side chain" with the term "receptor." During the time, new research results, specifically those published by Langley, had thrown light on "receptive substances," which probably led Ehrlich to make further modifications to his theory and concept [366, 367, 415].

Langley and Ehrlich's works on reception provided the background for subsequent fruitful investigation in the field of chemoreception [416, 417]. Around 1926–1937, Alfred Joseph Clark (1885–1941) observed that the binding of a ligand to its receptor could be quantitatively measured. Furthermore, he noted that there was a relationship between the ligand binding and the observed biological effect [418–420]. His success in quantifying the amount of acetylcholine needed to bind its receptor with the subsequent physiological response provided further insights into the mechanisms of receptor–ligand interaction and functioning. Alfred Joseph Clark (1885–1941) was a pioneer advocator of the reception theory and similarly helped to establish the receptor theory of drug action [415, 421]. Other contributors to this theory were Hans Walter Kosterlitz (1903–1996) and Walther Straub (1874–1944). Kosterlitz was a pioneer investigator of narcotic analgesics and multiple opioid receptors that led him to the discovery of endogenous opioid peptides. He also studied the mechanisms of enteric reflexes, carbohydrate and protein metabolism in normal and disease states [422]. The German pharmacologist Walther Straub (1874–1944) propounded a receptor theory that was physical in nature. His theory, called "potential-poison" theory, was not influenced by the initial receptor theory of his predecessors, Ehrlich and Langley. Straub investigated the antagonistic effects of muscarine and atropine on the heart of the sea snail (*Aplysia*) and the torpedo fish. He proposed that the effect of the substance (or poison muscarine) on the cell or tissue was due to a deformation or other physical disturbance of the cell membrane when the substances penetrated it. He further opined that the disturbance of the cell membrane was made possible due to the differences in concentration gradient or potential gradient between the outside and the inside of the cell. Straub noted that substances such as pilocarpine, physostigmine, nicotine, adrenalin function on the basis of his physical receptor theory. Supporters of this physical theory included Henry Dale (1875–1968), George Barger (1878–1939) [370, 423].

Sir Henry Hallett Dale (1875–1968) was an English pharmacologist and physiologist who studied the physiology of acetylcholine. For his study of this mediator as agent in the chemical transmission of nerve impulses (neurotransmission), he shared the 1936 Nobel Prize in Physiology or Medicine with Otto Loewi (1873–1961) [424]. Henry Dale, Marthe Louise Vogt (1903–2003), Wilhelm Siegmund Feldberg (1900–1993), and Otto Loewi (1873–1961) carried out subsequent works on receptor and its ligand (acetylcholine as the transmitter) that further made it possible to have a better understanding of the functions of this neurotransmitter [366, 424–426]. Following the initial conception of the receptor theory, scientists who worked in other areas of interest utilized the new information to investigate how the functions of receptors could be measured. A pioneer in this regard was Alfred Joseph Clark (1885–1941) who, using the equation established by Archibald Vivian Hill (1886–1977) in 1909 and 1910, first quantified the activation of receptors (in terms of physiological responses) caused by drug action [427, 428]. Another twentieth-century scientist, Raymond Perry Ahlquist (1914–1983) in 1948, described the action of adrenaline on its receptors [429]. In spite of the several pieces of researches and applications (of Ehrlich's ideas in medicine), only a few researchers believed that receptors existed in the body. During the 1950–60s, even eminent pharmacologists and scientists in Europe and other parts of the world ruled out the receptor concept. A research breakthrough was made when Sir James Whyte Black (1924–2010) discovered type 2 (H_2) receptor of histamine. This study eliminated any doubt about the existence of receptors, which was further strengthened following the design of the first receptor-specific treatment (histamine-H_2-receptor blockers for the treatment of stomach ulcers and beta-receptor blockers for the treatment of hypertension) [370].

Another research that represented breakthrough in physiology of receptor (receptology) was the first recordings of the electrical signs of single receptor channel openings by Erwin Neher (1944–) and Bert Sakmann (1942–) (Neher and Sakmann 1976). By this time, no one was in doubt that receptors are very important in physiological functions. As part of their contributions to science, the 1991 Nobel Prize in Physiology or Medicine was awarded jointly to Erwin Neher and Bert Sakmann "for their discoveries concerning the function of single ion channels in cells" (Details on the history of ion channel discovery have been discussed in Chap. 5).

Conclusively, therefore, receptors are the key regulators of traffic in the cell. Dysfunctions of receptors could lead to unimaginable aftermaths that may require stringent care of the sufferer. For instance, a mutation in one gene controlling a particular receptor can lead to damage of the functions of the affected tissue, and even multisystem disorders such as the case in cystic fibrosis [430–432].

As concepts in receptor physiology became widely acknowledged, understanding of epithelial transport led to a better understanding of the mechanism of regulation of GI functions in normal and disease states. A pioneer in studies of ion transport across epithelial membranes was the Danish scientist and physician Hans Henriksen Ussing (1911–2000) [433]. Further details on the history of GI epithelial ion transport are discussed in Chap. 12.

The Reception Theory as the Basis for Developing the First Antisecretory Pharmacotherapy Drugs

The development of the receptor theory as well as advancement on secretory function of the stomach and the role of stomach acid in pathology were important for the discovery of the cellular and molecular mechanisms of gastric acid secretion. The quest for cellular and molecular mechanisms of gastric acid secretion culminated in the development of drugs that act on the molecular determinants of acid secretion in the stomach. Secretory regulation of acid secretion dates as far back as 1906 when Edkins discovered the hormone gastrin that was later shown to stimulate gastric acid secretion. Histamine was isolated first by Sir Henry Dale, Professor Heinz Otto Schild (1906–1984), Sir James Black, and Professor Jean-Charles Schwartz. Subsequently in 1916, the GI source and functions of the hormone were discovered by Leon Popielski as a secretagogue that stimulates gastric acid. (Secretagogue is a substance that stimulates the release of another substance.) From 1910 to 1930s, research on histamine grew significantly and antihistamines were introduced. The wide distribution of histamine in body tissues was important for the increased interest in investigating this substance. The scientists Ungar Georges, Jean-Louis Parrot, and Daniel Bovet (1907–1992) who discovered the first antihistamine agent (an adrenolytic benzodioxan, piperoxan) were interested in finding an antagonistic agent that would be comparable to that exhibited by sympatholytic compounds toward epinephrine and by parasympatholytic compounds toward acetylcholine (Table 9.2) [434–436]. The 1957 Nobel Prize for Physiology or Medicine was awarded to Bovet in recognition of his work on antihistamines and curare. In 1942, the first antihistamine phenbenzamine was used in man. Subsequently mepyramine, pyrilamine, was developed and replaced phenbenzamine. Several antihistamines were thereafter developed, which were widely used to treat allergic disorders such as hay fever and allergic rhinitis [434].

Coupled with previously identified hydrochloric acid abundance in the stomach, the search for how acid secretion came about in the stomach was particularly heightened. Some scientists had long thought that the theory of chemoreception could provide insight into the acid secretory activity of the stomach. In the early 1900s, it was widely believed that excessive acid production in the stomach was responsible for some GI illnesses such as ulcers. But the mechanism for this remained unknown. A breakthrough was made in the early 1970s by the Scottish pharmacologist, Sir James Whyte Black (1924–2010). Black had a huge interest in physiological science and action of pharmacological agents on adrenergic and histamine receptors that were previously discovered and suspected to play a role in gastric acid secretion. His interest in physiology led him to establish the physiology department at the University of Glasgow, Scotland, UK. In the 1950s, Black had thought of formulating pharmacological agents that could modulate cell receptors to address human illnesses. After working to discover the beta-adrenergic receptor blocker (usually shortened as beta-blocker), propranolol in 1960, Black diverted research toward the histamine receptor. His research program on the histamine receptor launched in 1972 was similar to the design he had used for beta-adrenergic receptor. Antihistamines were widely known at this time as they were used to treat allergic reactions. The first histamine H_1 receptor and its antagonists (antihistamines) were already discovered back in the 1940s. Experimental investigations had indicated that there were possibly two types of histamine receptors: one that responds to antihistamines, the other did not. Black aimed at discovering the histamine that did not respond to antihistamines, but finding an antagonistic agent was a challenge. Interestingly, Black was interested whether or not the second agent was involved in gastric acid secretion. A bit later, the British–American chemist Graham J. Durant (1934–), British-born medicinal chemist, and Emeritus Smith Kline and French Professor of Medicinal Chemistry, Charon Robin Ganellin (1934–) and John Colin Emmett (1939–) and the pharmacologist Mike Parsons joined Black on the histamine project. The group tried to manipulate the chemical groups in histamine molecule in search for the antagonist that could affect the second proposed receptor. Prior to this time, there was no suitable chemical substance to define histamine H_2 receptors, not until the introduction of the compound burimamide (histamine antagonist) same year in 1972 [17, 437–442]. Further investigation led to the production of the drug metiamide, which was withdrawn during early clinical trials as it caused agranulocytosis, a life-threatening blood disorder. Thereafter, the researchers started substituting chemical groups with other groups in search for suitable antagonist. Subsequently, they discovered molecules that were antagonists and agonists. The agonist, 4-methylhistamine, stimulated acid secretion, but did not produce any of the known histamine responses. In 1973, Black and his team discovered the second receptor through the substitution of the thiourea group with a cyanoguanidine moiety. Almost the same time, groundbreaking research conducted by Allen L. Ganser (1942–) of Johann Wolfgang Goethe University Frankfurt, and John Gaetano Forte (1934–2012) of the University of California, Berkeley, in 1973 discovered proton pumps [443–446].

The first histamine H_2-receptor antagonist that revolutionized the treatment of peptic ulcer disease around the world—cimetidine was launched in 1977. The third receptor H_3 receptor was discovered by the French scientist Professor Jean-Charles Schwartz in 1983, and the antagonists of this receptor became widely available before 1990. The new H_3 antagonist non-imidazole has shown considerable promise for use in addressing memory, learning deficits, and epilepsy. Another H_3 antagonist pitolisant showed effectiveness for treating narcolepsy (a rare disabling sleep disorder characterized by excessive daytime sleepiness) and cataplexy (sudden loss of muscle tone) in adolescents who were refractory to traditionally available drugs such as modafinil, methylphenidate, mazindol, sodium oxybate, and D-amphetamine. The fourth H_4 receptor was discovered in 1999 and may play a role in inflammatory diseases including asthma [445, 446]. The 1988 Nobel Prize in Physiology or Medicine was awarded jointly to Sir James W. Black, Gertrude B. Elion (1918–1999), and George H. Hitchings (1905–1998) "for their discoveries of important principles for drug treatment" [447] (Fig. 11.17).

Further development of PPIs was made by Sachs and colleagues to address excessive production of gastric acid that characterizes peptic ulcer disease and GERD [463]. George Sachs and other researchers investigated proton pump inhibitors (PPI), which provided a good alternative, to gastric acid inhibition (H_2-receptor antagonists) for ulcer therapy. Sachs showed that acid secretion by the stomach is primarily due to an electroneutral ATP-dependent pump (H^+/K^+-ATPase or proton pump or potassium-activated ATPase or gastric hydrogen potassium ATPase) of the parietal cells [464–466]. In the early 1970s of the last century,

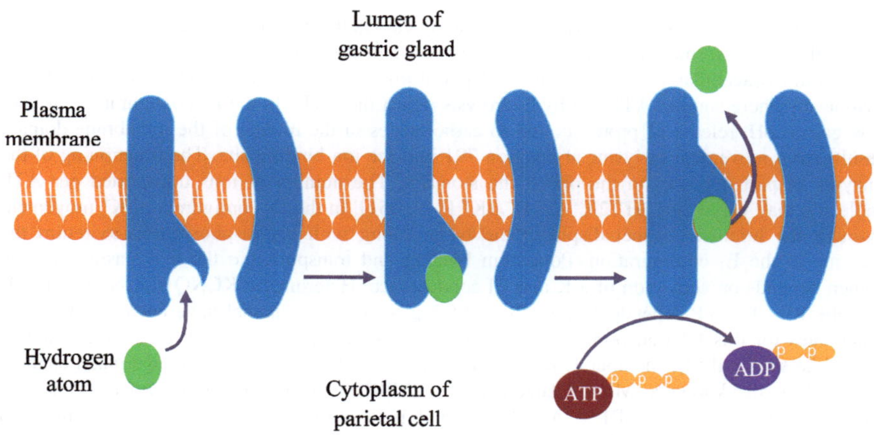

Fig. 11.17 Proton pump. This pump is an enzyme whose activities represent a key mechanism of secretion of the stomach acid—hydrochloric acid. The proton pump is responsible for the stomach acid secretion [443, 444, 448]. The H^+/K^+-ATPase of parietal cells of the inner cell lining of the stomach possesses an extensive secretory membrane system and is the major protein constituent of the gastric epithelium. It is a member of the eukaryotic class of P_2-type ATPase. Other members of this class include Ca^{2+}/ATPases and the Na^+/K^+ ATPases. Proton pump is the product of two genes. The protein is made up of two subunits: α consisting of 1000 amino acids and β which is made up of 300 amino acids. Hence, it is called a heterodimeric protein. The α-subunit is encoded by the gene ATP4A. The β-subunit is encoded by the gene ATP4B. The α-subunit contains the catalytic sites of the enzyme, forms the pore through the cell membrane, and allows the transport of ions. The catalytic α-subunit has ten transmembrane segments with a cluster of intramembrane carboxylic amino acids located in the middle of the transmembrane segments TM4, TM5, TM6, and TM8 and serving as the ion-binding domain [449]. Hydronium ions bind to two active sites present in the α-subunit. The α-subunit also has a phosphorylation site at amino acid Asp (aspartic acid) at position 385. The β-subunit has one transmembrane segment and forms an N-terminal cytoplasmic domain, a single transmembrane domain, and an extracellular domain containing 6–7 N-linked glycosylation sites. *N*-glycosylation is important for the enzyme assembly, maturation, and sorting. The β-subunit is responsible for stabilizing the pump by preventing it from functioning in the reverse direction. The release of the hormone gastrin causes enterochromaffin cells to release histamine, which directly binds to H_2 receptors on the parietal cell. This binding activates a cAMP-dependent pathway to stimulate KCl pathway which causes the enzyme to move from the cytoplasmic tubular membranes to deeply folded canaliculi of the stimulated parietal cell and secretes gastric acid by an electroneutral ATP-dependent hydrogen–potassium exchange. The proton pump exchanges potassium from the intestinal lumen with cytoplasmic hydrogen ion and is the enzyme primarily responsible for the acidification of the stomach contents and the activation of the digestive enzyme pepsin. The proton pump has two ATP-binding sites, both involved in enzyme activity, whose affinities vary as a function of the H^+ and K^+ concentrations [450]. To transport ions (potassium from the extracellular space and hydrogen ions from the cytoplasm) across the membrane, the enzyme alternates between two conformations, E_1 and E_2. The enzyme configuration depends on the binding of MgATP and subsequent phosphorylation. Similar to other P-type ATPases, a phosphate group is transferred from ATP to the proton pump. This phosphate transfer powers a conformational change in the enzyme that helps drive ion transport against a concentration gradient. In the E_1 conformation, a phosphate from ATP and hydrogen ion on the

cytoplasmic side binds to the enzyme. This binding causes the enzyme to change configuration to E_2. In the E_2 conformation, hydrogen ion is released in the lumen. The E_2 conformation binds potassium extracellularly and reverts to the E_1 conformation to release phosphate and K^+ into the cytoplasm where another ATP can by hydrolysis repeat the cycle. In the E_2 configuration, at very low gastric pH, release of proton bound to carboxylates in the middle of the membrane domain results from movement of lysine at position 791 into the ion-binding site. The functioning of the enzyme in part depends on the pH of the stomach. The number of ions transported per ATP molecule may vary from $2H^+/2K^+$ to $1H^+/1K^+$ [449, 451]. In the E_2 configuration, K^+ movement through the luminal channel displaces the lysine along with dephosphorylation to return the enzyme to the E_1 configuration. Potassium binding and transport into the cytoplasm from the lumen depends on activation of a K and Cl conductance. Heteromeric KCNQ1/KCNE2 channels and the Kir4.1 (ATP-dependent inwardly rectifying potassium) channel have been postulated as candidates for this K^+ transport. KCNQ1 is the gene symbol, voltage-gated potassium channel, KQT-like subfamily Q, member 1. KCNE2 is member 2 of the potassium voltage-gated channel subfamily E also known as MinK-related peptide 1 (MiRP1)—a protein that in humans is encoded by the *KCNE2* gene. MiRP1 is a voltage-gated potassium channel accessory subunit (beta subunit). Cl^- efflux across the luminal membrane is also necessary for gastric acid secretion. Cystic fibrosis transmembrane conductance regulator (CFTR), chloride intracellular channel protein 6 (CLIC-6), and solute carrier family 26 (anion exchanger), member 9 (SLC26A9) are candidates for Cl^- transport into the lumen. K^+ recycling across the luminal membrane is necessary to sustain the proton pump activity in gastric parietal cells. The K^+-Cl^- cotransporter type 4 (KCC4) is expressed in gastric parietal cells and is involved in gastric HCl secretion. It is located in the gastric parietal cells more abundantly at the luminal region of the gland than at the basal region. In addition to H^+/K^+-ATPase, KCC4 represents a new member for basal HCl secretion in the apical canalicular membrane of the resting parietal cell and contributes to increased acid secretion upon stimulation [452–455]. Apart from mechanisms involving gastrin, gastric acid secretion is regulated by neural (acetylcholine) and paracrine (histamine and somatostatin) mechanisms. Acetylcholine activates the M_3 subtype on the parietal cell. Like gastrin, stimulatory effect of acetylcholine is mediated via increase in cytosolic calcium. The effect of histamine is mediated via activation of adenylate cyclase and generation of cAMP. Somatostatin is the main inhibitor of gastric acid secretion. This inhibition is mediated via guanine nucleotide-binding proteins to inhibition of adenylate cyclase activity [456]. All these pathways modulate proton pump activity of the gastric parietal cell. Other studies have identified novel players in gastric acid secretion. Geibel et al. [457] identified another pathway-modulating gastric acid secretion through the stomach calcium-sensing receptor located on the basolateral membrane of gastric parietal cells. Activation of this receptor by Ca^{2+}, Mg^{2+}, or gadolinium ion (Gd^{3+}) leads to increase in the rate of acid secretion through the apical H^+/K^+-ATPase. The K^+ ionophores such as valinomycin, trinactin, nigericin, lasalocid, nystatin, salinomycin, and gramicidin A are experimentally used to stimulate the proton pump of oxyntic cell. Ionophores are lipophilic (lipid-soluble) complexing molecules that transport ions across a cell membrane. Thus, they can function as ion carriers in the living cell (plasma membrane) and synthetic vesicles (liposomes). Ionophores can reversibly bind ions. This property of ionophore is used by some microorganisms which produce the ionophore valinomycin (a dipeptide antibiotic which selectively translocate potassium across hydrophobic membrane of thickness $\sim$40 Å; a K^+ carrier) or channel former ionophore (gramicidin A) to increase the permeability of the membrane [458–461]. Nigericin exchanges K^+ for H^+. The ionophore calcimycin (A23187) is a calcium carrier. The activities of these ionophores can necessarily impair ion fluxes in the cell which results in disorders of energy transduction process in the cell [462] timoprazole, a substituted benzimidazole (called pyridylmethylsulfinyl benzimidazole), was discovered to have a substantial antisecretory activity [467, 468]. The antisecretory function of this agent did not depend on the stimulus type [449]. Unfortunately, however, timoprazole administration in laboratory animals resulted in the development of toxicity against the thyroid gland and thymus [469]. Another disadvantage about this agent was that it functioned only when the proton pump was transporting acid into the lumen, but also displayed a lag phase before it could inhibit the pump [465, 466, 469]. Around the turn of the 1970s of the last century, another benzimidazole derivative that did not cause thyroid dysfunction and at the same time had high antisecretory function was identified. The compound was named picoprazole—discovered in 1977 [467]. In strive for increased effectiveness of picoprazole against acid secretion, chemical modification of the pyridine ring of this agent led to the discovery of another agent which had greater time for accumulation in the parietal cell cytoplasm. This new compound was called omeprazole. Omeprazole became the first PPIs to be introduced into clinical practice in different parts of the world around the late 1980s for treatment of duodenal ulcer, gastric ulcer, reflux esophagitis, and Zollinger–Ellison syndrome [449, 467, 470]. Following ingestion of the drug, it is absorbed in the intestine, subsequently reaches the parietal cell via the bloodstream, and diffuses through the cytoplasm to affect the functions of the gastric proton pump. Omeprazole, upon activation in the parietal cytosol, is trapped as a sulfonamide in the acidic canaliculus, where it covalently binds to the proton pump to irreversibly block acid secretion [471].

The PPIs effectively inhibited proton pump and was more effective than the previously developed H_2-receptor antagonists. At a daily dose of 30 mg, lansoprazole produced faster relief of symptoms and superior healing rates in patients with gastric or duodenal ulcers or reflux esophagitis compared with H_2-receptor antagonists. Other members of this drug class are highly effective in addressing several gut disorders due to acid secretion and include pantoprazole, rabeprazole, and esomeprazole. The substituted benzimidazole covalently blocks the proton pump [449, 470].

In spite of the advancement in gastrology, gastritis and peptic ulcer erosive reflux esophagitis, active gastric ulcer, active duodenal ulcer, and the treatment of nonsteroidal anti-inflammatory drug (NSAID)-induced gastric and duodenal remained undefeated with pharmacotherapy [464]. Furthermore, continuous use of PPIs is associated with gastric anacidity, gastrin cell hyperplasia, hypergastrinemia, and a possible development of carcinoid tumors [471]. In 1983, the discovery of the spiral bacteria Campylobacter (*Helicobacter*) *pylori* by B. J. Marshall and R. J. Warren was globally considered a major breakthrough in pathophysiology of peptic ulcer disease and gastritis. The eradication of the *H. pylori* in the stomach was associated with improved gastroduodenal secretion and integrity [464]. A triple therapy with lansoprazole, clarithromycin, and amoxicillin or dual therapy with lansoprazole and amoxicillin were used to treat peptic ulcer diseases and related conditions that involved *H. pylori* infection [472].

Continuous search for more effective inhibition of stomach acid led to the development of potassium-competitive acid blockers also known as the prazans.

This group of drugs reversibly binds to proton pump and is fast-acting inhibitor of gastric proton pump; it is superior to the proton pump inhibitors [473]. A representative of this group is revaprazan. Revaprazan is a drug that reduces gastric acid secretion and is used for the treatment of gastritis and GERD. It acts as an acid pump antagonist (i.e., potassium-competitive acid blocker). Revaprazan is approved for use in Korea, but is yet to be approved in Europe or the USA [474].

11.4.4 Intestinal Secretions

Secretions of the intestine can be divided into small and large intestinal secretions. In small intestine, different regions have their peculiar secretory functions. Secretions of the small intestine include discharges from the epithelium of the intestines containing special glands (duodenal and ileal), bile, and pancreatic secretions. The large intestine secretes mainly mucus, various ions, enzymes, and other biologically active molecules [7, 475–477].

Duodenal Secretions

In the duodenum, the secretions are produced primarily by Brunner's glands (Fig. 11.18). The glands are named after the Swiss physician and physiologist, Johann Conrad Brunner (1653–1727), who is credited for accurately describing them. Brunner, who became the first to resect a pancreas, was born in Diessenhofen, a municipality in Frauenfeld District in the canton of Thurgau in Switzerland. At the age of 34 years, he became a professor of physiology and anatomy at the University of Heidelberg, Germany [478, 479].

The glands secrete approximately 1–2 L of juice per day and maintain a pH of about 7.4–7.8 units in the lumen of the upper small intestine. The maintenance of pH is achieved by their secretion of fluid rich in bicarbonate ions and mucus, thereby protecting the duodenal epithelium from the acidic chyme. Brunner's glands also secrete urogastrone (human epidermal growth factor), which inhibits acid secretion by the gastric parietal and chief cells. Urogastrone has been localized in other parts of the GI tract such as in acinar cells in the human submandibular gland [480–482].

Ileal Secretions

In small intestine, secretions are largely produced by the crypt of Lieberkühn named after the German physician Johann Nathanael Lieberkühn (1711–1756). Cells of the crypt include absorptive enterocytes, which express digestive enzymes that break down food nutrients and also mediate their absorption and transport through the epithelium into the blood or lymph; mucus-secreting goblet cells;

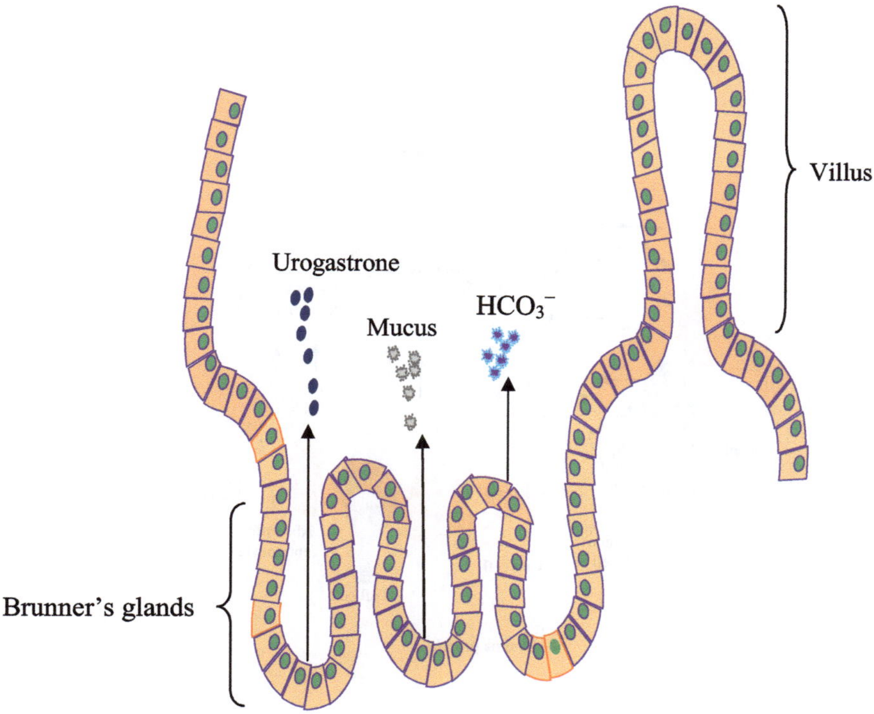

Fig. 11.18 Schematic diagram of Brunner's glands in the duodenum located close a villus. The gland secretes mainly urogastrone, bicarbonate ions, mucus that are responsible for maintaining the integrity of the gut

hormone-secreting enteroendocrine cells (1% of all epithelial cells); antimicrobial Paneth cells of the innate immune system—also regulate the gut microbiota and produce growth factors; wine glass-shaped cup cells (6% of the ileal epithelial cells); microfold or membranous (M) cells—cover the surface of the gut-associated lymphoid follicles and function as an interface between the luminal content and the underlying immune cells; and pear-shaped, barrel-shaped, or goblet-shaped tuft cells or brush cells (constitute less than 1% of ileal cells and are also found in the salivary duct system) (Fig. 11.19) [73, 340, 483–487].

Secretions by the intestinal glands (i.e., intestinal juice) contain hormones, digestive enzymes, mucus, and other substances that participate in the regulation of the functions of the gut. In addition, a large amount of fluid (water) is constantly secreted and circulated throughout the GI tract (GI fluid fluxes have been discussed above). These secretions are controlled by numerous factors including neural input from the central and enteral nervous systems, availability and types of nutrients, as well as self-secreted chemical components of the GI fluid [340, 485–487].

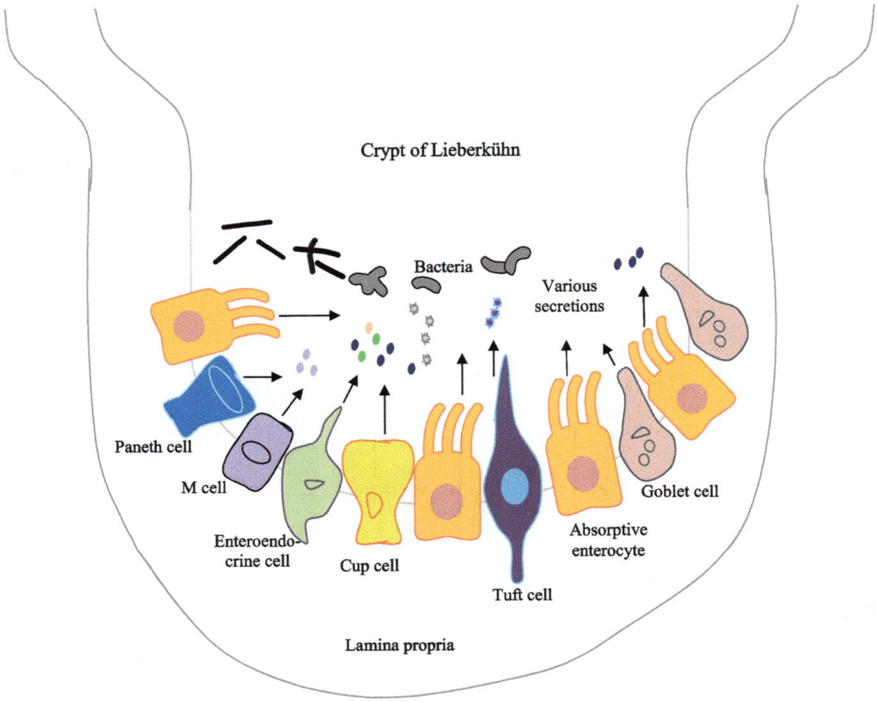

Fig. 11.19 Cell types of the crypt of Lieberkühn. The cells shown here are the currently known types present in the crypt. Some of their functions are not precisely known, but ongoing researches by many laboratories in the world are aimed at proving more information to understanding the functional roles and specifications of these cells. On the average, the life span of cells of the intestine is about 4–7 days. However, the life span of Paneth cells is about 9–15 times [73, 483, 484]. The average life span of intestinal cells is the approximate proliferative time and renewal of these cells. Proliferative cells reside in the epithelial invagination. New cells are formed in the crypts of Lieberkühn which then migrate upward and upon reaching the tip of the villus undergo apoptosis. Apoptotic cells are shed off into the intestinal lumen [488]

Pancreatic Secretions

Pancreatic secretion can be divided into exocrine and endocrine secretions. The endocrine secretions have been discussed in Chap. 8. Here, only the exocrine secretions will be discussed (Fig. 11.20). Over 80–90% of the volume of the pancreas is made of exocrine cells. It should be noted, however, that the activity of exocrine pancreas is influenced by the secretions of the endocrine pancreatic cells [489].

The exocrine pancreas comprises ducts that are arranged in clusters called acini (singular acinus) (Fig. 11.21). The exocrine pancreas secretes fluid that contains digestive enzymes and other substances that pass into the small intestine to digest ingested food. First, pancreatic secretions are secreted into the lumen of the acinus

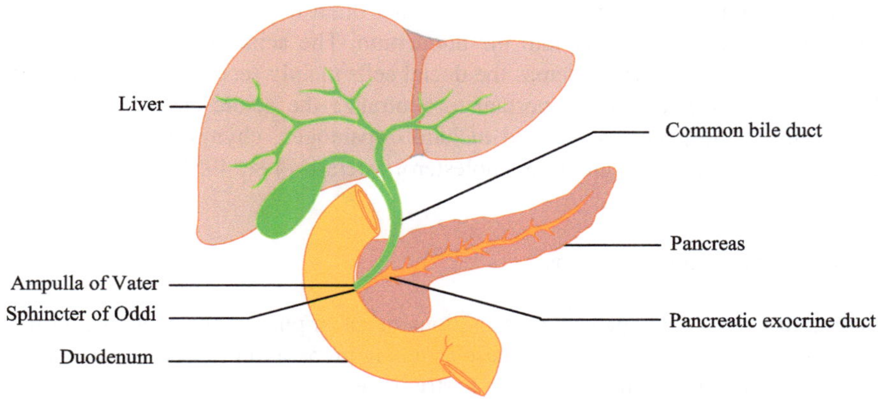

Fig. 11.20 Biliary tree showing major ducts leading to the ampulla of Vater through which secretion of both gallbladder and pancreas is emptied into the duodenum. The ampulla of Vater is the opening into the duodenum formed from the merging or confluence of main pancreatic duct and the common bile duct. The main pancreatic duct is formed from several intralobular ducts. The common bile duct is formed from the left and right hepatic bile ducts, which are the extensions of bile canaliculi. In tumors affecting the head of the pancreas, there is compression of the common bile duct leading a blockage in bile flow, which leads to hyperbilirubinemia, characterized by jaundice. The disease is also characterized by malabsorption and malnutrition. Consequently, it leads to a significant weight loss [490–493]

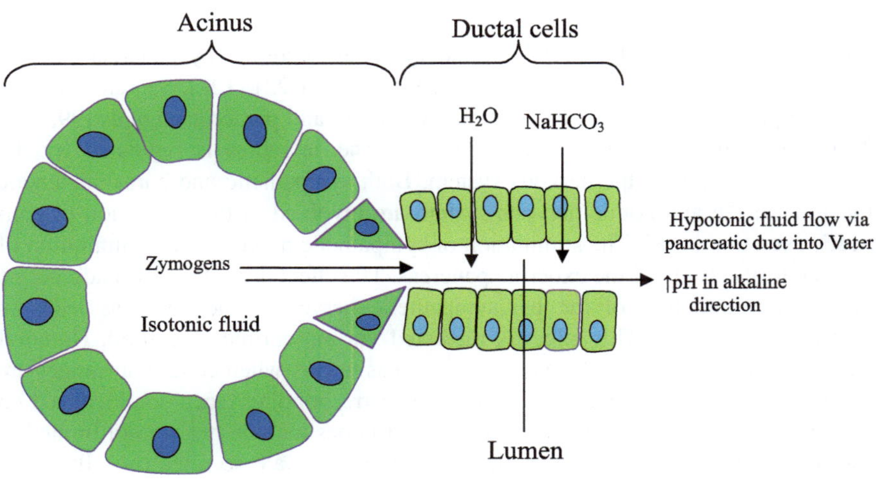

Fig. 11.21 Secretory unit of the exocrine pancreas and simplified schematic representation of pancreatic fluid flow. Acinar cells are the main secretory structures of zymogens, whereas the ductal cells secrete mainly bicarbonate ions. Both cells, however, secrete fluids, ions, and other biomolecules (see explanation in text)

and then accumulated in intralobular ducts that drain to the main pancreatic duct, which in turn drains directly into the duodenum. The acinar cells secrete mainly zymogens or proenzymes, whereas the ductal cells mainly secrete bicarbonate ions. These bicarbonate ions are important to neutralize the acidic chyme. The major zymogens of the pancreas secreted include trypsinogen, chymotrypsinogen, pancreatic lipase, pancreatic amylase, cholesterol esterase [494, 495].

Phases of Pancreatic Secretions

Like gastric secretions, pancreatic secretions occur in phases: the cephalic, gastric, and intestinal [496]. The cephalic phase occurs in preparation for food intake, and it is mediated via the vagus nerve. The gastric phase occurs immediately when food gets to stomach and is delivered to the duodenum. The phase is characterized by the action of hydrochloric acid and pepsin, as well as the effect exerted by nutrient flow into the duodenum. The intestinal phase takes place when food gets to the intestine (duodenum proper), and it is characterized by the secretion and action of CCK and secretin [496]. The intestinal phase is controlled by vago-vagal enteropancreatic reflexes, in which afferent neurons originating in the duodenal mucosa, and efferents mediating central input on the pancreatic ganglia, activate intrapancreatic postganglionic neurons [497].

Mechanism of Pancreatic Secretions

The mechanism of fluid, electrolyte, and zymogen secretion by pancreatic acinar cells is similar to that of salivary acinar cells (Fig. 11.22) [122]. Secretion by the exocrine pancreas is triggered by neural, hormonal, and paracrine signals [99, 489, 497]. The neural control of pancreatic exocrine secretion is regulated by the autonomic division of the nervous system. Both sympathetic and parasympathetic divisions of the autonomic nervous system are involved in the regulation of pancreatic secretion [498]. The stimulation of sympathetic nerves leads to inhibition of the secretory activity of the exocrine pancreas. The effect is mediated via adrenergic receptors. Stimulation of the parasympathetic nerves of the pancreas leads to increase in secretion (This effect is mediated via muscarinic receptors). Humoral control of the functions of the exocrine pancreas occurs when certain hormones are released via the bloodstream to activate their corresponding receptors, localized on the acinar and ductal cell membranes. The hormones include but are not limited to gastrin, cholecystokinin, and secretin. These hormones are secreted by cells of the stomach and duodenum in response to food and distension of the gut wall [496]. Paracrine factors released by the neighboring cells of the exocrine pancreas are also believed to influence the secretory functions of the pancreas. The neural, hormonal, and paracrine factors exert their activity through the corresponding receptors on the cells of the exocrine pancreas. Many functional receptors are expressed on the plasma membrane of pancreatic acinar and ductal cells. Activation of a subset of

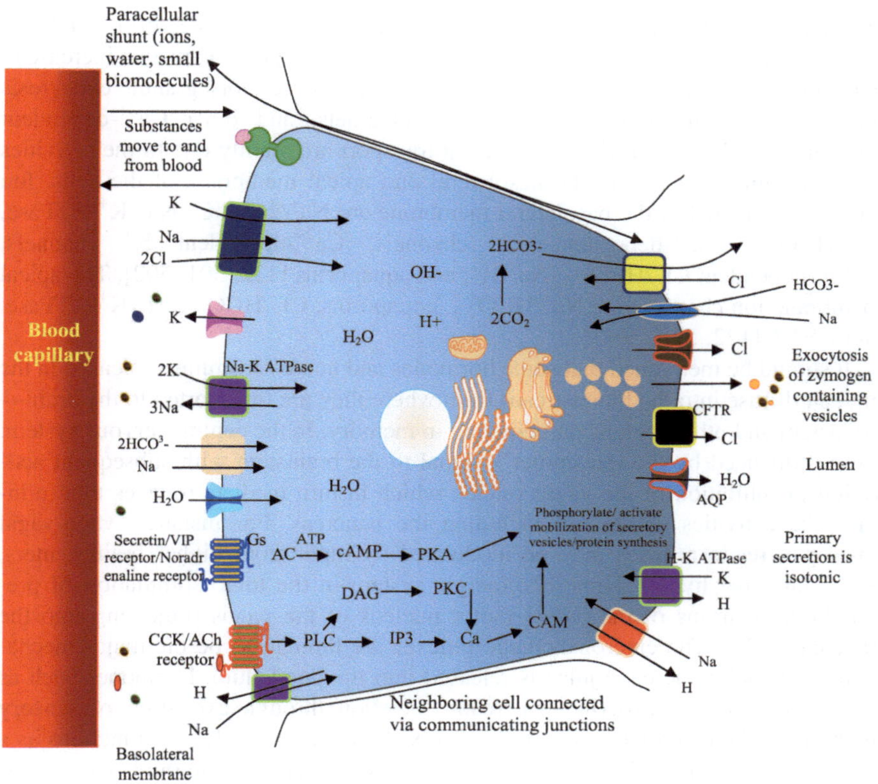

Fig. 11.22 Production and secretion of pancreatic components by acinar cells. The exocrine pancreatic cell expresses several receptors that control the activities of the cell. Activation of some of its receptors is related to secretion of fluid, ions, and zymogens (see further explanation in text)

these receptors leads to downstream signaling that culminates in increased intracellular calcium level, increased activity of certain cytosolic enzymes—which mediate the secretory process [499]. The signaling pathways not only lead to secretion of fluid, electrolytes, and enzymes but also *de novo* synthesis of biomolecules [500].

For instance, cholecystokinin (CCK) and secretin have been shown to substantially affect pancreatic secretion via CCK, secretin receptors. Similarly, VIP and noradrenaline also have substantial effect on pancreatic secretion through stimulation of VIP receptor and noradrenaline receptor, respectively. Acetylcholine released from vagal fibers can stimulate pancreatic secretion via activation of muscarinic receptors. The secretin/VIP receptor/noradrenaline receptor signaling is coupled to the stimulatory unit of the G protein. Upon activation, the stimulatory subunit dissociates to activate a membrane enzyme that relay the signal downstream, which leads to the increased production of cAMP with subsequent activation of PKA. The enzyme phosphorylates several protein targets including ion channels [122]. Other molecules such as CCK (via CCK-A receptors), ATP (via P2Y and P2X receptors), acetylcholine (via M1, M3 receptors) evoke an increase in cytosolic Ca^{2+} level. This rise in Ca^{2+} level mobilizes secretory granule exocytosis and activates intracellular enzymes, ion channels, and other Ca^{2+}-dependent molecules [122]. Pancreatic fluid and ion transport are mainly due to the activities of ion channels located in the basolateral and apical membrane of the cells. Ion channels activated in the basolateral membrane are $Na^+/2Cl^-/K^+$, Na^+/K^+-ATPase, $Na^+/HCO^-{}_3$, Ca^{2+}-dependent Cl^- channels, Ca^{2+}-dependent K^+ channels, cAMP-dependent $Cl^-/HCO^-{}_3$, Na^+/H^+, and aquaporins [122, 501, 502]. The apical membrane ion channels are $Na^+/HCO^-{}_3$, aquaporins, $Cl^-/HCO^-{}_3$, H^+/K^+-ATPase, and CFTR [122, 503, 504].

It should be mentioned that some hormones and neurotransmitters released in the gut can diffuse into the bloodstream from where they are transported to the brain—brainstem and other regions participating in memory. In the central nervous system, areas influenced by the transmitter's signal to the brainstem with subsequent activation or inhibition of the vagus nuclei, which in turn mediate reflexes that influence the activities of the gut including the pancreas. For instance, vago-vagal enteropancreatic reflexes may be modulated by input from higher brain centers, particularly the hypothalamic cholinergic system in the tonic stimulation of preganglionic neurons of the dorsal motor nucleus of the vagus projecting into the pancreas [497]. The enteropancreatic reflexes are thought to be the major mechanism by which pancreatic juice is released into the duodenum. Hormones such as CCK can stimulate exocrine pancreatic secretion through excitation of sensory afferents of the enteropancreatic reflexes [497]. However, CCK also functions as a paracrine factor [497]. The acinar cells secrete NaCl-rich isotonic fluid. The fluid is transported to the duct, where it is modified by volume and electrolyte composition —Cl^- is absorbed, whereas $HCO^-{}_3$ is secreted. The bulk of the fluid in the pancreatic juice is secreted by the ductal cells [56].

Gallbladder Secretions (Bile)

Bile is a greenish-yellow fluid secreted by liver cells (hepatocytes) and modified by the bile ductular system, stored in the gallbladder, and it is required for the digestion of fats, excretion of xenobiotics, and also display antimicrobial functions in the ileum (Figs. 11.23 and 11.24). Bile is composed of bile acids (about 50%), phospholipids (about 40%; and a level of 25–810 mg/100 mL), water, ions (Na^+, K^+, Ca^{2+}, $HCO^-{}_3$, Cl^-), and organic molecules. The organic molecules include proteins/peptides such as glutathione; bile acids; cholesterol (4%, and a level of 60–320 mg/100 mL); and bilirubin (2%). Bile acids comprise about 80% of the organic compounds. The concentration of the cations is slightly higher than the level in plasma, whereas for anions (Cl^-) the concentration is lower compared to plasma. The color of bile is mainly due to the presence of biliary pigments with concentration of about 50–200 mg/100 mL. In addition, bile may contain variable quantity of other molecules such as glucose [505–511].

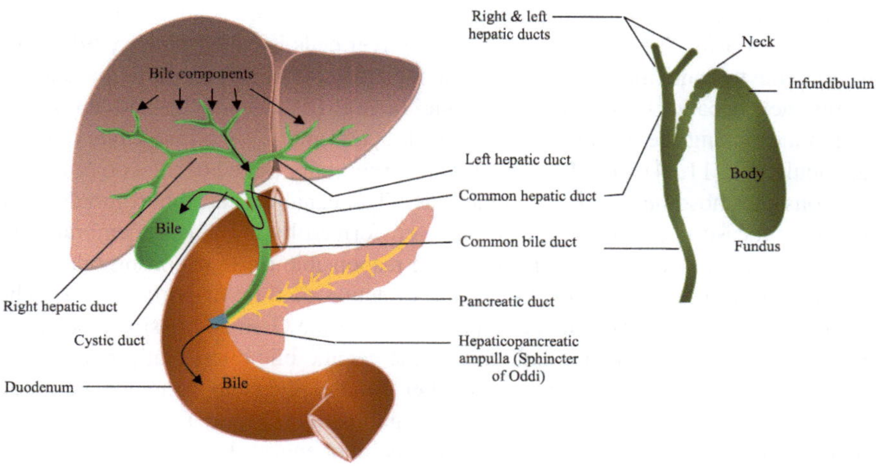

Fig. 11.23 Schematic diagram of the biliary tree (and associated structures), showing storage of bile in the gallbladder. After production by the cells of the liver, bile travels down into the hepatic bile ducts which merge with the cystic duct from the gallbladder to form the common bile duct (choledochus). The gallbladder stores bile until needed by the body for digestion of fats (The gallbladder is divided into fundus, body, and neck. It comprises the following layers of tissues—mucosa (epithelium and lamina propria) smooth muscle, perimuscular subserosa, and serosa). The primary function of the gallbladder is to store and concentrate bile for later secretion during digestion [512–514]. The common bile duct courses through the head of the pancreas and joins the pancreatic duct to form the ampulla of Vater which then empties into the duodenum. The flow of bile is indicated by the green arrows. The rate of bile flow is discussed in previous subsection of this chapter. However, for emphasis purpose, the average bile flow rate is about 620 mL per day. Of this, about 35% is due to the osmotic force exerted by bile acid secretion—often referred to as bile acid-dependent canalicular fraction. The fraction that is independent on bile acid secretion is estimated to be 35–40%—often referred to as bile acid-independent canalicular fraction. The fraction of bile due to secretions of the ducts is estimated at approximately 30% [508, 511]

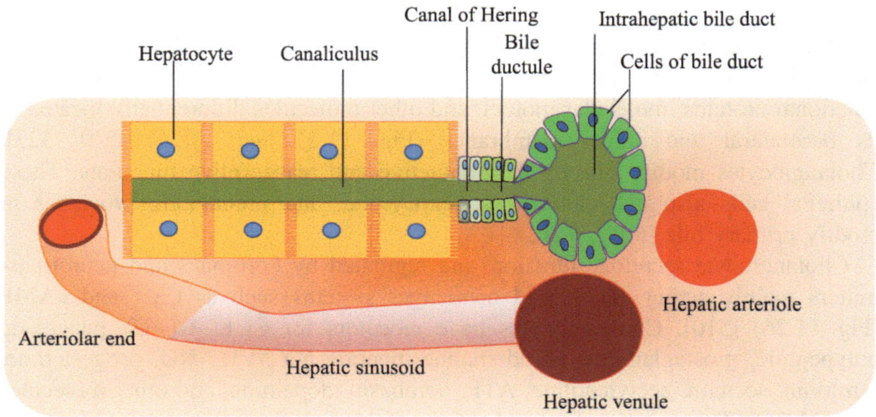

Fig. 11.24 Schematic diagram showing secretory unit of hepatobiliary system

Bile is produced by the hepatocytes which are the most abundant liver cell population (65%). The bile secreted by these liver cells is called primary bile. Like the enterocytes and many other cells of the body, hepatocytes have basolateral membrane (hepatic sinusoidal membrane facing Disse's space) and an apical membrane (facing the canaliculi). The hepatocytes empty the primary bile into the canaliculi (Fig. 11.24). (The hepatocytes are structurally linked to each other by gap junctions, desmosomes, and tight junctions.) The canaliculi are blind tubular tiny structures measuring about 1 μm in diameter, formed by the apical membranes of two adjacent hepatocytes. The canaliculi are responsible for transporting bile in the countercurrent direction (in respect to the direction of flow of portal blood). The canaliculi connect with Hering's canal (named after the German physiologist Ewald Hering, 1834–1918), which is linked with the hepatic bile ducts that progressively increase in diameter combining with other smaller ducts to form the common hepatic bile duct, which empties into the gallbladder. Upon stimulation by an adequate stimulus (such as entry of fat into the duodenum), bile is released via the common bile duct into the papilla of Vater and then finally into the duodenum (Fig. 11.23) [508, 511, 515–517].

The duct of the canaliculi is lined with cholangiocytes, which are formed from hepatoblasts [518]. Cholangiocytes and hepatocytes are the two types of epithelial cells of the liver. The cholangiocytes are the epithelial cells that line the intra- and extrahepatic ducts of the biliary tree, and they constitute about 5% of the liver cells (The ductular system responsible for the transport of bile from the liver to the duodenum is referred to as the biliary tree of the liver) [508, 512, 517, 519, 520]. The epithelial cells in small interlobular bile ducts are initially cuboidal in structure; however, they progressively become columnar and goblet in larger bile ducts including the extrahepatic bile ducts [521, 522]. Cholangiocytes have apical (facing the lumen) and basolateral (facing the interstitium) membranes [517, 518].

Mechanism of Gallbladder Secretions

Hepatocyte secretion of primary bile and its subsequent modification by the cholangiocytes is controlled by channels, transporters, exchangers, receptors, junctional proteins, molecular motors, and other molecules differentially located in its basolateral and apical membranes (Figs. 11.25 and 11.26) [520, 523]. Cholangiocytes modify bile by their secretive and reabsorptive functions. These epithelial cells also use endocytotic, exocytotic, and pinocytotic processes to modify primary bile [508, 512].

Cholangiocyte secretory functions are regulated by hormones and neurotransmitters mainly acting via second messenger systems such as Ca^{2+} and cAMP (Fig. 11.26) [510]. Cholangiocytes have receptors for CCK, CGRP, pancreatic polypeptide, prostaglandins, noradrenaline, peptide YY, VIP, NO, progesterone, serotonin, secretin, acetylcholine, ATP, bombesin, dopamine, and other molecules (Fig. 11.26). These molecules mediate the fluid and electrolyte secretion in the cholangiocyte [508, 519, 521, 522, 525, 535]. The hormone, secretin, for instance,

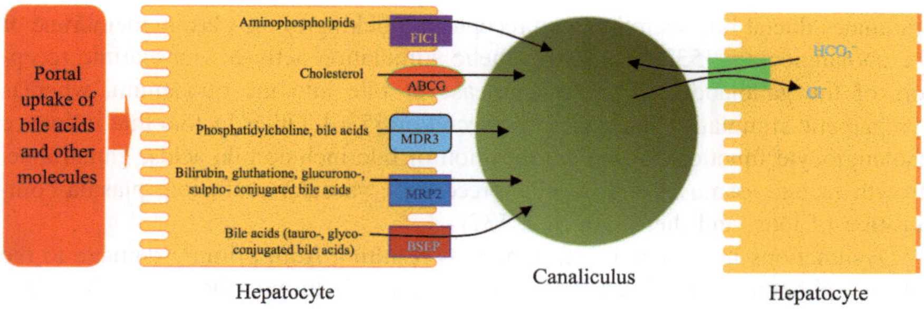

Fig. 11.25 Mechanism of transport of bile components into the canaliculus. Components used in the formation of bile are either imported into the hepatocytes from portal blood via the basolateral membrane of these hepatic epithelial cells or are synthesized in the cytoplasm of the cells. Basolateral hepatic transporters of bile components include Na^+ taurocholate co-transporting peptide and organic anion transport polypeptides [524]. Bile components are exported to the canaliculus via ATP-dependent bile salt export or excretory protein (BSEP); phosphatidylcholine is exported to the canaliculus via MDR3 (multi drug resistant protein or receptor type 3); cholesterol is exported to the canaliculus via the ABC (adenosine triphosphate-binding cassette)-G-5 and -8 transporters. Bilirubin, glutathione, glucurono-conjugated, and sulpho-conjugated bile acids are transported into the canaliculus via the multidrug-resistant protein type 2 (MRP2) receptor. Aminophospholipids are transported into the canaliculus via the familial intrahepatic cholestasis 1 (FIC1) receptor. These molecules combine to form micelles that are transported into gallbladder [525–531]. Intestinal bile transport and recirculation are reviewed in [524] and also discussed in Chap. 12

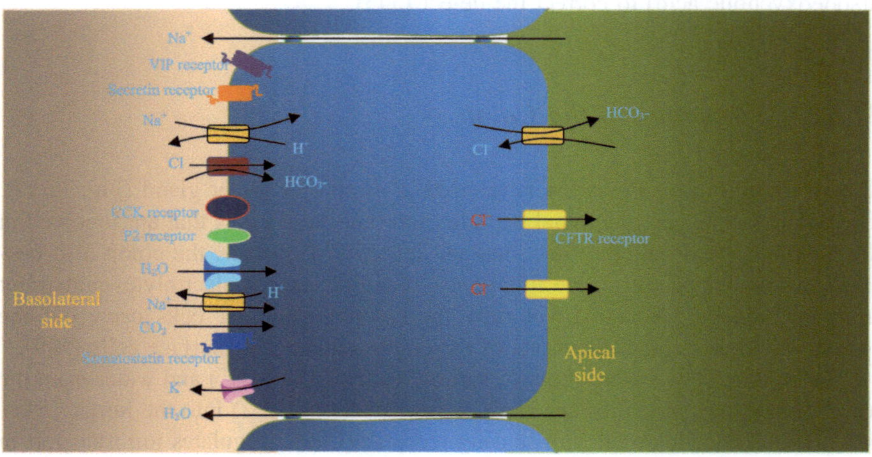

Fig. 11.26 Cholangiocyte functions. Cholangiocytes contribute to bile secretion via release of Cl^-, $HCO^-{}_3$, and water [521, 522, 532]. The released osmolytes control the fluidity and alkalinity of bile. The secretion of fluid and ions by the cholangiocytes is controlled by ion channels located in the basolateral and apical membranes. The basolateral Na^+/H^+ exchanger and $Na^+/HCO^-{}_3$ symporter are responsible for importing Na^+ and $HCO^-{}_3$ into the cytosol. The basolateral membrane also expresses a host of other ion channels including K^+ channels, aquaporins [510, 511, 533]. The apical $Cl^-/HCO^-{}_3$ exchanger, Cl^- conductance (CFTR), Ca^{2+}-activated Cl^- channel mediates the release of Cl^- and $HCO^-{}_3$ into the lumen [510, 511, 533]. The ion channels are activated by cAMP, Ca^{2+}, extracellular ATP, etc. [534]

stimulates ductal bile secretion, and receptor trafficking to the plasma membrane of the cholangiocytes [533]. Parasympathetic stimulation activates muscarinic receptors of the gallbladder to increase release of bile into the duodenum, whereas sympathetic stimulation decreases bile secretion [536]. Other factors that influence cholangiocyte functions or the composition of bile include bile acids, glutathione, growth factors, osmosensors, mechanoreceptors, vascular perfusion, plasma composition of ions, and diet [510, 512, 537].

Dysfunctions in cholangiocyte functions resulting from primary damage to the biliary epithelium have been implicated in several chronic cholestatic disorders, which are generally termed "vanishing bile duct syndromes" (cholangiopathies) [510, 512]. These diseases include primary biliary cirrhosis, AIDS cholangiopathy, chronic acalculous cholecystitis, cholestatic liver disease, Alagille's syndrome, cystic fibrosis, and biliary atresia [521, 522, 524, 538]. A subset of these diseases has no known causes and is termed "idiopathic cholangiopathy"—an example is primary sclerosing cholangitis [538, 539]. These diseases may be due to disorders in signaling pathways regulating cholangiocyte epithelial functions, inflammatory reactions, genetic mutations associated with bile formation and secretion [526, 538, 540–542]. Some gallbladder diseases are associated with gallbladder stasis, which may be due to gallstone formation or dysfunction in motility of the smooth muscle layer [543, 544]. Dysfunctions in bile functions, characterized by pathological reduction in the amount of bile acids, can be addressed through bile acid therapy, which involves supplemental administration of bile acids (such as ursodiol and ursodeoxycholic acid) to correct the defect [545].

11.4.5 Secretions of the Colon

The secretory unit of the colon is referred to as colonic pit or gland (Fig. 11.27). The colon secretes antimicrobial peptides and other molecules that are involved in the maintenance of colonic functions and integrity. However, the colon also possesses a number of absorptive functions—retaining mostly electrolytes and water from the lumen [546–548]. Not only the colonic glandular cells carry out the secretory and absorptive functions of the colon, but also the surface cells [549].

The colonic cells absorb mostly electrolytes (e.g., Na^+, Cl^-) and water from the lumen and at the same time secrete ions (especially $HCO^-{}_3$) into the lumen [546–548]. The extracellular calcium-sensing receptor (CaSR) regulates ion transport in the ileum and colon. CaSR stimulates Na^+ absorption in these regions of the gut [550]. The absorption of Na^+ is also stimulated by short-chain fatty acids (SCFA). SCFA are produced by the gut flora from nonabsorbable portion of ingested carbohydrate. Intake of resistant starch (i.e., starch that is relatively resistant to amylase digestion) can stimulate increased production of SCFA, which are absorbed by the enterocytes of the colon [551, 552]. The absorption of these ions due to the activities of CaSR and SCFA induces fluid absorption in these regions of the gut.

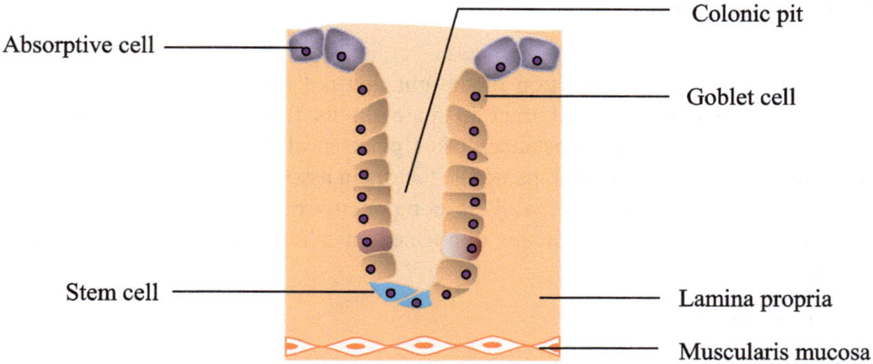

Fig. 11.27 Colonic pit or gland or crypt. In the large intestine, crypts are reduced in size and number. The crypts contain different types of cells—goblet cells, absorptive cells, stem cells. The goblet cells carry out most of the secretory functions of the colon. The number of goblet cells increases toward the rectum—this is necessary for enhancement of colonic wall lubrication, which is required not only to control sheer stress, but also for the smooth passage of feces. The absorptive cells are mainly concerned with absorption of water and ions. Stem cells are the progenitors, located at the base of the crypt [547, 552]

Reduced absorption of fluid can result in conditions of prolonged use of antibiotics, which substantially inhibit SCFA synthesis thereby resulting in diarrhea [551, 554].

The colon has high rate of secretion of glycoproteins that form the mucus. The thick colonic mucus acts as a physical barrier to shear stress [555]. Both soluble and insoluble fiber types increase mucus synthesis [555].

The secretion of HCO^-_3 via the apical Cl^-/HCO^-_3 exchanger is cAMP-dependent, Cl^--dependent, and SCFA-dependent. Cyclic AMP inhibits Cl^--dependent HCO^-_3 secretion. SCFA stimulates HCO^-_3 secretion via apical SCFA/HCO^-_3 exchanger. However, SCFA inhibits Cl^--dependent and cAMP-dependent luminal HCO^-_3 secretion [549, 556]. CaSR also regulates HCO^-_3 secretion in the colon [550, 556].

Recall that under normal conditions, an adult human colon absorbs about 1.5 L of fluid per day. This value varies among individuals, and under certain circumstances. For instance, increased secretion of the hormone aldosterone can increase fluid absorption in the colon by about 3–4 times [552]. Under physiological conditions, however, the absorption rate is usually balanced by the rate of secretion. When the secretion rate exceeds the rate of absorption, with impairment in colonic salvage, diarrhea consequently results [552].

11.5 Conclusions

GI secretion is a crucial function of the gut that not only regulates the series of activities occurring in the GI tract to maintain its functions, but also extragut activities. GI secretion of substances takes place in all regions of the gut and is controlled by hormonal, neural, paracrine factors in association with the activities of different cellular receptors as well as components of diets. Disorders in these secretory functions of the gut are implicated in a range of diseases that affect humans.

Recommended Readings

Original Articles

1. Gresz V, Kwon T-H, Gong H, Agre P, Steward MC, King LS, Nielsen S (2004) Immunolocalization of AQP-5 in rat parotid and submandibular salivary glands after stimulation or inhibition of secretion in vivo. Am J Physiol Gastrointest Liver Physiol 287(1):G151–G161
2. Kraaij S, Karagozoglu KH, Forouzanfar T, Veerman ECI, Brand HS (2014) Salivary stones: symptoms, aetiology, biochemical composition and treatment. Br Dent J 217:E23
3. Peacocke J, Lotz Z, de Beer C, Roux P, Mall AS (2012) The role of crude saliva and purified salivary mucins in the inhibition of the human immunodeficiency virus type 1. Virol J 9:177

Review Articles

1. Dent J (2006) Review article: from 1906 to 2006—a century of major evolution of understanding of gastro-oesophageal reflux disease. Aliment Pharmacol Ther 24(9):1269–1281
2. Hauser BR, Hoffman MP (2015) Regulatory mechanisms driving salivary gland organogenesis. Curr Top Dev Biol 115:111–130
3. Javitt NB (2014) History of hepatic bile formation: old problems, new approaches. Adv Physiol Educ 38:279–285
4. Lombaert I, Movahednia MM, Adine C, Ferreira JN (2017) Salivary gland regeneration: therapeutic approaches from stem cells to tissue organoids. Stem Cells 35(1):97–105
5. Rastogi R, Bhargava S, Mallarajapatna GJ, Singh SK (2012) Pictorial essay: salivary gland imaging. Indian J Radiol Imaging 22(4):325–333
6. Sami SS, Ragunath K (2013) The Los Angeles classification of gastroesophageal reflux disease. VideoGIE 1(1):103–104
7. Samuni Y, Baum BJ (2011) Gene delivery in salivary glands: from the bench to the clinic. Biochim Biophys Acta 1812(11):1515–1521
8. Sawaya RA, Macgill A, Parkman HP, Friedenberg FK (2012) Use of the Montreal global definition as an assessment of quality of life in reflux disease. Dis Esophagus 25(6):477–483

9. Spielmann N, Wong DT (2011) Saliva: diagnostics and therapeutic perspectives. Oral Dis 17(4):345–354
10. Tajiri K, Shimizu Y (2013) Liver physiology and liver diseases in the elderly. World J Gastroenterol 19(46):8459–8467
11. Tsuboi T (2008) Molecular mechanism of docking of dense-core vesicles to the plasma membrane in neuroendocrine cells. Med Mol Morphol 41:68

Books and Book Chapter

1. Del Valle J, Todisco A (2008) Gastric secretion. In: Yamada T (ed) Textbook of gastroenterology. Blackwell Publishing Ltd., Oxford
2. Keely SJ, Montrose MH, Barrett KE (2008) Electrolyte secretion and absorption: small intestine and colon. In: Yamada T (ed) Textbook of gastroenterology. Blackwell Publishing Ltd., Oxford
3. Owyang C, Williams JA (2008) Pancreatic secretion. In: Yamada T (ed) Textbook of gastroenterology. Blackwell Publishing Ltd., Oxford
4. Pandol SJ (2010) Regulation of whole-organ pancreatic secretion. The exocrine pancreas. Morgan & Claypool Life Sciences, San Rafael
5. Shnitka TK, Gilbert JAL, Harrison RC (eds) (1967) Gastric secretion: mechanisms and control. Pergamon Press, Oxford
6. Sturkie PD (1976) Secretion of gastric and pancreatic juice, pH of tract, digestion in alimentary canal, liver and bile, and absorption. In: Sturkie PD (ed) Avian physiology. Springer, Heidelberg
7. Weinman SA, Jalil S (2008) Bile secretion and cholestasis. In: Yamada T (ed) Textbook of gastroenterology. Blackwell Publishing Ltd., Oxford

Methods and Protocols

1. Gaffield MA, Betz WJ (2006) Imaging synaptic vesicle exocytosis and endocytosis with FM dyes. Nat Protoc 1(6):2916–2921

Bibliography

1. Fry C (2009) Secretions of the salivary glands and stomach. Surgery (Oxford) 27(12):503–506
2. Hendrix TR, Paulk HT (1977) Intestinal secretion. Int Rev Physiol 12:257–284
3. Bhattarai Y, Schmidt BA, Linden DR, Larson ED, Grover M, Beyder A et al (2017) Human derived gut microbiota modulates colonic secretion in mice by regulating 5-HT3 receptor expression via acetate production. Am J Physiol Gastrointest Liver Physiol 313(1):G80–G87
4. Smolka AJ, Schubert ML (2017) *Helicobacter pylori*-induced changes in gastric acid secretion and upper gastrointestinal disease. In: Tegtmeyer N, Backert S (eds) Molecular pathogenesis and signal transduction by helicobacter pylori, volume 400 of the series current topics in microbiology and immunology. Springer International Publishing AG, Cham
5. Davison JS (1989) Gastrointestinal secretion. Wright Publishing Company, London

6. Camilleri M (2004) Chronic diarrhea: a review on pathophysiology and management for the clinical gastroenterologist. Clin Gastroenterol Hepatol 2(3):198–206
7. Kiela PR, Ghishan FK (2016) Physiology of intestinal absorption and secretion. Best Pract Res Clin Gastroenterol 30(2):145–159
8. Iorgulescu G (2009) Saliva between normal and pathological. Important factors in determining systemic and oral health. J Med Life 2(3):303–307
9. Kvietys PR, Granger DN (2010) Role of intestinal lymphatics in interstitial volume regulation and transmucosal water transport. Ann N Y Acad Sci 1207(1):E29–E43
10. Whitehead WE (1989) Effects of psychological factors on gastrointestinal function. In: Snape WJ Jr (ed) Pathogenesis of functional bowel disease, part of the series topics in gastroenterology. Springer, New York
11. Schneeman B (2004) Food factors and gastrointestinal function: a critical interface. BioFactors 21(1–4):85–88
12. Collado MC, Cernada M, Neu J, Pérez-Martínez G, Gormaz M, Vento M (2015) Factors influencing gastrointestinal tract and microbiota immune interaction in preterm infants. Pediatr Res 77:726–731
13. Park HW, Lee MG (2012) Transepithelial bicarbonate secretion: lessons from the pancreas. Cold Spring Harb Perspect Med 2(10):a009571
14. Glass GBJ (1964) Proteins, mucosubstances, and biologically active components of gastric secretion. Adv Clin Chem 7:235–372
15. Krogdahl Å, Sundby A, Bakke AM (2011) Integrated function and control of the gut. Gut secretion and digestion. In: Anthony FP (ed) Encyclopedia of fish physiology—from genome to environment. Elsevier, MA
16. Claustre J, Toumi F, Trompette A, Jourdan G, Guignard H, Chayvialle JA, Plaisancié P (2002) Effects of peptides derived from dietary proteins on mucus secretion in rat jejunum. Am J Physiol Gastrointest Liver Physiol 283(3):G521–G528
17. Feldman M (2013) American journal of gastroenterology lecture: gastric acid secretion: still relevant? Am J Gastroenterol 108:347–352
18. Kumar S, Hoh JH (2001) Probing the machinery of intracellular trafficking with the atomic force microscope. Traffic 2:746–756
19. Nickel W, Rabouille C (2009) Mechanisms of regulated unconventional protein secretion. Nat Rev Mol Cell Biol 10:148–155
20. Farhan H, Rabouille C (2011) Signalling to and from the secretory pathway. J Cell Sci 124:171–180
21. Baron S, Vangheluwe P, Sepúlveda MR, Wuytack F, Raeymaekers L, Vanoevelen J (2010) The secretory pathway Ca(2+)-ATPase 1 is associated with cholesterol-rich microdomains of human colon adenocarcinoma cells. Biochim Biophys Acta 1798(8):1512–1521
22. Lodish H, Berk A, Zipursky SL, Matsudaira P, Baltimore D, Darnell J (2000) Molecular cell biology, 4th edn. W. H. Freeman, New York
23. Abdullah LH, Davis CW (2007) Regulation of airway goblet cell mucin secretion by tyrosine phosphorylation signaling pathways. Am J Physiol Lung Cell Mol Physiol 293: L591–L599
24. Shifflett DE, Jones SL, Moeser AJ, Blikslager AT (2004) Mitogen-activated protein kinases regulate COX-2 and mucosal recovery in ischemic-injured porcine ileum. Am J Physiol Gastrointest Liver Physiol 286(6):G906–G913
25. Takeuchi Y, Yamada J, Yamada T, Todisco A (1997) Functional role of extracellular signal-regulated protein kinases in gastric acid secretion. Am J Physiol Gastrointest Liver Physiol 273(6):G1263–G1272
26. Logsdon CD, Ji B (2013) The role of protein synthesis and digestive enzymes in acinar cell injury. Nat Rev Gastroenterol Hepatol 10:362–370
27. Capoccia BJ, Jin RU, Kong YY, Peek RM Jr, Fassan M, Rugge M, Mills JC (2013) The ubiquitin ligase Mindbomb 1 coordinates gastrointestinal secretory cell maturation. J Clin Invest 123(4):1475–1491

28. Nakamoto T, Srivastava A, Romanenko VG, Ovitt CE, Perez-Cornejo P, Arreola J et al (2007) Functional and molecular characterization of the fluid secretion mechanism in human parotid acinar cells. Am J Physiol Regul Integr Comp Physiol 292(6):R2380–R2390
29. Nagasawa J (1977) Exocytosis: the common release mechanism of secretory granules in glandular cells, neurosecretory cells, neurons and paraneurons. Arch Histol Jpn 40:31–47
30. Nagasawa J, Douglas WW, Schulz RA (1970) Ultrastructural evidence of secretion by exocytosis and of "synaptic vesicle" formation in posterior pituitary glands. Nature 227 (5256):407–409
31. Douglas WW, Nagasawa J, Schulz R (1971) Electron microscopic studies on the mechanism of secretion of posterior pituitary hormones and significance of microvesicles ("synaptic vesicles"): evidence of secretion by exocytosis and formation of microvesicles as a by-product of this process. Mem Soc Endocrinol 19:353–378
32. Halm DR, Halm ST (1999) Secretagogue response of goblet cells and columnar cells in human colonic crypts. Am J Physiol Cell Physiol 277(46):C501–C522
33. De Camilli P (1995) The eighth Datta lecture. Molecular mechanisms in synaptic vesicle recycling. FEBS Lett 369(1):3–12
34. Bonifacino JS, Glick BS (2004) The mechanisms of vesicle budding and fusion. Cell 116 (2):153–166
35. Wu L-G, Hamid E, Shin W, Chiang H-C (2014) Exocytosis and endocytosis: modes, functions, and coupling mechanisms. Annu Rev Physiol 76:301–331
36. Matthews G (2002) Synaptic vesicle exocytosis: does a lingering kiss lead to fusion? Neuron 35(6):1013–1014
37. Galli T, Haucke V, Gough NR (2003) Synaptic vesicle fusion followed by clathrin-mediated endocytosis. Sci STKE 2003(198):tr3
38. Castle AM, Huang AY, Castle JD (2002) The minor regulated pathway, a rapid component of salivary secretion, may provide docking/fusion sites for granule exocytosis at the apical surface of acinar cells. J Cell Sci 115:2963–2973
39. Koo SJ, Kochlamazashvili G, Rost B, Puchkov D, Gimber N, Lehmann M et al (2015) Vesicular synaptobrevin/VAMP2 levels guarded by AP180 control efficient neurotransmission. Neuron 88(2):330–344
40. Rajappa R, Gauthier-Kemper A, Böning D, Hüve J, Klingauf J (2016) Synaptophysin 1 clears synaptobrevin 2 from the presynaptic active zone to prevent short-term depression. Cell Rep 14(6):1369–1381
41. Tsuboi T, McMahon HT, Rutter GA (2004) Mechanisms of dense core vesicle recapture following "kiss and run" ("cavicapture") exocytosis in insulin-secreting cells. J Biol Chem 279:47115–47124
42. Oishi Y, Arakawa T, Tanimura A, Itakura M, Takahashi M, Tajima Y et al (2006) Role of VAMP-2, VAMP-7, and VAMP-8 in constitutive exocytosis from HSY cells. Histochem Cell Biol 125(3):273–281
43. Wang C-C, Shi H, Guo K, Ng CP, Li J, Gan BQ et al (2007) VAMP8/endobrevin as a general vesicular SNARE for regulated exocytosis of the exocrine system. Mol Biol Cell 18 (3):1056–1063
44. Barrera MJ, Sánchez M, Aguilera S, Alliende C, Bahamondes V, Molina C et al (2012) Aberrant localization of fusion receptors involved in regulated exocytosis in salivary glands of Sjögren's syndrome patients is linked to ectopic mucin secretion. J Autoimmun 39(1–2):83–92
45. Gaffield MA, Betz WJ (2007) Imaging synaptic vesicle exocytosis and endocytosis with FM dyes. Nat Protocols 1:2916–2921
46. Tsuboi T (2008) Molecular mechanism of docking of dense-core vesicles to the plasma membrane in neuroendocrine cells. Med Mol Morphol 41:68
47. Gong LW, Hafez I, Alvarez de Toledo G, Lindau M (2003) Secretory vesicles membrane area is regulated in tandem with quantal size in chromaffin cells. J Neurosci 23(21):7917–7921

48. Edgar WM (1992) Saliva: its secretion, composition and functions. Br Dent J 172(8):305–312
49. du Toit DF, Nortjé C (2004) Salivary glands: applied anatomy and clinical correlates. SADJ 59(2):65–66, 69–71, 73–74
50. Ellis H (2012) Anatomy of the salivary glands. Surgery 30(11):569–572
51. Humphrey SP, Williamson RT (2001) A review of saliva: normal composition, flow, and function. J Prosthet Dent 85(2):162–169
52. Ferguson DB (1999) The flow rate and composition of human labial gland saliva. Arch Oral Biol 44(1):S11–S14
53. Sonesson M (2011) On minor salivary gland secretion in children, adolescents and adults. Swed Dent J 215:9–64
54. Sonesson M, Ericson D, Kinnby B, Wickström C (2011) Glycoprotein 340 and sialic acid in minor-gland and whole saliva of children, adolescents, and adults. Eur J Oral Sci 119 (6):435–440
55. Varga G (2012) Physiology of the salivary glands. Surgery (Oxford) 30(11):578–583
56. Lee MG, Ohana E, Park HW, Yang D, Muallem S (2012) Molecular mechanism of pancreatic and salivary gland fluid and HCO3− secretion. Physiol Rev 92(1):39–74
57. Amano O, Mizobe K, Bando Y, Sakiyama K (2012) Anatomy and histology of rodent and human major salivary glands. Acta Histochem Cytochem 45(5):241–250
58. Hunter KD, Wilson WS (1995) The effects of antidepressant drugs on salivary flow and content of sodium and potassium ions in human parotid saliva. Arch Oral Biol 40:983–989
59. Silvers AR, Som PM (1998) Salivary glands. Radiol Clin North Am 36(5):941–966
60. Nadershah M, Salama A (2012) Removal of parotid, submandibular, and sublingual glands. Oral Maxillofac Surg Clin North Am 24(2):295–305
61. Lakraj AA, Moghimi N, Jabbari B (2013) Sialorrhea: anatomy, pathophysiology and treatment with emphasis on the role of botulinum toxins. Toxins (Basel) 5(5):1010–1031
62. Leung AKC, Kao CP (1999) Drooling in children. Paediatr Child Health 4(6):406–411
63. Zlotnik Y, Balash Y, Korczyn AD, Giladi N, Gurevich T (2015) Disorders of the oral cavity in parkinson's disease and parkinsonian syndromes. Parkinsons Dis 2015:379482
64. Restivo DA, Panebianco M, Casabona A, Lanza S, Marchese-Ragona R, Patti F, et al (2018) Botulinum toxin A for sialorrhoea associated with neurological disorders: evaluation of the relationship between effect of treatment and the number of glands treated. Toxins (Basel) 10 (2):E55
65. Dashtipour K, Bhidayasiri R, Chen JJ, Jabbari B, Lew M, Torres-Russotto D (2017) RimabotulinumtoxinB in sialorrhea: systematic review of clinical trials. J Clin Mov Disord 4:9
66. Petracca M, Guidubaldi A, Ricciardi L, Ialongo T, Del Grande A, Mulas D, et al (2015) Botulinum toxin A and B in sialorrhea: long-term data and literature overview. Toxicon 107 (Pt A):129–140
67. Kondo Y, Nakamoto T, Jaramillo Y, Choi S, Catalan MA, Melvin JE (2015) Functional differences in the acinar cells of the murine major salivary glands. J Dent Res 94(5):715–721
68. Delporte C, Bryla A, Perret J (2016) Aquaporins in salivary glands: from basic research to clinical applications. Int J Mol Sci 17(2):166
69. Holmberg KV, Hoffman MP (2014) Anatomy, biogenesis, and regeneration of salivary glands. Monogr Oral Sci 24:1–13
70. Shah AAK, Mulla AF, Mayank M (2016) Pathophysiology of myoepithelial cells in salivary glands. J Oral Maxillofac Pathol 20(3):480–490
71. Balachander N, Masthan KMK, Babu NA, Anbazhagan V (2015) Myoepithelial cells in pathology. J Pharm Bioallied Sci 7(1):S190–S193
72. Patterson K, Catalán MA, Melvin JE, Yule DI, Crampin EJ, Sneyd J (2012) A quantitative analysis of electrolyte exchange in the salivary duct. Am J Physiol Gastrointest Liver Physiol 303(10):G1153–G1163
73. Sato A (2007) Tuft cells. Anat Sci Int 82(4):187–199

74. Sato A, Miyoshi S (1996) Tuft cells in the main excretory duct epithelia of the three major rat salivary glands. Eur J Morphol 34(3):225–228
75. Segawa A, Loffredo F, Puxeddu R, Yamashina S, Testa Riva F, Riva A (1998) Exocytosis in human salivary glands visualized by high-resolution scanning electron microscopy. Cell Tissue Res 291(2):325–336
76. Hand AR, Pathmanathan D, Field RB (1999) Morphological features of the minor salivary glands. Arch Oral Biol 44(1):S3–S10
77. Stephens LC, King GK, Peters LJ, Ang KK, Schultheiss TE, Jardine JH (1986) Unique radiosensitivity of serous cells in rhesus monkey submandibular glands. Am J Pathol 124 (3):479–487
78. Davis PB (1987) Pathophysiology of cystic fibrosis with emphasis on salivary gland involvement. J Dent Res 66:667–671
79. Lydiatt DD, Bucher GS (2012) The historical evolution of the understanding of the submandibular and sublingual salivary glands. Clin Anat 25(1):2–11
80. Trivedi N (2015) Tumors of the parotid. Atlas of head and neck cancer surgery. Springer, New Delhi
81. Larian B (2016) Parotidectomy for benign parotid tumors. Otolaryngol Clin N Am 49:395–413
82. Carlson GW (2000) The salivary glands. Embryology, anatomy, and surgical applications. Surg Clin North Am 80(1):261–273, xii
83. Kuijper-Lenstra AH, Kramer MF (1975) Rate of protein synthesis in rat salivary gland cells after pilocarpine or feeding. I. Rate of (glyco) protein secretion from cells of mixed salivary glands. Cell Tissue Res 164(4):435–446
84. Ohshima H (2014) Dental and oral biology, anatomy. In: Caplan M, Bradshaw RA, Bylund DB, Carlson BM, Enna SJ, Hart GW et al (eds) Physiology of the digestive system. Human physiology. Reference module in biomedical sciences. Elsevier, MA
85. Rastogi R, Bhargava S, Mallarajapatna GJ, Singh SK (2012) Pictorial essay: salivary gland imaging. Indian J Radiol Imaging 22(4):325–333
86. Ekström J, Khosravani N, Castagnola M, Messana I (2011) Saliva and the control of its secretion. In: Ekberg O (ed) Dysphagia. Medical radiology. Springer, Heidelberg
87. Tandler B, Pinkstaff CA, Riva A (1994) Ultrastructure and histochemistry of human anterior lingual salivary glands (glands of Blandin and Nuhn). Anat Rec 240(2):167–177
88. Cheng SJ, Huang CF, Chen YC, Lee JJ, Chang HH, Chen HM et al (2009) Ultrastructural changes of posterior lingual glands after hypoglossal denervation in hamsters. J Anat 214 (1):163–170
89. Riva A, Puxeddu R, Uras L, Loy F, Serreli S, Testa Riva F (2000) A high resolution sem study of human minor salivary glands. Eur J Morphol 38(4):219–226
90. Kraitrakul S, Sirithunyaporn S, Yimtae K (2001) Distribution of minor salivary glands in the peritonsillar space. J Med Assoc Thai 84(3):371–378
91. Kumaresan R, Karthikeyan P, Mohammed F, Fairozekhan AT (2013) A novel technique for the management of Blandin-Nuhn mucocele: a case report. Int J Clin Pediatr Dent 6(3):201–204
92. Leon NG, Pardo GEM (2013) Mucocele of the glands of Blandin-Nuhn: a case report. Colomb Med (Cali) 44(1):46–47
93. Siqueira WL, Salih E, Wan DL, Helmerhorst EJ, Oppenheim FG (2008) Proteome of human minor salivary gland secretion. J Dent Res 87(5):445–450
94. Perez P, Rowzee AM, Zheng C, Adriaansen J, Baum BJ (2010) Salivary epithelial cells: an unassuming target site for gene therapeutics. Int J Biochem Cell Biol 42(6):773–777
95. Lantini MS, Proto E, Puxeddu P, Riva A, Testa Riva F (1990) Fine structure of excretory ducts of human salivary glands. J Submicrosc Cytol Pathol 22(3):465–475
96. Chitturi RT, Veeravarmal V, Nirmal RM, Reddy BVR (2015) Myoepithelial cells (MEC) of the salivary glands in health and tumours. J Clin Diagn Res 9(3):ZE14–ZE18

97. Ianez RF, Buim ME, Coutinho-Camillo CM, Schultz R, Soares FA, Lourenço SV (2010) Human salivary gland morphogenesis: myoepithelial cell maturation assessed by immunohistochemical markers. Histopathology 57(3):410–417
98. Nakamoto T, Romanenko V, Melvin JE (2007) The electrolyte and water secretion mechanism. J Oral Biosci 49(1):27–30
99. Cutler LS, Gremski W (1991) Epithelial-mesenchymal interactions in the development of salivary glands. Crit Rev Oral Biol Med 2(1):1–12
100. Chan YH, Huang TW, Young TH, Lou PJ (2011) Human salivary gland acinar cells spontaneously form three-dimensional structures and change the protein expression patterns. J Cell Physiol 226(11):3076–3085
101. Logsdon CD, Ji B (2013) The role of protein synthesis and digestive enzymes in acinar cell injury. Nat Rev Gastroenterol Hepatol 10(6):362–370
102. Turner JR, Sugiya H (2002) Understanding salivary fluid and protein secretion. Oral Dis 8:3–11
103. Kim SK, Cuzzort LM, McKean RK (1992) Amylase mRNA synthesis and ageing in rat parotid glands following isoproterenol-stimulated secretion. Arch Oral Biol 37(5):349–354
104. Horii A, Emi M, Tomita N, Nishide T, Ogawa M, Mori T, Matsubara K (1987) Primary structure of human pancreatic alpha-amylase gene: its comparison with human salivary alpha-amylase gene. Gene 60(1):57–64
105. Bonnefond A, Yengo L, Dechaume A, Canouil M, Castelain M, Roger E et al (2017) Relationship between salivary/pancreatic amylase and body mass index: a systems biology approach. BMC Med 15(1):37
106. Falchi M, El-Sayed Moustafa JS, Takousis P, Pesce F, Bonnefond A, Andersson-Assarsson JC et al (2014) Low copy number of the salivary amylase gene predisposes to obesity. Nat Genet 46:492–497
107. Castle D, Castle A (1998) Intracellular transport and secretion of salivary proteins. Crit Rev Oral Biol Med 9(l):4–22
108. Yokouchi H, Horii A, Emi M, Tomita N, Doi S, Ogawa M et al (1990) Cloning and characterization of a third type of human alpha-amylase gene, AMY2B. Gene 90(2):281–286
109. Samuelson LC, Wiebauer K, Gumucio DL, Meisler MH (1988) Expression of the human amylase genes: recent origin of a salivary amylase promoter from an actin pseudogene. Nucleic Acids Res 16(17):8261–8276
110. Carpenter D, Dhar S, Mitchell LM, Fu B, Tyson J, Shwan NA et al (2015) Obesity, starch digestion and amylase: association between copy number variants at human salivary (AMY1) and pancreatic (AMY2) amylase genes. Hum Mol Genet 24(12):3472–3480
111. Fábián TK, Hermann P, Beck A, Fejérdy P, Fábián G (2012) Salivary defense proteins: their network and role in innate and acquired oral immunity. Int J Mol Sci 13(4):4295–4320
112. Murakami M, Ohtake T, Dorschner RA, Gallo RL (2002) Cathelicidin antimicrobial peptides are expressed in salivary glands and saliva. J Dent Res 81(12):845–850
113. Storesund T, Schreurs O, Messelt EB, Kolltveit KM, Schenck K (2009) Trefoil factor family 3 expression in the oral cavity. Eur J Oral Sci 117(6):636–643
114. Jagla W, Wiede A, Hinz M, Dietzmann K, Gülicher D, Gerlach KL, Hoffmann W (1999) Secretion of TFF-peptides by human salivary glands. Cell Tissue Res 298(1):161–166
115. Taichman NS, Cruchley AT, Fletcher LM, Hagi-Pavli EP, Paleolog EM, Abrams WR et al (1998) Vascular endothelial growth factor in normal human salivary glands and saliva: a possible role in the maintenance of mucosal homeostasis. Lab Invest 78(7):869–875
116. Leclair EE (2003) Four BPI (bactericidal/permeability-increasing protein)-like genes expressed in the mouse nasal, oral, airway and digestive epithelia. Biochem Soc Trans 31 (Pt 4):801–805
117. Wheeler TT, Haigh BJ, McCracken JY, Wilkins RJ, Morris CA, Grigor MR (2002) The BSP30 salivary proteins from cattle, LUNX/PLUNC and von Ebner's minor salivary gland protein are members of the PSP/LBP superfamily of proteins. Biochim Biophys Acta 1579 (2–3):92–100

118. Wheeler TT, Hood KA, Maqbool NJ, McEwan JC, Bingle CD, Zhao S (2007) Expansion of the bactericidal/permeability increasing-like (BPI-like) protein locus in cattle. BMC Genom 8:75
119. Petersen OH, Tepikin AV (2008) Polarized calcium signaling in exocrine gland cells. Annu Rev Physiol 70:273–299
120. Salido GM, Sage SO, Rosado JA (2009) Biochemical and functional properties of the store-operated Ca^{2+} channels. Cell Signal 21(4):457–461
121. Chen KT, Ong HL, Liu X, Ambudkar IS (2011) Contribution of TRPC1 and Orai1 to Ca^{2+} entry activated by store depletion. Adv Exp Med Biol 704:435–449
122. Froehlich DA, Pangborn RM, Whitaker JR (1987) The effect of oral stimulation on human parotid salivary flow rate and alpha-amylase secretion. Physiol Behav 41(3):209–217
123. Williams JA (1981) Electrical correlates of secretion in endocrine and exocrine cells. Fed Proc 40(2):128–134
124. Park HS, Betzenhauser MJ, Zhang Y, Yule DI (2012) Regulation of Ca^{2+} release through inositol 1,4,5-trisphosphate receptors by adenine nucleotides in parotid acinar cells. Am J Physiol Gastrointest Liver Physiol 302(1):G97–G104
125. Benes C, Soltoff SP (2001) Modulation of PKCδ tyrosine phosphorylation and activity in salivary and PC-12 cells by Src kinases. Am J Physiol Cell Physiol 280(6):C1498–C1510
126. Ishikawa Y, Ishida H (2000) Aquaporin water channel in salivary glands. Jpn J Pharmacol 83(2):95–101
127. Putney JW Jr (1982) Inositol lipids and cell stimulation in mammalian salivary gland. Cell Calcium 3(4–5):369–383
128. Turner JT, Landon LA, Gibbons SJ, Talamo BR (1999) Salivary gland P2 nucleotide receptors. Crit Rev Oral Biol Med 10(2):210–224
129. Nauntofte B (1992) Regulation of electrolyte and fluid secretion in salivary acinar cells. Am J Physiol 263(6 Pt 1):G823–G837
130. Kim J-H, Park S-H, Moon YW, Hwang S, Kim D, Jo S-H et al (2009) Histamine H1 receptor induces cytosolic calcium increase and aquaporin translocation in human salivary gland cells. J Pharmacol Exp Ther 330(2):403–412
131. Ambudkar IS (2014) Ca^{2+} signaling and regulation of fluid secretion in salivary gland acinar cells. Cell Calcium 55(6):297–305
132. Beroukas D, Goodfellow R, Hiscock J, Jonsson R, Gordon TP, Waterman SA (2002) Up-regulation of M3-muscarinic receptors in labial salivary gland acini in primary Sjögren's syndrome. Lab Invest 82:203–210
133. Kajiya M, Ichimonji I, Min C, Zhu T, Jin J-O, Yu Q et al (2012) Muscarinic type 3 receptor induces cytoprotective signaling in salivary gland cells through epidermal growth factor receptor transactivation. Mol Pharmacol 82(1):115–124
134. Melvin JE, Yule D, Shuttleworth T, Begenisich T (2005) Regulation of fluid and electrolyte secretion in salivary gland acinar cells. Annu Rev Physiol 67:445–469
135. Maruyama Y, Gallacher DV, Petersen OH (1983) Voltage and Ca^{2+}-activated K^+ channel in baso-lateral acinar cell membranes of mammalian salivary glands. Nature 302(5911):827–829
136. Catalán MA, Peña-Munzenmayer G, Melvin JE (2014) Ca^{2+}-dependent K^+ channels in exocrine salivary glands. Cell Calcium 55(6):362–368
137. Ishikawa T, Murakami M, Seo Y (1994) Basolateral K^+ efflux is largely independent of maxi-K^+ channels in rat submandibular glands during secretion. Pflugers Arch 428(5–6):516–525
138. Morris AP, Gallacher DV, Fuller CM, Scott J (1987) Cholinergic receptor-regulation of potassium channels and potassium transport in human submandibular acinar cells. J Dent Res 66(2):541–546
139. Frizzell RA, Hanrahan JW (2012) Physiology of epithelial chloride and fluid secretion. Cold Spring Harb Perspect Med 2(6):a009563
140. Melvin JE (1999) Chloride channels and salivary gland function. Crit Rev Oral Biol Med 10 (2):199–209

141. Yang D, Shcheynikov N, Zeng W, Ohana E, So I, Ando H et al (2009) IRBIT coordinates epithelial fluid and HCO3− secretion by stimulating the transporters pNBC1 and CFTR in the murine pancreatic duct. J Clin Invest 119(1):193–202
142. Ando H, Kawaai K, Mikoshiba K (2014) IRBIT: a regulator of ion channels and ion transporters. Biochim Biophys Acta 1843(10):2195–2204
143. Yamaguchi S, Ishikawa T (2008) The electrogenic Na^{+}-HCO3− cotransporter NBCe1-B is regulated by intracellular $Mg2^{+}$. Biochem Biophys Res Commun 376(1):100–104
144. Lee S-K, Boron WF, Parker MD (2012) Relief of autoinhibition of the electrogenic Na-HCO_3 cotransporter NBCe1-B: role of IRBIT vs. amino-terminal truncation. Am J Physiol Cell Physiol 302(3):C518–C526
145. Jalali R, Guo J, Zandieh-Doulabi B, Bervoets TJM, Paine ML, Boron WF et al (2014) NBCe1 (SLC4A4) a potential pH regulator in enamel organ cells during enamel development in the mouse. Cell Tissue Res 358(2):433–442
146. Ando H, Kawaai K, Mikoshiba K (2014) IRBIT: a regulator of ion channels and ion transporters. Biochim Biophys Acta Mol Cell Res 1843(10):2195–2204
147. Castle D, Castle A (1998) Intracellular transport and secretion of salivary proteins. Crit Rev Oral Biol Med 9(1):4–22
148. Castle JD, Castle AM (1993) Sorting and secretion of salivary proteins. Crit Rev Oral Biol Med 4(3–4):393–398
149. Castle JD (1990) Sorting and secretory pathways in exocrine cells. Am J Respir Cell Mol Biol 2(2):119–126
150. Venkatesh SG, Tan J, Gorr SU, Darling DS (2007) Isoproterenol increases sorting of parotid gland cargo proteins to the basolateral pathway. Am J Physiol Cell Physiol 293(2):C558–C565
151. Fujita-Yoshigaki J, Matsuki-Fukushima M, Yokoyama M, Katsumata-Kato O (2013) Sorting of a HaloTag protein that has only a signal peptide sequence into exocrine secretory granules without protein aggregation. Am J Physiol Gastrointest Liver Physiol 305(10): G685–G696
152. von Zastrow M, Castle JD (1987) Protein sorting among two distinct export pathways occurs from the content of maturing exocrine storage granules. J Cell Biol 105(6 Pt 1):2675–2684
153. Carlo AS, Nykjaer A, Willnow TE (2014) Sorting receptor sortilin-a culprit in cardiovascular and neurological diseases. J Mol Med (Berl). 92(9):905–911
154. Padilla BE, Cottrell GS, Roosterman D, Pikios S, Muller L, Steinhoff M, Bunnett NW (2007) Endothelin-converting enzyme-1 regulates endosomal sorting of calcitonin receptor-like receptor and beta-arrestins. J Cell Biol 179(5):981–997
155. Katzmann DJ, Odorizzi G, Emr SD (2002) Receptor downregulation and multivesicular-body sorting. Nat Rev Mol Cell Biol 3:893–905
156. Chapman RE (1994) Vacuolar sorting: tracking down an elusive receptor. Curr Biol 4 (11):1019–1022
157. Varga G (2015) Physiology of the salivary glands. Surgery 33(12):581–586
158. Cannon RM, Lee CE (2013) Salivary gland physiology. In: Kountakis SE (ed) Encyclopedia of otolaryngology, head and neck surgery. Springer, Heidelberg
159. Hand AR (1990) The secretory process of salivary glands and pancreas. In: Riva A, Motta PM, Riva FT (eds) Ultrastructure of the extraparietal glands of the digestive tract. Springer, New York
160. Huang AY, Castle AM, Hinton BT, Castle JD (2001) Resting (basal) secretion of proteins is provided by the minor regulated and constitutive-like pathways and not granule exocytosis in parotid acinar cells. J Biol Chem 276(25):22296–22306
161. Wang J, Cawley NX, Voutetakis A, Rodriguez YM, Goldsmith CM, Nieman LK et al (2005) Partial redirection of transgenic human growth hormone secretion from rat salivary glands. Hum Gene Ther 16(5):571–583
162. Proctor GB (2016) The physiology of salivary secretion. Periodontol 2000 70(1):11–25

163. Okuma N, Saita M, Hoshi N, Soga T, Tomita M, Sugimoto M, Kimoto K (2017) Effect of masticatory stimulation on the quantity and quality of saliva and the salivary metabolomic profile. PLoS ONE 12(8):e0183109
164. Marquezin MC, Pedroni-Pereira A, Araujo DS, Rosar JV, Barbosa TS, Castelo PM (2016) Descriptive analysis of the masticatory and salivary functions and gustatory sensitivity in healthy children. Acta Odontol Scand 74(6):443–448
165. Catalán MA, Nakamoto T, Melvin JE (2009) The salivary gland fluid secretion mechanism. J Med Invest 56:192–196
166. Martinez JR (1987) Ion transport and water movement. J Dent Res 66:638–647
167. Thaysen JH, Thorn NA, Schwartz IL (1954) Excretion of sodium, potassium, chloride and carbon dioxide in human parotid saliva. Am J Physiol 178(1):155–159
168. Roussa E (2011) Channels and transporters in salivary glands. Cell Tissue Res 343(2):263–287
169. Ohana E (2015) Transepithelial ion transport across duct cells of the salivary gland. Oral Dis 21(7):826–835
170. Baker OJ (2016) Current trends in salivary gland tight junctions. Tissue Barriers 4(3): e1162348
171. Alam J, Choi YS, Koh JH, Kwok S-K, Park S-H, Song YW et al (2017) Detection of autoantibodies against aquaporin-1 in the sera of patients with primary Sjögren's syndrome. Immune Netw 17(2):103–109
172. Delporte C (2014) Aquaporins in salivary glands and pancreas. Biochim Biophys Acta 1840 (5):1524–1532
173. Delporte C, Steinfeld S (2006) Distribution and roles of aquaporins in salivary glands. Biochim Biophys Acta 1758(8):1061–1070
174. Gresz V, Kwon TH, Hurley PT, Varga G, Zelles T, Nielsen S et al (2001) Identification and localization of aquaporin water channels in human salivary glands. Am J Physiol Gastrointest Liver Physiol 281(1):G247–G254
175. Proctor GB, Carpenter GH (2007) Regulation of salivary gland function by autonomic nerves. Auton Neurosci 133(1):3–18
176. Lai Z, Yin H, Cabrera-Pérez J, Guimaro MC, Afione S, Michael DG et al (2016) Aquaporin gene therapy corrects Sjögren's syndrome phenotype in mice. Proc Natl Acad Sci U S A 113 (20):5694–5699
177. Carpenter GH (2013) The secretion, components, and properties of saliva. Annu Rev Food Sci Technol 4:267–276
178. Thie NM, Kato T, Bader G, Montplaisir JY, Lavigne GJ (2002) The significance of saliva during sleep and the relevance of oromotor movements. Sleep Med Rev 6(3):213–227
179. Esser D, Alvarez-Llamas G, de Vries MP, Weening D, Vonk RJ, Roelofsen H (2008) Sample stability and protein composition of saliva: implications for its use as a diagnostic fluid. Biomark Insights 3:25–27
180. de Almeida Pdel V, Grégio AM, Machado MA, de Lima AA, Azevedo LR (2008) Saliva composition and functions: a comprehensive review. J Contemp Dent Pract 9(3):72–80
181. Damle SG, Vidya I, Yadav R, Bhattal H, Loomba A (2012) Quantitative determination of inorganic constituents in saliva and their relationship with dental caries experience in children. Dentistry 2:131
182. Isenman L, Liebow C, Rothman S (1999) The endocrine secretion of mammalian digestive enzymes by exocrine glands. Am J Physiol Endocrinol Metab 276(2):E223–E232
183. Zolotukhin S (2013) Metabolic hormones in saliva: origins and functions. Oral Dis 19 (3):219–229
184. Tiwari M (2011) Science behind human saliva. J Nat Sci Biol Med 2(1):53–58
185. Granger DA, Shirtcliff EA, Booth A, Kivlighan KT, Schwartz EB (2004) The "trouble" with salivary testosterone. Psychoneuroendocrinology 29(10):1229–1240
186. Mathison RD, Davison JS, Befus AD, Gingerich DA (2010) Salivary gland derived peptides as a new class of anti-inflammatory agents: review of preclinical pharmacology of C-terminal peptides of SMR1 protein. J Inflamm 7:49

187. Leonora J, Tieche J-M, Celestin J (1987) Physiological factors affecting secretion of parotid hormone. Am J Physiol 252(4 Pt 1):E477–E484
188. Gröschl M (2009) The physiological role of hormones in saliva. BioEssays 31(8):843–852
189. Lindell SG, Suomi SJ, Shoaf S, Higley JD, Linnoila M (1999) Salivary prolactin as a marker for central serotonin turnover. Biol Psychiatry 46(4):568–572
190. Kanno T, Asada N, Yanase H, Iwanaga T, Ozaki T, Nishikawa Y et al (1999) Salivary secretion of highly concentrated chromogranin a in response to noradrenaline and acetylcholine in isolated and perfused rat submandibular glands. Exp Physiol 84(6):1073–1083
191. Gröschl M, Topf HG, Kratzsch J, Dötsch J, Rascher W, Rauh M (2005) Salivary leptin induces increased expression of growth factors in oral keratinocytes. J Mol Endocrinol 34 (2):353–366
192. Gröschl M, Topf HG, Bohlender J, Zenk J, Klussmann S, Dötsch J et al (2005) Identification of ghrelin in human saliva: production by the salivary glands and potential role in proliferation of oral keratinocytes. Clin Chem 51(6):997–1006
193. Gröschl M, Wendler O, Topf HG, Bohlender J, Köhler H (2009) Significance of salivary adrenomedullin in the maintenance of oral health: stimulation of oral cell proliferation and antibacterial properties. Regul Pept 154(1–3):16–22
194. Šimo L, Koči J, Žitňan D, Park Y (2011) Evidence for D1 dopamine receptor activation by a paracrine signal of dopamine in tick salivary glands. PLoS ONE 6(1):e16158
195. Sabbadini E, Berczi I (1995) The submandibular gland: a key organ in the neuro-immuno-regulatory network? NeuroImmunoModulation 2(4):184–202
196. Nederfors T, Dahlöf C (1996) Effects on salivary flow rate and composition of withdrawal of and re-exposure to the beta 1-selective antagonist metoprolol in a hypertensive patient population. Eur J Oral Sci 104(3):262–268
197. Arhakis A, Karagiannis V, Kalfas S (2013) Salivary alpha-amylase activity and salivary flow rate in young adults. Open Dent J 7:7–15
198. Gorr S-U, Venkatesh SG, Darling DS (2005) Parotid secretory granules: crossroads of secretory pathways and protein storage. J Dent Res 84(6):500–509
199. Messenger SW, Falkowski MA, Groblewski GE (2014) Ca^{2+}-regulated secretory granule exocytosis in pancreatic and parotid acinar cells. Cell Calcium 55(6):369–375
200. Cosen-Binker LI, Gaisano HY (2007) Recent insights into the cellular mechanisms of acute pancreatitis. Can J Gastroenterol 21(1):19–24
201. Leonora J, Tieche JM, Steinman RR (1993) Further evidence for a hypothalamus-parotid gland endocrine axis in the rat. Arch Oral Biol 38(10):911–916
202. Leonora J, Tieche JM, Celestin J (1987) Physiological factors affecting secretion of parotid hormone. Am J Physiol Endocrinol Metab 252(4):E477–E484
203. Anderson LC (1998) Hormonal regulation of salivary glands, with particular reference to experimental diabetes. Front Oral Biol 10:200–221
204. Yagi T, Ueda H, Amitani H, Asakawa A, Miyawaki S, Inui A (2012) The role of ghrelin, salivary secretions, and dental care in eating disorders. Nutrients 4(8):967–989
205. Nigro E, Piombino P, Scudiero O, Monaco ML, Schettino P, Chambery A, Daniele A (2015) Evaluation of salivary adiponectin profile in obese patients. Peptides 63:150–155
206. Rabe K, Lehrke M, Parhofer KG, Broedl UC (2008) Adipokines and insulin resistance. Mol Med 14(11–12):741–751
207. Mamali I, Roupas ND, Armeni AK, Theodoropoulou A, Markou KB, Georgopoulos NA (2012) Measurement of salivary resistin, visfatin and adiponectin levels. Peptides 33(1):120–124
208. Antuna-Puente B, Feve B, Fellahi S, Bastard JP (2008) Adipokines: the missing link between insulin resistance and obesity. Diabetes Metab 34(1):2–11
209. Cantarini L, Obici L, Simonini G, Cimaz R, Bacarelli MR, Merlini G et al (2012) Serum leptin, resistin, visfatin and adiponectin levels in tumor necrosis factor receptor-associated periodic syndrome (TRAPS). Clin Exp Rheumatol 30(3 Suppl 72):S108–S114

210. Tabari ZA, Ghaedi FB, Azadmehr A, Nohekhan A, Tabrizi MAA, Ardakani MRT et al (2015) . Salivary visfatin concentration in response to non-surgical periodontal therapy. J Clin Diagn Res 9:ZC05–ZC08
211. Özcan E, Saygun NI, Serdar MA, Kurt N (2015) Evaluation of the salivary levels of visfatin, chemerin, and progranulin in periodontal inflammation. Clin Oral Invest 19(4):921–928
212. Rao PV, Reddy AP, Lu X, Dasari S, Krishnaprasad A, Biggs E et al (2009) Proteomic identification of salivary biomarkers of type-2 diabetes. J Proteome Res 8(1):239–245
213. Loo JA, Yan W, Ramachandran P, Wong DT (2010) comparative human salivary and plasma proteomes. J Dent Res 89(10):1016–1023
214. Thomsson KA, Prakobphol A, Leffler H, Reddy MS, Levine MJ, Fisher SJ, Hansson GC (2002) The salivary mucin MG1 (MUC5B) carries a repertoire of unique oligosaccharides that is large and diverse. Glycobiology 12(1):1–14
215. Zalewska A, Zwierz K, Zółkowski K, Gindzieński A (2000) Structure and biosynthesis of human salivary mucins. Acta Biochim Pol 47(4):1067–1079
216. Mehrotra R, Thornton DJ, Sheehan JK (1998) Isolation and physical characterization of the MUC7 (MG2) mucin from saliva: evidence for self-association. Biochem J 334(Pt 2):415–422
217. Liu B, Rayment SA, Gyurko C, Oppenheim FG, Offner GD, Troxler RF (2000) The recombinant N-terminal region of human salivary mucin MG2 (MUC7) contains a binding domain for oral Streptococci and exhibits candidacidal activity. Biochem J 345(3):557–564
218. Liu B, Rayment S, Oppenheim FG, Troxler RF (1999) Isolation of human salivary mucin MG2 by a novel method and characterization of its interactions with oral bacteria. Arch Biochem Biophys 364(2):286–293
219. Park W-K, Chung J-W, Kim Y-K, Chung S-C, Kho H-S (2006) Influences of animal mucins on lysozyme activity in solution and on hydroxyapatite surfaces. Arch Oral Biol 51(10):861–869
220. Loomis RE, Prakobphol A, Levine MJ, Reddy MS, Jones PC (1987) Biochemical and biophysical comparison of two mucins from human submandibular-sublingual saliva. Arch Biochem Biophys 258(2):452–464
221. Fábián TK, Fejérdy P, Nguyen MT, Sőti C, Csermely P (2007) Potential immunological functions of salivary Hsp70 in mucosal and periodontal defense mechanisms. Arch Immunol Ther Exp 55:1–8
222. Soares RV, Siqueira CC, Bruno LS, Oppenheim FG, Offner GD, Troxler RF (2003) MG2 and lactoferrin form a heterotypic complex in salivary secretions. J Dent Res 82(6):471–475
223. de Sousa-Pereira P, Amado F, Abrantes J, Ferreira R, Esteves PJ, Vitorino R (2013) An evolutionary perspective of mammal salivary peptide families: cystatins, histatins, statherin and PRPs. Arch Oral Biol 58(5):451–458
224. Schlesinger DH, Hay DI, Levine MJ (1989) Complete primary structure of statherin, a potent inhibitor of calcium phosphate precipitation, from the saliva of the monkey, macaca arctoides. Int J Pept Protein Res 34(5):374–380
225. Schlesinger DH, Hay DI (1977) Complete covalent structure of statherin, a tyrosine-rich acidic peptide which inhibits calcium phosphate precipitation from human parotid saliva. J Biol Chem 252:1689–1695
226. Lamkin MS, Oppenheim FG (1993) Structural features of salivary function. Crit Rev Oral Biol Med 4(3–4):251–259
227. Hay DI, Smith DJ, Schluckebier SK, Moreno EC (1984) Relationship between concentration of human salivary statherin and inhibition of calcium phosphate precipitation in stimulated human parotid saliva. J Dent Res 63(6):857–863
228. Isola M, Cossu M, Diana M, Isola R, Loy F, Solinas P, Lantini MS (2012) Diabetes reduces statherin in human parotid: immunogold study and comparison with submandibular gland. Oral Dis 18(4):360–364
229. Isola M, Lantini M, Solinas P, Diana M, Isola R, Loy F, Cossu M (2011) Diabetes affects statherin expression in human labial glands. Oral Dis 17(7):685–689

230. Mednieks MI, Szczepanski A, Clark B, Hand AR (2009) Protein expression in salivary glands of rats with streptozotocin diabetes. Int J Exp Pathol 90(4):412–422
231. Li J, Helmerhorst EJ, Yao Y, Nunn ME, Troxler RF, Oppenheim FG (2004) Statherin is an in vivo pellicle constituent: identification and immuno-quantification. Arch Oral Biol 49 (5):379–385
232. Niemi LD, Johansson I (2004) Salivary statherin peptide-binding epitopes of commensal and potentially infectious *Actinomyces* spp. delineated by a hybrid peptide construct. Infect Immun 72(2):782–787
233. Tsai H, Bobek LA (1998) Human salivary histatins: promising anti-fungal therapeutic agents. Crit Rev Oral Biol Med 9(4):480–497
234. Piludu M, Lantini MS, Cossu M, Piras M, Oppenheim FG, Helmerhorst EJ et al (2006) Salivary histatins in human deep posterior lingual glands (of von Ebner). Arch Oral Biol 51 (11):967–973
235. Puri S, Edgerton M (2014) How does it kill?: understanding the candidacidal mechanism of salivary histatin 5. Eukaryot Cell 13(8):958–964
236. Bennick A (1982) Salivary proline-rich proteins. Mol Cell Biochem 45(2):83–99
237. Carlson DM (1988) Proline-rich proteins and glycoproteins: expression of salivary gland multigene families. Biochimie 70(11):1689–1695
238. Soares S, Mateus N, de Freitas V (2012) Interaction of different classes of salivary proteins with food tannins. Food Res Int 49(2):807–813
239. Carlson DM (1993) Salivary proline-rich proteins: biochemistry, molecular biology, and regulation of expression. Crit Rev Oral Biol Med 4(3–4):495–502
240. Rawlings ND, Barrett AJ (1990) Evolution of proteins of the cystatin superfamily. J Mol Evol 30(1):60–71
241. Dickinson DP (2002) Salivary (SD-type) cystatins: over one billion years in the making–but to what purpose? Crit Rev Oral Biol Med 13(6):485–508
242. Kordiš D, Turk V (2009) Phylogenomic analysis of the cystatin superfamily in eukaryotes and prokaryotes. BMC Evol Biol 9:266
243. Barrett AJ, Fritz H, Grubb A, Isemura S, Järvinen M, Katunuma N et al (1986) Nomenclature and classification of the proteins homologous with the cysteine-proteinase inhibitor chicken cystatin. Biochem J 236:312
244. Turk V, Stoka V, Turk D (2008) Cystatins: biochemical and structural properties, and medical relevance. Front Biosci 13:5406–5420
245. Abrahamson M, Alvarez-Fernandez M, Nathanson CM (2003) Cystatins. Biochem Soc Symp 70:179–199
246. Devos A, De Clercq N, Vercaeren I, Heyns W, Rombauts W, Peeters B (1993) Structure of rat genes encoding androgen-regulated cystatin-related proteins (CRPs): a new member of the cystatin superfamily. Gene 125(2):159–167
247. Parent AD, Cornwall GA, Liu LY, Smith CE, Hermo L (2011) Alterations in the testis and epididymis associated with loss of function of the cystatin-related epididymal spermatogenic (CRES) protein. J Androl 32(4):444–463
248. Wang L, Yuan Q, Chen S, Cai H, Lu M, Liu Y, Xu C (2012) Antimicrobial activity and molecular mechanism of the CRES protein. PLoS ONE 7(11):e48368
249. Lalmanach G, Naudin C, Lecaille F, Fritz H (2010) Kininogens: more than cysteine protease inhibitors and kinin precursors. Biochimie 92(11):1568–1579
250. Habermann E (1970) Kininogens, bradykinin, kallidin and kallikrein. Handb Exp Pharmacol 25:250–288
251. Khorchid A, Ikura M (2002) How calpain is activated by calcium. Nat Struct Mol Biol 9:239–241
252. Ono Y, Sorimachi H (2012) Calpains—an elaborate proteolytic system. Biochim Biophys Acta Proteins Proteom 1824(1):224–236
253. Schmaier AH (2000) Plasma kallikrein/kinin system: a revised hypothesis for its activation and its physiologic contributions. Curr Opin Hematol 7(5):261–265

254. Karlsrud TS, Buø L, Aasen AO, Johansen HT (1991) Characterization of kininogens in human malignant ascites. Thromb Res 63(6):641–650
255. Sharma JN, Al-Banoon A (2012) The role of inflammatory mediator bradykinin in cardiovascular and renal diseases. Sci Rep 1:142
256. Hojima Y, Maranda B, Moriwaki C, Schachter M (1977) Direct evidence for the location of kallikrein in the striated ducts of the cat's submandibular gland by the use of specific antibody. J Physiol 268(3):793–801
257. Heidland A, Röckel A, Schmid G (1979) Salivary kallikrein excretion in hypertension. Klin Wochenschr 57(19):1047–1052
258. Liu CY, Scott CF, Bagdasarian A, Pierce JV, Kaplan AP, Colman RW (1977) Potentiation of the function of Hageman factor fragments by high molecular weight kininogen. J Clin Invest 60(1):7–17
259. Paul M, Poyan Mehr A, Kreutz R (2006) Physiology of local renin-angiotensin systems. Physiol Rev 86(3):747–803
260. Basso N, Terragno NA (2001) History about the discovery of the renin-angiotensin system. Hypertension 38(6):1246–1249
261. Ian Phillips M, Schmidt-Ott KM (1999) The discovery of renin 100 years ago. Physiology 14(6):271–274
262. Van Epps HL (2005) Harry Goldblatt and the discovery of renin. J Exp Med 201(9):1351
263. Fournier D, Luft FC, Bader M, Ganten D, Andrade-Navarro MA (2012) Emergence and evolution of the renin–angiotensin–aldosterone system. J Mol Med (Berl) 90(5):495–508
264. Kon Y, Endoh D (1999) Renin in exocrine glands of different mouse strains. Anat Histol Embryol 28(4):239–242
265. Poulsen K, Jacobsen J (1983) The biochemistry of aggression-provoked renin. Clin Exp Hypertens A 5(7–8):969–973
266. Mercan R, Bıtık B, Tezcan ME, Kaya A, Tufan A, Özturk MA et al (2014) Minimally invasive minor salivary gland biopsy for the diagnosis of amyloidosis in a rheumatology clinic. ISRN Rheumatol 2014:354648
267. de Paula Eduardo F, de Mello Bezinelli L, de Carvalho DLC, Della-Guardia B, de Almeida MD, Marins LV, Corrêa L (2017) Minor salivary gland biopsy for the diagnosis of familial amyloid polyneuropathy. Neurol Sci 38(2):311–318
268. Sacsaquispe SJ, Antúnez-de Mayolo EA, Vicetti R, Delgado WA (2011) Detection of AA-type amyloid protein in labial salivary glands. Med Oral Patol Oral Cir Bucal 16(2): e149–e152
269. Yang GCH, Kuhel WI, Scognamiglio T (2014) Amyloid-rich low grade adenocarcinoma of the parotid: fine-needle aspiration cytology with histologic correlations. Diagn Cytopathol 42(9):798–801
270. Lechapt-Zalcman E, Authier FJ, Creange A, Voisin MC, Gherardi RK (1999) Labial salivary gland biopsy for diagnosis of amyloid polyneuropathy. Muscle Nerve 22(1):105–107
271. Bermejo-Pareja F, Antequera D, Vargas T, Molina JA, Carro E (2010) Saliva levels of Abeta1-42 as potential biomarker of Alzheimer's disease: a pilot study. BMC Neurol 10:108
272. Lee M, Guo JP, Kennedy K, McGeer EG, McGeer PL (2017) A method for diagnosing alzheimer's disease based on salivary amyloid-β protein 42 levels. J Alzheimers Dis 55 (3):1175–1182
273. Ralhan R, Desouza LV, Matta A, Chandra Tripathi S, Ghanny S, Dattagupta S et al (2009) iTRAQ-multidimensional liquid chromatography and tandem mass spectrometry-based identification of potential biomarkers of oral epithelial dysplasia and novel networks between inflammation and premalignancy. J Proteome Res 8(1):300–309
274. Gröschl M (2008) Current status of salivary hormone analysis. Clin Chem 54(11):1759–1769
275. Abikshyeet P, Ramesh V, Oza N (2012) Glucose estimation in the salivary secretion of diabetes mellitus patients. Diabetes Metab Syndr Obes 5:149–154

276. Nayak S, Bhad Patil WA, Doshi UH (2014) The relationship between salivary insulin-like growth factor I and quantitative cervical maturational stages of skeletal maturity. J Orthod 41 (3):170–174
277. Hassaneen M, Maron JL (2017) Salivary diagnostics in pediatrics: applicability, translatability, and limitations. Front Public Health 5:83
278. Estrada-Y-Martin RM, Orlander PR (2011) Salivary cortisol can replace free serum cortisol measurements in patients with septic shock. Chest 140(5):1216–1222
279. Perogamvros I, Owen LJ, Newell-Price J, Ray DW, Trainer PJ, Keevil BG (2009) Simultaneous measurement of cortisol and cortisone in human saliva using liquid chromatography–tandem mass spectrometry: application in basal and stimulated conditions. J Chromatogr B Analyt Technol Biomed Life Sci 877(29):3771–3775
280. Meulenberg PM, Hofman JA (1990) Differences between concentrations of salivary cortisol and cortisone and of free cortisol and cortisone in plasma during pregnancy and postpartum. Clin Chem 36(1):70–75
281. Malamud D, Rodriguez-Chavez IR (2011) Saliva as a diagnostic fluid. Dent Clin North Am 55(1):159–178
282. Lewis JG (2006) Steroid analysis in saliva: an overview. Clin Biochem Rev 27(3):139–146
283. Lee JM, Garon E, Wong DT (2009) Salivary diagnostics. Orthod Craniofac Res 12(3):206–211
284. Kolka CM, Bergman RN (2012) The barrier within: endothelial transport of hormones. Physiology (Bethesda) 27(4):237–247
285. Ekström J (1989) Autonomic control of salivary secretion. Proc Finn Dent Soc 85(4–5):323–331; 361–363
286. Garrett JR, Kidd A (1993) The innervation of salivary glands as revealed by morphological methods. Microsc Res Tech 26(1):75–91
287. Bhattacharya S, Verrill D, Giovannucci D (2011) The role of P2X4 receptors in calcium-mediated exocytosis in parotid acinar cells. Biophys J 100(3):p258a
288. Ferreira JN, Hoffman MP (2013) Interactions between developing nerves and salivary glands. Organogenesis 9(3):199–205
289. Eneroth CM, Hökfelt T, Norberg KA (1969) The role of the parasympathetic and sympathetic innervation for the secretion of human parotid and submandibular glands. Acta Otolaryngol 68(5):369–375
290. Norberg KA, Eneroth CM, Hökfelt T (1969) The significance of the autonomic innervation for the salivary secretion in the human parotid and submandibular glands. Acta Otolaryngol Suppl 263:193–194
291. Hettigoda NS, Fong AY, Badoer E, McKinley MJ, Oldfield BJ, Allen AM (2015) Identification of CNS neurons with polysynaptic connections to both the sympathetic and parasympathetic innervation of the submandibular gland. Brain Struct Funct 220(4):2103–2120
292. Sadi H, Finkelman M, Rosenberg M (2013) Salivary cortisol, salivary alpha amylase, and the dental anxiety scale. Anesth Prog 60(2):46–53
293. Ship JA, Fischer DJ (1997) The relationship between dehydration and parotid salivary gland function in young and older healthy adults. J Gerontol A Biol Sci Med Sci 52(5):M310–M319
294. Taubert M, Davies EMR, Back I (2007) Dry mouth. BMJ 334(7592):534
295. Ishikawa Y, Cho G, Yuan Z, Skowronski MT, Pan Y, Ishida H (2006) Water channels and zymogen granules in salivary glands. J Pharmacol Sci 100:495–512
296. Emmelin N (1987) Nerve interactions in salivary glands. J Dent Res 66(2):509–517
297. Proctor GB, Carpenter GH (2014) Salivary secretion: mechanism and neural regulation. Monogr Oral Sci 24:14–29
298. Chopra DP, Xue-Hu IC (1993) Secretion of alpha-amylase in human parotid gland epithelial cell culture. J Cell Physiol 155(2):223–233

299. Keating C (2004) The effects of dopamine agonists and antagonists on the secretory responses in the salivary glands of the locust (*Locusta migratoria*). J Insect Physiol 50 (1):17–23
300. Michalek R, Templeton D (1986) The role of dopamine in salivation in the rat parotid gland. Gen Pharmacol Vascular Syst 17(4):473–476
301. Evans AM, Green KL (1990) Characterization of the dopamine receptor mediating the hyperpolarization of cockroach salivary gland acinar cells in vitro. Br J Pharmacol 101 (1):103–108
302. Baumann O, Dames P, Kühnel D, Walz B (2002) Distribution of serotonergic and dopaminergic nerve fibers in the salivary gland complex of the cockroach Periplaneta americana. BMC Physiol 2:9
303. Just F, Walz B (1996) The effects of serotonin and dopamine on salivary secretion by isolated cockroach salivary glands. J Exp Biol 199(Pt 2):407–413
304. Ghafoor M (2012) Sjögren's before Sjögren: did Henrik Sjögren (1899–1986) really discover Sjögren's disease? J Maxillofac Oral Surg 11(3):373–374
305. Igoe A, Scofield RH (2013) Autoimmunity and infection in Sjögren's syndrome. Curr Opin Rheumatol 25(4):480–487
306. Baccaglini L, Baum BJ (2008) Hypothesis: Sjögren's syndrome: a possible pathogenetic mechanism involving somatostatin. Oral Dis 6(5):264–266
307. Abdulnour-Nakhoul S, Nakhoul HN, Kalliny MI, Gyftopoulos A, Rabon E, Doetjes R et al (2011) Ion transport mechanisms linked to bicarbonate secretion in the esophageal submucosal glands. Am J Physiol Regul Integr Comp Physiol 301(1):R83–R96
308. Joo NS, Krouse ME, Wu JV, Saenz Y, Jayaraman S, Verkman AS, Wine JJ (2001) HCO3− transport in relation to mucus secretion from submucosal glands. JOP 2(4):280–284
309. Orlando RC (2010) The integrity of the esophageal mucosa. Balance between offensive and defensive mechanisms. Best Pract Res Clin Gastroenterol 24(6):873–882
310. Sugimachi K, Sumiyoshi K, Nozoe T, Yasuda M, Watanabe M, Kitamura K et al (1995) Carcinogenesis and histogenesis of esophageal carcinoma. Cancer 75(6):1440–1445
311. Kuwano H, Ueo H, Sugimachi K, Inokuchi K, Toyoshima S, Enjoji M (1985) Glandular or mucus-secreting components in squamous cell carcinoma of the esophagus. Cancer 56 (3):514–518
312. Long JD, Orlando RC (1999) Esophageal submucosal glands: structure and function. Am J Gastroenterol 94(10):2818–2824
313. Shafik A, Shafik AA, El Sibai O, Mostafa RM (2004) Effect of straining on diaphragmatic crura with identification of the straining-crural reflex. The "reflex theory" in gastroesophageal competence. BMC Gastroenterol 4:24
314. Goyal RK, Chaudhury A (2008) Physiology of normal esophageal motility. J Clin Gastroenterol 42(5):610–619
315. Abdulnour-Nakhoul S, Nakhoul NL, Wheeler SA, Wang P, Swenson ER, Orlando RC (2005) HCO3− secretion in the esophageal submucosal glands. Am J Physiol Gastrointest Liver Physiol 288(4):G736–G744
316. Kongara KR, Soffer EE (1999) Saliva and esophageal protection. Am J Gastroenterol 94 (6):1446–1452
317. Akiba Y, Mizumori M, Kuo M, Ham M, Guth PH, Engel E, Kaunitz JD (2008) CO_2 chemosensing in rat oesophagus. Gut 57(12):1654–1664
318. Zhuang L, Peng J-B, Tou L, Takanaga H, Adam RM, Hediger MA, Freeman MR (2002) Calcium-selective ion channel, CaT1, is apically localized in gastrointestinal tract epithelia and is aberrantly expressed in human malignancies. Lab Invest 82:1755–1764
319. Lord RVN, Park JM, Wickramasinghe K, DeMeester SR, Oberg S, Salonga D et al (2003) Vascular endothelial growth factor and basic fibroblast growth factor expression in esophageal adenocarcinoma and Barrett esophagus. J Thorac Cardiovasc Surg 125(2):246–253

320. Fujiwara Y, Higuchi K, Takashima T, Hamaguchi M, Hayakawa T, Tominaga K et al (2006) Roles of epidermal growth factor and Na^+/H^+ exchanger-1 in esophageal epithelial defense against acid-induced injury. Am J Physiol Gastrointest Liver Physiol 290(4):G665–G673
321. Modlin IM, Lane G, Johnson SP, Schoenfeld PS, Allen J, Brill JV (2008) American gastroenterological association medical position statement on the management of gastroesophageal reflux disease. Gastroenterology 135:1383–1391
322. Odze RD (2005) Unraveling the mystery of the gastroesophageal junction: a pathologist's perspective. Am J Gastroenterol 100(8):1853–1867
323. Chandrasoma P (2005) Controversies of the cardiac mucosa and Barrett's oesophagus. Histopathology 46(4):361–373
324. Vakil N, van Zanten SV, Kahrilas P, Dent J, Jones R, Global Consensus Group (2006) The Montreal definition and classification of gastroesophageal reflux disease: a global evidence-based consensus. Am J Gastroenterol 101(8):1900–1920, 1943
325. Modlin IM, Hunt RH, Malfertheiner P, Moayyedi P, Quigley EM, Tytgat GNJ et al (2009) Diagnosis and management of non-erosive reflux disease—the Vevey NERD consensus group. Digestion 80(2):74–88
326. Ang D, Sifrim D, Tack J (2008) Mechanisms of heartburn. Nat Rev Gastroenterol Hepatol 5:383–392
327. Naini BV, Souza RF, Odze RD (2016) Barrett's esophagus: a comprehensive and contemporary review for pathologists. Am J Surg Pathol 40(5):e45–e66
328. Reid BJ, Li X, Galipeau PC, Vaughan T (2010) Barrett's oesophagus and oesophageal adenocarcinoma: time for a new synthesis. Nat Rev Cancer 10(2):87–101
329. Barbon C, Mungo B, Molena D, Yang SC (2015) Severe reflux-induced esophagitis. In: Pawlik TM, Maithel SK, Merchant NB (eds) Gastrointestinal surgery. Springer, New York
330. Hirano I (2011) Eosinophilic esophagitis and gastroesophageal reflux disease: there and back again. Clin Gastroenterol Hepatol 9(2):99–101
331. Hershcovici T, Fass R (2010) Nonerosive reflux disease (NERD)—an update. J Neurogastroenterol Motil 16(1):8–21
332. Gao F, Gao Y, Chen X, Qian J, Zhang J (2017) Comparison of oesophageal function tests between Chinese non-erosive reflux disease and reflux hypersensitivity patients. BMC Gastroenterol 17:67
333. Gindea C, Birla R, Hoara P, Caragui A, Constantinoiu S (2014) Barrett esophagus: history, definition and etiopathogeny. J Med Life 7(3):23–30
334. Van Eyken P (2000) Definition of Barrett's oesophagus. Acta Gastroenterol Belg 63(1):10–12
335. Schubert ML, Peura DA (2008) Control of gastric acid secretion in health and disease. Gastroenterology 134:1842
336. Rowland KJ, Choi PM, Warner BW (2013) The role of growth factors in intestinal regeneration and repair in necrotizing enterocolitis. Semin Pediatr Surg 22(2):101–111
337. Bimczok D, Kao JY, Zhang M, Cochrun S, Mannon P, Peter S et al (2015) Human gastric epithelial cells contribute to gastric immune regulation by providing retinoic acid to dendritic cells. Mucosal Immunol 8:533–544
338. Taupin D, Podolsky DK (2003) Trefoil factors: initiators of mucosal healing. Nat Rev Mol Cell Biol 4:721–732
339. Pakurar AS, Bigbee JW (2004) Digital histology: an interactive CD atlas with review text. Wiley, Hoboken
340. Wright NA (2000) Epithelial stem cell repertoire in the gut: clues to the origin of cell lineages, proliferative units and cancer. Int J Exp Path 81:117–143
341. Miyake K, Tanaka T, McNeil PL (2006) Disruption-induced mucus secretion: repair and protection. PLoS Biol 4(9):e276
342. Roberts NB (2006) Human pepsins—their multiplicity, function and role in reflux disease. Aliment Pharmacol Ther 24(2):2–9
343. Cui G, Waldum HL (2007) Physiological and clinical significance of enterochromaffin-like cell activation in the regulation of gastric acid secretion. World J Gastroenterol 28:493–496

344. Liu Y, Vosmaer GDC, Tytgat GNJ, S-d Xiao, Ten Kate FJW (2005) Gastrin (G) cells and somatostatin (D) cells in patients with dyspeptic symptoms: *Helicobacter pylori* associated and non-associated gastritis. J Clin Pathol 58(9):927–931
345. Sun F-P, Song Y-G, Cheng W, Zhao T, Yao Y-L (2002) Gastrin, somatostatin, G and D cells of gastric ulcer in rats. World J Gastroenterol 8(2):375–378
346. Park SM, Park HS (1993) G- and D-cell populations, serum and tissue concentrations of gastrin and somatostatin in patients with peptic ulcer diseases. Korean J Intern Med 8(1):1–7
347. Sekler I, Kobayashi S, Kopito RR (1996) A cluster of cytoplasmic histidine residues specifies pH dependence of the AE2 plasma membrane anion exchanger. Cell 86(6):929–935
348. Schreiber S, Garten D, Nguyen TH, Konradt M, Bücker R, Scheid P (2007) In situ measurement of pH in the secreting canaliculus of the gastric parietal cell and adjacent structures. Cell Tissue Res 329(2):313–320
349. Phan J, Benhammou JN, Pisegna JR (2015) Gastric hypersecretory states: investigation and management. Curr Treat Options Gastroenterol 13(4):386–397
350. Browning KN, Travagli RA (2014) Central nervous system control of gastrointestinal motility and secretion and modulation of gastrointestinal functions. Compr Physiol 4 (4):1339–1368
351. Konturek SJ, Kwiecien N, Obtulowicz W, Mikos E, Sito E, Oleksy J, Popiela T (1979) Cephalic phase of gastric secretion in healthy subjects and duodenal ulcer patients: role of vagal innervation. Gut 20(10):875–881
352. Håkanson R, Hedenbro J, Liedberg G, Sundler F, Vallgren S (1980) Mechanisms of gastric acid secretion after pylorus and oesophagus ligation in the rat. J Physiol 305:139–149
353. Sachs G, Zeng N, Prinz C (1997) Physiology of isolated gastric endocrine cells. Annu Rev Physiol 59:243–256
354. Fokina A, Konturek SJ, Kwiecien N, Radecki T (1979) Role of gastric antrum in gastric and intestinal phases of gastric secretion in dogs. J Physiol 295:229–239
355. Kester RC (1975) The intestinal phase of gastric secretion. Ann R Coll Surg Engl 56(5):231–245
356. Kelly KA, Nyhus LM, Harkins HN (1964) The vagal nerve and the intestinal phase of gastric secretion. Gastroenterology 46(2):163–166
357. Helander HF, Keeling DJ (1993) Cell biology of gastric acid secretion. Baillieres Clin Gastroenterol 7(1):1–21
358. Chew CS, Nakamura K, Ljungström M (1992) Calcium signaling mechanisms in the gastric parietal cell. Yale J Biol Med 65(6):561–623
359. Helander HF (1988) Physiology and pharmacology of the parietal cell. Baillieres Clin Gastroenterol 2(3):539–554
360. Yao X, Forte JG (2003) Cell biology of acid secretion by the parietal cell. Annu Rev Physiol 65:103–131
361. Hamada E, Nakajima T, Hata Y, Hazama H, Iwasawa K, Takahashi M et al (1997) Effect of caffeine on mucus secretion and agonist-dependent Ca^{2+} mobilization in human gastric mucus secreting cells. Biochim Biophys Acta Mol Cell Res 1356(2):198–206
362. Laddha SS, Wadodkar SG, Meghal SK (2009) cAMP-dependent phosphodiesterase inhibition and SAR studies on novel 6,8-disubstituted 2-phenyl-3-(substituted benzothiazole-2-yl)-4[3*H*]-quinazolinone. Med Chem Res 18(4):268–276
363. Sette C, Conti M (1996) Phosphorylation and activation of a cAMP-specific phosphodiesterase by the cAMP-dependent protein kinase—involvement of serine 54 in the enzyme activation. J Biol Chem 271:16526–16534
364. Koertge N (ed) (2007) New dictionary of scientific biography, 1st edn. Charles Scribners & Sons, Detroit
365. Bourgoin SM (2012) Encyclopedia of world biography, 2nd edn. Gale Research, Detroit
366. Bennett MR (2000) The concept of transmitter receptors: 100 years on. Neuropharmacology 39:523–546

367. Maehle A-H (2004) Receptive substances: John Newport Langley (1852–1925) and his Path to a receptor theory of drug action. Med Hist 48:153–174
368. Langley JN (1905) On the reaction of cells and of nerve-endings to certain poisons, chiefly as regards the reaction of striated muscle to nicotine and to curari. J Physiol 33:374–413
369. Rubin RP (2007) A brief history of great discoveries in pharmacology: in celebration of the centennial anniversary of the founding of the American Society of Pharmacology and Experimental Therapeutics. Pharmacol Rev 59(4):289–359
370. Maehle A-H (2009) A binding question: the evolution of the receptor concept. Endeavour 33 (4):135–140
371. Boes CJ (2014) Langley, John Newport. In: Aminoff MJ, Daroff RB (eds) Encyclopedia of the neurological sciences, 2nd edn. Elsevier, MA
372. Sawaya AC, Costa YD, Mazzafera P (2015) Unraveling the biosynthesis of pilocarpine in *Pilocarpus microphyllus*. Nat Prod Commun 10(5):721–724
373. Sawaya AC, Abreu IN, Andreazza NL, Eberlin MN, Mazzafera P (2008) HPLC-ESI-MS/MS of imidazole alkaloids in *Pilocarpus microphyllus*. Molecules 13(7):1518–1529
374. Dias DA, Urban S, Roessner U (2012) A historical overview of natural products in drug discovery. Metabolites 2(2):303–336
375. Richards D, Aronson J, Reynolds DJ, Coleman J (2011) Oxford handbook of practical drug therapy, 2nd edn. Oxford University Press, Oxford
376. Lidums I, Hebbard GS, Hollowaya RH (2000) Effect of atropine on proximal gastric motor and sensory function in normal subjects. Gut 47:30–36
377. Moulton BC, Fryer AD (2011) Muscarinic receptor antagonists, from folklore to pharmacology; finding drugs that actually work in asthma and COPD. Br J Pharmacol 163(1):44–52
378. O'Brien RD (1974) Atropine. In: Simpson LL, Curtis DR (eds) Poisons of plant origin. Plenum Press, New York
379. Schwarzenberger M, Stintzing F, Meyer U, Lindequist U (2012) Biochemical, microbiological and phytochemical studies on aqueous-fermented extracts from *Atropa belladonna* L. Part 2—phytochemistry. Pharmazie 67(5):460–466
380. Schwarzenberger M, Stintzing F, Meyer U, Lindequist U (2012) Biochemical, microbiological and phytochemical studies on aqueous-fermented extracts from *Atropa belladonna* L. Part 1—biochemistry and microbiology. Pharmazie 67(4):331–344
381. Lee MR (2007) *Solanaceae* IV: *Atropa belladonna*, deadly nightshade. J R Coll Physicians Edinb 37(1):77–84
382. Rajput H (2013) Effects of *Atropa belladonnas* an anti-cholinergic. Nat Prod Chem Res 1:104
383. Blackshaw LA (2008) New insights in the neural regulation of the lower oesophageal sphincter. Eur Rev Med Pharmacol Sci 12(1):33–39
384. Mittal R, Chiareli C, Liu J, Holloway R, Dixon W (1997) Atropine inhibits gastric distension and pharyngeal receptor mediated lower oesophageal sphincter relaxation. Gut 41(3):285–290
385. Zhang Q, Lehmann A, Rigda R, Dent J, Holloway RH (2002) Control of transient lower oesophageal sphincter relaxations and reflux by the GABAB agonist baclofen in patients with gastro-oesophageal reflux disease. Gut 50(1):19–24
386. Vakil N (2004) New pharmacological agents for the treatment of gastro-oesophageal reflux disease. Aliment Pharmacol Ther 19:1041–1049
387. Hirsch DP, Tytgat GNJ, Boeckxstaens GEE (2002) Transient lower oesophageal sphincterrelaxations—a pharmacological target for gastro-oesophageal reflux disease? Aliment Pharmacol Ther 16:17–26
388. Li S, Shi S, Chen F, Lin J (2014) The effects of baclofen for the treatment of gastroesophageal reflux disease: a meta-analysis of randomized controlled trials. Gastroenterol Res Pract 2014:307805
389. Tonini M, De Giorgio R, De Ponti F (2004) Progress with novel pharmacological strategies for gastro-oesophageal reflux disease. Drugs 64(4):347–361

390. Kahrilas PJ, Boeckxstaens G (2012) Failure of reflux inhibitors in clinical trials: bad drugs or wrong patients? Gut 61(10):1501–1509
391. Yamato S, Saha JK, Goyal RK (1992) Role of nitric oxide in lower esophageal sphincter relaxation to swallowing. Life Sci 50(17):1263–1272
392. Tomita R, Tanjoh K, Fujisaki S, Fukuzawa M (2003) Physiological studies on nitric oxide in the lower esophageal sphincter of patients with reflux esophagitis. Hepatogastroenterology 50(49):110–114
393. Baker DE (2007) Loperamide: a pharmacological review. Rev Gastroenterol Disord 7(3): S11–S18
394. Ehlert FJ, Pak KJ, Griffin MT (2012) Muscarinic agonists and antagonists: effects on gastrointestinal function. Handb Exp Pharmacol 208:343–374
395. Borella TL, De Luca LA, Jr Colombari DS, Menani JV (2008) Central muscarinic receptor subtypes involved in pilocarpine-induced salivation, hypertension and water intake. Br J Pharmacol 155(8):1256–1263
396. Carmine AA, Brogden RN (1985) Pirenzepine. A review of its pharmacodynamic and pharmacokinetic properties and therapeutic efficacy in peptic ulcer disease and other allied diseases. Drugs 30(2):85–126
397. Boulanger CM, Morrison KJ, Vanhoutte PM (1994) Mediation by M3-muscarinic receptors of both endothelium-dependent contraction and relaxation to acetylcholine in the aorta of the spontaneously hypertensive rat. Br J Pharmacol 112(2):519–524
398. Watson N, Barnes PJ, Maclagan J (1992) Actions of methoctramine, a muscarinic M2 receptor antagonist, on muscarinic and nicotinic cholinoceptors in guinea-pig airways in vivo and in vitro. Br J Pharmacol 105(1):107–112
399. Jakubík J, Zimčík P, Randáková A, Fuksová K, El-Fakahany EE, Doležal V (2014) Molecular mechanisms of methoctramine binding and selectivity at muscarinic acetylcholine receptors. Mol Pharmacol 86(2):180–192
400. Ehlert FJ (1996) The interaction of 4-DAMP mustard with subtypes of the muscarinic receptor. Life Sci 58(22):1971–1978
401. Holzer P (2009) Opioid receptors in the gastrointestinal tract. Regul Pept 155(1–3):11–17
402. Leppert W (2012) The impact of opioid analgesics on the gastrointestinal tract function and the current management possibilities. Contemp Oncol (Pozn) 16(2):125–131
403. Leppert W (2014) Oxycodone/naloxone in the management of patients with pain and opioid-induced bowel dysfunction. Curr Drug Targets 15(1):124–135
404. Gotfried J, Kataria R, Schey R (2017) The role of cannabinoids on esophageal function—what we know thus far. Cannabis Cannabinoid Res 2(1):252–258
405. Franklin JM, Carrasco GA (2012) Cannabinoid-induced enhanced interaction and protein levels of serotonin 5-HT(2A) and dopamine D_2 receptors in rat prefrontal cortex. J Psychopharmacol 26(10):1333–1347
406. Franklin JM, Carrasco GA (2013) Cannabinoid receptor agonists upregulate and enhance serotonin 2A (5-HT(2A)) receptor activity via ERK1/2 signaling. Synapse 67(3):145–159
407. Vandenberg RJ, Frencht CR, Barry PH, Shine J, Schofield PR (1992) Antagonism of ligand-gated ion channel receptors: two domains of the glycine receptor a subunit form the strychnine-binding site. Proc Natl Acad Sci U S A 89:1765–1769
408. Zhang M, Aguilera D, Das C, Vasquez H, Zage P, Gopalakrishnan V, Wolff J (2007) Measuring cytotoxicity: a new perspective on LC50. Anticancer Res 27(1A):35–38
409. Guo R, Wang T, Zhou G, Xu M, Yu X, Zhang X et al (2018) Botany, phytochemistry, pharmacology and toxicity of *Strychnos nux-vomica* L.: a review. Am J Chin Med 46(1):1–23
410. Gaddum JH (1957) John William Trevan, 1887-1956. Biogr Mem Fellows R Soc 3:273–288
411. Born S, Levit A, Niv MY, Meyerhof W, Behrens M (2013) The human bitter taste receptor TAS2R10 is tailored to accommodate numerous diverse ligands. J Neurosci 33(1):201–213
412. Rajendra S, Lynch JW, Schofield PR (1997) The glycine receptor. Pharmacol Ther 73 (2):121–146

413. Duan L, Yang J, Slaughter MM (2009) Caffeine inhibition of ionotropic glycine receptors. J Physiol 587(16):4063–4075
414. Kuijpers GA, Vergara LA, Calvo S, Yadid G (1994) Inhibitory effect of strychnine on acetylcholine receptor activation in bovine adrenal medullary chromaffin cells. Br J Pharmacol 113(2):471–478
415. Maehle AH, Prüll CR, Halliwell RF (2002) The emergence of the drug receptor theory. Nat Rev Drug Discov 1(8):637–641
416. Ehrlich P (1908) Nobel lecture on partial functions of the cell. In: Himmelweit F, Marquardt M, Dale HH (eds) The collected papers of P. Ehrlich, vol III. Pergamon Press, Oxford
417. Langley JN (1906) On nerve endings and on special excitable substances in cells. Proc R Soc B Biol Sci 70:170–194
418. Clark AJ (1926) The reaction between acetylcholine and muscle cells. J Physiol 61:530–547
419. Clark AJ (1926) The antagonism of acetylcholine by atropine. J Physiol 61:547–556
420. Clark AJ, Raventos J (1937) The antagonism of acetylcholine and of quaternary ammonium salts. Q J Exp Physiol 26:375–392
421. Verney EB, Barcroft J (1941) Alfred Joseph Clark. 1885–1941. Obit Not Fell R Soc 3 (10):969–984
422. Lees GM (1998) A tribute to the late Hans W. Kosterlitz: ploughing the lone furrow. Can J Physiol Pharmacol 76(3):244–251
423. Prüll CR (2006) Caught between the old and the new—Walther Straub (1874–1944), the question of drug receptors, and the rise of modern pharmacology. Bull Hist Med 80(3):465–489
424. Dale HH, Feldberg W, Vogt M (1936) Release of acetylcholine at voluntary motor nerve endings. J Physiol 86:353–380
425. Loewi O (1921) Über humorale Übertragbarkeit der Herznervenwirkung. I.Mittei-lung (About humoral transmissibility of the cardiac nervous system) Pflügers Arch Ges Physiol 189:239–242
426. Loewi O (1921) Über humorale Übertragbarkeit der Herznervenwirkung. II. Mittei-lung. Pflügers Arch Ges Physiol 193:201–213
427. Rang HP (2006) The receptor concept: pharmacology's big idea. Br J Pharmacol 147(S1): S9–S16
428. Limbird LE (2004) The receptor concept: a continuing evolution. Mol Interv 4(6):326–336
429. Ahlquist RP (1948) A study of the adrenotrophic receptors. Am J Physiol 155:586–600
430. Segal I, Yaakov Y, Adler SN, Blau H, Broide E, Santo M et al (2008) Cystic fibrosis transmembrane conductance regulator ion channel function testing in recurrent acute pancreatitis. J Clin Gastroenterol 42(7):810–814
431. Ashlock MA, Olson ER (2011) Therapeutics development for cystic fibrosis: a successful model for a multisystem genetic disease. Annu Rev Med 62:107–125
432. Elborn JS (2007) How can we prevent multisystem complications of cystic fibrosis? Semin Respir Crit Care Med 28(3):303–311
433. Larsen EH (2002) Hans H. Ussing—scientific work: contemporary significance and perspectives. Biochim Biophys Acta Biomembr 1566(1–2):2–15
434. Parsons ME, Ganellin CR (2006) Histamine and its receptors. Br J Pharmacol 147(1):S127–S135
435. Fogel WA (2015) Histamine. In: Parnham M (ed) Encyclopedia of inflammatory diseases. Springer Basel, Switzerland
436. Petersen H (1991) Histamine and the stomach: introduction. Scand J Gastroenterol 26 (180):2–3
437. Ganellin CR (2011) Personal reflections on Sir James Black (1924–2010) and histamine. Inflamm Res 60(1):103–110
438. Konturek SJ, Konturek PC, Konturek JW, Plonka M, Czesnikiewicz-Guzik M, Brzozowski T, Bielanski W (2006) Helicobacter pylori and its involvement in gastritis and peptic ulcer formation. J Physiol Pharmacol 57(3):29–50

439. Black RB, Hole D, Rhodes J (1971) Bile damage to the gastric mucosal barrier: the influence of pH and bile acid concentration. Gastroenterology 61(2):178–184
440. Black RB, Rhodes J, Hole D (1973) Measurement of bile damage to the gastric mucosa. The relation between the electrical potential difference and transmucosal movement of hydrogen and sodium ion. Am J Dig Dis 18(5):411–415
441. Black JW, Duncan WA, Durant CJ, Ganellin CR, Parsons EM (1972) Definition and antagonism of histamine H2-receptors. Nature 236:385–390
442. Black JW, Owen DA, Parsons ME (1975) An analysis of the depressor responses to histamine in the cat and dog: involvement of both H1- and H2-receptors. Br J Pharmacol 54 (3):319–324
443. Ganser AL, Forte JG (1973) K^+-stimulated ATPase in purified microsomes of bullfrog oxyntic cells. Biochim Biophys Acta 307:169–180
444. Ganser AL, Forte JG (1973) Ionophoretic stimulation of K^+-ATPase of oxyntic cell microsomes. Biochem Biophys Res Commun 54(2):690–696
445. Inocente C, Arnulf I, Bastuji H, Thibault-Stoll A, Raoux A, Reimão R et al (2012) Pitolisant, an inverse agonist of the histamine H3 receptor: an alternative stimulant for narcolepsy-cataplexy in teenagers with refractory sleepiness. Clin Neuropharmacol 35 (2):55–60
446. Black J (1988) Drugs from emasculated hormones: the principles of syntopic antagonism. In: Nobel lecture, physiology or medicine
447. Marx JL (1988) The 1988 nobel prize for physiology or medicine. Science 242(4878):516–517
448. Prinz C, Kajimura M, Scott D, Helander H, Shin J, Besancon M et al (1992) Acid secretion and the H, K ATPase of stomach. Yale J Biol Med 65(6):577–596
449. Shin JM, Munson K, Vagin O, Sachs G (2009) The gastric HK-ATPase: structure, function, and inhibition. Pflugers Arch 457(3):609–622
450. Soumarmon A, Lewin MJ (1986) Gastric (H^+, K^+)-ATPase. Biochimie 68(12):1287–1291
451. Sachs G, Shin JM, Vagin O, Lambrecht N, Yakubov I, Munson K (2007) The gastric H, K ATPase as a drug target: past, present, and future. J Clin Gastroenterol 41(2):S226–S242
452. Song P, Groos S, Riederer B, Feng Z, Krabbenhöft A, Smolka A, Seidler U (2009) KCNQ1 is the luminal K^+ recycling channel during stimulation of gastric acid secretion. J Physiol 587(15):3955–3965
453. Heitzmann D, Warth R (2008) Physiology and pathophysiology of potassium channels in gastrointestinal epithelia. Physiol Rev 88(3):1119–1182
454. Julio-Kalajzić F, Villanueva S, Burgos J, Ojeda M, Cid LP, Jentsch TJ, Sepúlveda FV (2018) K2P TASK-2 and KCNQ1-KCNE3K+ channels are major players contributing to intestinal anion and fluid secretion. J Physiol 596(3):393–407
455. Fujii T, Takahashi Y, Ikari A, Morii M, Tabuchi Y, Tsukada K et al (2009) Functional Association between K^+-Cl^- Cotransporter-4 and H^+, K^+-ATPase in the apical canalicular membrane of gastric parietal cells. J Biol Chem 284:619–629
456. Shamburek RD, Schubert ML (1993) Pharmacology of gastric acid inhibition. Baillieres Clin Gastroenterol 7(1):23–54
457. Geibel JP, Wagner CA, Caroppo R, Qureshi I, Gloeckner J, Manuelidis L et al (2001) The stomach divalent ion-sensing receptor scar is a modulator of gastric acid secretion. J Biol Chem 276(43):39549–39552
458. Bakker EP (1979) Ionophore antibiotics. Mechanism of action of antibacterial agents. Antibiotics 5(1):67–97
459. Daniele RP, Holian SK (1976) A potassium ionophore (valinomycin) inhibits lymphocyte proliferation by its effects on the cell membrane. Proc Natl Acad Sci U S A 73(10):3599–3602
460. Shamoo AE, Goldstein DA (1977) Isolation of ionophores from ion transport systems and their role in energy transduction. Biochim Biophys Acta Rev Biomembr 472(1):13–53

461. Bühlmann P, Chen LD (2012) Ion-selective electrodes with ionophore-doped sensing membranes. In: Gale PA, Steed JW (eds) Supramolecular chemistry: from molecules to nanomaterials. Wiley, Chichester
462. Daniele RP, Holian SK, Nowell PC (1978) A potassium ionophore (Nigericin) inhibits stimulation of human lymphocytes by mitogens. J Exp Med 147(2):571–581
463. Sachs G, Shin JM, Besancon M, Prinz C (1993) The continuing development of gastric acid pump inhibitors. Aliment Pharmacol Ther 7(1):4–12
464. Banić M, Malfertheiner P, Babić Z, Ostojić R, Kujundzic M, Fatović-Ferenčić S et al (2011) Historical impact to drive research in peptic ulcer disease. Dig Dis 29(5):444–453
465. Wallmark B, Sachs G, Mardh S, Fellenius E (1983) Inhibition of gastric (H^+-K^+)-ATPase by the substituted benzimidazole, picoprazole. Biochim Biophys Acta 728:31–38
466. Fellenius E, Berglindh T, Sachs G, Olbe L, Elander B, Sjostrand SE, Wallmark B (1981) Substituted benzimidazoles inhibit gastric acid secretion by blocking ($H^{++}K^+$)ATPase. Nature 290:159–161
467. Khan MOF, Deimling MJ, Philip A (2011) Medicinal chemistry and the pharmacy curriculum. Am J Pharm Educ 75(8):161
468. Olbe L, Carlsson E, Lindberg P (2003) A proton-pump inhibitor expedition: the case histories of omeprazole and esomeprazole. Nat Rev Drug Discov 2:132–139
469. Shin JM, Vagin O, Munson K, Kidd M, Modlin IM, Sachs G (2008) Molecular mechanisms in therapy of acid-related diseases. Cell Mol Life Sci 65(2):264–281
470. Lindberg P, Brändström A, Wallmark B, Mattsson H, Rikner L, Hoffmann KJ (1990) Omeprazole: the first proton pump inhibitor. Med Res Rev 10(1):1–54
471. Munson K, Garcia R, Sachs G (2005) Inhibitor and ion binding sites on the gastric H, K-ATPase. Biochemistry 44(14):5267–5284
472. Gremse DA (2001) Lansoprazole: pharmacokinetics, pharmacodynamics and clinical uses. Expert Opin Pharmacother 2(10):1663–1670
473. Simon WA, Herrmann M, Klein T, Shin JM, Huber R, Senn-Bilfinger J, Postius S (2007) Soraprazan: setting new standards in inhibition of gastric acid secretion. J Pharmacol Exp Ther 321(3):866–874
474. Kim HK, Park SH, Cheung DY, Cho YS, Kim JI, Kim SS et al (2010) Clinical trial: inhibitory effect of revaprazan on gastric acid secretion in healthy male subjects. J Gastroenterol Hepatol 25(10):1618–1625
475. Owyang C, Green L, Rader D (1983) Colonic inhibition of pancreatic and biliary secretion. Gastroenterology 84(3):470–475
476. Cumming JH (1975) Absorption and secretion by the colon. Gut 16(4):323–329
477. Matsumoto M, Ooga T, Kibe R, Aiba Y, Koga Y, Benno Y (2017) Colonic absorption of low-molecular-weight metabolites influenced by the intestinal microbiome: a pilot study. PLoS ONE 12(1):e0169207
478. Kuthmann E (1957) Johann Conrad Brunner as a university professor & therapeutist. Gesnerus 14(3–4):119–140
479. Bosmia AN, Tubbs RI, Clapp DC, Batzdorf U, Loukas M, Tubbs RS (2014) Johann Conrad Brunner (1653–1727) and the first description of syringomyelia. Childs Nerv Syst 30 (2):193–196
480. Heitz PU, Kasper M, van Noorden S, Polak JM, Gregory H, Pearse AG (1978) Immunohistochemical localisation of urogastrone to human duodenal and submandibular glands. Gut 19(5):408–413
481. Kasselberg AG, Orth DN, Gray ME, Stahlman MT (1985) Immunocytochemical localization of human epidermal growth factor/urogastrone in several human tissues. J Histochem Cytochem 33(4):315–322
482. Hirata Y, Moore GW, Bertagana C, Orth DN (1980) Plasma concentrations of immunoreactive human epidermal growth factor (urogastrone) in man. J Clin Endocrinol Metab 50 (3):440–444
483. Gerbe F, Legraverend C, Jay P (2012) The intestinal epithelium tuft cells: specification and function. Cell Mol Life Sci 69:2907–2917

484. May R, Qu D, Weygant N, Chandrakesan P, Ali N, Lightfoot SA et al (2014) Brief report: Dclk1 deletion in tuft cells results in impaired epithelial repair after radiation injury. Stem Cells 32(3):822–827
485. Humphries A, Wright NA (2008) Colonic crypt organization and tumorigenesis. Nat Rev Cancer 8:415–424
486. Potten CS (1998) Stem cells in gastrointestinal epithelium: numbers, characteristics and death. Philos Trans R Soc Lond B Biol Sci 353(1370):821–830
487. Baker AM, Cereser B, Melton S, Fletcher AG, Rodriguez-Justo M, Tadrous PJ et al (2014) Quantification of crypt and stem cell evolution in the normal and neoplastic human colon. Cell Rep 8(4):940–947
488. van der Flier LG, Clevers H (2009) Stem cells, self-renewal, and differentiation in the intestinal epithelium. Annu Rev Physiol 71(1):241–260
489. Leung PS (2010) Physiology of the pancreas. The renin-angiotensin system: current research progress in the pancreas volume 690 of the series advances in experimental medicine and biology. Springer, Dordrecht
490. Babu CSR, Sharma M (2014) Biliary tract anatomy and its relationship with venous drainage. J Clin Exp Hepatol 4(1):S18–S26
491. Allescher HD (1989) Papilla of vater: structure and function. Endoscopy 21(1):324–329
492. Behar J (2013) Physiology and pathophysiology of the biliary tract: the gallbladder and sphincter of Oddi–a review. ISRN Physiol 2013:837630
493. Ishiguro H, Yamamoto A, Nakakuki M, Yi L, Ishiguro M, Yamaguchi M et al (2012) Physiology and pathophysiology of bicarbonate secretion by pancreatic duct epithelium. Nagoya J Med Sci 74(1–2):1–18
494. Chang EB, Leung PS (2014) Pancreatic physiology. In: Leung PS (ed) The gastrointestinal system. Springer, Dordrecht
495. Adelson JW, Miller PE (1989) Heterogeneity of the exocrine pancreas. Am J Physiol Gastrointest Liver Physiol 256(5):G817–G825
496. Naruse S, Kitagawa M, Ishiguro H, Hayakawa T (2002) Feedback regulation of pancreatic secretion by peptide YY. Peptides 23(2):359–365
497. Singer MV, Niebergall-Roth E (2009) Secretion from acinar cells of the exocrine pancreas: role of enteropancreatic reflexes and cholecystokinin. Cell Biol Int 33(1):1–9
498. Verspohl EJ, Tacke R, Mutschler E, Lambrecht G (1990) Muscarinic receptor subtypes in rat pancreatic islets: binding and functional studies. Eur J Pharmacol 178(3):303–311
499. Low JT, Shukla A, Thorn P (2010) Pancreatic acinar cell: new insights into the control of secretion. Int J Biochem Cell Biol 42(10):1586–1589
500. Pandol SJ (2010) The exocrine pancreas. Morgan & Claypool Life Sciences, San Rafael
501. Gray MA, Greenwell JR, Argent BE (1990) The role of ion channels in the mechanism of pancreatic bicarbonate secretion. In: Young JA, Wong PYD (eds) Epithelial secretion of water and electrolytes. Springer, Heidelberg
502. Cotton CU (2000) Basolateral potassium channels and epithelial ion transport. Am J Respir Cell Mol Biol 23(3):270–272
503. Banales JM, Gradilone SA (2009) Primers on molecular pathways—ion channels: key regulators of pancreatic physiology. Pancreatology 9(5):556–559
504. Thévenod F (2005) Ion channels in secretory granules of the pancreas: molecular identification and their role in regulated secretion. In: Schultz C (ed) Defects of secretion in cystic fibrosis. Advances in experimental medicine and biology. Springer, Boston
505. Kosters A, Karpen SJ (2008) Bile acid transporters in health and disease. Xenobiotica 38(7–8):1043
506. Kuipers F, Bloks VW, Groen AK (2014) Beyond intestinal soap—bile acids in metabolic control. Nat Rev Endocrinol 10:488–498
507. Martínez-Augustin O, de Medina FS (2008) Intestinal bile acid physiology and pathophysiology. World J Gastroenterol 14(37):5630–5640
508. Esteller A (2008) Physiology of bile secretion. World J Gastroenterol 14(37):5641–5649

509. Rao RK, Samak G (2013) Bile duct epithelial tight junctions and barrier function. Tissue Barriers 1(4):e25718
510. Strazzabosco M (1997) Transport systems in cholangiocytes: their role in bile formation and cholestasis. Yale J Biol Med 70(4):427–434
511. Boyer JL (2013) Bile formation and secretion. Compr Physiol 3(3):1035–1078
512. Kanz MF (2010) Anatomy and physiology of the biliary epithelium. In: McQueen CA (ed) Comprehensive toxicology, 2nd edn. Elsevier, MA
513. Nagahashi M, Shirai Y, Wakai T, Sakata J, Ajioka Y, Hatakeyama K (2007) Perimuscular connective tissue contains more and larger lymphatic vessels than the shallower layers in human gallbladders. World J Gastroenterol 13(33):4480–4483
514. Okada K-I, Kijima H, Imaizumi T, Hirabayashi K, Matsuyama M, Yazawa N et al (2012) Clinical significance of wall invasion pattern of subserosa-invasive gallbladder carcinoma. Oncol Rep 28(5):1531–1536
515. Baumann C (2015) The Physiologist Ewald Hering (1834–1918): curriculum vitae. Strabismus 23(3):135–140
516. Ivy AC (1934) The physiology of the gall bladder. Physiol Rev 14(1):1–102
517. Tietz PS, Chen X-M, Gong A-Y, Huebert RC, Masyuk A, Masyuk T et al (2002) Experimental models to study cholangiocyte biology. World J Gastroenterol 8(1):1–4
518. Tanimizu N, Miyajima A, Mostov KE (2007) Liver progenitor cells develop cholangiocyte-type epithelial polarity in three-dimensional culture. Mol Biol Cell 18 (4):1472–1479
519. Tabibian JH, Masyuk AI, Masyuk TV, O'Hara SP, LaRusso NF (2013) Physiology of cholangiocytes. Compr Physiol 3(1):541–565
520. Tietz P, Levine S, Holman R, Fretham C, Larusso NF (1997) Characterization of apical and basolateral plasma membrane domains derived from cultured rat cholangiocytes. Analyt Biochem 254(2):192–199
521. Tietz PS, Larusso NF (2006) Cholangiocyte biology. Curr Opin Gastroenterol 22(3):279–287
522. Baiocchi L, LeSage G, Glaser S, Alpini G (1999) Regulation of cholangiocyte bile secretion. J Hepatol 31(1):179–191
523. Tormey JMcD, Diamond JM (1967) The ultrastructural route of fluid transport in rabbit gall bladder. J Gen Physiol 50(8):2031–2060
524. Alrefai WA, Gill RK (2007) Bile acid transporters: structure, function, regulation and pathophysiological implications. Pharm Res 24(10):1803–1823
525. Omenetti A, Yang L, Gainetdinov RR, Guy CD, Choi SS, Chen W et al (2011) Paracrine modulation of cholangiocyte serotonin synthesis orchestrates biliary remodeling in adults. Am J Physiol Gastrointest Liver Physiol 300(2):G303–G315
526. Syal G, Fausther M, Dranoff JA (2012) Advances in cholangiocyte immunobiology. Am J Physiol Gastrointest Liver Physiol 303(10):G1077–G1086
527. van Mil SWC, van Oort MM, van den Berg IET, Berger R, Houwen RHJ, Klomp LWJ (2004) D1FIC1 is expressed at apical membranes of different epithelial cells in the digestive tract and is induced in the small intestine during postnatal development of mice. Pediatr Res 56:981–987
528. Arrese M, Ananthananarayanan M, Suchy FJ (1998) Hepatobiliary transport: molecular mechanisms of development and cholestasis. Pediatr Res 44:141–147
529. Srivastava A (2014) Progressive familial intrahepatic cholestasis. J Clin Exp Hepatol 4 (1):25–36
530. Frankenberg T, Miloh T, Chen FY, Ananthanarayanan M, Sun AQ, Balasubramaniyan N et al (2008) The membrane protein ATPase class I type 8B member 1 signals through protein kinase C zeta to activate the farnesoid X receptor. Hepatology 48(6):1896–1905

531. Suchy FJ, Ananthanarayanan M (2006) Bile salt excretory pump: biology and pathobiology. J Pediatr Gastroenterol Nutr 43(1):S10–S16
532. Boyer J (1996) Bile duct epithelium: frontiers in transport physiology. Am J Physiol 270(1 Pt 1):G1–G5
533. Marinelli RA, Tietz PS, Pham LD, Rueckert L, Agre P, LaRusso NF (1999) Secretin induces the apical insertion of aquaporin-1 water channels in rat cholangiocytes. Am J Physiol 276(1 Pt 1):G280–G286
534. Melero S, Spirlì C, Zsembery A, Medina JF, Joplin RE, Duner E et al (2002) Defective regulation of cholangiocyte Cl^-/HCO3(−) and Na^+/H^+ exchanger activities in primary biliary cirrhosis. Hepatology 35(6):1513–1521
535. Mann R, Bhathal PS, Bell C (1991) Aminergic innervation of the gall bladder in man and dog. Clin Auton Res 1(3):205–213
536. Nelson DK, Glasbrenner B, Dahmen G, Riepl RL, Malfertheiner P, Adler G (1996) M1 muscarinic mechanisms regulate intestinal-phase gallbladder physiology in humans. Am J Physiol Gastrointest Liver Physiol 271(5):G824–G830
537. Jordan PH Jr (1964) Physiology of bile secretion. Am J Surg 107(2):367–370
538. Spirlì C, Fabris L, Duner E, Fiorotto R, Ballardini G, Roskams T et al (2003) Cytokine-stimulated nitric oxide production inhibits adenylyl cyclase and cAMP-dependent secretion in cholangiocytes. Gastroenterology 124(3):737–753
539. Tabibian JH, Trussoni CE, O'Hara SP, Splinter PL, Heimbach JK, LaRusso NF (2014) Characterization of cultured cholangiocytes isolated from livers of patients with primary sclerosing cholangitis. Lab Invest 94(10):1126–1133
540. Tabibian JH, O'Hara SP, Splinter PL, Trussoni CE, LaRusso NF (2014) Cholangiocyte senescence by way of N-ras activation is a characteristic of primary sclerosing cholangitis. Hepatology 59(6):2263–2275
541. Park J, Gores GJ, Patel T (1999) Lipopolysaccharide induces cholangiocyte proliferation via an interleukin-6-mediated activation of p44/p42 mitogen-activated protein kinase. Hepatology 29(4):1037–1043
542. Okado-Matsumoto A, Matsumoto A, Fujii J, Taniguchi N (2000) Effect of cAMP on inducible nitric oxide synthase gene expression: its dual and cell-specific functions. Antioxid Redox Signal 2(4):631–642
543. Utter A, Goss F (1997) Exercise and gall bladder function. Sports Med 23(4):218–227
544. Tierney S, Pitt HA, Lillemoe KD (1993) Physiology and pathophysiology of gallbladder motility. Surg Clin North Am 73(6):1267–1290
545. Hofmann AF (2007) Biliary secretion and excretion in health and disease: current concepts. Ann Hepatol 6(1):15–27
546. Li M, Ali SM, Umm-a-OmarahGilani S, Liu J, Li Y-Q, Zuo X-L (2014) Kudo's pit pattern classification for colorectal neoplasms: a meta-analysis. World J Gastroenterol 20 (35):12649–12656
547. Liu HH, Kudo SE, Juch JP (2003) Pit pattern analysis by magnifying chromoendoscopy for the diagnosis of colorectal polyps. J Formos Med Assoc 102(3):178–182
548. Gunduz-Demir C, Kandemir M, Tosun AB, Sokmensuer C (2010) Automatic segmentation of colon glands using object-graphs. Med Image Anal 14(1):1–12
549. Binder HJ, Rajendran V, Sadasivan V, Geibel JP (2005) Bicarbonate secretion: a neglected aspect of colonic ion transport. J Clin Gastroenterol 39(4 Suppl 2):S53–S58
550. Macleod RJ (2013) CaSR function in the intestine: hormone secretion, electrolyte absorption and secretion, paracrine non-canonical Wnt signaling and colonic crypt cell proliferation. Best Pract Res Clin Endocrinol Metab 27(3):385–402
551. Binder HJ (2010) Role of colonic short-chain fatty acid transport in diarrhea. Annu Rev Physiol 72:297–313

552. Milla PJ (2009) Advances in understanding colonic function. J Pediatr Gastroenterol Nutr 48 (2):S43–S45
553. Hedemann MS, Kristiansen E, Brunsgaard G (2002) Morphology of the large intestine of the pig: haustra versus taenia. Ann Anat 184(4):401–403
554. Tang L, Peng M, Liu L, Chang W, Binder HJ, Cheng SX (2015) Calcium-sensing receptor stimulates Cl(−)- and SCFA-dependent but inhibits cAMP-dependent HCO3(−) secretion in colon. Am J Physiol Gastrointest Liver Physiol 308(10):G874–G883
555. Brownlee IA, Havler ME, Dettmar PW, Allen A, Pearson JP (2003) Colonic mucus: secretion and turnover in relation to dietary fibre intake. Proc Nutr Soc 62(1):245–249
556. Vidyasagar S, Rajendran VM, Binder HJ (2004) Three distinct mechanisms of HCO3− secretion in rat distal colon. Am J Physiol Cell Physiol 287(3):C612–C621

Chapter 12
Chemical Digestion, Absorption, and Transport

Abstract Chemical digestion involves the catalytic processing of food in the gastrointestinal (GI) tract by digestive enzymes, aided by co-secreted substances, required to break down the food substances into simpler molecules for absorption. This process is necessary for transport and subsequent metabolic reactions that access the macroergic bonds in food molecules. The catalytic process of digestion starts from the mouth for lipids and carbohydrates. For proteins, the catalytic process begins from the stomach. The sources of the enzymes and co-secreted factors are the glands lining the GI tract, namely salivary, submucosal, gastric, and intestinal glands. This chapter provides up-to-date information on the course of discovery as well as the discoverers of major digestive enzymes in humans. The functions and mechanisms of action of all major digestive enzymes discovered up to the twenty-first century are discussed herein. This chapter is a useful reference source on the history and discovery of major digestive enzymes and the mechanisms of regulation of their functions. The clinical importance of the enzymes is systematically provided at strategic points of the discussion. This chapter provides cellular and molecular mechanisms on the absorption and transport of nutrients, ions, dietary elements, vitamins, toxic metals, and pharmacological drugs in the gut. Pathological implications of some of the components of GI metabolism are strategically outlined.

Keywords Chemical processing of food · Chemical digestion · Catalysis Carbohydrate · Lipid · Fatty acids · Protein · Amino acids · Protease Pepsin · Trypsin · Chymotrypsin · Duodenase · Enterokinase · Carbohydrase Amylase · Lipase · Lingual lipase · Gastric lipase · Pancreatic lipase Colipase · Alpha-glucosidases · Maltase-glucoamylase · Sucrase-isomaltase Lactase-phlorizin hydrolase · Trehalase · Disaccharidases · *Succus entericus* Trypsin inhibitor · Trypsin receptor · SWEET · Glucose chansporter TGA resynthesis · Chylomicrons · Basolateral exocytosis · Lipid absorption Dietary elements' gut epithelial ion transport · Calcium absorption and transport Iron absorption and transport · Magnesium absorption and transport

M. O. Welcome, *Gastrointestinal Physiology*,
https://doi.org/10.1007/978-3-319-91056-7_12

Zinc absorption and transport · Metal absorption and transport · Anion absorption and transport · Toxic metal absorption and transport Pharmacological drug · Vitamin absorption and transport · Water-soluble vitamins · Lipid-soluble vitamins · Bile acids · Enterohepatic recirculation Water absorption and transport · Cystic fibrosis · Pancreatic hyperenzymemia Gullo's syndrome · Variant pancreatic ducts · Pancreatic ductal system Alexander Mixailovich Ugolev · Alexander Yakovlevich (Jakulovich) Danilevsky or Alexander · Jakulowitsch Danilewsky · Anselme Payen Apollinaire Bouchardat · Claude Bernard · Erhard Friedrich Leuchs Gabriel Gustav Valentin · Hans Henriksen Ussing · Horace Middleton Vernon Jacques Loeb · James Batcheller Sumner · Jean-François Persoz Johann Nepomuk Eberle · Johannes Bohn · John Howard Northrop Julius Wohlgemuth · Lucio Gullo · Moses Kunitz · Nikolas Petrovich Shepovalnikov · Robert Robison · Roger moss Herriott · Rudolf P. H. Heidenhain Sigmund Rosenheim Otto · Theodor Schwann · Wendell Meredith Stanley Willy Kühne (Wilhelm Friedrich Kühne) · Zamolodchikova Tatyana Stepanovna

Abbreviations

5HT1B	5-Hydroxytryptamine type 1B
Asp	Aspartate
BACE-1	Beta-site APP-cleaving enzyme-1
Ca^{2+}	Calcium ion
CCK	Cholecystokinin
CD	Cluster of differentiation
CLPS	Colipase
CNS	Central nervous system
COX-1 & -2	Cyclooxygenases-1 and 2
CT	Computer tomography
Cu^{2+}	Copper ion
Cys	Cystine
F1-ATPase	F-type of adenosine triphosphate enzyme
HCl	Hydrochloric acid
Hg^{2+}	Mercury ion
kDa	Kilodalton
kg	Kilogram
mEq/L	Milliequivalent per liter
mg/dl	Milligram per deciliter
mL	Milliliter
mmol	Millimole
NBT-PABA	*N*-benzoyl-L-tyrosyl-p-aminobenzoic acid
NSAID	Non-steroid anti-inflammatory drug
PGs	Prostaglandins
SCFA	Short-chain fatty acids

α	Alpha
β	Beta
γ	Gamma

12.1 Introduction

Chemical digestion (biocatalytic processing of food) is the breakdown of food into small molecules that can be absorbed by special structures of the gastrointestinal (GI) epithelium. Chemical processing of food ensures, however, in concert with mechanical churning (processing) that the required substances or molecules in food are absorbed from the lumen of the GI tract. Chemical processing is accomplished by enzymes and other substances (e.g., acids), which are co-secreted into the lumen of the tract from salivary, submucosal, gastric, and intestinal glands. These glands produce the enzymes, which are usually stored in cytoplasmic vesicles, and released into the exterior of the glandular cell in response to specific stimuli. The enzymes function to cleave energy bonds in the food substances, whereas the co-secreted substances provide the required milieu needed for enzymatic action. The end products of the enzymatic reactions are then absorbed into the epithelial cells from where they are further broken down to simpler molecules to provide energy or packaged for transport or directly transported into the blood [1–6]. The basic principles of enzymatic cleavage are shown in Fig. 12.1.

Cleavage products of the enzymatic reactions as well as some molecules ingested together with food are absorbed in the intestine by transporters or diffuse into the enterocyte cytoplasm from where they are transported into the bloodstream to perform their respective functions [7–9].

12.2 Brief History of Chemical Digestion and Digestive Enzymes

Digestive enzymes generally may be divided into intracellular and extracellular types. The extracellular enzymes are the parietal (also known as membrane) and cavity types, which are responsible for breaking down food in the lumen of the intestines. Extracellular digestion mainly occurs in the membrane. Only a very small percentage of hydrolysis actually takes place in the lumen—cavity digestion [11].

The decomposition/fermentation theory was the dominating theory not until accumulating experimental evidences showed the role of specific enzymatic cleavage reaction as a necessary step in digestion [12]. In 1932, Rudolf Weidenhagen (1900–1979), a key personality in the physiology of digestion,

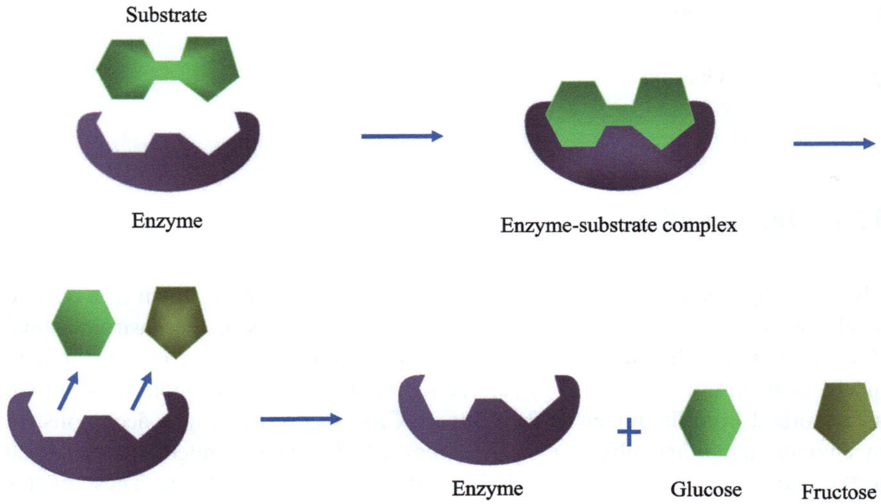

Fig. 12.1 Catalytic cycle of an enzyme. In the presence of a suitable substrate, the enzyme under favorable conditions, using its active sites, forms an association with the substrate, producing enzyme–substrate complex, which undergoes reactions to form the products, while the enzyme remains unchanged. The general equation for the enzymatic reaction is $E + S \leftrightarrow ES \leftrightarrow E + P$, where E is the enzyme, S is the substrate, ES is the enzyme–substrate complex, and P is the product [7–10]

became one of the first to extensively describe the lumenal catalysis of food including digestion of polysaccharides and disaccharides. The precise location of the disaccharide activity was first reported by Dahlqvist and co-authors in 1957 [13]. In the following year, the Russian physiologist Alexander Mixailovich Ugolev (1926–1991) discovered membrane (also known as contact or parietal) digestion or hydrolysis [11, 14, 15]. In one of his experiments, he placed raw and heat-treated (boiled) frog in different solutions containing gastric juice obtained from human, horse, and dog. The result indicated that the raw frog digested faster (and without residues) than the heat-treated frog. This experiment was reported in his book "The theory of adequate nutrition and trophology" published in 1980 [16, 17]. Other crucial contributors to parietal digestion include Arne Dahlqvist (1909–1995) and Bengt Borgström [13, 18–20]. Parietal digestion takes place on the membrane of the epithelial cells of the intestine. It occurs in the glycocalyx zone through the activities of fixed host intestinal enzymes resulting in the formation of absorbable units. These intestinal enzymes are synthesized in the enterocytes and transferred to the membrane of the microvilli where they carry out the process of hydrolysis [18]. Parietal digestion has directed arrangement of active centers of the participating enzymes, whose products of catalysis are coupled to the membrane transporters. Before absorption of nutrients takes place, the food must undergo sequential cavity and membrane digestion before it is transported into the enterocytes [21]. Further information on extracellular digestion is discussed in Tables 12.1, 12.2, and 12.3.

Table 12.1 Description of course of discovery of carbohydrate-digesting enzymes

Enzyme	Description	Discoverer	Year of discovery
Amylase	The hydrolase enzyme that catalyzes the breakdown (hydrolysis) of starch producing different lengths of sugars. Amylase refers to a group of enzymes of carbohydrate digestion—glycoside hydrolase, glycosidase, and transglucosidase. These enzymes hydrolyze or transglycosidize the glycosidic bonds of carbohydrates. About half of all enzymes of carbohydrate digestion are glycoside hydrolases, glycosidases, and transglucosidases [68]. All currently known carbohydrate enzymes are catalogued in a special database "Carbohydrate-Active Enzymes" (http://www.cazy.org). In 2009, the number of enzymes in the group of glycoside hydrolases was 113 members [68]. In 2013, the number increased to 133 members as more enzymes were identified. As at 2014, the current database number of carbohydrate-active enzymes in nature was 340,000 [69] Unlike other enzymes in the GI tract, amylase is secreted in the active form [72]. The enzyme is produced in the digestive tract of humans and other mammals, where it begins the chemical process of digestion. It occurs in almost all species of living things. In humans, the enzyme is produced in the salivary gland and pancreas. Amylase occurs as α-, β-, γ-amylases. The first two are referred to glycogenases [73–75] The α-amylases (also known as 1,4-α-D-glucan glucanohydrolase) are calcium metalloenzymes that randomly cleave the 1,4-alpha glycosidic linkages in carbohydrates to yield maltotriose and maltose from amylose, or maltose, glucose, and dextrin from amylopectin. The optimum pH at which the enzyme functions is around 6.7–7.0. Apart from animals, α-amylases are found in plants, fungi (e.g., ascomycetes, basidiomycetes) and bacteria (e.g., bacillus) [75, 76] Beta-amylase (also known as 1,4-α-D-glucan maltohydrolase, saccharogen amylase) is the enzyme that catalyzes the hydrolysis of the second α-1,4 glycosidic bond, cleaving off maltose at a time. This enzyme is widely distributed in nature. Beta-amylase is a very functional enzyme during the ripening of fruit; it breaks down starch into maltose. That is why ripe fruits have sweet flavor compared with unripe ones. Beside plants, the enzyme is also produced by bacteria and fungi. However, it is not synthesized in humans. Microbes use amylases to obtain nutrients by degrading starch. The enzymes produced by these microbes are released into the extracellular environ where they degrade the substrate. Some microbes in digestive tract produce these enzymes that break down some starch into simpler units [73–77] The optimum pH for β-amylase activity is around 4.0–5.0 [77] Gamma-amylase (also known as glucan 1,4-α-glucosidase, amyloglucosidase, exo-1,4-α-glucosidase, glucoamylase, lysosomal α-glucosidase, 1,4-α-D-glucan glucohydrolase) cleaves α(1–6) glycosidic bonds, as well as the last α(1–4) glycosidic	Erhard Friedrich Leuchs (1800–1837) first described the hydrolysis of starch by saliva and suggested that it was due to the presence of an enzyme, called "ptyalin." In 1833, Anselme Payen (1795–1871) and Jean-François Persoz (1805–1868) isolated an amylase complex from germinating barley and named it "diastase." In 1862, Alexander Jakulowitsch Danilewsky (1838–1923) separated pancreatic amylase from trypsin [70, 71] Almost a century after the discovery of amylase, the German Julius Wohlgemuth (1874–1948) in 1908 described a quantitative method (amyloclastic method) for determining the concentration of amylase (diastase) in the serum. The method can be used to assess amylase activity in body fluids including blood, urine, and duodenal fluid. Before the introduction of laparotomy in medical practice, amyloclastic method was extensively used for the diagnosis of pancreatitis [80, 81]. To execute the analysis of amylase in the test specimens, a known quantity of soluble starch is added in a series of test tubes containing known volumes of test specimens. If the test specimens contain amylase, the starch is digested stoichiometrically. The mixture is then incubated for 30 min at 37OC and immediately cooled to stop any ongoing reaction. Thereafter iodine is added into each test tube to assess the digestibility of the test specimens. If the starch in a given test tube has completely digested, it is evidenced by the complete absence of blue color. The presence of amylase in the test tubes is given as amylase units. Amylase units are calculated as mg of starch that would be completely digested by 1.0 cc of the test specimens [81–84]. With constant progress in technology, there are now new methods of evaluating amylase concentration in body fluids [85–89]	1831

(continued)

Table 12.1 (continued)

Enzyme	Description	Discoverer	Year of discovery
	bonds at the non-reducing end of amylose and amylopectin to yield glucose. The optimum pH for γ-amylase activity is around 3. γ-amylase is present in the livers of carp (Cyprinus carpio) and goldfish (Carassius auratus). The activities of this liver enzyme may be comparable with classic liver glycogen phosphorylase. Hepatic gamma-amylase may be an important glycogenolytic enzyme in carp but not in goldfish. This is because the activity of hepatic gamma-amylase is approximately one half that of glycogen phosphorylase in carp, whereas in goldfish, the activity of gamma-amylase is less than one-sixth that of the phosphorylase [78, 79]		
Disaccharidases	These are groups of enzymes classified as glycoside hydrolases that cleave the bonds found in disaccharides, into simpler sugars called monosaccharides. Disaccharides are synthesized in the epithelial cell and tethered to the lumenal plasma membrane of absorptive enterocytes. These enzymes are localized mostly in the brush border of the small intestine. They are further classified on the basis of the type of sugar cleaved as lactase, maltase, sucrase, among others (see below) [90, 91] Thus, lactase (also called lactase phlorizin hydrolase) catalyzes the hydrolysis or cleavage of lactose into glucose and galactose [94–96] Maltase is a brush border intestinal enzyme that cleaves maltose into two glucose units [90, 91] Sucrase cleaves sucrose into glucose and fructose [90] Alpha-glucosidases are intestinal brush border carbohydrate-hydrolase enzymes that act on 1,4-alpha bonds such as those seen in glycolipids and glycoproteins to release a single alpha-glucose molecule [104]. This is in contrast to beta-glucosidase. Alpha-glucosidase breaks down starch and disaccharides to glucose units. They carry out the final stage of starch hydrolysis in the intestine [67]. Examples of alpha-glucosidases are maltase-glucoamylase, sucrase-isomaltase, lactase phlorizin hydrolase, and trehalase [105–107] Sucrase-isomaltase is an intestinal brush border enzyme important for the digestion of carbohydrates [96]. The human sucrase-isomaltase is highly expressed in the brush border of the intestine and can hydrolyze α-1 → 2, α-1 → 4 and α-1 → 6, glucosidic bonds. This enzyme is responsible for 60–80% of maltase activity and virtually all sucrase activities in the intestine [108]. The usefulness of sucrase-isomaltase is seen in congenital deficiency of the enzyme. The disorder leads to the inability of the individual to break down sucrose and maltose [96, 105, 107]	Brown and Heron were the first to have observed that disaccharidases are present in the mucosa of the intestine [97], and this observation was confirmed by other investigators [92, 93]. However, a problem soon arose—whether or not the enzymes that hydrolyze disaccharides are localized to the intestinal mucosa (brush border) or found freely in the lumen. It was Mosenthal who suggested that these enzymes were localized in the lumen [98] This initial report of the hydrolysis of carbohydrates and proteins mainly in the lumen spurred a couple of researchers to believe that intestinal hydrolytic enzymes are secreted together with intestinal juice or succus entericus into the lumen [99]. However, there were scientists who continued to uphold the view that intestinal hydrolysis took place in the mucosa [19] In 1935, Cajori [100] experimentally showed that in dog intestine lactose was broken down by lactase. Following initial observation of Johnson & Kugler (1953) [101], the precise location of the disaccharide activity was first reported by Dahlqvist and co-authors in 1957 [13]. And almost the same time, similar results were reported by Alexander Mixailovich Ugolev (1926–1991). Ugolev who extensively studied digestion in the brush border of the gut invented the term “contact or membrane” digestion [11, 14–17] Sooner or later, immunohistochemical techniques enabled scientists to identify the localization of intestinal enzymes to the mucosa, precisely the brush border [102]. One of the most convincing results was provided by Miller and Crane who observed that the brush border possessed most of the alkaline phosphatase and disaccharidase activity of the intestinal mucosa [103]	ca. 1880

(continued)

Table 12.1 (continued)

Enzyme	Description	Discoverer	Year of discovery
	Maltase-glucoamylase (or just glucoamylase) is an intestinal brush border alpha-glucosidase digestive enzyme important for the digestion of carbohydrates, especially when there is reduced activity of lumenal alpha-amylase [96]. Such situation may occur in cases of malnutrition and immaturity. This enzyme is responsible for the hydrolysis of terminal $(1 \rightarrow 4)$-linked alpha-D-glucose residues successively from non-reducing ends of the molecules with release of beta-D-glucose [109, 110]. Maltase-glucoamylase consists of two subunits having different substrate specificity. This protein possesses both N-terminal and C-terminal catalytic domains. The N-terminal catalytic domain possesses the highest specificity and activity for maltose, whereas the C-terminal domain possesses the highest specificity and activity for glucose oligomers and dimers such as those in sucrose. The maltase-glucoamylase works in synergy with sucrase-isomaltase and alpha-amylase to digest carbohydrates in the gut. Deficiency of the glucoamylase is implicated in certain cases of chronic diarrhea [109, 110]. The activity of this enzyme is very important in infant who are deficient in pancreatic amylase [109]		
	Isomaltase is an enzyme that breaks the alpha $1 \rightarrow 6$ bonds in sugar, which cannot be cleaved by amylase or maltase. Isomaltase cleaves alpha-limit dextrin, a product of amylopectin digestion that retains its $1 \rightarrow 6$ linkage following digestion by alpha-amylase. The digestion of alpha-dextrin by isomaltase produces maltose. Isomaltase can break down the bonds in isomaltose, a natural disaccharide composed of glucose and fructose linked by alpha-1,6 bonds. Hence, the name is "isomaltase." Thus, isomaltose is similar to maltose, but instead of α-$(1 \rightarrow 4)$-bonds, it possesses α-$(1 \rightarrow 6)$-bond. Evidence suggests that intestinal disaccharidases from various species (including man) can hydrolyze isomaltulose (a sweetener that is similar to sucrose). However, the rate of digestion is slow [111–113]. Isomaltulose is also known by the trade name Palatinose, which is manufactured by enzymatic rearrangement (isomerization) from sucrose. This is a rearrangement of the alpha-1,2 bond between the glucose and the fructose molecule to an alpha-1,6 bond. The caramelization (pyrolysis, non-enzymatic browning of sugar) of sucrose produces fructose and glucose. It is used as sugar alternative [111]. It is one of the favorable sweeteners for diabetics and prediabetics. This carbohydrate is a derivative of a natural source of sucrose [90, 91]		
	Trehalase cleaves trehalose into two glucose units. Trehalose is a natural disaccharide formed by an α,α-1,1-glucoside bond between two α-D-glucose units. Trehalose functions in many organisms as an energy source or protectant against the effects of freezing or dehydration [114]		

(continued)

Table 12.1 (continued)

Enzyme	Description	Discoverer	Year of discovery
	Several factors seem to affect the activity of intestinal enzymes. Ethnic differences in the activity of intestinal enzymes have been reported [115]. Aging also affects the functionality of gut enzymes. Majority of the intestinal enzymes, including disaccharidases have been shown to decrease with aging [96]. Thus, aging is associated with reduced nutrient absorption, which may lead to malnutrition. Wallis et al. [104] previously reported reduced absorption of carbohydrate in the elderly. The authors suggested that this reduction was due to reduced activity of brush border disaccharidases or glucose transporter activities [104]. The results of Wallis et al. [104] also indicate that intestinal infection may reduce the amount of molecules absorbed. There was no gender difference in hydrolase activity and transport of molecules [104]. Ingestion of drugs may as well affect the activity of intestinal hydrolases [116]. The effect of chronic ethanol administration on the activity of intestinal brush border enzymes (sucrose, alkaline phosphatase as well as peptidases including enterokinase) was investigated and found to be significantly decreased [116] Deficiency of disaccharidase is implicated in carbohydrate malabsorption, intolerance, which may be a primary process or can be caused by a secondary phenomenon [91, 117–119]		

Note Apart from the amylase and disaccharidases, there are oligosaccharidases, which are enzymes that break down oligosaccharides into simpler molecules—disaccharides and monosaccharides. Oligosaccharides are carbohydrates that are composed of several monosaccharide residues joined through glycosidic linkages. Oligosaccharides consist of 2–10 monosaccharide residues [120]

Table 12.2 Description of characteristics of major protein-digesting enzymes, their mechanisms of action, and course of discovery

Enzyme	Brief description	Discoverer	Year of discovery
Pepsin	Pepsin is a monomeric proteolytic enzyme with the molecular weight 34.5 kDa and belongs to the group of protein enzymes called gastric aspartic proteinases. In this group of proteinases, the center of the cleft serves as the catalytic site and contains the amino acid aspartate. In pepsin, the aspartates are "Asp32 and Asp215" (the numbers denote the positions of the amino acid in the polypeptide) [158]. The catalytic activity of this protein is determined by the pH value. At a pH of 1.5, the catalytic activity of pepsin increases up to 90%. As the pH shifts toward the alkaline direction, the activity decreases. At a pH of 7, pepsin becomes irreversibly denatured. The optimum range of functioning for pepsin is at pH 1.0–4.0. At this pH range, there is usually higher number of catalytic sites available for reaction. This is due to the protonation of the aspartates in the catalytic site [158, 159] Pepsin is produced in the inactive form as a zymogen, called pepsinogen (with a molecular weight of 41.4 kDa), by the chief cells of the stomach mucosa [164, 165]. There are over 50 types of known pepsinogens for which the amino acid sequence has been determined. Extensively studied types of zymogens of pepsins are gastric digestive proteinases: pepsinogens A, B, C, F, progastricsin, and prochymosin (prorennin) [158, 163, 164]. All pepsins are produced from their corresponding pepsinogens [164, 165]. Pepsin A is the predominant gastric protease. The other types of pepsins are present in gastric juice but in trace amounts [166, 167]. But the precursors of pepsins occur at varying proportions in different mammals. The zymogens prochymosin and pepsinogens F are known to be fetus/infant-specific, especially in primates. In adult mammals, pepsinogen A and progastricsin are main types of the zymogens produced by the gastric mucosa [158, 168, 169] The secretion of pepsinogen and formation of its active form are controlled by many factors. Some hormones such as gastrin and neural inputs from the vagus nerve stimulate pepsinogen and HCl production. HCl released from parietal cells through intramolecular interactions activates this proenzyme causing it to unfold and cleaving itself by autocatalysis to form pepsin, a polypeptide of 44 amino acids. Pepsin is responsible for breaking down proteins into its constituent parts. The enzyme is a close relative of trypsin and chymotrypsin, which are the primary enzymes that degrade protein in the digestive tract [170] After discovery of the enzyme pepsin, a couple of years later, it was confirmed that the isolated gastric enzyme pepsin degraded proteins into peptones—water-soluble products of partial hydrolysis of proteins. Unfortunately, the purity of this enzyme could not be ascertained due to possible contamination with other gastric contents [1, 60]. The first crystallization of pepsin was achieved almost a century following the discovery of the enzyme. The first crystallization of an enzyme carried out in 1926 by the Harvard University	Theodor Schwann (1810–1882) coined the name "pepsin" from the Greek "pepsis," meaning "digestion" (from the Greek "peptein"—to digest) [1, 161] Schwann, a university of Bonn medical graduate, worked with Johannes Muller (1801–1858) and in 1836 discovered that a hydrophilic proteolytic factor in gastric juice digested egg white. It was Schwann who in 1836 reported the first digestive zymogen (pepsinogen) in pancreatic juice. However, the term zymogen was coined by Professor Rudolf P.H. Heidenhain in 1875 (1834–1897). Zymogen denotes a proenzyme. Heidenhain discovered that active enzymes in pancreas of living subjects are very small in quantity compared to postmortem pancreas. Consequently, he proposed that the enzymes existed as precursors in the living pancreas, which were converted to their active forms upon stimulation [162, 163]	1836

(continued)

Table 12.2 (continued)

Enzyme	Brief description	Discoverer	Year of discovery
	graduate James Batcheller Sumner (1887–1955) of Cornell University, Ithaca, New York. Sumner successfully obtained and crystallized the enzyme urease and showed convincingly that it was proteinous in nature. It was based on this work that John Howard Northrop (1891–1987), working in the famous Rockefeller Institute, USA, for the first time in 1929 obtained crystals of pepsin and showed convincingly that it was a protein [171]. Northrop who initially trained in chemistry developed interest in enzyme research after being introduced to the field of physiology by Jacques Loeb (1859–1924), a German physiologist who worked in Europe and the USA at the late nineteenth and early twentieth centuries [172–174] Thereafter, Northrop collaborated with other scientists to crystalize other enzymes. He also worked with renowned scientists notably Moses Kunitz (1887–1978), to isolate and crystallize trypsin [175, 176]. Subsequently, other gut enzymes including chymotrypsin, carboxypeptidase, and their precursors were crystalized. More importantly, Kunitz crystallized the enzymes that degrade nucleic acids, ribonuclease, and deoxyribonuclease, which provided future researchers with necessary information for investigating nucleic acids at a time when scientists were just beginning to understand the functions of cellular acids. Northrop also worked with Roger Moss Herriott (1908–1992), who after completing his PhD in 1932 at the Columbia University joined the Rockefeller Institute. He later gained a position at Johns Hopkins University (Baltimore, Maryland, USA) and became a Professor at the Johns Hopkins University School of Hygiene and Public Health and also served as chairman of biochemistry. Herriott studied pepsin and the biochemistry of other molecules. Herriott and Northrop obtained crystals of pepsin and its zymogen and showed the autocatalytic conversion of pepsinogen to pepsin. Their results were published in Science, one of the world leading journals in scientific research [177] In 1946 Northrop shared the Nobel Prize with Sumner (1887–1955) and another Rockefeller scientist Wendell Meredith Stanley (1904–1971). Stanley, while in 1935, applying Northrop's method of crystalizing protein, obtained crystals of tobacco mosaic virus, the organism that causes infection of many plants. Surprisingly, in crystal form, the organisms still remained infectious [178]. These discoveries raised much skepticism in the world. But as more and more laboratories all over the world began confirming these studies and synthesizing new and crystallizing many more proteins, the cognizance of the pioneers' work became widely recognized around the world. Further development of other techniques (such as chromatography) provided certain level of impetus to enzyme research. For instance, advancement in the area of X-ray diffraction techniques during the 70s of the last century granted scientists an opportunity to visualize the three-dimensional structure of pepsin. The results of dozens of experiments conducted in different regions of the world allowed for a better understanding of the catalytic reaction initiated by pepsin and other enzymes [179–183]		

(continued)

Table 12.2 (continued)

Enzyme	Brief description	Discoverer	Year of discovery
	Pepsin preferentially cleaves hydrophobic and aromatic amino acids such as phenylalanine, tyrosine, and tryptophan. These amino acids including other hydrophobic ones are preferred at the P1 and P'1 positions. The active-site cleft of the enzyme can accommodate at least seven residues of a substrate [184–187] A major point to note is that all enzymes are synthesized and released in their inactive form. Different types of gastric enzymes have different ways of activation. Moreover, the enzymes have different isozymes. When food is ingested, zymogens (inactive enzymes) are released into the gastric lumen and undergo conversion into active enzymes in the acidic environment of the stomach [7, 188, 189] The conversion of the zymogen into its active form is a complex process, and this involves a series of conformational changes and bond cleavage steps that lead to the unlocking of the active site and ultimately the removal and dissociation of the prosegment from the active center of the enzyme [190] The mechanism of pepsinogen activation to from pepsin is an autocalalytic process that involves intramolecular and intermolecular forces, modulated by the environmental conditions (acidity level and duration of exposure; for less than 2 min following release of pepsinogen, the active protein is formed) [191]. Cleavage of pepsinogen produces different pepsins through sequential, activation mechanisms [187] The gastric zymogens possess additional residues at the N terminus of the active enzyme, called a prosegment or intact activation segment that stabilizes the inactive form and thus prevents the substrate from getting access to the active site. Electrostatic interactions between the prosegment and the active enzyme moiety at a given pH regulate the stability of the inactive enzyme. Once the electrostatic forces are disrupted, zymogen activation begins [190] The first step in the activation of pepsinogen involves the initial release of the prosegment, and then subsequently steps involve the production of one or more pseudopepsins [158] During the activation, the prosegment and the active enzyme undergo changes in conformation, and the proteolytic cleavage of the prosegment can occur in a series of steps by intra- or intermolecular reactions [190]		

(continued)

Table 12.2 (continued)

Enzyme	Brief description	Discoverer	Year of discovery
	During intramolecular pepsinogen activation, gastric acidic environ exerts influences on the zymogen to be rapidly converted to the pepsinogen intermediate delta. This intermediate undergoes conformational change to bind the activation peptide portion of this same pepsinogen molecule in the active center to form an intramolecular enzyme–substrate complex (intermediate theta). This is followed by the intramolecular hydrolysis of the peptide bond between residues 44 and 45 of the pepsinogen molecule and the dissociation of the activation peptide from the pepsin. Intermediate delta apparently does not activate another pepsinogen molecule via an intermolecular process. Variable numbers of pepsinogen genes are found on the centromeric region of human chromosome 11 (p11–q13) [192] Inhibitors of pepsin include pepstatin and purified globin from hemoglobin. Pepstatin, a naturally occurring pentapeptide of Streptomyces origin, inhibits intramolecular pepsinogen activation, thus blocking the formation of pepsin. Purified globin from hemoglobin, a good pepsin substrate, binds to pepsinogen intermediate and inhibits pepsin formation [193]. Ascaris lumbricoides also produces a protein (Ascaris aspartic proteinase inhibitor-3) that has been shown to inhibit pepsin. The Ascaris pepsin inhibitor also inhibits gastricsin, and cathepsin E [194, 195] Pepsin-type enzymes belonging to the family of aspartic protease are produced by the malaria parasites and human immunodeficiency virus type 1 (HIV-1) [196]. In antiretroviral therapies for HIV/AIDS, HIV-1 protease inhibitors are the most potent drugs. These drugs inhibit aspartic proteases of the virus. The inhibitors compete with substrates for the active catalytic site of the enzymes. The protease inhibitors of aspartic protease used in HIV/AIDS therapy include ritonavir, lopinavir. Since the enzymes inhibited by these drugs are also produced by other pathogenic microbes, it could be expected that HIV/AIDS patients may not be prone to developing diseases caused by the corresponding microbes. Unfortunately, even though information is scanty, many factors actually modulate the occurrence of diseases. More importantly, there are reported cases of viral drug resistance to protease inhibitors, which reduces the effectiveness of the drugs and can even make the protease of other microbes more resistant to therapy [197]. An experimental investigation showed that HIV-1 protease inhibitors suppress Plasmodium falciparum by inhibiting P. falciparum aspartic protease plasmepsin II at pharmacological relevant concentration. Although their mechanism of action may differ, like pepstatin, HIV-1 protease inhibitors have been shown to inhibit aspartyl-type proteases produced by Candida albicans which causes the infection candidiasis [198–200]		

(continued)

Table 12.2 (continued)

Enzyme	Brief description	Discoverer	Year of discovery
	Alpha-secretase, beta-secretase (also known as memapsin-2 or beta-site amyloid precursor protein cleaving enzyme 1, beta-site APP-cleaving enzyme 1, membrane-associated aspartic protease 2, aspartyl protease 2), gamma-secretase, epsilon-secretase are also aspartic proteases produced in human cells responsible for processing and maturation of certain polypeptide molecules. Secretases are present in tissues of the pancreas, as well as in neurons [201–204]. They are produced mainly in the Golgi apparatus and also present in the endoplasmic reticulum and endosomes and are transported to their respective sites of activity. Following synthesis, beta-secretase and gamma-secretase are translocated to the plasma membrane where they function as transmembrane aspartyl proteases. Dysfunction in their activities has been implicated in a row of diseases affecting nervous system and associated organs. In this regards, substances known to inhibit the functions of such enzymes have shown promise for use in treating pathology resulting from their dysfunctions. For instance, inhibition of alpha-secretase, beta-secretase, gamma-secretase have shown considerable promise for treating brain degenerative diseases such as Alzheimer's disease. Unfortunately, the inhibitors are non-specific as they interact with and inhibit some signaling pathways and a subset of cell-surface receptors and proteins involved in embryonic development, hematopoiesis, cell adhesion, and cell-to-cell contact [205–208]. Fortunately, some NSAID analogues have been shown to preferentially inhibit beta-secretase without affecting notch signaling—one of the morphogen signaling pathways involved in development, hematopoiesis, cell adhesion, etc. Thus, investigation of the mechanisms of NSAID inhibition of beta-secretase may provide important information on the development of selective pharmacological agent for secretase inhibition to be used in addressing some pathological conditions [209]		
Trypsin	A proteolytic enzyme of digestive tract secreted by the exocrine pancreas in its inactive form—trypsinogen. (Enzymes of the pancreas where traditionally called ferments of the pancreas as it was believed that the enzymes caused food fermentation). There are three types of this zymogen—trypsinogen 1, 2, and 3—which are, respectively, cationic, anionic and mesotrypsinogen, which is also the order of their decreasing expression in the pancreas. The first two types are the main isozymes responsible for protein digestion [72]. The human pancreas secretes two major trypsinogen isoforms, cationic and anionic trypsinogen. There are at least 19 genetic variants of cationic trypsinogen gene. Certain mutations in this gene have been shown to contribute to hereditary, familial, or sporadic chronic pancreatitis. Cationic trypsinogen can be autoactivated for reasons not clearly understood. This variant of trypsinogen causes premature intrapancreatic digestive protease activation leading to increased risk of digestion of pancreatic tissue. Some mutations in the trypsinogen gene may inhibit autoactivation of cationic trypsinogen. It should be noted, however, that not all mutations may lead to a change in the catalytic activity of trypsin or its zymogen [210–212]	The German physiologist, Willy Kühne (1837–1900), who later changed his name to Wilhelm Friedrich Kühne, discovered the protein-digesting enzyme, trypsin. Kühne is also credited for introducing the term "enzyme" [213, 214]. Kühne also contributed to neuromuscular physiology, and the photochemical transduction of rhodopsin in retinal rod cells [214]. At one time, he studied under the guidance of the French physiologist Claude Bernard. Discovery of trypsin is sometimes credited to Alexander Yakovlevich (Jakulovich) Danilevsky (1838–1923); he must have identified the activity of the enzyme as far back as 1862, about a decade and half before Kühne's description of the enzyme [215]	1876

(continued)

Table 12.2 (continued)

Enzyme	Brief description	Discoverer	Year of discovery
	Trypsin is formed from the inactive trypsinogen. This 23.3 kDa peptide belongs to the family of serine protease, PA clan superfamily (Protease of mixed nucleophile, superfamily A). The PA clan is the largest group of proteases [216]. For details on classification of proteases, visit MEROPS database of proteolytic enzymes http://merops.sanger.ac.uk/		
	The family of protease enzymes has a long history of over 180 million years. In the course of evolution, a single ancestral gene is believed to have duplicated to produce at least two copies of genes: α-chymases; and granzymes B and H, cathepsin G, Mcpt8 and duodenases [217]		
	Trypsin cleaves peptide chains mainly at the carboxyl side of the amino acid lysine or arginine, except when either is followed by proline. The functionality of trypsin depends on many factors. It has optimal operating pH of about 7.5–8.5 and optimal temperature of about 37.1 °C [218, 219]		
	The formation of trypsin from trypsinogen is carried out by the duodenal/intestinal enzyme enterokinase. Trypsin activates pancreatic serine protease zymogens (chymotrypsinogen, proelastases), pancreatic metalloprotease zymogens (procarboxypeptidases) and prolipases to chymotrypsin, elastase, carboxypeptidase, and lipase, respectively [7].		
	As a conservative mechanism necessary to preserve the integrity of the intestine and pancreas, the activity of trypsin is properly suppressed in the pancreatic acinar cells under normal conditions [220]. The action of trypsin is regulated by a number of factors. Under normal conditions, conversion to trypsin occurs by stoichiometry, which is stimulus-dependent. Trypsin inhibitor and trypsin receptor in the pancreas aid in the regulation of trypsin activity. Other substances including the protein, Beta-site APP-cleaving enzyme-1 (BACE-1; also called Beta-secretase 1), a protease that is highly expressed in pancreatic acinar cells, and may have a protective role on the premature activation of trypsinogen in many pathological conditions occurring in the gut. For instance, duodenopancreatic reflux of the duodenal contents could result in premature activation of trypsinogen by enteropeptidase within the pancreas. This may lead to inflammatory process in the pancreas—known as pancreatitis [221]		
	The activity of trypsin is closely regulated by pancreatic secretory trypsin inhibitor (PSTI). The presence of this peptide was first reported in 1948 by Kazal and co-authors [222]. The peptide is also referred to serine protease inhibitor Kazal type I (SPINK1), which is one of the most widely distributed protease inhibitor families in the humans [223]. SPINK is a signal peptide composed of at least one Kazal domains which consist of about 50–60 amino acids [223]. It was also found that PSTI is analogous to the diazepam-binding inhibitor [224]		

(continued)

Table 12.2 (continued)

Enzyme	Brief description	Discoverer	Year of discovery
	previously, identified in the central nervous system [225, 226]. Subsequently it was identified in the peripheral tissues [227] and revealed identical amino acid sequence between that of diazepam-binding inhibitor and PSTI [224, 228]. Furthermore, their functional properties with regards to the regulation of pancreatic feedback were similar [229]. The protein also binds diazepam and, therefore, also known as endozepine or acyl-CoA-binding protein [230]. This protein (diazepam-binding inhibitor), a ligand of type A GABA (gamma-aminobutyric acid) receptors, displace natural ligand beta-carbolines and benzodiazepines [231]. GABA like benzodiazepines prolongs the opening of chloride ion channel thereby enhancing inhibitory responses [232]. In the GI tract PSTI has been detected in stomach, pancreas, and liver. The peptide has also been detected in extragut tissues including the heart, spleen, lung, kidney, and ovary [233] PSTI (in particular type I—see below) binds rapidly to trypsin to inhibit its activity, thus protecting the pancreas from premature activation of trypsinogen [223]. The peptide was purified from the pancreatic juice and was shown to stimulate pancreatic enzyme secretion [234] Furthermore, the peptide was found to exhibit cholecystokinin (CCK)-releasing activity in the rat intestine. Subsequently it was shown that this novel pancreatic peptide was trypsin-sensitive and mediates pancreatic secretion in the presence of protein-rich meal. The peptide competes with exogenous peptide, thus the greater the amount of protein present in the intestine, the greater the amount of the peptide released. It therefore functions as a monitoring peptide and is concerned with feedback regulation of pancreatic enzyme secretion [235–237]. It was previously thought that a dysbalance in the ratio of PSTI to trypsinogen could lead to the development of pancreatitis; however, a relatively recent study by Graf et al. [238] showed that this is unlikely, at least for chronic pancreatitis PSTI is secreted together with trypsin into pancreatic juice as PSTI-I and PSTI-II [238]. PSTI-I is a pancreatic monitor peptide that contains 61 amino acids, and possesses moderate affinity for trypsin. It functions by inducing a dose-dependent release of trypsinogen, and at same time, protecting the pancreas from premature activation of pancreatic proteases. The activity of PSTI-I and the control of trypsinogen is thought to be the first line of defense against pancreatitis [239]. PSTI-I inactivates trypsin to prevent pancreatic acinar cell damage [240]. PSTI-I also induces cholecystokinin release from the intestine in the presence of a protein-rich meal [239]		

(continued)

Table 12.2 (continued)

Enzyme	Brief description	Discoverer	Year of discovery
	However, the release of trypsin may exceed the capacity of PSTI and this may lead to activation of various pancreatic proteases, a condition that damages the pancreatic acinar cells. Trypsin produced in acinar cells activates the trypsin receptor—protease activated receptor-2, which is present at high densities on the lumenal surfaces of pancreatic acinar cells and duct cells [239]. Activation of this receptor by trypsin increases the production of cytokines, which subsequently regulates exocrine function via a negative feedback loop [240] The PTSI-II comprises 56 amino acid residues [235] and has no known influence on CCK release and pancreatic secretions [236, 241] Another peptide, known as lumenal CCK-releasing factor (LCRF) localized in the small intestine, play a role in feedback regulation of pancreatic functions [242]. Both LCRF and monitor peptide elicit CCK release in lumenal feedback regulation [242]. They serve as candidates for CCK-releasing peptides in the negative feedback process in the absence of pancreatic juice [243]. Evidence suggests that there is also secretin-releasing peptide or factor in the duodenum that is sensitive to acidic chyme, which upon stimulation mediates pancreatic secretion via a positive feedback mechanism [244–246]. Secretin-releasing peptide mediates the secretion of secretin into the bloodstream by the S cells of the duodenum. The activities of the releasing peptide are controlled by neurohumoral mechanisms involving vagal afferents. This released secretin then stimulates exocrine pancreatic secretion [246]		
Enterokinase (also called enteropeptidase)	Enterokinase is an enzyme secreted by the duodenum and mucosa of the small intestine (precisely in the intestinal brush border membrane), synthesized as a zymogen, called proenterokinase or proenteropeptidase. This zymogen requires activation by duodenase or trypsin (vide infra). The activation of proenterokinase make the enzyme membrane-bound through a conserved disulfide bond linking the catalytic sites [247–251]	Nikolas Petrovich Shepovalnikov [252]. Shepovalnikov is sometimes spelt Schepowalnikov or Shepovalnikow. Shepovalnikov was a student of Pavlov and worked closely with his mentor in the area of GI physiology. In his dissertation "Physiology of intestinal juice" required as partial fulfillment for the award of the degree of doctor of science in medicine, for the first time in 1899, Shepovalnikov reported that another enzyme was required to effect the action of trypsin in pure pancreatic juice [253]. Doctor of science in medicine is the highest academic degree awarded to an individual who had previously obtained a PhD (Doctor of Philosophy) in medicine, for which Shepovalnikov had worked for several years after completing his medical university education and while working with Pavlov. The degree is the equivalent of the Habilitation in Dutch-speaking countries. The degree work of Shepovalnikov was completed in the Military Medical Academy of Saint Petersburg, Russia. Both PhD and Doctor of science in medicine degrees are awarded on the basis of lay down requirements depending on the award institution or country and include exceptional research achievements and new findings not previously reported in the literature that have substantial practical applications.	1899

(continued)

Table 12.2 (continued)

Enzyme	Brief description	Discoverer	Year of discovery
	Enterokinase is responsible for the conversion of zymogens (e.g., trypsinogen) to its active product (e.g., trypsin). Its key role is to initiate the formation of proteolytic enzymes from their inactive precursors. This enzyme which was described by Pavlov as "ferment of ferments" decreases in expression from the duodenum distally to the proximal intestine and is undetectable in the midjejunum, ileum, and colon [247–250]. Enterokinase is found on villous enterocytes and maximal in the upper half of the villi, especially on the brush border. The enzyme is absent in the crypts. In situ mRNA hybridization localization of enterokinase mRNA to the enterocytes has shown the usefulness of the enzyme in aiding absorption. It should be noted, however, enterokinase has not been shown to be expressed in goblet or neuroendocrine cells. In the fetus, this enzyme develops from the 2 to 6 days of postnatal life [705] Enterokinase belongs to the family of protein enzymes called serine protease, a disulfide-linked heterodimer, consisting of a disulfide-linked heavy chain of molecular weight 82–140 kDa and a light chain. The heavy chain anchors enterokinase in the intestinal brush border membrane while the light chain of molecular weight 35–62 kDa contains the catalytic subunit. The heavy chain contains 784 amino acids, while the light chain contains 235 amino acids. Enterokinase is functionally and structurally similar to other serine proteases such as chymotrypsin [216, 254–256] The inactive proenteropeptidase produced by the villous enterocytes is converted by duodenase to the active form. The reaction is enhanced in the presence of calcium salts. The active enterokinase occurs in a soluble form in intestinal juice. The formation of enteropeptidase leads to the increased production of trypsin, which in turn further activates the formation of enteropeptidase [257] It is believed that the production of trypsin occurs via the activation by enterokinase or via autocatalysis by trypsin itself. Like enterokinase, the rate of rate of autocatalytic formation of trypsin is greatly enhanced in the presence of high concentrations of calcium salts. In addition magnesium sulfate also enhances the formation of trypsin. Certain anions also affect the conversion of trypsinogen to trypsin. The contribution to this conversion is greatest for acetate, and then followed by sulfate, oxalate, citrate, fluoride, and chloride ions followed in order by bromide, nitrate, and iodide anions [258]. Bile salts and bile itself enhance the autocatalytic activation of trypsinogen. However, this activation is also dependent on pH. The optimum pH for this reaction is observed at pH 5.4–7.8. Bile salts may activate the conversion of trypsinogen to trypsin by almost 55-fold [259]		

(continued)

Table 12.2 (continued)

Enzyme	Brief description	Discoverer	Year of discovery
	Enterokinase is the product of the serine protease-7 gene (PRSS7; also known as ENTK) located on chromosome 21q21. Certain types of nonsense (e.g., S712X/R857X; Q261X) and frame shift mutations (e.g., FsQ902) in this gene can lead to rare recessive disorders of enterokinase, resulting to enterokinase deficiency [260]. This rare recessively inherited/congenital deficiency of enterokinase leads to disorders characterized mainly by intestinal malabsorption, and severe failure to thrive [260]. The first case of congenital enteropeptidase deficiency was reported as far back as 1969. However, thanks to its rare rate of occurrence, not more than 20 cases have been confirmed worldwide. The disease is usually identified in the neonatal period and is characterized by failure to thrive, chronic diarrhea, low serum protein, generalized edema, low trypsin activity, and hypoproteinemia. Malabsorption syndrome in the infants is characterized by muscle wasting, failure to thrive, and hypoproteinemia (such as kwashiorkor), may include steatorrhea, severe anemia, which may be secondary to protein malabsorption and vitamin E deficiency [261, 262]. Some of the aforementioned symptoms may be found in other gut-related disorders such as sprue (psilosis characterized by non-absorption of nutrients, foul-smelling diarrhea, and emaciation), cystic fibrosis, protein-losing enteropathy, total villous atrophy. The use of enzyme supplements such as pancrelipase (creon) is integral to the management of congenital enterokinase deficiency [263–266]		
Duodenase	A duodenal serine protease that catalyzes the activation of enterokinase [267, 268]	The duodenase enzyme was first reported by the Russian scientist, Zamolodchikova Tatyana Stepanovna and colleagues	1992
Intestinal aminopeptidase (aminoacyl-peptide hydrolase)	Aminopeptidases are ubiquitous exoprotease-enzymes in living things (plants, eukaryotes, and prokaryotes) and tissues of different organisms. They are found in blood cells, kidney, CNS, components of intracellular organelles, depending on the pH of aminopeptidase optimal activity (basic, acidic, neutral). Aminopeptidases are one of the two major subclasses of the proteolytic enzymes (exopeptidases) that remove amino acids from the termini of peptides. It is a zinc metalloenzyme or zinc-binding metalloproteinase [269]. Ligands of aminopeptidases bind in a non-covalent mode by forming complexes with metal ions [270] Aminopeptidases attack their substrates exclusively from the amino-terminal end. It hydrolyzes the substrate peptide bond adjacent to an N-terminal of the amino acid (alanine) residue. There are different types of aminopeptidases: Aminopeptidase that cleaves one, two, and three peptide residues at a time are called peptidyl, dipeptidyl, and tripeptidyl aminopeptidases, respectively [272]. Aminopeptidases are also classified according to their cellular localization: membranous, cytosolic, and organellar types. Of particularly interest to	Linderstrøm-Lang K (−), a German scientist, who became the first to describe this intestinal enzymes [271]	1929

(continued)

Table 12.2 (continued)

Enzyme	Brief description	Discoverer	Year of discovery
	the discussion in this Chapter is the membranous type, found particularly as pancreatic/intestinal aminopeptidases localized on the membrane of enterocytes (e.g., alanyl-, arginyl-, leucyl-, methionyl-, phenylalanyl-, leucine-, aminopeptidase, aminopeptidase N or CD13). These intestinal brush border enzymes have huge role to play in amino acid production in the GI tract. They are generally involved in the terminal digestion of peptides generated from hydrolysis of proteins by the action of gastric and pancreatic proteases [104, 273–279] Aminopeptidases play considerable role in inflammation (e.g., multiple sclerosis, pancreatitis), cancer, acne, malaria, diabetes, and aging [270, 280, 281]. To this end, inhibitors of aminopeptidases have been found to be important in cancer therapy [282]. For example, aminopeptidase N inhibitor, Ubenimex, enhances radiation sensitivity in human cervical cancer [273] In contrast to the aminopeptidase, the intestinal carboxypeptidases function through the removal of C-terminal amino acids from dietary proteins and peptides. Carboxypeptidases also play a crucial role in the formation of neuropeptides and peptide hormones [283]. Examples of intestinal brush border carboxypeptidases include carboxypeptidase -A, -B, -O [284–286]. For further details on carboxypeptidases, see pancreatic enzymes		
Chymotrypsin (chymotrypsin A and B are the major types)	A digestive proteolytic enzyme (serine endopeptidase) secreted in its inactive form as chymotrypsinogen by the acinar cells of the exocrine pancreas. The zymogen is activated by trypsin in the duodenum. The chymotrypsin preferentially cleaves peptide–amide bonds where the carboxyl side of the amide bond (P1 position) is comprises hydrophobic amino acid (such as tyrosine, tryptophan, and phenylalanine). The side chain of these amino acids fits into the active site of chymotrypsin, referred to as a hydrophobic pocket (S1 position) [287–289]	Horace Middleton Vernon (1870–1951) of the Merton College, Oxford. He studied chemistry and physiology and later became a university lecturer of chemical physiology at Oxford [290–292]. During the early twentieth century, precisely in 1901, Vernon proposed that a pancreatic secretion has an intrinsic property of activating its own enzymes [293]. He noted that ferments of trypsin and rennet could be formed from their corresponding zymogens by the action of trypsin itself. Chymotrypsin was discovered by Vernon in 1913 as part of an isolate obtained from beef pancreas [294]. Unfortunately, his experimental results were not widely accepted until 1934 when Moses Kunitz and Northrop identified a novel enzyme, which was different from trypsin. They named it chymotrypsin [295, 296] Both the active and the inactive form of this enzyme and their subtypes were crystallized by the discoverers in the years 1935–1948. The different types of chymotrypsin, which Kunitz and Northrop isolated, were designated as alpha (A), beta (B), gamma, and delta [299–302] John Howard Northrop (1891–1987) was an American biochemist, professor of bacteriology and medical physics at the University of California, Berkeley. Moses Kunitz (1887–1978) was born in Slonim, Russia (now in Grodno region of Belarus) and later emigrated to the USA [303]	1913

(continued)

Table 12.2 (continued)

Enzyme	Brief description	Discoverer	Year of discovery
	Chymotrypsin cleaves peptide bonds through the action of its catalytic triad containing serine-195, histidine-57 and aspartic acid-102; they form the active site of the enzyme. The enzyme breaks down peptides by attacking the carbonyl group with the serine 195 residue, which is a powerful nucleophile. This results in a transient covalent linkage between the substrate and the enzyme, thus forming an enzyme–substrate intermediate. This intermediate is unstable and so under favorable conditions, the enzyme dissociates and the products are formed. Chymotrypsin catalyzes the reaction rate for the formation of products by a factor of 109 [289, 297, 298]	Northrop shared one half of the 1946 Nobel Prize in Chemistry with the American biochemist and virologist Wendell Meredith Stanley (1904–1971) "for their preparation of enzymes and virus proteins in a pure form" and the other half was awarded to the American chemist James Batcheller Sumner (1887–1955), "for his discovery that enzymes can be crystallized" [304] Further research by Joseph Stewart Fruton (1912–2007) and Max Bergmann (1886–1944) [305] led to the identification of other forms of chymotrypsin. Thereafter, in 1947, Jacobsen CF identified additional forms of chymotrypsin, designating them pi [306] Numerous forms of chymotrypsin have been reported [287]	
Pancreatic enzymes	These are digestive enzymes secreted by the pancreas into the intestinal lumen; they include amylase, trypsinogen, chymotrypsinogen, proelastase, procarboxypeptidases, pancreatic lipase, colipase, kallikreinogen, cholesterol ester hydrolase, ribonuclease, and deoxyribonuclease, among others [72, 307–313] The digestive role of pancreatic juice was first proposed by the German physician Johannes Bohn (1640–1718) in 1685. However, it was not until around the mid-nineteenth century when Eberle suggested that pancreatic juice has substances that possess catalytic activities. In 1834 Eberle suggested that pancreatic juice could emulsify fats and break down starch into simpler molecules. Subsequently, in 1838 Johannes Evangelista Purkinje demonstrated that pancreatic juice in the presence of bile not only emulsified, but also split fat. By 1845 Apollinaire Bouchardat (1806–1869), a leading pharmacist and diabetologist from France and Gabriel Gustav Valentin (1810–1883), a German physician and professor of physiology at the University of Bern, Switzerland, reported the digestion of potato by pancreatic juice. These studies laid further foundation for Alexander Yakovlevich (Jakulovich) Danilevsky (1838–1923), who after graduating from the Medical Faculty of Kharkiv National Medical University, Ukraine, began to conduct series of research in St. Petersburg and in many other cities. In 1862 Danilevsky conducted experiment to determine the role of some enzymes in metabolic processes. On the basis of his investigation, Danilevsky successfully isolated starch- and protein-digesting enzymes from the pancreas. Danilevsky later became a professor of therapeutics and also served as a rector of the Leipzig University, Leipzig, Germany [315–319]	Johann Nepomuk Eberle (1798–1834) [1, 215, 314]	1834

(continued)

Table 12.2 (continued)

Enzyme	Brief description	Discoverer	Year of discovery
	The discovery of pancreatic enzymes, their mechanisms of secretion and functions made it possible for future investigation of their role in health and disease. In 1926 Chiray M and colleagues became some of the pioneer scientists to study the action of secretin in pancreatic secretion in humans [320, 321]. Burton et al. (1960) studied changes in duodenal extracts following administration of both secretin and pancreozymin in humans [322]. A simultaneous report by Sun & Shay (1960) showed similar findings [323]. The scientists showed that the technique was useful in diagnosing pancreatic diseases. After obtaining a highly purified cholecystokinin, Erik Jorpes J, Viktor Mutt showed that not only pancreatic functions could be studied but also gallbladder evacuation. Thus they identified that the previous extracts used to study pancreatic functions also contained the hormone pancreozymin. The pancreozymin (cholecystokinin)-secretin test represents one of such tests that were introduced into clinical practice. The test is used to assess the functions of the pancreas and gallbladder [324, 325]. This method of investigation of secretory activity in normal and pathology [(pancreozymin (cholecystokinin)-secretin test)] was developed following the isolation of secretin in 1961, and cholecystokinin in 1964 [325]. In secretin-pancreozymin test, the hormone secretin is used to stimulate the production and secretion of pancreatic enzymes into the duodenum. The resulting fluid is then aspirated from the duodenum and evaluated for the expected components. The absence or level of the components being evaluated for is indicative of pancreatic or gallbladder pathology. The test can be used to measure the bicarbonate concentration or volume resulting from pancreatic secretion. Normal volume findings are in the range of about 2–6 mL bicarbonate per kg of body weight (or a bicarbonate concentration of 80–130 mEq/L). Abnormal levels (usually lower than normal) are suggestive of an obstructing malignancy such as cystic fibrosis. The test may identify dysfunction in asymptomatic cases of cystic fibrosis. Reduced bicarbonate and amylase concentration is usually diagnostic of chronic pancreatitis. The test is used in evaluating other conditions such as malabsorption and failure to thrive [326–330] Hormone stimulation tests correlates with imaging report (magnetic resonance imaging and magnetic resonance cholangiopancreatography) of pancreatic functions. Moreover, results of morphological alterations in pancreas (ductal dysfunction, parenchymal abnormalities, etc.) are comparable with results of hormonal stimulation tests [331, 332] Another test that can be used to analyze pancreatic functions is the Lundh test meal [333, 334]. Lundh test meal was invested by Lundh Göran in 1962 [335]. In Lundh test, trypsin concentration in the duodenum is measured after a test meal is administered, containing protein, fat, and sugar. The trypsin concentration in duodenal aspirates is then measured over several hours. A decrease in trypsin concentration indicates a low or abnormal pancreatic secretion. Over the years the Lundh test has been modified in many experimental conditions [336, 337]. The measurement of pancreatic enzymes in the duodenum following intravenous or		

(continued)

Table 12.2 (continued)

Enzyme	Brief description	Discoverer	Year of discovery
	oral stimulation is used for diagnosis in many settings. A reliable measure of pancreatic functions is the serum lipase activity investigation. Prior to Lundh meal test and 30 min after the test, serum lipase activity is evaluated. The results is a useful criterion for identifying pancreatic status, pancreatic insufficient in normal individuals and patients with cystic fibrosis [337]. For review of the Lundh test see: James [338]		
	In cystic fibrosis, pancreatic enzyme lipase, trypsin, chymotrypsin, and amylase) output may be normal or elevated and positively correlate with the volume of bicarbonate measurement [339]		
	Stimulation of pancreatic secretion followed by sequential serum lipase measurement was found to be a good indicator of exocrine pancreatic status. The use of glucose in Lundh meal test to evaluate the pancreatic endocrine axis functions can provide useful information on cystic fibrosis patients. It is believed that the traditional oral glucose tolerance test is not superior to Lundh meal test for identification people with cystic fibrosis and other related diseases. The glycemia post Lundh meal in healthy individuals peaks at 45 min (119–161 mg/dl) and then gradually declines, reaching the normal range at 120 min, whereas in patients with cystic fibrosis, glucose levels may be higher than 200 mg/dl 30–60 min post Lundh meal. Importantly, while Lundh meal test may be positive, the oral glucose tolerance test may show negative results [337]		
	There are other methods of evaluation of pancreatic functions. The protein CA 19-9 (cancer antigen 19-9) is a tumor-associated antigen that is elevated in some cystic fibrosis patients. It is thought that the protein originates from the respiratory system to cause increase in its serum level. The CA 19-9 levels may be associated with age, pulmonary function, pancreatic status, sweat chloride, pancreatic diseases, serum lipase, meconium ileus, distal intestinal obstruction—which are associated with cystic fibrosis [340]. Thus, CA 19-9 antigen levels are useful diagnostic measure of the disease. Truly, accumulating evidences have shown that CA 19-9 protein levels are associated with a couple of disease of the pancreas [341]		
	It should be noted that pancreatic enzymes may increase in the absence of organic pathology in the pancreas. In 1996, the Italian physician and pancreatologist, Professor Lucio Gullo (1938–2009) became the first to describe in adults a new syndrome characterized by a benign chronic increase in serum pancreatic enzymes in the absence of pancreatic or other pathologies. The condition is characterized by persistent fluctuations in serum amylase, pancreatic isoamylase, lipase, and trypsin, with no evidence of pancreatic disease as evidenced by imaging technologies. The use of abdominal CT, magnetic resonance cholangiopancreatography may be used to detect changes in the pancreas. In asymptomatic pancreatic hyperenzymemia, the pancreatic architectural structure is normal in almost 90% of individuals [342, 343]		

(continued)

Table 12.2 (continued)

Enzyme	Brief description	Discoverer	Year of discovery
	The fluctuations of serum pancreatic enzyme levels are marked by occasional transient normalization in serum enzymatic levels and may also be referred to as idiopathic pancreatic hyperenzymemia. It should be noted that some percentages of individuals with increased serum pancreatic level over a 5-day period of measurement may not have the syndrome. Over 60% of individuals with benign pancreatic hyperenzymemia have abnormal increase in lipase, amylase, and pancreatic isoamylase [344]. Some individuals may have increased salivary amylase with macroamylasemia and amylasuria [344, 345]. The condition was initially called chronic non-pathological pancreatic hyperenzymemia, but was renamed benign pancreatic hyperenzymemia. It is sometimes called Gullo's syndrome. The syndrome is usually familial, but may be sporadic. All pancreatic enzymes may be affected. However, there are conditions in which only one enzyme or two are affected—usually amylase or lipase. The condition has been reported in infants, adolescents, and adults [346]. In diagnosing the syndrome, an assay of serum pancreatic enzymes or urine is done over multiple times in one day. To qualify for familial condition, at least one family member must have the condition. It should be noted that when pancreatic hyperenzymemia is suspected in patients without any symptoms or pathologies, biochemical analysis should be done to rule out celiac disease, viral hepatitis, dyslipidemia, macroenzymemia, subclinical systemic disease, and malignancy. Increase in serum pancreatic enzymes may occur secondary to use of paracetamol, steroids, azathioprine, although, in such cases, the patient may not be apparently healthy [347–349] While in some cases, study of variant anatomy of the pancreas may be required, but which may not necessarily indicate pathological condition. Serum pancreatic enzymes have been shown to increase significantly even in such cases. The variant structures of the pancreas in these situations may include pancreas divisum, small intrapancreatic cyst, and anatomic variant of the Wirsung, secondary ducts, and mild dilatation [342]. In normal pancreatic anatomy (at least in the majority of cases), pancreatic fluid drains via a single duct, to the ampulla of Vater (named after Abraham Vater, 1684–1751) through the sphincter of Oddi. The main pancreatic duct drains at the ampulla of Vater and may connect with the duct of Santorini if present. The duct of Santorini (also called accessory pancreatic duct) is named after the Italian anatomist and professor of medicine, Giovanni Domenico Santorini (1681–1737), who was the first to describe the duct. This duct drains the anterior and superior portion of the pancreatic head and is found in about two-third of people, it drains at the minor papilla, in the rest of individuals it persists as a branch of the main pancreatic duct. The accessory pancreatic duct connects straight to the duodenum at the minor duodenal papilla. Thus this duct bypasses the Ampulla of Vater. In some cases, the accessory pancreatic duct may be non-functional. Pancreatic duct of Wirsung is continuous with the main pancreatic duct. The duct joins the pancreas to the common bile duct to supply		

(continued)

Table 12.2 (continued)

Enzyme	Brief description	Discoverer	Year of discovery
	pancreatic juices to the intestine for digestion. The duct of Wirsung was first reported by German-Italian anatomist Johann Georg Wirsung (1589–1643) [350–354]. Both ampulla of Vater and papilla of Santorini are sometimes called papilla duodeni. These variants are congenital in nature. In embryonic development, the pancreas is usually formed as a single organ during the seventh week via the fusion of two pancreatic primordia (one dorsal and one ventral), each containing a duct that drains into the duodenum. The ventral duct forms the distal part of the main duct of the pancreas (Wirsung's duct), whereas the dorsal primordium forms the proximal portion of the duct. The fusion process may be normal or abnormal resulting to the formation of variant ducts. Even when there is normal fusion, the formation of accessory duct remains a possibility as the distal portion of the embryonic dorsal pancreatic duct may not regress following normal fusion [355–357] Another pancreatic anomaly that may lead to abnormally high level of serum pancreatic enzyme level is pancreas divisum. Pancreas divisum is the most common congenital anomaly of the pancreatic ductal system, in which a single pancreatic duct is not formed, but rather remains as two distinct dorsal and ventral ducts. It results from failure of embryonic dorsal and ventral pancreas (primordia) to fusion thus leading to the formation of separate organs with two distinct ducts. The condition was first discovered by the Australian professor of anatomy, Joseph Hyrtl (1810–1894). In this pancreatic anomaly, the dorsal pancreatic duct drains most of the pancreatic glandular parenchyma via the minor papilla. The incidence of pancreas divisum is approximately 1–14%. This anomaly in some cases may remain asymptomatic and may be undetected [358, 359]. However, pancreas divisum may predispose an individual to pancreatitis and santorinicele. The later condition is a cystic dilatation of the distal dorsal duct, lying proximal to the minor papilla [359] For review on variant anatomy of pancreatic ductal system, see: Türkvatan et al. [350], Alexander [351], Borghei et al. [352] Proelastase: This is a pancreatic zymogen serine protease that is activated by trypsin to form elastase that hydrolyzes amides and esters, and other proteins including elastin. There are two forms of the enzyme—proelastase 1 and 2 [72]. Elastin is a type of protein found in connective tissue. Elastase can be detected in various fluids and waste of humans. It is also found in stool. A value greater 200 μg elastase/g stool is normal for pancreatic functioning [360–363] Procarboxypeptidases: These are protein zymogens, when activated cause hydrolysis of peptide bond at the carboxy-terminal (C-terminal) end of the protein. (recall that aminopeptidase cleaves bonds at the N-terminal). There are different types of carboxypeptidases with different functions in humans. These enzymes depending on the type are found in pancreas, different body fluids, lysosomes, endoplasmic reticulum.		

(continued)

Table 12.2 (continued)

Enzyme	Brief description	Discoverer	Year of discovery
	Carboxypeptidases are involved in numerous functions of the organism including maturation of proteins following synthesis, and regulation of physiological process, synthesis of peptides (e.g., insulin, growth factors). They play a role in blood clotting, wound healing, reproduction, etc. The types involved in digestion are pancreatic carboxypeptidases A1, A2, B1, and B2. They are synthesized in the pancreatic acinar cells together with other enzymes. These pancreatic enzymes may exist in three forms: the proenzyme, the active enzyme, and the activation peptide. Procarboxypeptidase A is produced exclusively in the pancreas and converted to its active form in the intestinal lumen [72, 364, 365]. Pancreatic enzymes can freely escape into blood during acute or chronic inflammation or neoplastic growth. The serum concentration of these enzymes has a high sensitivity and specificity in the diagnosis of cancer [366]		
	Carboxypeptidase B is an exoprotease synthesized as an inactive proenzyme procarboxypeptidase B by acinar cells together with other pancreatic proenzymes. During this activation a large N-terminal activation peptide (containing 81 amino acids with molecular weight 9398 Da), known as carboxypeptidase activation peptide is released [365]		
	A peptide enzyme (peptidase) is called an exopeptidase if it cleaves the terminal peptide bond in a protein. The reaction leads to release of a monomer or dimer from the protein. If the released amino acid monomer or dimer is from the amino-terminal, the exopeptidase is called aminopeptidase. If the released amino acid (s) is from the carboxy-terminal, the exopeptidase is called carboxypeptidase. Both enzymes have been discussed previously. In addition to their normal expression, these enzymes can occur pathologically in cancer cells. In contrast to exopeptidase, other peptidases cleave non-terminal peptide bonds, located within the molecule. These peptidases are called endopeptidases. Thus, the process cannot lead to the formation of amino acid monomers. Examples of endopeptidases include trypsin, chymotrypsins, oligopeptidases [277, 367–375]		
	Deoxyribonucleases and ribonuclease: These are types of nuclease enzymes that hydrolyze phosphodiester bonds that link nucleotides in nucleic acids. As with other proteases, nucleases depending on the type, may cleave residues at the ends of nucleic acids (exonuclease—exodeoxyribonucleases, -ribonucleases); cleave within the chain (endonucleases—endodeoxyribonucleases, -ribonucleases); topoisomerases, recombinases, ribozymes, or RNA splicing enzymes or other phosphodiesterases [376–379]		
	Ribonuclease pancreatic is the gene product of RNase1 (called ribonuclease I, endoribonuclease I, ribonucleic phosphatase, alkaline ribonuclease, oligonucleotidohydrolase). Pancreatic ribonuclease is an endonuclease that catalyzes the cleavage of RNA on the 3' side of pyrimidine nucleotides (cleavage of nucleoside 3′-phosphomononucleotides and 3′-phosphooligonucleotides ending in C-p or U-p with 2′,3′-cyclic phosphate intermediates) [380]		

(continued)

Table 12.2 (continued)

Enzyme	Brief description	Discoverer	Year of discovery
	Deoxyribonuclease pancreatic (pancreatic DNase) catalyzes the hydrolytic cleavage of phosphodiester linkages in the DNA backbone, which result to the breakdown of the DNA molecule [381, 382] Pancreatic kallikreinogen: This zymogen is highly expressed, together with trypsinogen, in the pancreas. Upon conversion to the active form, kallikrein, it cleaves inactive protein substrates, similar to the effect of trypsin on intestinal proenzymes [72, 383, 384]		
Alkaline phosphatase	Alkaline phosphatase is an intestinal brush border enzyme important for the digestion of nucleotides and proteins. The enzyme removes phosphates from nucleotides (DNA, RNA) and proteins, thus detoxifying free nucleotides and bacterial lipopolysaccharide. These enzymes are most active at alkaline pH, hence the name [104]. Apart from the intestine, the enzyme is found in the liver, bone, kidney, and placenta. Alkaline phosphatase is generally used as a marker of epithelial cell maturation [96]. However, the enzyme plays a variety of roles including growth and development of bones and teeth. The intestinal alkaline phosphatase is a component of the gut mucosal defense system and prevents bacterial invasion across the gut mucosal barrier [385]. It attenuates intestinal inflammation and modulates the gut microbiota. Intestinal alkaline phosphatase expression and function are preserved and enhanced during enteral feeding. However, their functions are lost with starvation. Thus, in critically ill patients whose mode of feeding is parenteral, this phosphatase activity is silenced leading to reduction in gut mucosal barrier function, thereby predisposing the individual to gut infection [385, 386]. Intestinal alkaline phosphatase is also involved in the regulation of intestinal pH, and absorption of lipids [386]. However the mechanism by which intestinal alkaline phosphatase regulates lipid absorption is not entirely understood	Robert Robison discovered alkaline phosphatase [387, 388]	1923

Table 12.3 Description of course of discovery of lipid-digesting enzymes and their characteristics

Enzyme	Brief description	Discoverer	Year of discovery
Lipase or triacylglycerol acylhydrolase or glycerol-ester hydrolase (lingual, gastric, pancreatic lipases)	The nineteenth century globally reckoned French physiologist was the first to discover that pancreatic secretions could emulsify and saponify fatty substances. Bernard attributed the reaction to the presence of an enzyme that was later called pancreatic lipase [427]. Relatively recently, extracellular lipases have been discovered in microbes (e.g., Acinetobacter sp.) [428–431] Lipases are secreted in the active form to hydrolyze dietary long-chain triacylglycerol to free fatty acids and monoacylglycerols in the intestinal lumen. However, pancreatic lipase can only function efficiently in the presence of bile acids, and colipase [see succeeding column]. Pancreatic lipase is made up of about 465 amino acids. The amino acid sequence of this enzyme bears some resemblance to the sequence in other mammals [72, 432, 433] Intestinal activity of lipase was not widely known until 1950–70s when it became recognized that lipases were produced not only by the pancreas but also glands in the oral cavity, stomach, and other body tissues. So, there are lingual, gastric, intestinal (pancreatic), hepatic, lipases. The activities of these lipases are all important in the metabolism of lipids [434]. Identification of lipases in different tissues of the body had a pretty long history. In 1953, Douglas et al. [435] became the first to publish the results of their study, which suggested a role of other regions of the GI tract, other than the pancreas, in digestion of food substances such as fat. Almost a decade later, Bank et al. [436] observed that gastric juice was capable of digesting fat. Consequently, there was increased interest in studying the lipolytic activity of gastric juice. In 1969, Clark SB, Brause B, Holt PR demonstrated lipolysis and absorption of fat in the gastric juice of rat. In the same year Darnton and Barrowman [437] investigated lipase activity in a range of tissues including pancreas, testis, cardiac, stomach, and liver, but reported true lipase activity only in the pancreas. A year later Hamosh et al. [438] successfully performed experimental investigation that led to the identification of lipase activity in other tissues. The authors identified a type of lipase (lipoprotein lipase) activity in adipose, mammary tissues, and serum of rat. This was a confirmation of the earlier report by Baskys et al. [439] on the presence of lipases in blood and other tissues. A year later, Cohen et al. [440] demonstrated lipase activity of human gastric and duodenal juice on fat. In a groundbreaking study, conducted by Barrowman & Darnton [441], it was identified beyond doubt that there is lipase activity in the stomach of mammal. In 1973, Margit Hamosh and Robert O. Scow discovered lingual lipase [442]. Lingual lipase is secreted by lingual serous glands (von Ebner glands). The glands that produce lipases are also found in the soft palate, anterior oral pharyngeal wall, and lateral oral pharyngeal region	Claude Bernard (1813–1878)	1846-48

(continued)

Table 12.3 (continued)

Enzyme	Brief description	Discoverer	Year of discovery
	of the oral cavity. Lipases are stored in these glands as zymogen granules and released on activation by different stimuli of varying intensity. In the presence of certain alimentary proteins and factors, lingual and gastric lipases are able to hydrolyze medium- and long-chain triglycerides. In conditions of high activity of these lipases, dysfunction in pancreatic lipases (due to inflammation or congenital diseases) may not be especially evident [443–446] Lingual lipases hydrolyze triglyceride to mostly diglyceride and free fatty acids. The pH at maximum activity is 3–6 [442]. It appears that lipid digestion, however, does not start in the mouth, but in the stomach by lingual/gastric lipases. The activity of these lipases is not hindered by low acidity of the stomach provided that the pH is not lower than 3 [447]. Gastric lipase is secreted together with other substances in the chief cells located in the mucosa of the gastric fundus [448, 449] The contribution of lingual and gastric lipases to lipid digestion varies from organism to organism [449, 450]. It has been suggested that gastric lipase may have higher contribution to lipid digestion than any other lipid secreted from other regions of the GI tract, whereas in rat lingual lipase is the predominating contributor to lipid digestion [442, 451–453]		
Colipase	Colipase is a protein cofactor of pancreatic lipase. It is a small protein with five disulfide bonds. This coenzyme enables lipase to anchor itself to the lipid-water interface. Bile salts have inhibitory effect on lipase, thus colipase helps to stabilize the association "lipase–colipase–lipid–water interface." Colipase is produced as a zymogen, procolipase, and it is activated by trypsin. The activation of procolipase also produces another peptide called enterostatin, which acts as a satiety signal peptide. Enterostatin is an enteropeptide that is made up of five amino acids—it has significant effects on peripheral and central processes. Peripherally, the protein inhibits fatty acid absorption. Its effects include reduction in food intake via mechanisms involving hypothalamic regulation downstream vagal efferents/afferents. Centrally, the peptide modulates serotonergic and opioidergic system to affect fat intake, bodyweight, and body fat. It has been shown that enterostatin also decreases insulin secretion, stimulates the secretion of steroids from the adrenal gland, increases the breakdown of brown adipose fats due to increase in the sympathicus activity. Although not exactly clear, enterostatin possibly acts via multiple receptor systems at different levels. The CCK-A, 5HT1B receptors may be candidates that mediate enterostatin activity. Another candidate receptor for enterostatin is F1-ATPase β-subunit (β-subunit of F-Type ATP synthase). This receptor is present in plasma, liver, amygdala, hypothalamus, and cortex [454–461]	Sigmund Rosenheim Otto (1871–1955) [462, 463]. However the name "colipase" was given by Maylie et al. [464] For further details, review Borgstrom et al. [465]	1910

(continued)

Table 12.3 (continued)

Enzyme	Brief description	Discoverer	Year of discovery
	Colipase binds to the C-terminal, non-catalytic domain of lipase to stabilize the active conformation and thus increases the overall hydrophobic binding site of the complex "lipase–colipase–protein micelle" [466]. The colipase coenzyme has both hydrophobic and hydrophilic surfaces. The hydrophobic surface of the colipase enzyme binds substrate leaving the hydrophilic surface positioned in the opposite plane [467]. This reaction is indeed modulated by bile salt (such as sodium taurodeoxycholate) micelles complex [468, 469]		
Pancreatic (cholesterol) esterase, also known as bile salt-activated lipase	Pancreatic cholesterol esterase is an enzyme that hydrolyzes dietary cholesterol esters, fat-soluble vitamins, triglycerides, and phospholipids to a form that is absorbable by the enterocyte. These enzymes are released by the exocrine pancreas into the intestinal lumen where they are activated by primary bile salts. The product of the hydrolysis includes cholesterol and free fatty acids [72, 470, 471]. Inhibition of the activity of this enzyme has shown considerable promise and as useful therapeutic option for controlling the absorption of dietary cholesterol in the intestine [472]		

Intracellular digestion occurs inside the cells of the GI tract, in which absorbed dimers or oligomers are further hydrolyzed inside the cytoplasm to form monomers, which are then transported by vessels into different body cells [21–23]. Following absorption of nutrients from the lumen of the intestine by the enterocytes, intraenterocytic digestion of absorbed dimers and oligomers occurs to efficiently carry out the process of transport to their site of utilization or storage. It is believed that intracellular digestion is directed toward partially digested molecules that are absorbed by the enterocytes. Some of the first researchers to report the absorption of partially digested food residues (e.g., disaccharides and oligosaccharides) were Chain, Mansford, and Pocchiari in 1960 [24]. However, it was not exactly clear whether or not these carbohydrates were absorbed in their original form as disaccharides or oligosaccharides. Around 1968, Hamilton and Mcmichael [25] set forth a hypothesis about the transport of disaccharides and how the microvilli and glycocalyx enhanced disaccharide transport. The scientists suggested that it was the hydrolytic products that were absorbed. In one report, during the same year, in which Hamilton and Mcmichael (1968) [25] formulated their hypothesis, Parsons and Prichard (1968) showed that disaccharides were absorbed in the intestine of *Rana temporaria, R. pipiens,* and *Bufo vulgaris* [26]. During the time, it was also reported that the absorption of maltose led to the increase of glucose in the lumen and blood. The experimenters concluded that the quantity of glucose that appeared in the intestine is not a direct measure of the activity of maltose hydrolysis. Whereas two sides of the pendulum could be swinging in this study about maltose hydrolysis (i.e., both glucose from maltose hydrolysis and maltose itself are absorbed), it could be that glucose molecules produced during hydrolysis were been absorbed in a faster rate. It is possible that the disaccharide was actually absorbed into the enterocyte. The Km (constant of Michaelis, i.e., [E][S]/[ES], where [E]—concentration of enzyme, [S]—concentration of substrate, [ES]—concentration of enzyme–substrate complex) for the transport of glucose is approximately one-fifth of the activity of disaccharides. The maximum speed of glucose transport by the glucose chansporter is lower than the maximal speed of disaccharide hydrolysis [7, 26–29]. This suggests the possibility that disaccharides may have been absorbed by the intestinal epithelium.

Recall that previously in 1961 Dahlqvist and Borgstrom had reported limited disaccharide hydrolysis in the intestine and low activity of glycosidase relative to the expected (calculated) value of carbohydrate absorbed, indicating that majority of disaccharides was absorbed unhydrolyzed, and thereafter underwent intracellular hydrolysis in the enterocyte [18].

Research data on the kinetic properties of absorption of hexoses and disaccharides were available already in the 1960–70s [30, 31]. However, because of the widespread of the theory of hydrolysis-related transport of glucose, it was still believed that disaccharides could not be transported and hence no intracellular hydrolysis takes place [32]. This view was in part driven by the fact that, as it was envisaged, transport of glucose was related to transport of disaccharides—i.e., they have the same mechanism of absorption. In 1984, Alvarado et al. reported that transport of hexoses is not associated with transport of disaccharides; they occur independently [33]. This single suggestion could have turned out to be an important path to the discovery of a disaccharide transporter. It should be noted that in protein digestion, dimers and oligomers of peptides are actively transported into the enterocyte where they undergo further hydrolysis to form monomers. In fact in microbes, polysaccharides are also transported. Consequently, intracellular digestion in such microbes represents an active process that ensures survival and continuity of life processes. A breakthrough on disaccharide transport was made in this century. In 2011, Meyer, et al. [34] reported the discovery of the first disaccharide transporter (SCRT—sucrose transporter) in *Drosophila melanogaster*. It is now thought that a disaccharide transporter may be functional part of transporter systems in higher animals. SCRT (made up of 599 amino acids) is the gene product of chromosome 67A3, functioning as carrier of sucrose in this animal. More recently, Likely et al. [35] reported that they have found the membrane transporter for the disaccharide sucrose in the intestine of *Homarus americanus*. This report confirms previous speculations about the role of disaccharide transporters' intestinal absorption.

Intracellular enzymes for synthesis or breakdown of relatively short or large chains of biomolecules are membrane bound [36, 37] and may be located in the cytosol, lysosome, Golgi complex, and endoplasmic reticulum [36, 38–41]. In these organelles, digestive processes are relatively frequent.

12.3 Sources of Enzymes for Chemical Digestion

Depending on the origin of the enzyme, digestion can be divided into three types: autolytic, symbiotic, and digestion proper. Digestive enzymes are produced in major regions of the digestive tract. These enzymes may be classified as artificial or natural. On the basis of the food type been digested, these biological catalysts may be carbohydrate-, protein-, lipid-digesting enzymes [42–46].

12.3.1 Autolytic Digestion

This type of digestion is carried out by the enzymes contained in food. Specific types of enzymes are added to food (enzyme supplementation) to aid digestive process for some individuals with pathologies such as exocrine pancreatic insufficiency, gastric and intestinal resection, and cystic fibrosis, which are characterized by malabsorption. The enzymes are derived from microbes, animals, and plants. Enzymes derived from animal sources (porcine or bovine sources) include trypsin, chymotrypsin, pancreatin, and pancrelipase [47–51]. Enzymes derived from microbes (similar to those in the gut) such as lipase have been successfully used as supplement to aid digestion. Microbes which are used as sources of enzymes are mostly fungi, and they include *Aspergillus oryzae* and *Rhizopus arrhizus*. The yeast

Kluyveromyces lactis serves as a source of lactase (beta-galactosidase). Some of these microbes are traditionally used for fermentation processes [47]. Enzymes derived from plants include bromelain, papain, which help in the in vivo digestion of protein as they mainly contain proteolytic enzymes. Papain is prepared from *Carica papaya* fruit. Papain is made up of many enzymes such as proteolytic enzymes, amylase, and lipase. However, the activity of lipase in papain is weak. Bromelain is a general name for the family of sulfhydryl-containing, proteolytic enzymes from the pineapple (*Ananas comosus*) fruit and stem. Bromelain contains such enzymes as peroxidase, acid phosphatase, and many proteases. It is used to supplement pepsin and/or trypsin. The enzymes may be used in combination with other substances that are also required for digestion. For instance bile, pancreatin, and bromelain may be used in combination as supplement to aid the process of intestinal digestion [47, 52–55].

12.3.2 Symbiotic Digestion

Under the activities of enzymes from symbiotic bacteria and protozoa, the macroorganism gets several benefits from the interaction. There are several millions of beneficial microbes residing in the gut beginning from the mouth to the anus. The metabolic and secretory products of these resident microbes control many physiological processes occurring in the macroorganism. Some of the functions of these microbes are dependent on the types of secreted enzymes. Enzymes of symbiotic bacteria, protozoa, and fungi include lipases, proteases, and different types of carbohydrases. These microbial enzymes are able to cleave certain food substances that are non-digested by the host enzymes. The products of the cleavage reactions by the microbes can serve different physiological roles for microbial activities and host. Thus humans benefit a lot from the association. In the same vein, the resident microbes have several benefits from the in vivo medium of the host and especially the dietary pattern/habit of the organism. Many undigested products of food taken into the alimentary canal are digested by the resident microbes. The products of this process provide a lot of benefit to the host [56–61].

12.3.3 Digestion Proper

This type of digestion is carried out by the enzymes produced by the enterocytes of the gut [45, 46]. This chapter is focused on this type of chemical digestion.

12.4 Digestion of Carbohydrate

Carbohydrates comprise the major constituents of the food intake of human. Carbohydrate is made up of chains of different monomers linked together by chemical bonds. To access the energy embedded in these bonds, carbohydrates must be broken down into simpler molecules and then oxidized to produce energy. The former is carried out by enzymes of the GI tract [62–65].

12.4.1 Carbohydrate-Digesting Enzymes

Carbohydrate-digesting enzymes are those biological catalysts that aid in the breakdown of carbohydrate food. There are two major types of digestive enzymes in humans that breakdown carbohydrates, namely α-amylase and mucosal α-glucosidase [66, 67]. These enzymes are responsible for the breakdown of carbohydrates to glucose that is absorbed in the intestine. Alpha-amylase may be classified as lingual, pancreatic (or intestinal) or as extracellular and intracellular or even as autolytic, microbial, and enterocytic. Table 12.1 gives a detailed description of the characteristics of major carbohydrate-digesting enzymes and their mechanisms of functioning.

12.4.2 Chemical Cleavage of Carbohydrate

Carbohydrate comprises about 50% of a typical Western diet [121, 122]. The fate of carbohydrate digestion is mainly to provide energy molecules for bodily functions. To obtain the energy molecules, carbohydrates must be first broken down, absorbed, and transported to the site of metabolism. Carbohydrate digestion starts from the mouth where food remains in the mouth for about 30s depending on the type and consistency of the ingested food. In the mouth, food is broken down (solid food) mechanically and chemically into smaller and simpler units. The major chemical cleavage in the mouth takes place on carbohydrate-containing products. If the carbohydrate is starch, it is hydrolyzed mainly to alpha-dextrin by salivary amylase [72, 90, 91]. However, it should be noted, generally, that hydrolysis of carbohydrates by salivary amylase split these macromolecules into a very small proportion of monosaccharides, but with increasing proportion of longer lengths of carbohydrate units such as disaccharides, trisaccharides, and oligosaccharides [72, 123]. Under the action of mucin and other salivary components, food particles are rolled to form a bolus, which is subsequently swallowed into the stomach. (The mechanisms and processes of swallowing have been discussed in Chap. 7). In the stomach, amylase is relatively inactive due to the high acidity [123]. The mechanical churning activity of the stomach results in the formation of chyme. Food stays in the stomach for approximately 2–3 h (depending on the type and consistency of the ingested food particles). Chyme travels into the initial region of the intestine upon relaxation of the antropyloric sphincter, also known as gastroduodenal pyloric sphincter [124, 125]. It is currently believed that at least four sphincters regulate the flow of chyme in the gastroduodenal region. Once in the duodenum, duodenal sphincters allow the thorough mixing of chyme with biliary and pancreatic secretions, which are required to ensure adequate digestion. Duodenal sphincters include the following—a distally located duodenal bulb sphincter; proximally located duodenal sphincter below the ampulla of Vater referred to as Ochsner muscle named after Albert J. Ochsner (1858–1925); and a physiological sphincter at the duodenojejunal junction [126, 127]. The first three sphincters coordinate the rate of gastric and duodenal solid carbohydrate food emptying [126]. The chyme stays in the duodenum for a few minutes depending on the type of ingested feeds [125]. The fourth physiological sphincter at the duodenojejunal junction coordinates food evacuation from the duodenum into the jejunum [124, 128]. In the upper intestine, alpha-dextrin is then digested by glucoamylase (alpha-dextrinases) to maltose and maltotriose [90, 91]. Here, other products of salivary activity such as trisaccharides and oligosaccharides are also broken down into smaller units by the pancreatic amylase [72, 123]. The food particles are emptied into the distal small bowel where disaccharides are further broken down to their monomers. The monomers—glucose, galactose, and fructose—produced are absorbed across the enterocytes of the upper half of the villus via special receptors–transporters [123]. The undigested disaccharides and oligosaccharides are then cleaved by their corresponding enzymes (maltase, isomaltase, sucrase, and lactase) and oligosaccharidases, respectively (see Table 12.1) [72, 90, 91]. For further information on α-glucosidases, see Table 12.1. It should be noted that the contributions of carbohydrases to the digestion of starch vary. The alpha-glucosidase mucosal maltase-glucoamylase contributes 20% of mucosal alpha-glucogenic activity, whereas sucrase-isomaltase contributes 65% of total alpha-glucogenesis. Human pancreatic alpha-amylase contributes less than 15% of alpha-glucogenesis [106]. Therefore, carbohydrate digestion is a combination of the catalytic products of different enzymes including pancreatic and salivary amylase as well as intestinal brush border enzymes, glucoamylase and sucrase-isomaltase. [109]. Food particles can remain in the distal half of the small intestine for more than 2 h [125].

12.4.3 Carbohydrate Absorption and Transport

Absorptive function of the GI tract can generally be referred to as the penetration of substances through the epithelium of the GI tract into the blood or lymph. Absorption of hexoses (glucose, galactose, and fructose) is the movement of these molecules from the intestinal lumen, across the epithelium into blood. Hexose sugar passes into the epithelial cell by membrane transporters of the epithelial cell. They are further transported from the cytosol to the basal side of the membrane and exit

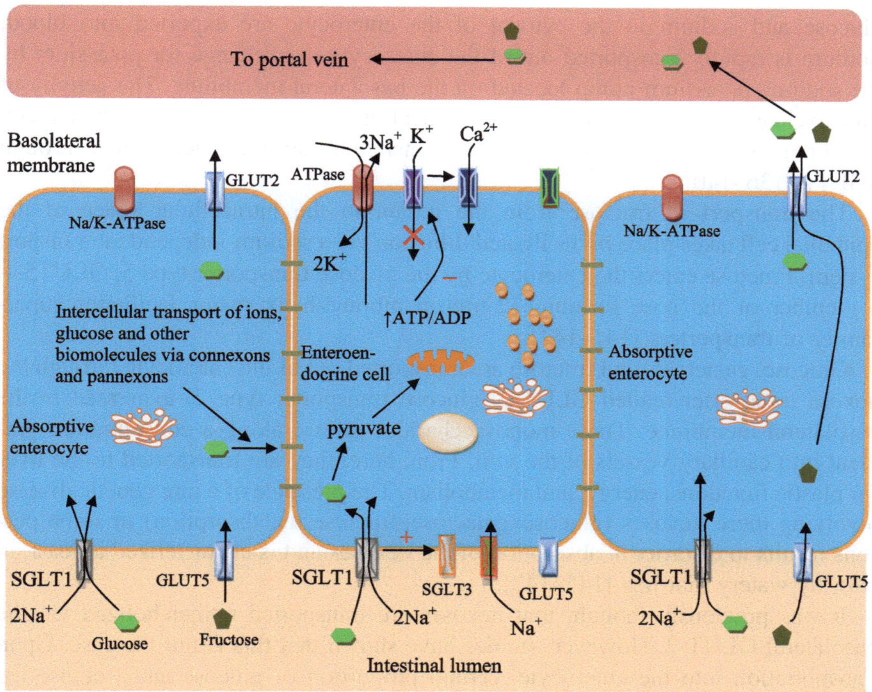

Fig. 12.2 Gut epithelial cells showing mechanism of absorption and transport of hexoses (such as glucose, fructose, galactose) through the enterocytes into bloodstream

the epithelial cell through receptors and channels according to their concentration gradient (Fig. 12.2) [123, 129, 130].

The transport of glucose and galactose from the lumen to the intracellular space of the epithelial cell occurs by co-transport mechanism using the sodium-dependent hexose transporter, which functions as a carrier molecule. Glucose and galactose are actively transported into the enterocyte by the Na^+-glucose co-transporter (SGLT1) via the transmembrane electrochemical Na^+ gradient [131]. This co-transporter was first suggested by Riklis and Quastel in 1958 [132]. Two years later, Robert Kellogg Crane (1919–2010) described the role of intestinal SGLT1 in glucose transport [133, 134]. In 1987, Ernest M. Wright and colleagues successfully cloned the transporter [135]. This carrier transports both glucose and sodium ion into the cell in a stepwise fashion. First, the carrier binds two molecules of sodium, which induces a conformational change, leading to the opening of the glucose-binding site. The binding of glucose molecule to the transporter causes reorientation of the transporter such that the binding sites with the sodium and glucose are moved to the interior of the cell. The two sodium molecules immediately dissociate into the cytoplasm, and glucose molecule is also detached. The transporter then returns to its original position, and it is ready to transport another molecule of glucose. Both glucose and sodium in the cytosol of the enterocyte are exported into blood. Sodium is rapidly transported out of the enterocyte in exchange for potassium by the sodium–potassium pump located on the basolateral membrane. The activity of this exchanger produces an electrochemical gradient across the epithelial cell membrane, which provides the energy responsible for driving glucose entry through SGLT1 [136–140].

The transport of fructose from the lumen to the intracellular space of the epithelial cell takes place by facilitated diffusion via a sodium-independent transport system. Fructose enters the enterocyte by the glucose transporter type 5, GLUT5—a member of the pore forming 12-transmembrane-helix major facilitator superfamily of transporters [141–144].

Glucose, galactose, and fructose are transported out of the enterocyte through the hexose transporter (called GLUT-2, glucose transporter type 2) expressed in the basolateral membrane. These monosaccharides diffuse along a concentration gradient into capillary vessels of the villi. From here, they are transported to the liver for plastic functions, energy, and metabolism. The presence of a rare genetic disease involving these hexoses (glucose–galactose–fructose malabsorption) in some persons results to disorder in absorption of the hexoses and is characterized at least, in part, by watery diarrhea [145–147].

It was previously thought that hexoses are transported unmetabolized via the basolateral GLUT-2. However, studies have shown that this is not the case. Upon transportation into the enterocyte, certain proportion of glucose and fructose are phosphorylated in the endoplasmic reticulum before it is released into the intracellular space to diffuse into the bloodstream. Fructose is converted to glucose. Remember that a small proportion of hexoses are transported unmetabolized via the GLUT-2. Exocytosis of glucose via the basal membrane cytoskeletal networks may be the main mechanism by which this hexose is transported out of the enterocyte. In 2001, Stümpel et al. [148] found that glucose transport pathway involves the structure of the cytoskeleton and the cell membrane of the basolateral membrane of epithelial cells. According to the study, the inactivation of gene GLUT2 did not change transepithelial transport of glucose. However, this glucose transport was inhibited by phlorizin, an inhibitor of SGLT-1. Furthermore, glucose transport was inhibited by an inhibitor of glucose-6-phosphate translocase (S4048) in the absence of GLUT2. Also, transport of [14C] 3-O-methylglucose terminated in the absence of GLUT2 (3-O-methylglucose is not metabolized in the body) [148]. In 2002, Masaya Hosokawa and Bernard Thorens reported that this type of transport of glucose involving the cytoskeleton is based on membrane invagination and may be responsible for about 30–40% of the glucose that enters the bloodstream from the intestine or hepatocytes [149]. The authors believed that this is the main pathway for glucose release into the bloodstream. In the following year, Santer et al. [150] reported that in congenital pathology of GLUT2, there is no observed disorder of glucose transport as the enterocyte membrane invagination mechanism is preserved. In patients with Fanconi–Bickel syndrome (disease characterized by GLUT2 deficiency), there is no observed disorder in the gut absorption of monosaccharides. This means there is another pathway other than glucose transport via GLUT2 [150].

Surprisingly, patients with such deficiency have a high expression of GLUT2-hexokinase enzyme (glucose-6-phosphate translocase type 1). Patients with a deficiency of glucose-6-phosphate translocase type 1 have disrupted transport of monosaccharides [150]. Thus, the main pathway glucose is transported into bloodstream, at least for the enterocyte and hepatocyte, which is based on the structure of the basolateral membrane of the cell. This means a decrease in blood glucose (or its inappropriate increase) after consuming carbohydrates indicates a possible disorder of the structure of microskeleton of the epithelial cells or enzyme activity in the gut lumen or epithelial cell, rather than the traditional transporter of glucose, GLUT-2 [150].

Apart from the GLUTs and SGLT, more recently another glucose chansporter (known as SWEET) has been characterized and found to be expressed in microbes, plants, and animals [151]. In contrast to plants where many SWEETs are present, only a few SWEETs have been characterized in animals. In humans, for example, only one SWEET has been identified—SWEET1 [151]. SWEET1 is a bidirectional 7-transmembrane spanning helical sugar transporter that mediates influx and efflux of hexoses and sucrose across the plasma, vacuolar membranes, or endoplasmic reticulum [144, 152–155].

12.4.4 The Fate of Absorbed Hexoses

Hexoses absorbed are transported via the portal vein to the liver, where it is stored as glycogen or transported to tissues for energy production. Major percentage of the glucose absorbed is transported to the brain. Of the total amount of 160 g of glucose required for daily consumption of the whole organism, a typical human brain weighing 1.5 kg uses 100–150 g of glucose per day and 20% of the total consumption of oxygen. In conditions of active brain activity, consumption of glucose by the brain cells, which is required to maintain adequate brain functions, increases to ~90% (at rest, the contribution of glucose is about 40% to brain functions) [156, 157]. Amount of glucose present in body fluids is about 20 g, and the quantity readily available as glycogen reserves is about 190 g [157].

12.5 Protein-Digesting Enzymes and Chemical Processing of Proteins

Protein forms up to 30% of a typical Western human diet—the total energy obtained from protein food [122]. Before the energy in protein can be obtained, it must be broken down, first, into smaller units. Like the carbohydrates, the cleavage is carried out by enzymes, referred to as protein-digesting enzymes (see Table 12.2).

12.5.1 Protein-Digesting Enzymes

Protein-digesting enzymes are those biological catalysts that aid in the breakdown of protein. They may be classified as gastric or intestinal in origin. Table 12.2 gives a list of the characteristics and description of major protein-digesting enzymes and their mechanisms of action.

12.5.2 Chemical Cleavage of Protein

Digestion of ingested protein begins in the stomach where it is broken down by the proteolytic enzyme pepsin secreted as pepsinogen by oxyntic cells. The cleavage products include amino acids and peptides of varying length—di-, tri-, and oligopeptides. Further digestion of protein takes place in the intestine where undigested residues of varying lengths or partially digested protein residues are further broken down to smaller units by mainly trypsin, chymotrypsin, and carboxypeptidases. The hydrolyzed peptides—amino acids or di-, tri-, and oligopeptides—are absorbed into the enterocytes. The amino acids are released to the basolateral side from where they are transported by the portal system to the liver (for details on the digestion proteins by different intestinal enzymes and their mechanisms of regulation, review Table 12.2). It was previously believed that the di-, tri-, and oligopeptides are completely hydrolyzed to individual amino acids before they are transported via the basolateral membrane to the portal system. However, some authors have suggested that some peptides may be transported via the basolateral membrane into the portal vein without undergoing complete hydrolysis [7, 389].

12.5.3 Amino Acid and Peptide Absorption and Transport in the Gut

Amino Acid Transporters and Transport Systems

Amino acids are absorbed via lumenal (membrane) amino acid transporters, whereas peptides are absorbed via peptide transporter. The amino acid transporters are mainly solute carriers (SLC) and may be classified according to different criteria which include their chemical and functional properties and transport mechanisms. The transporters are generally classified as amino acid transport systems. Amino acid transport systems are functionally identified transport activity present in a variety of cells. These systems of amino acid transport are classified based on different nomenclatures that use letters (and numbers or signs) to designate the transporter. Amino acid transporters may be classified according to the HUGO

(Human Genome Organization) nomenclature, which is based on the gene responsible for the production of that transporter. There are 11 families of amino acid transporters including the sodium-dependent neutral amino acid transporters (SLC38), proton-coupled amino acid transporters (SLC36), vesicular inhibitory amino acid transporters (SLC32), multifunctional anion exchanger (SLC26), some mitochondrial carriers (SLC25), proton-oligopeptide co-transporter (SLC15), cationic amino acid transporter (SLC7), among others [390–397]. There is also Cl-dependent amino acid transporter, which is also present in the intestine [398]. Various GI glands and organs possess specific amino acid transporters [399].

The Christensen nomenclature classifies amino acid transport systems into "L," "A," "ASC." System L is Leucine and other large hydrophobic neutral as well as branched chain amino acid transporter [400, 401]. This system is Na^+-independent transporter that participates in amino acid exchange rather than net uptake [401–403]. System A is **a**lanine and other small and polar neutral amino acid preferring transporter. System A is Na^+-dependent and catalyzes amino acid net uptake, particularly of neutral amino acids, including glutamine. System A transporters can substitute Li^+ for Na^+. System A can also transport synthetic amino acids [401]. System ASC is **a**lanine, **s**erine, and **c**ysteine preferring transporter [401]. System ASC also recognizes other aliphatic amino acids. The transport mechanism of system ASC is Na^+-dependent and appears not to mediate amino acid uptake, rather it participate in amino acid exchange [401]. This classification was pioneered by Oxender and Christensen [404].

Another nomenclature uses the chemical properties of amino acids to classify their transporters. The letters "x" for anionic, "y" for cationic, and "z" for neutral are used to designate amino acid transport of cationic amino acids (system y^+), anionic amino acids (system x^-_{AG}), and neutral amino acids (system z). In this system of nomenclature, lowercase acronyms indicate Na^+-independent transporters, whereas uppercase acronyms are used for Na^+-dependent transporters [400, 402, 403].

Still, a fourth nomenclature classifies amino acid transporters on the basis of functional studies in kidney and intestine and the amino acid profile in the urine of individuals with different aminoacidurias. This system classifies amino acid transporters into five types: "neutral system, which is methionine preferring and transports all neutral amino acids; acidic system transports glutamate and aspartate; basic system transports cystine and cationic amino acids; iminoglycine system transports proline, hydroxyproline, and glycine; β-amino acid system transports unnatural (also called non-coded, non-proteinogenic) amino acids such as beta-alanine and other several other synthesized amino acids". The acidic, basic, and neutral amino acid transporters belong to the sodium-dependent amino acid transporters, also known as system A and N transporters (**a**mino acid and **N**a^+) [400]. The sodium-dependent amino acid transporters bind amino acids only after binding sodium. The transporter then undergoes a conformational change, leading to transport of both sodium and the amino acid into the cytosol of the enterocyte. The transporter returns to its original conformation and is ready to transport another molecule of amino acid. The movement of sodium across the epithelial membrane creates an electrochemical gradient across the epithelium. This gradient which is coupled to the activity of basolateral ATPase (Na^+/K^+ pump) drives the activity of the sodium-dependent amino acid transporter. Importantly, this energy is also used to drive water absorption into the cell. The basolateral membrane of the enterocyte contains transporters which export amino acids from the cell into the bloodstream [405–409] (see Fig. 12.3).

The mechanism of absorption of di- and tripeptides in the small intestine is shown in Fig. 12.3. Peptides are absorbed into epithelial cell by the H^+ co-transporter, PepT1. In the cytosol of the enterocyte, di- and tripeptides are digested into amino acids by cytoplasmic peptidases and the amino acids are exported from the cell into the bloodstream [410]. The expression of PepT1 is dependent on stage of development, diet, hormones, diurnal rhythm, and health state. In general, intake of protein diet upregulates the expression of peptide transporter, whereas starvation reduces its expression [410]. The expression of these amino acid transporters are also regulated by hormones and may be influenced by the range of extracellular pH [401].

Fig. 12.3 Epithelial cell of the ileum, showing mechanism of absorption and transport of amino acids and peptides through the enterocyte. Amino acids are transported into the enterocyte by sodium-dependent symporters, and peptides are transported into the enterocyte via the proton-dependent PePT1, which uses the proton-motive force and the gradient of other amino acids to efficiently absorb amino acids/peptides from the lumen of the gut. Also, there is excitatory amino acid transporter (EAAT) transporter that moves L-glutamate into the cytosol of the enterocyte [400, 411, 412]. Inside the enterocyte, the peptides are further broken down into amino acids. The amino acids in the enterocyte are transported through the basolateral membrane via special transporters. Because of the usefulness of amino acids even to the enterocyte functioning, antiporters located in the basolateral membrane cooperate with amino acid transport facilitators to release amino acids without depleting cells of nutrients [400]. The widely studied basolateral amino acid transporters are TAT1, LAT1, and LAT2. T-type amino acid transporter (TAT1) is a type of basolateral aromatic amino acid uniporter—an aromatic amino acid-facilitated diffusion transporter, responsible for homeostasis of amino acids in the cell. It does this by balancing the concentration of the extra- and intracellular levels of the amino acids [413]. TAT1 can import and export L-Phe (phenylalanine) in and out of the cell [413]. TAT1 regulate amino acid concentration in cell by exporting these biomolecules via the LAT2-4F2hc exchanger—this way the transporter participates in recycling of its imported amino acids [413, 414]. LAT2-4F2hc is the heterodimerization product of L-type amino acid transporter (LAT2) and 4F2 heavy chain (4F2hc)/CD98. (LAT1 can also heterodimerize). LAT2 is involved in the reabsorption of amino acids [414, 415]. The heterodimeric form of LAT2 (y+LAT2-4F2hc) also imports amino acids such as glutamine (Gln), arginine (Arg), and leucine (Leu) [416]. But y+LAT2-4F2hc transporter has been shown to mediate import of amino acids in exchange for glutamine [416]. The y +LAT1-4F2hc heterodimer can also import amino acids (cationic types) [417]. The LAT-type transporters belong to the family of permease-related proteins, glycoprotein-associated amino acid transporters. Both LAT1 and LAT2 are antiporters (1–1 ratio transporter) and require heterodimerization to execute their functions [417, 418]. They consist of membrane spanning light chain (designated "y+") and heavy chain (4F2hc) or rBAT (related to basic amino acid transport), which are glycosylated proteins [417, 419, 420]. The "y+" chain is responsible for carrying out the transport process of the antiporter, while the heavy chain is responsible for trafficking of the complex to the plasma membrane [419]

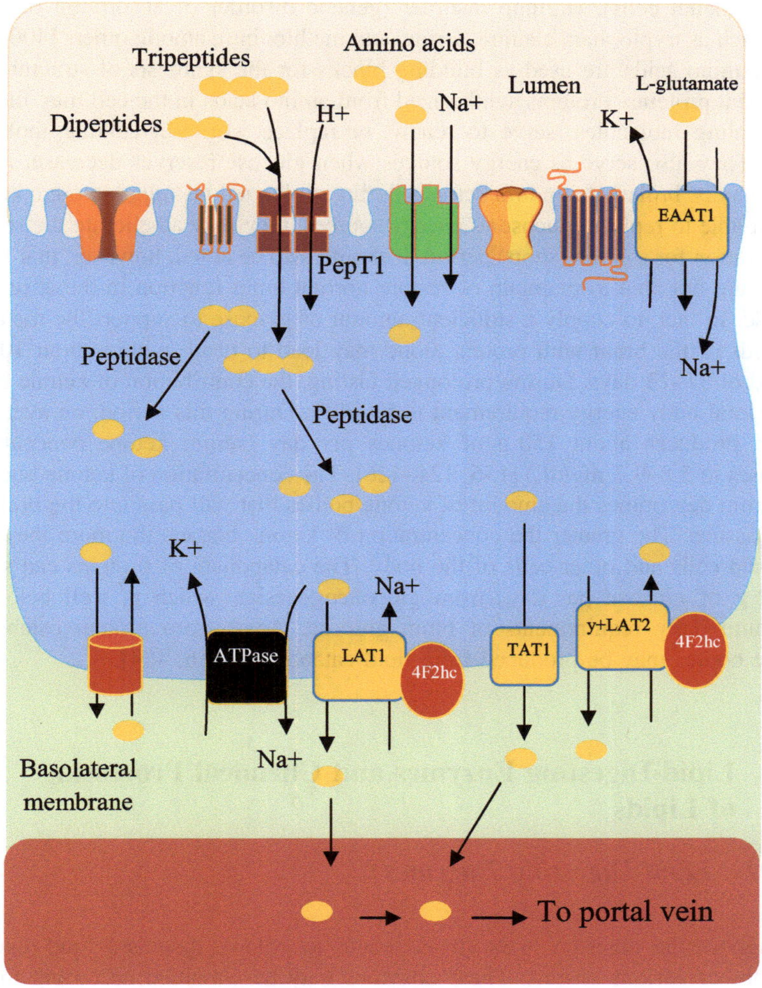

Fate of Absorbed Amino Acids

Amino acids that pass to the extracellular space of the basolateral membrane diffuse into the capillary from where they are transported to the liver. In the liver, some are used for maintenance of liver functioning; others are transported to various tissues for cellular renewal and functioning. The transport of amino acids in intestine is critical for the supply of amino acids to all tissues and the homeostasis of plasma amino acid levels. The usefulness of the gut amino acid transport is seen in inherited disorders of amino acid transport including amino acid intolerance, cystinuria (a genetic disorder of reabsorption of cystine, arginine, lysine, and ornithine in the brush border membrane of the proximal renal tubule and in the GI tract epithelial cells), Hartnup disorder (genetic disorder of absorption of amino acids such as tryptophan, alanine, serine, and methionine), among others [400, 421–423]. Amino acids are used as building blocks for the synthesis of structural and functional proteins. Proteins synthesized from amino acids in the cell may function as signaling molecules, serve to renew, or replace worn-out cellular molecules [413]. They also serve as energy sources when glucose reserves decrease. Amino acids may be broken down into ketones during prolonged fasting. Although ketones are not able to replace glucose as energy sources, they may ameliorate the severity of starvation for a given (short) period. It should also be noted, however, that ketone bodies are not able to maintain or restore normal brain function in the absence of glucose. In fact, to supply a sufficient amount of glucose to support the metabolic demands of the brain with protein alone may lead to death within about 10 days instead of 57–73 days. During prolonged fasting, the contribution of ketone bodies to the total body energy requirement is 50–70%. During this period, on average, a person produces about 150 g of ketones per day (serum ketone concentration increases to 5.8–9.7 mg/dL) [156, 424–426]. The concentration of ketone bodies in the serum determines the amount of ketone bodies that will pass into the brain and other tissues. The greater the concentration of ketone bodies, the more they enter the brain cells and other cells of the body. The catabolism of proteins can supply 17–32 g of glucose per day (from gluconeogenesis), which is well below the minimum daily requirements of brain glucose. Apart from protein catabolism, ketone bodies may be produced from lipid catabolism [156, 424].

12.6 Lipid-Digesting Enzymes and Chemical Processing of Lipids

12.6.1 Lipid-Digesting Enzymes

Throughout the digestive tract (from mouth to colon), there are lipid-digesting enzymes of various origins. These enzymes may be classified according to their region of synthesis in the GI tract: lingual, gastric, intestinal lipid-digesting enzymes. They may be classified as artificial or natural; microbial or intrinsic human lipid-digesting enzymes [72]. The level of pancreatic enzymes (lipases) in pancreatic fluid is affected by many factors including lipid content of ingested food, age, gender of the individual [72].

12.6.2 Lipid Digestion

In a typical diet in developed countries, approximately 30–40% of calories are obtained from dietary lipids, mostly in the form of triglycerides [121, 122, 473].

Triglycerides are the major components found in simple lipid—called fats [474]. Dietary fats may be animal or plant in origin. Examples of animal fats are cheese, milk fat, butter, and pork lard. Examples of fats obtained from plants (also known as vegetable fats) are coconut oil, palm oil, olive oil, grape oil, wheat germ oil, pumpkin seed oil, sesame oil, rice bran oil, almond oil, rapeseed oil, peanut oil, corn oil, sunflower oil, and safflower oil [475–478]. The animal fats contain mainly saturated, monounsaturated, and polyunsaturated fatty acids, cholesterol, as well as vitamin E. The vegetable fats are composed of different proportions of saturated, unsaturated, monounsaturated, and polyunsaturated fatty acids and vitamin E, but do not contain cholesterol [479–482]. Further information on the types and the clinical implication of these fatty acids can be found in Klonoff [483].

The daily lipid intake contains about 11–12 g of phospholipid, mainly phosphatidylcholine. The sterol (also a type of lipid) in a typical diet contains mainly cholesterol, derived from animal fat and accounts for about 80% of dietary sterol. Sterol derived from plant may constitute about 20% of the total sterol in a typical diet [474]. To gain the usefulness of fats and oils, they must be digested by the respective enzymes in the digestive tract. Digestion of lipids starts in the stomach by the action of both lingual and gastric lipases. However, an insignificant amount of lipid may be digested in the mouth. It is believed that a substantial amount of lipid is digested in downstream regions of the gut—stomach; in addition, pancreatic lipase with other cofactors is also produced in the upper intestine to participate in lipid digestion. The major component in fats of human diet is triacylglycerol (TAG). TAG consists of three fatty acids linked to glycerol. In the GI tract, lipases hydrolyze TAG to release free fatty acids and monoglycerides [484, 485]. Inhibition of this enzyme slows down lipid digestion and thus cholesterol absorption. This fact has been used for GI lipases, especially pancreatic lipase, which is inhibited by the drug orlistat (tetrahydrolipstatin)—which is used for treatment of obesity [486–488].

The activity of lipase on fats is low since the enzyme is water-soluble, whereas fat is water-insoluble. Thus, lipase can only act on the surface of fats. The poor solubility of fats in the gut prevents adequate digestion. The digestion of lipids is substantially enhanced by emulsification, a process by which fat globules are broken down into smaller droplets through the mechanical activity of the GI tract (GI motility). Furthermore, emulsification increases the surface area for the action of lipase to digest TAG. Another agent that promotes emulsification is bile acid, a derivative of cholesterol. Bile acids aid emulsification through the intercalation of the hydrophobic portions of bile acids into the lipid (recall that bile acids possess both hydrophilic and hydrophobic domains; i.e., they are amphipathic). This way, bile acids participate in coating lipid droplets. The coenzyme, colipase, also an amphipathic protein, is responsible for binding and anchoring lipase to the surface of the emulsion droplet. Following the completion of lipid digestion, monoglycerides and fatty acids associate with bile salts and phospholipids and/or other fat-soluble components to form micelles in the lumen of the small intestine. Micelles are hundreds of times smaller than emulsion droplets. Micelles are responsible for transporting monoglycerides, fatty acids, and other lipid-soluble

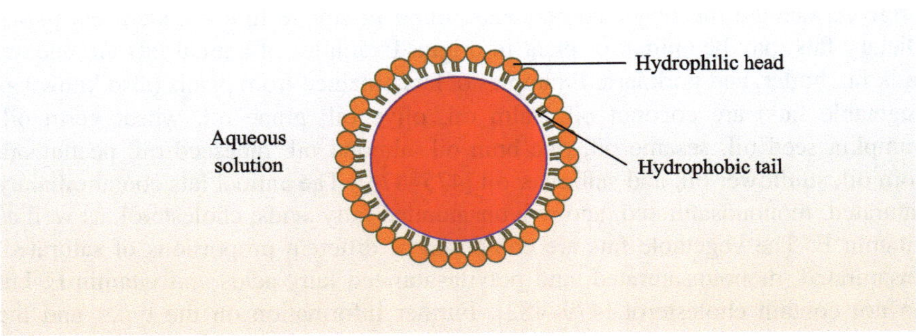

Fig. 12.4 Micelle

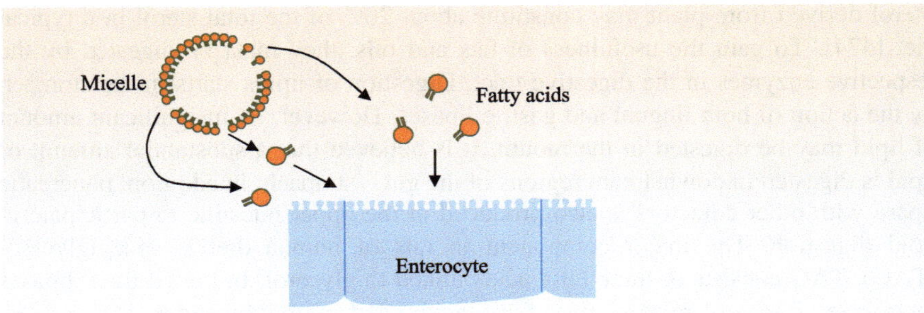

Fig. 12.5 Diffusion of micelle components to the membrane of the ileal enterocyte. Free fatty acids are usually classified according to their chain length. The most important for health and well-being are the short-chain free fatty acids, followed by the medium-chain fatty acids and lastly the long-chain fatty acids. The short- and medium-chain triglycerides are transported directly into the portal venous system [496–500]

molecules (e.g., certain vitamins and cholesterol) to the surface of the enterocyte where they can be absorbed through the apical membrane of these cells (Figs. 12.4 and 12.5) [489–495].

12.6.3 Lipid Absorption and Transport Mechanism

The products of lipid digestion may diffuse or are transported into the enterocyte via membrane carriers or transporters. Absorption of fatty acids depends on the chain lengths and involves fatty acid transporter and diffusion via the plasma membrane (flip-flop mechanism) (Fig. 12.6) [485, 491, 501].

There are membrane protein-transporters of fatty acids, which are distributed in different ways in the cells. These transporters regulate the concentration of fatty

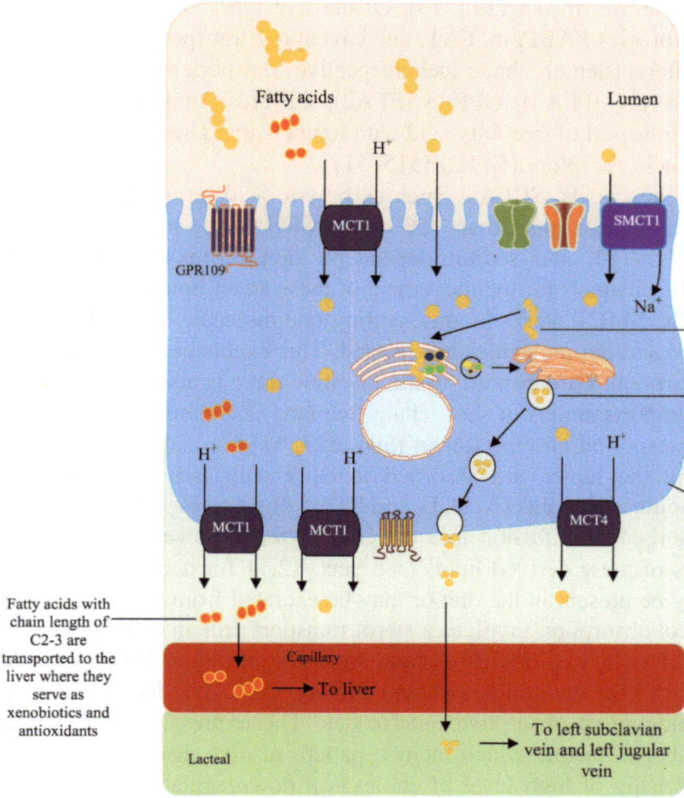

Fig. 12.6 Epithelial cells showing mechanism of absorption and transport of fatty acids through the enterocyte. Fatty acid transporters are located in both small and large intestinal enterocytes [502, 503]. Short-chain fatty acids may be transported into the blood via the monocarboxylate transporter 1, MCT1 [504]. While short-chain fatty acid may be produced by digestion of dietary lipid, they are usually produced by the anaerobic gut microbiota via fermentation of dietary (resistant) fibers in the large intestine required for energy and numerous cellular processes in health and disease. Short-chain fatty acids (e.g., butyrate) are known to prevent the development of cancer [505, 506]. The long-chain length fatty acids (as a constituent of TAG) combine with other lipid components in the ER to form chylomicrons—which are the main transport vehicles for the lipid components [507]. However, some of the absorbed products of digestion of dietary fats, following resynthesis into TAG, are stored as lipid droplets in the cytoplasm of the enterocyte [508, 509]

acids in the cell by translocations of intracellular fatty acid transporters to the membrane (receptor number changes) [510]. The fatty acid transporters include plasma membrane fatty acid-binding protein (FABPpm); fatty acid translocase (FAT); caveolin, 56-kDa renal fatty acid-binding protein, and fatty acid transport protein (FATP) [473, 511]. FATP belongs to the family of fatty acid transport proteins (SLC27A1-6) [512]. Transporter FABPpm is similar to the mitochondrial aspartate aminotransferase (mAspAT) [513]. Of the five types of fatty acid transporters, intestine expresses FABPpm, FAT, and caveolin-1 transporters [514]. Free fatty acids with different lengths have their respective transporters. The receptors GPR40 (FFA1), GPR41 (FFA3), GPR43 (FFA2), GPR120, and GPR109A are responsible for the transport of free fatty acid with long chains. These free fatty acid receptors are G protein receptors (GPRs) [515–517].

For short-chain fatty acids, SCFAs, (such as butyric acid, propionic acid, and acetic acid, which comprise more than 95% of all SCFA), the monocarboxylate transporter, MCT-1, 2, 4, and sodium-dependent monocarboxylate transporter (SMCT) serve as the channels for the movement of these small lipid molecules into the bloodstream. The SMCT works by transporting one molecule of free fatty acid anion for every two sodium molecules transported. This exchanger also stimulates Cl^- and water absorption. Furthermore, in the intestine there is also SCFA/HCO_3^- exchanger, which imports anions of short-chain free fatty acids into the enterocyte, while HCO_3^- is transported into the lumen [506, 517, 518].

Phospholipids in the lumen are absorbed together with fatty acids into the enterocyte. Lipid-soluble vitamins (A, D, E, and K) are digested together with lipids in the gut and absorbed by diffusion into the enterocyte. However, there may be special transporters of these derived lipids (see Sect. 12.11 for details).

Cholesterol may be present in the diet or may be excreted from the liver in bile. Intestinal cholesterol absorption involves a sterol transport protein "Niemann–Pick C1 Like 1" (NPC1L1) that carries cholesterol from the intestinal lumen into the enterocyte. This 145-kDa integral membrane protein, NPC1L1, localizes to the brush border membrane of absorptive enterocytes. The expression of NPC1L1 depends on a number of factors which include pattern of diet, disease, and inheritance. Diets comprising of high level of cholesterol downregulates the NPC1L1 transport [519]. Polymorphisms in the gene responsible for the production of this sterol transporter affect cholesterol absorption and plasma low-density lipoprotein levels, hence influencing whole-body cholesterol homeostasis [519]. Animal studies indicate that mice lacking the NPC1L1 gene exhibit a significant reduction in the intestinal uptake and absorption of cholesterol and phytosterols. Painstaking years, involved in unraveling the physiology of this transporter, led to important application of the results in medicine. Ezetimibe is a specific inhibitor of intestinal cholesterol absorption recently introduced into medical practice. The drug was identified as the molecular target for cholesterol absorption inhibitors in the enterocyte brush border membrane [520]. Ezetimibe binds to a portion in the extracellular domain of the NPC1L1 transporter to inhibit the transport function of this sterol transporter inhibiting intestinal absorption of cholesterol, phytosterols, and certain oxysterols [519, 521]. The cholesterol absorption inhibitor ezetimibe has been shown to be effective at reducing levels of LDL cholesterol, particularly when combined with a statin, a drug that inhibits cholesterol synthesis in the liver. Statins function as HMG CoA (3-hydroxy-3-methyl-glutaryl-coenzyme A) reductase inhibitors. The combination of ezetimibe and a statin (e.g., ezetimibe/atorvastatin, ezetimibe/simvastatin) provides a further lowering of LDL cholesterol (in comparison with statin monotherapy), which is associated with reduction of the risk of myocardial infarction, stroke, and diabetes mellitus [491, 522, 523].

Apart from the NPC1L1 transporter, the ATP-binding cassette (ABC) proteins, ABCG5 and ABCG8 present in the apical membrane, also participate in cholesterol efflux [491]. ABCG5 and ABCG8 channels are very important in maintaining cholesterol homeostasis. Other cholesterol transporters include scavenger receptor class B type I (SR-BI), cluster determinant 36 (CD36), and ABCA1 [524].

12.6.4 Resynthesis of TGA and the Synthesis of Chylomicrons in the Enterocyte

After absorption of the components (fatty acids and monoglycerides, phospholipids, cholesterol, fat-soluble vitamins) of micelle into the enterocyte, they are transported to the ER. In the ER, fatty acids and monoglycerides are resynthesized into TAG by lipid re-esterification enzymes [485].

Cholesterol is conveyed to the ER via active transport and re-esterified into neutral lipids and utilized within the ER to form lipoproteins [485, 491]. Lipoproteins are secreted from the basolateral side of the enterocyte [485, 491]. Apolipoprotein B (apoB) and microsomal triglyceride transfer protein (MTP) are involved in the assembly and secretion of lipoproteins as well as synthesis of chylomicrons in the intestine [491, 525]. It was previously thought that cholesterol, phospholipids, and vitamin E are all packed into chylomicrons and transported to the liver; however, accumulating evidences suggest that cholesterol, phospholipids, and vitamin E can also be secreted from enterocytes as components of high-density lipoproteins (HDL). Thus, lipoproteins may be secreted free of apoB and free of vitamin E [491]. ApoB is the major lipoprotein found in chylomicrons (also known as ultra-low-density lipoprotein), very low-density lipoprotein (VLDL), intermediate-density lipoprotein (IDL), and low-density lipoprotein (LDL) particles. It serves as a receptor for lipid molecules (including cholesterol, fats) and as a ligand for VLDL and LDL receptors in cells of the body. MTP is a carrier of triglyceride essential for the assembly of apoB-containing lipoproteins by the liver and the small intestine [525–527]. Adipose tissue MTP possibly plays a role in triglyceride formation from fatty acids [526]. MTP helps to link apoB, TAG, and cholesterol esters in the ER to form vesicles. The vesicles bud off and are transported to the Golgi for further lipidation to form chylomicrons. The chylomicrons are composed of a core of triglycerides and cholesterol esters, which are enclosed by cholesterol, phospholipids, and proteins. The molecules of the chylomicrons are positioned in such a way that hydrophobic structures are directed toward the core, whereas hydrophilic ones are directed to the surface [485, 491, 501, 515, 528].

12.6.5 Basolateral Exocytosis and Transport of Chylomicrons

Recall that upon synthesis, chylomicrons are secreted by the Golgi complex in the form of vesicles and then transported into the basolateral side of the enterocyte. The vesicles fuse with the plasma membrane of the enterocyte and undergo exocytosis to release the chylomicrons into the extracellular space [485, 491, 501].

Chylomicrons are relatively large particles, so they do not pass into blood capillaries. Instead chylomicral particles enter lacteals—lymphatic capillaries formed at the center of each villus. In the lymphatic vessels, lymph appears milky, especially when large numbers of the chylomicral particles are absorbed. Through the lymphatic vessels, chylomicrons drain into the general circulation (vena cava) via the thoracic duct in the chest (Fig. 12.7, also see Figs. 2.12 and 2.13). From the general circulation, the chylomicrons reach the liver [485, 491, 501].

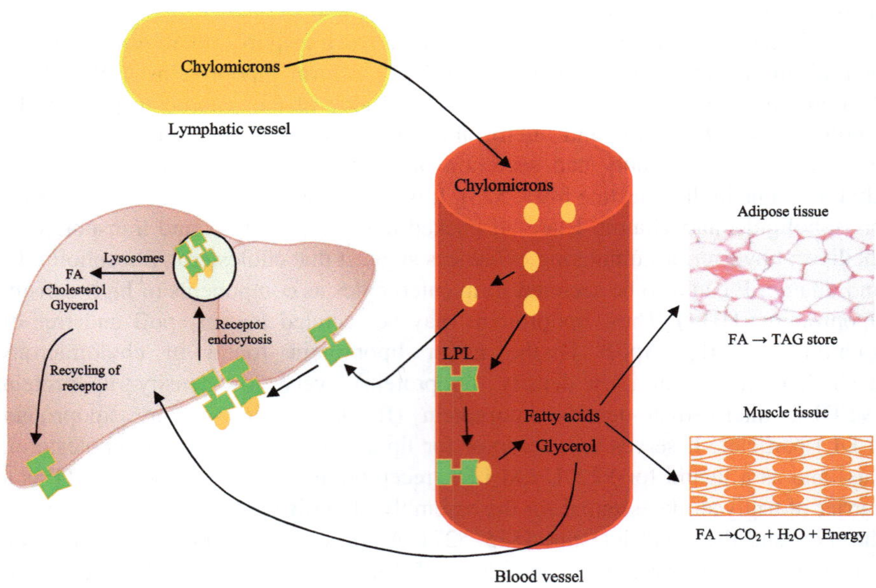

Fig. 12.7 Transport of chylomicrons to the liver and tissues of the body. Chylomicrons in the lymphatic vessel empty into the blood vessel. In blood vessel, chylomicrons may be broken down by lipoprotein lipase (LPL) located on the endothelium of capillaries to produce fatty acids (FA) and glycerol. Fatty acids are transported into the adipose tissue for storage and energy production when necessary. Fatty acids are also transported into muscles where they are mainly utilized for energy. Glycerol is transported to the liver, where it is used for metabolism. A major part of the chylomicrons is transported to the liver. In the liver, chylomicrons bind with its receptors and the receptor–ligand complex is endocytosed into the hepatocyte to form vesicle known as endosome. The endosome formed is degraded by lysosomal enzymes to produce remnants of the receptors, fatty acids, glycerol, cholesterol, and other lipid molecules depending on the constituents of the chylomicrons [529–531]

Chylomicrons deliver absorbed TAG to various cells and tissues of the body (Fig. 12.7). Some TAG and lipoproteins in chylomicrons undergo hydrolysis which is catalyzed by lipoprotein lipase, localized in the endothelium of capillary blood vessels (Fig. 12.7). Fatty acids and monoglycerides (glycerol) released from the hydrolysis of TAG diffuse into various cells of the body [485, 491, 501].

12.6.6 Fate of Absorbed Lipids

The absorbed lipid components are used for a variety of physiological processes in the body. Lipids and fats form major part of the membrane and many components of the cell. They also serve as energy sources during prolonged starvation. Certain types of fats are essential nutrients that perform critical functions in the body. Essential fats are required as a carrier of lipid-soluble vitamins, serve as substrates for the synthesis of steroid hormones, and also enhance the bioavailability of fat-soluble micronutrients [532, 533]. Some lipids serve as second messengers in cellular signaling processes or participate in signal transduction process. Lipids of plasma membranes play a critical role in sensing and responding to stress stimuli [533]. Several cellular functions of lipids in the cell have been discussed in Chap. 3. These benefits of fats notwithstanding, high lipid intake can predispose an individual to the development of coronary artery disease, hypertension, obesity, and other related diseases [532, 533].

12.7 Absorption and Transport of Dietary Elements

Dietary or nutritive (essential) metals (elements) such as zinc (Zn), calcium (Ca), phosphorus (P), copper (Cu), selenium (Se), magnesium (Mg), iron (Fe), potassium (K), sodium (Na) constitute a significant fraction of all elements absorbed in the gut. These elements are required for normal cellular processes. These elements at recommended range are inevitable substances in the normal functioning of the body [534].

12.7.1 Historical Background of Intestinal Epithelial Ion Transport

Some ions may move passively across the lipid bilayer, whereas the movement of other ions is regulated by stimuli impinging on the cell. Such stimuli may include the activation of neurotransmitter receptors such as acetylcholine receptors by acetylcholine. Movement of ions in and out of the cells of the lining of the GI tract is crucial in the regulation of GI functions. A pioneer in the investigation of transport of ions across the epithelium of the GI tract is Hans Henriksen Ussing (1911–2000). Ussing became acquainted with the Danish professor Schack August Steenberg Krogh (1874–1949), Hungarian radiochemist and Nobel Laureate George Charles de Hevesy (1885–1966), the Danish physicist Niels Henrik David Bohr (or just Niels Bohr) (1885–1962) when the latter three had planned to use artificial radioactive isotopes to study the dynamic state of the living organism. August Krogh won the 1920 Nobel Prize in Physiology or Medicine 1920 "for his discovery of the capillary motor regulating mechanism." His interest majored in respiratory physiology, and he investigated the mechanism of transport of gases across the capillary of the lungs. Niels Bohr won the 1922 Nobel Prize in Physics "for his contribution to unraveling the structure of atoms and the radiation emanating from them." The interest of Niels Bohr in studying physiological systems with the tracer could have been stimulated not only by his friends Krogh and others, but also his father, Christian Bohr (1855–1911), who was a Professor of Physiology, Copenhagen University, Copenhagen, Denmark. Hevesy (1885–1966) was recognized in 1943 for his key role in the development of isotopes as tracers in the study of chemical processes. Hevesy is credited for discovering radioactive tracing [535–537].

This collaboration that Ussing had with top scientists with deep background in radioisotope labeling was important for his future career. Ussing made important findings in the physiology of epithelial transport, especially exchange diffusion, unidirectional fluxes, flux-ratio equation, and solvent drag. It was Ussing who introduced the concepts of "short-circuit current," "active transport," and "paracellular shunt pathway." He showed that movement of sodium ions in cells is regulated by active transport mechanisms. Ussing's model provided further opportunities in investigating the mechanism of transport of hormones and drugs using radioisotope techniques. He was a pioneer in solute-coupled water transport, a major pathway regulating movement of water in the GI tract. Using the frog's intestine, Ussing showed that there was sodium recirculation. Ussing's experimental results in epithelial membrane transport, provided insights into the application for studying intestinal diffusion. His methods were applied to study the physiology of other cells, tissues, and organs such as kidney, respiratory epithelia, and exocrine organs [535]. The Ussing chamber has provided useful information on the mechanism of epithelial transport [538].

12.7.2 Absorption and Transport of Calcium

Calcium is absorbed by calcium-binding protein (CaBP) via active transport process. The synthesis of this protein needs vitamin D (1,25 dihydroxyvitamin D_3 (1,25 $(OH)_2D_3$)). Lumenal calcium binds to an integral protein of the brush border membrane, which transport calcium ions into the enterocyte down its concentration gradient [539]. Calcium-binding protein (CaBP) serves as an intracellular calcium buffer that allows large amounts of calcium to pass into the cytosol while keeping the free calcium concentration in the cell low. There are at least two different basolateral transporters of calcium in the enterocyte (see Fig. 12.8). Calcium is needed in all cells of the body for signaling purposes and is involved in a variety of cellular functioning. Unabsorbed calcium in the intestinal lumen is excreted in the stool [539]. The GI tract is one of the major routes of calcium excretion in the body (in addition to urinary calcium excretion) [540].

12.7.3 Absorption and Transport of Iron

Iron is a metal found in some key proteins of the body such as hemoglobin, myoglobin, neuroglobin, and many enzymes including cytochrome P-450. This iron is found in heme, a prosthetic group present in these proteins. Free heme can initiate the formation of free radical and cause damage to the cells [548, 549].

The major source of iron in the body is hemoglobin (contains about 75% of body iron), the oxygen carrying protein of red blood cell. However, red blood cells have a life span of about 120 days following which the hemoglobin is degraded in the reticuloendothelial systems and in the liver. Iron and amino acids from the globin are recycled, while the porphyrin is degraded. Bilirubin is the end product of heme metabolism. Because iron is lost from the body, a given range of quantity must be taken in daily to prevent the reduction in its level and to maintain homeostasis. Dietary iron may be in the form of heme or iron III (Fe^{+3}). Both ferric (Fe^{+3}) and ferrous (Fe^{+2}) ions form insoluble complexes with many proteins and ions such as phosphate, hydroxide, and bicarbonate [550–553]. In anemia prophylaxis or treatment, complexes of iron such as ferrous gluconate, ferrous sulfate, and ferrous fumarate are administered to replenish the loss [554, 555].

At low pH concentration, such as in the stomach, the complexes formed are more soluble, promoting absorption of iron. In the brush border of the intestine, Fe^{3+} in foods is converted to Fe^{2+}, which is more soluble, hence better absorbed into the enterocyte. This conversion of Fe^{3+} to Fe^{2+} is enhanced by ascorbic acid. This conversion of Fe^{3+} to Fe^{2+} is carried out by the brush border protein, duodenal cytochrome B (Dcytb), a ferrireductase that feeds enterocyte membrane iron transporter with Fe^{2+} in the duodenal enterocytes (Fig. 12.9) [556–561].

Fe^{2+} is transported via the H^+-coupled divalent metal-ion transporter-1 (DMT1), also called divalent cation transporter (DCT1), which was formerly called Nramp2 (Natural-resistance-associated macrophage protein 1) (Fig. 12.9). DMT1 is an apical transmembrane transporter of enterocytes, responsible for major uptake of iron in the duodenum [562–564]. DMT1 is also the primary transporter in other cells including erythroid precursor cells in which iron-transferrin complex is mobilized from endosomes to cytosol [564]. The activities of DMT1 are coupled to proton electrochemical potential gradient to drive active transport of Fe^{2+} into the enterocyte [559]. It should be noted that the activities of divalent metal-ion transporter may be dependent on the level cellular pH [564].

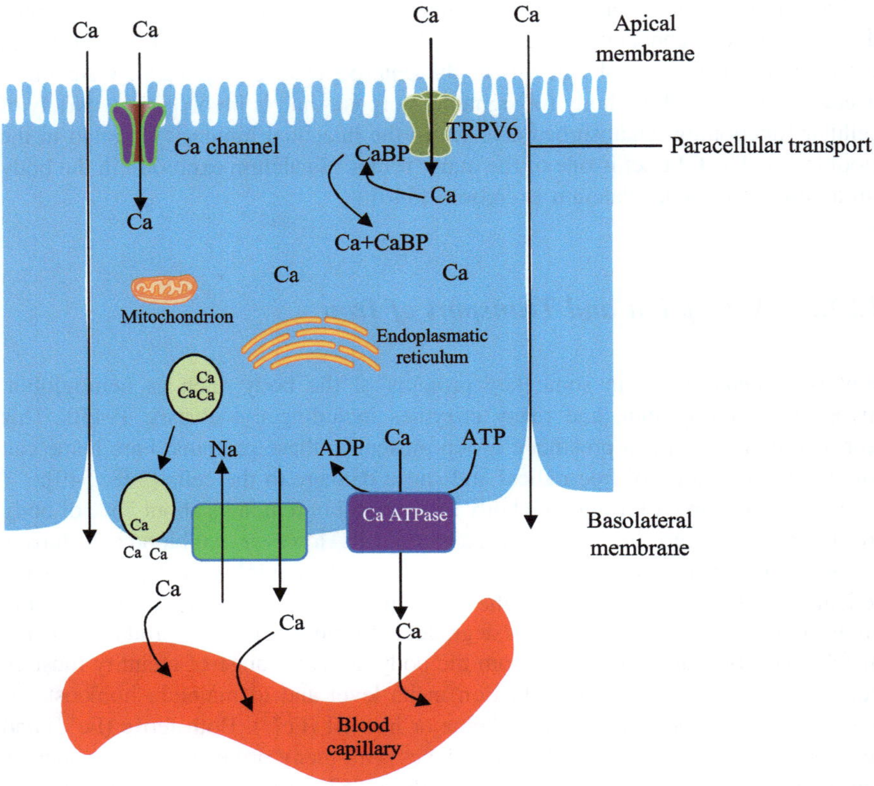

Fig. 12.8 Mechanism of absorption and transport of calcium (Ca) in the enterocyte. Absorption and transport of Ca involves both transcellular and paracellular pathways. Transcellular transport is the transfer of Ca through the enterocyte, while paracellular shunt pathway is the transfer of Ca through a very tiny passage between enterocytes. The transcellular pathway is also called the active saturable pathway, while the paracellular is passively non-saturable [541–543]. The transcellular transport is active largely in the duodenum and upper jejunum, while the paracellular shunt is active throughout the intestine [544]. Thus, at low lumenal calcium level, transcellular flux accounts for a larger fraction of the absorbed Ca. However, at high calcium intake, paracellular shunt participates to account for a high quantity of ion influx across the cell. In the transcellular transport, Ca diffuses through CaT1 or the calcium channel (TRPV6, transient receptor vallinoid potential type 6) into the cytosol, where it may bind with CaBPs (such as calmodulin, calretinin, and calbindin-D_{9k}). Calcium levels in the cytosol are also regulated by the ER and mitochondria. These organelles are responsible for Ca sequestration [544]. The basolateral membrane export of Ca into circulatory system involves the ATP-dependent Ca^{2+} pump and a Ca^{2+}/Na^{+} exchanger, as well as exocytosis as the terminal event in a proposed vesicular transport mechanism [541, 544, 545]. The paracellular and transcellular fluxes of Ca are vitamin D-dependent [543]. The expression of the rate-limiting proteins in Ca transport, CaT1 and CaBP, is significantly reduced at high lumenal Ca levels [544]. This is because the vitamin D3, the hormone, that regulates the biosynthesis and functions of CaBP and CaT1 is significantly reduced when lumenal Ca level increases [544]. Vitamin D3 regulates cytosolic CaBPs, and Ca channel receptors, including TRPV6 (previously known as CaT1 or epithelial Ca channel), which independently regulates active intestinal Ca absorption. The paracellular involvement of vitamin D3 in Ca shunt is based on vitamin D regulation of junctional proteins of epithelial cells including cadherin and claudins [546, 547]

There are at least four isoforms of DMT1. However, their physiological characteristics and role in different anemic conditions are still been investigated [563]. DMT1 has been found to play an important role in iron metabolism and hemogenesis. DMT1 was significantly decreased in inflammatory bowel disease (IBD) such as Crohn's disease [568]. Recent investigation has observed a pathological role of DMT1 in anemia, hereditary hemochromatosis, and other iron-overload disorders [563]. Certain mutations in human DMT1 can lead to severe microcytic–hypochromic anemia [563]. Thus, DMT1 is a key therapeutic target for these illnesses.

Once transported by the DMT1 into the enterocyte, Fe^{2+} binds to a cytosolic globular protein, ferritin, the primary intracellular iron-storage protein and forms Fe^{2+}/ferritin complex which acts as a buffer of free intracellular iron levels. Ferritin helps to regulate the concentration of intracellular iron to prevent iron-induced toxicity, iron deficiency, and iron overload. Ferritin that is not bound to iron is called apoferritin [556].

The heme from diet may be imported into the enterocyte via the apical transporter, HCP1 (Fig. 12.9). In the enterocyte, the enzyme, heme oxygenase which catalyzes the conversion of heme to bilirubin in the liver and spleen, also breaks down heme or the heme that is not catalyzed by this enzyme is transported via the basolateral carriers, FLVCR, a putative heme transporter or BCRP (Fig. 12.9) [569–571]. The activity of heme oxygenase is regulated by many factors including fasting, hypoglycemia, hormones (e.g., epinephrine, glucagon), cyclic AMP [570].

The basolateral transporters of iron include the duodenal iron export protein-transporter—iron-responsive element (IRE) IREG1, the transmembrane protein—ferroportin [556, 572]. IREG1 is responsible for iron efflux from the basolateral membrane of the enterocyte [556]. Ferroportin transports iron from the inside of the enterocyte to the outside [572].

Ferroportin, the major ion transporter on the basolateral membrane, is regulated by the iron regulatory hormone, called hepcidin (hepatic bactericidal protein), 25-amino acid peptide exclusively synthesized by the liver [573] named based on site of synthesis (hep-) and its in vitro antibacterial properties (-cidin). But the hormone was first isolated from human urine [573, 574]. The hormone is synthesized from the 84 amino acid preprohormone through an initial cleavage, which produces the 60-amino acid prohormone. There are numerous chain lengths of hepcidin. Hepcidin inhibits iron transport by binding to ferroportin. The ferroportin–hepcidin complex is then degraded in the lysosome of the enterocyte. Inhibiting ferroportin prevents iron from being exported, and the iron is sequestered in the cells. This way, iron export to the portal system is reduced. Hepcidin also regulates ferroportin activity in the reticuloendothelial system cells such as macrophages. In infection and inflammatory reactions, hepcidin level significantly increases, thus causing decrease in serum iron due to reduced iron export in the enterocyte, iron trapping within macrophages and liver cells. In condition such as hemochromatosis, hepcidin level is very low, which results in iron overload. This is due to excessive ferroportin-mediated iron transport [573, 574]. Dysfunction of this protein leads to excessive intestinal absorption of iron and increased iron release by

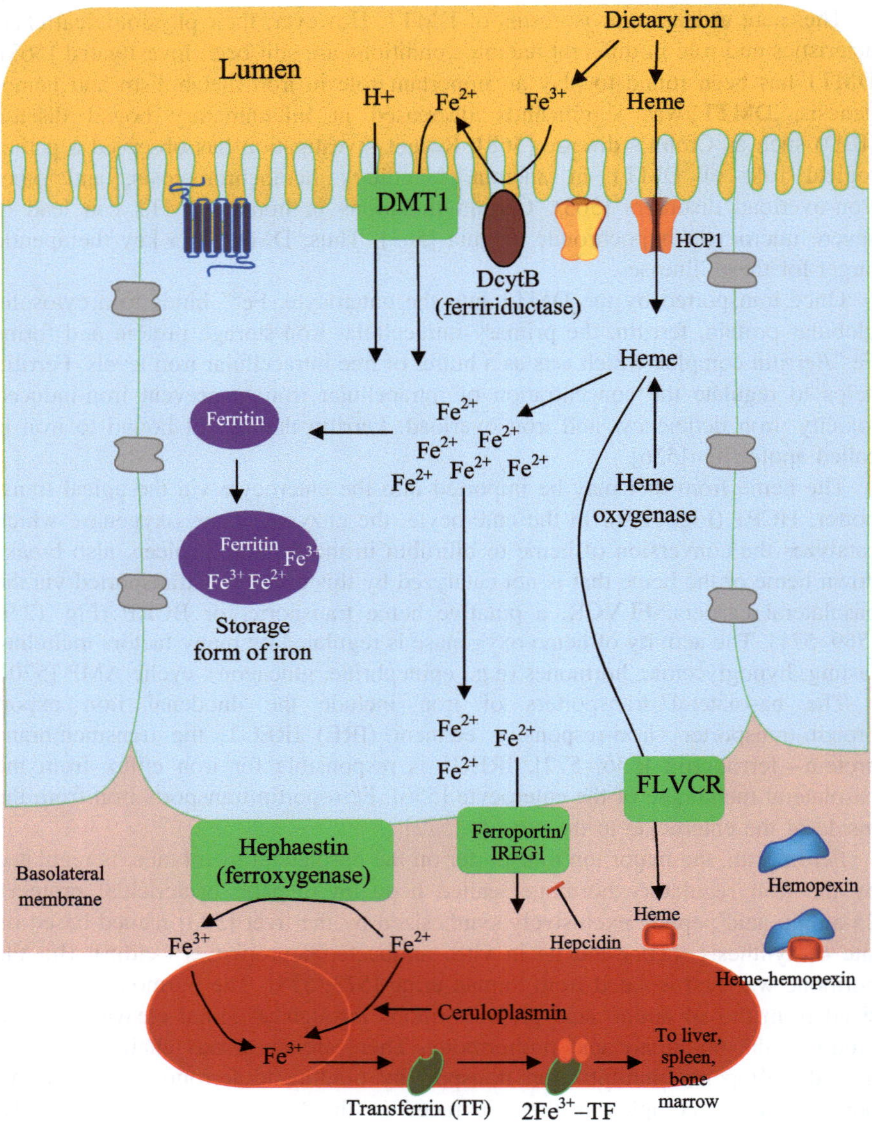
Dietary iron
Lumen
H+
Fe^{2+}
Fe^{3+}
Heme
DMT1
DcytB
(ferriridúctase)
HCP1
Heme
Ferritin
Fe^{2+}
Fe^{2+} Fe^{2+}
Fe^{2+} Fe^{2+} Fe^{2+}
Heme
oxygenase
Ferritin
Fe^{3+}
Fe^{3+} Fe^{2+}
Storage
form of iron
Fe^{2+} Fe^{2+}
Fe^{2+}
FLVCR
Ferroportin/
IREG1
Hephaestin
(ferroxygenase)
Basolateral
membrane
Hemopexin
Heme
Hepcidin
Fe^{3+}
Fe^{2+}
Heme-hemopexin
Ceruloplasmin
Fe^{3+}
To liver,
spleen,
bone
marrow
Transferrin (TF)
$2Fe^{3+}$–TF

◂**Fig. 12.9** Schematic representation of absorption and transport of iron in the enterocyte. Iron in dietary iron is converted to iron II and transported via a ferrireductase protein called the duodenal cytochrome B (DcytB). The iron II is readily transported into the enterocyte by the divalent metal transporter, DMT1. DMT1 mediates both H^+-coupled Fe^{2+} transport and uncoupled fluxes. In the enterocyte, excess iron II may be stored by binding to ferritin, an intracellular iron-storage protein. Iron II in the enterocyte is also transported to the basolateral membrane from where it is exported by ferroportin or IREG1 (the iron permease) localized to the basolateral membrane [565]). Hephaestin cooperates with ferroportin to convert iron II to iron III, which is readily transported by apo-transferrin. Ceruloplasmin can also convert iron II to iron III. Apo-transferrin binds with iron II to form iron III-transferrin complex. Dietary heme is transported into the enterocyte via the heme carrier protein 1 (HCP1). Heme in the enterocyte may be transported via *anemia-inducing feline leukemia virus subgroup C receptor* (FLVCR) or breast cancer resistance protein (BCRP/ABCG2) transporter through the basolateral membrane to the extracellular space. The heme binds with another heme-transport protein called hemopexin to form heme–hemopexin, which is transported to the liver and cells of the reticuloendothelial system for degradation [552, 566, 567]

macrophages, a condition that leads to iron overload. Hepcidin level is regulated by the plasma level of iron. Increase in serum iron increases the expression of hepcidin [573].

Before iron is transported into the circulation, it must be converted from Fe^{2+} to Fe^{3+}. The conversion is mediated by a basolateral transmembrane copper-dependent ferroxidase, called hephaestin, discovered in 1999 by Christopher D. Vulpe and co-workers [560]. This ferroxidase closely cooperates with ferroportin. While the main site of expression of hephaestin is the small intestine, the peptide has been discovered in bone trabecular cells, large intestine, kidney, spleen, breast, and placenta. However, the role of this peptide in these cells and tissues is not exactly known [565, 575]. The conversion of Fe^{2+} to Fe^{3+} allows the binding of iron to its carrier (transferrin) in the blood. The blood carrier of iron can only bind ferric iron (Fe^{3+}) under physiological conditions [575]. Mutation of hephaestin has been implicated in sex-linked anemia, which results in iron deficiency in laboratory animals [565, 575].

Hephaestin is noted for its homology to ceruloplasmin, a serum dehydrogenase protein involved in copper detoxification and storage. Ceruloplasmin, the major copper-carrying protein in the blood, participates in both iron and copper metabolism. Ceruloplasmin exhibits a copper-dependent oxidase activity, oxidizing Fe^{2+} to Fe^{3+}. Thus, this protein aids in transporting Fe^{3+} to transferrin [575–577].

When Fe^{2+} has been converted to Fe^{3+}, it is ready to bind to transferrin, the iron-binding blood plasma glycoprotein. In the blood, Fe^{3+} readily binds to transferrin for transport and delivery to the liver, spleen, and bone marrow. Transferrin has two specific high-affinity Fe^{3+}-binding sites. The complex formed from the reaction between transferrin and iron is reversible. Less than 0.1% (4 mg) of total body iron binds to transferrin. Notwithstanding, however, iron-transferrin complex forms the most vital iron pool having the highest rate of turnover (about 25 mg per day) in the body. The affinity of transferrin for Fe^{3+} is dependent on pH. At a pH 7.4, the affinity is highest. The affinity of transferrin for Fe^{3+} decreases with decreasing pH. Fe^{3+}-transferrin complex is transported to the liver, spleen, and bone marrow, where they bind to the transferrin receptor on the respective cells. The ligand–receptor association is transported into the cell in a vesicle by receptor-mediated endocytosis [578, 579].

Recall that heme is exported from the basolateral membrane of the enterocyte into the circulation. Because heme is highly toxic, it must be transported to the site of degradation [557]. Many proteins including albumin, haptoglobin, hemopexin, cooperate to trap the plasma heme. Of particular note is hemopexin, a heme-binding plasma glycoprotein which, after haptoglobin, forms the second line of defense against hemoglobin-mediated oxidative damage during intravascular hemolysis [580, 581]. Hemopexin has the highest binding affinity to heme among known proteins [581]. Hemopexin binds heme to form heme–hemopexin complexes, which are delivered to hepatocytes by CD91-receptor-mediated endocytosis after which hemopexin is recycled back to the circulation and heme is delivered to the hepatic parenchymal cells [580–582].

12.7.4 Absorption and Transport of Magnesium

Magnesium (Mg) is an essential (macro-) mineral with many physiological roles in the body. This element is involved in activation of enzymes, metabolic pathways, regulation of membrane channels, and muscle contraction [583]. Mg is usually ingested together with food. Mg is absorbed throughout the intestine, but the distal part of the ileum is the predominate site of absorption. Mg absorption occurs by passive diffusion and active transport. The former takes place by paracellular shunt, while the later occurs by transcellular pathway. The paracellular shunt is important only in leaky epithelia and accounts for the majority of Mg absorbed into the circulation [583, 584]. The transcellular pathway Mg involves lumenal uptake through the Mg^{2+} transporter with subsequent basolateral extrusion, which possibly involves Na^+/Mg^{2+} exchanger [583, 585].

12.7.5 Absorption and Transport of Zinc

Zinc (Zn) is a micronutrient essential for many physiological and biochemical processes occurring in the body. Zn is a catalytic cofactor for several hundred metalloproteins. Zn is ingested in the dietary form and transported by the apical brush border transmembrane proteins into the cytosol of the gut epithelial cells [586–588]. These transporters of zinc are composed of two different families of proteins—Zn transporter (ZnT) and the Zrt-(zinc-regulated transporters), Irt (iron-regulated transporter)-like protein (ZIP) transporters [589, 590]. There are at least 10 ZnT members and 15 ZIP member transporters in human cells. The ones expressed in the enterocyte include ZnT1, ZnT2, ZnT4, ZnT5, ZnT6, ZnT7, ZIP4, and ZIP5. These proteins have varying localizations and functional roles. In general, ZnT family transporters function to lower the cellular zinc concentration by promoting zinc efflux from cells or into intracellular vesicles, while the ZIP proteins act to increase cellular zinc by promoting extracellular zinc uptake and vesicular Zn release into the cytosol [591–594]. The accessory organs of the GI tracts also express a range of Zn transporters [592].

Zn transporters expressed in the apical membrane of the enterocyte are ZIP4 and ZnT5 [592]. Notable intracellular zinc transporters in the enterocyte are ZnT2,4–7. These intracellular Zn transporters are localized in vesicles, secretory granules, endosomes, and Golgi complex [592, 594, 595]. Importantly, the secretory vesicles and granules may be involved in trafficking of Zn to the basolateral membrane for export. Metallothionein, an intracellular metal buffer, plays a role in the release of Zn into the portal circulation [592, 594]. However, there are basolateral Zn transporters, which include ZnT1 and ZIP5. The former is thought to be the main transporter involved in cellular Zn efflux from enterocytes to the bloodstream [592, 594]. Zn transporters are believed to function by facilitated diffusion, secondary active transport, or as symporters. The process of Zn transport is ATP-independent [591]. Zinc transporters are involved in a range of disorders including cancers, immune, and metabolic diseases.

12.7.6 Absorption and Transport of Other Metals

The DMT1 is a transporter for multiple metals. This is probably due to the fact that that DMT1 has many isoforms [596]. DMT1 is involved in the transport of iron (Fe^{2+}), and cobalt (Co^{2+}), cadmium (Cd^{2+}), copper (Cu^{2+}), zinc (Zn^{2+}), nickel (Ni^{2+}), and manganese (Mn^{2+}) [562, 596]. It is possible that some of these metals have their specific transporter responsible for transport of the metal in the intestinal epithelium [597]. For instance, copper transporter-1 and -2 are involved in apical copper transport, while Cu-ATPases (ATP7A and ATP7B) function as copper efflux pumps [598, 599]. The ATP7A, for instance, can bind and expel seven copper ions at a time, thereby regulating the intracellular concentration of this metal [600]. The usefulness of metal transporters is observed in a variety of human diseases. For instance, Wilson's disease involves biallelic ATP7B gene mutation. The disease is characterized by pathological accumulation of copper in the liver, brain, and other organs [597, 601, 602].

12.8 Absorption and Transport of Anions

Dietary derived sulfate, phosphate, chloride, and oxalate are absorbed in the intestines, and their circulating levels are controlled by renal tubular mechanisms. The process of anion absorption is facilitated by sulfate, phosphate, chloride, and oxalate transporters that exist in the epithelial cells of the kidneys and intestines

[603, 604]. The mechanism controlling the absorption of anions depends on the type of anion involved [605]. For instance, absorption of anions, such as phosphate, Cl^-, and $SCFA^-$ depends on the activity of their respective transporters. Putative anion transporter (PAT1, SLC26A6) transports organic anions such as formate and oxalate, as well as inorganic ions. This transporter may be dependent, in part, by the outward movement of Cl^- [605]. Phosphate ion (PO_4^-) is transported by Na^+/PO_4^- co-transporter that moves both sodium and phosphate into the enterocyte. For each phosphate transported, one sodium molecule is also transported. There are different types of Na^+-coupled phosphate transporters, which are grouped as type I (SLC17), type II (SLC34), and type III (SLC20) [603, 606]. Sulfate transporters are Na^+-coupled sulfate carrier proteins (SLC13) and sulfate anion exchangers (SLC26) [606]. The transport of oxalate ions (salt or ester of oxalic acid) occurs via oxalate transporter SLC26A6. The transport of oxalate can also occur by paracellular pathway [607]. It is believed that certain kidney disease such as calcium oxalate nephrolithiasis may be due to dysfunction in intestinal transport of oxalate [608]. The monocarboxylate ions are transported by the proton-coupled monocarboxylate ($H^+/RCOO^-$) transporter, which moves one molecule each of proton and monocarboxylate into the enterocyte. For detailed information on the absorption and transport of inorganic anions in the enterocyte, review [609]. Transport of Cl^- and HCO_3^- has been discussed in Chap. 11.

12.9 Absorption and Transport of Toxic Metals

Toxic (nonessential) heavy metals (type D heavy metals) such as mercury (Hg), cadmium (Cd), lead (Pb), arsenic (As), mercury (Hg), and selenium (Se), and some oxy-anions or organic ions (e.g., methylmercury, MeHg) in target organs and tissues have no known nutritive value. Rather they cause substantial toxicity (harm) to the organism [610–614].

Absorption and transport of nonessential (toxic) heavy metals (type D heavy metals) depends on the properties of the metal and the physiology of the absorbing membrane [615]. Toxic metals gain entry into target cells, through mechanisms of ionic or molecular mimicry, at the site of transporters of essential elements [610, 616, 617]. The absorption of the toxic metals into enterocyte by molecular or ionic mimicry involves a phenomenon, whereby the bonding of metal ions to certain biomolecules results in the formation of metal complexes that can behave as homologues of endogenous biomolecules or the toxic metal species mimic an essential element that allows it gain access into the cells of the gut [610, 616, 617].

A number of carrier proteins have been implicated in the transport of some toxic metals, in particular, amino acid transporters (i.e., system $b^{0,+}$, system L) and organic anion transporters (i.e., OAT1 and OAT3) [610].

12.10 Absorption and Transport of Pharmacological Drugs

Several drugs administered orally pass through many barriers before they reach their target site. The pharmacokinetic behavior of drugs largely depends on the structural architecture and physiological properties of the intestine and liver [618, 619]. The epithelium of the intestine represents a major barrier that drugs must overcome to be absorbed. This is to a greater extent true at least for drugs that have not overcome the first bypass metabolism (by organs such as the intestine, liver, and lungs) and especially for non-lipophilic drugs. Even though lipophilic drugs may readily diffuse across the apical membrane of the enterocyte, their subsequent passage into circulation may remain a challenge [619, 620]. In particular, for orally ingested drugs, absorption also depends on the drug properties as well as patient's properties [620].

Passage of drugs across the intestinal epithelium occurs by diffusion and carrier-mediated uptake in the apical membrane of enterocytes [618]. The transporters are ATP-binding cassette (ABC) transporters which are expressed in the apical membrane of the enterocyte and hepatocyte [621, 622]. ABC transporters, such as P-glycoprotein (multidrug resistance protein MDR-1/ABC-B1), MDR-2/ABC-C2, breast cancer resistance protein (BCRP/ABCG2), substantially influence the pharmacokinetics of a large number of drugs and modulate the effectiveness of drug therapy [619, 623, 624]. The steroid, xenobiotic, and endobiotic receptor, pregnane X receptor (PXR), is a nuclear receptor that senses toxic substances and regulates their level of production and activity by gene expression. This nuclear receptor PXR is responsible for transcriptional regulation that results to detoxification of toxic substances. The receptor is involved in the metabolism of drugs, bile acid, fatty acids, lipids, glucose, etc. [625]. The intestinal PepT1 transporter plays a predominant role in the carrier-mediated intestinal absorption of beta-lactam antibiotics and native oligopeptides [626]. For details on drug absorption, see El-Kattan and Varma [627].

It should be noted that not all components of ingested feeds may be digested. Indigestible components of the ingesta are usually passed out as feces, but in some cases, they can conglomerate to form a mass that obstructs the intestinal lumen, thereby compromising the state of health of the individual. For a review of indigestible residues as potential cause of intestinal obstruction, see Clinical Correlate 12.1.

Clinical Correlate 12.1

Non-digestible Mass as Potential Cause of Gastrointestinal Obstruction—Gastrointestinal Bezoars

Introduction

The term bezoar was used in the ancient times to denote a substance that served as a remedy to control the effect of a poison. So, the term "bezoar" simply means "antidote" (from Arabic word "badzehr" or the Persian word "panzehr"). During the time, bezoars from animals were considered as antidotes to several poisons and diseases that affected mankind. However, with progress in science and technology, the use of bezoars as antidotes gradually faded away until the eighteenth century [628, 629].

Bezoar is an uncommon result of intentionally or accidentally ingestion of non-digestible substances, which subsequently accumulate and become trapped in the lumen of the GI tract [630, 631]. The incidence of GI bezoars is variable and ranges from about 0.3 to 4.8% of all cases of GI obstruction [631].

Classification

The classification of bezoar is based on the site of formation and the composition of the bezoars. On the basis of site of formation, bezoars can be classified as gastric, small intestinal (duodenal, jejunal, ileal), colonic, and rectosigmoidal bezoars [631]. The stomach is the most common location of bezoars [631]. On the basis of composition, bezoars can be classified as phytobezoars, trichobezoars, lactobezoars, and pharmacobezoars [629]. Phytobezoars are formed from non-digestible components (fibers, seeds, skins) of vegetable or fruit such as celery, pumpkins, grape, and especially persimmons [629]. Phytobezoars are the most frequent type of bezoars, and they represent about 40% of all reported cases of bezoars [631]. Trichobezoars are formed from ingested hair and food particles [629, 631]. Trichobezoar formation may be related to psychological disorders such as trichophagia (eating of hairs), which in turn is associated with trichotillomania (pull out of one's own hairs) [631, 632]. Pharmacobezoars are formed due to accumulation of laxative medications such as psyllium, and guar gum in the GI tract. Lactobezoars are formed from milk protein [629, 631]. The formation of lactobezoar occurs almost exclusively in infants who are fed with synthetic milk [631]. Apart from the above-mentioned causes of bezoar formation, worms (e.g., ascaris) and ingested materials such as plastic and toilet paper have been reported to cause bezoar formation [631].

Mechanism of formation

The mechanism of formation of bezoar depends on the type of bezoar. The non-digestible material or foreign bodies can mix with mucus and partially digested food to form a conglomeration [631, 633, 634]. In the formation of phytobezoars for instance, food containing high quantity of cellulose, lignin, and tannins are likely to form a non-digestible mass with mucus and other contents of the GI lumen [631, 633]. The tannins (e.g., catechins), in particular, can react with gastric acid and polymerize to form a mass with cellulose and other indigestible plant components [631, 634]. The formation of pharmacobezoars occurs due to the absorbent properties of bulk-forming medications [631]. Extended release drugs can also cause the formation of pharmacobezoars [635].

Risk factors

Risk factors for bezoar formation include illnesses that compromise gut motility, partial gastrectomy, diabetes mellitus with gastroparesis, cystic fibrosis, psychological disorders, and excessive fiber ingestion [629, 631, 632].

Diagnosis

For diagnosis of bezoar, it is advised to first obtain a detailed patient history. Physical examination may occasionally reveal a palpable abdominal mass and abdominal distension due to gastric outlet obstruction [629]. In addition, the patient may have one or more of the following: vomiting, nausea, early satiety, epigastric pain, ulceration, GI bleeding, or even perforation [629]. The gold standard for instrumental diagnosis of bezoars is endoscopy. This method allows for visualization, tissue sampling of bezoar, and can also be used for therapeutic purposes. The use of computed tomography is highly accurate and is confirming the diagnosis of GI bezoar. However, ultrasound, plain abdominal radiography, and barium study can also be used [629, 636].

Treatment

The treatment depends on the bezoar type that is formed and the presence of complications. Trichobezoars and phytobezoars with associated GI obstruction are usually removed by laparotomy and enterotomy [629–631, 636–638]. However, fragmentation with colonoscopy has been successfully used to treat phytobezoars [630]. Endoscopy can be used to remove these bezoars [629, 639]. More so, administration of cellulase enzyme (3–5 g dissolved in 300–500 mL of water and given *per os* daily for 2–5 days) has been reported to dissolve phytobezoars [629]. Lactobezoars can be managed conservatively. The patient is placed on *nil per os*, intravenous infusion, and gastric lavage. However, endoscopic or surgical treatment remains an option if conservative treatment fails [629]. The treatment for pharmacobezoars can substantially vary depending on the ingested medication and may involve gastric decontamination, dissolution therapy, whole bowel irrigation, endoscopy, or surgery [629, 635, 640].

Complication

Complications from phytobezoars can include ulceration, bleeding, gastric outlet obstruction, ileus, perforation [629, 631, 637]. Even though gastric bezoars are the most common, terminal ileal obstruction and ileus are the most frequent complications [630]. Ulceration and bleeding are usually associated with gastric bezoars [631].

12.11 Absorption and Transport of Vitamins

Vitamins are substances needed in minute quantity for the normal functioning of the body cells (see Table 12.4). These vitamins must be taken in as diet to avoid occurrence of diseases associated with their abnormal decrease such as birth defects, cancers, inflammatory, and cardiovascular diseases. They are important in many enzymatic reactions and redox homeostasis and may serve as cofactors. However, certain vitamins including thiamin, riboflavin, pantothenate, biotin, folate, phylloquinone can be synthesized by the gut microbiota in the large intestine and subsequently absorbed by the colonocytes [641–643]. The mechanism of absorption and transport of vitamins, in part, depends on the solubility of the vitamin in the aqueous medium of the intestinal lumen. Thus, absorption and transport of water-soluble vitamins are different from lipid-soluble vitamins [644, 645].

12.11.1 Water-Soluble Vitamins

Water-soluble vitamins (see Table 12.4) are absorbed by sodium-dependent secondary transport [604, 646].

Absorption and transport of water-soluble vitamins involves specialized carrier-mediated mechanisms (i.e., facilitated diffusion). However, some water-soluble vitamins can be absorbed by diffusion or secondary active transport which is sodium-dependent [647]. Although the precise mechanism of transport of water-soluble vitamins is yet to be unraveled, carrier-mediated transport of water-soluble vitamins has been shown to involve the regulation of intraenterocytic molecules including PKA, PKC, and calmodulin [647–649]. The vitamins involved besides initiating intracellular signaling may be transported to the basolateral membrane from where they are effluxed to the extracellular space to diffuse to the blood. The transport of these vitamins depends not only on the physiology of the GI tract, but also on factors such as surrounding temperature and cellular energy, and pH [650–653].

Table 12.4 Vitamins and their common names

Group		
Lipid-soluble	Vitamin A	Retinol
	Vitamin D	Calciferol
	Vitamin E	Alpha-tocopherol
	Vitamin K	Phylloquinone (vitamin K1), menaquinone, menadione
Water-soluble	Vitamin B1	Thiamin
	Vitamin B2	Riboflavin
	Vitamin B3	Niacin or nicotinic acid
	Vitamin B4	Choline
	Vitamin B5	Pantothenic acid
	Vitamin B6	Pyridoxine
	Vitamin B7 or vitamin H	Biotin
	Vitamin B9	Folic acid (folate, folacin)
	Vitamin B10	Para amino benzoic acid
	Vitamin B11	Pteryl-hepta-glutamic acid, salicylic acid
	Vitamin B12	Cobalamin
	Vitamin B13	Orotic acid (produced by gut microbiota)
	Vitamin B14	No name
	Vitamin B15	Pangamic acid
	Vitamin B16	Dimethylglycine, dimethylamino acetic acid
	Vitamin B17	Amygdalin or laetrile
	Vitamin B18, 19, 20	Carnitine
	Vitamin B21	
	Vitamin B22	
	Vitamin C	Ascorbic acid (ascorbate)

Note This is not a complete list of currently identified (known) vitamins. References to Table 12.4 [PubChem Search Beta https://pubchem.ncbi.nlm.nih.gov; 604, 660]

The water-soluble vitamins ascorbic acid, biotin, folate, pantothenic acid, thiamine, riboflavin, thiamin, cobalamin, and pyridoxine (see Table 12.4) use intestinal carrier-mediated transport to absorb vitamins from the diet [649, 654]. But these vitamins may have multiple transporters or receptors that mediate their movement into the cell. For instance, the major transporters of folate in the intestine (small and large bowel) are proton-coupled folate transporter (PCFT) and reduced folate carrier (RFC) which are all solute carriers [655]. PCFT and RFC are responsible for the movement of folate across the apical membrane of the enterocyte and basolateral membrane of the choroid plexus into the cerebrospinal fluid, and basolateral membrane of renal proximal tubules. The two transporters differ by their optimal level of functioning at a given pH. The optimal level of functioning of RFC occurs at pH of about 7.4, whereas for PCFT it is around 5.5–6.5 [655–657].

Genetic mutations in these transporters have been associated with hereditary folate malabsorption syndrome. The disease is characterized by failure to thrive, macrocytic anemia, immune deficiency, developmental delays, and other related disorders [656, 658, 659].

The transport of vitamin B1 is carrier-mediated, though Na^+-independent, but it is saturable [648, 649]. Vitamins B2–B6 also are absorbed and transported via carrier-mediated mechanisms [647].

The absorption and transport of vitamin H is dependent on pantothenate, Na^+-dependent, carrier-mediated at the apical membrane of human colonocytes [649]. Biotin absorption into the enterocytes of the small intestine of humans and other mammals occurs via carrier-mediated and Na^+-dependent mechanisms. Efflux of biotin from the basolateral membrane of the enterocyte also occurs via a carrier-mediated process, but the process is Na^+-independent and electrogenic [654].

Cobalamine (vitamin B12) absorption and transport are based on the secretion of gastric intrinsic factor and the receptor located in the brush border of the epithelium (Fig. 12.10) [649]. The absorption of cobalamin involves a specific facilitated transport mechanism. Vitamin B12 molecules readily form complexes with dietary proteins. Vitamin B12 molecules that are released by digestion of protein in the stomach bind to a glycoprotein found in saliva and gastric juice called R protein (also known as haptocorrin, transcobalamin-1 or cobalophilin). When the B12–R protein complex reaches the duodenum, the R protein is digested by pancreatic proteases, releasing B12 to bind to another protein found in gastric juice called intrinsic factor (IF) of Castle, named after William Bosworth Castle (1897–1990), an American physician and physiologist who first reported in 1929 the presence of a substance (an intrinsic factor) which acted on a dietary substance (extrinsic factor) to produce a material required for the normal maturation of erythrocytes. Vitamin B12 binds with the parietal-cell-secreted glycoprotein, gastric intrinsic factor, and is absorbed in the jejunum [661–664]. The B12–IF complex cannot be cleaved by proteolytic enzymes. In the brush boarder of the distal ileum, in the presence of calcium ions, the complex dimerizes and binds to its cognate receptor. The B12–IF–receptor complex is internalized, and the vitamin dissociates from the complex to combine with transcobalamin II. The B12–transcobalamin II complex diffuses into the bloodstream from where the B12 is delivered to the tissues of the body [643, 663].

Dietary ascorbic acid is readily absorbed in the GI tract and circulated in free form in the blood. It is transported against concentration gradient into many tissues, reabsorbed in the kidney tubule system [642]. Vitamin C can enter cells both in its reduced and oxidized forms—ascorbic acid and dehydroascorbate [665]. The apical transporters of ascorbate are glucose transporters (GLUT1–3), **s**odium-dependent **v**itamin **C** co-**t**ransporters (SVCT), and surface glycoproteins [666–668]. There are two types of SVCT. SVCT1 is expressed predominantly on the apical membrane, whereas SVCT2 is mainly localized at the basolateral membrane [642, 669, 670]. The basolateral efflux of ascorbate is also mediated via volume-sensitive and Ca^{2+}-dependent anion channels, hemichannels of gap junctions, and exocytosis of vesicles containing the vitamin [665].

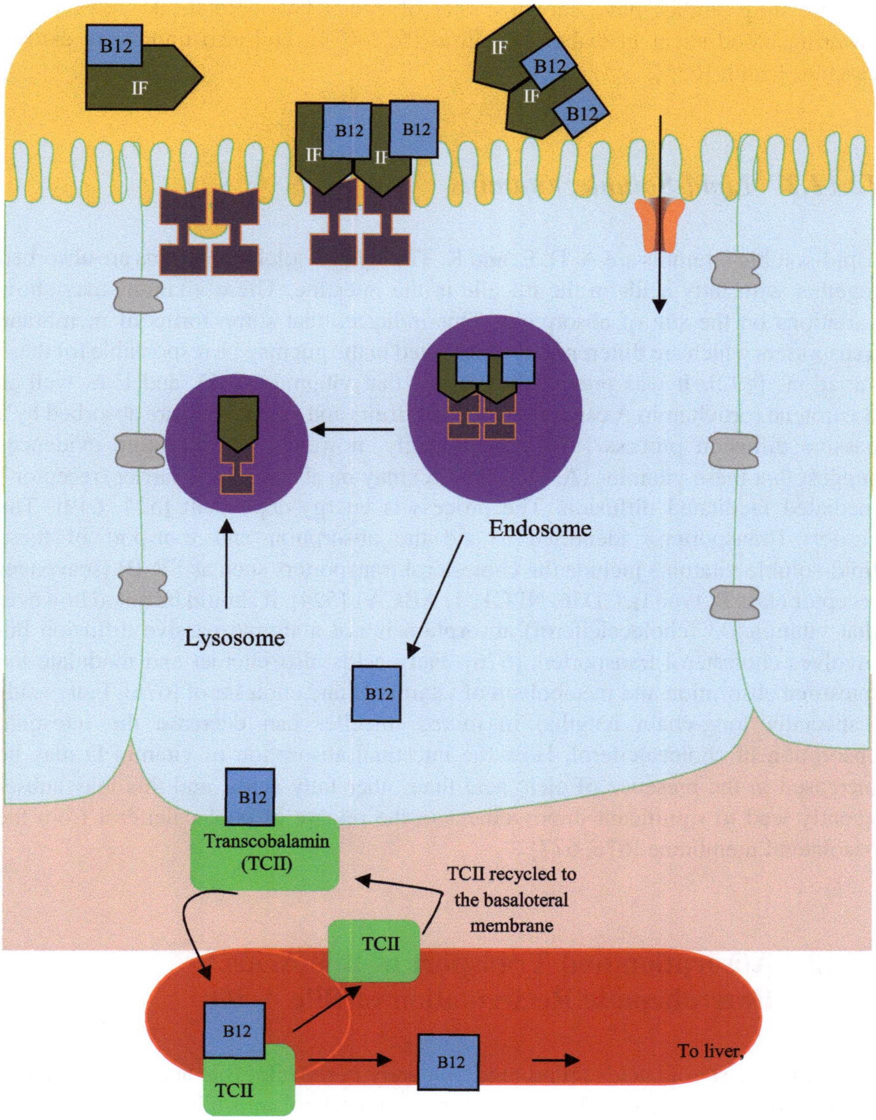

Fig. 12.10 Schematic representation of intestinal absorption of vitamin B12

Diets abundant with fruits and vegetables contain vitamin C and flavonoids, as well as other components, required for the normal functioning of the body. However, flavonoids are known to inhibit the absorption of vitamin C and glucose [671]. Thus, flavonoids may serve as alternatives for lowering blood sugar following a mixed meal of flavonoid-rich vegetables and carbohydrates. Indeed cocoa accumulating studies have shown a role of cocoa flavonoid drink (flavanol) in lowering blood sugar in diabetes mellitus [672–674], and also improving cardiovascular health [675].

12.11.2 *Lipid-Soluble Vitamins*

Lipid-soluble vitamins are A, D, E, and K. These lipid-soluble vitamins are absorbed together with fatty acids in the micelle in the intestine. These vitamins may show variations on the site of absorption. This indicates that some forms of membrane transporters which are differentially distributed in the gut may be responsible for these variations [492]. It was previously thought that vitamins A, D, and E as well as β-carotene (provitamin A carotenoid of some fruits and vegetables) are absorbed by a passive diffusion process [524]. Interestingly, however, accumulating evidences suggest that these vitamins (A, D, E, and K) may be absorbed via carrier (receptor-) mediated facilitated diffusion. The process is energy-dependent [524, 649]. The carriers (transporters) identified to aid the absorption and transport of these lipid-soluble vitamins include the cholesterol transporters such as SR-BI (scavenger receptor class B, type I), CD36, NPC1L1, ABCA1 [524]. It should be noted however that vitamin D3 (cholecalciferol) absorption is not a simple passive diffusion but involves cholesterol transporters [676]. Fatty acids also interact and modulate the intestinal absorption and metabolism of vitamin D and cholesterol [676]. Fatty acids (especially long-chain lengths) in mixed micelles can decrease the intestinal absorption of cholecalciferol. How the intestinal absorption of vitamin D may be increased in the presence of oleic acid than other fatty acids, and this may subsequently lead to significant improvement in the release of cholecalciferol from the basolateral membrane [676, 677].

12.12 Absorption and Transport of Bile Acids. Enterohepatic Recirculation of Bile Acids

Bile acids are steroid acids found predominantly in the bile. Bile acids are produced in vivo from cholesterol. The conversion of cholesterol into bile acids in the liver involves multiple enzymatic steps. The rate-limiting step is catalyzed by the enzyme cholesterol 7 alpha-hydroxylase. This enzyme converts cholesterol to produce primary bile acids (CA, cholic acid—3α,7α,12α-trihydroxy-5β-cholanoic acid; CDCA, chenodeoxycholic acid—3α,7α-dihydroxy-5β-cholanoic acid) (Fig. 12.11) [678]. In humans, cholic acid (taurocholic and glycocholic acid) and chenodeoxycholic acid (taurochenodeoxycholic and glycochenodeoxycholic acid) are the major bile salts. The expression of cholesterol 7 alpha-hydroxylase increases with primary bile acids and cholesterol and is inhibited by the actions of the ileal

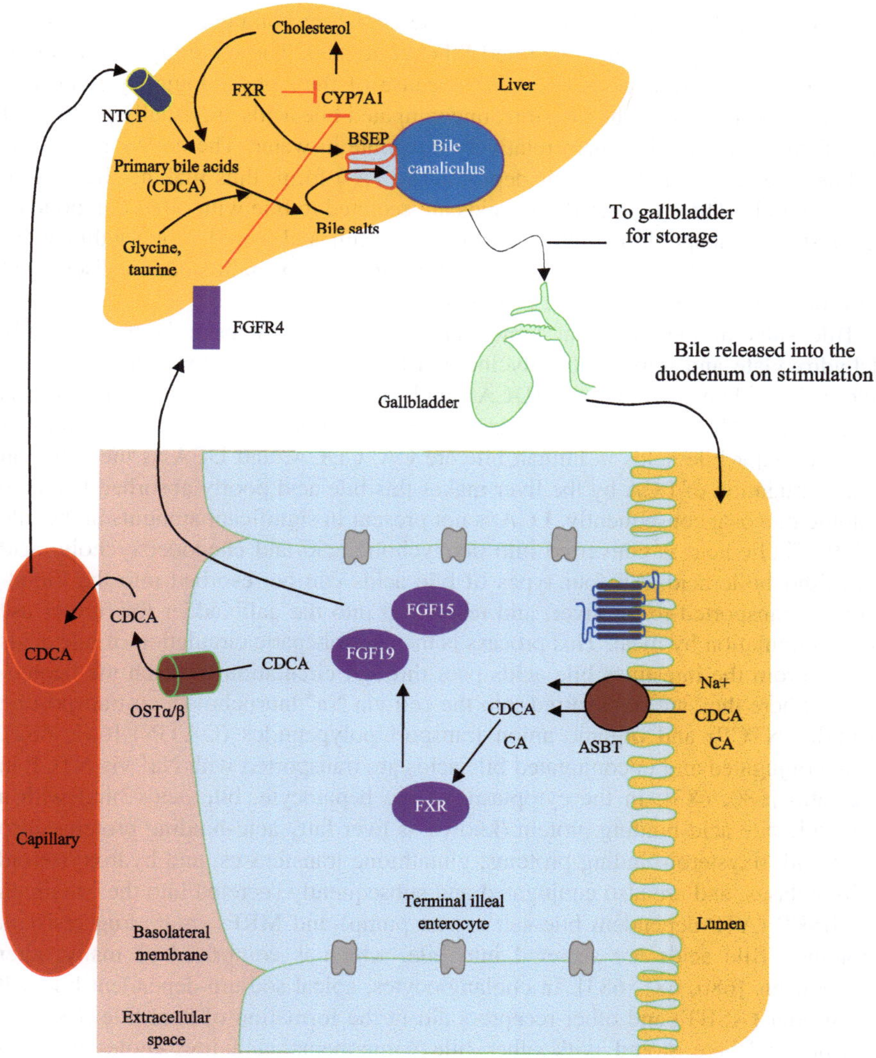

Fig. 12.11 Reabsorption of bile acids (CA, CDCA) into the enterocyte of the terminal ileum. About 500–600 mg of bile acids is synthesized daily [692]. In the liver, the bile acids are then conjugated with taurine or glycine to form bile salts. In addition, both conjugated and unconjugated bile acids are transported to the liver via the enterohepatic pathway. Bile salts in the liver are subsequently transported to the gallbladder for storage and are released into the intestine upon meal stimulation. Bacteria in the terminal ileum deconjugate the bile salts to form bile acids and also dehydroxylate the primary bile acids to form secondary bile acids—deoxycholic and lithocholic acids. Bile acids may be passively absorbed in the upper intestine; conjugated bile acids are absorbed in the terminal ileum. Approximately 500–600 mg of bile is passed into the intestine per day, and almost the same quantity of bile acids (500–600 mg) is lost via feces. Of these, about 300–600 mg of bile acids is lost daily in feces. Fecal losses comprise about 5% of the total bile acid pool in humans, which is estimated at approximately 3 g. The remaining 95% of bile acids are returned to the liver through the enterohepatic recirculation [692–694]

hormones FGF15 and FGF19. Primary bile acids are conjugated with taurine or glycine to form deoxycholic acid and lithocholic acid. These conjugated forms of bile acids are referred to as bile salts because of their physiologically important acid–base properties. In the intestine, unconjugated bile acids are usually protonated (HA form), thus making them relatively insoluble in water. The conjugated forms of bile acids are usually in their deprotonated (A-) form, thus making them more water-soluble. The conjugated bile acids are secreted to bile with other components (cholesterol, phospholipids, etc.) via bile canaliculi and stored in the gallbladder. Upon stimulation by diet or other factors, bile stored in the gallbladder is discharged into the intestinal lumen (Fig. 12.11) [678–680].

Bile salts in the intestinal lumen are acted upon by bacteria that partially dehydroxylate and remove the glycine and taurine groups to form the secondary bile acids—deoxycholic acid (DCA) and lithocholic acid (LCA) in humans (Fig. 12.11). The process is called deconjugation of bile salts [678, 680]. It appears that the major bile acids in human bile are CA, CDCA, and DCA as the sulfation and conjugation of LCA by the liver makes this bile acid poorly absorbed from the colonic mucosa; consequently, LCA is not present in significant amounts in the bile [679]. Cholic acid is converted into deoxycholic acid and chenodeoxycholic acid into lithocholic acid. All four types of bile acids can be resorbed into the bloodstream, transported to the liver, and resecreted into the gallbladder for further use upon stimulation by meal. This process is the enterohepatic circulation of bile acids [681]. From the intestines bile acids pass through circulation to reach the hepatocytes, where they are transported into the cell via Na^+ taurocholate co-transporting peptide (NTCP) and organic anion transport polypeptides (OATPs) [681, 682]. Both conjugated and unconjugated bile acids are transported with Na^+ via NTCP in 1:2 ratio [680, 683]. In the cytoplasm of the hepatocyte, bile acids bind with a cytosolic bile acid-binding protein (known as liver fatty acid-binding protein), also react with oxysterol-binding proteins, glutathione transferases, and hydroxysteroid dehydratases, and are also conjugated and subsequently secreted into the canaliculi via BSEP (ATP-dependent bile salt export pump) and MRPs (multidrug resistant proteins). Bile acids are secreted into bile, which is comprised of many other components [680, 681, 683]. In cholangiocytes, apical sodium-dependent bile salt transporter (ASBT) and other receptors aid in the formation of final bile [683].

Bile acids are stored with other bile components including cholesterol and phospholipids in the gallbladder as mixed micelles [680]. The release of bile into the duodenum is stimulated by GI hormones whose secretion is activated, especially by fat-rich meal. The result is gallbladder contraction and release of bile into the intestinal lumen [678].

Because of the amphipathic nature of bile acids (hydrophobic core and a number of attached hydrophilic hydroxyl groups), they facilitate the digestion of fat- and lipid-soluble vitamins by the formation of micelles in the intestine. Thus, bile acids act as surfactants that emulsify lipids into micelles. Bile acids also behave like hormones by signaling through the farnesoid X receptor and GPBAR1 (also known as Takeda G protein-coupled receptor 5, TGR5) [678, 680].

Recall that a portion of the pool of primary bile acids is modified by the gut microbiota into secondary bile acids in intestinal lumen. This modification increases the hydrophobicity of bile acids. It also increases the amount of bile acids that are passively absorbed into the enterocyte. However, this modification makes bile acids potentially more harmful to cause cancer and cholesterol gallstone disease. Diets rich in fat and poor in fiber increase the amount of conjugated bile acids in the lumen. The higher the conjugation, the lower the absorption of bile in the ileum. Diets rich in meat and seafood increase the level of taurine, and thus, more conjugated bile acids are produced [678, 684, 685]. Bile acid metabolism is also affected by a host of other factors including duration of sleep, alcohol, and drugs [686].

Bile acids are absorbed in the intestine by passive diffusion, especially in the jejunum. However, major part of the bile acid pool (primary and secondary) is absorbed by active transport mechanism in the terminal ileal enterocytes from where they are transported to the blood and return to the liver. Approximately 95% of bile acids are taken up by hepatocytes. The proportion of bile acid uptake by the liver depends on a number of factors which include dietary pattern, GI transit time, drug ingestion, and diseases such as cancer, IBD, and diarrhea [678, 687, 688].

In the ileum, absorption of bile acids from the lumen into cytosol of the epithelial cells is carried out by the apical Na^+ bile salt transporter (ASBT). The efflux rate across the basolateral membrane into the portal blood is carried out by the Na^+-independent organic solute transporter α/β (OSTα/β) heterodimer via a facilitated diffusion mechanism [680, 681, 689–691]. Bile acid transport by ASBT is similar to NTCP; sodium-dependent transport of Na^+/BA in a 2:1 ratio [680].

The process of bile acid synthesis is regulated by feedback and feed-forward loops by transcriptional and posttranscriptional mechanisms. A key regulator of bile salt homeostasis is the bile acid receptor—farnesoid X receptor, which activates transcription of the genes responsible for the synthesis of BSEP, OATP, and the small heterodimer partner 1 (SHP). SHP is a transcriptional repressor that mediates bile acid-induced inhibition of the bile salt uptake by OATP [683, 695].

The cholestatic drugs, such as rifampicin, glibenclamide, and cyclosporin A, are known to reduce the apical efflux rate of taurocholate across NTCP and BSEP [682]. Dysfunctions of the hepatic bile acid transporters and receptors are implicated in the pathophysiology of cholestatic liver disease [681, 696, 697]. Recent discovery of pharmacological agents that influence bile acid metabolism may have functional implication on pathogenesis in metabolic disorders, obesity, and other gut diseases [696, 697].

12.13 Absorption and Transport of Water

The GI tract, especially the ileum, absorbs a lot of water every day, which is required to ensure adequate and ongoing functions of the GI tract and the body as a whole. Receptors and channels involved in water transport include aquaporins, intercellular junctions (e.g., gap junctions as well as hemichannels), glucose transporters (e.g., GLUTs, SGLT1). While water can passively flow through traditional water channels (aquaporins) and through paracellular pathway (e.g., gap junctions), it is now believed that substantial quantity of water absorption and transport in the GI tract is associated with solute and electrolyte transport [698–701]. The passive transport of water occurs due to osmotic pressure gradient [701]. Approximately 30% of water transported in the GI tract is believed to occur by this passive transport via the plasma membrane [136, 702]. Solute and electrolyte transport in the GI tract is associated with water transport. The process is known as secondary active transport. An example of a transporter involved in secondary active transport of water is the Na^+/glucose co-transporter (SGLT1). For every transport cycle, the SGLT1 transports one glucose molecule to two Na^+ ions to 220–400 ($\sim$260) water molecules [699, 701]. Therefore, this transporter can be designated as Na^+/glucose/water transporter or Na^+/glucose co-transporter of water. About 35% of water is transported through the transporter. Furthermore, SGLT1 is believed to be responsible for about 35% of flow by osmosis [136]. The Na^+ that enters the enterocyte is pumped out via basolateral membrane by $3Na^+/2K^+$ pump, while the glucose passes out through the basolateral membrane via the GLUT-2 or exocytosis [703]. Other ion transporters that facilitate water transport include the carbonic anhydrase type 2, apical CFTR-Cl-channel, $Na^+/K^+/2Cl^-$ co-transporter, Na^+/H^+ exchanger type 3, epithelial sodium channel, Cl/HCO_3^- exchangers, $2Na^+$/SCFA, and SCFA-HCO_3^- exchanger [506, 517, 518, 704].

12.14 Conclusion

Digestive enzymes and receptors including carbohydrases, proteinases, lipases, nucleases, and transporters/channels represent crucial components of the machinery of digestion in the gut. The transport of ions, nutrients, and products of gut metabolism is based on complex mechanisms involving active and passive transport via transcellular and paracellular fluxes as well as exocytosis. The mechanism of GI functioning is based on regulated signaling events at different levels of cellular and molecular organization. While some mechanisms of transport have evolved novel and multiple ways in counteracting imbalance, dysfunction of the major components of GI transport has been implicated in a variety of diseases.

Recommended Readings

Original Articles

1. Akiyama H, Okamura Y, Nagashima T, Yokoi A, Muraji T, Uetani Y (2006) Intracranial hemorrhage and vitamin K deficiency associated with biliary atresia: summary of 15 cases and review of the literature. Pediatr Neurosurg 42(6):362–367

2. Alatas FS, Hayashida M, Matsuura T, Saeki I, Yanagi Y, Taguchi T (2012) Intracranial hemorrhage associated with vitamin K-deficiency bleeding in patients with biliary atresia: focus on long-term outcomes. J Pediatr Gastroenterol Nutr 54(4):552–557
3. Chakhtoura MT, Nakhoul NN, Shawwa K, Mantzoros C, El Hajj Fuleihan GA (2016) Hypovitaminosis D in bariatric surgery: A systematic review of observational studies. Metabolism 65(4):574–85
4. D'Aquila T, Sirohi D, Grabowski JM, Hedrick VE, Paul LN, Greenberg AS, et al (2015) Characterization of the Proteome of Cytoplasmic Lipid Droplets in Mouse Enterocytes after a Dietary Fat Challenge. PLoS One 10(5):e0126823
5. Dong R, Sun S, Liu XZ, Shen Z, Chen G, Zheng S (2017) Fat-soluble vitamin deficiency in pediatric patients with biliary atresia. Gastroenterol Res Pract 2017:7496860
6. Huang Q, Li N, Zhang W, Zhu W, Li Q, Wang B, Li J (2008) Na $^{+}$-dependent neutral amino acid transporter ASCT2 is downregulated in seriously traumatized human intestinal epithelial cells. J Pediatr Gastroenterol Nutr 46(1):71–9

Review Articles

1. Albahrani AA, Greaves RF (2016) Fat-soluble vitamins: clinical indications and current challenges for chromatographic measurement. Clin Biochem Rev 37 (1):27–47
2. Bröer S (2008) Amino acid transport across mammalian intestinal and renal epithelia. Physiol Rev 88(1):249–286
3. Ceranowicz P, Cieszkowski J, Warzecha Z, Kuśnierz-Cabala B, Dembiński A (2015) The beginnings of pancreatology as a field of experimental and clinical medicine. Biomed Res Int 2015:128095
4. Cruz-Mora J, Martínez-Hernández NE, del Campo-López FM, Viramontes-Hörner D, Vizmanos-Lamotte B, Muñoz-Valle JF et al (2014) Effects of a symbiotic on gut microbiota in mexican patients with end-stage renal disease. J Ren Nutr 24(5):330–335
5. Ianiro G, Pecere S, Giorgio V, Gasbarrini A, Cammarota G (2016) Digestive enzyme supplementation in gastrointestinal diseases. Curr Drug Metab 17 (2):187–193
6. Struyvenberg MR, Martin CR, Freedman SD (2017) Practical guide to exocrine pancreatic insufficiency—Breaking the myths. BMC Medicine 15:29
7. Williams JA (2013) The Nobel pancreas: a historical perspective. Gastroenterology 144(6):1166–1169

Bibliography

1. Fruton JS (2002) A history of pepsin and related enzymes. Q Rev Biol 77(2):127–147
2. Sampath-Kumar PS, Fruton JS (1974) Studies on the extended active sites of acid proteinases. Proc Natl Acad Sci U S A 71(4):1070–1072

3. Klockars M, Reitamo S (1975) Tissue distribution of lysozyme in man. J Histochem Cytochem 23(12):932–940
4. Isenman L, Liebow C, Rothman S (1999) The endocrine secretion of mammalian digestive enzymes by exocrine glands. Am J Physiol 276(2 Pt 1):E223–E232
5. Pelaseyed T, Bergström JH, Gustafsson JK, Ermund A, Birchenough GMH, Schütte A et al (2014) The mucus and mucins of the goblet cells and enterocytes provide the first defense line of the gastrointestinal tract and interact with the immune system. Immunol Rev 260 (1):8–20
6. Hamosh M (1990) Lingual and gastric lipases. Nutrition 6(6):421–428
7. Berg JM, Tymoczko JL, Stryer L (2002) Biochemistry, 5th edn. WH Freeman, New York
8. Rabin BR (1970) The mechanism of enzyme action. J Clin Pathol Suppl (Assoc Clin Pathol) 4:1–7
9. Gao J, Ma S, Major DT, Nam K, Pu J, Truhlar DG (2006) Mechanisms and free energies of enzymatic reactions. Chem Rev 106(8):3188–3209
10. Antonov VK, Ginodman LM, Rumsh LD, Kapitannikov YV, Barshevskaya TN, Yavashev LP et al (1981) Studies on the mechanisms of action of proteolytic enzymes using heavy oxygen exchange. Eur J Biochem 117(1):195–200
11. Ugolev AM, Iezuitova NN, Smirnova LF (1984) Role of digestive enzymes in the permeability of the enterocyte. In: Csáky TZ (ed) Pharmacology of intestinal permeation II, vol 70. Springer, Heidelberg
12. Clericuzio A (2012) Chemical and mechanical theories of digestion in early modern medicine. Stud Hist Philos Biol Biomed Sci 43(2):329–337
13. Borgstrom B, Dahlqvist A, Lundh G, Sjovall J (1957) Studies of intestinal digestion and absorption in the human. J Clin Invest 36(10):1521–1536
14. Ugolev AM (1965) Membrane (contact) digestion. Physiol Rev 45:555–595
15. Ugolev AM, Kooshuck RI (1966) Hydrolysis of dipeptides in cells of the small intestine. Nature 212:859–860
16. Ugolev AM (1980) Trophic ecology, a new interdisciplinary science. Vestnik Acad Sci USSR 1:50–61
17. Ugolev AM (1991) The theory of adequate nutrition and trophic ecology. Nauka, St. Petersburg
18. Dahlqvist A, Borgstrom B (1961) Digestion and absorption of disaccharides in man. Biochem J 81(2):411–418
19. Cajori FA (1933) The enzyme activity of dogs' intestinal juice and its relation to intestinal digestion. Amer. J. Physiol 104:659–668
20. Miller D, Crane RK (1961) The digestive function of the epithelium of the small intestine. I. An intracellular locus of disaccharide and sugar phosphate hydrolysis. Biochim Biophys Acta 52:281–293
21. Parfenov AI (2010) On the occasion of the 50th anniversary of A. M. Ugolev's current theory of digestion. Ter Arkh 82(2):5–10
22. Raymond DA (1996) Glycobiology: toward understanding the function of sugars. Chem Rev 96(2):683–720
23. McNaught AD, Wilkinson A; International Union of Pure and Applied Chemistry IUPAC (2014) Compendium of chemical terminology (the gold book). Blackwell Scientific Publications, Oxford
24. Chain EB, Mansford KRL, Pocchiari F (1960) The absorption of sucrose, maltose and higher oligosaccharides from the isolated rat small intestine. J Physiol 154:39–51
25. Hamilton JD, McMichael HB (1968) Role of the microvillus in the absorption of disaccharides. Lancet 2(7560):154–157
26. Parsons DS, Prichard JS (1968) Disaccharide absorption by amphibian small intestine in vitro. J Physiol 199(137–150):137
27. Runge SW, Hill BJF, Moran WM (2006) A simple classroom teaching technique to help students understand Michaelis-Menten kinetics. CBE Life Sci Educ 5(4):348–352

28. Johnson KA, Goody RS (2011) The original Michaelis constant: translation of the 1913 Michaelis-Menten paper. Biochemistry 50(39):8264–8269
29. Abbott GW (2017) Chansporter complexes in cell signaling. FEBS Lett 591(17):2556–2576
30. Dahlqvist A, Thomson DL (1963) The digestion and absorption of maltose and trehalose by the intact rat. Acta Physiol Scand 59(1–2):111–125
31. Malathi P, Ramaswamy K, Caspary WF, Crane RK (1973) Studies on the transport of glucose from disaccharides by hamster small intestine in vitro. I. Evidence for a disaccharidase-related transport system. Biochim Biophys Acta Biomembr 307(3):613–626
32. Warden DA, Fannin FF, Evans JO, Hanke DW, Diedrich DF (1980) A hydrolase-related transport system is not required to explain the intestinal uptake of glucose liberated from phlorizin. Biochim Biophys Acta 599(2):664–672
33. Alvarado F, Lherminier M, Phan H-H (1984) Hamster intestinal disaccharide absorption: extracellular hydrolysis precedes transport of the monosaccharide products. J Physiol 355:493–507
34. Meyer H, Vitavska O, Wieczorek H (2011) Identification of an animal sucrose transporter. J Cell Sci 124:1984–1991
35. Likely R, Johnson E, Ahearn GA (2015) Functional characterization of a putative disaccharide membrane transporter in crustacean intestine. J Comp Physiol B 185(2):173–183
36. Haeuw JF, Michalski JC, Strecker G, Spik G, Montreuil J (1991) Cytosolic glycosidases: do they exist? Glycobiology 1(5):487–492
37. De Gasperi R, Daniel PF, Warren CD (1992) A human lysosomal alpha-mannosidase specific for the core of complex glycans. J Biol Chem 267(14):9706–9712
38. Herscovics A (1999) Importance of glycosidases in mammalian glycoprotein biosynthesis. Biochim Biophys Acta 1473(1):96–107
39. Grard T, Herman V, Saint-Pol A, Kmiecik D, Labiau O, Mir AM et al (1996) Oligomannosides or oligosaccharide-lipids as potential substrates for rat liver cytosolic alpha-D-mannosidase. Biochem J 316(3):787–792
40. Daniel PF, Winchester B, Warren CD (1994) Mammalian alpha-mannosidases–multiple forms but a common purpose? Glycobiology 4(5):551–566
41. Moremen KW (2002) Golgi alpha-mannosidase II deficiency in vertebrate systems: implications for asparagine-linked oligosaccharide processing in mammals. Biochim Biophys Acta 1573(3):225–235
42. Karasov WH, Douglas AE (2013) Comparative digestive physiology. Compr Physiol 3 (2):741–783
43. Wang Z, Xu S, Du K, Huang F, Chen Z, Zhou K et al (2016) Evolution of digestive enzymes and RNASE1 provides insights into dietary switch of cetaceans. Mol Biol Evol 33 (12):3144–3157
44. Uneyama H, San Gabriel A, Kawai M, Tomoe M, Torii K (2008) Physiological role of dietary free glutamate in the food digestion. Asia Pac J Clin Nutr 17(1):372–375
45. Pereverzev VA, Lobanok LM (2014) Physiology of digestion. In: Kubarko AI (ed) Normal physiology. Visheishaya Shkola, Minsk, Belarus
46. Nikitina OS, Welcome MO, Pereverzev VA (2016) Human anatomy and physiology. In: 2 parts. Part 1. Belarusian State Medical University Press, Minsk
47. Roxas M (2008) The role of enzyme supplementation in digestive disorders. Altern Med Rev 13(4):307–314
48. Ianiro G, Pecere S, Giorgio V, Gasbarrini A, Cammarota G (2016) Digestive enzyme supplementation in gastrointestinal diseases. Curr Drug Metab 17(2):187–193
49. Layer P, Keller J (2003) Lipase supplementation therapy: standards, alternatives, and perspectives. Pancreas 26(1):1–7
50. Mößeler A, Kamphues J (2017) Black-box gastrointestinal tract-needs and prospects of gaining insights of fate of fat, protein, and starch in case of exocrine pancreatic insufficiency by using fistulated pigs. Nutrients 9(2):E150

51. Nakajima K, Oshida H, Muneyuki T, Kakei M (2012) Pancrelipase: an evidence-based review of its use for treating pancreatic exocrine insufficiency. Core Evid 7:77–91
52. Lorkowski G (2012) Gastrointestinal absorption and biological activities of serine and cysteine proteases of animal and plant origin: review on absorption of serine and cysteine proteases. Int J Physiol Pathophysiol Pharmacol 4(1):10–27
53. Iwamuro M, Okada H, Matsueda K, Inaba T, Kusumoto C, Imagawa A, Yamamoto K (2015) Review of the diagnosis and management of gastrointestinal bezoars. World J Gastrointest Endosc 7(4):336–345
54. Pavan R, Jain S, Shraddha Kumar A (2012) Properties and therapeutic application of bromelain: a review. Biotechnol Res Int 2012:976203
55. Rathnavelu V, Alitheen NB, Sohila S, Kanagesan S, Ramesh R (2016) Potential role of bromelain in clinical and therapeutic applications. Biomed Rep 5(3):283–288
56. Ushakova NA, Nekrasov RV, Pravdin IV, Sverchkova NV, Kolomiyets EI, Pavlov DS (2015) Mechanisms of the effects of probiotics on symbiotic digestion. Biol Bull Russ Acad Sci 42(5):394–400
57. Czjzek M (2017) Biochemistry: a wine-induced breakdown. Nature 544:45–46
58. Cruz-Mora J, Martínez-Hernández NE, del Campo-López FM, Viramontes-Hörner D, Vizmanos-Lamotte B, Muñoz-Valle JF, García-García G, Parra-Rojas I, Castro-Alarcón N (2014) Effects of a symbiotic on gut microbiota in mexican patients with end-stage renal disease. J Ren Nutr 24(5):330–335
59. Derrien M, Veiga P (2017) Rethinking diet to aid human-microbe symbiosis. Trends Microbiol 25(2):100–112
60. Hutter T, Gimbert C, Bouchard F, Lapointe F-J (2015) Being human is a gut feeling. Microbiome 3:9
61. Pudlo NA, Urs K, Kumar SS, German JB, Mills DA, Martens EC (2015) Symbiotic human gut bacteria with variable metabolic priorities for host mucosal glycans. mBio 6(6):e01282–15
62. Austin GL, Ogden LG, Hill JO (2011) Trends in carbohydrate, fat, and protein intakes and association with energy intake in normal-weight, overweight, and obese individuals: 1971–2006. Am J Clin Nutr 93(4):836–843
63. Medagama A, Fernando D, Widanapathirana H (2015) Energy and nutrient intakes of Sri Lankan patients with type 2 diabetes mellitus: a cross-sectional survey. BMC Res Notes 8:753
64. Reynolds RD, Lickteig JA, Howard MP, Deuster PA (1998) Intakes of high fat and high carbohydrate foods by humans increased with exposure to increasing altitude during an expedition to Mt. Everest. J Nutr 128(1):50–55
65. Wismann J, Willoughby D (2006) Gender differences in carbohydrate metabolism and carbohydrate loading. J Int Soc Sports Nutr 3:28
66. Lin AH-M, Lee B-H, Nichols BL, Quezada-Calvillo R, Rose DR, Naim HY, Hamaker BR (2012) Starch source influences dietary glucose generation at the mucosal α-glucosidase level. J Biol Chem 287:36917–36921
67. Lin AH-M, Lee B-H, Chang W-J (2016) Small intestine mucosal α-glucosidase: a missing feature of in vitro starch digestibility. Food Hydrocoll 53:163–171
68. Cantare BL, Coutinho PM, Rancurel C, Bernard T, Lombard V, Henrissat B (2009) The carbohydrate-active enzymes database (CAZy): an expert resource for glycogenomics. Nucleic Acids Res 37:D233–D238
69. Lombard V, Golaconda Ramulu H, Drula E, Coutinho PM, Henrissat B (2014) The carbohydrate-active enzymes database (CAZy) in 2013. Nucleic Acids Res 42:D490–D495
70. Velmurugan C, Natarajan R (2015) Research productivity of amylase in microbiology in Indian perspective: a scientometric analysis. In: Manimaran D, Velmurugan C, Elangovan N (eds) Microbial production of amylase in Bacillus Cereus Sp. Lambert Academic Publishing, Saarbrücken
71. Wisniak J (2004) Anselme Payen. Educ Quím 16(4):114–126

72. Fieker A, Philpott J, Armand M (2011) Enzyme replacement therapy for pancreatic insufficiency: present and future. Clin Exp Gastroenterol 4:55–73
73. Mehta D, Satyanarayana T (2016) Bacterial and archaeal α-amylases: diversity and amelioration of the desirable characteristics for industrial applications. Front Microbiol 7:1129
74. Gopinath SCB, Anbu P, Md Arshad MK, Lakshmipriya T, Voon CH, Hashim U, Chinni SV (2017) Biotechnological processes in microbial amylase production. Biomed Res Int 2017:1272193
75. Wu AC, Ral J-P, Morell MK, Gilbert RG (2014) New perspectives on the role of α- and β-amylases in transient starch synthesis. PLoS ONE 9(6):e100498
76. Oudjeriouat N, Moreau Y, Santimone M, Svensson B, Marchis-Mouren G, Desseaux V (2003) On the mechanism of alpha-amylase. Eur J Biochem 270(19):3871–3879
77. Todaka D, Matsushima H, Morohashi Y (2000) Water stress enhances beta-amylase activity in cucumber cotyledons. J Exp Bot 51(345):739–745
78. Picukans I, Umminger BL (1979) Comparative activities of glycogen phosphorylase and gamma-amylase in livers of carp (Cyprinus carpio) and goldfish (Carassius auratus). Comp Biochem Physiol B 62(4):455–457
79. Simon K, Gładysz A, Bakońska-Pacoń E, Sobiech KA, Dzik T (1991) The activity of alpha-amylase and gamma-amylase in serum and pancreatic homogenate of rats with experimental liver damage treated with colchicine. Mater Med Pol 23(2):103–106
80. Soininen K, Ceska M, Adlercreutz H (1972) Comparison between a new chromogenic α-amylase test (phadebas) and the wohlgemuth amyloclastic method in urine. Scand J Clin Lab Investig 30(3):291–297
81. Chua KS, Tan IK, Vengadiswaran R, Peiris JT (1979) An assessment of four methods for the assay of amylase activity. Ann Acad Med Singapore 8(2):187–192
82. Hathaway JA, Hunter DT, Berrett CR (1970) An automated method for the determination of amylase. Clin Biochem 3(3):217–224
83. Moriyama T, Ikeda H (1996) Hydrolases acting on glycosidic bonds: chromatographic and electrophoretic separations. J Chromatogr B Biomed Appl 684(1–2):201–216
84. Lehane DP, Wissert PJ, Lum G, Levy AL (1977) Amylase activity in serum and urine: comparison of results by the amyloclastic, dyed-starch, and nephelometric techniques. Clin Chem 23(6):1061–1065
85. Fenton J, Foery R, Piatt L, Geschwindt K (1982) A new chromogenic amylase method compared with two established methods. Clin Chem 28(4 Pt 1):704–706
86. Behringer V, Borchers C, Deschner T, Möstl E, Selzer D, Hohmann G (2013) Measurements of salivary alpha amylase and salivary cortisol in hominoid primates reveal within-species consistency and between-species differences. PLoS ONE 8(4):e60773
87. Foo AY, Rosalki SB (1986) Measurement of plasma amylase activity. Ann Clin Biochem 23:624–637
88. Rohleder N, Nater UM (2009) Determinants of salivary alpha-amylase in humans and methodological considerations. Psychoneuroendocrinology 34(4):469–485
89. Branca P, Rodriguez RM, Rogers JT, Ayo DS, Moyers JP, Light RW (2001) Routine measurement of pleural fluid amylase is not indicated. Arch Intern Med 161(2):228–232
90. Mones RL, Yankah A, Duelfer D, Bustami R, Mercer G (2011) Disaccharidase deficiency in pediatric patients with celiac disease and intact villi. Scand J Gastroenterol 46(12):1429–1434
91. Sandeep GK, Sonny CKF, Joseph FF (1999) Disaccharidase activities in children: normal values and comparison based on symptoms and histologic changes. J Pediatric Gastroenterol Nutr 28(3):246–251
92. Reid EW (1901) Intestinal absorption of maltose. J Physiol (Lond) 26:427–435
93. Plimmer RHA (1907) On the presence of lactase in the intestines of animals and on the adaptation of the intestine to lactose. J Physiol (Lond) 35:20–31

94. Stevens JA, Kidder DE (1972) The distribution of trehalase, sucrase, -amylase, glucoamylase and lactase (-galactosidase) along the small intestine of five pigs. Br J Nutr 28(1):129–137
95. Dahlqvist A (1970) Assay of intestinal disaccharidases. Enzymol Biol Clin (Basel) 11 (1):52–66
96. Varljen J, Detel D, Batičić L, Erakovic VH, Štrbo N, Ćuk M, Milin Č (2005) Age dependent activity of brush-border enzymes in BALB/c mice. Croat Chem Acta 78(3):379–384
97. Brown HT, Heron J (1880) Uber die hydrolytischen Wirkungen des Pankreas und des Dunndarmes. Ann Chem Pharmacol 204:228–251
98. Mosenthal HO (1911) Observations on the succus entericus. J Exp Med 13(3):319–327
99. Dyck WP, Bonnet D, Lasater J, Stinson C, Hall FF (1974) Hormonal stimulation of intestinal disaccharidase release in the dog. Gastroenterology 66(4):533–538
100. Cajori FA (1935) The lactase activity of the intestinal mucosa of the dog and some characteristics of intestinal lactase. J Biol Chem 109:159–168
101. Johnson FR, Kugler JH (1953) The distribution of alkaline phosphatase in the mucosal cells of the small intestine of the rat, cat and dog. J Anat (Lond) 87:247–256
102. Nachlas MM, Monis B, Rosenblatt D, Seligman AM (1960) Improvement in the histochemical localization of leucine aminopeptidase with a new substrate, L-leucyl-4-methoxy-2-naphthylamide. J Biophys Biochem Cytol 7:261–264
103. Miller D, Crane RK (1961) The digestive function of the epithelium of the small intestine. II. Localization of disaccharide hydrolysis in the isolated brush border portion of intestinal epithelial cells. Biochim Biophys Acta (Amst) 52:293–298
104. Wallis JL, Lipski PS, Mathers JC, James OFW, Hirst BH (1993) Duodenal brush-border mucosal glucose transport and enzyme activities in aging man and effect of bacterial contamination of the small intestine. Digest Dis Sci 38(3):403–409
105. Gudmand-Høyer E, Skovbjerg H (1996) Disaccharide digestion and maldigestion. Scand J Gastroenterol Suppl 216:111–121
106. Jones K, Sim L, Mohan S, Kumarasamy J, Liu H, Avery S et al (2011) Mapping the intestinal alpha-glucogenic enzyme specificities of starch digesting maltase-glucoamylase and sucrase-isomaltase. Bioorg Med Chem 19(13):3929–3934
107. Keller J, Layer P (2014) The pathophysiology of malabsorption. Viszeralmedizin 30(3):150–154
108. Gericke B, Amiri M, Naim HY (2016) The multiple roles of sucrase-isomaltase in the intestinal physiology. Mol Cell Pediatr 3:2
109. Lee PC, Werlin S, Trost B, Struve M (2004) Glucoamylase activity in infants and children: normal values and relationship to symptoms and histological findings. J Pediatr Gastroenterol Nutr 39(2):161–165
110. Lebenthal E, Khin-Maung-U Zheng BY, Lu RB, Lerner A (1994) Small intestinal glucoamylase deficiency and starch malabsorption: a newly recognized alpha-glucosidase deficiency in children. J Pediatr 124(4):541–546
111. Lina BA, Jonker D, Kozianowski G (2002) Isomaltulose (Palatinose): a review of biological and toxicological studies. Food Chem Toxicol 40(10):1375–1381
112. Suraphad P, Suklaew PO, Ngamukote S, Adisakwattana S, Mäkynen K (2017) The effect of isomaltulose together with green tea on glycemic response and antioxidant capacity: a single-blind, crossover study in healthy subjects. Nutrients 9(5):E464
113. Takazoe I, Frostell G, Ohta K, Topitsoglou V, Sasaki N (1985) Palatinose—a sucrose substitute. Pilot studies. Swed Dent J 9(2):81–87
114. Richards AB, Krakowka S, Dexter LB, Schmid H, Wolterbeek AP, Waalkens-Berendsen DH et al (2002) Trehalose: a review of properties, history of use and human tolerance, and results of multiple safety studies. Food Chem Toxicol 40(7):871–898
115. Kolho KL, Savilahti E (2000) Ethnic differences in intestinal disaccharidase values in children in Finland. J Pediatr Gastroenterol Nutr 30(3):283–287

116. Eloy R, Battinger F, Bignon JY, Ananna A, Grenier JF (1979) Experimental study in rats. Intestinal brush border enzymes and chronic alcohol ingestion. Res Exp Med 175(3):257–269
117. Neale G (1971) Disaccharidase deficiencies. J Clin Pathol Suppl (R Coll Pathol) 5:22–28
118. Herber R (1972) Disaccharidase deficiency in health and disease. Calif Med 116(6):23–37
119. Dahlqvist A (1962) Specificity of the human intestinal disaccharidases and implications for hereditary disaccharide intolerance. J Clin Invest 41(3):463–470
120. Semenza G (1976) Small intestinal disaccharidases: their properties and role as sugar translocators across natural and artificial membranes. In: Martonosi AN (ed) The enzymes of biological membranes. Plenum Press, New York
121. Allen LR, Stephen WA (2006) Low-carbohydrate diets. Am Fam Phys 73(11):1942–1948
122. Food and Nutrition Board. Institute of Medicine (2005) Dietary reference intakes for energy, carbohydrate, fiber, fat, fatty acids, cholesterol, protein, and amino acids. National Academies Press, Washington
123. Levin RJ (1994) Digestion and absorption of carbohydrates—from molecules and membranes to humans. Am J Clin Nutr 59(3):690S–698S
124. Shafik A, El Sibai O, Shafik AA, Shafik IA (2006) Demonstration of a physiologic sphincter at duodeno-jejunal junction. Front Biosci 11:2790–2794
125. Bronner F (2003) Mechanisms of intestinal calcium absorption. J Cell Biochem 88(2):387–393
126. Shafik A, Shafik IA, Sibai OE, Shafik AA (2007) Duodeno-jejunal junction dyssynergia: description of a novel syndrome. World J Gastroenterol 13(30):4112–4116
127. Kellogg EL, Kellogg WA (1921) Chronic duodenal obstruction with duodeno-jejunostomy as a method of treatment. Ann Surg 73(5):578–608
128. Virchenko SB, Sayenko VF, Kucherenko TL, Tsedik NN, Elbrønd H, Djurhuus JC, Funch-Jensen P (1993) The duodenojejunal junction and Treitz ligament in the regulation of duodenal emptying. Scand J Gastroenterol 28(9):753–759
129. Cheeseman CI (2002) Intestinal hexose absorption: transcellular or paracellular fluxes. J Physiol 544(Pt 2):336
130. Hewitt JA (1924) The metabolism of carbohydrates. Part III. The absorption of glucose, fructose and galactose from the small intestine. Biochem J 18(1):161–170
131. Hamilton KL, Butt AG (2013) Glucose transport into everted sacs of the small intestine of mice. Adv Physiol Educ 37(4):415–426
132. Riklis E, Quastel JH (1958) Effects of cations on sugar absorption by isolated surviving guinea pig intestine. Can J Biochem Physiol 36:347–362
133. Crane RK (1962) Hypothesis for mechanism of intestinal active transport of sugars. Fed Proc 21:891–895
134. Crane RK (1960) Intestinal absorption of sugars. Physiol Rev 40:789–825
135. Hediger MA, Coady MJ, Ikeda TS, Wright EM (1987) Expression cloning and cDNA sequencing of the Na^+/glucose co-transporter. Nature 330:379–381
136. Wright EM, Loo DDF, Hirayama BA, Turk E (2004) Surprising versatility of Na^+-glucose cotransporters: SLC5. Physiology 19(6):370–376
137. Röder PV, Geillinger KE, Zietek TS, Thorens B, Koepsell H, Daniel H (2014) The role of SGLT1 and GLUT2 in intestinal glucose transport and sensing. PLoS ONE 9(2):e89977
138. Dyer J, Vayro S, King TP, Shirazi-Beechey SP (2003) Glucose sensing in the intestinal epithelium. Eur J Biochem 270(16):3377–3388
139. Tyagi NK, Puntheeranurak T, Raja M, Kumar A, Wimmer B, Neundlinger I, et al (2011) A biophysical glance at the outer surface of the membrane transporter SGLT1. Biochim Biophys Acta 1808(1):1–18
140. Krofchick D, Silverman M (2003) Investigating the conformational states of the rabbit Na^+/glucose cotransporter. Biophys J 84(6):3690–3702
141. Wilder-Smith CH, Li X, Ho SS, Leong SM, Wong RK, Koay ES, Ferraris RP (2014) Fructose transporters GLUT5 and GLUT2 expression in adult patients with fructose intolerance. United Eur Gastroenterol J 2(1):14–21

142. Douard V, Ferraris RP (2008) Regulation of the fructose transporter GLUT5 in health and disease. Am J Physiol Endocrinol Metab 295(2):E227–E237
143. Jones HF, Butler RN, Brooks DA (2011) Intestinal fructose transport and malabsorption in humans. Am J Physiol Gastrointest Liver Physiol 300(2):G202–G206
144. Xu Y, Tao Y, Cheung LS, Fan C, Chen L-Q, Xu S et al (2014) Structures of bacterial homologues of SWEET transporters in two distinct conformations. Nature 515:448–452
145. Naftalin RJ (2014) Does apical membrane GLUT2 have a role in intestinal glucose uptake? Version 1. F1000Res 3:304
146. Drozdowski LA, Thomson ABR (2006) Intestinal sugar transport. World J Gastroenterol 12 (11):1657–1670
147. Sundler F (2004) GI tract, general anatomy (cells). In: Martini L (ed) Encyclopedia of endocrine diseases. Elsevier, MA, USA
148. Stümpel F, Burcelin R, Jungermann K, Thorens B (2001) Normal kinetics of intestinal glucose absorption in the absence of GLUT2: evidence for a transport pathway requiring glucose phosphorylation and transfer into the endoplasmic reticulum. PNAS 98(20):11330–11335
149. Hosokawa M, Thorens B (2002) Glucose release from GLUT2-null hepatocytes: characterization of a major and a minor pathway. Am J Physiol Endocrinol Metab 282:E794–E801
150. Santer R, Hillebrand G, Steinmann B, Schaub J (2003) Intestinal glucose transport: evidence for a membrane traffic–based pathway in humans. Gastroenterology 124:34–39
151. Deng D, Yan N (2016) GLUT, SGLT, and SWEET: structural and mechanistic investigations of the glucose transporters. Protein Sci 25(3):546–558
152. Tao Y, Cheung LS, Li S, Eom JS, Chen LQ, Xu Y et al (2015) Structure of a eukaryotic SWEET transporter in a homotrimeric complex. Nature 527(7577):259–263
153. Eom JS, Chen LQ, Sosso D, Julius BT, Lin IW, Qu XQ et al (2015) SWEETs, transporters for intracellular and intercellular sugar translocation. Curr Opin Plant Biol 25:53–62
154. Li-Qing C, Bi-Huei H, Sylvie L, Hitomi T, Mara LH, Xiao-Qing Q et al (2010) Sugar transporters for intercellular exchange and nutrition of pathogens. Nature 468(7323):527–532
155. Hitomi T, Wolf BF (2010) Facilitative plasma membrane transporters function during ER transit. FASEB J 24(8):2849–2858
156. White H, Venkatesh B (2011) Clinical review: ketones and brain injury. Crit Care 15(2):219
157. Welcome MO, Razvodovsky YE, Pereverzeva EV, Pereverzev VA (2013) State of cognitive functions of students-medics with different relationship to alcohol use. Belarusian State Medical University Press, Minsk, Belarus
158. Kageyama T (2002) Pepsinogens, progastricsins, and prochymosins: structure, function, evolution, and development. Cell Mol Life Sci 59(2):288–306
159. Roberts NB (2006) Review article: human pepsins—their multiplicity, function and role in reflux disease. Aliment Pharmacol Ther 24(2):2–9
160. McFarlane J, Dunbar VE, Borsook H, Wasteneys H (1927) The stages of the peptic hydrolysis of egg albumin. J Gen Physiol 10(3):437–450
161. Kousoulis AA, Tsoucalas G, Armenis I, Marineli F, Karamanou M, Androutsos G (2012) From the "hungry acid" to pepsinogen: a journey through time in quest for the stomach's secretion. Ann Gastroenterol 25(2):119–122
162. Matlin KS, Caplan MJ (2017) The secretory pathway at 50: a golden anniversary for some momentous grains of silver. Mol Biol Cell 28(2):229–232
163. Hajdu SI, Tang P (2008) The saga of carcinoid and oat-cell carcinoma. Ann Clin Lab Sci 38 (4):414–417
164. Lee D, Ryle AP (1967) Pepsin D: A minor component of commercial pepsin preparations. Biochem J 104(3):742–748
165. Lee D, Ryle AP (1967) Pepsinogen D. A fourth proteolytic zymogen from pig gastric mucosa. Biochem J 104(3):735–741

166. Thompson JD, Higgins DG, Gibson TJ (1994) CLUSTAL W: improving the sensitivity of progressive multiple sequence alignment through sequence weighting, position-specific gap penalties and weight-matrix choice. Nucleic Acids Res 22:4673–4680
167. Ward PH, Neumann VK, Chiang L (1978) Partial characterization of pepsins and gastricsins and their zymogens from human and toad gastric mucosae. Comp Biochem Physiol Part B Comp Biochem 61(4):491–498
168. Majewska M, Lipka A, Panasiewicz G, Gowkielewicz M, Jozwik M, Majewski MK, Szafranska B (2017) Identification of novel placentally expressed aspartic proteinase in humans. Int J Mol Sci 18(6):1227
169. Shen S, Jiang J, Yuan Y (2017) Pepsinogen C expression, regulation and its relationship with cancer. Cancer Cell Int 17:57
170. Florkin M (1957) Discovery of pepsin by Theodor Schwann. Rev Med Liege 12(5):139–144
171. Northrop JH (1929) Crystalline pepsin. Science 69:580
172. Simoni RD, Hill RH, Vaughan M (2002) Urease, the first crystalline enzyme and the proof that enzymes are proteins: the work of James B. Sumner. J Biol Chem 277(35):23e
173. Hasnain S (2016) Impact and influence of crystallography across the sciences. IUCrJ 3(Pt 6):389–390
174. Andersen OS (2005) A brief history of the journal of general physiology. J Gen Physiol 125 (1):3–12
175. Northrop JH, Kunitz M (1932) Crystalline trypsin. I. Isolation and tests of purity. J Gen Physiol 16:267–294
176. Kunitz M, Northrop JH (1934) The isolation of crystalline trypsinogen and its conversion into crystalline trypsin. Science 80:505–506
177. Herriott RM, Northrop JH (1936) Isolation of crystalline pepsinogen from swine gastric mucosae and its autocatalytic conversion into pepsin. Science 83:469–470
178. Stanley W (1935) Isolation of a crystalline protein possessing the properties of tobacco-mosaic virus. Science 81:644–645
179. Eisenstein M (2016) The field that came in from the cold. Nat Methods 13:19–22
180. McPherson A, Gavira JA (2014) Introduction to protein crystallization. Acta Crystallogr F Struct Biol Commun 70(Pt 1):2–20
181. McPherson A (2017) Protein crystallization. Methods Mol Biol 1607:17–50
182. Coskun O (2016) Separation techniques: chromatography. North Clin Istanb 3(2):156–160
183. Alberts B, Johnson A, Lewis J, Raff M, Roberts K, Walter P (2002) Molecular biology of the cell, 4th edn. Garland Science, New York
184. Ehren J, Govindarajan S, Moron B, Minshull J, Khosla C (2008) Protein engineering of improved prolyl endopeptidases for celiac sprue therapy. Protein Eng Des Sel 21(12):699–707
185. Moose RE, Clemente JC, Jackson LR, Ngo M, Wooten K, Chang R et al (2007) Analysis of binding interactions of pepsin inhibitor-3 to mammalian and malarial aspartic proteases. Biochemistry 46(49):14198–14205
186. Sampath-Kumar PS, Fruton JS (1974) Studies on the extended active sites of acid proteinases. Proc Natl Acad Sci U S A 71(4):1070–1072
187. Kay J, Dykes CW (1977) The first cleavage site in pepsinogen activation. Adv Exp Med Biol 95:103–127
188. Khan AR, James MN (1998) Molecular mechanisms for the conversion of zymogens to active proteolytic enzymes. Protein Sci 7(4):815–836
189. Pizauro JM Jr, Ferro JA, de Lima ACF, Routman KS, Portella MC (2004) The zymogen-enteropeptidase system: a practical approach to study the regulation of enzyme activity by proteolytic cleavage. Biochem Mol Biol Educ 32(1):45–48
190. Richter C, Tanaka T, Yada RY (1998) Mechanism of activation of the gastric aspartic proteinases: pepsinogen, progastricsin and prochymosin. Biochem J 335(3):481–490
191. Sanny CG, Hartsuck JA, Tang J (1975) Conversion of pepsinogen to pepsin. Further evidence for intramolecular and pepsin-catalyzed activation. J Biol Chem 250(7):2635–2639

192. Taggart RT, Mohandas TK, Shows TB, Bell GI (1985) Variable numbers of pepsinogen genes are located in the centromeric region of human chromosome 11 and determine the high-frequency electrophoretic polymorphism. PNAS 82(18):6240–6244
193. Marciniszyn J Jr, Huang JS, Hartsuck JA, Tang J (1976) Mechanism of intramolecular activation of pepsinogen. Evidence for an intermediate delta and the involvement of the active site of pepsin in the intramolecular activation of pepsinogen. J Biol Chem 251 (22):7095–7102
194. Kageyama T (1998) Molecular cloning, expression and characterization of an Ascaris inhibitor for pepsin and cathepsin E. Eur J Biochem 253(3):804–809
195. Zalatoris J, Rao-Naik C, Fecho G, Girdwood K, Kay J, Dunn BM (1998) Expression, purification, and characterization of the recombinant pepsin inhibitor from Ascaris suum. Adv Exp Med Biol 436:387–389
196. Parikh S, Gut J, Istvan E, Goldberg DE, Havlir DV, Rosenthal PJ (2005) Antimalarial activity of human immunodeficiency virus type 1 protease inhibitors. Antimicrob Agents Chemother 49(7):2983–2985
197. Yang H, Nkeze J, Zhao RY (2012) Effects of HIV-1 protease on cellular functions and their potential applications in antiretroviral therapy. Cell Biosci 2:32
198. Parikh S, Liu J, Sijwali P, Gut J, Goldberg DE, Rosenthal PJ (2006) Antimalarial effects of human immunodeficiency virus type 1 protease inhibitors differ from those of the aspartic protease inhibitor pepstatin. Antimicrob Agents Chemother 50(6):2207–2209
199. Boss C, Richard-Bildstein S, Weller T, Fischli W, Meyer S, Binkert C (2003) Inhibitors of the Plasmodium falciparum parasite aspartic protease plasmepsin II as potential antimalarial agents. Curr Med Chem 10(11):883–907
200. dos Santos AL (2010) HIV aspartyl protease inhibitors as promising compounds against Candida albicans André Luis Souza dos Santos. World J Biol Chem 1(2):21–30
201. Venugopal C, Demos CM, Rao KS, Pappolla MA, Sambamurti K (2008) Beta-secretase: structure, function, and evolution. CNS Neurol Disord: Drug Targets 7(3):278–294
202. Wolfe MS (2010) Structure, mechanism and inhibition of gamma-secretase and presenilin-like proteases. Biol Chem 391(8):839–847
203. Epis R, Marcello E, Gardoni F, Di Luca M (2012) Alpha, beta-and gamma-secretases in Alzheimer's disease. Front Biosci (Schol Ed) 4:1126–1150
204. Kametani F (2008) Epsilon-secretase: reduction of amyloid precursor protein epsilon-site cleavage in Alzheimer's disease. Curr Alzheimer Res 5(2):165–171
205. Vassar R (2007) Beta-secretase (BACE) as a drug target for Alzheimer's disease. Adv Drug Deliv Rev 54(12):1589–1602
206. Tomita T (2009) Secretase inhibitors and modulators for Alzheimer's disease treatment. Expert Rev Neurother 9(5):661–679
207. Evin G, Sernee MF, Masters CL (2006) Inhibition of gamma-secretase as a therapeutic intervention for Alzheimer's disease: prospects, limitations and strategies. CNS Drugs 20 (5):351–372
208. Schlenk F (1985) Early research on fermentation—a story of missed opportunities. Trends in Biochem Sci 10(6):252–254
209. Kukar T, Golde TE (2008) Possible mechanisms of action of NSAIDs and related compounds that modulate γ-secretase cleavage. Curr Top Med Chem 8(1):47–53
210. Teich N, Le Maréchal C, Kukor Z, Caca K, Witzigmann H, Chen JM et al (2004) Interaction between trypsinogen isoforms in genetically determined pancreatitis: mutation E79K in cationic trypsin (PRSS1) causes increased transactivation of anionic trypsinogen (PRSS2). Hum Mutat 23(1):22–31
211. Hegyi E, Sahin-Tóth M (2017) Genetic risk in chronic pancreatitis: the trypsin-dependent pathway. Dig Dis Sci 62(7):1692–1701
212. Zator Z, Whitcomb DC (2017) Insights into the genetic risk factors for the development of pancreatic disease. Therap Adv Gastroenterol 10(3):323–336
213. Kühne (1877) Über das Verhalten verschiedener organisirter und sog. Ungeformter Fermente, Verhandlungen des Heidelb. Naturhist Med Vereins, Neue Folge (On the

behavior of various organized and so-called unformed ferments, Negotiations of the Heidelb. Naturhist Med Assoc, New Series) 1(3):190–193
214. Lanska DJ (2014) Kühne, Wilhelm (Willy) Friedrich. In: MJ Aminoff, RB Daroff (eds) Encyclopedia of the neurological sciences, 2nd ed. Elsevier, MA, USA
215. Howard JM, Hess W (2002) History of the pancreas: mysteries of a hidden organ. Springer, New York
216. Rawlings ND, Barrett AJ (1993) Evolutionary families of peptidases. Biochem J 290 (1):205–218
217. Maike G (2006) Sculpted through time: evolution and function of serine proteases from the mast cell chymase locus. Uppsala University, Uppsala
218. Murthy SN, Kostman J, Dinoso VP Jr (1980) Effect of pH, substrate, and temperature on tryptic activity of duodenal samples. Dig Dis Sci 25(4):289–294
219. Cheison CS, Brand J, Leeb E, Kulozik U (2011) Analysis of the effect of temperature changes combined with different alkaline pH on the β-lactoglobulin trypsin hydrolysis pattern using MALDI-TOF-MS/MS. J Agric Food Chem 59(5):1572–1581
220. Hirota M, Ohmuraya M, Baba H (2006) The role of trypsin, trypsin inhibitor, and trypsin receptor in the onset and aggravation of pancreatitis. J Gastroenterol 41(9):832–836
221. Hoffmeister A, Dietz G, Zeitschel U, Mössner J, Rossner S, Stahl T (2009) BACE1 is a newly discovered protein secreted by the pancreas which cleaves enteropeptidase in vitro. JOP 10(5):501–506
222. Kazal LA, Spicer DS, Brahinsky RA (1948) Isolation of a crystalline trypsin inhibitor-anticoagulant protein from pancreas. J Am Chem Soc 70:3034–3040
223. Wang G-P, Xu C-S (2010) Pancreatic secretory trypsin inhibitor: more than a trypsin inhibitor. World J Gastrointest Pathophysiol 1(2):85–90
224. Herzig KH, Schon I, Tatemoto K, Ohe Y, Li Y, Folsch UR, Owyang C (1996) Diazepam binding inhibitor is a potent cholecystokinin-releasing peptide in the intestine. Proc Nat Acad Sci 93:7927–7932
225. Li Y, Hao Y, Owyang C (2000) Diazepam-binding inhibitor mediates feedback regulation of pancreatic secretion and postprandial release of cholecystokinin. J Clin Invest 105(3):351–359
226. Costa E, Guidotti A (1991) Diazepam binding inhibitor (DBI): a peptide with multiple biological actions. Life Sci 49(5):325–344
227. Anholt RRH, Pederson PL, Desouza EB, Snyder SH (1986) The peripheral-type benzodiazepine receptor: localization to the mitochondrial outer membrane. J Biol Chem 261:576–583
228. Gersuk VH, Rose TM, Todaro GJ (1995) Molecular cloning and chromosomal localization of a pseudogene related to the human acyl-CoA binding protein/diazepam binding inhibitor. Genomics 25:469–476
229. Li Y, Hao Y, Owyang C (2000) Diazepam-binding inhibitor mediates feedback regulation of pancreatic secretion and postprandial release of cholecystokinin. J Clin Invest 105:351–359
230. Rose TM, Schultz ER, Todaro GJ (1992) Molecular cloning of the gene for the yeast homolog (ACB) of diazepam binding inhibitor/endozepine/acyl-CoA-binding protein. Proc Nat Acad Sci 89:11287–11291
231. Gray PW, Glaister D, Seeburg PH, Guidotti A, Costa E (1986) Cloning and expression of cDNA for human diazepam binding inhibitor, a natural ligand of an allosteric regulatory site of the gamma-aminobutyric acid type A receptor. Proc Nat Acad Sci 83:7547–7551
232. Haefely W (1984) Benzodiazepine interactions with GABA receptors. Neurosci Lett 47 (3):201–206
233. Shibata T, Ogawa M, Takata N, Matsuda K, Niinobu T, Uda K et al (1987) Distribution of pancreatic secretory trypsin inhibitor in various human tissues and its inactivation in the gastric mucosa. Res Commun Chem Pathol Pharmacol 55(2):243–248
234. Fushiki T, Fukuoka S, Iwai K (1984) Stimulatory effect of an endogenous peptide in rat pancreatic juice on pancreatic enzyme secretion in the presence of atropine. Biochem Biophys Res Commun 118:532–537

235. Iwai K, Fukuoka S, Fushiki T, Tsujikawa M, Hirose M, Tsunasawa S, Sakiyama F (1987) Purification and sequencing of a trypsin-sensitive cholecystokinin-releasing peptide from rat pancreatic juice. Its homology with pancreatic secretory trypsin inhibitor. J Biol Chem 262:8956–8959
236. Kyoko M, Rieko N, Akihiro F, Kenichi K (1989) Stimulatory effect of monitor peptide and human pancreatic secretory trypsin inhibitor on pancreatic secretion and cholecystokinin release in conscious rats. Pancreas 4(2):139–144
237. Miyasaka K, Funakoshi A, Nakamura R, Kitani K, Uda K, Murata A, Ogawa M (1989) Differences in stimulatory effects between rat pancreatic secretory trypsin inhibitor-61 and -56 on rat pancreas. Japan J Physiol 39(6):891–899
238. Graf R, Klauser S, Fukuoka SI, Schiesser M, Bimmler D (2003) The bifunctional rat pancreatic secretory trypsin inhibitor/monitor peptide provides protection against premature activation of pancreatic juice. Pancreatology 3(3):195–206
239. Hirota M, Ohmuraya M, Baba H (2006) The role of trypsin, trypsin inhibitor, and trypsin receptor in the onset and aggravation of pancreatitis. J Gastroenterol 41(9):832–836
240. Hoffmeister A, Dietz G, Zeitschel U, Mössner J, Rossner S, Stahl T (2009) BACE1 is a newly discovered protein secreted by the pancreas which cleaves enteropeptidase in vitro. JOP 10(5):501–506
241. Uda K-I, Murata A, Nishijima J-I, Doi S, Tomita N, Ogawa M, Mori T (1994) Elevation of circulating monitor peptide/pancreatic secretory trypsin inhibitor-I (PSTI-61) after turpentine-induced inflammation in rats: hepatocytes produce it as an acute phase reactant. J Surg Res 57(5):563–568
242. Miyasaka K, Funakoshi A (1998) Luminal feedback regulation, monitor peptide, CCK-releasing peptide, and CCK receptors. Pancreas 16(3):277–283
243. Herzig K-H (1998) Cholecystokinin- and secretin-releasing peptides in the intestine—a new regulatory interendocrine mechanism in the gastrointestinal tract. Regul Pept 73(2):89–94
244. Song Y, Li P, Lee KY, Chang T, Chey WY (1999) Canine pancreatic juice stimulates the release of secretin and pancreatic secretion in the dog. Am J Physiol 277(3 Pt 1):G731–G735
245. Li P, Lee KY, Chang T-M, Chey WY (1990) Mechanism of acid-induced release of secretin in ratspresence of a secretin-releasing peptide. J Clin Invest 86:1474–1479
246. Chey WY, Chang TM (2003) Neural control of the release and action of secretin. J Physiol Pharmacol 54(4):105–112
247. Mellanby J, Woolley VJ (1912) The ferments of the pancreas: Part I. The generation of trypsin from trypsinogen by enterokinase. J Physiol 45(5):370–388
248. Mellanby J, Woolley VJ (1913) The ferments of the pancreas. Part II. The action of calcium salts in the generation of trypsin from trypsinogen. J Physiol 46(2):159–172
249. Zamolodchikova TS, Sokolova EA, Aleksandrov SL, Mirgorodskaia OA, Morozov IA, Vorotyntseva TI (1998) Duodenase—a potential activator of cascade of digestive proteases. Bioorg Khim 24(4):300–305
250. Pletnev VZ, Zamolodchikova TS, Pangborn WA, Duax WL (2000) Crystal structure of bovine duodenase, a serine protease, with dual trypsin and chymotrypsin-like specificities. Proteins 41(1):8–16
251. Gaspariana ME, Bobika TV, Kima YV, Ponomarenkoa NA, Dolgikha DA, Gabibova AG, Kirpichnikova MP (2013) Heterogeneous catalysis on the phage surface: display of active human enteropeptidase. Biochimie 95(11):2076–2081
252. Shlygin GK (1956) The importance of determining enterokinase and alkaline phosphatase for the assessment of the state of human intestines. Clin Chim Acta 1(5):421–433
253. Shabalova NP (ed) (2011) Pediatric gastroenterology: a guide for physicians. MEDPress-Inform, Moscow, Russia
254. Kunitz M (1939) Formation of trypsin from crystalline trypsinogen by means of enterokinase. J Gen Physiol 22(4):429–446
255. Light A, Janska H (1989) Enterokinase (enteropeptidase): comparative aspects. Trends Biochem Sci 14(3):110–112

256. Hooper JD, Clements JA, Quiqley JP, Antalis TM (2001) Type II transmembrane serine proteases. Insights into an emerging class of cell surface proteolytic enzymes. J Biol Chem 276(2):857–860
257. Lu D, Yuan X, Zheng X, Sadler JE (1997) Bovine proenteropeptidase is activated by trypsin, and the specificity of enteropeptidase depends on the heavy chain. J Biol Chem 272:31293–31300
258. McDonald MR, Kunitz M (1941) The effect of calcium and other ions on the autocatalytic formation of trypsin from trypsinogen. J Gen Physiol 25(1):53–73
259. Sarkany RP, Moreland BH (1985) Enhancement of the autocatalytic activation of trypsinogen to trypsin by bile and bile acids. Biochim Biophys Acta 839(3):262–267
260. Holzinger A, Maier EM, Bück C, Mayerhofer PU, Kappler M, Haworth JC et al (2002) Mutations in the proenteropeptidase gene are the molecular cause of congenital enteropeptidase deficiency. Am J Hum Genet 70(1):20–25
261. Hadorn B, Tarlow MJ, Lloyd JK, Wolff OH (1969) Intestinal enterokinase deficiency. Lancet 1(7599):812–813
262. Holzinger A, Maier EM, Bück C, Mayerhofer PU, Kappler M, Haworth JC et al (2002) Mutations in the proenteropeptidase gene are the molecular cause of congenital enteropeptidase deficiency. Am J Hum Genet 70(1):20–25
263. Ferrone M, Raimondo M, Scolapio JS (2007) Pancreatic enzyme pharmacotherapy. Pharmacotherapy 27(6):910–920
264. Kuhn RJ, Gelrud A, Munck A, Caras S (2010) 0 CREON (pancrelipase delayed-release capsules) for the treatment of exocrine pancreatic insufficiency. Adv Ther 27(12):895–916
265. Fieker A, Philpott J, Armand M (2011) Enzyme replacement therapy for pancreatic insufficiency: present and future. Clin Exp Gastroenterol 4:55–73
266. David WC, Amit B, Katrin B, Suntje S-S, Shufang L, Mahesh F et al (2016) Efficacy and safety of pancrelipase/pancreatin in patients with exocrine pancreatic insufficiency and a medical history of diabetes mellitus. Pancreas 45(5):679–686
267. Antonov VK, Vorotyntseva TI, Zamolodchikova TS (1992) Duodenase—a new serine proteinase with unusual specificity. Dokl Akad Nauk 324(6):1318–1322
268. Zamolodchikova TS, Sokolova EA, Lu D, Sadler JE (2000) Activation of recombinant proenteropeptidase by duodenase. FEBS Lett 466(2–3):295–299
269. Sanderink GJ, Artur Y, Siest G (1989) Human aminopeptidases: a review of the literature. J Clin Chem Clin Biochem 26(12):795–807
270. Mucha A, Drag M, Dalton JP, Kafarski P (2010) Metallo-aminopeptidase inhibitors. Biochimie 92(11):1509–1529
271. Linderstrøm-Lang K (1929) Über Darmerepsin. Hoppe Seylers Z Physiol Chem 182:151–174
272. Bradshaw RA (2013) Aminopeptidases. In: Lennarz W, Lane M (eds) Encyclopedia of biological chemistry, 2nd edn. Elsevier, MA, USA
273. Tsukamoto H, Shibata K, Kajiyama H, Terauchi M, Nawa A, Kikkawa F (2008) BMC Cancer 8:74
274. Grdiša M, Vitale L (1991) Types and localization of aminopeptidases in different human blood cells. Int J Biochem 23(3):339–345
275. Sidorowicz W, Jackson GC, Behal FJ (1980) Multiple molecular forms of human pancreas alanine aminopeptidase. Clin Chim Acta 104(2):169–179
276. Maroux S, Louvard D, Barath J (1973) The aminopeptidase from hog intestinal brush border. Biochim Biophys Acta Enzymol 321(1):282–295
277. Magee AI, Grant DA, Hermon-Taylor J (1977) The apparent molecular weights of human intestinal aminopeptidase, enterokinase and maltase in native duodenal fluid. Biochem J 165 (3):583–585
278. Allen T (1993) Aminopeptidases: structure and function. FASEB J 7(2):290–298
279. Wickström M, Larsson R, Nygren P, Gullbo J (2011) Aminopeptidase N (CD13) as a target for cancer chemotherapy. Cancer Sci 102(3):501–508

280. Gu Y-Q, Chao WS, Walling LL (1996) Localization and post-translational processing of the wound-induced leucine aminopeptidase proteins of tomato. J Biological Chem 271:25880–25887
281. Thielitz A, Reinhold D, Vetter R, Bank U, Helmuth M, Hartig R et al (2007) Inhibitors of dipeptidyl peptidase IV and aminopeptidase N target major pathogenetic steps in acne initiation. J Invest Dermatol 127(5):1042–1051
282. Bauvois B, Dauzonne D (2006) Aminopeptidase-N/CD13 (EC 3.4.11.2) inhibitors: chemistry, biological evaluations, and therapeutic prospects. Med Res Rev 26(1):88–130
283. Sapio MR, Fricker LD (2014) Carboxypeptidases in disease: insights from peptidomic studies. Proteomics Clin Appl 8:327–337
284. Lyons PJ, Fricker LD (2011) Carboxypeptidase O is a glycosylphosphatidy linositol-anchored intestinal peptidase with acidic amino acid specificity. J Biol Chem 286 (45):39023–39032
285. Yoshioka M, Erickson RH, Kim YS (1988) Digestion and assimilation of proline-containing peptides by rat intestinal brush border membrane carboxypeptidases. Role of the combined action of angiotensin-converting enzyme and carboxypeptidase P. J Clin Invest 81(4):1090–1095
286. Berger J, Schneeman BO (1988) Intestinal zinc and carboxypeptidase A and B activity in response to consumption of test meals containing various proteins by rats. J Nutr 118 (6):723–728
287. Wilcox PE (1970) Chymotrypsinogens – chymotrypsins. Methods Enzymol 19:64–108
288. Appel W (1986) Chymotrypsin: molecular and catalytic properties. Clin Biochem 19 (6):317–322
289. Gráf L, Szilágyi L, Venekei I (2013) Chymotrypsin. In: Neil RD, Salvesen G (eds) Handbook of proteolytic enzymes. Elsevier, San Diego, CA, USA
290. Fleming TC, Riddel GH (1961) Studies of the antigenic properties of alpha chymotrypsin. Am J Ophthalmol 51(5):1104/232–1107/235
291. Vernon HM (1914) The activation of trypsinogen. Biochem J 8(5):494–529
292. Vernon HM (1901) The conditions of action of pancreatic rennin and diastase. J Physiol 27 (3):174–199
293. Bedford T (1951) H. M. Vernon, M.A., M.D. Br J Ind Med 8(2):96–97
294. Vernon HM (1913) The auto-catalysis of trypsinogen. J Physiol 47(4–5):325–338
295. Kunitz M, Northrop JH (1934) The isolation of crystalline trypsinogen and its conversion into crystalline trypsin. Science 80(2083):505–506
296. Kunitz M, Northrop JH (1934) Autocatalytic activation of trypsinogen in the presence of concentrated ammonium or magnesium sulfate. Science 80(2069):190
297. Yoosuke T, Shinichi F, Tomoaki M, Kenichi M (1986) Structure of human cholecystokinin gene and its chromosomal location. Gene 50(1–3):353–360
298. Manchester KL (2004) The crystallization of enzymes and virus proteins: laying to rest the colloidal concept of living systems. Endeavour 28(1):25–29
299. Kunitz M (1948) Crystallization of salt-free chymotrypsinogen and chymotrypsin from solution in dilute ethyl alcohol. J Gen Physiol 32:265–269
300. Kunitz M (1938) Formation of new crystalline enzymes from chymotrypsin, isolation of beta and gamma chymotrypsin. J Gen Physiol 32:207–237
301. Kunitz M, Northrop J (1935) Crystalline chymotrypsin and chymotrypsinogen. Isolation, crystallization and general properties of a new proteolytic enzyme and its precursor. J Gen Physiol 18:433–458
302. Northrop J, Kunitz M, Herriott R (1948) Crystalline enzymes, 2nd edn. Columbia Univ Press, NY, USA
303. Herriott RM (1989) Moses Kunitz: December 19, 1887–April 20, 1978. Biogr Mem Natl Acad Sci 58:305–317
304. Koertge N (ed) (2007) New dictionary of scientific biography, 1st edn. Charles Scribners & Sons, Detroit

305. Fruton GW, Bergmann M (1942) The multiple specificity of chymotrypsin. J Biol Chem 145:253–265
306. Jacobsen CF (1947) The activation of chymotrypsin. Compt Rend Tray Lab Carlsberg Ser Chim 25:325–337
307. Kato M, Hayakawa S, Naruse S, Kitagawa M, Ishiguro H, Nakae Y, Hayakawa T (1997) Change of pancreatic enzymes, pancreatic stone protein (PSP), and plasma alpha(2)-macroglobulin-trypsin complex-like substance (MTLS) in the activation of pancreatic juice. Pancreas 15(4):345–349
308. Vendrell J, Guasch A, Coll M, Villegas V, Billeter M, Wider G et al (1992) Pancreatic procarboxypeptidases: their activation processes related to the structural features of the zymogens and activation segments. Biol Chem Hoppe Seyler 373(7):387–392
309. Ventura S, Gomis-Rüth FX, Puigserver A, Avilés FX, Vendrell J (1997) Pancreatic procarboxypeptidases: oligomeric structures and activation processes revisited. Biol Chem 378(3–4):161–165
310. Okazaki H, Igarashi M, Nishi M, Sekiya M, Tajima M, Takase S et al (2008) Identification of neutral cholesterol ester hydrolase, a key enzyme removing cholesterol from macrophages. J Biol Chem 283(48):33357–33364
311. Andersson E-L, Hernell O, Bläckberg L, Fält H, Lindquist S (2011) BSSL and PLRP2: key enzymes for lipid digestion in the newborn examined using the Caco-2 cell line. J Lipid Res 52(11):1949–1956
312. Omar A, Krebs A (1975) An analysis of pancreatic enzymes used in epidermal separation. Arch Dermatol Res 253(2):203–212
313. Donà F, Houseley J (2014) Unexpected DNA loss mediated by the DNA Binding activity of ribonuclease A. PLoS ONE 9(12):e115008
314. Busnardo AC, DiDio LJA, Tidrick RT, Thomford NR (1983) History of the pancreas. Am J Surg 146(5):539–550
315. Beger HG, Buchler M, Kozarek R, Lerch M, Neoptolemos JP, Warshaw A et al (eds) (2008) The pancreas: an integrated textbook of basic science, medicine, and surgery. Wiley-Blackwell, Oxford, UK
316. Navarro S (2014) A brief history of the anatomy and physiology of a mysterious and hidden gland called the pancreas. Gastroenterol Hepatol 37:527–534
317. Trevisani F (1983) Ratio and experimentum: Johannes Bohn (1640–1718) and Italian experimental physiology. Clio Med 17(4):199–206
318. Karamanou M, Koutsilieris M, Laios K, Marineli F, Androutsos G (2014) Apollinaire Bouchardat (1806–1886): founder of modern Diabetology. Hormones 13(2):296–300
319. Chast F (2000) Apollinaire Bouchardat, pharmacist, nutritionist. Ann Pharm Fr 58(6):435–442
320. Chiray M, Salmon AR, Mercier A (1926) Action of purified secretin on external secretion of pancreas in man. Bull Soc Med Hop Paris 50:1417
321. Goldstein F, Wirts CW, Cozzolino HJ, Menduke H (1964) Secretin tests of pancreatic and biliary tract disease. Arch Intern Med 114(1):124–131
322. Burton P, Evans DG, Harper AA, Howath T, Oleesky S, Scott JE, Varley H (1960) A test of pancreatic function in man based on the analysis of duodenal contents after administration of secretin and pancreozymin. Gut 1:111–124
323. Sun DC, Shay H (1960) Pancreozymin-secretin test. The combined study of serum enzymes and duodenal contents in the diagnosis of pancreatic disease. Gastroenterology 38:570–581
324. McGillivray DC, Stordy SN, Bogoch A (1966) The pancreozymin-secretin test after partial gastrectomy. Can Med Assoc J 94(24):1261–1263
325. Jorpes EJ, Mutt V (1973) Secretin and cholecystokinin (CCK). In: Jorpes EJ, Mutt V (eds) Secretin, cholecystokinin, pancreozymin and gastrin. Handbook of experimental pharmacology. Springer, Heidelberg
326. Wong LT, Turtle S, Davidson AG (1982) Secretin pancreozymin stimulation test and confirmation of the diagnosis of cystic fibrosis. Gut 23(9):744–750

327. Mee AS, Girdwood AH, Walker E, Gilinsky NH, Kottler RE, Marks IN (1985) Comparison of the oral (PABA) pancreatic function test, the secretin-pancreozymin test and endoscopic retrograde pancreatography in chronic alcohol induced pancreatitis. Gut 26(11):1257–1262
328. Raimondo M, Imoto M, DiMagno EP (2003) Rapid endoscopic secretin stimulation test and discrimination of chronic pancreatitis and pancreatic cancer from disease controls. Clin Gastroenterol Hepatol 1(5):397–403
329. John G, Lieb II, Draganov Peter V (2008) Pancreatic function testing: here to stay for the 21st century. World J Gastroenterol 14(20):3149–3158
330. Chowdhury RS, Forsmark CE (2003) Pancreatic function testing. Aliment Pharmacol Ther 17:733–750
331. Malfertheiner P, Büchler M (1989) Correlation of imaging and function in chronic pancreatitis. Radiol Clin North Am 27(1):51–64
332. Alkaade S, Cem Balci N, Momtahen AJ, Burton F (2008) Normal pancreatic exocrine function does not exclude MRI/MRCP chronic pancreatitis findings. J Clin Gastroenterol 42 (8):950–955
333. Moeller DD, Dunn GD, Klotz AP (1972) Comparison of the pancreozymin-secretin test and the Lundh test meal. Am J Digest Dis 17(9):799–805
334. Burton P, Harper AA, Howat HT, Scott JE, Varley H (1960) The use of cholecystokinin to test gall bladder function in man. Gut 1:193–204
335. Lundh G (1962) Pancreatic exocrine function in neoplastic and inflammatory disease; a simple and reliable new test. Gastroenterology 42:275–280
336. Czakó L, Hajnal F, Németh J, Lonovics J (2000) Assessment of pancreatic enzyme secretory capacity by a modified Lundh test. Int J Pancreatol 27(1):13–19
337. Augarten A, Dubenbaum L, Yahav Y, Katznelson D, Szeinberg A, Blank A, Sack J (1999) Lundh meal: a single non-invasive challenge test for evaluation of exocrine and endocrine pancreatic function in cystic fibrosis patients. Int J Clin Lab Res 29(3):114–116
338. James O (1973) Progress report: the Lundh test. Gut 14:582–591
339. Wong LT, Turtle S, Davidson AG (1982) Secretin pancreozymin stimulation test and confirmation of the diagnosis of cystic fibrosis. Gut 23(9):744–750
340. Augarten A, Berman H, Aviram M, Diver-Habber A, Akons H, Ben Tur L et al (2003) Serum CA 19-9 levels as a diagnostic marker in cystic fibrosis patients with borderline sweat tests. Clin Exp Med 3(2):119–123
341. Fong ZV, Winter JM (2012) Biomarkers in pancreatic cancer: diagnostic, prognostic, and predictive. Cancer J 18(6):530–538
342. Gullo L, Lucrezio L, Calculli L, Salizzoni E, Coe M, Migliori M et al (2009) Magnetic resonance cholangiopancreatography in asymptomatic pancreatic hyperenzymemia. Pancreas 38(4):396–400
343. Paola T, Raffaele P, Marina M, De Roberto G (2010) A tribute to Lucio Gullo, MD (1938–2009). Pancreas 39(4):423–424
344. Galassi E, Birtolo C, Migliori M, Bastagli L, Gabusi V, Stanghellini V, De Giorgio R (2014) A 5-year experience of benign pancreatic hyperenzymemia. Pancreas 43(6):874–878
345. Gullo L (2000) Familial pancreatic hyperenzymemia. Pancreas 20(2):158–160
346. Mariani A (2010) Chronic asymptomatic pancreatic hyperenzymemia: is it a benign anomaly or a disease? JOP. J Pancreas 11(2):95–98
347. Gullo L, Lucrezio L, Migliori M, Bassi M, Nesticò V, Costa PL (2008) Benign pancreatic hyperenzymemia or Gullo's syndrome. Adv Med Sci 53(1):1–5
348. Gullo L, Migliori M (2007) Benign pancreatic hyperenzymemia in children. Eur J Pediatr 166:125–129
349. Gullo L (2007) Day-to-day variations of serum pancreatic enzymes in benign pancreatic hyperenzymemia. Clin Gastroenterol Hepathol 5(1):70–74
350. Türkvatan A, Erden A, Türkoğlu MA, Yener Ö (2013) Congenital variants and anomalies of the pancreas and pancreatic duct: imaging by magnetic resonance cholangiopancreaticography and multidetector computed tomography. Korean J Radiol 14(6):905–913

351. Alexander LF (2012) Congenital pancreatic anomalies, variants, and conditions. Radiol Clin North Am 50(3):487–498
352. Borghei P, Sokhandon F, Shirkhoda A, Morgan DE (2013) Anomalies, anatomic variants, and sources of diagnostic pitfalls in pancreatic imaging. Radiology 266(1):28–36
353. Dinter D, Löhr JM, Neff KW (2007) Bifid tail of the pancreas: benign bifurcation anomaly. AJR Am J Roentgenol 189(5):W251–W253
354. Kanne JP, Rohrmann CA, Lichtenstein JE (2006) Eponyms in radiology of the digestive tract: historical perspectives and imaging appearances. Part 2. Liver, biliary system, pancreas, peritoneum, and systemic disease. Radiographics 26(2):465–480
355. Avisse C, Flament J-B, Delattre J-F (2000) Ampulla of Vater: anatomic, embryologic, and surgical aspects. Surg Clin North Am 80(1):201–212
356. Kamisawa T, Koike M, Okamoto A (1999) Embryology of the pancreatic duct system. Digestion 60(2):161–165
357. Kamisawa T, Takuma K, Egawa N, Tsuruta K, Sasaki T (2010) A new embryological theory of the pancreatic duct system. Dig Surg 27(2):132–136
358. Stern CD (1986) A historical perspective on the discovery of the accessory duct of the pancreas, the ampulla of Vater and pancreas divisum. Gut 27:203–212
359. Wang D-B, Yu J, Fulcher AS, Turner MA (2013) Pancreatitis in patients with pancreas divisum: Imaging features at MRI and MRCP. World J Gastroenterol 19(30):4907–4916
360. Largman C, Brodrick JW, Geokas MC (1976) Purification and characterization of two human pancreatic elastases. Biochemistry 15(11):2491–2500
361. Gonzales AC, Vieira SM, Maurer RL, Silva FA, Silveira TR (2011) Use of monoclonal faecal elastase-1 concentration for pancreatic status assessment in cystic fibrosis patients. J Pediatr (Rio J) 87(2):157–162
362. Löser C, Möllgaard A, Fölsch UR (1996) Faecal elastase 1: a novel, highly sensitive, and specific tubeless pancreatic function test. Gut 39(4):580–586
363. Allan BJ, Zager NI, Keller PJ (1970) Human pancreatic proteins: Amylase, proelastase, and trypsinogen. Archiv Biochem Biophys 136(2):529–540
364. Shamamian P, Goldberg JD, Ye XY, Stewart JD, White PJ, Gilvarg C (2006) Evaluation of pro-carboxypeptidase A and carboxypeptidase A as serologic markers for adenocarcinoma of the pancreas. HPB (Oxford) 8(6):451–457
365. Müller CA, Appelros S, Uhl W, Büchler W, Borgström A (2002) Serum levels of procarboxypeptidase B and its activation peptide in patients with acute pancreatitis and non-pancreatic diseases. Gut 51(2):229–235
366. Kemik O, Kemik AS, Sumer A, Beğenik H, Dülger AC, Purisa S, Tuzun S (2012) Serum procarboxypeptidase A and carboxypeptidase A levels in pancreatic disease. Hum Exp Toxicol 31(5):447–451
367. Marlow VL, Cianfanelli FR, Porter M, Cairns LS, Dale JK, Stanley-Wall NR (2014) The prevalence and origin of exoprotease-producing cells in the Bacillus subtilis biofilm. Microbiology 160(Pt 1):56–66
368. Block H, Maertens B, Spriestersbach A, Kubicek J, Schäfer F (2015) Proteolytic Affinity Tag Cleavage. Methods Enzymol 559:71–97
369. Villanueva J, Shaffer DR, Philip J, Chaparro CA, Erdjument-Bromage H, Olshen AB et al (2006) Differential exoprotease activities confer tumor-specific serum peptidome patterns. J Clin Invest 116(1):271–284
370. Finlay M-F, McCloud L (1990) Intestinal leucine aminopeptidase and alkaline phosphatase: Genetic regulation and development in mice. Biochem Genet 28(5–6):267–281
371. Stefanovic V, Vlahovic P, Ardaillou N, Ronco P, Ardaillou R (1992) Cell surface aminopeptidase A and N activities in human glomerular epithelial cells. Kidney Int 41:1571–1580
372. Shapiro LH, Ashmun RA, Roberts WM, Look AT (1991) Separate promoters control transcription of the human aminopeptidase N gene in myeloid and intestinal epithelial cells. J Biol Chem 266(18):11999–12007

373. Nitta S, Komatsu A, Ishii T, Iwamoto H, Numata K (2016) Synthesis of peptides with narrow molecular weight distributions via exopeptidase-catalyzed aminolysis of hydrophobic amino-acid alkyl esters. Polym J 48:955–961
374. Stressler T, Ewert J, Merz M, Funk J, Claaßen W, Lutz-Wahl S et al (2016) A novel glutamyl (aspartyl)-specific aminopeptidase a from Lactobacillus delbrueckii with promising properties for application. PLoS ONE 11(3):e0152139
375. Gomez S, Gluschankof P, Lepage A, Cohen P (1988) Relationship between endo- and exopeptidases in a processing enzyme system: activation of an endoprotease by the aminopeptidase B-like activity in somatostatin-28 convertase. Proc Natl Acad Sci U S A 85 (15):5468–5472
376. Yang W (2011) Nucleases: diversity of structure, function and mechanism. Q Rev Biophys 44(1):1–93
377. Dhananjaya BL, Souza DCJ (2010) An overview on nucleases (DNase, RNase, and phosphodiesterase) in snake venoms. Biochemistry (Mosc) 75(1):1–6
378. Tolun G, Myers RS (2003) A real-time DNase assay (ReDA) based on PicoGreen® fluorescence. Nucleic Acids Res 31(18):e111
379. Liu Y, Zhang Y, Dong P, An R, Xue C, Ge Y et al (2015) Digestion of nucleic acids starts in the stomach. Sci Rep 5:11936
380. Seno M, Futami J, Kosaka M, Seno S, Yamada H (1994) Nucleotide sequence encoding human pancreatic ribonuclease. Biochim Biophys Acta 1218(3):466–468
381. Baril E, Mitchener J, Lee L, Baril B (1977) Action of pancreatic DNase: requirements for activation of DNA as a template-primer for DNA polymerase. Nucleic Acids Res 4(8):2641–2653
382. Zhu B, Zhang L, Zhang Y-Y, Wang L, Li X-G, Liu T et al (2016) DNase I aggravates islet β-cell apoptosis in type 2 diabetes. Mol Med Rep 13(6):4577–4584
383. Fernández-Zapico ME (2010) Never conclude with a negative result, explore all possibilities before changing your hypothesis: an interview with Dr. Catherine Figarella, Former Director Groupe de Recherche sur lesGlandes Exocrines, Faculté de Médecine, Marseille, France; Active Member of the Board ofthe French Cystic Fibrosis Association: Vaincre la Mucoviscidose. Pancreatology 10:109–111
384. Ryan JW, Moffat JG, Thompson AG (1964) Role of bradykinin in the development of acute pancreatitis. Nature 204:1212–1213
385. Goldberg RF, Austen WG Jr, Zhang X, Munene G, Mostafa G, Biswas S et al (2008) Intestinal alkaline phosphatase is a gut mucosal defense factor maintained by enteral nutrition. Proc Natl Acad Sci U S A 105(9):3551–3556
386. Lallès J-P (2014) Intestinal alkaline phosphatase: novel functions and protective effects. Nutr Rev 72(2):82–94
387. Whyte MP (2010) Physiological role of alkaline phosphatase explored in hypophosphatasia. Ann NY Acad Sci 1192:190–200
388. Millán JL, Whyte MP (2016) Alkaline phosphatase and hypophosphatasia. Calcif Tissue Int 98:398–416
389. Argiles JM, Lopez-Soriano FJ (1990) Intestinal amino acid transport: an overview. Int J Biochem 22(9):931–937
390. Daniel H, Kottra G (2004) The proton oligopeptide cotransporter family SLC15 in physiology and pharmacology. Pflugers Arch 447(5):610–618
391. Reimer RJ, Edwards RH (2004) Organic anion transport is the primary function of the SLC17/type I phosphate transporter family. Pflugers Arch 447(5):629–635
392. Eiden LE, Schafer MK, Weihe E, Schutz B (2004) The vesicular amine transporter family (SLC18): amine/proton antiporters required for vesicular accumulation and regulated exocytotic secretion of monoamines and acetylcholine. Pflugers Arch 447(5):636–640
393. Palmieri F (2004) The mitochondrial transporter family (SLC25): physiological and pathological implications. Pflugers Arch 447(5):689–709
394. Mount DB, Romero MF (2004) The SLC26 gene family of multifunctional anion exchangers. Pflugers Arch 447(5):710–721

395. Gasnier B (2004) The SLC32 transporter, a key protein for the synaptic release of inhibitory amino acids. Pflugers Arch 447(5):752–755
396. Boll M, Daniel H, Gasnier B (2004) The SLC36 family: proton-coupled transporters for the absorption of selected amino acids from extracellular and intracellular proteolysis family. Pflugers Arch 447(5):776–779
397. Mackenzie B, Erickson JD (2004) Sodium-coupled neutral amino acid (System N/A) transporters of the SLC38 gene family. Pflugers Arch 447(5):784–795
398. Munck LK (1995) Chloride dependent amino acid transport in the human small intestine. Gut 36:215–219
399. Jenstad M, Chaudhry FA (2013) The amino acid transporters of the glutamate/GABA-glutamine cycle and their impact on insulin and glucagon secretion. Front Endocrinol 4:199
400. Bröer S (2008) Amino acid transport across mammalian intestinal and renal epithelia. Physiol Rev 88(1):249–286
401. Reimer RJ, Chaudhry FA, Gray AT, Edwards RH (2000) Amino acid transport system A resembles system N in sequence but differs in mechanism. PNAS 97(14):7715–7720
402. Boudko DY (2012) Molecular basis of essential amino acid transport from studies of insect nutrient amino acid transporters of the SLC6 family (NAT-SLC6). J Insect Physiol 58 (4):433–449
403. Wagner CA, Lang F, Bröer S (2001) Function and structure of heterodimeric amino acid transporters. Am J Physiol Cell Physiol 281(4):C1077–C1093
404. Oxender DL, Christensen HN (1963) Distinct mediating systems for the transport of neutral amino acids by the Ehrlich cell. J Biol Chem 238:3686–3699
405. Kekuda R, Prasad PD, Fei YJ, Torres-Zamorano V, Sinha S, Yang-Feng TL et al (1996) Cloning of the sodium-dependent, broad-scope, neutral amino acid transporter Bo from a human placental choriocarcinoma cell line. J Biol Chem 271(31):18657–18661
406. Kekuda R, Torres-Zamorano V, Fei YJ, Prasad PD, Li HW, Mader LD et al (1997) Molecular and functional characterization of intestinal Na(+)-dependent neutral amino acid transporter B0. Am J Physiol 272(6 Pt 1):G1463–G1472
407. Pochini L, Scalise M, Galluccio M, Indiveri C (2014) Membrane transporters for the special amino acid glutamine: structure/function relationships and relevance to human health. Front Chem 2:61
408. Baerlocher KE, Scriver CR, Mohyuddin F (1970) Ontogeny of iminoglycine transport in mammalian kidney. Proc Natl Acad Sci U S A 65(4):1009–1016
409. Revsin B, Morrow G 3rd (1979) Imino acid transport in human diploid fibroblasts. Exp Cell Res 119(1):55–61
410. Gilbert ER, Wong EA, Webb KE Jr (2008) Board-invited review: Peptide absorption and utilization: implications for animal nutrition and health. J Anim Sci 86(9):2135–2155
411. Mordrelle A, Jullian E, Costa C, Cormet-Boyaka E, Benamouzig R, Tomé D, Huneau J-F (2000) EAAT1 is involved in transport of L-glutamate during differentiation of the Caco-2 cell line. Am J Physiol Gastrointest Liver Physiol 279(2):G366–G373
412. Vermeulen MAR, de Jong J, Vaessen MJ, van Leeuwen PAM, Houdijk APJ (2011) Glutamate reduces experimental intestinal hyperpermeability and facilitates glutamine support of gut integrity. World J Gastroenterol 17(12):1569–1573
413. Mariotta L, Ramadan T, Singer D, Guetg A, Herzog B, Stoeger C et al (2012) T-type amino acid transporter TAT1 (Slc16a10) is essential for extracellular aromatic amino acid homeostasis control. J Physiol 590(24):6413–6424
414. Ramadan T, Camargo SM, Herzog B, Bordin M, Pos KM, Verrey F (2007) Recycling of aromatic amino acids via TAT1 allows efflux of neutral amino acids via LAT2-4F2hc exchanger. Pflugers Arch 454(3):507–516
415. Fraga S, Pinho M, Soares-da-Silva P (2005) Expression of LAT1 and LAT2 amino acid transporters in human and rat intestinal epithelial cells. Amino Acids 29(3):229–233

416. Bröer A, Wagner CA, Lang F, Bröer S (2000) The heterodimeric amino acid transporter 4F2hc/y+LAT2 mediates arginine efflux in exchange with glutamine. Biochem J 349(3):787–795
417. Pfeiffer R, Rossier G, Spindler B, Meier C, Kühn L, Verrey F (1999) Amino acid transport of y+L-type by heterodimers of 4F2hc/CD98 and members of the glycoprotein-associated amino acid transporter family. EMBO J 18(1):49–57
418. Rossier G, Meier C, Bauch C, Summa V, Sordat B, Verrey F, Kühn LC (1999) LAT2, a new basolateral 4F2hc/CD98-associated amino acid transporter of kidney and intestine. J Biol Chem 274(49):34948–34954
419. Bröer A, Friedrich B, Wagner CA, Fillon S, Ganapathy V, Lang F, Bröer S (2001) Association of 4F2hc with light chains LAT1, LAT2 or y+LAT2 requires different domains. Biochem J 355(3):725–731
420. del Amo EM, Urtti A, Yliperttula M (2008) Pharmacokinetic role of L-type amino acid transporters LAT1 and LAT2. Eur J Pharm Sci 35(3):161–174
421. Eggermann T, Venghaus A, Zerres K (2012) Cystinuria: an inborn cause of urolithiasis. Orphanet J Rare Dis 7:19
422. Claes DJ, Jackson E (2012) Cystinuria: mechanisms and management. Pediatr Nephrol 27(11):2031–2038
423. Tahmoush AJ, Alpers DH, Feigin RD, Armbrustmacher V, Prensky AL (1976) Hartnup disease. Clinical, pathological, and biochemical observations. Arch Neurol 33(12):797–807
424. Garcia ML, Benavides J, Valdivieso F (1980) Ketone body transport in renal brush border membrane vesicles. Biochim Biophys Acta Biomembr 600(3):922–930
425. Ding F, Yao J, Rettberg JR, Chen S, Brinton RD (2013) Early decline in glucose transport and metabolism precedes shift to ketogenic system in female aging and Alzheimer's mouse brain: implication for bioenergetic intervention. PLoS ONE 8(11):e79977
426. Prins ML (2012) Cerebral ketone metabolism during development and injury. Epilepsy Res 100(3):218–223
427. de Romo AC (1989) Tallow and the time capsule: Claude Bernard's discovery of the pancreatic digestion of fat. Hist Philos Life Sci 11(2):253–274
428. Kim S-H, Park I-H, Lee S-C, Lee Y-S, Yi Z, Kim TO et al (2008) Discovery of three novel lipase (lipA1, lipA2, and lipA3) and lipase-specific chaperone (lipB) genes present in Acinetobacter sp. DYL129. Appl Microbiol Biotechnol 77(5):1041–1051
429. Kim H-K, Lee J-K, Kim H, Oh T-K (1996) Characterization of an alkaline lipase from Proteus vulgaris K80 and the DNA sequence of the encoding gene. FEMS Microbiol Lett 135(1):117–121
430. Kok RG, van Thor JJ, Nugteren-Roodzant IM, Brouwer MBW, Egmond MR, Nudel CB et al (1995) Characterization of the extracellular lipase, LipA, of Acinetobacter calcoaceticus BD413 and sequence analysis of the cloned structural gene. Mol Microbiol 15(5):803–818
431. Johnston FA Jr, Sell HM (1944) Changes in chemical composition of tung kernels during germination. Plant Physiol 19(4):694–698
432. Lowe ME, Rosenblum JL, Strauss AW (1989) Cloning and characterization of human pancreatic lipase cDNA. J Biol Chem 264(33):20042–20048
433. Lowe ME (1994) Pancreatic triglyceride lipase and colipase: insights into dietary fat digestion. Gastroenterology 107(5):1524–1536
434. Gargouri Y, Pieroni G, Moreau H, Ferrato F, Rivière C, Saunière JF et al (1986) Lipases of the digestive system. Reprod Nutr Dev 26(5B):1163–1176
435. Douglas GJ Jr, Reinauer AJ, Brooks WC, Pratt JH (1953) The effect on digestion and absorption of excluding the pancreatic juice from the intestine. Gastroenterology 23(3):452–459
436. Bank S, Krut LH, Marks IN, Bronte-Stewart B, Uys PJ (1964) Hydrolysis of fat by human gastric juice. Gut 5(5):480–484
437. Darnton SJ, Barrowman J (1969) A specific histochemical method for the determination of pancreatic lipase. Histochem J 1(6):551–557

438. Hamosh M, Clary TR, Chernick SS, Scow RO (1970) Lipoprotein lipase activity of adipose and mammary tissue and plasma triglyceride in pregnant and lactating rats. Biochim Biophys Acta 210(3):473–482
439. Baskys B, Klein E, Lever WF (1963) Lipases of blood and tissue. Purification and properties of pancreatic lipase. Arch Biochem Biophys 102:201–209
440. Cohen M, Morgan RG, Hofmann AF (1971) Lipolytic activity of human gastric and duodenal juice against medium and long chain triglycerides. Gastroenterology 60(1):1–15
441. Barrowman JA, Darnton SJ (1970) The lipase of rat gastric mucosa. A histochemical demonstration of the enzymatic activity against a medium chain triglyceride. Gastroenterology 59:13–21
442. Hamosh M, Scow RO (1973) Lingual lipase and its role in the digestion of dietary lipid. J Clin Invest 52:88–95
443. Newman L, Haryono R, Keast R (2013) Functionality of fatty acid chemoreception: a potential factor in the development of obesity? Nutrients 5(4):1287–1300
444. Hayes JR, Pence DH, Scheinbach S, D'Amelia RP, Klemann LP, Wilson NH, Finley JW (1994) Review of triacylglycerol digestion, absorption, and metabolism with respect to Salatrim triacylglycerols. J Agric Food Chem 42(2):474–483
445. Voigt N, Stein J, Galindo MM, Dunkel A, Raguse JD, Meyerhof W et al (2014) The role of lipolysis in human orosensory fat perception. J Lipid Res 55(5):870–882
446. Turki S, Kallel H (2012) Emerging approaches for the treatment of fat malabsorption due toexocrine pancreatic insufficiency. In: Brzozowski T (ed) New advances in the basic and clinical gastroenterology. InTech, Rijeka, Croatia
447. Fink CS, Hamosh P, Hamosh M (1984) Fat digestion in the stomach: stability of lingual lipase in the gastric environment. Pediatr Res 18(3):248–254
448. Moreau H, Bernadac A, Gargouri Y, Benkouka F, Laugier R, Verger R (1989) Immunocytolocalization of human gastric lipase in chief cells of the fundic mucosa. Histochemistry 91(5):419–423
449. Hamosh M (1990) Lingual and gastric lipases. Nutrition 6(6):421–428
450. Kawai T, Fushiki T (2003) Importance of lipolysis in oral cavity for orosensory detection of fat. Am J Physiol Regul Integr Comp Physiol 285(2):R447–R454
451. Sedlakova A, Kohút A, Sarisský M (2001) Changes of gastric lipase activity after ethanol and indomethacin administration: influence of pretreatment with allopurinol, pentoxifylline and L-DOPA. Physiol Res 50:299–307
452. DeNigris SJ, Hamosh M, Kasbekar DK, Lee TC, Hamosh P (1988) Lingual and gastric lipases: species differences in the origin of prepancreatic digestive lipases and in the localization of gastric lipase. Biochim Biophys Acta Lipids Lipid Metab 959(1):38–45
453. Carrière F, Grandval P, Gregory PC, Renou C, Henniges F, Sander-Struckmeier S, Laugier R (2005) Does the pancreas really produce much more lipase than required for fat digestion? JOP 6(3):206–215
454. Park M-J, Lin L, Thomas S, Braymer HD, Smith PM, Harrison DHT, York DA (2004) The F1-ATPase β-subunit is the putative enterostatin receptor. Peptides 25(12):2127–2133
455. Dittrich M, Schulten K (2005) Zooming in on ATP hydrolysis in F1. J Bioenerg Biomembr 37:441–444
456. Dittrich M, Hayashi S, Schulten K (2004) ATP hydrolysis in the bTP and bDP catalytic sites of F1-ATPase. Biophys J 87:2954–2967
457. Dittrich M, Hayashi S, Schulten K (2003) On the mechanism of ATP hydrolysis in F1-ATPase. Biophys J 85:2253–2266
458. Lin L, Park M, York DA (2007) Enterostatin inhibition of dietary fat intake is modulated through the melanocortin system. Peptides 28(3):643–649
459. Park M-J, Farrell J, Lemmon K, York DA (2009) Enterostatin alters protein trafficking to inhibit insulin secretion in Beta-TC6 cells. Peptides 30(10):1866–1873
460. Berger K, Winzell MS, Mei J, Erlanson-Albertsson C (2004) Enterostatin and its target mechanisms during regulation of fat intake. Physiol Behav 83(4):623–630

461. Davis RC, Xia YR, Mohandas T, Schotz MC, Lusis AJ (1991) Assignment of the human pancreatic colipase gene to chromosome 6p21.1 to pter. Genomics 10(1):262–265
462. Rosenheim O (1910) On pancreatic lipase. The separation of lipase from its co-enzyme. J Physiol 15:14–16
463. King H (1956) Sigmund Otto Rosenheim. 1871–1955. Biogr Mem Fellows R Soc 2:256–267
464. Maylie MF, Charles M, Cache C, Desnuelle P (1971) Isolation and partial identification of a pancreatic colipase. Biochim Biophys Acta 229:286–289
465. Borgstrom B, Erlanson-Albertsson C, Wieloch T (1979) Pancreatic colipase: chemistry and physiology. J Lipid Res 20(7):805–816
466. van Tilbeurgh H, Bezzine S, Cambillau C, Verger R, Carrière F (1999) Colipase: structure and interaction with pancreatic lipase. Biochim Biophys Acta Mol Cell Biol Lipids 1441(2–3):173–184
467. Crandall WV, Lowe ME (2001) Colipase residues Glu64 and Arg65 are essential for normal lipase-mediated fat digestion in the presence of bile salt micelles. J Biol Chem 276 (16):12505–12512
468. Chapus C, Rovery M, Sarda L, Verger R (1988) Minireview on pancreatic lipase and colipase. Biochimie 70(9):1223–1234
469. Bezzine S, Ferrato F, Ivanova MG, Lopez V, Verger R, Carrière F (1999) Human pancreatic lipase: colipase dependence and interfacial binding of lid domain mutants. Biochemistry 38 (17):5499–5510
470. John S, Thangapandian S, Sakkiah S, Lee KW (2010) Discovery of potential pancreatic cholesterol esterase inhibitors using pharmacophore modelling, virtual screening, and optimization studies. J Enzyme Inhib Med Chem 26(4):535–545
471. Karmanskiĭ IM, Pichugin AL, Iusupova GI, Sysoev IuA (1981) Isolation of cholesterol esterase from the pancreatic juice of dogs. Biull Eksp Biol Med 91(3):330–332
472. Heidrich JE, Contos LM, Hunsaker LA, Deck LM, Vander Jagt DL (2004) Inhibition of pancreatic cholesterol esterase reduces cholesterol absorption in the hamster. BMC Pharmacol 4:5
473. Hui TY, Bernlohr DA (1997) Fatty acid transporters in animal cells. Front Biosci 2:d222–d231
474. Tso P, Fujimoto K (1991) The absorption and transport of lipids by the small intestine. Brain Res Bull 27(3–4):477–482
475. Orsavova J, Misurcova L, Ambrozova J, Vicha R, Mlcek J (2015) Fatty acids composition of vegetable oils and its contribution to dietary energy intake and dependence of cardiovascular mortality on dietary intake of fatty acids. Int J Mol Sci 16:12871–12890
476. Food and Agriculture Organization of the United Nations. Food and Agriculture Organization of the United Nations (FAO) (2017) Oilcrops complex: policy changes and industry measures, Annual compendium 2016. FAO, Roma
477. Food and Agriculture Organization of the United Nations (FAO) (2017) Food outlook: biannual report on global food markets. FAO, Rome
478. US Department of Agriculture, Agricultural Research Service, Nutrient Data Laboratory (2015) USDA national nutrient database for standard reference. United States, Washington, DC
479. McNaught AD, Wilkinson A (1997) IUPAC compendium of chemical terminology (the gold book), 2nd edn. International Union of Pure and Applied Chemistry (IUPAC) Compendium of Chemical Terminology. Blackwell Scientific Publications, Oxford
480. Park J-M, Kim N-K, Yang C-Y, Moon K-W, Kim J-M (2014) Determination of the authenticity of dairy products on the basis of fatty acids and triacylglycerols content using GC analysis. Korean J Food Sci Anim Resour 34(3):316–324
481. Gupta SV, Yamada N, Fungwe TV, Khosla P (2003) Replacing 40% of dietary animal fat with vegetable oil is associated with lower HDL cholesterol and higher cholesterol ester transfer protein in cynomolgus monkeys fed sufficient linoleic acid. J Nutr 133(8):2600–2606

482. Kim JM, Kim HJ, Park JM (2015) Determination of milk fat adulteration with vegetable oils and animal fats by gas chromatographic analysis. J Food Sci 80(9):C1945–C1951
483. Klonoff DC (2007) Replacements for trans fats—will there be an oil shortage? J Diabetes Sci Technol 1(3):415–422
484. Liao TH, Hamosh P, Hamosh M (1984) Fat digestion by lingual lipase: mechanism of lipolysis in the stomach and upper small intestine. Pediatr Res 18(5):402–409
485. Phan CT, Tso P (2001) Intestinal lipid absorption and transport. Front Biosci 6:D299–D319
486. Drent ML, van der Veen EA (1993) Lipase inhibition: a novel concept in the treatment of obesity. Int J Obes Relat Metab Disord 17(4):241–244
487. Nelson RH, Miles JM (2005) The use of orlistat in the treatment of obesity, dyslipidaemia and Type 2 diabetes. Expert Opin Pharmacother 6(14):2483–2491
488. Dickerson LM, Carek PJ (2000) Drug therapy for obesity. Am Fam Physician 61(7):2131–2138
489. Wang TY, Liu M, Portincasa P, Wang DQ-H (2013) New insights into the molecular mechanism of intestinal fatty acid absorption. Eur J Clin Invest 43(11):1203–1223
490. Harrison EH (2012) Mechanisms involved in the intestinal absorption of dietary vitamin A and provitamin A carotenoids. Biochim Biophys Acta 1821(1):70–77
491. Iqbal J, Hussain MM (2009) Intestinal lipid absorption. Am J Physiol Endocrinol Metab 296:E1183–E1194
492. Goncalves A, Gontero B, Nowicki M, Margier M, Masset G, Amiot M-J, Reboul E (2015) Micellar lipid composition affects micelle interaction with class B scavenger receptor extracellular loops. J Lipid Res 56(6):1123–1133
493. Mansbach CM 2nd, Cohen RS, Leff PB (1975) Isolation and properties of the mixed lipid micelles present in intestinal content during fat digestion in man. J Clin Invest 56(4):781–791
494. Abrahamse E, Minekus M, van Aken GA, van de Heijning B, Knol J, Bartke N et al (2012) Development of the digestive system—experimental challenges and approaches of infant lipid digestion. Food Dig 3(1–3):63–77
495. Dejgaard SY, Presley JF (2014) New automated single-cell technique for segmentation and quantitation of lipid droplets. J Histochem Cytochem 62(12):889–901
496. Mowat AM, Agace WW (2014) Regional specialization within the intestinal immune system. Nat Rev Immunol 14:667–685
497. St-Onge M-P, Jones PJH (2002) Physiological effects of medium-chain triglycerides: potential agents in the prevention of obesity. J Nutr 132(3):329–332
498. Hardy SC, Tanpowpong P, Kleinman RE (2014) Nutrition in children with liver disease: evaluation and management. In: Murray K, Horslen S (eds) Diseases of the liver in children. Springer, New York
499. Ziarnik E, Nesbitt J (2015) Chyle leak after esophageal surgery. In: Pawlik T, Maithel S, Merchant N (eds) Gastrointestinal surgery. Springer, New York
500. Mattes RD (2009) Oral detection of short-, medium-, and long-chain free fatty acids in humans. Chem Senses 34(2):145–150
501. Senior JR (1964) Intestinal absorption of fats. J Lipid Res 5:495–521
502. Kirat D, Kato S (2006) Monocarboxylate transporter 1 (MCT1) mediates transport of short-chain fatty acids in bovine caecum. Exp Physiol 91(5):835–844
503. Sellin JH, De Soignie R (1998) Short-chain fatty acids have polarized effects on sodium transport and intracellular pH in rabbit proximal colon. Gastroenterology 114(4):737–747
504. Kirat D, Masuoka J, Hayashi H, Iwano H, Yokota H, Taniyama H, Kato S (2006) Monocarboxylate transporter 1 (MCT1) plays a direct role in short-chain fatty acids absorption in caprine rumen. J Physiol 576(2):635–647
505. Nedjadi T (2014) Trans-cellular transport of short chain fatty acids in the large intestine. BMC Genom 15(2):P4
506. den Besten G, van Eunen K, Groen AK, Venema K, Reijngoud DJ, Bakker BM (2013) The role of short-chain fatty acids in the interplay between diet, gut microbiota, and host energy metabolism. J Lipid Res 54(9):2325–2340

507. Abumrad NA, Davidson NO (2012) Role of the gut in lipid homeostasis. Physiol Rev 92 (3):1061–1085
508. D'Aquila T, Hung Y-H, Carreiro A, Buhman KK (2016) Recent discoveries on absorption of dietary fat: presence, synthesis, and metabolism of cytoplasmic lipid droplets within enterocytes. Biochim Biophys Acta 1861(8):730–747
509. Uchida A, Lee HJ, Cheng JX, Buhman KK (2013) Imaging cytoplasmic lipid droplets in enterocytes and assessing dietary fat absorption. Methods Cell Biol 116:151–166
510. Schwenk RW, Dirkx E, Coumans WA, Bonen A, Klip A, Glatz JFC, Luiken JJFP (2010) Requirement for distinct vesicle-associated membrane proteins in insulin- and AMP-activated protein kinase (AMPK)-induced translocation of GLUT4 and CD36 in cultured cardiomyocytes. Diabetologia 53:2209–2219
511. Abumrad N, Harmon C, Ibrahimi A (1998) 0 Membrane transport of long-chain fatty acids: evidence for a facilitated process. J Lipid Res 39:2309–2318
512. Kazantzis M, Stahl A (2012) Fatty acid transport proteins, implications in physiology and disease. Biochim Biophys Acta Mol Cell Biol Lipids 1821(5): 852–857
513. Glatz JFC, Luiken JJFP, Bonen A (2010) Membrane fatty acid transporters as regulators of lipid metabolism: implications for metabolic disease. Physiol Rev 90:367–417
514. Shim J, Moulson CL, Newberry EP, Lin M-H, Xie Y, Kennedy SM et al (2009) Fatty acid transport protein 4 is dispensable for intestinal lipid absorption in mice. J Lipid Res 50 (3):491–500
515. Tazoe H, Otomo Y, Karaki S, Kato I, Fukami Y, Terasaki M, Kuwahara A (2009) Expression of short-chain fatty acid receptor GPR41 in the human colon. Biomed Res 30 (3):149–156
516. Tazoe H, Otomo Y, Kaji I, Tanaka R, Karaki SI, Kuwahara A (2008) Roles of short-chain fatty acids receptors, GPR41 and GPR43 on colonic functions. J Physiol Pharmacol 59 (2):251–262
517. Ganapathy V, Thangaraju M, Prasad PD, Martin PM, Singh N (2013) Transporters and receptors for short-chain fatty acids as the molecular link between colonic bacteria and the host. Curr Opin Pharmacol 13(6):869–874
518. Moschen I, Bröer A, Galić S, Lang F, Bröer S (2012) Significance of short chain fatty acid transport by members of the monocarboxylate transporter family (MCT). Neurochem Res 37 (11):2562–2568
519. Davis HR Jr, Altmann SW (2009) Niemann-Pick C1 Like 1 (NPC1L1) an intestinal sterol transporter. Biochim Biophys Acta Mol Cell Biol Lipids 1791(7):679–683
520. Kramer W, Girbig F, Corsiero D, Pfenninger A, Frick W, Jähne G et al (2005) Aminopeptidase N (CD13) is a molecular target of the cholesterol absorption inhibitor ezetimibe in the enterocyte brush border membrane. J Biol Chem 280:1306–1320
521. Bays HE, Neff D, Tomassini JE, Tershakovec AM (2008) Ezetimibe: cholesterol lowering and beyond. Expert Rev Cardiovasc Ther 6(4):447–470
522. Fraunberger P, Gröne E, Gröne HJ, Drexel H, Walli AK (2017) Ezetimibe reduces cholesterol content and NF-kappaB activation in liver but not in intestinal tissue in guinea pigs. J Inflamm (Lond) 14:3
523. Le NA, Tomassini JE, Tershakovec AM, Neff DR, Wilson PW (2015) Effect of switching from statin monotherapy to ezetimibe/simvastatin combination therapy compared with other intensified lipid-lowering strategies on lipoprotein subclasses in diabetic patients with symptomatic cardiovascular disease. J Am Heart Assoc 4(10):e001675
524. Reboul E, Borel P (2011) Proteins involved in uptake, intracellular transport and basolateral secretion of fat-soluble vitamins and carotenoids by mammalian enterocytes. Prog Lipid Res 50(4):388–402
525. Ueshima K, Akihisa-Umeno H, Sawada M, Nagayoshi A, Ozaki T, Takakura S, Manda T, Mutoh S (2004) Possible involvement of enhanced intestinal microsomal triglyceride transfer protein (MTP) gene expression in acceleration of lipid absorption by a western-type diet in apolipoprotein E knockout mice. Life Sci 76(2):179–190

526. Swift LL, Kakkad B, Boone C, Jovanovska A, Jerome WG, Mohler PJ, Ong DE (2005) Microsomal triglyceride transfer protein expression in adipocytes: a new component in fat metabolism. FEBS Lett 579(14):3183–3189
527. Beierfuß A, Dietrich H, Kremser C, Hunjadi M, Ritsch A, Rülicke T et al (2017) Knockout of apolipoprotein E in rabbit promotes premature intervertebral disc degeneration: a new in vivo model for therapeutic approaches of spinal disc disorders. PLoS ONE 12(11): e0187564
528. Wetterau JR, Lin MC, Jamil H (1997) Microsomal triglyceride transfer protein. Biochim Biophys Acta 1345:136–150
529. Havel RJ (1995) Chylomicron remnants: hepatic receptors and metabolism. Curr Opin Lipidol 6(5):312–316
530. Feingold KR, Grunfeld C (2000) Introduction to lipids and lipoproteins. In: De Groot LJ, Chrousos G, Dungan K, Feingold KR, Grossman A, Hershman JM et al (eds) Endotext [Internet]. MDText.com Inc, South Dartmouth
531. Röhrl C (1831) Stangl H (2013) HDL endocytosis and resecretion. Biochim Biophys Acta 11:1626–1633
532. Boateng L, Ansong R, Owusu WB, Steiner-Asiedu M (2016) Coconut oil and palm oil's role in nutrition, health and national development: a review. Ghana Med J 50(3):189–196
533. Escribá PV, González-Ros JM, Goñi FM, Kinnunen PKJ, Vigh L, Sánchez-Magraner L et al (2008) Membranes: a meeting point for lipids, proteins and therapies. J Cell Mol Med 12 (3):829–875
534. Bridges CC, Zalups RK (2005) Molecular and ionic mimicry and the transport of toxic metals. Toxicol Appl Pharmacol 204(3):274–308
535. Larsen EH (2002) Hans H. Ussing–scientific work: contemporary significance and perspectives. Biochim Biophys Acta Biomembr 1566(1–2):2–15
536. Lindemann B (2001) Hans Ussing, experiments and models. J Membr Biol 184(3):203–210
537. Hamilton KL (2011) Ussing's "little chamber": 60 years + old and counting. Front Physiol 2:6
538. Larsen EH (2009) Hans Henriksen Ussing. 30 December 1911–22 December 2000. Biogr Mems Fellows R Soc 55:305–335
539. Coetzer H, Claassen N, van Papendorp DH, Kruger MC (1994) Calcium transport by isolated brush border and basolateral membrane vesicles: role of essential fatty acid supplementation. Prostaglandins Leukot Essent Fatty Acids 50(5):257–266
540. Ross AC, Taylor CL, Yaktine AL, Del Valle HB (eds) (2011) Dietary reference intakes for calcium and vitamin D. National Academies Press, Washington
541. Wasserman RH, Chandler JS, Meyer SA, Smith CA, Brindak ME, Fullmer CS et al (1992) Intestinal calcium transport and calcium extrusion processes at the basolateral membrane. J Nutr 122(3):662–671
542. Wasserman RH, Fullmer CS (1995) Vitamin D and intestinal calcium transport: facts, speculations and hypotheses. J Nutr 125(7):1971S–1979S
543. Christakos S (2012) Mechanism of action of 1,25-dihydroxyvitamin D3 on intestinal calcium absorption. Rev Endocr Metab Disord 13(1):39–44
544. Bronner F (2003) Mechanisms of intestinal calcium absorption. J Cell Biochem 88(2):387–393
545. Ghishan FK, Dannan G, Arab N, Kikuchi K (1987) Intestinal maturation: calcium transport by basolateral membranes. Pediatr Res 21(3):257–260
546. Bronner F (2009) Recent developments in intestinal calcium absorption. Nutr Rev 67 (2):109–113
547. Christakos S, Dhawan P, Ajibade D, Benn BS, Feng J, Joshi SS (2010) Mechanisms involved in vitamin D mediated intestinal calcium absorption and in non-classical actions of vitamin D. J Steroid Biochem Mol Biol 121(1–2):183–187
548. Trent JT III, Watts RA, Hargrove MS (2001) Human neuroglobin, a hexacoordinate hemoglobin that reversibly binds oxygen. J Biol Chem 276:30106–30110

549. Tiso M, Tejero J, Basu S, Azarov I, Wang X, Simplaceanu V et al (2011) Human neuroglobin functions as a redox regulated nitrite reductase. J Biol Chem 286(20):18277–18289
550. Abbaspour N, Hurrell R, Kelishadi R (2014) Review on iron and its importance for human health. J Res Med Sci 19(2):164–174
551. Linder MC (2013) Mobilization of stored iron in mammals: a review. Nutrients 5(10):4022–4050
552. Muñoz M, Villar I, García-Erce JA (2009) An update on iron physiology. World J Gastroenterol 15(37):4617–4626
553. Sheftel AD, Mason AB, Ponka P (2012) The long history of iron in the universe and in health and disease. Biochim Biophys Acta 1820(3):161–187
554. Santiago P (2012) Ferrous versus ferric oral iron formulations for the treatment of iron deficiency: a clinical overview. Sci World J 2012:846824
555. Short MW, Domagalski JE (2013) Iron deficiency anemia: evaluation and management. Am Fam Physician 87(2):98–104
556. McKie AT, Marciani P, Rolfs A, Brennan K, Wehr K, Barrow D et al (2000) A novel duodenal iron-regulated transporter, IREG1, implicated in the basolateral transfer of iron to the circulation. Mol Cell 5(2):299–309
557. Ascenzi P, Bocedi A, Visca P, Altruda F, Tolosano E, Beringhelli T, Fasano M (2005) Hemoglobin and heme scavenging. IUBMB Life 57(11):749–759
558. Latunde-Dada GO, Simpson RJ, McKie AT (2008) Duodenal cytochrome B expression stimulates iron uptake by human intestinal epithelial cells. J Nutr 138(6):991–995
559. Mackenzie B, Garrick MD (2005) Iron imports. II. Iron uptake at the apical membrane in the intestine. Am J Physiol Gastrointest Liver Physiol 289(6):G981–G986
560. Vulpe CD, Kuo Y-M, Murphy TL, Cowley L, Askwith C, Libina N et al (1999) Hephaestin, a ceruloplasmin homologue implicated in intestinal iron transport, is defective in the sla mouse. Nat Genet 21:195–199
561. Avivi A, Gerlach F, Joel A, Reuss S, Burmester T, Nevo E, Hankeln T (2010) Neuroglobin, cytoglobin, and myoglobin contribute to hypoxia adaptation of the subterranean mole rat Spalax. Proc Natl Acad Sci U S A 107(50):21570–21575
562. Tallkvist J, Bowlus CL, Lönnerdal B (2003) Effect of iron treatment on nickel absorption and gene expression of the divalent metal transporter (DMT1) by human intestinal Caco-2 cells. Pharmacol Toxicol 92(3):121–124
563. Shawki A, Knight PB, Maliken BD, Niespodzany EJ, Mackenzie B (2012) H(+)-coupled divalent metal-ion transporter-1: functional properties, physiological roles and therapeutics. Curr Top Membr 70:169–214
564. Mackenzie B, Ujwal ML, Chang MH, Romero MF, Hediger MA (2006) Divalent metal-ion transporter DMT1 mediates both H^+-coupled $Fe2^+$ transport and uncoupled fluxes. Pflugers Arch 451(4):544–558
565. Kuo YM, Su T, Chen H, Attieh Z, Syed BA, McKie AT et al (2004) Mislocalisation of hephaestin, a multicopper ferroxidase involved in basolateral intestinal iron transport, in the sex linked anaemia mouse. Gut 53(2):201–206
566. Krishnamurthya P, Xieb T, Schuetza JD (2007) The role of transporters in cellular heme and porphyrin homeostasis. Pharmacol Ther 114(3):345–358
567. Fiorito V, Geninatti Crich S, Silengo L, Aime S, Altruda F, Tolosano E (2013) Lack of plasma protein hemopexin results in increased duodenal iron uptake. PLoS ONE 8(6): e68146
568. Wu W, Song Y, He C, Liu C, Wu R, Fang L et al (2015) Divalent metal-ion transporter 1 is decreased in intestinal epithelial cells and contributes to the anemia in inflammatory bowel disease. Sci Rep 5:16344
569. Philip M, Zaballa A, Phelps S, Abkowitz J (2011) FLVCR, a heme export protein, is required for T cell development beyond the CD4+ CD8+ double-positive stage. J Immunol 186:18–64

570. Bakken AF, Thaler MM, Schmid R (1972) Metabolic regulation of heme catabolism and bilirubin production. I. Hormonal control of hepatic heme oxygenase activity. J Clin Invest 51(3):530–536
571. Przybyszewska J, Żekanowska E (2014) The role of hepcidin, ferroportin, HCP1, and DMT1 protein in iron absorption in the human digestive tract. Prz Gastroenterol 9(4):208–213
572. Tailor CS, Willett BJ, Kabat D (1999) A putative cell surface receptor for anemia-inducing feline leukemia virus subgroup C is a member of a transporter superfamily. J Virol 73 (8):6500–6505
573. D'Angelo G (2013) Role of hepcidin in the pathophysiology and diagnosis of anemia. Blood Res 48(1):10–15
574. Rossi E (2005) Hepcidin—the iron regulatory hormone. Clin Biochem Rev 26(3):47–49
575. Fuqua BK, Lu Y, Darshan D, Frazer DM, Wilkins SJ, Wolkow N et al (2014) The multicopper ferroxidase hephaestin enhances intestinal iron absorption in mice. PLoS ONE 9 (6):e98792
576. Vashchenko G, MacGillivray RTA (2013) Multi-copper oxidases and human iron metabolism. Nutrients 5:2289–2313
577. Ramírez-Cárdenas L, Costa NMB, Reis FP (2005) Copper-iron metabolism interaction in rats. Nutr Res 25(1):79–92
578. Aaseth J, Gerhardsson L, Skaug MA, Alexander J (2016) General chemistry of metal toxicity and basis for metal complexation. In: Aaseth J, Crisponi G, Anderson O (eds) Chelation therapy in the treatment of metal intoxication. Elsevier, Oxford
579. Ponka P, Lok CN (1999) 0 The transferrin receptor: role in health and disease. Int J Biochem Cell Biol 31(10):1111–1137
580. Delanghe JR, Langlois MR (2001) Hemopexin: a review of biological aspects and the role in laboratory medicine. Clin Chim Acta 312(1–2):13–23
581. Tolosano E, Altruda F (2004) Hemopexin: structure, function, and regulation. DNA Cell Biol 21(4):297–306
582. Smith A, Morgan WT (1978) Transport of heme by hemopexin to the liver: evidence for receptor-mediated uptake. Biochem Biophys Res Commun 84(1):151–157
583. Schweigel M, Martens H (2000) Magnesium transport in the gastrointestinal tract. Front Biosci 5:D666–D677
584. Hardwick LL, Jones MR, Brautbar N, Lee DB (1990) Site and mechanism of intestinal magnesium absorption. Miner Electrolyte Metab 16(2–3):174–180
585. Leonhard-Marek S, Stumpff F, Brinkmann I, Breves G, Martens H (2005) Basolateral Mg^{2+}/Na^{+} exchange regulates apical nonselective cation channel in sheep rumen epithelium via cytosolic Mg^{2+}. Am J Physiol Gastrointest Liver Physiol 288(4):G630–G645
586. Wang X, Zhou B (2010) Dietary zinc absorption: A play of Zips and ZnTs in the gut. IUBMB Life 62(3):176–182
587. Sekler I, Sensi SL, Hershfinkel M, Silverman WF (2007) Mechanism and regulation of cellular zinc transport. Mol Med 13(7–8):337–343
588. Cragg RA, Phillips SR, Piper JM, Varma JS, Campbell FC, Mathers JC, Ford D (2005) Homeostatic regulation of zinc transporters in the human small intestine by dietary zinc supplementation. Gut 54(4):469–478
589. Kambe T, Hashimoto A, Fujimoto S (2014) Current understanding of ZIP and ZnT zinc transporters in human health and diseases. Cell Mol Life Sci 71(17):3281–3295
590. Liuzzi JP, Cousins RJ (2004) Mammalian zinc transporters. Annu Rev Nutr 24:151–172
591. Cousins RJ, Liuzzi JP, Lichten LA (2006) Mammalian zinc transport, trafficking, and signals. J Biol Chem 281:24085–24089
592. Cousins RJ (2010) Gastrointestinal factors influencing zinc absorption and homeostasis. Int J Vitam Nutr Res 80:243–248
593. Krebs NF (2000) Overview of zinc absorption and excretion in the human gastrointestinal tract. J Nutr 130(5):1374S–1377S
594. Myers S, Shastri MD, Adulcikas J, Sohal SS, Norouzi S (2017) Zinc and gastrointestinal disorders: a role for the zinc transporters Zips and ZnTs. Curr Pharm Des 23(16):2328–2332

595. Maywald M, Wessels I, Rink L (2017) Zinc signals and immunity. Int J Mol Sci 18(10): E2222
596. Garrick MD, Dolan KG, Horbinski C, Ghio AJ, Higgins D, Porubcin M et al (2003) DMT1: a mammalian transporter for multiple metals. Biometals 16(1):41–54
597. Przybyłkowski A, Gromadzka G, Wawer A, Grygorowicz T, Cybulska A, Członkowska A (2013) Intestinal expression of metal transporters in Wilson's disease. Biometals 26(6):925–934
598. Lutsenko S, Barnes NL, Bartee MY, Dmitriev OY (2007) Function and regulation of human copper-transporting ATPases. Physiol Rev 87(3):1011–1046
599. Mercer SW, Wang J, Burke R (2017) In vivo modeling of the pathogenic effect of copper transporter mutations that cause menkes and wilson diseases, motor neuropathy, and susceptibility to alzheimer's disease. J Biol Chem 292(10):4113–4122
600. Ariöz C, Li Y, Wittung-Stafshede P (2017) The six metal binding domains in human copper transporter, ATP7B: molecular biophysics and disease-causing mutations. Biometals 30 (6):823–840
601. Gupta A, Lutsenko S (2009) Human copper transporters: mechanism, role in human diseases and therapeutic potential. Future Med Chem 1(6):1125–1142
602. Jayakanthan S, Braiterman LT, Hasan NM, Unger VM, Lutsenko S (201) Human copper transporter ATP7B (Wilson disease protein) forms stable dimers in vitro and in cells. J Biol Chem 292(46):18760–18774
603. Markovich D (2010) Sulfate and phosphate transporters in mammalian renal and gastrointestinal systems. In: Gerencser GA (ed) Epithelial transport physiology. Humana Press, New York
604. Kiela PR, Ghishan FK (2016) Physiology of intestinal absorption and secretion. Best Pract Res Clin Gastroenterol 30(2):145–159
605. Nozawa T, Sugiurac S, Hashino Y, Tsuji A, Tamai I (2004) Role of anion exchange transporter PAT1 (SLC26A6) in intestinal absorption of organic anions. J Drug Target 12 (2):97–104
606. Hediger MA, Clémençon B, Burrier RE, Bruford EA (2013) The ABCs of membrane transporters in health and disease (SLC series): introduction. Mol Aspects Med 34(2–3):95–107
607. Knauf F, Ko N, Jiang Z, Robertson WG, Van Itallie CM, Anderson JM, Aronson PS (2011) Net intestinal transport of oxalate reflects passive absorption and SLC26A6-mediated secretion. J Am Soc Nephrol 22(12):2247–2255
608. Hatch M, Freel RW (2008) The roles and mechanisms of intestinal oxalate transport in oxalate homeostasis. Sem Nephrol 28(2):143–151
609. Field M (2003) Intestinal ion transport and the pathophysiology of diarrhea. J Clin Invest 111(7):931–943
610. Bridges CC, Zalups RK (2005) Molecular and ionic mimicry and the transport of toxic metals. Toxicol Appl Pharmacol 204(3):274–308
611. Kerper LE, Ballatori N, Clarkson TW (1992) Methylmercury transport across the blood-brain barrier by an amino acid carrier. Am J Physiol 262(5 Pt 2):R761–R765
612. Bridges CC, Zalups RK (2017) Mechanisms involved in the transport of mercuric ions in target tissues. Arch Toxicol 91(1):63–81
613. Chmielowska-Bąk J, Izbiańska K, Deckert J (2013) The toxic Doppelganger: on the ionic and molecular mimicry of cadmium. Acta Biochim Pol 60(3):369–374
614. Clarkson TW (1993) Molecular and ionic mimicry of toxic metals. Annu Rev Pharmacol Toxicol 33:545–571
615. Foulkes EC (2000) Transport of toxic heavy metals across cell membranes. Proc Soc Exp Biol Med 223(3):234–240
616. Ballatori N (2002) Transport of toxic metals by molecular mimicry. Environ Health Perspect 110(5):689–694
617. Simmons-Willis TA, Koh AS, Clarkson TW, Ballatori N (2002) Transport of a neurotoxicant by molecular mimicry: the methylmercury-L-cysteine complex is a substrate

for human L-type large neutral amino acid transporter (LAT) 1 and LAT2. Biochem J 367 (1):239–246
618. Oude Elferink RP, de Waart R (2007) Transporters in the intestine limiting drug and toxin absorption. J Physiol Biochem 63(1):75–81
619. Chan LM, Lowes S, Hirst BH (2004) The ABCs of drug transport in intestine and liver: efflux proteins limiting drug absorption and bioavailability. Eur J Pharm Sci 21(1):25–51
620. Pang KS (2003) Modeling of intestinal drug absorption: roles of transporters and metabolic enzymes (for the Gillette review series). DMD 31(12):1507–1519
621. Lemos C, Jansen G, Peters GJ (2008) Drug transporters: recent advances concerning BCRP and tyrosine kinase inhibitors. Br J Cancer 98(5):857–862
622. Dietrich CG, Geier A, Oude Elferink RPJ (2003) ABC of oral bioavailability: transporters as gatekeepers in the gut. Gut 52(12):1788–1795
623. Albermann N, Schmitz-Winnenthal FH, Z'graggen K, Volk C, Hoffmann MM, Haefeli WE, Weiss J (2005) Expression of the drug transporters MDR1/ABCB1, MRP1/ABCC1, MRP2/ ABCC2, BCRP/ABCG2, and PXR in peripheral blood mononuclear cells and their relationship with the expression in intestine and liver. Biochem Pharmacol 70(6):949–58
624. Tsuji A, Tamai I (1996) Carrier-mediated intestinal transport of drugs. Pharm Res 13 (7):963–977
625. Hukkanen J, Hakkola J, Rysä J (2014) Pregnane X receptor (PXR)—a contributor to the diabetes epidemic? Drug Metab Drug Interact 29(1):3–15
626. Tamai I, Nakanishi T, Hayashi K, Terao T, Sai Y, Shiraga T et al (1997) The predominant contribution of oligopeptide transporter PepT1 to intestinal absorption of beta-lactam antibiotics in the rat small intestine. J Pharm Pharmacol 49(8):796–801
627. El-Kattan A, Varma M (2012) Oral absorption, intestinal metabolism and human oral bioavailability. In: Paxton J (ed) Topics on drug metabolism. InTech, China
628. Kajal P, Bhutani N, Tyagi N, Arya P (2017) Trichobezoar with and without Rapunzel syndrome in paediatric population: a case series from a tertiary care centre of Northern India. Int J Surg Case Rep 40:23–26
629. Eng K, Kay M (2012) Gastrointestinal bezoars: history and current treatment paradigms. Gastroenterol Hepatol (N Y) 8(11):776–778
630. Jao SW, Wang LT, Wu CC, Hsiao CW (2015) Removal of a giant rectosigmoid phytobezoar without Laparotomy. Indian J Surg 77(1):69–71
631. Iwamuro M, Okada H, Matsueda K, Inaba T, Kusumoto C, Imagawa A, Yamamoto K (2015) Review of the diagnosis and management of gastrointestinal bezoars. World J Gastrointest Endosc 7(4):336–345
632. Koulas SG, Zikos N, Charalampous C, Christodoulou K, Sakkas L, Katsamakis N (2008) Management of gastrointestinal bezoars: an analysis of 23 cases. Int Surg 93(2):95–98
633. Kim Y, Park BJ, Kim MJ, Sung DJ, Kim DS, Yu YD, Lee JH (2013) Biliary phytobezoar resulting in intestinal obstruction. World J Gastroenterol 19(1):133–136
634. Andrus CH, Ponsky JL (1988) Bezoars: classification, pathophysiology, and treatment. Am J Gastroenterol 83(5):476–478
635. Guillermo PTJ, Carlos PHJ, Ivonne BAM, Herminio TF, Rubén RP (2014) Extended release potassium salts overdose and endoscopic removal of a pharmacobezoar: a case report. Toxicol Rep 1:209–213
636. Arda K, Yilmaz S, Calikoglu U, Olçer T (1995) Duodenal phytobezoar: a case report and review of the literature. Acta Gastroenterol Belg 58(5–6):470–474
637. Corona-Cruz JF, Sánchez-Lozada R, Gracida-Mancilla NI, Palomeque-López A, Vega Chavaje GR (2005) Intestinal trichobezoar as a cause of small bowel obstruction. Case report and review of the literature. Gac Med Mex 141(5):417–419
638. Balogun OS, Osinowo AO, Afolayan MO, Adesanya AA (2017) An obstructing small bowel phytobezoar in an elderly female nigerian: a case report and literature review. Case Rep Surg 2017:6962876
639. Ugenti I, Travaglio E, Lagouvardou E, Iambrenghi OC, Martines G (2017) Successful endoscopic treatment of gastric phytobezoar: a case report. Int J Surg Case Rep 37:45–47

640. Lung D, Cuevas C, Zaid U, Ancock B (2011) Venlafaxine pharmacobezoar causing intestinal ischemia requiring emergent hemicolectomy. J Med Toxicol 7(3):232–235
641. Said HM, Mohammed ZM (2006) Intestinal absorption of water-soluble vitamins: an update. Curr Opin Gastroenterol 22(2):140–146
642. Eck P, Kwon O, Chen S, Mian O, Levine M (2013) The human sodium-dependent ascorbic acid transporters SLC23A1 and SLC23A2 do not mediate ascorbic acid release in the proximal renal epithelial cell. Physiol Rep 1(6):e00136
643. Kozyraki R, Cases O (2013) Vitamin B12 absorption: mammalian physiology and acquired and inherited disorders. Biochimie 95(5):1002–1007
644. Cooperstone JL, Goetz HJ, Riedl KM, Harrison EH, Schwartz SJ, Kopec RE (2017) Relative contribution of α-carotene to postprandial vitamin A concentrations in healthy humans after carrot consumption. Am J Clin Nutr 106(1):59–66
645. Goncalves A, Roi S, Nowicki M, Dhaussy A, Huertas A, Amiot MJ, Reboul E (2015) Fat-soluble vitamin intestinal absorption: absorption sites in the intestine and interactions for absorption. Food Chem 172:155–160
646. Said HM (2011) Intestinal absorption of water-soluble vitamins in health and disease. Biochem J 437(3):357–372
647. Said HM, Ortiz A, Ma TY (2003) A carrier-mediated mechanism for pyridoxine uptake by human intestinal epithelial Caco-2 cells: regulation by a PKA-mediated pathway. Am J Physiol Cell Physiol 285(5):C1219–C1225
648. Said HM, Ortiz A, Kumar CK, Chatterjee N, Dudeja PK, Rubin S (1999) Transport of thiamine in human intestine: mechanism and regulation in intestinal epithelial cell model Caco-2. Am J Physiol 277(4 Pt 1):C645–C651
649. Said HM, Kumar C (1999) Intestinal absorption of vitamins. Curr Opin Gastroenterol 15 (2):172–176
650. Sahoo S, Aurich MK, Jonsson JJ, Thiele I (20140 Membrane transporters in a human genome-scale metabolic knowledgebase and their implications for disease. Front Physiol 5:91
651. Said HM (2011) Intestinal absorption of water-soluble vitamins in health and disease. Biochem J 437(3):357–372
652. Bifari F, Ruocco C, Decimo I, Fumagalli G, Valerio A, Nisoli E (2017) Amino acid supplements and metabolic health: a potential interplay between intestinal microbiota and systems control. Genes Nutr 12:27
653. Lodish H, Berk A, Zipursky SL, Matsudaira P, Baltimore D, Darnell J (2000) Molecular cell biology, 4th edn. WH Freeman, New York
654. Said HM (2004) Recent advances in carrier-mediated intestinal absorption of water-soluble vitamins. Annu Rev Physiol 66:419–446
655. Zhao R, Goldman ID (2013) The proton-coupled folate transporter: physiological and pharmacological roles. Curr Opin Pharmacol 13(6):875–880
656. Desmoulin SK, Hou Z, Gangjee A, Matherly LH (2012) The human proton-coupled folate transporter—biology and therapeutic applications to cancer. Cancer Biol Ther 13(14):1355–1373
657. Matherly LH, Hou Z (2008) Structure and function of the reduced folate carrier: a paradigm of a major facilitator superfamily mammalian nutrient transporter. Vitam Horm 79:145–184
658. Geller J, Kronn D, Jayabose S, Sandoval C (2002) Hereditary folate malabsorption: family report and review of the literature. Medicine (Baltimore) 81(1):51–68
659. Aluri S, Zhao R, Lubout C, Goorden SMI, Fiser A, Goldman ID (2018) Hereditary folate malabsorption due to a mutation in the external gate of the proton-coupled folate transporter SLC46A1. Blood Adv 2(1):61–68
660. Thomas DR (2006) Vitamins in aging, health, and longevity. Clin Interv Aging 1(1):81–91
661. Hall BE (1950) Nature of the intrinsic factor of Castle. Br Med J 2(4679):585–589
662. Festen HP (1991) Intrinsic factor secretion and cobalamin absorption. Physiology and pathophysiology in the gastrointestinal tract. Scand J Gastroenterol Suppl 188:1–7

663. Gräsbeck R (2006) Imerslund-Gräsbeck syndrome (selective vitamin B12 malabsorption with proteinuria). Orphanet J Rare Dis 1:17
664. Adkins Y, Lönnerdal B (2001) High affinity binding of the transcobalamin II-cobalamin complex and mRNA expression of haptocorrin by human mammary epithelial cells. Biochim Biophys Acta 1528(1):43–48
665. Corti A, Casini AF, Pompella A (2010) Cellular pathways for transport and efflux of ascorbate and dehydroascorbate. Archiv Biochem Biophys 500(2):107–115
666. Subramanian VS, Sabui S, Moradi H, Marchant JS, Said HM (2018) Inhibition of intestinal ascorbic acid uptake by lipopolysaccharide is mediated via transcriptional mechanisms. Biochim Biophys Acta 1860(2):556–565
667. Rivas CI, Zúñiga FA, Salas-Burgos A, Mardones L, Ormazabal V, Vera JC (2008) Vitamin C transporters. J Physiol Biochem 64(4):357–375
668. May JM (2011) The SLC23 family of ascorbate transporters: ensuring that you get and keep your daily dose of vitamin C. Br J Pharmacol 164(7):1793–1801
669. Boyer JC, Campbell CE, Sigurdson WJ, Kuo S-M (2005) Polarized localization of vitamin C transporters, SVCT1 and SVCT2, in epithelial cells. Biochem Biophys Res Commun 334 (1):150–156
670. Savini I, Rossi A, Pierro C, Avigliano L, Catani MV (2008) SVCT1 and SVCT2: key proteins for vitamin C uptake. Amino Acids 34(3):347–355
671. Song J, Kwon O, Chen S, Daruwala R, Eck P, Park JB, Levine M (2002) Flavonoid inhibition of sodium-dependent vitamin C transporter 1 (SVCT1) and glucose transporter isoform 2 (GLUT2), intestinal transporters for vitamin C and Glucose. J Biol Chem 277 (18):15252–15260
672. Babu PV, Liu D, Gilbert ER (2013) Recent advances in understanding the anti-diabetic actions of dietary flavonoids. J Nutr Biochem 24(11):1777–1789
673. Cordero-Herrera I, Martín MÁ, Goya L, Ramos S (2014) Cocoa flavonoids attenuate high glucose-induced insulin signalling blockade and modulate glucose uptake and production in human HepG2 cells. Food Chem Toxicol 64:10–19
674. Petyaev IM, Bashmakov YK (2017) Dark chocolate: opportunity for an alliance between medical science and the food industry? Front Nutr 4:43
675. Vlachojannis J, Erne P, Zimmermann B, Chrubasik-Hausmann S (2016) The impact of cocoa flavanols on cardiovascular health. Phytother Res 30(10):1641–1657
676. Goncalves A, Gleize B, Roi S, Nowicki M, Dhaussy A, Huertas A et al (2013) Fatty acids affect micellar properties and modulate vitamin D uptake and basolateral efflux in Caco-2 cells. J Nutr Biochem 24(10):1751–1757
677. Reboul E (2015) Intestinal absorption of vitamin D: from the meal to the enterocyte. Food Funct 6:356–362
678. Kuipers F, Bloks VW, Groen AK (2014) Beyond intestinal soap—bile acids in metabolic control. Nat Rev Endocrinol 10:488–498
679. Martínez-Augustin O, de Medina FS (2008) Intestinal bile acid physiology and pathophysiology. World J Gastroenterol 14(37):5630–5640
680. Kosters A, Karpen SJ (2008) Bile acid transporters in health and disease. Xenobiotica 38 (7–8): 1043
681. Alrefai WA, Gill RK (2007) Bile acid transporters: structure, function, regulation and pathophysiological implications. Pharm Res 24(10):1803–1823
682. Mita S, Suzuki H, Akita H, Hayashi H, Onuki R, Hofmann AF, Sugiyama Y (2006) Inhibition of bile acid transport across Na^+/taurocholate cotransporting polypeptide (SLC10A1) and bile salt export pump (ABCB 11)-coexpressing LLC-PK1 cells by cholestasis-inducing drugs. Drug Metab Dispos 34(9):1575–1581
683. Kullak-Ublick GA, Stieger B, Meier PJ (2004) Enterohepatic bile salt transporters in normal physiology and liver disease. Gastroenterology 126(1):322–342
684. Ridlon JM, Kang DJ, Hylemon PB, Bile Bajaj JS (2014) Acids and the gut microbiome. Curr Opin Gastroenterol 30(3):332–338

685. Ramírez-Pérez O, Cruz-Ramón V, Chinchilla-López P, Méndez-Sánchez N (2017) The role of the gut microbiota in bile acid metabolism. Ann Hepatol 16(1):S15–S20
686. Chiang JYL, Ferrell JM (2018) Bile acid metabolism in liver pathobiology. Gene Expr. https://doi.org/10.3727/105221618X15156018385515
687. Staels B, Fonseca VA (2009) Bile acids and metabolic regulation mechanisms and clinical responses to bile acid sequestration. Diabetes Care 32(2):S237–S245
688. Swanson HI, Wada T, Xie W, Renga B, Zampella A, Distrutti E et al (2013) Role of nuclear receptors in lipid dysfunction and obesity-related diseases. Drug Metab Dispos 41(1):1–11
689. Hofmann AF (1999) Bile acids: the good, the bad, and the ugly. News Physiol Sci 14:24–29
690. Ballatori N, Christian WV, Lee JY, Dawson PA, Soroka CJ, Boyer JL et al (2005) OSTalpha-OSTbeta: a major basolateral bile acid and steroid transporter in human intestinal, renal, and biliary epithelia. Hepatology 42(6):1270–1279
691. Dawson PA, Hubbert M, Haywood J, Craddock AL, Zerangue N, Christian WV, Ballatori N (2005) The heteromeric organic solute transporter α-β, Ostα-Ostβ, is an ileal basolateral bile acid transporter. J Biol Chem 280(8):6960–6968
692. Kakiyama G, Muto A, Takei H, Nittono H, Murai T, Kurosawa T et al (2014) A simple and accurate HPLC method for fecal bile acid profile in healthy and cirrhotic subjects: validation by GC-MS and LC-MS. J Lipid Res 55:978–990
693. Sauter GH, Munzing W, Von RC, Paumgartner G (1999) Bile acid malabsorption as a cause of chronic diarrhea: diagnostic value of 7alpha-hydroxy-4-cholesten-3-one in serum. Dig Dis Sci 44:14–19
694. Li T, Chiang JY (2014) Bile acid signaling in metabolic disease and drug therapy. Pharmacol Rev 66(4):948–983
695. Dawson PA, Lan T, Rao A (2009) Bile acid transporters. J Lipid Res 50(12):2340–2357
696. Festa C, De Marino S, Carino A, Sepe V, Marchianò S, Cipriani S et al (2017) Targeting bile acid receptors: discovery of a potent and selective farnesoid X receptor agonist as a new lead in the pharmacological approach to liver diseases. Front Pharmacol 8:162
697. Roda A, Pellicciari R, Gioiello A, Neri F, Camborata C, Passeri D et al (2014) Semisynthetic bile acid FXR and TGR5 agonists: physicochemical properties, pharmacokinetics, and metabolism in the rat. J Pharmacol Exp Ther 350(1):56–68
698. Duquette PP, Bissonnette P, Lapointe JY (2001) Local osmotic gradients drive the water flux associated with Na(+)/glucose cotransport. Proc Natl Acad Sci U S A 98(7):3796–3801
699. Loo DD, Wright EM, Zeuthen T (2002) 0 Water pumps. J Physiol 542(Pt 1):53–60
700. Sidorov JJ (1976) Intestinal absorption of water and electrolytes. Clin Biochem 9:117–120
701. Lapointe JY, Gagnon MP, Gagnon DG, Bissonnette P (2002) Controversy regarding the secondary active water transport hypothesis. Biochem Cell Biol 80(5):525–533
702. Wright EM, Turk E (2004) The sodium glucose cotransport family SLC5. Pflügers Arch 447(5):813–815
703. Wright EM, Loo DD (2000) Coupling between Na+, sugar, and water transport across the intestine. Ann N Y Acad Sci 915:54–66
704. Sullivan SK, Field M (1991) Ion transport across mammalian small intestine. In: Schultz SG, Field M, Frizzell RA (eds) Handbook of physiologv. Section 6. The gastrointestinal system, vol 4. Intestinal absorption and secretion. Oxford University Press, New York
705. Yuan X, Zheng X, Lu D, Rubin DC, Pung CY, Sadler JE (1998) Structure of murine enterokinase (enteropeptidase) and expression in small intestine during development. Am J Physiol 274(2 Pt 1):G342–G349

Chapter 13
Excretory Functions of the Gastrointestinal Tract. Defecation

Abstract The gastrointestinal (GI) represents a critical hub where certain by-products of metabolism are channeled for removal from the body. Disorders that affect the excretion of metabolic waste products from the gut can have serious consequences to health. Furthermore, the undigested products of metabolism are also removed from the distal gut. This ensures continuous and adequate functioning of the GI tract, which are all required to maintain ongoing life processes. This chapter deals with the metabolic products produced in the gut that are channeled for removal as well as the evacuation of the residual wastes of digestion and the associated health implications.

Keywords Ammonia · Ammonium transporters · Anorectal afferents Anorectal afferents · Anorectal disorders · Anorectal innervation Anorectal physiology · Bilirubin diglucuronide · Bilirubin · Bilirubinemia Hyperbilirubinemia · Conjugated bilirubin · Unconjugated bilirubin Constipation · Defecation · Defecation reflex · Dual incontinence External anal sphincter · Fecal incontinence · Gut microbiota · Hyperuraemia Hyperammonemia · Internal anal sphincter · Internal anal sphincter achalasia Lactulose · Linaclotide · Liver function tests · Nutrietics · Urea cycle Plecanatide · Prebiotics · Probiotics · Reticuloendothelial system Stercobilin · Synbiotics · Ultrashort-segment Hirschsprung's disease Urea · Urobilin · Urobilinogen · Valves of Houston

Abbreviations

GI	Gastrointestinal
ALAT	Alanine transaminase
ASAT	Aspartate aminotransferases
gammaGT	Gamma-glutamyltransferase
ALP	Alkaline phosphatase
GC-C	Guanylate cyclase C

M. O. Welcome, *Gastrointestinal Physiology*,
https://doi.org/10.1007/978-3-319-91056-7_13

13.1 Introduction

The GI tract functions as an excretory tract by removing waste products of metabolism. The metabolic waste products (e.g., ammonia, urea, bilirubin) are released by the digestive glands into the lumen of the GI tract [1–4]. To maintain the functioning of the gut and related organs, these metabolites must be actively managed and excreted from the body. The salts of heavy metals, drugs as well as calcium ions are also excreted from the digestive tract. In addition, the removal of residual contents of digestion, a useful function of the distal gut, is critical to health [4–6]. This chapter discusses the gut as an excretory organ and the evacuation of the residual wastes of digestion, as well as the associated pathophysiological implications.

13.2 Ammonia Handling in the Gut

Ammonia is a product of metabolism in many cells and tissues of the body. The GI tract is believed to be the main source of ammonia in the body. However, other tissues of the body including the liver, kidney skin, central nervous system are also involved in ammonia handling [7, 8]. In the GI tract, ammonia is produced by the metabolic reactions of the body cells and gut microbes [9, 10]. GI tract resident microbes involved in ammonia production include gram-negative anaerobes (e.g., *Clostridia, enterobacteria*, and *Bacillus* species), gram-positive anaerobes (e.g., *streptococci*). Of the gut resident bacteria, it is believed that *Saccharomyces* and *Lactobacillus* species produce the least quantity of ammonia in the gut [10]. Consequently, these low ammonia output agents (*Lactobacillus*, *Saccharomyces boulardii*, etc.) form major constituents of probiotics (see below).

The epithelial cells of the stomach, duodenum, jejunum, ileum, and colon produce substantial quantity of ammonia [11]. Gut ammonia production increases with increase in protein ingestion or an upper GI bleeding via the fermentative action of the resident bacteria [8, 11–13]. Luminal production of ammonia may be higher in the colon than other regions of the body, which is due to the higher number of commensal microbes in this region of the gut [13]. The synthesized ammonia is transported by ammonium transporters **Rh B g**lycoprotein (RhBG) and **Rh C g**lycoprotein (RhCG) [8]. The RhCG is an apical transporter, while the RhBG is the basolateral transporter of ammonia in the gut [7, 8, 14]. The class A of this family of transporters (RhAG) is expressed only in erythrocytes and their precursors where it mediates the transport of ammonium ion [8]. In the GI tract, RhCG ammonium transporter secretes ammonium NH_4^+ into the lumen, whereas RhBG mediates its release into the portal system [7, 8, 14]. But NH_4^+ can be transported by other ion channels such as $Na^+/K^+/2Cl^-$ co-transporter, H^+/K^+-ATPase, Na^+/K^+-ATPase, K^+ channels. The NH_4^+ is substituted for K^+ in these ion channels. The nonionic form of ammonia NH_3 is transported mainly by diffusion [8, 14].

The ion channels release ammonia into the bloodstream. The ammonia is transported to the liver via the portal vein. In a normal individual, it is believed that almost all ammonia produced by the gut is transported to the liver for detoxification. Consequently, the gut contributes little or no ammonia to the systemic ammonia level [11]. The liver receives excess circulating alanine and glutamine from the GI tract to produce less toxic substances that can be excreted via urine [15]. Increase in ammonia concentration above normal (i.e., systemic hyperammonemia) has been associated with impaired liver functions as observed in cirrhosis [11]. Systemic hyperammonemia in cirrhosis is mainly due to increased activity of the intestinal glutaminase which deaminates glutamine [16]. Although normal concentration of ammonia varies around the world, a value less than 35 μmol/L (60 mcg/dL) is usually considered normal. But slightly higher values below 60 μmol/L may not necessarily result in visible symptoms. Above 61 μmol/L, the individual may experience seizure and vomiting. Such a concentration of ammonia can lead to lethargy and subsequently to coma. A level above 250 μmol/L is characterized for hyperammonemic crisis, resulting in coma and death [17]. Considerable increase in systemic ammonia level can predispose an individual to the development of encephalopathy—a condition that may result from cirrhosis—known as hepatic encephalopathy [16]. The neurotoxicity of ammonia is due to excessive activation of brain N-methyl-D-aspartate receptors and involves dysfunctions of multiple pathways and neurotransmitters as well as receptors including glutamine and nitric oxide [9, 18].

13.3 Urea Handling in the Gut

Certain microbes such as the autochthonous bacteria can utilize NH_3 as a source of nitrogen to fuel amino acid and protein synthesis [13]. However, ammonia in humans is toxic—as its accumulation in the body can lead to serious health consequences [9]. Thus, ammonia is converted to less toxic substrates. This conversion process mainly occurs in the liver where ammonia is metabolized to urea ($(NH_2)_2CO$) or glutamine through multistep biochemical reactions via the ornithine–urea cycle [9, 11]. Urea synthesis occurs through a cycle of enzymatic reactions called urea cycle (also known as the ornithine cycle) discovered by the German physician Sir Hans Adolf Krebs (1900–1981) and his medical student Kurt Henseleit in 1932 [19–21]. For details on the biochemical steps of the urea cycle, review Berg et al. [22] and Randall [9]. The urea that is produced in ornithine–urea cycle is released into the bloodstream where it travels to the kidneys and is excreted in urine. Even though urea is a much less toxic substance than ammonia, it can cause serious health problems when its rate of excretion is decreased. Decreased excretion of urea, which consequently leads to increase in serum urea, can occur in some diseases. Increase in serum urea level (hyperuraemia) can lead to uremia and renal failure. The increase in serum urea can result in its diffusion from blood vessels to the intestine. Agents like montmorillonite can be used for urea adsorption to enhance the removal of urea from the body [23–26].

Inborn errors of metabolism can occur in the urea cycle—simply referred to as urea cycle disorders. These disorders are rare congenital disorders due to deficiency of the enzymes or transporters that are responsible for the removal of ammonia from the body. The result is accumulation of ammonia [18, 27, 28].

13.4 Prebiotics, Probiotics, Synbiotics and Gut Ammonia and Urea Handling

While carbohydrate fermentation by gut microbes leads to the generation of short-chain fatty acids and is thus beneficial for the host, protein fermentation is associated with formation of toxic metabolites such as phenol- and sulfur-containing compounds as well as ammonia [13]. Diseases of the gut and the accessory organs are associated with increase in these toxic metabolites. To repress the production of these toxic agents, certain agents are used. Widely used agents include lactulose, *Bifidobacterium*, *Saccharomyces boulardii*, *Lactobacilli* [10, 13]. *Bifidobacterium* and *S. boulardii* form the main constituents of nutritional supplements that exert an immense health benefit to the host. These nutritional supplements are called probiotics. An example of a prebiotic agent is lactulose. Prebiotics are non-digestible food components that confer considerable benefit to the host by selectively stimulating growth and activity of the gut resident bacteria [13]. (Note that this nutrietic that is non-digestible by the host enzymes is digestible by the enzymes of the gut flora.) When used in their combination (i.e., probiotics plus prebiotics), the nutrietics are called synbiotics. Probiotics, prebiotics, and synbiotics substantially enhance the removal of toxic substances from the body via urine and feces [13, 29, 30]. These agents reduce ammonia production, in part, by their activities on the resident microbes and also reduce the activity of glutaminase, thus decreasing consumption of glutamine by the gut, so that ammonia production is significantly reduced [31, 32].

13.5 Excretion of Bilirubin

Bilirubin is the end product of heme metabolism by cells of the reticuloendothelial system (macrophages and Kupffer cells), and it is secreted together with other components of bile into the intestine following stimulation by meal [33]. Bilirubin is degraded by cells of the reticuloendothelial system and liver to form bilirubin diglucuronide, which is transported into bile. Upon release of bile into the intestine, it is degraded into urobilinogen. Intestinal microflora acts on urobilinogen and reduces it to stercobilin. Some quantities of urobilinogen are transported to the liver and excreted in the urine (Fig. 13.1) [34–37].

13.5.1 Diagnostic Usefulness of Conjugated and Unconjugated Bilirubin

Bilirubin level in the blood is a useful measure of the functions of the liver, level of breakdown of red blood cells or biliary functions. Impairment of the liver functions leads to increase in conjugated bilirubin, which may leak out into circulation to increase the level of serum bilirubin. However, the liver may also lose the ability to conjugate bilirubin, thus increasing the level of serum unconjugated bilirubin. Consequently, both conjugated and unconjugated bilirubin of the serum are increased. Such a situation may be seen in hepatitis, a condition characterized by inflammation of the liver cells [41, 42]. Drugs such as paracetamol may induce liver injury and cause serum bilirubin level to increase [43, 44].

In conditions of increased breakdown of red blood cells, such as in hemolytic anemia, the serum level of unconjugated bilirubin increases above normal. However, urine bilirubin will not increase since unconjugated bilirubin is not water-soluble. If the liver and biliary tree are functioning normally, then unconjugated bilirubin returns to the liver where it is conjugated, excreted in bile, metabolized to urobilinogen, and reabsorbed before it will show as an increase in urine urobilinogen. Also, at very high level of conjugated bilirubin in the blood, some quantities may be secreted in the urine. This may indicate an extrahepatic pathology. In hemolytic anemia, in contrast to jaundice of hepatic origin, liver function tests (ALAT—alanine transaminase, ASAT—aspartate aminotransferases, gammaGT—gamma-glutamyltransferase, and ALP—alkaline phosphatase) are normal. Furthermore, there may be splenomegaly due to the increased activity of the cells of the reticuloendothelial system [41, 42].

Since conjugated bilirubin may be secreted in bile and stored in gallbladder, an obstruction of the bile duct results in the inability of the liver to excrete bile even though it can conjugate the bilirubin. Thus, conjugated bilirubin significantly increases. This type of condition is cholestatic disease or obstructive liver disease, and the type of jaundice is called obstructive jaundice. The urine level of bilirubin and urobilinogen helps to differentiate obstructive liver disease from other causes of jaundice [45–47].

Another type of increased level of bilirubin in blood naturally occurs shortly after birth in newborns. This type of bilirubinemia is called physiological neonatal jaundice and may resolve after about one week. It should be noted, however, that the condition is more frequent and worst in infants born prematurely. Neonatal jaundice may be due to immaturity of glucuronyl transferase and rapid breakdown of fetal red blood cells and their gradual conversion to adult type [48–50].

Certain genetic mutations affecting the enzymes that catalyze the breakdown of heme may lead to a type of increased level of bilirubinemia referred to as familial hyperbilirubinemia [51, 52].

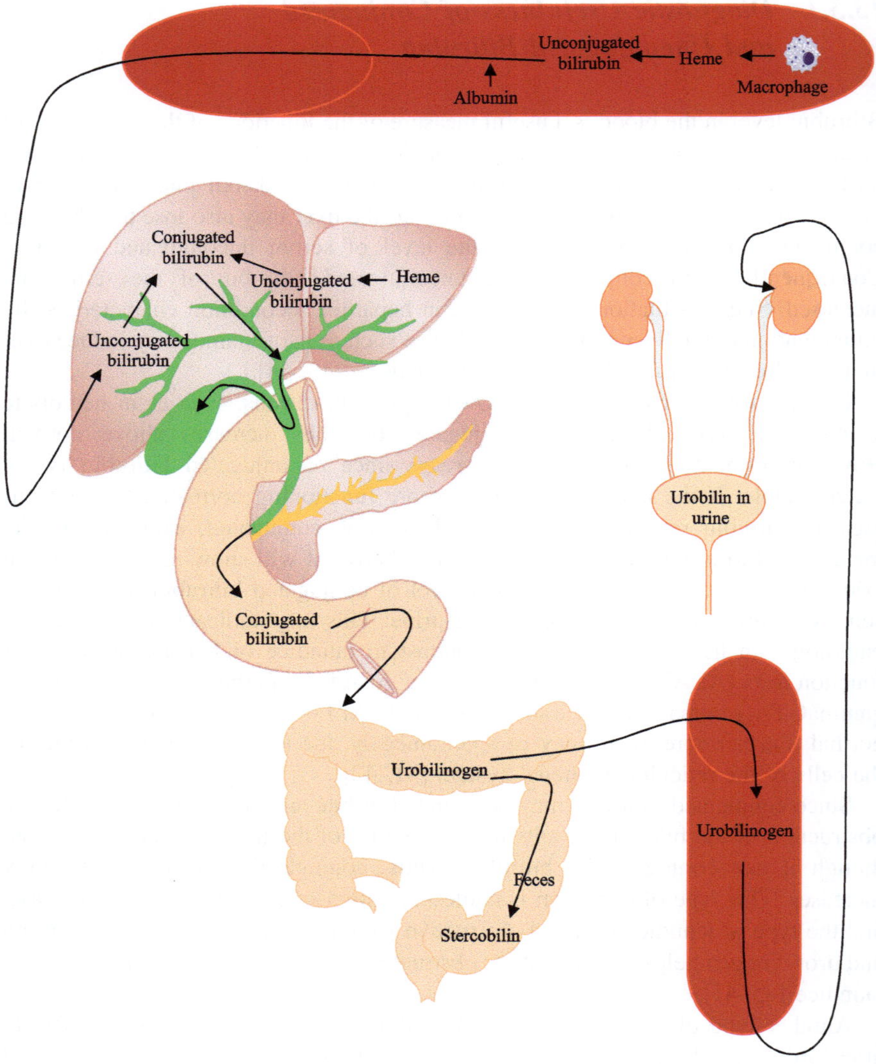
Unconjugated
bilirubin
Heme
Macrophage
Albumin
Conjugated
bilirubin
Unconjugated
bilirubin
Heme
Unconjugated
bilirubin
Urobilin in
urine
Conjugated
bilirubin
Urobilinogen
Urobilinogen
Feces
Stercobilin

◂**Fig. 13.1** Metabolism/excretion of bilirubin. The heme in hemoglobin is degraded by macrophages circulating in the bloodstream to produce bilirubin. First, heme is acted upon by heme oxygenase producing methemoglobin, which further forms biliverdin. The enzyme biliverdin reductase converts biliverdin, producing bilirubin. In the plasma, albumin may associate with bilirubin, to form a complex that is transported to the liver. In the liver, bilirubin is conjugated to glucuronic acid by the enzyme glucuronyl transferase to produce conjugated bilirubin, which is water-soluble. Conjugated bilirubin is released together in bile into the intestine where it is acted upon by gut bacterial proteases (especially of *Clostridium* spp.) deconjugating and metabolizing bilirubin to produce urobilinogen, a colorless compound. Urobilinogen can be reduced to stercobilinogen, which can be oxidized to stercobilin. Stercobilin is what gives stool its brown color. Some quantities of urobilinogen may be reabsorbed into circulation and transported to the kidney as urobilin—one of the oxidation products of urobilinogen in the intestine. Urobilin is the substance responsible for yellow color of urine. Depending on the functioning of gut microbiota, certain quantity of conjugated bilirubin that escapes metabolism by gut microbiota may be transported to the liver via enterohepatic circulation from where it is reexcreted in the bile. Unconjugated bilirubin is also called indirect bilirubin. The conjugated form is also called direct bilirubin. Bilirubin is what gives the yellow discoloration in jaundice [34–40]

13.6 Gastrointestinal Excretion of Some Chemicals and Drugs

Some chemicals including drugs can be excreted through the GI tract and expelled as a constituent of feces. This occurs by biliary and intestinal excretion [53, 54]. For instance, hexachlorobenzene is excreted in feces by the intestine. Intestinal excretion involves the diffusion of the substance from blood into the lumen of the intestine (especially the colon) where it is eliminated with feces [54, 55]. Biliary excretion is one of the major pathways of elimination of drugs and other metabolites [56]. However, certain drugs or metabolites that are excreted in bile may be reabsorbed from the GI tract and returned to the liver and subsequently enter the general circulation. In some cases, the gut commensals may metabolize drugs or chemicals before it is excreted in feces or recirculated to the liver [57, 58].

13.7 Defecation

Defecation can be defined as the elimination or expulsion of residual undigested food and wastes from the body itself in the form of feces through the anus. The anus is an extension from the rectum. The rectum and the anus represent the most distal regions of the GI tract (Fig. 13.2) [59, 60].

The rectum functions as a reservoir for feces and as a pump for emptying stool in a socially appropriate environment [59, 60]. The spiral folds of the wall of the rectum (called valves of Houston) provide a mechanical barrier that prevents the progression of feces. The accumulation of feces in this region increases the barrier effect of these transverse valves (Fig. 13.2) [60]. The valves are named after the

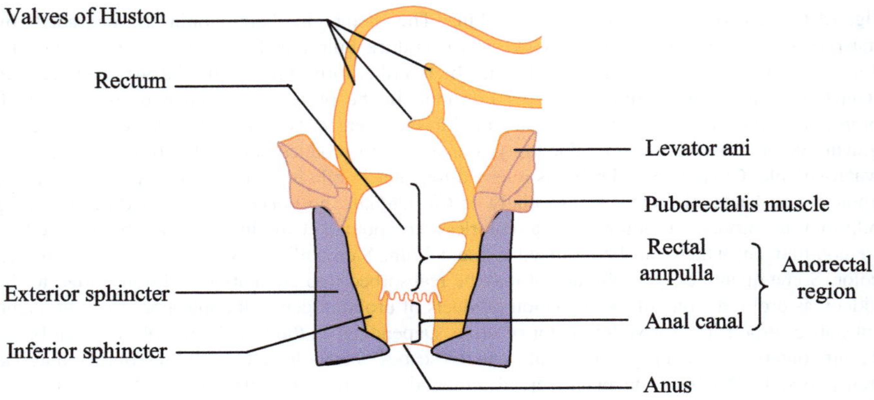

Fig. 13.2 Rectum

Irish–British anatomist and surgeon John Houston (1802–1845), who was the first to describe them [61].

The anus is a muscular tube measuring about 2–4 cm in length. In the luminal membrane, the anus is covered by epithelium. The upper aspect of the anal canal is covered by columnar epithelium, which gradually changes to stratified squamous epithelium (having sebaceous and apocrine sweat glands) through a transitional zone. The wall of the anus is composed of both skeletal and smooth muscle cells. The smooth muscle cells occur in two layers—inner circular and outer longitudinal. The former expands to form the internal anal sphincter. The tonicity of the internal anal sphincter ensures that the anus is closed at rest, which is required to maintain continence [59, 62, 63]. The longitudinal muscle of the anal canal terminates as septa, penetrating the striated muscles of the anus [64]. The skeletal muscles (levator ani muscles) expand to form the external anal sphincter [59]. These striated muscles are composed of pubococcygeus, iliococcygeus, and puborectalis—which also form the pelvic floor and are also involved in the maintenance of continence [60]. The striated muscles are also capable of tonic activity at rest, but they can rapidly contract during voluntary exertion to push down fecal matter [64].

The anus forms an angle in relation to the axis of the rectum, known as the anorectal angle. It is the angle between the longitudinal axis of anal canal and the posterior line of the rectum, parallel to the longitudinal rectal axis. The angle is difficult to measure. That is why measurements are usually subjective, and thus, the reported values may vary and also depend on the technique of investigation [60, 65, 66]. This angle is used as indicator of the activity of the puborectal muscle. The anorectal angle changes in many pathological conditions involving the anorectal region. Under physiological resting condition, the average value is estimated to be 90°–96° (the range is 65°–100°, but can reach 117° in some normal subjects) [65]. Upon squeezing (contraction), the angle becomes more acute, around 70°–98° [60, 67]. Upon straining, which occurs during defecation (relaxation), the angle becomes more obtuse, at 110°–155° [60, 65, 67]. The anorectal angle does not have gender differences. In anorectal disorders, this angle has been shown to increase substantially [67].

The anal canal is primarily responsible for the regulation of continence and defecation. These complex physiological events occur through regulated activity of autonomic nervous system (sympathetic and parasympathetic divisions), intrinsic neurons of the anal wall, in addition to the skeletal and smooth muscles. Environmental factors also play a crucial role in modulating the activities of the anorectal region [59, 68, 69].

The intrinsic nervous system of the anus is represented by the enteric nervous plexuses [69]. The myenteric plexus of the anorectal region form contacts (synapses) with the extrinsic nerves of this region [64]. The extrinsic nervous system is mainly represented by the parasympathetic pelvic nerves [68, 69]. But the rectal motility is also controlled by sympathetic hypogastric nerves [68]. Both divisions of the autonomic nervous system innervate the internal anal sphincter [60, 64]. The parasympathetic nerves originate from the sacral cord S2–4 to form the inferior hypogastric plexus, which gives rise to the rectal nerves (superior, middle, and inferior) that innervate the rectum and anus [64, 70]. Stimulation of the parasympathetic nerves inhibits the activity of the internal anal sphincter [60]. The sympathetic nerves originate from the lower thoracic ganglia to form the superior hypogastric plexus [64, 71]. Stimulation of the sympathetic nerves results in contraction and relaxation of the internal anal sphincter [60].

The sensory information from the anus is sent to the dorsal aspect of spinal cord S2–4 via the dorsal root ganglia [72, 73].

The motor nerves that innervate the external anal sphincter originate in ventral sacral spinal cord S2–4 [Onuf's nucleus, named after the Polish scientist Bronislaw Onufrowicz (1863–1928)]. The neurons of this nucleus give rise to pudendal nerves (right and left divisions), which in turn give rise to the inferior rectal branches (also known as inferior hemorrhoidal branches) that innervate this region of the anus [60, 64]. It is believed that the branches of the pudendal nerve innervate the puborectalis muscle, whereas the pubococcygeus and iliococcygeus muscles are innervated by fibers of sacral nerve S3 and S4 [64].

13.7.1 Mechanism of Defecation: Defecation Reflex

Accumulation of fecal matter in the rectal ampulla induces distention of the rectal wall, stimulating stretch (sensory) receptors that begin signaling processes that culminate in the release of tonic activity of the rectal muscles. This ultimately leads to defecation [60, 74, 75]. The process of defecation is reflexive in nature and can be controlled at will [6].

The defecation reflex involves intrinsic and extrinsic reflexes. The intrinsic reflex is mediated via the myenteric plexus and mediates the activity of the anorectal smooth muscle layer and internal anal sphincter. In the extrinsic reflex, sensory nerves of the parasympathetic division transmit afferent impulses to the neurons of the dorsal spinal cord S2–4. The signal is processed, and the information (i.e., response) is relayed back via the parasympathetic motor fibers that innervate the anorectum. The motor response arriving at the anorectum is inhibitory, which is required to release the tonic activity of the internal anal sphincter, causing muscle relaxation [69, 76–78]. Though both the internal anal sphincter and external anal sphincter contribute to maintenance of continence, the external anal sphincter plays a major role in defecation [64]. But the actions of the two different sphincters are complementary [79]. The voluntary control of defecation occurs via cortical influences on the lower motor neurons that innervate the external anal sphincter [76]. Under normal condition, during defecation, the contraction of respiratory diaphragm and abdominal wall results in an increase in the intra-abdominal and rectal pressure. At almost the same time, the pelvic floor muscles and anal sphincter muscles relax [64]. During relaxation of these muscles, the rectum contracts to expel the fecal mass [69, 76, 80, 81].

The control of continence involves three anal reflexes, which are mediated by the pelvic and pudendal nerves. The rectoanal inhibitory reflex is required for releasing the tonic activity of the muscles of the anal sphincters. The rectoanal contractile reflex is a reflex that prevents unexpected release of fecal mass from the rectum. The sensorimotor response is a transient reflex initiated by puborectalis contraction in response to sensation resulting from the desire to defecate [60, 82, 83].

Anorectal structure and functions can be studied using magnetic resonance imaging (MRI), video X-ray, computed tomography (CT) scans, manometry, endoanal ultrasound (anal endosonography), scintigraphy, electromyographic recordings using anal plug electrodes or cutaneous electrodes, X-ray fluoroscopy (with barium), and radiopaque markers or radioisotopes. For instance, processes of defecation can be studied by recording defecating proctograms or video myogram defecography [64, 80, 84–87].

13.7.2 Pathological Conditions that Are Associated with Defecation

The anorectal region is implicated in a range of disorders, which may be functional and/or structural. The functional anorectal disorders include fecal incontinence, constipation, levator ani syndrome (a type of chronic proctalgia, which also includes unspecified anorectal pain), proctalgia fugax (fleeting pain), fecal impaction, dyssynergic defecation [60, 88–90]. Structural disorders include rectocele, cystocele, hemorrhoids, rectal intussusception, prolapse, anal fissure [64, 88, 91]. Here, only some functional disorders will be briefly discussed. These disorders of the anorectal region are associated with considerable psychological distress and social stigma [60].

Constipation: This refers to difficulty in evacuation of fecal mass through the anus. The condition can be due to slow colonic transit and increased colonic water absorption leading to formation of hard fecal mass, thereby causing the individual to excessive straining during passage of feces. Constipation can be categorized into slow transit type and outlet obstruction type. The slow transit type of constipation is characterized by reduced movement of fecal mass in the GI tract. The outlet obstruction type is due to blockade or narrowing of the lumen, resulting in difficulty in passage of fecal mass [60, 90]. Predisposing factors to the occurrence of constipation mostly involve decreased intake of fluid and low fiber diet. Consequently, individuals with constipation are advised to increase fluid and fiber intake. The use of laxatives is helpful. However, if constipation becomes chronic especially in the elderly people, colorectal cancer should be suspected [60, 90, 92]. Recent clinical trials of linaclotide and plecanatide have shown high efficacy and considerable promise for the treatment of chronic idiopathic constipation [93, 94]. These pharmacological agents function by activating GI mucosal guanylate cyclase C (GC-C) receptors. The GC-C receptors via interaction with the hormone uroguanylin regulate GI fluid and electrolyte balance [93, 94].

Fecal Incontinence: This is the involuntary passage of fecal mass via the anus [60]. The diagnosis of fecal incontinence can be made when this condition is experienced at least 3 months in a person of at least 4 years of age. Functional fecal incontinence is associated with abnormal functioning in the absence of structural detection in the neuromuscular unit of the anorectum [90, 95].

The etiology of fecal incontinence is multifactorial and can be caused by diarrhea, functional disability, cognitive impairment, traumatic anal injury, neurologic deficits, inflammatory conditions, disordered rectal compliance or accommodation, constipation, impaired anorectal sensation, neuropathy of the nerves supplying the muscles of the anorectum, and other anorectal dysfunctions [60, 64]. Fecal incontinence is more frequent in the elderly, especially above 80 years, and it may occur together with urinary incontinence (known as dual incontinence) [60].

The condition may be due to internal anal sphincter incompetence, decreased anorectal sensation, structural or neuromuscular dysfunction of the pelvic floor and anorectum and possibly due to surgery, irradiation, or other causes [60].

Internal Anal Sphincter Achalasia: Internal anal sphincter achalasia (also known as ultrashort-segment Hirschsprung's disease) is defined as the inability of the internal anal sphincter to relax [96–98]. The result is obstruction of the anal outlet [99]. The cause of the condition is not exactly clear, but it is believed that dysfunctions in neurotransmitter (e.g., NO) signaling, neuromuscular signaling, and interstitial cells of Cajal may be possible factors that contribute to the disorder [100]. Diagnosis of this clinical condition can be made with anorectal manometry, which shows the absence of rectosphincteric reflex on rectal balloon inflation [98]. The condition is usually treated with botulinum toxin [97].

Annoted Review Texts on Anorectal Disorders

1. Bharucha AE, Wald AM (2010) Anorectal disorders. Am J Gastroenterol 105: 786–794
2. Foxx-Orenstein AE, Umar SB, Crowell MD (2014) Common anorectal disorders. Gastroenterol Hepatol (N Y) 10(5): 294–301

3. Henderson PK, Cash BD (2014) Common anorectal conditions: evaluation and treatment. Curr Gastroenterol Rep 16:408
4. Lacy BE, Weiser K (2009) Common anorectal disorders: diagnosis and treatment. Curr Gastroenterol Rep 11(5):413–419
5. Pfenninger JL, Zainea GG (2001) Common anorectal conditions: part I. Symptoms and complaints. Am Fam Physician 63(12):2391–2398
6. Pfenninger JL, Zainea GG (2001) Common anorectal conditions: part II. Lesions. Am Fam Physician 64(1):77–89
7. Remes-Troche JM, Satish SC Rao (2008) Neurophysiological testing in anorectal disorders. Expert Rev Gastroenterol Hepatol 2(3): 323–335

13.8 Conclusions

The GI tract represents a crucial hub for the mobilization of excretory wastes as well as evacuation of residual products of digestion. Substances such as ammonia and other metabolic wastes are continuously channeled for removal to ensure normal physiological functioning of the gut and the organism as a whole. Disorders in these functions of the gut can result in serious health consequences.

Recommended Readings
Original Articles

1. Cappabianca S, Reginelli A, Iacobellis F, Granata V, Urciuoli L, Alabiso ME, et al (2011) Dynamic MRI defecography vs. entero-colpo-cysto-defecography in the evaluation of midline pelvic floor hernias in female pelvic floor disorders. Int J Colorectal Dis 26:1191
2. El Sayed RF, Alt CD, Maccioni F, Meissnitzer M, Masselli G, Manganaro L, et al (2017) Magnetic resonance imaging of pelvic floor dysfunction—joint recommendations of the ESUR and ESGAR Pelvic Floor Working Group. Eur Radiol 27(5):2067–2085
3. Flusberg M, Sahni VA, Erturk SM, Mortele KJ (2011) Dynamic MR defecography: assessment of the usefulness of the defecation phase. AJR 196: W394–W399
4. Reginelli A, Di Grezia G, Gatta G, Iacobellis F, Rossi C, Giganti M, et al (2013) Role of conventional radiology and MRi defecography of pelvic floor hernias. BMC Surg 13(2):S53
5. Roos JE, Weishaupt D, Wildermuth S, Willmann JK, Marincek B, Hilfiker PR (2002) Experience of 4 years with open MR defecography: pictorial review of anorectal anatomy and disease. Radio-Graphics 22(4):817–832
6. Schreyer AG, Paetzel C, Fürst A, Dendl LM, Hutzel E, Müller-Wille R, et al (2012) Dynamic magnetic resonance defecography in 10 asymptomatic volunteers. World J Gastroenterol 18(46):6836–6842

Review Articles

1. Burton BK (2000) Urea cycle disorders. Clin Liver Dis 4(4):815–830
2. Watford M (2003) The urea cycle: teaching intermediary metabolism in a physiological setting. Biochem Mol Biol Educ 31(5):289–297
3. Yu SWB, Rao SSC (2014) Anorectal physiology/pathophysiology in the elderly. Clin Geriatr Med 30(1):95–106

Books

1. Berg JM, Tymoczko JL, Stryer L (2002) Biochemistry, 5th edn. WH Freeman, New York
2. Bernstein LE, Rohr F, Helm JR, eds. (2015) Nutrition management of inherited metabolic diseases—lessons from metabolic university. Springer International Publishing, Cham, Switzerland
3. National Collaborating Centre for Women's and Children's Health (UK) (2010) Neonatal jaundice. NICE clinical guidelines, no. 98. RCOG Press, London, UK

Bibliography

1. Bergen WG, Wu G (2009) Intestinal nitrogen recycling and utilization in health and disease. J Nutr 139(5):821–825
2. Ross AC, Taylor CL, Yaktine AL, Del Valle HB (eds) (2011) Dietary reference intakes for calcium and vitamin D. National Academies Press, Washington, DC, USA
3. Fried DE, Watson RE, Robson SC, Gulbransen BD (2017) Ammonia modifies enteric neuromuscular transmission through glial γ-aminobutyric acid signaling. Am J Physiol Gastrointest Liver Physiol 313(6):G570–G580
4. Pereverzev VA, Lobanok LM (2014) Physiology of digestion. In: Kubarko AI (ed) Normal physiology. Visheishaya Shkola, Minsk, Belarus
5. Nikitina OS, Welcome MO, Pereverzev VA (2016) Human anatomy and physiology. Belarusian State Medical University Press, Minsk
6. Bassotti G, Antonelli E, Villanacci V, Baldoni M, Dore MP (2014) Colonic motility in ulcerative colitis. United Eur Gastroenterol J 2(6):457–462
7. Weiner ID, Verlander JW (2003) Renal and hepatic expression of the ammonium transporter proteins, Rh B Glycoprotein and Rh C Glycoprotein. Acta Physiol Scand 179(4):331–338
8. Handlogten ME, Hong S-P, Zhang L, Vander AW, Steinbaum ML, Campbell-Thompson M, Weiner ID (2005) Expression of the ammonia transporter proteins Rh B glycoprotein and Rh C glycoprotein in the intestinal tract. Am J Physiol Gastrointest Liver Physiol 288(5): G1036–G1047
9. Randall DJ (2011) Nitrogenous-waste balance: excretion of ammonia. In: Farrell A (ed) Encyclopedia of fish physiology: from genome to environment. Academic Press, MA, USA
10. Vince AJ, Burridge SM (1980) Ammonia production by intestinal bacteria: the effects of lactose, lactulose and glucose. J Med Microbiol 13(2):177–191
11. van de Poll MCG, Ligthart-Melis GC, Olde Damink SWM, van Leeuwen PAM, Beets-Tan RGH, Deutz NEP et al (2008) The gut does not contribute to systemic ammonia release in humans without portosystemic shunting. Am J Physiol Gastrointest Liver Physiol 295(4): G760–G765

12. Sugarbaker SP, Revhaug A, Wilmore DW (1987) The role of the small intestine in ammonia production after gastric blood administration. Ann Surg 206(1):5–17
13. De Preter V, Vanhoutte T, Huys G, Swings J, Rutgeerts P, Verbeke K (2006) Effect of lactulose and *Saccharomyces boulardii* administration on the colonic urea-nitrogen metabolism and the bifidobacteria concentration in healthy human subjects. Aliment Pharmacol Ther 23(7):963–974
14. Weiner ID, Verlander JW (2011) Role of NH_3 and NH_4+ transporters in renal acid-base transport. Am J Physiol Renal Physiol 300(1):F11–23
15. Phromphetcharat V, Jackson A, Dass PD, Welbourne TC (1981) Ammonia partitioning between glutamine and urea: Interorgan participation in metabolic acidosis. Kidney Int 20:598–605
16. Romero-Gómez M, Jover M, Galán JJ, Ruiz A (2009) Gut ammonia production and its modulation. Metab Brain Dis 24(1):147–157
17. Rohr F (2015) Nutrition management of urea cycle disorders. In: Bernstein L, Rohr F, Helm J (eds) Nutrition management of inherited metabolic diseases. Springer, Cham, Switzerland
18. Gropman AL, Summar M, Leonard JV (2007) Neurological implications of urea cycle disorders. J Inherit Metab Dis 30(6):865–879
19. Kinne-Saffran E, Kinne RK (1999) Vitalism and synthesis of urea. From Friedrich Wöhler to Hans A. Krebs. Am J Nephrol 19(2):290–294
20. Krebs HA (1973) The discovery of the ornithine cycle of urea synthesis. Biochem Educ 1 (2):19–23
21. Nickelsen K, Graßhoff G (2009) Concepts from the bench: Hans Krebs, Kurt Henseleit and the urea cycle. In: Hon G, Schickore J, Steinle F (eds) Going amiss in experimental research. Springer, Dordrecht, Netherlands
22. Berg JM, Tymoczko JL, Stryer L (2002) Biochemistry, 5th edn. WH Freeman, New York
23. Long L-H, Zhang Y-T, Wang X-F, Cao Y-X (2009) Montmorillonite adsorbs urea and accelerates urea excretion from the intestine. Appl Clay Sci 46(1):57–62
24. Thadhani R, Pascual M, Bonventre JV (1996) Acute renal failure. N Engl J Med 334:1448–1460
25. Weiner ID, Mitch WE, Sands JM (2015) Urea and ammonia metabolism and the control of renal nitrogen excretion. Clin J Am Soc Nephrol 10(8):1444–1458
26. Murea M (2012) Advanced kidney failure and hyperuricemia. ACKD 19(6):419–424
27. Goodman MW, Zieve L, Konstantinides FN, Cerra FB (1984) Mechanism of arginine protection against ammonia intoxication in the rat. Am J Physiol Gastrointest Liver Physiol 247(3):G290–G295
28. Brosnan ME, Brosnan JT (2007) Orotic acid excretion and arginine metabolism. J Nutr 137 (6):1656S–1661S
29. Cheng W, Lu J, Li B, Lin W, Zhang Z, Wei X et al (2017) Effect of functional oligosaccharides and ordinary dietary fiber on intestinal microbiota diversity. Front Microbiol 8:1750
30. Markowiak P, Śliżewska K (2017) Effects of probiotics, prebiotics, and synbiotics on human health. Nutrients 9(9):E1021
31. Hawkins RA, Jessy J, Mans AM, Chedid A, De Joseph MR (1994) Neomycin reduces the intestinal production of ammonia from glutamine. Adv Exp Med Biol 368:125–134
32. van Berlo CLH, van Leeuwen PAM, Soeters PB (1988) Porcine intestinal ammonia liberation: influence of food intake, lactulose and neomycin treatment. J Hepatol 7(2):250–257
33. Ascenzi P, Bocedi A, Visca P, Altruda F, Tolosano E, Beringhelli T, Fasano M (2005) Hemoglobin and heme scavenging. IUBMB Life 57(11):749–759
34. Quinn KD, Nguyen NQT, Wach MM, Wood TD (2012) Tandem mass spectrometry of bilin tetrapyrroles by electrospray ionization and collision induced dissociation. Rapid Commun Mass Spectrom 26(16):1767–1775

35. Kotal P, Van Der Veere CN, Sinaasappel M, Elferink RO, Vítek L, Brodanová M et al (1997) Intestinal excretion of unconjugated bilirubin in man and rats with inherited unconjugated hyperbilirubinemia. Pediatr Res 42:195–200
36. Tiribelli C, Ostrow JD (2005) Intestinal flora and bilirubin. J Hepatol 42(2):170–172
37. Fargo MV, Grogan SP, Saguil A (2017) Evaluation of jaundice in adults. Am Fam Physician 95(3):164–168
38. Daniel H, Gholami AM, Berry D, Desmarchelier C, Hahne H, Loh G et al (2014) High-fat diet alters gut microbiota physiology in mice. ISME J 8(2):295–308
39. Jeschke MG (2009) The hepatic response to thermal injury: is the liver important for postburn outcomes? Mol Med 15(9–10):337–351
40. Nakamura T, Sato K, Akiba M, Ohnishi M (2006) Urobilinogen, as a bile pigment metabolite, has an antioxidant function. J Oleo Sci 55(4):191–197
41. Kasper D, Fauci A, Hauser S, Longo D, Jameson J, Loscalzo J (eds) (2015) Harrison's principles of internal medicine, 19th edn. McGraw Hill, New York, USA
42. Warrell DA, Cox TM, Firth JD (2010) The Oxford textbook of medicine, 5th edn. Oxford University Press, London, UK
43. Sabaté M, Ibáñez L, Pérez E, Vidal X, Buti M, Xiol X et al (2011) Paracetamol in therapeutic dosages and acute liver injury: causality assessment in a prospective case series. BMC Gastroenterol 11:80
44. Mahadevan SBK, McKiernan PJ, Davies P, Kelly DA (2006) Paracetamol induced hepatotoxicity. Arch Dis Child 91(7):598–603
45. Ryder SD, Beckingham IJ (2001) Other causes of parenchymal liver disease. BMJ 322 (7281):290–292
46. Beckingham IJ, Ryder SD (2001) Investigation of liver and biliary disease. BMJ 322 (7277):33–36
47. Assy N, Jacob G, Spira G, Edoute Y (1999) Diagnostic approach to patients with cholestatic jaundice. World J Gastroenterol 5(3):252–262
48. Vitek L (2003) Intestinal metabolism of bilirubin in the pathogenesis of neonatal jaundice. J Paediatr 143(6):810
49. Stokowski LA (2006) Fundamentals of phototherapy for neonatal jaundice. Adv Neonatal Care 6(6):303–312
50. Sadeghnia A, Ganji M, Armanian AM (2014) A comparison between the effect of fluorescent lamps and quartz halogen incandescent filament lamps on the treatment of hyperbilirobinemia in newborns with the gestational age of 35 weeks or more. Int J Prev Med 5(9):1186–1191
51. Berthelot P, Dhumeaux D (1978) New insights into the classification and mechanisms of hereditary, chronic, non-haemolytic hyperbilirubinaemias. Gut 19(6):474–480
52. Nambu M, Namihisa T (1996) Hepatic transport of serum bilirubin, bromsulfophthalein, and indocyanine green in patients with congenital non-hemolytic hyperbilirubinemia and patients with constitutional indocyanine green excretory defect. J Gastroenterol 31(2):228–236
53. Rollins DE, Klaassen CD (1979) Biliary excretion of drugs in man. Clin Pharmacokinet 4 (5):368–379
54. Rozman K (1985) Intestinal excretion of toxic substances. Arch Toxicol Suppl 8:87–93
55. Jandacek RJ, Genuis SJ (2013) An assessment of the intestinal lumen as a site for intervention in reducing body burdens of organochlorine compounds. Sci World J 2013:205621
56. Sharifi M, Ghafourian T (2014) Estimation of biliary excretion of foreign compounds using properties of molecular structure. AAPS J 16(1):65–78
57. Mikov M (1994) The metabolism of drugs by the gut flora. Eur J Drug Metab Pharmacokinet 19(3):201–207
58. Roberts MS, Magnusson BM, Burczynski FJ, Weiss M (2002) Enterohepatic circulation: physiological, pharmacokinetic and clinical implications. Clin Pharmacokinet 41(10): 751–790

59. Barleben A, Mills S (2010) Anorectal anatomy and physiology. Surg Clin North Am 90 (1):1–15
60. Yu SW, Rao SS (2014) Anorectal physiology and pathophysiology in the elderly. Clin Geriatr Med 30(1):95–106
61. Ikard RW (2015) Spiral rectal valves: anatomy, eponyms, and clinical significance. Clin Anat 28(4):436–441
62. Jorge JM, Wexner SD (1997) Anatomy and physiology of the rectum and anus. Eur J Surg 163(10):723–731
63. Irving MH, Catchpole B (1992) ABC of colorectal diseases. Anatomy and physiology of the colon, rectum, and anus. BMJ 304(6834):1106–1108
64. Raizada V, Mittal RK (2008) Pelvic floor anatomy and applied physiology. Gastroenterol Clin North Am 37(3):493–vii
65. Kim AY (2011) How to interpret a functional or motility test—defecography. J Neurogastroenterol Motil 17(4):416–420
66. Penninckx F, Debruyne C, Lestar B, Kerremans R (1991) Intraobserver variation in the radiological measurement of the anorectal angle. Gastrointest Radiol 16(1):73–76
67. Piloni V, Fioravanti P, Spazzafumo L, Rossi B (1999) Measurement of the anorectal angle by defecography for the diagnosis of fecal incontinence. Int J Colorectal Dis 14(2):131–135
68. Ridolfi TJ, Tong W-D, Takahashi T, Kosinski L, Ludwig KA (2009) Sympathetic and parasympathetic regulation of rectal motility in rats. J Gastrointest Surg 13(11):2027–2033
69. Shafik A, El-Sibai O, Ahmed I (2002) Parasympathetic extrinsic reflex: role in defecation mechanism. World J Surg 26(6):737–741
70. Mauroy B, Demondion X, Bizet B, Claret A, Mestdagh P, Hurt C (2007) The female inferior hypogastric (=pelvic) plexus: anatomical and radiological description of the plexus and its afferences—applications to pelvic surgery. Surg Radiol Anat 29(1):55–66
71. Gallachera K, Santosa LC, Campoy L, Bezuidenhout AJ, Gilbert RO (2016) Development of a peripheral nerve stimulator-guided technique for equine pudendal nerve blockade. Vet J 217:72–77
72. Origoni M, Maggiore ULR, Salvatore S, Candiani M (2014) Neurobiological mechanisms of pelvic pain. Biomed Res Int 2014:903848
73. Vermeulen W, De Man JG, Pelckmans PA, De Winter BY (2014) Neuroanatomy of lower gastrointestinal pain disorders. World J Gastroenterol 20(4):1005–1020
74. Palit S, Lunniss PJ, Scott SM (2012) The physiology of human defecation. Dig Dis Sci 57 (6):1445–1464
75. Katsui R, Kuniyasu H, Matsuyoshi H, Fujii H, Nakajima Y, Takaki M (2009) The plasticity of the defecation reflex pathway in the enteric nervous system of guinea pigs. J Smooth Muscle Res 45(1):1–13
76. Bajwa A, Emmanuel A (2009) The physiology of continence and evacuation. Best Pract Res Clin Gastroenterol 23(4):477–485
77. Winge K, Rasmussen D, Werdelin LM (2003) Constipation in neurological diseases. J Neurol Neurosurg Psychiatry 74:13–19
78. Krier J (1989) Motor function of anorectum and pelvic floor musculature. In: Wood JD (ed) Handbook of physiology. The gastrointestinal system: motility and circulation. American Physiological Society, Bethesda, Maryland, United States
79. Bajwa A, Emmanuel A (2009) The physiology of continence and evacuation. Best Pract Res Clin Gastroenterol 23(4):477–485
80. Petros P, Swash M, Bush M, Fernandez M, Gunnemann A, Zimmer M (2012) Defecation 1: testing a hypothesis for pelvic striated muscle action to open the anorectum. Tech Coloproctol 16(6):437–443
81. Shafik A, Shafik IA, El Sibai O, Shafik AA (2006) The "opening time" and "pelvic floor electromyographic lag time": two novel tools in the assessment of the anorectal evacuation time. J Invest Surg 19(5):307–311
82. Remes-Troche JM, De-Ocampo S, Valestin J, Rao SS (2010) Rectoanal reflexes and sensorimotor response in rectal hyposensitivity. Dis Colon Rectum 53(7):1047–1054

83. De Ocampo S, Remes-Troche JM, Miller MJ, Rao SS (2007) Rectoanal sensorimotor response in humans during rectal distension. Dis Colon Rectum 50(10):1639–1646
84. Bush M, Petros P, Swash M, Fernandez M, Gunnemann A (2012) Defecation 2: internal anorectal resistance is a critical factor in defecatory disorders. Tech Coloproctol 16(6):445–450
85. Fletcher JG, Busse RF, Riederer SJ, Hough D, Gluecker T, Harper CM, Bharucha AE (2003) Magnetic resonance imaging of anatomic and dynamic defects of the pelvic floor in defecatory disorders. Am J Gastroenterol 98(2):399–411
86. Wald A (1994) Colonic and anorectal motility testing in clinical practice. Am J Gastroenterol 89(12):2109–2115
87. Yabunaka K, Nakagami G, Komagata K, Sanada H (2017) Ultrasonographic follow-up of functional chronic constipation in adults: a report of two cases. SAGE Open Med Case Rep 5:2050313X17694234
88. Ko CY, Tong J, Lehman RE, Shelton AA, Schrock TR, Welton ML (1997) Biofeedback is effective therapy for fecal incontinence and constipation. Arch Surg 132:829–834
89. Whitehead WE (1996) Functional anorectal disorders. Semin Gastrointest Dis 7(4):230–236
90. Bharucha AE, Wald A, Enck P, Rao S (2006) Functional anorectal disorders. Gastroenterology 130(5):1510–1518
91. Shafik A (1982) A new concept of the anatomy of the anal sphincter mechanism and the physiology of defecation. XV. Chronic anal fissure: a new theory of pathogenesis. Am J Surg 144(2):262–268
92. Chokhavatia S, John ES, Bridgeman MB, Dixit D (2016) Constipation in elderly patients with noncancer pain: focus on opioid-induced constipation. Drugs Aging 33(8):557–574
93. Miner PB Jr (2018) Efficacy and safety of plecanatide in treating constipation predominant irritable bowel syndrome. Expert Opin Pharmacother 19(2):177–183
94. Shah ED, Kim HM, Schoenfeld P (2018) Efficacy and tolerability of guanylate cyclase-C agonists for irritable bowel syndrome with constipation and chronic idiopathic constipation: a systematic review and meta-analysis. Am J Gastroenterol 113(3):329–338
95. Kinugasa Y, Arakawa T, Murakami G, Fujimiya M, Sugihara K (2014) Nerve supply to the internal anal sphincter differs from that to the distal rectum: an immunohistochemical study of cadavers. Int J Colorectal Dis 29(4):429–436
96. Holschneider AM, Kunst M (2008) Anal sphincter achalasia and ultrashort hirschsprung's disease. In: Holschneider A, Puri P (eds) Hirschsprung's disease and allied disorders. Springer, Heidelberg
97. Chumpitazi BP, Fishman SJ, Nurko S (2009) Long-term clinical outcome after botulinum toxin injection in children with nonrelaxing internal anal sphincter children with nonrelaxing internal anal sphincter. Am J Gastroenterol 104:976–983
98. De Caluwé D, Yoneda A, Akl U, Puri P (2001) Internal anal sphincter achalasia: outcome after internal sphincter myectomy. J Pediatr Surg 36:736–738
99. Koivusalo AI, Pakarinen MP, Rintala RJ (2009) Botox injection treatment for anal outlet obstruction in patients with internal anal sphincter achalasia and Hirschsprung's disease. Pediatr Surg Int 25(10):873–876
100. Doodnath R, Puri P (2009) Internal anal sphincter achalasia. Semin Pediatr Surg 18(4): 246–248

Chapter 14
Helicobacter Pylori

Abstract *Helicobacter pylori* (*H. pylori*) is a gastro-infective agent that colonizes the stomach of nearly half of humans in the world. The infective nature of this agent is due to its ability to change the physiology of the immediate gastric environment. Several factors are responsible for determining the ability of *H. pylori* to cause infection. This chapter deals with the pathophysiology of *H. pylori*.

Keywords *Helicobacter pylori* · *H. pylori* · Peptic ulcer disease
Gastritis · Gastric ulcer · Duodenal ulcer · Gastric cancer · Urease
Rapid urease test · *H. pylori* and pregnancy · Cytotoxin-associated antigen
Cytotoxin-associated gene · *H. pylori* chemoreceptors · Eradication therapy
H. pylori therapy

Abbreviations

Cag	Cytotoxin-associated gene
CagA	Cytotoxin-associated antigen
GI	Gastrointestinal
H. pylori	*Helicobacter pylori*
MALT	Mucosa-associated lymphoid tissue
NOD	Nucleotide-binding oligomerization domain

14.1 Introduction

Helicobacter pylori (*H. pylori*) is a gram-negative spiral or curved bacillus, discovered in 1982 by the Australian scientists Robin Warren (1937-present) and Barry Marshall (1951–present), who won the 2005 Nobel Prize for their discovery of the bacteria and its role in the development of gastritis, peptic ulcer disease, and duodenal ulcer [1–3]. *H. pylori* is now believed to cause other diseases of the stomach including cancer and mucosa-associated lymphoid tissue (MALT) lymphoma [4].

M. O. Welcome, *Gastrointestinal Physiology*,
https://doi.org/10.1007/978-3-319-91056-7_14

Emerging studies indicate possible association of *H. pylori* infection and immune-mediated gastric and extragastric diseases such as colorectal cancer, pancreatic cancers, iron deficiency anemia, vitamin B12 deficiency, idiopathic thrombocytopenic purpura, obesity, diabetes mellitus, pulmonary, cardiovascular, neurological/psychological, dermatological, ocular disorders [5–11]. This gastroduodenal disease-causing microbe was first cultured in 1983 [12]. *H. pylori* is one of the most frequent bacterial infections in the world [13, 14]. *H. pylori* influences several aspects of gastric physiology to its benefits and causes detrimental effects to the host. This chapter deals with the pathophysiology of *H. pylori*.

14.2 Pathogenicity of *H. Pylori*: Cytotoxin-Associated Gene Products

H. pylori usually colonizes humans during the early childhood age [15]. The pathogenicity of this bacterium depends on many factors but the cytotoxin-associated gene (Cag) island, and the enzyme urease has been shown to be crucial to the development of diseases associated with colonization of the bacterium. Cag encodes proteins that are incorporated to form a functional type IV secretion system. This secretory system is responsible for transporting the products synthesized by the bacterium into the epithelial cells of the stomach of the host [16, 17]. One of the substances transported via this secretion system is cytotoxin-associated antigen (CagA). This protein is an immune response oncoprotein with a molecular weight of approximately 120–145 kDa, influencing several signaling cascades that ensure maintenance of the actin-cytoskeletal structure, which may eventually lead to the development of diseases [16, 18]. Some Cag gene products (CagL, CagI, and other adhesins) are used to bind the host receptor $\alpha 5\beta 1$ integrin, which ensures tight contact between *H. pylori* and epithelial cells of the stomach [13, 16].

14.3 *H. Pylori* Urease, Urease Transporter, and Physiology of Gastric Microenvironment

H. pylori produces substantial quantity of urease (urea amidohydrolase), a 550-kDa acid-sensitive nickel-containing enzyme, which is utilized by the bacterium for successful colonization of the host [19]. The amount of urease synthesized by *H. pylori* is estimated to be about 10–15% of its total protein synthesis [20]. The enzyme urease occurs in two locations in the bacterium—cytoplasm and bacterial (extracellular) surface [21, 22]. Both locations of urease are important for resistance of *H. pylori* to acid, enhancing the survival of the organism in the gastric microenvironment [19, 22]. While the extracellular urease can easily gain access to urea, the cytoplasmic urease requires a urea transporter protein. This transporter is called acid-activated urea channel, UreI. The channel transports urea into the bacterial cytoplasm, leading to the activation of cytoplasmic urease at acidic pH [20, 23].

Urease catalyzes the hydrolysis of urea to generate ammonia and carbon dioxide [21]. The ammonia released by the activities of this enzyme can change mucosal pH of the microenvironment of the stomach. The growth and motility of *H. pylori* depend on changes in the surrounding pH of the stomach [22]. Evidence indicates that generation of ammonia enhances survival of the bacterium in the mucosa of the stomach through the neutralization of gastric acid [21, 24]. The enzyme urease is thought to control *H. pylori* overgrowth by raising the mucus pH above 8 units, thereby suppressing the activities of the microbe [22]. The optimum range of pH for *H. pylori* is in the range 4–8 units. The optimal level of activity of the enzyme is a pH range of ~2–7, at which urease activity is optimal [20, 22, 25].

14.4 *H. Pylori* Acid Chemoreceptor Sensing

Emerging data indicates that *H. pylori* senses acidity in the stomach via multiple pathways. This chemotactic behavior is mediated through the chemotaxis protein (also known as transducer-like proteins, Tlp) TlpA, TlpB, TlpC, and TlpD [26–28]. The chemoreceptors TlpA and TlpD are acid sensors that control the movement of the bacterium away from areas of high acid concentration (hydrogen ion concentration ~100 nM). The chemoreceptor TlpD mediates the bacterial movement to areas of basic pH. The chemoreceptor TlpB represses motility of *H. pylori* toward the basic pH [27]. Thus, the activities of these receptors are inescapable for colonization and persistence *H. pylori* in the gastric mucus as the receptors regulate the movement and behavior of the bacterium and obviate eradication detrimental changes in gastric pH or mucus [26, 29]. Indeed majority of *H. pylori* are found within approximately 30 μm deep in the mucus layer of the stomach [26]. It should be mentioned that *H. pylori* also utilizes the chemoreceptors to sense other substances such as ammonia, urea, zinc, nickel, bicarbonate, arginine [30, 31].

14.5 *H. Pylori* and Gastritis

Gastritis is the inflammation of the mucosa of the stomach [32]. *H. pylori* causes chronic inflammation of the gastric mucosa in infected individuals [33]. The type of gastric inflammation caused by the bacterium is called atrophic gastritis—gastric inflammation that involves loss of mucosal glands due to long-standing *H. pylori* infection [32, 34]. There are currently different clinical classifications and staging of gastritis, which can be assessed at Sugimoto et al. [35], Rugge et al. [32], Rugge et al. [36].

The inflammatory reaction of the gastric mucosa is largely due to the immune responses which *H. pylori* and its enzyme urease elicit in the gastric mucosa. These immune responses are associated with large accumulation of IgG, IgA, IgM secreting cells in the gastric mucosa, recruitment of lymphocytes and neutrophils to the site of microbial aggression, and formation of tertiary lymphoid tissues at the site of aggression [21, 25, 37]. Importantly, these responses correlate with CagA antibodies [38].

Tests used for identification of *H. pylori* infection are described in Clinical Correlate 14.1.

Clinical Correlate 14.1
Tests for *H. Pylori*

Tests for *H. pylori* can broadly be divided into invasive (direct) and noninvasive (indirect). The invasive tests are those that require endoscopic excision of gastric mucosa for the examination. The noninvasive tests require blood, breath, or stool [12, 14, 39, 40]. In terms of specificity and sensitivity, the invasive tests are not superior to the noninvasive tests [39].

Invasive methods

Invasive tests include rapid urease test, histology, smear (cytology), bacterial culture, and polymerase chain reaction [40, 41]. Gastric tissues (biopsies) for these tests are excised from the corpus and antrum (distal aspect of the stomach, about 2 cm from the pylorus) [40, 42]. These sites are most implicated in the development of gastric cancer. However, the report of Kim et al. [42] shows that the sensitivity and specificity of biopsy from the upper body greater curvature are higher compared to the antrum, lesser curvature, and upper body lesser curvature [42].

Rapid urease test: The rapid urease test a rapid, simple, and inexpensive diagnostic test that detects the presence of urease in gastric mucosa [12, 41, 43]. This is one of the first tests that is done following gastric endoscopy for *H. pylori* identification [41]. The rapid urease test is used to detect the presence of an active infection. To carry out the test, a sample of gastric mucosa collected during endoscopy is placed in a tube, gel, or other media containing urea and a pH indicator. The presence of urease (presumably from *H. pylori*) will result to the hydrolysis of urea to carbon dioxide and ammonia. The products of urea hydrolysis increase the pH of the medium resulting to color change in the pH indicator. The color changes and the value of the pH depend on the quantity of bacteria in the sample [4, 12]. For an agar gel, for instance, a positive test with color change requires about 10^5 *H. pylori* in the tissue sample [12]. Rapid urease test results can be obtained in minutes to 24 h. Note that this duration, in part, depends on the number of microbes present in the sample [4]. The sensitivity of the rapid urease test is about 80–100% and specificity is 97–100% [12, 39, 40, 44].

Histology: Histological examination of biopsy sample obtained from the gastric mucosa provides information on whether or not there is tissue inflammation. It also gives information on cancerous and precancerous changes in the gastric tissue [41]. Histological examination should be carried out by an experienced specialist as sensitivity depends on experience as well as area of tissue collection. The sensitivity and specificity of histology are around 90–95% [40, 44].

Culture: Part of the biopsy sample obtained during endoscopy may be used for culture and sensitivity testing. This is expected to improve management of the patient's condition [41]. Culture and sensitivity testing is indicated for patients who have had failure of at least two eradication therapies. Failure of treatment may be caused by many factors but susceptibility and resistance of the bacteria are important factors that determine success of therapy. The specificity of culture is 100%, but the sensitivity depends on experience of the microbiologist [40, 45–47].

Polymerase chain reaction: The polymerase chain reaction (PCR) for the identification of *H. pylori* is usually used for research purposes, but also has immense clinical use [40, 48, 49]. The sensitivity of PCR is approximately 94%, and the specificity is ~96% [39]. A new method (e.g., loop-mediated isothermal amplification, LAMP) of genetic detection of *H. pylori* has been developed [49, 50].

Noninvasive tests

The noninvasive tests include the urea breath test, stool test, and serology [41, 44]. These tests are normally used for post-treatment monitoring of patients and are performed at 4–6th week following completion of eradication therapy [4, 41].

Urea breath test: The breath test uses ^{13}C- or ^{14}C-labeled urea to determine the presence of *H. pylori* infection. The ^{13}C- or ^{14}C-labeled urea is ingested orally. If the bacterium is present, then the ingested substance will be broken down into ammonia and carbon dioxide as $^{13}CO_2$ or $^{14}CO_2$. The carbon dioxide diffuses into the bloodstream and is transported to the lungs, where it is expired [4, 51, 52]. Carbon-13 isotope in $^{13}CO_2$ is non-poisonous and non-radioactive and can be measured in the breath with a special mass spectrometer. Carbon-14 isotope in $^{14}CO_2$ is radioactive and thus requires a nuclear machine (e.g., scintillograph) for the analysis [4, 53]. The breath test is a noninvasive method recommended for monitoring *H. pylori* eradication therapy [14, 54]. The sensitivity of this test is approximately 98.9–100% and specificity is about 98.9–99.5% [4, 15, 54]. Most studies report a higher sensitivity for the ^{13}C-urea breath test compared to serological assay and stool antigen test [14].

Stool antigen test: The test uses an enzyme immunoassay to detect the presence of antigens against *H. pylori* in stool samples [4, 15]. The enzyme immunoassay is either based on the use of polyclonal or monoclonal antibodies. The use of monoclonal enzyme-linked immunosorbent assay (ELISA) shows a superior result compared with the use of polyclonal antibodies [4, 54]. The sensitivity and specificity of monoclonal ELISA are 94–97% and

100%, respectively, whereas for polyclonal ELISA, the sensitivity is 88%, and specificity is 97% [4, 55]. The *H. pylori* stool antigen test is a reliable method to diagnose an active infection. It is also recommended for use in investigating the success of eradication therapy [4].

Serological test: The test involves collection of serum samples for detecting IgA and IgG for *H. pylori* infection [56–59]. The *H. pylori* IgG and IgA levels can be used to monitor the success of eradication therapy [57, 60]. The serological tests show high accuracy. For instance, at a cutoff greater than 10 U/ml, the sensitivity and specificity of IgG antibody test are 96.5 and 93%, respectively [56].

Rapid urine *H. pylori* antibody test: Urine antibody test for *H. pylori* was recently introduced [61, 62]. The test has a high degree of accuracy and shows promise for use in detecting *H. pylori* infection and monitoring the progress of eradication therapy [62]. The sensitivity of the test is 85.7–97.6% and the specificity is 95.7–100% [61, 62].

14.6 *H. Pylori* and Gastric Ulcer

Gastric ulcer is a break (erosion or ulceration) in the lining of the gastric mucosal observed during endoscopic examination [63, 64]. In specific terms, an ulcer can be defined as mucosal break of at least 3 mm deep [65]. One of the most common upper GI ulceration resulting to substantial loss of per capita income and reduction of quality of life, peptic ulcer [66–68], is described in Clinical Correlate 14.2.

Clinical Correlate 14.2
Peptic Ulcer Disease

Peptic ulcer disease is characterized by erosion of the mucosa of the stomach and proximal duodenum, due to multiple etiological factors but *H. pylori* infection and use of non-steroidal anti-inflammatory drugs (NSAID) are major causes [69, 70]. About 70% of gastric ulcers and 95% of duodenal are associated with *H. pylori* infection [71]. Thus, the prevalent types of peptic ulcer disease are H. pylori-associated and NSAID-associated peptic ulcer diseases. The other types are either non-*H. pylori*-associated or non-NSAID-associated disease and they may be classified according to different schemes [72]. For example, idiopathic and refractory peptic ulcer diseases may be classified under non-*H. pylori*-associated or non-NSAID-associated disease [73]. An idiopathic peptic ulcer disease is a clinical condition with an unknown cause [70, 73]. The mucosal breaks in refractory peptic ulcer disease do not heal completely after 8–12 weeks of standard therapy. While the causes of idiopathic peptic ulcer disease are not known, refractory peptic ulcer disease is mainly due to NSAID use, persistent *H. pylori* infection, resistant or highly virulence *H. pylori* strains, antibiotic resistance, presence of hypersecretory syndrome, malignancy, or even large size ulcers [12, 74]. It is believed that certain factors may predispose an individual to the development of peptic ulcer disease. These factors include alcohol consumption, smoking, psychological stress, disordered mucosal immune response, genetic predisposition or family history, disordered acid secretion and rapid gastric emptying, use of NSAID, aspirin or antiplatelet agents, central as well as total obesity [70, 75–80]. Protective factors against the development of the disease may be consumption of chili peppers, concurrent parasite infestation, moderate exercise [78, 79].

H. pylori infection is usually acquired during the childhood age from infected person through fecal-to-oral or oral-to-oral transmission [81]. A study by Mhaskar et al. [78] reported risk factors for *H. pylori* infection as smoking, lower socioeconomic status, consumption of restaurant food, meat, and non-filtered water [78].

Following ingestion, the bacteria localizes to the stomach through a process that involves interaction with epithelial cell membrane glycolipid receptors [82]. The attachment of *H. pylori* to the mucous layer involves certain structures on the surface of the epithelial cell, including lactosylceramide sulfate and GM3 ganglioside [83]. These membrane structures of the epithelial cell show high expression in the antral region of the stomach compared to the fundal region. This differential expression of these structures is one of the reasons why the number of *H. pylori* varies in different regions of the stomach [83].

The response of the host immune system substantially determines *H. pylori* colonization and subsequent infection of the stomach. The nucleotide-binding oligomerization domain (NOD) protein-1 and -2 (NOD-1 and NOD-2) are intracellular pathogen recognition signaling molecules that are implicated in innate immune response. The NOD proteins interact with certain components of the bacterial peptidoglycan cell wall that has been transferred to the host epithelial cell cytoplasm upon colonization [16, 84–87]. The bacterial muramyl dipeptide is one of the components transferred to the host epithelial cell cytoplasm to interact with intracellular signaling molecules. Adequate response by NOD will result to synthesis of interferon-β and other cytokines (e.g., tumor necrosis factor, interleukin-1), which subdue the activities of *H. pylori* [16, 87–90]. It is imperative to mention that the immune response resulting from *H. pylori* colonization and infection is due to multiple factors which include CagA and other related bacterial peptides as well as host immune components [16]. Inadequate immune response to *H. pylori* colonization of the gut induces damage to the mucosa of the stomach. This damage initially presents as an inflammation, weakening the mucus layer, and may subsequently lead to peptic ulcer disease or other associated conditions [81, 91].

H. pylori colonization of the stomach alone is believed to be associated with 18% reduction in the mucus layer independent of underlying atrophy (Fig. 14.1) [91, 92]. The mucosal changes are due to the action of *H. pylori* enzymes (e.g., urease), which disintegrate mucosal structures (polymeric mucin and lipid), thus leading to loss of mucosal surface [83]. The result is further erosion by the gastric juice [92].

Peptic ulcer disease may be either symptomatic or asymptomatic [93]. The symptoms of peptic ulcer disease include epigastric pain or discomfort, loss of appetite, loss of weight, nausea or vomiting, feeling of fullness. There may be alarm symptoms. The pain experienced in peptic ulcer disease is relieved by ingestion of food or antacids, but may resurface between periods of meal intake. The epigastric pain in peptic ulcer disease usually causes awakening at night. The alarm symptoms point to the presence of complication or malignancy and require immediate medical attention [69, 94].

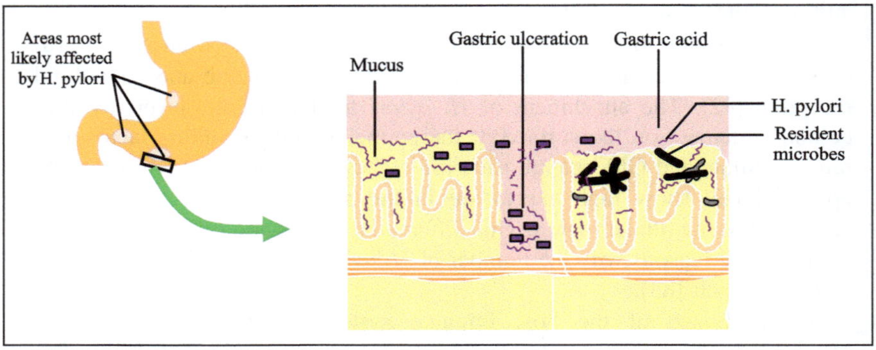

Fig. 14.1 Gastric ulcers caused by *H. pylori*

Diagnosis: For diagnosis of *H. pylori*-associated peptic ulcer disease, it is important to have a detailed history of the patient with appropriate investigations. The required investigations include endoscopy, histology, culture, PCR, and rapid urease test. The urea breath test, serology, stool antigen, and rapid urine tests are alternative tests [4, 95].

Treatment of *H. pylori*-associated peptic ulcer disease: Treatment of the patient depends on the history/state of the patient and presenting complains. If *H. pylori* infection is diagnosed, the infection should be eradicated and antisecretory therapy (such as proton pump inhibitor, PPI) initiated and given for four weeks [69]. The standard triple therapy is recommended for treatment of the infection, and it contains bismuth, metronidazole, and tetracycline. The addition of PPI gives a better result [4, 96]. However, other recommendations, which have shown high effectiveness, include quadruple, sequential, and concomitant therapies [4]. The treatment involves a combination of different antibiotics such as clarithromycin, metronidazole, tetracycline, amoxicillin, fluoroquinolones, and tinidazole. The antibiotics are used in combination with bismuth and PPIs [4, 97–99]. On the basis of resistance to clarithromycin, these antibiotics are usually outlined as first-, second-, and third-line treatments [4]. For a detailed treatment outline, review Garza-González et al. [4].

Note that patients on NSAIDs and aspirin have to discontinue the use of these drugs upon presentation [69].

Treatment of *H. pylori*-associated peptic ulcer disease in pregnancy: The use of bismuth, tetracyclines, and quinolones are contraindicated in pregnancy [4]. For *H. pylori* eradication therapy in pregnancy, amoxicillin and clarithromycin are recommended. The PPI used is omeprazole. Drugs should be taken for 14 days. In addition, iron and folic acid should be given in therapeutic doses [4, 100].

Probiotics: The use of probiotics is encouraged in the treatment of *H. pylori* infection. The microbes in probiotics are believed to interfere with *H. pylori* activity on the host's mucus membrane [4, 101, 102]. Although the mechanisms are not clear, emerging evidences indicate that *Lactobacillus*- and *Bifidobacterium*-containing probiotics are effective in *H. pylori* eradication therapy [4, 101–103].

Surgery: Surgery is indicated for complications associated with *H. pylori* infection/peptic ulcer disease. Surgery should be considered in cases when the ulcer is unresponsive to treatment [69].

Complications of peptic ulcer disease: Possible complications of peptic ulcer disease include bleeding, perforation, and gastric outlet obstruction. The first is the most common complications of the disease [69, 104, 105].

Lifestyle changes: Patients with the disease are advised to quit alcohol consumption, smoking, and stressful work or conditions that may necessarily aggravate the condition. Moderate level of physical activity is beneficial [64].

14.7 *H. Pylori* and Gastric Cancer

In 1994, the International Agency for Research on Cancer of the World Health Organization classified *H. pylori* as a group I or definite carcinogen [106]. *H. pylori*-associated inflammation (pangastritis) leads to loss of gastric glands (i.e., atrophic gastritis) in the majority of people. Atrophic gastritis is associated with intestinal metaplasia and dysplasia [107]. The intestinal metaplasia/dysplasia weakens mucosal defense resulting to peptic ulceration [108]. The precancerous changes associated with *H. pylori* infection develop in several decades following colonization [16]. These changes increase the risk of the development of cancer (carcinoma) of the distal stomach (antrum or body) by about 5–90-fold [16, 107, 109]. Chronic pangastritis due to *H. pylori* infection is also associated with lymphoma of gastric mucosal-associated lymphoid tissue (MALT). A substantial proportion of gastric MALT lymphoma is related to *H. pylori* infection [16]. About 70% of all gastric cancer cases worldwide are directly attributable to prior *H. pylori* infection [16].

CagA also acts as a bacterial oncoprotein as it is capable of inducing malignancy. This protein is transported into the gastric epithelial cell via the type IV secretory system. CagA deranges multiple host signaling pathways implicated in the pathogenesis of gastric cancer. It causes dysfunctions of many genetic and epigenetic molecules that predispose the individual to the formation of cancer [110–112]. CagA positivity in addition to the pattern and intensity of inflammation and acid secretion determines the differences associated with *H. pylori*-induced gastric disorders [113].

Certain products of *H. pylori* activities such as ammonia, reactive oxygen, and nitrogen species among others have cytotoxic effects on the gastric epithelium and are capable of inducing cellular damage that may result to the development of cancer in the stomach [16, 21, 24, 108]. These molecules induce oxidative stress that result to cellular and DNA damage. In these conditions, though, a repair process is instituted by the cell, but in certain cases especially when the immune responses are not adequate, antioxidant and repair system are depleted, the damage usually accumulate and the likelihood of genetic errors increases, which may result to carcinogenesis [16].

14.8 Conclusions

H. pylori-associated infection is one of the most prevalent infections in the world. The bacterium colonizes the gut of about half of the world inhabitants. Its pathogenicity depends on many factors including Cag island, host factors such as immunity, genetic predisposition, excessive alcohol consumption, and smoking. *H. pylori* is one of the most frequent causes of gastritis, peptic ulcer disease, and gastric cancer. Early diagnosis of the condition is useful for patient's management.

Recommended Readings

Original Articles

1. Hu HM, Kuo CH, Lo YC, Wu MT, Wu IC, Lu CY et al (2007) Evaluation of the two immunochromatographic methods for detecting urine and serum IgG antibodies to helicobacter pylori and comparison of accuracy and clinical utility. Hepatogastroenterology 54(73):119–123
2. Jalalypour F, Farajnia S, Somi MH, Hojabri Z, Yousefzadeh R, Saeedi N (2016) Comparative evaluation of RUT, PCR and ELISA tests for detection of infection with cytotoxigenic *H. pylori*. Adv Pharm Bull 6(2):261–266
3. Liu J, Liu H, Zhang T, Ren X, Nadolny C, Dong X et al (2014) Serum *Helicobacter pylori* NapA antibody as a potential biomarker for gastric cancer. Sci Rep 4:4143

4. Okuda M, Kamiya S, Booka M, Kikuchi S, Osaki T, Hiwatani T et al (2013) Diagnostic accuracy of urine-based kits for detection of *Helicobacter pylori* antibody in children. Pediatr Int 55(3):337–341
5. Patnayak R, Reddy V, Jena A, Gavini S, Thota A, Nandyala R, Chowhan AK (2016) *Helicobacter pylori* in cholecystectomy specimens-morphological and immunohistochemical assessment. J Clin Diagn Res 10(5):EC01–EC03
6. Wang JW, Hsu CN, Tai WC, Ku MK, Hung TH, Tseng KL et al (2016) The association of *Helicobacter pylori* eradication with the occurrences of chronic kidney diseases in patients with peptic ulcer diseases. PLoS One 11(10): e0164824
7. Wu DC, Kuo CH, Lu CY, Su YC, Yu FJ, Lee YC et al (2001) Evaluation of an office-based urine test for detecting *Helicobacter pylori*: a prospective pilot study. Hepatogastroenterology 48(39):614–617

Review Articles

1. Ricci C, Holton J, Vaira D (2007) Diagnosis of *Helicobacter pylori*: invasive and non-invasive tests. Best Pract Res Clin Gastroenterol 21(2):299–313
2. Singhal AV, Sepulveda AR (2005) *Helicobacter heilmannii* gastritis: a case study with review of literature. Am J Surg Pathol 29(11):1537–1539
3. Tiszlavicz L, Suták J, Wittmann T, Cserni G (1999) *Helicobacter heilmannii* (*Gastrospirillum hominis*)-associated gastritis: first reported cases in Hungary. Orv Hetil 140(16):887–890
4. Yang HR (2016) Updates on the diagnosis of *Helicobacter pylori* infection in children: what are the differences between adults and children? Pediatr Gastroenterol Hepatol Nutr 19(2):96–103
5. Yu LY, Sun LN, Zhang XH, Li YQ, Yu L, Yuan ZQ et al (2017) A review of the novel application and potential adverse effects of proton pump inhibitors. Adv Ther 34(5):1070–1086

Guidelines

1. De Francesco V, Bellesia A, Ridola L, Manta R, Zullo A (2017) First-line therapies for *Helicobacter pylori* eradication: a critical reappraisal of updated guidelines. Ann Gastroenterol 30(4):373–379
2. Dzieniszewski J, Jarosz M (2006) Guidelines in the medical treatment of *Helicobacter pylori* infection. J Physiol Pharmacol 57(3):143–154
3. Fock KM, Katelaris P, Sugano K, Ang TL, Hunt R, Talley NJ et al (2009) Second Asia-pacific consensus guidelines for *Helicobacter pylori* infection. J Gastroenterol Hepatol 24(10):1587–1600
4. Malfertheiner P, Megraud F, O'Morain CA, Gisbert JP, Kuipers EJ, Axon AT et al (2017) Management of *Helicobacter pylori* infection-the Maastricht V/ Florence Consensus Report Gut 66(1):6–30

Books

1. Hunt RH, Tytgat GNJ eds (2000) *Helicobacter pylori*: basic mechanisms to clinical cure 2000. Springer, Dordrecht, Netherlands

2. Kim N (ed) (2016) *Helicobacter pylori.* Springer, Singapore
3. Suzuki H, Warren R, Marshall B (eds) (2016) *Helicobacter pylori.* Springer, Tokyo, Japan

Bibliography

1. Warren JR, Marshall B (1983) Unidentified curved bacilli on gastric epithelium in active chronic gastritis. Lancet 1:1273–1275
2. Marshall BJ, Warren JR (1984) Unidentified curved bacilli in the stomach of patients with gastritis and peptic ulceration. Lancet 1(8390):1311–1315
3. Konturek SJ, Konturek PC, Konturek JW, Plonka M, Czesnikiewicz-Guzik M, Brzozowski T, Bielanski W (2006) *Helicobacter pylori* and its involvement in gastritis and peptic ulcer formation. J Physiol Pharmacol 57(3):29–50
4. Garza-González E, Perez-Perez GI, Maldonado-Garza HJ, Bosques-Padilla FJ (2014) A review of *Helicobacter pylori* diagnosis, treatment, and methods to detect eradication. World J Gastroenterol 20(6):1438–1449
5. Franceschi F, Tortora A, Gasbarrini G, Gasbarrini A (2014) *Helicobacter pylori* and extragastric diseases. Helicobacter 19(1):52–58
6. Campuzano-Maya G (2011) Cure of alopecia areata after eradication of Helicobacter pylori: a new association? World J Gastroenterol 17(26):3165–3170
7. Hershko C, Ianculovich M, Souroujon M (2007) A hematologist's view of unexplained iron deficiency anemia in males: impact of Helicobacter pylori eradication. Blood Cells Mol Dis 38(1):45–53
8. Kabeer KK, Ananthakrishnan N, Anand C, Balasundaram S (2017) Prevalence of *Helicobacter Pylori* infection and stress, anxiety or depression in functional dyspepsia and outcome after appropriate intervention. J Clin Diagn Res 11(8):VC11–VC15
9. Guarneri C, Lotti J, Fioranelli M, Roccia MG, Lotti T, Guarneri F (2017) Possible role of *Helicobacter pylori* in diseases of dermatological interest. J Biol Regul Homeost Agents 31 (2 Suppl 2):57–77
10. Goni E, Franceschi F (2016) *Helicobacter pylori* and extragastric diseases. Helicobacter 21 (1):45–48
11. Testerman TL, Morris J (2014) Beyond the stomach: an updated view of Helicobacter pylori pathogenesis, diagnosis, and treatment. World J Gastroenterol 20(36):12781–12808
12. Uotani T, Graham DY (2015) Diagnosis of *Helicobacter pylori* using the rapid urease test. Ann Transl Med 3(1):9
13. Shaffer CL, Gaddy JA, Loh JT, Johnson EM, Hill S, Hennig EE et al (2011) *Helicobacter pylori* exploits a unique repertoire of type IV secretion system components for pilus assembly at the bacteria-host cell interface. PLoS Pathog 7(9):e1002237
14. Nocon M, Kuhlmann A, Leodolter A, Roll S, Vauth C, Willich SN, Greiner W (2009) Efficacy and cost-effectiveness of the 13C-urea breath test as the primary diagnostic investigation for the detection of *Helicobacter pylori* infection compared to invasive and non-invasive diagnostic tests. GMS Health Technol Assess 5:Doc14
15. Konstantopoulos N, Rüssmann H, Tasch C, Sauerwald T, Demmelmair H, Autenrieth I, Koletzko S (2001) Evaluation of the *Helicobacter pylori* stool antigen test (HpSA) for detection of Helicobacter pylori infection in children. Am J Gastroenterol 96(3):677–683
16. Kim SS, Ruiz VE, Carroll JD, Moss SF (2011) *Helicobacter pylori* in the pathogenesis of gastric cancer and gastric lymphoma. Cancer Lett 305(2):228–238
17. Tseng T-T, Tyler BM, Setubal JC (2009) Protein secretion systems in bacterial-host associations, and their description in the gene ontology. BMC Microbiol 9(1):S2

18. Jiménez-Soto LF, Clausen S, Sprenger A, Ertl C, Haas R (2013) Dynamics of the cag-type IV secretion system of *Helicobacter pylori* as studied by bacterial co-infections. Cell Microbiol 15(11):1924–1937
19. Krishnamurthy P, Parlow M, Zitzer JB, Vakil NB, Mobley HL, Levy M et al (1998) *Helicobacter pylori* containing only cytoplasmic urease is susceptible to acid. Infect Immun 66(11):5060–5066
20. Sachs G, Scott DR, Weeks DL, Rektorscheck M, Melchers K (2001) Regulation of urease for acid habitation. In: Mobley HLT, Mendz GL, Hazell SL (eds) *Helicobacter pylori*: physiology and genetics. ASM Press, Washington, DC, USA
21. Mobley HL (1996) The role of Helicobacter pylori urease in the pathogenesis of gastritis and peptic ulceration. Aliment Pharmacol Ther 10(1):57–64
22. Sidebotham RL, Worku ML, Karim QN, Dhir NK, Baron JH (2003) How Helicobacter pylori urease may affect external pH and influence growth and motility in the mucus environment: evidence from in-vitro studies. Eur J Gastroenterol Hepatol 15(4):395–401
23. Weeks DL, Eskandari S, Scott DR, Sachs G (2000) A H + -gated urea channel: the link between *Helicobacter pylori* urease and gastric colonization. Science 287(5452):482–485
24. Mobley HL, Hu LT, Foxal PA (1991) *Helicobacter pylori* urease: properties and role in pathogenesis. Scand J Gastroenterol Suppl 187:39–46
25. Sjöström JE, Larsson H (1996) Factors affecting growth and antibiotic susceptibility of *Helicobacter pylori*: effect of pH and urea on the survival of a wild-type strain and a urease-deficient mutant. J Med Microbiol 44(6):425–433
26. Croxen MA, Sisson G, Melano R, Hoffman PS (2006) The *Helicobacter pylori* chemotaxis receptor TlpB (HP0103) is required for pH taxis and for colonization of the gastric mucosa. J Bacteriol 188(7):2656–2665
27. Huang JY, Goers Sweeney E, Guillemin K, Amieva MR (2017) Multiple acid sensors control *Helicobacter pylori* colonization of the stomach. PLoS Pathog 13(1):e1006118
28. Machuca MA, Johnson KS, Liu YC, Steer DL, Ottemann KM, Roujeinikova A (2017) *Helicobacter pylori* chemoreceptor TlpC mediates chemotaxis to lactate. Sci Rep 7(1):14089
29. Williams SM, Chen YT, Andermann TM, Carter JE, McGee DJ, Ottemann KM (2007) *Helicobacter pylori* chemotaxis modulates inflammation and bacterium-gastric epithelium interactions in infected mice. Infect Immun 75(8):3747–3757
30. Aihara E, Closson C, Matthis AL, Schumacher MA, Engevik AC, Zavros Y et al (2014) Motility and chemotaxis mediate the preferential colonization of gastric injury sites by *Helicobacter pylori*. PLoS Pathog 10(7):e1004275
31. Behrens W, Schweinitzer T, McMurry JL, Loewen PC, Buettner FFR, Menz S, Josenhans C (2016) Localisation and protein-protein interactions of the *Helicobacter pylori* taxis sensor TlpD and their connection to metabolic functions. Sci Rep 6:23582
32. Rugge M, Pennelli G, Pilozzi E, Fassan M, Ingravallo G, Russo VM et al (2011) Gastritis: the histology report. Dig Liver Dis 43(4):S373–S384
33. Kuipers EJ (1998) Relationship between *Helicobacter pylori*, atrophic gastritis and gastric cancer. Aliment Pharmacol Ther 12(1):25–36
34. Rugge M, Correa P, Di Mario F, El-Omar E, Fiocca R, Geboes K et al (2008) OLGA staging for gastritis: a tutorial. Dig Liver Dis 40(8):650–658
35. Sugimoto M, Ban H, Ichikawa H, Sahara S, Otsuka T, Inatomi O et al (2017) Efficacy of the Kyoto classification of gastritis in identifying patients at high risk for gastric cancer. Intern Med 56(6):579–586
36. Rugge M, Meggio A, Pennelli G, Piscioli F, Giacomelli L, De Pretis G, Graham DY (2007) Gastritis staging in clinical practice: the OLGA staging system. Gut 56(5):631–636
37. Quiding-Järbrink M, Ahlstedt I, Lindholm C, Johansson E-L, Lönroth H (2001) Homing commitment of lymphocytes activated in the human gastric and intestinal mucosa. Gut 49:519–525

38. Vorobjova T, Ren Z, Dunkley M, Clancy R, Maaroos HI, Labotkin R et al (2006) Response of IgG1 and IgG2 subclasses to *Helicobacter pylori* in subjects with chronic inflammation of the gastric mucosa, atrophy and gastric cancer in a country with high Helicobacter pylori infection prevalence. APMIS 114(5):372–380
39. Khalifehgholi M, Shamsipour F, Ajhdarkosh H, Ebrahimi Daryani N, Pourmand MR, Hosseini M et al (2013) Comparison of five diagnostic methods for *Helicobacter pylori*. Iran J Microbiol 5(4):396–401
40. Pajares-García JM (1998) Diagnosis of *Helicobacter pylori*: invasive methods. Ital J Gastroenterol Hepatol 30(3):S320–S323
41. Burette A (1998) How (who?) and when to test or retest for H. pylori. Acta Gastroenterol Belg 61(3):336–343
42. Kim CG, Choi IJ, Lee JY, Cho SJ, Nam BH, Kook MC et al (2009) Biopsy site for detecting Helicobacter pylori infection in patients with gastric cancer. J Gastroenterol Hepatol 24 (3):469–474
43. Wong A, Ching SS, Long AS (2014) The use of a second biopsy from the gastric body for the detection of Helicobacter pylori using rapid urease test. Singapore Med J 55(12):644–647
44. Bermejo F, Boixeda D, Gisbert JP, Defarges V, Sanz JM, Redondo C et al (2002) Rapid urease test utility for *Helicobacter pylori* infection diagnosis in gastric ulcer disease. Hepatogastroenterology 49(44):572–575
45. Arslan N, Yılmaz Ö, Demiray-Gürbüz E (2017) Importance of antimicrobial susceptibility testing for the management of eradication in Helicobacter pylori infection. World J Gastroenterol 23(16):2854–2869
46. Zamani M, Rahbar A, Shokri-Shirvani J (2017) Resistance of *Helicobacter pylori* to furazolidone and levofloxacin: a viewpoint. World J Gastroenterol 23(37):6920–6922
47. Chen YL, Mo XQ, Huang GR, Huang YQ, Xiao J, Zhao LJ et al (2016) Gene polymorphisms of pathogenic Helicobacter pylori in patients with different types of gastrointestinal diseases. World J Gastroenterol 22(44):9718–9726
48. Ho GY, Windsor HM (2000) Accurate diagnosis of *Helicobacter pylori*: polymerase chain reaction tests. Gastroenterol Clin North Am 29(4):903–915
49. Bakhtiari S, Alvandi A, Pajavand H, Navabi J, Najafi F, Abiri R (2016) Development and diagnostic evaluation of loop-mediated isothermal amplification using a new gene target for rapid detection of *Helicobacter pylori*. Jundishapur J Microbiol 9(5):e28831
50. Yari F, Abiri R, Aryan E, Ahmadi Jouybari T, Navabi J, Alvandi A (2016) Loop-mediated isothermal amplification as a fast noninvasive method of *Helicobacter pylori* diagnosis. J Clin Lab Anal 30(5):464–470
51. Zhou Q, Li L, Ai Y, Pan Z, Guo M, Han J (2017) Diagnostic accuracy of the 14C-urea breath test in *Helicobacter pylori* infections: a meta-analysis. Wien Klin Wochenschr 129(1–2):38–45
52. Ling D (2013) Carbon-13 urea breath test for *Helicobacter pylori* infection in patients with uninvestigated ulcer-like dyspepsia: an evidence-based analysis. Ont Health Technol Assess Ser 13(19):1–30
53. Goddard AF, Logan RPH (1997) Urea breath tests for detecting *Helicobacter pylori*. Aliment Pharmacol Ther 11:641–649
54. Manes G, Zanetti MV, Piccirillo MM, Lombardi G, Balzano A, Pieramico O (2005) Accuracy of a new monoclonal stool antigen test in post-eradication assessment of Helicobacter pylori infection: comparison with the polyclonal stool antigen test and urea breath test. Dig Liver Dis 37(10):751–755
55. Paimela HM, Oksala NK, Kääriäinen IP, Carlson PJ, Kostiala AA, Sipponen PI (2006) Faecal antigen tests in the confirmation of the effect of Helicobacter eradication therapy. Ann Med 38(5):352–356
56. Shady MMA, Fathy HA, Ali A, Galal EM, Fathy GA, Sibaii H (2015) Comparison of serum IgG antibody test with gastric biopsy for the detection of *Helicobacter pylori* infection among Egyptian children. Open Access Maced J Med Sci 3(2):303–306

57. Pandya HB, Patel JS, Agravat HH, Singh NKR (2014) Non-invasive diagnosis of *Helicobacter pylori*: evaluation of two enzyme immunoassays, testing serum IgG and IgA response in the Anand district of Central Gujarat, India. J Clin Diagn Res 8(6):DC12–DC15
58. Shuto M, Fujioka T, Matsunari O, Okamoto K, Mizukami K, Okimoto T et al (2017) Association between gastric cancer risk and serum *Helicobacter pylori* antibody titers. Gastroenterol Res Pract 2017:1286198
59. She RC, Wilson AR, Litwin CM (2009) Evaluation of *Helicobacter pylori* immunoglobulin G (IgG), IgA, and IgM serologic testing compared to stool antigen testing. Clin Vaccine Immunol 16(8):1253–1255
60. Li S, Lu AP, Zhang L, Li YD (2003) Anti-*Helicobacter pylori* immunoglobulin G (IgG) and IgA antibody responses and the value of clinical presentations in diagnosis of H. pylori infection in patients with precancerous lesions. World J Gastroenterol 9(4):755–758
61. Okuda M, Mabe K, Lin Y, Chaochen W, Taniguchi Y, Kato M, Kikuchi S (2017) Rapid urine antibody test for *Helicobacter pylori* infection in adolescents. Pediatr Int 59(7):798–802
62. Mabe K, Kikuchi S, Okuda M, Takamasa M, Kato M, Asaka M (2017) Diagnostic accuracy of urine *Helicobacter pylori* antibody test in junior and senior high school students in Japan. Helicobacter 22(1):e12329
63. Desai JC, Goo T, Fukata M, Sanyal S, Dikman A, Miller K et al (2009) NSAID-induced antral ulcers are associated with distinct changes in mucosal gene expression. Aliment Pharmacol Ther 30(1):71–81
64. Shephard RJ (2017) Peptic ulcer and exercise. Sports Med 47(1):33–40
65. Yeomans NDNJ (2008) Systematic review: ulcer definition in NSAID ulcer prevention trials. Aliment Pharmacol Ther 15(27):465–472
66. Henke CJ, Levin TR, Henning JM, Potter LP (2000) Work loss costs due to peptic ulcer disease and gastroesophageal reflux disease in a health maintenance organization. Am J Gastroenterol 95(3):788–792
67. Joish VN, Donaldson G, Stockdale W, Oderda GM, Crawley J, Sasane R et al (2005) The economic impact of GERD and PUD: examination of direct and indirect costs using a large integrated employer claims database. Curr Med Res Opin 21(4):535–544
68. Wen Z, Li X, Lu Q, Brunson J, Zhao M, Tan J et al (2014) Health related quality of life in patients with chronic gastritis and peptic ulcer and factors with impact: a longitudinal study. BMC Gastroenterol 14:149
69. Ramakrishnan K, Salinas RC (2007) Peptic ulcer disease. Am Fam Physician 76(7):1005–1012
70. Quan C, Talley NJ (2002) Management of peptic ulcer disease not related to *Helicobacter pylori* or NSAIDs. Am J Gastroenterol 97(12):2950–2961
71. Ford AC, Gurusamy KS, Delaney B, Forman D, Moayyedi P (2016) Eradication therapy for peptic ulcer disease in *Helicobacter pylori*-positive people. Cochrane Database Syst Rev 4: CD003840
72. Yuan Y, Padol IT, Hunt RH (2006) Peptic ulcer disease today. Nat Clin Pract Gastroenterol Hepatol 3(2):80–89
73. Chung C-S, Chiang T-H, Lee Y-C (2015) A systematic approach for the diagnosis and treatment of idiopathic peptic ulcers. Korean J Intern Med 30(5):559–570
74. Kim HU (2015) Diagnostic and treatment approaches for refractory peptic ulcers. Clin Endosc 48(4):285–290
75. Rosenstock S, Jørgensen T, Bonnevie O, Andersen L (2003) Risk factors for peptic ulcer disease: a population based prospective cohort study comprising 2416 danish adults. Gut 52 (2):186–193
76. Najm WI (2011) Peptic ulcer disease. Prim Care Clin Office Pract 38(3):383–394
77. Kang JM, Kim N, Lee BH, Park HK, Jo HJ, Shin CM et al (2011) Risk factors for peptic ulcer bleeding in terms of *Helicobacter pylori*, NSAIDs, and antiplatelet agents. Scand J Gastroenterol 46(11):1295–1301

78. Mhaskar RS, Ricardo I, Azliyati A, Laxminarayan R, Amol B, Santosh W, Boo K (2013) Assessment of risk factors of *Helicobacter pylori* infection and peptic ulcer disease. J Glob Infect Dis 5(2):60–67
79. Levenstein S, Rosenstock S, Jacobsen RK, Jorgensen T (2015) Psychological stress increases risk for peptic ulcer, regardless of Helicobacter pylori infection or use of nonsteroidal anti-inflammatory drugs. Clin Gastroenterol Hepatol 13(3):498–506
80. Boylan MR, Khalili H, Huang ES, Chan AT (2014) Measures of adiposity are associated with increased risk of peptic ulcer. Clin Gastroenterol Hepatol 12(10):1688–1694
81. Abadi AT, Kusters JG (2016) Management of Helicobacter pylori infections. BMC Gastroenterol 16(1):94
82. Slomiany BL, Murty VL, Piotrowski J, Slomiany A (1994) Gastroprotective agents in mucosal defense against Helicobacter pylori. Gen Pharmacol 25(5):833–841
83. Slomiany BL, Slomiany A (1992) Mechanism of Helicobacter pylori pathogenesis: focus on mucus. J Clin Gastroenterol 14(1):S114–S121
84. Caruso R, Warner N, Inohara N, Núñez G (2014) NOD1 and NOD2: signaling, host defense, and inflammatory disease. Immunity 41(6):898–908
85. Watanabe T, Asano N, Kudo M, Strober W (2017) Nucleotide-binding oligomerization domain 1 and gastrointestinal disorders. Proc Jpn Acad Ser B Phys Biol Sci 93(8):578–599
86. Mogensen TH (2009) Pathogen recognition and inflammatory signaling in innate immune defenses. Clin Microbiol Rev 22(2):240–273
87. Patel MK, Trombly MI, Kurt-Jones EA (2012) Innate immune responses to *Helicobacter pylori* infection: an overview. In: Houghton J (ed) Helicobacter species. Methods in molecular biology (methods and protocols), vol 921. Humana Press, Totowa, New Jersey, United States
88. Kim HS, Kim J, Im HN, An DR, Lee M, Hesek D et al (2014) Structural basis for the recognition of muramyltripeptide by Helicobacter pylori Csd4, a d, l-carboxypeptidase controlling the helical cell shape. Acta Crystallogr D Biol Crystallogr 70(Pt 11):2800–2812
89. Barnich N, Aguirre JE, Reinecker H-C, Xavier R, Podolsky DK (2005) Membrane recruitment of NOD2 in intestinal epithelial cells is essential for nuclear factor–κB activation in muramyl dipeptide recognition. J Cell Biol 170(1):21–26
90. Semper RP, Mejías-Luque R, Groß C, Anderl F, Müller A, Vieth M et al (2014) *Helicobacter pylori*-induced IL-1β secretion in innate immune cells is regulated by the NLRP3 inflammasome and requires the cag pathogenicity island. J Immunol 193(7):3566–3576
91. Newton JL, Jordan N, Oliver L, Strugala V, Pearson J, James OF, Allen A (1998) Helicobacter pylori in vivo causes structural changes in the adherent gastric mucus layer but barrier thickness is not compromised. Gut 43(4):470–475
92. Allen A, Newton J, Oliver L, Jordan N, Strugala V, Pearson JP, Dettmar PW (1997) Mucus and H. pylori. J Physiol Pharmacol 48(3):297–305
93. Lee SP, Sung IK, Kim JH, Lee SY, Park HS, Shim CS (2017) Risk factors for the presence of symptoms in peptic ulcer disease. Clin Endosc 50(6):578–584
94. Khademi H, Radmard A-R, Malekzadeh F, Kamangar F, Nasseri-Moghaddam S, Johansson M et al (2012) Diagnostic accuracy of age and alarm symptoms for upper GI malignancy in patients with dyspepsia in a GI clinic: a 7-year cross-sectional study. PLoS ONE 7(6):e39173
95. Fashner J, Gitu AC (2015) Diagnosis and treatment of peptic ulcer disease and *H. pylori* infection. Am Fam Physician 91(4):236–242
96. Hunt RH (1997) Peptic ulcer disease: defining the treatment strategies in the era of Helicobacter pylori. Am J Gastroenterol 92(4):36S–43S
97. Ji Y, Lu H (2018) Meta-analysis: high-dose vs. low-dose metronidazole-containing therapies for Helicobacter pylori eradication treatment. PLoS ONE 13(1):e0189888
98. Safavi M, Sabourian R, Foroumadi A (2016) Treatment of *Helicobacter pylori* infection: current and future insights. World J Clin Cases 4(1):5–19

99. Almeida N, Donato MM, Romãozinho JM, Luxo C, Cardoso O, Cipriano MA et al (2015) Beyond maastricht IV: are standard empiric triple therapies for Helicobacter pylori still useful in a South-European country? BMC Gastroenterol 15:23
100. Davis S (2014) Eradication of *Helicobacter pylori.* S Afr Fam Pract 56(3):1–4
101. Feng JR, Wang F, Qiu X, McFarland LV, Chen PF, Zhou R et al (2017) Efficacy and safety of probiotic-supplemented triple therapy for eradication of Helicobacter pylori in children: a systematic review and network meta-analysis. Eur J Clin Pharmacol 73(10):1199–1208
102. Lau CS, Ward A, Chamberlain RS (2016) Probiotics improve the efficacy of standard triple therapy in the eradication of Helicobacter pylori: a meta-analysis. Infect Drug Resist 9:275–289
103. Wen J, Peng P, Chen P, Zeng L, Pan Q, Wei W, He J (2017) Probiotics in 14-day triple therapy for Asian pediatric patients with Helicobacter pylori infection: a network meta-analysis. Oncotarget 8(56):96409–96418
104. Milosavljevic T, Kostić-Milosavljević M, Jovanović Krstić M (2011) Complications of peptic ulcer disease. Dig Dis 29(5):491–493
105. Tonolini M, Ierardi AM, Bracchi E, Magistrelli P, Vella A, Carrafiello G (2017) Non-perforated peptic ulcer disease: multidetector CT findings, complications, and differential diagnosis. Insights Imaging 8(5):455–469
106. International Agency for Research on Cancer (IARC) (1994) Evaluation of carcinogenic risks to humans. Lyon, Frnace
107. Sipponen P, Hyvärinen H (1993) Role of *Helicobacter pylori* in the pathogenesis of gastritis, peptic ulcer and gastric cancer. Scand J Gastroenterol Suppl 196:3–6
108. Dixon MF (1994) Pathophysiology of *Helicobacter pylori* infection. Scand J Gastroenterol Suppl 201:7–10
109. Kuipers EJ (1999) Exploring the link between Helicobacter pylori and gastric cancer. Aliment Pharmacol Ther 13(1):3–11
110. Saber T, Ghonaim MM, Yousef AR, Khalifa A, Al Qurashi H, Shaqhan M, Samaha M (2015) Association of *Helicobacter pylori* cagA gene with gastric cancer and peptic ulcer in saudi patients. J Microbiol Biotechnol 25(7):1146–1153
111. Hatakeyama M (2006) The role of *Helicobacter pylori* CagA in gastric carcinogenesis. Int J Hematol 84(4):301–308
112. Hatakeyama M (2014) *Helicobacter pylori* CagA and gastric cancer: a paradigm for hit-and-run carcinogenesis. Cell Host Microbe 15(3):306–316
113. Graham DY, Yamaoka Y (1998) *H. pylori* and cagA: relationships with gastric cancer, duodenal ulcer, and reflux esophagitis and its complications. Helicobacter 3(3):145–151

Chapter 15
Functional Relationship Between the Gut and Other Tissues/Organs of the Body

Abstract The enteral system provides a pivotal functional connectivity with other organs and tissues of the body, referred to as gut–extraenteric tissue axis. This functional relationship regulates several functions of the body including higher mental (memory, cognition), cardiac, renal, pulmonary, hepatic, pancreatic functions. Examples of such relationship include gut–brain (enterocerebral) axis, gut–liver (enterohepatic) axis, gut–pancreas (enteropancreatic) axis, gut–liver–pancreas (enterohepatopancreatic/enteropancreatohepatic) axis or triangle, gut–heart (enterocardiac) axis, gut–kidney (enterorenal, enteronephric) axis, gut–lung (enteropulmonary) axis, gut–bone (entero-osseous) axis, gut–skin (enterocutaneous) axis, and gut–brain–skin (enterocerebrocutaneous) triangle. Other axes (usually referred to as systems) include hepatobiliary, pancreatobiliary, and pancreatohepatobiliary systems. These axes, triangles, and systems play a crucial role in the maintenance of homeostasis not only in the GI tract, but also in extraenteric tissues/organs. It is important to note that apart from the functional relationship, some of these axes, triangles, and systems are structurally connected to the gut. This chapter is concerned with this functional and structural connectivity.

Keywords Gut–liver (enterohepatic) axis · Gut–brain (enterocerebral) axis · Gut–pancreas (enteropancreatic) axis · Gut–liver-pancreas (enterohepatopancreatic/enteropancreatohepatic) axis or triangle · Gut–heart (enterocardiac) axis · Gut–bone (entero-osseous) axis · Gut–kidney (enterorenal, enteronephric) axis · Gut–lung (enteropulmonary) axis · Gut–skin (enterocutaneous) axis · Gut–brain–skin (enterocerebrocutaneous) triangle · Hepatobiliary · Pancreatobiliary · Pancreatohepatobiliary system · Small intestinal bacterial overgrowth

Abbreviations

cAMP	Cyclic adenosine monophosphate
CCK	Cholecystokinin
cGMP	Cyclic guanosine monophosphate
GDP	Guanosine diphosphate
GI	Gastrointestinal

M. O. Welcome, *Gastrointestinal Physiology*,
https://doi.org/10.1007/978-3-319-91056-7_15

GIP	Glucose-dependent insulinotropic polypeptide/gastro-inhibitory polypeptide
GLP-1	Glucagon-like peptide 1
GPCR	G protein-coupled receptor
IGF-1	Insulin-like growth factor-1
OXM	Oxyntomodulin
PP	Pancreatic polypeptide
SBP	Spontaneous bacterial peritonitis

15.1 Introduction

Apart from the traditional functions of the gut discussed in previous chapters of this book, the gastrointestinal (GI) tract regulates several functions of extraenteric tissues/organs. The enteral system (gut) provides functional connectivity with other tissues and organs of the body. This functional link is usually referred to as axis or triangle. Through this functional linkage, the gut regulates almost all activities of the body. Some of the axes or triangles include gut–brain (enterocerebral) axis, gut–liver (enterohepatic) axis, gut–pancreas (enteropancreatic) axis, gut–liver–pancreas (enterohepatopancreatic/enteropancreatohepatic) axis or triangle, gut–heart (enterocardiac) axis, gut–kidney (enteronephric) axis, gut–lung (enteropulmonary) axis, gut–skin (enterocutaneous) axis, and gut–brain–skin (enterocerebrocutaneous) triangle. Other axes (also referred to as systems) include hepatobiliary, pancreatobiliary, and pancreatohepatobiliary systems. It is important to note that "system" is normally used when the relationship also involves a structural connectivity, in addition to the functional linkage. The structural connectivity allows direct signaling of molecules from one organ or tissue to the other via anatomical passages or lumen. The functional connectivity involves the transmission of molecules secreted by the gut to exert their effects in a distant tissue or organ. Functional connectivity is neurally or hormonally mediated. Many aspects of this functional connectivity are mediated by the activities of the gut microbiota [1–8]. The gut microbiota is currently considered as a metabolic organ that modulates adiposity, local and systemic metabolism, including cardiac and hepatic metabolism [9]. To distinguish the effect of molecules released by the epithelial cells of the gut from those secreted by the gut commensals, some authors refer to the gut microbiota–extraenteric tissue/organ axis [1, 2]. The chapter deals with the functional and structural connectivity between the gut and other tissues/organs of the body and the implication for disease development.

15.2 Composition and Classification of the Gut Microbiota

Microbes colonize every part of the human body, but are especially found in the mucosa of the respiratory, urinary, genital, and GI tracts as well as on the skin. However, the gut, due to availability of nutrients, in the course of evolution has always had a higher proportion (more than 90%) of the total microbes that inhabit the human body [10]. The gut microbiota consists of several trillions of microbes ($\sim 10^{14}$), particularly bacteria and archaea with a fewer number of viruses, fungi, and protozoa, from at least 500–1000 species [11–13]. But a study has indicated that the total number of species may be far more than 1000 [14]. Both the number of microorganisms and their genes exceeds that of the human cells by several folds. The gut microbiota exceeds the number of cells in the human body by tenfold and the genes by 150-fold [10, 11, 15]. But a recent study reported revised estimates of the number of gut microbes showing that the number of bacteria in the body is almost the same as the number of human cells [16].

Over 90% of the resident bacteria constitute *Firmicutes* and *Bacteroidetes* species and comprise mainly anaerobes [11]. The remaining microbes are composed of *Actinobacteria*, *Cyanobacteria,* *Fusobacteria*, *Proteobacteria*, and *Verrucomicrobia* [10, 12]. The gut microbiota is populated on the outer layer of the intestinal mucus, whereas the inner intestinal mucus layer has a fewer microbes, but with immense protective features on the gut [12]. The bacterial composition of these two layers also substantially differs [10]. The number of microbes in the gut increases from proximal to the distal region of the GI tract [11]. There are about 500–700 or more bacterial species, especially anaerobes, in the oral cavity [17, 18]. An intimate kiss lasting for about 10s can transfer as much as 80 million bacteria, mostly comprising *Lactobacillus* and *Bifidobacterium* in healthy individuals [18]. But the oral microbiota also contains *Actinomyces*, *Fusobacterium*, *Granulicatella*, *Gemella*, *Haemophilus*, *Neisseria*, *Porphyromonas*, *Rothia*, *Streptococcus*, *Veillonella* [18]. In the esophagus, a wide range of commensal microbes have been identified, especially in the distal region. Though the composition of microbes in the esophagus may vary and depend on the state of health of the individual, esophageal microbiota mainly consists of *Actinobacteria*, *Bacteroidetes*, *Firmicutes*, and *Proteobacteria* [19, 20]. The number of bacteria in the stomach is estimated at about 10^1–10^2 per gram of gastric contents; 10^3 in the duodenum; 10^4–10^7 in the small intestine; and 10^{11}–10^{12} in the large intestine [10]. The composition of bacteria in these regions of the gut also differs. For further review, see Sekirov et al. (2010) [10]. It should be mentioned that the composition of the gut microbiota depends on many factors including the individual's age, dietary pattern, mode of delivery/birth, antibiotic administration, diseases, and a range of environmental factors including potentially poisonous fumes [11, 13].

15.3 Gut–Liver Axis

Gut–liver axis refers to the functional and structural connectivity between the gut and the liver that ensures regulated bidirectional movement of substances between the two organs for maintenance of homeostasis. This connectivity involves the gut microbiota, intestinal barrier, and liver. The homeostasis of this connectivity is maintained by the gut microbiota [12, 21]. Structural and functional details on the intestinal/mucosal barrier and permeability have been discussed in Chap. 4. The integrity of the intestinal mucosa depends on such factors as nutrition (pattern of diet, food types), lifestyle, antibiotic use, mucosal receptors, luminal pH, and immune response. Importantly, the non-digestible portion of carbohydrates is known to act as prebiotics, which can considerably change the anatomy and physiology of the GI mucosa [12, 22, 23]. Furthermore, fermentation of non-digestible portion of carbohydrates by the gut microbiota can provide about 5–15% of the daily energy needs [12].

The functions of intestinal barrier (intestinal permeability) and liver are controlled in part by the activities of the gut microbiota. The metabolism of bile acid by the gut microbiota also has substantial influence on the gut–liver axis. The controlled transfer of certain substances synthesized by the gut commensals substantially affects the functions of the liver. At the same time, the liver also controls the functions of the gut microbiota through secretion of certain proteins (e.g., factors of the immune system) as well as bile acids [24]. Therefore, the connectivity between the gut and liver can be referred to as a bidirectional relationship.

Disorders of any of these components of the gut–liver axis can result in serious health consequences. Such disorders have been implicated in inflammatory bowel disease, alcoholic and non-alcoholic fatty liver disease, celiac disease, obesity, diabetes, cardiovascular diseases, allergic diseases, irritable bowel syndrome, rheumatoid arthritis, cancer, autism, steatosis, hepatitis (e.g., autoimmune type), liver cirrhosis, cholestatic liver diseases, and liver failure. The gut microbiota also plays a crucial role in complications of liver cirrhosis—spontaneous bacterial peritonitis (SBP), portal hypertension, and hepatic encephalopathy [12, 13, 25–28]. About 20–75% of cases of chronic liver diseases have been shown to involve derangement of the gut flora (e.g., small intestinal bacterial overgrowth) [29, 30]. The pathogenesis of some of these disorders has been described [31]. Here, only a few of the disorders will be discussed.

SBP is an acute bacterial infection of ascitic fluid due to prolonged bacteremia, occurring on account of bacterial translocation into the bloodstream [32–34]. SBP is one of the features of liver cirrhosis, and it is characterized by bacterascites, fever, abdominal pain, rebound, abdominal tenderness, and altered mental status [33].

Portal hypertension can be defined as elevation of pressures in the portal system to about 5–10 mmHg or higher [35, 36]. The pressure in the hepatic venous system is usually less than 5 mmHg. However, in disease such as cirrhosis, the value may increase to 10 mmHg. Portal blood pressure of 5–10 mmHg is termed mild portal hypertension. A clinically significant pressure in the portal system is equal to or greater than 10 mmHg [35]. Portal hypertension is a serious clinical condition characterized by GI bleeding (mostly esophageal varices, but also gastric varices), splenomegaly, ascites, portosystemic encephalopathy, and liver failure [36].

Hepatic encephalopathy, also known as portosystemic encephalopathy, is a reversible syndrome of impaired brain functions occurring in persons with advanced liver disease [37, 38]. Portosystemic encephalopathy is an integral sign of liver decompensation in cirrhotic patients [36]. The syndrome is due to brain cell metabolic derangement, brain edema, systemic inflammatory response, or other conditions involving extensive functional loss of brain cells and dysfunctions in the blood–brain barrier [37–39]. This condition causes clinically significant dysfunctions in neurotransmission and gliotransmission. Although the severity and duration of symptoms may vary, overt portosystemic encephalopathy is characterized by spatiotemporal disorientation, disorder of consciousness, personality changes, and drowsiness. The end stages of portosystemic encephalopathy are characterized by stupor and coma [37–40]. Portosystemic encephalopathy is diagnosed with the aid of psychometric tests [37].

15.4 Gut–Pancreas Axis

Gut–pancreas axis is a structural and functional cross talk between the gut and the pancreas that ensures a regulated functioning of both organs. This axis was named “enteroinsular axis” in 1969 by Unger H. Roger (1924–) and Eisentraut M. Anna (–), referring to the functional link between the gut and the endocrine pancreas, whereby food substances trigger the release of certain molecules from the small intestine to influence the endocrine pancreatic functions. In turn, substances released from the endocrine pancreas also modulate the exocrine pancreatic secretory activity as well as metabolic functions [41].

In 1979, Prof. Dr. Werner Creutzfeldt (1924–2006) made an outstanding contribution by suggesting that the enteroinsular axis comprises input and output signals (nutrient, neural, hormonal) between the gut and islets [42–45]. So, food in the GI tract stimulates insulin secretion via humoral and neural mechanisms [46]. The hormones secreted by the gut upon stimulation by food (especially carbohydrate) are the incretins (glucose-dependent insulinotropic polypeptide, GIP, and glucagon-like peptide 1, GLP-1). The incretins stimulate the secretion of insulin and inhibit glucagon as well as somatostatin secretion [47, 48]. Pancreatic hormones and neurotransmitters released upon incretin stimulation also include ghrelin, pancreastatin, nitric oxide, substance P, peptide YY [49]. Details on incretins have been discussed in Chap. 8.

Further, research has showed that the exocrine pancreas is also functionally related to the endocrine pancreas via the pancreatic islet–acinar axis. Recent studies indicate that insulin secreted by the islet cells directly regulates the functions of exocrine pancreas in particular zymogen secretion [49, 50].

15.5 Gut–Brain Axis

The functional relationship between the gut and brain is referred to as the gut–brain or enterocerebral axis [51]. The gut–brain axis is controlled by the gut microbiota, gut hormones, enteric nervous system, central nervous system, and nutrients [52]. These components of the gut–brain axis interact bidirectionally through neurohumoral pathways and direct connections between the vagus nerve and the brainstem, and between the spinal nerves and the spinal cord [52, 53]. Many hormones and peptides are secreted in the gut in response to food ingestion. These hormones have been discussed extensively in Chap. 8. The hormones not only act on the GI tract (enteric nervous system), but also on the cells of the brainstem and hypothalamus [47, 52–54].

Diets containing dietary fibers (e.g., oligofructose and raffinose), dairy products, unsaturated fatty acids, or mixed nutrients can induce secretion of gut hormones such as GLP-1, GIP, cholecystokinin (CCK), pancreatic polypeptide (PP), peptide YY, and oxyntomodulin (OXM). GLP-1, CCK, PP, peptide YY, and OXM act to increase satiety, thereby decreasing food ingestion. In contrast, ghrelin increases appetite, thereby increasing food ingestion [52]. These hormones, by signaling to the brain, control appetite and satiety, thus regulating feeding behavior. The brain also controls other aspects of gut functioning such as GI secretion, motility, in part through the vago-vagal reflexes [53]. But it should be mentioned that the hormones and peptides not only control the feeding behavior, secretion, and motility, but also cortical functioning including attention, memory, emotion, cognition [55–57].

Small molecules such as cytokines (tumor necrosis factor alpha, interferon gamma, interleukin-6,-8) released in the gut in response to inflammation can affect neurotransmission or mediate other processes in the gut–brain axis [52].

The gut microbiota can affect the brain through its metabolic products. In particularly, the short-chain fatty acids (e.g., butyric acid and propionic acid) produced by the gut microbiota via neurohumoral, neuroimmune, and autonomic pathways influence brain functioning and represent a crucial pathogenic basis for some neurodevelopmental disorders involving social and cognitive deficits [52, 58–60]. Consequently, nutritional therapy has been used to address many gut–brain disorders including autism and inflammatory diseases [52]. The prebiotic oligofructose and arabinoxylan oligosaccharides not only increase short-chain fatty acids in the gut, but also increase the number of beneficial microbes (e.g., *Lactobacillus* and *Bifidobacterium* spp.) and enhance extraenteric tissue and organ functions [52, 61]. The short-chain fatty acids and other products of metabolism of these gut commensals can also increase the expression of GLP-1 and decrease that of ghrelin [52, 62].

The gut–brain axis is involved in diseases such as inflammatory bowel disease, obesity, type 2 diabetes mellitus, behavioral and psychiatric disorders [52, 63, 64].

15.6 Gut–Heart Axis

The gut–heart axis, also known as enterocardiac axis, refers to the functional relationship between the gut and the heart (Fig. 15.1). Certain hormones released by the gut influence distant tissues such as the heart in many ways. Such hormones include incretins, amylin, adrenomedullin, glucagon, insulin [65, 66].

Incretin (GLP-1): The GLP-1 is the most widely studied incretin that affects cardiovascular functions. GLP-1 receptors are expressed on cardiomyocytes. Following stimulation of the intestinal mucosal cells by meal, GLP-1 is released and transported via the bloodstream to the heart where it localizes and activates the cardiac GLP-1 receptor. The receptor is coupled to GDP-associated G protein, which, upon stimulation, mediates changes in cAMP level (Fig. 15.1). GLP-1 has positive inotropic and chronotropic effects on the heart [67–69]. GLP-1 receptor activation promotes the secretion of atrial natriuretic peptide, which mediates decrease in arterial blood pressure. Thus, GLP-1 receptor agonists can be used as antihypertensive agents [66].

Glucagon: Glucagon is a peptide hormone, mostly produced by alpha cells of the endocrine pancreas. The hormone is antagonistic in function to that of insulin. Glucagon functions via activation of its cognate receptors, which are coupled to GPCR. The stimulation of this receptor increases intracellular cAMP level, which in turn mediates several signaling pathways culminating in the activation of ion channels and gene expression in atrial and ventricular cardiac muscle cells. The

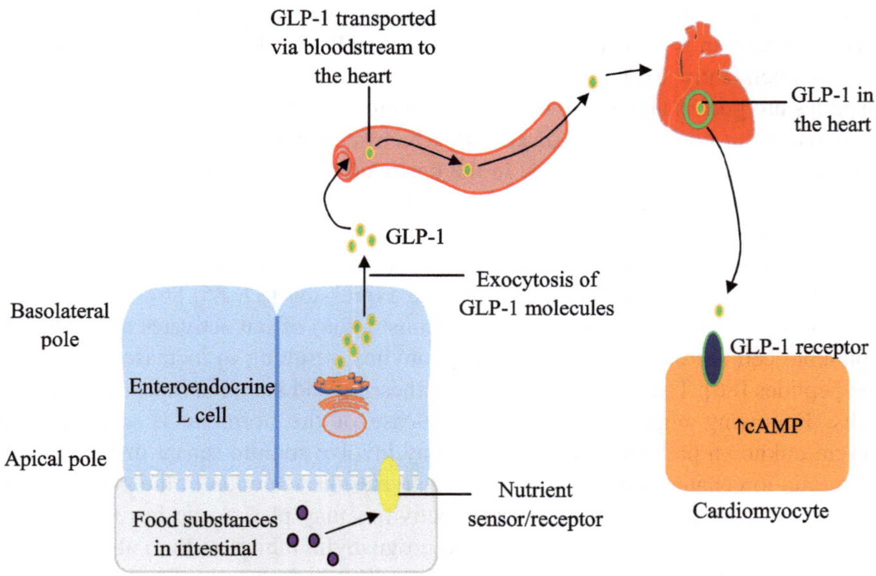

Fig. 15.1 Gut–heart axis

signaling pathways mediated by the action of glucagon lead to positive inotropic and chronotropic effects on the heart [70–73].

Insulin: Insulin is a peptide hormone mainly produced by the alpha cells of the endocrine pancreas [74]. Upon release from the endocrine pancreas, insulin is transported to several tissues and organs of the body including the heart, where it exerts a variety of effects. Insulin has a positive inotropic effect on the heart [74, 75]. The positive inotropic effect of insulin on myocardial cells is due to increase in intracellular calcium fluxes [75–77]. A similar positive inotropic effect is exerted by insulin-like growth factor-1 (IGF-1) [78].

Amylin is a peptide hormone mainly produced by a certain cell of the endocrine pancreas and enteroendocrine cells [79]. Amylin has positive inotropic and chronotropic as well as coronary vasodilatory effects on the heart [79–81].

Adrenomedullin: Though evidences are weak, research indicates that adrenomedullin exerts positive chronotropic and negative inotropic effect on the heart [79].

15.7 Gut–Kidney Axis

The gut–kidney axis, also known as enterorenal or enteronephric axis, is the functional linkage between the gut and the kidney. Some hormones and peptides produced by intestinal mucosal cells in response to salt intake have been implicated in the regulation of renal functions. These peptide hormones are called guanylins—guanylin and uroguanylin [82]. Other members of the guanylins include renoguanylin and lymphoguanylin [83]. Renoguanylin is structurally and functionally similar to guanylin and uroguanylin and highly expressed in the kidney [83]. Lymphoguanylin is also structurally and functionally similar to the other guanylins and is abundantly expressed in lymphoid tissues [84]. The receptors of these peptides are guanylate cyclase C. The stimulation of this receptor leads to increase in intracellular concentration of cGMP, which mediates a couple of downstream and upstream signaling pathways. In the upstream pathway, cGMP mainly inhibits Na^+/H^+ exchanger and stimulates the ion channel—cystic fibrosis transmembrane regulator [82, 85, 86]. However, the hormone can function upstream via PLA_2 pathway to inhibit K^+ channels [86]. The downstream pathways stimulate or inhibit transcription factors that play a role in gene expression [82, 85, 86].

Oral ingestion (but not intravenous administration) of salt activates the intestinal epithelial cell secretory pathway of the guanylins, resulting in increased release of the peptides [86]. The prohormones are synthesized and stored in intestinal mucosal cells. Following ingestion of salt, the release of the hormone is activated via yet-an-unknown pathway. The pathway may involve sodium sensor or the activity of certain ion channels that subsequently activates the secretion of the peptides into the bloodstream [87]. GI taste receptor activities may play a role in regulation of intestinal salt signaling [2]. Guanylin and uroguanylin inhibit sodium absorption in both the intestines and stimulate electrolyte (sodium and potassium) excretion in the kidney, thereby inducing electrolyte excretion. They also induce anion (Cl^-, HCO_3^-) and water secretion. Therefore, the hormones can be referred to as natriuretics, kaliuretics, and diuretics [82]. The guanylins can affect renal secretory functions via influences on aldosterone and angiotensin II. The renal effect of guanylin peptides is also related to their effects on atrial natriuretic peptide [86, 88]. Guanylin signaling in the kidney can be mediated via pendrin, an ion channel that is widely expressed in aldosterone, angiotensin II, and atrial natriuretic peptide—sensitive nephrons. The hormones aldosterone, angiotensin II, and atrial natriuretic peptide can upregulate pendrin expression. The guanylin peptides downregulate pendrin-mediated Cl^- reabsorption and bicarbonate secretion via the ion channel Cl^-/HCO_3^- exchanger [86, 88, 89].

The gut peptide hormones can also exert their influences on the kidney through the brain, referred to as the gut–brain–kidney axis or triangle. While some of the peptides produced in the gut may act directly on renal tissues, others may act on the CNS to indirectly affect renal functions [2]. For instance, the hormone amylin synthesized by the pancreas acts centrally to stimulate the renin–angiotensin system to increase sodium excretion. The resultant effect is increased in renin secretion. The effect is largely mediated via angiotensin AT_1 receptor [90].

Research indicates that many aspects of the link between the gut and the kidney are mediated by the gut microbiota. The composition of gut microbiota in patients with chronic kidney diseases has been shown to be different from healthy individuals [91]. Dysfunctions of the gut microbiota (dysbiosis) can result in increased production of uremic toxins such as *p*-cresol sulfate, indoxyl sulfate, indole-3 acetic and trimethylamine *N*-oxide, resulting from the fermentative action of the gut microbiota on protein nutrients and other nitrogen-containing molecules [91–94]. Increased production of these toxins by the gut microbiota can result in disorders of metabolism and local and systemic inflammatory and immune responses [91–93]. Furthermore, these toxins can damage the intestinal barrier and composition of the gut microbiota, leading to translocation of potentially pathogenic bacteria into the bloodstream [95]. These disorders can culminate in kidney injury and are currently considered as crucial pathogenic mechanisms involved in the progression of chronic kidney disease and kidney failure. Emerging evidences suggest that these gut–kidney disorders may underlie the development of some cardiovascular diseases such as hypertension and atherosclerosis [96, 97]. More so, the metabolic products of the gut microbiota (e.g., short-chain fatty acids) serve as positive regulators of blood pressure, metabolism, and immunoinflammatory response [97]. Accumulating studies have showed that nutritional intervention using probiotics, prebiotics, and synbiotics in the diseases of the gut and kidney mentioned above can lead to positive outcome [91–94].

15.8 Gut–Lung Axis

Gut–lung axis is the functional relationship between the gut and the lungs. This association is mediated by the activities of the microbiota of the gut and respiratory tract, which are believed to influence each other, modulating several functions of the organism including immune response and the development of certain diseases [13, 98, 99]. Though data are scanty, accumulating evidences suggest that diseases of the gut such as inflammatory bowel disease are associated with pulmonary dysfunctions. Patients with gut disorders also develop pulmonary impairments [13, 100]. A couple of factors modulate the gut–lung functional relationship, and they include age, gender, diet, lifestyle, and use of antibiotics. This suggests that the development of lung diseases such as asthma, chronic obstructive pulmonary disease, and cystic fibrosis may be influenced by intestinal dysbiosis [13].

The mechanisms of the functional relationship between the gut and lungs are not completely understood, but bidirectional cross talks between the microbial community of the respiratory tract and gut appear to be crucial. Though the number of microbes is far more less in the lungs/respiratory tract than in the gut, except for the nasal mucosa, they contain the same species of bacteria—*Bacteroidetes*, *Firmicutes*, and *Proteobacteria*. The number of species of lung microbiota is estimated at about 500. The lung microbiota is more or less similar to the oral microbiota. New data indicate that disorder in oral microbiota is associated with several diseases including those involving pulmonary impairment. Changes in oral microbiota species can easily translocate lung–oral cavity boundaries and induce immune responses that underlie the onset of some diseases [13, 98, 100–103]. However, disorder of intestinal mucosa due to gut diseases may increase translocation of bacteria to the bloodstream, resulting in bacteremia that subsequently may lead to pulmonary impairment. A similar condition may occur in the epithelium of the lungs [104–107].

15.9 Gut–Bone Axis

The gut–bone axis, also known as entero-osseous axis, represents a functional cross talk between epithelial cells of the gut and the bone cells—osteoblasts and osteoclasts [108]. Several studies have documented pathways that may likely connect the gut to the bone [109, 110]. The intestinal hormones involved in bone homeostasis include GIP, GLP-1, and GLP-2 and are the main hormones that affect bone functions and structure. The receptors for these and other gut-derived hormones are expressed in bone cells [108]. But the endocrine pancreas and the adipose tissue also release hormones that regulate bone functions. Such hormones include insulin, amylin, adrenomedullin, leptin, resistin, adiponectin [108, 111, 112]. The liver also synthesizes certain proteins such as IGF-1 and immune factors that affect bone mass [113].

Bone cells express receptors for GIP. Activation of the GIP receptors leads to increased synthesis and proliferation of bone components and also inhibits apoptosis, thereby increasing bone mass. Thus, the hormone is anabolic. Furthermore, the hormone GIP inhibits PTH-induced bone resorption; i.e., GIP is antiresorptive [108, 114].

GLP-2 receptors are expressed in bone cells. Activation of the GLP-2 receptor inhibits bone resorption [108, 115].

GLP-1 receptors are not present in bone cells; hence, the effect of activation of these receptors on bone is indirect [108, 116, 117].

A huge part of the effect exerted by the gut on bone is mediated by the gut microbiota. The mechanisms by which the gut microbiota regulates bone remodeling and homeostasis are believed to occur mainly through the immune system [113].

15.10 Gut–Skin Axis and Gut–Brain–Skin Triangle

The gut–skin axis, also known as enterocutaneous axis, is a functional relationship between the gut and skin that ensures adequate regulation of cutaneous functions [118–120]. The gut microbiota modulates the functions of the skin via its metabolic capacity and immune responses [121]. Bacteriotherapy involving the ingestion *Lactobacillus* strains has been shown to reduce skin infection [6]. The use of supplements such as oral probiotics, prebiotics, and synbiotics has been shown not only to reduce skin inflammation, but also reduce stress. Disorders of the gut microbiota have been implicated in the pathogenesis of acne, seborrheic dermatitis, and atopic dermatitis [118, 119].

Acne is a chronic inflammatory skin disease characterized by *Propionibacterium acnes* colonization, inflammation of the sebaceous glands resulting in increased keratin formation and sebum production [122]. Seborrheic dermatitis is a chronic relapsing, pruritic inflammatory skin disease characterized by scaling and poorly defined erythematous patches [123]. The itch in seborrheic dermatitis mostly affects areas with high sebaceous glands such as the face, scalp, upper chest, and back. The etiopathogenesis of the disease is not fully understood, but it has been suggested that Malassezia yeast may be involved as a causative agent. However, lifestyle and environmental factors play essential role in the development of the disease [124, 125]. Atopic dermatitis, also known as eczema, is a non-contagious, chronic relapsing, pruritic inflammation of the epidermis and dermis, having typical distribution and morphology of rash, primarily prevalent in families with atopic diseases such as bronchial asthma, allergic rhinoconjunctivitis, and eczema [126, 127]. For further review of these disorders, see Johnson and Nunley [124], Berk and Scheinfeld [125], Darsow et al. [126], Correale et al. [127], and Simpson et al. [128].

The skin diseases mentioned above also involve disorders in nervous system functions [6]. Patients with such skin infections have higher score in mental tests on depression and anxiety, compared to other patients with diabetes for instance [119]. In addition, such patients also experience GI disorders such as constipation, abdominal bloating, halitosis, and reflux of gastric contents [119]. These findings point to the presence of a functional relationship involving the gut, brain, and skin, often referred to as gut–brain–skin axis or triangle. Although the link between the gut and skin health was suggested since antiquity, John H. Stokes and Donald M. Pillsbury were some of the pioneers that suggested that a link between the nervous system and the gut was responsible for some skin diseases such as acne [129]. Patients with such skin disease usually have lower numbers of beneficial bacteria in the gut [119]. Consequently, oral administration of beneficial bacteria (*Lactobacillus* species, *Saccharomyces cerevisiae*), probiotics, has been shown to improve both skin inflammation and nervous system functions [62, 119, 130].

Reduction in gastric acid contents (hypochlorhydria) has been shown to be associated with skin infections [118]. In addition to emotional stress, depression, and anxiety, hypochlorhydria is a risk factor for small intestinal bacterial overgrowth [119]. Small intestinal bacterial overgrowth refers to an increase in the number or alteration in the composition of bacteria in the small intestine. This clinical condition is characterized by increase in small intestinal bacterial number above 10^5–10^6 per mL. In this circumstance, substantial part of the types of normal bacteria in the small intestine may be replaced with gram-negative ones instead of predominantly gram-positive bacteria that is found in the tract [131]. In small intestinal bacterial overgrowth, the increased number of bacteria above normal competes with the cells of the body for available nutrients, leading to malabsorption, weight loss, and nutritional deficiencies. There is also increased production of toxic metabolites that may cause injury to the epithelial cells, which may in turn result in increased intestinal permeability [119]. The signs and symptoms of the condition include bloating, chronic diarrhea, abdominal pain, and constipation [119]. For small intestinal bacterial overgrowth therapy, antibiotics are usually prescribed, but prokinetics, probiotics, prebiotics, and fecal transplant have been shown to restore normal gut microbiota and intestinal permeability [118, 119, 131, 132].

15.11 Conclusions

The gut–extraenteric functional link provides extensive interaction between the gut and almost all tissues and organs of the body. This functional connectivity (in addition to structural relationship in some cases) regulates a wide range of activities in the entire organism including higher mental, cardiorenal, respiratory, cutaneous functions. Dysfunctions in the connectivity can lead to serious health consequences in the affected axis or system.

Recommended Readings

Original Articles

1. Eriksson L, Lif Holgerson P, Johansson I (2017) Saliva and tooth biofilm bacterial microbiota in adolescents in a low caries community. Sci Rep 7 (1):5861
2. Labus JS, Hollister EB, Jacobs J, Kirbach K, Oezguen N, Gupta A et al (2017) Differences in gut microbial composition correlate with regional brain volumes in irritable bowel syndrome. Microbiome 5:49
3. McDonald D, Ackermann G, Khailova L, Baird C, Heyland D, Kozar R et al (2016) Extreme dysbiosis of the microbiome in critical illness. mSphere 1(4). pii: e00199–16
4. Rogers MB, Firek B, Shi M, Yeh A, Brower-Sinning R, Aveson V et al (2016) Disruption of the microbiota across multiple body sites in critically ill children. Microbiome 4(1):66
5. Saito Y, Fujii M, Watanabe T, Maruyama K, Kowatari Y, Ogata H, Kumagai T (2017) Randomized, double-blind, placebo-controlled, parallel-group study of the effect of Lactobacillus paracasei K71 intake on salivary release of secretory immunoglobulin A. Biosci Microbiota Food Health 36(2):55–63
6. Saito Y, Mihara T, Maruyama K, Saito J, Ikeda M, Tomonaga A, Kumagai T (2017) Effects of intake of *Lactobacillus casei* subsp. *casei* 327 on skin conditions: a randomized, double-blind, placebo-controlled, parallel-group study in women. Biosci Microbiota Food Health 36(3):111–120
7. Tang WHW, Wang Z, Kennedy DJ, Wu Y, Buffa JA, Agatisa-Boyle B et al (2015) Gut microbiota-dependent trimethylamine N-oxide (TMAO) pathway contributes to both development of renal insufficiency and mortality risk in chronic kidney disease. Circ Res 116:448–455

Review Articles

1. Dicembrini I, Mannucci E, Rotella CM (2012) Bone: incretin hormones perceiver or receiver? Exp Diabetes Res 2012:519784
2. Hemarajata P, Versalovic J (2013) Effects of probiotics on gut microbiota: mechanisms of intestinal immunomodulation and neuromodulation. Therap Adv Gastroenterol 6(1):39–51
3. Ho P, Ross DA (2017) More than a gut feeling: the implications of the gut microbiota in psychiatry. Biol Psychiatry 81(5):e35–e37
4. Jose PA, Raj D (2015) Gut microbiota in hypertension. Curr Opin Nephrol Hypertens 24(5):403–409
5. Kilian M, Chapple IL, Hannig M, Marsh PD, Meuric V, Pedersen AM et al (2016) The oral microbiome—an update for oral healthcare professionals. Br Dent J 221:657–666
6. Konturek PC, Brzozowski T, Konturek SJ (2011) Stress and the gut: pathophysiology, clinical consequences, diagnostic approach and treatment options. J Physiol Pharmacol 62(6):591–9

7. Leung HYC, Leong PK, Chen J, Ko KM (2017) Inter-organ relationships among gut, lung and skin beyond the pathogenesis of allergies: relevance to the Zang-Fu theory in chinese medicine. Chinese Med 8(3):78205
8. Quigley EMM (2013) Gut bacteria in health and disease. Gastroenterol Hepatol (N Y) 9(9):560–569
9. Slattery J, MacFabe DF, Frye RE (2016) The significance of the enteric microbiome on the development of childhood disease: a review of prebiotic and probiotic therapies in disorders of childhood. Clin Med Insights Pediatr 10:91–107
10. Wade WG (2013) The oral microbiome in health and disease. Pharmacol Res 69(1):137–43

Books

1. Fong IW (2014) The role of microbes in common non-infectious diseases. Springer, New York
2. Mackie R, White B, Isaacson RE (1997) Gastrointestinal microbiology: gastrointestinal microbes and host interactions. Springer, New York
3. Rosenberg E, Gophna U (eds) (2011) Beneficial microorganisms in multicellular life forms. Springer, Heidelberg

Databases

1. Human Oral Microbiome Database (HOMD). http://www.homd.org.
2. National Institutes of Health (NIH) Human Microbiome Project (HMP). http://hmpdacc.org

Bibliography

1. Fenske W, Athanasiou T, Harling L, Drechsler C, Darzi A, Ashrafian H (2013) Obesity-related cardiorenal disease: the benefits of bariatric surgery. Nat Rev Nephrol 9:539–551
2. Michell AR, Debnam ES, Unwin RJ (2008) Regulation of renal function by the gastrointestinal tract: potential role of gut-derived peptides and hormones. Annu Rev Physiol 70:379–403
3. Pereverzev VA, Lobanok LM (2014) Physiology of digestion. In: Kubarko AI (ed) Normal physiology. Visheishaya Shkola, Minsk, Belarus
4. Nikitina OS, Welcome MO, Pereverzev VA (2016) Human anatomy and physiology. Belarusian State Medical University Press, Minsk
5. Vitetta L, Manuel R, Zhou JY, Linnane AW, Hall S, Coulson S (2014) The overarching influence of the gut microbiome on end-organ function: the role of live probiotic cultures. Pharmaceuticals (Basel) 7(9):954–989
6. Arck P, Handjiski B, Hagen E, Pincus M, Bruenahl C, Bienenstock J, Paus R (2010) Is there a 'gut-brain-skin axis'? Exp Dermatol 19(5):401–405
7. Benson MD, Gandhi MR (2000) Ultrasound of the hepatobiliary-pancreatic system. World J Surg 24(2):166–170

8. Olivier AK, Gibson-Corley KN, Meyerholz DK (2015) Animal models of gastrointestinal and liver diseases. Animal models of cystic fibrosis: gastrointestinal, pancreatic, and hepatobiliary disease and pathophysiology. Am J Physiol Gastrointest Liver Physiol 308(6): G459–G471
9. Serino M, Blasco-Baque V, Nicolas S, Burcelin R (2014) Managing the manager: gut microbes, stem cells and metabolism. Diabetes Metab 40(3):186–190
10. Sekirov I, Russell SL, Antunes LCM, Finlay BB (2010) Gut microbiota in health and disease. Physiol Rev 90(3):859–904
11. Vajro P, Paolella G, Fasano A (2013) Microbiota and gut-liver axis: a mini-review on their influences on obesity and obesity related liver disease. J Pediatr Gastroenterol Nutr 56 (5):461–468
12. Paolella G, Mandato C, Pierri L, Poeta M, Di Stasi M, Vajro P (2014) Gut-liver axis and probiotics: their role in non-alcoholic fatty liver disease. World J Gastroenterol 20 (42):15518–15531
13. Shukla SD, Budden KF, Neal R, Hansbro PM (2017) Microbiome effects on immunity, health and disease in the lung. Clin Transl Immunol 6:e133
14. Frank DN, St Amand AL, Feldman RA, Boedeker EC, Harpaz N, Pace NR (2007) Molecular-phylogenetic characterization of microbial community imbalances in human inflammatory bowel diseases. Proc Natl Acad Sci USA 104:13780–13785
15. Zhu B, Wang X, Li L (2010) Human gut microbiome: the second genome of human body. Protein Cell 1(8):718–725
16. Sender R, Fuchs S, Milo R (2016) Revised estimates for the number of human and bacteria cells in the body. PLoS Biol 14(8):e1002533
17. Dewhirst FE, Chen T, Izard J, Paster BJ, Tanner ACR, Yu W-H et al (2010) The human oral microbiome. J Bacteriol 192(19):5002–5017
18. Kort R, Caspers M, van de Graaf A, van Egmond W, Keijser B, Roeselers G (2014) Shaping the oral microbiota through intimate kissing. Microbiome 2:41
19. Liu N, Ando T, Ishiguro K, Maeda O, Watanabe O, Funasaka K et al (2013) Characterization of bacterial biota in the distal esophagus of Japanese patients with reflux esophagitis and Barrett's esophagus. BMC Infect Dis 13:130
20. Harris JK, Fang R, Wagner BD, Choe HN, Kelly CJ, Schroeder S et al (2015) Esophageal microbiome in eosinophilic esophagitis. PLoS ONE 10(5):e0128346
21. Acharya C, Sahingur SE, Bajaj JS (2017) Microbiota, cirrhosis, and the emerging oral-gut-liver axis. JCI Insight 2(19). pii: 94416
22. Odamaki T, Bottacini F, Kato K, Mitsuyama E, Yoshida K, Horigome A et al (2018) Genomic diversity and distribution of *Bifidobacterium longum* subsp. *longum* across the human lifespan. Sci Rep 8:85
23. Basson A, Trotter A, Rodriguez-Palacios A, Cominelli F (2016) Mucosal interactions between genetics, diet, and microbiome in inflammatory bowel disease. Front Immunol 7:290
24. Brandl K, Kumar V, Eckmann L (2017) Gut-liver axis at the frontier of host-microbial interactions. Am J Physiol Gastrointest Liver Physiol 312(5):G413–G419
25. Wiest R, Albillos A, Trauner M, Bajaj J, Jalan R (2017) Targeting the gut-liver axis in liver disease. J Hepatol pii: S0168-8278(17)32016-0
26. Arab JP, Martin-Mateos RM, Shah VH (2018) Gut-liver axis, cirrhosis and portal hypertension: the chicken and the egg. Hepatol Int 12(1):24–33
27. Fukui H (2015) Gut-liver axis in liver cirrhosis: how to manage leaky gut and endotoxemia. World J Hepatol 7(3):425–442
28. Kabeerdoss J, Jayakanthan P, Pugazhendhi S, Ramakrishna BS (2015) Alterations of mucosal microbiota in the colon of patients with inflammatory bowel disease revealed by real time polymerase chain reaction amplification of 16S ribosomal ribonucleic acid. Indian J Med Res 142(1):23–32

29. Compare D, Coccoli P, Rocco A, Nardone OM, De Maria S, Cartenì M, Nardone G (2012) Gut–liver axis: the impact of gut microbiota on non alcoholic fatty liver disease. Nutr Metab Cardiovasc Dis 22(6):471–476
30. Kalaitzakis E (2014) Gastrointestinal dysfunction in liver cirrhosis. World J Gastroenterol 20 (40):14686–14695
31. Visschers RG, Luyer MD, Schaap FG, Olde Damink SW, Soeters PB (2013) The gut-liver axis. Curr Opin Clin Nutr Metab Care 16(5):576–581
32. Koulaouzidis A, Bhat S, Karagiannidis A, Tan WC, Linaker BD (2007) Spontaneous bacterial peritonitis. Postgrad Med J 83(980):379–383
33. Koulaouzidis A, Bhat S, Saeed AA (2009) Spontaneous bacterial peritonitis. World J Gastroenterol 15(9):1042–1049
34. Lata J, Stiburek O, Kopacova M (2009) Spontaneous bacterial peritonitis: a severe complication of liver cirrhosis. World J Gastroenterol 15(44):5505–5510
35. Procopet B, Berzigotti A (2017) Diagnosis of cirrhosis and portal hypertension: imaging, non-invasive markers of fibrosis and liver biopsy. Gastroenterol Rep 5(2):79–89
36. Banerjee JK (2012) Portal hypertension. Med J Armed Forces India 68(3):276–279
37. Ferenci P (2017) Hepatic encephalopathy. Gastroenterol Rep (Oxf) 5(2):138–147
38. Dharel N, Bajaj JS (2015) Definition and nomenclature of hepatic encephalopathy. J Clin Exp Hepatol 5(1):S37–S41
39. Pantham G, Mullen KD (2017) Practical issues in the management of overt hepatic encephalopathy. Gastroenterol Hepatol (N Y) 13(11):659–665
40. Shawcross DL (2018) Diagnosis and management of hepatic encephalopathy. Br J Nurs 27 (3):S7–S13
41. Unger RH, Eisentraut AM (1969) Entero-insular axis. Arch Intern Med 123:261–266
42. Creutzfeldt W (1979) The incretin concept today. Diabetologia 16:75–85
43. Schmidt WE, Meier JJ (2006) Werner Creutzfeldt May 11, 1924–August 30, 2006. Regul Pept 137(3):105–106
44. Creutzfeldt W, Nauck M (1992) Gut hormones and diabetes mellitus. Diabetes/Metab Rev 8 (2):149–177
45. Göke B (2006) Obituary for Prof. Dr. med. Dr. h.c. Werner Creutzfeldt, FRCP. May 11, 1924 to August 30, 2006. Digestion 74:55–56
46. Schönfeld JV, Goebell H, Mütter MK (1994) The islet-acinar axis of the pancreas. Int J Pancreatol 16(2–3):131–140
47. Ranganath LR (2007) The entero-insular axis: implications for human metabolism. Clin Chem Lab Med 46(1):43–56
48. Ørskov C (1992) Glucagon-like peptide-1, a new hormone of the entero-insular axis. Diabetologia 35(8):701–711
49. Barreto SG, Carati CJ, Toouli J, Saccone GT (2010) The islet-acinar axis of the pancreas: more than just insulin. Am J Physiol Gastrointest Liver Physiol 299(1):G10–G22
50. Williams JA, Goldfine ID (1985) The insulin-pancreatic acinar axis. Diabetes 34(10):980–986
51. Foster JA (2013) Gut feelings: bacteria and the brain. Cerebrum 2013:9
52. Pimentel GD, Micheletti TO, Pace F, Rosa JC, Santos RVT, Lira FS (2012) Gut-central nervous system axis is a target for nutritional therapies. Nutr J 11:22
53. Furness JB, Callaghan BP, Rivera LR, Cho HJ (2014) The enteric nervous system and gastrointestinal innervation: integrated local and central control. Adv Exp Med Biol 817:39–71
54. Chaudhri O, Small C, Bloom S (2006) Gastrointestinal hormones regulating appetite. Philos Trans R Soc Lond B Biol Sci 361(1471):1187–1209
55. Nyberg F, Hallberg M (2013) Growth hormone and cognitive function. Nat Rev Endocrinol 9(6):357–365
56. Mayer EA, Knight R, Mazmanian SK, Cryan JF, Tillisch K (2014) Gut microbes and the brain: paradigm shift in neuroscience. J Neurosci 34(46):15490–15496
57. Galland L (2014) The gut microbiome and the brain. J Med Food 17(12):1261–1272

58. Li Q, Zhou JM (2016) The microbiota-gut-brain axis and its potential therapeutic role in autism spectrum disorder. Neuroscience 324:131–139
59. MacFabe DF (2015) Enteric short-chain fatty acids: microbial messengers of metabolism, mitochondria, and mind: implications in autism spectrum disorders. Microb Ecol Health Dis 26:28177
60. Petra AI, Panagiotidou S, Hatziagelaki E, Stewart JM, Conti P, Theoharides TC (2015) Gut-microbiota-brain axis and its effect on neuropsychiatric disorders with suspected immune dysregulation. Clin Ther 37(5):984–995
61. Grootaert C, Van den Abbeele P, Marzorati M, Broekaert WF, Courtin CM, Delcour JA et al (2009) Comparison of prebiotic effects of arabinoxylan oligosaccharides and inulin in a simulator of the human intestinal microbial ecosystem. FEMS Microbiol Ecol 69(2):231–242
62. Geurts L, Neyrinck AM, Delzenne NM, Knauf C, Cani PD (2014) Gut microbiota controls adipose tissue expansion, gut barrier and glucose metabolism: novel insights into molecular targets and interventions using prebiotics. Benef Microbes 5(1):3–17
63. Carabotti M, Scirocco A, Maselli MA, Severia C (2015) The gut-brain axis: interactions between enteric microbiota, central and enteric nervous systems. Ann Gastroenterol 28 (2):203–209
64. Hyland N, Stanton C (eds) (2016) The gut-brain axis: dietary, probiotic, and prebiotic interventions on the microbiota. Academic Press, Cambridge, MA, USA
65. Sollazzo R, Sanges M (2016) The second brain and possible interactions with the heart. In: Roncella A, Pristipino C (eds) Psychotherapy for ischemic heart disease. Springer, Cham, Switzerland
66. Kim M, Platt MJ, Shibasaki T, Quaggin SE, Backx PH, Seino S et al (2013) GLP-1 receptor activation and Epac2 link atrial natriuretic peptide secretion to control of blood pressure. Nat Med 19:567–575
67. Ruiz-Hurtado G, Morel E, Domínguez-Rodríguez A, Llach A, Lezoualc'h F, Benitah JP, Gomez AM (2013) Epac in cardiac calcium signaling. J Mol Cell Cardiol 58:162–171
68. Holz GG, Kang G, Harbeck M, Roe MW, Chepurny OG (2006) Cell physiology of cAMP sensor Epac. J Physiol 577(1):5–15
69. Sheikh A (2013) Direct cardiovascular effects of glucagon like peptide-1. Diabetol Metab Syndr 5:47
70. Méry P-F, Brechler V, Pavoine C, Pecker F, Fischmeister R (1990) Glucagon stimulates the cardiac Ca^{2+} current by activation of adenylyl cyclase and inhibition of phosphodiesterase. Nature 345:158–161
71. Farah AE (1983) Glucagon and the heart. In: Lefebvre PJ (ed) Glucagon II. Handbook of experimental pharmacology, vol. 66 of the series. Springer, Heidelberg
72. Williams JF Jr, Childress RH, Chip JN, Border JF (1969) Hemodynamic effects of glucagon in patients with heart disease. Circulation 39:38–47
73. Entman ML, Levey GS, Epstein SE (1969) Mechanism of action of epinephrine and glucagon on the canine heart: evidence for increase in sarcotubular calcium stores mediated by cyclic 3′,5′-AMP. Circ Res 25:429–438
74. Rieker RP, Lee JC, Downing SE (1975) Positive inotropic action of insulin on piglet heart. Yale J Biol Med 48(5):353–359
75. Hsu C-H, Wei J, Chen Y-C, Yang S-P, Tsai C-S, Lin C-I (2006) Cellular mechanisms responsible for the inotropic action of insulin on failing human myocardium. J Heart Lung Transpl 25(9):1126–1134
76. von Lewinski D, Bruns S, Walther S, Kögler H, Pieske B (2005) Insulin causes [Ca2 +] i-dependent and [Ca2 +]i-independent positive inotropic effects in failing human myocardium. Circulation 111(20):2588–2595
77. Engebretsen KM, Kaczmarek KM, Morgan J, Holger JS (2011) High-dose insulin therapy in beta-blocker and calcium channel-blocker poisoning. Clin Toxicol (Phila) 49(4):277–283

78. von Lewinski D, Voss K, Hülsmann S, Kögler H, Pieske B (2003) Insulin-like growth factor-1 exerts Ca2 + -dependent positive inotropic effects in failing human myocardium. Circ Res 92(2):169–176
79. Kaygisiz Z, Ozden H, Erkasap N, Köken T, Gündüz T, Ikizler M, Kural T (2009) Effects of proadrenomedullin N-terminal 20 peptide and calcitonin on isolated perfused rat hearts. Anadolu Kardiyol Derg 9(3):176–182
80. Kaygisiz Z, Ozden H, Erkasap N, Koken T, Gunduz T, Ikizler M, Kural T (2010) Positive inotropic, positive chronotropic and coronary vasodilatory effects of rat amylin: mechanisms of amylin-induced positive inotropy. Acta Physiol Hung 97(4):362–374
81. Opgaard SO, de Vries R, Tom B, Edvinsson L, Saxena PR (1999) Positive inotropy of calcitonin gene-related peptide and amylin on porcine isolated myocardium. Eur J Pharmacol 385(2–3):147–154
82. Sindić A, Schlatter E (2006) Cellular effects of guanylin and uroguanylin. J Am Soc Nephrol 17(3):607–616
83. Yuge S, Inoue K, Hyodo S, Takei Y (2003) A novel guanylin family (guanylin, uroguanylin, and renoguanylin) in eels: possible osmoregulatory hormones in intestine and kidney. J Biol Chem 278:22726–22733
84. Forte LR, Eber SL, Fan X, London RM, Wang Y, Rowland LM et al (1999) Lymphoguanylin: cloning and characterization of a unique member of the guanylin peptide family. Endocrinology 140(4):1800–1806
85. Sindic A (2013) Current understanding of guanylin peptides actions. ISRN Nephrol 2013:813648
86. Rozenfeld J, Tal O, Kladnitsky O, Adler L, Efrati E, Carrithers SL et al (2012) The pendrin anion exchanger gene is transcriptionally regulated by uroguanylin: a novel enterorenal link. Am J Physiol Renal Physiol 302(5):F614–F624
87. Preston RA, Afshartous D, Forte LR, Rodco R, Alonso AB, Garg D, Raij L (2012) Sodium challenge does not support an acute gastrointestinal–renal natriuretic signaling axis in humans. Kidney Int 82:1313–1320
88. Wall SM, Lazo-Fernandez Y (2015) The role of pendrin in renal physiology. Annu Rev Physiol 77:363–378
89. Wall SM (2017) Renal intercalated cells and blood pressure regulation. Kidney Res Clin Pract 36(4):305–317
90. Mathai ML, Sosa Leon LA, May CN, Thomson CE, McKinley MJ (2005) Amylin induces natriuresis by a central angiotensin-dependent mechanism. Regul Pept 130(1–2):91–96
91. Rossi M, Johnson DW, Campbell KL (2015) The kidney-gut axis: implications for nutrition care. J Ren Nutr 25(5):399–403
92. Coppo R (2018) The gut-kidney axis in IgA nephropathy: role of microbiota and diet on genetic predisposition. Pediatr Nephrol 33(1):53–61
93. Wing MR, Patel SS, Ramezani A, Raj DS (2016) Gut microbiome in chronic kidney disease. Exp Physiol 101(4):471–477
94. Fernandez-Prado R, Esteras R, Perez-Gomez MV, Gracia-Iguacel C, Gonzalez-Parra E, Sanz AB et al (2017) Nutrients turned into toxins: microbiota modulation of nutrient properties in chronic kidney disease. Nutrients 9(5). pii: E489
95. Sabatino A, Regolisti G, Brusasco I, Cabassi A, Morabito S, Fiaccadori E (2015) Alterations of intestinal barrier and microbiota in chronic kidney disease. Nephrol Dial Transplant 30 (6):924–933
96. Ito S, Yoshida M (2014) Protein-bound uremic toxins: new culprits of cardiovascular events in chronic kidney disease patients. Toxins (Basel) 6(2):665–678
97. Huang W, Zhou L, Guo H, Xu Y, Xu Y (2017) The role of short-chain fatty acids in kidney injury induced by gut-derived inflammatory response. Metabolism 68:20–30
98. Marsland BJ, Trompette A, Gollwitzer ES (2015) The gut-lung axis in respiratory disease. Ann Am Thorac Soc 12(2):S150–S156
99. Keely S, Talley NJ, Hansbro PM (2012) Pulmonary-intestinal cross-talk in mucosal inflammatory disease. Mucosal Immunol 5:7–18

100. Wang H, Liu J-S, Peng S-H, Deng X-Y, Zhu D-M, Javidiparsijani S et al (2013) Gut-lung crosstalk in pulmonary involvement with inflammatory bowel diseases. World J Gastroenterol 19(40):6794–6804
101. Bingula R, Filaire M, Radosevic-Robin N, Bey M, Berthon JY, Bernalier-Donadille A et al (2017) Desired turbulence? Gut-lung axis, immunity, and lung cancer. J Oncol 2017:5035371
102. Zhao Y, Wang J, Liu Z, Lin H, Shi Y, Sun X (2017) Pulmonary dysfunction in 114 patients with inflammatory bowel disease. Medicine (Baltimore) 96(18):e6808
103. Dickson RP, Huffnagle GB (2015) The lung microbiome: new principles for respiratory bacteriology in health and disease. PLoS Pathog 11(7):e1004923. https://doi.org/10.1371/journal.ppat.1004923
104. Berg RD (1999) Bacterial translocation from the gastrointestinal tract. Adv Exp Med Biol 473:11–30
105. Vaishnavi C (2013) Translocation of gut flora and its role in sepsis. Indian J Med Microbiol 31(4):334–342
106. Dickson RP, Singer BH, Newstead MW, Falkowski NR, Erb-Downward JR, Standiford TJ, Huffnagle GB (2016) Enrichment of the lung microbiome with gut bacteria in sepsis and the acute respiratory distress syndrome. Nat Microbiol 1(10):16113
107. García-Tsao G (2001) Bacterial translocation: cause or consequence of decompensation in cirrhosis? J Hepatol 34(1):150–155
108. Isales CM, Hamrick M (2008) Nutritional hormones and the entero-osseous axis. J Musculoskelet Neuronal Interact 8(4):348–350
109. Levasseur R, Lacombe D, de Vernejoul MC (2005) LRP5 mutations in osteoporosis-pseudoglioma syndrome and high-bone-mass disorders. Jt Bone Spine 72 (3):207–214
110. Chen D, Zhao CM (2011) The possible existence of a gut-bone axis suggested by studies of genetically manipulated mouse models? Curr Pharm Des 17(16):1552–1555
111. Lerner A, Matthias T (2016) Gut-bone cross talks and implications in celiac disease. Int J Celiac Dis 4(1):19–23
112. Zhao LJ, Jiang H, Papasian CJ, Maulik D, Drees B, Hamilton J, Deng HW (2008) Correlation of obesity and osteoporosis: effect of fat mass on the determination of osteoporosis. J Bone Miner Res 23(1):17–29
113. Novince CM, Whittow CR, Aartun JD, Hathaway JD, Poulides N, Chavez MB et al (2017) Commensal gut microbiota immunomodulatory actions in bone marrow and liver have catabolic effects on skeletal homeostasis in health. Sci Rep 7(1):5747
114. Christensen MB, Lund A, Calanna S, Jørgensen NR, Holst JJ, Vilsbøll T, Knop FK (2018) Glucose-dependent insulinotropic polypeptide (GIP) inhibits bone resorption independently of insulin and glycemia. J Clin Endocrinol Metab 103(1):288–294
115. Pacheco-Pantoja EL, Ranganath LR, Gallagher JA, Wilson PJ, Fraser WD (2011) Receptors and effects of gut hormones in three osteoblastic cell lines. BMC Physiol 11:12
116. Zhao C, Liang J, Yang Y, Yu M, Qu X (2017) The impact of glucagon-like peptide-1 on bone metabolism and its possible mechanisms. Front Endocrinol (Lausanne) 8:98
117. Meng J, Ma X, Wang N, Jia M, Bi L, Wang Y et al (2016) Activation of GLP-1 receptor promotes bone marrow stromal cell osteogenic differentiation through β-catenin. Stem Cell Rep 6(4):579–591
118. Craig MJ (2016) Atopic dermatitis and the intestinal microbiota in humans and dogs. Vet Med Sci 2(2):95–105
119. Bowe WP, Logan AC (2011) Acne vulgaris, probiotics and the gut-brain-skin axis—back to the future? Gut Pathogens 3:1
120. Zákostelská Z, Málková J, Klimešová K, Rossmann P, Hornová M, Novosádová I et al (2016) Intestinal microbiota promotes psoriasis-like skin inflammation by enhancing Th17 response. PLoS ONE 11(7):e0159539
121. O'Neill CA, Monteleone G, McLaughlin JT, Paus R (2016) The gut-skin axis in health and disease: a paradigm with therapeutic implications. BioEssays 38(11):1167–1176

122. Titus S, Hodge J (2012) Diagnosis and treatment of acne. Am Fam Physician 86(8):734–740
123. Mokos BZ, Kralj M, Basta-Juzbašić A, Jukić LI (2012) Seborrheic dermatitis: an update. Acta Dermatovenerol Croat 20(2):98–104
124. Johnson BA, Nunley JR (2000) Treatment of seborrheic dermatitis. Am Fam Physician 61 (9):2703–2710
125. Berk T, Scheinfeld N (2010) Seborrheic dermatitis. PT 35(6):348–352
126. Darsow U, Raap U, Ständer S (2014) Atopic dermatitis. In: Carstens E, Akiyama T (eds) Itch: mechanisms and treatment. CRC Press/Taylor & Francis, Boca Raton, Florida, USA
127. Correale CE, Walker C, Murphy L, Craig TJ (1999) Atopic dermatitis: a review of diagnosis and treatment. Am Fam Physician 60(4):1191–1198
128. Simpson EL, Keck LE, Chalmers JR, Williams HC (2012) How should an incident case of atopic dermatitis be defined? A systematic review of primary prevention studies. J Allergy Clin Immunol 130(1):137–144
129. Stokes JH, Pillsbury DM (1930) The effect on the skin of emotional and nervous states iii. Theoretical and practical consideration of a gastro-intestinal mechanism. Arch Derm Syphilol 22(6):962–993
130. Sarkar A, Lehto SM, Harty S, Dinan TG, Cryan JF, Burnet PWJ (2016) Psychobiotics and the manipulation of bacteria–gut–brain signals. Trends Neurosci 39(11):763–781
131. Bures J, Cyrany J, Kohoutova D, Förstl M, Rejchrt S, Kvetina J et al (2010) Small intestinal bacterial overgrowth syndrome. World J Gastroenterol 16(24):2978–2990
132. Dukowicz AC, Lacy BE, Levine GM (2007) Small intestinal bacterial overgrowth: a comprehensive review. Gastroenterol Hepatol (N Y) 3(2):112–122

MIX
Papier aus verantwortungsvollen Quellen
Paper from responsible sources
FSC® C105338

If you have any concerns about our products,
you can contact us on
ProductSafety@springernature.com

In case Publisher is established outside the EU,
the EU authorized representative is:
Springer Nature Customer Service Center GmbH
Europaplatz 3, 69115 Heidelberg, Germany

Printed by Libri Plureos GmbH
in Hamburg, Germany